D0215293

Discovering Statistics Using SPSS

ISM INTRODUCING STATISTICAL METHODS

This series provides accessible but in-depth introductions to statistical methods that are not covered in any detail in standard introductory courses. The books are aimed at both the beginning researcher who needs to know how to use a particular technique and the established researcher who wishes to keep up to date with recent developments in statistical data analysis.

Editorial board

Other titles in this series

Introducing Multilevel Modeling
Ita Kreft and Jan de Leeuw

Introducing Social Networks
Alain Degenne and Michel Forsé

Introducing ANOVA and ANCOVA
Andrew Rutherford

Introducing LISREL: A Guide for the Uninitiated
Adamantios Diamantopoulos and Judy Siguaw

Discovering Statistics Using SPSS

SECOND EDITION

(and sex, drugs and rock 'n' roll)

ANDY FIELD

First edition published 2000
This edition first published 2005

SAGE Publications Ltd
1 Oliver's Yard
55 City Road
London EC1Y 1SP

SAGE Publications Inc.
2455 Teller Road
Thousand Oaks, California 91320

SAGE Publications India Pvt Ltd
B-42, Panchsheel Enclave
Post Box 4109
New Delhi 110 017

British Library Cataloguing in Publication data

A catalogue record for this book is available from the British Library

ISBN 0 7619 4451 6
ISBN 0 7619 4452 4 (pbk)

Library of Congress Control Number: 2004099538

Typeset by C&M Digitals (P) Ltd, Chennai, India
Printed in Great Britain by The Alden Press, Oxford

CONTENTS

PREFACE

Karma Police, arrest this man, he talks in maths, he buzzes like a fridge, he's like a detuned radio.

Radiohead (1997)

INTRODUCTION

Since time immemorial, social science students have despised statistics. For one thing, most have a non-mathematical background, which makes understanding complex statistical equations very difficult. The major advantage in being taught statistics in the early 1990s (as I was) compared to the 1960s was the development of computer software to do all of the hard work. The advantage of learning statistics now rather than 10 years ago is that these packages are now considerably easier to use thanks to Windows™/MacOS™. SPSS is, in my opinion, the best of the commercially available statistical packages and is commonly used in many universities. So what possessed me to write a book about statistics and SPSS?

Many good textbooks already exist that describe statistical theory. Howell (2002), Stevens (1992) and Wright (2002) have all written wonderful and clear books but use computer examples only as addenda to the theory. Likewise, several excellent books on SPSS already exist (Kinnear & Gray, 2000; Foster, 2001), but these concentrate on 'doing the test'. Using SPSS without any statistical knowledge can be a dangerous thing (it is only a tool, not a divine source of wisdom). As such, I want to use SPSS as a tool for teaching statistical concepts. In doing so I hope that the reader will gain a better understanding of both theory and practice.

Primarily, I want to answer the kinds of questions that I found myself asking while learning statistics and using SPSS as an undergraduate (things like 'How can I understand how this statistical test works without knowing too much about the maths behind it?' 'What does that button do?' 'What the hell does this output mean?'). SPSS has a complex set of options for each test, many of which are overlooked by books and tutors alike. I hope to be able to explain what these options actually do and why you might use them. Related to this point I want to be non-prescriptive. Too many books tell the reader what to do ('click on this button', 'do this', 'do that', etc.) and this can create the impression that statistics and SPSS are inflexible. SPSS has many options designed to allow you to tailor a given test to your particular needs. Therefore, although I make recommendations, I hope to give the readers enough background in theory to enable them to make their own decisions about which options are appropriate for the analysis they want to do.

A second aim was to have a book that could be read at several levels (see the next section for more guidance). There are chapters for first-year undergraduates (1, 2, 3, 4, 7 and 13), chapters for second-year undergraduates (5, 8, 9, 10, 11 and 12) and chapters on more advanced topics that postgraduates might use (6, 14, 15 and 16). All of these chapters should be accessible to everyone, and I hope to achieve this by flagging the level of each section (see the next section).

WHAT'S NEW?

I have a sticky history with maths because I used to be terrible at it. At 13, I was almost bottom of my class, yet 12 years later I'd written a statistics textbook (and 17 years later I've finished the second edition). The difference between the 13 year old who failed his exam and the 15 year old who did quite well was a good teacher: my brother, Paul. In fact I owe my life as an academic to Paul's ability to do what my teachers couldn't: teach me stuff in an engaging way. To this day he still pops up in times of need to teach me things (a crash course in computer programming a few Christmases ago springs to mind). Anyway, the reason he's a great teacher is because he's able to make things interesting and relevant to me. Sadly he seems to have got the 'good teaching' genes in the family (and he doesn't even work as a bloody teacher so they're wasted!), but I have tried to take his approach to my lectures and books. One thing I have learnt is that people appreciate the human touch, and so in the first edition I tried to inject a lot of my own personality (or should that be 'lack of personality'?). So, there were some light-hearted (some have said 'smutty' but I prefer 'light-hearted') examples and there was a large dose of my humour thrown in—which might have been a bad thing because I am a lecturer and, therefore, have no sense of humour!

When I wrote the first edition all I really wanted to do was write the kind of stats book that I would enjoy reading. Pretty selfish I know, but I just thought if I had a reference book that had a few examples in it that amused me then it would make life a lot easier when I needed to look something up. I honestly didn't think anyone would buy the thing (well, apart from my Mum and Dad) and I anticipated a glut of feedback along the lines of 'the whole of chapter X is completely wrong and you're an arrant fool', or 'think how many trees have died in the name of this rubbish, you should be ashamed of yourself you brainless idiot'. In fact, even the publishers didn't think it would sell (they have only revealed this subsequently I might add). What I absolutely didn't expect was to receive loads of extremely nice emails from people who liked the book. To this day it still amazes me that anyone reads it, let alone takes the time to write me a nice email (and at the risk of sounding corny, knowing that the book has helped people never ceases to put a smile on my face). Anyway, why am I telling you all this? Well, seeing as at least some people appreciated the style of the first edition I've taken this as a green light to include even more stupid examples, more smut and more bad taste. In short, lots more sex, drugs and rock 'n' roll. To those who hate all of this, I apologise, but if nothing else it keeps *me* amused.

Aside from adding more smut, I was forced reluctantly to expand the academic content! Most of the expansions have resulted from someone (often several people) emailing me to ask how to do something. So, in theory, this edition should answer any question anyone has asked me over the past 4 years! The general changes in each chapter are:

- **Examples**: Each chapter now has several examples at the end for you to work through, with answers supplied.
- **More data**: Because of the above, you have more data sets!
- **Reporting your analysis**: There are sections on how to report the analysis in most chapters. I've based these on the APA publication style manual (American Psychological Association, 2001).
- **Glossary**: Key terms are now identified in the text and there is a glossary of all of these terms.
- **Effect sizes**: Most topics now have a discussion of how to calculate effect sizes.
- **Summaries**: There's a lot of information in the text and I have a rambling (I prefer 'chatty') writing style, so there are now boxed summaries of the key points (see next section on how to use this book).
- **Boxes**: I've added numerous boxes that highlight interesting issues.
- **Syntax**: There is now some discussion of using SPSS syntax scattered throughout the book.
- **SPSS 13 Compliant**: Although this was the least amount of change because SPSS 13 is really not much different to version 9, I have made everything up to date with the latest version of SPSS.

The specific changes in each chapter are:

- **Chapter 1 (Basics)**: Has been expanded to include more background on statistical theory (e.g. confidence intervals, significance testing, power, effect sizes).
- **Chapter 2 (SPSS)**: Used to be squashed into Chapter 1 but has been updated to deal with SPSS 13 and now includes some discussion of the dreaded syntax window.
- **Chapter 3 (Exploring data)**: Bears little resemblance to what was Chapter 2 in the first edition. It now covers screening data, transforming data, checking assumptions of data and much more.
- **Chapter 4 (Correlation)**: Used to be Chapter 3 and hasn't changed much actually.
- **Chapter 5 (Regression)**: Used to be Chapter 4 but now has some more advanced material such as a section on using categorical predictors, as well as discussions of eigenvalues/vectors. I've also expanded the original sections in subtle ways too (e.g. greater discussion of sample size).
- **Chapter 6 (Logistic regression)**: The former Chapter 5 has been expanded in subtle ways (the theory part mainly), but I have also reorganized some of the material so that the theory and SPSS bits are more clearly separated.
- **Chapter 7 (*t*-tests)**: Hasn't changed in any specific ways from the old Chapter 6 apart from rewriting some stuff that I decided I didn't like!
- **Chapter 8 (GLM 1)**: Is basically the same as the old Chapter 7 except it covers the Welch and Brown–Forsythe versions of F for when homogeneity of variance can't be assumed.

- **Chapter 9 (GLM 2)**: This covers analysis of covariance (ANCOVA), which used to share a chapter with factorial ANOVA but now has a chapter of its own. I've got some extra material on planned contrasts and have corrected some embarrassing mistakes from the first edition☺
- **Chapter 10 (GLM 3)**: Covers factorial ANOVA (which used to share a chapter with ANCOVA). There are entirely new sections on the theory of two-way independent ANOVA, doing contrasts (with syntax), doing simple effects analysis, interpreting interaction graphs, and how we can conceptualize two-way ANOVA as a GLM!
- **Chapter 11 (GLM 4)**: Used to be Chapter 9 on repeated-measures ANOVA. I've added a section on the theory of repeated-measures ANOVA, and on conducting simple effects analysis in two-way repeated-measures ANOVA.
- **Chapter 12 (GLM 5)**: Covers mixed ANOVA (used to be in with repeated-measures ANOVA in the old Chapter 9). There is a completely new example, and with the luxury of a whole chapter I've expanded the interpretation sections a bit.
- **Chapter 13 (Non-parametric statistics)**: This is a completely new chapter (although borrows a few things from the old Chapter 2). I've added theory sections for the Mann–Whitney and Wilcoxon tests; I've also added completely new sections that cover Friedman's ANOVA, Jonckheere's test and the Kruskal–Wallis test.
- **Chapter 14 (MANOVA)**: hasn't changed much from the old Chapter 10.
- **Chapter 15 (Factor analysis)**: I've rearranged some of the material so that bits that were hidden away in the interpretation section are now clearly placed in the theory section. There is an entirely new section on reliability analysis.
- **Chapter 16 (Categorical data)**: This is basically a completely new chapter (although the chi-square example is old). I've included a theory section for the chi-square test, the likelihood ratio and Yates' correction, discussion of odds ratios, and loglinear analysis, none of which were in the last edition!

GOODBYE

I wrote at the beginning of the first edition that 'This book is the result of 2 years (give or take a few weeks to write up my Ph.D.) of trying to fulfil these aims. It isn't perfect and so I'd love to have feedback (good or bad) from the people who really matter: you, the readers'. This sentiment still very much applies except that it's now the result of 2½ years' work. Over the past 4 years I've become very attached to this book: it began life as a labour of love and it still is. What with the unexpected success of the first edition, and having done such a huge update for the second edition (300 pages compared to the 50 pages that the publishers actually wanted me to write!), I am hugely anxious that I've stuffed it all up by changing it! So, whether it'll live to see a third edition is anyone's guess, but in the meantime I'd still love to have feedback from the people who still matter: you, the readers.

Andy

Email: discoveringstatistics@sussex.ac.uk
Web: http://www.sussex.ac.uk/Users/andyf/

HOW TO USE THIS BOOK

When the publishers asked me to write a section on 'How to use this book' it was obviously tempting to write 'Buy a large bottle of Olay anti-wrinkle cream (which you'll need to fend off the effects of ageing while you read), find a comfy chair, sit down, fold back the front cover, begin reading and stop when you reach the back cover.' However, I think they wanted something more useful☺

WHAT BACKGROUND KNOWLEDGE DO I NEED?

In essence, I assume you know nothing about statistics, but I do assume you have some very basic grasp of computers (I won't be telling you how to switch them on, for example) and maths (although I have included a quick revision of some very basic concepts so I really don't assume anything).

DO THE CHAPTERS GET MORE DIFFICULT AS I GO THROUGH THE BOOK?

In a sense they do (Chapter 14 on MANOVA is more difficult than Chapter 1) but in other ways they don't (Chapter 13 on non-parametric statistics is arguably less complex than Chapter 12, and Chapter 7 on the *t*-test is definitely less complex than Chapter 6 on logistic regression). Why have I done this? Well, I've ordered the chapters to make statistical sense (to me, at least). Many books teach different tests in isolation and never really give you a grip of the similarities between them; this, I think, creates an unnecessary mystery. Most of the tests in this book are the same thing expressed in slightly different ways. So, I wanted the book to tell this story. To do this I have to do certain things such as explain regression fairly early on because it's the foundation on which nearly everything else is built!

However, to help you through I've coded each section with an icon. These icons are designed to give you an idea of the difficulty of the section. It doesn't necessarily mean you can skip the sections (but see Smart Alex in the next section), but it will let you know whether a section is about your level, or whether it's going to push you. I've based the icons on my own teaching so they may not be entirely accurate for everyone (especially as systems vary in different countries!):

① This means 'level 1' and I equate this to first-year undergraduate in the UK. These are sections that everyone should be able to understand.
② This is the next level and I equate this to second-year undergraduate in the UK. These are topics that I teach my second years and so anyone with a bit of background in statistics should be able to get to grips with them. However, some of these sections will be quite challenging even for second years. These are intermediate sections.
③ This is 'level 3' and represents difficult topics. I'd expect third-year (final-year) UK undergraduates and recent postgraduate students to be able to tackle these sections.
④ This is the highest level and represents very difficult topics. I would expect these sections to be very challenging to undergraduates and recent postgraduates, but postgraduates with a reasonable background in research methods shouldn't find them too much of a problem.

WHY DO I KEEP SEEING STUPID FACES EVERYWHERE?

Smart Alex: Alex is a very important character because he appears when things get particularly difficult. He's basically a bit of a smart alec and so whenever you see his face you know that something scary is about to be explained. When the hard stuff is over he reappears to let you know that it's safe to continue. Now, this is not to say that all of the rest of the material in the book is easy; he just let's you know the bits of the book that you can skip if you've got better things to do with your life than read all 800 pages! So, if you see Smart Alex then you can *skip the section* entirely and still understand what's going on. You'll also find that Alex pops up at the end of each chapter to give you some tasks to do to see whether you're as smart as he is. Incidentally, any physical similarity between Smart Alex and my editor is entirely coincidental!

Cramming Samantha: Samantha hates statistics. In fact, she thinks it's all a boring waste of time and she just wants to pass her exam and forget that she ever had to know anything about normal distributions. So, she appears and gives you a summary of the key points that you need to know. So if, like Samantha, you're cramming for an exam, she will tell you the essential information to save you having to trawl through hundreds of pages of my drivel.

Brian Haemorrhage: Brian's job is to pop up to ask questions and look permanently confused. It's no surprise to note, therefore, that he doesn't look entirely different from the author. As the book progresses he becomes increasingly despondent. Read into that what you will.

Curious Cat: He also pops up and asks questions (because he's curious). Actually the only reason he's here is because I wanted a cat in the book … and preferably one that looks like mine. Of course the educational specialists think he needs a specific role, and so his role is to look cute and make bad cat-related jokes.

WHAT ARE THE BOXES FOR?

You'll notice boxes throughout the book. These boxes can be ignored if you chose to, because they are generally asides to the main body of the book. They either try to explain very complicated things, or illustrate interesting points about statistics with which you can impress your friends.

WHAT IS ON THE CD-ROM?

The CD-ROM contains many interesting things:

- **Data files**: Mostly you'll use it to work through examples because it contains all of the data files for all of the examples in the book (the data for each chapter are contained in a separate folder).
- **Answers to Smart Alex tasks**: Because I don't want to kill any more trees than necessary, in addition to the book there are seemingly hundreds of pages of extra material on the CD-ROM. The most important documents are the answers to the tasks at the end of each chapter. For each task the document contains the SPSS output and a brief explanation of it so you can check your answers.
- **Additional material**: For some topics I have written extensive documents on the CD-ROM. These are mainly advanced topics or topics for which a detailed explanation in the main body of the book would have been a distraction. However, in case anyone wants to know more, there are descriptions on the CD. For example, I have some files on running contrasts using syntax, on calculating Welch's *F*-ratio, Jonckheere's test and so on. Most of you can ignore these files, but they're there in case anyone is interested.
- **Appendix material**: Some of the appendix material (such as mathematical calculations) has also been put in files on the CD-ROM to save paper.
- **Programs**: I have put a copy of G*Power on the CD-ROM, which is a useful program for calculating power.

THE WEBSITE

A companion website containing useful material for both lecturers and students is available on www.sagepub.co.uk/field

Happy reading!

ACKNOWLEDGEMENTS

Thanks to SPSS Inc. for allowing me use of their screen images. SPSS can be contacted at 444 North Michigan Avenue, Chicago, Illinois 60611 (USA) or First Floor, St Andrew's House, West Street, Woking, GU21 1EB (UK). Also check out their web pages (http://www.spss.com) for support and information. Thanks also to Axel Buchner for granting permission to put his program G*Power on the CD-ROM.

The first edition of this book wouldn't have happened if it hadn't been for Dan Wright, who not only had an unwarranted faith in a then-postgraduate to write the book, but also read and commented on many draft chapters. This new edition has benefited from many people sending me emails offering suggestions based on the first edition: I'm really grateful for everyone who has taken time out of their busy lives to point out mistakes, suggest improvements and provide wonderful and detailed feedback, so my sincere thanks go to Peter de Heus, Tilly Houtmans, Don Hunt, Paul Tinsley, Keith Tolfrey, Jaap Dronkers and Nick Smith (to name but a few!). Sage also commissioned various anonymous reviews, and, again, these threw up some very helpful suggestions, so I'm grateful to these people, whoever they may be (Tony Cassidy, having admitted to being one, gets a personal thanks!). Jeremy Miles stopped me making a complete and utter arse of myself (in the book—sadly his powers don't extend to everyday life) by pointing out many glaring errors; he's also been a very nice person to know over the past few years (apart from when he's saying that draft sections of my books are, and I quote, 'bollocks'!). Gareth Williams and Lynne Slocombe offered some inspired pedagogic suggestions; however, I will never forgive them for suggesting a glossary of key terms. Laura Murray also road-tested the first edition as both an undergraduate and then as a postgraduate tutor on my course; she then took it upon herself to take time away from her Ph.D. to comment on chapters in this new edition and suggested numerous improvements. Particular thanks for this edition go to David Hitchin, the unsung hero of statistics at the University of Sussex. He spent vast amounts of his time providing feedback on the first edition and on draft chapters of this new edition and teaching me numerous things about statistics over email. All of these people have read the first edition in its entirety, or have read vast chunks (if not all) of it and taken time out of their busy lives to spot errors, suggest improvements, or just simply tell me stuff I didn't previously know. I'm not sure what that says about their mental states, but they are all responsible for a great many improvements and I'm eternally grateful to their good natures: may they live long and their data sets be normal☺

At the risk of this becoming an Oscar-ceremony-type thing I am also very grateful to the following people for writing nice reviews of the first edition on either Amazon.com or Amazon.co.uk: George H Marshall (Scotland), Andrew (Italy), Dr Simon Marshall (University of Loughborough, UK), a reader from Royal Holloway University of London, Dr Keith Tolfrey (Manchester Metropolitan University, UK), James MacCabe (Institute of Psychiatry, London, UK), Geoff Bird (University College London, UK), abare@telcel (UK), Jeremy Miles (although slightly biased by knowing me), Ken Kolosh (IL, USA), Mehmet Yusuf Yahyagil (Turkey), Sandra Casillas (CA, USA), a reader from Missoula (MT, USA), Paulo Ferreira Leite (Brazil), a reader from Fair Lawn (NJ, USA), lts 1102 from Cresco (PA, USA), Shaun Galloway (Hungary), smcol from USA, a reader from Bowling Green (KY, USA), Denis E. Hommrich (KY, USA), Hsi Lung Wu (Taiwan), Iwan Wahyu (Singapore) and Mark Gray. I really appreciate you all taking the time to write reviews and am bowled over by some of the nice things you wrote. If you're ever in Brighton, I owe you a pint!

The people at Sage are without doubt some of the most hardened drinkers I have ever had the misfortune to be taken out by. Since the first edition I have been very fortunate to work with Michael Carmichael, who despite his failings on the football field (!) has provided me with some truly memorable nights out. In his spare time he's also a superb editor and all-round lovely person and deserves a medal for putting up with me. To be fair, though, I deserve a medal for humouring his obsessions with glossaries and such like.

If you own any of my books, you'll know I always write listening to music. For the first edition I owed my sanity to the following for providing great sounds to which to write: Fugazi, Beck, Busta Rhymes, Abba, The Cardiacs, Mercury Rev, Ben & Jason, Plug, Roni Size, Supergrass, Massive Attack, Elvis Costello, The Smashing Pumpkins, Radiohead, Placebo, Money Mark, Love, Hefner, Nick Cave, DJ Shadow, Elliott Smith, Muse, Arvo Pärt, AC/DC and Quasi. For this second edition, I think my playlist is somewhat indicative of the depths of mental anguish to which I sunk☺ The delightful tunes of the following saw me through: Emperor, Cradle of Filth, Frank Black and the Catholics, Blondie, Fugazi, Radiohead, Peter Gabriel, Genesis (Peter Gabriel era—had a bit of progressive rock phase around Chapter 5), Metallica, The White Stripes, Sevara Nazarkhan, Nusrat Fateh Ali Khan, Killing Joke, The Beyond, Jane's Addiction, Nevermore, The French and Hefner, Iron Maiden, The Mars Volta, Morrissey, Slipknot, Granddaddy, Mark Lanegan, PJ Harvey.

Finally, all this book-writing nonsense requires many lonely hours (often late at night) of typing. Without some wonderful friends to drag me out of my dimly lit room from time to time I'd be even more of a gibbering cabbage than I already am. In particular I'm grateful to Graham and Benie, Martin, Doug, Paul, Darren, Helen Liddle and Mark for reminding me that there is more to life than statistics and to Leonora for thinking that geeky psychologists who like statistics make good husband material☺

Dedication

Like the first edition, this book is dedicated to my brother Paul and my cat Fuzzy (who is the same cat as in the last edition but has reverted from his nick-name 'Beana' to the original name I gave him!), because one of them is a constant source of intellectual inspiration and the other wakes me up in the morning by sitting on me and purring in my face until I give him cat food: mornings will be considerably more pleasant when my brother gets over his love of cat food for breakfast☺

SYMBOLS USED IN THIS BOOK

Mathematical operators

Σ This symbol (called sigma) means 'add everything up'. So, if you see something like Σx_i it just means 'add up all of the scores you've collected'.

Π This symbol means 'multiply everything'. So, if you see something like Πx_i it just means 'multiply all of the scores you've collected'.

$\sqrt{x}$ This means 'take the square root of x'.

Greek symbols

α The probability of making a Type I error
β The probability of making a Type II error
β_i Standardized regression coefficient
χ^2 Chi-square test statistic
χ^2_F Friedman's ANOVA test statistic
ε Usually stands for 'error'
η^2 Eta squared (a measure of the size of an effect)
μ The mean of a population of scores
ρ The correlation in the population
σ^2 The variance in a population of data
σ The standard deviation in a population of data
σ_X The standard error of the mean
τ Kendall's tau (non-parametric correlation coefficient)
ω^2 Omega squared (effect size measure)

English symbols

b_i The regression coefficient (unstandardized)
df Degrees of freedom
e_i The error associated with the ith person
F F-ratio (test statistic used in ANOVA)

H	Kruskal–Wallis test statistic
k	The number of levels of a variable (i.e. the number of treatment conditions), or the number of predictors in a regression model
ln	Natural logarithm
MS	The mean squared error (mean square). The average variability in the data.
N, n, n_i	The sample size. N usually denotes the total sample size, whereas n usually denotes the size of a particular group
P	Probability (the probability value, p-value or significance of a test are usually denoted by p)
r	Pearson's correlation coefficient
r_b, r_{pb}	Biserial correlation coefficient and point-biserial correlation coefficient respectively
r_s	Spearman's rank correlation coefficient
R	The multiple correlation coefficient
R^2	The coefficient of determination (i.e. the proportion of variance within some data explained by the model)
s	The standard deviation of a sample of data
s^2	The variance of a sample of data
SS	The sum of squares or sum of squared errors to give it its full title
SS_A	The sum of squares for variable A
SS_M	The model sum of squares (i.e. the variability explained by the model fitted to the data)
SS_R	The residual sum of squares (i.e. the variability that the model can't explain—the error in the model)
SS_T	The total sum of squares (i.e. the total variability within the data)
t	Test statistic for student's t-test
T	Test statistic for Wilcoxon's matched-pairs signed-ranks test
U	Test statistic for the Mann–Whitney test
Ws	Test statistic for Wilcoxon's rank-sum test
$\bar{X}$ or $\bar{x}$	The mean of a sample of scores
z	A data point expressed in standard deviation units

Some maths revision

1. **Two negatives make a positive**: although in life two wrongs don't make a right, in mathematics they do! When we multiply a negative number with another negative number, the result is a positive number. For example, $-2 \times -4 = 8$.
2. **A negative number multiplied by a positive one make a negative number**: if you multiply a positive number by a negative number then the result is another negative number. For example, $2 \times -4 = -8$, or $-2 \times 6 = -12$.
3. **BODMAS**: this is an acronym for the order in which mathematical operations are performed. It stands for Brackets, Order, Division, Multiplication, Addition, Subtraction and this is the order in which you should carry out operations within an

equation. Mostly these operations are self-explanatory (e.g. always calculate things within brackets first) except for order, which actually refers to power terms such as squares. Four squared, or 4^2, used to be called four raised to the order of 2, hence why these terms are called 'order' in BODMAS (also, if we called it power, we'd end up with BPDMAS, which doesn't roll off the tongue quite so nicely). Let's look at an example of BODMAS: what would be the result of $1 + 3 \times 5^2$? The answer is 76 (not 100 as some of you might have thought). There are no brackets so the first thing is to deal with the order term: 5^2 is 25, so the expression becomes $1 + 3 \times 25$. There is no division, so we can move on to multiplication: 3×25, which gives us 75. BODMAS tells us to deal with addition next: $1 + 75$, which gives us 76 and the result is obtained. If I'd written the original expression as $(1 + 3) \times 5^2$, then the answer would have been 100 because we deal with the brackets first: $(1 + 3) = 4$, so the expression becomes 4×5^2. We then deal with the order term, so the expression becomes $4 \times 25 = 100$!

4. http://www.easymaths.com is a good site for revising basic maths.

PRAISE FOR THE FIRST EDITION

FROM STUDENTS:

Without sounding like I'm going to set up a church in your name, I think you ought to know how well written it is. It's so clear it's like seeing all the stats I've been supposedly studying for such a long time through a clean pair of glasses.
Megan Gray, University of Sussex

Your combination of the technical with the conceptual is just great. Your sense of humour doesn't hurt either!
Abigail Levy

You have an unusual ability to explain complex things in a simple, understandable way.
Ruth Mann

I would not have been able to finish my studies this easily if it weren't for your book.
Marleen Smits, University of Maastricht

Your book has made me a much happier student.
Lisa Oliver, University of British Columbia

I've never seen the concepts explained so well or illustrated so clearly.
Joel Philip

I defended my dissertation in psychology in March and your book SAVED MY LIFE!
Noelle Leonard

I think you'll go down in history as the first person ever to inject humour into a statistics textbook!
Carol McSweeney, Brookes University

You are the epitome of a good teacher! Although I haven't met you or sat in one of your lectures, your empathy and genuine desire to help students resonates in your book.
Johannah Sirkka

FROM ACADEMICS:

Last year I discovered your web pages (and thereby your books), and it completely transformed my teaching of basic statistics to the graduate students from 'okay' to 'great'. As a result, this year (and for the foreseeable future), I will be using your book as the chief text in the class.
Michael Marsiske, University of Florida

I have prescribed your book for one of my courses this year, and can't wait for my students to arrive. I have been teaching statistics for 25 years and used a lot of different prescribed books. But this one is the best. You gave me a new insight into things I had trouble with over all these years.
Henry Steel, University of Stellenbosch, South Africa

The way you've managed to bring the reader so quickly into the world of advanced statistics without a lot of technical jargon is quite remarkable.
Michael A. Karchmer, Director, Gallaudet Research Institute, Gallaudet University

Many thanks for writing such an excellent, inspiring book on a topic that is almost impossible to make interesting – something you have worked miracles on!
Dr Keith Tolfrey, Senior Lecturer, Department of Exercise & Sports Science, Manchester Metropolitan University

Since I discovered your discovery of statistics, I recommended your book to my colleagues and I guess that at least half of the department has your book on the shelf now. When somebody doesn't know what to do with his/her data we just say, look it up in Andy!
Froukje Dijk, University of Maastricht

CHAPTER 1

EVERYTHING YOU EVER WANTED TO KNOW ABOUT STATISTICS (WELL, SORT OF)

1.1. WHAT WILL THIS CHAPTER TELL US? ①

I realize that many people will use this book to 'dip in' to chapters that describe the statistical test that they've suddenly discovered they have to use. However, for any of those chapters to make sense you might find it useful to know a few important things about statistics. This chapter is an attempt to give you a brief overview of some important statistical concepts such as how we use statistical models to answer scientific questions.

1.2. BUILDING STATISTICAL MODELS ①

In the social sciences we are usually interested in discovering something about a phenomenon that we assume actually exists (a 'real-world' phenomenon). These real-world phenomena can be anything from the behaviour of interest rates in the economic market to the behaviour of undergraduates at the end-of-exam party. Whatever the phenomenon we desire to explain, we seek to explain it by collecting data from the real world, and then using these data to draw conclusions about what is being studied. As statisticians our job is to take the available data and to use them in a meaningful way and this often involves building statistical models of the phenomenon of interest.

The reason for building statistical models of real-world data is best explained by analogy. Imagine an engineer wishes to build a bridge across a river. That engineer would be pretty daft if she just built any old bridge, because the chances are that it would fall down.

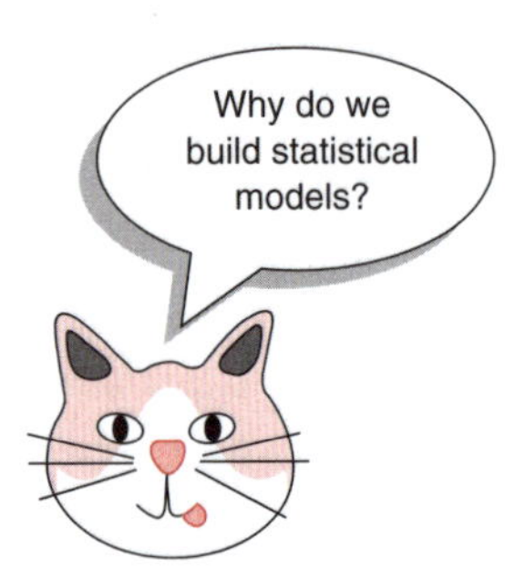

Instead, the engineer collects data from the real world: she looks at bridges in the real world and sees what materials they are made from, what structures they use and so on (she might even collect data about whether these bridges are damaged!). She then uses this information to construct a model. She builds a scaled-down version of the real-world bridge because it is impractical, not to mention expensive, to build the actual bridge itself. The model may differ from reality in several ways—it will be smaller for a start—but the engineer will try to build a model that best fits the situation of interest based on the data available. Once the model has been built, it can be used to predict things about the real world: for example, the engineer might test whether the bridge can withstand strong winds by placing the model in a wind tunnel. It seems obvious that it is important that the model is an accurate representation of the real world. Social scientists do much the same thing as engineers: we build models of real-world processes in an attempt to predict how these processes operate under certain conditions. We don't have direct access to the processes, so we collect data that represent the processes and then use these data to build statistical models (we reduce the process to a statistical model). We then use this statistical model to make predictions about the real-world phenomenon. Just like the engineer, we want our models to be as accurate as possible so that we can be confident that the predictions we make are also accurate. However, unlike engineers we don't have access to the real-world situation and so we can only ever *infer* things about psychological, societal or economic processes based upon the models we build. If we want our inferences to be accurate then the statistical model we build must represent the data collected (the *observed data*) as closely as possible. The degree to which a statistical model represents the data collected is known as the *fit* of the model and this is a term you will frequently come across.

Figure 1.1 illustrates the kinds of models that an engineer might build to represent the real-world bridge that she wants to create. The first model (a) is an excellent representation of the real-world situation and is said to be a *good fit* (i.e. there are a few small differences but the model is basically a very good replica of reality). If this model is used to make predictions about the real world, then the engineer can be confident that these predictions will be very accurate, because the model so closely resembles reality. So, if the model collapses in a strong wind, then there is a good chance that the real bridge would collapse also. The second model (b) has some similarities to the real world: the model includes some of the basic structural features, but there are some big differences from the real-world bridge (i.e. the absence of one of the supporting towers). This is what we might term a *moderate fit* (i.e. there are some differences between the model and the data but there are also some great similarities). If the engineer uses this model to make predictions about the real world then these predictions may be inaccurate and possibly catastrophic (e.g. if the bridge collapses in strong winds this could be due to the absence of a second supporting tower). So, using this model results in predictions that we can have some confidence in but not complete confidence. The final model (c) is completely different to the real-world situation. This model bears no structural similarities to the real bridge and so could be termed a poor fit (in fact, it might more accurately be described as an abysmal fit!). As such, any predictions based on this model are likely to be completely inaccurate. Extending this analogy to the social sciences we can say that it is important when we fit a

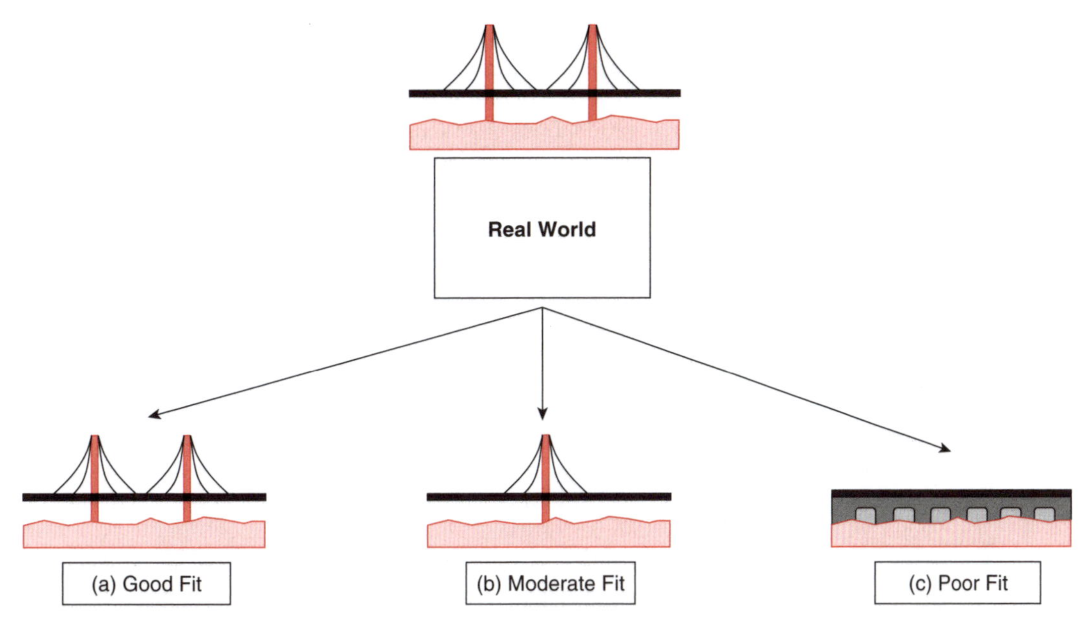

Figure 1.1 Fitting models to real-world data (see text for details)

statistical model to a set of data that this model fits the data well. If our model is a poor fit of the observed data then the predictions we make from it will be equally poor.

1.3. POPULATIONS AND SAMPLES ①

As researchers, we are interested in finding results that apply to an entire population of people or things. For example, psychologists want to discover processes that occur in all humans, biologists might be interested in processes that occur in all cells, economists want to build models that apply to all salaries and so on. A population can be very general (all human beings) or very narrow (all male ginger cats called Bob), but in either case scientists rarely, if ever, have access to every member of a population. Psychologists cannot collect data from every human being and ecologists cannot observe every male ginger cat called Bob. Therefore, we collect data from a small subset of the population (known as a sample) and use these data to infer things about the population as a whole. The bridge-building engineer cannot make a full-size model of the bridge she wants to build and so she builds a small-scale model and tests this model under various conditions. From the results obtained from the small-scale model the engineer infers things about how the full-sized bridge will respond. The small-scale model may respond differently to a full-sized version of the bridge, but the larger the model, the more likely it is to behave in the same way as the full-sized bridge. This metaphor can be extended to social scientists. We never have access to the entire population (the real-size bridge) and so we collect smaller samples (the scaled-down bridge) and use the behaviour within the sample to infer things about the behaviour in the population. The bigger the sample, the more likely it is to reflect the whole population. If we take several random samples from the population, each of these samples will give us slightly different results. However, on average, results from large samples should be fairly similar.

1.4. SIMPLE STATISTICAL MODELS ①

1.4.1. The mean, sums of squares, variance and standard deviations ①

One of the simplest models used in statistics is the mean. Some of you may have trouble thinking of the mean as a model, but in fact it is because it represents a summary of data. The **mean** is a hypothetical value that can be calculated for any data set; it doesn't have to be a value that is actually observed in the data set. For example, if we took five statistics lecturers and measured the number of friends that they had, we might find the following data: 1, 2, 3, 3 and 4. If we take the mean number of friends, this can be calculated by adding the values we obtained, and dividing by the number of values measured: $(1 + 2 + 3 + 3 + 4)/5 = 2.6$. Now, we know that it is impossible to have 2.6 friends (unless you chop someone up with a chainsaw and befriend their arm) so the mean value is a *hypothetical* value. As such, the mean is a model created to summarize our data. Now, we can determine whether this is an accurate model by looking at how different our real data are from the model that we have created. One way to do this is to look at the difference between the data we observed and the model fitted. Figure 1.2 shows the number of friends that each statistics lecturer had, and also the mean number that we calculated earlier on. The line representing the mean can be thought of as our model, and the circles are the observed data. The diagram also has a series of vertical lines that connect each observed value to the mean value. These lines represent the **deviance** between the observed data and our model and can be thought of as the error in the model. We can calculate the magnitude of these deviances by simply subtracting the mean value ($\bar{x}$) from each of the observed values (x_i).[1] For example, lecturer 1 had only 1 friend and so the difference is $x_1 - \bar{x} = 1 - 2.6 = -1.6$. You might notice that the deviance is a minus number, and this represents the fact that our model *overestimates* this lecturer's popularity: it predicts that he will have 2.6 friends yet in reality he has only 1 friend (bless him!). Now, how can we use these deviances to estimate the accuracy of the model? One possibility is to add up the deviances (this would give us an estimate of the total error). If we were to do this we would find that:

$$\begin{aligned}\text{total error} &= \text{sum of deviances}\\ &= \sum(x_i - \bar{x}) = (-1.6) + (-0.6) + (0.4) + (0.4) + (1.4) = 0\end{aligned}$$

1 The x_i simply refers to the observed score for the *i*th person (so, the *i* can be replaced with a number that represents a particular individual). For these data: for lecturer 1, $x_i = x_1 = 1$; for lecturer 3, $x_i = x_3 = 3$; for lecturer 5, $x_i = x_5 = 4$.

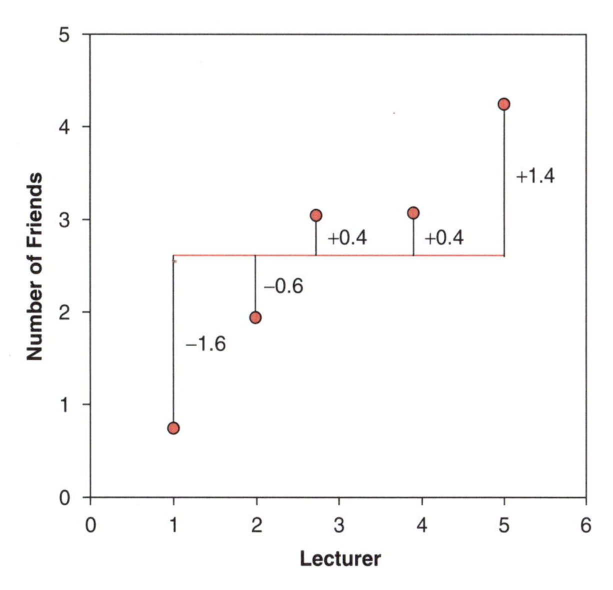

Figure 1.2 Graph showing the difference between the observed number of friends that each statistics lecturer had, and the mean number of friends (the red line)

In effect the result tells us that there is no total error between our model and the observed data, so the mean is a perfect representation of the data. Now, this clearly isn't true: there were errors but some of them were positive, some were negative and they have simply cancelled each other out. It is clear that we need to avoid the problem of which direction the error is in and one mathematical way to do this is to square each error,[2] that is multiply each error by itself. So, rather than calculating the sum of errors, we calculate the sum of squared errors. In this example:

$$\begin{aligned}\text{sum of squared errors (SS)} &= \sum (x_i - \bar{x})(x_i - \bar{x}) \\ &= (-1.6)^2 + (-0.6)^2 + (0.4)^2 + (0.4)^2 + (1.4)^2 \\ &= 2.56 + 0.36 + 0.16 + 0.16 + 1.96 \\ &= 5.20\end{aligned}$$

The sum of squared errors (SS) is a good measure of the accuracy of our model. However, it is fairly obvious that the sum of squared errors is dependent upon the amount of data that has been collected—the more data points the higher the SS. To overcome this problem we calculate the average error by dividing the SS by the number of observations (*N*). If we are interested only in the average error for the sample, then we can divide by

2 When you multiply a negative number by itself it becomes positive.

N alone. However, we are generally interested in using the error in the sample to estimate the error in the population and so we divide the SS by the number of observations minus 1 (the reason why is explained in Box 8.2). This measure is known as the ***variance*** and is a measure that we will come across a great deal:

$$\text{variance } (s^2) = \frac{\text{SS}}{N-1} = \frac{\sum(x_i - \bar{x})^2}{N-1} = \frac{5.20}{4} = 1.3$$

The variance is, therefore, the average error between the mean and the observations made (and so is a measure of how well the model fits the actual data). There is one problem with the variance as a measure: it gives us a measure in units squared (because we squared each error in the calculation). In our example we would have to say that the average error in our data (the variance) was 1.3 friends squared. It makes little enough sense to talk about 1.3 friends, but it makes even less to talk about friends squared! For this reason, we often take the square root of the variance (which ensures that the measure of average error is in the same units as the original measure). This measure is known as the standard deviation and is simply the square root of the variance. In this example the standard deviation is:

$$\begin{aligned} s &= \sqrt{\frac{\sum(x_i - \bar{x})^2}{N-1}} \\ &= \sqrt{1.3} \\ &= 1.14 \end{aligned}$$

The standard deviation is, therefore, a measure of how well the mean represents the data. Small standard deviations (relative to the value of the mean itself) indicate that data points are close to the mean. A large standard deviation (relative to the mean) indicates that the data points are distant from the mean (i.e. the mean is not an accurate representation of the data). A standard deviation of 0 would mean that all of the scores were the same. Figure 1.3 shows the overall ratings (on a five-point scale) of two lecturers after each of five different lectures. Both lecturers had an average rating of 2.6 out of 5 across the lectures. However, the first lecturer had a standard deviation of 0.55 (relatively small compared to the mean). It should be clear from the graph that ratings for this lecturer were consistently close to the mean rating. There was a small fluctuation, but generally his lectures did not vary in popularity. As such, the mean is an accurate representation of his ratings. The mean is a good fit of the data. The second lecturer, however, had a standard deviation of 1.82 (relatively high compared to the mean). The ratings for this lecturer are clearly more spread from the mean; that is, for some lectures he received very high ratings, and for others his ratings were appalling. Therefore, the mean is not such an accurate representation of his performance because there was a lot of variability in the popularity of his lectures. The mean is a poor fit of the data. This illustration should hopefully make clear why the standard deviation is a measure of how well the mean represents the data.

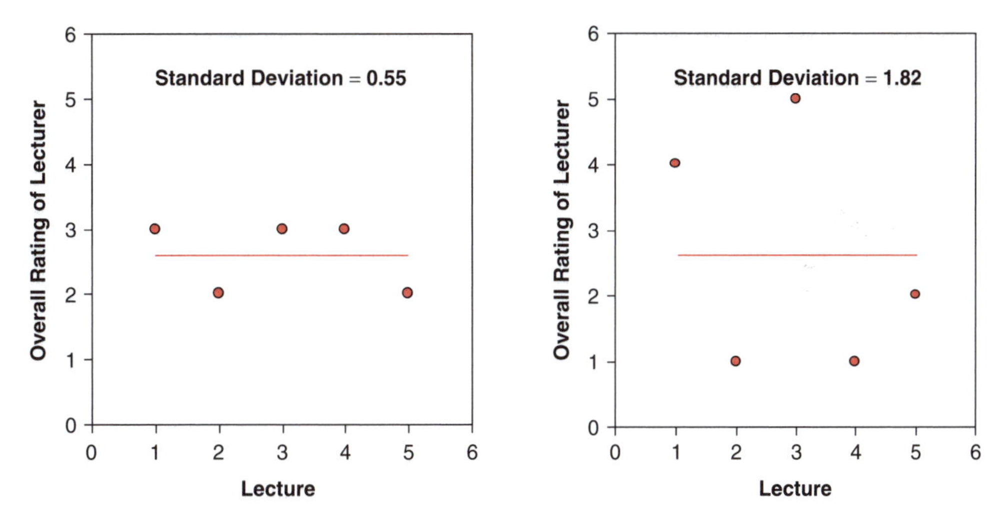

Figure 1.3 Graphs illustrating data that have the same mean but different standard deviations

The discussion of means, sums of squares and variance may seem a side-track from the initial point, but in fact the mean is probably one of the simplest statistical models that can be fitted to data. What do I mean by this? Well, everything in statistics essentially boils down to one equation:

$$\text{Outcome}_i = (\text{Model}_i) + \text{error}_i \tag{1.1}$$

This just means that the data we observe can be predicted from the model we choose to fit to the data plus some amount of error. With the mean, when I say that the mean is a simple statistical model, then all I mean is that we can replace the word 'Model' with the word 'mean' in that equation. If we return to our example involving the number of friends that statistics lecturers have and look at lecturer 1, for example, we observed that he had one friend and the mean of all lecturers was 2.6. So, the equation becomes:

$$\begin{aligned} \text{Outcome}_{\text{lecturer 1}} &= \overline{X} + \varepsilon_{\text{lecturer 1}} \\ 1 &= 2.6 + \varepsilon_{\text{lecturer 1}} \end{aligned}$$

From this we can work out that the error is 1 – 2.6, or –1.6. If we replace this value into the equation we get 1 = 2.6 – 1.6 or 1 = 1. Although it probably seems like I'm stating the obvious, it is worth bearing this general equation in mind throughout this book because if you do you'll discover that most things ultimately boil down to this one simple idea!

Likewise, the variance and standard deviation illustrate another fundamental concept: how the goodness-of-fit of a model can be measured. If we're looking at how well a model fits the data (in this case our model is the mean) then we generally look at deviation from the model, we look at the sum of squared error, and in general terms we can write this as:

$$\text{deviation} = \sum(\text{observed} - \text{model})^2 \tag{1.2}$$

Put another way, we assess models by comparing the data we observe to the model we've fitted to the data, and then square these differences. Again, you'll come across this fundamental idea time and time again throughout this book.

1.5. FREQUENCY DISTRIBUTIONS ①

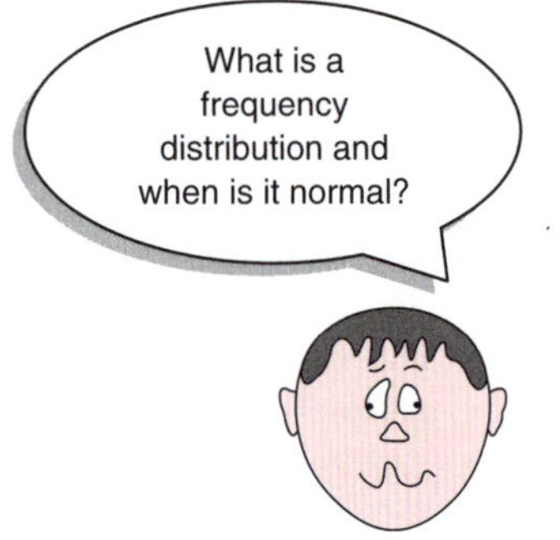

Once you've collected some data a very useful thing to do is to plot a graph of how many times each score occurs. This is known as a frequency distribution, or histogram, which is simply a graph plotting values of observations on the horizontal axis, with a bar showing how many times each value occurred in the data set. Frequency distributions can be very useful for assessing properties of the distribution of scores. For one thing, by looking at which score has the tallest bar, we can immediately see the mode, which is simply the score that occurs most frequently in the data set.

Frequency distributions come in many different shapes and sizes. It is quite important, therefore, to have some general descriptions for common types of distributions. In an ideal world our data would be distributed symmetrically around the centre of all scores. As such, if we drew a vertical line through the centre of the distribution then it should look the same on both sides. This is known as a normal distribution and is characterized by the bell-shaped curve with which you might already be familiar. This shape basically implies that the majority of scores lie around the centre of the distribution (so the largest bars on the histogram are all around the central value). Also, as we get further away from the centre the bars get smaller, implying that as scores start to deviate from the centre their frequency is decreasing. As we move still further away from the centre our scores become very infrequent (the bars are very short). An example of a normal distribution is shown in Figure 1.4.

1.5.1. Properties of frequency distributions ①

There are two main ways in which a distribution can deviate from normal: (1) lack of symmetry (called skewness) and (2) pointyness (called kurtosis). Skewed distributions are not symmetrical and instead the most frequent scores (the tall bars on the graph) are clustered at one end of the scale. So, the typical pattern is a cluster of frequent scores at one end of the scale and the frequency of scores tailing off towards the other end of

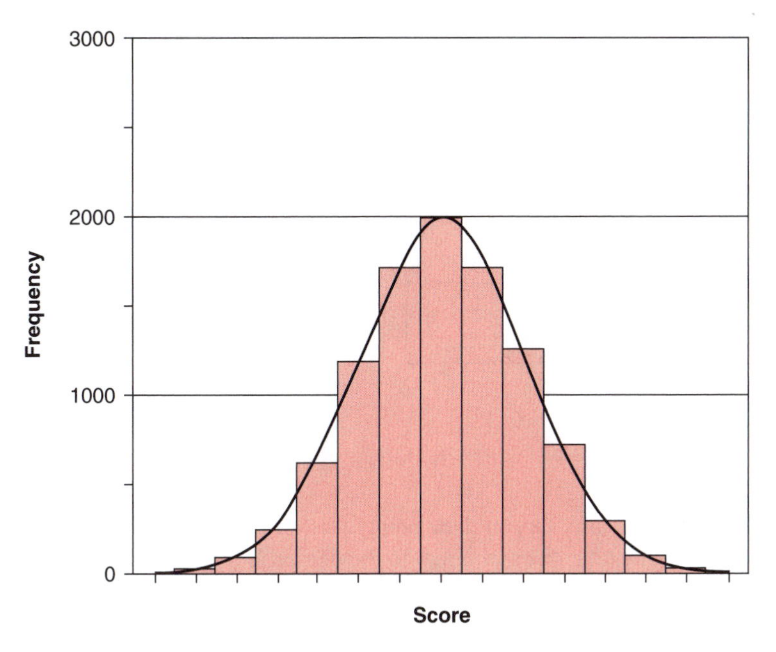

Figure 1.4 A 'normal' distribution (the curve shows the idealized shape)

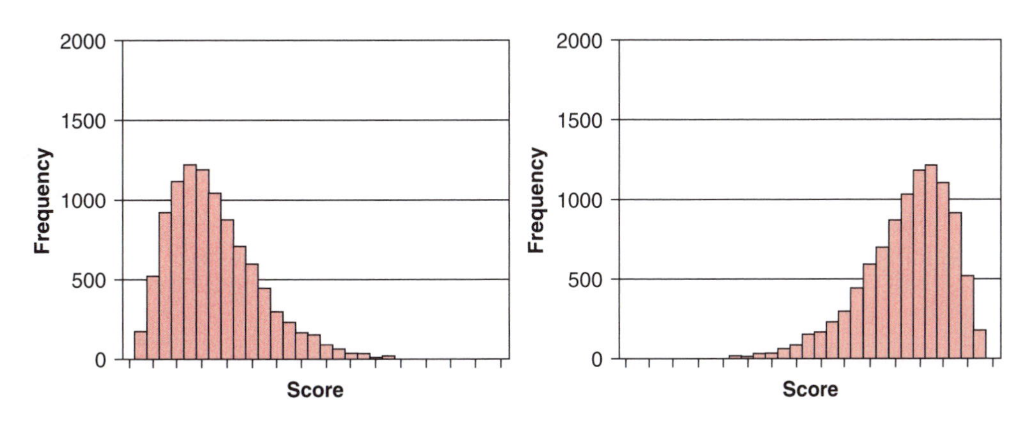

Figure 1.5 A positively (left) and negatively (right) skewed distribution

the scale. A skewed distribution can be either *positively skewed* (the frequent scores are clustered at the lower end and the tail points towards the higher or more positive scores) or *negatively skewed* (the frequent scores are clustered at the higher end of and the tail points towards the lower more negative scores). Figure 1.5 shows examples of these distributions.

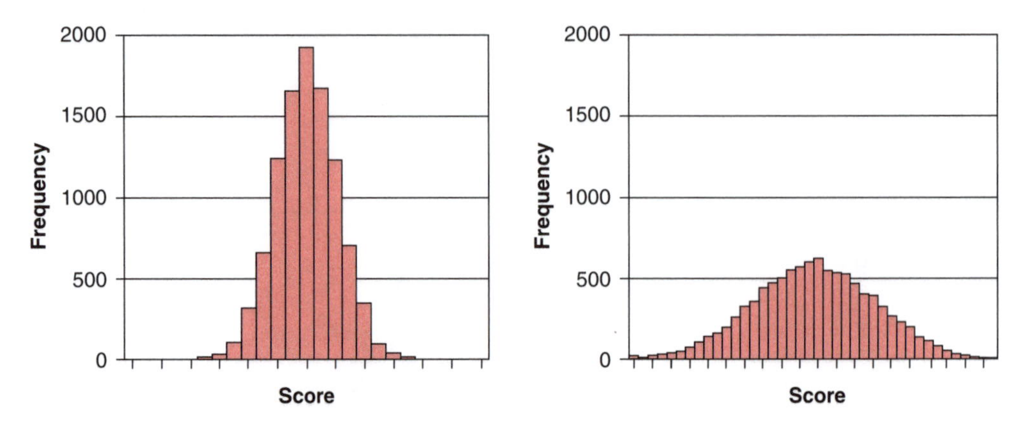

Figure 1.6 A leptokurtic (left) and platykurtic (right) distribution

Distributions also vary in their pointyness, or kurtosis. Kurtosis, despite sounding like some kind of exotic disease, refers to the degree to which scores cluster in the tails of the distribution. This characteristic is usually shown up by how flat or pointy a distribution is. A platykurtic distribution is one that has many scores in the tails (a so-called heavy-tailed distribution) and so is quite flat. In contrast, leptokurtic distributions are relatively thin in the tails and so look quite pointy. You can remember this by 'the *lept*okurtic distribution *leapt* up in the air, and the *plat*ykurtic distribution was flat like a *plat*eau' (Figure 1.6), or the surrealists amongst you can remember it as 'the platykurtic distribution had a duck-bill and the leptokurtic distribution compulsively stole things'. Ideally, we want our data to be normally distributed (i.e. not too skewed, and not too pointy or flat!).

In a normal distribution the values of skew and kurtosis are 0 (i.e. the distribution is neither too pointy, nor too flat, and is perfectly symmetrical). If a distribution has values of skew or kurtosis above or below 0 then this indicates a deviation from normal (see Chapter 3).

1.5.2. The standard deviation and the shape of the distribution ①

As well as telling us about the accuracy of the mean as a model of our data set, the variance and standard deviation also tell us about the shape of the distribution of scores. If the mean represents the data well then most of the scores will cluster close to the mean and the resulting standard deviation is small relative to the mean. When the mean is a worse representation of the data, the scores cluster more widely around the mean (think back to Figure 1.3) and the standard deviation is larger. Figure 1.7 shows two distributions that have the same mean (50) but different standard deviations. One has a large standard deviation relative to the mean (SD = 25) and this results in a flatter distribution that is more spread out, whereas the other has a small standard deviation relative to the mean (SD = 15) resulting in a more pointy distribution in which scores close to the mean are very frequent

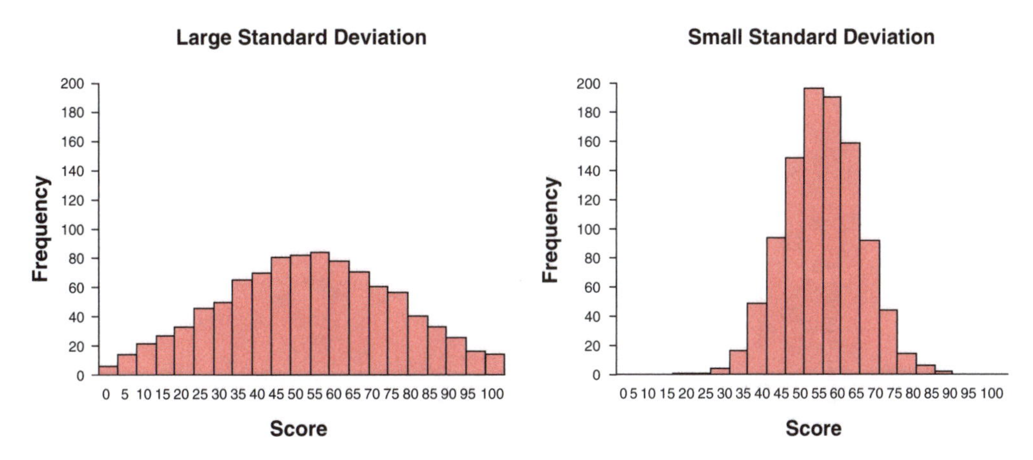

Figure 1.7 Two distributions with the same mean, but large and small standard deviations

but scores further from the mean become increasingly infrequent. The main message is that as the standard deviation gets larger, the distribution gets fatter.

1.5.3. What is the standard normal distribution? ①

Another way to look at frequency distributions is in terms of probability: they give us some idea of how likely a given score is to occur. Beachy Head is a large, windy cliff on the Sussex coast (not far from where I live) that has something of a reputation for attracting suicidal people, who seem to like throwing themselves off of it (and after several months of rewriting this book I find my thoughts drawn towards that peaceful chalky cliff top more and more often). Figure 1.8 shows a frequency distribution of some completely made up data of the number of suicides at Beachy Head in a year by people of different ages (although I made these data up they are roughly based on general suicide statistics such as those in Williams, 2001). There were 171 suicides in total and you can see that the suicidal lemmings were most frequently aged between about 25 and 30 (the highest bar), which is quite worrying for me (although it does mean that if I can just keep going for this year the urges will decrease next year!). The graph also tells us that, for example, very few people aged above 70 committed suicide at Beachy Head.

I said earlier that we could think of frequency distributions in terms of probability. To explain this, imagine that someone asked you 'how likely is it that a 70 year old committed suicide at Beachy Head?'; what would your answer be? The chances are that if you looked at the frequency distribution you might respond 'not very likely' because you can see a couple of people out of the 171 suicides were aged around 70. What about if someone asked 'how likely is it that a 30 year old committed suicide?' Again, by looking at the graph, you might say 'it's actually quite likely' because 32 out of the 171 suicides were by people aged around 30 (that's more than one in every five people that committed suicide).

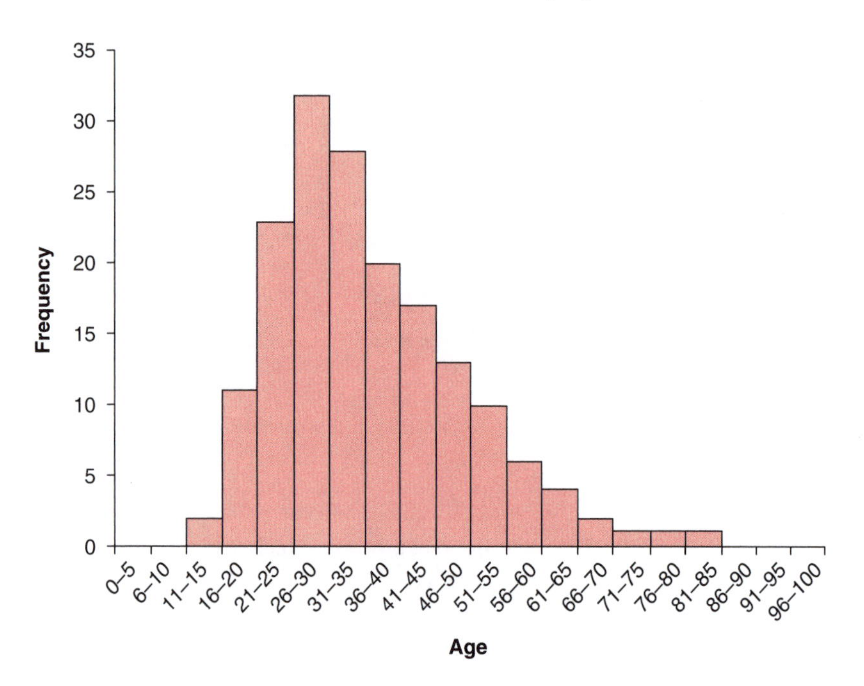

Figure 1.8 Frequency distribution showing the number of suicides at Beachy Head in year by age

So based on the frequencies of different scores it should start to become clear that we could use this information to estimate the probability that a particular score will occur. So, we could ask, based on our data, what's the probability of a suicide victim being aged 16–20? A probability value can range from 0 (there's no chance whatsoever of the event happening) to 1 (the event will definitely happen). So, for example, when I talk to my publishers I tell them there's a probability of 1 that I will have completed the revisions to this book by June 2003. However, when I talk to anyone else, I might, more realistically, tell them that there's a 0.1 probability of me finishing the revisions on time (or put another way, a 10% chance, or a 1 in 10 chance that I'll complete the book in time). In reality, the probability of me meeting the deadline is 0 (not a chance in hell) because I never manage to meet publishers' deadlines! If probabilities don't make sense to you then just ignore the decimal point and think of them as percentages instead (i.e. 0.1 probability that something will happen = 10% chance that something will happen).

I've talked in vague terms about how frequency distributions can be used to get a rough idea of the probability of a score occurring. However, we can be precise. For any distribution of scores we could, in theory, calculate the probability of obtaining a score of a certain size—it would be incredibly tedious and complex to do it, but we could. To spare our sanity, statisticians have identified several common distributions. For each one they have worked out mathematical formulae that specify idealized versions of these distributions (they are specified in terms of a curved line). These idealized distributions are known as probability distributions and from these distributions it is possible to calculate the probability of getting particular scores based on the frequencies with which a particular score occurs in a distribution with these common shapes. One of these 'common' distributions is the normal distribution, which I've already mentioned. Statisticians have calculated the probability of certain scores occurring in a normal distribution with a mean of 0 and a standard deviation of 1. Therefore, if we have any data that are shaped like a normal distribution then if the mean and standard deviation are 0 and 1 respectively we can use the tables of probabilities for the normal distribution to see how likely it is that a particular score will occur in the data (I've produced such a table in the Appendix to this book).

The obvious problem is that not all of the data we collect will have a mean of 0 and standard deviation of 1! For example, we might have a data set that has a mean of 567 and a standard deviation of 52.98! Luckily any data set can be converted into a data set that has a mean of 0 and a standard deviation of 1. First, to centre the data at zero, we take each score and subtract from it the mean of all scores. Then, we divide the resulting score by the standard deviation to ensure the data have a standard deviation of 1. The resulting scores are known as z-scores and, in equation form, the conversion that I've just described is:

$$z = \frac{X - \bar{X}}{s}$$

The table of probability values that have been calculated for the standard normal distribution is shown in the Appendix. Why is this table important? Well, if we look at our suicide data, we can answer the question 'what's the probability that someone who threw themselves off of Beachy Head was aged 70 or older?' First we convert 70 into a z-score. Say the mean of the suicide scores was 36 and the standard deviation 13; the 70 will become $(70 - 36)/13 = 2.62$. We then look up this value in the column labelled 'smaller portion' (i.e. the area above the value 2.62). We should find that the probability is .0044, or, put another way, only a 0.44% chance that a suicide victim would be 70 years old or more. By looking at the column labelled 'bigger portion' we can also see the probability that a suicide victim was aged 70 or less! This probability is .9956, or, put another way, there's a 99.56% chance that a suicide victim was less than 70 years old!

Hopefully you can see from these examples that the normal distribution and z-scores allow us to go a first step beyond our data in that from a set of scores we can calculate the probability that a particular score will occur. So, we can see whether scores of a certain size are likely or unlikely to occur in a distribution of a particular kind. You'll see just how useful this is in due course, but it is worth mentioning at this stage that certain z-scores are particularly important. This is because their value cuts off certain important percentages of the distribution. The first important value of z is 1.96 because this cuts off the top 2.5% of the distribution, and its counterpart at the opposite end (−1.96) cuts off the bottom 2.5% of the distribution. As such, taken together, these values cut off 5%

of scores, or, put another way, 95% of *z*-scores lie between −1.96 and 1.96. The other two important benchmarks are ±2.58 and ±3.29, which cut off 1% and 0.1% of scores respectively. Put another way, 99% of *z*-scores lie between −2.58 and 2.58, and 99.9% of them lie between −3.29 and 3.29. Remember these values because they'll crop up time and time again!

1.6. IS MY SAMPLE REPRESENTATIVE OF THE POPULATION? ①

1.6.1. The standard error ①

We've seen that the standard deviation and frequency distributions tell us something about how well the mean represents that data, but I mentioned earlier on that usually we collect data from samples rather than the entire population. I also mentioned that if you take several samples from a population, then these samples would differ slightly. Therefore, it's also important to know how well a particular sample represents the population. This is where we use the standard error. Many students get confused about the difference between the standard deviation and the standard error (usually because the difference is never explained clearly). However, the standard error is an important concept to grasp, so I'll do my best to explain it to you.

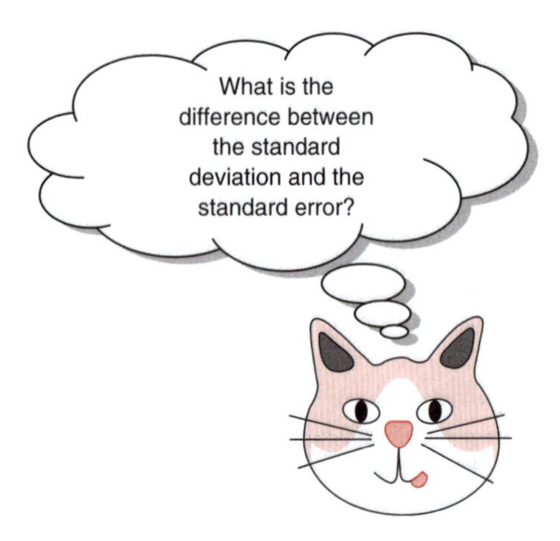

We have already learnt that social scientists use samples as a way of estimating the behaviour in a population. Imagine that we were interested in the ratings of all lecturers (so, lecturers in general were the population). We could take a sample from this population. When someone takes a sample from a population, they are taking one of many possible samples. If we take several samples from the same population, then each sample has its own mean, and some of these sample means will be different (not every sample will have the same mean).

Figure 1.9 illustrates the process of taking samples from a population. Imagine that we could get ratings of all lecturers on the planet and that on average, the rating is 3 (this is the *population mean*). Of course, we can't collect ratings of all lecturers, so we use a sample. For each of these samples we can calculate the average, or *sample mean*. Let's imagine we took nine different samples (as in the diagram); you can see that some of the samples have the same mean as the population but some have different means: the first sample of lecturers were rated as 3, but the second sample were, on average, rated only 2. This illustrates *sampling variation*: that is, samples will vary because they contain different members of the population; a sample that by chance includes some very good lecturers will have a higher average than a sample that, by chance, includes some awful lecturers! We can actually plot the sample means as a frequency distribution, or histogram,[3] just like I have done in the diagram. This distribution shows that there were three samples that had a mean of 3, means of 2 and 4 occurred in two samples each, and means of 1 and 5 occurred in only one sample each. The end result is a nice symmetrical distribution known as a sampling distribution. A sampling distribution is simply the frequency distribution of sample means from the same population. In theory we'd take hundreds or thousands of samples to construct a sampling distribution, but I'm just using nine to keep the diagram simple! The sampling distribution tells us about the behaviour of samples from the population, and you'll notice that it is centred at the same value as the mean of the population (i.e. 3). This means that if we took the average of all sample means we'd get the value of the population mean. Now, if the average of the sample means is the same value as the population mean, then if we know the accuracy of that average then we'd know something about how likely it is that a given sample is representative of the population. So how do we determine the accuracy of the population mean?

Think back to the discussion of the standard deviation. We used the standard deviation as a measure of how representative the mean was of the observed data. Small standard deviations represented a scenario in which most data points were close to the mean, a large standard deviation represented a situation in which data points were widely spread from the mean. If you were to calculate the standard deviation between *sample means* then this too would give you a measure of how much variability there was between the means of different samples. The standard deviation of sample means is known as the *standard error of the mean (SE)*. Therefore, the standard error could be calculated by taking the difference between each sample mean and the overall mean, squaring these differences, adding them up, and then dividing by the number of samples.

Of course, in reality we cannot collect hundreds of samples and so we rely on approximations of the standard error (luckily for us lots of clever statisticians have calculated ways in which the standard error can be worked out from the sample standard deviation). Luckily, we don't need to understand why this approximation works. We can just trust that these people are ridiculously clever and know what they're talking about. The standard error can be calculated by dividing the sample standard deviation (s) by the square root of the sample size (N):[4]

$$\sigma_{\overline{X}} = \frac{s}{\sqrt{N}}$$

---

3 This is just a graph of each sample mean plotted against the number of samples that has that mean—see Chapter 2 for more details.

4 In fact it should be the *population* standard deviation (σ) that is divided by the square root of the sample size; however, for large samples this is a reasonable approximation.

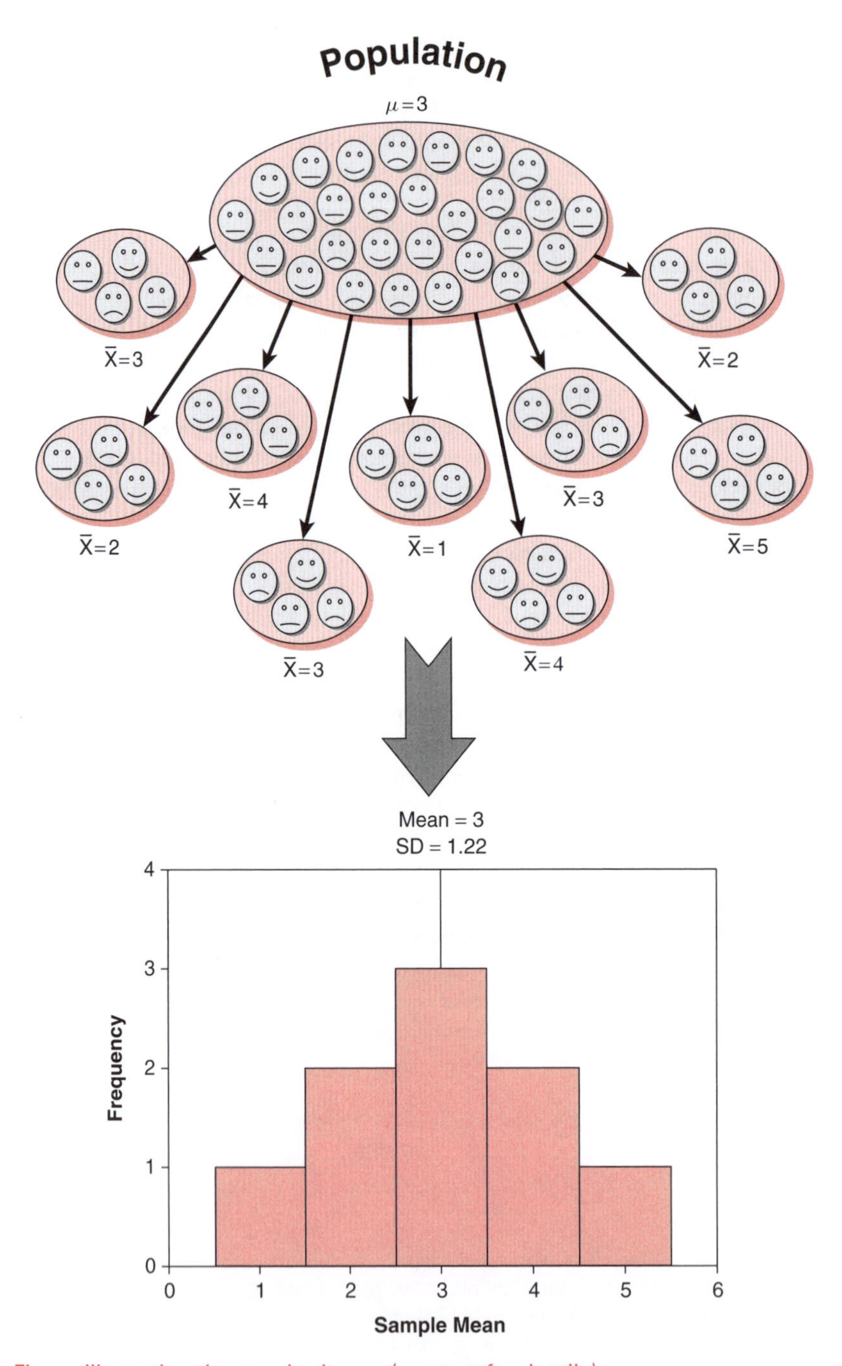

Figure 1.9 Figure illustrating the standard error (see text for details)

Cramming Samantha's Tip

To sum up, the standard error is the standard deviation of sample means. As such, it is a measure of how representative a sample is likely to be of the population. A large standard error (relative to the sample mean) means that there is a lot of variability between the means of different samples and so the sample we have might not be representative of the population. A small standard error indicates that most sample means are similar to the population mean and so our sample is likely to be an accurate reflection of the population.

1.6.2. Confidence intervals ②

Remember that usually we're interested in using the sample mean as an estimate of the true value of the mean (i.e. the value in the population). We've just seen that different samples will give rise to different values of the mean, and we can use the standard error to get some idea of the extent to which sample means differ. A different approach to assessing the accuracy of the sample mean as an estimate of the mean in the population is to calculate boundaries within which we believe the true value of the mean will fall. Such boundaries are called confidence intervals. The basic idea behind confidence intervals is to construct a range of values within which we think the population value falls.

Let's imagine an example: Domjan, Blesbois & Williams (1998) examined the learnt release of sperm in Japanese quail. The basic idea is that if a quail is allowed to copulate with a female quail in a certain context (an experimental chamber) then this context will serve as a cue to copulation and this in turn will affect semen release (although during the test phase the poor quail were tricked into copulating with a terry cloth with an embalmed female quail head stuck on top — sick eh?). Anyway, if we look at the mean amount of sperm released in the experimental chamber, there is a true mean (the mean in the population); let's imagine it's 15 million sperm. Now, in our actual sample, we might find the mean amount of sperm released was 17 million. Because we don't know the true mean, we don't really know whether our sample value of 17 million is a good or bad estimate of this value. What we can do instead is use an interval estimate: we use our sample value as the mid-point, but set a lower and upper limit as well. So, we might say, we think the true value of the mean sperm release is somewhere between 12 million and 22 million spermatozoa (note 17 million falls exactly between these values). Of course, in this case the true value (15 million) does fall within these limits. However, what if we'd set smaller limits, what if we'd said we think the true value falls between 16 and 18 million (again, note that 17 million is in the middle)? In this case the interval does not contain the true value of the mean. Let's now imagine that you were particularly fixated with Japanese quail sperm, and you repeated the experiment 50 times using different samples. Each time you did the experiment again you constructed an interval around the sample mean as I've just described. Figure 1.10 shows this scenario: the circles represent the mean for each sample with the lines sticking out of them representing the intervals for these means. The true value of the mean (the mean in the population) is 15 million and is shown by a vertical line. The first thing to note is that most of the sample means are different from the true mean (this is because of sampling variation as described in the previous section). Second, although most of the intervals do contain the true mean (they cross the vertical line, meaning that 15 million spermatozoa falls somewhere between the lower and upper boundaries), a few do not.

Up until now I've avoided the issue of how we might calculate the intervals. The crucial thing with confidence intervals is to construct them in such a way that they tell us something useful. Therefore, we calculate them so that they have certain properties: in particular they tell us the likelihood that they contain the true value of the thing we're trying to estimate (in this case, the mean).

Typically we look at 95% confidence intervals, and sometimes 99% confidence intervals, but they all have a similar interpretation: they are limits constructed such that a certain percentage of the time (be that 95% or 99%) the true value of the population mean will fall within these limits. So, when you see a 95% confidence interval for a mean, think of it like this: if we'd collected 100 samples, calculated the mean and then calculated a confidence interval for that mean (a bit like in Figure 1.10) then for 95 of these samples, the confidence intervals we constructed would contain the true value of the mean in the population.

To calculate the confidence interval, we need to know the limits within which 95% of means will fall. How do we calculate these limits? Remember back in section 1.5.3 that I said that 1.96 was an important value of z (a score from a normal distribution with a mean of 0 and standard deviation of 1) because 95% of z-scores fall between −1.96 and 1.96. This means that if our sample means had a mean of 0 and a standard error of 1, then the limits of our confidence interval would be −1.96 and +1.96. You might also remember that we can convert scores into z-scores using this equation:

$$z = \frac{X - \overline{X}}{s}$$

If we know that our limits would be −1.96 and 1.96 in z-scores, then what are the corresponding scores in our raw data? To find this out, we can replace z in the equation (because there are two values, we get two equations):

$$1.96 = \frac{X - \overline{X}}{s} \qquad -1.96 = \frac{X - \overline{X}}{s}$$

What we need to know is the value of X in these equations, and to discover this we simply rearrange them:

$$\begin{aligned} 1.96 \times s &= X - \overline{X} & \qquad -1.96 \times s &= X - \overline{X} \\ (1.96 \times s) + \overline{X} &= X & \qquad (-1.96 \times s) + \overline{X} &= X \end{aligned}$$

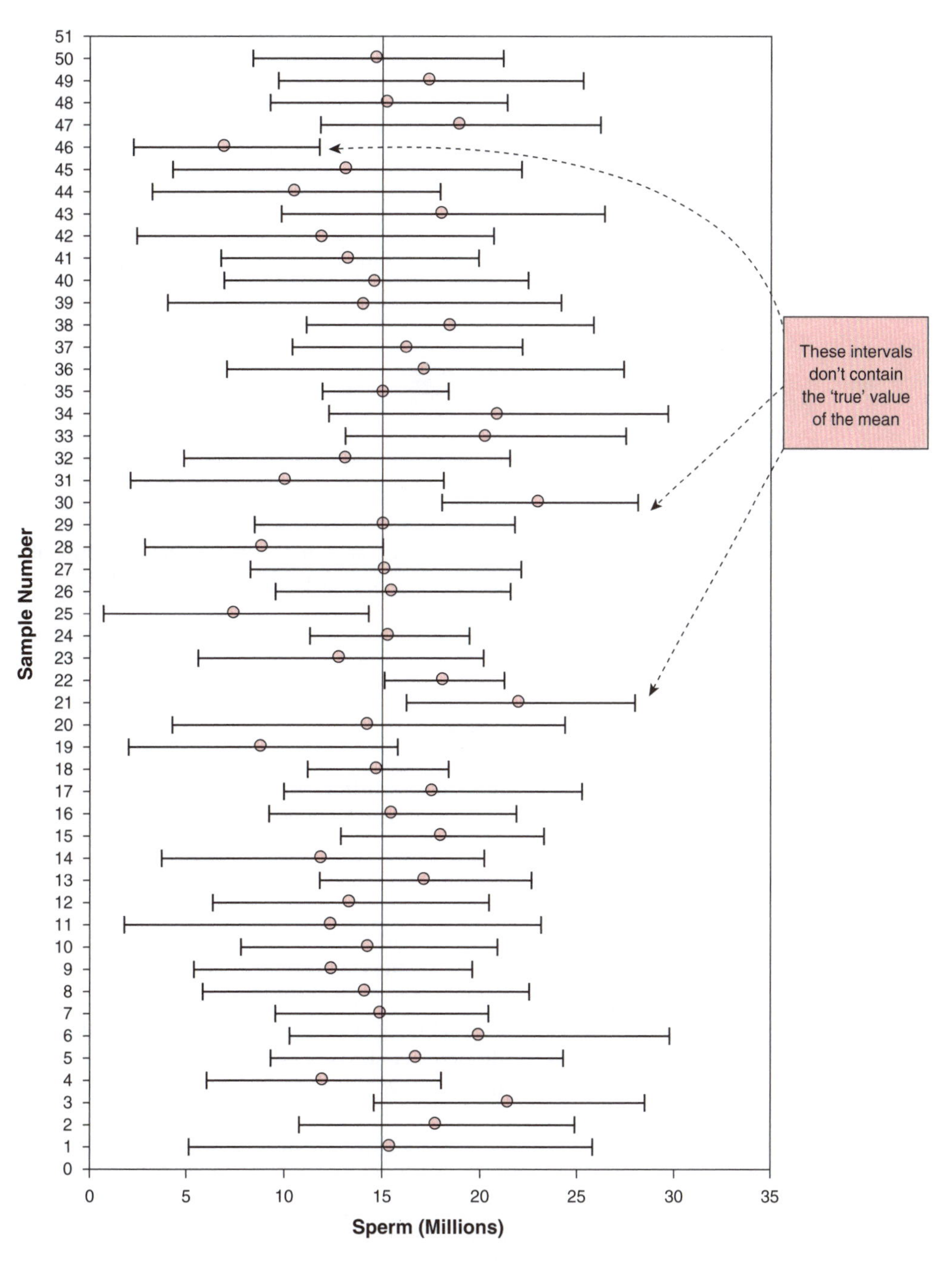

Figure 1.10 Figure showing confidence intervals of the sperm counts of Japanese quail (horizontal axis) for 50 different samples (vertical axis)

Therefore, the confidence interval can easily be calculated once the standard deviation (s in the equations above) and mean ($\overline{X}$ in the equations) are known. However, in fact we use the standard error and not the standard deviation because we're interested in the variability of sample means, not the variability in observations within the sample. The lower boundary of the confidence interval is, therefore, the mean minus 1.96 times the standard error, and the upper boundary is the mean plus 1.96 standard errors.

$$\text{Lower boundary of confidence interval} = \overline{X} - (1.96 \times \text{SE})$$

$$\text{Upper boundary of confidence interval} = \overline{X} + (1.96 \times \text{SE})$$

As such, the mean is always in the centre of the confidence interval. If the mean accurately represents the true mean, then the confidence interval of that mean should be small: because 95% of confidence intervals contain the true mean, we can assume that this confidence interval contains the true mean; therefore, if the interval is small, the sample mean must be very close to the true mean. Conversely, if the confidence interval is very wide then the sample mean could be very different from the true mean, indicating that it is a bad representation of the population. You'll find that confidence intervals will come up time and time again throughout this book.

1.7. LINEAR MODELS ①

The mean is an example of a statistical model, but you may well ask what other kinds of statistical models can be built. Well, if the truth is known there is only one model that is generally used, and this is known as the linear model. To some social scientists it may not be entirely obvious that my previous statement is correct, yet a statistician would acknowledge my sentiment much more readily. The reason for this is that there are a variety of different names given to statistical procedures that are based on the linear model. A classic example is that analysis of variance (ANOVA) and regression are identical systems (Cohen, 1968), yet they have different names and are used largely in different contexts (due to a divide in methodological philosophies—see Cronbach, 1957).

The word *linear* literally means 'relating to a line' but in statistical terms the line to which it refers is a straight one. A linear model is, therefore, a model that is based upon a straight line; this means that we are usually trying to summarize our observed data in terms of a straight line. For example, in the chapter describing regression, it will become clear that two variables can be negatively related (this just means that as values of one variable increase, values of the other variable decrease). In such circumstances, the relationship may be summarized by a straight line.

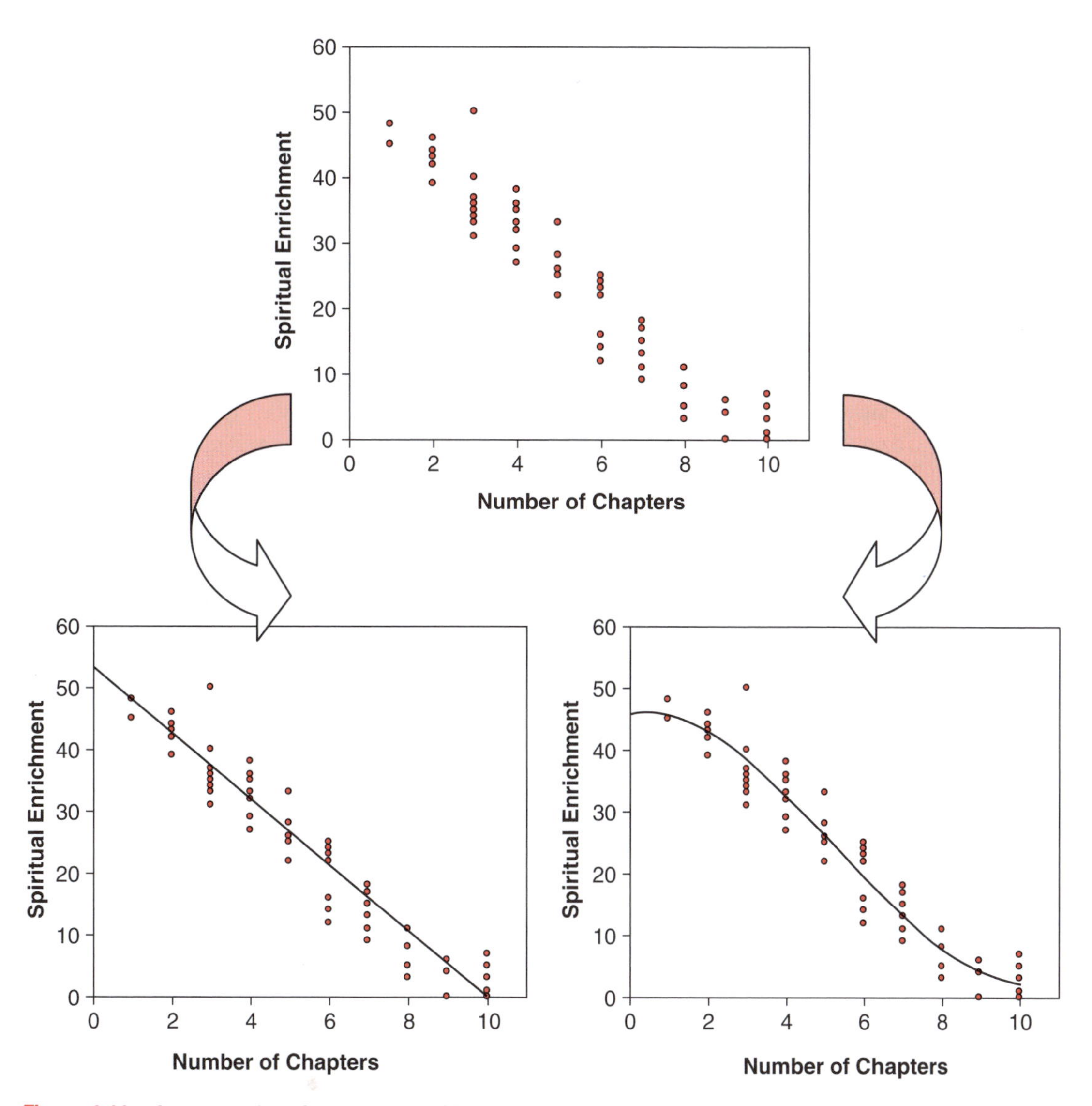

Figure 1.11 A scatterplot of some data with no model fitted to the data, with a linear model fitted, and with a non-linear model fitted

Suppose we measured how many chapters of this book a person had read, and then measured their spiritual enrichment; we could represent these hypothetical data in the form of a scatterplot in which each dot represents an individual's score on both variables. Figure 1.11 shows such a graph, and also shows the same graph but with a line that summarizes the pattern of these data. A third version of the scatterplot is also included but has a curved line to summarize the general pattern of the data. As such, Figure 1.11 illustrates how we can fit different types of models to the same data. In this case we can use a straight line to represent our data and it shows that the more chapters a person reads, the less their spiritual enrichment. However, we can also use a curved line to summarize the data and this shows that when most, or all, of the chapters have been read, spiritual enrichment seems to increase slightly (presumably because once the book is read everything suddenly makes sense—yeah, as if!). Neither of the two types of model is necessarily correct, but it will be the case that one model fits the data better than another and this is why when we use statistical models it is important for us to assess how well a given model fits the data.

Most of the statistics used in the social sciences are based on linear models, which means that we try to fit straight line models to the data collected. This is interesting because most published scientific studies are ones with statistically significant results. Given that most social scientists are only ever taught how to use techniques based on the linear model, published results will be those that have successfully used linear models. Data that fit a non-linear pattern are likely to be wrongly ignored (because the wrong model will have been applied to the data, leading to non-significant results). This is why good researchers plot their data first: plots tell you a great deal about what models should be applied to data. It is possible, therefore, that some areas of science are progressing in a biased way, so, if you collect data that look non-linear, why not try redressing the balance and investigating different statistical techniques (which is easy for me to say when I don't include such techniques in this book!)?

1.8. HOW CAN WE TELL IF OUR MODEL REPRESENTS THE REAL WORLD? ①

So far, we've seen that we use samples to estimate what's happening in a larger population to which we don't have access. We've also seen that it's important to establish whether or not a model is a good or bad fit of the data, and whether it is representative of the population. I've explained all of this using the mean as an example, but what happens when our model is more complicated? Scientists are usually interested in more complex scenarios such as 'is there a relationship between the amount of gibberish that people speak and the amount of vodka jelly they've eaten?', or 'is the mean amount of chocolate I eat higher when I'm writing statistics books than when I'm not?'. In these cases, we fit models that are more complex: we're actually looking at detecting effects in our population and quantifying these effects. This section explains some of the ways in which we quantify effects and decide whether they are meaningful. Essentially, as we'll see, this is a four-stage process (but see Box 1.1):

1. Generate a hypothesis (or hypotheses)—this will usually be a prediction that some kind of effect exists in the population.
2. Collect some useful data.

3. Fit a statistical model to the data—this model will test your original predictions.
4. Assess this model to see whether it supports your initial predictions.

Scientists are usually interested in testing hypotheses; that is, testing the scientific questions that they generate. Within these questions, there is usually a prediction that the researcher has made. This prediction is called an experimental hypothesis (it is the prediction that your experimental manipulation will have some effect or that certain variables will relate to each other). The reverse possibility—that your prediction is wrong and that the predicted effect doesn't exist—is called the null hypothesis. A couple of examples are:

1. *Hamburgers make you fat*: the experimental hypothesis is that the more hamburgers you eat, the more you start to resemble a beached whale; the null hypothesis is that people will be equally fat regardless of how many hamburgers they eat.
2. *Cheese gives you nightmares*: the experimental hypothesis would be that those who eat cheese before bedtime have more nightmares than those who don't; the null hypothesis would be that those who eat cheese before bedtime have about the same number of nightmares as those who don't. (In case you're interested, I frequently eat cheese before bedtime, and most of the rest of the day in fact, and I don't have that many nightmares—except when I'm writing stats books and then I dream that I'm being taunted by a mean number.)

Box 1.1

Cheating in research ①

The four-stage process I describe in this chapter works only if you follow the steps without any cheating. Imagine I wanted to place a bet on who would win the Rugby World Cup (which is happening as I type). Being an Englishman, I might want to bet on England to win the tournament. To do this I'd:

1. Place my bet, choosing my team (England) and odds available at the betting shop (currently 6/4).
2. See which team wins the tournament.
3. Collect my winnings (if England do the decent thing and actually win).

To keep everyone happy, this process needs to be equitable: the betting shops set their odds such that they're not paying out too much money (which keeps them happy), but so that they do pay out sometimes (to keep the customers happy). The betting shop can offer any odds before the tournament has ended, but it can't change them once the tournament is over (or the last game has started). Similarly, I can choose any team before the tournament, but I can't then change my mind halfway through, or after the final game!

(Continued)

Box 1.1 (Continued)

The situation in research is similar: we can choose any hypothesis (rugby team) we like before the data are collected, but we can't change our minds halfway through data collection (or after data collection). Likewise we have to decide on our probability level (or betting odds) before we collect data. If we do this, the four-stage process works. However, researchers sometimes cheat. They don't write down their hypotheses before they conduct their experiments, sometimes they change them when the data are collected (like me changing my team after the World Cup is over), or worse still decide on them after the data are collected! With the exception of some complicated procedures called *post hoc* tests, this is cheating. Similarly, researchers can be guilty of choosing which significance level to use after the data are collected and analysed, like a betting shop changing the odds after the tournament.

Every time that you change your hypothesis or the details of your analysis you appear to increase the chance of finding a significant result, but in fact you are making it more and more likely that you will publish results that other researchers can't reproduce (which is very embarrassing!). If, however, you follow the rules carefully and do your significance testing at the 5% level you at least know that in the long run at most only one result out of every 20 will risk this public humiliation.

(With thanks to David Hitchin for this box, and with apologies to him for turning it into a rugby example!)

Most of this book deals with *inferential statistics*, which tell us whether the experimental hypothesis is likely to be true—they help us to confirm or reject our predictions. Crudely put, we fit a statistical model to our data and see how well it fits the data (in terms of the variance it explains). If it fits the data well (i.e. explains a lot of the variation in scores) then we assume our initial prediction is true: we accept the experimental hypothesis. Of course, we can never be completely sure that either hypothesis is correct, and so we work with probabilities instead. Specifically we calculate the probability that the results we have obtained occurred by chance—as this probability decreases, we gain greater confidence that the experimental hypothesis is actually correct and that the null hypothesis can be rejected.

To illustrate this idea, Fisher[5] (1925) describes an experiment designed to test a claim by a woman that she could determine, by tasting a cup of tea, whether the milk or the tea was added first to the cup. Fisher thought that he should give the woman some cups of tea, some of which had the milk added first and some of which had the milk added last, and see whether she could correctly identify them. The woman would know that there are an equal number of cups in which milk was added first or last but wouldn't know in which order the cups were placed. If we take the simplest situation in which there are only two cups then the woman has 50% chance of guessing correctly. If she did guess correctly we wouldn't be that confident in concluding that she can tell the difference between cups in

5 To find out more about Fisher visit http://www.economics.soton.ac.uk/staff/aldrich/fisherguide/rafreader.htm

Figure 1.12 Ronald Fisher contemplating the consequences of setting $p = .05$

which the milk was added first from those in which it was added last, because most of us could perform fairly well on this task just by guessing. However, what about if we complicated things by having six cups? There are 20 orders in which these cups can be arranged and the woman would only guess the correct order one time in 20 (or 5% of the time). If she got the correct order, we would probably be very confident that she could genuinely tell the difference (and bow down in awe of her finely tuned palette). If you'd like to know more about Fisher and his tea-tasting antics see Field & Hole (2003), but for our purposes the take-home point is that only when there was a very small probability that the woman could complete the tea-task by luck alone would we conclude that she had genuine skill in detecting whether milk was poured into a cup before or after the tea was added.

It's no coincidence that I chose the example of six cups above (where the tea-taster had a 5% chance of getting the task right by guessing), because Fisher suggested that only when we are 95% certain that a result is genuine (i.e. not a chance finding) should we accept it as being true.[6] The opposite way to look at this is to say that if there is only a 5% probability of something occurring by chance then we can accept that it is a true finding: we say it is a *statistically significant* finding (see Box 1.2 for a more specific discussion of what statistically significant actually means). This criterion of 95% confidence forms the basis of modern statistics and yet there is very little justification for it other than Fisher said so and he was a ridiculously clever bloke so we trust his judgement (Figure 1.12). Nevertheless, sometimes when I'm staring at my filing cabinet full of experimental results that were significant at only a 93% probability, I wonder how different my career would be had Fisher woken up that day in a 90% kind of a mood.

6 Of course, in reality, it might not be true—we're just prepared to believe that it is!

Research journals tend to have a bias towards publishing positive results (in which the experimental hypothesis is supported) and so had Fisher woken up in a 90% mood that morning, I'd have had a lot more 'successful' experiments and probably be vice-chancellor of my university by now (hmm, so maybe it's just as well for my university that he woke up in a 95% mood!).

1.8.1. Test statistics ①

So, how do we tell whether our model is a good representation of what's going on in the real world? Well, when we collect data, the data we get will vary (in fact we saw how to measure this variance in section 1.4.1). There are two types of variance (see also Chapter 7):

- Systematic variation: This variation is due to some genuine effect (be that the effect of an experimenter doing something to all of the participants in one sample but not in other samples, or natural variation between sets of variables). You can think of this as variation that can be explained by the model that we've fitted to the data.
- Unsystematic variation: This is variation that isn't due to the effect in which we're interested (so could be due to natural differences between people in different samples such as differences in intelligence or motivation). You can think of this as variation that can't be explained by the model that we've fitted to the data.

If we're trying to establish whether a model is a reasonable representation of what's happening in the population then we often calculate a test statistic. A test statistic is simply a statistic that has known properties; specifically we know how frequently different values of this statistic occur. By knowing this, we can calculate the probability of obtaining a particular value. An analogy used by Field and Hole (2003) is the age at which people die. Past data have told us the distribution of the age of death. For example, we know that on average men die at about 75 years old, and that this distribution is top heavy; that is, most people die above the age of about 50 and it's fairly unusual to die in your twenties (he says with a sigh of relief!). So, the frequencies of the age of demise at older ages are very high but are lower at younger ages. From these data, it would be possible to calculate the probability of someone dying at a certain age. If we randomly picked someone and asked them their age, and it was 53, we could tell them how likely it is that they will die before their next birthday (at which point they'd probably punch us!). Also, say we met a man of 110; we could calculate how probable it was that he would have lived that long (it would be a very small probability because most people die before they reach that age). The way we use test statistics is rather similar: we know their distributions and this allows us, once we've calculated the test statistic, to discover the probability of having found a value as big as we have. So, if we calculated a test statistic and its value was 110 (rather like our old man) we can then calculate the probability of obtaining a value that large.

So, how do we calculate these test statistics? This depends on which statistic you're using (and as you'll discover later in the book there are lots of them: t, F and χ^2 to name only three). However, most of these statistics represent essentially the same thing:

$$\text{test statistic} = \frac{\text{variance explained by the model}}{\text{variance not explained by the model}}$$

The exact form of this equation changes from test to test, but essentially we're always comparing the amount of variance explained by the model we've fitted to the data against the variance that can't be explained by the model (see Chapters 5 and 7 in particular for a more detailed explanation). The reason why this ratio is so useful is intuitive really: if our model is good then we'd expect it to be able to explain more variance than it can't explain. In this case, the test statistic will always be greater than 1 (but not necessarily significant).

Given that we know how frequently different values of different test statistics occur by chance, once we've calculated a particular test statistic we can calculate the probability of getting this value. As we saw in the previous section, these probability values tell us how likely it is that our model, or effect, is genuine and not just a chance result. The more variation our model explains (compared to the variance it can't explain), the bigger the test statistic will be, and the more unlikely it is to occur by chance (like our 110 year old man). So, as test statistics get bigger, the probability of them occurring becomes smaller. When this probability falls below .05 (Fisher's criterion), we accept this as giving us enough confidence to assume that the test statistic is as large as it is because our model explains a sufficient amount of variation to reflect what's genuinely happening in the real world (the population). Put another way, we accept our experimental hypothesis and reject our null hypothesis—however, Box 1.2 explains some common misconceptions about this process.

Box 1.2

What we can and can't conclude from a significant test statistic ②

1. **The importance of an effect**: We've seen already that the basic idea behind hypothesis testing involves us generating an experimental hypothesis and a null hypothesis, fitting a statistical model to the data, and assessing that model with a test statistic. If the probability of obtaining the value of our test statistic by chance is less than .05 then we generally accept the experimental hypothesis as true: there is an effect in the population. Normally we say 'there is a *significant* effect of …'. However, don't be fooled by that word 'significant', because even if the probability of our effect being a chance result is small (less than .05) it doesn't necessarily follow that the effect is important. Very small and unimportant effects can turn out to be statistically

(Continued)

Box 1.2 (Continued)

significant just because huge numbers of people have been used in the experiment (see Field & Hole, page 60).

2. **Non-significant results**: Once you've calculated your test statistic, you calculate the probability of that test statistic occurring by chance; if this probability is greater than .05 you reject your experimental hypothesis. However, this does *not* mean that the null hypothesis is true. Remember that the null hypothesis is that there is no effect in the population. All that a non-significant result tells us is that the effect is not big enough to be anything other than a chance finding—it doesn't tell us that the effect is 0. As Cohen (1990) points out, a non-significant result should never be interpreted (despite the fact it often is) as 'no difference between means' or 'no relationship between variables'. Cohen also points out that the null hypothesis is *never* true because we know from sampling distributions (see section 1.6) that two random samples will have slightly different means, and even though these differences can be very small (e.g. one mean might be 10 and another might be 10.00001) they are nevertheless different. In fact, even such a small difference would be deemed as statistically significant if a big enough sample were used. So, significance testing can ever tell us that the null hypothesis is true, because it never is!
3. **Significant results**: OK, we may not be able to accept the null hypothesis as being true, but we can at least conclude that it is false when our results are significant, right? Wrong! A significant test statistic is based on probabilistic reasoning, which severely limits what we can conclude. Again, Cohen (1994), who was an incredibly lucid writer on statistics, points out that formal reasoning relies on an initial statement of fact followed by a statement about the current state of affairs, and an inferred conclusion. This syllogism illustrates what I mean:

   - If a man has no arms then he can't play guitar.
     - This man plays guitar.
     - Therefore, this man has arms.

The syllogism starts with a statement of fact that allows the end conclusion to be reached because you can deny the man has no arms (the antecedent) by denying that he can't play guitar (the consequent). A comparable version of the null hypothesis is:

- If the null hypothesis is correct, then this test statistic cannot occur.
  - This test statistic has occurred.
  - Therefore, the null hypothesis is false.

This is all very nice except that the null hypothesis is not represented in this way because it is based on probabilities. Instead it should be stated as follows:

- If the null hypothesis is correct, then this test statistic is highly unlikely.
  - This test statistic has occurred.
  - Therefore, the null hypothesis is highly unlikely.

Box 1.2 (Continued)

If we go back to the guitar example we could get a similar statement:

- If a man plays guitar then he probably doesn't play for Fugazi (this is true because there are thousands of people who play guitar but only two who play guitar in the band Fugazi!).
  - Guy Picciotto plays for Fugazi.
  - Therefore, Guy Picciotto probably doesn't play guitar.

This should hopefully seem completely ridiculous—the conclusion is wrong because Guy Picciotto does play guitar. This illustrates a common fallacy in hypothesis testing. In fact significance testing allows us to say very little about the null hypothesis.

1.8.2. One- and two-tailed tests ①

When using statistical tests we either have a specific prediction about what will happen, such as 'the more someone reads this book, the more they want to kill its author', or we don't really know what will happen, such as 'reading more of this book could increase or decrease the reader's desire to kill its author'. The former example is directional: we've explicitly said that people will want to kill me more as they read more of this book. If we tested this hypothesis statistically, the test is called a one-tailed test. The second hypothesis is non-directional: we stated that the desire to kill me will change because of reading more of this book, but we haven't said whether that desire will increase or decrease. If we tested this hypothesis statistically, the test would be a two-tailed test.

Imagine we wanted to discover whether reading this book increased or decreased the desire to kill me. We could do this either (experimentally) by taking two groups, one who had read this book and one who hadn't, or (correlationally) by measuring the amount of this book that had been read and the corresponding desire to kill me. If we have no directional hypothesis then there are three possibilities. (1) People who read this book want to kill me more than those who don't so the difference (mean for those reading the book minus the mean for non-readers) is positive. Correlationally, the more of the book you read, the more you want to kill me—a positive relationship. (2) People who read this book want to kill me less than those who don't so the difference (mean for those reading the book minus the mean for non-readers) is negative. Correlationally, the more of the book you read, the less you want to kill me—a negative relationship. (3) There is no difference between readers and non-readers in their desire to kill me—the mean for readers minus the mean for non-readers is exactly zero. Correlationally, there is no relationship between reading this book and wanting to kill me. This final option is the null hypothesis. The direction of the test statistic (i.e. whether it is positive or negative) depends on whether the difference is positive or negative. Assuming there is a positive difference or relationship (reading this book makes you want to kill me), then to detect this difference we have to take account of the fact that

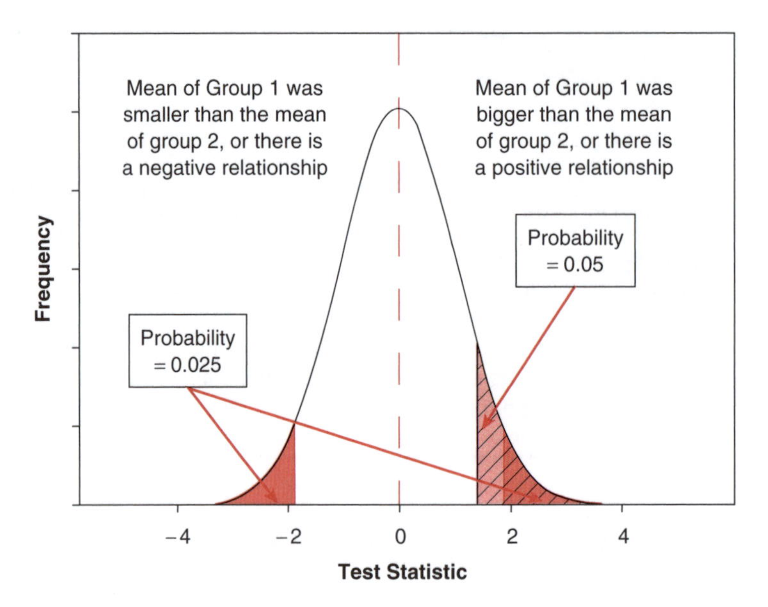

Figure 1.13 Diagram to show the difference between one- and two-tailed tests

the mean for readers is bigger than for non-readers (and so derive a positive test statistic). However, if we've predicted incorrectly and actually reading this book makes readers want to kill me less then the test statistic will actually be negative.

What are the consequences of this? Well, if, at the .05 level, we needed to get a test statistic bigger than say 10 and the one we get is actually −12, then we would reject the hypothesis even though a difference does exist. To avoid this we can look at both ends (or tails) of the distribution of possible test statistics. This means we will catch both positive and negative test statistics. However, doing this has a price because to keep our criterion probability of .05 we have to split this probability across the two tails: so we have .025 at the positive end of the distribution and .025 at the negative end. Figure 1.13 shows this situation—the tinted areas are the areas above the test statistic needed at a .025 level of significance. Combine the probabilities at both ends and we get .05—our criterion value. Now if we have made a prediction, then we put all our eggs in one basket and look only at one end of the distribution (either the positive or the negative end depending on the direction of the prediction we make). Consequently, we can just look for the value of the test statistic that would occur by chance with a probability of .05. In Figure 1.13, the diagonal lines show the area above the positive test statistic needed at a .05 level of significance. The important point is that if we make a specific prediction then we need a smaller test statistic to find a significant result (because we are looking in only one tail), but if our prediction happens to be in the wrong direction then we'll miss detecting the effect that does exist! In this context, it's important to remember what I said in Box 1.1: you can't place a bet or change your bet when the tournament is over—if you didn't make a prediction of direction before you collected the data, you are too late to predict the direction and claim the advantages of a one-tailed test.

1.8.3. Type I and Type II errors ①

We have seen that we use test statistics to tell us about the true state of the world (to a certain degree of confidence). Specifically, we're trying to see whether there is an effect in our population. There are two possibilities in the real world: there is, in reality, an effect in the population, or there is, in reality, no effect in the population. We have no way of knowing which of these possibilities is true; however, we can look at test statistics and their associated probability to tell us which of the two is more likely. Obviously, it is important that we're as accurate as possible, which is why Fisher originally said that we should be very conservative and only believe that a result is genuine when we are 95% confident that it is—or when there is only a 5% chance that the results could occur by chance. However, even if we're 95% confident there is still a small chance that we get it wrong. There are two mistakes we can make: a Type I and a Type II error. A Type I error occurs when we believe that there is a genuine effect in our population, when in fact there isn't. If we use Fisher's criterion then the probability of this error is .05 (or 5%) when there is no effect in the population—this value is known as the α-level. Assuming there is no effect in our population, if we replicated our data collection 100 times we could expect that on five occasions we would obtain a test statistic large enough to make us think that there is a genuine effect in the population even though there isn't. The opposite is a Type II error, which occurs when we believe that there is no effect in the population when, in reality, there is. This would occur when we obtain a small test statistic (perhaps because there is a lot of natural variation between our samples). In an ideal world, we want the probability of this error to be very small (if there is an effect in the population then it's important that we can detect it). Cohen (1992) suggests that the maximum acceptable probability of a Type II error would be .2 (or 20%)—this is called the β-level. That would mean that if we took 100 samples of data from a population in which an effect exists, we would fail to detect that effect in 20 of those samples (so we'd miss 1 in 5 genuine effects).

There is obviously a trade-off between these two errors: if we lower the probability of accepting an effect as genuine (i.e. make α smaller) then we increase the probability that we'll reject an effect that does genuinely exist (because we've been so strict about the level at which we'll accept that an effect is genuine). The exact relationship between the Type I and Type II error is not straightforward because they are based on different assumptions: to make a Type I error there has to be no effect in the population, whereas to make a Type II error the opposite is true (there has to be an effect that we've missed). So, although we know that as the probability of making a Type I error decreases, the probability of making a Type II error increases, the exact nature of the relationship is usually left for the researcher to make an educated guess (Howell, 2002, pp. 104–107 gives a great explanation of the trade-off between errors).

1.8.4. Effect sizes ②

The framework for testing whether effects are genuine that I've just presented has a few problems, most of which have been briefly explained in Box 1.2. The first problem we encountered was knowing how important an effect is: just because a test statistic is significant doesn't mean that the effect it measures is meaningful or important. The solution to this criticism is to measure the size of the effect that we're testing in a standardized way. When we measure the size of an effect (be that an experimental manipulation or the strength of relationship between variables) it is known as an effect size. An effect size is simply an objective and standardized measure of the magnitude of the observed effect. The fact that the measure is standardized just means that we can compare effect sizes across different studies that have measured different variables, or have used different scales of measurement (so an effect size based on speed in milliseconds could be compared to an effect size based on heart rates). Many measures of effect size have been proposed, the most common of which are Cohen's d and Pearson's correlation coefficient r (see Field, 2001, and Chapter 4). Many of you will be familiar with the correlation coefficient as a measure of the strength of relationship between two variables (see Chapter 4 if you're not); however, it is also a very versatile measure of the strength of an experimental effect. It's a bit difficult to reconcile how the humble correlation coefficient can also be used in this way; however, this is only because students are typically taught about it within the context of non-experimental research. I don't want to get into it now, but as you read through Chapters 4, 7 and 8 it will (I hope!) become clear what I mean. Personally, I prefer Pearson's correlation coefficient r as an effect size measure because it is constrained to lie between 0 (no effect) and 1 (a perfect effect).[7]

Effect sizes are useful because they provide an objective measure of the importance of an effect. So, it doesn't matter what effect you're looking for, what variables have been measured, or how those variables have been measured: we know that a correlation coefficient of 0 means there is no effect, and a value of 1 means that there is a perfect effect. Cohen (1988, 1992) has also made some widely accepted suggestions about what constitutes a large or small effect:

- **r = .10 (small effect)**: in this case, the effect explains 1% of the total variance.
- **r = .30 (medium effect)**: the effect accounts for 9% of the total variance.
- **r = .50 (large effect)**: the effect accounts for 25% of the variance.

---

7 The correlation coefficient can also take on minus values (but not below −1), which is useful when we're measuring a relationship between two variables because the sign of r tells us about the direction of the relationship, but in experimental research the sign of r merely reflects the way in which the experimenters coded their groups (see Chapter 4).

We can use these guidelines to assess the importance of our effects (regardless of the significance of the test statistic). However, r is not measured on a linear scale so an effect with $r = .6$ isn't twice as big as one with $r = .3$! Such is the utility of effect size estimates that the American Psychological Association is now recommending that all psychologists report these effect sizes in the results of any published work. So it's a habit well worth getting into.

A final thing to mention is that when we calculate effect sizes we calculate them for a given sample. When we looked at means in a sample, we saw that we used them to draw inferences about the mean of the entire population (which is the value in which we're actually interested). The same is true of effect sizes: the size of the effect in the population is the value in which we're interested, but because we don't have access to this value, we use the effect size in the sample to estimate the likely size of the effect in the population (see Field, 2001).

1.8.5. Statistical power ②

We've seen that effect sizes are an invaluable way to express the importance of a research finding. The effect size in a population is intrinsically linked to three other statistical properties: (1) the sample size on which the sample effect size is based; (2) the probability level at which we will accept an effect as being statistically significant (the α-level); and (3) the ability of a test to detect an effect of that size (known as the statistical power, not to be confused with statistical powder, which is an illegal substance that makes you understand statistics better). As such, once we know three of these properties, then we can always calculate the remaining one. It will also depend on whether the test is one- or two-tailed (see section 1.8.2). Typically, in psychology we use an α-level of .05 (see earlier) so we know this value already. The power of a test is the probability that a given test will find an effect assuming that one exists in the population. If you think back you might recall that we've already come across the probability of failing to detect an effect when one genuinely exists (β, the probability of a Type II error). It follows that the probability of detecting an effect if one exists must be the opposite of the probability of not detecting that effect (i.e. $1 - \beta$). I've also mentioned that Cohen (1988, 1992) suggests that we would hope to have a .2 probability of failing to detect a genuine effect, and so the corresponding level of power that he recommended was $1 - .2$, or .8. We should aim to achieve a power of .8, or an 80% chance of detecting an effect if one genuinely exists. The effect size in the population can be estimated from the effect size in the sample, and the sample size is determined by the experimenter anyway so that value is easy to calculate. Now, there are two useful things we can do knowing that these four variables are related:

1. **Calculate the power of a test**: Given that we've conducted our experiment, we will have already selected a value of α, we can estimate the effect size based on our sample, and we will know how many participants we used. Therefore, we can use these values to calculate β, the power of our test. If this value turns out to be .8 or more

we can be confident that we achieved sufficient power to detect any effects that might have existed, but if the resulting value is less, then we might want to replicate the experiment using more participants to increase the power.

2. **Calculate the sample size necessary to achieve a given level of power**: Given that we know the value of α and β, we can use past research to estimate the size of effect that we would hope to detect in an experiment. Even if no one had previously done the exact experiment that you intend to do, we can still estimate the likely effect size based on similar experiments. We can use this estimated effect size to calculate how many participants we would need to detect that effect (based on the values of α and β that we've chosen).

The latter use is the most common: to determine how many participants should be used to achieve the desired level of power. The actual computations are very cumbersome, but fortunately, there are now computer programs available that will do them for you (one example is G*Power, which is free and can be found on the CD-ROM, another is nQuery Adviser—see Field (1998b) for a review—but this has to be bought!). Also, Cohen (1988) provides extensive tables for calculating the number of participants for a given level of power (and vice versa). Based on Cohen (1992) we can use the following guidelines: if we take the standard α-level of .05 and require the recommended power of .8, then we need 783 participants to detect a small effect size (r = .1), 85 participants to detect a medium effect size (r = .3) and 28 participants to detect a large effect size (r = .5).

1.9. SOME CONCLUDING ADVICE

I just want to end this chapter with some general advice. Statistical procedures are just a way of number crunching and so even if you put rubbish into an analysis you will still reach conclusions that are statistically meaningful, but are unlikely to be empirically meaningful. There is a temptation to see statistics as some God-like way of determining the real truth, but statistics are merely a tool. My statistics lecturer used to say, quite rightly, 'if you put garbage in, you get garbage out' and to be honest I never really understood what he was going on about until people started presenting me with data sets containing millions of seemingly randomly connected variables that they wanted me to help them interpret. In a statistical analysis there is no substitute for empirical thinking! Bear that in mind throughout this book. Oh, and some final advice: never trust anything a psychologist says about statistics☺

1.10. WHAT HAVE WE DISCOVERED ABOUT STATISTICS? ①

OK, that's been your crash course in statistical theory! Hopefully your brain is still relatively intact. The key point I want you to understand is that when you carry out research you're trying to see whether some effect genuinely exists in your population (the effect you're interested in will depend on your research interests and your specific predictions). You won't be able to collect data from the entire population (unless you want to spend your entire life, and probably several after-lives, collecting data) so you use a sample instead. Using the data from this sample, you fit a statistical model to test your predictions, or, put another way, detect the effect you're looking for. Statistics boils down to one simple idea: observed data can be predicted from some kind of model and an error associated with that model. You use that model (and usually the error associated with it) to calculate a test statistic. If that model can explain a lot of the variation in the data collected (the probability of obtaining that test statistic is less than .05) then you infer that the effect you're looking for genuinely exists in the population. If the probability of obtaining that test statistic is more than .05, then you conclude that the effect was too small to be detected. Rather than rely on significance, you can also quantify the effect in your sample in a standard way as an *effect size* and this can be helpful in gauging the importance of that effect. Now, onto SPSS!

1.11. KEY TERMS THAT WE'VE DISCOVERED

- α-level
- β-level
- Confidence interval
- Deviance
- Effect size
- Experimental hypothesis
  - Frequency distribution
  - Histogram
- Kurtosis
- Leptokurtic
- Linear model
- Mean
- Mode
- Normal distribution
- Null hypothesis
- One-tailed test
- Platykurtic
- Population
- Probability distribution
- Sample
- Sampling distribution
- Skewness
- Standard deviation
- Standard error
- Statistical power
- Sum of squared errors (SS)
- Systematic variation
- Test statistic
- Two-tailed test
- Type I error
- Type II error
- Unsystematic variation
- Variance
- z-score

1.12. SMART ALEX'S STATS QUIZ

Smart Alex knows everything there is to know about statistics and SPSS. He also likes nothing more than to ask people stats questions just so that he can be smug about how much he knows. So, why not really annoy him and get all of the answers right?

1. Why do we use samples?①
2. What is the mean and how do we tell if it's representative of our data?①

3. What's the difference between the standard deviation and the standard error?①
4. What is a test statistic and what does it tell us?①
5. What are Type I and Type II errors?①
6. What is an effect size and how is it measured?②
7. What is statistical power?②

Some brief answers can be found in the file **Answers(Chapter 1).pdf** on the CD-ROM.

1.13. FURTHER READING

Cohen, J. (1990). Things I have learned (so far). *American Psychologist*, *45*, 1304–1312.

Cohen, J. (1994). The earth is round ($p < .05$). *American Psychologist*, *49*, 997–1003. A couple of beautiful articles by the best modern writer of statistics that we've had.

Field, A. P. & Hole, G. J. (2003). *How to design and report experiments*. London: Sage. I am rather biased, but I think this is a good overview of basic statistical theory.

Wright, D. B. (2002). *First steps in statistics*. London: Sage. Chapters 1, 4 and 5 are very clear introductions to sampling, confidence intervals and other important statistical ideas.

CHAPTER 2

THE SPSS ENVIRONMENT

2.1. WHAT WILL THIS CHAPTER TELL US? ①

Hopefully Chapter 1 hasn't put you off of this book. If not, Chapter 2 deals with the other background information that you need to know: the SPSS environment. There are several excellent texts that give introductions to the general environment within which SPSS operates. The best ones include Kinnear & Gray (2000) and Foster (2001). These texts are well worth reading if you are unfamiliar with computers and SPSS generally. However, I appreciate the limited funds of most students and so to make this text usable for those inexperienced with SPSS this section provides a guide to the SPSS environment—but if you need a more thorough account then see the previously cited texts and the SPSS manuals (you might also find the SPSS help files useful). This chapter looks at the key windows in SPSS (the *data editor* and *viewer*) and also looks at how to create variables, enter data and adjust properties of your variables. We finish off by looking at how to load files and save them!

2.2. VERSIONS OF SPSS ①

This book is based primarily on version 13 of SPSS (at least in terms of the diagrams); however, don't be fooled too much by version numbers because SPSS have a habit of releasing 'new' versions fairly regularly. Although this makes them a lot of money and creates a nice market for people writing books on SPSS, there are in reality few differences in these new releases that most of us would actually notice. Occasionally they have a major overhaul (version 7.0 saw a dramatic change from version 6.0, and version 10 altered the way in which variables are entered), but most of the time you can get by with a book that doesn't explicitly cover the version you're using (the first edition of this book was based on version 9, but could be easily used with versions 7, 8, 10, 11 and 12 without

much difficulty!). So, this revised book, although dealing with version 13, will happily cater for earlier versions (7.0, 7.5 and 8.0—there are few differences between versions 7.0, 8.0 and 9.0 but any obvious differences are highlighted where relevant). I also suspect it'll be useful with versions 14, 15 and 16 when they appear (although SPSS may decide to change everything just to spite me!). There are various differences in terms of data entry with versions before version 10, so if you're using version 9 or earlier then I've included a file called **Field2000(Chapter1).pdf** on the CD-ROM in the back of this book. This file is a copy of Chapter 1 from the first edition of this book and will tell you how to enter data for these earlier versions.

2.3. GETTING STARTED ①

SPSS mainly uses two windows: the data editor (this is where you input your data and carry out statistical functions) and the viewer (this is where the results of any analysis appear).[1] There are several additional windows that can be activated such as the *SPSS* syntax editor (see section 2.6), which allows you to enter SPSS commands manually (rather than using the window-based menus). At most levels of expertise, the syntax window is redundant because you can carry out most analyses by clicking merrily with your mouse. However, there are various additional functions that can be accessed using syntax and sick individuals who enjoy statistics can find numerous uses for it! Just to prove that I am the sickest of them all, there are a few sections where I'll force you to use it☺

Once SPSS has been activated, a start-up window will appear (see Figure 2.1), which allows you to select various options.[2] If you already have a data file on disk that you would like to open then select *Open an existing data source* by clicking on the ○ so that it looks like ◉: this is the default option. In the space underneath this option there will be a list of recently used data files that you can select with the mouse. To open a selected file click on OK. If you want to open a data file that isn't in the list then simply select *More Files...* with the mouse and click on OK. This will open a standard explorer window that allows you to browse your computer and find the file you want (see section 2.8). Now it might be the case that you want to open something other than a data file, for example a *viewer* document containing the results of your last analysis. You can do this by selecting *Open another type of file* by clicking on the ○ (so that it looks like ◉) and either

1 In versions of SPSS earlier than version 7.0, graphs appear in a separate window known as the *chart carousel*; however, versions 7.0 and after include graphs in the output window, which is called the *output navigator* (version 7.0) and the *output viewer* (version 8.0 and after).

2 In fact, this window doesn't appear in versions earlier than SPSS 10. Instead a blank data editor window is loaded and you can load files using the menus (see section 2.8 or the file **Field2000 (Chapter1).pdf** on the CD-ROM).

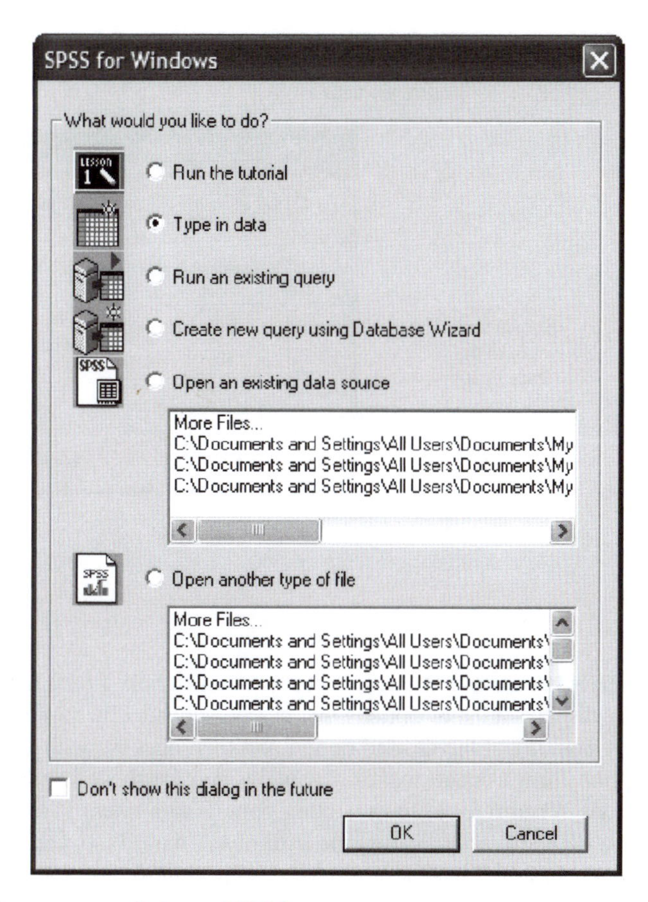

Figure 2.1 The start-up window of SPSS

selecting a file from the list or selecting *More Files...* and browsing your computer. If you're starting a new analysis (as we are here) then we want to type our data into a new data editor. Therefore, we need to select *Type in data* (by again clicking on the appropriate) and then clicking on . This will load a blank *data editor* window.

2.4. THE DATA EDITOR ①

The main SPSS window includes a data editor for entering data. This window is where most of the action happens. At the top of this screen is a menu bar similar to the ones you might have seen in other programs (such as Microsoft Word). Figure 2.2 shows this menu bar and the data editor. There are several menus at the top of the screen (e.g. *File*, *Edit*, etc.) that can be activated by using the computer mouse to move the on-screen arrow onto the desired menu and then pressing the left mouse button once (pressing this button is

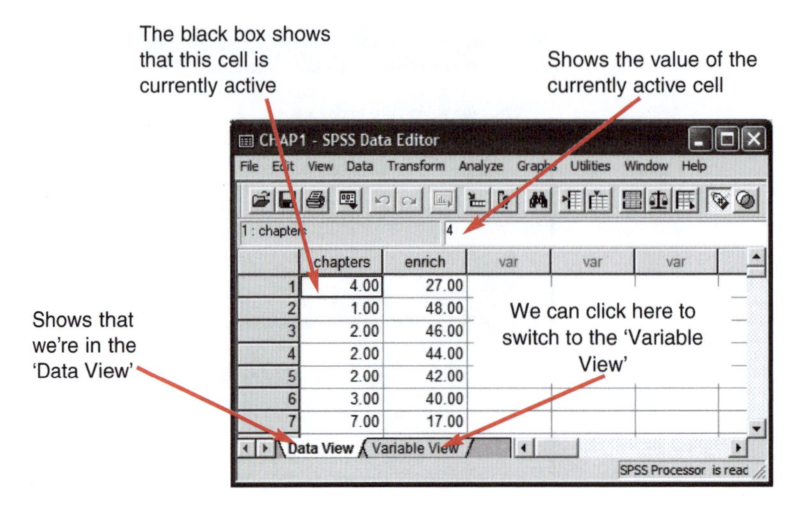

Figure 2.2 The SPSS Data Editor (version 10 onwards)

usually known as *clicking*). When you have clicked on a menu, a menu box will appear that displays a list of options that can be activated by moving the on-screen arrow so that it is pointing at the desired option and then clicking with the mouse. Often, selecting an option from a menu makes a window appear; these windows are referred to as *dialog boxes*. When referring to selecting options in a menu I will notate the action using bold type with arrows indicating the path of the mouse (so, each arrow represents placing the on-screen arrow over a word and clicking the mouse's left button). So, for example, if I were to say that you should select the *Save As...* option in the *File* menu, I would write this as select **File⇒Save As...**.

The data editor has two views: the data view and the variable view. The data view is for entering data into the data editor, and the variable view allows us to define various characteristics of the variables within the data editor.[3] At the bottom of the data editor, you should notice that there are two tabs labelled 'Data View' and 'Variable View' (Data View / Variable View) and all we do to switch between these two views is click on these tabs. If we're in the data view then this tab will have a white background and the variable view will have a shaded background, and if we're in the variable view then the opposite is true. Let's first look at some general features of the data editor; these are features that don't change when we switch between the data view and the variable view. First off, let's look at the menus.

In many computer packages you'll find that within the menus some letters are underlined: these underlined letters represent the *keyboard shortcut* for accessing that function. It is possible to select many functions without using the mouse, and the experienced keyboard

3 Before SPSS version 10, data entry was done using a single view (see Field, 2000, Chapter 1, which is included on the CD-ROM as **Field2000(Chapter1).pdf**).

user may find these shortcuts faster than manoeuvring the mouse arrow to the appropriate place on the screen. The letters underlined in the menus indicate that the option can be obtained by simultaneously pressing *Alt* on the keyboard and the underlined letter. So, to access the *Save As...* option, using only the keyboard, you should press *Alt* and *F* on the keyboard simultaneously (which activates the *File* menu); then, keeping your finger on the *Alt* key, press A (which is the underlined letter). In SPSS 13 (at least when running under Windows XP), these underlined letters are not visible; however, you can still use keyboard shortcuts. If you press *Alt* then the underlined letters become visible, and having pressed *Alt* once you can just press the underlined letters to navigate the menus. So, throughout this book I'll include the underlined letters for people who want to use the keyboard shortcuts, but for those of you that don't, try not to be put off by the underlined letters in this book!

Below is a brief reference guide to each of the menus and some of the options that they contain. This is merely a summary and we will discover the wonders of each menu as we progress through the book.

- **File**: This menu allows you to do general things such as saving data, graphs, or output. Likewise, you can open previously saved files and print graphs, data or output. In essence, it contains all of the options that are customarily found in *File* menus.
- **Edit**: This menu contains edit functions for the data editor. In SPSS for Windows it is possible to *cut* and *paste* blocks of numbers from one part of the data editor to another (which can be very handy when you realize that you've entered lots of numbers in the wrong place). You can also use *Options* to select various preferences such as the font that is used for the output. The default preferences are fine for most purposes; the only thing you might want to change (for the sake of the environment) is to set the text output page size length of the viewer to infinite (this saves hundreds of trees when you come to print things). To do this you'll need to select the 'Viewer' tab at the top of the window.
- **View**: This menu deals with system specifications such as whether you have grid lines on the data editor, or whether you display value labels (exactly what value labels are will become clear later).
- **Data**: This menu allows you to make changes to the data editor. The important features are *insert variable*, which is used to insert a new variable into the data editor (i.e. add a column); *insert case,* which is used to add a new row of data between two existing rows of data; *split file*, which is used to split the file by a grouping variable (see section 3.4) and *select cases*, which is used to run analyses on only a selected sample of cases.
- **Transform**: You should use this menu if you want to manipulate one of your variables in some way. For example, you can use *recode* to change the values of certain variables (e.g. if you wanted to adopt a slightly different coding scheme for some reason)—see Box 3.2. The *compute* function is also useful for transforming data (e.g. you can create a new variable that is the average of two existing variables). This function allows you to carry out any number of calculations on your variables (see section 3.3.4).
- **Analyze**: The fun begins here, because the statistical procedures lurk in this menu.[4] Below is a brief guide to the options in the statistics menu that will be used during the course of this book (this is only a small portion of what is available):

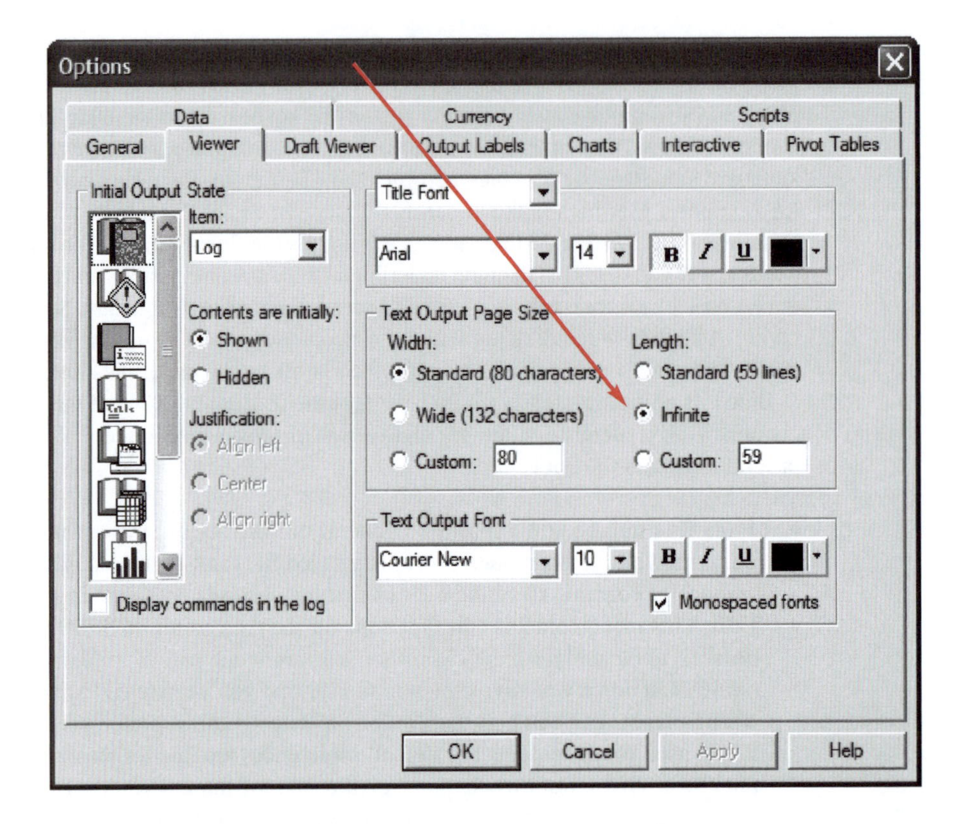

(a) **Descriptive Statistics**:[5] This menu is for conducting descriptive statistics (mean, mode, median, etc.), frequencies and general data exploration. There is also a command called *crosstabs* that is useful for exploring frequency data and performing tests such as chi-square, Fisher's exact test and Cohen's kappa.

(b) **Compare Means**: This is where you can find *t*-tests (related and unrelated—Chapter 7) and one-way independent ANOVA (Chapter 8).

(c) **General Linear Model**:[6] This menu is for complex ANOVA such as two-way (unrelated, related or mixed), one-way ANOVA with repeated measures and multivariate analysis of variance (MANOVA)—see Chapters 9, 10, 11, 12 and 14.

(d) **Correlate**: It doesn't take a genius to work out that this is where the correlation techniques are kept! You can do bivariate correlations such as Pearson's *R*, Spearman's rho (ρ) and Kendall's tau (τ) as well as partial correlations (see Chapter 4).

(e) **Regression**: There are a variety of regression techniques available in SPSS. You can do simple linear regression, multiple linear regression (Chapter 5) and more advanced techniques such as logistic regression (Chapter 6).

---

4 This menu is called **Statistics** in version 8.0 and earlier.

5 This menu is called **Summarize** in version 8.0 and earlier.

6 This is called **ANOVA Models** in version 6 of SPSS.

(f) **Loglinear**: Loglinear analysis is hiding in this menu, waiting for you, and ready to pounce like a tarantula from its burrow (Chapter 16).
(g) **Data Reduction**: You find factor analysis here (Chapter 15).
(h) **Scale**: Here you'll find reliability analysis (Chapter 15).
(i) **Nonparametric Tests**: There are a variety of non-parametric statistics available such the chi-square goodness-of-fit statistic, the binomial test, the Mann–Whitney test, the Kruskal–Wallis test, Wilcoxon's test and Friedman's ANOVA (Chapter 13).

- **Graphs**: SPSS comes with its own, fairly versatile, graphing facilities. The types of graphs you can do include: bar charts, histograms, scatterplots, box–whisker plots, pie charts and error bar graphs to name but a few. There is also the facility to edit any graphs to make them look snazzy—which is pretty smart if you ask me.
- **Window**: This allows you to switch from window to window. So, if you're looking at the output and you wish to switch back to your data sheet, you can do so using this menu. There are icons as shortcuts for most of the options in this menu so it isn't particularly useful.
- **Help**: This is an invaluable menu because it offers you on-line help on both the system itself and the statistical tests. Although the statistics help files are fairly useless at times (after all, the program is not supposed to teach you statistics) and certainly no substitute for acquiring a good knowledge of your own, they can sometimes get you out of a sticky situation.

As well as the menus there is also a set of *icons* at the top of the data editor window (see Figure 2.2) that are shortcuts to specific, frequently used, facilities. All of these facilities can be accessed via the menu system but using the icons will save you time. Below is a brief list of these icons and their function:

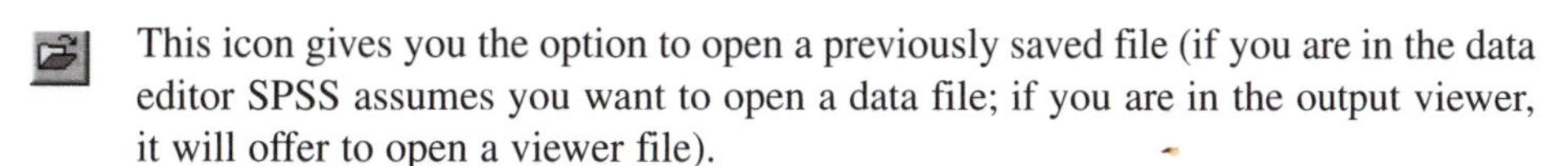

This icon gives you the option to open a previously saved file (if you are in the data editor SPSS assumes you want to open a data file; if you are in the output viewer, it will offer to open a viewer file).

This icon allows you to save files. It will save the file you are currently working on (be it data or output). If the file hasn't already been saved it will produce the *save data as* dialog box.

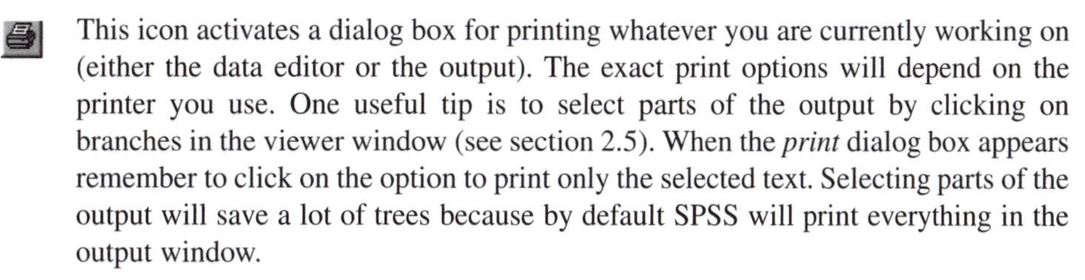

This icon activates a dialog box for printing whatever you are currently working on (either the data editor or the output). The exact print options will depend on the printer you use. One useful tip is to select parts of the output by clicking on branches in the viewer window (see section 2.5). When the *print* dialog box appears remember to click on the option to print only the selected text. Selecting parts of the output will save a lot of trees because by default SPSS will print everything in the output window.

Clicking on this icon will activate a list of the last 12 dialog boxes that were used. From this list you can select any box from the list and it will appear on the screen. This icon makes it easy for you to repeat parts of an analysis.

This icon allows you to go directly to a case (i.e. a participant). This is useful if you are working on large data files. For example, if you were analysing a survey with 3000 respondents it would get pretty tedious scrolling down the data sheet to find a particular participant's responses. This icon can be used to skip directly to a case (e.g. case 2407). Clicking on this icon activates a dialog box that requires you to type in the case number required.

Clicking on this icon will give you information about a specified variable in the data editor (a dialog box allows you to choose which variable you want summary information about).

This icon doesn't allow you to go bird watching, but it does enable you to search for words or numbers in your data file and output window.

Clicking on this icon inserts a new case in the data editor (so, it creates a blank row at the point that is currently highlighted in the data editor). This function is very useful if you need to add new data or if you forget to put a particular participant's data in the data editor.

Clicking on this icon creates a new variable to the left of the variable that is currently active (to activate a variable simply click once on the name at the top of the column).

Clicking on this icon is a shortcut to the **Data⇒Split File...** function (see section 3.4). Social scientists often conduct experiments on different groups of people. In SPSS we differentiate groups of people by using a coding variable (see section 2.4.4), and this function lets us divide our output by such a variable. For example, we might test males and females on their statistical ability. We can code each participant with a number that represents their gender (e.g. 1 = female, 0 = male). If we then want to know the mean statistical ability of each gender we simply ask the computer to split the file by the variable **gender**. Any subsequent analyses will be performed on the men and women separately.

This icon is a shortcut to the **Data⇒Weight Cases...** function. This function is necessary when we come to input frequency data (see section 16.4.2) and is useful for some advanced issues in survey sampling.

This icon is a shortcut to the **Data⇒Select Cases...** function. If you want to analyse only a portion of your data, this is the option for you! This function allows you to specify what cases you want to include in the analysis.

Clicking on this icon will either display, or hide, the value labels of any coding variables. We often group people together and use a coding variable to let the computer know that a certain participant belongs to a certain group. For example, if we coded gender as 1 = female, 0 = male then the computer knows that every time it comes across the value 1 in the **gender** column, that person is a female. If you press this icon, the coding will appear on the data editor rather than the numerical values; so, you will see the words *male* and *female* in the **gender** column rather than a series of numbers. This idea will become clear in section 2.4.4.

2.4.1. Entering data into the data editor ①

When you first load SPSS it will provide a blank data editor with the title *Untitled* (this of course is daft because once it has been given the title 'untitled' it ceases to be untitled!). When inputting a new set of data, you must input your data in a logical way. The SPSS data editor is arranged such that *each row represents data from one individual while each column represents a variable*. There is no discrimination between independent and dependent variables: both types should be placed in a separate column. The key point is that each row represents one person's data. Therefore, any information about that case should be entered across the data editor. For example, imagine you were interested in sex differences in perceptions of pain created by hot and cold stimuli. You could place some people's hands in a bucket of very cold water for a minute and ask them to rate how painful they thought the experience was on a scale of 1 to 10. You could then ask them to hold a hot potato and again measure their perception of pain. Imagine I was a participant. You would have a single row representing my data, so there would be a different column for my name, my age, my gender, my pain perception for cold water, and my pain perception for a hot potato: Andy, 31 (by the time this comes out), male, 7, 10.

The column with the information about my gender is a grouping variable: I can belong to either the group of males or the group of females, but not both. As such, this variable is a between-group variable (different people belong to different groups). Rather than representing groups with words, in SPSS we have to use numbers. This involves assigning each group a number, and then telling SPSS which number represents which group. Therefore, between-group variables are represented by a single column in which the group to which the person belonged is defined using a number (see section 2.4.4). For example, we might decide that if a person is male then we give them the number 0, and if they're female we give them the number 1. We then have to tell SPSS that every time it sees a 1 in a particular column the person is a female, and every time it sees a 0 the person is a male. Variables that specify to which of several groups a person belongs can be used to split up data files (so, in the pain example you could run an analysis on the male and female participants separately—see section 3.4).

Finally, the two measures of pain are a repeated measure (all participants were subjected to hot and cold stimuli). Therefore, levels of this variable can be entered in separate columns (one for pain to a hot stimulus and one for pain to a cold stimulus).

Cramming Samantha's Tip

In summary, any variable measured with the same participants (a repeated measure) should be represented by several columns (each column representing one level of the repeated-measures variable). However, any variable that defines different groups of people (such as when a between-group design is used and different participants are assigned to different levels of the independent variable) is defined using a single column. This idea will become clearer as you learn about how to carry out specific procedures.

The data editor is made up of lots of *cells*, which are just boxes in which data values can be placed. When a cell is active it becomes highlighted with a black surrounding box (as in Figure 2.2). You can move around the data editor, from cell to cell, using the arrow keys ← ↑ ↓ → (found on the right of the keyboard) or by clicking the mouse on the cell that you wish to activate. To enter a number into the data editor simply move to the cell in which you want to place the data value, type the value, then press the appropriate arrow button for the direction in which you wish to move. So, to enter a row of data, move to the far left of the row, type the value and then press → (this process inputs the value and then moves you into the next cell on the right).

2.4.2. Creating a variable ①

The first step in entering your data is to create some variables using the 'Variable View' of the data editor, and then to input your data using the 'Data View' of the data editor. We'll go through these two steps by working through an example. Imagine we were interested in looking at the differences between lecturers and students. We took a random sample of five psychology lecturers from the University of Sussex and five psychology students and then measured how many friends they had, their weekly alcohol consumption (in units), their yearly income and how neurotic they were (higher score is more neurotic). These data are in Table 2.1.

2.4.3. The 'Variable View' ①

Before we actually input any data into the data editor, we need to create the variables. To create variables we use the variable view of the data editor. To access this view click on

Table 2.1 Some data with which to play

	No. of Friends	Alcohol	Income (p.a.)	Neuroticism
Lecturer	5	10	20000	10
Lecturer	2	15	40000	17
Lecturer	0	20	35000	14
Lecturer	4	5	22000	13
Lecturer	1	30	50000	21
Student	10	25	5000	7
Student	12	20	100	13
Student	15	16	3000	9
Student	12	17	10000	14
Student	17	18	10	13

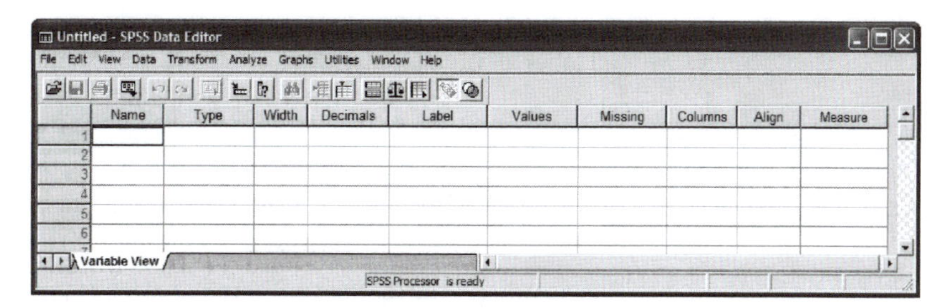

Figure 2.3 The variable view of the SPSS data editor

the 'Variable View' tab at the bottom of the data editor (Data View / Variable View); the contents of the window will change and the variable view tab will turn white (see Figure 2.3).

Every row of the variable view represents a variable, and you set characteristics of a particular variable by entering things into the labelled columns. You can change various characteristics such as the width of the column, how many decimal places are displayed, and whether values are aligned to the left, right or centre (play around and you'll get the hang of it). The most important characteristics are probably the following:

Name You can enter a name in this column for each variable. This name will appear at the top of a column when you're in the data view, and is really only to help you identify variables in the data view. There are some general rules about variables' names such as that they must be eight characters or less, you can use only lower-case letters, and you cannot use a blank space. If you violate any of these rules the computer will tell you that the variable name is invalid when you click on a different cell, or try to move off the cell using the arrow keys.

Type You can have different types of data. The most common are numeric (this is the default that SPSS assumes, and just means that it is a variable containing numbers). The other one you will come across is a *string* variable, which is short for 'string of letters'. So, if you wanted to type in people's names, for example, you would need to change the variable type to be string rather than numeric.

Label The name of the variable (see above) is restricted to eight characters, so SPSS provides us with a facility to give our variable a more meaningful title (and this label can also have capital letters and space characters too—great!). This may seem pointless, but is actually one of the best habits you can get into. If you have a variable called 'number of times I wanted to shoot myself during Andy Field's statistics lecture', then you might have called it 'shoot'. If you don't add a label, then SPSS will use the variable name (in this case shoot) in all of the output from an analysis. That's all well and good, but believe me, a few months down the line when you return to the data and analysis you'll probably think 'what the bugger did shoot stand for?' or 'what the hell is variable sftg45c?' So, get into a good habit and label all of your variables!

Values — This column is for assigning numbers to represent groups of people (see section 2.4.4 below).

Missing — This column is for assigning numbers to missing data (see section 2.4.6 below).

Measure — This is where you define the level at which a variable was measured (*Nominal*, *Ordinal* or *Scale*).

To enter the data above into the SPSS data editor we need to create several variables. If we begin with the variable **No. of friends**, we should follow these steps:

1. Move the on-screen arrow (using the mouse) to the first white cell in the column labelled *Name*.
2. Type the word *friends* (remember we can only use a name shorter than eight characters so we can't type in *number of friends*).
3. Move off this cell using the arrow keys on the keyboard (you can also just click on a different cell, but this is actually a very slow way of doing it).

You've just created your first variable! You should notice that once you've typed a name, SPSS creates default settings for the variable (such as assuming its numeric and assigning 2 decimal places). Now because I want you to get into good habits, move to the cell in the column labelled *Label* and type 'Number of Friends'. Finally, we can specify the level at which a variable was measured (see Box 2.1) by going to the column labelled *Measure* and selecting *Nominal*, *Ordinal* or *Scale* from the drop-down list (see Box 2.1).

Scale
Ordinal
Nominal

Box 2.1

Levels of measurement ①

There are various levels at which variables can be measured: *nominal*, *ordinal*, *interval* and *ratio*. The lowest level is nominal data, where numbers merely represent names. For example, if we asked people whether reading this chapter bores them they will answer *yes* or *no*. Therefore, people fall into two categories: bored and not bored. There is no indication as to exactly how bored the bored people are and therefore the data are merely labels, or categories into which people can be placed. If we assign a number to 'bored' and 'not bored' then obviously the value of the number is irrelevant: it merely represents a category (so, if you are using the variable as a coding variable (next section) then the data are *nominal*). Ordinal data give us more information

(Continued)

Box 2.1 (Continued)

than nominal data. If we use an ordinal scale to measure something, we can tell not only that things have occurred, but also the order in which they occurred. However, these data tell us nothing about the differences between values. For example, if we ordered three people according to how bored they were—most bored, middle and least bored. These labels do tell us something about the level of boredom. In using ordered categories we now know that the most bored person was more bored than the least bored person! Interval data are scores that are measured on a scale along the whole of which intervals are equal. For example, rather than asking people if they are bored we could measure boredom along a 10-point scale (0 being very interested and 10 being very bored). For data to be interval it should be true that the increase in boredom represented by a change from 3 to 4 along the scale should be the same as the change in boredom represented by a change from 9 to 10. Ratio data have this property, but in addition we should be able to say that someone who had a score of 8 was twice as bored as someone who scored only 4. These two types of data are represented by the *Scale* option in SPSS. It should be obvious that in some social sciences (notably psychology) it is extremely difficult to establish whether data are interval (can we really tell whether a change on the boredom scale represents a genuine change in the experience of boredom?)—see Field & Hole (2003).

Once the variable has been created, you can return to the data view by clicking on the 'Data View' tab at the bottom of the data editor (Data View Variable View). The contents of the window will change, and you'll notice that the first column now has the label *friends*. To enter the data, click on the white cell at the top of the column labelled 'friends' and type the first value, 5. To register this value in this cell, we have to move to a different cell and because we are entering data down a column, the most sensible way to do this is to press the ↓ key on the keyboard. This action moves you down to the next cell, and the number 5.00 should appear in the cell above. Enter the next number, 2, and then press ↓ to move down to the next cell, and so on.

2.4.4. Creating coding variables ①

In the previous sections I have mentioned coding variables and this section is dedicated to a fuller description of this kind of variable (it is a type of variable that you will use a lot). A coding variable (also known as a grouping variable) is a variable consisting of a series of numbers that represent levels of a treatment variable or describe different groups of people. In experiments, coding variables are used to represent independent variables that have been measured between groups (i.e. different participants were assigned to different groups). So, if you were to run an experiment with one group of participants in an experimental condition and a different group of participants in a control group, you might assign the experimental group a code of 1, and the control group a code of 0. When you come to put the data into the data editor, then you would create a variable (which you might call **group**) and type in the value 1 for any participants in the experimental group, and 0 for any participant in the control group. These codes tell the computer that all of the cases that have been assigned the value 1 should be treated as belonging to the same group, and likewise for the cases assigned the value 0. In situations other than experiments, you might simply use codes to distinguish naturally occurring groups of people (e.g. you might give students a code of 1 and lecturers a code of 0).

There is a simple rule for how variables should be placed in the SPSS data editor: levels of the between-group variables go down the data editor whereas levels of within-subject (repeated measures) variables go across the data editor. We shall see exactly how to put this rule into operation in Chapter 6.

We have a coding variable in our data: the one describing whether a person was a lecturer or student. To create this coding variable, we follow the steps for creating a normal variable, but we also have to tell the computer which numeric codes have been assigned to which groups. So, first of all, return to the variable view if you're not already in it and then move to the cell in the second row of the data editor and in the column labelled *Name* type a name (let's call it **group**). I'm still trying to instil good habits, so move along the second row to the column called *Label* and give the variable a full description such as 'Is the person a lecturer or a student?' Then to define the group codes, move along the second row to the column labelled Values. You'll see that this cell says *None* and there's a button (see Figure 2.4). Click on this button to access the *value labels* dialog box (see Figure 2.4).

The *value labels* dialog box is used to specify group codes. This can be done in three easy steps. First, click with the mouse in the white space next to where it says *Value* (or press *Alt* and *u* at the same time) and type in a code (e.g. 1). These codes are completely arbitrary; for the sake of convention people typically use 0, 1, 2, 3, etc., but in practice you could have a code of 495 if you were feeling particularly arbitrary. The second step is to click the mouse in the white space below, next to where it says *Value Label* (or press *tab*, or *Alt* and *e* at the same time) and type in an appropriate label for that group. In Figure 2.4 I have typed in 2 as my code and given this a label of *Student*. The third step is to add this coding to the list by clicking on Add. In Figure 2.4 I have already defined a code of 1 for the lecturer group; to add the coding for the student group click on Add. When you have defined all of your coding values simply click on OK; if you click on OK and have forgotten to add your final coding to the list, SPSS will display a message warning you that any pending changes will be lost. In plain English this simply tells you to go back and click on Add before continuing. Finally, coding variables always represent categories and so the level at which they are measured is nominal (or ordinal if the categories have a meaningful order). Therefore, you should specify the level at which the variable was measured by going to the column labelled *Measure* and selecting *Nominal* (or *Ordinal* if the groups have a meaningful order) from the drop-down list (see earlier).

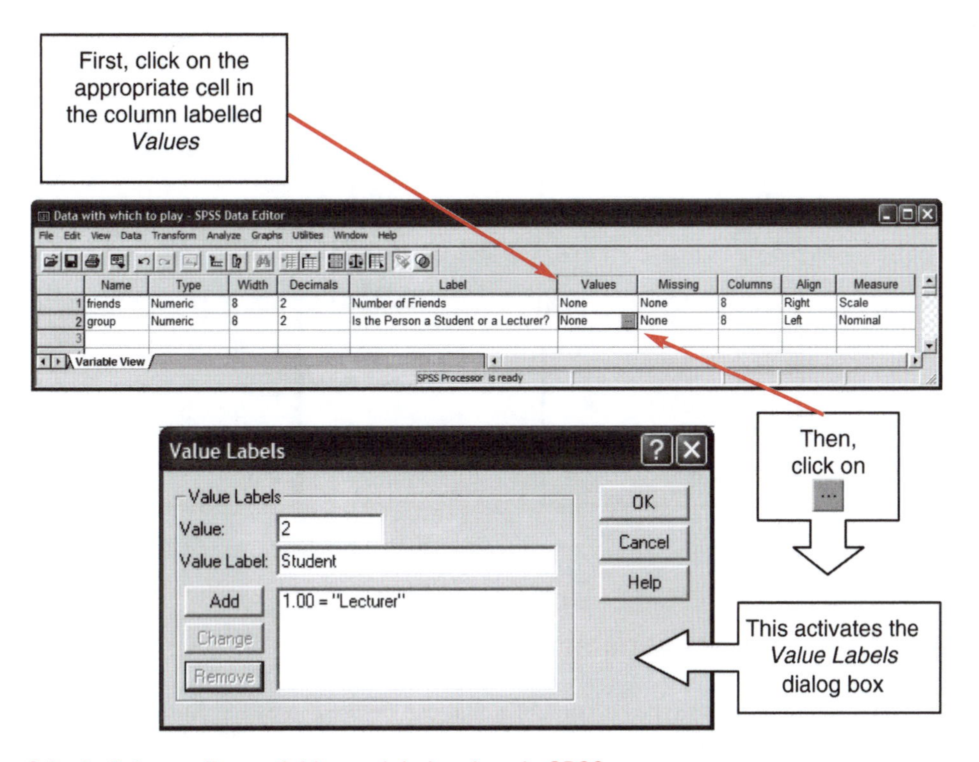

Figure 2.4 Defining coding variables and their values in SPSS

Having defined your codes, you can then switch to the data view and type these numerical values into the appropriate column (so if a person was a lecturer, type a value of 1; if they were a student, type the value 2). What is really groovy is that you can get the computer to display the codes themselves, or the value labels that you gave them by clicking on [icon] (see Figure 2.5). Figure 2.5 shows how the data should be arranged for a coding variable. Now remember that each row of the data editor represents one participant's data and so in this example it is clear that the first five participants were lecturers whereas participants 6–10 were students. This example also demonstrates why grouping variables are used for variables that have been measured between subjects: because by using a coding variable it is impossible for a participant to belong to more than one group. This situation should occur in a between-group design (i.e. a participant should not be tested in both the experimental and the control group). However, in repeated-measures designs (within subjects) each participant is tested in every condition and so we would not use this sort of coding variable (because each participant does take part in every experimental condition).

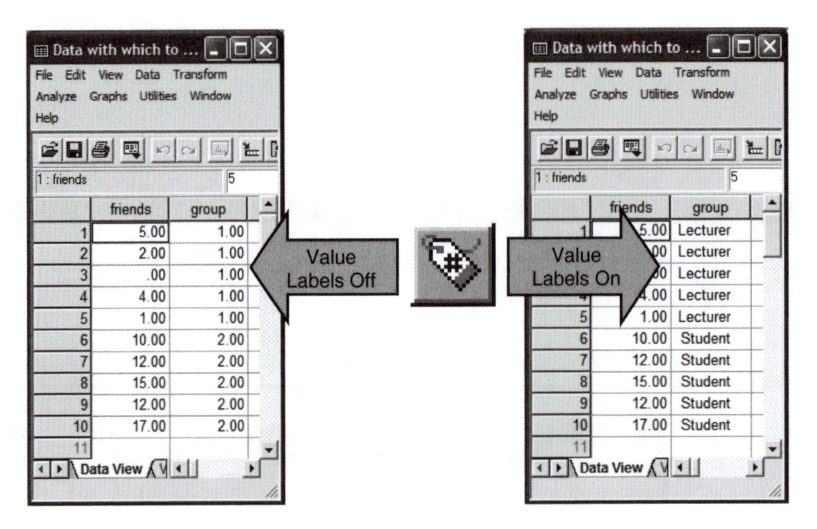

Figure 2.5 Coding values in the data editor with the value labels switched off and on

2.4.5. Types of variables ①

There are different types of variables that can be used in SPSS. In the majority of cases you will find yourself using numeric variables. These variables are ones that contain numbers and include the type of coding variables that have just been described. However, one of the other options when you create a variable is to specify the type of variable and this is done by clicking in the column labelled Type in the variable view and then clicking on ... (rather like when we specified value labels in Figure 2.4). This will activate the *variable type* dialog box in Figure 2.6, which shows the default settings. By default, a variable is set up to be *numeric* and store eight digits, but you can change this value by typing a new number in the space labelled *Width* in the dialog box. Under normal circumstances you wouldn't require SPSS to retain any more than eight characters unless you were doing calculations that need to be particularly precise. Another default setting is to have 2 decimal places displayed (in fact, you'll notice by default that when you type in whole numbers SPSS will add a decimal place with two zeros after it—this can be disconcerting initially!). It is easy enough to change the number of decimal places for a given variable by simply replacing the 2 with a new value depending on the level of precision you require.

The *variable type* dialog box also allows you to specify other types of variable. For the most part you will use numeric values. However, the other variable type of use is a string variable. A string variable is simply a line of text and could represent comments about a certain participant, or other information that you don't wish to analyse as a grouping variable (such as the participant's name). If you select the string variable option, SPSS lets you specify the width of the string variable (which by default is eight characters) so that you can insert longer strings of text if necessary.

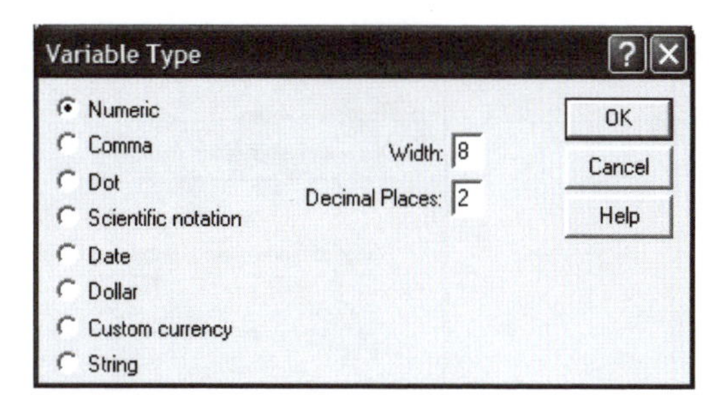

Figure 2.6 Defining the type of variable being used

2.4.6. Missing values ①

Although as researchers we strive to collect complete sets of data, it is often the case that we have missing data. Missing data can occur for a variety of reasons: in long questionnaires participants accidentally miss out questions; in experimental procedures mechanical faults can lead to a datum not being recorded; and in research on delicate topics (e.g. sexual behaviour) participants may exert their right not to answer a question. However, just because we have missed out on some data for a participant doesn't mean that we have to ignore the data we do have (although it sometimes creates statistical difficulties). However, we do need to tell the computer that a value is missing for a particular person. The principle behind missing values is quite similar to that of coding variables in that we choose a numeric value to represent the missing data point. This value simply tells the computer that there is no recorded value for a participant for a certain variable. The computer then ignores that cell of the data editor (it does not use the value you select in the analysis). You need to be careful that the chosen code doesn't correspond with any naturally occurring data value. For example, if we tell the computer to regard the value 9 as a missing value and several participants genuinely scored 9, then the computer will treat their data as missing when, in reality, it is not.

To specify missing values you simply click in the column labelled Missing in the variable view and then click on ... to activate the *missing values* dialog box in Figure 2.7. By default SPSS assumes that no missing values exist but if you do have data with missing values you can choose to define them in one of three ways. The first is to select discrete values (by clicking on the circle next to where it says *Discrete missing values*) which are single values that represent missing data. SPSS allows you to specify up to three discrete values to represent missing data. The reason why you might choose to have several numbers to represent missing values is that you can assign a different meaning to each discrete value. For example, you could have the number 8 representing a response of 'not applicable', a code of 9 representing a 'don't know' response, and a code of 99 meaning that the participant failed to give any response. As far as the computer is concerned it will ignore any data cell containing these values; however, using different codes may be a useful way to remind you of why a particular score is missing. Usually, one discrete value is enough and in an experiment in which attitudes are measured on a 100-point scale

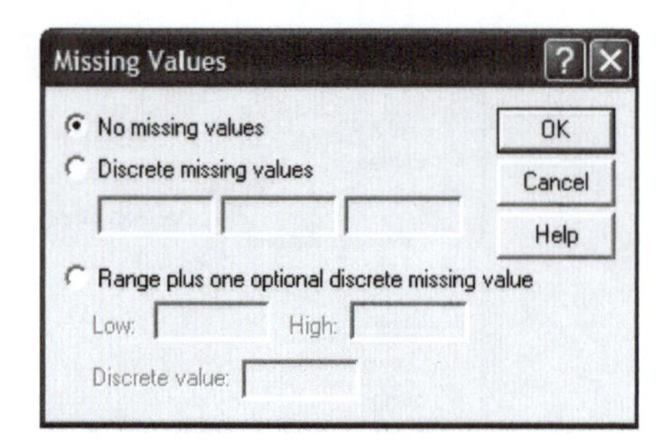

Figure 2.7 Defining missing values

(so scores vary from 1 to 100) you might choose 666 to represent missing values because (1) this value cannot occur in the data that have been collected and (2) because missing data create statistical problems, you will regard the people who haven't given you responses as children of Satan! The second option is to select a range of values to represent missing data and this is useful in situations in which it is necessary to exclude data falling between two points. So, we could exclude all scores between 5 and 10. The final option is to have a range of values and one discrete value.

2.4.7. Changing the column format ①

The final option available to us when we define a variable is to adjust the formatting of the column within the data editor. The default option is to have a column that is eight characters wide with all numbers and text aligned to the right-hand side of the column. Both of these defaults can be changed. In the variable view, you can increase or decrease the default value of 8 by clicking the up and down arrows ( ). To change the alignment of the column move to the column labelled Align and click in the cell for the variable you want to change. You can then activate the drop-down list by clicking on and select one of the options: *Left*, *Right* or *Center*. It is particularly useful to adjust the column width when you have a coding variable with value labels that exceed eight characters in length.

Right
Left
Right
Center

2.4.8. Quick test ①

Having created the variables **friends** and **type** with a bit of guidance, try to enter the rest of the variables in Table 2.1 yourself. The finished data and variable views should look like those in Figure 2.8 (more or less!).

2.5. THE OUTPUT VIEWER ①

Alongside the data editor window, there is a second window known as the *viewer* (or the output viewer as I prefer to call it). In days of old, the output window (as it was called then)

Hooray, I've entered my first set of data! - SPSS Data Editor

File Edit View Data Transform Analyze Graphs Utilities Window Help

1 : friends 5

	friends	group	alcohol	income	nerotic
1	5.00	1.00	10.00	20000.00	10.00
2	2.00	1.00	15.00	40000.00	17.00
3	.00	1.00	20.00	35000.00	14.00
4	4.00	1.00	5.00	22000.00	13.00
5	1.00	1.00	30.00	50000.00	21.00
6	10.00	2.00	25.00	5000.00	7.00
7	12.00	2.00	20.00	100.00	13.00
8	15.00	2.00	16.00	3000.00	9.00
9	12.00	2.00	17.00	10000.00	14.00
10	17.00	2.00	18.00	10.00	13.00
11					

Data View / Variable View

SPSS Processor is ready

Hooray, I've entered my first set of data! - SPSS Data Editor

File Edit View Data Transform Analyze Graphs Utilities Window Help

	Name	Type	Width	Decimals	Label	Values	Missing	Columns	Align	Measure
1	friends	Numeric	8	2	Number of Frie	None	None	8	Right	Scale
2	group	Numeric	8	2	Is the Person	{1.00, Lecturer	None	8	Right	Nominal
3	alcohol	Numeric	8	2	Alcohol Consu	None	None	8	Right	Scale
4	income	Numeric	8	2	Income (Per A	None	None	8	Right	Scale
5	nerotic	Numeric	8	2	Neurotisim	None	None	8	Right	Scale
6										

Data View / Variable View

SPSS Processor is ready

Figure 2.8 Finished data view and variable view for the data in Table 2.1

displayed only statistical results in a rather bland font, and any graphs were displayed in a separate window.[7] The new, improved and generally amazing output viewer happily displays graphs, tables and statistical results in the same window (and all in a much nicer font and nicely arranged tables). My prediction in the last edition of this book that future versions of SPSS will include a tea-making facility in the output viewer have sadly not come to fruition (SPSS Inc. take note!!).

Figure 2.9 shows the basic layout of the output viewer. On the right-hand side there is a large space in which the output is displayed. SPSS displays both graphs and the results of statistical analyses in this part of the viewer. It is also possible to edit graphs and to do this you simply double-click on the graph you wish to edit (this creates a new window in which the graph can be edited). On the left-hand side of the output viewer there is a tree diagram illustrating the structure of the output. This tree diagram is useful when you have conducted several analyses because it provides an easy way of accessing specific parts of the

7 It became the *output navigator* in versions 7.0 and 7.5, but changed its name to *viewer* in version 8.

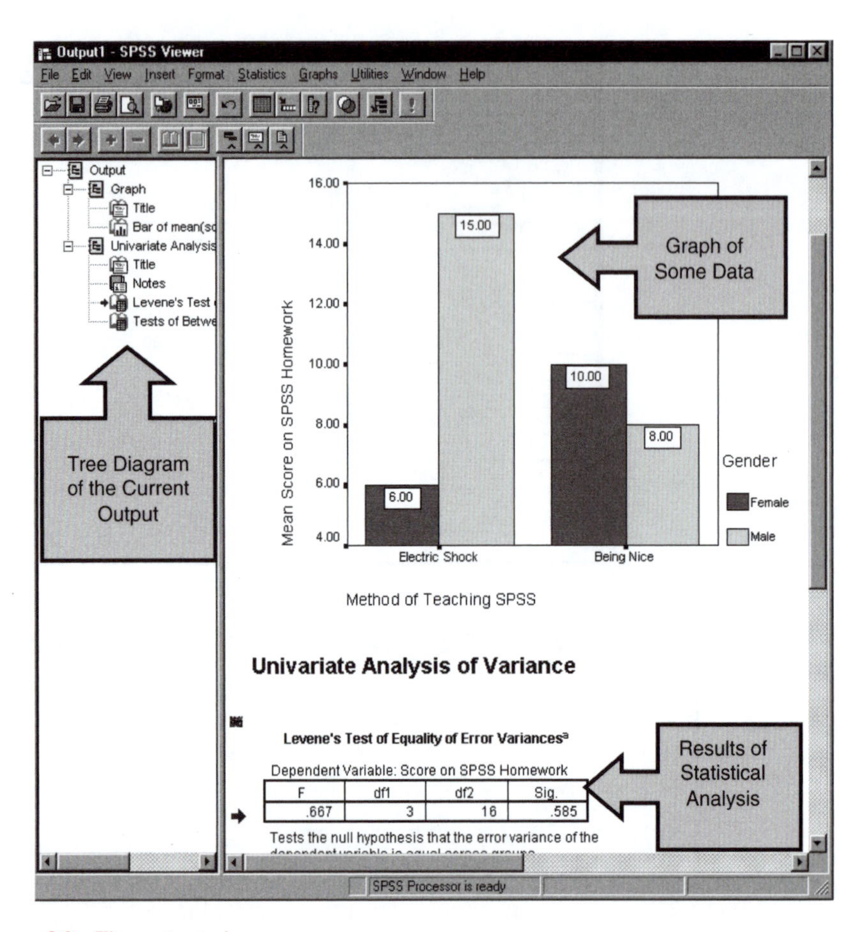

Figure 2.9 The output viewer

output. The tree structure is fairly self-explanatory in that every time you conduct a procedure (such as drawing a graph or running a statistical procedure), SPSS lists this procedure as a main heading.

In Figure 2.9 I conducted a graphing procedure and then conducted a univariate analysis of variance (ANOVA) and so these names appear as main headings. For each procedure there are a series of sub-procedures, and these are listed as branches under the main headings. For example, in the ANOVA procedure there are several sections to the output such as a Levene's test (which tests the assumption of homogeneity of variance) and the between-group effects (i.e. the F-test of whether the means are significantly different). You can skip to any one of these sub-components of the ANOVA output by clicking on the appropriate branch of the tree diagram. So, if you wanted to skip straight to the between-group effects you should move the on-screen arrow to the left-hand portion of the window and click where it says *Tests of Between-Subjects Effects*. This action will highlight this part of the output in the main part of the viewer. You can also use this tree diagram to select parts of the output (which is useful for printing). For example, if you decided that you wanted to print out a graph but you didn't want to print the whole output, you can click on the word *Graph* in the tree structure and that graph will become highlighted in the output. It is then possible through the print menu to select to print only the selected part of the output. In this context it is worth noting that if you click on a main heading (such as *Univariate Analysis of Variance*) then SPSS will highlight not only that main heading but all of the sub-components as well. This is extremely useful when you want to print the results of a single statistical procedure.

There are several icons in the output viewer window that help you to do things quickly without using the drop-down menus. Some of these icons are the same as those described for the data editor window so I will concentrate mainly on the icons that are unique to the viewer window.

As with the data editor window, this icon activates the print menu. However, when this icon is pressed in the viewer window it activates a menu for printing the output. When the print menu is activated you are given the default option of printing the whole output, or you can choose to select an option for printing the output currently visible on the screen, or most useful is an option to print a selection of the output. To choose this last option you must have already selected part of the output (see above).

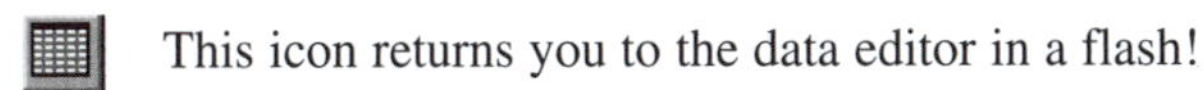

This icon returns you to the data editor in a flash!

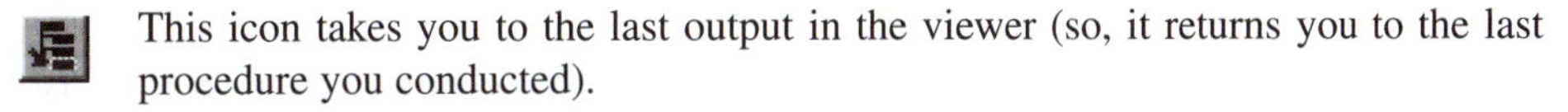

This icon takes you to the last output in the viewer (so, it returns you to the last procedure you conducted).

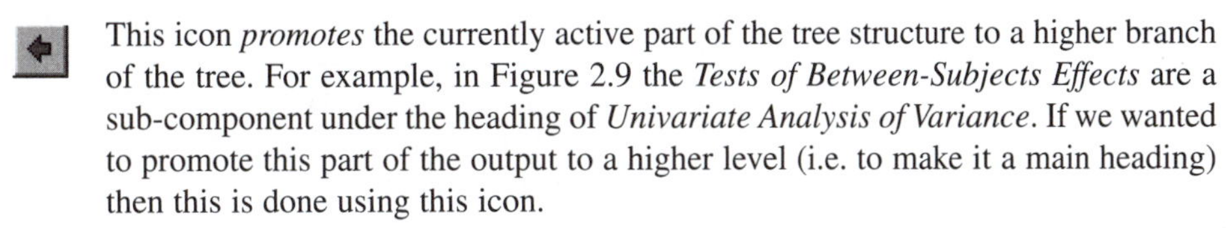

This icon *promotes* the currently active part of the tree structure to a higher branch of the tree. For example, in Figure 2.9 the *Tests of Between-Subjects Effects* are a sub-component under the heading of *Univariate Analysis of Variance*. If we wanted to promote this part of the output to a higher level (i.e. to make it a main heading) then this is done using this icon.

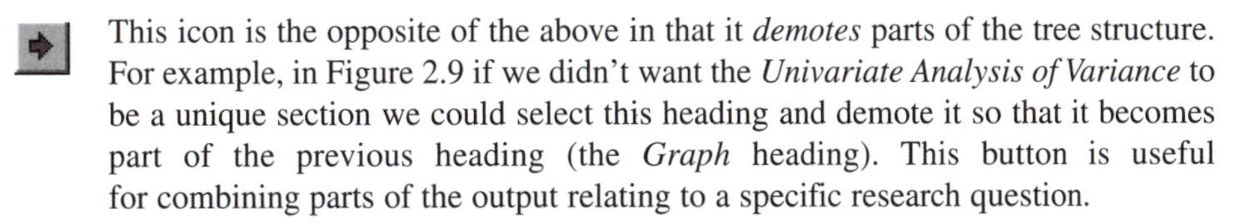

This icon is the opposite of the above in that it *demotes* parts of the tree structure. For example, in Figure 2.9 if we didn't want the *Univariate Analysis of Variance* to be a unique section we could select this heading and demote it so that it becomes part of the previous heading (the *Graph* heading). This button is useful for combining parts of the output relating to a specific research question.

This icon collapses parts of the tree structure, which simply means that it hides the sub-components under a particular heading. For example, in Figure 2.9 if we selected the heading *Univariate Analysis of Variance* and pressed this icon, all of the sub-headings would disappear. The sections that disappear from the tree structure don't disappear from the output itself; the tree structure is merely condensed. This can be useful when you have been conducting lots of analyses and the tree diagram is becoming very complex.

This icon expands any collapsed sections. By default all of the main headings are displayed in the tree diagram in their expanded form. If, however, you have opted to collapse part of the tree diagram (using the icon above) then you can use this icon to undo your dirty work.

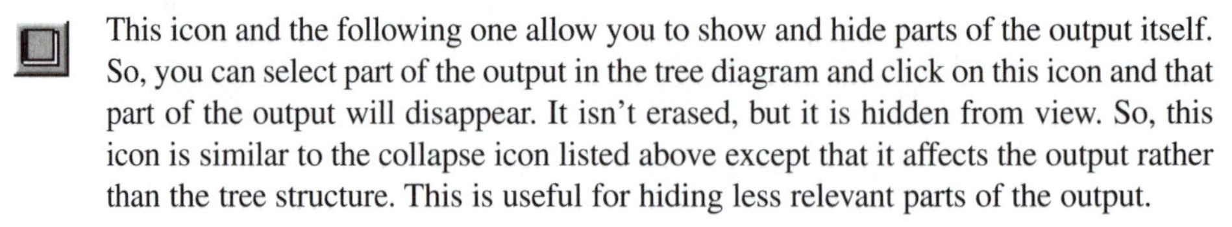

This icon and the following one allow you to show and hide parts of the output itself. So, you can select part of the output in the tree diagram and click on this icon and that part of the output will disappear. It isn't erased, but it is hidden from view. So, this icon is similar to the collapse icon listed above except that it affects the output rather than the tree structure. This is useful for hiding less relevant parts of the output.

This icon undoes the previous one, so if you have hidden a selected part of the output from view and you click on this icon, that part of the output will reappear. By default, all parts of the output are shown and so this icon is not active: it will become active only when you have hidden part of the output.

Although this icon looks rather like a paint roller, it unfortunately does not paint the house for you. What it does do is to insert a new heading into the tree diagram. For example, if you had several statistical tests that related to one of many research questions you could insert a main heading and then demote the headings of the relevant analyses so that they all fall under this new heading.

Assuming you had done the above, you can use this icon to provide your new heading with a title. The title you type in will actually appear in your output. So, you might have a heading like 'Research Question number 1' which tells you that the analyses under this heading relate to your first research question.

This final icon is used to place a text box in the output window. You can type anything into this box. In the context of the previous two icons, you might use a text box to explain what your first research question is (e.g. 'My first research question is whether or not boredom has set in by the end of the first chapter of my book. The following analyses test the hypothesis that boredom levels will be significantly higher at the end of the first chapter than at the beginning.').

2.6. THE SYNTAX WINDOW ③

I've mentioned earlier that sometimes it's useful to use SPSS syntax. This is a language of commands for carrying out statistical analyses and data manipulations. Most of the time you'll do the things you need to using SPSS dialog boxes; however, SPSS syntax can be useful. For one thing there are certain things you can do with syntax that you can't do through dialog boxes (admittedly most of these things are fairly advanced, but there will be a few places in this book where I show you some nice tricks using syntax). The second reason for using syntax is if you often carry out very similar analyses on data sets. In these situations it is often quicker to do the analysis and save the syntax as you go along. Fortunately this is easily done because many dialog boxes in SPSS have a Paste button. When you've specified your analysis using the dialog box, if you click on this button it will

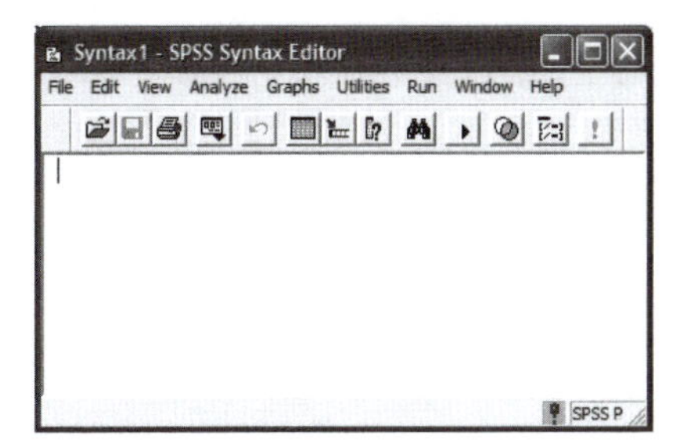

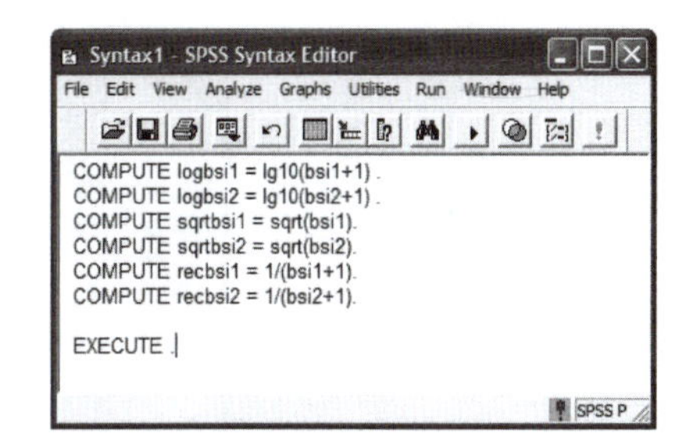

Figure 2.10 A new syntax window (left) and a syntax window with some syntax in it (right)

paste the syntax into a syntax window for you. To open a syntax window simply use the menus **File⇒New⇒Syntax...** and a blank syntax window will appear as in Figure 2.10. In this window you can type your syntax commands. The rules of SPSS syntax can be quite frustrating (e.g. each line has to end with a full stop and if you forget this full stop you'll get an error message), and way beyond the scope of this book. However, as we go through I'll show you a few things that will give you a flavour of how syntax can be used. Most of you won't have to use it, but for those that do this flavour will hopefully be enough to start you on your way. Once you've typed in your syntax you have to run it using the **Run** menu. **Run⇒All** will run all of the syntax in the window (clicking on will also do this), or you can highlight a selection of your syntax using the mouse and use **Run⇒Selection** to process the selected syntax. You can also run the current command by using **Run⇒Current** (or press *Ctrl* and *R* on the keyboard), or run all the syntax from the cursor to the end of the syntax window using **Run⇒To End**.

2.7. SAVING FILES ①

Although most of you should be familiar with how to save files in Windows it is a vital thing to know and so I will briefly describe what to do. To save files simply use the icon (or use the menus **File⇒Save** or **File⇒Save As...**). If the file is a new file, then clicking on this icon will activate the *Save Data As* dialog box (see Figure 2.11). If you are in the data editor when you select *Save As...* then SPSS will save the data file you are currently working on, but if you are in the viewer window then it will save the current output.

There are several features of the dialog box in Figure 2.11. First, you need to select a location at which to store the file. Typically, there are two types of locations where you can save data: the hard drive (or drives) and the floppy drive (and with the advent of rewritable CD-ROM drives, zip drives, jaz drives and the like you may have many other choices of location on your particular computer). The first thing to do is select either the floppy drive, by double clicking on , or the hard drive, by double clicking on . Once you have

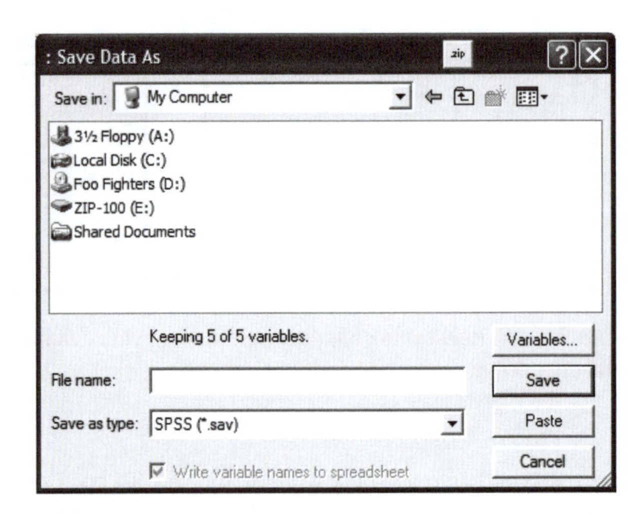

Figure 2.11 The *Save Data As* dialog box

chosen a main location the dialog box will display all of the available folders on that particular device (you may not have any folders on your floppy disk in which case you can create a folder by clicking on). Once you have selected a folder in which to save your file, you need to give your file a name. If you click in the space next to where it says *File name*, a cursor will appear and you can type a name of up to 10 letters. By default, the file will be saved in an SPSS format, so if it is a data file it will have the file extension *.sav*, if it is a viewer document it will have the file extension *.spo*, and if it is a syntax file it will have the file extension *.sps*. However, you can save data in different formats such as Microsoft Excel files and tab-delimited text. To do this just click on where it says *Save as type* and a list of possible file formats will be displayed. Click on the file type you require. Once a file has previously been saved, it can be saved again (updated) by clicking on . This icon appears in both the data editor and the viewer, and the file saved depends on the window that is currently active. The file will be saved in the location at which it is currently stored.

2.8. RETRIEVING A FILE ①

Throughout this book you will work with data files that have been provided on a CD-ROM. It is, therefore, important that you know how to load these data files into SPSS. The procedure is very simple. To open a file, simply use the icon (or use the menu **File⇒Open⇒Data...**) to activate the dialog box in Figure 2.12. First, you need to find the location at which the file is stored. If you are loading a file from the CD-ROM then access the CD-ROM drive by clicking on where it says *Look in* and a list of possible location drives will be displayed. Once the CD-ROM drive has been accessed (by clicking on) you should see a list of files and folders that can be opened. As with saving a file, if you are currently in the data editor then SPSS

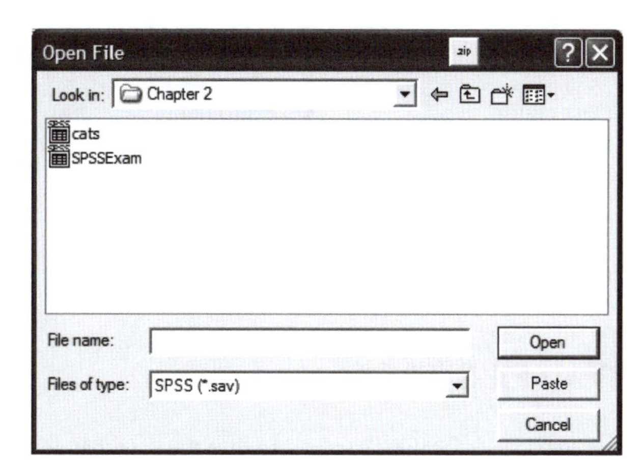

Figure 2.12 Dialog box to open a file

will display only SPSS data files to be opened (if you are in the viewer window then only output files will be displayed). If you use the menus and used the path **File⇒Open⇒Data...** then data files will be displayed, but if you used the path **File⇒Open⇒Output...**, then viewer files will be displayed and if you used **File⇒Open⇒Syntax...** then syntax files will be displayed (you get the general idea). You can open a folder by double-clicking on the folder icon. Once you have tracked down the required file you can open it either by selecting it with the mouse and then clicking on [Open], or by double-clicking on the icon next to the file you want (e.g. double-clicking on [icon]). The data/output will then appear in the appropriate window. If you are in the data editor and you want to open a viewer file, then click on [▼] where it says *Files of type* and a list of alternative file formats will be displayed. Click on the appropriate file type (viewer document (**.spo*), syntax file (**.sps*), Excel file (**.xls*), text file (**.dat, *.txt*)) and any files of that type will be displayed for you to open.

2.9. WHAT HAVE WE DISCOVERED ABOUT STATISTICS? ①

This chapter has provided a basic introduction to the SPSS environment. We've seen that SPSS uses two main windows: the data editor and the viewer. The data editor has both a data view (where you input the raw scores) and a variable view (where you define variables and their properties). The viewer is a window in which any output appears, such as tables, statistics and graphs. We also created our first data set by creating some variables and inputting some data. In doing so we discovered that we can code groups of people using numbers (coding variables) and discovered that rows in the data editor represent people (or cases of data) and columns represent different variables. Finally, we had a look at the syntax window and were told how to open and save files.

2.10. KEY TERMS THAT WE'VE DISCOVERED

- Data editor
- Data view
- Interval data
- Nominal data
- Numeric variables
- Ordinal data
- Ratio data
- String variables
- Syntax editor
- Variable view
- Viewer

2.11. SMART ALEX'S TASK

Smart Alex's task for this chapter is to save the data that you've entered in this chapter. Save it somewhere on the hard drive of your computer (or a floppy disk if you're not working on your own computer). Give it a sensible title and save it somewhere easy to find (perhaps create a folder called 'My Data Files' in the 'My Documents' part of Windows if you're using a PC).①

2.12. FURTHER READING

Einspruch, E. L. (1998). *An introductory guide to SPSS for Windows*. Thousand Oaks, CA: Sage.

Foster, J. J. (2001). *Data analysis using SPSS for Windows: a beginner's guide* (2nd edition). London: Sage.

Kinnear, P. R. & Gray, C. D. (2000). *SPSS for Windows made simple (Release 10)*. Hove: Psychology Press.

CHAPTER 3

EXPLORING DATA

3.1. WHAT WILL THIS CHAPTER TELL US? ①

As the title suggests, this chapter will teach you the first steps in analysing data: exploring it. Wright (2003) uses a quote by Rosenthal who said that researchers should 'make friends with their data'. Now, this wasn't meant to imply that people who use statistics may as well befriend their data because it's the only friend they'll have; instead Rosenthal meant that many researchers often rush their analysis. Wright makes the analogy of a fine wine: you should savour the bouquet and delicate flavours to truly enjoy the experience. That's perhaps overstating the joys of data analysis; however, rushing your analysis is, I suppose, a bit like gulping down a bottle of wine: the outcome is messy and incoherent! This chapter looks at some ways in which you can explore your data. We start off by looking at how we can screen data, check some basic assumptions about the data, and then look at some descriptive statistics and plot some graphs.

3.2. PARAMETRIC DATA ①

3.2.1. Assumptions of parametric data ①

Many of the statistical procedures described in this book are parametric tests based on the normal distribution (which is described in section 1.5). A parametric test is one that requires data from one of the large catalogue of distributions that statisticians have described and for data to be parametric certain assumptions must be true. If you use a parametric test when your data are not parametric then the results are likely to be inaccurate. Therefore, it is very important that you check the assumptions before deciding which statistical test is appropriate. Throughout this book you will become aware of my obsession with assumptions and checking them. Most parametric tests based on the normal distribution have four basic assumptions that must be met for the test to be accurate. Many

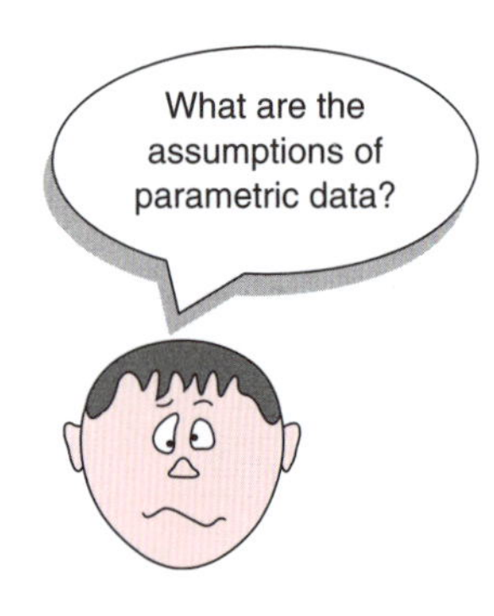

students find checking assumptions a pretty tedious affair, and often get confused about how to tell whether or not an assumption has been met. Therefore, this chapter is designed to take you on a step-by-step tour of the world of parametric assumptions (wow, how exciting!). Now, you may think that assumptions are not very exciting, but they can have great benefits: for one thing you can impress your supervisor/lecturer by spotting all of the test assumptions that they have violated throughout their career. You can then rubbish, on statistical grounds, the theories they have spent their lifetime developing—and they can't argue with you.[1] Well, as I always say, 'if you're going to go through the agony of doing statistics, you may as well do them properly'. The assumptions of parametric tests are as follows:

1. **Normally distributed data**: It is assumed that the data are from one or more normally distributed populations. The rationale behind hypothesis testing relies on having normally distributed populations and so if this assumption is not met then the logic behind hypothesis testing is flawed (we came across these principles in Chapter 1 and they're explained again in Chapter 6). Most researchers eyeball their sample data (using a histogram) and if the sample data look roughly normal, then the researchers assume that the populations are also. We shall see in this chapter that you can go a step beyond this approach and test whether your sample data differ significantly from normal.
2. **Homogeneity of variance**: This assumption means that the variances should be the same throughout the data. In designs in which you test several groups of participants this assumption means that each of these samples comes from populations with the same variance. In correlational designs, this assumption means that the variance of one variable should be stable at all levels of the other variable (see section 3.6).
3. **Interval data**: Data should be measured at least at the interval level (see Box 2.1 in Chapter 2). This means that the distance between points of your scale should be equal at all parts along the scale. For example, if you had a 10-point anxiety scale, then the difference in anxiety represented by a change in score from 2 to 3 should be the same as that represented by a change in score from 9 to 10.
4. **Independence**: This assumption is that data from different participants are independent, which means that the behaviour of one participant does not influence the behaviour of another. In repeated-measures designs (in which participants are measured in more than one experimental condition), we expect scores in the experimental conditions to be non-independent for a given participant, but behaviour between different participants should be independent. As an example, imagine two people, Paul and Julie, were participants in an experiment where they had to indicate whether they

1 When I was doing my Ph.D., we were set a task by our statistics lecturer in which we had to find some published papers and criticize the statistical methods in them. I chose one of my supervisor's papers and proceeded to slate completely every aspect of the data analysis (and I was being *very* pedantic about it all). Imagine my horror when my supervisor came bounding down the corridor with a big grin on his face and declared that, unbeknownst to me, he was the second marker of my essay. Luckily, he had a sense of humour and I got a good mark☺

remembered having seen particular photos earlier on in the experiment. If Paul and Julie were to confer about whether they'd seen certain pictures then their answers would *not* be independent: Julie's response to a given question would depend on Paul's answer, and this would violate the assumption of independence. If Paul and Julie were unable to confer (if they were in different rooms) then their responses should be independent (unless they're telepathic!): Paul's responses should not be influenced by Julie's.

The assumptions of interval data and independent measurements are, unfortunately, tested only by common sense. The assumption of homogeneity of variance is tested in different ways for different procedures and so although we'll discuss this assumption in section 3.6 we will also discuss these tests as the need arises. This leaves us with only the assumption of normality to check and in many respects this is the most important of the assumptions anyway. The easiest way to check this assumption is to look at the distribution of the sample data. If the sample data are normally distributed then we tend to assume that they came from a normally distributed population. Likewise if our sample is not normally distributed then we assume that it came from a non-normal population. So, to test this assumption we can start by plotting how frequently each score occurs (a *frequency distribution*) and see if it looks normal. Let's see how to do this.

3.3. GRAPHING AND SCREENING DATA ①

We'll now have a look at how we can use frequency distributions, other graphs and descriptive statistics to screen our data. We'll use an example to illustrate what to do. A biologist was worried about the potential health effects of music festivals. So, one year she went to the Glastonbury music festival (for those of you outside the UK, you can pretend it is Roskilde festival, Ozzfest, Lollapalooza or something!) and measured the hygiene of 810 concert goers over the three days of the festival. In theory each person was measured on each day but because it was difficult to track people down, there were some missing data on days 2 and 3. Hygiene was measured using a standardized technique (don't worry it *wasn't* licking the person's armpit) that results in a score ranging between 0 (you smell like a rotting corpse that's hiding up a skunk's backside) and 5 (you smell of sweet roses on a fresh spring day). Now I know from bitter experience that sanitation is not always great at these places (Reading festival seems particularly bad ...) and so this researcher predicted that personal hygiene would go down dramatically over the three days of the festival. The data file can be found on the CD-ROM and is called **GlastonburyFestival.sav** (see section 2.8 for a reminder of how to open a file).

3.3.1. Step 1: spot the obvious mistakes using histograms ①

The first thing to do is to plot a histogram to look at the distribution of data. SPSS has the facilities to plot numerous different types of graph. If you click on the *Graphs* menu, a menu bar should drop down showing you the variety of graphs on offer. You may be familiar with some of these graphs (e.g. pie charts, bar charts, etc.):

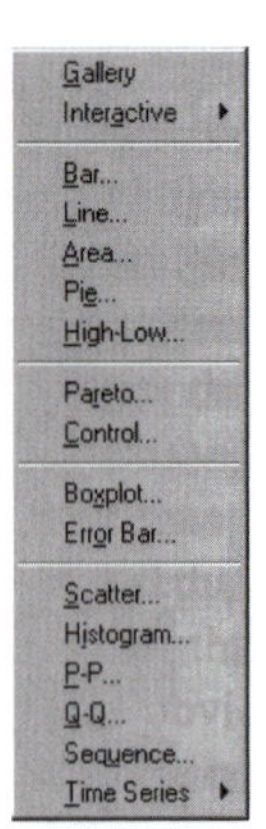

- **Bar**: Usually used to plot means of different groups of people (*Summaries for groups of cases*), or means of different variables (*Summaries of separate variables*).
- **Line**: Also used for plotting means.
- **Pie**: Used for plotting frequencies and percentages.
- **Boxplot**: Box–whisker diagram used for showing the median, spread and interquartile range of scores.
- **Error Bar**: Shows the mean and the 95% confidence interval for that mean (see Chapter 7).
- **Scatter**: Shows relationships between two variables (see Chapter 5).
- **Histogram**: Shows the frequency of different scores (useful for checking the distribution of scores).

You'll notice at the top there is also an option labelled *Interactive* and this opens up another menu that lists much the same graphs as on this menu. The interactive graphs are actually a lot more aesthetically pleasing (by and large) and are supposed to change interactively when you change the data (hence the name!). So, if we want to plot a histogram, we could select **Graphs⇒Interactive⇒Histogram** to open the dialog box in Figure 3.1.

You should note that the variables in the data editor are listed on the left-hand side and you can drag any of these variables into any of the blank spaces. The two axes are represented by arrows and you simply drag the variable into the space. On the vertical axis it already contains a variable called *Count*. This is because we want to plot the frequency of each score and so SPSS will count how often each score occurs. To begin with let's draw a histogram of the hygiene data for day 1 of the festival. Select this variable from the list and drag it into the space by the horizontal axis. The final dialog box should look like Figure 3.1. You might notice a space labelled *Panel variables* and if

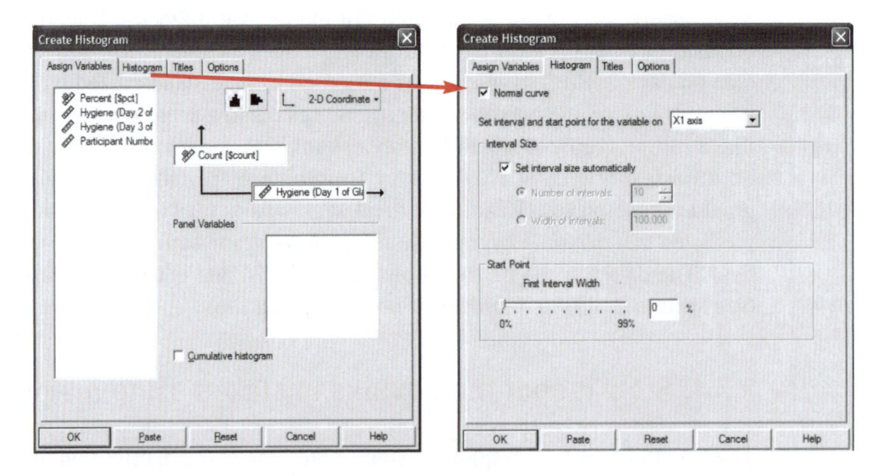

Figure 3.1 Dialog box for an interactive histogram (clicking on the tabs at the top of the dialog box changes the content of the box)

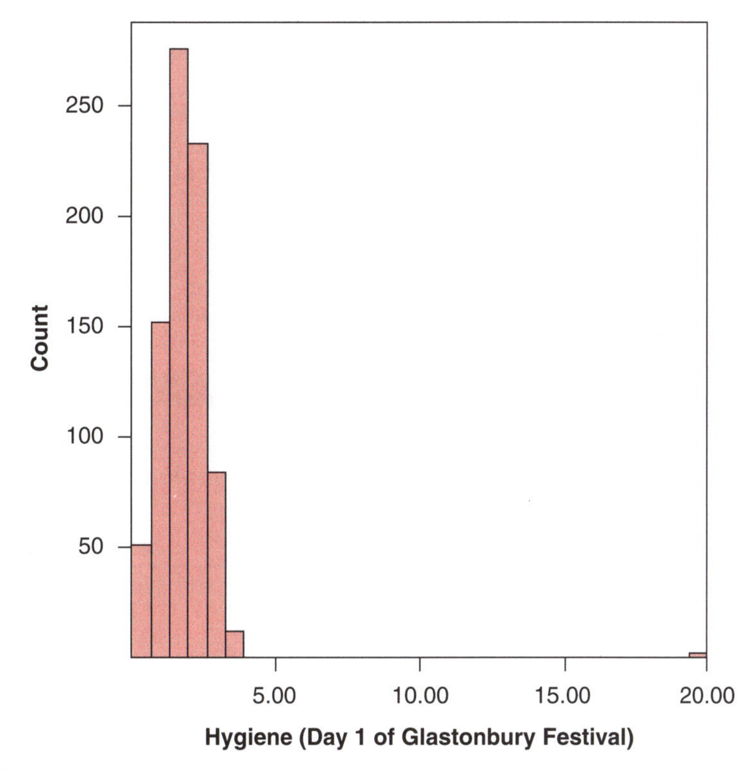

Figure 3.2 Histogram of day 1 of the Glastonbury festival hygiene scores

you drag a variable into that space, SPSS will create a histogram for each level of that variable. So if, for example, you had a coding variable for gender and you placed it in that space, you'd get two histograms: one for women and one for men. The default options are fine so you can now just click on OK (if you want to plot a normal curve over the top you should select the *Histogram* tab at the top and select the option for *Normal curve*).

The resulting histogram is shown in Figure 3.2 and the first thing that should leap out at you is that there appears to be one case that is very different to the others. All of the scores appear to be squashed up at one end of the distribution because they are all less than 5 (yielding a very leptokurtic distribution!) except for one, which has a value of 20. This is an outlier: a score very different to the rest. Outliers bias the mean and inflate the standard deviation (see Field & Hole, 2003) and screening data is an important way to detect them. What's odd about this outlier is that it has a score of 20, which is above the top of our scale (remember our hygiene scale ranged only from 0 to 5) and so it must be a mistake (or the person had obsessive compulsive disorder and had washed themselves into a state of extreme cleanliness). However, with 810 cases, how on earth do we find out which case it

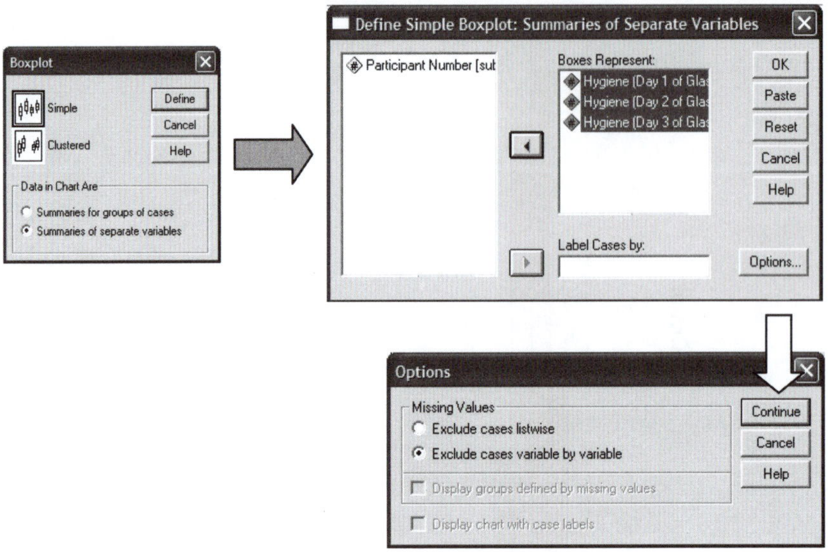

Figure 3.3 Dialog boxes to define a boxplot for several variables

was? We could just look through the data, but that would certainly give us a headache and so instead we can use a boxplot.

I'll talk more about boxplots later in this chapter, so for now just trust me and select **Graphs⇒Boxplot** to access the dialog box. Within that just select *Summaries of separate variables* and click on Define. This opens another dialog box that lists all of the variables in the data editor on the left-hand side and has another box labelled *Boxes Represent*. Select all three variables (representing the three days of the festival) from the list and transfer them to the box labelled *Boxes Represent* by clicking on ▸. Now click on Options... and in that dialog box select *Exclude cases variable by variable*. We have to select this option because some of our variables have missing data and if we *Exclude cases listwise* then if a case has missing values on any of the three variables then data for that case won't be used at all. Now obviously we don't want that; we only want to exclude cases for the variable on which that case has missing values: *Excluding cases variable by variable* does this for us. The final dialog box should look like Figure 3.3. Click on OK to plot the graphs.

The resulting boxplots are shown in Figure 3.4. The important thing to note is that the outlier that we detected in the histogram is shown up as an asterisk on the boxplot and next to it is a number. That number relates to the case of data that's producing this outlier. It's actually case 611. If you go to the data editor (data view), you can quickly locate this case by clicking on the Go to Case button and typing 611 in the dialog box that appears. That takes you straight to case 611. Looking at this case reveals a score of 20.02, which is probably a mis-typing of 2.02. You'd have to go back to the raw data and check. We'll assume you've checked the raw data and it should be 2.02, so replace the value 20.02 with the value 2.02 before continuing this example.

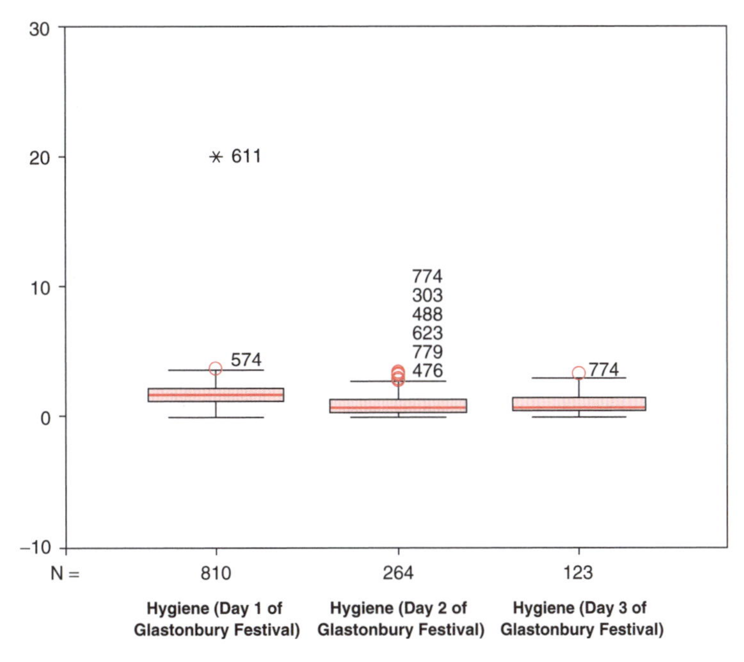

Figure 3.4 Boxplots of the hygiene scores over the three days of the Glastonbury festival

Now that we've removed the mistake let's re-plot the histogram. While we're at it we should plot histograms for the data from day 2 and day 3 of the festival as well. Figure 3.5 shows the resulting three histograms. The first thing to note is that the data from day 1 look a lot more healthy now we've removed the data point that was mis-typed. In fact the distribution is amazingly normal looking: it is nicely symmetrical and doesn't seem too pointy or flat—these are good things! However, the distributions for days 2 and 3 are not nearly so symmetrical. In fact, they both look positively skewed. In general, what this seems to suggest is that by days 2 and 3, hygiene scores were much more clustered around the low end of the scale. Remember that the lower the score the less hygienic the person is, so this suggests that generally people became smellier as the festival progressed. The skew occurs because a substantial minority insisted on upholding their levels of hygiene (against all odds!) over the course of the festival (baby wet-wipes are indispensable I find). However, these skewed distributions might cause us a problem if we want to use parametric tests. In the next section we'll look at ways to try to quantify the skewness and kurtosis of these distributions.

3.3.2. Step 2: descriptive statistics and boxplots ①

Having had a first glance at the distributions of hygiene scores and detected and corrected an incorrectly entered value, we should move on to look at ways to quantify the shape of

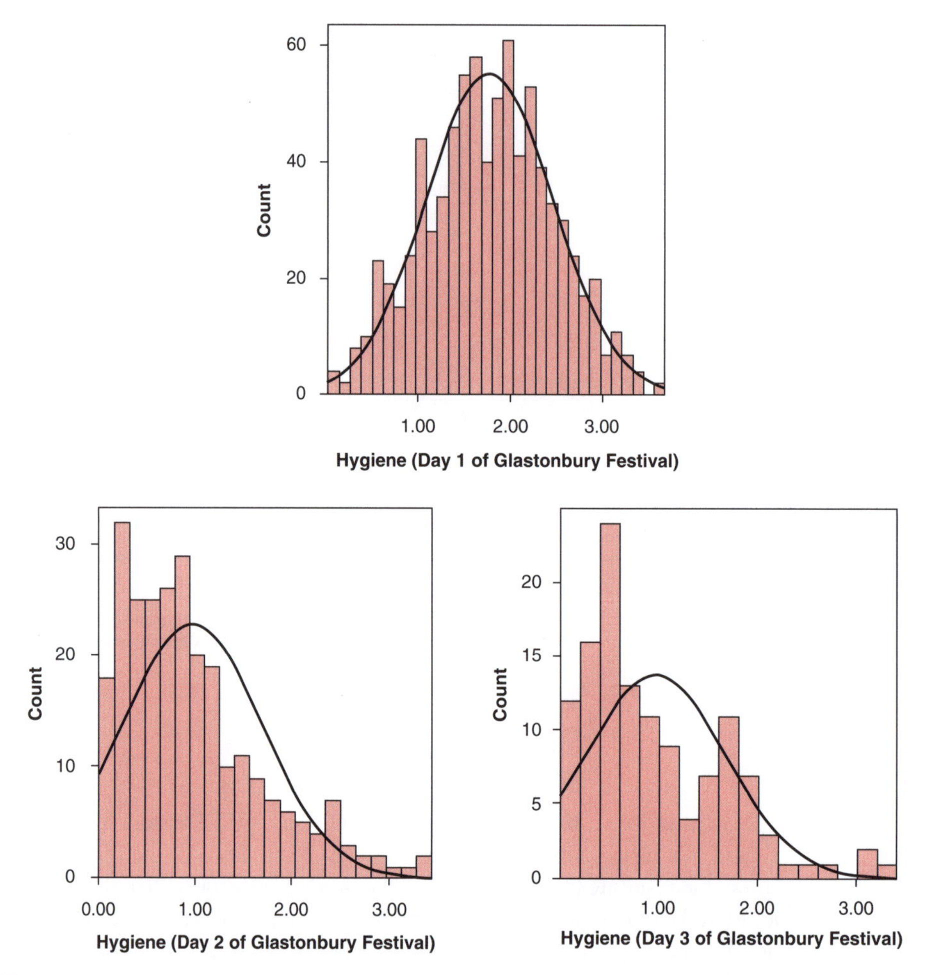

Figure 3.5 Histograms of the hygiene scores over the three days of Glastonbury

the distributions and to look for outliers. To explore the distribution of the variables further, we can use the *frequencies* command, which is accessed using the file path **Analyze⇒Descriptive Statistics⇒Frequencies...**.[2] The main dialog box is shown in Figure 3.6. The variables in the data editor should be listed on the left-hand side, and they can be transferred to the box labelled *Variable(s)* by clicking on a variable (or highlighting several with the mouse) and then clicking on ▶. Any analyses you choose to do will be done on every variable listed in the *Variable(s)* box. If a variable listed in the *Variable(s)* box is

2 Remember that this menu path would be **Statistics⇒Summarize⇒Frequencies...** in version 8.0 and earlier.

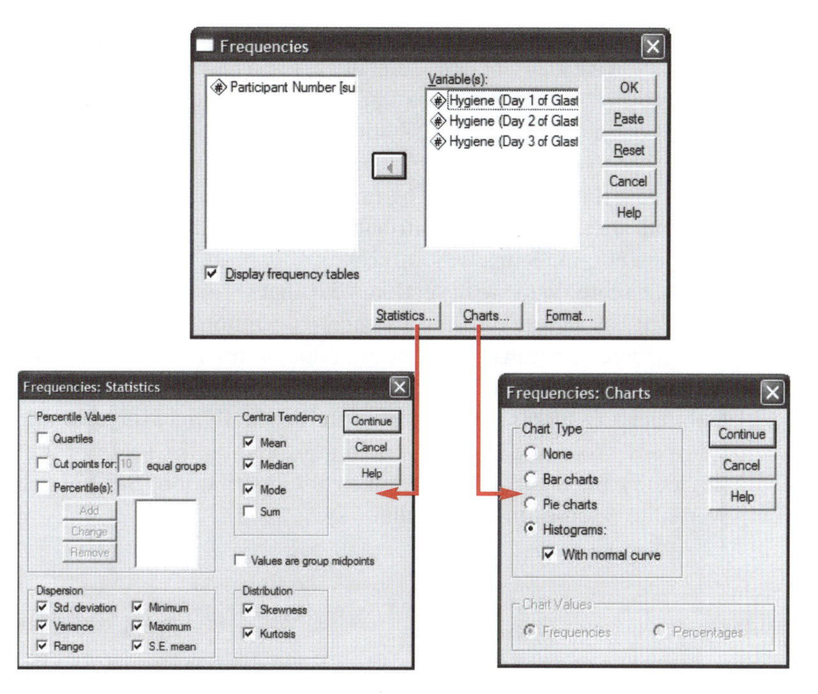

Figure 3.6 Dialog boxes for the *frequencies* command

selected using the mouse, it can be transferred back to the variable list by clicking on the arrow button (which should now be pointing in the opposite direction). By default, SPSS produces a frequency distribution of all scores in table form. However, there are two other dialog boxes that can be selected that provide other options. The *statistics* dialog box is accessed by clicking on Statistics..., and the *charts* dialog box is accessed by clicking on Charts....

The *statistics* dialog box allows you to select several options of ways in which a distribution of scores can be described, such as measures of central tendency (mean, mode, median), measures of variability (range, standard deviation, variance, quartile splits), measures of shape (kurtosis and skewness). To describe the characteristics of the data we should select the mean, mode, median, standard deviation, variance and range. To check that a distribution of scores is normal, we need to look at the values of kurtosis and skewness. The *charts* option provides a simple way to plot the frequency distribution of scores (as a bar chart, a pie chart or a histogram). We've already plotted histograms of our data so we don't need to select these options, but you could use these options in future analyses. When you have selected the appropriate options, return to the main dialog box by clicking on Continue. Once in the main dialog box, click on OK to run the analysis.

SPSS Output 3.1 shows the table of descriptive statistics for the three variables in this example. From this table, we can see that, on average, hygiene scores were 1.77 (out of 5) on day 1 of the festival, but went down to .96 and .98 on day 2 and 3 respectively.

The other important measures for our purposes are the skewness and the kurtosis, both of which have an associated standard error. The values of skewness and kurtosis should be zero in a normal distribution. Positive values of skewness indicate a pile-up of scores on the left of the distribution, whereas negative values indicate a pile-up on the right. Positive values of kurtosis indicate a pointy distribution whereas negative values indicate a flat distribution. The further the value is from zero, the more likely it is that the data are not normally distributed. However, the actual values of skewness and kurtosis are not, in themselves, informative. Instead, we need to take the value and convert it to a *z*-score. We saw in section 1.5.3 that a *z*-score is simply a score from a distribution that has a mean of 0 and a standard deviation of 1. The reason for converting scores to a *z*-score is because it is a way of standardizing them. So, we can take any variable measured in any units and convert it to a *z*-score. By converting to a *z*-score we can compare any scores even if they were originally measured in different units. To transform any score to a *z*-score you simply subtract the mean of the distribution and then divide by the standard deviation of the distribution. Skewness and kurtosis are converted to *z*-scores in exactly this way.

$$z_{\text{skewness}} = \frac{S - 0}{SE_{\text{skewness}}} \qquad z_{\text{kurtosis}} = \frac{K - 0}{SE_{\text{kurtosis}}}$$

In the above equations, the values of S (skewness) and K (kurtosis) and their respective standard errors are produced by SPSS. These *z*-scores can be compared against values that you would expect to get by chance alone (i.e. known values for the normal distribution shown in the Appendix). So, an absolute value greater than 1.96 is significant at $p < .05$, above 2.58 is significant at $p < .01$ and absolute values above about 3.29 are significant at $p < .001$. Large samples will give rise to small standard errors and so when sample sizes are big, significant values arise from even small deviations from normality. In most samples it's OK to look for values above 1.96; however, in small samples this criterion should be increased to the 2.58 one and in very large samples, because of the problem of small standard errors that I've described, no criterion should be applied! If you have a large sample (200 or more) it is more important to look at the shape of the distribution visually and to look at the value of the skewness and kurtosis statistics rather than calculate their significance.

For the hygiene scores, the *z*-score of skewness is –.004/.086 = .047 on day 1, 1.095/.150 = 7.300 on day 2, and 1.033/.218 = 4.739 on day 3. It is pretty clear then that although on day 1 scores are not at all skewed, on days 2 and 3 there is a very significant positive skew shown by z-scores greater than 1.96 (as was evident from the histogram). However, bear in mind what I just said about large samples! The kurtosis *z*-scores are: – .410/.172 = – 2.38 on day 1, .822/.299 = 2.75 on day 2, and .732/.433 = 1.69 on day 3. These values indicate significant kurtosis (at $p < .05$) for all three days, but because of the large sample, this isn't surprising and so we could take comfort in the fact that all values of kurtosis are below our upper threshold of 3.29.

Cramming Samantha's Tips

- To check that the distribution of scores is normal, we need to look at the values of *skewness* and *kurtosis* in the SPSS output.
- Positive values of kurtosis indicate a pointy distribution whereas negative values indicate a flat distribution.
- The further the value is from zero, the more likely it is that the data are not normally distributed.

Statistics

		Hygiene (Day 1 of Glastonbury Festival)	Hygiene (Day 2 of Glastonbury Festival)	Hygiene (Day 3 of Glastonbury Festival)
N	Valid	810	264	123
	Missing	0	546	687
Mean		1.7711	.9609	.9765
Std. Error of Mean		.02437	.04436	.06404
Median		1.7900	.7900	.7600
Mode		2.00	.23	.44[a]
Std. Deviation		.69354	.72078	.71028
Variance		.48100	.51952	.50449
Skewness		−.004	1.095	1.033
Std. Error of Skewness		.086	.150	.218
Kurtosis		−.410	.822	.732
Std. Error of Kurtosis		.172	.299	.433
Range		3.67	3.44	3.39
Minimum		.02	.00	.02
Maximum		3.69	3.44	3.41

a Multiple modes exist. The smallest value is shown

SPSS Output 3.1

The output provides tabulated frequency distributions of each variable (not reproduced here). These tables list each score and the number of times that it is found within the data set. In addition, each frequency value is expressed as a percentage of the sample. Also, the cumulative percentage is given, which tells us how many cases (as a percentage) fell below a certain score. So, for example, we can see that only 15.4% of hygiene scores were below 1 on the first day of the festival. Compare this to the table for day 2: 63.3% of scores were less than 1!

Having checked the shape of the distribution, we need to have a look for outliers (see Box 3.1). There are two ways to do this: (1) look at a boxplot, and (2) look at *z*-scores. I'll leave Smart Alex to deal with *z*-scores in Box 3.2, and I'll stick with the pretty pictures—they're more my kind of level. We have already said how to get SPSS to do boxplots in

Figure 3.3 and so repeat this process for the hygiene data. The resulting boxplots should now look different (remember we changed that incorrect value) and are shown in Figure 3.7.

Box 3.1

Outliers ①

An outlier is a score very different from the rest of the data. When we analyse data we have to be aware of such values because they bias the model we fit to the data. A good example of this bias can be seen by looking at a simple statistical model such as the mean. Field & Hole (2003) describe the example of the customer ratings of the first edition of this book on Amazon.co.uk. These ratings can range from 1 to 5 stars. At the time of writing, seven people had reviewed and rated the first edition with ratings (in the order given) of 2, 5, 4, 5, 5, 5, 5. All but one of these ratings are fairly similar (mainly 5 and 4) but the first rating was quite different from the rest—it was a rating of 2. On the graph there are the seven reviewers on the horizontal axis and their ratings on the vertical axis. On this graph there is a horizontal line that represents the mean of all seven scores (4.43 as it happens). It should be clear that all of the scores except one lie close to this line. The score of 2 is very different and lies some way below the mean. This score is an example of an outlier. The dashed horizontal line represents the mean of the scores when the outlier is not included (4.83). This line is higher than the original mean indicating that by ignoring this score the mean increases (it increases by 0.4). This shows how a single score can bias the mean; in this case the first rating (of 2) drags the average down. In practical terms this has a bigger implication because Amazon round off to half numbers, so that single score has made a difference between the average rating reported by Amazon as a generally glowing 5 stars and the less impressive 4.5 stars (although obviously I am consumed with bitterness about this whole affair, I am actually quietly pleased because it's given me a great example of an outlier!).

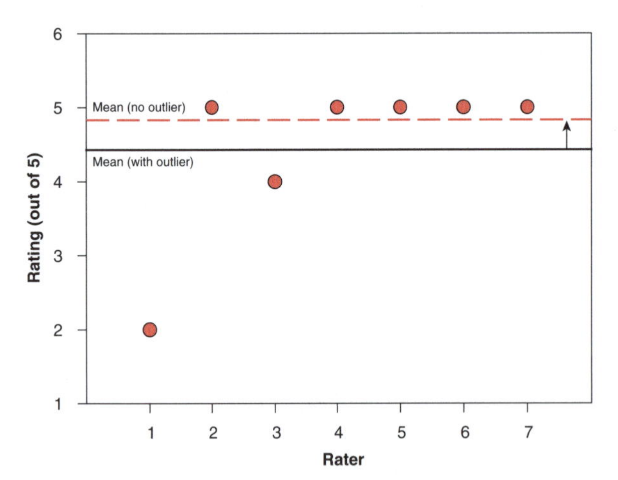

(Data for this example from http://www.amazon.co.uk/.)

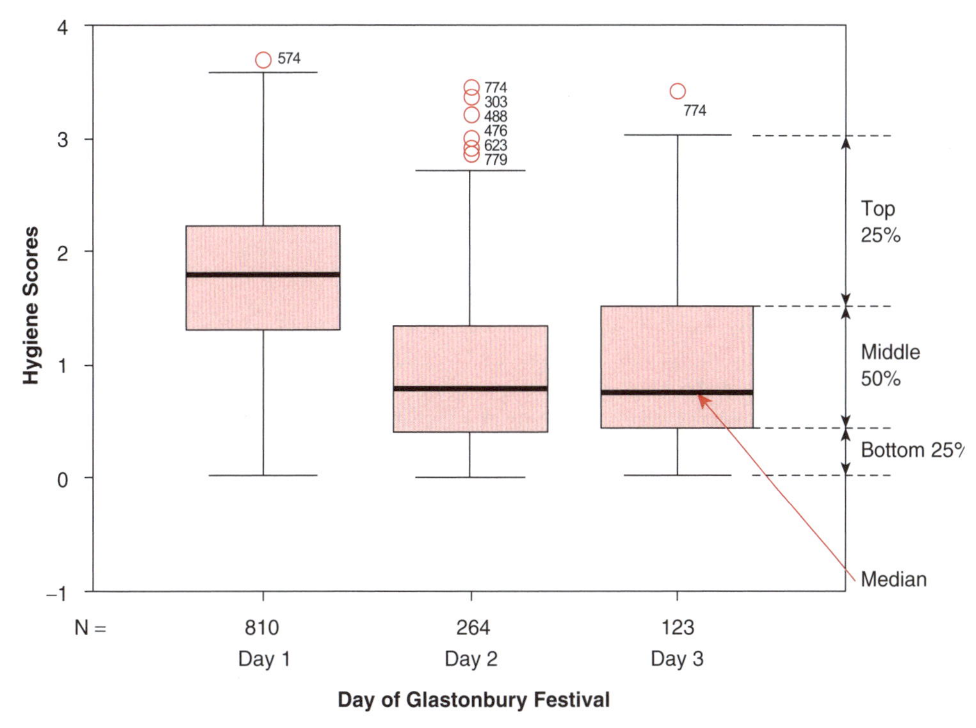

Figure 3.7 Boxplots of hygiene scores for the three days of the Glastonbury festival

What is a boxplot?

The boxplots in Figure 3.7 tell us something about the distributions of the hygiene scores on the three days of the Glastonbury festival. The boxplots (also called *box–whisker diagrams*) show us the lowest score (the bottom horizontal line on each plot) and the highest (the top horizontal line of each plot). The distance between the lowest horizontal line and the lowest edge of the tinted box is the range between which the lowest 25% of scores fall (called the *bottom quartile*). The box (the tinted area) shows the middle 50% of scores (known as the interquartile range): that is, 50% of the scores are bigger than the lowest part of the tinted area but smaller than the top part of the tinted area. The distance between the top edge of the tinted box and the top horizontal line shows the range between which the top 25% of scores fall (the *top quartile*). In the middle of the tinted box is a slightly thicker horizontal line. This represents the value of the median, which if you arranged the hygiene scores in order would be the middle score. So, these diagrams show us the range of scores, the range between which the middle 50% of scores fall, and the median score. Like histograms, they also tell us whether the distribution is symmetrical or skewed. On day 1, which is symmetrical, the whiskers on either side of the box are of equal length (the range of the top and bottom 25% of scores is the same) however, on days 2 and 3 the whisker coming out of the top of the box is longer than that at the bottom, which shows that the distribution is skewed (i.e. the top 25% of scores are spread out over a wider range than the bottom 25%). Finally, you'll notice some circles above each boxplot. These are cases that are deemed to be outliers. Each circle has a number next to it that tells us in which row of the data editor to find that case. In the next section we'll see what can be done about these outliers.

Box 3.2

Using *z*-scores to find outliers ③

To check for outliers we can look at *z*-scores. These are simply a way of *standardizing* your data set. By this I simply mean expressing your scores in terms of a distribution that has a known mean and standard deviation. Specifically, we express scores in terms of a distribution with a mean of 0 and a standard deviation of 1 (see section 1.5). The benefit of this is that we can establish benchmarks that we can apply to any data set (regardless of what its original mean and standard deviation were). To convert a data set into *z*-scores is easy: you simply take each score (X) and convert it to a *z*-score by subtracting the mean of all scores ($\bar{X}$) from it and dividing by the standard deviation of all scores (s):

$$z = \frac{X - \bar{X}}{s}$$

Rather than do this for each score manually, we can get SPSS to do it for us using the **Analyze⇒Descriptive Statistics⇒Descriptives...** dialog box, selecting a variable (such as day 2 of the hygiene data as in the diagram), or several variables, and selecting *Save standardized values as variables* before clicking on OK. SPSS will now create a new variable in the data editor (with the same name prefixed with the letter *z*). We can then use these scores and count how many fall within certain important limits. If we take the absolute value (i.e. we ignore whether the *z*-score is positive or negative) then in a normal distribution we'd expect about 5% to have absolute values greater than 1.96 (we often use 2 for convenience), and 1% to have absolute values greater than 2.58, and none to be greater than about 3.29.

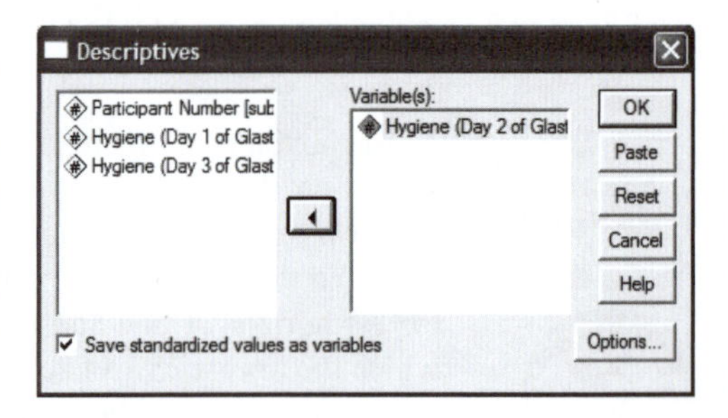

Alternatively, you could use some SPSS syntax in an SPSS syntax window to create the *z*-scores and count them for you. The CD-ROM contains a syntax file, **Outliers (Percentage of Z-scores).sps**, which I've written to produce a table for day 2 of the Glastonbury hygiene data. Load this file and run the syntax, or open a syntax window (see section 2.6) and type the following (remembering all of the full stops)—the explanations of the code are surrounded by *s and don't need to be typed!:

(Continued)

Box 3.2 (Continued)

```
DESCRIPTIVES
VARIABLES= day2/SAVE.
```

*This uses SPSS's descriptives function on the variable day2 (instead of using the dialog box) to save the *z*-scores in the data editor (these will be saved as a variable called zday2).*

```
COMPUTE outlier1 = abs(zday2).
EXECUTE.
```

*This creates a new variable in the data editor called *outlier1*, which contains the absolute values of the *z*-scores that we just created.*

```
RECODE
outlier1 (3.29 thru Highest = 4) (2.58 thru Highest = 3) (1.96 thru Highest = 2) (Lowest thru 2 = 1).
EXECUTE.
```

This recodes the variable called outlier1 according to the benchmarks I've described. So, if a value is greater than 3.29 it's assigned a code of 4, if it's between 2.58 and 3.29 then it's assigned a code of 3, if it's between 1.96 and 2.58 it's assigned a code of 2 and if it's less than 1.96 it gets a code of 1.

```
VALUE LABELS outlier1
1 'Absolute z-score less than 2' 2 'Absolute z-score greater than 1.96' 3 'Absolute z-score greater than 2.58'
4 'Absolute z-score greater than 3.29'.
```

This assigns appropriate labels to the codes we defined above.

```
FREQUENCIES
VARIABLES=outlier1
/ORDER= ANALYSIS .
```

*Finally, this syntax uses the *frequencies* facility of SPSS to produce a table telling us the percentage of ones, twos, threes and fours found in the variable outlier1.*

The table produced by this syntax is shown. Look at the column labelled 'Valid Percent'. We would expect to see 95% of cases with absolute value less than 1.96, 5% (or less) with an absolute value greater than 1.96, and 1% (or less) with an absolute value greater than 2.58. Finally, we'd expect no cases above 3.29 (well, these cases are significant outliers). For hygiene scores on day 2 of Glastonbury, 93.2% of values had *z*-scores less than 1.96. Put another way, 6.8% were above (looking at the table we get this figure by adding 4.5% + 1.5% + 0.8%). This is slightly more than the 5% we would expect in a normal distribution. Looking at values above 2.58, we would expect to find only

(Continued)

Box 3.2 (Continued)

OUTLIER 1

		Frequency	Percent	Valid Percent	Cumulative Percent
Valid	Absolute *z*-score less than 2	246	30.4	93.2	93.2
	Absolute *z*-score greater than 1.96	12	1.5	4.5	97.7
	Absolute *z*-score greater than 2.58	4	.5	1.5	99.2
	Absolute *z*-score greater than 3.29	2	.2	.8	100.0
	Total	264	32.6	100.0	
Missing	System	546	67.4		
Total		810	100.0		

1%, but again here we have a higher value of 2.3% (1.5% + 0.8%). Finally, we find that 0.8% of cases were above 3.29 (so, 0.8% are significant outliers). This suggests that there may be too many outliers in this data set and we might want to do something about them!

3.3.3. Step 3: correcting problems in the data ②

The last section showed us various ways to explore our data; we saw how to look for problems with our distribution of scores and how to detect outliers. The next question is what do we do about these problems? If you detect outliers in the data there are several options for reducing the impact of these values. However, before you do any of these things, it's worth checking that the data have been entered correctly for the problem cases. If the data are correct then the three main options you have are:

1. *Remove the case*: This entails deleting the data from the person who contributed the outlier. However, this should be done only if you have good reason to believe that this case is not from the population that you intended to sample. For example, if you were investigating factors that affected how much cats purr and one cat didn't purr at all, this would likely be an outlier (all cats purr). Upon inspection, if you discovered that this cat was actually a dog (hence why it didn't purr) wearing a cat costume, then you'd have grounds to exclude this case because it comes from a different population (dogs who like to dress as cats) than your target population (cats).
2. *Transform the data*: If you have a non-normal distribution then this should be done anyway (and skewed distributions will by their nature generally have outliers because it's these outliers that skew the distribution!). Such transformations (see below) should reduce the impact of these outliers.

3. *Change the score*: If transformation fails, then you can consider replacing the score. This on the face of it may seem like cheating (you're changing the data from what was actually collected); however, if the score you're changing is very unrepresentative and biases your statistical model anyway then changing the score is the lesser of two evils! There are several options for how to change the score:

   (a) *The next highest score plus one*: Change the score to be one unit above the next highest score in the data set.
   (b) *Convert back from a z-score*: A z-score of 3.29 constitutes an outlier (see Box 3.2) so we can calculate what score would give rise to a z-score of 3.29 (or perhaps 3) by rearranging the z-score equation in Box 3.2, which gives us: $X = (z \times s) + \bar{X}$. All this means is that we calculate the mean ($\bar{X}$) and standard deviation (s) of the data, we know the z is 3 (or 3.29 if you want to be exact) so we just add three times the standard deviation to the mean, and replace our outliers with that score.
   (c) *The mean plus two standard deviations*: A variation on the above method is to use the mean plus two times the standard deviation (rather than three times the standard deviation).

The next section is quite hair-raising so don't worry if it doesn't make much sense—many undergraduate courses won't cover transforming data so feel free to ignore this section if you want to!

Of all of the options above, transforming the data is perhaps the best because rather than changing a single score, you carry out the same transformation on all scores (so you're not just singling out a score to be changed, but doing something that reduces the impact of extreme scores). So, the idea behind transformations is that you transform all of your data to correct for distributional problems or outliers. Although some students often (understandably) think that transforming data sounds dodgy (the phrase 'fudging your results' springs to some people's minds!), in fact it isn't because you do the same thing to all of your data.[3] As such, transforming the data won't change the relationships between variables (the relative differences between people for a given variable stay the same), but it does change the differences between different variables (because it changes the units of measurement). Therefore, even if you only have one variable that has a skewed distribution, you should still transform any other variables in your data set if you're going to compare differences between that variable and others that you intend to transform. So, for example, with our hygiene data, we might want to look at how hygiene levels changed across the three days (i.e. compare the mean on day 1 to the means on days 2 and 3). The data for days 2 and 3 were skewed and need to be transformed, but because we might later compare the data to scores on day 1, we would also have to transform the day 1 data

3 Although there aren't statistical consequences of transforming data, as Grayson (2004) points out, there may be empirical or scientific implications that outweigh the statistical benefits. Specifically, transformation means that we're now addressing a different construct to the one originally measured, and this has obvious implications for interpreting those data.

(even though they aren't skewed). If we don't change the day 1 data as well, then any differences in hygiene scores we find from day 1 to day 2 or 3 will simply be due to us transforming one variable and not the others.

When you have positively skewed data (as we do) that need correcting, there are three transformations that can be useful (and as you'll see, these can be adapted to deal with negatively skewed data too):[4]

1. **Log transformation ($\log(X_i)$)**: Taking the logarithm of a set of numbers squashes the right tail of the distribution. As such it's a good way to reduce positive skew. However, you can't get a log value of zero or negative numbers, so if your data tend to zero or produce negative numbers you need to add a constant to all of the data before you do the transformation. For example, if you have zeros in the data then do $\log(X_i + 1)$, or if you have negative numbers add whatever value makes the smallest number in the data set positive.
2. **Square root transformation ($\sqrt{X_i}$)**: Taking the square root of large values has more of an effect than taking the square root of small values. Consequently, taking the square root of each of your scores will bring any large scores closer to the centre—rather like the log transformation. As such, this can be a useful way to reduce positively skewed data; however, you still have the same problem with negative numbers (negative numbers don't have a square root).
3. **Reciprocal transformation ($1/X_i$)**: Dividing 1 by each score also reduces the impact of large scores. The transformed variable will have a lower limit of 0 (very large numbers will become close to zero). One thing to bear in mind with this transformation is that it reverses the scores in the sense that scores that were originally large in the data set become small (close to zero) after the transformation, but scores that were originally small become big after the transformation. For example, imagine two scores of 1 and 10; after the transformation they become $1/1 = 1$ and $1/10 = 0.1$: the small score becomes bigger than the large score after the transformation. However, you can avoid this by reversing the scores before the transformation, by finding the highest score and changing each score to the highest score minus the score you're looking at. So, you do a transformation $1/(X_{\text{Highest}} - X_i)$.

Any one of these transformations can also be used to correct negatively skewed data, but first you'll have to reverse the scores (i.e. subtract each score from the highest score obtained)—if you do this, don't forget to reverse the scores back afterwards!

3.3.4. Step 4: transforming the data using SPSS ②

The next issue is how we do these transformations on SPSS, and this entails the *compute* command. The *compute* command enables us to carry out various functions on columns

4 You'll notice in this section that I keep writing X_i. We saw in Chapter 1 that this refers to the observed score for the *i*th person (so, the *i* could be replaced with the name of a particular person, e.g. for Graham, $X_i = X_{\text{Graham}}$ = Graham's score, and for Carol, $X_i = X_{\text{Carol}}$ = Carol's score).

of data in the data editor. Some typical functions are adding scores across several columns, taking the square root of the scores in a column, or calculating the mean of several variables. To access the *compute* dialog box, use the mouse to specify **Transform⇒Compute...**. The resulting dialog box is shown in Figure 3.8. In this dialog box there is a list of functions on the right-hand side, and a calculator-like keyboard in the centre. Most of the functions on the calculator are obvious but the most common are listed below.

Addition: This button places a plus sign in the command area. For example, with our hygiene data, 'day1 + day2' creates a column in which each row contains the hygiene score from the column labelled *day1* added to the score from the column labelled *day2* (e.g. for participant 1: 2.65 + 1.35 = 4).

Subtraction: This button places a minus sign in the command area. For example, if we wanted to calculate the change in hygiene from day 1 to day 2 we could type 'day2-day1'. This creates a column in which each row contains the score from the column labelled *day1* subtracted from the score from the column labelled *day2* (e.g. for participant 1: 2.65 − 1.35 = −1.30). Therefore, this person's hygiene went down by 1.30 (on our five-point scale) from day 1 to day 2 of the festival.

Multiply: This button places a multiplication sign in the command area. For example, 'day1 * day2' creates a column that contains the score from the column labelled *day1* multiplied by the score from the column labelled *day2* (e.g. for participant 1: 2.65 × 1.35 = 3.58).

Divide: This button places a division sign in the command area. For example, 'day1/day2' creates a column that contains the score from the column labelled *day1* divided by the score from the column labelled *day2* (e.g. for participant 1: 2.65/1.35 = 1.96).

Exponentiation: This button is used to raise the preceding term by the power of the succeeding term. So, 'day1**2' creates a column that contains the scores in the *day1* column raised to the power of 2 (i.e. the square of each number in the *day1* column: for participant 1, $(2.65)^2 = 7.02$). Likewise, 'day1**3' creates a column with values of **day1** cubed.

Less than: This operation is usually used for 'include case' functions. If you click on the button, a dialog box appears that allows you to select certain cases on which to carry out the operation. So, if you typed 'day1 < 1', then SPSS would carry out the *compute* function only for those participants whose hygiene score on day 1 of the festival was less than 1 (i.e. if **day1** was 0.99 or less). So, we might use this if we only wanted to look at the people who were already smelly on the first day of the festival!

Less than or equal to: This operation is the same as above except that cases that are exactly 1 would be included as well.

More than: This operation is generally used to include cases above a certain value. So, if you clicked on and then typed 'day1 > 1' then SPSS will carry out any analysis only on cases for which hygiene scores on day 1 of the festival were greater than 1 (i.e. 1.01 and above). This could be used if we wanted to exclude people who were already smelly at the start of the festival (as we'd be interested in the effects of the festival on hygiene and these people would contaminate the data (not to mention our nostrils) by reeking of putrefaction in the first place!).

More than or equal to: This operation is the same as above but will include cases that are exactly 1 as well.

Equal to: You can use this operation to include cases for which participants have a specific value. So, 'day1 = 1' typed in the *if* dialog box ensures that only cases that have a value of 1 for the **day1** variable are included. This is most useful when you have a coding variable and you want to look at only one of the groups. For example, if we had a variable **music** that coded the musical affiliations of the festival goers (see section 5.10) as 1 = Indie Kid, 2 = Metaller, 3 = Crusty and 4 = no affiliation, and we typed 'music = 2', then the analysis would be carried out only on the people who identify themselves as 'metallers' (people who listen to heavy metal).

Not equal to: This operation will include all cases except those with a specific value. So, 'music ~= 2' will include all cases except those that identified themselves as metallers.

Variable type: This button opens a dialog box that allows you to give the new variable a full name, and to specify what type of variable it is (e.g. numeric).

The results of any compute function will be produced in a new column in the data editor and so the first thing to do is to type in a label for this new variable (in the box labelled *Target Variable*). If you type in a variable name that already exists in the data editor then SPSS will tell you and ask you whether you want to replace this existing variable. If you respond with 'Yes' then SPSS will replace the data in the existing column with the result of the compute function. If you respond with 'No' then nothing will happen and you will need to rename the target variable. If you click on another dialog box appears, where you can give the variable a descriptive label, and where you can specify whether it is a numeric or string variable (see section 2.4.5). The box labelled *Numeric Expression* is the space in which arithmetic commands are typed (I've called this space the command area). You can enter variable names into the command area by selecting the variable required from the variables list and then clicking on . Likewise, you can select certain functions from the list of available functions and enter them into the command area by clicking on .

Some of the most useful functions are listed in Table 3.1, which shows the standard form of the function, the name of the function, an example of how the function can be used and what SPSS would output if that example were used. There are several basic functions for calculating means, standard deviations and sums of columns. There are also functions such as the square root and logarithm functions that are useful for transforming data that

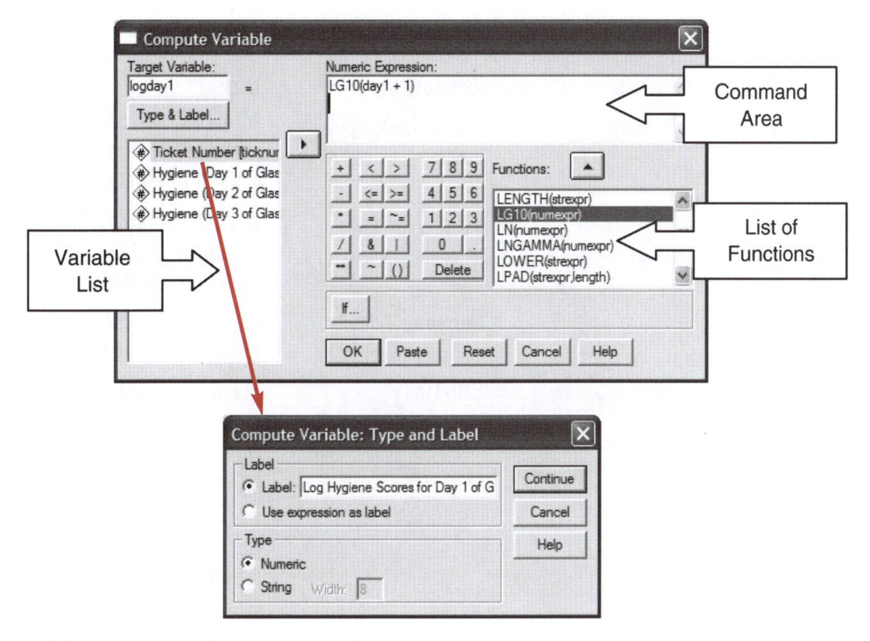

Figure 3.8 Dialog box for the compute function

are skewed and these are the functions we'll use now. For the interested reader, the SPSS base systems user's guide (and the SPSS help files) has details of all of the functions available through the *compute* dialog box (Norušis, 1997; see also SPSS Inc., 1997).

Now we've found out some basic information about the compute function, let's use it to transform our data. First open the main compute dialog box by using the **Transform⇒Compute...** menu path. Enter the name **logday1** into the box labelled *Target Variable* and then click on [Type&Label...] and give the variable a more descriptive name such as *Log transformed hygiene scores for day 1 of Glastonbury festival*. Then scroll down the list of functions until you find the one called LG10(numexpr). Highlight this function and transfer it to the command area by clicking on [▲]. When the command is transferred, it appears in the command area as 'LG10(?)' and the question mark should be replaced with a variable name (which can be typed manually or transferred from the variables list). So replace the question mark with the variable **day1**. Now, for the day 2 hygiene scores there is a value of 0 in the original data, and there is no logarithm of the value 0. To overcome this we should add a constant to our original scores before we take the log of those scores. Any constant will do, provided that it makes all of the scores greater than 0. In this case our lowest score is 0 in the data set so we can simply add 1 to all of the scores and that will ensure that all scores are greater than zero. To do this, make sure the cursor is still inside the brackets and click on [+] and then [1]. The final dialog box should look like Figure 3.8, and note that the expression reads LG10(day1 + 1), that is SPSS will add 1 to each of the day1 scores and then take the log of the resulting values. Click on [OK] to create a new variable **logday1** containing the transformed

Table 3.1 Some useful *compute* functions

Function	Name	Example Input	Output
Mean(?,?, ..)	Mean	Mean(day1, day2, day3)	For each row, SPSS calculates the average hygiene score across the three days of the festival
SD(?,?, ..)	Standard Deviation	SD(day1, day2, day3)	Across each row, SPSS calculates the standard deviation of the values in the columns labelled *day1*, *day2* and *day3*
SUM(?,?, ..)	Sum	Sum(day1, day2)	For each row, SPSS adds the values in the columns labelled *day1* and *day2*
SQRT(?)	Square Root	SQRT(day2)	Produces a column that contains the square root of each value in the column labelled *day2*. Useful for transforming skewed data or data with heterogeneous variances
ABS(?)	Absolute Value	ABS(day1)	Produces a variable that contains the absolute value of the values in the column labelled *day1* (absolute values are ones where the signs are ignored: so −5 becomes +5 and +5 stays as +5)
LG10(?)	Base 10 Logarithm	LG10(day1)	Produces a variable that contains the logarithmic (to base 10) values of *day1*. This is useful for transforming positively skewed data
Normal(stddev)	Normal Random Numbers	Normal(5)	Produces a variable containing pseudo-random numbers from a normal distribution with a mean of 0 and a standard deviation of 5

values. Now have a go at creating similar variables **logday2** and **logday3** for the day 2 and day 3 data!

Now we know how to log-transform our data, we can have a look at other types of transformation. For example, if we wanted to use the square root transformation, we could run through the same process, by using a name such as **sqrtday1** and selecting the *SQRT (numexpr)* function from the list. This will appear in the box labelled *Numeric Expression* as SQRT(?), and you can simply replace the question mark with the variable you want to change—in this case **day1**. The final expression will read *SQRT(day1)*. Again, try repeating this for **day2** and **day3** to create variables called **sqrtday2** and **sqrtday3**.

Finally, if we wanted to use a reciprocal transformation on the data from day 1, we could use a name such as **recday1** and then simply click on 1 and then /. Ordinarily you would select the variable name that you want to transform from the list and transfer it across by clicking on ▸ (or just type the name of the variable). However, the day 2 data contain a zero value and if we try to divide 1 by 0 then we'll get an error message (you can't divide by 0).

As such we need to add a constant to our variable just as we did for the log transformation. Any constant will do, but 1 is a convenient number for these data. So, instead of selecting the variable we want to transform, click on (). This places a pair of brackets in the box labelled *Numeric Expression*, then make sure the cursor is between these two brackets and select the variable you want to transform from the list and transfer it across by clicking on ► (or type the name of the variable manually). Now click on + and then 1 (or type + *1* using your keyboard). The box labelled *Numeric Expression* should now contain the text *1/(day1 + 1)*. Click on OK to create a new variable containing the transformed values. Repeat this process yourself for the data for the remaining two days. If you're computing lots of variables it may be quicker to use the syntax windows as explained in Box 3.3.

Box 3.3

Using syntax to compute new variables ③

If you're computing a lot of new variables it can be quicker to use syntax—for example, to create the transformed data in the example in this chapter. The CD-ROM contains a syntax file, **Transformations.sps**, which I've written to do all nine of the transformations we've discussed. Load this file and run the syntax, or open a syntax window (see section 2.6) and type the following:

```
COMPUTE logday1 = LG10 (day1 + 1).
COMPUTE logday2 = LG10 (day2 + 1).
COMPUTE logday3 = LG10 (day3 + 1).
COMPUTE sqrtday1 = SQRT(day1).
COMPUTE sqrtday2 = SQRT(day2).
COMPUTE sqrtday3 = SQRT(day3).
COMPUTE recday1 = 1/(day1+1).
COMPUTE recday2 = 1/(day2+1).
COMPUTE recday3 = 1/(day3+1).
EXECUTE.
```

Each COMPUTE command is doing the equivalent of what you'd do using the compute dialog box in Figure 3.8. So, the first three lines just ask SPSS to create three new variables (**logday1**, **logday2** and **logday3**), which are just the log transformations of the variables day1, day2 and day3 plus 1. The next three lines do much the same but use the SQRT function, and so take the square root of **day1**, **day2** and **day3** to create new variables called **sqrtday1**, **sqrtday2** and **sqrtday3** respectively. The next three lines do the reciprocal transformation in much the same way. The final line has the command EXECUTE without which none of the compute commands beforehand will be excuted! Note also that every line ends will a fullstop.

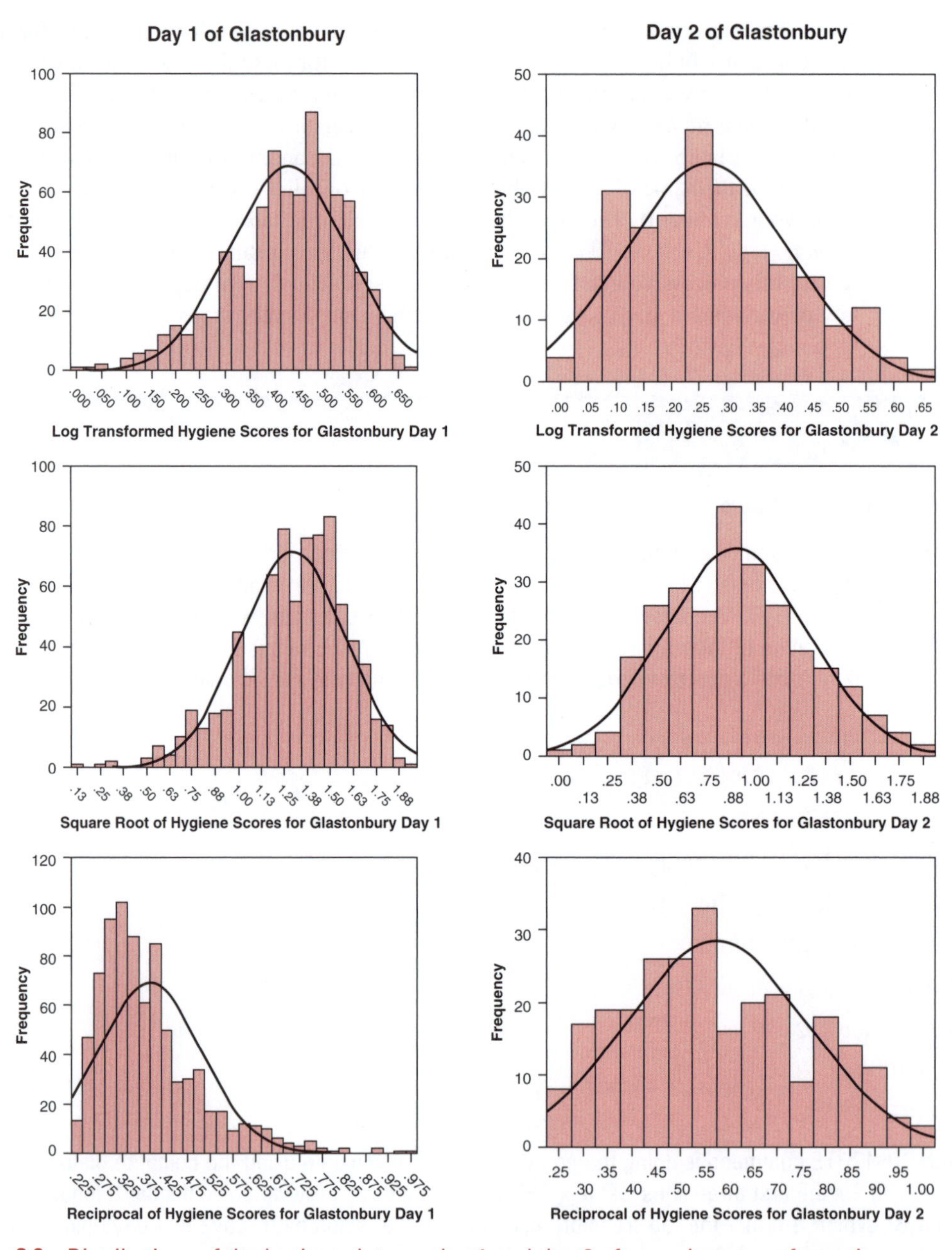

Figure 3.9 Distributions of the hygiene data on day 1 and day 2 after various transformations

Let's have a look at our new distributions (Figure 3.9). Compare these with the untransformed distributions in Figure 3.5. Now, you can see that all three transformations have cleaned up the hygiene scores for day 2: the positive skew is gone (the square root transformation in particular has been useful). However, because our hygiene scores on day 1 were more or less symmetrical to begin with, they have now become slightly negatively skewed for the log and square root transformations, and positively skewed for the reciprocal transformation![5] If we're using scores from day 2 alone then we could use the transformed scores; however, if we want to look at the change in scores then we'd have to weigh up whether the benefits of the transformation for the day 2 scores outweigh the problems it creates in the day 1 scores—data analysis can be frustrating sometimes!

If the transformation corrects the problems in your distribution then run whatever statistical analysis you need to run on the transformed scores. If it doesn't correct the problem then you should consider using a test that does not rely on the assumption of normally-distributed data and as you go through the various chapters of this book I'll point out these tests.[6]

3.4. EXPLORING GROUPS OF DATA ①

Sometimes we have data in which there are different groups of people (men and women, different universities, people with depression and people without, for example). There are several ways to produce basic descriptive statistics for separate groups of people (and we will come across some of these methods in the next section). However, I intend to use this opportunity to introduce you to a function called *split file*, which allows you to repeat any analysis on several groups of cases. The *split file* function allows you to specify a grouping variable (remember these variables are used to specify categories of people). Any subsequent procedure in SPSS will then be carried out, in turn, on each category specified by that grouping variable.

You're probably getting sick of the hygiene data from the Glastonbury festival so let's use the data in the file **SPSSExam.sav**. This file contains data regarding students' performance on an SPSS exam. Four variables were measured: **exam** (first-year SPSS exam scores as a percentage), **computer** (measure of computer literacy in per cent), **lecture** (percentage of SPSS lectures attended) and **numeracy** (a measure of numerical ability out of 15). There is a variable called **uni** indicating whether the student attended Sussex University or Duncetown University. To begin with open the file **SPSSExam.sav** (see section 2.8 for a reminder of how to open a file). Let's begin by looking at the data as a whole.

---

5 The reversal of the skew for the reciprocal transformation is because, as I mentioned earlier, the reciprocal has the effect of reversing the scores.

6 For convenience a lot of textbooks refer to these tests as *non-parametric tests* or *assumption-free* tests and stick them in a separate chapter. Actually neither of these terms are particularly accurate (e.g. none of these tests are assumption free) but in keeping with tradition I've put them in a chapter (13) on their own, ostracized from their 'parametric' counterparts and feeling lonely.

3.4.1. Running the analysis for all data ①

To see the distribution of the variables, we can use the *frequencies* command, which we came across in the previous section (see Figure 3.6). Use this dialog box and place all four variables (**exam**, **computer**, **lecture** and **numeracy**) in the *Variable(s)* box. Then click on Statistics... to select the *statistics* dialog box and select some measures of central tendency (mean, mode, median), measures of variability (range, standard deviation, variance, quartile splits) and measures of shape (kurtosis and skewness). Also click on Charts... to access the *charts* dialog box and select a frequency distribution of scores with a normal curve (see Figure 3.6 if you need any help with any of these options). Return to the main dialog box by clicking on Continue and once in the main dialog box, click OK to run the analysis.

3.4.2. SPSS output for all data ①

SPSS Output 3.2 shows the table of descriptive statistics for the four variables in this example. From this table, we can see that, on average, students attended nearly 60% of lectures, obtained 58% in their SPSS exam, scored only 51% on the computer literacy test and only 5 out of 15 on the numeracy test. In addition, the standard deviation for computer literacy was relatively small compared to that of the percentage of lectures attended and exam scores. These latter two variables had several modes (multimodal). The other important measures are the skewness and the kurtosis, both of which have an associated standard error. We came across these measures earlier on and found that we can convert these values to *z*-scores by dividing by their standard errors. For the SPSS exam scores, the *z*-score of skewness is $-.107/.241 = -.44$. For numeracy, the *z*-score of skewness is $.961/.241 = 3.99$. It is pretty clear then that the numeracy scores are positively skewed, indicating a pile-up of scores on the left of the distribution (so, most students got low scores). Try calculating the *z*-scores for the other variables.

The output provides tabulated frequency distributions of each variable (not reproduced here). These tables list each score and the number of times that it is found within the data set. In addition, each frequency value is expressed as a percentage of the sample (in this case the frequencies and percentages are the same because the sample size was 100). Also, the cumulative percentage is given, which tells us how many cases (as a percentage) fell below a certain score. So, for example, we can see that 66% of numeracy scores were 5 or less, 74% were 6 or less, and so on. Looking in the other direction, we can work out that only 8% (100 – 92%) got scores greater than 8.

Finally, we are given histograms of each variable with the normal distribution overlaid. These graphs are displayed in Figure 3.10 and show us several things. First, it looks as though computer literacy is fairly normally distributed (a few people are very good with computers and a few are very bad, but the majority of people have a similar degree of knowledge). The exam scores are very interesting because this distribution is quite clearly

Statistics

		Computer literacy	Percentage on SPSS exam	Percentage of lectures attended	Numeracy
N	Valid	100	100	100	100
	Missing	0	0	0	0
Mean		50.7100	58.1000	59.7650	4.8500
Std. Error of Mean		.8260	2.1316	2.1685	.2706
Median		51.5000	60.0000	62.0000	4.0000
Mode		54.00	72.00[a]	48.50[a]	4.00
Std. Deviation		8.2600	21.3156	21.6848	2.7057
Variance		68.2282	454.3535	470.2296	7.3207
Skewness		−.174	−.107	−.422	.961
Std. Error of Skewness		.241	.241	.241	.241
Kurtosis		.364	−1.105	−.179	.946
Std. Error of Kurtosis		.478	.478	.478	.478
Range		46.00	84.00	92.00	13.00
Minimum		27.00	15.00	8.00	1.00
Maximum		73.00	99.00	100.00	14.00

a Multiple modes exist. The smallest value is shown

SPSS Output 3.2

not normal; in fact, it looks suspiciously bimodal (there are two peaks indicative of two modes). This observation corresponds to the earlier information from the table of descriptive statistics. Lecture attendance is generally quite normal, but the tails of the distribution are quite heavy (i.e. although most people attend the majority of lectures, there are a reasonable number of dedicated souls who attend them all and a larger than 'normal' proportion who attend very few). This is why there are high frequencies at the two ends of the distribution. Finally, the numeracy test has produced very positively skewed data (i.e. the majority of people did very badly on this test and only a few did well). This corresponds to what the skewness statistic indicated.

Descriptive statistics and histograms are a good way of getting an instant picture of the distribution of your data. This snapshot can be very useful: for example, the bimodal distribution of SPSS exam scores instantly indicates a trend that students are typically either very good at statistics or struggle with it (there are relatively few who fall in between these extremes). Intuitively, this finding fits with the nature of the subject: statistics is very easy once everything falls into place, but before that enlightenment occurs it all seems hopelessly difficult! Although there is a lot of information that we can obtain from histograms and descriptive information about a distribution, there are more objective ways in which we can assess the degree of normality in a set of data (see section 3.5).

3.4.3. Running the analysis for different groups ①

If we wanted to obtain separate descriptive statistics for each of the universities, we could split the file, and then proceed using the *frequencies* command described in the previous

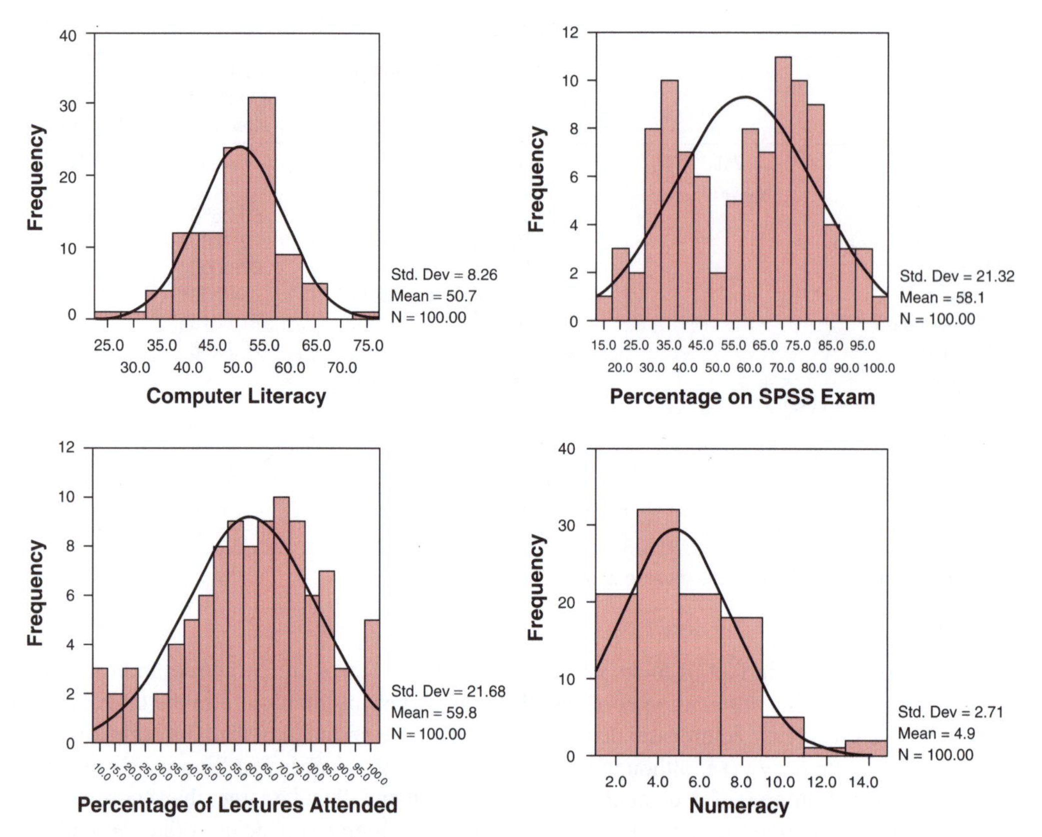

Figure 3.10 Histograms of the SPSS exam data

section. To split the file, simply use the menu path **Data⇒Split File...** or click on . In the resulting dialog box (Figure 3.11) select the option *Organize output by groups*. Once this option is selected, the *Groups Based on* box will activate. Select the variable containing the group codes by which you wish to repeat the analysis (in this example select **Uni**), and transfer it to the box by clicking on . By default, SPSS will then sort the file by these groups (i.e. it will list one category followed by the other in the data editor window). Once you have split the file, use the *frequencies* command (see previous section). I have requested statistics for only **numeracy** and **exam** scores; you can select the other variables too if you want.

3.4.4. Output for different groups ①

The SPSS output is split into two sections: first the results for students at Duncetown University, then the results for those attending Sussex University. SPSS Output 3.3 shows

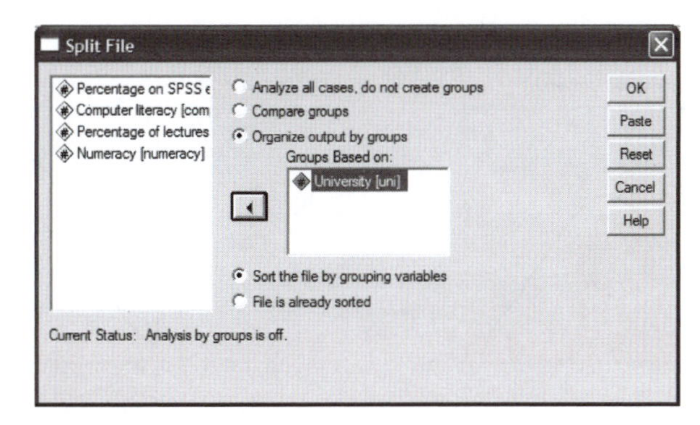

Figure 3.11 Dialog box for the *split file* command

Duncetown University

Statistics[a]

		Percentage on SPSS exam	Numeracy
N	Valid	50	50
	Missing	0	0
Mean		54.4400	3.1800
Std. Error of Mean		2.7779	.2094
Median		53.0000	3.0000
Mode		47.00	2.00
Std. Deviation		19.6429	1.4803
Variance		385.8433	2.1914
Skewness		.259	.621
Std. Error of Skewness		.337	.337
Kurtosis		−.893	−.100
Std. Error of Kurtosis		.662	.652
Range		77.00	6.00
Minimum		22.00	1.00
Maximum		99.00	7.00

a University = Duncetown University

Sussex University

Statistics[a]

		Percentage on SPSS exam	Numeracy
N	Valid	50	50
	Missing	0	0
Mean		61.7600	6.5200
Std. Error of Mean		3.1774	.3717
Median		67.5000	6.5000
Mode		77.00	5.00
Std. Deviation		22.4677	2.6283
Variance		504.7984	6.9078
Skewness		−.482	.697
Std. Error of Skewness		.337	.337
Kurtosis		−.931	.648
Std. Error of Kurtosis		.662	.662
Range		82.00	12.00
Minimum		15.00	2.00
Maximum		97.00	14.00

a University = Sussex University

SPSS Output 3.3

the two main summary tables. From these tables it is clear that Sussex students scored higher on both their SPSS exam and the numeracy test than their Duncetown counterparts. In fact, looking at the means reveals that, on average, Sussex students scored 6% more on the SPSS exam than Duncetown students, and had numeracy scores twice as high. The standard deviations for both variables are slightly higher for Sussex, but not greatly so.

Figure 3.12 shows the histograms of these variables split according to the university attended. The first interesting thing to note is that for exam marks, the distributions are

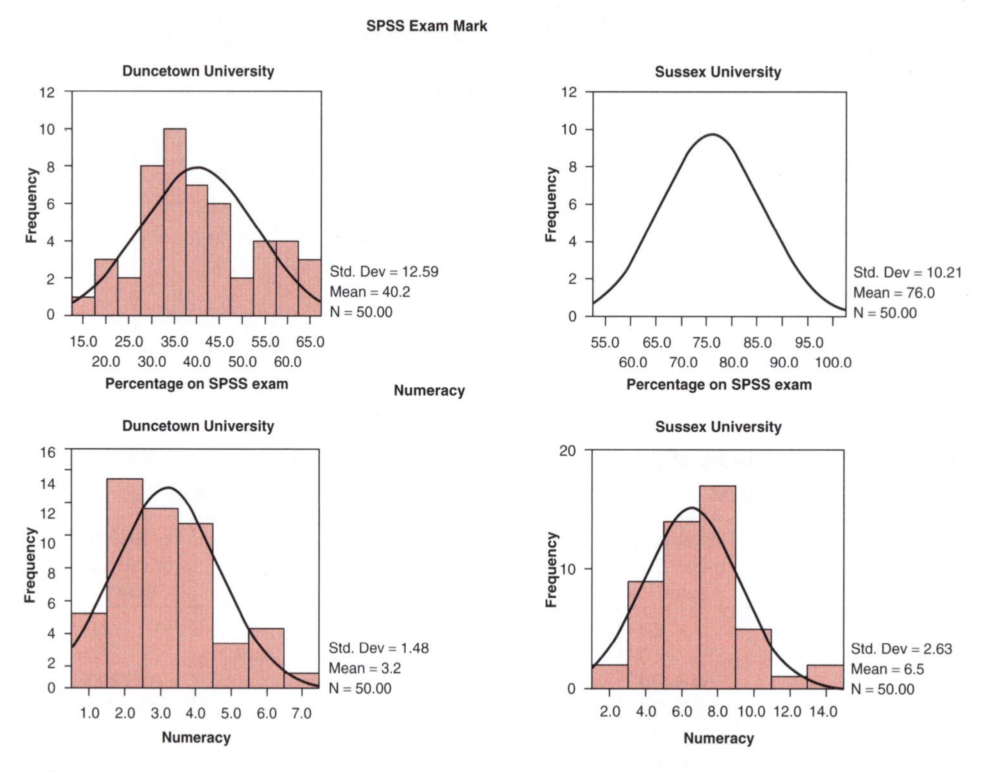

Figure 3.12 Distributions of exam and numeracy scores for Duncetown University and Sussex University students

both fairly normal. This seems odd because the overall distribution was bimodal. However, it starts to make sense when you consider that for Duncetown the distribution is centred around a mark of about 35%, but for Sussex the distribution is centred around a mark of around 75%. This illustrates how important it is to look at distributions within groups. If we were interested in comparing Duncetown to Sussex it wouldn't matter that overall the distribution of scores was bimodal; all that's important is that each group comes from a normal distribution, and in this case it appears to be true. When the two samples are combined, these two normal distributions create a bimodal one (one of the modes being around the centre of the Duncetown distribution, and the other being around the centre of the Sussex data!). For numeracy scores, the distribution is slightly positively skewed in the Duncetown group (there is a larger concentration at the lower end of scores) whereas Sussex students are fairly normally distributed around a mean of 7. Therefore, the overall positive skew observed previously is due to the mixture of universities (the Duncetown students contaminate Sussex's normally distributed scores!). When you have finished with the *split file* command, remember to *switch it off* (otherwise SPSS will carry on doing every analysis on each group separately). To switch this function off, return to the *split file* dialog box and select *Analyze all cases, do not create groups*.

3.5. TESTING WHETHER A DISTRIBUTION IS NORMAL ①

It is all very well to look at histograms, but they tell us little about whether a distribution is close enough to normality to be useful. Looking at histograms is subjective and open to abuse (I can imagine researchers sitting looking at a completely distorted distribution and saying 'yep, well Bob, that looks normal to me', and Bob replying 'yep, sure does'). What is needed is an objective test to decide whether or not a distribution is normal. The skewness and kurtosis statistics that we saw earlier tell us a bit about deviations from normality, but they deal with only one aspect of non-normality each. Another way of looking at the problem is to see whether the distribution as a whole deviates from a comparable normal distribution. The Kolmogorov–Smirnov (Figure 3.13) and Shapiro–Wilk tests do just this: they compare the scores in the sample to a normally distributed set of scores with the same mean and standard deviation. If the test is non-significant ($p > .05$) it tells us that the distribution of the sample is not significantly different from a normal distribution (i.e. it is probably normal). If, however, the test is significant ($p < .05$) then the distribution in question is significantly different from a normal distribution (i.e. it is non-normal). These tests seem great: in one easy procedure they tell us whether our scores are normally distributed (nice!). However, they have their limitations because with large sample sizes it is very easy to get significant results from small deviations from normality, and so a significant test doesn't necessarily tell us whether the deviation from normality is enough to bias any statistical procedures that we apply to the data. I guess the take-home message is: by all means use these tests, but plot your data as well and try to make an informed decision about the extent of non-normality.

Figure 3.13 Andrei Kolmogorov, wishing he had a Smirnov

3.5.1. Doing the Kolmogorov–Smirnov test on SPSS ①

The Kolmogorov–Smirnov (K–S from now on) test can be accessed through the *explore* command (**Analyze⇒Descriptive Statistics⇒Explore…**).[7] Figure 3.14 shows the dialog boxes for the *explore* command. First, enter any variables of interest in the box labelled *Dependent List* by highlighting them on the left-hand side and transferring them by clicking on ▶. For this example, just select the exam scores and numeracy scores. It is also possible to select a factor (or grouping variable) by which to split the output (so, if you select **uni** and transfer it to the box labelled *Factor List,* SPSS will produce exploratory analysis for each group—a bit like the *split file* command). If you click on Statistics... a dialog box appears, but the default option is fine (it will produce means, standard deviations and so on). The more interesting option for our purposes is accessed by clicking on Plots... . In this dialog box select the option ☑ Normality plots with tests, and this will produce both the K–S test and some graphs called *normal Q–Q plots* for all of the variables selected. By default, SPSS will produce boxplots (split according to group if a factor has been specified) and stem and leaf diagrams as well. Click on Continue to return to the main dialog box and then click OK to run the analysis.

3.5.2. Output from the *explore* procedure ①

The first table produced by SPSS contains descriptive statistics (mean etc.) and should have the same values as the tables obtained using the frequencies procedure. The important table is that of the K–S test.

Tests of Normality

	Kolmogorov–Smirnov[a]		
	Statistic	df	Sig.
Percentage on SPSS exam	.102	100	.012
Numeracy	.153	100	.000

a Lilliefors Significance Correction

This table includes the test statistic itself, the degrees of freedom (which should equal the sample size) and the significance value of this test. Remember that a significant value (*Sig.* less than .05) indicates a deviation from normality. For both numeracy and SPSS exam, the K–S test is highly significant, indicating that both distributions are not normal. This result is likely to reflect the bimodal distribution found for exam scores, and the positively skewed distribution observed in the numeracy scores. However, these tests confirm that

7 This menu path would be **Statistics⇒Summarize⇒Explore…** in version 8.0 and earlier.

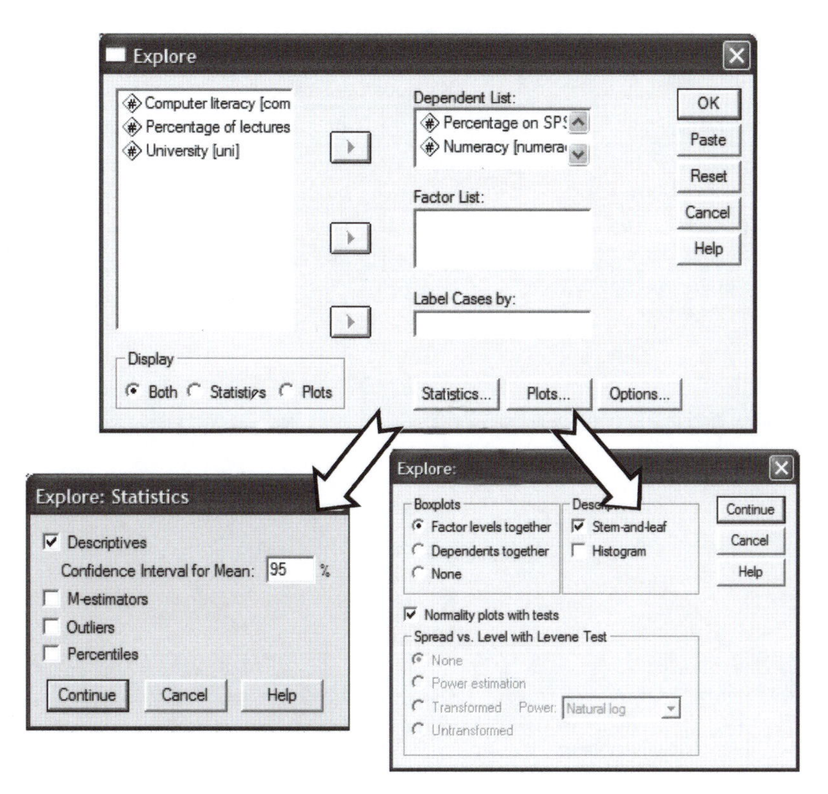

Figure 3.14 Dialog boxes for the *explore* command

these deviations were *significant*. This finding is important because the histograms tell us only that our sample distributions deviate from normal; they do not tell us whether this deviation is large enough to be important. The test statistic for the K–S test is denoted by *D* and so we can report these results in the following way. The percentage on the SPSS exam, $D(100) = 0.10$, $p < .05$, and the numeracy scores, $D(100) = 0.15$, $p < .001$, were both significantly non-normal. The numbers in brackets are the degrees of freedom (*df*) from the table.

As a final point, bear in mind that when we looked at the exam scores for separate groups, the distributions seemed quite normal; now if we'd asked for separate tests for the two universities (by placing **uni** in the box labelled *Factor List* as in Figure 3.17) the K–S test might not have been significant. In fact if you try this out, you'll get the table in SPSS Output 3.4, which shows that the percentages on the SPSS exam are indeed normal within the two groups (the values in the column labelled *Sig.* are greater than .05). This is important because if our analysis involves comparing groups then what's important is not the overall distribution, but the distribution in each group.

Tests of Normality

	University	Kolmogorov–Smirnov[a]			Shapiro–Wilk		
		Statistic	df	Sig.	Statistic	df	Sig.
Percentage on SPSS exam	Duncetown University	.106	50	.200*	.972	50	.283
	Sussex University	.073	50	.200*	.984	50	.715
Numeracy	Duncetown University	.183	50	.000	.941	50	.015
	Sussex University	.155	50	.004	.932	50	.007

* This is a lower bound of the true significance.
a Lilliefors Significance Correction

SPSS Output 3.4

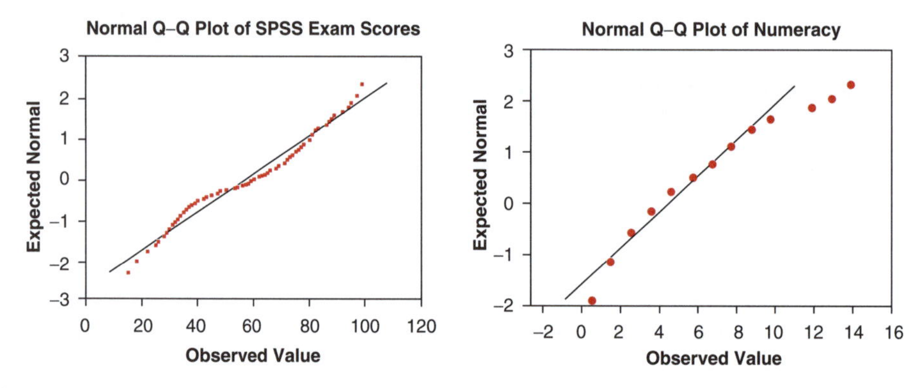

Figure 3.15 Normal Q–Q plots of numeracy and SPSS exam scores

SPSS also produces a normal Q–Q plot for any variables specified (see Figure 3.15). The normal Q–Q chart plots the values you would expect to get if the distribution were normal (expected values) against the values actually seen in the data set (observed values). The expected values are a straight diagonal line, whereas the observed values are plotted as individual points. If the data are normally distributed, then the observed values (the dots on the chart) should fall exactly along the straight line (meaning that the observed values are the same as you would expect to get from a normally distributed data set). Any deviation of the dots from the line represents a deviation from normality. So, if the Q–Q plot looks like a straight line with a wiggly snake wrapped around it then you have some deviation from normality! Specifically, when the line sags consistently below the diagonal, or consistently rises above it, then this shows that the kurtosis differs from a normal distribution, and when the curve is S-shaped, the problem is skewness.

In both of the variables analysed we already know that the data are not normal, and these plots confirm this observation because the dots deviate substantially from the line. It is noteworthy that the deviation is greater for the numeracy scores, and this is consistent with the higher significance value of this variable on the K–S test. A deviation from normality such as this tells us that we cannot use a parametric test, because the assumption of normality is not tenable. In these circumstances we can sometimes turn to non-parametric tests as a means of testing the hypothesis of interest (see Chapter 4).

3.6. TESTING FOR HOMOGENEITY OF VARIANCE ①

So far I've concentrated on the assumption of normally distributed data; however, at the beginning of this chapter I mentioned another assumption: homogeneity of variance. This assumption means that as you go through levels of one variable, the variance of the other should not change. If you've collected groups of data then this means that the variance of your outcome variable or variables should be the same in each of these groups. If you've collected continuous data (such as in correlational designs), this assumption means that the variance of one variable should be stable at all levels of the other variable. Let's illustrate this with an example. An audiologist was interested in the effects of loud concerts on people's hearing. So, she decided to send 10 people on tour with the loudest band she could find, Motörhead. These people went to concerts in Brixton (London), Brighton, Bristol, Edinburgh, Newcastle, Cardiff and Dublin and after each concert the audiologist measured the numbers of hours after the concert that these people had ringing in their ears.

Figure 3.16 shows the number of hours that each person had ringing in their ears after each concert (each person is represented by a circle). The horizontal lines represent the average number of hours that there was ringing in the ears after each concert and these means are connected by a line so that we can see the general trend of the data. Remember that for each concert, the circles are the scores from which the mean is calculated. Now, we can see in both graphs that the means increase as the people go to more concerts. So, after the first concert their ears ring for about 12 hours, but after the second they ring for about 15–20 hours, and by the final night of the tour, they ring for about 45–50 hours (2 days). So, there is a cumulative effect of the concerts on ringing in the ears. This pattern is found in both graphs; the difference between the graphs is not in terms of the means (which are roughly the same), but in terms of the spread of scores around the mean. If you look at the left-hand graph, the spread of scores around the mean stays the same after each concert (the scores are fairly tightly packed around the mean). Put another way, if you measured the vertical distance between the lowest score and the highest score after the Brixton concert, and then did the same after the other concerts, all of these distances would be fairly similar. Although the means increase, the spread of scores for hearing loss is the same at each level of the concert variable (the spread of scores is the same after Brixton, Brighton, Bristol, Edinburgh, Newcastle, Cardiff and Dublin). This is what we mean by *homogeneity of variance*. The right-hand graph shows a different picture: if you look at the spread of scores after the Brixton concert, they are quite tightly packed around the mean (the vertical distance from the lowest score to the highest score is small), but after the Dublin show (for example) the scores are very spread out around the mean (the vertical distance from the lowest score to the highest score is large). This is an example of *heterogeneity of variance*: that is, at some levels of the concert variable the variance of scores is different to other levels (in this case the vertical distance from the lowest to highest score is different after different concerts).

Hopefully you have got a grip of what homogeneity of variance actually means. Now, how do we test for it? Many statistical procedures have their own unique way to look for it: in correlational analysis of such a regression we tend to use graphs (see section 5.8.7) and for groups of data we tend to use a test called Levene's test. Levene's test tests the

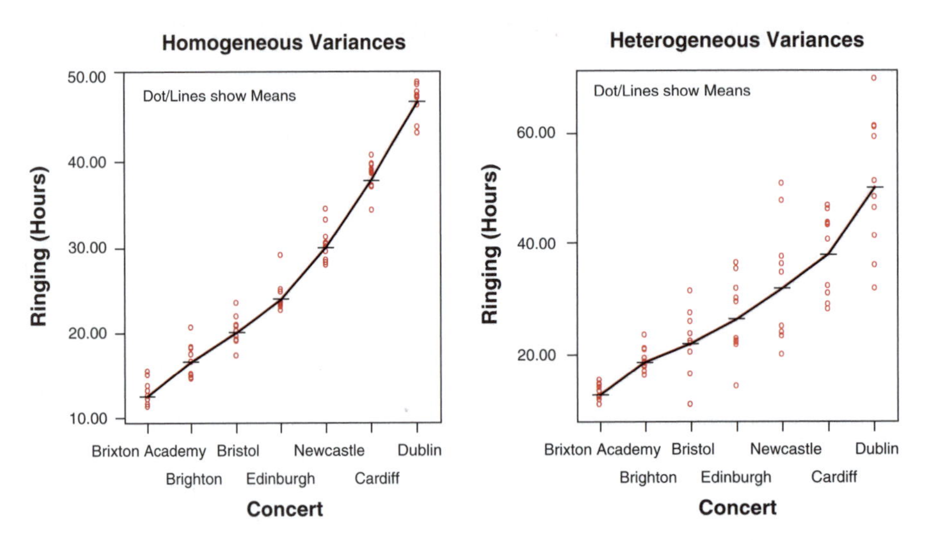

Figure 3.16 Graphs illustrating data with homogeneous and heterogeneous variances

hypothesis that the variances in the groups are equal (i.e. the difference between the variances is zero). Therefore, if Levene's test is significant at $p \leq .05$ then we can conclude that the null hypothesis is incorrect and that the variances are significantly different—therefore, the assumption of homogeneity of variances has been violated. If, however, Levene's test is non-significant (i.e. $p > .05$) then we must accept the null hypothesis that the difference between the variances is zero—the variances are roughly equal and the assumption is tenable. Although Levene's test can be selected as an option in many of the statistical tests that require it, it can also be examined when you're exploring data (and strictly speaking it's better to examine it now than wait until your main analysis).

As with the K–S test (and other tests of normality), when the sample size is large, small differences in group variances can produce a Levene's test that is significant (because, as we saw in Chapter 1, the power of the test is improved). A useful double check is to look at the variance ratio. This is the ratio of the variances between the group with the biggest variance and the group with the smallest variance. So, if you look at the variances in different groups, take the highest value you find and divide it by the smallest value you find. If this ratio is less than 2, then it's safe to assume homogeneity of variance.

We can get Levene's test using the *explore* menu that we used in the previous section. For this example, we'll use the SPSS exam data that we used in the previous section (in the file **SPSSExam.sav**). Once the data are loaded, use **Analyze⇒Descriptive Statistics⇒Explore...** to open the dialog box in Figure 3.17. To keep things simple we'll just look at the SPSS exam scores and the numeracy scores from this file, so transfer these two variables from the list on the left-hand side to the box labelled *Dependent List* by clicking on ▶ next to this box, and because we want to split the output by the grouping variable to compare the variances, select the variable **uni** and transfer it to the box labelled

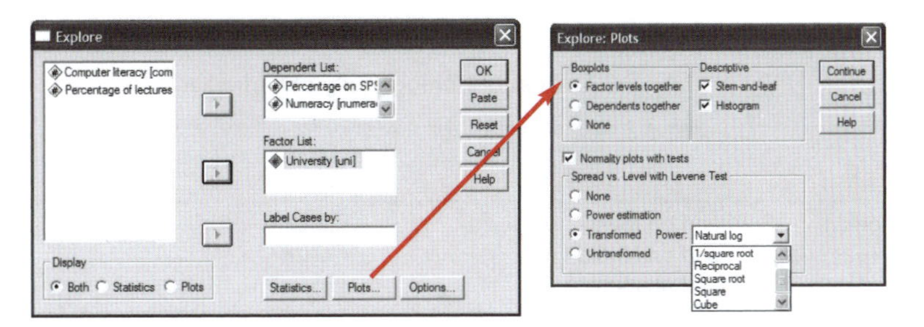

Figure 3.17 Exploring groups of data and obtaining Levene's test

Factor List by clicking on the appropriate . Then click on Plots... to open the other dialog box in Figure 3.17. To get Levene's test we need to select one of the options where it says *Spread vs. Level with Levene's test*. If you select Untransformed Levene's test is carried out on the raw data (a good place to start). However, if the variances turn out to be unequal you can attempt to correct this inequality by using one of the transformations we've already seen in this chapter. Using the same dialog box if we now select Transformed you should notice a drop-down list that becomes active (see Figure 3.17) and if we select one of the transformations from this list and run the analysis, SPSS will calculate what Levene's test would be if we were to transform the data using this method. This can save you a lot of time trying out different transformations. So, to begin with run the analysis with Untransformed selected, and then go back and run it with Transformed selected (try out a log transformation to begin with). When you've finished with this dialog box click on Continue to return to the main *explore* dialog box and then click OK to run the analysis.

SPSS Output 3.5 shows the table for Levene's test first when the analysis was run with Untransformed selected, and then when it was run with Transformed selected (natural log). You should read the statistics based on the mean. For the untransformed data it appears that Levene's test is non-significant for the SPSS exam scores (values in the column labelled *Sig.* are more than .05) indicating that the variances are not significantly different (i.e. they are similar and the homogeneity of variance assumption is tenable). However, for the numeracy scores, Levene's test is significant (values in the column labelled *Sig.* are less than .05) indicating that the variances are significantly different (i.e. they are not the same and the homogeneity of variance assumption has been violated). So, this might give us reason to attempt a transformation on the numeracy scores. Let's see what happens when we do. Well, for the log transformed scores, the problem has been reversed: Levene's test is now significant for the SPSS exam scores (values in the column labelled *Sig.* are less than .05) but is no longer significant for the numeracy scores (values in the column labelled *Sig.* are less than .05). In this situation, provided we were comparing our two universities on their numeracy and SPSS exam scores separately, we should transform numeracy scores, but not the exam scores. Levene's test can be denoted by the letter F and there are two different degrees of freedom. As such you can report it, in general form, as F(df1, df2) = value, *Sig*. So, for the percentage on the SPSS exam, we could say the variances are equal, $F(1, 98) = 2.58$, *ns*, but for numeracy scores the variances are significantly different, $F(1, 98) = 7.37$, $p < .01$. Try reporting the results for the log transformed scores.

Untransformed Data

Test of Homogeneity of Variance

		Levene Statistic	df1	df2	Sig.
Percentage on SPSS exam	Based on Mean	2.584	1	98	.111
	Based on Median	2.089	1	98	.152
	Based on Median and with adjusted df	2.089	1	94.024	.152
	Based on trimmed mean	2.523	1	98	.115
Numeracy	Based on Mean	7.368	1	98	.008
	Based on Median	5.366	1	98	.023
	Based on Median and with adjusted df	5.366	1	83.920	.023
	Based on trimmed mean	6.766	1	98	.011

Log Transformed Data

Test of Homogeneity of Variance

		Levene Statistic	df1	df2	Sig.
Percentage on SPSS exam	Based on Mean	25.055	1	98	.000
	Based on Median	24.960	1	98	.000
	Based on Median and with adjusted df	24.960	1	64.454	.000
	Based on trimmed mean	25.284	1	98	.000
Numeracy	Based on Mean	.211	1	98	.647
	Based on Median	.279	1	98	.598
	Based on Median and with adjusted df	.279	1	97.664	.598
	Based on trimmed mean	.238	1	98	.627

SPSS Output 3.5

3.7. GRAPHING MEANS ①

Having done all of this data exploration, you might want to do some graphs of the means. This section shows you how to use some of the graphing procedures in SPSS to create graphs of means (this is useful when writing up your data or presenting it to others). How you create these graphs in SPSS depends largely on how you collect your data (whether you use different people or the same people to measure different levels of a variable). To begin with, imagine that a film company director was interested in whether there was really such a thing as a 'chick flick' (a film that typically appeals to women more than men).

He took 20 men and 20 women and showed half of each sample a film that was supposed to be a 'chick flick' (*Bridget Jones' Diary*), and the other half of each sample a film that didn't fall into the category of 'chick flick' (*Memento*—a brilliant film by the way). In all cases he measured their physiological arousal as a measure of how much they enjoyed the film. The data are in a file called **ChickFlick.sav** on the CD-ROM. Load this file now.

To draw a graph of the means, we'll use the interactive graphs in SPSS. The main advantage of these graphs is that you can put error bars on your graph (see Chapter 8) and they can be updated even when the data file is closed (see Box 3.4). You can also plot graphs using the non-interactive graphing procedures in SPSS, but to save trees I've described how to do this on a file on the CD-ROM (**Answers(Chapter 3).pdf**) and in Chapter 4. Let's look at a simple plot: first simply click on **Graphs⇒Interactive⇒Bar...** to activate the dialog box in Figure 3.18. Notice that there is a schematic representation of the horizontal and the vertical axes and both have a gap into which you can drag a variable from the list on the left-hand side. The *y*-axis needs to be the dependent variable, or the thing you've measured, or more simply the thing for which you want to display the mean. In this case it would be **arousal**, so select arousal from the list and drag it into the space on the vertical axis. The *x*-axis should be the variable by which we want to split the arousal data. To keep things simple, let's just plot the means for the two films. To do this select the variable **film** from the list on the left-hand side and drag it into the space on the horizontal axis.[8] At the bottom of the dialog box there is a section labelled *Bars Represent* ... and this becomes active only after you've specified a variable on the vertical axis (this is why it says *Bars Represent Arousal (arousal)* because we specified arousal on the vertical axis). By default, bars will represent the mean (which is generally what you want), but if you select the drop-down list by clicking on ▾ you'll see that you can plot virtually any other descriptive statistic from other measures of central tendency (such as the median and mode), measuresof dispersion (variance or standard deviation) and measures such as skewness and kurtosis.

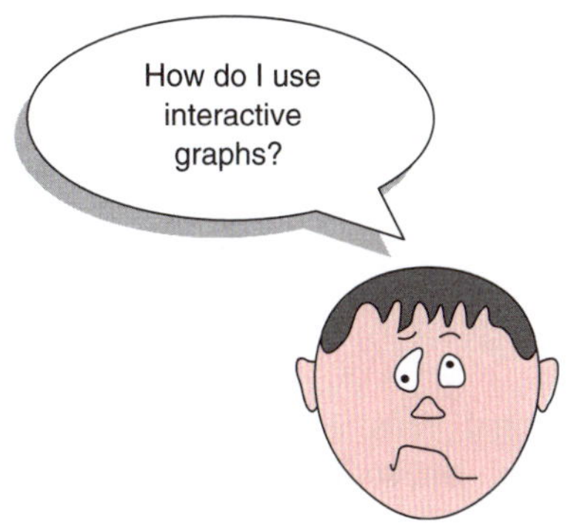

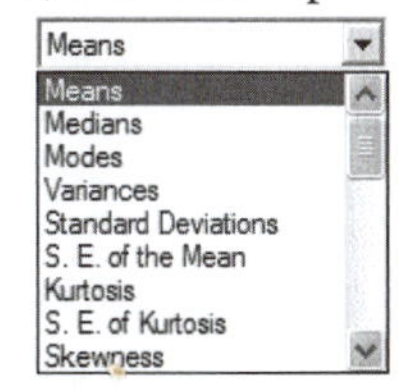

If we want to add error bars (see section 7.3), we should click on the *Error Bars* tab at the top of the dialog box and select ☑ Display Error Bars. By default, SPSS will display a confidence interval of the mean (see section 1.6.2), but if you select the drop-down list by clicking on ▾ next to where it says *Units* then you can choose to plot the standard deviation (see section 1.4.1) or the standard error (see section 1.6) of the mean instead. Finally, by default SPSS plots error bars that stick out both sides of the mean. However, you can choose a different display by

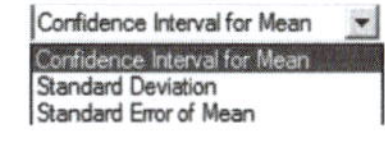

8 For this to work it's important that you've specified the variable as a nominal variable (see section 2.4.3).

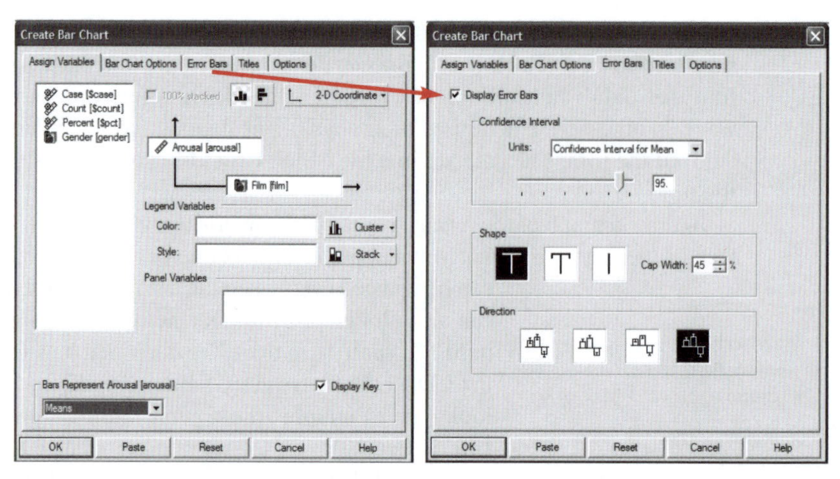

Figure 3.18 Dialog boxes for an interactive bar chart with error bars

selecting one of the four buttons at the bottom. My personal preference is the final button which just has the bars pointing away from the top of each bar, so, I generally select this too.

Figure 3.19 shows the resulting bar chart. This just simply displays the mean (and the confidence interval of those means). This graph shows us that, on average, people were more aroused by *Memento* than they were by *Bridget Jones' Diary*. However, we originally wanted to look for gender effects, so this graph isn't really telling us what we need to know. The graph we need is a *clustered graph* and this is easily created by using the same dialog box as before (see Figure 3.18), but specifying a second variable (in this case gender) in the box where it says *Legend Variable* next to where it says *Color*. Drag the variable **gender** into this space. The finished dialog box is shown in Figure 3.20 and the resulting bar chart is shown in Figure 3.21. This graph is more informative. It tells us what the last graph did: that is, arousal was overall higher for *Memento* than *Bridget Jones' Diary*, but it also splits this by gender. Look first at the mean arousal for *Bridget Jones' Diary*; this shows that males were actually more aroused during this film than females. This indicates they enjoyed the film more than the women did! Contrast this with *Memento*, for which arousal levels are comparable in males and females. On the face of it, this contradicts the idea of a 'chick flick': it actually seems that men enjoy chick flicks more than the chicks (probably because it's the only help we get to understand the complex workings of the female mind!).

You can edit almost every aspect of the graph by double-clicking on the graph and then clicking on the thing you want to change. You can change the bar colours, the axis titles, the scale of each axis and so on. You can also do things like make the bars three-dimensional. However, tempting as this may be (it can look quite pretty) try to resist the temptation of 3-D charts (unless you're actually plotting three variables): the extra dimension doesn't tell us anything useful and usually only makes the graph more difficult to read (see Wright, 2003).

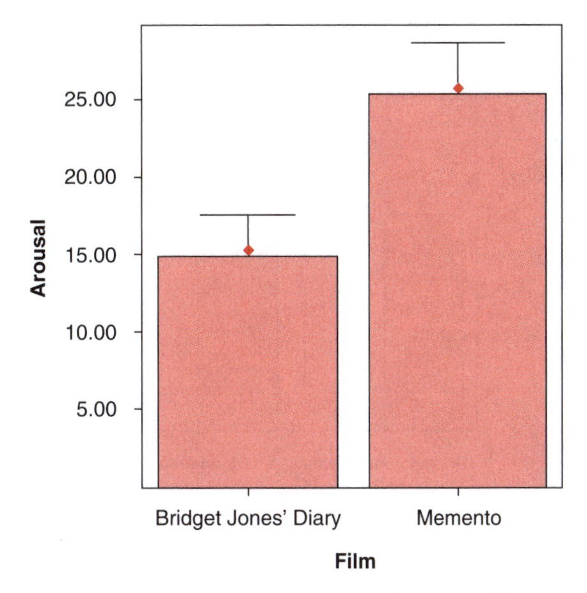

Figure 3.19 Bar chart of the mean arousal for each of the two films

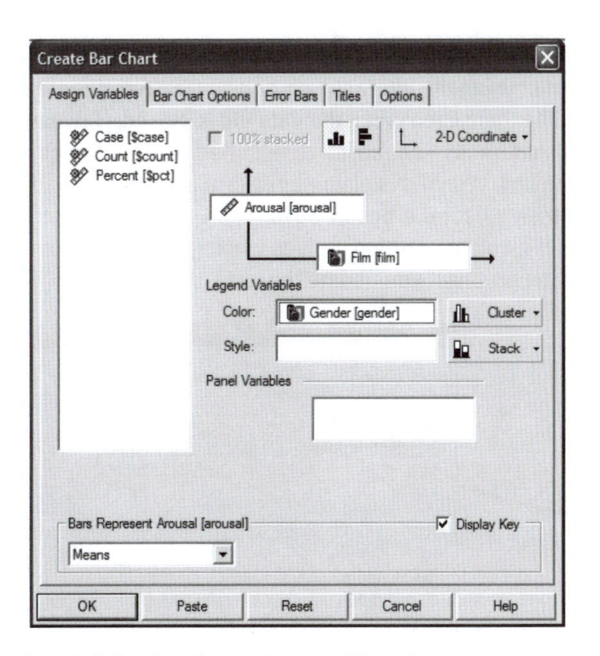

Figure 3.20 Completed dialog box for a clustered bar chart

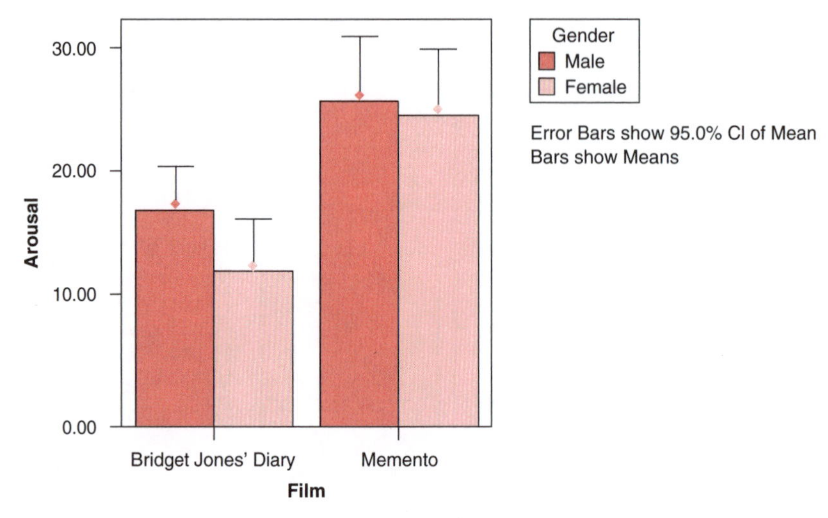

Figure 3.21 Clustered graph showing the mean arousal during the two films for males and females

Box 3.4

Making interactive graphs interactive! ①

The major benefit of using the interactive graphs in SPSS (rather than the non-interactive ones) is that you can update them even when the data file is closed. However, for this to happen, you need to check that this interactive capability is switched on. Use the **Edit⇒Options**... menu to access the *options* dialog box. Within this dialog box there is a tab at the top labelled *Interactive*, so click on this tab and then select the option *Save data with the chart*. This will make your interactive charts interactive! If you use large data files you might find it better to switch this option off again and just redraw the graphs if you update the data.

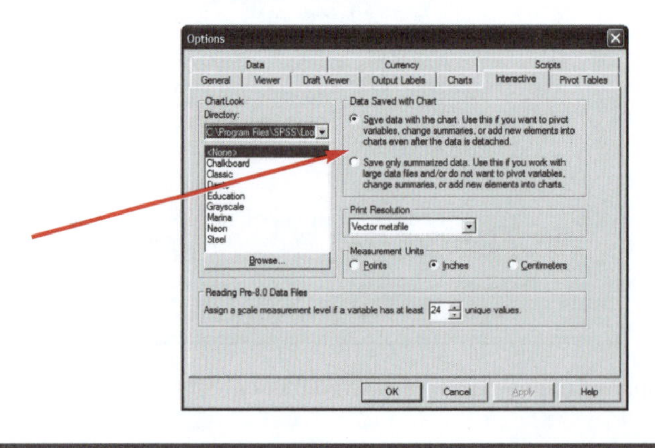

3.8. WHAT HAVE WE DISCOVERED ABOUT STATISTICS? ①

This chapter has covered a lot of ground very quickly! I've tried to show you the importance of examining your data properly before you embark on the main statistical analysis. To begin with, have a look at the distributions of your data and try to spot any data-entry errors, or outliers. If your distributions are not normal then see whether the data can be transformed to correct the problem, or look at replacing outliers. If you're looking at data that compare different groups then look at the distributions within each group. We can also apply statistical tests to see whether our data are normal (the *Kolmogorov–Smirnov* test) and look at other graphs such as the *Q–Q plot*. On the way we discovered some of the uses of the *transform* function of SPSS and the *split file* command. Finally we had a brief look at how to plot means using interactive graphs.

3.9. KEY TERMS THAT WE'VE DISCOVERED

- Bimodal
- Boxplot (box–whisker plot)
- Homogeneity of variance
- Independence
- Interquartile range
- Kolmogorov–Smirnov test
- Levene's test
- Median
- Multimodal
- Normally distributed data
- Outlier
- Parametric test
- Shapiro–Wilk test
- Transformation
- Variance ratio

3.10. SMART ALEX'S TASKS

- Using the data from Chapter 2 (which you should have saved, but if you didn't, re-enter it from Table 2.1) plot some graphs of the mean number of friends, alcohol consumption, income and neuroticism of students and lecturers. Which have the most friends, drink the most, earn the most and are the most neurotic?①
- Using the **ChickFlick.sav** data, check the distributions for the two films (ignore gender): are they normally distributed?①
- Using the **SPSSExam.sav** data (remember that numeracy scores appear positively skewed (see Figure 3.10)), transform these data using one of the transformations described in this chapter: do the data become normal?②

Some brief answers can be found in the file **Answers(Chapter 3).pdf** on the CD-ROM.

3.11. FURTHER READING

Tabachnick, B. G. & Fidell, L. S. (2001). *Using multivariate statistics* (4th edition). Boston: Allyn & Bacon. Chapter 4 is the definitive guide to screening data!

Wright, D. B. (2002). *First steps in statistics*. London: Sage. Chapter 2 is a good introduction to graphing data.

CHAPTER 4

CORRELATION

4.1. WHAT WILL THIS CHAPTER TELL US? ①

It is often interesting for researchers to know what relationship exists, if any, between two or more variables. A correlation is a measure of the linear relationship between variables. For example, I might be interested in whether there is a relationship between the amount of time spent reading this book and the reader's understanding of statistics and SPSS. There are several ways in which these two variables could be related: (1) they could be *positively related*, which would mean that the more time a person spent reading this book, the better their understanding of statistics and SPSS; (2) they could be not related at all, which would mean that a person's understanding of statistics and SPSS remained the same regardless of how much time they spend reading this book; or (3) they could be *negatively related*, which would mean that the more a person reads this book, the worse their understanding of statistics and SPSS gets. How can we tell if two variables are related? This chapter looks first at how we can express the relationships between variables statistically by looking at two measures: *covariance* and the *correlation coefficient*. We then move on to look at how we can represent relationships graphically, before discovering how to carry out and interpret correlations in SPSS. The chapter ends by looking at more complex measures of relationships; in doing so it acts as a precursor to the chapter on multiple regression.

4.2. HOW DO WE MEASURE RELATIONSHIPS? ①

4.2.1. A detour into the world of covariance ①

The simplest way to look at whether two variables are associated is to look at whether they *covary*. To understand what covariance is, we first need to think back to the concept of variance that we met in Chapter 1. Remember that the variance of a single variable represents the average amount that the data vary from the mean. Numerically, it is described by equation (4.1):

$$\text{variance}(s^2) = \frac{\sum(x_i - \bar{x})^2}{N-1} = \frac{\sum(x_i - \bar{x})(x_i - \bar{x})}{N-1} \tag{4.1}$$

The mean of the sample is represented by $\bar{x}$, x_i is the data point in question and N is the number of observations (see section 1.4.1). If we are interested in whether two variables are related, then we are interested in whether changes in one variable are met with similar changes in the other variable. **Therefore, when one variable deviates from its mean we would expect the other variable to deviate from its mean in a similar way**. To illustrate what I mean, imagine we took five people and subjected them to a certain number of advertisements promoting toffee sweets, and then measured how many packets of those sweets each person bought during the next week. The data are given in Table 4.1 as well as the mean and standard deviation (s) of each variable.

Table 4.1

Subject:	1	2	3	4	5	Mean	*s*
Adverts Watched	5	4	4	6	8	5.4	1.67
Packets Bought	8	9	10	13	15	11.0	2.92

If there were a relationship between these two variables, then as one variable deviates from its mean, the other variable should deviate from its mean in the same or directly opposite way. Figure 4.1 shows the data for each participant (squares represent the number of packets bought and circles represent the number of adverts watched); the dashed line is the average number of packets bought and the full line is the average number of adverts watched. The vertical lines represent the differences between the observed values and the mean of the relevant variable. The first thing to notice about Figure 4.1 is that there is a very similar pattern of differences for both variables. For the first three participants the observed values are below the mean for both variables, and for the last two people the observed values are above the mean for both variables. This pattern is indicative of a potential relationship between the two variables (because it seems that if a person's score is below the mean for one variable then their score for the other will also be below the mean).

So, how do we calculate the exact similarity between the pattern of differences of the two variables displayed in Figure 4.1? One possibility is to calculate the total amount of difference but we would have the same problem as in the single variable case (see section 1.4.1). Also, by simply adding the differences, we would gain little insight into the *relationship* between the variables. Now, in the single variable case, we squared the differences to eliminate the problem of positive and negative differences cancelling out each other. When there are two variables, rather than squaring each difference, we can multiply

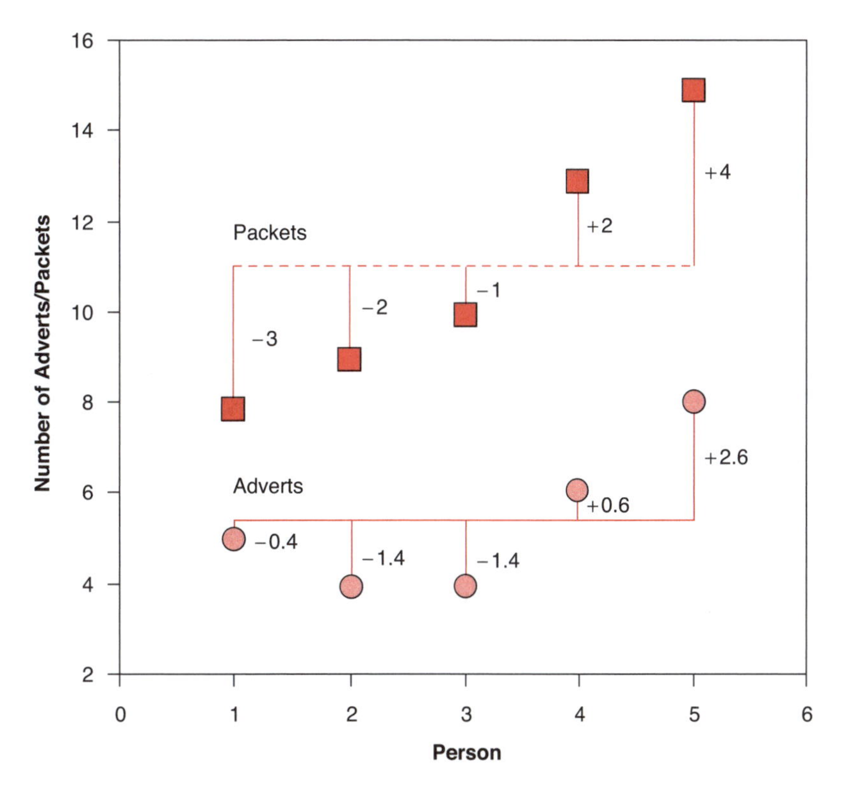

Figure 4.1 Graphical display of the differences between the observed data and the means of two variables

the difference for one variable by the corresponding difference for the second variable. If both errors are positive or negative then this will give us a positive value (indicative of the errors being in the same direction), but if one error is positive and one negative then the resulting product will be negative (indicative of the errors being opposite in direction). When we multiply the differences of one variable by the corresponding differences of the second variable, we get what is known as the cross-product deviations. As with the variance, if we want an average value of the combined differences for the two variables, we must divide by the number of observations (we actually divide by $N - 1$). This averaged sum of combined differences is known as the *covariance*. We can write the covariance in equation form as in equation (4.2)—you will notice that the equation is the same as the equation for variance, except that instead of squaring the differences, we multiply them by the corresponding difference of the second variable:

$$\text{cov}(x, y) = \frac{\sum(x_i - \bar{x})(y_i - \bar{y})}{N - 1} \tag{4.2}$$

For the data in Table 4.1 and Figure 4.1 we reach the following value:

$$
\begin{aligned}
\operatorname{cov}(x, y) &= \frac{\sum(x_i - \bar{x})(y_i - \bar{y})}{N-1} \\
&= \frac{(-0.4)(-3) + (-1.4)(-2) + (-1.4)(-1) + (-0.6)(2) + (2.6)(4)}{4} \\
&= \frac{1.2 + 2.8 + 1.4 + 1.2 + 10.4}{4} \\
&= \frac{17}{4} \\
&= 4.25
\end{aligned}
$$

Calculating the covariance is a good way to assess whether two variables are related to each other. A positive covariance indicates that as one variable deviates from the mean, the other variable deviates in the same direction. On the other hand, a negative covariance indicates that as one variable deviates from the mean (e.g. increases), the other deviates from the mean in the opposite direction (e.g. decreases).

There is, however, one problem with covariance as a measure of the relationship between variables and that is that it depends upon the scales of measurement used. So, covariance is not a standardized measure. For example, if we use the data above and assume that they represented two variables measured in miles then the covariance is 4.25 (as calculated above). If we then convert these data into kilometres (by multiplying all values by 1.609) and calculate the covariance again then we should find that it increases to 11. This dependence on the scale of measurement is a problem because it means that we cannot compare covariances in an objective way—so, we cannot say whether a covariance is particularly large or small relative to another data set unless both data sets were measured in the same units.

4.2.2. Standardization and the correlation coefficient ①

To overcome the problem of dependence on the measurement scale, we need to convert the covariance into a standard set of units. This process is known as standardization. A very basic form of standardization would be to insist that all experiments use the same units of measurement, say metres—that way, all results could be easily compared. However, what happens if you want to measure attitudes—you'd be hard pushed to measure them in metres! Therefore, we need a unit of measurement into which any scale of measurement can be converted. The unit of measurement we use is the *standard deviation*. We came across this measure in section 1.4.1 and saw that, like the variance, it is a measure of the average deviation from the mean. If we divide any distance from the mean by the standard deviation, it gives us that distance in standard deviation units. For example, for the data in Table 4.1, the standard deviation for the number of packets bought is approximately 3.0 (the exact value is 2.91). In Figure 4.1 we can see that the observed value for participant 1 was three packets less than the mean (so, there was an error of –3 packets of sweets). If we divide this deviation, −3, by the standard deviation, which is approximately 3, then we get a value of −1. This tells us that the difference between participant 1's score and the mean was −1 standard deviation. So, we can express the deviation from the mean for a participant in standard units by dividing the observed deviation by the standard deviation.

It follows from this logic that if we want to express the covariance in a standard unit of measurement we can simply divide by the standard deviation. However, there are two variables and, hence, two standard deviations. Now, when we calculate the covariance we actually calculate two deviations (one for each variable) and then multiply them. Therefore, we do the same for the standard deviations: we multiply them and divide by the product of this multiplication. The standardized covariance is known as a *correlation coefficient* and is defined by equation (4.3) in which s_x is the standard deviation of the first variable and s_y is the standard deviation of the second variable (all other letters are the same as in the equation defining covariance):

$$r = \frac{\text{cov}_{xy}}{s_x s_y} = \frac{\sum(x_i - \bar{x})(y_i - \bar{y})}{(N-1)s_x s_y} \tag{4.3}$$

The coefficient in equation (4.3) is known as the Pearson product-moment correlation coefficient (or Pearson correlation coefficient as it tends to be known) and was invented by Pearson (what a surprise!).[1] If we look back at Table 4.1 we see that the standard deviation for the number of adverts watched (s_x) was 1.67, and for the number of packets of crisps bought (s_y) was 2.92. If we multiply these together we get $1.67 \times 2.92 = 4.88$. Now, all we need to do is take the covariance, which we calculated a few pages ago as being 4.25, and divide by these multiplied standard deviations. This gives us $r = 4.25/4.88 = .87$ (this is the same value that you'll come across later in SPSS Output 4.1).

By standardizing the covariance we end up with a value that has to lie between −1 and +1 (if you find a correlation coefficient less than −1 or more than +1 you can be sure that something has gone hideously wrong!). A coefficient of +1 indicates that the two variables are perfectly positively correlated, so as one variable increases the other increases by a proportionate amount. Conversely, a coefficient of −1 indicates a perfect negative relationship: if one variable increases the other decreases by a proportionate amount. A coefficient of 0 indicates no linear relationship at all and so if one variable changes the other stays the same. We also saw in section 1.8.4 that because the correlation coefficient is a standardized measure of an observed effect, it is a commonly used measure of the size of an effect and that values of ± .1 represent a small effect, ± .3 is a medium effect and ± .5 is a large effect.

---

1 You will find Pearson's product-moment correlation coefficient denoted by both r and R. Typically, the upper-case form is used in the context of regression because it represents the multiple correlation coefficient; however, for some reason, when we square r (as in section 4.5.3) an upper-case R is used. Don't ask me why—it's just to confuse me I suspect.

Cramming Samantha's Tips

- A crude measure of the relationship between variances is the *covariance*.
- If we standardize this value we get *Pearson's correlation coefficient, r.*
- The correlation coefficient has to lie between −1 and +1.
- A coefficient of +1 indicates a perfect positive relationship, a coefficient of −1 indicates a perfect negative relationship, a coefficient of 0 indicates no linear relationship at all.
- The correlation coefficient is a commonly used measure of the size of an effect: values of ± 0.1 represent a small effect, ± 0.3 is a medium effect and ± 0.5 is a large effect.

4.3. DATA ENTRY FOR CORRELATION ANALYSIS USING SPSS ①

Data entry for correlation, regression and multiple regression is straightforward because each variable is entered in a separate column. So, for each variable you have measured, create a variable in the data editor with an appropriate name, and enter a participant's scores across one row of the data editor. There may be occasions on which you have one or more categorical variables (such as gender) and these variables can also be entered in a column (but remember to define appropriate value labels). As an example, if we wanted to calculate the correlation between the two variables in Table 4.1 we would enter these data as in Figure 4.2. Throughout this chapter we are going to analyse a data set based on undergraduate exam performance. The data for several examples are stored in a single file on the sample disk called **ExamAnxiety.sav**. If you open this data file you will see that these data are laid out in the data editor as separate columns and that **gender** has been coded appropriately. We will discover to what each of the variables refers as we progress through this chapter.

	adverts	packets
1	5.00	8.00
2	4.00	9.00
3	4.00	10.00
4	6.00	13.00
5	8.00	15.00
6		

Figure 4.2 Data entry for correlation. The data editor tells us that participant 1 was shown five adverts and subsequently purchased eight packets of sweets

4.4. GRAPHING RELATIONSHIPS: THE SCATTERPLOT ①

How do I draw a graph of the relationship between two variables?

Before conducting any correlational analysis it is *essential* to plot a scatterplot to look at the general trend of the data. A scatterplot is simply a graph that plots each person's score on one variable against their score on another (and their score on a third variable can also be included on a 3-D scatterplot). A scatterplot tells us several things about the data such as whether there seems to be a relationship between the variables, what kind of relationship it is and whether any cases are markedly different from the others. A case that differs substantially from the general trend of the data is known as an *outlier* and such cases can severely bias the correlation coefficient (see Box 3.1 and section 5.6.1.1 for more detail). We can use a scatterplot to show us if any cases look like outliers.

Drawing a scatterplot using SPSS is dead easy. You can draw it using the interactive graphs menu, or using non-interactive graphs. We'll have a look at both because there are some useful scatterplots that you can do only using non-interactive graphs.

4.4.1. Simple scatterplot ①

This type of scatterplot is for looking at just two variables. For example, a psychologist was interested in the effects of exam stress on exam performance. So, she devised and validated a questionnaire to assess state anxiety relating to exams (called the Exam Anxiety Questionnaire, or EAQ). This scale produced a measure of anxiety scored out of 100. Anxiety was measured before an exam, and the percentage mark of each student on the exam was used to assess the exam performance. Before seeing if these variables were correlated, the psychologist would draw a scatterplot of the two variables (her data are in the file **Exam Anxiety.sav** and you should already have this file loaded into SPSS). To plot these two variables you can use an interactive graph so select **Graphs⇒Interactive⇒Scatterplot…** to open the dialog box in Figure 4.3. In this dialog box all of the variables in the data editor are displayed on the left-hand side and there are several empty spaces on the right-hand side. You simply click on a variable from the list on the left and drag it to move it to the appropriate box. On any graph the vertical axis is known as the *y*-axis (or *ordinate*) of the graph. This variable should be the dependent variable (the outcome that was measured),[2] which in this case is **exam performance (%) [exam]**. The horizontal axis is known as the *X*-axis (or *abscissa*) of the graph. This variable should be the independent variable, which in this case is **exam anxiety [anxiety]**. The dialog box in Figure 4.3 has a schematic representation of the *y*-axis (the arrow pointing upwards) and the *x*-axis (the arrow pointing right), and each has a space. To specify these variables we need to drag them into the appropriate

2 In experimental research it is customary to plot the independent variable on the horizontal axis and the dependent variable on the vertical axis. In this form of controlled research, the implication is that changes in the independent variable (the variable that the experimenter has manipulated) cause changes in the dependent variable. In correlational research, variables are measured simultaneously and so no cause and effect relationship can be established. As such, these terms are used loosely!

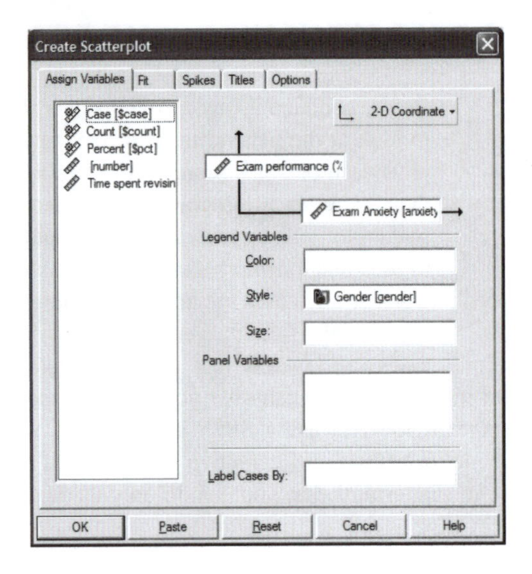

Figure 4.3 Dialog box for a simple scatterplot

space. So, drag **exam performance (%) [exam]** into the space for the *y*-axis and drag **exam anxiety [anxiety]** into the space for the *x*-axis.

Lower down in the dialog box there are several other spaces into which you can drag variables. You can use a grouping variable to define different categories on the scatterplot (it will display each category using a different colour, symbol or graph). This function is useful for looking at the relationship between two variables for groups of people. In the current example, we have data relating to whether the student was male or female, so we can use these other options in the following way:

- **Color**: If you select **Gender [gender]** from the list and drag it to this space the resulting scatterplot would show male data points in a different colour to female data points.
- **Style**: If you select **Gender [gender]** from the list and drag it to this space the resulting scatterplot would use different symbols for male and female data points. This is what I've chosen to do and the resulting dialog box should look like Figure 4.3.
- **Size**: If you select **Gender [gender]** from the list and drag it to this space the symbols for male and female data points would be different sizes in the resulting scatterplot.
- **Panel Variables**: If you select **Gender [gender]** from the list and drag it to this space SPSS will produce two scatterplots (side by side)—one for males and one for females.
- **Label Cases By**: If you select **Gender [gender]** from the list and drag it to this space SPSS will put a label next to each data point on the scatterplot indicating whether that data point came from a male or a female. With lots of data points this can make the plot hard to read! One use might be if you had a variable with a participant's name, or a code, in which case each point on the scatterplot will be labelled with a code that identifies the person, which could be useful for identifying outliers and then actually being able to fathom out which case is the outlier!

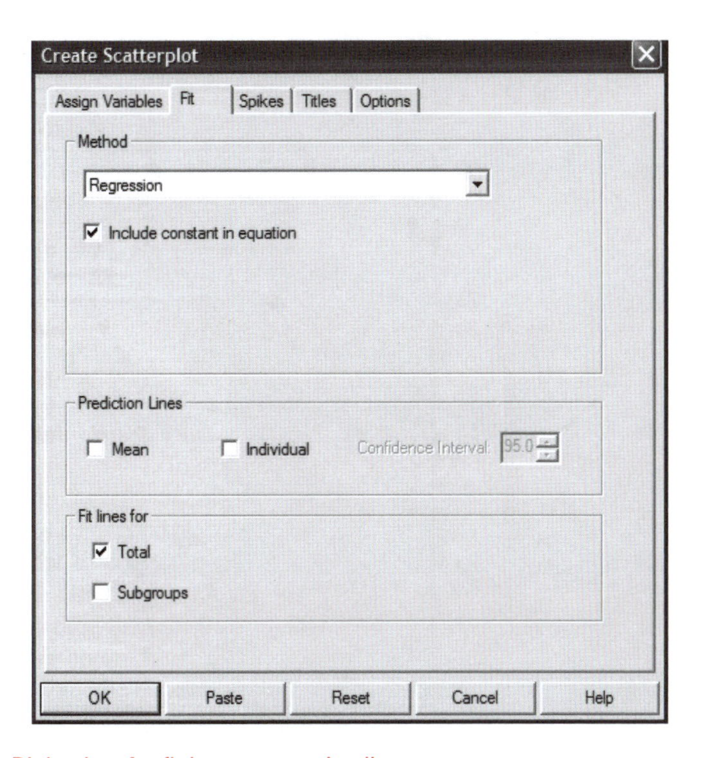

Figure 4.4 Dialog box for fitting a regression line

At the top of the dialog box there are several tabs that give you access to other options. If you click on *Titles* for example you'll see a dialog box that gives you space to type in a title for the scatterplot. One tab that might be useful (especially when you've read the next chapter) is *Fit* (Figure 4.4). Within this dialog box you can choose a line to plot by selecting a *Method* from the drop-down list. If you chose *regression* then SPSS will fit the straight line that best represents the relationship between the variables in your scatterplot (see Chapter 5). If you choose *mean* then SPSS plots a horizontal line representing the mean, and if you choose *smoother* then it fits the curved line that best describes the relationship between the variables on your scatterplot. If you select ☑ Total then SPSS plots these lines for the data overall, and if you select ☑ Subgroups then it will plot separate lines for different groups specified by any grouping variable that you may have used in the original scatterplot dialog box (these options can both be selected at the same time so you can have an overall line and subgroup lines on the same plot). In our example, we used gender as a grouping variable so we'd get a line for males and a line for females. For the time being don't select any fit options, but as you're reading through Chapter 5 bear in mind that this is how you plot a regression line on a scatterplot!

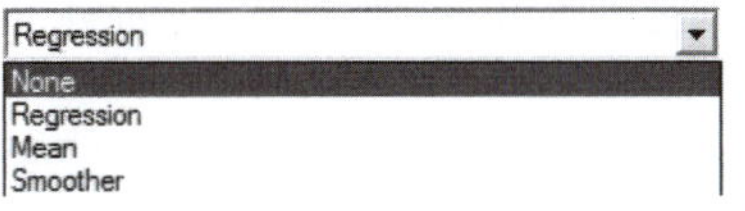

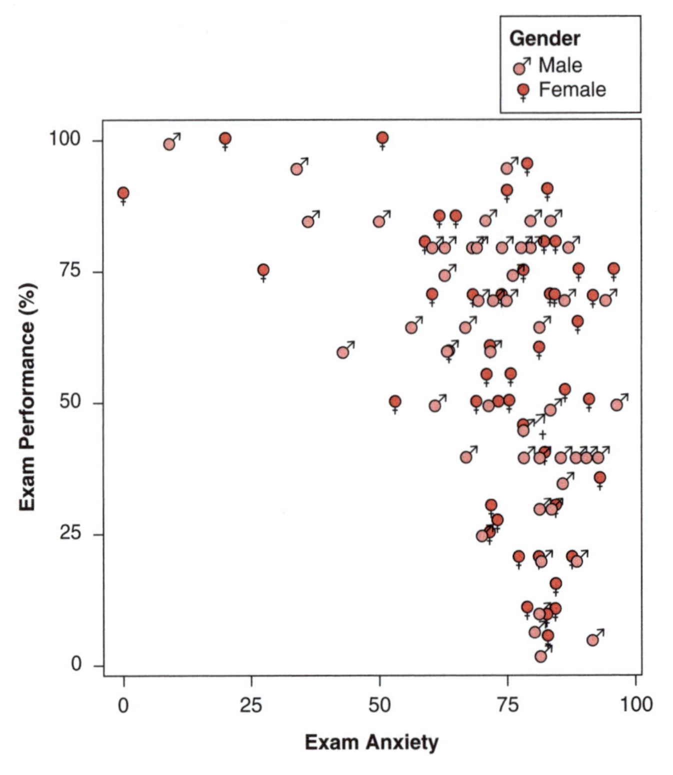

Figure 4.5 Scatterplot of exam performance against exam anxiety

The resulting scatterplot is shown in Figure 4.5. The scatterplot on your screen will look slightly different in that the male and female data will be displayed using different symbols (triangles and circles). I have replaced the male and female markers with different symbols (this provided the opportunity for me to show off by using gender-appropriate symbols!). The scatterplot shows that the majority of students suffered from high levels of anxiety (there are very few cases that had anxiety levels below 60). Also, there are no obvious outliers in that most points seem to fall within the vicinity of other points. There also seems to be some general trend in the data such that higher levels of anxiety are associated with lower exam scores and low levels of anxiety are almost always associated with high examination marks. The gender markers show that anxiety seems to affect males and females in the same way (because the different symbols are fairly evenly interspersed). Another noticeable trend in these data is that there were no cases having low anxiety and low exam performance—in fact, most of the data are clustered in the upper region of the anxiety scale.

Had there been any data points which obviously didn't fit the general trend of the data then it would be necessary to try to establish if there was a good reason why these participants responded so differently. Sometimes outliers are just errors of data entry (i.e. you mis-typed a value) and so it is wise to double-check the data in the editor window for any case that looks unusual. If an outlier can't be explained by incorrect data entry, then it is important to try to establish whether there might be a third variable affecting this person's score. For example, a person could be experiencing anxiety about something other than the exam and their score on the anxiety questionnaire might have picked up on this anxiety, but it may be specific anxiety about the exam that interferes with performance. Hence, this person's unrelated anxiety did not affect their performance, but resulted in a high anxiety score. If there is a good reason for a participant responding differently to everyone else then you can consider eliminating that participant from the analysis in the interest of building an accurate model. However, participants' data should not be eliminated because they don't fit with your hypotheses—only if there is a good explanation of why they behaved so oddly.

4.4.2. The 3-D scatterplot ①

I'm now going to show you one of the few times it's appropriate to use a 3-D graph! A 3-D scatterplot is used to display the relationship between three variables. The reason why it's all right to use a 3-D graph here is because the third dimension is actually telling us something useful (and isn't just there to look flash).[3] As an example, imagine our researcher decided that exam anxiety might not be the only factor contributing to exam performance. So, she also asked participants to keep a revision diary from which she calculated the number of hours spent revising for the exam. She wanted to look at the relationships between these variables simultaneously. To create a 3-D scatterplot, use the same dialog box as in Figure 4.3 above (access this dialog box in the same way). In this dialog box there is a button with a drop-down menu labelled *2-D Co-ordinate*. Click on the arrow to access the drop-down menu and you'll see a second option appear labelled *3-D Co-ordinate*. Click on this option and the schematic representation of the axis will change to include a third axis: the *z*-axis. Using Figure 4.3 as a guide, complete the dialog box as before (try putting the gender variable in the colour box this time) but select *3-D Co-ordinate* and then grab and drag **Time spent revising [revise]** into the space for the *z*-axis. The resulting scatterplot is shown in Figure 4.6. Usually 3-D scatterplots can be quite difficult to interpret and so their usefulness in exploring data can be limited; however, they are great for displaying data in a concise way. Now, if you double-click on this graph you'll be able to rotate the axis to get a better idea of the relationship between the three variables—this is helpful for you; however, ultimately you'll have to decide which angle best represents the relationship for when you want

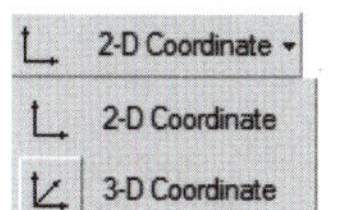

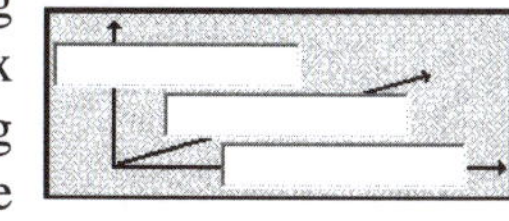

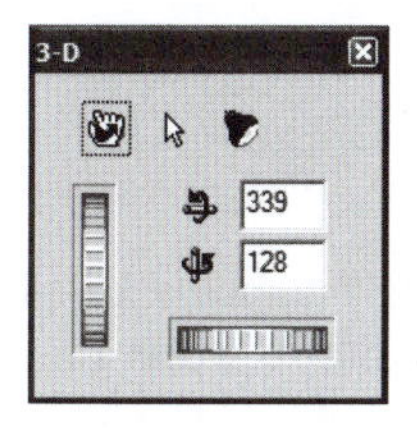

3 It's not actually wrong to use 3-D plots when you're only plotting two variables. I mean it's not like anything bad will happen to you (unless you show it to certain statisticians who will lock you in a room and make you write I must not do 3-D graphs 75,172 times on the blackboard), it's just not a clear way to present data.

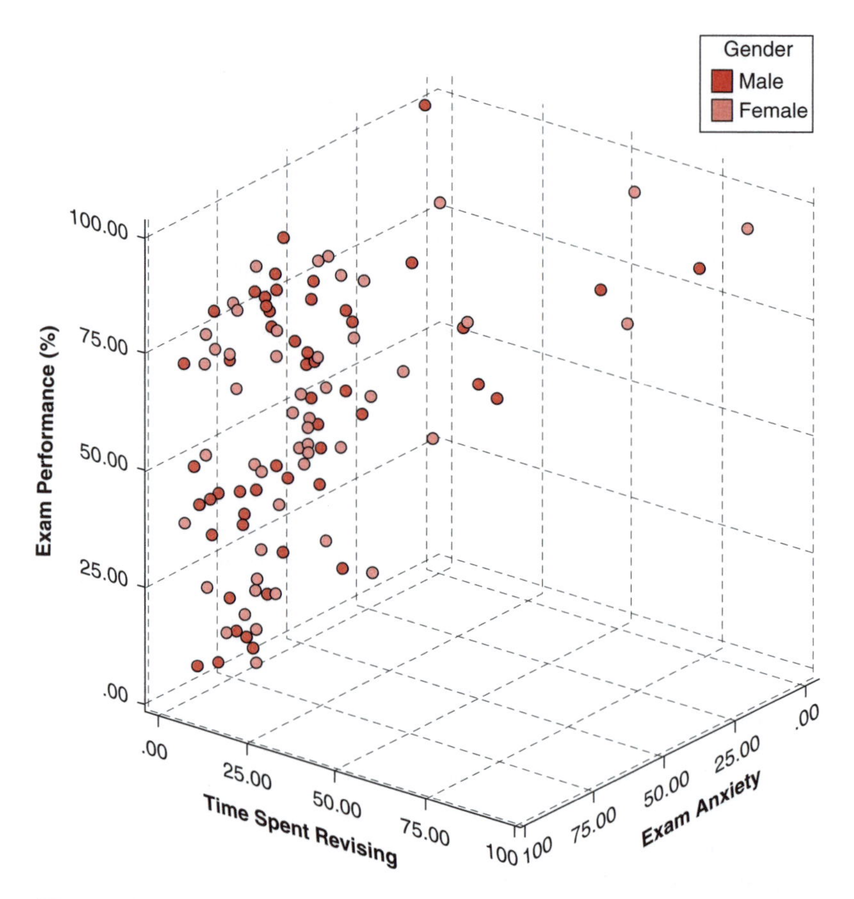

Figure 4.6 A 3-D scatterplot of exam performance plotted against exam anxiety and the amount of time spent revising

to put the graph on a 2-D piece of paper! Have a play with the graph and see if you can get the same picture as me (the screenshot near this text will help!).

4.4.3. Overlay scatterplot ①

There are two types of scatterplot that you can't do using interactive graphs: an overlay scatterplot and a matrix scatterplot. Both are very useful and so worth finding out about. An overlay scatterplot is one in which several pairs of variables are plotted on the same axes. Imagine our researcher wanted to look at the role of both exam anxiety and revision time on exam performance (but not the relationship between exam anxiety and revision); it would be useful to plot the relationships between revision time and exam performance and between exam anxiety and performance simultaneously. To get the best use out of this

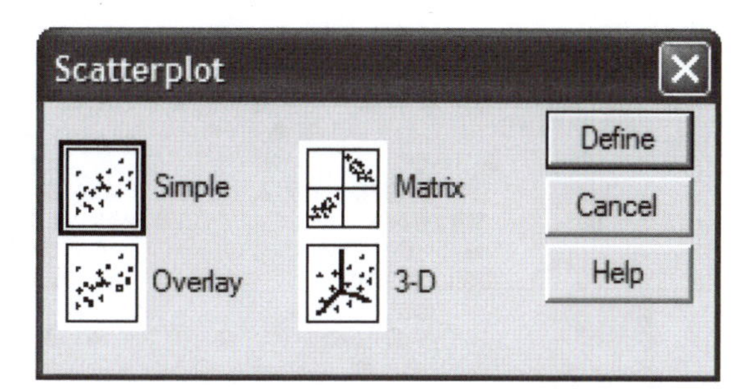

Figure 4.7 Main *scatterplot* dialog box

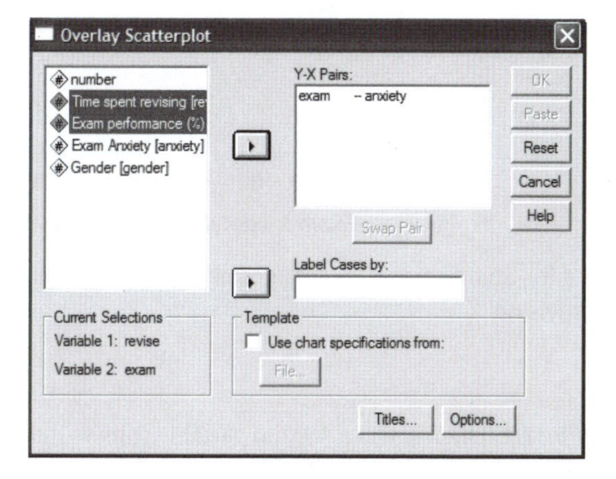

Figure 4.8 Dialog box for creating an overlay scatterplot. The first pair of variables has already been specified and the second pair has been selected but not transferred to the pair list

type of scatterplot keep one variable constant and plot it against several others. In the examination example, in which the effect of both anxiety and revision time on exam performance is of interest, you should plot **anxiety** (*X*) against **exam** (*Y*), and then overlay **revise** (*X*) against **exam** performance (*Y*).

To plot these combinations, simply use the menus as follows: **Graphs⇒Scatter**.... This activates the dialog box in Figure 4.7, which in turn gives you four options for the different types of scatterplot available. By default a simple scatterplot is selected as is shown by the black rim around the picture. If you wish to draw a different scatterplot then move the on-screen arrow over one of the other pictures and click with the left button of the mouse. When you have selected a scatterplot click on Define.

To plot an overlay scatterplot, click on *Overlay* in the main *scatterplot* dialog box (see Figure 4.7) and the dialog box in Figure 4.8 appears. To select a pair of variables, click on one variable from the list (this will appear as *Variable 1* in the section labelled *Current*

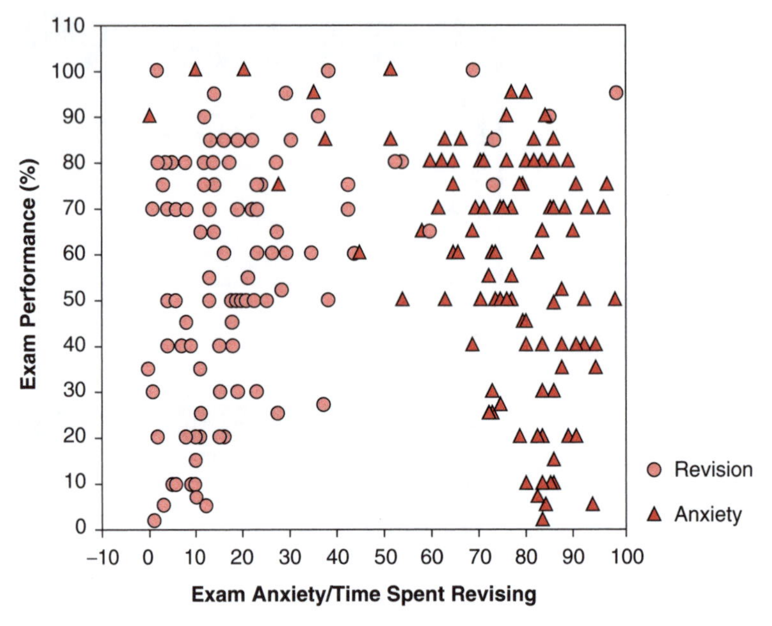

Figure 4.9 Scatterplot of exam scores against both exam anxiety and time spent revising

Selections) and then select a second variable (this will be listed as *Variable 2*). Transfer the variable pair by clicking on ▸. The pair will appear in the space labelled *Y-X Pairs*. The order of variables relates to the axis on which they will be plotted, so in Figure 4.8 **exam** will appear on the *Y*-axis and **anxiety** on the *X*-axis. The second pairing of **exam** and **revise** should also be transferred to the *Y–X* pair list such that **exam** is listed first (and so is plotted on the *Y*-axis) and **revise** second (so will be plotted on the *X*-axis). If, when you transfer two variables, they appear the wrong way round to how you want to plot them, you can swap the order of variables (and hence the axis on which they will be plotted) by clicking on Swap Pair.

From Figure 4.9 it is clear that although anxiety is negatively related to exam performance, it looks as though exam performance is positively related to revision time. So, as revision time increases, exam performance increases also, but as anxiety increases, exam performance decreases. The overlay scatterplot clearly shows these different relationships.

4.4.4. Matrix scatterplot ①

Instead of plotting several variables on the same axes on a 3-D scatterplot (which can be difficult to interpret with several variable pairs), it is possible to plot a matrix of scatterplots. This type of plot allows you to see the relationship between all combinations

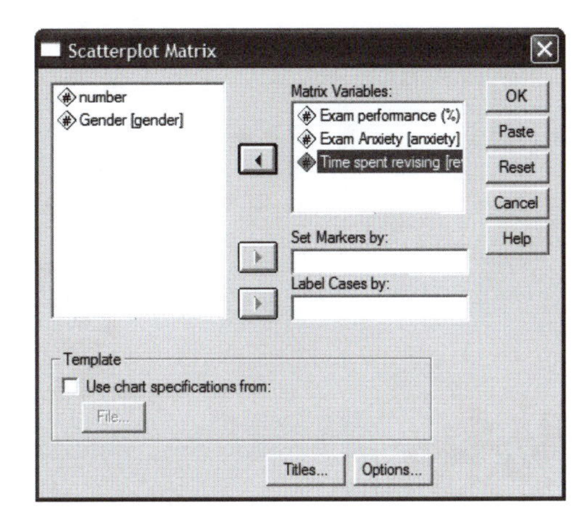

Figure 4.10 Dialog box for a matrix scatterplot

of many different pairs of variables. To conduct a matrix scatterplot for the same data as was used for the overlay scatterplot, select the *Matrix* scatterplot option in the main *scatterplot* dialog box (Figure 4.7). The completed dialog box is shown in Figure 4.10: variables are listed on the left and any you want plotted should be transferred to the box labelled *Matrix Variables* using the ▶ button. For our data, select and transfer **Exam performance (%) [exam], Exam Anxiety [anxiety]** and **Time spent revising [revise]** to the box labelled *Matrix Variables*. As with the simple scatterplot there is an option to split the scatterplot by a selected grouping variable (*Set Markers by …*) but there is no need to use this option for these data. The resulting scatterplot of exam performance against exam anxiety and revision time is shown in Figure 4.11.

The six scatterplots in Figure 4.11 represent the various combinations of each variable plotted against each other variable. So, the grid references represent the following plots:

- **B1**: exam performance (Y) vs. anxiety (X)
- **C1**: exam performance (Y) vs. revision time (X)
- **A2**: anxiety (Y) vs. exam performance (X)
- **C2**: anxiety (Y) vs. revision time (X)
- **A3**: revision time (Y) vs. exam performance (X)
- **B3**: revision time (Y) vs. anxiety (X)

Thus, the three scatterplots below the diagonal of the matrix are the same plots as the ones above the diagonal but with the axes reversed. From this matrix we can see that revision time and anxiety are inversely related (so, the more time spent revising the less anxiety the participant had about the exam). Also, in the scatterplot of revision time against anxiety (grids C2 and B3) it looks as though there is one possible outlier—there is a single participant who spent very little time revising yet suffered very little anxiety about the exam. As all participants who had low anxiety scored highly on the exam we can deduce that this

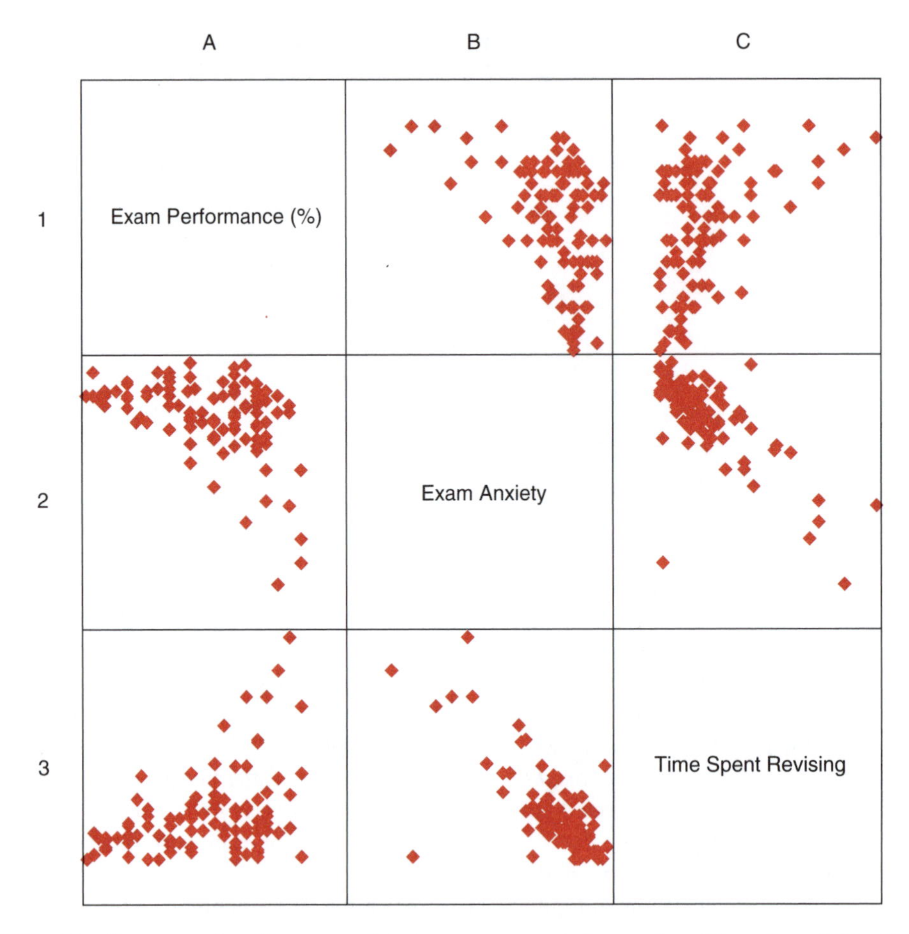

Figure 4.11 Matrix scatterplot of exam performance, exam anxiety and revision time. Grid references have been added for clarity

person also did well on the exam (don't you just hate a smart alec!). We could choose to examine this case more closely if we believed that their behaviour was caused by some external factor (such as taking brain-pills!). Matrix scatterplots are very convenient for examining pairs of relationships between variables. However, I don't recommend plotting them for more than three or four variables because they become very confusing indeed.

4.5. BIVARIATE CORRELATION ①

Having taken a preliminary glance at the data, we can proceed to conduct the correlation analysis. For this provisional look at correlations we will return to the data in Table 4.1, which looked at whether there was a relationship between the number of adverts watched

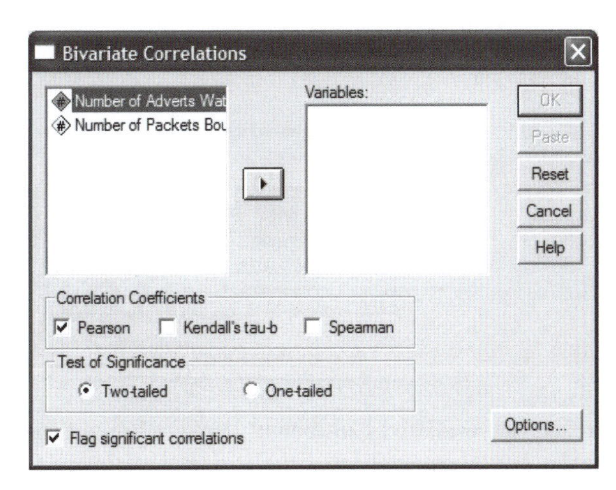

Figure 4.12 Dialog box for conducting a bivariate correlation

and the number of packets of toffees subsequently purchased. Create two columns in the data editor and input these data (the lazy among you can access the file **Advert.sav** on the disk).

There are two types of correlation: *bivariate* and *partial*. A bivariate correlation is a correlation between two variables (as described at the beginning of this chapter) whereas a partial correlation looks at the relationship between two variables while 'controlling' the effect of one or more additional variables. Pearson's product-moment correlation coefficient (described earlier) and Spearman's rho (see section 4.5.4) are examples of bivariate correlation coefficients. To conduct a bivariate correlation you need to find the *Correlate* option of the *Analyze* menu. The main dialog box is accessed by the menu path **Analyze⇒Correlate⇒Bivariate...** and is shown in Figure 4.12. Using the dialog box it is possible to select which of three correlation statistics you wish to perform. The default setting is Pearson's product-moment correlation, but you can also calculate Spearman's correlation and Kendall's correlation—we shall see the differences between these correlation coefficients in due course. In addition, it is possible to specify whether or not the test is one- or two-tailed (see section 1.8.2). A one-tailed test should be selected when you have a directional hypothesis (e.g. 'the more adverts a person watches, the more packets of sweets they will have bought'). A two-tailed test (the default) should be used when you cannot predict the nature of the relationship (i.e. 'I'm not sure whether watching more adverts will be associated with an increase or a decrease in the number of packets of sweets a person buys').

Having accessed the main dialog box, you should find that the variables in the data editor are listed on the left-hand side of the dialog box (Figure 4.12). There is an empty box labelled *Variables* on the right-hand side. You can select any variables from the list using the mouse and transfer them to the *Variables* box by clicking on ▶. SPSS will create a table of correlation coefficients for all of the combinations of variables. This table is called a correlation matrix. For our current example, select the variables **Number of Adverts**

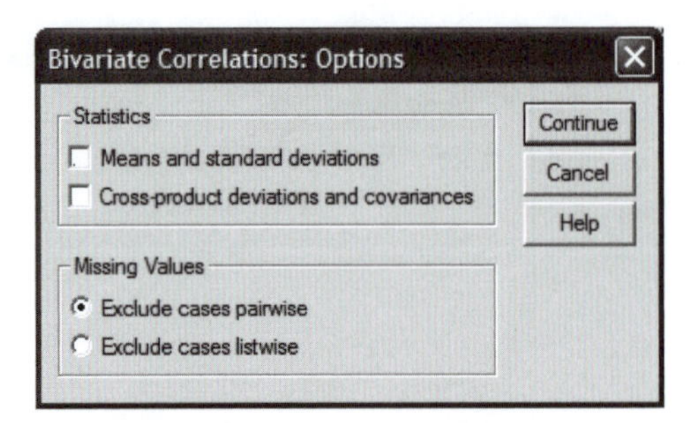

Figure 4.13 Dialog box for bivariate correlation options

Watched [adverts] and **Number of Packets Bought [packets]** and transfer them to the variables list. Having selected the variables of interest you can choose between three correlation coefficients: Pearson's product-moment correlation coefficient, Spearman's rho and Kendall's tau. Any of these can be selected by clicking on the appropriate tick-box with a mouse.

If you click on Options... then another dialog box appears with two *Statistics* options and two options for missing values (Figure 4.13). The *Statistics* options are enabled only when Pearson's correlation is selected; if Pearson's correlation is not selected then these options are disabled (they appear in a light grey rather than black and you can't activate them). This deactivation occurs because these two options are meaningful only for parametric data and Pearson's correlation is used with those kinds of data. If you select the tick-box labelled *Means and standard deviations* then SPSS will produce the mean and standard deviation of all of the variables selected for correlation. If you activate the tick-box labelled *Cross-product deviations and covariances* then SPSS will give you the values of these statistics for each of the variables being correlated. The cross-product deviations tell us the sum of the products of mean corrected variables, which is simply the numerator (top half) of equation (4.2). The covariances option gives us values of the covariance between variables, which could be calculated manually using equation (4.2). In other words, these covariance values are the cross-product deviations divided by (N – 1) and represent the unstandardized correlation coefficient. In most instances, you will not need to use these options but they occasionally come in handy!

To illustrate what these options do, select them for the advert data. Leave the default options in the main dialog box as they are. The resulting output from SPSS is shown in SPSS Output 4.1. The section labelled *Sum of Squares and Cross-products* shows us the cross-product (17 in this example) that we calculated from equation (4.2), and the sums of squares for each variable. The sums of squares are calculated from the top half of equation (4.1). The value of the covariance between the two variables is 4.25, which is the same value as was calculated from equation (4.2). The covariance within each variable is the same as the variance for each variable (so, the variance for the number of adverts seen is 2.8, and the variance for the number of packets bought is 8.5). These variances can be calculated manually from equation (4.1). Also note that Pearson's correlation coefficient between the two variables is .871, which is the same as the value we calculated in section 4.2.2.

Correlations

		Adverts Watched	Number of Packets
Adverts Watched	Pearson Correlation	1.000	.871
	Sig. (2-tailed)	.	.054
	Sum of Squares and Cross-products	11.200	17.000
	Covariance	2.800	4.250
	N	5	5
Number of Packets	Pearson Correlation	.871	1.000
	Sig. (2-tailed)	.054	.
	Sum of Squares and Cross-products	17.000	34.000
	Covariance	4.250	8.500
	N	5	5

SPSS Output 4.1 Output from the SPSS correlation procedure

4.5.1. Pearson's correlation coefficient ①

Pearson's correlation coefficient was described in full at the beginning of this chapter. For those of you unfamiliar with basic statistics, it is not meaningful to talk about means unless we have data measured at an interval or ratio level (for revision see Box 2.1 or Field & Hole, 2003). Pearson's correlation requires only that data are interval for it to be an accurate measure of the linear relationship between two variables. However, if you want to establish whether the correlation coefficient is significant, then more assumptions are required: for the test statistic to be valid data have to be normally distributed. Although typically we would want both variables to be normally distributed, there is one exception to this rule: one of the variables can be a categorical variable provided there are only two categories (in fact if you look at section 4.5.6 you'll see that this is the same as doing a *t*-test, but I'm jumping the gun a bit). In any case, if your data are non-normal (see Chapter 3) or are not measured at the interval level then you should deselect the Pearson tick-box (Figure 4.14).

Reload the data from the file **ExamAnxiety.sav** so that we can calculate the correlation between exam performance, revision time and exam anxiety. The data from the exam performance study are parametric and so a Pearson's correlation can be applied. Access the main *bivariate correlations* dialog box (**Analyze⇒Correlate⇒Bivariate...**) and transfer **exam**, **anxiety** and **revise** to the box labelled *Variables* (Figure 4.15). The dialog box also allows you to specify whether the test should be one- or two-tailed. One-tailed tests should be used when there is a specific direction to the hypothesis being tested, and two-tailed tests should be used when a relationship is expected, but the direction of the relationship is not predicted. Our researcher predicted that at higher levels of anxiety, exam performance would be poor. Therefore, the test for these variables should be one-tailed because before the data were collected the researcher predicted a specific kind

Figure 4.14 Karl Pearson

of relationship. What's more, a positive correlation between revision time and exam performance was predicted so this test too should be one-tailed. To ensure that the output displays the one-tailed significance values select [One-tailed] and then click on [OK].

SPSS Output 4.2 provides a matrix of the correlation coefficients for the three variables. Underneath each correlation coefficient both the significance value of the correlation and the sample size (*N*) on which it is based, are displayed. Each variable is perfectly correlated with itself (obviously) and so $r = 1$ along the diagonal of the table. Exam performance is negatively related to exam anxiety with a Pearson correlation coefficient of $r = -.441$ and there is a less than .001 probability that a correlation coefficient this big would have occurred by chance in a sample of 103 people (as indicated by the two asterisks after the coefficient). This significance value tells us that the probability of this correlation being a 'fluke' is very low (close to zero in fact). Hence, we can have confidence that the relationship between exam performance and anxiety is genuine. Usually, social scientists accept any probability value below .05 as being statistically meaningful and so any probability value below .05 is regarded as indicative of genuine effect (and SPSS will mark any correlation coefficient significant at this level with an asterisk). The output also shows that exam performance is positively related to the amount of time spent revising, with a coefficient of $r = .397$, which is also significant at $p < .001$. Finally, exam anxiety appears to be negatively related to the time spent revising ($r = -.709$, $p < .001$).

In psychological terms, this all means that as anxiety about an exam increases, the percentage mark obtained in that exam decreases. Conversely, as the amount of time revising

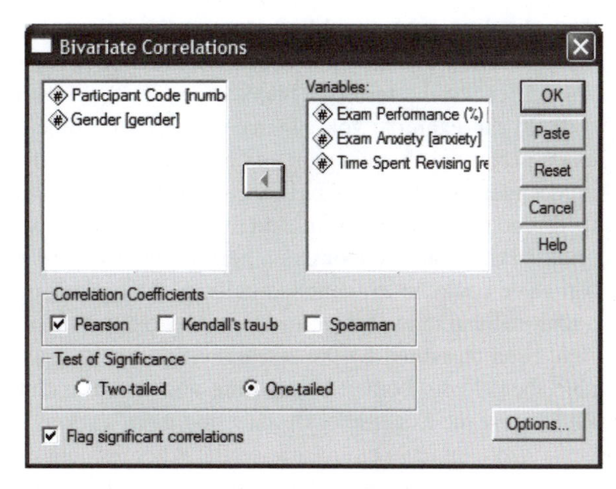

Figure 4.15 Completed dialog box for the exam performance data

Correlations

		Exam performance (%)	Exam Anxiety	Time spent revising
Exam performance (%)	Pearson Correlation	1.000	−.441**	.397**
	Sig. (1-tailed)	.	.000	.000
	N	103	103	103
Exam Anxiety	Pearson Correlation	−.441**	1.000	−.709**
	Sig. (1-tailed)	.000	.	.000
	N	103	103	103
Time spent revising	Pearson Correlation	.397**	−.709**	1.000
	Sig. (1-tailed)	.000	.000	.
	N	103	103	103

**Correlation is significant at the .01 level (1-tailed)

SPSS Output 4.2 Output from SPSS for a Pearson's correlation

increases, the percentage obtained in the exam increases. Finally, as revision time increases the student's anxiety about the exam decreases. So there is a complex interrelationship between the three variables.

4.5.2. A word of warning about interpretation: causality ①

Considerable caution must be taken when interpreting correlation coefficients because they give no indication of the direction of *causality*. So, in our example, although we can conclude that exam performance goes down as anxiety about that exam goes up, we cannot say that high exam anxiety *causes* bad exam performance. This caution is for two reasons:

- **The third-variable problem**: In any bivariate correlation causality between two variables cannot be assumed because there may be other measured or unmeasured variables affecting the results. This is known as the *third-variable* problem or the *tertium quid*. In our example you can see that revision time does relate significantly to both exam performance and exam anxiety and there is no way of telling which of the two independent variables, if either, causes exam performance to change. So, if we had measured only exam anxiety and exam performance we might have assumed that high exam anxiety caused poor exam performance. However, it is clear that poor exam performance could be explained equally well by a lack of revision. There may be several additional variables that influence the correlated variables and these variables may not have been measured by the researcher. So, there could be another, unmeasured, variable that affects both revision time and exam anxiety. For example, Field & Hole (2003) give the example of Broca's finding of a strong relationship between gender and brain size: women's brains are, on the whole, smaller than men. This was, at the time, used as evidence to argue that women were intellectually inferior to men. The obvious problem is that this relationship takes no account of body size: people with bigger bodies have bigger brains irrespective of intellectual ability. Gould (1981) famously reanalysed Broca's data and demonstrated that the strong relationship between gender and brain size disappeared when you accounted for body size!
- **Direction of causality**: Correlation coefficients say nothing about which variable causes the other to change. Even if we could ignore the third-variable problem described above, and we could assume that the two correlated variables were the only important ones, the correlation coefficient doesn't indicate in which direction causality operates. So, although it is intuitively appealing to conclude that exam anxiety causes exam performance to change, there is no *statistical* reason why exam performance cannot cause exam anxiety to change. Although the latter conclusion makes no human sense (because anxiety was measured before exam performance), the correlation does not tell us that it isn't true.

4.5.3. Using R^2 for interpretation ①

Although we cannot make direct conclusions about causality, we can take the correlation coefficient a step further by squaring it. The correlation coefficient squared (known as the coefficient of determinatfion, R^2) is a measure of the amount of variability in one variable that is explained by the other. For example, we may look at the relationship between exam anxiety and exam performance. Exam performances vary from person to person because of any number of factors (different ability, different levels of preparation and so on). If we were to add up all of this variability (rather like when we calculated the sum of squares in section 1.4.1) then we would have an estimate of how much variability exists in exam performances. We can then use R^2 to tell us how much of this variability is accounted for by exam anxiety. These two variables had a correlation of −.4410 and so the value of R^2 will be $(-.4410)^2 =$ .194. This value tells us how much of the variability in exam performance can be explained by exam anxiety. If we convert this value into a percentage (multiply by 100) we can say that exam anxiety accounts for 19.4% of the variability in exam performance. So, although exam anxiety was highly correlated with exam performance, it can account for only 19.4% of variation in exam scores. To put this value into perspective, this leaves 80.6% of the variability still to be accounted for by other variables. I should note at this point that although R^2 is an extremely useful measure of the substantive importance of an effect, it cannot be used to infer causal relationships. Although we usually talk in terms of 'the variance in y accounted for by x' or even the variation in one variable explained by the other, this still says nothing of which way causality runs. So, although exam anxiety can account for 19.4% of the variation in exam scores, it does not necessarily cause this variation.

4.5.4. Spearman's correlation coefficient ①

Spearman's correlation coefficient, r_s, is a non-parametric statistic and so can be used when the data have violated parametric assumptions such as non-normally distributed data (see Chapter 3). Spearman's test works by first ranking the data, and then applying Pearson's equation (equation (4.3)) to those ranks. As a statistics lecturer I am always interested in the factors that determine whether a student will do well on a statistics course. One potentially important factor is their previous expertise with mathematics (at the very least past experience will determine whether an equation makes sense!). Imagine I took 25 students and looked at their degree grades for the statistics course at the end of their first year at university. In the UK, a student can get a first-class mark, an upper second-class mark, a lower second, a third, a pass or a fail. These grades are categories, but they have an order to them (an upper second is better than a lower second). I could also ask these students what grade they got in their GCSE maths exams. In the UK GCSEs are school exams taken at age 16 that are graded A, B, C, D, E or F. Again, these grades are categories that have an order of importance (an A grade is better than all of the lower grades). When you have categories like these that can be ordered in a meaningful way, the data are said to be *ordinal*. The data are not interval, because a first-class degree encompasses a 30% range (70–100%) whereas an upper second only covers a 10% range (60–70%). When data have been measured at only the ordinal level they are said to be non-parametric and Pearson's correlation is not appropriate. Therefore, Spearman's correlation coefficient is used. The data for this study are in the file **grades.sav**. The data are in two columns: one labelled **stats** and one labelled **gcse**. Each of the categories described above has been coded with a numeric value. In both cases, the highest grade (first class or A grade) has been coded with the value 1, with subsequent categories being labelled 2, 3 and so on. Note that for each numeric code I have provided a value label (just like we did for coding variables).

The procedure for doing the Spearman correlation (Figure 4.16) is the same as for the Pearson correlation except that in the *bivariate correlations* dialog box (Figure 4.15), we need to select ☑ Spearman and deselect the option for a Pearson correlation. At this stage, you should also specify whether you require a one- or two-tailed test. For the example above, I predicted that better grades in GCSE maths would correlate with better degree grades for my statistics course. This hypothesis is directional and so a one-tailed test should be selected.

Figure 4.16 Charles Spearman, ranking furiously

SPSS Output 4.3 shows the output for a Spearman correlation on the variables **stats** and **gcse**. The output is very similar to that of the Pearson correlation: a matrix is displayed giving the correlation coefficient between the two variables (.455), underneath is the significance value of this coefficient (.011) and finally the sample size (25). The significance value for this correlation coefficient is less than .05; therefore, it can be concluded that there is a significant relationship between a student's grade in GCSE maths and their degree grade for their statistics course. The correlation itself is positive: therefore, we can conclude that as GCSE grades improve, there is a corresponding improvement in degree grades for statistics. As such, the hypothesis was supported. Finally, it is good to check that the value of *N* corresponds to the number of observations that were made. If it doesn't then data may have been excluded for some reason.

Correlations

			Statistics Grade	GCSE Maths Grade
Spearman's rho	Statistics Grade	Correlation Coefficient	1.000	.455*
		Sig. (1-tailed)	.	.011
		N	25	25
	GCSE Maths Grade	Correlation Coefficient	.455*	1.000
		Sig. (1-tailed)	.011	.
		N	25	25

*Correlation is significant at the .05 level (1-tailed).

SPSS Output 4.3

4.5.5. Kendall's tau (non-parametric) ①

Kendall's tau, τ, is another non-parametric correlation and it should be used rather than Spearman's coefficient when you have a small data set with a large number of tied ranks. This means that if you rank all of the scores and many scores have the same rank, Kendall's tau should be used. Although Spearman's statistic is the more popular of the two coefficients, there is much to suggest that Kendall's statistic is actually a better estimate of the correlation in the population (see Howell, 1997, p. 293). As such, we can draw more accurate generalizations from Kendall's statistic than from Spearman's. To carry out Kendall's correlation on the statistics degree grades data simply follow the same steps as for the Pearson and Spearman correlation but select ☑ Kendall's tau-b and deselect the Pearson option. The output is much the same as for Spearman's correlation.

You'll notice from SPSS Output 4.4 that the actual value of the correlation coefficient is less than the Spearman correlation (it has decreased from .455 to .354). Despite the difference in the correlation coefficients we can still interpret this result as being a highly significant positive relationship (because the significance value of .015 is less than .05). However, Kendall's value is a more accurate gauge of what the correlation in the population would be. As with the Pearson correlation we cannot assume that the GCSE grades caused the degree students to do better in their statistics course.

Correlations

			Statistics Grade	GCSE Maths Grade
Kendall's tau_b	Statistics Grade	Correlation Coefficient	1.000	.354*
		Sig. (1-tailed)	.	.015
		N	25	25
	GCSE Maths Grade	Correlation Coefficient	.354*	1.000
		Sig. (1-tailed)	.015	.
		N	25	25

*Correlation is significant at the .05 level (1-tailed).

SPSS Output 4.4

4.5.6. Biserial and point–biserial correlations ②

The biserial and point–biserial correlation coefficients are distinguished by only a conceptual difference yet their statistical calculation is quite different. These correlation coefficients are used when one of the two variables is dichotomous (i.e. it is categorical with only two categories). An example of a dichotomous variable is being pregnant, because a woman can be either pregnant or not (she cannot be 'a bit pregnant'). Often it is necessary to investigate relationships between two variables when one of the variables is dichotomous. The difference between the use of biserial and point–biserial correlations depends on whether the dichotomous variable is discrete or continuous. This difference is very subtle. A discrete, or true, dichotomy is one for which there is no underlying continuum between the categories. An example of this is whether someone is dead or alive: a person can be only dead or alive, they can't be 'a bit dead'. Although you might describe a person as being 'half-dead'—especially after a heavy drinking session—they are clearly still alive if they are still breathing! Therefore, there is no continuum between the two categories. However, it is possible to have a dichotomy for which a continuum does exist. An example is passing or failing a statistics test: some people will only just fail whilst others will fail by a large margin; likewise some people will scrape a pass whilst others clearly excel. So although participants fall into only two categories there is clearly an underlying continuum along which people lie. Hopefully, it is clear that in this case there is some kind of continuum underlying the dichotomy, because some people passed or failed more dramatically than others. *The point–biserial correlation coefficient (r_{pb}) is used when one variable is a discrete dichotomy (e.g. pregnancy), whereas the biserial correlation coefficient (r_b) is used when one variable is a continuous dichotomy (e.g. passing or failing an exam).* The biserial correlation coefficient cannot be calculated directly in SPSS: first you must calculate the point–biserial correlation coefficient and then use an equation to adjust that figure.

Imagine that I was interested in the relationship between the gender of a cat and how much time it spent away from home (what can I say? I love cats so these things interest me). I had heard that male cats disappeared for substantial amounts of time on long-distance roams around the neighbourhood (something about hormones driving them to find mates) whereas female cats tended to be more homebound. So, I used this as a purrfect (sorry!) excuse to go and visit lots of my friends and their cats. I took a note of the gender of the cat and then asked the owners to note down the number of hours that their cat was absent from home over a week. Clearly the time spent away from home is measured at an interval level—and let's assume it meets the other assumptions of parametric data—whilst the gender of the cat is discrete dichotomy. A point–biserial correlation has to be calculated and this is simply a Pearson correlation when the dichotomous variable is coded with 0 for one category and 1 for the other (actually you can use any values and SPSS will change the lower one to 0 and the higher one to 1 when it does the calculations). So, to conduct these correlations in SPSS assign the **gender** variable a coding scheme as described in section 2.4.4 (in the saved data the coding is 1 for a male and 0 for a female). The **time** variable simply has time in hours recorded as normal. These data are in the file **pbcorr.sav**. Carry out Pearson's correlation (as in section 4.5.1).

SPSS Output 4.5 shows the correlation matrix of **time** and **gender**. The point–biserial correlation coefficient is r_{pb} = .378, which has a one-tailed significance value of .001. The significance test for this correlation is actually the same as performing an independent samples *t*-test on the data (see Chapter 7). The sign of the correlation (i.e. whether the relationship was positive or negative) will depend entirely on which way round the coding of the dichotomous variable was made. To prove that this is the case, the data file **pbcorr.sav** has an extra variable called **recode** which is the same as the variable **gender** except that the coding is reversed (1 = female, 0 = male). If you repeat the Pearson correlation using **recode** instead of **gender** you will find that the correlation coefficient becomes −.378. The sign of the coefficient is completely dependent on which category you assign to

Correlations

		Time away from home (hours)	Gender of cat
Time Away from Home (Hours)	Pearson Correlation	1.000	.378**
	Sig. (1-tailed)	.	.001
	N	60	60
Gender of Cat	Pearson Correlation	.378**	1.000
	Sig. (1-tailed)	.001	.
	N	60	60

**Correlation is significant at the .01 level (1-tailed).

SPSS Output 4.5

which code and so we must ignore all information about the direction of the relationship. However, we can still interpret R^2 as before. So in this example, $R^2 = (.378)^2 = .143$. Hence, we can conclude that gender accounts for 14.3% of the variability in time spent away from home.

Imagine now that we wanted to convert the point–biserial correlation into the biserial correlation coefficient (r_b) (because some of the male cats were neutered and so there might be a continuum of maleness that underlies the gender variable). We must use equation (4.4) in which P_1 is the proportion of cases that fell into category 1 (the number of male cats) and P_2 is the proportion of cases that fell into category 2 (the number of female cats). In this equation y is the ordinate of the normal distribution at the point where there is P_1% of the area on one side and P_2% on the other (this will become clearer as we do an example):

$$r_{\text{b}} = \frac{r_{\text{pb}}\sqrt{(P_1 P_2)}}{y} \tag{4.4}$$

To calculate P_1 and P_2 simply use the menus **Analyze⇒Descriptive Statistics ⇒ Frequencies** and select the variable **gender**. There is no need to click on any further options as the defaults will give you what you need to know (namely, the percentage of male and female cats). It turns out that 53.3% (.533 as a proportion) of the sample were female (this is P_2) whilst the remaining 46.7% (.467 as a proportion) were male (this is P_1). To calculate *y*, we use these values and the values of the normal distribution displayed in Appendix A.1. Find the ordinate when the normal curve is split with .467 as the smaller portion and .533 as the larger portion (in fact we will have to use the nearest values to those, which are .4681 and .5319 respectively). The ordinate value is .3977. If we replace these values into equation (4.4) we get .475 (see below), which is quite a lot higher than the value of the point–biserial correlation (.378). Therefore, the choice of correlation coefficient can make a substantial difference to the result. You should, therefore, be careful to decide whether your dichotomous variable has an underlying continuum, or whether it is a truly discrete variable.

$$r_{\mathrm{b}} = \frac{r_{\mathrm{pb}}\sqrt{(P_1 P_2)}}{y}$$
$$= \frac{(.378)\sqrt{(.533 \times .467)}}{.3977}$$
$$= .475$$

4.6. PARTIAL CORRELATION ②

4.6.1. The theory behind part and partial correlation ②

I mentioned earlier that there is a type of correlation that can be done that allows you to look at the relationship between two variables when the effects of a third variable are held constant. For example, analyses of the exam anxiety data (in the file **ExamAnxiety.sav**) showed that exam performance was negatively related to exam anxiety, but positively related to revision time, and revision time itself was negatively related to exam anxiety. This scenario is complex, but given that we know that revision time is related to both exam anxiety and exam performance, then if we want a pure measure of the relationship between exam anxiety and exam performance we need to take account of the influence of revision time. Using the values of R^2 for these relationships, we know that exam anxiety accounts for 19.4% of the variance in exam performance, that revision time accounts for 15.7% of the variance in exam performance, and that revision time accounts for 50.2% of the variance in exam anxiety. If revision time accounts for half of the variance in exam anxiety, then it seems feasible that at least some of the 19.4% of variance in exam performance that is accounted for by anxiety is the same variance that is accounted for by revision time. As such, some of the variance in exam performance explained by exam anxiety is not *unique* and can be accounted for by revision time. **A correlation between two variables in which the effects of other variables are held constant is known as a partial correlation.**

Figure 4.17 illustrates the principle behind partial correlation. In part 1 of the diagram there is a box labelled 'Exam Performance' that represents the total variation in exam scores (this value would be the variance of exam performance). There is also a box that represents the variation in exam anxiety (again, this is the variance of that variable). We know already that exam anxiety and exam performance share 19.4% of their variation (this value is the correlation coefficient squared). Therefore, the variations of these two variables overlap (because they share variance) creating a third box (the one with diagonal lines). The overlap of the boxes representing exam performance and exam anxiety is the common variance. Likewise, in part 2 of the diagram the shared variation between exam performance and revision time is illustrated. Revision time shares 15.7% of the variation in exam scores. This shared variation is represented by the area of overlap (filled with diagonal lines). We know that revision time and exam anxiety also share 50% of their

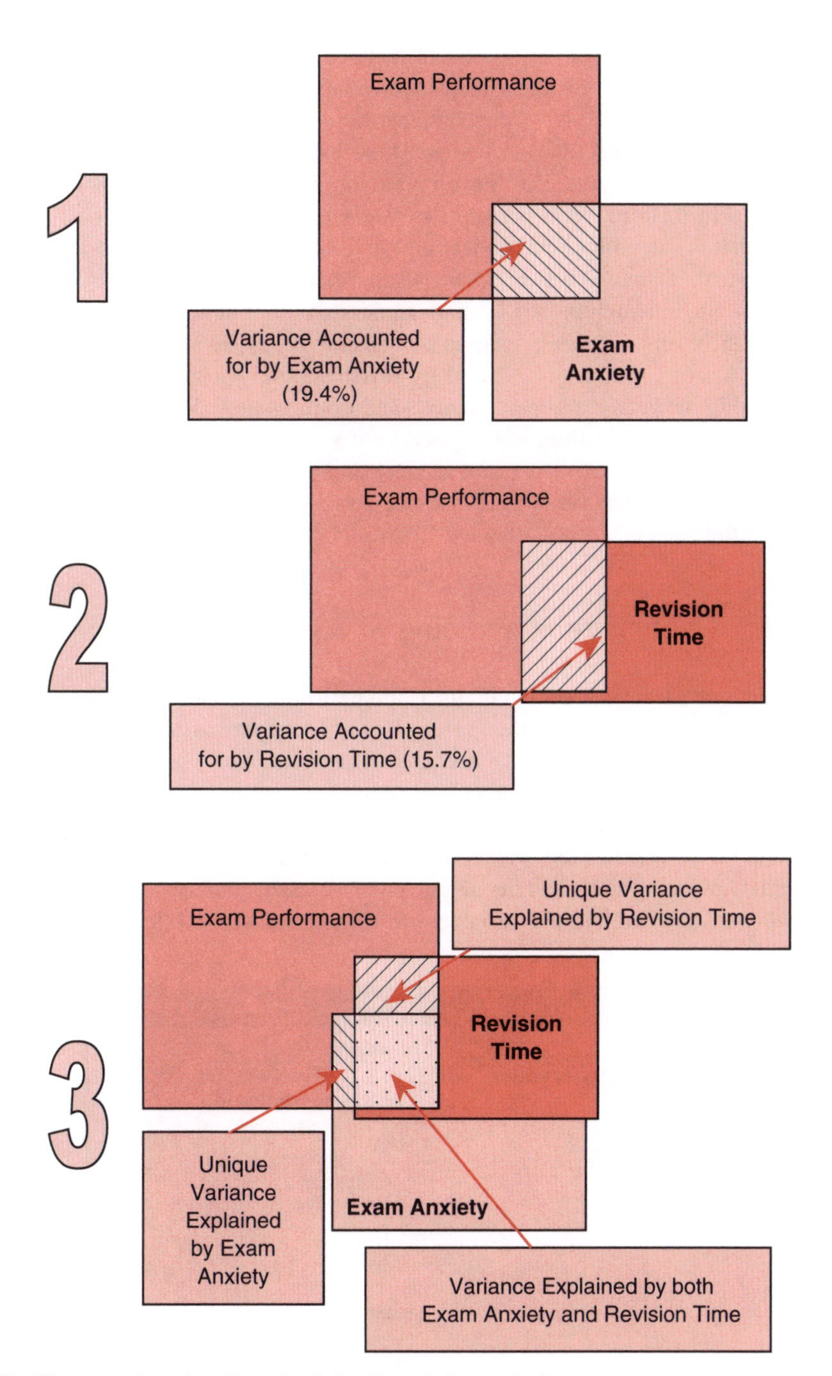

Figure 4.17 Diagram showing the principle of partial correlation

variation: therefore, it is very probable that some of the variation in exam performance shared by exam anxiety is the same as the variance shared by revision time.

Part 3 of the diagram shows the complete picture. The first thing to note is that the boxes representing exam anxiety and revision time have a large overlap (this is because they share 50% of their variation). More important, when we look at how revision time and anxiety contribute to exam performance we see that there is a portion of exam performance that is shared by both anxiety and revision time (the dotted area). However, there are still small chunks of the variance in exam performance that are unique to the other two variables. So, although in part 1, exam anxiety shared a large chunk of variation in exam performance, some of this overlap is also shared by revision time. If we remove the portion of variation that is also shared by revision time, we get a measure of the unique relationship between exam performance and exam anxiety. **We use partial correlations to find out the size of the unique portion of variance. Therefore, we could conduct a partial correlation between exam anxiety and exam performance while 'controlling' the effect of revision time**. Likewise, we could carry out a partial correlation between revision time and exam performance 'controlling' for the effects of exam anxiety.

4.6.2. Partial correlation using SPSS ②

To conduct a partial correlation on the exam performance data select the *Correlate* option from the *Analyze* menu and then select *Partial* (**Analyze⇒Correlate⇒Partial**) and the dialog box in Figure 4.18 will be activated. This dialog box lists all of the variables in the data editor on the left-hand side and there are two empty spaces on the right-hand side. The first space is for listing the variables that you want to correlate and the second is for declaring any variables the effects of which you want to control. In the example I have described, we want to look at the unique effect of exam anxiety on exam performance and so we want to correlate the variables **exam** and **anxiety**, while controlling for **revise**.

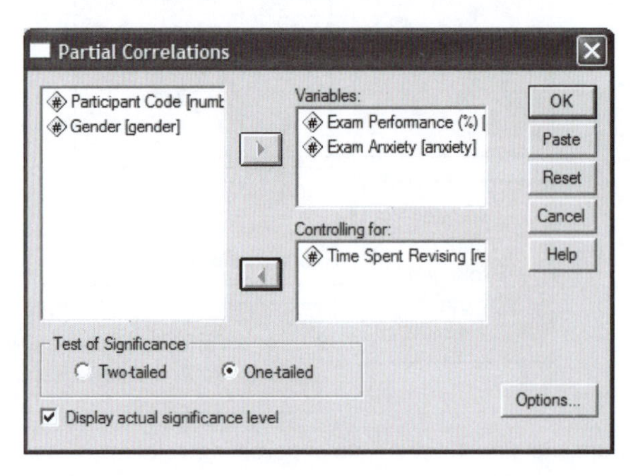

Figure 4.18 Main dialog box for conducting a partial correlation

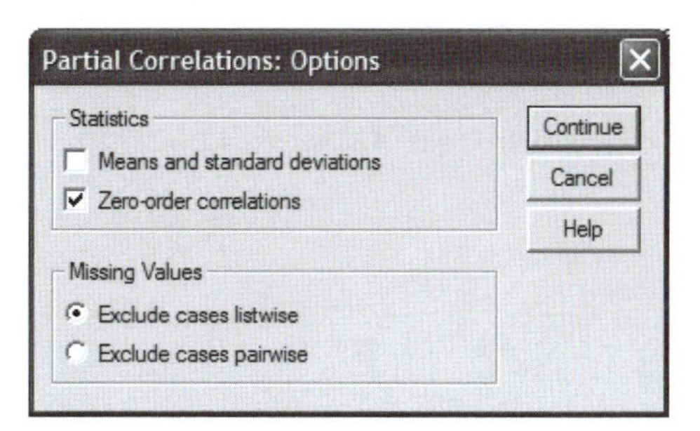

Figure 4.19 Options for partial correlation

Figure 4.18 shows the completed dialog box. If you click on Options... then another dialog box appears as shown in Figure 4.19.

These further options are similar to those in bivariate correlation except that you can choose to display zero-order correlations. Zero-order correlations are the bivariate correlation coefficients without controlling for any other variables (i.e. they're the Pearson correlation coefficient). So, in our example, if we select the tick-box for zero-order correlations SPSS will produce a correlation matrix of **anxiety**, **exam** and **revise**. If you haven't conducted bivariate correlations before the partial correlation then this is a useful way to compare the correlations that haven't been controlled against those that have. This comparison gives you some insight into the contribution of different variables. Tick the box for zero-order correlations but leave the rest of the options as they are.

SPSS Output 4.6 shows the output for the partial correlation of exam anxiety and exam performance controlling for revision time. The first thing to notice is the matrix of zero-order correlations, which we asked for using the *options* dialog box. The correlations displayed here are identical to those obtained from the Pearson correlation procedure (compare this matrix with the one in SPSS Output 4.2). Underneath the zero-order correlations is a matrix of correlations for the variables **anxiety** and **exam** but controlling for the effect of revision. **In this instance we have controlled for one variable and so this is known as a first-order partial correlation**. It is possible to control for the effects of two variables at the same time (a second-order partial correlation) or control three variables (a third-order partial correlation) and so on. First, notice that the partial correlation between exam performance and exam anxiety is –.2467, which is considerably less than the correlation when the effect of revision time is not controlled for ($r = -.4410$). In fact, the correlation coefficient is nearly half what it was before. Although this correlation is still statistically significant (its *p*-value is still below .05), the relationship is diminished. In terms of variance, the value of R^2 for the partial correlation is .06, which means that exam anxiety can now account for only 6% of the variance in exam performance. When the effects of revision time were not controlled for, exam anxiety shared 19.4% of the variation in exam scores and so the inclusion of revision time has severely diminished the amount of variation in exam scores shared by anxiety. As such, a truer measure of the role of exam anxiety has been obtained. Running this analysis has shown us that exam anxiety alone does explain much of the variation in exam scores, and we have discovered a complex relationship between anxiety and revision that might otherwise

```
PARTIAL CORRELATION COEFFICIENTS

Zero Order Partials

              EXAM      ANXIETY     REVISE

EXAM        1.0000      -.4410       .3967
          (      0)  (     101)  (     101)
            P = .     P = .000    P = .000

ANXIETY     -.4410      1.0000      -.7092
          (    101)  (       0)  (     101)
          P = .000     P = .      P = .000

REVISE       .3967      -.7092      1.0000
          (    101)  (     101)  (       0)
          P = .000    P = .000      P = .


(Coefficient / (D.F.) / 1-tailed Significance)

PARTIAL CORRELATION COEFFICIENTS

Controlling for..    REVISE

              EXAM      ANXIETY

EXAM        1.0000      -.2467
          (      0)  (     100)
            P = .     P = .006

ANXIETY     -.2467      1.0000
          (    100)  (       0)
          P = .006     P = .

(Coefficient / (D.F.) / 1-tailed Significance)

" . " is printed if a coefficient cannot be computed
```

SPSS Output 4.6 Output from a partial correlation

have been ignored. Although causality is still not certain, because relevant variables are being included, the third-variable problem is, at least, being addressed in some form.

These partial correlations can be done when variables are dichotomous. So, to use an earlier example, we could examine the relationship between brain size and gender (which is dichotomous) controlling for body size. It's also fine for the 'third' variable to be dichotomous, so, for example, we could look at the relationship between bladder relaxation and the number of large tarantulas crawling up your leg (both continuous variables) controlling for gender (a dichotomous variable).[4]

4.6.3. Semi-partial (or part) correlations ②

In the next chapter, we will come across another form of correlation known as a semi-partial correlation (also referred to as a part correlation). While I'm babbling on about partial correlations it is worth me explaining the difference between this type of correlation and a semi-partial correlation. **When we do a partial correlation between two variables, we control for the effects of a third variable. Specifically, the effect that the third variable has on *both* variables in the correlation is controlled. In a semi-partial correlation we control for the effect that the third variable has on only one of the variables in the correlation**. Figure 4.20 illustrates this principle for the exam performance data. The partial correlation that we calculated took account not only of the effect of revision on exam performance, but also of the effect of revision on anxiety. If we were to calculate the semi-partial correlation for the same data, then this would control for only the effect of revision on exam performance (the effect of revision on exam anxiety is ignored). Partial correlations are most useful for looking at the unique relationship between two variables when other variables are ruled out. Semi-partial correlations are, therefore, useful when trying to explain the variance in one particular variable (an outcome) from a set of predictor variables. This idea leads us nicely towards Chapter 5.

Figure 4.20 The difference between a partial and a semi-partial correlation

4 Both these examples are, in fact, simple cases of hierarchical regression (see the next chapter) and the first example is also an example of analysis of covariance. This may be confusing now, but as we progress through the book I hope it'll become clearer that virtually all of the statistics that you use are actually the same things dressed up in different names.

4.7. HOW TO REPORT CORRELATION COEFFICIENTS ①

Reporting correlation coefficients is pretty easy: you just have to say how big they are and what their significance value was (although the significance value isn't *that* important because the correlation coefficient is an effect size in its own right!). Four things to note are that (1) there should be no zero before the decimal point for the correlation coefficient or the probability value (because neither can exceed 1); (2) if you are quoting a one-tailed probability, you should say so; (3) each correlation coefficient is represented by a different letter (and some of them are Greek!); and (4) there are standard criteria of probabilities that we use (.05, .01 and .001). Let's take a few examples from this chapter:

- ✓ There was a significant relationship between the number of adverts watched and the number of packets of sweets purchased, $r = .87$, p (one-tailed) $< .05$.
- ✓ Exam performance was significantly correlated with exam anxiety, $r = -.44$, and time spent revising, $r = .40$; the time spent revising was also correlated with exam anxiety, $r = -.71$ (all ps $< .001$).
- ✓ There was a positive relationship between a person's statistics grade, and their GCSE maths grade, $r_s = .46$, $p < .05$.
- ✓ There was a positive relationship between a person's statistics grade, and their GCSE maths grade, $\tau = .35$, $p < .05$. (Note that I've quoted Kendall's tau here.)
- ✓ The gender of the cat was significantly related to the time the cat spent away from home, $r_{pb} = .38$, $p < .01$.

A final point to note is that in the social sciences we have several *standard* levels of statistical significance. Primarily, the most important criterion is that the significance value is less than .05; however, if the exact significance value is much lower then we can be much more confident about the strength of the experimental effect. In these circumstances we like to make a big song and dance about the fact that our result isn't just significant at .05, but is significant at a much lower level as well (hooray!). The values we use are .05, .01, .001 and .0001. You are rarely ever going to be in the fortunate position of being able to report an effect that is significant at a level less than .0001!

4.8. WHAT HAVE WE DISCOVERED ABOUT STATISTICS? ①

This chapter has looked at ways to study relationships between variables. We began by looking at how we might measure relationships statistically by developing what we already know about variance (from Chapter 1) to look at variance shared between variables. This shared variance is known as *covariance*. Next we looked at how to represent relationships graphically by using *scatterplots*. In their simplest form, these graphs plot values of one variable against another. However, you can plot more than two variables on a 3-D scatterplot, or an overlay scatterplot. It's also possible to distinguish groups of people on a scatterplot using different symbols or colours. We then discovered that when data are parametric we can measure the strength of a relationship using Pearson's correlation coefficient, r. When data violate the assumptions of parametric tests we can use Spearman's r_s, or for small data sets Kendall's τ may be more accurate. We also saw that correlations can be calculated between two variables when one of those variables is a dichotomy (i.e. comprised of two categories); when the categories have no underlying continuum then we use the point–biserial correlation, r_{pb}, but when the categories do have an underlying continuum we use the biserial correlation, r_b. Finally, we looked at the difference between *partial correlations*, in which the relationship between two variables is measured controlling for the effect that one or more variable has on both of those variables, and *semi-partial correlations*, in which the relationship between two variables is measured controlling for the effect that one or more variable has on only one of those variables.

4.9. KEY TERMS THAT WE'VE DISCOVERED

- Biserial correlation
- Bivariate correlation
- Coefficient of determination
- Covariance
- Cross-product deviations
- Kendall's tau
- Partial correlation
- Pearson correlation coefficient
- Point–biserial correlation
- Scatterplot
- Semi-partial correlation
- Spearman's correlation coefficient
- Standardization

4.10. SMART ALEX'S TASKS

The answers for these tasks can be found on the CD-ROM in the file **Answers(Chapter 4).pdf**.

- A student was interested in whether there was a positive relationship between the time spent doing an essay and the mark received. He got 45 of his friends and timed how long they spent writing an essay (**hours**) and the percentage they got in the essay (**essay**). He also translated these grades into their degree classifications (**grade**): first, upper second, lower second and third class. Using the data in the file **EssayMarks.sav** find out what the relationship was between the time spent doing an essay and the eventual mark in terms of percentage and degree class (draw a scatterplot too!).①
- Using the **ChickFlick.sav** data from Chapter 3, is there a relationship between gender and arousal? Using the same data, is there a relationship between the film watched and arousal?①

4.11. FURTHER READING

Howell, D. C. (2002). *Statistical methods for psychology* (5th edition). Belmont, CA: Duxbury. Chapter 9 provides more detailed coverage of correlation and regression than Wright but is less reader-friendly. Chapter 10 is great for biserial and point–biserial correlation.

Wright, D. B. (2002). *First steps in statistics*. London: Sage. This book has a very clear introduction to the concept of correlation and regression (Chapter 8).

CHAPTER 5

REGRESSION

5.1. WHAT WILL THIS CHAPTER TELL US? ①

In the last chapter we looked at how to measure (and draw) relationships between two variables. However, we can take this process a step further and look at predicting one variable from another. A simple example might be to try to predict levels of stress from the amount of time until you have to give a talk. You'd expect this to be a negative relationship (the smaller the amount of time until the talk, the larger the anxiety). We could then extend this basic relationship to answer a question such as 'if there's 10 minutes to go until someone has to give a talk, how anxious will they be?' This is the essence of regression analysis: it's a way of predicting some kind of outcome from one or more predictor variables. This chapter looks at regression analysis in a fair bit of depth. I'll start off with an introduction to the basic principles in regression using the example of predicting an outcome from a single predictor (simple regression). In doing so, I'll explain what the regression model is, how we estimate it, how we assess how well it fits the data, how all of these things are done on SPSS, and how the SPSS output is interpreted. I'll then look at the more complex case of when an outcome is predicted from several predictor variables (multiple regression). Although I don't go into how the model is estimated, instead I'll describe it and the various methods for specifying which variables are entered into the model. You'll also find out some of the things that can effect the accuracy of the model, and how to asses them before moving on to doing an analysis on SPSS, interpreting the output, and finding out how well the model fits the data and can be generalized. If you're still awake at this point, the chapter ends by looking at how we can incorporate categorical variables into a regression analysis using something called dummy variables (whoopee!).

5.2. AN INTRODUCTION TO REGRESSION ①

Correlations can be a very useful research tool but they tell us nothing about the predictive power of variables. In regression analysis we fit a predictive model to our data and use that model to predict values of the dependent variable (DV) from one or more independent variables (IVs).[1] Simple regression seeks to predict an outcome variable from a single predictor variable whereas multiple regression seeks to predict an outcome from several predictors. This is an incredibly useful tool because it allows us to go a step beyond the data that we actually possess. In section 1.4.1 I introduced you to the idea that we can predict any data using the following general equation:

$$\text{Outcome}_i = (\text{Model}_i) + \text{error}_i \tag{5.1}$$

This just means that the outcome we're trying to predict for a particular person can be predicted by whatever model we fit to the data plus some kind of error. In regression the model we fit is a linear model ('linear model' just means 'model based on a straight line') and you can imagine it as trying to summarize a data set with a straight line (think back to Figure 1.11). As such, the word 'Model' in the equation above simply gets replaced by some things that define the line that we fit to the data (see next section).

With any data set there are several lines that could be used to summarize the general trend and so we need a way to decide which of many possible lines to choose. For the sake of drawing accurate conclusions we want to fit a model that *best* describes the data. There are several ways to fit a straight line to the data you have collected. The simplest way would be to use your eye to gauge a line that looks as though it summarizes the data well. However, the 'eyeball' method is very subjective and so offers no assurance that the model is the best one that could have been chosen. Instead, we use a mathematical technique to establish the line that best describes the data collected. This method is called the *method of least squares*.

5.2.1. Some important information about straight lines ①

I mentioned above that in our general equation the word model gets replaced by 'some things that define the line that we fit to the data'. In fact, any straight line can be defined by two things: (1) the slope (or gradient) of the line (usually denoted by b_1), and (2) the point at which the line crosses the vertical axis of the graph (known as the *intercept* of the

---

1 Unfortunately, you will come across people (and SPSS for that matter) referring to regression variables as dependent and independent variables (as in controlled experiments). However, correlational research by its nature seldom controls the independent variables to measure the effect on a dependent variable. Instead, variables are measured simultaneously and without strict control. It is, therefore, inaccurate to label regression variables in this way. For this reason I label 'independent variables' as *predictors*, and the 'dependent variable' as the *outcome*.

line, b_0). In fact our general model becomes equation (5.2) in which Y_i is the outcome that we want to predict and X_i is the *i*th participant's score on the predictor variable.[2] b_1 is the gradient of the straight line fitted to the data and b_0 is the intercept of that line. These parameters b_1 and b_0 are known as the *regression coefficients* and will crop up time and time again in this book, where you may see them referred to generally as *b* (without any subscript) or b_i (meaning the *b* associated with variable *i*). There is a residual term, ε_i, which represents the difference between the score predicted by the line for participant *i* and the score that participant *i* actually obtained. The equation is often conceptualized without this residual term (so, ignore it if it's upsetting you); however, it is worth knowing that this term represents the fact that our model will not fit perfectly the data collected:

$$Y_i = (b_0 + b_1 X_i) + \varepsilon_i \tag{5.2}$$

A particular line has a specific intercept and gradient. Figure 5.1 shows a set of lines that have the same intercept but different gradients, and a set of lines that have the same gradient but different intercepts. Figure 5.1 also illustrates another useful point: that the gradient of the line tells us something about the nature of the relationship being described. In Chapter 4 we saw how relationships can be either positive or negative (and I don't mean the difference between getting on well with your girlfriend and arguing all the time!). A line that has a gradient with a positive value describes a positive relationship, whereas a line with a negative gradient describes a negative relationship. So, if you look at the graph in Figure 5.1 in which the gradients differ but the intercepts are the same, then the thicker

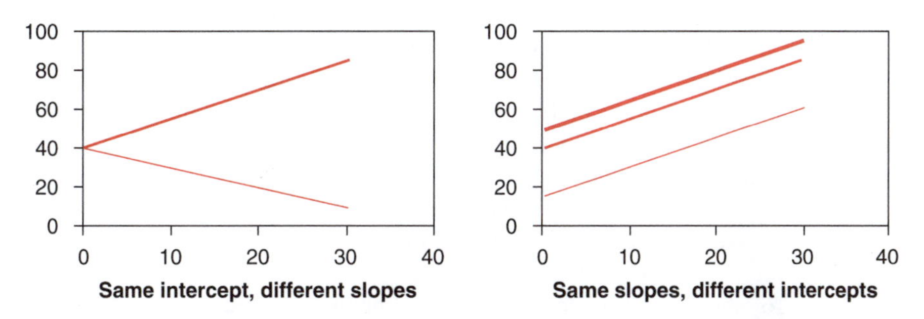

Figure 5.1 Lines with the same gradients but different intercepts, and lines that share the same intercept but have different gradients

---

2 You'll sometimes see this equation written as:

$$Y_i = (\beta_0 + \beta_1 X_i) + \varepsilon_i$$

The only difference is that this equation has got βs in it instead of *b*s and in fact both versions are the same thing—they just use different letters to represent the coefficients. I actually prefer the β-notation (and used it in the previous edition); however, when SPSS estimates the coefficients in this equation, it labels them *b*. To be consistent with the SPSS output that you'll end up looking at, I've therefore used *b*s instead of βs.

line describes a positive relationship whereas the thinner line describes a negative relationship.

If it is possible to describe a line knowing only the gradient and the intercept of that line, then we can use these values to describe our model (because in linear regression the model we use is a straight line). So, the model that we fit to our data in linear regression can be conceptualized as a straight line that can be described mathematically by equation (5.2). With regression we strive to find the line that best describes the data collected, then estimate the gradient and intercept of that line. Having defined these values, we can insert different values of our predictor variable into the model to estimate the value of the outcome variable.

5.2.2. The method of least squares ①

I have already mentioned that the method of least squares is a way of finding the line that best fits the data (i.e. finding a line that goes through, or as close to, as many of the data points as possible). This 'line of best fit' is found by ascertaining which line, of all of the possible lines that could be drawn, results in the least amount of difference between the observed data points and the line. Figure 5.2 shows that when any line is fitted to a set of data, there will be small differences between the values predicted by the line and the data that were actually observed. We are interested in the vertical differences between the line and the actual data because we are using the line to predict values of Y from values of the X-variable. Although some data points fall exactly on the line, others lie above and below the line, indicating that there is a difference between the model fitted to these data and the data collected. Some of these differences are positive (they are above the line, indicating that the model underestimates their value) and some are negative (they are below the line, indicating that the model overestimates their value). These differences are usually called residuals. In the discussion of variance in section 1.4.1 I explained that if we sum positive and negative differences then they tend to cancel each other out. To avoid this problem we square the differences before adding them up. These squared differences provide a gauge of how well a particular line fits the data: if the squared differences are large, the line is not representative of the data; if the squared differences are small then the line is representative. The sum of squared differences (or SS for short) can be calculated for any line that is fitted to some data; the 'goodness-of-fit' of each line can then be compared by looking at the sum of squares for each. The method of least squares works by selecting the line that has the lowest sum of squared differences (so it chooses the line that best represents the observed data). One way to select this optimal line would be to fit every possible line to a set of data, calculate the sum of squared differences for each line, and then choose the line for which this value is smallest. This would take quite a long time to do! Fortunately, there is a mathematical technique for finding maxima and minima and this technique (calculus) is used to find the line that minimizes the sum of squared differences. The end result is that the value of the slope and intercept of the 'line of best fit' can be estimated. Social scientists generally refer to this line of best fit as a regression line.

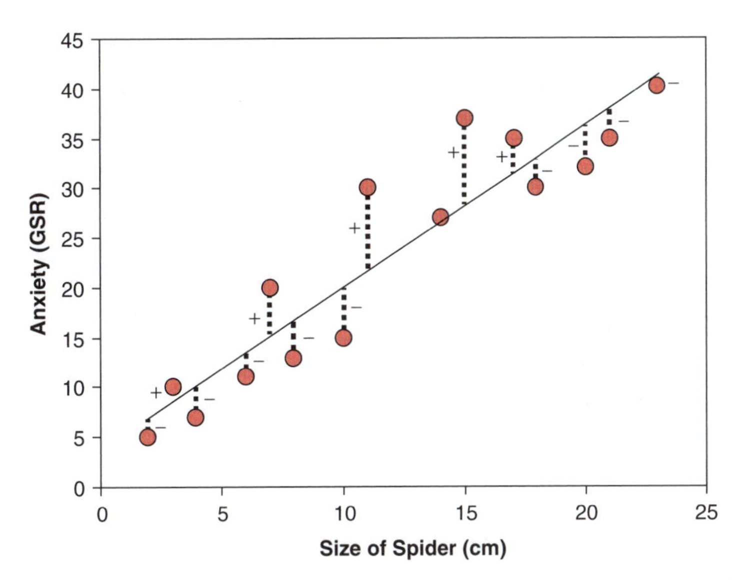

Figure 5.2 This graph shows a scatterplot of some data with a line representing the general trend. The vertical lines (dashed) represent the differences (or residuals) between the line and the actual data

5.2.3. Assessing the goodness-of-fit: sums of squares, R and R^2 ①

Once we have found the line of best fit it is important that we assess how well this line fits the actual data (we assess the goodness-of-fit of the model). We do this because even though this line is the best one available, it can still be a lousy fit of the data! In section 1.4.1 we saw that one measure of the adequacy of a model is the sum of squared differences (or more generally we assess models using equation (5.3)). If we want to assess the line of best fit, we need to compare it against something, and the thing we choose is the most basic model we can find. So we use equation (5.3) to calculate the fit of the most basic model, and then the fit of the best model (the line of best fit), and basically if the best model is any good then it should fit the data significantly better than our basic model. This is all quite abstract so let's look at an example:

$$\text{deviation} = \sum(\text{observed} - \text{model})^2 \tag{5.3}$$

Imagine that I was interested in predicting record sales (Y) from the amount of money spent advertising that record (X). One day my boss came in to my office and said 'Andy, how many records will we sell if we spend £100,000 on advertising?' If I didn't have an accurate model of the relationship between record sales and advertising, what would my

best guess be? Well, probably the best answer I could give would be the mean number of record sales (say, 200,000) because on average that's how many records we expect to sell. This response might well satisfy a brainless record company executive. However, what if he had asked 'How many records will we sell if we spend £1 on advertising?' Again, in the absence of any accurate information, my best guess would be to give the average number of sales (200,000). There is a problem: whatever amount of money is spent on advertising I always predict the same levels of sales. It should be pretty clear then that the mean is fairly useless as a model of a relationship between two variables—but it is the simplest model available.

So, as a basic strategy for predicting the outcome, we might choose to use the mean, because on average (*sic*) it will be a fairly good guess of an outcome. Using the mean as a model, we can calculate the difference between the observed values and the values predicted by the mean (equation (5.3)). We saw in section 1.4.1 that we square all of these differences to give us the sum of squared differences. This sum of squared differences is known as the total sum of squares (denoted SS_T) because it is the total amount of differences present when the most basic model is applied to the data. This value represents how good the mean is as a model of the observed data. Now, if we fit the more sophisticated model to the data, such as a line of best fit, we can again work out the differences between this new model and the observed data, again using (equation (5.3)). In the previous section we saw that the method of least squares finds the best possible line to describe a set of data by minimizing the difference between the model fitted to the data and the data themselves. However, even with this optimal model there is still some inaccuracy, which is represented by the differences between each observed data point and the value predicted by the regression line. As before, these differences are squared before they are added up so that the directions of the differences do not cancel out. The result is known as the *sum of squared residuals* or residual sum of squares (SS_R). This value represents the degree of inaccuracy when the best model is fitted to the data. We can use these two values to calculate how much better the regression line (the line of best fit) is than just using the mean as a model (i.e. how much better is the best possible model than the worst model?). The improvement in prediction resulting from using the regression model rather than the mean is calculated by calculating the difference between SS_T and SS_R. This difference shows us the reduction in the inaccuracy of the model resulting from fitting the regression model to the data. This improvement is the model sum of squares (SS_M). Figure 5.3 shows each sum of squares graphically.

If the value of SS_M is large then the regression model is very different from using the mean to predict the outcome variable. This implies that the regression model has made a big improvement to how well the outcome variable can be predicted. However, if SS_M is small then using the regression model is little better than using the mean (i.e. the regression model is no better than taking our 'best guess'). A useful measure arising from these sums of squares is the proportion of improvement due to the model. This is easily calculated by dividing the sum of squares for the model by the total sum of squares. The resulting value is called R^2 and to express this value as a percentage you should multiply it by 100. So, R^2 represents the amount of variance in the outcome explained by the model (SS_M)

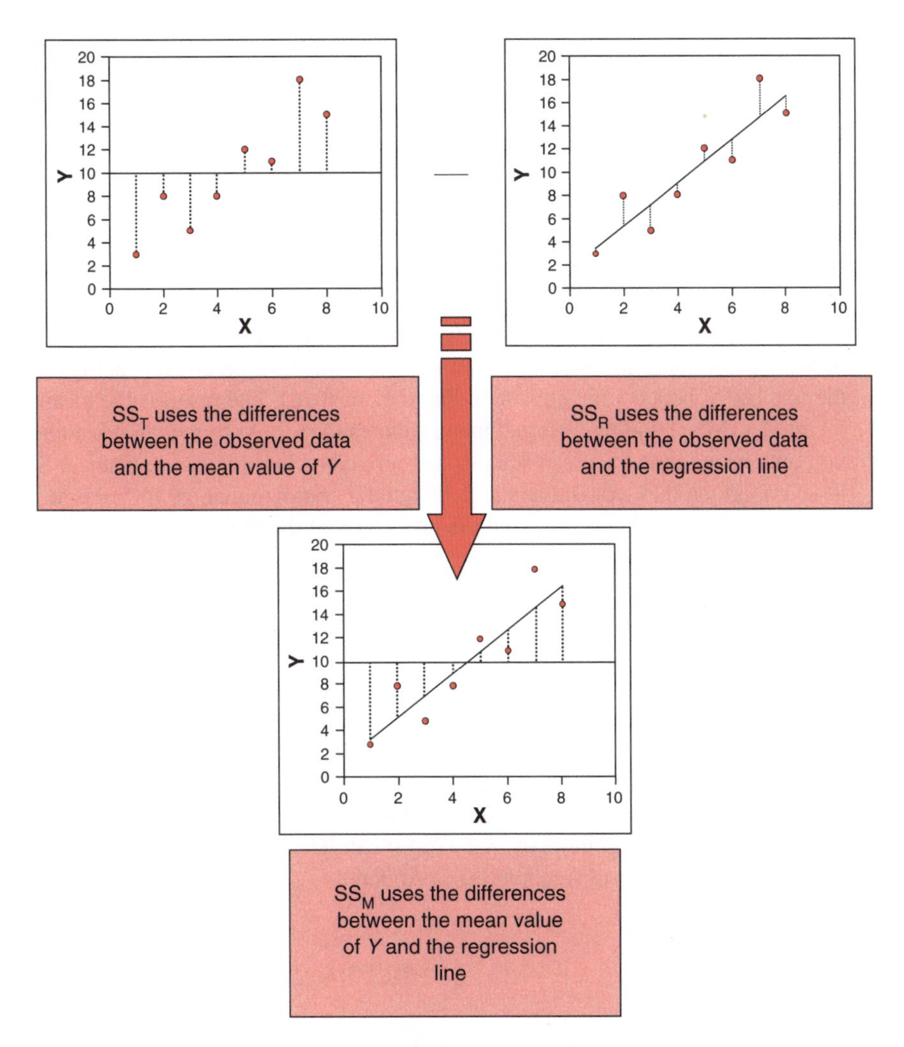

Figure 5.3 Diagram showing from where the regression sums of squares derive

relative to how much variation there was to explain in the first place (SS_T). Therefore, as a percentage, it represents the percentage of the variation in the outcome that can be explained by the model:

$$R^2 = \frac{SS_M}{SS_T} \tag{5.4}$$

Interestingly, this value is the same as the R^2 we met in Chapter 4 (section 4.5.3) and you'll notice that it is interpreted in the same way. Therefore, in simple regression we can take the square root of this value to obtain the Pearson correlation coefficient. As such, the correlation coefficient provides us with a good estimate of the overall fit of the regression model, and R^2 provides us with a good gauge of the substantive size of the relationship.

A second use of the sums of squares in assessing the model is through the *F*-test. The *F*-test is something we will cover in greater depth in Chapter 8, but briefly this test is based upon the ratio of the improvement due to the model (SS_M) and the difference between the model and the observed data (SS_R). In fact, rather than using the sums of squares themselves, we take the mean sums of squares (referred to as the mean squares or MS). To work out the mean sums of squares it is necessary to divide by the degrees of freedom (this is comparable to calculating the variance from the sums of squares—see section 1.4.1). For SS_M the degrees of freedom are simply the number of variables in the model, and for SS_R they are the number of observations minus the number of parameters being estimated (i.e. the number of beta coefficients including the constant). The result is the mean squares for the model (MS_M) and the residual mean squares (MS_R). At this stage it isn't essential that you understand how the mean squares are derived (it is explained in Chapter 8). However, it is important that you understand that the *F*-ratio (equation (5.5)) is a measure of how much the model has improved the prediction of the outcome compared to the level of inaccuracy of the model:

$$F = \frac{MS_M}{MS_R} \tag{5.5}$$

If a model is good, then we expect the improvement in prediction due to the model to be large (so, MS_M will be large) and the difference between the model and the observed data to be small (so, MS_R will be small). In short, a good model should have a large *F*-ratio (greater than 1 at least) because the top half of equation (5.5) will be bigger than the bottom. The exact magnitude of this *F*-ratio can be assessed using critical values for the corresponding degrees of freedom (as in Appendix A.3).

5.2.4. Assessing individual predictors ①

We've seen that the predictor in a regression model has a coefficient (b_1), which in simple regression represents the gradient of the regression line. The value of *b* represents the change in the outcome resulting from a unit change in the predictor. If the model was useless at predicting the outcome, then if the value of the predictor changes, what might we expect the change in the outcome to be? Well, if the model is very bad then we would expect the change in the outcome to be zero. Think back to Figure 5.3 (see the panel representing SS_T) in which we saw that using the mean was a very bad way of predicting the outcome. In fact, the line representing the mean is flat, which means that as the predictor variable changes, the value of the outcome does *not* change (because for each level of the predictor variable, we predict that the outcome will equal the mean value). The important point here is that a bad model (such as the mean) will have regression coefficients of 0 for the predictors. A regression coefficient of 0 means: (1) a unit change in the predictor variable results in no change in the predicted value of the outcome (the predicted value of the outcome does not change at all), and (2) the gradient of the regression line is zero, meaning that the regression line is flat. Hopefully, you'll see that it logically follows that if a variable significantly predicts an outcome, then it should have a b-value significantly different from zero. This hypothesis is tested using a t-test (see Chapter 7). The **t-statistic** tests the null hypothesis that the value of b is zero: therefore, if it is significant we accept the hypothesis that the b-value is significantly different from zero and that the predictor variable contributes significantly to our ability to estimate values of the outcome.

One problem with testing whether the b-values are different from zero is that their magnitude depends on the units of measurement. Therefore, the t-test is calculated by taking account of the standard error. The standard error tells us something about how different b-values would be if we took lots and lots of samples of data regarding record sales and advertising budgets and calculated the b-values for each sample. We could plot a frequency distribution of these samples to discover whether the b-values from all samples would be relatively similar, or whether they would be very different (think back to section 1.6). We can use the standard deviation of this distribution (known as the *standard error*) as a measure of the similarity of b-values across samples. If the standard error is very small, then it means that most samples are likely have a b-value similar to the one in the sample collected (because there is little variation across samples). The t-test tells us whether the b-value is different from zero relative to the variation in b-values for similar samples. When the standard error is small even a small deviation from zero can reflect a meaningful difference because b is representative of the majority of possible samples.

Equation (5.6) shows how the t-test is calculated and you'll find a general version of this equation in Chapter 7 (equation (7.1)). The b_{expected} is simply the value of b that we would expect to obtain if the null hypothesis were true. I mentioned earlier that the null hypothesis is that b is zero and so this value can be replaced by zero. The equation simplifies to become the observed value of b divided by the standard error with which it is associated:[3]

$$
\begin{aligned}
t &= \frac{b_{\text{observed}} - b_{\text{expected}}}{\text{SE}_b} \\
&= \frac{b_{\text{observed}}}{\text{SE}_b}
\end{aligned}
\tag{5.6}
$$

3 To see that this is true you can use the values from SPSS Output 5.3 below to calculate t for the constant (t = 134.140/7.537 = 17.79) and advertising budget. For the advertising budget, you'll get a different value of t if you use the values as reported in the SPSS output. This is because SPSS rounds values 3 decimal places in the table, but calculates t using unrounded values (usually this doesn't make too much difference but in this case it does!). To obtain the unrounded values, double-click on the table in the SPSS output and then double-click on the value that you wish to see in full. You should find that t = .096124/.009632 = 9.979.

The values of t have a special distribution that differs according to the degrees of freedom for the test. In regression, the degrees of freedom are simply $N - p - 1$, where N is the total sample size and p is the number of predictors. In simple regression when we have only one predictor, this reduces down to $N - 2$. Having established which t-distribution needs to be used, the observed value of t can then be compared to the values that we would expect to find by chance alone: if t is very large then it is unlikely to have occurred by chance (these values can be found in Appendix A.2). SPSS provides the exact probability that the observed value of t would occur if the value of b was, in fact, zero. As a general rule, if this observed significance is less than .05, then social scientists agree that b is significantly different from zero; put another way, the predictor makes a significant contribution to predicting the outcome (see Chapter 1).

5.3. DOING SIMPLE REGRESSION ON SPSS ①

So far, we have seen a little of the theory behind regression, albeit restricted to the situation in which there is only one predictor. To help clarify what we have learnt so far, we will go through an example of a simple regression on SPSS. Earlier on I asked you to imagine that I worked for a record company and that my boss was interested in predicting record sales from advertising. There are some data for this example in the file **Record1.sav**. This data file has 200 rows, each one representing a different record. There are also two columns, one representing the sales of each record in the week after release and the other representing the amount (in pounds) spent promoting the record before release. This is the format for entering regression data: the outcome variable and any predictors should be entered in different columns, and each row should represent independent values of those variables. The pattern of the data is shown in Figure 5.4 and it should be clear that a positive relationship exists: so, the more money spent advertising the record, the more it is likely to sell. Of course there are some records that sell well regardless of advertising (top left of scatterplot), but there are none that sell badly when advertising levels are high (bottom right of scatterplot). The scatterplot also shows the line of best fit for these data: bearing in mind that the mean would be represented by a flat line at around the 200,000 sales mark, the regression line is noticeably different.

To find out the parameters that describe the regression line, and to see whether this line is a useful model, we need to run a regression analysis. To do the analysis you need to access the main dialog box by using the **Analyze⇒Regression⇒Linear...** menu path. Figure 5.5 shows the resulting dialog box. There is a space labelled *Dependent* in which you should place the outcome variable (in this example **sales**). So, select **sales** from the list on the left-hand side, and transfer it by clicking on ▸. There is another space labelled *Independent(s)* in which any predictor variable should be placed. In simple regression we use only one predictor (in this example **adverts**) and so you should select **adverts** from the list and click on ▸ to transfer it to the list of predictors. There are a variety of options available, but these will be explored within the context of multiple regression (see section 5.5). For the time being just click on OK to run the basic analysis.

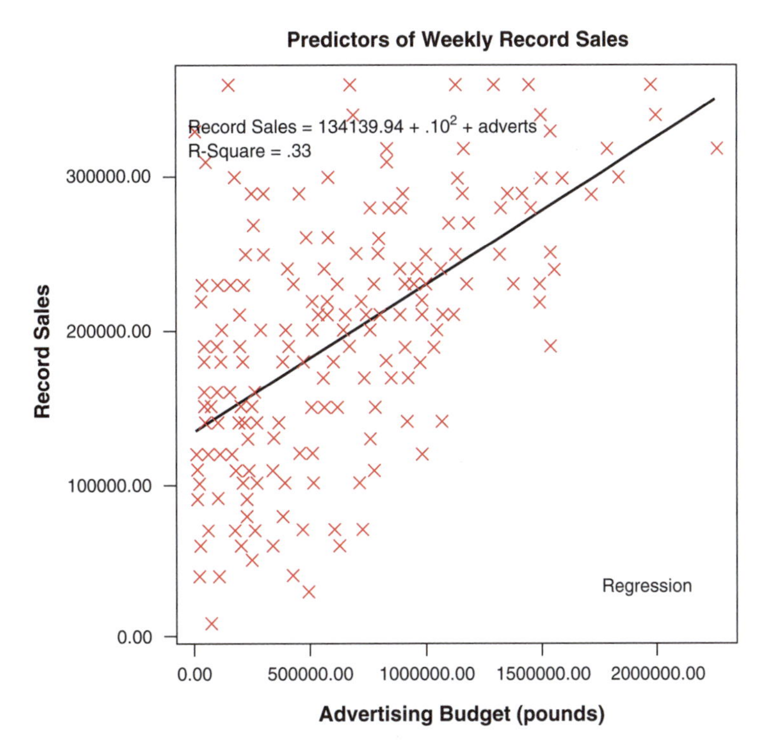

Figure 5.4 Scatterplot showing the relationship between record sales and the amount spent promoting the record

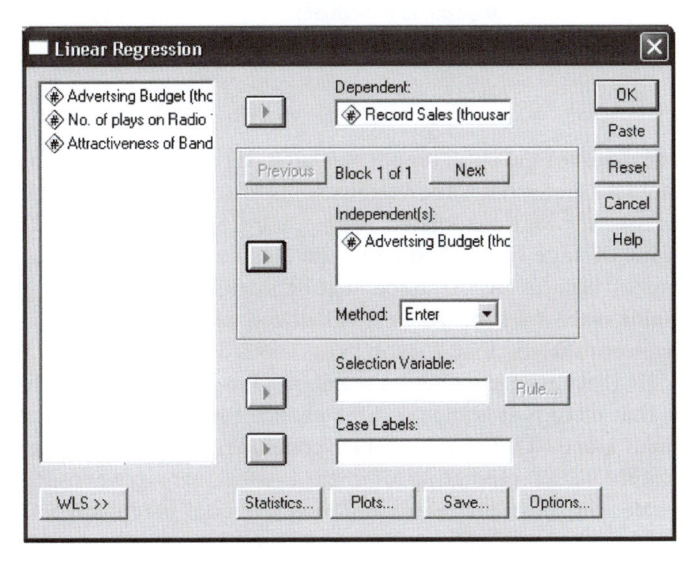

Figure 5.5 Main dialog box for regression

5.4. INTERPRETING A SIMPLE REGRESSION ①

5.4.1. Overall fit of the model ①

The first table provided by SPSS is a summary of the model. This summary table (SPSS Output 5.1) provides the value of R and R^2 for the model that has been derived. For these data, R has a value of .578 and because there is only one predictor, this value represents the simple correlation between advertising and record sales (you can confirm this by running a correlation using what you were taught in Chapter 4). The value of R^2 is .335, which tells us that advertising expenditure can account for 33.5% of the variation in record sales. In other words, if we are trying to explain why some records sell more than others, we can look at the variation in sales of different records. There might be many factors that can explain this variation, but our model, which includes only advertising expenditure, can explain 33% of it. This means that 66% of the variation in record sales cannot be explained by advertising alone. Therefore, there must be other variables that have an influence also.

Model Summary

Model	R	R Square	Adjusted R Square	Std. Error of the Estimate
1	.578[a]	.335	.331	65.9914

a Predictors: (Constant), Advertising Budget (thousands of pounds)

SPSS Output 5.1

The next part of the output reports an analysis of variance (ANOVA—see Chapter 8). The summary table shows the various sums of squares described in Figure 5.3 and the degrees of freedom associated with each (SPSS Output 5.2). From these two values, the average sums of squares (the mean squares) can be calculated by dividing the sums of squares by the associated degrees of freedom. The most important part of the table is the F-ratio, which is calculated using equation (5.5), and the associated significance value of that F-ratio. For these data, F is 99.59, which is significant at $p < .001$ (because the value in the column labelled *Sig.* is less than .001). This result tells us that there is less than a 0.1% chance that an F-ratio this large would happen by chance alone. Therefore, we can conclude that our regression model results in significantly better prediction of record sales than if we used the mean value of record sales. In short, the regression model overall predicts record sales significantly well.

ANOVA[b]

Model		Sum of Squares	df	Mean Square	F	Sig.
1	Regression	433687.833	1	433687.833	99.587	.000[a]
	Residual	862264.167	198	4354.870		
	Total	1295952.0	199			

a Predictors: (Constant), Advertising Budget (thousands)
b Dependent Variable: Record Sales (thousands)

SPSS Output 5.2

5.4.2. Model parameters ①

The ANOVA tells us whether the model, overall, results in a significantly good degree of prediction of the outcome variable. However, the ANOVA doesn't tell us about the individual contribution of variables in the model (although in this simple case there is only one variable in the model and so we can infer that this variable is a good predictor). The table in SPSS Output 5.3 provides details of the model parameters (the beta values) and the significance of these values. We saw in equation (5.2) that b_0 was the Y-intercept and this value is the value B (in the SPSS output) for the constant. So, from the table, we can say that b_0 is 134.14, and this can be interpreted as meaning that when no money is spent on advertising (when $X = 0$), the model predicts that 134,140 records will be sold (remember that our unit of measurement was thousands of records). We can also read off the value of b_1 from the table and this value represents the gradient of the regression line. It is 9.612E-02, which in unabbreviated form is .09612.[4] Although this value is the slope of the regression line, it is more useful to think of this value as representing *the change in the outcome associated with a unit change in the*

4 You might have noticed that this value is reported by SPSS as 9.612E-02 and many students find this notation confusing. Well, this notation simply means 9.61×10^{-2} (which might be a more familiar notation). OK, some of you are still confused. Well think of E-02 as meaning 'move the decimal place 2 steps to the left', so 9.612E-02 becomes .09612. If the notation read 9.612E-01, then that would be .9612, and if it read 9.612E-03, that would be .009612. Likewise, think of E+02 (notice the minus sign has changed) as meaning 'move the decimal place 2 places to the right'. So 9.612E+02 becomes 961.2.

Coefficients[a]

Model		Unstandardized Coefficients		Standardized Coefficients		
		B	Std. Error	Beta	t	Sig.
1	(Constant)	134.140	7.537		17.799	.000
	Advertising Budget (thousands of pounds)	9.612E-02	.010	.578	9.979	.000

a Dependent Variable: Record Sales (thousands)

SPSS Output 5.3

predictor. Therefore, if our predictor variable is increased by one unit (if the advertising budget is increased by 1), then our model predicts that .096 extra records will be sold. Our units of measurement were thousands of pounds and thousands of records sold, so we can say that for an increase in advertising of £1000 the model predicts 96 ($.096 \times 1000 = 96$) extra record sales. As you might imagine, this investment is pretty bad for the record company: they invest £1000 and get only 96 extra sales! Fortunately, as we already know, advertising accounts for only one-third of record sales!

We saw earlier that, in general, values of the regression coefficient b represent the change in the outcome resulting from a unit change in the predictor and that if a predictor is having a significant impact on our ability to predict the outcome then this b should be different from 0 (and big relative to its standard error). We also saw that the t-test tells us whether the b-value is different from 0. SPSS provides the exact probability that the observed value of t would occur if the value of b were 0. As a general rule, if this observed significance is less than .05, then social scientists agree that the result reflects a genuine effect (see Chapter 1). For these two values, the probabilities are .000 (zero to 3 decimal places) and so we can say that the probability of these t-values occurring if the values of b were 0 is less than .001. Therefore, the bs are different from 0 and we can conclude that the advertising budget makes a significant contribution ($p < .001$) to predicting record sales.

5.4.3. Using the model ①

So far, we have discovered that we have a useful model, one that significantly improves our ability to predict record sales. However, the next stage is often to use that model to make some predictions. The first stage is to define the model by replacing the b-values in equation (5.2) with the values from SPSS Output 5.3. In addition, we can replace the X and Y with the variable names so that the model becomes:

$$\begin{aligned}\text{Record Sales}_i &= b_0 + b_1\text{Advertising Budget}_i \\ &= 134.14 + (.09612 \times \text{Advertising Budget}_i)\end{aligned} \tag{5.7}$$

It is now possible to make a prediction about record sales, by replacing the advertising budget with a value of interest. For example, imagine a record executive wanted to spend £100,000 on advertising a new record. Remembering that our units are already in thousands of pounds, we can simply replace the advertising budget with 100. He would discover that record sales should be around 144,000 for the first week of sales:

$$\begin{aligned}\text{Record Sales}_i &= 134.14 + (.09612 \times \text{Advertising Budget}_i) \\ &= 134.14 + (.09612 \times 100) \\ &= 143.75\end{aligned} \tag{5.8}$$

5.5. MULTIPLE REGRESSION: THE BASICS ②

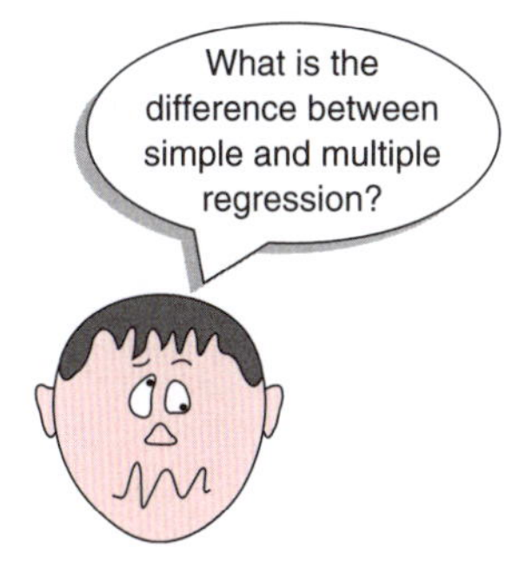

To summarize what we have learnt so far, in simple linear regression the outcome variable Y is predicted using the equation of a straight line (equation (5.2)). Given that we have collected several values of Y and X, the unknown parameters in the equation can be calculated. They are calculated by fitting a model to the data (in this case a straight line) for which the sum of the squared differences between the line and the actual data points is minimized. This method is called the method of least squares. Multiple regression is a logical extension of these principles to situations in which there are several predictors. Again, we still use our basic equation of:

$$\text{Outcome}_i = (\text{Model}_i) + \text{error}_i$$

But this time the model is slightly more complex. It is basically the same as for simple regression except that for every extra predictor you include, you have to add a coefficient; so, each predictor variable has its own coefficient, and the outcome variable is predicted from a combination of all the variables multiplied by their respective coefficients plus a residual term (see equation (5.9)—the brackets aren't necessary, they're just to make the connection to the general equation):

$$Y_i = (b_0 + b_1X_1 + b_2X_2 + \cdots + b_nX_n) + \varepsilon_i \tag{5.9}$$

Y is the outcome variable, b_1 is the coefficient of the first predictor (X_1), b_2 is the coefficient of the second predictor (X_2), b_n is the coefficient of the nth predictor (X_n), and ε_i is the difference between the predicted and the observed value of Y for the ith participant. In this case, the model fitted is more complicated, but the basic principle is the same as simple regression. That is, we seek to find the linear combination of predictors that correlate maximally with the outcome variable. Therefore, when we refer to the regression model in multiple regression, we are talking about a model in the form of equation (5.9).

5.5.1. An example of a multiple regression model ②

Imagine that our record company executive was interested in extending his model of record sales to incorporate another variable. We know already that advertising accounts for 33% of variation in record sales, but a much larger 67% remains unexplained. The record executive could measure a new predictor to the model in an attempt to explain some of the unexplained variation in record sales. He decides to measure the number of times the record is played on Radio 1 (Britain's national radio station) during the week prior to release. The existing model that we derived using SPSS (see equation (5.7)) can now be extended to include this new variable (**airplay**):

$$\text{Record Sales}_i = b_0 + b_1\text{Advertising Budget}_i + b_2\text{Airplay}_i + \varepsilon_i \qquad (5.10)$$

The new model is based on equation (5.9) and includes a b-value for both predictors (and, of course, the constant). If we calculate the b-values, we could make predictions about record sales based not only on the amount spent on advertising but also in terms of radio play. There are only two predictors in this model and so we could display this model graphically in three dimensions (Figure 5.6).

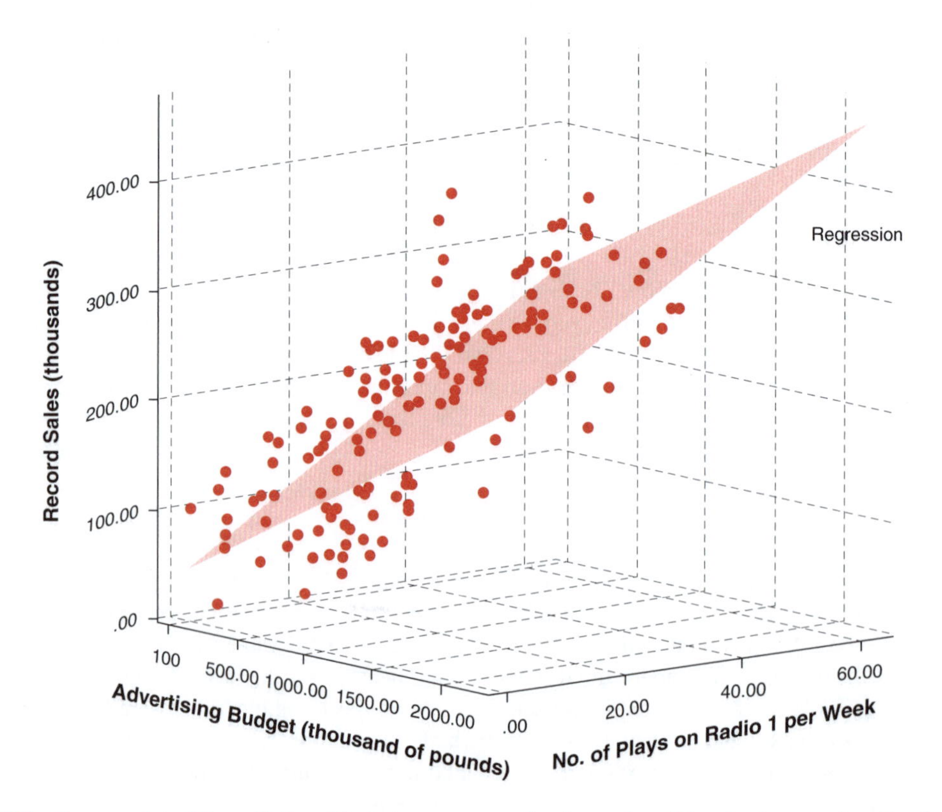

Figure 5.6 Scatterplot of the relationship between record sales, advertising budget and radio play

Equation (5.9) describes the tinted trapezium in the diagram (this is known as the regression *plane*) and the red dots represent the observed data points. Like simple regression, the plane fitted to the data aims to best predict the observed data. However, there are invariably some differences between the model and the real-life data (this fact is evident because some of the dots do not lie exactly on the tinted area of the graph). The *b*-value for advertising describes the slope of the top and bottom of the regression plane whereas the *b*-value for airplay describes the slope of the left and right sides of the regression plane. Knowledge of these two slopes allows us to place the regression plane in space.

It is fairly easy to visualize a regression model with two predictors, because it is possible to plot the regression plane using a 3-D scatterplot. However, multiple regression can be used with three, four or even ten or more predictors. Although you can't immediately visualize what such complex models look like, or know what the *b*-values represent, you should be able to apply the principles of these basic models to more complex scenarios.

5.5.2. Sums of squares, *R* and R^2 ②

When we have several predictors, the partitioning of sums of squares is the same as in the single variable case except that the model we refer to takes the form of equation (5.9) rather than simply being a two-dimensional straight line. Therefore, an SS_T can be calculated that represents the difference between the observed values and the mean value of the outcome variable. SS_R still represents the difference between the values of *Y* predicted by the model and the observed values. Finally, SS_M can still be calculated and represents the difference between the values of *Y* predicted by the model and the mean value. Although the computation of these values is much more complex than in simple regression, conceptually these values are the same.

When there are several predictors it does not make sense to look at the simple correlation coefficient and instead SPSS produces a multiple correlation coefficient (labelled Multiple *R*). Multiple *R* is the correlation between the observed values of *Y* and the values of *Y* predicted by the multiple regression model. Therefore, large values of the multiple *R* represent a large correlation between the predicted and observed values of the outcome. A multiple *R* of 1 represents a situation in which the model perfectly predicts the observed data. As such, multiple *R* is a gauge of how well the model predicts the observed data. It follows that the resulting R^2 can be interpreted in the same way as simple regression: it is the amount of variation in the outcome variable that is accounted for by the model.

5.5.3. Methods of regression ②

If we are interested in constructing a complex model with several predictors, then how do we decide which predictors to use? A great deal of care should be taken in selecting predictors for a model because the values of the regression coefficients depend upon the variables in the model. Therefore, the predictors included and the way in which they are entered into the model can have a great impact. In an ideal world, predictors should be selected based on past research.[5] If new predictors are being added to existing models then select these new variables based on the substantive *theoretical* importance of these variables. One thing *not* to do is select hundreds of random predictors, bung them all into a regression analysis and hope for the best. In addition to the problem of selecting predictors, there are several ways in which variables can be entered into a model. When predictors are all completely uncorrelated the order of variable entry has very little effect on the parameters calculated; however, in social science research we rarely have uncorrelated predictors and so the method of predictor selection is crucial.

5.5.3.1. Hierarchical (Blockwise Entry) ②

In hierarchical regression predictors are selected based on past work and the experimenter decides in which order to enter predictors into the model. As a general rule, known predictors (from other research) should be entered into the model first in order of their importance in predicting the outcome. After known predictors have been entered, the experimenter can add any new predictors into the model. New predictors can be entered all in one go, in a stepwise manner, or hierarchically (such that the new predictor suspected to be the most important is entered first).

5.5.3.2. Forced entry ②

Forced entry (or *Enter* as it is known in SPSS) is a method in which all predictors are forced into the model simultaneously. Like hierarchical, this method relies on good theoretical reasons for including the chosen predictors, but unlike hierarchical the experimenter makes no decision about the order in which variables are entered.

5.5.3.3. Stepwise methods ②

In stepwise regressions decisions about the order in which predictors are entered into the model are based on a purely mathematical criterion. In the *forward* method, an initial model is defined that contains only the constant (b_0). The computer then searches for the predictor (out of the ones available) that best predicts the outcome variable—it does this by selecting the predictor that has the highest simple correlation with the outcome. If this predictor significantly improves the ability of the model to predict the outcome, then this predictor is retained in the model and the computer searches for a second predictor. The criterion used for selecting this second predictor is that it is the variable that has the largest semi-partial correlation with the outcome. Let me explain this in plain English. Imagine that the first predictor can explain 40% of the variation in the outcome variable; then there is still 60% left unexplained. The computer searches for the predictor that can

---

5 I, rather cynically, qualify this suggestion by proposing that predictors be chosen based on past research that has utilized good methodology. If basing such decisions on regression analyses, select predictors based only on past research that has used regression appropriately and yielded reliable, generalizable models!

explain the biggest part of the remaining 60% (so, it is not interested in the 40% that is already explained). In statistical terms you can think of this like a partial correlation in that the computer correlates each of the predictors with the outcome while controlling for the effect of the first predictor. The reason that it is called a *semi*-partial correlation is because the effects of the first predictor are partialled out of only the remaining predictors, and are not controlled for in the outcome itself. This semi-partial correlation gives a measure of how much 'new variance' in the outcome can be explained by each remaining predictor (see section 4.6). The predictor that accounts for the most new variance is added to the model and, if it makes a significant contribution to the predictive power of the model, it is retained and another predictor is considered.

The *stepwise* method in SPSS is the same as the forward method, except that each time a predictor is added to the equation, a removal test is made of the least useful predictor. As such the regression equation is constantly being reassessed to see whether any redundant predictors can be removed. The *backward* method is the opposite of the forward method in that the computer begins by placing all predictors in the model and then calculating the contribution of each one by looking at the significance value of the *t*-test for each predictor. This significance value is compared against a removal criterion (which can be either an absolute value of the test statistic or a probability value for that test statistic). If a predictor meets the removal criterion (i.e. if it is not making a statistically significant contribution to how well the model predicts the outcome variable) it is removed from the model and the model is re-estimated for the remaining predictors. The contribution of the remaining predictors is then reassessed.

If you do decide to use a stepwise method then the backward method is preferable to the forward method. This is because of suppressor effects, which occur when a predictor has a significant effect but only when another variable is held constant. Forward selection is more likely than backward elimination to exclude predictors involved in suppressor effects. As such, the forward method runs a higher risk of making a Type II error (i.e. missing a predictor that does in fact predict the outcome).

5.5.3.4. Choosing a method ②

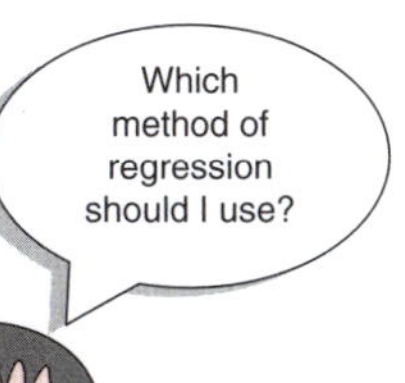

SPSS allows you to opt for any one of these methods and it is important to select an appropriate one. The forward, backward and stepwise methods all come under the general heading of *stepwise methods* because they all rely on the computer selecting variables based upon mathematical criteria. Many writers argue that this takes many important methodological decisions out of the hands of the researcher. What's more, the models derived by computer often take advantage of random sampling variation and so decisions about which variables should be included will be based upon slight differences in their semi-partial correlation. However, these slight statistical differences may contrast dramatically with the theoretical importance of a predictor to the model. For this reason stepwise methods are best avoided except for exploratory model building (see Wright, 1997, p. 181). When there is a sound theoretical literature available, then base your model upon what past research tells you. Include any meaningful variables in the model in their order of importance. After this initial analysis, repeat the regression but exclude any variables that were statistically redundant the first time around. There are important considerations in deciding which predictors should be included. It is important not to include too many predictors. As a general rule, the fewer predictors the better, and certainly include only predictors for which you have a good theoretical grounding (it is meaningless to measure hundreds of variables and then put them all into a regression model). So, be selective and remember you should have a decent sample size (some suggest at least 15 participants per predictor—but see section 5.6.2.3).

5.6. HOW ACCURATE IS MY REGRESSION MODEL? ②

When we have produced a model based on a sample of data there are two important questions to ask: (1) does the model fit the observed data well, or is it influenced by a small number of cases, and (2) can my model generalize to other samples? These questions are vital to ask because they affect how we use the model that has been constructed. These questions are also, in some sense, hierarchical because we wouldn't want to generalize a bad model. However, it is a mistake to think that because a model fits the observed data well we can draw conclusions beyond our sample. Generalization is a critical additional step and if we find that our model is not generalizable, then we must restrict any conclusions based on the model to the sample used. First, we will look at how we establish whether a model is an accurate representation of the actual data, and then in section 5.6.2 we move on to look at how we assess whether a model can be used to make inferences beyond the sample of data that has been collected.

5.6.1. Assessing the regression model I: diagnostics ②

To answer the question of whether the model fits the observed data well, or if it is influenced by a small number of cases, we can look for outliers and influential cases (the difference is explained in Box 5.1). We will look at these in turn.

5.6.1.1. Outliers and residuals ②

An outlier is a case that differs substantially from the main trend of the data (see Box 3.1). Figure 5.7 shows an example of such a case in regression. Outliers can cause your model to be biased because they affect the values of the estimated regression coefficients. For example, Figure 5.7 uses the same data as Figure 5.2 except that the score of one participant has been changed to be an outlier (in this case a person who was very calm in the presence of a very big spider). The change in this one point has had a dramatic effect on the regression model chosen to fit the data. With the outlier present, the regression model changes: its gradient is reduced (the line becomes flatter) and the intercept increases (the new line will cross the Y-axis at a higher point). It should be clear from this diagram that it is important to try to detect outliers to see whether the model is biased in this way.

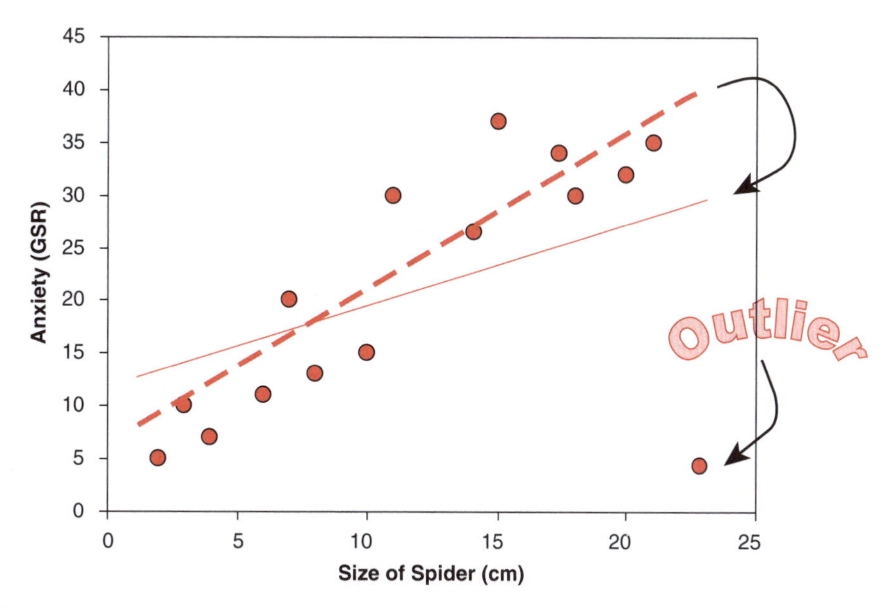

Figure 5.7 Graph demonstrating the effect of an outlier. The dashed line represents the original regression line for these data (see Figure 5.2), whereas the full line represents the regression line when an outlier is present

If you think about how you might detect an outlier, how do you think you might do it? Well, we know that an outlier, by its nature, is very different from all of the other scores. This being true, do you think that the model will predict that person's score very accurately? The answer is clearly *no*: looking at Figure 5.7 it is evident that even though the outlier has biased the model, the model still predicts that one value very badly (the regression line is a long way from the outlier). Therefore, if we were to work out the differences between the data values that were collected, and the values predicted by the model, we could detect an outlier by looking for large differences. This process is the same as looking for cases that the model predicts inaccurately. The differences between the values of the outcome predicted by the model and the values of the outcome observed in the sample are known as *residuals*. These residuals effectively represent the error present in the model. If a model fits the sample data well then all residuals will be small (if the model was a perfect fit of the sample data—all data points fall on the regression line—then all residuals would be zero). If a model is a poor fit of the sample data then the residuals will be large. Also, if any cases stand out as having a large residual, then they could be outliers.

The *normal* or unstandardized residuals described above are measured in the same units as the outcome variable and so are difficult to interpret across different models. What we can do is to look for residuals that stand out as being particularly large. However, we cannot define a universal cut-off point for what constitutes a large residual. To overcome this problem, we use standardized residuals, which are the residuals divided by an estimate of their standard deviation. We came across standardization in section 1.4.1 as a means of converting variables into a standard unit of measurement (the standard deviation); we also came across *z*-scores (see Box 3.2) in which variables are converted into standard deviation units (i.e. they're converted into scores that are distributed around a mean of 0 with a standard deviation of 1). By converting residuals into *z*-scores (standardized residuals) we can compare residuals from different models and use what we know about the properties of *z*-scores to devise universal guidelines for what constitutes an acceptable (or unacceptable) value. For example, we know from Chapter 3 that in a normally distributed sample, 95% of *z*-scores should lie between −1.96 and +1.96, 99% should lie between −2.58 and +2.58, and 99.9% (i.e. nearly all of them) should lie between −3.29 and +3.29. Some general rules for standardized residuals are derived from these facts: (1) standardized residuals with an absolute value greater than 3.29 (actually we usually just use 3) are cause for concern because in an average sample a value this high is unlikely to happen by chance; (2) if more than 1% of our sample has standardized residuals with an absolute value greater than 2.58 (we usually just say 2.5) there is evidence that the level of error within our model is unacceptable (the model is a fairly poor fit of the sample data); and (3) if more than 5% of cases have standardized residuals with an absolute value greater than 1.96 (we usually use 2 for convenience) then there is also evidence that the model is a poor representation of the actual data.

A third form of residual is the Studentized residual, which is the unstandardized residual divided by an estimate of its standard deviation that varies point by point. These residuals have the same properties as the standardized residuals but usually provide a more precise estimate of the error variance of a specific case.

5.6.1.2. Influential cases ③

As well as testing for outliers by looking at the error in the model, it is also possible to look at whether certain cases exert undue influence over the parameters of the model. So, if we were to delete a certain case, would we obtain different regression coefficients? This type of analysis can help to determine whether the regression model is stable across the sample, or whether it is biased by a few influential cases. Again, this process will unveil outliers.

There are several residual statistics that can be used to assess the influence of a particular case. One statistic is the adjusted predicted value for a case when that case is excluded from the analysis. In effect, the computer calculates a new model without a particular case and then uses this new model to predict the value of the outcome variable for the case that was excluded. If a case does not exert a large influence over the model then we would expect the adjusted predicted value to be very similar to the predicted value when the case is included. Put simply, if the model is stable then the predicted value of a case should be the same regardless of whether or not that case was used to calculate the model. The difference between the adjusted predicted value and the original predicted value is known as DFFit (see below). We can also look at the residual based on the adjusted predicted value: that is, the difference between the adjusted predicted value and the original observed value. This is the deleted residual. The deleted residual can be divided by the standard error to give a standardized value known as the Studentized deleted residual. This residual can be compared across different regression analyses because it is measured in standard units.

The deleted residuals are very useful to assess the influence of a case on the ability of the model to predict that case. However, they do not provide any information about how a case influences the model as a whole (i.e. the impact that a case has on the model's ability to predict *all* cases). One statistic that does consider the effect of a single case on the model as a whole is Cook's distance. Cook's distance is a measure of the overall influence of a case on the model and Cook & Weisberg (1982) have suggested that values greater than 1 may be cause for concern.

A second measure of influence is leverage (sometimes called hat values), which gauges the influence of the observed value of the outcome variable over the predicted values. The average leverage value is defined as $(k + 1)/n$ in which k is the number of predictors in the model and n is the number of participants.[6] Leverage values can lie between 0 (indicating that the case has no influence whatsoever) and 1 (indicating that the case has complete influence over prediction). If no cases exert undue influence over the model then we would expect all of the leverage value to be close to the average value ($(k + 1)/n$). Hoaglin & Welsch (1978) recommend investigating cases with values greater than twice the average ($2(k + 1)/n$) and Stevens (1992) recommends using three times the average ($3(k + 1)/n$) as a cut-off point for identifying cases having undue influence. We shall see how to use these cut-off points in section 5.8.6. However, cases with large leverage values will not necessarily have a large influence on the regression coefficients because they are measured on the outcome variables rather than the predictors.

Related to the leverage values are the Mahalanobis distances (Figure 5.8), which measure the distance of cases from the mean(s) of the predictor variable(s). You need to look for the cases with the highest values. It is not easy to establish a cut-off point at which to worry, although Barnett & Lewis (1978) have produced a table of critical values dependent on the number of predictors and the sample size. From their work it is clear that even with large samples ($N = 500$) and five predictors, values above 25 are cause for concern. In smaller samples ($N = 100$) and with fewer predictors (namely three) values greater than 15 are problematic, and in very small samples ($N = 30$) with only two predictors values greater than 11 should be examined. However, for more specific advice, refer to Barnett & Lewis's (1978) table.

It is possible to run the regression analysis with a case included and then rerun the analysis with that same case excluded. If we did this, undoubtedly there would be some difference between the b-coefficients in the two regression equations. This difference would tell us how much influence a particular case has on the parameters of the regression model. To take a hypothetical example, imagine two variables that had a perfect negative relationship except for a single case (case 30). If a regression analysis was done on the 29 cases that were perfectly linearly related then we would get a model in which the predictor variable X perfectly predicts

---

6 You may come across the average leverage denoted as p/n in which p is the number of parameters being estimated. In multiple regression, we estimate parameters for each predictor and also for a constant and so p is equivalent to the number of predictors plus one $(k + 1)$.

Figure 5.8 Prasanta Chandra Mahalanobis staring into his distances

the outcome variable *Y*, and there are no errors. If we then run the analysis but this time include the case that didn't conform (case 30), then the resulting model has different parameters. Some data are stored in the file **dfbeta.sav** which illustrate such a situation. Try running a simple regression first with all the cases included and then with case 30 deleted. The results are summarized in Table 5.1, which shows (1) the parameters for the regression model when the extreme case is included or excluded; (2) the resulting regression equations; and (3) the value of *Y* predicted from participant 30's score on the *X*-variable (which is obtained by replacing the *X* in the regression equation with participant 30's score for *X*, which was 1).

When case 30 is excluded, these data have a perfect negative relationship: hence the coefficient for the predictor (b_1) is −1 (remember that in simple regression this term is the same as the Pearson correlation coefficient), and the coefficient for the constant (the intercept, b_0) is 31. However, when case 30 is included, both parameters are reduced[7] and the difference between the parameters is also displayed. The difference between a parameter estimated using all cases and estimated when one case is excluded is known as the DFBeta in SPSS. DFBeta is calculated for every case and for each of the parameters in the model. So, in our hypothetical example, the DFBeta for the constant is –2, and the DFBeta for the predictor variable is 0.1. By looking at the values of the DFBetas, it is possible to identify cases that have a large influence on the parameters of the regression model. Again, the units of measurement used will affect these values and so SPSS produces a standardized DFBeta. These standardized values are easier to use because universal cut-off points can be applied. In this case absolute values above 1 indicate cases that substantially influence the model parameters (although Stevens, 1992, suggests looking at cases with absolute values greater than 2).

7 The value of b_1 is reduced because the data no longer have a perfect linear relationship and so there is now variance that the model cannot explain.

Table 5.1 The difference in the parameters of the regression model when one case is excluded

Parameter (*b*)	Case 30 Included	Case 30 Excluded	Difference
Constant (intercept)	29.00	31.00	−2.00
Predictor (gradient)	−0.90	−1.00	0.10
Model (regression line)	Y = (−0.9)X + 29	Y = (−1)X + 31	
Predicted *Y*	28.10	30.00	−1.09

A related statistic is the DFFit, which is the difference between the predicted value for a case when the model is calculated including that case and when the model is calculated excluding that case: in this example the value is −1.09 (see Table 5.1). If a case is not influential then its DFFit should be zero—hence, we expect non-influential cases to have small DFFit values. However, we have the problem that this statistic depends on the units of measurement of the outcome and so a DFFit of 0.5 will be very small if the outcome ranges from 1 to 100, but very large if the outcome varies from 0 to 1. Therefore, SPSS also produces standardized versions of the DFFit values (standardized DFFit). A final measure is that of the covariance ratio (CVR), which is a measure of whether a case influences the variance of the regression parameters. A description of the computation of this statistic would leave most readers dazed and confused, so suffice to say that when this ratio is close to 1 the case has very little influence on the variances of the model parameters. Belsey, Kuh & Welsch (1980) recommend the following:

- If $CVR_i > 1 + [3(k + 1)/n]$ then deleting the *i*th case will damage the precision of some of the model's parameters.
- If $CVR_i < 1 - [3(k + 1)/n]$ then deleting the *i*th case will improve the precision of some of the model's parameters.

In both expressions, k is the number of predictors, CVR_i is the covariance ratio for the *i*th participant, and n is the sample size.

Box 5.1

The difference between residuals and influence statistics ③

In this section I've described two ways to look for cases that might bias the model: residual and influence statistics. To illustrate how these measures differ, imagine that the Mayor of London at the turn of the last century was interested in how drinking affected mortality. London is divided up into different regions called boroughs, and so he might measure the number of pubs and the number of deaths over a period of time in eight of his boroughs. The data are in a file called **pubs.sav**

(Continued)

Box 5.1 (Continued)

The scatterplot of these data reveals that without the last case there is a perfect linear relationship (the dashed straight line). However, the presence of the last case (case 8) changes the line of best fit dramatically (although this line is still a significant fit of the data—do the regression analysis and see for yourself).

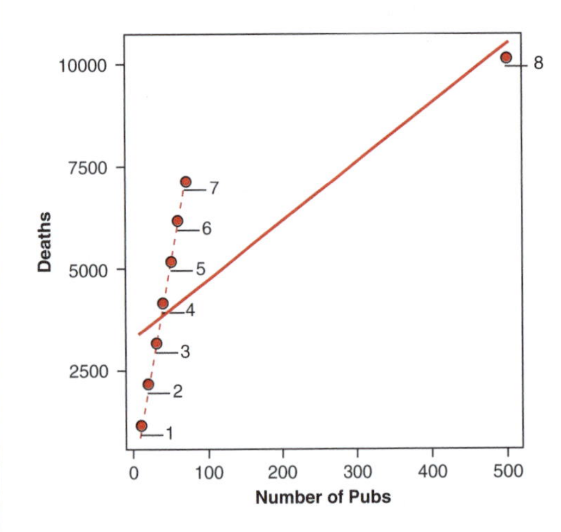

What's interesting about these data is when we look at the residuals and influence statistics. The standardized residual for case 8 is the second *smallest*: this outlier produces a very small residual (most of the non-outliers have larger residuals) because it sits very close to the line that has been fitted to the data. How can this be? Look at the influence statistics and you'll see that they're massive for case 8: case 8 exerts a huge influence over the model.

As always when you see a statistical oddity you should ask what was happening in the real world. The last data point represents the City of London, a tiny area of only 1 square mile in the centre of London where very few people lived but where thousands of commuters (even then) came to work and had lunch in the pubs. Hence the pubs didn't rely on the resident population for their business and the residents didn't consume all of their beer! Therefore, there were a massive number of pubs.

This illustrates that a case exerting a massive influence can produce a small residual—so look at both! (I'm very grateful to David Hitchin for this example, and he in turn got it from Dr Richard Roberts.)

Case Summaries[a]

		Standardized Residual	Centred Leverage Value	Standardized DFFIT	Standardized DFBETA Intercept	Standardized DFBETA PUBS
1		−1.33839	.04074	−.74402	−.74317	.36886
2		−.87895	.03196	−.40964	−.40766	.18484
3		−.41950	.02424	−.17697	−.17494	.07132
4		.03995	.01759	.01606	.01572	−.00564
5		.49940	.01200	.20042	.19337	−.05933
6		.95885	.00748	.40473	.38333	−.09618
7		1.41830	.00402	.68084	.62996	−.12023
8		−.27966	.86196	−460379232.7	92676016.019	−430238878.2
Total	N	8	8	8	8	8

a Limited to first 100 cases

5.6.1.3. A final comment on diagnostic statistics ②

There are a lot of diagnostic statistics that should be examined after a regression analysis, and it is difficult to summarize this wealth of material into a concise conclusion. However, one thing I would like to stress is a point made by Belsey et al. (1980) who noted the dangers inherent in these procedures. The point is that diagnostics are tools that enable you to see how good or bad your model is in terms of fitting the sampled data. They are a way of assessing your model. They are *not*, however, a way of justifying the removal of data points to effect some desirable change in the regression parameters (e.g. deleting a case that changes a non-significant *b*-value into a significant one). Stevens (1992), as ever, offers excellent advice:

> If a point is a significant outlier on *Y*, but its Cook's distance is < 1, there is no real need to delete that point since it does not have a large effect on the regression analysis. However, one should still be interested in studying such points further to understand why they did not fit the model. (p. 118)

5.6.2. Assessing the regression model II: generalization ②

When a regression analysis is done, an equation can be produced that is correct for the sample of observed values. However, in the social sciences we are usually interested in generalizing our findings outside of the sample. So, although it can be useful to draw conclusions about a particular sample of people, it is usually more interesting if we can then assume that our conclusions are true for a wider population. For a regression model to generalize we must be sure that underlying assumptions have been met, and to test whether the model does generalize we can look at cross-validating it.

5.6.2.1. Checking assumptions ②

To draw conclusions about a population based on a regression analysis done on a sample, several assumptions must be true (see Berry, 1993).

- **Variable types**: All predictor variables must be quantitative or categorical (with two categories), and the outcome variable must be quantitative, continuous and unbounded. By quantitative I mean that they should be measured at the interval level and by unbounded I mean that there should be no constraints on the variability of the outcome. If the outcome is a measure ranging from 1 to 10 yet the data collected vary between 3 and 7, then these data are constrained.
- **Non-zero variance**: The predictors should have some variation in value (i.e. they do not have variances of 0).

- **No perfect** multicollinearity: There should be no perfect linear relationship between two or more of the predictors. So, the predictor variables should not correlate too highly (see section 5.6.2.4).
- **Predictors are uncorrelated with 'external variables'**: *External variables* are variables that haven't been included in the regression model which influence the outcome variable.[8] These variables can be thought of as similar to the 'third variable' that was discussed with reference to correlation. This assumption means that there should be no external variables that correlate with any of the variables included in the regression model. Obviously, if external variables do correlate with the predictors, then the conclusions we draw from the model become unreliable (because other variables exist that can predict the outcome just as well).
- **Homoscedasticity**: At each level of the predictor variable(s), the variance of the residual terms should be constant. This just means that the residuals at each level of the predictor(s) should have the same variance (homoscedasticity); when the variances are very unequal there is said to be heteroscedasticity (see section 3.6 as well).
- Independent errors: For any two observations the residual terms should be uncorrelated (or independent). This eventuality is sometimes described as a lack of autocorrelation. This assumption can be tested with the Durbin–Watson test, which tests for serial correlations between errors. Specifically, it tests whether adjacent residuals are correlated. The test statistic can vary between 0 and 4 with a value of 2 meaning that the residuals are uncorrelated. A value greater than 2 indicates a negative correlation between adjacent residuals, whereas a value below 2 indicates a positive correlation. The size of the Durbin–Watson statistic depends upon the number of predictors in the model, and the number of observations. For accuracy, you should look up the exact acceptable values in Durbin & Watson's (1951) original paper. As a very conservative rule of thumb, values less than 1 or greater than 3 are definitely cause for concern; however, values closer to 2 may still be problematic depending on your sample and model.
- **Normally distributed errors**: It is assumed that the residuals in the model are random, normally distributed variables with a mean of 0. This assumption simply means that the differences between the model and the observed data are most frequently zero or very close to zero, and that differences much greater than zero happen only occasionally. Some people confuse this assumption with the idea that predictors have to be normally distributed. In fact, predictors do not need to be normally distributed (see section 5.10).
- **Independence**: It is assumed that all of the values of the outcome variable are independent (in other words, each value of the outcome variable comes from a separate entity).
- **Linearity**: The mean values of the outcome variable for each increment of the predictor(s) lie along a straight line. In plain English this means that it is assumed that the relationship we are modelling is a linear one. If we model a non-linear relationship using a linear model then this obviously limits the generalizability of the findings.

---

8 Some authors choose to refer to these external variables as part of an error term that includes any random factor in the way in which the outcome varies. However, to avoid confusion with the residual terms in the regression equations I have chosen the label 'external variables'. Although this term implicitly washes over any random factors, I acknowledge their presence here!

This list of assumptions probably seems pretty daunting and in fact most undergraduates (and some academics for that matter) tend to regard assumptions as rather tedious things about which no one really need worry. In fact, when I mention statistical assumptions to most psychologists they tend to give me that 'you really are a bit of a pedant' look and then ignore me. However, regardless of my status as a pedant, there are good reasons for taking assumptions seriously. Imagine that I go over to a friend's house, the lights are on and it's obvious that someone is at home. I ring the doorbell and no one answers. From that experience, I conclude that my friend hates me and that I am a terrible, unlovable, person. How tenable is this conclusion? Well, there is a reality that I am trying to tap (i.e. whether my friend likes or hates me), and I have collected data about that reality (I've gone to his house, seen that he's at home, rang the doorbell and got no response). Imagine that in reality my friend likes me (he never was a good judge of character!); in this scenario, my conclusion is false. Why have my data led me to the wrong conclusion? The answer is simple: I had assumed that my friend's doorbell was working and under this assumption the conclusion that I made from my data was accurate (my friend heard the bell but chose to ignore it because he hates me). However, this assumption was not true—his doorbell was not working, which is why he didn't answer the door—and as a consequence the conclusion I drew about reality was completely false.

Enough about doorbells, friends and my social life: the point to remember is that when assumptions are broken we stop being able to draw accurate conclusions about reality. In terms of regression, when the assumptions are met, the model that we get for a sample can be accurately applied to the population of interest (the coefficients and parameters of the regression equation are said to be *unbiased*). Some people assume that this means that when the assumptions are met the regression model from a sample is always identical to the model that would have been obtained had we been able to test the entire population. Unfortunately, this belief isn't true. What an unbiased model does tell us is that *on average* the regression model from the sample is the same as the population model. However, you should be clear that even when the assumptions are met, it is possible that a model obtained from a sample may not be the same as the population model—but the likelihood of them being the same is increased.

5.6.2.2. Cross-validation of the model ③

Even if we can't be confident that the model derived from our sample accurately represents the entire population, there are ways in which we can assess how well our model can predict the outcome in a different sample. Assessing the accuracy of a model across different samples is known as cross-validation. If a model can be generalized, then it must be capable of accurately predicting the same outcome variable from the same set of predictors in a different group of people. If the model is applied to a different sample and there is a severe drop in its predictive power, then the model clearly does *not* generalize. As a first rule of thumb, we should aim to collect enough data to obtain a reliable regression model (see the next section). Once we have a regression model there are two main methods of cross-validation.

- **Adjusted R^2**: In SPSS, not only are the values of R and R^2 calculated, but so is an adjusted R^2. This adjusted value indicates the loss of predictive power or *shrinkage*.

Whereas R^2 tells us how much of the variance in Y is accounted for by the regression model from our sample, the adjusted value tells us how much variance in Y would be accounted for if the model had been derived from the population from which the sample was taken. SPSS derives the adjusted R^2 using Wherry's equation. However, this equation has been criticized because it tells us nothing about how well the regression model would predict an entirely different set of data (how well can the model predict scores of a different sample of data from the same population?). One version of R^2 that does tell us how well the model cross-validates uses Stein's formula which is shown in equation (5.11) (see Stevens, 1992):

$$\text{adjusted } R^2 = 1 - \left[\left(\frac{n-1}{n-k-1}\right)\left(\frac{n-2}{n-k-2}\right)\left(\frac{n+1}{n}\right)\right](1-R^2) \qquad (5.11)$$

In Stein's equation, R^2 is the unadjusted value, n is the number of cases and k is the number of predictors in the model. For the more mathematically minded of you, it is worth using this equation to cross-validate a regression model.

- **Data splitting**: This approach involves randomly splitting your data set in half, computing a regression equation on both halves of the data and then comparing the resulting models. However, researchers rarely have large enough data sets to perform this kind of analysis.

5.6.2.3. Sample size in regression ③

In the previous section I said that it's important to collect enough data to obtain a reliable regression model. Well, how much is enough? You'll find a lot of rules of thumb floating about, the two most common being that you should have 10 cases of data for each predictor in the model, or 15 cases of data per predictor. So, with five predictors, you'd need 50 or 75 cases respectively (depending on the rule you use). These rules are very pervasive (even I used the 15 cases per predictor rule in the first edition of this book) but they oversimplify the issue considerably. In fact, the sample size required will depend on the size of effect that we're trying to detect (i.e. how strong is the relationship we're trying to measure?) and how much power we want to detect these effects (see Chapter 1). The simplest rule of thumb is that the bigger the sample size the better! The reason is that the estimate of R that we get from regression is dependent on the number of predictors, k, and the sample size, N. In fact the expected R for random data is $k/(N-1)$ and so with small sample sizes random data can appear to show a strong effect: for example, with six predictors and 21 cases of data, $R = 6/(21-1) = .3$ (a medium effect size by Cohen's criteria described in section 4.2.2). Obviously for random data we'd want the expected R to be 0 (no effect) and for this to be true we need large samples (to take the previous example, if we had 100 cases, not 21, then the expected R would be a more acceptable 0.06).

It's all very well knowing that larger is better, but researchers usually need some more concrete guidelines (much as we'd all love to collect 1000 cases of data it isn't always practical!). Green (1991) gives two rules of thumb for the *minimum* acceptable sample size, the first based on whether you want to test the overall fit of your regression model (i.e. test the R^2), and the second based on whether you want to test the individual predictors within the model (i.e. test b-values of the model). If you want to test the model overall, then he recommends a minimum sample size of $50 + 8k$, where k is the number of predictors. So, with five predictors, you'd need a sample size of $50 + 40 = 90$. If you want to test the individual predictors then he suggests a minimum sample size of $104 + k$, so again taking the example of five predictors you'd need a sample size of $104 + 5 = 109$. Of course, in most cases we're interested both in the overall fit and in the contribution of individual predictors, and in this situation Green recommends you calculate both of the minimum sample sizes I've just described, and use the one that has the largest value (so, in the five-predictor example, we'd use 109 because it is bigger than 90).

Now, these guidelines are all right as a rough and ready guide, but they still oversimplify the problem. As I've mentioned, the sample size required actually depends on the size of the effect (i.e. how well our predictors predict the outcome) and how much statistical power we want to detect these effects. Miles and Shevlin (2001) produce some extremely useful graphs that illustrate the sample sizes needed to achieve different levels of power, for different effect sizes, as the number of predictors vary. For precise estimates of the sample size you should be using, I recommend using these graphs. I've summarized some general findings from Miles and Shevlin in Figure 5.9. This diagram shows the sample size required to achieve a high level of power (I've taken Cohen's 1988 benchmark of .8) depending on the

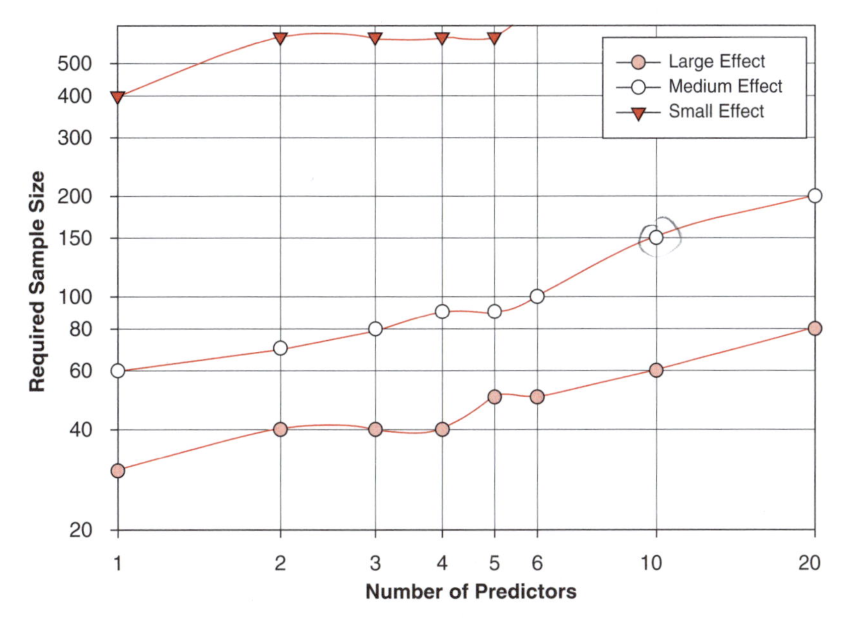

Figure 5.9 Graph to show the sample size required in regression depending on the number of predictors and the size of expected effect

number of predictors and the size of expected effect. Some general findings from this are: (1) if you expect to find a large effect then a sample size of 80 will always suffice (with up to 20 predictors) and if there are fewer predictors then you can afford to have a smaller sample; (2) if you're expecting a medium effect, then a sample size of 200 will always suffice (up to 20 predictors), but you should always have a sample size above 60, and with six or less predictors you'll be fine with a sample of 100; and (3) if you're expecting a small effect size then just don't bother unless you have the time and resources to collect at least 600 cases of data (and many more if you have six or more predictors!).

5.6.2.4. Multicollinearity ②

Multicollinearity exists when there is a strong correlation between two or more predictors in a regression model. Multicollinearity poses a problem only for multiple regression because (without wishing to state the obvious) simple regression requires only one predictor. Perfect collinearity exists when at least one predictor is a perfect linear combination of the others (the simplest example being two predictors that are perfectly correlated—they have a correlation coefficient of 1). If there is perfect collinearity between predictors it becomes impossible to obtain unique estimates of the regression coefficients because there are an infinite number of combinations of coefficients that would work equally well. Put simply, if we have two predictors that are perfectly correlated, then the values of *b* for each variable are interchangeable. The good news is that perfect collinearity is rare in real-life data. The bad news is that less than perfect collinearity is virtually unavoidable. Low levels of collinearity pose little threat to the models generated by SPSS, but as collinearity increases so do the standard errors of the *b*-coefficients, which in turn affects whether these coefficients are found to be statistically significant. In short, high levels of collinearity increase the probability that a good predictor of the outcome will be found non-significant and rejected from the model (a Type II error). There are three other reasons why the presence of multicollinearity poses a threat to the validity of multiple regression analysis:

- **It limits the size of *R***: Remember that R is a measure of the multiple correlation between the predictors and the outcome and that R^2 indicates the variance in the outcome for which the predictors account. Imagine a situation in which a single variable predicts the outcome variable fairly successfully (e.g. $R = .80$) and a second predictor variable is then added to the model. This second variable might account for a lot of the variance in the outcome (which is why it is included in the model), but the variance it accounts for is the same variance accounted for by the first variable. In other words, once the variance accounted for by the first predictor has been removed, the second predictor accounts for very little of the remaining variance (the second variable accounts for very little *unique variance*). Hence, the overall variance in the outcome accounted for by the two predictors is little more than when only one predictor is used (so R might increase from .80 to .82). This idea is connected to the notion of partial correlation that was explained in Chapter 3. If, however, the two predictors are completely uncorrelated, then the second predictor is likely to account for different variance in the outcome to that accounted for by the first predictor. So, although in itself the second predictor might account for only a little of the variance in the outcome, the variance it does account for is different to that of the other predictor (and so when both predictors are included, R is substantially larger, say .95). Therefore, having uncorrelated predictors is beneficial.

- **Importance of predictors**: Multicollinearity between predictors makes it difficult to assess the individual importance of a predictor. If the predictors are highly correlated, and each accounts for similar variance in the outcome, then how can we know which of the two variables is important? Quite simply we can't tell which variable is important—the model could include either one, interchangeably.
- **Unstable predictor equations**: I have described how multicollinearity increases the variances of the regression coefficients, resulting in unstable predictor equations. This means that the estimated values of the regression coefficients (the b-values) will be unstable from sample to sample.

One way of identifying multicollinearity is to scan a correlation matrix of all of the predictor variables and see if any correlate very highly (by very highly I mean correlations of above .80 or .90). This is a good 'ball park' method but misses more subtle forms of multicollinearity. Luckily, SPSS produces various collinearity diagnostics, one of which is the variance inflation factor (VIF). The VIF indicates whether a predictor has a strong linear relationship with the other predictor(s). Although there are no hard and fast rules about what value of the VIF should be cause for concern, Myers (1990) suggests that a value of 10 is a good value at which to worry. What's more, Bowerman & O'Connell (1990) suggest that if the average VIF is greater than 1, then multicollinearity may be biasing the regression model. Related to the VIF is the tolerance statistic, which is its reciprocal (1/VIF). As such, values below .1 indicate serious problems, although Menard (1995) suggests that values below .2 are worthy of concern.

Other measures that are useful in discovering whether predictors are dependent are the *eigenvalues of the scaled, uncentred cross-products matrix*, the *condition indexes* and the *variance proportions*. These statistics are extremely complex and will be covered as part of the interpretation of SPSS output (see section 5.8.5). If none of this has made any sense then have a look at Hutcheson & Sofroniou (1999, pp. 78–85) who give a really clear explanation of multicollinearity.

5.7. HOW TO DO MULTIPLE REGRESSION USING SPSS ②

5.7.1. Main options ②

Imagine that the record company executive was now interested in extending the model of record sales to incorporate other variables. He decides to measure two new variables: (1) the number of times the record is played on Radio 1 (Britain's largest national radio station) during the week prior to release (**airplay**), and (2) the attractiveness of the band (**attract**). Before a record is released, the executive notes the amount spent on advertising,

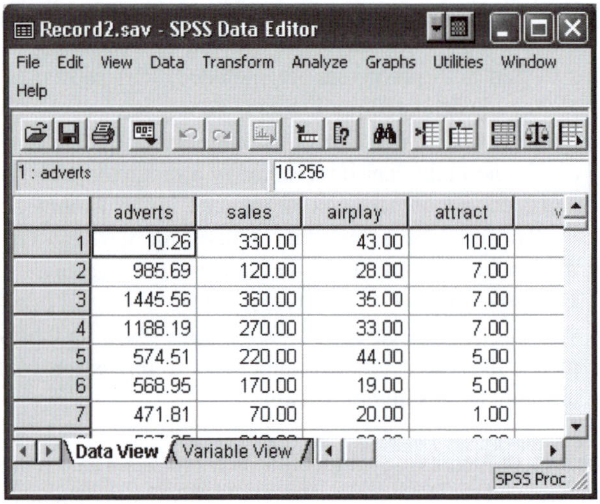

Record2.sav - SPSS Data Editor

File Edit View Data Transform Analyze Graphs Utilities Window Help

1 : adverts 10.256

	adverts	sales	airplay	attract
1	10.26	330.00	43.00	10.00
2	985.69	120.00	28.00	7.00
3	1445.56	360.00	35.00	7.00
4	1188.19	270.00	33.00	7.00
5	574.51	220.00	44.00	5.00
6	568.95	170.00	19.00	5.00
7	471.81	70.00	20.00	1.00

Data View / Variable View

SPSS Proc

Figure 5.10 Data layout for multiple regression

the number of times the record is played on radio the week before release, and the attractiveness of the band. He does this for 200 different records (each made by a different band). Attractiveness was measured by asking a random sample of the target audience to rate the attractiveness of each band on a scale from 0 (hideous potato-heads) to 10 (gorgeous sex objects). The mode attractiveness given by the sample was used in the regression (because he was interested in what the majority of people thought, rather than the average of people's opinions).

These data are in the file **Record2.sav** and you should note that each variable has its own column (the same layout as for correlation), and each row represents a different record. So, the first record had £10,260 spent advertising it, it sold 330,000 copies, it received 43 plays on Radio 1 the week before release and it was made by a band that the majority of people rated as gorgeous sex objects (Figure 5.10).

The executive has past research indicating that the advertising budget is a significant predictor of record sales, and so he should include this variable in the model first. His new variables (**airplay** and **attract**) should, therefore, be entered into the model *after* advertising budget. This method is hierarchical (the researcher decides in which order to enter variables into the model based on past research). To do a hierarchical regression in SPSS we have to enter the variables in blocks (each block representing one step in the hierarchy). To get to the main *regression* dialog box you must go to the *Analyze* menu and select *Regression* and then *Linear* (**Analyze⇒Regression⇒Linear**). The main dialog box is shown in Figure 5.11 and is the same as when we encountered it for simple regression.

The main dialog box is fairly self-explanatory in that there is a space to specify the dependent variable (outcome), and a space to place one or more independent variables (predictor variables). As usual, the variables in the data editor are listed on the left-hand

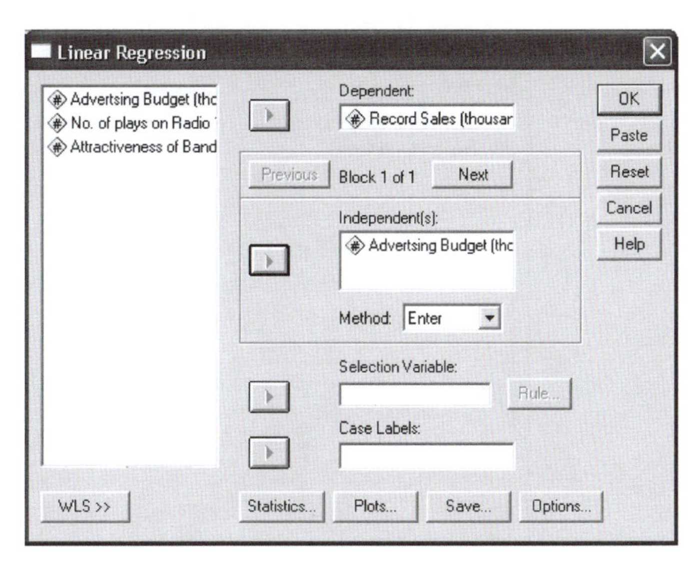

Figure 5.11 Main dialog box for block 1 of the multiple regression

side of the box. Highlight the outcome variable (record sales) in this list by clicking on it and then transfer it to the box labelled *Dependent* by clicking on ▶. We also need to specify the predictor variable for the first block. We decided that advertising budget should be entered into the model first (because past research indicates that it is an important predictor), so highlight this variable in the list and transfer it to the box labelled *Independent(s)* by clicking on ▶. Underneath the *Independent(s)* box, there is a drop-down menu for specifying the *Method* of regression (see section 5.5.3). You can select a different method of variable entry for each block by clicking on ▼, next to where it says *Method*. The default option is forced entry, and this is the option we want, but if you were carrying out more exploratory work, you might decide to use one of the stepwise methods (forward, backward, stepwise or remove).

Enter
Stepwise
Remove
Backward
Forward

Having specified the first block in the hierarchy, we need to move on to to the second. To tell the computer that you want to specify a new block of predictors you must click on Next. This process clears the *Independent(s)* box so that you can enter the new predictors (you should also note that above this box it now reads *Block 2 of 2*, indicating that you are in the second block of the two that you have so far specified). We decided that the second block would contain both of the new predictors and so you should click on **airplay** and **attract** in the variables list and transfer them, one by one, to the *Independent(s)* box by clicking on ▶. The dialog box should now look like Figure 5.12. To move between blocks use the Previous and Next buttons (so, for example, to move back to block 1, click on Previous).

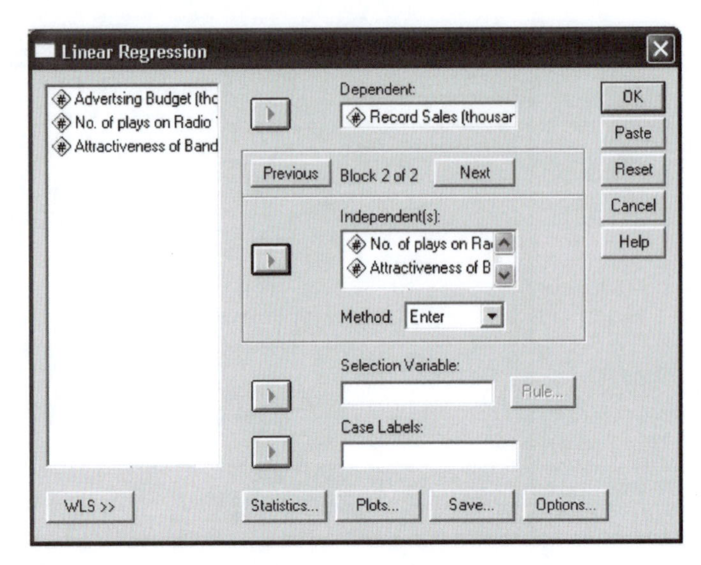

Figure 5.12 Main dialog box for block 2 of the multiple regression

It is possible to select different methods of variable entry for different blocks in a hierarchy. So, although we specified forced entry for the first block, we could now specify a stepwise method for the second. Given that we have no previous research regarding the effects of attractiveness and airplay on record sales, we might be justified in requesting a stepwise method for this block. However, because of the problems with stepwise methods, I am going to stick with forced entry for both blocks in this example.

5.7.2. Statistics ②

In the main *regression* dialog box click on Statistics... to open a dialog box for selecting various important options relating to the model (Figure 5.13). Most of these options relate to the parameters of the model; however, there are procedures available for checking the assumptions of no multicollinearity (collinearity diagnostics) and serial independence of errors (Durbin–Watson). When you have selected the statistics you require (I recommend all but the covariance matrix as a general rule) click on Continue to return to the main dialog box.

- **Estimates**: This option is selected by default because it gives us the estimated coefficients of the regression model (i.e. the estimated *b*-values). Test statistics and their significance are produced for each regression coefficient: a *t*-test is used to see whether each *b* differs significantly from zero (see section 5.2.4).[9]

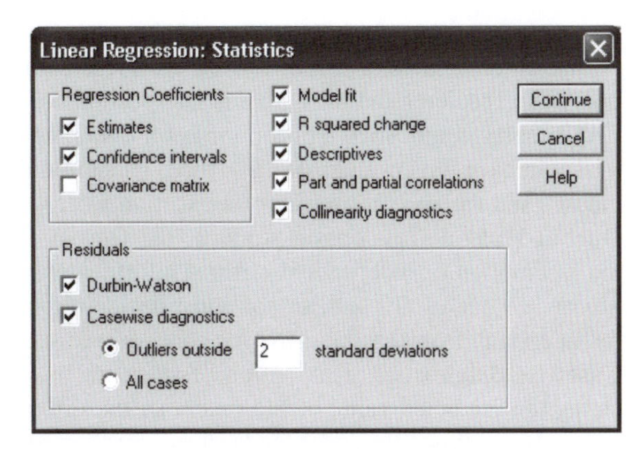

Figure 5.13 *Statistics* dialog box for regression analysis

- **Confidence intervals**: This option, if selected, produces confidence intervals for each of the unstandardized regression coefficients. Confidence intervals can be a very useful tool in assessing the likely value of the regression coefficients in the population—I shall describe their exact interpretation later.
- **Covariance matrix**: If selected, this option will display a matrix of the covariances, correlation coefficients and variances between the regression coefficients of each variable in the model. A variance–covariance matrix is produced with variances displayed along the diagonal and covariances displayed as off-diagonal elements. The correlations are produced in a separate matrix.
- **Model fit**: This option is vital and so is selected by default. It provides not only a statistical test of the model's ability to predict the outcome variable (the *F*-test—see section 5.2.3), but also the value of *R* (or multiple *R*), the corresponding R^2, and the adjusted R^2.
- **R squared change**: This option displays the change in R^2 resulting from the inclusion of a new predictor (or block of predictors). This measure is a useful way to assess the contribution of new predictors (or blocks) to explaining variance in the outcome.
- **Descriptives**: If selected, this option displays a table of the mean, standard deviation and number of observations of all of the variables included in the analysis. A correlation matrix is also displayed showing the correlation between all of the variables and the one-tailed probability for each correlation coefficient. This option is extremely useful because the correlation matrix can be used to assess whether predictors are interrelated (which can be used to establish whether there is multicollinearity).
- **Part and partial correlations**: This option produces the zero-order correlation (the Pearson correlation) between each predictor and the outcome variable. It also produces the partial correlation between each predictor and the outcome, controlling for all other predictors in the model. Finally, it produces the part correlation (or semi-partial correlation) between each predictor and the outcome. This correlation represents the

relationship between each predictor and the part of the outcome that is not explained by the other predictors in the model. As such, it measures the unique relationship between a predictor and the outcome (see section 4.6).

- **Collinearity diagnostics**: This option is for obtaining collinearity statistics such as the VIF, tolerance, eigenvalues of the scaled, uncentred cross-products matrix, condition indexes and variance proportions (see section 5.6.2.4).
- **Durbin-Watson**: This option produces the Durbin–Watson test statistic, which tests the assumption of independent errors. Unfortunately, SPSS does not provide the significance value of this test, so you must decide for yourself whether the value is different enough from 2 to be cause for concern (see section 5.6.2.1).
- **Casewise diagnostics**: This option, if selected, lists the observed value of the outcome, the predicted value of the outcome, the difference between these values (the residual) and this difference standardized. Furthermore, it will list these values either for all cases, or just for cases for which the standardized residual is greater than 3 (when the ± sign is ignored). This criterion value of 3 can be changed, and I recommend changing it to 2 for reasons that will become apparent. A summary table of residual statistics indicating the minimum, maximum, mean and standard deviation of both the values predicted by the model and the residuals (see section 5.8.6) is also produced.

5.7.3. Regression plots ②

Once you are back in the main dialog box, click on Plots... to activate the regression *plots* dialog box shown in Figure 5.14. This dialog box provides the means to specify several graphs, which can help to establish the validity of some regression assumptions. Most of these plots involve various *residual* values, which will be described in more detail in section 5.7.4.

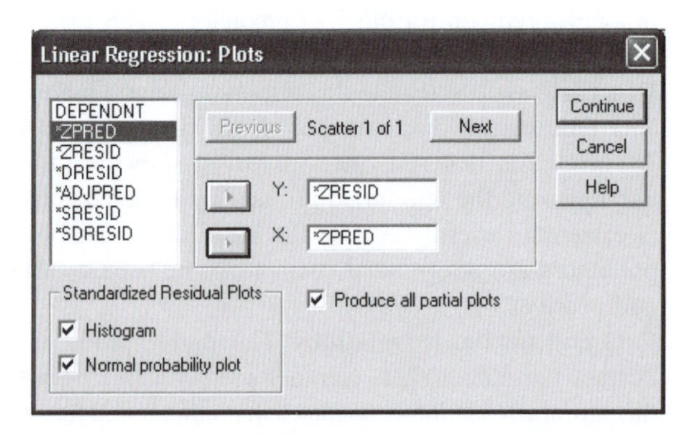

Figure 5.14 Linear regression: *plots* dialog box

On the left-hand side of the dialog box is a list of several variables.

- **DEPENDNT** (*the outcome variable*).
- ***ZPRED** (*the standardized predicted values* of the dependent variable based on the model). These values are standardized forms of the values predicted by the model.
- ***ZRESID** (*the standardized residuals*, or errors). These values are the standardized differences between the observed data and the values that the model predicts.
- ***DRESID** (*the deleted residuals*). See section 5.6.1.1 for details.
- ***ADJPRED** (the *adjusted predicted values*). See section 5.6.1.2 for details.
- ***SRESID** (*the Studentized residual*).
- ***SDRESID** (the *Studentized deleted residual*). This value is the deleted residual divided by its standard error.

The variables listed in this dialog box all come under the general heading of residuals, and were discussed in detail in section 5.6.1.1. For a basic analysis it is worth plotting *ZRESID (y-axis) against *ZPRED (x-axis), because this plot is useful to determine whether the assumptions of random errors and homoscedasticity have been met. A plot of *SRESID (y-axis) against *ZPRED (x-axis) will show up any heteroscedasticity also. Although often these two plots are virtually identical, the latter is more sensitive on a case-by-case basis. To create these plots simply select a variable from the list, and transfer it to the space labelled either X or Y (which refer to the axes) by clicking on ▸. When you have selected two variables for the first plot (as is the case in Figure 5.14) you can specify a new plot by clicking on Next. This process clears the spaces in which variables are specified. If you click on Next and would like to return to the plot that you last specified, then simply click on Previous. You can specify up to nine plots.

You can also select the tick-box labelled *Produce all partial plots* which will produce scatterplots of the residuals of the outcome variable and each of the predictors when both variables are regressed separately on the remaining predictors. Regardless of whether the previous sentence made any sense to you, these plots have several important characteristics that make them worth inspecting. First, the gradient of the regression line between the two residual variables is equivalent to the coefficient of the predictor in the regression equation. As such, any obvious outliers on a partial plot represent cases that might have undue influence on a predictor's regression coefficient. Second, non-linear relationships between a predictor and the outcome variable are much more detectable using these plots. Finally, they are a useful way of detecting collinearity. For these reasons, I recommend requesting them.

There are several options for plots of the standardized residuals. First, you can select a histogram of the standardized residuals (this is extremely useful for checking the assumption of normality of errors). Second, you can ask for a normal probability plot, which also provides information about whether the residuals in the model are normally distributed. When you have selected the options you require, click on Continue to take you back to the main *regression* dialog box.

5.7.4. Saving regression diagnostics ②

In section 5.6 we met two types of regression diagnostics: those that help us assess how well our model fits our sample and those that help us detect cases that have a large influence on the model generated. In SPSS we can choose to save these diagnostic variables in the data editor (so, SPSS will calculate them and then create new columns in the data editor in which the values are placed).

To save regression diagnostics you need to click on Save... in the main *regression* dialog box. This process activates the *save* new variables dialog box (see Figure 5.15). Once this dialog box is active, it is a simple matter to tick the boxes next to the required statistics. Most of the available options were explained in section 5.6 and Figure 5.15 shows what I consider to be a fairly basic set of diagnostic statistics. Standardized (and Studentized) versions of these diagnostics are generally easier to interpret and so I suggest selecting them in preference to the unstandardized versions. Once the regression has been run, SPSS creates a column in your data editor for each statistic requested and it has a standard set of variable names to describe each one. After the name, there will be a number that refers to the analysis that has been run. So, for the first regression run on a data set the variable names will be followed by a 1, if you carry out a second regression it will create a new set of variables with names followed by a 2 and so on. The names of the variables in the data editor are displayed below. When you have selected the diagnostics you require

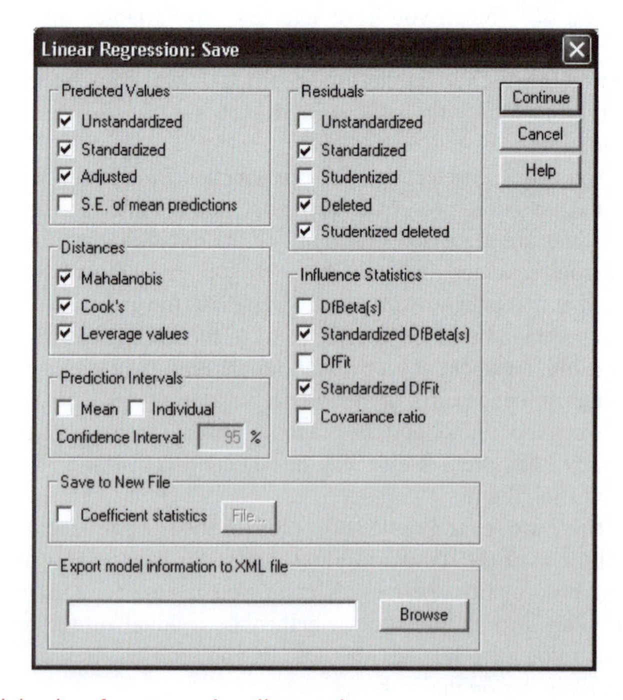

Figure 5.15 Dialog box for regression diagnostics

(by clicking in the appropriate boxes), click on Continue to return to the main *regression* dialog box.

- **pre_1**: unstandardized predicted value.
- **zpr_1**: standardized predicted value.
- **adj_1**: adjusted predicted value.
- **sep_1**: standard error of predicted value.
- **res_1**: unstandardized residual.
- **zre_1**: standardized residual.
- **sre_1**: Studentized residual.
- **dre_1**: deleted residual.
- **sdr_1**: Studentized deleted residual.
- **mah_1**: Mahalanobis distance.
- **coo_1**: Cook's distance.
- **lev_1**: centred leverage value.
- **sdb0_1**: standardized DFBETA (intercept).
- **sdb1_1**: standardized DFBETA (predictor 1).
- **sdb2_1**: standardized DFBETA (predictor 2).
- **sdf_1**: standardized DFFIT.
- **cov_1**: covariance ratio.

5.7.5. Further options ②

As a final step in the analysis, you can click on Options... to take you to the *options* dialog box (Figure 5.16). The first set of options allows you to change the criteria used for entering variables in a stepwise regression. If you insist on doing stepwise regression, then it's probably best that you leave the default criterion of .05 probability for entry alone. However, you can make this criterion more stringent (.01). There is also the option to build a model that doesn't include a constant (i.e. has no *Y*-intercept). This option should also be left alone! Finally, you can select a method for dealing with missing data points. By default, SPSS excludes cases listwise, which means that if a person has a missing value for any variable, then they are excluded from the whole analysis. So, for example, if our record company boss didn't have an attractiveness score for one of his bands, their data would not be used in the regression model. Another option is to exclude cases on a pairwise basis, which means that if a participant has a score missing for a particular variable, then their data are excluded only from calculations involving the variable for which they have no score. So, data for the band for which there was no attractiveness rating would still be used to calculate the relationships between advertising budget, airplay and record sales. However, if you do this, many of your variables may not make sense, and you can end up with absurdities such as R^2 either negative or greater than 1.0. So, it's not a good option.

Another possibility is to replace the missing score with the average score for this variable and then include that case in the analysis (so, our example band would be given an attractiveness rating equal to the average attractiveness of all bands). The problem

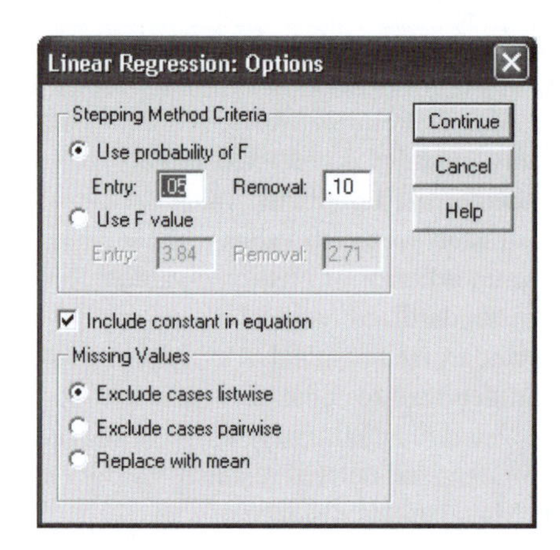

Figure 5.16 Options for linear regression

with this final choice is that it is likely to suppress the true value of the standard deviation (and more importantly the standard error). The standard deviation will be suppressed because for any replaced case there will be no difference between the mean and the score, whereas if data had been collected for that case there would, almost certainly, have been some difference between the score and the mean. Obviously, if the sample is large and the number of missing values small then this is not a serious consideration. However, if there are many missing values this choice is potentially dangerous because smaller standard errors are more likely to lead to significant results that are a product of the data replacement rather than a genuine effect. The final option is to use the 'Missing Value Analysis' routine in SPSS. This is for experts. It makes use of the fact that if two or more variables are present and correlated for most cases in the file, and an occasional value is missing, you can replace the missing values with estimates far better than the mean (Tabachnick & Fidell, 2001, Chapter 4, describe some of these procedures).

5.8. INTERPRETING MULTIPLE REGRESSION ②

A good strategy to adopt with regression is to measure predictor variables for which there are sound theoretical reasons for expecting them to predict the outcome. Run a regression analysis in which all predictors are entered into the model and examine the output to see which predictors contribute substantially to the model's ability to predict the outcome. Once you have established which variables are important, rerun the analysis including only the important predictors and use the resulting parameter estimates to define your regression model. If the initial analysis reveals that there are two or more significant predictors then you could consider running a forward stepwise analysis (rather than forced entry) to find out the individual contribution of each predictor.

I have spent a lot of time explaining the theory behind regression and some of the diagnostic tools necessary to gauge the accuracy of a regression model. It is important to remember that SPSS may appear to be very clever, but in fact it is not. Admittedly, it can do lots of complex calculations in a matter of seconds, but what it can't do is control the quality of the model that is generated—to do this requires a human brain (and preferably a trained one). SPSS will happily generate output based on any garbage you decide to feed into the data editor and SPSS will not judge the results or give any indication of whether the model can be generalized or if it is valid. However, SPSS provides the statistics necessary to judge these things, and at this point our brains must take over the job—which is slightly worrying (especially if your brain is as small as mine)!

Having selected all of the relevant options and returned to the main dialog box, we need to click on OK to run the analysis. SPSS will spew out copious amounts of output in the viewer window, and we now turn to look at how to make sense of this information.

5.8.1. Descriptives ②

The output described in this section is produced using the options in the linear regression *statistics* dialog box (see Figure 5.13). To begin with, if you selected the *Descriptives* option, SPSS will produce the table seen in SPSS Output 5.4. This table tells us the mean and standard deviation of each variable in our data set, so we now know that the average number of record sales was 193,200. This table isn't necessary for interpreting the regression model, but it is a useful summary of the data. In addition to the descriptive statistics, selecting this option produces a correlation matrix too. The table shows three things. First, it shows the value of the Pearson correlation coefficient between every pair of variables (e.g. we can see that the advertising budget had a large positive correlation with record sales, $R = .578$). Second, the one-tailed significance of each correlation is displayed (e.g. the correlation above is significant, $p < .001$). Finally, the number of cases contributing to each correlation ($N = 200$) is shown.

You might notice that along the diagonal of the matrix the values for the correlation coefficients are all 1.00 (i.e. a perfect positive correlation). The reason for this is that these values represent the correlation of each variable with itself, so obviously the resulting values are 1. The correlation matrix is extremely useful for getting a rough idea of the relationships between predictors and the outcome, and for a preliminary look for multicollinearity. If there is no multicollinearity in the data then there should be no substantial correlations ($R > .9$) between predictors.

If we look only at the predictors (ignore record sales) then the highest correlation is between the attractiveness of the band and the amount of airplay which is significant at a .01 level ($R = .182$, $p = .005$). Despite the significance of this correlation, the coefficient is small and so it looks as though our predictors are measuring different things (there is no collinearity). We can see also that of all of the predictors, the number of plays on Radio 1 correlates best with the outcome ($R = .599$, $p < .001$) and so it is likely that this variable will best predict record sales.

Cramming Samantha's Tip

Use the descriptive statistics to check the correlation matrix for multicollinearity; that is, predictors that correlate too highly with each other, $R > .9$.

Descriptive Statistics

	Mean	Std. Deviation	N
Record Sales (thousands)	193.2000	80.6990	200
Advertising Budget (thousands of pounds)	614.4123	485.6552	200
No. of plays on Radio 1 per week	27.5000	12.2696	200
Attractiveness of Band	6.7700	1.3953	200

Correlations

		Record Sales (thousands)	Advertising Budget (thousands of pounds)	No. of plays on Radio 1 per week	Attractiveness of Band
Pearson Correlation	Record Sales (thousands)	1.000	.578	.599	.326
	Advertising Budget (thousands of pounds)	.578	1.000	.102	.081
	No. of plays on Radio 1 per week	.599	.102	1.000	.182
	Attractiveness of Band	.326	.081	.182	1.000
Sig. (1-tailed)	Record Sales (thousands)	.	.000	.000	.000
	Advertising Budget (thousands of pounds)	.000	.	.076	.128
	No. of plays on Radio 1 per week	.000	.076	.	.005
	Attractiveness of Band	.000	.128	.005	.
N	Record Sales (thousands)	200	200	200	200
	Advertising Budget (thousands of pounds)	200	200	200	200
	No. of plays on Radio 1 per week	200	200	200	200
	Attractiveness of Band	200	200	200	200

SPSS Output 5.4 Descriptive statistics for regression analysis

5.8.2. Summary of model ②

The next section of output describes the overall model (so it tells us whether the model is successful in predicting record sales). Remember that we chose a hierarchical method and so each set of summary statistics is repeated for each stage in the hierarchy. In SPSS Output 5.5 you should note that there are two models. Model 1 refers to the first stage in the hierarchy when only advertising budget is used as a predictor. Model 2 refers to when all three predictors are used. SPSS Output 5.5 is the *model summary* and this table was produced using the *Model fit* option. This option is selected by default in SPSS because it provides us with some very important information about the model: the values of R, R^2 and the adjusted R^2. If the *R squared change* and *Durbin-Watson* options were selected, then these values are included also (if they weren't selected you'll find that you have a smaller table).

The model summary table is shown in SPSS Output 5.5 and you should notice that under this table SPSS tells us what the dependent variable (outcome) was and what the predictors were in each of the two models. In the column labelled R are the values of the multiple correlation coefficient between the predictors and the outcome. When only advertising budget is used as a predictor, this is the simple correlation between advertising and record sales (.578). In fact all of the statistics for model 1 are the same as the simple regression model earlier (see section 5.4). The next column gives us a value of R^2, which we already know is a measure of how much of the variability in the outcome is accounted for by the predictors. For the first model its value is .335, which means that advertising budget accounts for 33.5% of the variation in record sales. However, when the other two predictors are included as well (model 2), this value increases to .665 or 66.5% of the variance in record sales. Therefore, if advertising accounts for 33.5%, we can tell that attractiveness and radio play account for an additional 33%.[9] So, the inclusion of the two new predictors has explained quite a large amount of the variation in record sales.

Model Summary[c]

					Change Statistics					
Model	R	R Square	Adjusted R Square	Std. Error of the Estimate	R Square Change	F Change	df1	df2	Sig. F Change	Durbin–Watson
1	.578[a]	.335	.331	65.9914	.335	99.587	1	198	.000	
2	.815[b]	.665	.660	47.0873	.330	96.447	2	196	.000	1.950

a Predictors: (Constant), Advertising Budget (thousands of pounds)
b Predictors: (Constant), Advertising Budget (thousands of pounds), Attractiveness of Band, No. of plays on Radio 1 per week
c Dependent Variable: Record sales (thousands)

SPSS Output 5.5 Regression model summary

9 That is, 33% = 66.5% – 33.5% (this value is the *R Square Change* in the table).

The adjusted R^2 gives us some idea of how well our model generalizes and ideally we would like its value to be the same, or very close to, the value of R^2. In this example the difference for the final model is small (in fact the difference between the values is .665 – .660 = .005, about 0.5%). This shrinkage means that if the model were derived from the population rather than a sample it would account for approximately 0.5% less variance in the outcome. Advanced students might like to apply Stein's formula to the R^2 to get some idea of the likely value of it in different samples. Stein's formula was given in equation (5.11) and can be applied by replacing n with the sample size (200) and k with the number of predictors (3):

$$\begin{aligned}\text{adjusted } R^2 &= 1 - \left[\left(\frac{200-1}{200-3-1}\right)\left(\frac{200-2}{200-3-2}\right)\left(\frac{200+1}{200}\right)\right](1-.665)\\ &= 1 - [(1.015)(1.015)(1.005)](.335)\\ &= 1 - .347\\ &= .653\end{aligned}$$

This value is very similar to the observed value of R^2 (.665) indicating that the cross-validity of this model is very good.

The change statistics are provided only if requested and these tell us whether the change in R^2 is significant. The significance of R^2 can actually be tested using an F-ratio, and this F is calculated from the following equation (in which N is the number of cases or participants, and k is the number of predictors in the model):

$$F = \frac{(N-k-1)R^2}{k(1-R^2)}$$

In SPSS Output 5.5, the change in this F is reported for each block of the hierarchy. So, model 1 causes R^2 to change from zero to .335, and this change in the amount of variance explained gives rise to an F-ratio of 99.587, which is significant with a probability less than .001. Bearing in mind, for this first model, that we have only one predictor (so $k = 1$) and 200 cases ($N = 200$), this F comes from the equation above:[10]

$$F_{\text{Model 1}} = \frac{(200-1-1).334648}{1(1-.334648)} = 99.587$$

The addition of the new predictors (model 2) causes R^2 to increase by 0.330 (see above). We can calculate the F-ratio for this change using the same equation, but because we're

10 To get the same values as SPSS we have to use the exact value of R^2, which is .3346480676231 (if you don't believe me double-click on the table in the SPSS output that reports this value, then double-click on the cell of the table containing the value of R^2 and you'll see that .335 becomes the value that I've just typed!).

looking at the change in models we use the change in, R^2_{Change}, and the R^2 in the new model (model 2 in this case so I've called it R_2^2) and we also use the change in the number of predictors, k_{Change} (model 1 had one predictor and model 2 had three predictors, so the change in the number of predictors is 3 – 1 = 2), and the number of predictors in the new model, k_2 (in this case because we're looking at model 2). Again, if we use a few more decimal places than in the SPSS table, we get approximately the same answer as SPSS:

$$
\begin{aligned}
F_{\text{Change}} &= \frac{(N - k_2 - 1)R^2_{\text{Change}}}{k_{\text{Change}}(1 - R_2^2)} \\
&= \frac{(200 - 3 - 1) \times .330}{2(1 - .664668)} \\
&= 96.44
\end{aligned}
$$

As such, the change in the amount of variance that can be explained gives rise to an F-ratio of 96.44, which is again significant ($p < .001$). The change statistics therefore tell us about the difference made by adding new predictors to the model.

Finally, if you requested the Durbin–Watson statistic it will be found in the last column of the table in SPSS Output 5.5. This statistic informs us about whether the assumption of independent errors is tenable (see section 5.6.2.1). As a conservative rule I suggested that values less than 1 or greater than 3 should definitely raise alarm bells (although I urge you to look up precise values for the situation of interest). The closer to 2 that the value is, the better, and for these data the value is 1.950, which is so close to 2 that the assumption has almost certainly been met.

SPSS Output 5.6 shows the next part of the output, which contains an analysis of variance (ANOVA) that tests whether the model is significantly better at predicting the outcome than using the mean as a 'best guess'. Specifically, the F-ratio represents the ratio of the improvement in prediction that results from fitting the model, relative to the inaccuracy that still exists

ANOVA[c]

Model		Sum of Squares	df	Mean Square	F	Sig.
1	Regression	433687.833	1	433687.833	99.587	.000[a]
	Residual	862264.167	198	4354.870		
	Total	1295952.0	199			
2	Regression	861377.418	3	287125.806	129.498	.000[b]
	Residual	434574.582	196	2217.217		
	Total	1295952.0	199			

a Predictors: (Constant), Advertising Budget (thousands of pounds)
b Predictors: (Constant), Advertising Budget (thousands of pounds), Attractiveness of Band, No. of Plays on Radio 1 per Week
c Dependent Variable: Record Sales (thousands)

SPSS Output 5.6

in the model (see section 5.2.3). This table is again split into two sections: one for each model. We are told the value of the sum of squares for the model (this value is SS_M in section 5.2.3 and represents the improvement in prediction resulting from fitting a regression line to the data rather than using the mean as an estimate of the outcome). We are also told the residual sum of squares (this value is SS_R in section 5.2.3 and represents the total difference between the model and the observed data). We are also told the degrees of freedom (*df*) for each term. In the case of the improvement due to the model, this value is equal to the number of predictors (one for the first model and three for the second), and for SS_R it is the number of observations (200) minus the number of coefficients in the regression model. The first model has two coefficients (one for the predictor and one for the constant) whereas the second has four (one for each of the three predictors and one for the constant). Therefore, model 1 has 198 degrees of freedom whereas model 2 has 196. The average sum of squares (MS) is then calculated for each term by dividing the SS by the *df*. The *F*-ratio is calculated by dividing the average improvement in prediction by the model (MS_M) by the average difference between the model and the observed data (MS_R). If the improvement due to fitting the regression model is much greater than the inaccuracy within the model then the value of *F* will be greater than 1 and SPSS calculates the exact probability of obtaining the value of *F* by chance. For the initial model the *F*-ratio is 99.587, which is very unlikely to have happened by chance ($p < .001$). For the second model the value of *F* is even higher (129.498), which is also highly significant ($p < .001$). We can interpret these results as meaning that the initial model significantly improved our ability to predict the outcome variable, but that the new model (with the extra predictors) was even better (because the *F*-ratio is more significant).

Cramming Samantha's Tip

The fit of the regression model can be assessed using the **Model Summary** and **ANOVA** tables from SPSS. Look for the R^2 to tell you the proportion of variance explained by the model. If you have done a hierarchical regression then you can assess the improvement of the model at each stage of the analysis by looking at the change in R^2 and whether this change is significant (look for values less than .05 in the column labelled *Sig. F Change*). The ANOVA also tells us whether the model is a significant fit of the data overall (look for values less than .05 in the column labelled *Sig.*). Finally, there is an assumption that errors in regression are independent; this assumption is likely to be met if the Durbin–Watson statistic is close to 2 (and between 1 and 3).

5.8.3. Model parameters ②

So far we have looked at several summary statistics telling us whether or not the model has improved our ability to predict the outcome variable. The next part of the output is concerned with the parameters of the model. SPSS Output 5.7 shows the model parameters for

Coefficients[a]

Model		Unstandardized Coefficients		Standardized Coefficients	t	Sig.	95% Confidence Interval for B	
		B	Std. Error	Beta			Lower Bound	Upper Bound
1	(Constant)	134.140	7.537		17.799	.000	119.278	149.002
	Advertising Budget (thousands of pounds)	9.612E–02	.010	.578	9.979	.000	.077	.115
2	(Constant)	–26.613	17.350		–1.534	.127	–60.830	7.604
	Advertising Budget (thousands of pounds)	8.488E–02	.007	.511	12.261	.000	.071	.099
	No. of plays on Radio 1 per week	3.367	.278	.512	12.123	.000	2.820	3.915
	Attractiveness of Band	11.086	2.438	.192	4.548	.000	6.279	15.894

a Dependent Variable: Record Sales (thousands)

Coefficients[a]

Model		Correlations			Collinearity Statistics	
		Zero-order	Partial	Part	Tolerance	VIF
1	Advertising Budget (thousands of pounds)	.578	.578	.578	1.000	1.000
2	Advertising Budget (thousands of pounds)	.578	.659	.507	.986	1.015
	No. of plays on Radio 1 per week	.599	.655	.501	.959	1.043
	Attractiveness of Band	.326	.309	.188	.963	1.038

a Dependent Variable: Record Sales (thousands)

SPSS Output 5.7 Coefficients of the regression model[11]

both steps in the hierarchy. Now, the first step in our hierarchy was to include advertising budget (as we did for the simple regression earlier in this chapter) and so the parameters for the first model are identical to the parameters obtained in SPSS Output 5.3. Therefore, we will be concerned only with the parameters for the final model (in which all predictors were included). The format of the table of coefficients will depend on the options selected. The confidence interval for the *b*-values, collinearity diagnostics and the part and partial correlations will be present only if selected in the dialog box in Figure 5.13.

Remember that in multiple regression the model takes the form of equation (5.9) and in that equation there are several unknown quantities (the *b*-values). The first part of the table

11 To spare your eyesight I have split this part of the output into two tables; however, it should appear as one long table in the SPSS viewer.

gives us estimates for these b-values and these values indicate the individual contribution of each predictor to the model. If we replace the b-values into equation (5.9) we find that we can define the model as in equation (5.12):

$$\begin{aligned}\text{Sales}_i &= b_0 + b_1\text{Advertising}_i + b_2\text{Airplay}_i + b_3\text{Attractiveness}_i \\ &= -26.61 + (0.08\text{Advertising}_i) + (3.37\text{Airplay}_i) \\ &\quad + (11.09\text{Attractiveness}_i) \end{aligned} \tag{5.12}$$

The b-values tell us about the relationship between record sales and each predictor. If the value is positive we can tell that there is a positive relationship between the predictor and the outcome whereas a negative coefficient represents a negative relationship. For these data all three predictors have positive b-values indicating positive relationships. So, as advertising budget increases, record sales increase; as plays on the radio increase so do record sales; and finally more attractive bands will sell more records. The b-values tell us more than this though. They tell us to what degree each predictor affects the outcome *if the effects of all other predictors are held constant.*

- **Advertising budget** (b = 0.085): This value indicates that as advertising budget increases by one unit, record sales increase by .085 units. Both variables were measured in thousands; therefore, for every £1000 more spent on advertising, an extra .085 thousand records (85 records) are sold. This interpretation is true only if the effects of attractiveness of the band and airplay are held constant.
- **Airplay** (b = 3.367): This value indicates that as the number of plays on radio in the week before release increases by one, record sales increase by 3.367 units. Therefore, every additional play of a song on radio (in the week before release) is associated with an extra 3.367 thousand records (3367 records) being sold. This interpretation is true only if the effects of attractiveness of the band and advertising are held constant.
- **Attractiveness** (b = 11.086): This value indicates that a band rated one higher on the attractiveness scale can expect additional record sales of 11.086 units. Therefore, every unit increase in the attractiveness of the band is associated with an extra 11.086 thousand records (11,086 records) being sold. This interpretation is true only if the effects of radio airplay and advertising are held constant.

Each of these beta values has an associated standard error indicating to what extent these values would vary across different samples, and these standard errors are used to determine whether or not the b-value differs significantly from zero. As we saw in section 5.2.4, a t-statistic can be derived that tests whether a b-value is significantly different from zero. In simple regression, a significant value of t indicates that the slope of the regression line is significantly different from horizontal, but in multiple regression, it is not so easy to visualize what the value tells us. Well, it is easiest to conceptualize the t-tests as measures of whether the predictor is making a significant contribution to the model. Therefore, if the t-test associated with a b-value is significant (if the value in the column labelled *Sig.* is less than .05) then the predictor is making a significant contribution to the model. The smaller the value of *Sig.* (and the larger the value of t) the greater the contribution of that predictor. For this model, the advertising budget ($t(196) = 12.26$, $p < .001$), the amount of radio play prior to release ($t(196) = 12.12$, $p < .001$) and attractiveness of the band ($t(196) = 4.55$, $p < .001$) are all significant predictors of record sales.[12] From the magnitude of the t-statistics we can see that the advertising budget and radio play had a similar impact whereas the attractiveness of the band had less impact.

The b-values and their significance are important statistics to look at; however, the standardized versions of the b-values are in many ways easier to interpret (because they are not dependent on the units of measurement of the variables). The standardized beta values are provided by SPSS (labelled as Beta, β_i) and they tell us the number of standard deviations that the outcome will change as a result of one standard deviation change in the predictor. The standardized beta values are all measured in standard deviation units and so are directly comparable; therefore, they provide a better insight into the 'importance' of a predictor in the model. The standardized beta values for airplay and advertising budget are virtually identical (.512 and .511 respectively) indicating that both variables have a comparable degree of importance in the model (this concurs with what the magnitude of the t-statistics told us)[13]. To interpret these values literally, we need to know the standard deviations of all of the variables and these values can be found in SPSS Output 5.4.

- **Advertising budget** (*standardized* $\beta = .511$): This value indicates that as advertising budget increases by one standard deviation (£485,655), record sales increase by .511 standard deviations. The standard deviation for record sales is 80,699 and so this constitutes a change of 41,240 sales ($.511 \times 80{,}699$). Therefore, for every £485,655 more spent on advertising, an extra 41,240 records are sold. This interpretation is true only if the effects of attractiveness of the band and airplay are held constant.
- **Airplay** (*standardized* $\beta = .512$): This value indicates that as the number of plays on radio in the week before release increases by one standard deviation (12.27), record sales increase by .512 standard deviations. The standard deviation for record sales is 80,699 and so this constitutes a change of 41,320 sales ($.512 \times 80{,}699$). Therefore, if Radio 1 plays the song an extra 12.27 times in the week before release, 41,320 extra record sales can be expected. This interpretation is true only if the effects of attractiveness of the band and advertising are held constant.

12 For all of these predictors I wrote $t(196)$. The number in brackets is the degrees of freedom. We saw in section 5.2.4 that in regression the degrees of freedom are $N - p - 1$, where N is the total sample size (in this case 200) and p is the number of predictors (in this case three). For these data we get $200 - 3 - 1 = 196$.

13 The reason why the relative magnitudes of the t-statistics are comparable to the standardized beta values is because they are derived by dividing by the standard error and so they represent a *Studentized* version of the beta values. The standardized beta values are calculated by dividing by the standard deviation. In Chapter 1 we saw that the standard deviation and standard error are closely related: therefore, standardized and Studentized statistics are, in some sense, comparable.

- **Attractiveness** (*standardized* β = .192): This value indicates that a band rated one standard deviation (1.40 units) higher on the attractiveness scale can expect additional record sales of .192 standard deviations units. This constitutes a change of 15,490 sales (.192 × 80,699). Therefore, a band with an attractiveness rating 1.40 higher than another band can expect 15,490 additional sales. This interpretation is true only if the effects of radio airplay and advertising are held constant.

Imagine that we collected 100 samples of data measuring the same variables as our current model. For each sample we could create a regression model to represent the data. If the model is reliable then we hope to find very similar parameters in both. Therefore, each sample should produce approximately the same *b*-values. The confidence interval of the unstandardized beta values are boundaries constructed such that in 95% of these samples these boundaries will contain the true value of *b* (see section 1.6.2). Therefore, if we'd collected 100 samples, and calculated the confidence intervals for *b*, we are saying that 95% of these confidence intervals would contain the true value of *b*. Therefore, we can be fairly confident that the confidence interval we have constructed for this sample will contain the true value of *b* in the population. This being so, a good model will have small confidence intervals, indicating that the value of *b* in this sample is close to the true value of *b* in the population. The sign (positive or negative) of the *b*-values tells us about the direction of the relationship between the predictor and the outcome. Therefore, we would expect a very bad model to have confidence intervals that cross zero, indicating that in some samples the predictor has a negative relationship to the outcome whereas in others it has a positive relationship. In this model, the two best predictors (advertising and airplay) have very tight confidence intervals indicating that the estimates for the current model are likely to be representative of the true population values. The interval for attractiveness is wider (but still does not cross zero) indicating that the parameter for this variable is less representative, but nevertheless significant.

If you asked for part and partial correlations, then they will appear in the output in separate columns of the table. The zero-order correlations are the simple Pearson correlation coefficients (and so correspond to the values in SPSS Output 5.4). The partial correlations represent the relationships between each predictor and the outcome variable, controlling for the effects of the other two predictors. The part correlations represent the relationship between each predictor and the outcome, controlling for the effect that the other two variables have on the outcome. In effect, these part correlations represent the unique relationship that each predictor has with the outcome. If you opt to do a stepwise regression, you would find that variable entry is based initially on the variable with the largest zero-order correlation and then on the part correlations of the remaining variables. Therefore, airplay would be entered first (because it has the largest zero-order correlation), then advertising budget (because its part correlation is bigger than attractiveness) and then finally attractiveness. Try running a forward stepwise regression on these data to see if I'm right! Finally, we are given details of the collinearity statistics, but these will be discussed in section 5.8.5.

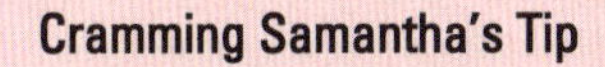

Cramming Samantha's Tip

The individual contribution of variables to the regression model can be found in the **Coefficients** table from SPSS. If you have done a hierarchical regression then look at the values for the final model. For each predictor variable, you can see if it has made a significant contribution to predicting the outcome by looking at the column labelled *Sig.* (values less than .05 are significant). You should also look at the standardized beta values because these tell you the importance of each predictor (bigger absolute value = more important). The tolerance and VIF values will also come in handy later on so make a note of them!

5.8.4. Excluded variables ②

At each stage of a regression analysis SPSS provides a summary of any variables that have not yet been entered into the model. In a hierarchical model, this summary has details of the variables that have been specified to be entered in subsequent steps, and in stepwise regression this table (SPSS Output 5.8) contains summaries of the variables that SPSS is considering entering into the model. For this example, there is a summary of the excluded variables for the first stage of the hierarchy (there is no summary for the second stage because all predictors are in the model). The summary gives an estimate of each predictor's beta value if it was entered into the equation at this point and calculates a *t*-test for this value. In a stepwise regression, SPSS should enter the predictor with the highest *t*-statistic and will continue entering predictors until there are none left with *t*-statistics that have significance values less than .05. The partial correlation also provides some indication as to what contribution (if any) an excluded predictor would make if it were entered into the model.

Excluded Variables[b]

Model		Beta In	t	Sig.	Partial Correlation	Collinearity Statistics		
						Tolerance	VIF	Minimum Tolerance
1	No. of plays on Radio 1 per week	.546[a]	12.513	.000	.665	.990	1.010	.990
	Attractiveness of Band	.281[a]	5.136	.000	.344	.993	1.007	.993

a Predictors in the Model: (Constant), Advertising Budget (thousands of pounds)
b Dependent Variable: Record Sales (thousands)

SPSS Output 5.8

5.8.5. Assessing the assumption of no multicollinearity ②

SPSS Output 5.7 provided some measures of whether there is collinearity in the data. Specifically, it provides the VIF and tolerance statistics (with tolerance being 1 divided by the VIF). There are a few guidelines from section 5.6.2.3 that can be applied here:

- If the largest VIF is greater than 10 then there is cause for concern (Myers, 1990; Bowerman & O'Connell, 1990).
- If the average VIF is substantially greater than 1 then the regression may be biased (Bowerman & O'Connell, 1990).
- Tolerance below .1 indicates a serious problem.
- Tolerance below .2 indicates a potential problem (Menard, 1995).

For our current model the VIF values are all well below 10 and the tolerance statistics all well above .2; therefore, we can safely conclude that there is no collinearity within our data. To calculate the average VIF we simply add the VIF values for each predictor and divide by the number of predictors (k):

$$\overline{\text{VIF}} = \frac{\sum_{i=1}^{k} \text{VIF}_i}{k} = \frac{1.015 + 1.043 + 1.038}{3} = 1.032$$

The average VIF is very close to 1 and this confirms that collinearity is not a problem for this model. SPSS also produces a table of eigenvalues of the scaled, uncentred cross-products matrix, condition indexes and variance proportions. There is a lengthy discussion, and an example, of collinearity in section 6.8 and how to detect it using variance proportions and so I will limit myself now to saying that we are looking for large variance proportions on the same *small* eigenvalues. Therefore, in SPSS Output 5.9 we look at the bottom few rows of the table (these are the small eigenvalues) and look for any variables that both have high-variance proportions for that eigenvalue. The variance proportions vary

Collinearity Diagnostics[a]

Model	Dimension	Eigenvalue	Condition Index	Variance Proportions			
				(Constant)	Advertising Budget (thousands of pounds)	No. of plays on Radio 1 per week	Attractiveness of Band
1	1	1.785	1.000	.11	.11		
	2	.215	2.883	.89	.89		
2	1	3.562	1.000	.00	.02	.01	.00
	2	.308	3.401	.01	**.96**	.05	.01
	3	.109	5.704	.05	.02	**.93**	.07
	4	2.039E–02	13.219	.94	.00	.00	**.92**

a Dependent Variable: Record Sales (thousands)

SPSS Output 5.9

between 0 and 1, and for each predictor should be distributed across different dimensions (or eigenvalues). For this model, you can see that each predictor has most of its variance loading onto a different dimension (advertising has 96% of variance on dimension 2, airplay has 93% of variance on dimension 3 and attractiveness has 92% of variance on dimension 4). These data represent a classic example of no multicollinearity. For an example of when collinearity exists in the data and some suggestions about what can be done, see Chapters 5 and 11 (section 6.8).

Cramming Samantha's Tip

To check the assumption of no multicollinearity, use the VIF values from the table labelled **Coefficients** in the SPSS output. If these values are less than 10 then that indicates that there probably isn't cause for concern. If you take the average of VIF values, and this average is not substantially greater than 1, then that also indicates that there's no cause for concern.

Box 5.2

What are eigenvectors and eigenvalues? ④

The definitions and mathematics of eigenvalues and eigenvectors are very complicated and most of us need not worry about them (although they do crop up again in Chapters 14 and 15). However, although the mathematics of them is hard, they are quite easy to visualize! Imagine we have two variables: the salary a supermodel earns in a year, and how attractive they are. Also imagine these two variables are normally distributed and so can be considered together as a bivariate normal distribution. If these variables are correlated, then their scatterplot forms an ellipse. This is shown in the scatterplots below: if we draw a dashed line around the outer values of the scatterplot we get something oval shaped. Now, we can draw two lines to measure the length and height of this ellipse. These lines are the *eigenvectors* of the original correlation matrix for these two variables (a vector is just a set of numbers that tells us the location of a line in geometric space). Note that the two lines we've drawn (one for height and one for width of the oval) are perpendicular; that is, they are at 90 degrees, which means that they are independent from one another. So, with two variables, eigenvectors are just lines measuring the length and height of the ellipse that surrounds the scatterplot of data for those variables. If we add a third variable (e.g. experience of the supermodel) then all that happens is our scatterplot gets a third dimension, the ellipse turns into something shaped like a rugby ball (or American football), and because we now have a third dimension (height, width and depth) we get an extra eigenvector to measure this extra dimension.

(Continued)

Box 5.2 (Continued)

If we add a fourth variable, a similar logic applies (although it's harder to visualize): we get an extra dimension, and an eigenvector to measure that dimension. Now, each eigenvector has an *eigenvalue* that tells us its length (i.e. the distance from one end of the eigenvector to the other). So, by looking at all of the eigenvalues for a data set, we know the dimensions of the ellipse or rugby ball: put more generally, we know the dimensions of the data. Therefore, the eigenvalues show how evenly (or otherwise) the variances of the matrix are distributed.

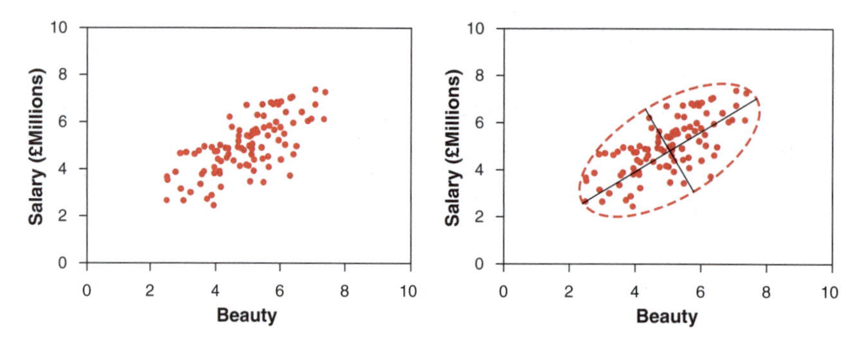

In the case of two variables, the *condition* of the data is related to the ratio of the larger eigenvalue to the smaller. Let's look at the two extremes: when there is no relationship at all between variables, and when there is a perfect relationship. When there is no relationship, the scatterplot will, more or less, be contained within a circle (or a sphere if we had three variables). If we again draw lines that measure the height and width of this circle we'll find that these lines are the same length. The eigenvalues measure the length; therefore, the eigenvalues will also be the same. So, when we divide the largest eigenvalue by the smallest we'll get a value of 1 (because the eigenvalues are the same). When the variables are perfectly correlated (i.e. there is perfect collinearity) then the scatterplot forms a straight line and the ellipse surrounding it will also collapse to a straight line. Therefore, the height of the ellipse will be very small indeed (it will approach zero). Consequently, when we divide the largest eigenvalue by the smallest we'll get a value that tends to infinity (because the smallest eigenvalue is close to zero). Therefore, an infinite condition index is a sign of deep trouble.

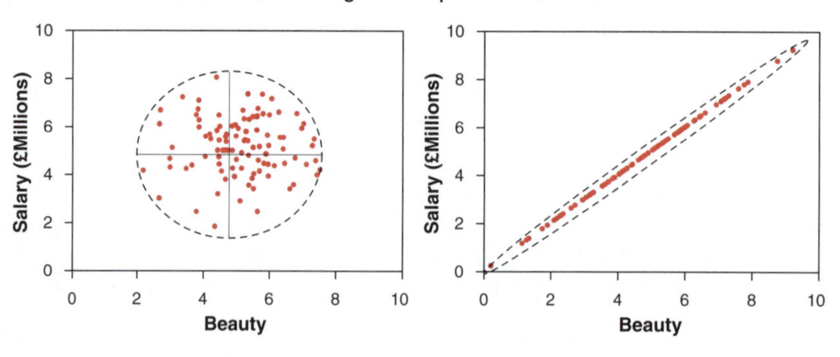

5.8.6. Casewise diagnostics ②

SPSS produces a summary table of the residual statistics and these should be examined for extreme cases. SPSS Output 5.10 shows any cases that have a standardized residual less than –2 or greater than 2 (remember that we changed the default criterion from 3 to 2 in Figure 5.13). I mentioned in section 5.6.1.1 that in an ordinary sample we would expect 95% of cases to have standardized residuals within about ±2. We have a sample of 200, so it is reasonable to expect about 10 cases (5%) to have standardized residuals outside of these limits. From SPSS Output 5.10 we can see that we have 12 cases (6%) that are outside of the limits: therefore, our sample is within 1% of what we would expect. In addition, 99% of cases should lie within ±2.5 and so we would expect only 1% of cases to lie outside of these limits. From the cases listed here, it is clear that two cases (1%) lie outside of the limits (cases 164 and 169). Therefore, our sample appears to conform to what we would expect for a fairly accurate model. These diagnostics give us no real cause for concern except that case 169 has a standardized residual greater than 3, which is probably large enough for us to investigate this case further.

You may remember that in section 5.7.4 we asked SPSS to save various diagnostic statistics. You should find that the data editor now contains columns for these variables. It is perfectly acceptable to check these values in the data editor, but you can also get SPSS to list the values in your viewer window too. To list variables you need to use the *Case Summaries* command, which can be found by using the **Analyze⇒Reports⇒Case Summaries...** menu path.[15] Figure 5.17 shows the dialog box for this function. Simply

Casewise Diagnostics[a]

Case Number	Std. Residual	Record Sales (thousands)	Predicted Value	Residual
1	2.125	330.00	229.9203	100.0797
2	–2.314	120.00	228.9490	–108.9490
10	2.114	300.00	200.4662	99.5338
47	–2.442	40.00	154.9698	–114.9698
52	2.069	190.00	92.5973	97.4027
55	–2.424	190.00	304.1231	–114.1231
61	2.098	300.00	201.1897	98.8103
68	–2.345	70.00	180.4156	–110.4156
100	2.066	250.00	152.7133	97.2867
164	–2.577	120.00	241.3240	–121.3240
169	3.061	360.00	215.8675	144.1325
200	–2.064	110.00	207.2061	-97.2061

a Dependent Variable: Record Sales (thousands)

SPSS Output 5.10

15 **Statistics⇒Summarize⇒Case Summaries...** in version 8.0 and earlier.

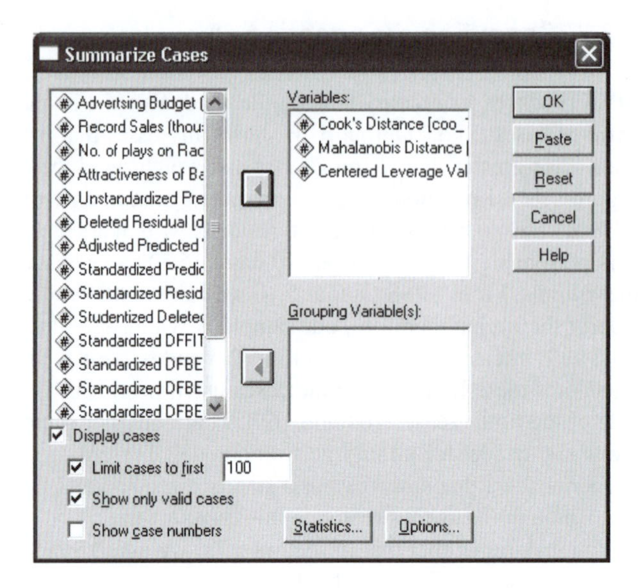

Figure 5.17

select the variables that you want to list and transfer them to the box labelled *Variables* by clicking on ▶. By default, SPSS will limit the output to the first 100 cases, but if you want to list all of your cases then simply deselect this option. It is also very important to select the *Show case numbers* option because otherwise you might not be able to identify a problem case.

One useful strategy is to use the casewise diagnostics to identify cases that you want to investigate further. So, to save space, I created a coding variable (1 = include, 0 = exclude) so that I could specify the 12 cases listed in SPSS Output 5.11 in one group, and all other cases in the other. By using this coding variable and specifying it as a grouping variable in the *summarize cases* dialog box, I could look at those 12 cases together and discard all others.

SPSS Output 5.11 shows the influence statistics for the 12 cases that I selected. None of them have a Cook's distance greater than 1 (even case 169 is well below this criterion) and so none of the cases is having an undue influence on the model. The average leverage can be calculated as .02 ($k + 1/n = 4/200$) and so we are looking for values either twice as large as this (.04) or three times as large (.06) depending on which statistician you trust most (see section 5.6.1.2)! All cases are within the boundary of three times the average and only case 1 is close to two times the average. Finally, from our guidelines for the Mahalanobis distance we saw that with a sample of 100 and three predictors, values greater than 15 were problematic. We have three predictors and a larger sample size, so this value will be a conservative cut-off, yet none of our cases come close to exceeding this criterion. The evidence suggests that there are no influential cases within our data (although all cases would need to be examined to confirm this fact).

Case Summaries

	Case Number	Standardized DFBETA Intercept	Standardized DFBETA ADVERTS	Standardized DFBETA AIRPLAY	Standardized DFBETA ATTRACT	Standardized DFFIT	COVRATIO
1	1	−.31554	−.24235	.15774	.35329	.48929	.97127
2	2	.01259	−.12637	.00942	−.01868	−.21110	.92018
3	10	−.01256	−.15612	.16772	.00672	.26896	.94392
4	47	.06645	.19602	.04829	−.17857	−.31469	.91458
5	52	.35291	−.02881	−.13667	−.26965	.36742	.95995
6	55	.17427	−.32649	−.02307	−.12435	−.40736	.92486
7	61	.00082	−.01539	.02793	.02054	.15562	.93654
8	68	−.00281	.21146	−.14766	−.01760	−.30216	.92370
9	100	.06113	.14523	−.29984	.06766	.35732	.95888
10	164	.17983	.28988	−.40088	−.11706	−.54029	.92037
11	169	−.16819	−.25765	.25739	.16968	.46132	.85325
12	200	.16633	−.04639	.14213	−.25907	−.31985	.95435
Total N		12	12	12	12	12	12

Case Summaries

	Case Number	Cook's Distance	Mahalanobis Distance	Centered Leverage Value
1	1	.05870	8.39591	.04219
2	2	.01089	.59830	.00301
3	10	.01776	2.07154	.01041
4	47	.02412	2.12475	.01068
5	52	.03316	4.81841	.02421
6	55	.04042	4.19960	.02110
7	61	.00595	.06880	.00035
8	68	.02229	2.13106	.01071
9	100	.03136	4.53310	.02278
10	164	.07077	6.83538	.03435
11	169	.05087	3.14841	.01582
12	200	.02513	3.49043	.01754
Total N		12	12	12

SPSS Output 5.11

We can look also at the DFBeta statistics to see whether any case would have a large influence on the regression parameters. An absolute value greater than 1 is a problem and in all cases the values lie within ±1, which shows that these cases have no undue influence over the regression parameters. There is also a column for the covariance ratio. We saw in section 5.6.1.2 that we need to use the following criteria:

- $CVR_i > 1 + [3(k + 1)/n] = 1 + [3(3 + 1)/200] = 1.06$.
- $CVR_i < 1 - [3(k + 1)/n] = 1 - [3(3 + 1)/200] = 0.94$.

Therefore, we are looking for any cases that deviate substantially from these boundaries. Most of our 12 potential outliers have CVR values within or just outside these boundaries. The only case that causes concern is case 169 (again!) whose CVR is some way below the bottom limit. However, given Cook's distance for this case, there is probably little cause for alarm.

You would have requested other diagnostic statistics and from what you know from the earlier discussion of them you are well advised to glance over them in case of any unusual cases in the data. However, from this minimal set of diagnostics we appear to have a fairly reliable model that has not been unduly influenced by any subset of cases.

Cramming Samantha's Tips

You need to look for cases that might be influencing the regression model.

- Look at standardized residuals and check that no more than 5% of cases have absolute values above 2, and that no more than about 1% have absolute values above 2.5. Any case with a value above about 3 could be an outlier.
- Look in the data editor for the values of Cook's distance: any value above 1 indicates a case that might be influencing the model.
- Calculate the average leverage (the number of predictors plus 1, divided by the sample size) and then look for values greater than twice or three times this average value.
- For Mahalanobis distance, a crude check is to look for values above 25 in large samples (500) and values above 15 in smaller samples (100). However, Barnett & Lewis (1978) should be consulted for more detailed analysis.
- Look for absolute values of DFBeta greater than 1.
- Calculate the upper and lower limit of acceptable values for the covariance ratio, CVR. The upper limit is 1 plus three times the average leverage, whereas the lower limit is 1 minus three times the average leverage. Cases that have a CVR that falls outside of these limits may be problematic.

5.8.7. Checking assumptions ②

As a final stage in the analysis, you should check the assumptions of the model. We have already looked for collinearity within the data and used Durbin–Watson to check whether the residuals in the model are independent. In section 5.7.3 we asked for a plot of *ZRESID against *ZPRED and for a histogram and normal probability plot of the residuals. The graph of *ZRESID and *ZPRED should look like a random array of dots evenly dispersed around zero. If this graph funnels out, then the chances are that there is heteroscedasticity in the data. If there is any sort of curve in this graph then the chances are that the data have broken the assumption of linearity. Figure 5.18 shows several examples of the plot of standardized residuals against standardized predicted values. Panel (a) shows the graph for the data in our record sales example. Note how the points are randomly and evenly dispersed throughout the plot. This pattern is indicative of a situation in which the assumptions of linearity and homoscedasticity have been met. Panel (b) shows a similar plot for a data set that violates the assumption of homoscedasticity. Note that the points form the shape of a funnel so they become more spread out across the graph. This funnel shape is typical of heteroscedasticity and indicates increasing variance across the residuals. Panel (c) shows a plot of some data in which there is a non-linear relationship between the outcome and the predictor. This pattern is shown up by the residuals. A line illustrating the curvilinear relationship has been drawn over the top of the graph to illustrate the trend in the data. Finally, panel (d) represents a situation in which the data not only represent a non-linear relationship but also show heteroscedasticity. Note first the curved trend in the data, and then also note that at one end of the plot the points are very close together, whereas at the other end they are widely dispersed. When these assumptions have been

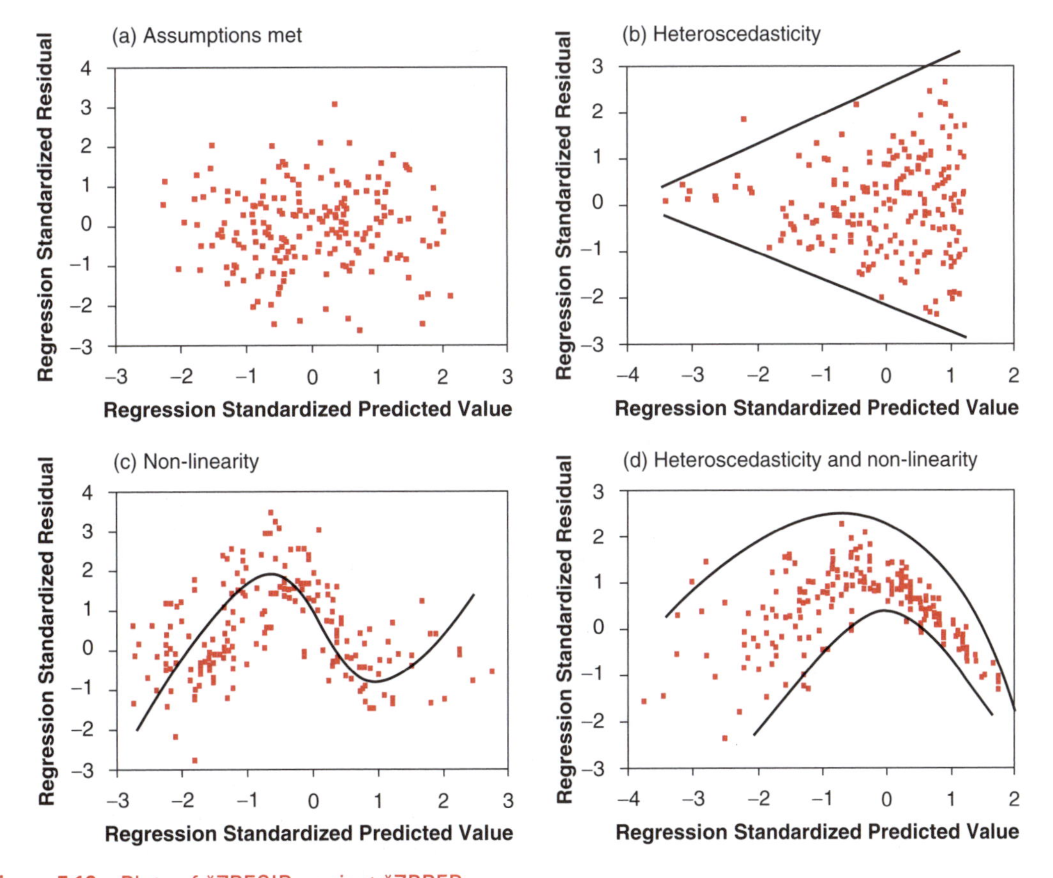

Figure 5.18 Plots of *ZRESID against *ZPRED

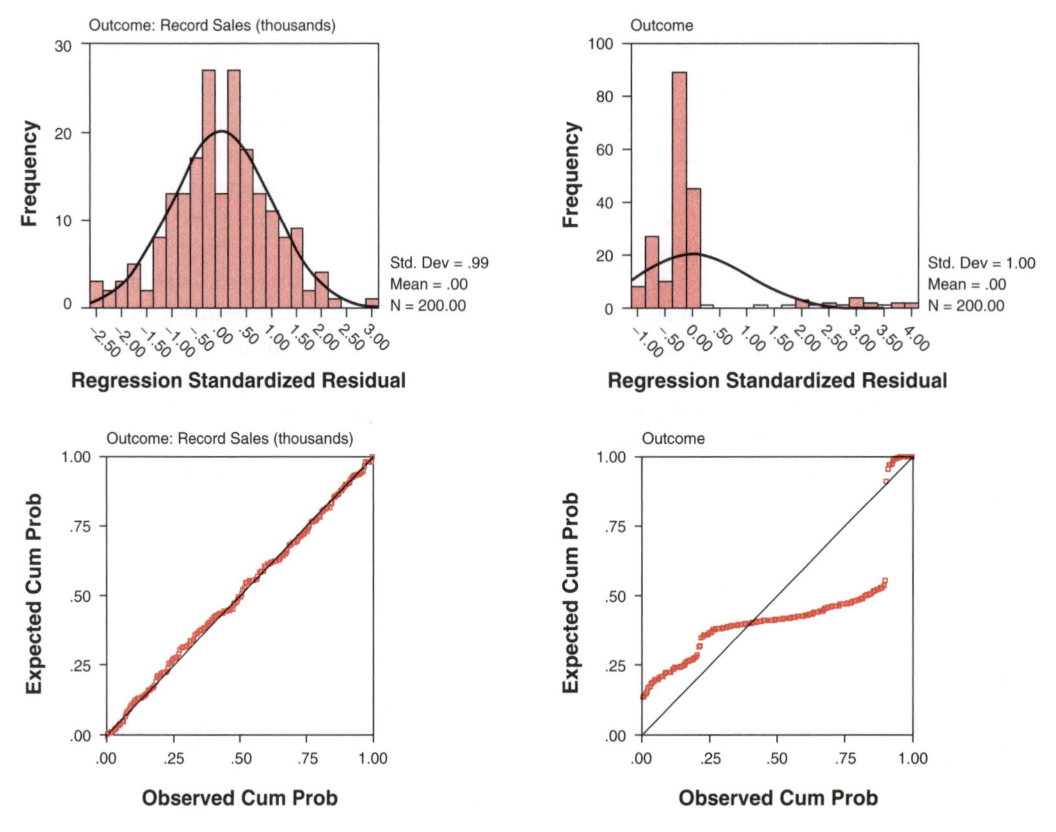

Figure 5.19 Histograms and normal P–P plots of normally distributed residuals (left-hand side) and non-normally distributed residuals (right-hand side)

broken you will not see these exact patterns, but hopefully these plots will help you to understand the types of anomalies you should look out for.

To test the normality of residuals, we must look at the histogram and normal probability plot selected in Figure 5.14. Figure 5.19 shows the histogram and normal probability plot of the data for the current example (left-hand side). The histogram should look like a normal distribution (a bell-shaped curve). SPSS draws a curve on the histogram to show the shape of the distribution. For the record company data, the distribution is roughly normal (although there is a slight deficiency of residuals exactly on zero). Compare this histogram with the extremely non-normal histogram next to it and it should be clear that the non-normal distribution is extremely skewed (unsymmetrical). So, you should look for a curve that has the same shape as the one for the record sales data: any deviation from this curve is a sign of non-normality and the greater the deviation the more non-normally distributed the residuals. The normal probability plot also shows up deviations from normality (see Chapter 2). The straight line in this plot represents a normal distribution, and the points represent the observed residuals.

Therefore, in a perfectly normally distributed data set, all points will lie on the line. This is pretty much what we see for the record sales data. However, next to the normal probability plot of the record sales data is an example of an extreme deviation from normality. In this plot, the dots are very distant from the line, which indicates a large deviation from normality. For both plots, the non-normal data are extreme cases and you should be aware that the deviations from normality are likely to be subtler. Of course, you can use what you learnt in Chapter 2 to do a K–S test on the standardized residuals to see whether they deviate significantly from normality.

A final set of plots specified in Figure 5.14 was the partial plots. These plots are scatterplots of the residuals of the outcome variable and each of the predictors when both variables are regressed separately on the remaining predictors. I mentioned earlier that obvious outliers on a partial plot represent cases that might have undue influence on a predictor's regression coefficient and that non-linear relationships and heteroscedasticity can be detected using these plots as well.

For advertising budget the partial plot shows the strong positive relationship to record sales. The gradient of the line is b for advertising in the model (this line does not appear

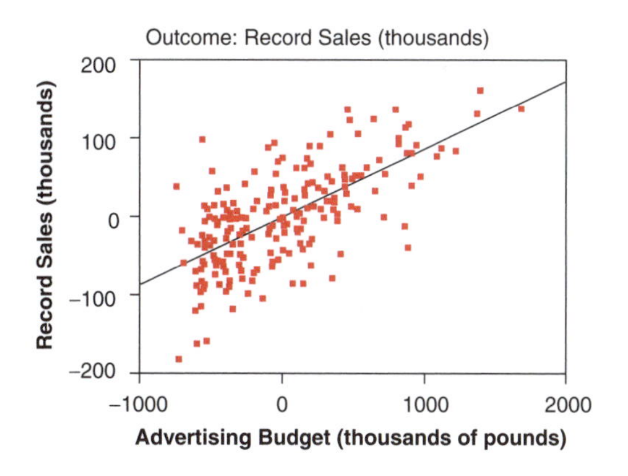

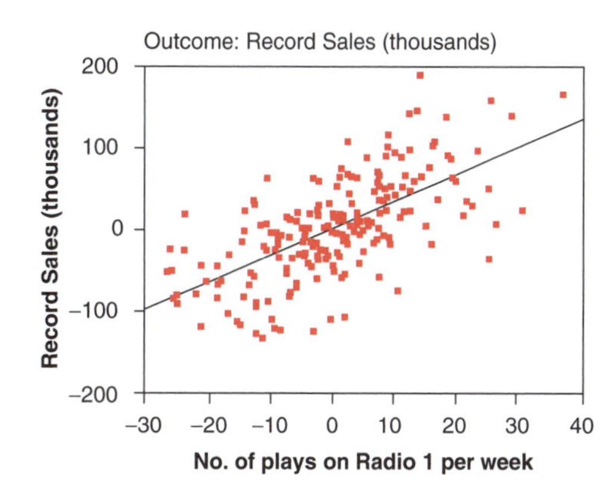

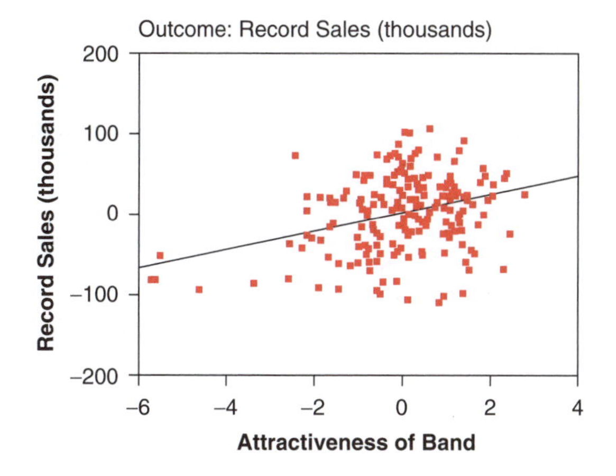

by default). There are no obvious outliers on this plot, and the cloud of dots is evenly spaced out around the line, indicating homoscedasticity.

For airplay the partial plot shows a strong positive relationship to record sales. The gradient and pattern of the data are similar to advertising (which would be expected given the similarity of the standardized betas of these predictors). There are no obvious outliers on this plot, and the cloud of dots is evenly spaced around the line, indicating homoscedasticity.

For attractiveness the partial plot shows a positive relationship to record sales. The relationship looks less linear than the other predictors do and the dots seem to funnel out indicating greater variance at high levels of attractiveness. There are no obvious outliers on this plot, but the funnel-shaped cloud of dots might indicate a violation of the assumption of homoscedasticity. We would be well advised to collect some more data for unattractive bands to verify the current model.

We could summarize by saying that the model appears, in most senses, to be both accurate for the sample and generalizable to the population. The only slight glitch is some concern over whether attractiveness ratings had violated the assumption of homoscedasticity. Therefore, we could conclude that in our sample advertising budget and airplay are fairly equally important in predicting record sales. Attractiveness of the band is a significant predictor of record sales but is less important than the other two predictors (and probably needs verification because of possible heteroscedasticity). The assumptions seem to have been met and so we can probably assume that this model would generalize to any record being released.

Cramming Samantha's Tips

You need to check some of the assumptions of regression to make sure your model generalizes beyond your sample.

- Look at the graph of ZRESID* plotted against ZPRED*. If it looks like a random array of dots then this is good. If the dots seem to get more or less spread out over the graph (look like a funnel) then this is probably a violation of the assumption of homogeneity of variance. If the dots have a pattern to them (i.e. a curved shape) then this is probably a violation of the assumption of linearity. If the dots seem to have a pattern and are more spread out at some points on the plot than others then this probably reflects violations of both homogeneity of variance *and* linearity. Any of these scenarios puts the validity of your model into question. Repeat the above for all partial plots too.
- Look at histograms and P–P plots. If the histograms look like normal distributions (and the P–P plot looks like a diagonal line), then all is well. If the histogram looks non-normal and the P–P plot looks like a wiggly snake curving around a diagonal line then things are less good! Be warned, though: distributions can look very non-normal in small samples even when they are!

5.9. HOW TO REPORT MULTIPLE REGRESSION ②

If you follow the American Psychological Association guidelines for reporting multiple regression then the implication seems to be that tabulated results are the way forward. They also seem in favour of reporting, as a bare minimum, the standardized betas, their significance value and some general statistics about the model (such as the R^2). If you do decide to do a table then the beta values and their standard errors are also very useful and personally I'd like to see the constant as well because then readers of your work can construct the full regression model if they need to. Also, if you've done a hierarchical regression you should report these values at each stage of the hierarchy. So, basically, you want to reproduce the table labelled *coefficients* from the SPSS output and omit some of the non-essential information. For the example in this chapter we might produce a table like that in Table 5.2.

Table 5.2 How to report multiple regression

	B	*SE B*	β
Step 1			
Constant	134.14	7.54	
Advertising Budget	0.10	0.01	.58*
Step 2			
Constant	–26.61	17.35	
Advertising Budget	0.09	0.01	.51*
Plays on BBC Radio 1	3.37	0.28	.51*
Attractiveness	11.09	2.44	.19*

Note $R^2 = .34$ for Step 1; $\Delta R^2 = .33$ for Step 2 ($ps < .001$). * $p < .001$.

See if you can look back through the SPSS output in this chapter and work out from where the values came. Things to note are: (1) I've rounded off to 2 decimal places throughout; (2) for the standardized betas there is no zero before the decimal point (because these values cannot exceed 1) but for all other values less than one the zero is present; (3) the significance of the variable is denoted by an asterisk with a footnote to indicate the significance level being used (if there's more than one level of significance being used you can denote this with multiple asterisks, such as *$p < .05$, **$p < .01$, and ***$p < .001$); and (4) the R^2 for the initial model and the change in R^2 (denoted as ΔR^2) for each subsequent step of the model are reported below the table.

5.10. CATEGORICAL PREDICTORS AND MULTIPLE REGRESSION ③

Often in regression analysis you'll collect data about groups of people (some examples from the social sciences might be ethnic group, gender, socio-economic status, diagnostic category). You might want to include these groups as predictors in the regression model; however, we saw from our assumptions that variables need to be continuous or categorical with only two categories. We saw in section 4.5.6 that a point–biserial correlation is Pearson's *r* between two variables when one is continuous and the other has two categories coded as 0 and 1. We've also learnt that simple regression is based on Pearson's *r*, so it shouldn't take a great deal of imagination to see that, like the point–biserial correlation, we could construct a regression model with a predictor that has two categories (e.g. gender). Likewise, it shouldn't be too inconceivable that we could then extend this model to incorporate several predictors that had two categories. All that is important is that we code the two categories with the values of 0 and 1. Why is it important that there are only two categories and that they're coded 0 and 1? Actually, I don't want to get into this here because this chapter is already too long, the publishers are going to break my legs if it gets any longer, and I explain it anyway later in the book (sections 7.8 and 8.2.2). So, for the time being, just trust me!

5.10.1. Dummy coding ③

The obvious problem with wanting to use categorical variables as predictors is that often you'll have more than two categories. For example, if you'd measured religiosity you might have categories of Muslim, Jewish, Hindu, Catholic, Buddhist, Protestant, Jedi (for those of you not in the UK, we had a census here a few years back in which a significant portion of people put down Jedi as their religion). Clearly these groups cannot be distinguished using a single variable coded with zeros and ones. In these cases we have to use what's called dummy variables. Dummy coding is a way of representing groups of people using only zeros and ones. To do it, we have to create several variables—in fact, the number of variables we need is one less than the number of groups we're recoding. There are eight basic steps:

1. Count the number of groups you want to recode and subtract 1.
2. Create as many new variables as the value you calculated in step 1. These are your dummy variables.
3. Choose one of your groups as a baseline (i.e. a group against which all other groups should be compared). This should usually be a control group, or, if you don't have a specific hypothesis, it should be the group that represents the majority of people (because it might be interesting to compare other groups against the majority).
4. Having chosen a baseline group, assign that group values of 0 for all of your dummy variables.

5. For your first dummy variable, assign the value 1 to the first group that you want to compare against the baseline group. Assign all other groups 0 for this variable.
6. For the second dummy variable assign the value 1 to the second group that you want to compare against the baseline group. Assign all other groups 0 for this variable.
7. Repeat this until you run out of dummy variables.
8. Place all of your dummy variables into the regression analysis!

Let's try this out using an example. In Chapter 3 we came across an example in which a biologist was worried about the potential health effects of music festivals. She went to the Glastonbury music festival and measured the hygiene of concert goers over the three days of the festival using a technique that results in a score ranging between 0 (you smell like you've bathed in sewage) and 5 (you smell of freshly baked bread). Now, in Chapter 3, we just looked at the distribution of scores for the three days of the festival; but imagine if the biologist wanted to look at predictors of whether hygiene would decrease over the festival. The data file called **GlastonburyFestivalRegression.sav** can be found on the CD-ROM. This contains the hygiene scores for each of three days of the festival, but it also contains a variable called *change* which is the change in hygiene over the three days of the festival (so it's the change from day 1 to day 3).[16] Finally, the biologist categorized people according to their musical affiliation: if they mainly liked alternative music she called them 'indie kid', if they mainly liked heavy metal she called them a 'metaller' and if they mainly liked sort of hippy/folky/ambient type of stuff (you know—the levellers, eat static, that sort of thing) then she labelled them a 'crusty'. Anyone not falling into these categories was labelled 'no musical affiliation'. In the data file she coded these groups 1, 2, 3 and 4 respectively.

The first thing we should do is calculate the number of dummy variables. We have four groups, and so there will be three dummy variables (one less than the number of groups). Next we need to choose a baseline group. We're interested in comparing those that have different musical affiliations against those that don't, so our baseline category will be 'no musical affiliation'. We give this group a code of 0 for all of our dummy variables. For our first dummy variable, we could look at the 'crusty' group, and to do this we give anyone that was a crusty a code of 1, and everyone else a code of 0. For our second dummy variable, we could look at the 'metaller' group, and to do this we give anyone that was a metaller a code of 1, and everyone else a code of 0. We have one dummy variable left and this will have to look at our final category: 'indie kid'. To do this we give anyone that was an indie kid a code of 1, and everyone else a code of 0. The resulting coding scheme is shown in Table 5.3. The thing to note is that each group has a code of 1 on only one of the dummy variables (except the base category that is always coded as 0).

As I said, we'll look at why dummy coding works in sections 7.8 and 8.2.2, but for the time being let's look at how to recode our grouping variable into these dummy

16 Remember from Chapter 3 that not everyone could be measured on day 3, so there is only a change score for a subset of the original sample.

Table 5.3 Dummy coding for the Glastonbury festival data

	Dummy Variable 1	Dummy Variable 2	Dummy Variable 3
No Affiliation	0	0	0
Indie Kid	0	0	1
Metaller	0	1	0
Crusty	1	0	0

variables using SPSS. To recode variables you need to use the *Recode* function. Use the **Transform⇒ Recode⇒Into Different Variables...** to access the dialog box in Figure 5.20. The first dialog box lists all of the variables in the data editor, and you need to select the one you want to recode (in this case **music**) and transfer it to the box labelled *Numeric Variable → Output Variable* by clicking on ▶. You then need to name the new variable (the output variable as SPSS calls it), so go to the part that says *Output Variable* and in the space below where it says *Name* write a name for your first dummy variable (you might call it **music1**). You can also give this variable a more descriptive name by typing something in the space labelled *Label* (for this first dummy variable I've called it No Affiliation vs. Crusty). When you've done this click on Change to transfer this new variable to the box labelled *Numeric Variable → Output Variable* (this box should now say *music → Music1*).

Having defined the first dummy variable, we need to tell SPSS how to recode the values of the variable **music** into the values that we want for the new variable, **music1**. To do this click on Old and New Values... to access the second dialog box in Figure 5.20. This dialog box is used to change values of the original variable into different values for the new variable. For our first dummy variable, we want anyone who was a crusty to get a code of 1 and everyone else to get a code of 0. Now, crusty was coded with the value 3 in the original variable, so you need to type the value 3 in the section labelled *Old Value* in the space labelled *Value*. The new value we want is 1, so we need to type the value 1 in the section labelled *New Value* in the space labelled *Value*. When you've done this, click on Add to add this change to the list of changes (the list is displayed in the box labelled *Old → New*, which should now say *3 → 1* as in the diagram). The next thing we need to do is to change the remaining groups to have a value of 0 for the first dummy variable. To do this just select All other values and type the value 0 in the section labelled *New Value* in the space labelled *Value*. When you've done this, click on Add to add this change to the list of changes (this list will now also say *ELSE → 0*). When you've done this click on Continue to return to the main dialog box, and then click on OK to create the first dummy variable. This variable will appear as a new column in the data editor, and you should notice it will have a value of 1 for anyone originally classified as a crusty and a value of 0 for everyone else. Now try creating the remaining two dummy variables (call them **Music2** and **Music3**) using the same principles.

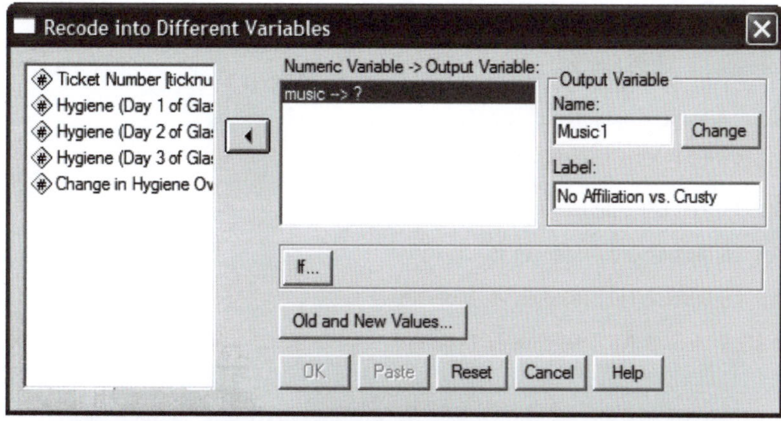

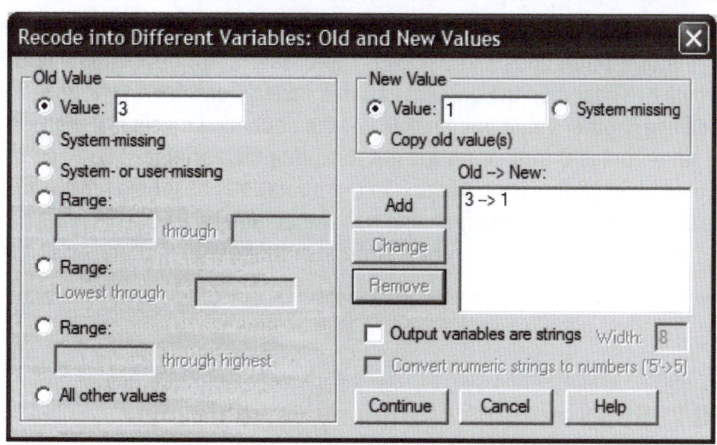

Figure 5.20 Dialog boxes for the *Recode* function

Box 5.3

Using syntax to recode ③

If you're doing a lot of recoding it soon becomes pretty tedious using the dialog boxes all of the time. The CD-ROM contains a syntax file, **RecodeGlastonburyData.sps**, which I've written to create all of the dummy variables we've discussed. Load this file and run the syntax, or open a syntax window (see section 2.6) and type the following:

(Continued)

Box 5.3 (Continued)

```
RECODE music (3 = 1) (ELSE = 0) INTO Music1.
RECODE music (2 = 1) (ELSE = 0) INTO Music2.
RECODE music (1 = 1) (ELSE = 0) INTO Music3.
VARIABLE LABELS Music1 'No Affiliation vs. Crusty'.
VARIABLE LABELS Music2 'No Affiliation vs. Metaller'.
VARIABLE LABELS Music3 'No Affiliation vs. Indie Kid'.
EXECUTE .
```

Each RECODE command is doing the equivalent of what you'd do using the compute dialog box in Figure 5.20. So, the first three lines just ask SPSS to create three new variables (**Music1**, **Music2** and **Music3**), which are based on the original variable **music**. For the first variable, if **music** is 3 then it becomes 1, and every other value becomes 0. For the second, if **music** is 2 then it becomes 1, and every other value becomes 0 and so on for the third dummy variable. The lines beginning VARIABLE LABELS just tell SPSS to assign the text in the quotations as labels for the variables **Music1**, **Music2** and **Music3** respectively. The final line has the command EXECUTE without which none of the commands beforehand will be executed! Note also that every line ends will a full stop.

5.10.2. SPSS output for dummy variables ③

Let's assume you've created the three dummy coding variables (if you're stuck there is a data file called **GlastonburyDummy.sav** (the 'Dummy' refers to the fact it has dummy variables in it—I'm not implying that if you need to use this file you're a dummy!). With dummy variables, you have to enter all related dummy variables in the same block (so use the *Enter* method). So, in this case we have to enter our dummy variables in the same block; however, if we'd had another variable (e.g. socio-economic status) that had been transformed into dummy variables, we could have entered these dummy variables in a different block (so, it's only dummy variables that have recoded the same variable that need to be entered in the same block). Use what you've learnt in this chapter to run a multiple regression using the change scores as the outcome, and the three dummy variables (entered in the same block) as predictors. Let's have a look at the output.

SPSS Output 5.12 shows the model statistics. This shows that by entering the three dummy variables we can explain 7.6% of the variance in the change in hygiene scores (the R^2-value × 100). In other words, 7.6% of the variance in the change in hygiene can be explained by the musical affiliation of the person. The ANOVA (which shows the same thing as the R^2 change statistic because there is only one step in this regression) tells us

Model Summary[b]

Model	R	R Square	Adjusted R Square	Std. Error of the Estimate	Change Statistics					Durbin–Watson
					R Square Change	F Change	df1	df2	Sig. F Change	
1	.276[a]	.076	.053	.68818	.076	3.270	3	119	.024	1.893

a Predictors: (Constant), No Affiliation vs. Indie Kid, No Affiliation vs. Crusty, No Affiliation vs. Metaller
b Dependent Variable: Change in Hygiene Over The Festival

ANOVA[b]

Model		Sum of Squares	df	Mean Square	F	Sig.
1	Regression	4.646	3	1.549	3.270	.024[a]
	Residual	56.358	119	.474		
	Total	61.004	122			

a Predictors: (Constant), No Affiliation vs. Indie Kid, No Affiliation vs. Crusty, No Affiliation vs. Metaller
b Dependent Variable: Change in Hygiene Over The Festival

SPSS Output 5.12

that the model is significantly better at predicting the change in hygiene scores than having no model (or, put another way, the 7.6% of variance that can be explained is a significant amount). Most of this should be clear from what you've read in this chapter already; what's more interesting is how we interpret the individual dummy variables.

SPSS Output 5.13 shows a basic *coefficients* table for the dummy variables (I've excluded the confidence intervals and collinearity diagnostics). The first thing to notice is that each dummy variable appears in the table with a useful label (such as Crusty vs. No Affiliation). This has happened because when we recoded our variables we gave each variable a label; if we hadn't done this then the table would contain the rather less helpful variable names (music1, music2 and music3). The labels I suggested giving each variable give us a real hint about what each dummy variable represents. The first dummy variable (No Affiliation vs. Crusty) shows the difference between the change in hygiene scores for the no affiliation group and the crusty group. Remember that the beta value tells us the change in the outcome due to a unit change in the predictor. In this case, a unit change in the predictor is the change from 0 to 1. As such it shows the shift in the change in hygiene scores that results from the dummy variable changing from 0 to 1 (crusty). By including all three dummy variables at the same time, our baseline category is always 0, so this actually represents the shift in the change in hygiene scores if a person has no musical affiliation, compared to someone who is a crusty. This shift is actually the difference between the two

Coefficients[a]

Model		Unstandardized Coefficients		Standardized Coefficients		
		B	Std. Error	Beta	t	Sig.
1	(Constant)	−.554	.090		−6.134	.000
	No Affiliation vs. Crusty	−.412	.167	−.232	−2.464	.015
	No Affiliation vs. Metaller	.028	.160	.017	.177	.860
	No Affiliation vs. Indie Kid	−.410	.205	−.185	−2.001	.048

a Dependent Variable: Change in Hygiene Over The Festival

SPSS Output 5.13

group means. To illustrate this, I've produced a table of the group means for each of the four groups. These means represent the average change in hygiene scores for the three groups (i.e. the man of each group on our outcome variable). If we calculate the difference in these means for the no affiliation group and the crusty group we get: crusty—no affiliation = (−.9658) − (−.5543) = −.412. In other words, the change in hygiene scores is greater for the crusty group than it is for the no affiliation group (crusties' hygiene decreases more over the festival than those with no musical affiliation). This value is the same as the beta value in SPSS Output 5.13! So, the beta values tell us the relative difference between each group and the group that we chose as a baseline category. This beta value is converted to a *t*-statistic and the significance of this *t* reported. This *t*-statistic is testing, as we've seen before, whether the beta value is 0 and when we have two categories coded with 0 and 1, that means it's testing whether the difference between groups is zero. If it is significant then it means that the group coded with 1 is significantly different from the baseline category—so, it's testing the difference between two means, which is the context in which you'll be more familiar with the *t*-statistic (see Chapter 7). For our first dummy variable, the *t*-test is significant, and the beta value has a negative value so we could say that the change in hygiene scores goes down as a person changes from having no affiliation to being a crusty. Bear in mind that a decrease in hygiene scores represents more change (you're becoming smellier) so what this actually means is that hygiene decreased significantly more in crusties compared to those with no musical affiliation!

Report

Change in Hygiene Over The Festival

Musical Affiliation	Mean	N	Std. Deviation
Indie Kid	−.9643	14	.67020
Metaller	−.5259	27	.57583
Crusty	−.9658	24	.76038
No Musical Affiliation	−.5543	58	.70834
Total	−.6750	123	.70713

Moving onto our next dummy variable, this compares metallers with those who have no musical affiliation. The beta value again represents the shift in the change in hygiene scores if a person has no musical affiliation, compared to someone who is a metaller. If we calculate the difference in the group means for the no affiliation group and the metaller group we get: metaller—no affiliation = (−.5259) − (−.5543) = .028. This value is again the same as the beta value in SPSS Output 5.13! For this second dummy variable, the *t*-test is not significant, so we could say that the change in hygiene scores is the same if a person changes from having no affiliation to being a metaller. In other words, the change in hygiene scores is not predicted by whether someone is a metaller compared to if they have no musical affiliation.

For the final dummy variable, we're comparing indie kids with those that have no musical affiliation. The beta value again represents the shift in the change in hygiene scores if a person has no musical affiliation, compared to someone who is an indie kid. If we calculate the difference in the group means for the no affiliation group and the indie kid group we get: indie kid—no affiliation = (−.9643) − (−.5543) = −.410. It should be no surprise to you by now that this is the beta value in SPSS Output 5.13! The *t*-test is significant, and the beta value has a negative value, so, as with the first dummy variable, we could say that the change in hygiene scores goes down as a person changes from having no affiliation to being an indie kid. Bear in mind that a decrease in hygiene scores represents more change (you're becoming smellier) so what this actually means is that hygiene decreased significantly more in indie kids compared to those with no musical affiliation!

So, overall, this analysis has shown that compared to having no musical affiliation, crusties and indie kids get significantly smellier across the three days of the festival, but metallers don't. This section has introduced some really complex ideas that I expand upon in Chapters 7 and 8. It might all be a bit much to take in, and so if you're confused or want to know more about why dummy coding works in this way I suggest reading sections 7.8 and 8.2.2 and then coming back here. Alternatively, read Hardy's (1993) excellent monograph!

5.11. WHAT HAVE WE DISCOVERED ABOUT STATISTICS? ①

This chapter is possibly the longest book chapter ever written, and if you feel like you aged several years while reading it then, well, you probably have (look around, there are cobwebs in the room, you have a long beard, and when you go outside you'll discover a second ice age has been and gone leaving only you and a few woolly mammoths to populate the planet). However, on the plus side, you hopefully now know an awful lot about regression, which is the foundation of virtually all statistics! The rest of what you'll learn in this book is merely variations on the themes of this chapter.

We started the chapter by looking at the case of when you have one predictor and one outcome. This allowed us to look at some basic principles such as the equation of a straight line, the method of least squares, and how to assess how well our model fits the data using some important quantities that you'll come across in future chapters: the model sum of squares, SS_M, the residual sum of squares, SS_R, and the total sum of squares, SS_T. We used these values to calculate several important statistics such as R^2 and the *F*-ratio. We also learnt how to do a regression on SPSS, and how we can plug the resulting beta values into the equation of a straight line to make predictions about our outcome.

Next, we saw that the question of a straight line can be extended to include several predictors and looked at different methods of placing these predictors in the model (hierarchical, forced entry, stepwise). Next, we looked at factors that can affect the accuracy of a modal (outliers and influential cases) and ways to identify these factors. We then moved on to look at the assumptions necessary to generalize our model beyond the sample of data we've collected before discovering how to do the analysis on SPSS, and how to interpret the output, create our multiple regression model and test its reliability and generalizability. I finished the chapter by looking at how we can use categorical predictors in regression (and in passing we discovered more about the recode function). In general, multiple regression is a long process and should be done with care and attention to detail. There are a lot of important things to consider and you should approach the analysis in a systematic fashion. I hope this chapter helps you to do that! One thing we haven't looked at is whether we can use a categorical variable as an outcome (rather than a predictor) and the next chapter looks at a special kind of regression in which you can do just that! Before then, I need to take vast quantities of Valium.

5.12. KEY TERMS THAT WE'VE DISCOVERED

- Adjusted predicted value
- Adjusted R^2
- Autocorrelation
- b_i
- β_i
- Cook's distance
- Covariance ratio (CVR)
- Cross-validation
- Deleted residual
- DFBeta
- DFFit
- Dummy variables
- Durbin–Watson test
- F-ratio
- Generalization
- Goodness-of-fit
- Hat values
- Homoscedasticity
- Heteroscedasticity
- Independent errors
- Leverage
- Mahalanobis distances
- Mean squares
- Model sum of squares
- Multiple R
- Multiple regression
- Multicollinearity
- Outcome variable
- Perfect collinearity
- Predictor variable
- Residual
- Residual sum of squares
- Shrinkage
- Simple regression
- Standardized DFBeta
- Standardized DFFit
- Standardized residuals
- Stepwise regression
- Studentized deleted residuals
- Studentized residuals
- Suppressor effects
- t-statistic
- Tolerance
- Total sum of squares
- Unstandardized residuals
- Variance inflation factor (VIF)

5.13. SMART ALEX'S TASKS

The answers to these tasks can be found on the CD-ROM in the file called **Answers (Chapter 5).pdf**.

- A fashion student was interested in factors that predicted the salaries of catwalk models. She collected data from 231 models. For each model she asked them their salary per day on days when they were working (**salary**), their age (**age**), how many years they had worked as a model (**years**), and then got a panel of experts from modelling agencies to rate the attractiveness of each model as a percentage with 100% being perfectly attractive (**beauty**). The data are on the CD-ROM in the file **Supermodel.sav**. Unfortunately, this fashion student bought some substandard statistics text and so doesn't know how to analyse her data☺ Can you help her out by conducting a multiple regression to see which factor predicts a model's salary? How valid is the regression model?②
- Using the Glastonbury data from this chapter (with the dummy coding in **GlastonburyDummy.sav**), which you should've already analysed, comment on whether you think the model is reliable and generalizable.③

5.14. FURTHER READING

Bowerman, B. L. & O'Connell, R. T. (1990). *Linear statistical models: an applied approach* (2nd edition). Belmont, CA: Duxbury. This text is only for the mathematically minded or postgraduate students but provides an extremely thorough exposition of regression analysis.

Hardy, M. A. (1993). *Regression with dummy variables*. Sage university paper series on quantitative applications in the Social Sciences, 07-093. Newbury Park, CA: Sage.

Howell, D. C. (2002). *Statistical methods for psychology* (5th edition). Belmont, CA: Duxbury. Chapters 9 and 15 are excellent introductions to the mathematics behind regression analysis.

Miles, J. & Shevlin, M. (2001). *Applying regression and correlation: a guide for students and researchers*. London: Sage. This is an extremely readable text that covers regression in loads of detail but with minimum pain—highly recommended.

Stevens, J. (1992). *Applied multivariate statistics for the social sciences* (2nd edition). Hillsdale, NJ: Erlbaum. Chapter 3.

CHAPTER 6

LOGISTIC REGRESSION

6.1. WHAT WILL THIS CHAPTER TELL US? ①

In the previous chapter we looked at how to model the relationship between one or more predictor variables and an outcome. This chapter extends these ideas to look at how we can predict outcome variables that are categorical using logistic regression. It's a tough chapter (in fact I could've surrounded the whole thing with a Smart Alex, but I haven't!) and logistic regression isn't one of my strong points, so it's probably a bit rubbish. Nevertheless, we begin the chapter with a brief bit of theory to help you understand what logistic regression is all about, before moving straight into an example. You'll learn how to carry out logistic regression in SPSS and how to interpret the resulting output. We end the chapter with a second example, and use this example as a way to look at how we can combat multicollinearity. In the process we also find out why the England football (soccer for the Americans) team can't take penalties.

6.2. BACKGROUND TO LOGISTIC REGRESSION ①

In a nutshell, logistic regression is multiple regression but with an outcome variable that is a categorical dichotomy and predictor variables that are continuous or categorical. In plain English, this simply means that we can predict which of two categories a person is likely to belong to given certain other information. A trivial example is to look at which variables predict whether a person is male or female. We might measure laziness, pig-headedness, alcohol consumption and number of burps that a person does in a day. Using logistic regression, we might find that all of these variables predict the gender of the person, but the technique will also allow us to predict whether a certain person is likely to be male or female. So, if we picked a random person and discovered they scored highly on laziness, pig-headedness, alcohol consumption and the number of burps, then the regression model might tell us that, based on this information, this person is likely to be male.

Admittedly, it is unlikely that a researcher would ever be interested in the relationship between flatulence and gender (it is probably too well established by experience to warrant research!), but logistic regression can have life-saving applications. In the biomedical literature (i.e. medical research) logistic regression has applications such as formulating models about the sorts of factors that might determine whether a tumour is cancerous or benign (for example). A database of patients can be used to establish which variables are influential in predicting the malignancy of a tumour. These variables can then be measured for a new patient and their values placed in a logistic regression model, from which a probability of malignancy could be estimated. If the probability value of the tumour being malignant is suitably low then the doctor may decide not to carry out expensive and painful surgery that in all likelihood is unnecessary.

In the social sciences we rarely face such life-threatening decisions yet logistic regression is still a very useful tool. For this reason, it is tragic that many textbooks often overlook it. In this chapter I hope to redress the balance by explaining the principles behind logistic regression and how to carry out the procedure on SPSS.

6.3. WHAT ARE THE PRINCIPLES BEHIND LOGISTIC REGRESSION? ③

I don't wish to dwell on the underlying principles of logistic regression because they aren't necessary to understand the test (I am living proof of this fact). However, I do wish to draw a few parallels to normal regression in an attempt to couch logistic regression in a framework that will be familiar to everyone who has got this far in the book (what do you mean you haven't read the regression chapter yet?!). Now would be a good time for the equation-phobes to look away. In simple linear regression, we saw that the outcome variable Y is predicted from the equation of a straight line:

$$Y_i = b_0 + b_1 X_1 + \varepsilon_i \tag{6.1}$$

in which b_0 is the Y-intercept, b_1 is the gradient of the straight line, X_1 is the value of the predictor variable and ε is a residual term. Given the values of Y and X_1, the unknown parameters in the equation can be estimated by finding a solution for which the squared distance between the observed and predicted values of the dependent variable is minimized (the method of least squares).

This stuff should all be pretty familiar. In multiple regression, in which there are several predictors, a similar equation is derived in which each predictor has its own coefficient. As such, Y is predicted from a combination of each predictor variable multiplied by its respective regression coefficient:

$$Y_i = b_0 + b_1 X_1 + b_2 X_2 + \cdots + b_n X_n + \varepsilon_i \tag{6.2}$$

in which b_n is the regression coefficient of the corresponding variable X_n. In logistic regression, instead of predicting the value of a variable Y from a predictor variable X_1 or several predictor variables (Xs), we predict the *probability* of Y occurring given known values of X_1 (or Xs). The logistic regression equation bears many similarities to the regression equations just described. In its simplest form, when there is only one predictor variable X_1, the logistic regression equation from which the probability of Y is predicted is given by equation (6.3):

$$P(Y) = \frac{1}{1 + e^{-(b_0 + b_1 X_1 + \varepsilon_i)}} \tag{6.3}$$

in which $P(Y)$ is the probability of Y occurring, e is the base of natural logarithms, and the other coefficients form a linear combination much the same as in simple regression. In fact, you might notice that the bracketed portion of the equation is identical to the linear regression equation in that there is a constant (b_0), a predictor variable (X_1) and a coefficient (or weight) attached to that predictor (b_1). Just like linear regression, it is possible to extend this equation so as to include several predictors. When there are several predictors the equation becomes:

$$P(Y) = \frac{1}{1 + e^{-(b_0 + b_1 X_1 + b_2 X_2 + \cdots + b_n X_n + \varepsilon_i)}} \tag{6.4}$$

Equation (6.4) is the same as the equation used when there is only one predictor except that the linear combination has been extended to include any number of predictors. So, whereas the one-predictor version of the logistic regression equation contained the simple linear regression equation within it, the multiple-predictor version contains the multiple regression equation.

Despite the similarities between linear regression and logistic regression, there is a good reason why we cannot apply linear regression directly to a situation in which the outcome variable is dichotomous. The reason is that one of the assumptions of linear regression is that the relationship between variables is linear (see section 5.6.2.1). In that section we saw how important it is that the assumptions of a model are met for that model to be accurate. Therefore, for linear regression to be a valid model, the observed data should contain a linear relationship. When the outcome variable is dichotomous, this assumption is usually violated (see Berry, 1993). One way around this problem is to transform the data using the logarithmic transformation (see Berry & Feldman, 1985, and Chapter 3). This has the effect of making the *form* of the relationship linear whilst leaving the relationship itself as non-linear (so it is a way of expressing a non-linear relationship in a linear way). The logistic regression equation described above is based on this principle: it expresses the multiple linear regression equation in logarithmic terms and thus overcomes the problem of violating the assumption of linearity.

The exact form of the equation can be arranged in several ways but the version I have chosen expresses the equation in terms of the probability of Y occurring (i.e. the probability that a case belongs in a certain category). As such, the resulting value from the equation is a probability value that varies between 0 and 1. A value close to 0 means that Y is very unlikely to have occurred, and a value close to 1 means that Y is very likely to have occurred. Also, just like linear regression, each predictor variable in the logistic regression equation has its own coefficient. When we run the analysis we need to estimate the value of these coefficients so that we can solve the equation. These parameters are estimated by fitting models, based on the available predictors, to the observed data. The chosen model will be the one that, when values of the predictor variables are placed in it, results in values of Y closest to the observed values. Specifically, the values of the parameters are estimated using maximum-likelihood estimation, which selects coefficients that make the observed values most likely to have occurred. So, as with multiple regression, we try to fit a model to our data that allows us to estimate values of the outcome variable from known values of the predictor variable or variables.

6.3.1. Assessing the model: the log-likelihood statistic ③

We've seen that the logistic regression model predicts the probability of an event occurring for a given person (we would denote this as $P(Y_i)$ the probability that Y occurs for the ith person), based on observations of whether or not the event did occur for that person (we could denote this as Y_i, the actual outcome for the ith person). So, for a given person, Y will be either 0 (the outcome didn't occur) or 1 (the outcome did occur), and the predicted value, $P(Y)$, will be a value between 0 (there is no chance that the outcome will occur) and 1 (the outcome will occur). We saw in multiple regression (see the previous chapter) that if we want to assess whether a model fits the data we can compare the observed and predicted values of the outcome (if you remember, we use R^2, which is the Pearson correlation between observed values of the outcome and the values predicted by the regression model). Likewise, in logistic regression, we can use the observed and predicted values to assess the fit of the model. The measure we use is the log-likelihood:

$$\text{log-likelihood} = \sum_{i=1}^{N} \{Y_i \text{In}(P(Y_i)) + (1 - Y_i)\text{In}[1 - P(Y_i)]\} \tag{6.5}$$

The log-likelihood is therefore based on summing the probabilities associated with the predicted and actual outcomes (see Tabachnick & Fidell, 2001). The log-likelihood statistic is analogous to the residual sum of squares in multiple regression in the sense that it is an indicator of how much unexplained information there is after the model has been fitted. It follows, therefore, that large values of the log-likelihood statistic indicate poorly fitting statistical models, because the larger the value of the log-likelihood, the more unexplained observations there are.

Now, it's possible to calculate a log-likelihood for different models and to compare these models by looking at the difference between their log-likelihoods. One use of this is to compare the state of a logistic regression model against some kind of baseline state. The baseline state that's usually used is the model when only the constant is included. In multiple regression, the baseline model we use is the mean of all scores (this is our best guess of the outcome when we have no other information). In logistic regression, if we want to predict the outcome what would our best guess be? Well, we can't use the mean score because our outcome is made of zeros and ones and so the mean is meaningless! However, if we know the frequency of zeros and ones, then the best guess will be the category with the largest number of cases. So, if an outcome occurs 107 times, and doesn't occur only 72 times, then out best guess of the outcome will be that it occurs (because it occurs 107 compared to 72 times when it doesn't occur). As such, as in multiple regression, our baseline model is the model that gives us the best prediction when we know nothing other than the values of the outcome: in logistic regression this will be to predict the outcome that occurs most often. This is the logistic regression model when only the constant is included. If we then add one or more predictors to the model, we can compute the improvement of the model as follows:

$$\chi^2 = 2[\text{LL(New)} - \text{LL(Baseline)}] \quad (df = k_{\text{new}} - k_{\text{baseline}}) \tag{6.6}$$

So, we merely take the new model and subtract from it the baseline model (the model when only the constant is included). You'll notice that we multiply this value by 2; this is because it gives the result a chi-square distribution (see Chapter 16 and Appendix A4) and so makes it easy to calculate the significance of the value. The chi-square distribution we use has degrees of freedom equal to the number of parameters in the new model minus the number of parameters in the baseline model. The number of parameters, *k*, in the baseline model will always be 1 (the constant is the only parameter to be estimated), and any subsequent model will have degrees of freedom equal to the number of predictors plus 1 (i.e. the number of predictors plus one parameter representing the constant).

6.3.2. Assessing the model: *R* and R^2 ③

When we talked about linear regression, we saw that the multiple correlation coefficient *R* and the corresponding R^2 were useful measures of how well the model fits the data. We've also just seen that the likelihood ratio is similar in the respect that it is based on the level of correspondence between predicted and actual values of the outcome. However, you can calculate a more literal version of the multiple correlation in logistic regression known as the *R*-statistic. This *R*-statistic is the partial correlation between the outcome variable and each of the predictor variables and it can vary between –1 and 1. A positive value indicates that as the predictor variable increases so does the likelihood of the event occurring. A negative value implies that as the predictor variable increases the likelihood of the outcome occurring decreases. If a variable has a small value of *R* then it contributes only a small amount to the model.

The equation for R is given in equation (6.7). The −2LL is the −2 log-likelihood for the original model, the Wald statistic is calculated as described in the next section, and the degrees of freedom can be read from the summary table for the variables in the equation. However, because this value of R is dependent upon the Wald statistic it is by no means an accurate measure (we'll see in the next section that the Wald statistic can be inaccurate under certain circumstances). For this reason the value of R should be treated with some caution, and it is invalid to square this value and interpret it as you would in linear regression.

$$R = \pm\sqrt{\left(\frac{\text{Wald} - (2 \times df)}{-2\text{LL(Original)}}\right)} \tag{6.7}$$

There is some controversy over what would make a good analogue to the R^2-value in linear regression, but one measure described by Hosmer & Lemeshow (1989) can be easily calculated. In SPSS terminology, Hosmer and Lemeshow's R^2_{L} is calculated as in equation (6.8):

$$R^2_{\text{L}} = \frac{-2\text{LL(Model)}}{-2\text{LL(Original)}} \tag{6.8}$$

As such, R^2_{L} is calculated by dividing the model chi-square (based on the log-likelihood) by the *original* −2LL (the log-likelihood of the model before any predictors were entered). R^2_{L} is the proportional reduction in the absolute value of the log-likelihood measure and as such it is a measure of how much the badness-of-fit improves as a result of the inclusion of the predictor variables. It can vary between 0 (indicating that the predictors are useless at predicting the outcome variable) and 1 (indicating that the model predicts the outcome variable perfectly).

However, this is not the measure used by SPSS. SPSS uses Cox & Snell's R^2_{CS} (1989), which is based on the log-likelihood of the model (*LL*(*New*)) and the log-likelihood of the original model (*LL*(*Baseline*)), and the sample size, *n*:

$$R^2_{CS} = 1 - e^{\left[-\frac{2}{n}(\text{LL(New)} - \text{LL(Baseline)})\right]} \tag{6.9}$$

However, this statistic never reaches its theoretical maximum of 1. Therefore, Nagelkerke (1991) suggested the following amendment (Nagelkerke's R^2_{N}):

$$R^2_{\text{N}} = \frac{R^2_{CS}}{1 - e^{\left[\frac{2(\text{LL(Baseline)})}{n}\right]}} \tag{6.10}$$

Although all of these measures differ in their computation (and the answers you get), collectively they can be seen as somewhat the same. So, in terms of interpretation they can be seen as similar to the R^2 in linear regression in that they provide a gauge of the substantive significance of the model.

6.3.3. Assessing the contribution of predictors: the Wald statistic ②

As in linear regression, we want to know not only how well the model fits the data overall, but also the individual contribution of predictors. In linear regression, we used the estimated regression coefficients (*b*) and their standard errors to compute a *t*-statistic. In logistic regression there is an analogous statistic know as the Wald statistic, which has a special distribution known as the chi-square distribution. Like the *t*-test in linear regression, the Wald statistic tells us whether the *b*-coefficient for that predictor is significantly different from zero. If the coefficient is significantly different from zero then we can assume that the predictor is making a significant contribution to the prediction of the outcome (*Y*). Equation (6.11) shows how the Wald statistic is calculated and you can see it's basically identical to the *t*-statistic in linear regression (see equation (5.6)): it is the value of the regression coefficient divided by its associated standard error. The Wald statistic is usually used to ascertain whether a variable is a significant predictor of the outcome: however, it is probably more accurate to examine the likelihood ratio statistics. The reason why the Wald statistic (Figure 6.1) should be used cautiously is because when the regression coefficient (*b*) is large, the standard error tends to become inflated, resulting in the Wald statistic being underestimated (see Menard, 1995). The inflation of the standard error increases the probability of rejecting a predictor as being significant when in reality it is making a significant contribution to the model (i.e. you are more likely to make a Type II error).

$$\text{Wald} = \frac{b}{\text{SE}_b} \tag{6.11}$$

Figure 6.1 Abraham Wald writing 'I must not devise test statistics prone to having inflated standard errors' on the blackboard 100 times

6.3.4. Exp *b* ③

More crucial to the *interpretation* of logistic regression is the value of exp b (*Exp(B)* in the SPSS output), which is an indicator of the change in odds resulting from a unit change in the predictor. As such, it is similar to the b-coefficient in logistic regression but easier to understand (because it doesn't require a logarithmic transformation). When the predictor variable is categorical exp b is easier to explain, so imagine we had a simple example in which we were trying to predict whether or not someone got pregnant from whether or not their partner used a condom the last time they made love. The odds of an event occurring are defined as the probability of an event occurring divided by the probability of that event not occurring (see equation (6.12)) and should not be confused with the more colloquial usage of the word to refer to probability. So, for example, the odds of becoming pregnant are the probability of becoming pregnant divided by the probability of not becoming pregnant:

$$\begin{aligned} \text{odds} &= \frac{P(\text{event})}{P(\text{no event})} \\ P(\text{event } Y) &= \frac{1}{1 + e^{-(b_0 + b_1 X_1)}} \\ P(\text{no event } Y) &= 1 - P(\text{event } Y) \end{aligned} \tag{6.12}$$

To calculate the change in odds that results from a unit change in the predictor, we must first calculate the odds of becoming pregnant given that a condom *wasn't* used. We then calculate the odds of becoming pregnant given that a condom *was* used. Finally, we calculate the proportionate change in these two odds.

To calculate the first set of odds, we need to use equation (6.3) to calculate the probability of becoming pregnant given that a condom wasn't used. If we had more than one predictor we would use equation (6.4). There are three unknown quantities in this equation: the coefficient of the constant (b_0), the coefficient for the predictor (b_1) and the value of the predictor itself (X). We'll know the value of X from how we coded the condom use variable (chances are we would've used 0 = condom wasn't used and 1 = condom was used). The values of b_1 and b_0 will be estimated for us. We can calculate the odds as in equation (6.12).

Next, we calculate the same thing after the predictor variable has changed by one unit. In this case, because the predictor variable is dichotomous, we need to calculate the odds of getting pregnant, given that they *did* use a condom. So, the value of X is now 1 (rather than 0).

We now know the odds before and after a unit change in the predictor variable. It is a simple matter to calculate the proportionate change in odds by dividing the odds after a unit change in the predictor by the odds before that change:

$$\Delta\text{odds} = \frac{\text{odds after a unit change in the predictor}}{\text{original odds}} \tag{6.13}$$

This proportionate change in odds is exp b, so we can interpret exp b in terms of the change in odds: if the value is greater than 1 then it indicates that as the predictor increases, the odds of the outcome occurring increase. Conversely, a value less than 1 indicates that as the predictor increases, the odds of the outcome occurring decrease. We'll see how this works with a real example shortly.

6.3.5. Methods of logistic regression ②

As with multiple regression, there are several different methods that can be used in logistic regression.

6.3.5.1. The forced entry method

The default method of conducting the regression is 'enter'. This is the same as forced entry in multiple regression in that all of the covariates are placed into the regression model in one block, and parameter estimates are calculated for each block. Some researchers believe that this method is the only appropriate method for theory testing (Studenmund & Cassidy, 1987) because stepwise techniques are influenced by random variation in the data and so seldom give replicable results if the model is retested within the same sample.

6.3.5.2. Stepwise methods

If you are undeterred by the criticisms of stepwise methods in the previous chapter, then you can select either a forward or a backward stepwise method. When the forward method is employed the computer begins with a model that includes only a constant and then adds single predictors into the model based on a specific criterion. This criterion is the value of the *score* statistic: the variable with the most significant score statistic is added to the model. The computer proceeds until none of the remaining predictors have a significant score statistic (the cut-off point for significance being .05). At each step, the computer also examines the variables in the model to see whether any should be removed. It does this in one of three ways. The first way is to use the likelihood ratio statistic described in 16.2.2 (the *Forward:LR* method) in which case the current model is compared to the model when that predictor is removed. If the removal of that predictor makes a significant difference to how well the model fits the observed data, then the computer retains that predictor (because the model is better if the predictor is included). If, however, the removal of the predictor makes little difference to the model then the computer rejects that predictor. Rather than using the likelihood ratio statistic, which estimates how well the model fits the observed data, the computer could use the conditional statistic as a removal criterion (*Forward:Conditional*). This statistic is an arithmetically less intense version of the likelihood ratio statistic and so there is little to recommend it over the likelihood ratio method. The final criterion is the Wald statistic, in which case any predictors in the model that have significance values of the Wald statistic (above the default removal criterion of .1) will be removed. Of these methods the likelihood ratio method is the best removal criterion because the Wald statistic can, at times, be unreliable (see section 6.3.3).

The opposite of the forward method is the backward method. This method uses the same three removal criteria, but instead of starting the model with only a constant, it begins the model with all predictors included. The computer then tests whether any of these predictors can be removed from the model without having a substantial effect on how well the model fits the observed data. The first predictor to be removed will be the one that has the least impact on how the model fits the data.

6.3.5.3. How do we select a method? ②

As with ordinary regression (previous chapter), the method of regression chosen will depend on several things. The main consideration is whether you are testing a theory or merely carrying out exploratory work. As noted earlier, some people believe that stepwise methods have no value for theory testing. However, stepwise methods are defensible when used in situations in which no previous research exists on which to base hypotheses for testing, and in situations where causality is not of interest and you merely wish to find a model to fit your data (Menard, 1995; Agresti & Finlay, 1986). Also, as I mentioned for ordinary regression, if you do decide to use a stepwise method then the backward method is preferable to the forward method. This is because of suppressor effects, which occur when a predictor has a significant effect but only when another variable is held constant. Forward selection is more likely than backward elimination to exclude predictors involved in suppressor effects. As such, the forward method runs a higher risk of making a Type II error. In terms of the test statistic used in stepwise methods, the Wald statistic has a tendency to be inaccurate in certain circumstances (see section 6.3.3) and so the likelihood ratio method is best.

6.4. RUNNING THE ANALYSIS: A RESEARCH EXAMPLE ②

As my first research example I am going to look at a simple example from developmental psychology. A good example of a dichotomy is the passing or failing of a test, which in developmental psychology terms often relates to the possession (or not) of a cognitive ability. This example looks at children's understanding of display rules based on the child's age, and whether the child possesses a theory of mind. Put simplistically, a display rule is a convention of displaying an appropriate emotion in a given situation. For example, if you receive a Christmas present that you don't like, the appropriate emotional display is to smile politely and say 'thank you Auntie Kate, I've always wanted a rotting cabbage'. The inappropriate emotional display is to start crying and scream 'why did you buy me a rotting cabbage you selfish old bag?' There is some evidence that young children lack an understanding of appropriate display rules and this has been linked to the acquisition of a theory of mind (which is simply the ability to understand what another person might be thinking).

For this example, our researchers are interested in whether the understanding of emotional display rules was linked to having a theory of mind. The rationale is that it might be necessary for a child to understand how another person thinks to realize how their emotional displays will affect that person: if you can't put yourself in Aunt Kate's mind, then you won't realize that she might be upset by you calling her an old bag. To test this theory, several children were given a standard false belief task (a task used to measure whether someone has a theory of mind) that they could either pass or fail and their age in months was also measured. In addition, each child was given a display rule task, which they could either pass or fail. So, the following variables were measured:

- **Outcome (dependent variable)**: Possession of display rule understanding (Did the child pass the test: Yes/No?).
- **Predictor (independent variable)**: Possession of a theory of mind (Did the child pass the false belief task: Yes/No?).
- **Predictor (independent variable)**: Age in months.

In this experiment, there is a dichotomous outcome variable, a categorical predictor and a continuous predictor. This scenario is ideal for logistic regression.

6.4.1. The main analysis ②

To carry out logistic regression, the data must be entered as for normal regression: they are arranged in the data editor in three columns (one representing each variable). The data can be found in the file **display.sav** in the Chapter 6 folder of the data disk. Looking at the data editor you should notice that both of the categorical variables have been entered as coding variables (see section 2.4.4); that is, numbers have been specified to represent categories. For ease of interpretation, the outcome variable should be coded 1 (event occurred) and 0 (event did not occur); in this case, 1 represents having display rule understanding, and 0 represents an absence of display rule understanding. For the false belief task a similar coding has been used (1 = passed the false belief task, 2 = failed the false belief task). Logistic regression is located in the *Regression* menu accessed via the *Analyze* menu: **Analyze⇒Regression⇒Binary Logistic**. Following this menu path activates the main *logistic regression* dialog box shown in Figure 6.2.

The main dialog box is very similar to the standard *regression* option box. There is a space to place a dependent variable (or outcome variable). In this example, the outcome was the display rule task, so we can simply click on **display** and transfer it to the *Dependent* box by clicking on . There is also a box for specifying the covariates (the predictor variables). It is possible to specify both main effects and interactions in logistic regression. To specify a main effect, simply select one predictor (e.g. **age**) and then transfer this variable to the *Covariates* box by clicking on . To input an interaction, click on more than one variable on the left-hand side of the dialog box (i.e. highlight two or more variables) and then click on >a*b> to move them to the *Covariates* box. In this example

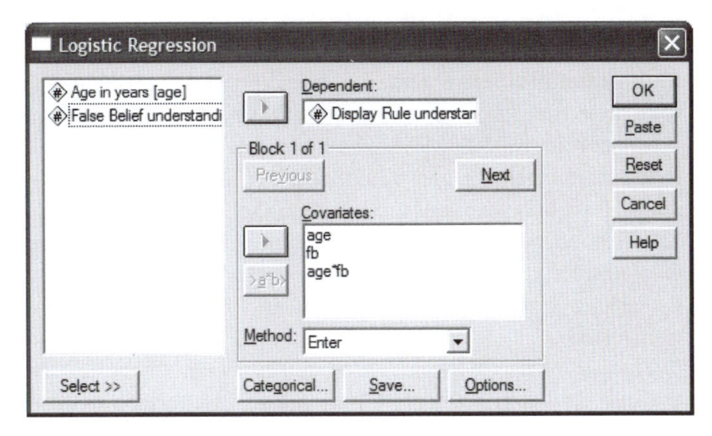

Figure 6.2 *Logistic regression* main dialog box

there are only two predictors and therefore there is only one possible interaction (the **age** × **fb** interaction), but if you have three predictors then you can select several two-way interactions and the three-way interaction as well.

6.4.2. Method of regression ②

As with multiple regression, there are several different methods that can be used in logistic regression (see section 6.3.5). You can select a particular method of regression by clicking on the down arrow next to the box labelled *Method*. Figure 6.3 shows part of the logistic regression menu when the methods of regression are activated. For this analysis select a *Forward:LR* method of regression. Having spent a vast amount of time telling you never to do stepwise analyses, it's probably a bit strange to hear me suggest doing forward regression here. Well, for one thing this study is the first in the field and so we have no past research to tell us which variables to expect to be reliable predictors. Second, I didn't show you a stepwise example in the regression chapter and so this will be a useful way to demonstrate how a stepwise procedure operates!

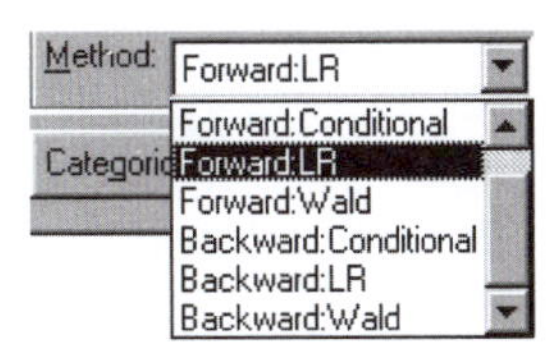

Figure 6.3 Method of regression

6.4.3. Categorical predictors ②

In this example there is one categorical predictor variable. One of the great things about logistic regression is that it is quite happy to accept categorical predictors. However, it is necessary to 'tell' SPSS which variables, if any, are categorical by clicking on Categorical... in the main *logistic regression* dialog box to activate the dialog box in Figure 6.4. In this dialog box, the covariates are listed on the left-hand side, and there is a box on the right-hand side in which categorical covariates can be placed. Simply highlight any categorical variables you have (in this example click on **fb**) and transfer them to the *Categorical Covariates* box by clicking on ▸. In Figure 6.4, **fb** has already been selected and transferred.

There are many ways in which you can treat categorical predictors: by default, SPSS will use deviation contrasts on all categorical variables. In section 5.10 in the previous chapter we saw that categorical predictors could be incorporated into regression by recoding them using zeros and ones (known as dummy coding). Now, actually, there are different ways you can arrange this coding depending on what you want to compare, and SPSS has several 'standard' ways built into it that you can select (we'll come across these methods in more detail in Chapter 8). By default SPSS uses *Indicator* coding, which is the standard dummy variable coding that I explained in section 5.10 (and you can choose to have either the first or last category as your baseline). To change to a different kind of contrast click on the down arrow in the *Change Contrast* box. Figure 6.5 shows that it is possible to select simple contrasts, difference contrasts, Helmert contrasts, repeated contrasts, polynomial contrasts and deviation contrasts just as in ANOVA (see Table 8.6). These techniques will be discussed in detail in Chapter 8 and so there I'll explain what these contrasts actually do. However, you won't come across indicator contrasts in that chapter and so we'll use them here (as in Figure 6.4). To reiterate, when an indicator contrast is used, levels of the categorical variable are recoded using standard dummy variable coding (see sections 5.10 and 7.8).

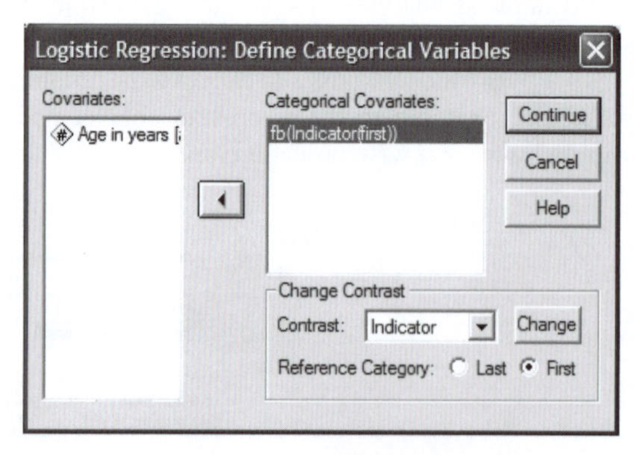

Figure 6.4 Defining categorical variables in logistic regression

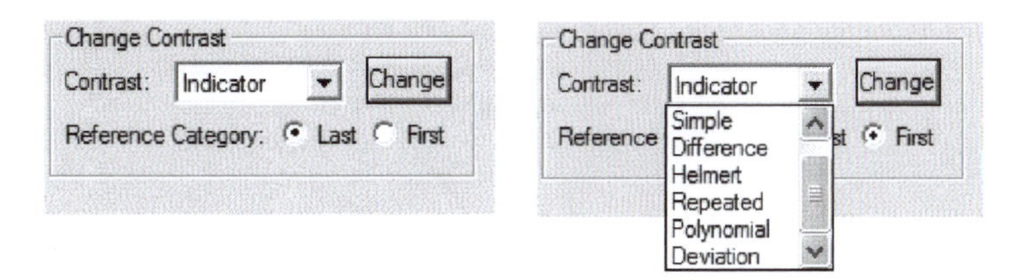

Figure 6.5 Selecting contrasts for categorical predictors

6.4.4. Obtaining residuals ②

As with linear regression it is possible to ask SPSS to save a set of residuals as new variables in the data editor. These residual variables can then be examined to see how well the model fits the observed data. To save residuals click on Save... in the main *logistic regression* dialog box (Figure 6.2). SPSS saves each of the selected variables into the data editor but they can be listed in the output viewer by using the *Case Summaries* command (see section 5.8.6) and selecting the residual variables of interest. The residuals dialog box in Figure 6.6 gives us several options and most of these are the same as those in multiple regression (refer to section 5.7.4). Two residuals that are unique to logistic regression are the *predicted probabilities* and the *predicted group memberships*. The predicted probabilities are the probabilities of *Y* occurring given the values of each predictor for a given participant. As such, they are derived from equation (6.4) for a given case. The predicted group membership is self-explanatory in that it predicts to which of the two categories of *Y* a participant is most likely to belong based on the model. The group memberships are based on the predicted probabilities and I will explain these values in more detail when we consider how to interpret the residuals. It is worth selecting all of the available options, or as a bare minimum select the same options as in Figure 6.6.

I cannot stress enough the importance of examining residuals after any analysis. One of the many benefits of computer packages over analysing data by hand is that it is very easy to obtain residual diagnostics which would otherwise be incredibly time consuming to obtain and would require a fairly substantial understanding of algebra. Nevertheless, the advancement of computing facilities has probably not been met with a proportionate advancement in the consideration of residuals. Sadly, social scientists, who have spent far too long being trained only to hunt out probability values below .05, often ignore the examination of residuals! I have stressed the importance of building accurate models, and that running a regression without checking how well the model fits the data is like buying a new pair of trousers without trying them on—they might look fine on the hanger but get them home and you find you're Johnny-tight-pants. The trousers might do their job (they cover your legs and keep you warm) but they have no real-life value (because they cut off the blood circulation to your legs and other important appendages). Likewise, regression does its job regardless of the data—it will create a model—but the real-life value of the model may be limited.

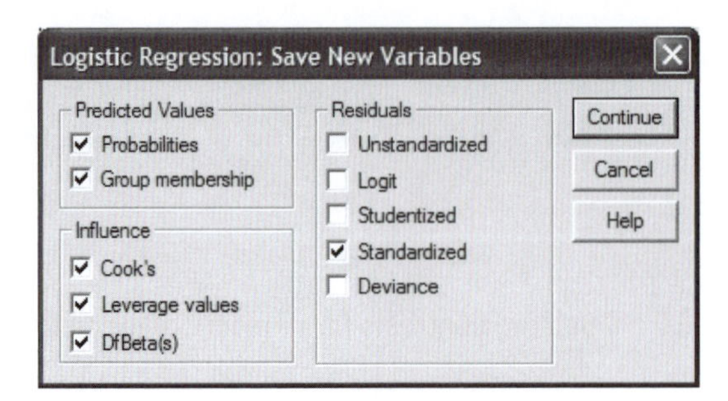

Figure 6.6 Dialog box for obtaining residuals for logistic regression

Testing the degree to which a model fits the data collected is the essence of diagnostic statistics. If the model fits the data well, then we can have more confidence that the coefficients of the model are accurate and are not being influenced by a few stray data points. At best, failure to examine residuals can lead to ignorance of inaccurate coefficients in a model. At worst, this inaccurate model could then be used to accept a theoretical hypothesis that is, in fact, not true.

6.4.5. Further options ②

There is a final dialog box that offers further options. This box is shown in Figure 6.7 and is accessed by clicking on Options... in the main *logistic regression* dialog box. For the most part, the default settings in this dialog box are fine. I mentioned in section 6.4.2 that when a stepwise method is used there are default criteria for selecting and removing predictors from the model. These default settings are displayed in the *options* dialog box under *Probability for Stepwise*. The probability thresholds can be changed, but there is really no need unless you have a good reason for wanting harsher criteria for variable selection. Another default is to arrive at a model after a maximum of 20 iterations. When we fit a model to our data we try to fit the best model possible and the maximum iterations can be conceived of as the maximum number of attempts that the computer will make to find the best model. Unless you have a very complex model, 20 iterations will be more than adequate. We saw in Chapter 5 that regression equations contain a constant that represents the Y-intercept (i.e. the value of Y when the value of the predictors is zero). By default SPSS includes this constant in the model, but it is possible to run the analysis without this constant and this has the effect of making the model pass through the origin (i.e. Y is zero when X is zero). Given that we are usually interested in producing a model that best fits the data we have collected, there is little point in running the analysis without the constant included.

One option that is very useful is a classification plot, which is a histogram of the actual and predicted values of the outcome variable. This plot is useful for assessing the fit of the model to the observed data. It is also possible to do a *Casewise listing of residuals* either

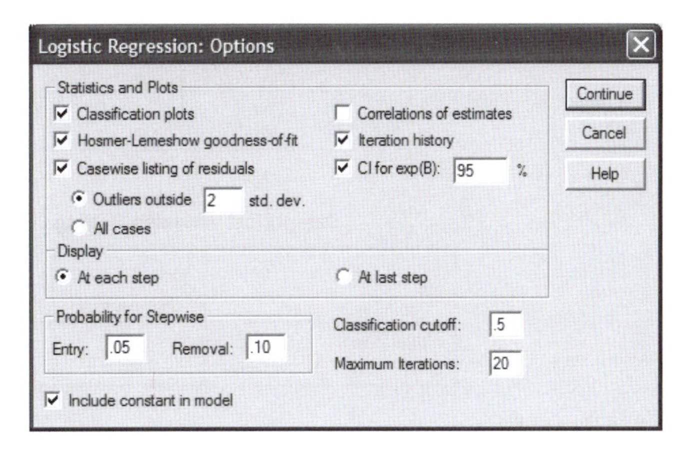

Figure 6.7 Dialog box for logistic regression options

for any cases for which the standardized residual is greater than 2 standard deviations (this value can be changed but the default is sensible), or for all cases. I recommend a more thorough examination of residuals but this option can be useful for a quick inspection. You can ask SPSS to display a confidence interval (see section 1.6.2) for the exp *b* statistic and SPSS version 12 displays a 95% confidence interval and this is appropriate (if it says anything else then change it to 95%) and it is a useful statistic to have. More important, you can request the *Hosmer-Lemeshow goodness-of-fit* statistic, which can be used to assess how well the chosen model fits the data. The remaining options are fairly unimportant: you can choose to display all statistics and graphs at each stage of an analysis (the default), or only after the final model has been fitted. Finally, you can display a correlation matrix of parameter estimates for the terms in the model, and you can display coefficients and log-likelihood values at each iteration of the parameter estimation process—the practical function of doing this is lost on most of us mere mortals!

6.5. INTERPRETING LOGISTIC REGRESSION ②

When you have selected all of the options I've just described, click on OK and watch the output spew out in the viewer window. I shall explain each part of the analysis in turn. In earlier versions of SPSS the output from logistic regression was unformatted (it just appeared as text—see the first edition of this book if you're really that interested!) but these days the output is formatted in nice tables!

6.5.1. The initial model ②

SPSS Output 6.1 tells us the parameter codings given to the categorical predictor variable. Indicator coding was chosen with two categories, and so the coding is the same as the

Dependent Variable Encoding

Original Value	Internal Value
No	0
Yes	1

Categorical Variables Codings

		Frequency	Parameter (1)
False Belief understanding	No	29	.000
	Yes	41	1.000

SPSS Output 6.1

values in the data editor. If *deviation* coding had been chosen then the coding would have been −1 (**fb** Yes) and 1 (**fb** No). With a *simple* contrast the codings would have been − 0.5 (**fb** No) and .5 (**fb** Yes) if the latter category was selected as the reference category, or vice versa if the former category was selected as the reference category. The parameter codings are important for calculating the probability of the outcome variable ($P(Y)$), but, we shall come to that later.

For this first analysis we requested a forward stepwise method and so the initial model is derived using only the constant in the regression equation. SPSS Output 6.2 tells us about the model when only the constant is included (i.e. all predictor variables are omitted). Although SPSS doesn't display this value, the log-likelihood of this baseline model (see section 6.3.1), is 96.124 (trust me for the time being!). This represents the fit of the model when the most basic model is fitted to the data. When including only the constant, the computer bases the model on assigning every participant to a single category of the outcome variable. In this example, SPSS can decide either to predict that every child has display rule understanding, or to predict that all children do not have display rule understanding. It could make this decision arbitrarily, but because it is crucial to try to maximize how well the model predicts the observed data SPSS will predict that every child belongs to the category in which most observed cases fell. In this example there were 39 children who had display rule understanding and only 31 who did not. Therefore, if SPSS predicts that every child has display rule understanding then this prediction will be correct 39 times out of 70 (i.e. 56% approx.). However, if SPSS predicted that every child did not have display rule understanding, then this prediction would be correct only 31 times out of 70 (44% approx.). As such, of the two available options it is better to predict that all children had display rule understanding because this results in a greater number of correct predictions. The output shows a contingency table for the model in this basic state. You can see that SPSS has predicted that all children have display rule understanding, which results in 0% accuracy for the children who were observed to have no display rule understanding, and 100% accuracy for those children observed to have passed the display rule task. Overall, the model

Classification Table[a,b]

			Predicted		
			Display Rule understanding		Percentage
Observed			No	Yes	Correct
Step 0	Display Rule understanding	No	0	31	.0
		Yes	0	39	100.0
	Overall Percentage				55.7

a Constant is included in the model
b The cut value is .500

Variables in the Equation

		B	S.E.	Wald	df	Sig.	Exp(B)
Step 0	Constant	.230	.241	.910	1	.340	1.258

Variables not in the Equation

			Score	df	Sig.
Step 0	Variables	AGE	15.956	1	.000
		FB(1)	24.617	1	.000
		AGE by FB(1)	23.987	1	.000
	Overall Statistics		26.257	3	.000

SPSS Output 6.2

correctly classifies 55.71% of children. The next part of the output summarizes the model, and at this stage this entails quoting the value of the constant (b_0), which is equal to 0.23.

The final table of the output is labelled *Variables not in the Equation*. The bottom line of this table reports the residual chi-square statistic as 26.257 which is significant at $p < .0001$ (it labels this statistic *Overall Statistics*). This statistic tells us that the coefficients for the variables not in the model are significantly different from zero—in other words, that the addition of one or more of these variables to the model will significantly affect its predictive power. If the probability for the residual chi-square had been greater than .05 it would have meant that none of the variables excluded from the model could make a significant contribution to the predictive power of the model. As such, the analysis would have terminated at this stage.

The remainder of this table lists each of the predictors in turn with a value of Roa's efficient score statistic for each one (column labelled *Score*). In large samples when the null hypothesis is true, the score statistic is identical to the Wald statistic and the likelihood ratio statistic. It is used at this stage of the analysis because it is computationally less intensive than the Wald statistic and so can still be calculated in situations when the Wald statistic would prove prohibitive. Like any test statistic, Roa's score statistic has a specific distribution from which statistical significance can be obtained. In this example, all excluded variables have significant score statistics at $p < .001$ and so all three could potentially make a contribution to the model. However, as mentioned in section 6.4.2, the stepwise calculations are relative and so the variable that will be selected for inclusion is the one with the highest value for the score statistic that is significant at a .05 level of significance. In this example, that variable will be **fb** because it has the highest value of the score statistic. The next part of the output deals with the model after this predictor has been added.

6.5.2. Step 1: false belief understanding ③

So, in the first step, false belief understanding (**fb**) is added to the model as a predictor. As such a child is now classified as having display rule understanding based on whether they passed or failed the false belief task. Now, this can be easily explained if we look at the crosstabulation for the variables **fb** and **display**.[1] The model will use false belief understanding to predict whether a child has display rule understanding by applying the crosstabulation table shown in Table 6.1. So, the model predicts that all of the children who showed false belief understanding will have display rule understanding. There were 41 children with false belief understanding, so the model predicts that these 41 children had display rule understanding (as such it is correct 33 times out of 41, and incorrect 8 times out of 41). In addition, this new model predicts that all of the 29 children who didn't show false belief understanding did not show display rule understanding (as such it is correct 23 times out of 29 and incorrect 6 times out of 29).

SPSS Output 6.3 shows summary statistics about the new model (which we've already seen contains **fb**). The overall fit of the new model is assessed using the log-likelihood statistic (see section 6.3.1). In SPSS, rather than reporting the log-likelihood itself, the value is multiplied by –2 (and sometimes referred to as –2LL): this multiplication is done because –2LL has an approximately chi-square distribution and so makes it possible to compare values against those that we might expect to get by chance alone. Remember that large values of the log-likelihood statistic indicate poorly fitting statistical models.

Table 6.1 Crosstabulation of display rule understanding with false belief understanding

		False Belief Understanding (fb)	
		No	Yes
Display Rule	No	23	8
Understanding (display)	Yes	6	33

1 The dialog box to produce this table can be obtained through the menus by **Analyze⇒Descriptive Statistics⇒Crosstabs**.

Omnibus Tests of Model Coefficients

		Chi-square	df	Sig.
Step 1	Step	26.083	1	.000
	Block	26.083	1	.000
	Model	26.083	1	.000

Model Summary

Step	−2 Log likelihood	Cox & Snell R Square	Nagelkerke R Square
1	70.042	.311	.417

Classification Table[a]

Observed			Predicted		
			Display Rule understanding		Percentage Correct
			No	Yes	
Step 1	Display Rule understanding	No	23	8	74.2
		Yes	6	33	84.6
	Overall Percentage				80.0

a The cut value is .500

SPSS Output 6.3

At this stage of the analysis the value of −2 × log-likelihood should be less than the value when only the constant was included in the model (because lower values of −2LL indicate that the model is predicting the outcome variable more accurately). When only the constant was included, −2LL = 96.124, but now **fb** has been included this value has been reduced to 70.042. This reduction tells us that the model is better at predicting display rule understanding than it was before **fb** was added. The question of how much better the model predicts the outcome variable can be assessed using the *model chi-square statistic*, which measures the difference between the model as it currently stands and the model when only the constant was included. We saw in section 6.3.1 that we could assess the significance of the change in a model by taking the log-likelihood of the new model and subtracting the log-likelihood of the baseline model from it. The value of the model chi-square statistic works on this principle and is, therefore, equal to −2LL with **fb** included minus the value of −2LL when only the constant was in the model (96.124 − 70.042 = 26.083). This value has a chi-square distribution and so its statistical significance can be easily calculated.[2]

2 The degrees of freedom will be the number of parameters in the new model (the number of predictors plus 1, which in this case, with one predictor, means 2) minus the number of parameters in the baseline model (which is 1, the constant). So, in this case, $df = 2 - 1 = 1$.

In this example, the value is significant at a .05 level and so we can say that overall the model is predicting display rule understanding significantly better than it was with only the constant included. The model chi-square is an analogue of the *F*-test for the linear regression sum of squares (see Chapter 5). In an ideal world we would like to see a non-significant −2LL (indicating that the amount of unexplained data is minimal) and a highly significant model chi-square statistic (indicating that the model including the predictors is significantly better than without those predictors). However, in reality it is possible for both statistics to be highly significant.

There is a second statistic called the *step* statistic that indicates the improvement in the predictive power of the model since the last stage. At this stage there has been only one step in the analysis and so the value of the improvement statistic is the same as the model chi-square. However, in more complex models in which there are three or four stages, this statistic gives you a measure of the improvement of the predictive power of the model since the last step. Its value is equal to −2LL at the current step minus −2LL at the previous step. If the improvement statistic is significant then it indicates that the model now predicts the outcome significantly better than it did at the last step, and in a forward regression this can be taken as an indication of the contribution of a predictor to the predictive power of the model. Similarly, the *block* statistic provides the change in −2LL since the last block (for use in hierarchical or blockwise analyses).

Finally, the classification table at the end of this section of the output indicates how well the model predicts group membership. Because the model is using false belief understanding to predict the outcome variable, this classification table is the same as Table 6.1. The current model correctly classifies 23 children who don't have display rule understanding but misclassifies 8 others (i.e. it correctly classifies 74.19% of cases). For children who do have display rule understanding, the model correctly classifies 33 and misclassifies 6 cases (i.e. correctly classifies 84.62% of cases). The overall accuracy of classification is, therefore, the weighted average of these two values (80%). So, when only the constant was included, the model correctly classified 56% of children, but now, with the inclusion of **fb** as a predictor, this has risen to 80%.

The next part of the output (SPSS Output 6.4) is crucial because it tells us the estimates for the coefficients for the predictors included in the model. This section of the output gives us the coefficients and statistics for the variables that have been included in the model at this point (namely, **fb** and the constant). The *b*-value is the same as the *b*-value in linear regression: they are the values that we need to replace in equation (6.4) to establish

Variables in the Equation

								95.0% C.I. for EXP(B)	
		B	S.E.	Wald	df	Sig.	Exp(B)	Lower	Upper
Step 1[a]	FB(1)	2.761	.605	20.856	1	.000	15.812	4.835	51.706
	Constant	−1.344	.458	8.592	1	.003	.261		

a Variable(s) entered on step 1: FB.

SPSS Output 6.4

the probability that a case falls into a certain category. We saw in linear regression that the value of b represents the change in the outcome resulting from a unit change in the predictor variable. The interpretation of this coefficient in logistic regression is very similar in that it represents the change in the *logit* of the outcome variable associated with a one-unit change in the predictor variable. The logit of the outcome is simply the natural logarithm of the odds of Y occurring.

The crucial statistic is the Wald statistic,[3] which has a chi-square distribution and tells us whether the b-coefficient for that predictor is significantly different from zero. If the coefficient is significantly different from zero then we can assume that the predictor is making a significant contribution to the prediction of the outcome (Y). We came across the Wald statistic in section 6.3.3 and saw that it should be used cautiously because, when the regression coefficient (b) is large, the standard error tends to become inflated, resulting in the Wald statistic being underestimated (see Menard, 1995). However, for these data it seems to indicate that false belief understanding is a significant predictor of display rule understanding (note that the significance of the Wald statistic is less than .05).

In section 6.3.2 we saw that we could calculate an analogue of R using equation (6.7). For these data, the Wald statistic and its *df* can be read from SPSS Output 6.4 (20.856 and 1 respectively), and the original −2LL was 96.12. Therefore, R can be calculated as:

$$\begin{aligned} R &= \pm\sqrt{\left(\frac{20.856 - (2 \times 1)}{96.124}\right)} \\ &= .4429 \end{aligned} \tag{6.14}$$

In the same section we saw that Hosmer and Lemeshow's measure (R^2_L) is calculated by dividing the model chi-square by the *original* −2LL. In this example the model chi-square after all variables have been entered into the model is 26.083, and the original −2LL (before any variables were entered) was 96.124. So, $R^2_L = 26.083/96.124 = .271$, which is different to the value we would get by squaring the value of R given above ($R^2 = .4429^2 = .196$).

Earlier on in SPSS Output 6.3, SPSS gave us two other measures of R^2 that were described in section 6.3.2. The first is Cox and Snell's measure,[4] which SPSS reports

3 SPSS actually quotes the Wald statistic squared, so for these data it would be $(2.761/.605)^2 = 20.8$ as reported in the table—see equation (6.11).

4 This is calculated from equation (6.9). Remember that this equation uses the log-likelihood, whereas SPSS reports −2 × log-likelihood. *LL(New)* is, therefore, 70.042/−2 = −35.021, and the *LL(Baseline)* = 96.124/−2 = −48.062. The sample size, n, is 70. Thus:

$$\begin{aligned} R^2_{CS} &= 1 - e^{\{-\frac{2}{70}[-35.021-(-48.062)]\}} \\ &= 1 - e^{-0.3726} \\ &= 1 - 0.6889 \\ &= .311 \end{aligned}$$

as .311, and the second is Nagelkerke's adjusted value,[5] which SPSS reports as .417. As you can see, there's a fairly substantial difference between the two values!

The final thing we need to look at is exp *b* (*Exp(B)* in the SPSS output), which was described in section 6.3.4. To calculate the change in odds that results from a unit change in the predictor for this example, we must first calculate the odds of a child having display rule understanding given that they *don't* have second-order false belief task understanding. We then calculate the odds of a child having display rule understanding given that they *do* have false belief understanding. Finally, we calculate the proportionate change in these two odds.

To calculate the first set of odds, we need to use equation (6.12) to calculate the probability of a child having display rule understanding given that they failed the false belief task. The parameter coding at the beginning of the output told us that children who failed the false belief task were coded with a 0, so we can use this value in place of *X*. The value of b_1 has been estimated for us as 2.7607 (see *Variables in the Equation* in SPSS Output 6.4), and the coefficient for the constant can be taken from the same table and is −1.3437. We can calculate the odds as in equation (6.15):
Now, we calculate the same thing after the predictor variable has changed by one unit. In this case, because the predictor variable is dichotomous, we need to calculate the odds of

$$\begin{aligned} P(\text{event } Y) &= \frac{1}{1+e^{-(b_0+b_1X_1)}} \\ &= \frac{1}{1+e^{-[-1.3437+(2.7607\times 0)]}} \\ &= .2069 \end{aligned}$$

$$\begin{aligned} P(\text{no event } Y) &= 1 - P(\text{event } Y) \\ &= 1 - .2069 \\ &= .7931 \end{aligned} \tag{6.15}$$

$$\begin{aligned} \text{odds} &= \frac{.2069}{.7931} \\ &= .2609 \end{aligned}$$

5 This is calculated from equation (6.10)—there's a slight difference because I've used values rounded to 3 decimal places:

$$\begin{aligned} R_N^2 &= \frac{.311}{1-e^{[\frac{2(-48.062)}{70}]}} \\ &= \frac{.311}{1-e^{-1.3732}} \\ &= \frac{.311}{1-0.2533} \\ &= .416 \end{aligned}$$

a child passing the display rule task, given that they have *passed* the false belief task. So, the value of the false belief variable, X, is now 1 (rather than 0). The resulting calculations are shown in equation (6.16):

$$\begin{aligned}
P(\text{event } Y) &= \frac{1}{1+e^{-(b_0+b_1X_1)}} \\
&= \frac{1}{1+e^{-[-1.3437+(2.7607\times 1)]}} \\
&= .8049 \\[1em]
P(\text{no event } Y) &= 1 - P(\text{event } Y) \\
&= 1 - .8049 \\
&= .1951 \\[1em]
\text{odds} &= \frac{.8049}{.1951} \\
&= 4.1256
\end{aligned} \tag{6.16}$$

We now know the odds before and after a unit change in the predictor variable. It is now a simple matter to calculate the proportionate change in odds by dividing the odds after a unit change in the predictor by the odds before that change:

$$\begin{aligned}
\Delta\text{odds} &= \frac{\text{odds after a unit change in the predictor}}{\text{original odds}} \\
&= \frac{4.1256}{.2609} \\
&= 15.8129
\end{aligned} \tag{6.17}$$

You should notice that the value of the proportionate change in odds is the same as the value that SPSS reports for exp b (allowing for differences in rounding). We can interpret exp b in terms of the change in odds. If the value is greater than 1 then it indicates that as the predictor increases, the odds of the outcome occurring increase. Conversely, a value less than 1 indicates that as the predictor increases, the odds of the outcome occurring decrease. In this example, we can say that the odds of a child who has false belief understanding also having display rule understanding are 15 times higher than those of a child who does not have false belief understanding.

In the options (see section 6.4.5), we requested a confidence interval for exp b and it can also be found in the output. The way to interpret this confidence interval is to say that if we ran 100 experiments and calculated confidence intervals for the value of exp b, then these intervals would encompass the actual value of exp b in the population (rather than the sample) on 95 occasions. So, in this case, we can be fairly confident that the population

Model if Term Removed

Variable		Model Log Likelihood	Change in −2 Log Likelihood	df	Sig. of the Change
Step 1	FB	−48.062	26.083	1	.000

Variables not in the Equation

			Score	df	Sig.
Step 1	Variables	AGE	2.313	1	.128
		AGE by FB(1)	1.261	1	.261
	Overall Statistics		2.521	2	.283

SPSS Output 6.5

value of exp b lies between 4.84 and 51.71. However, there is a 5% chance that a sample could give a confidence interval that 'misses' the true value.

The test statistics for **fb** if it were removed from the model are in SPSS Output 6.5. Now, recall earlier on I said how the regression would place variables into the equation and then test whether they then met a removal criterion. Well, the *Model if Term Removed* part of the output tells us the effects of removal. The important thing to note is the significance value of the log-likelihood ratio (log LR). The log LR for this model is highly significant ($p < .0001$) which tells us that removing **fb** from the model would have a significant effect on the predictive ability of the model—in other words, it would be a very bad idea to remove it!

Finally, we are told about the variables currently not in the model. First of all, the residual chi-square (labelled *Overall Statistics* in the output), which is non-significant, tells us that none of the remaining variables have coefficients significantly different from zero. Furthermore, each variable is listed with its score statistic and significance value, and for both variables their coefficients are not significantly different from zero (as can be seen from the significance values of .128 for age and .261 for the interaction of age and false belief understanding). Therefore, no further variables will be added to the equation.

The next part of the output (SPSS Output 6.6) displays the classification plot that we requested in the *options* dialog box. This plot is a histogram of the predicted probabilities of a child passing the display rule task. If the model perfectly fits the data, then this histogram should show all of the cases for which the event has occurred on the right-hand side, and all the cases for which the event hasn't occurred on the left-hand side. In other words, all the children who passed the display rule task should appear on the right and all those who failed should appear on the left. In this example, the only significant predictor is dichotomous and so there are only two columns of cases on the plot. If the predictor is

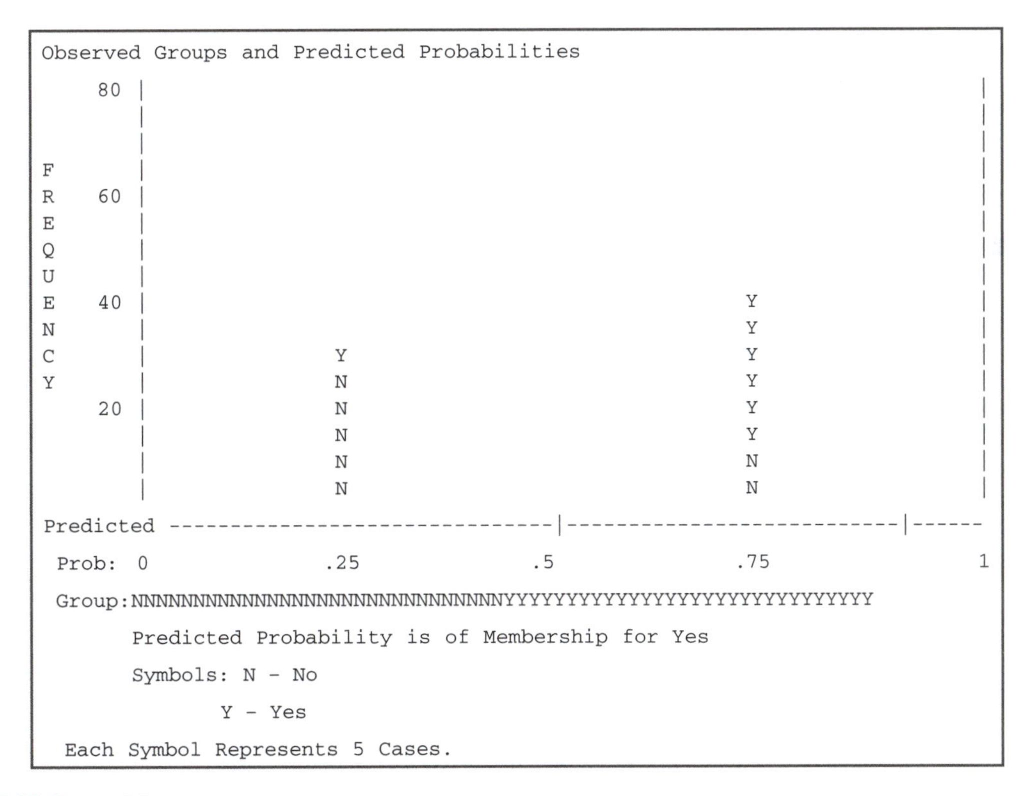

```
Observed Groups and Predicted Probabilities
     80 |                                                              |
        |                                                              |
        |                                                              |
F       |                                                              |
R    60 |                                                              |
E       |                                                              |
Q       |                                                              |
U       |                                                              |
E    40 |                                       Y                      |
N       |                                       Y                      |
C       |            Y                          Y                      |
Y       |            N                          Y                      |
     20 |            N                          Y                      |
        |            N                          Y                      |
        |            N                          N                      |
        |            N                          N                      |
Predicted -----------------------------|--------------------------|------
 Prob: 0            .25              .5               .75             1
 Group:NNNNNNNNNNNNNNNNNNNNNNNNNNNNNNYYYYYYYYYYYYYYYYYYYYYYYYYYYYYY
       Predicted Probability is of Membership for Yes
       Symbols: N - No
                Y - Yes
  Each Symbol Represents 5 Cases.
```

SPSS Output 6.6

a continuous variable, the cases are spread out across many columns. As a rule of thumb, the more that the cases cluster at each end of the graph, the better. This statement is true because such a plot would show that when the outcome did actually occur (i.e. the child did pass the display rule task) the predicted probability of the event occurring is also high (i.e. close to 1). Likewise, at the other end of the plot it would show that when the event didn't occur (i.e. when the child failed the display rule task) the predicted probability of the event occurring is also low (i.e. close to 0). This situation represents a model that is correctly predicting the observed outcome data. If, however, there are a lot of points clustered in the centre of the plot then it shows that for many cases the model is predicting a probability of .5 that the event will occur. In other words, for these cases there is little more than a 50:50 chance that the data are correctly predicted—as such the model could predict these cases just as accurately by simply tossing a coin! Also, a good model will ensure that few cases are misclassified; in this example there are two *N*s on the right of the model and one *Y* on the left of the model. These are misclassified cases, and the fewer of these there are, the better the model.

Cramming Samantha's Tips

- The overall fit of the final model is shown by the −2 × log-likelihood statistic and its associated chi-square statistic. If the significance of the chi-square statistic is less than .05, then the model is a significant fit of the data.
- Check the table labelled *Variables in the Equation* to see which variables significantly predict the outcome.
- For each variable in the model, look at the Wald statistic and its significance (which again should be below .05). More important, though, use the value of *Exp(B)* for interpretation. If the value is greater than 1 then as the predictor increases, the odds of the outcome occurring increase. Conversely, a value less than 1 indicates that as the predictor increases, the odds of the outcome occurring decrease. For the aforementioned interpretation to be reliable the confidence interval of *Exp(B)* should not cross 1!
- Check the table labelled *Variables not in the Equation* to see which variables did not significantly predict the outcome.

6.5.3. Listing predicted probabilities ②

It is possible to list the expected probability of the outcome variable occurring based on the final model. In section 6.4.4 we saw that SPSS could save residuals and also predicted probabilities. SPSS saves these predicted probabilities and predicted group memberships as variables in the data editor and names them PRE_1 and PGR_1 respectively. These probabilities can be listed using the **Analyze⇒Reports⇒Case Summaries...** dialog box (see section 5.8.6). SPSS Output 6.7 shows a selection of the predicted probabilities (because the only significant predictor was a dichotomous variable, there will be only two different probability values). It is also worth listing the predictor variables as well to clarify from where the predicted probabilities come.

We found from the model that the only significant predictor of display rule understanding was false belief understanding. This could have a value of either 1 (pass the false belief task) or 0 (fail the false belief task). If these two values are placed into equation (6.4) with the respective regression coefficients, then the two probability values are derived. In fact, we calculated these values as part of equation (6.15) and equation (6.16) and you should note that the calculated probabilities correspond to the values in PRE_1. These values tells us that when a child doesn't possess second-order false belief understanding (**fb** = 0, No), there is a probability of .2069 that they will pass the display rule task—approximately a 21% chance (1 out of 5 children). However, if the child does pass the false belief task (**fb** = 1, yes), there is a probability of .8049 that they will pass the display rule task—an 80.5% chance (4 out of 5 children). Consider that a probability of 0 indicates no chance of the child passing the display rule task, and a probability of 1 indicates that the child will definitely pass the display rule task. Therefore, the values

Case Summaries[a]

	Case Number	Age in years	False Belief understanding	Display Rule understanding	Predicted probability	Predicted group
1	1	24.00	No	No	.20690	No
2	5	36.00	No	No	.20690	No
3	9	34.00	No	Yes	.20690	No
4	10	31.00	No	No	.20690	No
5	11	32.00	No	No	.20690	No
6	12	30.00	Yes	Yes	.80488	Yes
7	20	26.00	No	No	.20690	No
8	21	29.00	No	No	.20690	No
9	29	45.00	Yes	Yes	.80488	Yes
10	31	41.00	No	Yes	.20690	No
11	32	32.00	No	No	.20690	No
12	43	56.00	Yes	Yes	.80488	Yes
13	60	63.00	No	Yes	.20690	No
14	66	79.00	Yes	Yes	.80488	Yes
Total N		14	14	14	14	14

a Limited to first 100 cases

SPSS Output 6.7

obtained provide strong evidence for the role of false belief understanding as a prerequisite for display rule understanding.

Assuming we are content that the model is accurate and that false belief understanding has some substantive significance, then we could conclude that false belief understanding is the single best predictor of display rule understanding. Furthermore age and the interaction of age and false belief understanding in no way predict display rule understanding. As a homework task, why not rerun this analysis using the forced entry method of analysis—how do your conclusions differ?

This conclusion is fine in itself, but to be sure that the model is a good one, it is important to examine the residuals. In section 6.4.4 we saw how to get SPSS to save various residuals in the data editor. We can now list these residuals using the **Analyze⇒Reports⇒Case Summaries** dialog box and interpret them.

6.5.4. Interpreting residuals ②

The main purpose of examining residuals in logistic regression is (1) to isolate points for which the model fits poorly, and (2) to isolate points that exert an undue influence on the model. To assess the former we examine the residuals, especially the Studentized residual, standardized residual and deviance statistics. All of these statistics have the common property that 95% of cases in an average, normally distributed sample should have values which lie within ±1.96, and 99% of cases should have values that lie within ±2.58. Therefore, any values outside of ±3 are cause for concern and any outside of about ±2.5 should be examined more closely. To assess the influence of individual cases we use

Table 6.2 Summary of residual statistics saved by SPSS

Label	Name	Comment
PRE_1	Predicted Value	
PGR_1	Predicted Group	
COO_1	Cook's Distance	Should be less than 1
LEV_1	Leverage	Lies between 0 (no influence) and 1 (complete influence). The expected leverage is $(k+1)/N$, where k is the number of predictors and N is the sample size. In this case it would be $2/70 = 0.03$
RES_1	Residual	
LRE_1	Logit Residual	
SRE_1	Studentized Residual	5% should lie between ±1.96, and 1% should lie between ±2.58.
ZRE_1	Standardized Residual	Cases above 3 are cause for concern and cases close to 3
DEV_1	Deviance	warrant inspection
DFB0_1	DFBeta for the Constant	Should be less than 1
DFB1_1	DFBeta for the first predictor (**fb**)	

influence statistics such as Cook's distance (which is interpreted in the same way as for linear regression: as a measure of the change in the regression coefficient if a case is deleted from the model). Also, the value of DFBeta, which is a standardized version of Cook's statistic, tells us something of the influence of certain cases—any values greater than 1 indicate possible influential cases. Additionally, leverage statistics or hat values, which should lie between 0 (the case has no influence whatsoever) and 1 (the case exerts complete influence over the model), tell us about whether certain cases are wielding undue influence over the model. The expected value of leverage is defined as for linear regression. These statistics were explained in more detail in Chapter 5.

If you request these residual statistics, SPSS saves them in as new columns in the data editor; Table 6.2 lists each of the variable names and what they represent. There are additional comments to summarize the expected values of some of the statistics.

SPSS Output 6.8 shows the basic residual statistics for this example (Cook's distance, leverage, standardized residuals and DFBeta values). Note that all cases have DFBetas less than 1, and leverage statistics (LEV_1) close to the calculated expected value of 0.03. There are also no unusually high values of Cook's distance (COO_1) which, all in all, means that there are no influential cases having an effect on the model. Cook's distance is an unstandardized measure and so there is no absolute value at which you can say that a case is having an influence. Instead, you should look for values of Cook's distance which are particularly high compared to the other cases in the sample. However, Stevens (1992) suggests that a value greater than 1 is problematic. About half of the leverage values are a little high but given that the other statistics are fine, this is probably no cause for concern.

Case Summaries[a]

	Case Number	Analog of Cook's influence statistics	Leverage value	Normalized residual	DFBETA for constant	DFBETA for FB(1)
1	1	.00932	.03448	−.51075	−.04503	.04503
2	2	.00932	.03448	−.51075	−.04503	.04503
3	3	.00932	.03448	−.51075	−.04503	.04503
4	4	.00932	.03448	−.51075	−.04503	.04503
5	5	.00932	.03448	−.51075	−.04503	.04503
6	6	.13690	.03448	1.95789	.17262	−.17262
7	7	.00932	.03448	−.51075	−.04503	.04503
8	8	.00932	.03448	−.51075	−.04503	.04503
9	9	.13690	.03448	1.95789	.17262	−.17262
10	10	.00932	.03448	−.51075	−.04503	.04503
11	11	.00932	.03448	−.51075	−.04503	.04503
12	12	.00606	.02439	.49237	.00000	.03106
13	13	.00932	.03448	−.51075	−.04503	.04503
14	14	.10312	.02439	−2.03101	.00000	−.12812
15	15	.00932	.03448	−.51075	−.04503	.04503
16	16	.00932	.03448	−.51075	−.04503	.04503
17	17	.13690	.03448	1.95789	.17262	−.17262
18	18	.00932	.03448	−.51075	−.04503	.04503
19	19	.00606	.02439	.49237	.00000	.03106
20	20	.00932	.03448	−.51075	−.04503	.04503
21	21	.00932	.03448	−.51075	−.04503	.04503
22	22	.00606	.02439	.49237	.00000	.03106
23	23	.00606	.02439	.49237	.00000	.03106
24	24	.10312	.02439	−2.03101	.00000	−.12812
25	25	.00932	.03448	−.51075	−.04503	.04503
26	26	.00932	.03448	−.51075	−.04503	.04503
27	27	.00932	.03448	−.51075	−.04503	.04503
28	28	.00932	.03448	−.51075	−.04503	.04503
29	29	.00606	.02439	.49237	.00000	.03106
30	30	.00932	.03448	−.51075	−.04503	.04503
31	31	.13690	.03448	1.95789	.17262	−.17262
32	32	.00932	.03448	−.51075	−.04503	.04503
33	33	.00606	.02439	.49237	.00000	.03106
34	34	.00606	.02439	.49237	.00000	.03106
35	35	.00606	.02439	.49237	.00000	.03106
36	36	.00606	.02439	.49237	.00000	.03106
37	37	.10312	.02439	−2.03101	.00000	−.12812
38	38	.00606	.02439	.49237	.00000	.03106
39	39	.13690	.03448	1.95789	.17262	−.17262
40	40	.10312	.02439	−2.03101	.00000	−.12812
41	41	.00606	.02439	.49237	.00000	.03106
42	42	.00606	.02439	.49237	.00000	.03106
43	43	.00606	.02439	.49237	.00000	.03106
44	44	.00606	.02439	.49237	.00000	.03106
45	45	.00606	.02439	.49237	.00000	.03106
Total N		45	45	45	45	45

a Limited to first 100 cases

Case Summaries[a]

	Case Number	Analog of Cook's influence statistics	Leverage value	Normalized residual	DFBETA for constant	DFBETA for FB(1)
1	46	.10312	.02439	−2.03101	.00000	−.12812
2	47	.00606	.02439	.49237	.00000	.03106
3	48	.00932	.03448	−.51075	−.04503	.04503
4	49	.00932	.03448	−.51075	−.04503	.04503
5	50	.10312	.02439	−2.03101	.00000	−.12812
6	51	.00606	.02439	.49237	.00000	.03106
7	52	.00606	.02439	.49237	.00000	.03106
8	53	.00606	.02439	.49237	.00000	.03106
9	54	.00606	.02439	.49237	.00000	.03106
10	55	.00606	.02439	.49237	.00000	.03106
11	56	.00606	.02439	.49237	.00000	.03106
12	57	.10312	.02439	−2.03101	.00000	−.12812
13	58	.00606	.02439	.49237	.00000	.03106
14	59	.00606	.02439	.49237	.00000	.03106
15	60	.13690	.03448	1.95789	.17262	−.17262
16	61	.00606	.02439	.49237	.00000	.03106
17	62	.00606	.02439	.49237	.00000	.03106
18	63	.00606	.02439	.49237	.00000	.03106
19	64	.00606	.02439	.49237	.00000	.03106
20	65	.00606	.02439	.49237	.00000	.03106
21	66	.00606	.02439	.49237	.00000	.03106
22	67	.00606	.02439	.49237	.00000	.03106
23	68	.10312	.02439	−2.03101	.00000	−.12812
24	69	.00606	.02439	.49237	.00000	.03106
25	70	.00606	.02439	.49237	.00000	.03106
Total N		25	25	25	25	25

a Limited to first 100 cases

SPSS Output 6.8

The standardized residuals all have values of less than ±2.5 and predominantly have values less than ±2 and so there seems to be very little here to concern us.

You should note that these residuals are slightly unusual because they are based on a single predictor that is categorical. This is why there isn't a lot of variability in the values of the residuals. Also, if substantial outliers or influential cases had been isolated, you are not justified in eliminating these cases to make the model fit better. Instead these cases should be inspected closely to try to isolate a good reason why they were unusual. It might simply be an error in inputting data, or it could be that the case was one which had a special reason for being unusual: for example, the child had found it hard to pay attention to the false belief task and you had noted this at the time of the experiment. In such a case, you may have good reason to exclude the case and duly note the reasons why.

Cramming Samantha's Tips

You need to look for cases that might be influencing the logistic regression model.

- Look at standardized residuals and check that no more than 5% of cases have absolute values above 2, and that no more than about 1% have absolute values above 2.5. Any case with a value above about 3 could be an outlier.
- Look in the data editor for the values of Cook's distance: any value above 1 indicates a case that might be influencing the model.
- Calculate the average leverage (the number of predictors plus 1, divided by the sample size) and then look for values greater than twice or three times this average value.
- Look for absolute values of DFBeta greater than 1.

6.5.5. Calculating the effect size ②

We've already seen (section 6.3.2) that SPSS produces a value of *r* for each predictor, based on that Wald statistic. This is your effect size measure; however, it's worth bearing in mind what I've already said about it: it won't be accurate when the Wald statistic is inaccurate!

6.6. HOW TO REPORT LOGISTIC REGRESSION ②

To be honest, because logistic regression is fairly rarely used (in my discipline of psychology at any rate), it's difficult to find any concrete guidelines about how to report one! So, this is just my personal view on what should be reported based on the American Psychological Association guidelines for reporting multiple regression (see section 5.9). As with multiple regression I'd be in favour of tabulating the results, unless it's a very simple model. As a bare minimum, the beta values and their standard errors and significance value and some general statistics about the model (such as the R^2 and goodness-of-fit statistics) should be reported. I'd also highly recommend reporting *Exp(B)* and its confidence interval. I'd like to see the constant as well because then readers of your work can construct the full regression model if they need to. You might also consider reporting the variables that were not significant predictors as this can be as valuable as knowing which predictors were significant. For the example in this chapter we might produce a table like that in Table 6.3.

Hopefully you can work out from where the values came by looking back through the chapter so far. As with multiple regression, I've rounded off to 2 decimal places throughout; for the R^2 and standardized betas there is no zero before the decimal point (because these values cannot exceed 1) but for all other values less than 1 the zero is present; and the significance of the variable is denoted by asterisks with a footnote to indicate the significance level being used.

Table 6.3 How to report logistic regression

		95% CI for exp *b*		
	B (SE)	Lower	exp *b*	Upper
Included				
Constant	−1.34* (0.46)	4.84	15.81	51.71
False Belief Understanding	2.76** (0.60)			

Note $R^2 = .27$ (Hosmer & Lemeshow), .31 (Cox & Snell), .42 (Nagelkerke). Model $\chi^2(1) = 26.08$, $p < .001$. * $p < .01$. ** $p < .001$.

6.7. ANOTHER EXAMPLE ②

This example was originally inspired by events in the soccer World Cup finals of 1998 (a long time ago now, but such crushing disappointments are not easily forgotten!). Unfortunately for me (being an Englishman), I was subjected to watching England get knocked out of the championships by losing a penalty shootout. Reassuringly, 6 years later I have just watched (yesterday) England get knocked out of the European championships in another penalty shootout! I am consoled by the fact they are consistent in their ineptitude.

Now, if I were the England coach, I might be interested in finding out what factors predict whether or not a player will score a penalty. Those of you who hate football can read this example as being factors that predict success in a free-throw in basketball or netball, a penalty in hockey or a penalty kick in rugby[6] or a field goal in American football. Now, this research question is perfect for logistic regression because our outcome variable is a dichotomy: a penalty can be either scored or missed. Imagine that past research (Eriksson, Beckham & Vassell, 2004; Hoddle, Batty & Ince, 1998) had shown that there are two factors that reliably predict whether a penalty kick will be saved or scored. The first factor is whether the player taking the kick is a worrier (this factor can be measured using a measure such as the Penn State Worry Questionnaire, PSWQ). The second factor is the player's past success rate at scoring (so, whether the player has a good track record

6 Although this would be an unrealistic example because our rugby team, unlike their football counterparts, have Jonny Wilkinson who is the lord of penalty kicks and we bow at his great left foot in wonderment (well, I do …)!

of scoring penalty kicks). It is fairly well accepted that anxiety has detrimental effects on performance of a variety of tasks and so it was also predicted that state anxiety might be able to account for some of the unexplained variance in penalty success.

This example is a classic case of building on a well-established model, because two predictors are already known and we want to test the effect of a new one. So, 75 football players were selected at random and before taking a penalty kick in competition they were given a state anxiety questionnaire to complete (to assess anxiety before the kick was taken). These players were also asked to complete the PSWQ to give a measure of how much they worried about things generally, and their past success rate was obtained from a database. Finally, a note was made of whether the penalty was scored or missed. The data can be found in the file **penalty.sav** which contains four variables, each in a separate column.

- **Scored**: This variable is our outcome and it is coded such that 0 = penalty missed, and 1 = penalty scored.
- **Pswq**: This variable is the first predictor variable and it gives us a measure of the degree to which a player worries.
- **Previous**: This variable is the percentage of penalties scored by a particular player in their career. As such, it represents previous success at scoring penalties.
- **Anxious**: This variable is our third predictor and it is a variable that has not previously been used to predict penalty success. **Anxious** is a measure of state anxiety before taking the penalty.

6.7.1. Running the analysis: block entry regression ②

To run the analysis, we must first select the main *logistic regression* dialog box, and using the mouse to specify **Analyze⇒Regression⇒Binary Logistic** can do this. In this example, we know of two previously established predictors and so it is a good idea to enter these predictors into the model in a single block. Then we can add the new predictor in a second block (by doing this we effectively examine an old model and then add a new variable to this model to see whether the model is improved). This method is known as block entry and Figure 6.8 shows how it is specified.

It is easy to do block entry regression. First, you should use the mouse to select the variable **scored** from the variables list and then transfer it to the box labelled *Dependent* by clicking on ▸. Second, you should select the two previously established predictors. So, select **pswq** and **previous** from the variables list and transfer them to the box labelled *Covariates* by clicking on ▸. Our first block of variables is now specified. To specify the second block, click on Next to clear the *Covariates* box, which should now be labelled *Block 2 of 2*. Now select **anxious** from the variables list and transfer it to the box labelled *Covariates* by clicking on ▸. We could at this stage select some interactions to be included in the model, but unless there is a sound theoretical reason for believing that the predictors should interact there is no need. Make sure that *Enter* is selected as the method of regression (this method is the default and so should be selected already).

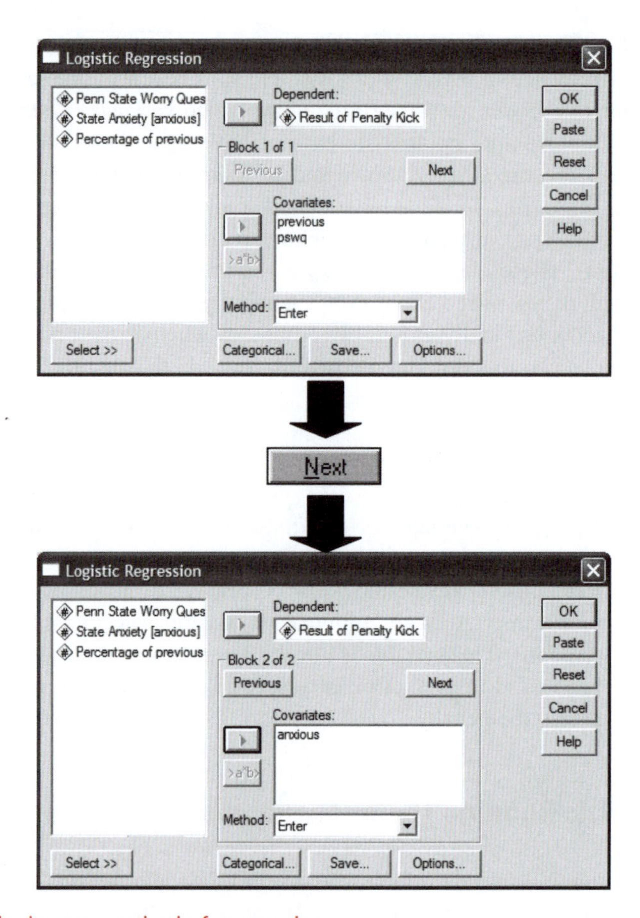

Figure 6.8 Block entry method of regression

Once the variables have been specified, you should select the options described in sections 6.4.4 and 6.4.5, but because none of the predictors are categorical there is no need to use the *Categorical...* option. When you have selected the options and residuals that you want you can return to the main *logistic regression* dialog box and click on OK.

6.7.2. Interpreting output ③

The output of the logistic regression will be arranged in terms of the blocks that were specified. In other words, SPSS will produce a regression model for the variables specified in block 1, and then produce a second model that contains the variables from both blocks 1 and 2. First, SPSS Output 6.9 shows the results from block 0: the output tells us that 75 cases have been accepted, and that the dependent variable has been coded 0 and 1

Dependent Variable Encoding

Original Value	Internal Value
Missed Penalty	0
Scored Penalty	1

Block 0: Beginning Block

Classification Table[a,b]

Observed			Predicted		
			Result of Penalty Kick		Percentage Correct
			Missed Penalty	Scored Penalty	
Step 0	Result of Penalty Kick	Missed Penalty	0	35	.0
		Scored Penalty	0	40	100.0
	Overall Percentage				53.3

a Constant is included in the model
b The cut value is .500

Variables in the Equation

		B	S.E.	Wald	df	Sig.	Exp(B)
Step 0	Constant	.134	.231	.333	1	.564	1.143

Variables not in the Equation

			Score	df	Sig.
Step 0	Variables	PREVIOUS	34.109	1	.000
		PSWQ	34.193	1	.000
	Overall Statistics		41.558	2	.000

SPSS Output 6.9

(because this variable was coded as 0 and 1 in the data editor, these codings correspond exactly to the data in SPSS). We are then told about the variables that are in and out of the equation. At this point only the constant is included in the model, and so to be perfectly honest none of this information is particularly interesting!

The results from block 1 are shown in SPSS Output 6.10 and in this analysis we forced SPSS to enter **previous** and **pswq** into the regression model. Therefore, this part of the output provides information about the model after the variables **previous** and **pswq** have been added. The first thing to note is that the −2LL is 48.66, which is a change of 54.98 (which is the value given by the *model chi-square*). This value tells us about the model as a whole whereas the *block* tells us how the model has improved since the last block. The change in the amount of information explained by the model is significant ($p < .0001$) and so using previous experience and worry as predictors significantly improves our ability to predict penalty success. A bit further down, the classification table shows us that 84% of cases can be correctly classified using **pswq** and **previous**.

In the display rule example, SPSS did not produce Hosmer & Lemeshow's goodness-of-fit test. The reason is that this test can't be calculated when there is only one predictor and that predictor is a categorical dichotomy! However, for this example the test can be calculated. The important part of this test is the test statistic itself (7.93) and the significance value (.339). This statistic tests the hypothesis that the observed data are significantly different from the predicted values from the model. So, in effect, we want a non-significant value for this test (because this would indicate that the model does not differ significantly from the observed data). We have a non-significant value here, which is indicative of a model that is predicting the real-world data fairly well.

The part of SPSS Output 6.10 labelled *Variables in the Equation* then tells us the parameters of the model when **previous** and **pswq** are used as predictors. The significance values of the Wald statistics for each predictor indicate that both **pswq** and **previous** significantly predict penalty success ($p < .01$). The values of exp b for **previous** indicates that if the percentage of previous penalties scored goes up by one, then the odds of scoring a penalty also increase (because exp b is greater than 1). The confidence interval for this value ranges from 1.02 to 1.11 so we can be very confident that the value of exp b in the population lies somewhere between these two values. What's more, because both values are greater than 1 we can also be confident that the relationship between **previous** and penalty success found in this sample is true of the whole population of footballers. The values of exp b for **pswq** indicate that if the level of worry increases by one point along the Penn State worry scale, then the odds of scoring a penalty decrease (because exp b is less than 1). The confidence interval for this value ranges from .68 to .93 so we can be very confident that the value of exp b in the population lies somewhere between these two values. In addition, because both values are less than 1 we can be confident that the relationship between **pswq** and penalty success found in this sample is true of the whole population of footballers. If we had found that the confidence interval ranged from less than 1 to more than 1, then this would limit the generalizability of our findings because the value exp b in the population could indicate either a positive (*Exp(B)* > 1) or negative (*Exp(B)* < 1) relationship.

A glance at the classification plot (SPSS Output 6.11) also brings us good news because most cases are clustered at the ends of the plots and few cases lie in the middle of the plot. This reiterates what we know already: that the model is correctly classifying most cases. We can, at this point, also calculate R^2 (see section 6.3.2) by dividing the model chi-square by the original value of –2LL. The result is shown in equation (6.18) and we can interpret the result as meaning that the model can account for 53% of the variance in penalty success (so, roughly half of what makes a penalty kick successful is still unknown):

$$\begin{aligned} R^2 &= \frac{\text{model chi-square}}{\text{original} - 2\text{LL}} \\ &= \frac{54.977}{103.6385} \\ &= .53 \end{aligned} \tag{6.18}$$

Block 1: Method = Enter

Omnibus Tests of Model Coefficients

		Chi-square	df	Sig.
Step 1	Step	54.977	2	.000
	Block	54.977	2	.000
	Model	54.977	2	.000

Model Summary

Step	−2 Log likelihood	Cox & Snell R Square	Nagelkerke R Square
1	48.662	.520	.694

Hosmer and Lemeshow Test

Step	Chi-square	df	Sig.
1	7.931	7	.339

Contingency Table for Hosmer and Lemeshow Test

		Result of Penalty Kick = Missed Penalty		Result of Penalty Kick = Scored Penalty		
		Observed	Expected	Observed	Expected	Total
Step 1	1	8	7.904	0	.096	8
	2	8	7.779	0	.221	8
	3	8	6.705	0	1.295	8
	4	4	5.438	4	2.562	8
	5	2	3.945	6	4.055	8
	6	2	1.820	6	6.180	8
	7	2	1.004	6	6.996	8
	8	1	.298	7	7.702	8
	9	0	.108	11	10.892	11

Classification Table[a]

			Predicted		
			Result of Penalty Kick		
	Observed		Missed Penalty	Scored Penalty	Percentage Correct
Step 1	Result of Penalty Kick	Missed Penalty	30	5	85.7
		Scored Penalty	7	33	82.5
	Overall Percentage				84.0

a The cut value is .500

(Continued)

Variables in the Equation

		B	S.E.	Wald	df	Sig.	Exp(B)	95.0% C.I. for EXP(B)	
								Lower	Upper
Step 1[a]	PREVIOUS	.065	.022	8.609	1	.003	1.067	1.022	1.114
	PSWQ	−.230	.080	8.309	1	.004	.794	.679	.929
	Constant	1.280	1.670	.588	1	.443	3.598		

a. Variable(s) entered on step 1: PREVIOUS, PSWQ

SPSS Output 6.10

```
                     Observed Groups and Predicted Probabilities

        16 +                                                                  +
           I                                                                  I
           I                                                                  I
F          I                                                                  I
R       12 +                                                                  +
E          I                                                                 SI
Q          I                                                                 SI
U          I                                                                 SI
E        8 +                                                                 S+
N          IMM                                                               SI
C          IMM                                                               SI
Y          IMM                                                               SI
         4 +MM                                                              SS+
           IMM          M           SS                 S         S          SSI
           IMMM         M    M  S   SM  S           S SS    SM  SSS        SSSI
           IMMMM MM  M  M   MMS M   MM  S  S        S SM   MSM  SSM       SSSSI
Predicted ---------------+--------------+--------------+---------------
  Prob: 0               .25            .5             .75               1
  Group:    MMMMMMMMMMMMMMMMMMMMMMMMMMMMMMSSSSSSSSSSSSSSSSSSSSSSSSSSSSSS

          Predicted Probability is of Membership for Scored Penalty
          The Cut Value is .50
          Symbols: M - Missed Penalty
                   S - Scored Penalty
          Each Symbol Represents 1 Case.
```

SPSS Output 6.11

SPSS Output 6.12 shows what happens to the model when our new predictor is added (**anxious**). This part of the output describes block 2, which is just the model described in block 1 but with a new predictor added. So, we begin with the model that we had in block 1 and we then add **anxious** to it. The effect of adding anxious to the model is to reduce the –2LL to 47.416 (a reduction of 1.246 from the model in block 1 as shown in the *model chi-square* and *block* statistics). This improvement is non-significant, which tells us that including **anxious** in the model has not significantly improved our ability to predict whether a penalty will be scored or missed. The classification table tells us that the model is now correctly classifying 85.33% of cases. Remember that in block 1 there were 84% correctly classified and so an extra 1.33% of cases are now classified (not a great deal more—in fact, examining the table shows us that only one extra case has now been correctly classified).

The section labelled *Variables in the Equation* now contains all three predictors and something very interesting has happened: **pswq** is still a significant predictor of penalty success, but **previous** experience no longer significantly predicts penalty success. In addition, state anxiety appears not to make a significant contribution to the prediction of penalty success.

Block 2: Method = Enter

Omnibus Tests of Model Coefficients

		Chi-square	df	Sig.
Step 1	Step	1.246	1	.264
	Block	1.246	1	.264
	Model	56.223	3	.000

Model Summary

Step	–2 Log likelihood	Cox & Snell R Square	Nagelkerke R Square
1	47.416	.527	.704

Hosmer and Lemeshow Test

Step	Chi-square	df	Sig.
1	9.937	7	.192

Contingency Table for Hosmer and Lemeshow Test

		Result of Penalty Kick = Missed Penalty		Result of Penalty Kick = Scored Penalty		
		Observed	Expected	Observed	Expected	Total
Step 1	1	8	7.926	0	.074	8
	2	8	7.769	0	.231	8
	3	9	7.649	0	1.351	9
	4	4	5.425	4	2.575	8
	5	1	3.210	7	4.790	8
	6	4	1.684	4	6.316	8
	7	1	1.049	7	6.951	8
	8	0	.222	8	7.778	8
	9	0	.067	10	9.933	10

Classification Table[a]

Observed			Predicted: Result of Penalty Kick: Missed Penalty	Predicted: Result of Penalty Kick: Scored Penalty	Predicted: Percentage Correct
Step1	Result of Penalty Kick	Missed Penalty	30	5	85.7
		Scored Penalty	6	34	85.0
	Overall Percentage				85.3

a The cut value is .500

Variables in the Equation

		B	S.E.	Wald	df	Sig.	Exp(B)	95.0% C.I. for EXP(B) Lower	95.0% C.I. for EXP(B) Upper
Step 1[a]	PREVIOUS	.203	.129	2.454	1	.117	1.225	.950	1.578
	PSWQ	−.251	.084	8.954	1	.003	.778	.660	.917
	ANXIOUS	.276	.253	1.193	1	.275	1.318	.803	2.162
	Constant	−11.493	11.802	.948	1	.330	.000		

a Variable(s) entered on step 1: ANXIOUS

SPSS Output 6.12

How can it be that previous experience no longer predicts penalty success, and neither does anxiety, yet the ability of the model to predict penalty success has improved slightly?

The classification plot (SPSS Output 6.13) is similar to before and the contribution of **pswq** to predicting penalty success is relatively unchanged. What has changed is the contribution of previous experience. If we examine the values of exp *b* for both **previous** and **anxious** it is clear that they both potentially have a positive relationship to penalty success (i.e. as they increase by a unit, the odds of scoring improve). However, the confidence intervals for these values cross 1, which indicates that the direction of this relationship may be unstable in the population as a whole (i.e. the value of exp *b* in our sample may be quite different to the value if we had data from the entire population).

You may be tempted to use this final model to say that although worry is a significant predictor of penalty success, the previous finding that experience plays a role is incorrect. This would be a dangerous conclusion to make and now we shall see why.

6.8. TESTING FOR MULTICOLLINEARITY ③

In section 5.6.2.4 we saw how multicollinearity can affect the parameters of a regression model. Logistic regression is equally as prone to the biasing effect of collinearity and it is

```
                       Observed Groups and Predicted Probabilities

      16 +                                                                             +
         I                                                                             I
         I                                                                             I
F        I                                                                             I
R     12 +                                                                             +
E        I                                                                            SI
Q        I                                                                            SI
U        I                                                                            SI
E      8 +M                                                                           S+
N        IM                                                                           SI
C        IM                                                                           SI
Y        IMM                                                                         SSI
       4 +MM              M                                                S         SS+
         IMM              M     S                                          SS        SSI
         IMMMM            M     S              S                S  MS     SS        SSSI
         IMMMMMM  MM   M  M     S  S     MM   MMS    S     S     SSMSMM   SSSS   M  SSSSI
Predicted  -----------------+------------------+------------------+-----------------
   Prob:   0               .25                 .5                .75                1
   Group:    MMMMMMMMMMMMMMMMMMMMMMMMMMMMMMSSSSSSSSSSSSSSSSSSSSSSSSSSSSSSS

       Predicted Probability is of Membership for Scored Penalty
       The Cut Value is .50
       Symbols: M - Missed Penalty
                S - Scored Penalty
       Each Symbol Represents 1 Case.

-

   CASE Observed
         SCORED        Pred       PGroup        Resid        ZResid
        58  S  M**     .9312        S          -.9312       -3.6790

       S = Selected U=Unselected cases
       ** = Misclassified cases

* Cases with Studentized residuals greater than 2 are listed.
        The Cut Value is .50
```

SPSS Output 6.13

essential to test for collinearity following a logistic regression analysis. Unfortunately, SPSS does not have an option for producing collinearity diagnostics in logistic regression (which can create the illusion that it is unnecessary to test for it!). However, you can obtain

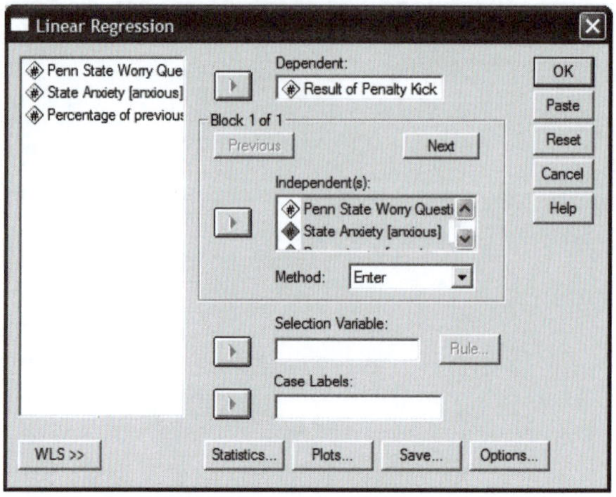

Figure 6.9 *Linear regression* dialog box for penalty data

statistics such as the tolerance and VIF by simply running a linear regression analysis using the same outcome and predictors. So, for the current example, access the *linear regression* dialog box by using the mouse to specify **Analyze⇒Regression⇒Linear**. The completed dialog box is shown in Figure 6.9. It is unnecessary to specify lots of options (we are using this technique only to obtain tests of collinearity) but it is essential that you click on Statistics... and then select *Collinearity diagnostics* in the dialog box. Once you have selected Collinearity diagnostics, switch off all of the default options, click on Continue to return you to the *linear regression* dialog box, and then click on OK to run the analysis.

The results of the linear regression analysis are shown in SPSS Output 6.14. From the first table we can see that the tolerance values are .014 for **previous** and **anxious** and .575 for **pswq**. In Chapter 5 we saw various criteria for assessing collinearity. Menard (1995) suggests that a tolerance value less than .1 almost certainly indicates a serious collinearity problem. Myers (1990) also suggests that a VIF value greater than 10 is cause for concern and in these data the values are over 70 for both **anxious** and **previous**. It seems from these values that there is an issue of collinearity between the predictor variables. We can investigate this issue further by examining the collinearity diagnostics.

SPSS Output 6.14 also shows a table labelled *Collinearity Diagnostics*. In this table, we are given the eigenvalues of the scaled, uncentred cross-products matrix, the condition index and the variance proportions for each predictor. If any of the eigenvalues in this table are much larger than others then the uncentred cross-products matrix is said to be ill-conditioned, which means that the solutions of the regression parameters can be greatly affected by small changes in the predictors or outcome. In plain English, these values give us some idea as to how accurate our regression model is: if the eigenvalues are fairly similar then the derived model is likely to be unchanged by small changes in the measured variables. The *condition*

Coefficients[a]

Model		Collinearity Statistics	
		Tolerance	VIF
1	State Anxiety	.014	71.764
	Percentage of previous penalties scored	.014	70.479
	Penn State Worry Questionnaire	.575	1.741

a Dependent Variable: Result of Penalty Kick

Collinearity Diagnostics[a]

Model	Dimension	Eigenvalue	Condition Index	Variance Proportions			
				(Constant)	State Anxiety	Percentage of previous penalties scored	Penn State Worry Questionnaire
1	1	3.434	1.000	.00	.00	.00	.01
	2	.492	2.641	.00	.00	.00	.04
	3	7.274E-02	6.871	.00	.01	.00	.95
	4	5.195E-04	81.303	1.00	.99	.99	.00

a Dependent Variable: Result of Penalty Kick

SPSS Output 6.14 Collinearity diagnostics for penalty data

indexes are another way of expressing these eigenvalues and represent the square root of the ratio of the largest eigenvalue to the eigenvalue of interest (so, for the dimension with the largest eigenvalue, the condition index will always be 1). For these data the final dimension has a condition index of 81.3, which is massive compared to the other dimensions. Although there are no hard and fast rules about how much larger a condition index needs to be to indicate collinearity problems, this case clearly shows that a problem exists.

The final step in analysing this table is to look at the variance proportions. The variance of each regression coefficient can be broken down across the eigenvalues and the variance proportions tell us the proportion of the variance of each predictor's regression coefficient that is attributed to each eigenvalue. These proportions can be converted to percentages by multiplying them by 100 (to make them more easily understood). So, for example, for **pswq** 95% of the variance of the regression coefficient is associated with eigenvalue number 3, 4% is associated with eigenvalue number 2, and 1% is associated with eigenvalue number 1. In terms of collinearity, we are looking for predictors that have high proportions on the same *small* eigenvalue, because this would indicate that the variances of their regression coefficients are dependent. So we are interested mainly in the bottom few rows of the table (which represent small eigenvalues). In this example, 99% of the variance in the regression

Correlations

		Result of Penalty Kick	State Anxiety	Percentage of previous penalties scored	Penn State Worry Questionnaire
Result of Penalty Kick	Pearson Correlation	1.000	−.668**	.674**	−.675**
	Sig. (2-tailed)	.	.000	.000	.000
	N	75	75	75	75
State Anxiety	Pearson Correlation	−.668**	1.000	**−.993****	.652**
	Sig. (2-tailed)	.000	.	.000	.000
	N	75	75	75	75
Percentage of previous penalties scored	Pearson Correlation	.674**	**−.993****	1.000	−.644**
	Sig. (2-tailed)	.000	.000	.	.000
	N	75	75	75	75
Penn State Worry Questionnaire	Pearson Correlation	−.675**	.652**	−.644**	1.000
	Sig. (2-tailed)	.000	.000	.000	.
	N	75	75	75	75

**Correlation is significant at the .01 level (2-tailed).

SPSS Output 6.15

coefficients of both **anxiety** and **previous** is associated with eigenvalue number 4 (the smallest eigenvalue), which clearly indicates dependency between these variables.

The result of this analysis is pretty clear cut: there is collinearity between state anxiety and previous experience of taking penalties and this dependency results in the model becoming biased. To illustrate from where this collinearity stems, SPSS Output 6.15 shows the result of a Pearson correlation between all of the variables in this regression analysis (you can run such an analysis yourself). From this output we can see that **anxious** and **previous** are highly negatively correlated ($r = -.99$); in fact they are nearly perfectly correlated. Both **previous** and **anxious** correlate with penalty success[7] but because they are correlated so highly with each other, it is unclear which of the two variables predicts penalty success in the regression.

This discussion begs the question of what to do when you have identified collinearity. Well, put simply, there's not much you can do. One obvious solution is to omit one of the variables (so, for example, we might stick with the model from block 1 that ignored state anxiety). The problem with this should be obvious: there is no way of knowing which variable to omit. The resulting theoretical conclusions are, therefore, meaningless because, statistically speaking, any of the variables involved in collinearity could be omitted. In short,

7 As a point of interest, some may question whether it is legitimate to do a Pearson correlation on a dichotomous variable such as penalty success; however, this is simply doing a point–biserial correlation as described in Chapter 4.

there are no statistical grounds for omitting one variable over another. Even if a predictor is removed, Bowerman & O'Connell (1990) recommend that another equally important predictor that does not have such strong multicollinearity should replace it. They go on to suggest collecting more data to see whether the multicollinearity can be lessened. Another possibility when there are several predictors involved in the multicollinearity is to run a factor analysis on these predictors and to use the resulting factor scores as a predictor (see Chapter 15). The safest (although unsatisfactory) remedy is to acknowledge the unreliability of the model. So, if we were to report the analysis of which factors predict penalty success, we might acknowledge that previous experience significantly predicted penalty success in the first model, but propose that this experience might affect penalty taking by increasing state anxiety. This statement would be highly speculative because the correlation between **anxious** and **previous** tells us nothing of the direction of causality, but it would acknowledge the inexplicable link between the two predictors. I'm sure that many of you may find the lack of remedy for collinearity grossly unsatisfying—unfortunately statistics is frustrating sometimes!

6.9. THINGS THAT CAN GO WRONG ④

SPSS solves logistic regression problems by an iterative procedure; it guesses a likely solution, checks to see how well that solution fits the data, and then repeats the process to try to improve the fit. It stops when it thinks that its guess is close enough to the true solution. Sometimes, instead of pouncing on the correct solution quickly, you'll notice nothing happening: SPSS begins to move infinitely slowly, or appears to have just got fed up with you asking it to do stuff and has gone on strike. If it can't find a correct solution, then sometimes it actually does give up, quietly offering you (without any apology) a result which is completely incorrect. Usually this is revealed by an implausibly large standard error. Two situations can provoke this.

6.9.1. Incomplete information from the predictors ④

Imagine you're trying to predict lung cancer from smoking (foul habit believed to increase the risk of cancer) and whether or not you eat tomatoes (which are believed to reduce the

Do you smoke?	Do you eat tomatoes?	Do you have Cancer?
Yes	No	Yes
Yes	Yes	Yes
No	No	Yes
No	Yes	??????

risk of cancer). You collect data from people who do and don't smoke, and from people who do and don't eat tomatoes; however, this isn't sufficient unless you collect data from all combinations of smoking and tomato eating. Imagine you ended up with the following data:

Observing only the first three possibilities does not prepare you for the outcome of the fourth. We have no way of knowing whether this last person will have cancer or not based on the other data we've collected. Therefore, SPSS will have problems unless you've collected data from all combinations of your variables. This should be checked before you run the analysis using a crosstabulation table, and I describe how to do this in Chapter 16.

This point applies not only to categorical variables, but also to continuous ones. Suppose that you wanted to investigate factors related to human happiness. These might include age, gender, sexual orientation, religious beliefs, levels of anxiety and even whether a person is right-handed. You interview 1000 people, record their characteristics, and whether they are happy ('yes' or 'no'). Although a sample of 1000 seems quite large, is it likely to include an 80 year old highly anxious left-handed Buddhist lesbian? If you found one such person and she was happy, should you conclude that everyone else in the same category is happy? It would, obviously, be better to have several more people in this category to confirm that this combination of characteristics causes happiness.

As a general point, whenever samples are broken down into categories and one or more combinations are empty it creates problems. These will probably be signalled by coefficients that have unreasonably large standard errors. Conscientious researchers produce and check multiway crosstabulations of all categorical independent variables. Lazy but cautious ones don't bother with crosstabulations, but look carefully at the standard errors. Those who don't bother with either should expect trouble.

6.9.2. Complete separation ④

A second situation in which logistic regression collapses might surprise you: it's when the outcome variable can be perfectly predicted by one variable or a combination of variables! This is known as complete separation.

Let's look at an example: imagine you placed a pressure pad under your door mat and connected it to your security system so that you could detect burglars when they creep in at night. However, because your teenage children (which you would have if you're old enough and rich enough to have security systems and pressure pads) and their friends are often coming home in the middle of the night. When they tread on the pad you want it to work out the probability that the person is a burglar and not your teenager. Therefore, you could measure the weight of some burglars and some teenagers and use logistic regression to predict the outcome (teenager or burglar) from the weight. The graph would show a line of triangles at 0 (the data points for all of the teenagers you weighed) and a line of triangles at 1 (the data points for burglars you weighed). Note that these lines of triangles

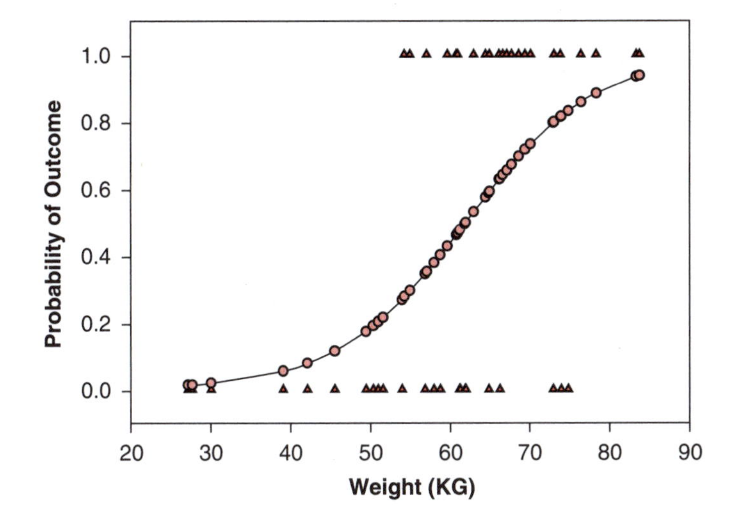

overlap (some teenagers are as heavy as burglars). We've seen that in logistic regression, SPSS tries to predict the probability of the outcome given a value of the predictor. In this case, at low weights the fitted probability follows the bottom line of the plot, and at high weights it follows the top line. At intermediate values it tries to follow the probability as it changes.

Imagine that we had the same pressure pad, but our teenage children had left home to go to university. We're now interested in distinguishing burglars from our pet cat based on weight. Again, we can weigh some cats and weigh some burglars. This time the graph still has a row of triangles at 0 (the cats we weighed) and a row at 1 (the burglars) but now the rows of triangles do not overlap: there is no burglar who weighs the same as a cat—obviously there were no cat burglars in the sample (groan now at that sorry excuse for a joke!). This is known as perfect separation: the outcome (cats and burglars) can be perfectly predicted from weight (anything less than 15 kg is a cat, anything more than 40 kg is a burglar). If we try to calculate the probabilities of the outcome given a certain weight then we run into trouble. When the weight is low, the probability is 0, and when the weight is high the probability is 1, but what happens inbetween? We have no data between 15 and 40 kg on which to base these probabilities. The figure shows two possible probability curves that we could fit to these data, one much steeper than the other. Either one of these curves is valid based on the data we have available. The lack of data means that SPSS will be uncertain how steep it should make the intermediate slope, it will try to bring the centre as close to vertical as possible, and its estimates veer unsteadily towards infinity (hence large standard errors).

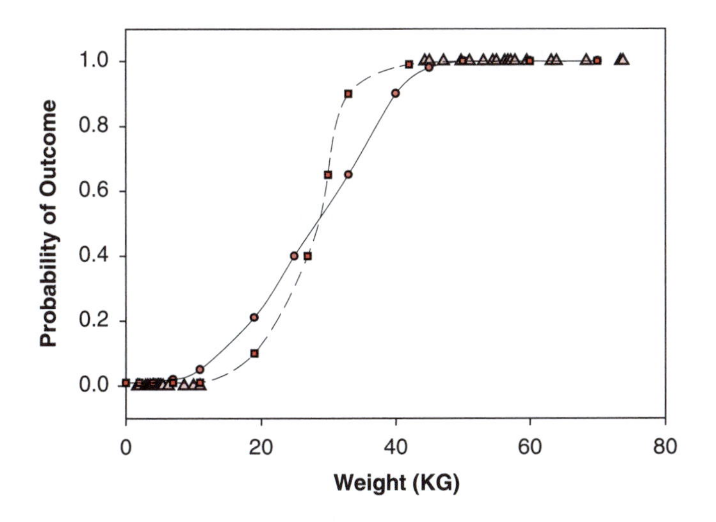

This problem often arises when too many variables are fitted to too few cases. Often the only satisfactory solution is to collect more data, but sometimes a neat answer is found by adopting a simpler model.

6.10. WHAT HAVE WE DISCOVERED ABOUT STATISTICS? ①

Writing this chapter taught me an awful lot (like to ignore my brain when it thinks 'I know what would be really good in my book, a chapter on logistic regression'!). Still, I hope I've covered up my ineptitude at least most of the time. We began the chapter by looking at why we can't use linear regression (Chapter 5) when we have a dichotomous outcome. We then looked into some of theory of logistic regression by looking at the regression equation and what it means. Then we moved on to assessing the model and talked about the log-likelihood statistic and the associated chi-square test. I talked about different methods of obtaining equivalents to R^2 in regression (Hosmer & Lemeshow, Cox & Snell, and Nagelkerke). We also discovered the Wald statistic and *Exp(B)*. The rest of the chapter looked at two examples using SPSS to carry out logistic regression. So, hopefully, you should have a pretty good idea of how to conduct and interpret a logistic regression by now.

6.11. KEY TERMS THAT WE'VE DISCOVERED

- –2LL
- Chi-square distribution
- Complete separation
- Cox & Snell's R^2_{CS}
- *Exp(B)*
- Hosmer & Lemeshow's R^2_L
- Logistic regression
- Log-likelihood
- Maximum-likelihood estimation
- Nagelkerke's R^2_N
- Odds
- Roa's efficient score statistic
- Wald statistic

6.12. SMART ALEX'S TASKS

The answers to these tasks can be found on the CD-ROM in the file called **Answers(Chapter 6).pdf**.

- Recent research has shown that lecturers are among the most stressed workers. A researcher wanted to know exactly what it was about being a lecturer that created this stress and subsequent burnout. She took 467 lecturers and administered several questionnaires to them that measured: **Burnout** (burnt out or not), **Perceived Control** (high score = low perceived control), **Coping Style** (high score = high ability to cope with stress), **Stress from Teaching** (high score = teaching creates a lot of stress for the person), **Stress from Research** (high score = research creates a lot of stress for the person), and **Stress from Providing Pastoral Care** (high score = providing pastoral care creates a lot of stress for the person). The outcome of interest was burnout, and Cooper, Sloan & Williams' (1988) model of stress indicates that perceived control and coping style are important predictors of this variable. The remaining predictors were measured to see the unique contribution of different aspects of a lecturer's work to their burnout—can you help her out by conducting a logistic regression to see which factor predicts burnout? The data are in **Burnout.sav**.③
- A health psychologist interested in research into HIV wanted to know the factors that influenced condom use with a new partner (relationship less than 1 month old). The outcome measure was whether a condom was used (**Use**: condom used = 1, Not used = 0). The predictor variables were mainly scales from the Condom Attitude Scale (CAS) by Sacco, Levine, Reed & Thompson (*Psychological Assessment: A journal of Consulting and Clinical Psychology*, 1991). **Gender** (gender of the person); **Safety** (relationship safety, measured out of 5, indicates the degree to which the person views this relationship as 'safe' from sexually transmitted disease); **Sexexp** (sexual experience, measured out of 10, indicates the degree to which previous experience influences attitudes towards condom use); **Previous** (a measure not from the CAS, this variable measures whether or not the couple used a condom in their previous encounter, 1 = condom used, 0 = not used, 2 = no previous encounter with this partner); **selfcon** (self-control, measured out of 9, indicates the degree of self-control that a person has when it comes to condom use: that is, do they get carried away with the heat of the moment, or do they exert control?); **Perceive** (perceived risk, measured out of 6, indicates the degree to which the person feels at risk from unprotected sex). Previous research (Sacco, Rickman, Thompson, Levine & Reed, in *Aids Education and Prevention*, 1993) has shown that **gender**, **relationship safety** and **perceived risk** predict condom use. Carry out an appropriate analysis to verify these previous findings, and to test whether self-control, previous usage and sexual experience can predict any of the remaining variance in condom use. (1) Interpret all important parts of the SPSS output. (2) How reliable is the final model? (3) What are the probabilities that participants 12, 53 and 75 will use a condom? (4) A female, who used a condom in her previous encounter with her new partner, scores 2 on all variables except perceived risk (for which she scores 6). Use the model to estimate the probability that she will use a condom in her next encounter. Data are in the file **condom.sav**.③

6.13. FURTHER READING

Hutcheson, G. & Sofroniou, N. (1999). *The multivariate social scientist*. London: Sage. Chapter 4.

Menard, S. (1995). *Applied logistic regression analysis*. Sage university paper series on quantitative applications in the social sciences, 07–106. Thousand Oaks, CA: Sage. This is a fairly advanced text, but great nevertheless. Unfortunately, few basic-level texts include logistic regression so you'll have to rely on what I've written!

Miles, J. & Shevlin, M. (2001). *Applying regression and correlation: a guide for students and researchers*. London: Sage. Chapter 6 is a nice introduction to logistic regression.

CHAPTER 7

COMPARING TWO MEANS

7.1. WHAT WILL THIS CHAPTER TELL US? ①

Rather than looking at relationships between variables, researchers are sometimes interested in looking at differences between groups of people. In particular, in experimental research we often want to manipulate what happens to people so that we can make causal inferences. For example, if we take two groups of people and randomly assign one group a programme of dieting pills, and the other group a programme of sugar pills (which they think will help them lose weight) then if the people who take the dieting pills lose more weight than those on the sugar pills we can infer that the diet pills caused the weight loss. This is a powerful research tool because it goes one step beyond merely observing variables and looking for relationships (as in correlation and regression).[1] This chapter is the first of many that looks at this kind of research scenario, and we start with the simplest scenario: when we have two groups, or, to be more specific, when we want to compare two means. As we'll see, there are two different ways of collecting data: we can either expose different people to different experimental manipulations (*between-group* or *independent* design), or take a single group of people and expose them to different experimental manipulations at different points in time (a *repeated-measures* design). First we have a look at how to graph these different types of data, and then we look at statistical procedures for analysing independent designs (the *independent t-test*) and then move on to repeated-measures designs (the *related t-test*).

---

1 People sometimes get confused and think that certain statistical procedures allow causal inferences and others don't. This isn't true; it's the fact that in experiments we manipulate one variable systematically to see its effect on another that allows the causal inference. In other forms of research we merely observe changes in variables without any intervention from an experimenter, and so we cannot say which variable causes a change in the other but merely say that they change in a particular way. As you'll discover, the statistical procedures for analysing these different types of data are, in fact, mathematically identical!

7.2. REVISION OF EXPERIMENTAL RESEARCH ①

Often in the social sciences we are not just interested in looking at which variables covary, or predict an outcome. Instead, we might want to look at the effect of one variable on another by systematically changing some aspect of that variable. So, rather than collecting naturally occurring data as in correlation and regression, we manipulate one variable to observe its effect on another. As a simple case, we might want to see what the effect of positive encouragement has on learning about statistics. I might, therefore, randomly split some students into two different groups:

- **Group 1 (positive reinforcement)**: During seminars I congratulate all students in this group on their hard work and success. Even when people get things wrong, I am supportive and say things like 'that was very nearly the right answer, you're coming along really well' and then give the student a nice piece of chocolate.
- **Group 2 (negative reinforcement)**: This group receives a normal university-style seminar, so during seminars I give relentless verbal abuse to all of the students even when they give the correct answer. I demean their contributions and am patronizing and dismissive of everything they say. I tell students that they are stupid, worthless and shouldn't be doing the course at all.

The thing that I have manipulated is the teaching method (positive reinforcement versus negative reinforcement). This variable is known as the independent variable (IV) and in this situation it is said to have two levels, because it has been manipulated in two ways (i.e. reinforcement has been split into two types: positive and negative). Once I have carried out this manipulation I must have some kind of outcome that I am interested in measuring. In this case it is statistical ability, and I could measure this variable using the end-of-year statistics exam results. This outcome variable is known as the dependent variable, or DV, because we assume that these scores will depend upon the type of teaching method used (the independent variable).

7.2.1. Two methods of data collection ①

When we collect data, we can choose between two methods of data collection. The first is to manipulate the independent variable using different participants. This method is the one described above, in which different groups of people take part in each experimental condition (a between-group, between-subjects or independent design). The second method is to manipulate the independent variable using the same participants. Simplistically, this method means that we give a group of students positive reinforcement for a few weeks and test their statistical abilities and then begin to give this same group negative reinforcement for a few weeks before testing them again (a within-subject or repeated-measures design).

The way in which the data are collected determines the type of test that is used to analyse the data.

7.2.2. Two types of variation ①

To begin with, imagine you've used a repeated-measures design in an experiment that has two conditions. Therefore, the same participants participate in both conditions, or, put another way, we measure participants' behaviour in condition 1 *and* in condition 2. For example, imagine we were trying to see whether you could train chimpanzees to run the economy. In one training phase they are sat in front of a chimp-friendly computer and press buttons which change various parameters of the economy; once these parameters have been changed a figure appears on the screen indicating the economic growth resulting from those parameters. Now, chimps can't read (I don't think) so this feedback is meaningless. A second training phase is the same except that if the economic growth is good, they get a banana (if growth is bad they do not)—this feedback is valuable to the average chimp.

Let's take a step back and think what would happen if we did *not* introduce an experimental manipulation (i.e. there were no bananas in the second training phase so condition 1 and condition 2 were identical). If there is no experimental manipulation then we expect a chimp's behaviour to be the same in both conditions. We expect this because external factors such as age, gender, IQ, motivation and arousal will be the same for both conditions (a chimp's gender etc. will not change from when they are tested in condition 1 to when they are tested in condition 2). If the performance measure is reliable (i.e. our test of how well they run the economy), and the variable or characteristic that we are measuring (in this case, ability to run an economy) remains stable over time, then a participant's performance in condition 1 should be very highly related to their performance in condition 2. So, chimps who score highly in condition 1 will also score highly in condition 2, and those who have low scores for condition 1 will have low scores in condition 2. However, performance won't be *identical*; there will be small differences in performance created by unknown factors. This variation in performance is known as *unsystematic variation*.

If we introduce an experimental manipulation (i.e. provide bananas as feedback in one of the training sessions), then we do something different to participants in condition 1 from what we do to them in condition 2. So, the *only* difference between conditions 1 and 2 is the manipulation that the experimenter has made (in this case that the chimps get bananas as a positive reward in one condition but not in the other). Therefore, any difference between the means of the two conditions is probably due to the experimental manipulation. So, if the chimps perform better in one training phase than the other then this *has* to be due to the fact that bananas were used to provide feedback in one training phase but not the other. Differences in performance created by a specific experimental manipulation are known as *systematic variation*.

Now let's have a think about what happens when we use different participants—an independent design. In this design we still have two conditions, but this time different participants participate in each condition. Going back to our example, one group of chimps receive training without feedback, whereas a second group of different chimps do receive feedback on their performance via bananas.[2] Imagine again that we didn't have an experimental manipulation. If we did nothing to the groups, then we would still find some variation between behaviour between the groups because they contain different chimps who will vary in their ability, motivation, IQ and other factors. In short, the type of factors that were held constant in the repeated-measures design are free to vary in the independent design. So, the unsystematic variation will be bigger than for a repeated-measures design. If we again introduce a manipulation (i.e. bananas) then, again, we will see additional variation created by this manipulation. As such, in both the repeated-measures design and the independent design there are always two sources of variation:

- **Systematic variation**: This variation is due to the experimenter doing something to all of the participants in one condition but not in the other condition.
- **Unsystematic variation**: This variation results from random factors that exist between the experimental conditions (such as natural differences in ability).

The role of statistics is to discover how much variation there is in performance, and then to work out how much of this is systematic and how much is unsystematic.

In a repeated-measures design, differences between two conditions can be caused by only two things: (1) the manipulation that was carried out on the participants, or (2) any other factor that might affect the way in which a person performs from one time to the next. The latter factor is likely to be fairly minor compared to the influence of the experimental manipulation. In an independent design, differences between the two conditions can also be caused by one of two things: (1) the manipulation that was carried out on the participants, or (2) differences between the characteristics of the people allocated to each of the groups. The latter factor in this instance is likely to create considerable random variation both within each condition and between them. Therefore, the effect of our experimental manipulation is likely to be more apparent in a repeated-measures design than in a between-group design because in the former unsystematic variation can be caused only by differences in the way in which someone behaves at different times. In between-group designs we have differences in innate ability contributing to the unsystematic variation. Therefore, this error variation will almost always be much larger than if the same participants had been used. When we look at the effect of our experimental manipulation, it is always against a background of 'noise' caused by random, uncontrollable differences between our conditions. In a repeated-measures design this 'noise' is kept to a minimum and so the effect of the experiment is more likely to show up. This means that repeated-measures designs have more power to detect effects that genuinely exist than independent designs.

---

2 When I say 'via' I don't mean that the bananas developed little banana mouths that opened up and said 'well done old chap, the economy grew that time' in chimp language. I mean that when they got something right they received a banana as a reward for their correct response.

7.2.3. Randomization ①

In both repeated-measures and independent designs it is important to try to keep the unsystematic variation to a minimum. By keeping the unsystematic variation as small as possible we get a more sensitive measure of the experimental manipulation. Generally, scientists use the randomization of participants to achieve this goal. Many statistical tests (e.g. the *t*-test) work by identifying the systematic and unsystematic sources of variation and then comparing them. This comparison allows us to see whether the experiment has generated considerably more variation than we would have got had we just tested participants without the experimental manipulation. Randomization is important because it eliminates most other sources of systematic variation, which allows us to be sure that any systematic variation between experimental conditions is due to the manipulation of the independent variable.

Let's look at a repeated-measures design first. When the same people participate in more than one experimental condition this obviously violates the notion of randomly assigning people to groups. Although they are naive during the first experimental condition they come to the second experimental condition with prior experience of what is expected of them. At the very least they will be familiar with the dependent measure (e.g. the task they're performing). The two most important sources of systematic variation in this type of design are:

- Practice effects: Participants may perform differently in the second condition because of familiarity with the experimental situation and/or the measures being used.
- Boredom effects: Participants may perform differently in the second condition because they are tired or bored from having completed the first condition.

Although these effects are impossible to eliminate completely, we can ensure that they produce no systematic variation between our conditions. We can do this by counterbalancing the order in which a person participates in a condition. We can randomly decide either that a participant does condition 1 before condition 2, or that they have condition 2 before condition 1. If we look back at the teaching method example at the beginning of this chapter, if the same participants were used in both conditions then we might find that statistical ability was higher after the negative reinforcement condition. However, every student experienced the negative reinforcement after the positive reinforcement and so went into the negative reinforcement condition already having a better knowledge of statistics than when they began the positive reinforcement condition. So, the apparent improvement after negative reinforcement is not due to the experimental manipulation (i.e. it's not because negative reinforcement works), it is because participants had attended more statistics seminars by the time they had finished the negative reinforcement condition. We can ensure that the number of statistics seminars does not introduce a systematic bias by randomizing the order of conditions, so we could randomly take half of the students and give them negative reinforcement first while the remainder have positive reinforcement first.

If we turn our attention to independent designs, a similar argument can be applied. We know that different participants participate in different experimental conditions and that these participants will differ in many respects (their IQ, attention span, etc.). Although we know that these factors (known as confounding variables) contribute to the variation between conditions, we need to make sure that these variables contribute to the unsystematic variation and *not* the systematic variation. The way to ensure that confounding variables are unlikely to contribute systematically to the variation between experimental conditions is to allocate participants randomly to a particular experimental condition. This should ensure that these confounding variables are evenly distributed across conditions.

A good example is the effects of alcohol on personality. You might give one group of people 5 pints, and keep a second group sober, and then count how many fights each person gets into. The effect that alcohol has on people can be very variable because of different tolerance levels: teetotal people can become very drunk on a small amount, whilst alcoholics need to consume vast quantities before the alcohol affects them. Now, if you allocated a bunch of teetotal participants to the condition that consumed alcohol, then you might find no difference between them and the sober group (because the teetotal participants are all unconscious after the first glass and so can't become involved in any fights). As such, the person's prior experiences with alcohol will create systematic variation that cannot be dissociated from the effect of the experimental manipulation. The best way to reduce this eventuality is to allocate participants randomly to conditions.

7.3. INPUTTING DATA AND DISPLAYING MEANS WITH ERROR BAR CHARTS ①

As I tried to stress in Chapter 3, it is always a good idea to look at your data with graphs before you begin the analysis. Graphs help us to understand what is going on within the data and what kinds of results we might expect to obtain. We've already seen how to plot bar charts of means in section 3.7. Bar charts are often reported in academic papers and are a nice way to summarize data but they give a very narrow view of the data (e.g. bar charts give no indication of the number of scores that contributed to each mean, nor do they tell us how much variation there was in scores). I mentioned back in section 3.7 something called an error bar and these are a better way to examine data. An error bar chart displays not only the mean, but also the 95% confidence interval (see section 1.6.2) of the mean of each experimental condition.

Hopefully you're becoming familiar with confidence intervals because we have come across them in Chapters 1, 5 and 6. Let's go back to my sperm example from section 1.6.2, in which we were trying to measure the average amount of sperm (in millions) produced by Japanese quail. I explained that if we were to take 100 samples from a population and calculate the mean of each, and construct the 95% confidence interval for each, then this means that 95 of those 100 confidence intervals would contain the true value of the mean (the value in the population). So, if a confidence interval has a lower limit of 2 and an upper limit of 16, then this means that we're 95% confident that this interval is one that contains the population mean (i.e. the true amount of sperm produced by quail). Therefore, an error bar graph displays the limits within which we think the true value of the mean lies.

You'll discover shortly that the *t*-test works on the principle that if two samples are taken from the same population then they should have fairly similar means. We know from Chapter 1 that it is possible that any two samples could have slightly different means (and the standard error will tell us a little about how different we can expect sample means to be). Now, the confidence interval tells us the limits within which the population mean is likely to fall (in fact, the size of the confidence interval will depend on the size of the standard error). In the example above we know that the population mean is probably between 2 and 16 million sperms. What if we now took a second sample of quail and found the confidence interval ranged from 4 to 15? This interval overlaps a lot with our first sample:

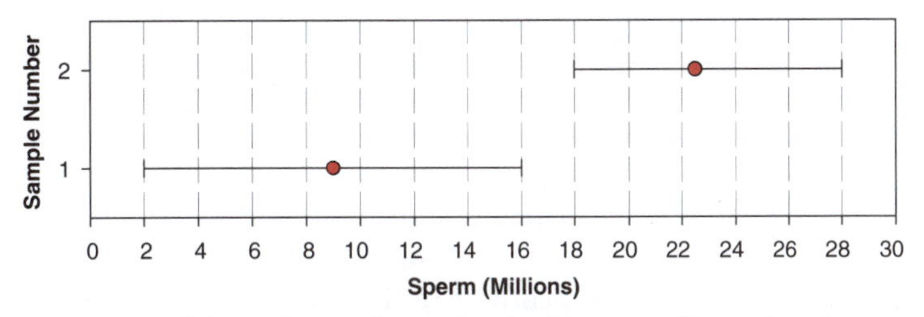

The fact that the confidence intervals overlap in this way tells us that these means could plausibly come from the same population: in both cases we're 95% confident that the intervals contain the true value of the mean, and both intervals overlap considerably, so they contain many similar values. What if the confidence interval for our second sample ranges from 18 to 28? If we compare this with out first sample we'd get:

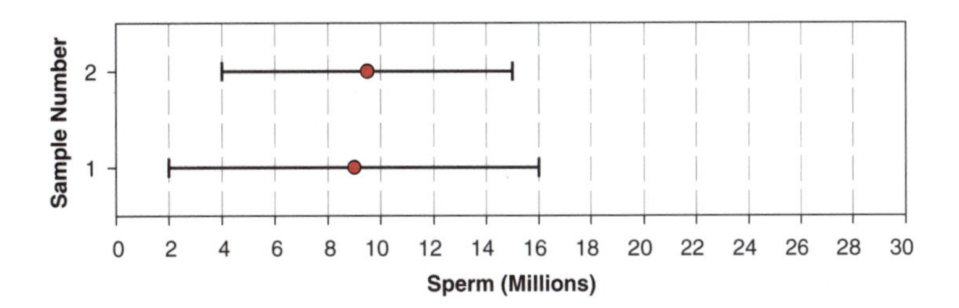

Now, these confidence intervals don't overlap at all. So, one confidence interval (which we're 95% sure contains the true value of the mean) tells us that the population mean is somewhere between 2 and 16 million, whereas the other confidence (which we're also 95% sure contains the true value of the mean) tells us that the population mean is somewhere between 18 and 28. This suggests that either our samples must come from different populations, or both samples come from the same population but one of the confidence intervals doesn't contain the population mean. If we've used 95% confidence intervals then we know that the second possibility is unlikely (this happens only 5 times in 100 or 5% of the time), so the first explanation is more likely.

OK, I can hear you all thinking 'so what if the samples come from a different population?' Well, if we have taken two random samples of people, and we have tested them on some measure (e.g. fear of statistics textbooks), then we expect these people to belong to the same population. If their sample means are so different as to suggest that, in fact, they come from different populations, why might this be? The answer is that our experimental manipulation has induced a difference between the samples.

To reiterate, when an experimental manipulation is successful, we expect to find that our samples have come from different populations. If the manipulation is unsuccessful, then we expect to find that the samples came from the same population (e.g. the sample means should be fairly similar). Now, the 95% confidence interval tells us something about the likely value of the population mean. If we take samples from two populations, then we expect the confidence intervals to be different (in fact, to be sure that the samples were from different populations we would not expect the two confidence intervals to overlap). If we take two samples from the same population, then we would expect, if our measure is reliable, the confidence intervals to be very similar (i.e. they should overlap completely with each other). You may well ask where this diversion into the theory of hypothesis testing is leading us; well it follows from what I have just said that if our experimental manipulation is successful, then the confidence intervals of the experimental groups should not overlap. If the manipulation is unsuccessful then the confidence intervals will overlap. In terms of the error bar graph, we should find that **if the bars on our error bar graph do not overlap this is indicative of a significant difference between groups.** It's also worth mentioning at this point that error bars from SPSS are suitable only for normally distributed data.

7.3.1. Error bar graphs for between-group designs ①

When data are collected using different participants in each group, we need to input the data using a coding variable (see section 2.4.4). So, the data editor will have two columns of data. The first column is a coding variable (called something like **group**) which, when we have only two groups, will obviously have two codes (for convenience I suggest 0 = group 1, and 1 = group 2). The second will have values for the dependent variable.

Throughout this chapter I use the same data set, not because I am too lazy to think up different data sets, but because it allows me to illustrate various things. The example relates to whether arachnophobia (fear of spiders) is specific to real spiders or whether pictures of spiders can evoke similar levels of anxiety. Twenty-four arachnophobes were used in all. Twelve were asked to play with a big hairy tarantula with big fangs and an evil look in its eight eyes. Their subsequent anxiety was measured. The remaining 12 were shown only pictures of the same big hairy tarantula and again their anxiety was measured. The data are in the file **spiderBG.sav**, but for those who want some practice in inputting data, they are also presented in Table 7.1. Remember that each row in the data editor represents a different participant's data. Therefore, you need a column representing the group to which they belonged and a second column representing their anxiety. The data in Table 7.1 show only the group codes and not the corresponding label. When you enter the data into SPSS remember to tell the computer that a code of 0 represents the group that were shown the picture, and that a code of 1 represents the group that saw the real spider (see section 2.4.4).

Table 7.1 Data from spiderBG.sav

Participant	Group	Anxiety
1	0	30
2	0	35
3	0	45
4	0	40
5	0	50
6	0	35
7	0	55
8	0	25
9	0	30
10	0	45
11	0	40
12	0	50
13	1	40
14	1	35
15	1	50
16	1	55
17	1	65
18	1	55
19	1	50
20	1	35
21	1	30
22	1	50
23	1	60
24	1	39

When you have entered the data (or accessed the file **spiderBG.sav**) you should access the *error bar* dialog box by using the mouse to click on **Graphs⇒Error Bar…**. The initial dialog box is shown in Figure 7.1. There are two choices of error bar graph. The first is a simple error bar chart (for plotting levels of a single independent variable) and the second is a clustered error bar graph. The clustered graph can be used when a second between-group independent variable has been measured (such as gender). We will stick to a simple error bar graph. The next choice is whether the data in the chart summarize groups of cases or separate variables. There is a simple rule here: select *Summaries for groups of cases* when the data were collected using different participants (as is the case here) and select

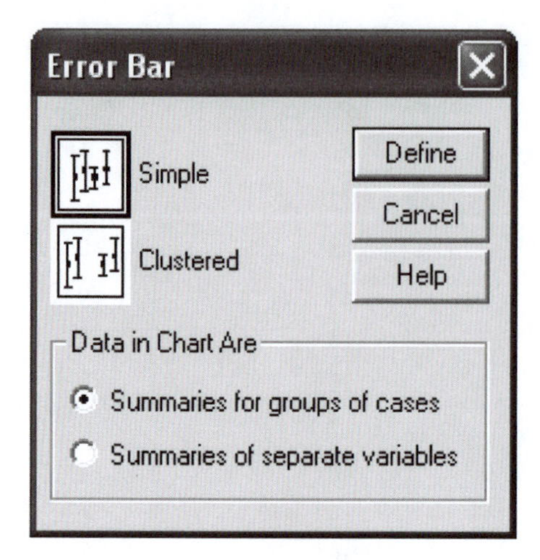

Figure 7.1 Initial dialog box for error bar graphs

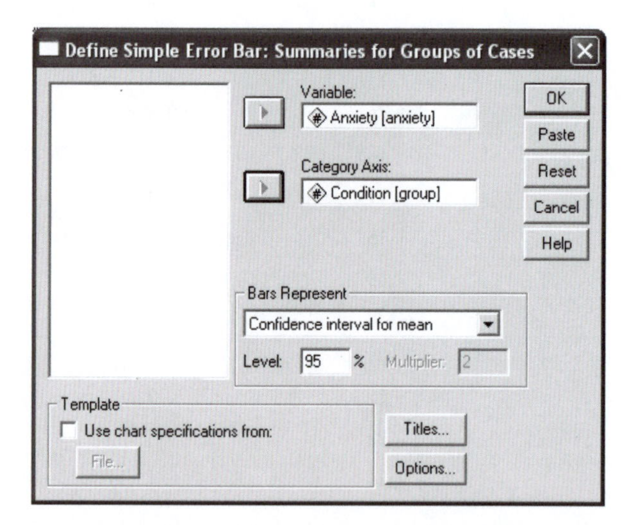

Figure 7.2 Main *error bar* dialog box for summaries for groups of cases

Summaries of separate variables when the data were collected using the same participants. When you have selected the appropriate options, click on Define.

When you click on Define a new dialog box appears (Figure 7.2) that allows you to specify which variables you want to plot. In this example we have only two variables: **group** and **anxiety**. So, using the mouse, select **anxiety** from the variables list and insert it into the space labelled *Variable* by clicking on ▶. Then, highlight **group** in the variables list and transfer it to the space labelled *Category Axis* by clicking on ▶. You can plot numerous things using this type of graph and the default option is to plot the 95% confidence

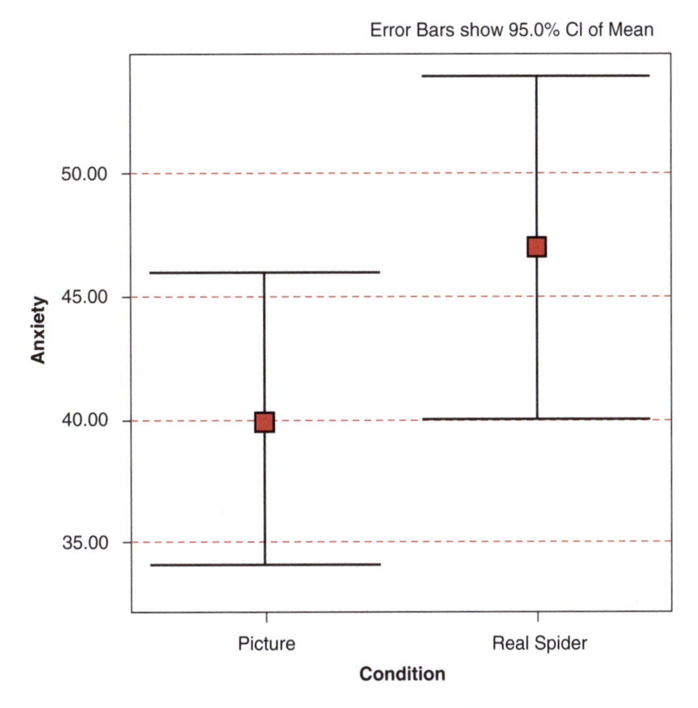

Figure 7.3 (Nicely edited!) error bar graph of **spiderBG.sav**

interval. This is the most useful type of graph and so the default options can be left as they are.

The resulting error bar graph is shown in Figure 7.3 and, as you can see, it looks like two Is. In the middle of each of the two bars is a square that represents the mean of each group. The vertical bar shows the confidence interval around that mean. So, from the graph of these data we can see that when a picture of a spider was used, the average level of anxiety was 40 and that the population mean is likely to fall between about 34 and 46. However, when a real spider was used, the average level of anxiety was 47, and the population mean is likely to fall between about 40 and 54. More important, the error bars overlap considerably, indicating that these samples are unlikely to be from different populations (so the experimental manipulation was unsuccessful). To see whether this last statement is true, you'll have to wait until section 7.6.3.

7.3.2. Error bar graphs for repeated-measures designs ②

If we repeated the study just described but using the same participants in each condition, we could produce an error bar graph identical to that shown in Figure 7.3. The problem

Table 7.2 Data from **spiderRM.sav**

Subject	Picture (Anxiety score)	Real (Anxiety score)
1	30	40
2	35	35
3	45	50
4	40	55
5	50	65
6	35	55
7	55	50
8	25	35
9	30	30
10	45	50
11	40	60
12	50	39

with creating an error bar graph of repeated-measures data is that SPSS treats the data as though different groups of participants were used. To prove that I'm not talking rubbish, the data for this study are included in a file called **spiderRM.sav**. In this file, the values of **anxiety** are identical to the between-group data. However, the data are arranged as if the same participants were used in each condition (so, each participant was exposed to a picture of a spider and their anxiety was measured, and at some other time the same participants were exposed to the real spider and their anxiety was measured again). The data are arranged differently now because the same participants were used. In SPSS, each row of the data editor represents a single participant, and so with this design the data are arranged in two columns (one representing the **picture** condition and one representing the **real** condition). The data are displayed in Table 7.2 and I recommend that you try inputting these data into a new data editor file rather than accessing the file on disk.

To plot an error bar graph of these data simply select the *error bar* dialog box by using the mouse to select **Graphs⇒Error Bar...** and then in the dialog box (see Figure 7.1) click on [Summaries of separate variables] and then [Define]. This process will bring up the main dialog box. Once the dialog box is activated you simply select the two variables of interest (**picture** and **real**) and then click on [OK]. The resulting graph should be identical to the one that you plotted for the between-subject data. Now, this is a problem because I spent an awful lot of time telling you how repeated-measures designs eliminated a lot of the unsystematic variance from the data, and by plotting repeated-measures data in this way we ignore the repeated-measures component of the data. Loftus & Masson (1994) suggest that one way to overcome this problem is to eliminate the between-subject variability by normalizing participant means (this just means ensuring that all participants have the same mean across

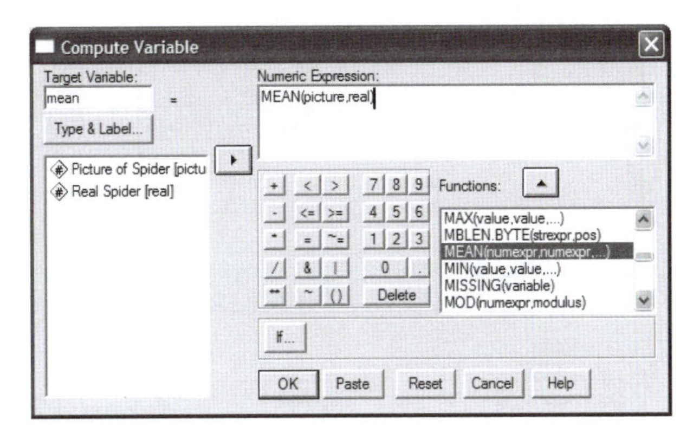

Figure 7.4 Using the *compute* function to calculate the mean of two columns

conditions). To normalize the means, we need to use the *compute* function of SPSS and carry out several steps (all of this was just a sneaky ploy to get you to use the *compute* function). We came across this function in section 3.3.4.

In section 3.3.4 we discovered that the *compute* command allows us to carry out various functions on columns of data (e.g. adding, or multiplying columns). Before we go on, go back to this section and refresh you memory about the wonders of this function!

7.3.2.1. Step 1: Calculate the mean for each participant ②

Assuming you've revised the *compute* command, we can use it to produce more accurate within-subject error bar charts. To begin with, we need to calculate the average anxiety for each participant and so we use the *mean* function. Access the main *compute* dialog box by using the **Transform⇒Compute...** menu path. Enter the name **mean** into the box labelled *Target Variable* and then scroll down the list of functions until you find the one called *MEAN(numexpr, numexpr,..)*. Highlight this function and transfer it to the command area by clicking on ▲. When the command is transferred, it appears in the command area as 'MEAN(?,?)' and the question marks should be replaced with variable names (which can be typed manually or transferred from the variables list). So replace the first question mark with the variable **picture**, and the second one with the variable **real**. The completed dialog box should look like the one in Figure 7.4.

7.3.2.2. Step 2: Calculate the grand mean ②

The grand mean is the mean of all scores (regardless of which condition the score comes from) and so for the current data this value will be the mean of all 24 scores. One way to calculate this is by hand (i.e. add up all of the scores and divide by 24); however, an easier way is to use the means that we have just calculated. These means represent the average score for each participant and so if we take the average of those mean scores, we will have the mean of all participants (i.e. the grand mean)—phew, there were a lot of means in that

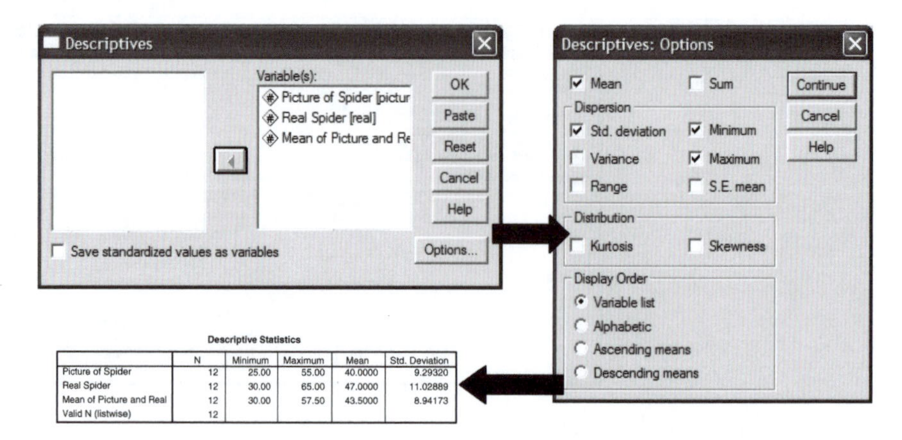

Descriptive Statistics

	N	Minimum	Maximum	Mean	Std. Deviation
Picture of Spider	12	25.00	55.00	40.0000	9.29320
Real Spider	12	30.00	65.00	47.0000	11.02889
Mean of Picture and Real	12	30.00	57.50	43.5000	8.94173
Valid N (listwise)	12				

Figure 7.5 Dialog boxes and output for descriptive statistics

sentence! OK, to do this we can use a useful little gadget called the *descriptives* command (you could also use the *Explore* or *Frequencies* functions that we came across in Chapter 3, but as I've already covered those we'll try something different). Access the *descriptives* function using the **Analyze⇒Descriptive Statistics⇒Descriptives...** menu path. The dialog box in Figure 7.5 should appear. The *descriptives* command is used to get basic descriptive statistics for variables and by clicking on [Options...] a second dialog box is activated. Select the variable **mean** from the list and transfer it to the box labelled *Variable(s)* by clicking on [▸]. Then, use the *options* dialog box to specify only the mean (you can leave the default settings as they are, but it is only the mean in which we are interested). If you run this analysis the output should provide you with some self-explanatory descriptive statistics for each of the three variables (assuming you selected all three). You should see that we get the mean for the picture condition, and the mean of the real spider condition, but it's actually the final variable we're interested in: the mean of the picture and spider condition. The mean of this variable is the grand mean, and you can see from the summary table that its value is 43.50. We will use this grand mean in the following calculations.

7.3.2.3. Step 3: Calculate the adjustment factor ②

If you look at the variable labelled **mean**, you should notice that the values for each participant are different, which tells us that some people had greater anxiety than others did across the conditions. The fact that participants' mean anxiety scores differ represents individual differences between different people (so, it represents the fact that some of the participants are generally more scared of spiders than others). These differences in natural anxiety to spiders contaminate the error bar graphs, which is why if we don't adjust the values that we plot, we will get the same graph as if a between-subjects design had been used. Loftus & Masson (1994) argue that to eliminate this contamination we should equalize the means between participants (i.e. adjust the scores in each condition such that when we take the mean score across conditions, it is the same for all participants). To do this,

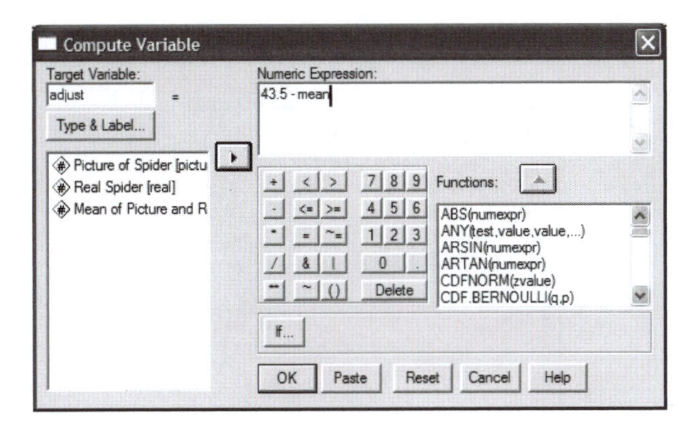

Figure 7.6 Calculating the adjustment factor

we need to calculate an adjustment factor by subtracting each participant's mean score from the grand mean. We can use the *compute* function to do this calculation for us. Activate the *compute* dialog box, give the target variable a name (I suggest **adjust**) and then use the command '43.5 - mean'. This command will take the grand mean (43.5) and subtract from it each participant's average anxiety level (see Figure 7.6).

This process creates a new variable in the data editor called **adjust**. The scores in the column **adjust** represent the difference between each participant's mean anxiety and the mean anxiety level across all participants. You'll notice that some of the values are positive and these participants are one's who were less anxious than average. Other participants were more anxious than average and they have negative adjustment scores. We can now use these adjustment values to eliminate the between-subject differences in anxiety.

7.3.2.4. Step 4: Create adjusted values for each variable ②

So far, we have calculated the difference between each participant's mean score and the mean score of all participants (the grand mean). This difference can be used to adjust the existing scores for each participant. First we need to adjust the scores in the **picture** condition. Once again, we can use the *compute* command to make the adjustment. Activate the *compute* dialog box in the same way as before, and then title our new variable **picture2** (you can then click on Type&Label... and give this variable a label such as 'adjusted values for the picture condition'). All we are going to do is to add each participant's score in the **picture** condition to their adjustment value. Select the variable **picture** and transfer it to the command area by clicking on ▶, then click on + and select the variable **adjust** and transfer it to the command area by clicking on ▶. The completed dialog box is shown in Figure 7.7. Now do the same thing for the variable **real**: create a variable called **real2** that contains the values of **real** added to the value in the **adjust** column.

Now, the variables **real2** and **picture2** represent the anxiety experienced in each condition, adjusted so as to eliminate any between-subject differences. If you don't believe me, then use the *compute* command to create a variable **mean2**, that is the average of **picture2** and **real2** (just like we did in section 7.3.2.1). You should find that the value in this

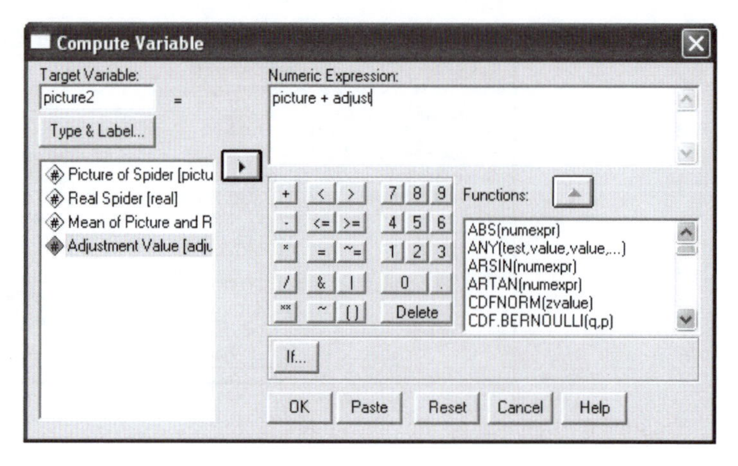

Figure 7.7 Adjusting the values of **picture**

column is the same for every participant, thus proving that the between-subject variability in means is gone: the value will be 43.50—the grand mean.

7.3.2.5. Drawing the error bar graph ②

Drawing the graph itself is similar to the between-group scenario. First, access the main dialog box through the **Graphs⇒Error Bar...** menus. Then in this dialog box (see Figure 7.1) click on Summaries of separate variables and then Define. This process will bring up the main dialog box. Once the dialog box is activated you simply select the two variables of interest (we want the adjusted values and so we need to select **picture2** and **real2**) and then click on OK (see Figure 7.8).

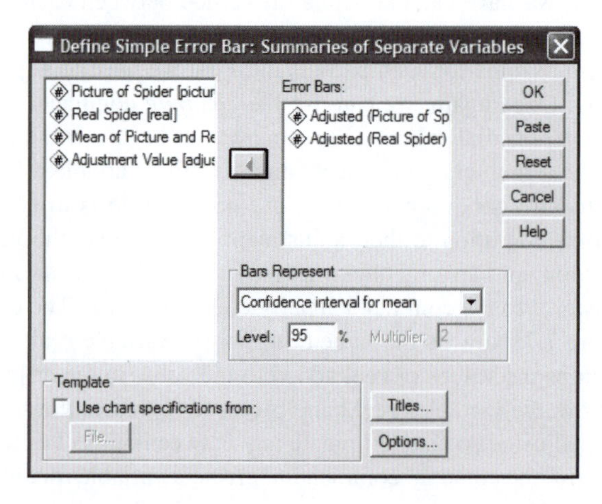

Figure 7.8 Dialog box for drawing an error bar graph of a repeated-measures variable

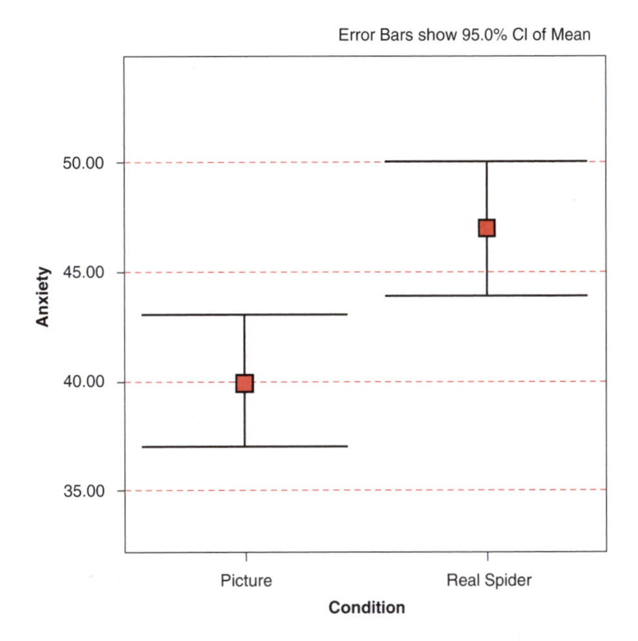

Figure 7.9 Error bar graph of the adjusted values of **spiderRM.sav**

The resulting error bar graph is shown in Figure 7.9. You'll notice that the error bar graph of the adjusted values is somewhat different to the error bar graph of the unadjusted values (look back to Figure 7.3). The first thing to notice is that the error bars do not overlap. Earlier I said that when error bars do not overlap we can be confident that our samples have not come from the same population (and so our experimental manipulation has been successful). Therefore, the graph of the repeated-measures version of these data indicates that real spiders may have evoked significantly greater (because the mean is higher) degrees of anxiety than just a picture. This example illustrates how different conclusions can be reached depending upon whether data were collected using the same or different participants (I expand upon this point in section 7.7).

7.4. TESTING DIFFERENCES BETWEEN MEANS: THE *T*-TEST ①

The simplest form of experiment that can be done is one with only one independent variable that is manipulated in only two ways and only one outcome is measured. More often than not the manipulation of the independent variable involves having an experimental condition and a control group (see Field & Hole, 2003). Some examples of this kind of design are:

- Is the movie *Scream 2* scarier than the original *Scream*? We could measure heart rates (which indicate anxiety) during both films and compare them.
- Does listening to music while you work improve your work? You could get some people to write an essay (or book!) listening to their favourite music, and then write a different essay when working in silence (this is a control group). You could then compare the essay grades!
- Does listening to Andy's favourite music improve your work? You could repeat the above but rather than letting people work with their favourite music, you could play them some of my favourite music (as listed in the acknowledgements) and watch the quality of their work plummet!

The *t*-test was designed to analyse these sorts of scenarios. Of course, there are more complex experimental designs and we will look at these in subsequent chapters. There are, in fact, two different *t*-tests and the one you use depends on whether the independent variable was manipulated using the same participants or different:

- **Independent means *t*-test**: This test is used when there are two experimental conditions and different participants were assigned to each condition (this is sometimes called the *independent measures* or *independent samples t*-test).
- **Dependent means *t*-test**: This test is used when there are two experimental conditions and the same participants took part in both conditions of the experiment (this test is sometimes referred to as the *matched-pairs* or *paired-samples t*-test).

7.4.1. Rationale for the *t*-test ①

Both *t*-tests have a similar rationale:

- Two samples of data are collected and the sample means calculated. These means might differ by either a little or a lot.
- If the samples come from the same population, then we expect their means to be roughly equal (see section 1.6). Although it is possible for their means to differ by chance alone, we would expect large differences between sample means to occur very infrequently. Under what's known as *the null hypothesis* we assume that the experimental manipulation has no effect on the participants: therefore, we expect the sample means to be very similar.
- We compare the difference between the sample means that we collected to the difference between the sample means that we would expect to obtain by chance. We use the standard error (see section 1.6) as a gauge of the variability between sample means. If the standard error is small, then we expect most samples to have very similar means. When the standard error is large we expect to obtain large differences in sample means by chance alone. If the difference between the samples we have collected is larger than what we would expect based on the standard error then we can assume one of two things:

(a) That sample means in our population fluctuate a lot by chance alone and we have, by chance, collected two samples that are atypical of the population from which they came.
(b) The two samples come from different populations but are typical of their respective parent population. In this scenario, the difference between samples represents a genuine difference between the samples (and so the null hypothesis is incorrect).

- As the observed difference between the sample means gets larger, the more confident we become that the second explanation is correct (i.e. that the null hypothesis should be rejected). If the null hypothesis is incorrect, then we accept what is known as *the experimental hypothesis* (also sometimes called the *alternative hypothesis* or something similar): that is, that the two sample means differ because of the different experimental manipulation imposed on each sample.

In essence, we calculate the *t*-test using equation (7.1), but the exact form that this equation takes will depend on how the data were collected (i.e. whether the same or different participants were used in each experimental condition).

$$t = \frac{\begin{array}{c}\text{observed difference}\\ \text{between sample}\\ \text{means}\end{array} - \begin{array}{c}\text{expected difference}\\ \text{between population means}\\ \text{(if null hypothesis is true)}\end{array}}{\begin{array}{c}\text{estimate of the standard error of the difference between}\\ \text{two sample means}\end{array}} \qquad (7.1)$$

7.4.2. Assumptions of the *t*-test ①

Both the independent and dependent *t*-tests are *parametric tests* based on the normal distribution (see Chapter 3). Therefore, they assume:

- Data are from normally distributed populations.
- Data are measured at least at the interval level.

The independent *t*-test, because it is used to test different groups of people, also assumes:

- Variances in these populations are roughly equal (*homogeneity of variance*).
- Scores are independent (because they come from different people).

These assumptions were explained in detail in Chapter 3 and in that chapter I emphasized the need to check these assumptions before you reach the point of carrying out your statistical test. As such, I won't go into them again, but it does mean that if you have ignored my advice and haven't checked these assumptions then you need to do it now! SPSS also incorporates some procedures into the *t*-test (e.g. Levene's test, see section 3.6, can be done at the same time as the *t*-test). Let's now look at each of the two *t*-tests in turn.

7.5. THE DEPENDENT *T*-TEST ①

Now we know how to graph means, we can move on to analysing whether differences between group means are statistically meaningful. If we stay with our repeated-measures data for the time being we can look at the dependent *t*-test, or paired-samples *t*-test. The dependent *t*-test is easy to calculate. In effect, we use a numeric version of equation (7.1):

$$t = \frac{\overline{D} - \mu_D}{s_D/\sqrt{N}} \tag{7.2}$$

Equation (7.2) compares the mean difference between our samples ($\bar{D}$) with the difference that we would expect to find between population means (μ_D), and then takes into account the standard error of the differences ($S_D/\sqrt{N}$). If the null hypothesis is true, then we expect there to be no difference between the population means (hence $\mu_D = 0$).

7.5.1. Sampling distributions and the standard error ①

In equation (7.1) I referred to the lower half of the fraction as the standard error of differences. The standard error was introduced in section 1.6 and is simply the standard deviation of what's known as a *sampling distribution*. Have a look back at this section now to refresh your memory about sampling distributions and the standard error. Sampling distributions have several properties that are important. For one thing, if the population is normally distributed then so is the sampling distribution; in fact, if the samples contain more than about 50 scores the sampling distribution will always be normally distributed. The mean of the sampling distribution is equal to the mean of the population—so, if you calculated the average of all of the sample means then the value you obtain should be the same as the population mean. This property makes sense because if a sample is representative of the population then you would expect its mean to be equal to that of the population. However, sometimes samples are unrepresentative and their mean differs from the population mean. On average, though, a sample mean will be very close to the population mean and only rarely will the sample mean be substantially different from the population. A final property of a sampling distribution is that its standard deviation is equal to the standard deviation of the population divided by the square root of the number of observations in the sample. As I mentioned before, this standard deviation is known as the standard error.

We can extend this idea to look at the *differences* between sample means. If you were to take several pairs of samples from a population and calculate their means, then you could also calculate the difference between their means. I mentioned earlier that *on average* sample means will be very similar to the population mean: as such, most samples will have very similar means. Therefore, most of the time the difference between sample means from the same population will be zero, or close to zero. However, sometimes one or both of the samples could have a mean very deviant from the population mean and so it is possible to obtain large differences between sample means by chance alone. However, this would happen less frequently.

In fact, if you plotted these differences between sample means as a histogram, you would again have a sampling distribution with all of the properties previously described. The standard deviation of this sampling distribution is called the standard error of differences. A small standard error tells us that most pairs of samples from a population will have very similar means (i.e. the difference between sample means should normally be very small). A large standard error tells us that sample means can deviate quite a lot from the population mean and so differences between pairs of samples can be quite large by chance alone.

7.5.2. The dependent *t*-test equation explained ①

In an experiment, a person's score in condition 1 will be different to their score in condition 2, and this difference could be very large or very small. If we were to calculate the differences between each person's score in each condition and add up these differences we would get the total amount of difference. If we then divide this total by the number of participants we get the average difference (so how much, on average, a person's score differed in condition 1 compared to condition 2). This average difference is $\bar{D}$ in equation (7.2) and it is an indicator of the systematic variation in the data (i.e. it represents the experimental effect). We know that if we had taken two random samples from a population (and not done anything to these samples) then the means could be different just by chance. As we've just seen, the degree to which two sample means are likely to differ is determined by the size of the standard error of differences. We need to be sure that the observed difference between our samples is due to our experimental manipulation (and not a chance result). Knowing the mean difference alone is not useful because it depends upon the scale of measurement and so we standardize the value. One way to standardize the sum of group differences would be to divide by the sample standard deviation of those differences (see section 1.4.1). If you think about what the standard deviation represents, it is a measure of the average deviation from the mean, and so the standard deviation of the differences between conditions represents the average deviation from the mean difference. As such, the standard deviation is a measure of how much variation there is between participants' difference scores. Thus, the standard deviation of differences represents the *unsystematic* variation in the experiment.

To clarify, imagine that an alien came down and cloned me (because I am petrified of spiders) 12 times (heaven forbid that there should ever be 12 of me running around the planet!). All of my clones would be the same as me, and would behave in an identical way to me. Therefore, we would all be quite scared of the picture of the spider and might all have an anxiety score of 40. What's more, we would all soil our underpants at the sight of real spiders and so would have anxiety scores of 50. Remember that we are clones so we have behaved identically. If you calculated the difference between our anxiety scores for the picture and the real spider, the difference would be 10 for each of us. If we then calculate the standard deviation of these difference scores it will be 0 (because we all got the same scores and so there is no variation at all between scores). Therefore, if all of our participants were the same, the standard deviation would be 0; that is, there would be no unsystematic variation. In other words, all of the differences in anxiety result from showing us real spiders instead of pictures and none of the differences can be explained by other factors (such as individual differences).

Although dividing by the standard deviation would be useful as a means of standardizing the average difference between conditions, we are interested in knowing how the difference between sample means compares to what we would expect to find had we not imposed an experimental manipulation. We can use the properties of the sampling distribution: instead of dividing the average difference between conditions by the standard deviation of differences, we could divide it by the standard error of differences. Dividing by the standard error not only standardizes the average difference between conditions, but also tells us how the difference between the two sample means compares in magnitude to what would be expected by chance alone. If the standard error is large, then large differences between samples are more common (because the distribution of differences is more spread out). Conversely, if the standard error is small, then large differences between sample means are uncommon (because the distribution is very narrow and centred around zero). Therefore, if the average difference between our samples is large, and the standard error of differences is small, then we can be confident that the difference we observed in our sample is not a chance result. If the difference is not a chance result then it must have been caused by the experimental manipulation.

In a perfect world, we could calculate the standard error by taking all possible pairs of samples from a population, calculating the differences between their means, and then working out the standard deviation of these differences. However, in reality this is impossible. Therefore, we estimate the standard error from the standard deviation of differences obtained within the sample (S_D) and the sample size (N). Think back to section 1.6 where we saw that the standard error is simply the standard deviation divided by the square root of the sample size; likewise the standard error of differences ($\sigma_{\bar{D}}$) is simply the standard deviation of differences divided by the square root of the sample size:

$$\sigma_{\bar{D}} = \frac{s_D}{\sqrt{N}}$$

If the standard error of differences is a measure of the unsystematic variation within the data, and the sum of difference scores represents the systematic variation, then it should be clear that **the *t*-statistic is simply the ratio of the systematic variation in the experiment to the unsystematic variation**. If the experimental manipulation creates any kind of effect, then we would expect the systematic variation to be much greater than the unsystematic variation (so at the very least, t should be greater than 1). If the experimental manipulation is unsuccessful then we might expect the variation caused by individual differences to be much greater than that caused by the experiment (and so t will be less than 1). We can compare the obtained value of t against the maximum value we would expect to get by chance alone in a t-distribution with the same degrees of freedom (these values can be found in Appendix A.2); if the value we obtain exceeds this critical value we can be confident that this reflects an effect of our independent variable.

7.5.3. Dependent *t*-tests using SPSS ①

Using our spider data (**spiderRM.sav**), we have 12 spider-phobes who were exposed to a picture of a spider (**picture**) and on a separate occasion a real live tarantula (**real**). Their anxiety was measured in each condition (half of the participants were exposed to the picture before the real spider while the other half were exposed to the real spider first). I have already described how the data are arranged, and so we can move straight on to doing the test itself. First, we need to access the main dialog box by using the **Analyze⇒Compare Means⇒Paired-Samples T Test...** menu pathway (Figure 7.10). Once the dialog box is activated, select two variables from the list (click on the first variable with the mouse and then the second and note that the variable names appear in the box labelled *Current Selections*) and transfer them to the box labelled *Paired Variables* by clicking on ▶. If you want to carry out several t-tests then you can select another pair of variables, transfer them to the variables list, and then select another pair and so on. In this case, we want only one test. If you click on Options... then another dialog box appears that gives you

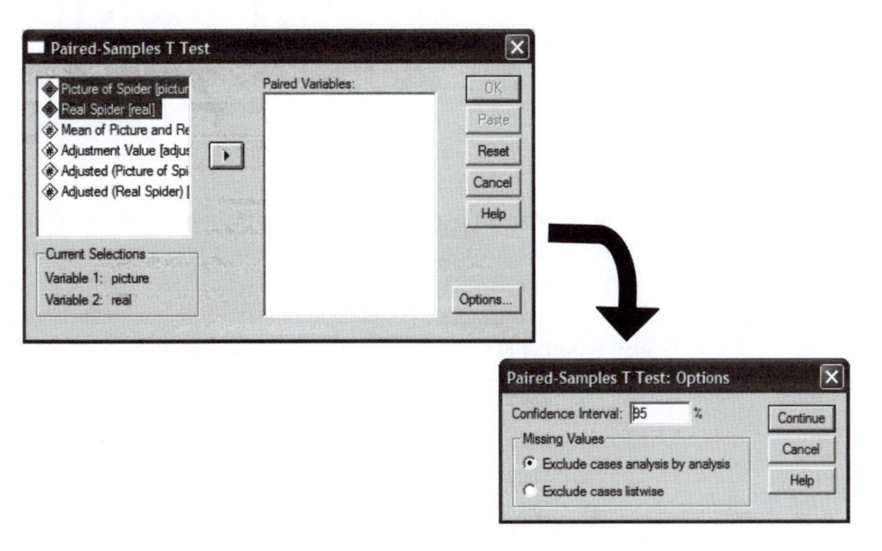

Figure 7.10 Main dialog box for paired-samples t-test

the chance to change the width of the confidence interval that is calculated. The default setting is for a 95% confidence interval and this is fine; however, if you want to be stricter about your analysis you could choose a 99% confidence interval but you run a higher risk of failing to detect a genuine effect (a Type II error). To run the analysis click on OK.

7.5.4. Output from the dependent *t*-test ①

The resulting output produces three tables. SPSS Output 7.1 shows a table of summary statistics for the two experimental conditions. For each condition we are told the mean, the number of participants (*N*) and the standard deviation of the sample. In the final column we are told the standard error (see section 7.5.1), which is the sample standard deviation divided by the square root of the sample size ($SE = s/\sqrt{N}$), so for the picture condition $SE = 9.2932/\sqrt{12} = 9.2932/3.4641 = 2.68$.

SPSS Output 7.1 also shows the Pearson correlation between the two conditions. When repeated measures are used it is possible that the experimental conditions will correlate (because the data in each condition come from the same people and so there could be some constancy in their responses). SPSS provides the value of Pearson's *r* and the two-tailed significance value (see Chapter 4). For these data the experimental conditions yield a fairly large correlation coefficient ($r = .545$) but are not significantly correlated because $p > .05$.

Paired Samples Statistics

		Mean	N	Std. Deviation	Std. Error Mean
Pair 1	Picture of Spider	40.0000	12	9.2932	2.6827
	Real Spider	47.0000	12	11.0289	3.1838

Paired Samples Correlations

		N	Correlation	Sig.
Pair 1	Picture of Spider & Real Spider	12	.545	.067

SPSS Output 7.1

SPSS Output 7.2 shows the most important of the tables: the one that tells us whether the difference between the means of the two conditions was large enough *not* to be a chance result. First, the table tells us the mean difference between scores (this value—$\bar{D}$ in equation (7.2)—is the difference between the mean scores of each condition: $40 - 47 = -7$). The table also reports the standard deviation of the differences between the means and more importantly the standard error of the differences between participants' scores in each condition (see section 7.5.1). The test statistic, *t*, is calculated by dividing the mean of differences by the standard error of differences (see equation (7.2): $t = -7/2.8311 = -2.47$). The size of *t* is compared against known values based on the degrees of freedom. When the same participants have been used, the degrees of freedom are simply the sample size minus 1 ($df = N - 1 = 11$). SPSS uses the degrees of freedom to calculate the exact probability that a value of *t* as big as the one obtained could occur by chance. This probability value is in

Paired Samples Test

		Paired Differences					t	df	Sig. (2-tailed)
					95% Confidence Interval of the Differences				
		Mean	Std. Deviation	Std. Error Mean	Lower	Upper			
Pair 1	Picture of Spider – Real Spider	–7.0000	9.8072	2.8311	–13.2312	–.7688	–2.473	11	.031

SPSS Output 7.2

the column labelled *Sig.* By default, SPSS provides only the two-tailed probability, which is the probability when no prediction was made about the direction of group differences. If a specific prediction was made (e.g. we might predict that anxiety will be higher when a real spider is used) then the one-tailed probability should be reported and this value is obtained by dividing the two-tailed probability by 2 (more on this later). The two-tailed probability for the spider data is very low ($p = .031$) and in fact it tells us that there is only a 3.1% chance that a value of t this big could happen by chance alone. As social scientists we are prepared to accept as statistically meaningful anything that has less than a 5% chance of occurring by chance. The fact that the t-value is a minus number tells us that condition 1 (the **picture** condition) had a smaller mean than the second (the **real** condition) and so the real spider led to greater anxiety than the picture. When we report a t-test we always include the degrees of freedom (in this case 11), the value of the t-statistic, and the level at which this value is significant. Therefore, we can conclude that exposure to a real spider caused significantly more reported anxiety in spider-phobes than exposure to a picture ($t\ (11) = -2.47$, $p < .05$). This result was predicted by the error bar chart in Figure 7.9.

The final thing that this output provides is a 95% confidence interval for the mean difference. Imagine we took 100 samples from a population of difference scores and calculated their means ($\bar{D}$) and a confidence interval for that mean. In 95 of those samples the constructed confidence interval contains the true value of the mean difference. The confidence interval tells us the boundaries within which the true mean difference is likely to lie.[3] So, assuming this sample's confidence interval is one of the 95 out of 100 that contains the population value, we can say that the true mean difference lies between –13.23

3 We saw in section 1.6.2 that these intervals represent the value of two (well, 1.96 to be precise) standard errors either side of the mean of the sampling distribution. For these data, in which the mean difference was –7 and the standard error was 2.8311, these limits will be $-7 \pm (1.96 \times 2.8311)$. However, because we're using the t-distribution, not the normal distribution, we use the critical value of t to compute the confidence intervals. This value is (with 11 degrees of freedom as in this example) 2.201 (two-tailed), which gives us $-7 \pm (2.201 \times 2.8311)$.

and −0.77. The importance of this interval is that it does not contain zero (i.e. both limits are negative) because this tells us that the true value of the mean difference is unlikely to be zero. Crucially, if we were to compare pairs of random samples from a population we expect most of the differences between sample means to be zero. This interval tells us that, based on our two samples, the true value of the difference between means is unlikely to be zero. Therefore, we can be confident that our two samples do not represent random samples from the same population. Instead they represent samples from different populations induced by the experimental manipulation.

7.5.5. Calculating the effect size ②

Even though our *t*-statistic is statistically significant, this doesn't mean our effect is important in practical terms. To discover whether the effect is substantive we need to use what we know about effect sizes (see section 1.8.4). I'm going to stick with the effect size *r* because it's widely understood, frequently used, and yes, I'll admit it, I actually like it! Converting a *t*-value into an *r*-value is actually really easy; we can use the following equation (from Rosenthal, 1991, p. 19, and Rosnow & Rosenthal, 2005, p. 328):

$$r = \sqrt{\frac{t^2}{t^2 + df}}$$

We know the value of *t* and the *df* from the SPSS output and so we can compute *r* as follows:

$$\begin{aligned} r &= \sqrt{\frac{-2.473^2}{-2.473^2 + 11}} \\ &= \sqrt{\frac{6.116}{17.116}} \\ &= .60 \end{aligned}$$

If you think back to our benchmarks for effect sizes this represents a very large effect (it is above .5, the threshold for a large effect). Therefore, as well as being statistically significant, this effect is large and so represents a substantive finding.

7.5.6. Reporting the dependent *t*-test ①

There is a fairly standard way to report any test statistic: you usually state the finding to which the test relates and then report the test statistic, its degrees of freedom, and the probability value of that test statistic. There has also been a recent move (by the American

Psychological Association amongst others) to recommend that an estimate of the effect size is routinely reported. Although effect sizes are still rather sporadically used, I want to get you into good habits so we'll start thinking about effect sizes now. In this example the SPSS output tells us that the value of t was −2.47, that the degrees of freedom on which this was based was 11, and that it was significant at $p = .031$. We can also see the means for each group. We could write this as:

✓ On average, participants experienced significantly greater anxiety to real spiders ($M = 47.00$, $SE = 3.18$), than to pictures of spiders ($M = 40.00$, $SE = 2.68$, $t(11) = -2.43$, $p < .05$, $r = .60$).

Note how we've reported the means in each group (and standard errors) in the standard format. For the test statistic note that we've used an italic t to denote the fact that we've calculated a t-statistic, then in brackets we've put the degrees of freedom and then stated the value of the test statistic. The probability can be expressed in several ways: often people report things to a standard level of significance (such as .05) as I have done here, but sometimes people will report the exact significance. Finally, note that I've reported the effect size at the end—you won't always see this in published papers but that's no excuse for you not to report it!

Try to avoid writing vague, unsubstantiated things like this:

✗ People were more scared of real spiders ($t = -2.47$).

More scared than what? Where are the *df*? Was the result statistically significant? Was the effect important (what was the effect size)?

Cramming Samantha's Tips

- The dependent *t*-test compares two means, when those means have come from the same group of people; for example, if you have two experimental conditions and have used the same participants in each condition.
- Look at the column labelled *Sig.* If the value is less than .05 then the means of the two groups are significantly different.
- Look at the values of the means to tell you how the groups differ.
- SPSS only provides the two-tailed significance value; if you want the one-tailed significance just divide the value by 2.
- Report the *t*-statistic, the degrees of freedom and the significance value. Also report the means and their corresponding standard errors (or draw an error bar chart).
- If you're feeling brave, calculate the effect size and report this too!

7.6. THE INDEPENDENT *T*-TEST ①

7.6.1. The independent *t*-test equation explained ①

The independent *t*-test is used in situations in which there are two experimental conditions and different participants have been used in each condition. There are two different equations that can be used to calculate the *t*-statistic depending on whether the samples contain an equal number of people. As with the dependent *t*-test we can calculate the *t*-statistic by using a numerical version of equation (7.1). With the dependent *t*-test we could look at differences between pairs of scores, because the scores came from the same participants and so individual differences between conditions were eliminated. Hence, the difference in scores should reflect only the effect of the experimental manipulation. Now, when different participants participate in different conditions then pairs of scores will differ not just because of the experimental manipulation, but also because of other sources of variance (such as individual differences between participants' motivation and IQ etc.). If we cannot investigate differences between conditions on a *per participant* basis (by comparing pairs of scores as we did for the dependent *t*-test) then we must make comparisons on a *per condition* basis (by looking at the overall effect in a condition—see equation (7.3)):

$$t = \frac{(\overline{X}_1 - \overline{X}_2) - (\mu_1 - \mu_2)}{\text{estimate of the standard error}} \tag{7.3}$$

Instead of looking at differences between pairs of scores, we now look at differences between the overall means of the two samples and compare them to the differences we would expect to get between the means of the two populations from which the samples come. If the null hypothesis is true then the samples have been drawn from the same population. Therefore, under the null hypothesis $\mu_1 = \mu_2$ and $\mu_1 - \mu_2 = 0$. Therefore, under the null hypothesis the equation becomes:

$$t = \frac{\overline{X}_1 - \overline{X}_2}{\text{estimate of the standard error}} \tag{7.4}$$

In the dependent *t*-test we divided the mean difference between pairs of scores by the standard error of these differences. For the independent *t*-test we are looking at differences between groups and so we need to divide by the standard deviation of differences between groups. We can still apply the logic of sampling distributions to this situation. Now, imagine we took several pairs of samples—each pair containing one sample from the two different populations—and compared the means of these samples. From what we have learnt about sampling distributions, we know that the majority of samples from a population will have fairly similar means. Therefore, if we took several pairs of samples (from different populations), the differences between the sample means will be similar across pairs. However, often the difference between a pair of sample means will deviate by a small amount and very occasionally it will deviate by a large amount. If we could plot a sampling distribution of the differences between every pair of sample means that could be taken from two populations, then we would find that it had a normal distribution with a mean equal to the difference between population means ($\mu_1 - \mu_2$). The sampling distribution would tell us by how much we can expect the means of two (or more) samples to differ. As before, the standard deviation of the sampling distribution (the standard error) tells us how variable the differences between sample means are by chance alone. If the standard deviation is high then large differences between sample means can occur by chance; if it is small then only small differences between sample means are expected. It therefore makes sense that we use the standard error of the sampling distribution to assess whether the difference between two sample means is statistically meaningful or simply a chance result. Specifically, we divide the difference between sample means by the standard deviation of the sampling distribution.

So, how do we obtain the standard deviation of the sampling distribution of differences between sample means? Well, we use the variance sum law, which states that the variance of a difference between two independent variables is equal to the sum of their variances (see Howell, 2002, Chapter 7). This statement means that the variance of the sampling distribution is equal to the sum of the variances of the two populations from which the samples were taken. We saw earlier that the standard error is the standard deviation of the sampling distribution of a population. We can use the sample standard deviations to calculate the standard error of each population's sampling distribution:

$$\text{SE of sampling distribution of population 1} = \frac{s_1}{\sqrt{N_1}}$$
$$\text{SE of sampling distribution of population 2} = \frac{s_2}{\sqrt{N_2}}$$

Therefore, remembering that the variance is simply the standard deviation squared, we can calculate the variance of each sampling distribution:

$$\text{variance of sampling distribution of population 1} = \left(\frac{s_1}{\sqrt{N_1}}\right)^2 = \frac{s_1^2}{N_1}$$
$$\text{variance of sampling distribution of population 2} = \left(\frac{s_2}{\sqrt{N_2}}\right)^2 = \frac{s_2^2}{N_2}$$

The variance sum law means that to find the variance of the sampling distribution of differences we merely add together the variances of the sampling distributions of the two populations:

$$\text{variance of the sampling distribution of differences} = \frac{s_1^2}{N_1} + \frac{s_2^2}{N_2}$$

To find out the standard error of the sampling distribution of differences we merely take the square root of the variance (because variance is the standard deviation squared):

$$\text{SE of the sampling distribution of differences} = \sqrt{\left(\frac{s_1^2}{N_1} + \frac{s_2^2}{N_2}\right)}$$

Therefore, equation (7.4) becomes equation (7.5):

$$t = \frac{\overline{X}_1 - \overline{X}_2}{\sqrt{\left(\frac{s_1^2}{N_1} + \frac{s_2^2}{N_2}\right)}} \tag{7.5}$$

Equation (7.5) is true only when the sample sizes are equal. Often in the social sciences it is not possible to collect samples of equal size (because, for example, people may not complete an experiment). When we want to compare two groups that contain different numbers of participants then equation (7.5) is not appropriate. Instead the pooled variance estimate *t*-test is used which takes account of the difference in sample size by *weighting* the variance of each sample. Intuitively, you might realize that large samples are better than small ones because they more closely approximate the population; therefore, we weight the variance by the size of sample on which it's based (we actually weight by the degrees of freedom, which are the sample size minus 1). Therefore, the pooled variance estimate is:

$$s_p^2 = \frac{(n_1 - 1)s_1^2 + (n_2 - 1)s_2^2}{n_1 + n_2 - 2}$$

This is simply a weighted average in which each variance is multiplied (weighted) by its degrees of freedom, and then we divide by the sum of weights (or sum of the two degrees of freedom). The resulting weighted average variance is then just replaced in the *t*-test equation:

$$t = \frac{\overline{X}_1 - \overline{X}_2}{\sqrt{\frac{s_p^2}{n_1} + \frac{s_p^2}{n_2}}}$$

As with the dependent *t*-test we can compare the obtained value of *t* against the maximum value we would expect to get by chance alone in a *t*-distribution with the same degrees of freedom (these values can be found in Appendix A.2); if the value we obtain exceeds this critical value we can be confident that this reflects an effect of our independent variable.

The derivation of the *t*-statistic is merely to provide a conceptual grasp of what we are doing when we carry out a *t*-test on SPSS. Therefore, if you don't know what on earth I'm babbling on about then don't worry about it (just spare a thought for my cat: he has to listen to this rubbish all the time!) because SPSS knows how to do it and that's all that matters!

7.6.2. The independent *t*-test using SPSS ①

I have probably bored most of you to the point of wanting to eat your own legs by now. Equations are boring and that is why SPSS was invented to help us minimize our contact with them. Using our spider data again (**spiderBG.sav**), we have 12 spider-phobes who were exposed to a picture of a spider and 12 different spider-phobes who were exposed to a real-life tarantula (the groups are coded using the variable **group**). Their anxiety was measured in each condition (**anxiety**). I have already described how the data are arranged (see section 7.3.1), and so we can move straight on to doing the test itself. First, we need to access the main dialog box by using the **Analyze⇒Compare Means⇒Independent-Samples T Test...** menu pathway (see Figure 7.11). Once the dialog box is activated, select the dependent variable from the list (click on **anxiety**) and transfer it to the box labelled *Test Variable(s)* by clicking on ▶. If you want to carry out *t*-tests on several dependent variables then you can select other dependent variables and transfer them to the variables list. However, there are good reasons why it is not a good idea to carry out lots of tests (see Chapter 8).

Next, we need to select an independent variable (the grouping variable). In this case, we need to select **group** and then transfer it to the box labelled *Grouping Variable*. When your grouping variable has been selected the Define Groups... button will become active and you should click on

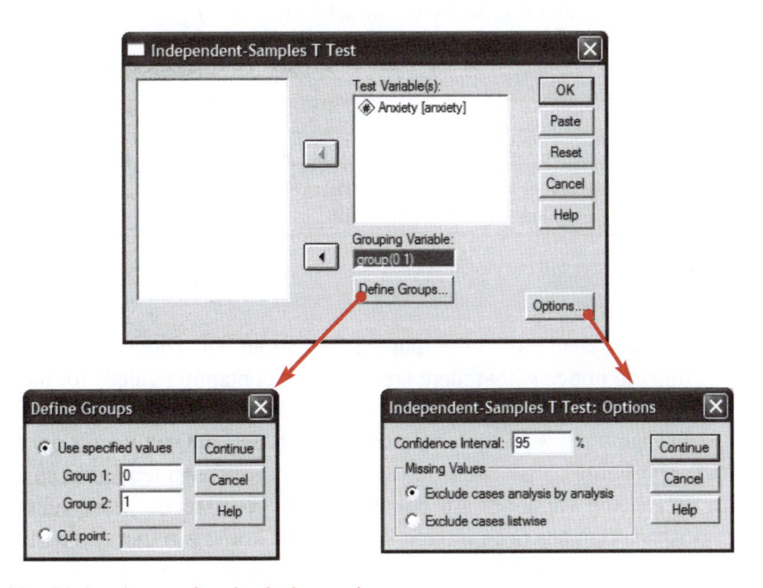

Figure 7.11 Dialog boxes for the independent means *t*-test

it to activate the *define groups* dialog box. SPSS needs to know what numeric codes you assigned to your two groups, and there is a space for you to type the codes. In this example, we coded our picture group as 0 and our real group as 1, and so these are the codes that we type. Alternatively you can specify a *Cut point*, in which case SPSS will assign all cases greater than or equal to that value to one group and all the values below the cut point to the second group. This facility is useful if you are testing different groups of participants based on something like a median split. So, you might want to classify people as spider-phobes or non-spider-phobes, and so you measure their score on a spider phobia questionnaire and calculate the median. You then classify anyone with a score above the median as phobic, and those below the median as non-phobes. Rather than recoding all of your participants and creating a coding variable, you would simply type the median value in the box labelled *Cut point*. When you have defined the groups, click on Continue to return to the main dialog box. If you click on Options... then another dialog box appears that gives you the same options as for the dependent *t*-test. To run the analysis click on OK.

7.6.3. Output from the independent *t*-test ①

The output from the independent *t*-test contains only two tables. The first table (SPSS Output 7.3) provides summary statistics for the two experimental conditions. From this table, we can see that both groups had 12 participants (column labelled *N*). The group who saw the picture of the spider had a mean anxiety of 40, with a standard deviation of 9.29. What's more, the standard error of that group (the standard deviation of the sampling distribution) is 2.68 ($SE = 9.2932/\sqrt{12} = 9.2932/3.4641 = 2.68$). In addition, the table tells us that the average anxiety level in participants who were shown a real spider was 47, with a standard deviation of 11.03, and a standard error of 3.18 ($SE = 11.03/\sqrt{12} = 11.03/3.4641 = 3.18$).

Group Statistics

	Condition	N	Mean	Std. Deviation	Std. Error Mean
Anxiety	Picture	12	40.0000	9.2932	2.6827
	Real Spider	12	47.0000	11.0289	3.1838

SPSS Output 7.3

The second table of output (SPSS Output 7.4) contains the main test statistics. The first thing to notice is that there are two rows containing values for the test statistics: one row is labelled *Equal variances assumed*, while the other is labelled *Equal variances not assumed*. In Chapter 3, we saw that parametric tests assume that the variances in experimental groups are roughly equal. Well, in reality there are adjustments that can be made in situations in which the variances are not equal. The rows of the table relate to whether or not this assumption has been broken. How do we know whether this assumption has been broken?

Well, we could just look at the values of the variances and see whether they are similar (e.g. we know that the standard deviations of the two groups are 9.29 and 11.03 and if we

Independent Samples Test

		Levene's Test for Equality of Variances		t-test for Equality of Means						
									95% Confidence Interval of the Mean	
		F	Sig.	t	df	Sig. (2-tailed)	Mean Difference	Std. Error Difference	Lower	Upper
Anxiety	Equal variances assumed	.782	.386	−1.681	22	.107	−7.0000	4.1633	−15.6342	1.6342
	Equal variances not assumed			−1.681	21.385	.107	−7.0000	4.1633	−15.6486	1.6486

SPSS Output 7.4

square these values then we get variances of 86.30 and 121.66). However, this measure would be very subjective and probably prone to academics thinking 'ooh look, the variance in group 1 is only 3000 times larger than the variance in group 2: that's roughly equal'. Fortunately, there is a test that can be performed to see whether the variances are different enough to cause concern. Levene's test (see section 3.6) is similar to a *t*-test in that it tests the hypothesis that the variances in the two groups are equal (i.e. the difference between the variances is zero). Therefore, if Levene's test is significant at $p \leq .05$ then we can conclude that the null hypothesis is incorrect and that the variances are significantly different—therefore, the assumption of homogeneity of variances has been violated. If, however, Levene's test is non-significant (i.e. $p > .05$) then we must accept the null hypothesis that the difference between the variances is zero—the variances are roughly equal and the assumption is tenable. For these data, Levene's test is non-significant (because $p = .386$, which is greater than .05) and so we should read the test statistics in the row labelled *Equal variances assumed*. Had Levene's test been significant, then we would have read the test statistics from the row labelled *Equal variances not assumed*.

Having established that the assumption of homogeneity of variances is met, we can move on to look at the *t*-test itself. We are told the mean difference ($\bar{X}_1 - \bar{X}_2 = 40 - 47 = -7$) and the standard error of the sampling distribution of differences, which is calculated using the lower half of equation (7.5):

$$\sqrt{\left(\frac{s_1^2}{N_1} + \frac{s_2^2}{N_2}\right)} = \sqrt{\left(\frac{9.29^2}{12} + \frac{11.03^2}{12}\right)}$$
$$= \sqrt{(7.19 + 10.14)}$$
$$= \sqrt{17.33}$$
$$= 4.16$$

The *t*-statistic is calculated by dividing the mean difference by the standard error of the sampling distribution of differences ($t = -7/4.16 = -1.68$). The value of *t* is then assessed against the value of *t* you might expect to get by chance when you have certain degrees of freedom. For the independent *t*-test, degrees of freedom are calculated by adding the two sample sizes and then subtracting the number of samples ($df = N_1 + N_2 - 2 = 12 + 12 - 2 = 22$). SPSS produces the exact significance value of *t*, and we are interested in whether this value is less than or greater than .05. In this case the two-tailed value of *p* is .107, which is greater than .05, and so we would have to conclude that there was no significant difference between the means of these two samples. In terms of the experiment, we can infer that spider-phobes are made equally anxious by pictures of spiders as they are by the real thing.

Now, we use the two-tailed probability when we have made no specific prediction about the direction of our effect (see section 1.8.2). For example, if we were unsure whether a real spider would induce more or less anxiety, then we would have to use a two-tailed test. However, often in research we can make specific predictions about which group has the highest mean. In this example, it is likely that we would have predicted that a real spider would induce greater anxiety than a picture and so we predict that the mean of the real group would be greater than the mean of the picture group. In this case, we can use a one-tailed test (for more discussion of this issue see section 1.8.2, or Rowntree, 1981, Chapter 7). Some students get very upset by the fact that SPSS produces only the two-tailed significance and are confused by why there isn't an option that can be selected to produce the one-tailed significance. The answer is simple: there is no need for an option because the one-tailed probability can be ascertained by dividing the two-tailed significance value by 2. In this case, the two-tailed probability was .107, therefore the one-tailed probability is .054 (.107/2). The one-tailed probability is still greater than .05 (albeit by a small margin) and so we would still have to conclude that spider-phobes were equally as anxious when presented with a real spider as spider-phobes who were presented with a picture of the same spider. This result was predicted by the error bar chart in Figure 7.3

7.6.4. Calculating the effect size ②

To discover whether our effect is substantive we can use the same equation as in section 7.5.5 to convert the *t*-statistics into a value of *r*. We know the value of *t* and the *df* from the SPSS output and so we can compute *r* as follows:

$$\begin{aligned} r &= \sqrt{\frac{-1.681^2}{-1.681^2 + 22}} \\ &= \sqrt{\frac{2.826}{24.826}} \\ &= .34 \end{aligned}$$

If you think back to our benchmarks for effect sizes this represents a medium effect (it is around .3, the threshold for a medium effect). Therefore, even though the effect was non-significant, it still represented a fairly substantial effect. You may also notice that the effect has shrunk, which may seem slightly odd given that we used exactly the same data (but see section 7.7)!

7.6.5. Reporting the independent *t*-test ①

The rules that I made up, erm, I mean, reported for the dependent *t*-test pretty much apply for the independent *t*-test. The SPSS output tells us that the value of *t* was –1.68, that the degrees of freedom on which this was based was 22, and that it was not significant at $p < .05$. We can also see the means for each group. We could write this as:

✓ On average, participants experienced greater anxiety to real spiders ($M = 47.00$, $SE = 3.18$), than to pictures of spiders ($M = 40.00$, $SE = 2.68$). This difference was not significant $t(22) = -1.68$, $p > .05$; however, it did represent a medium sized effect $r = .34$.

Note how we've reported the means in each group (and standard errors) as before. For the test statistic everything is much the same as before except that I've had to report that *p* was greater than (>) .05 rather than less than (<). Finally, note that I've commented on the effect size at the end.

Cramming Samantha's Tips

- The independent *t*-test compares two means, when those means have come from different groups of people; for example, if you have two experimental conditions and have used different participants in each condition.
- Look at the column labelled *Levene's Test for Equality of Variance*. If the *Sig.* value is less than .05 then the assumption of homogeneity of variance has been broken and you should only look at the row in the table labelled *Equal variances not assumed*. If the *Sig.* value of Levene's test is bigger than .05 then you should look at the row in the table labelled *Equal variances assumed*.
- Look at the column labelled *Sig.*: if the value is less than .05 then the means of the two groups are significantly different.
- Look at the values of the means to tell you how the groups differ.
- SPSS only provides the two-tailed significance value; if you want the one-tailed significance just divide the value by 2.
- Report the *t*-statistic, the degrees of freedom and the significance value. Also report the means and their corresponding standard errors (or draw an error bar chart).
- If you're feeling brave, calculate the effect size and report this too!

7.7. BETWEEN GROUPS OR REPEATED MEASURES? ①

The two examples in this chapter are interesting (honestly!) because they illustrate the difference between data collected using the same participants and data collected using different participants. The two examples in this chapter use the same scores in each condition. When analysed as though the data came from the same participants the result was a significant difference between means, but when analysed as though the data came from different participants there was no significant difference between group means. This may seem like a puzzling finding—after all, the numbers were identical in both examples. What this illustrates is the relative *power* of repeated-measures designs. When the same participants are used across conditions the unsystematic variance (often called the error variance) is reduced dramatically, making it easier to detect any systematic variance. It is often assumed that the way in which you collect data is irrelevant, but I hope to have illustrated that it can make the difference between detecting a difference and not detecting one. In fact, researchers have carried out studies using the same participants in experimental conditions, then repeated the study using different participants in experimental conditions, and then used the method of data collection as an independent variable in the analysis. Typically, they have found that the method of data collection interacts significantly with the results found (see Erlebacher, 1977).

7.8. THE *T*-TEST AS A GENERAL LINEAR MODEL ②

A lot of you might think it's odd that I've chosen to represent the effect size for my *t*-tests using *r*, the correlation coefficient. In fact you might well be thinking 'but correlations show relationships, not differences between means'. I used to think this too until I read a fantastic paper by Cohen (1968), which made me realize what I'd been missing: the complex, thorny, weed-infested and large Andy-eating tarantula-inhabited world of statistics suddenly turned into a beautiful meadow filled with tulips and little bleating lambs all jumping for joy at the wonder of life. Well, actually it didn't, I'm still a bumbling fool trying desperately to avoid having the life blood sucked from my flaccid corpse by the tarantulas of statistics; but it was a good paper! What I'm about to say will either make no sense at all, or might help you to appreciate that all statistical procedures are basically the same: they're just more or less elaborate versions of the correlation coefficient!

In Chapter 5 we saw that the *t*-test was used to test whether the regression coefficient of a predictor was equal to 0. The experimental design for which the independent *t*-test is used can be conceptualized as a regression equation (after all, there is one independent variable (predictor) and one dependent variable (outcome)). If we want to predict our outcome, then we can use the general equation that I've mentioned at various points:

If we want to use a linear model, then we saw that this general equation becomes equation (5.2) in which the model is defined by the slope and intercept of a straight line.

$$\text{Outcome}_i = (\text{Model}_i) + \text{error}_i$$

Equation (7.6) is a very similar equation in which A_i is the dependent variable (outcome), b_0 is the intercept, b_1 is the weighting of the predictor, and G_i is the independent variable (predictor). Now, I've also included the same equation but with some of the letters replaced with what they represent in the spider experiment (so, A = **anxiety**, G = **group**). When we run an experiment with two conditions, the independent variable has only two values (group 1 or group 2). There are several ways in which these groups can be coded (in the spider example we coded group 1 with the value 0 and group 2 with the value 1). This coding variable is known as a *dummy variable* and values of this variable represent groups of people. The brave ones amongst you will have come across this coding in section 5.10:

$$A_i = b_0 + b_1 G_i + \varepsilon_i$$
$$\text{Anxiety}_i = b_0 + b_1 \text{Group}_i + \varepsilon_i \tag{7.6}$$

Using the spider example, we know that the mean **anxiety** of the picture group was 40, and that the **group** variable is equal to 0 for this condition. Look at what happens when the **group** variable is equal to 0 (the picture condition): equation (7.6) becomes (if we ignore the residual term):

$$40 = b_0 + (b_1 \times 0)$$
$$b_0 = 40$$

Therefore, b_0 (the intercept) is equal to the mean of the picture group (i.e. it is the mean of the group coded as 0). Now let's look at what happens when the **group** variable is equal to 1. This condition is the one in which a real spider was used, and the mean **anxiety** ($\bar{X}_{\text{real}}$) of this condition was 47. Remembering that we have just found out that b_0 is equal to the mean of the picture group ($\bar{X}_{\text{picture}}$), equation (7.6) becomes:

$$47 = b_0 + (b_1 \times 1)$$
$$47 = 40 + b_1$$
$$b_1 = 47 - 40$$
$$b_1 = \bar{X}_{\text{real}} - \bar{X}_{\text{picture}}$$

b_1, therefore, represents the difference between the group means. As such, we can represent a two-group experiment as a regression equation in which the coefficient of the independent variable (b_1) is equal to the difference between group means, and the intercept (b_0) is equal to the mean of the group coded as 0. In regression, the t-test is used to ascertain whether the regression coefficient (b_1) is equal to 0, and when we carry out a t-test on grouped data we therefore test whether the difference between group means is equal to 0.

To prove that I'm not making it up as I go along, use the data in **spiderBG.sav** and run a simple linear regression using **group** as the predictor and **anxiety** as the outcome. **Group** is coded using a 0 and 1 coding scheme and so represents the dummy variable

Coefficients[a]

Model		Unstandardized Coefficients		Standardized Coefficients		
		B	Std. Error	Beta	t	Sig.
1	(Constant)	40.000	2.944		13.587	.000
	Condition	7.000	4.163	.337	1.681	.107

a Dependent Variable: Anxiety

SPSS Output 7.5 Regression analysis of between-group spider data

described above. The resulting SPSS output should contain the regression summary table shown in SPSS Output 7.5. The first thing to notice is the value of the constant (b_0): its value is 40, the same as the mean of the base category (the picture group). The second thing to notice is that the value of the regression coefficient b_1 is 7, which is the difference between the two group means (47 – 40 = 7). Finally, the *t*-statistic, which tests whether b_1 is significantly different from 0, is the same as for the independent *t*-test (see SPSS Output 7.4) and so is the significance value.[4]

This section has demonstrated that experiments can be represented in terms of linear models and this concept is essential in understanding the following chapters on the general linear model.

7.9. WHAT IF OUR DATA ARE NOT NORMALLY DISTRIBUTED? ②

We've seen in this chapter that there are adjustments that can be made to the *t*-test when the assumption of homogeneity of variance is broken, but what about when you have non-normally distributed data? One possibility is to try to correct the distribution using a transformation (see Chapter 3), but this doesn't always work. Another useful solution is to use one of a group of tests commonly referred to as *non-parametric tests*. These tests have fewer assumptions than their parametric counterparts and so are useful when your data violate the assumptions of parametric data described in section 3.2.1. Some of these tests are described in Chapter 13. The non-parametric counterpart of the *dependent t-test* is called the *Wilcoxon signed-rank test* (section 13.3), and the independent *t*-test has two non-parametric counterparts (both extremely similar) called the *Wilcoxon rank-sum test* and the *Mann–Whitney test* (section 13.2). I'd recommend reading these sections before moving on.

---

4 In fact, the value of the *t*-statistic is the same but has a positive sign rather than negative. You'll remember from the discussion of the point–biserial correlation in section 4.5.6 that when you correlate a dichotomous variable the direction of the correlation coefficient depends entirely upon which cases are assigned to which groups. Therefore, the direction of the *t*-statistic here is similarly influenced by which group we select to be the base category (the category coded as 0).

7.10. WHAT HAVE WE DISCOVERED ABOUT STATISTICS? ①

We've covered a lot of ground in this chapter, and as always seems to be the case what started out as a 'quick chapter on comparing two means' has ended up as a book in it's own right. These perfectionist tendencies are the cross I have to bear!

We started the chapter, about five decades ago, by looking at different ways in which data can be collected and why it might be interesting to compare two means. We looked at some important issues such as randomization. I then talked about error bar charts, first of all for situations in which you've collected data using different participants, and went on to look at some of the issues and adjustments necessary when you've used the same participants. We then had a look at some general conceptual features of the *t*-test, a parametric test that's used to test differences between two means. After this general taster, we moved on to look specifically at the dependent *t*-test (used when your conditions involve the same people). I explained how it was calculated, how to do it on SPSS and how to interpret the results. We then discovered much the same for the independent *t*-test (used when your conditions involve different people). After this I droned on excitedly about how a situation with two conditions can be conceptualized as a general linear model, by which point those of you who have a life had gone to the pub for a stiff drink. Finally, I mentioned the non-parametric equivalents of the *t*-tests: the Mann–Whitney and Wilcoxon rank-sum tests (used when you are comparing two conditions in which different people participated) and the Wilcoxon signed-rank test (used when you want to compare two conditions in which the same people participated). These are described in Chapter 13. Having done all that, I'm off to have a cup of Darjeeling tea and very large bar of chocolate.

7.11. KEY TERMS THAT WE'VE DISCOVERED

- Between-group design
- Between-subjects design
- Boredom effects
- Confounding variables
- Counterbalancing
- Dependent *t*-test
- Dependent variable
- Error bar chart
- Grand mean
- Independent design
- Independent *t*-test
- Independent variable
- Practice effects
- Randomization
- Repeated-measures design
- Standard error of differences
- Variance sum law
- Within-subject design

7.12. SMART ALEX'S TASKS

The following two scenarios are taken from Field & Hole (2003). In each case analyse the data on SPSS. Detailed answers to these tasks can be found in Field & Hole, or in shorter form on the CD-ROM in the file called **Answers (Chapter 7).pdf**.

- **Task 1**: One of my pet hates is 'pop psychology' books. Along with banishing Freud from all bookshops, it is my vowed ambition to rid the world of these rancid putrefaction-ridden wastes of trees. Not only do they give psychology a very bad name by stating the obvious and charging people for the privilege, but also they are considerably less enjoyable to look at than the trees killed to produce them (admittedly the same could be said for the turgid tripe that I produce in the name of education but let's not go there just for now!). Anyway, as part of my plan to rid the world of popular psychology I did a little experiment. I took two groups of people who were in relationships and randomly assigned them to one of two conditions. One group read the famous popular psychology book *Women are from Bras and men are from Penis*, whereas another group read *Marie Claire*. I tested only 10 people in each of these groups, and the dependent variable was an objective measure of their happiness with their relationship after reading the book. I didn't make any specific prediction about which reading material would improve relationship happiness. The data are in the file **Penis.sav**; analyse them with the appropriate *t*-test.①
- **Task 2**: Imagine Twaddle and Sons, the publishers of *Women are from Bras and men are from Penis*, were upset about my claims that their book was about as useful as a paper umbrella. They decided to take me to task and design their own experiment in which participants read their book, and one of my books (Field & Hole) at different times. Relationship happiness was measured after reading each book. To maximize their chances of finding a difference they used a sample of 500 participants, but got each participant to take part in both conditions (they read both books). The order in which books were read was counterbalanced and there was a delay of 6 months between reading the books. They predicted that reading their wonderful contribution to popular psychology would lead to greater relationship happiness than reading some dull and tedious book about experiments. The data are in **Field&Hole.sav**; analyse them using the appropriate *t*-test.①

7.13. FURTHER READING

Field, A. P. & Hole, G. (2003). *How to design and report experiments*. London: Sage. In my completely unbiased opinion this is a useful book to get some more background on experimental methods.

Rosnow, R. L. & Rosenthal, R. (2005). *Beginning behavioural research: a conceptual primer* (5th edition). Englewood Cliffs, NJ: Pearson/Prentice Hall.

Rowntree, D. (1981). *Statistics without tears: a primer for non-mathematicians*. London: Penguin. Chapters 4, 5, 6 and 7 provide an excellent and understandable introduction to the ideas of sampling distributions, hypothesis testing and statistical inference.

Wright, D. B. (2002). *First steps in statistics*. London: Sage. An excellent and clear chapter on the *t*-test (Chapter 6).

CHAPTER 8

COMPARING SEVERAL MEANS: ANOVA (GLM 1)

8.1. WHAT WILL THIS CHAPTER TELL US? ①

The previous chapter showed us how we could look at differences between two conditions or groups of people. The tests we explored in that chapter are useful tools in social science research; however, they are limited to situations in which there are only two levels of the independent variable (i.e. two experimental groups). It is common to run experiments in which there are three, four or even five levels of the independent variable and in these cases the tests in Chapter 7 are inappropriate. Instead, a technique called analysis of variance (or ANOVA to its friends) is used. ANOVA has the advantage that it can be used to analyse situations in which there are several independent variables. In these situations, ANOVA tells us how these independent variables interact with each other and what effects these interactions have on the dependent variable. So, this chapter will begin by explaining the theory of ANOVA when different participants are used (independent ANOVA). We'll then look at how to carry out the analysis on SPSS and interpret the results.

8.2. THE THEORY BEHIND ANOVA ②

8.2.1. Inflated error rates ②

Before explaining how ANOVA works, it is worth mentioning why we don't simply carry out several *t*-tests to compare all combinations of groups that have been tested. Imagine

a situation in which there were three experimental conditions and we were interested in differences between these three groups. If we were to carry out *t*-tests on every pair of groups, then we would have to carry out three separate tests: one to compare groups 1 and 2, one to compare groups 1 and 3, and one to compare groups 2 and 3. If each of these *t*-tests uses a .05 level of significance then for each test the probability of falsely rejecting the null hypothesis (known as a Type I error) is only 5%. Therefore, the probability of no Type I errors is .95 (95%) for each test. If we assume that each test is independent (hence, we can multiply the probabilities) then the overall probability of no Type I errors is $(.95)^3 = .95 \times .95 \times .95 = .857$, because the probability of no Type I errors is .95 for each test and there are three tests. Given that the probability of no Type I errors is .857, then we can calculate the probability of making at least one Type I error by subtracting this number from 1 (remember that the maximum probability of any event occurring is 1). So, the probability of at least one Type I error is $1 - .857 = .143$, or 14.3%. Therefore, across this group of tests, the probability of making a Type I error has increased from 5% to 14.3%, a value greater than the criterion accepted by social scientists. This error rate across statistical tests conducted on the same experimental data is known as the familywise or experimentwise error rate. An experiment with three conditions is a relatively simple design, and so the effect of carrying out several tests is not severe. If you imagine that we now increase the number of experimental conditions from three to five (which is only two more groups) then the number of *t*-tests that would need to be done increases to 10.[1] The familywise error rate can be calculated using the general equation (8.1), in which n is the number of tests carried out on the data. With 10 tests carried out, the familywise error rate is .40 ($1 - .95^{10} = .40$), which means that there is a 40% chance of having made at least one Type I error. For this reason we use ANOVA rather than conducting lots of *t*-tests:

$$\text{familywise error} = 1 - (.95)^n \tag{8.1}$$

When we perform a *t*-test, we test the hypothesis that the two samples have the same mean. Similarly, ANOVA tells us whether three or more means are the same, so it tests the

---

1 These comparisons are group 1 vs. 2, 1 vs. 3, 1 vs. 4, 1 vs. 5, 2 vs. 3, 2 vs. 4, 2 vs. 5, 3 vs. 4, 3 vs. 5 and 4 vs. 5. The number of tests required is calculated using this equation:

$$\text{number of comparisons, } C = \frac{k!}{2(k-2)!}$$

in which k is the number of experimental conditions. The ! symbol stands for *factorial*, which means that you multiply the value preceding the symbol by all of the whole numbers between zero and that value (so $5! = 5 \times 4 \times 3 \times 2 \times 1 = 120$). So, with five conditions we find that:

$$C = \frac{5!}{2(5-2)!} = \frac{120}{2 \times 6} = 10$$

hypothesis that all group means are equal. An ANOVA produces an *F-statistic* or *F-ratio*, which is similar to the *t*-statistic in that it compares the amount of systematic variance in the data to the amount of unsystematic variance. However, ANOVA is an *omnibus* test, which means that it tests for an overall experimental effect: so, there are things that an ANOVA cannot tell us. Although ANOVA tells us whether the experimental manipulation was generally successful, it does not provide specific information about which groups were affected. Assuming an experiment was conducted with three different groups, the *F*-ratio simply tells us that the means of these three samples are not equal (i.e. that $\bar{X}_1 = \bar{X}_2 = \bar{X}_3$ is *not* true). However, there are several ways in which the means can differ. The first possibility is that all three sample means are significantly different ($\bar{X}_1 \neq \bar{X}_2 \neq \bar{X}_3$). A second possibility is that the means of group 1 and 2 are the same but group 3 has a significantly different mean from both of the other groups ($\bar{X}_1 = \bar{X}_2 \neq \bar{X}_3$). Another possibility is that groups 2 and 3 have similar means but group 1 has a significantly different mean ($\bar{X}_1 \neq \bar{X}_2 = \bar{X}_3$). Finally, groups 1 and 3 could have similar means but group 2 has a significantly different mean from both ($\bar{X}_1 = \bar{X}_3 \neq \bar{X}_2$). So, the *F*-ratio tells us only that the experimental manipulation has had some effect, but it doesn't tell us specifically what the effect was.

8.2.2. ANOVA as regression ②

Many social scientists are unaware that ANOVA and regression are conceptually the same procedure. The reason is largely historical in that two distinct branches of methodology developed in the social sciences: correlational research and experimental research. Researchers interested in controlled experiments adopted ANOVA as their flagship statistic whereas those looking for real-world relationships adopted multiple regression. As we all know, scientists are intelligent, mature and rational people and so neither group was tempted to slag off the other and claim that their own choice of methodology was far superior to the other (yeah right!). With the divide in methodologies came a chasm between the statistical methods adopted by the two opposing camps (Cronbach, 1957, documents this divide in a lovely article). This divide has lasted many decades to the extent that now students are generally taught regression and ANOVA in very different contexts and most textbooks teach ANOVA in an entirely different way to regression.

Although many considerably more intelligent people than I have attempted to redress the balance (notably the great Jacob Cohen, 1968), I am passionate about making my own small, feeble-minded attempt to enlighten you (and set the ball rolling in sections 5.10 and 7.8). There are many good reasons why I think ANOVA should be taught within the context of regression. First, it provides a familiar context: I wasted many trees trying to explain regression, so why not use this base of knowledge to explain a new concept (it should make it easier to understand). Second, the traditional method of teaching ANOVA (known as the variance-ratio method) is fine for simple designs, but becomes impossibly cumbersome in more complex situations (such as analysis of covariance). The regression model extends very logically to these more complex designs without anyone needing to get bogged down in mathematics. Finally, the variance-ratio method becomes extremely unmanageable in unusual circumstances such as when you have unequal sample sizes.[2] The regression method makes these situations considerably simpler. Although these reasons are good enough, it is also the case that SPSS has moved away from the variance-ratio method of ANOVA and progressed towards solely using the regression model (known as the general linear model, or GLM). The end result of the two approaches to ANOVA are the same and most textbooks already detail the variance-ratio approach (I recommend Howell, 2002), so it makes sense for me to tackle the regression approach.

ANOVA is a way of comparing the ratio of systematic variance to unsystematic variance in an experimental study. The ratio of these variances is known as the *F*-ratio. Any of you who have read Chapter 5 should recognize the *F*-ratio (see section 5.2.3) because it was used to assess how well a regression model can predict an outcome compared to the error within that model. If you haven't read Chapter 5 (surely not!) have a look before you carry on (it should only take you a couple of weeks to read). The *F*-ratio in ANOVA is exactly the same as in regression, except that the regression model contains only categorical predictors (i.e. grouping variables). So, just as the *t*-test could be represented by the linear regression equation (see section 7.8), ANOVA can be represented by the multiple regression equation in which the number of predictors is one less than the number of categories of the independent variable.

Let's take an example. There was a lot of controversy, when I wrote the first edition of this book, surrounding the drug Viagra. Admittedly there's less controversy now, but the controversy has been replaced by an alarming number of spam emails on the subject (for which I'll no doubt be grateful in 20 years' time), so I'm going to stick with the example. Viagra is a sexual stimulant (used to treat impotence) that broke into the black market under the belief that it will make someone a better lover (oddly enough there were a glut of journalists taking the stuff at the time in the name of 'investigative journalism'… hmmm!). Suppose we tested this belief by taking three groups of participants and administering one group with a placebo (such as a sugar pill), one group with a low-dose of Viagra and one with a high dose. The dependent variable was an objective measure of libido (I will tell you only that it was measured over the course of a week—the rest I shall leave to your own imagination). The data can be found in the file **Viagra.sav** (which is described in detail later in this chapter) and are in Table 8.1.

If we want to predict levels of libido from the different levels of Viagra then we can use the general equation that keeps popping up:

$$\text{Outcome}_i = (\text{Model}_i) + \text{error}_i$$

If we want to use a linear model, then we saw in section 7.8 that when there are only two groups we could replace the 'model' in this equation with a linear regression equation with one dummy variable to describe two experimental groups. This dummy variable was a categorical variable with two numeric codes (a 0 for one group and a 1 for the other). With

2 Having said this, it is well worth the effort trying to obtain equal sample sizes in your different conditions because unbalanced designs do cause statistical complications (see section 8.2.9).

Table 8.1 Data in **Viagra.sav**

	Placebo	Low Dose	High Dose
	3	5	7
	2	2	4
	1	4	5
	1	2	3
	4	3	6
$\bar{X}$	2.20	3.20	5.00
s	1.30	1.30	1.58
s^2	1.70	1.70	2.50
Grand Mean = **3.467** Grand SD = **1.767** Grand Variance = **3.124**			

three groups, however, we can extend this idea and use a multiple regression model with two dummy variables. In fact, as a general rule we can extend the model to any number of groups and the number of dummy variables needed will be one less than the number of categories of the independent variable. In the two-group case, we assigned one category as a base category (remember in section 7.8 we chose the picture condition to act as a base) and this category was coded with a 0. When there are three categories we also need a base category and you should choose the condition with which you intend to compare the other groups. Usually this category will be the control group. In most well-designed social science experiments there will be a group of participants who act as a baseline for other categories. This baseline group should act as the reference or base category, although the group you choose will depend upon the particular hypotheses that you want to test. In unbalanced designs (in which the group sizes are unequal) it is important that the base category contains a fairly large number of cases to ensure that the estimates of the regression coefficients are reliable. In the Viagra example, we can take the placebo group as the base category because this group was a placebo control. We are interested in comparing both the high- and low-dose groups to the group who received no Viagra at all. If the placebo group is the base category then the two dummy variables that we have to create represent the other two conditions: so, we should have one dummy variable called High and the other one called Low. The resulting equation is described in equation (8.2):

$$\text{Libido}_i = b_0 + b_2\text{High}_i + b_1\text{Low}_i + \varepsilon_i \tag{8.2}$$

In equation (8.2), a person's libido can be predicted from knowing their group code (i.e. the code for the high- and low-dose dummy variables) and the intercept (b_0) of the model. The dummy variables in equation (8.2) can be coded in several ways, but the simplest way is to use a similar technique to that of the t-test. The base category is always coded as 0. If a participant was given a high dose of Viagra then they are coded with a 1 for the High

Table 8.2 Dummy coding for the three-group experimental design

Group	Dummy Variable 1 (High y)	Dummy Variable 2 (Low y)
Placebo	0	0
Low-Dose Viagra	0	1
High-Dose Viagra	1	0

dummy variable, and 0 for all other variables. If a participant was given a low dose of Viagra then they are coded with the value 1 for the Low dummy variable, and coded with a 0 for all other variables (this is the same type of scheme we used in section 5.10). Using this coding scheme we can express each group by combining the codes of the two dummy variables (see Table 8.2).

Placebo group: Let's examine the model for the placebo group. In the placebo group both the High and Low dummy variables are coded as 0. Therefore, if we ignore the error term (ε_i), the regression equation becomes:

$$\begin{aligned}\text{Libido}_i &= b_0 + (b_2 \times 0) + (b_1 \times 0)\\ \text{Libido}_i &= b_0\\ \overline{X}_{\text{Placebo}} &= b_0\end{aligned}$$

This is a situation in which the high- and low-dose groups have both been excluded (because they are coded with a 0). We are looking at predicting the level of libido when both doses of Viagra are ignored, and so the predicted value will be the mean of the placebo group (because this group is the only one included in the model). Hence, the intercept of the regression model, b_0, is always the mean of the base category (in this case the mean of the placebo group).

High-dose group: If we examine the high-dose group, the dummy variable High will be coded as 1 and the dummy variable Low will be coded as 0. If we replace the values of these codes into equation (8.2) the model becomes:

$$\begin{aligned}\text{Libido}_i &= b_0 + (b_2 \times 1) + (b_1 \times 0)\\ \text{Libido}_i &= b_0 + b_2\end{aligned}$$

We know already that b_0 is the mean of the placebo group. If we are interested in only the high-dose group then the model should predict that the value of Libido for a given participant equals the mean of the high-dose group. Given this information, the equation becomes:

$$\begin{aligned}\text{Libido}_i &= b_0 + b_2\\ \overline{X}_{\text{High}} &= \overline{X}_{\text{Placebo}} + b_2\\ b_2 &= \overline{X}_{\text{High}} - \overline{X}_{\text{Placebo}}\end{aligned}$$

Hence, b_2 represents the difference between the means of the high-dose group and the placebo group.

Low-dose group: Finally, if we look at the model when a low dose of Viagra has been taken, the dummy variable Low is coded as 1 (and hence High is coded as 0). Therefore, the regression equation becomes:

$$\text{Libido}_i = b_0 + (b_2 \times 0) + (b_1 \times 1)$$
$$\text{Libido}_i = b_0 + b_1$$

We know that the intercept is equal to the mean of the base category and that for the low-dose group the predicted value should be the mean libido for a low dose. Therefore the model can be reduced down to:

$$\text{Libido}_i = b_0 + b_1$$
$$\overline{X}_{\text{Low}} = \overline{X}_{\text{Placebo}} + b_1$$
$$b_1 = \overline{X}_{\text{Low}} - \overline{X}_{\text{Placebo}}$$

Hence, b_1 represents the difference between the means of the low-dose group and the placebo group. This form of dummy variable coding is the simplest form, but as we shall see later, there are other ways in which variables can be coded to test specific hypotheses. These alternative coding schemes are known as *contrasts* (see section 8.2.10.2). The idea behind contrasts is that you code the dummy variables in such a way that the *b*-values represent differences between groups that you are interested in testing.

To illustrate exactly what is going on I have created a file called **dummy.sav** in the Chapter 8 folder of the CD-ROM. This file contains the Viagra data but with two additional variables (**dummy1** and **dummy2**) that specify to which group a data point belongs (as in Table 8.2). Access this file and run multiple regression analysis using **libido** as the outcome and **dummy1** and **dummy2** as the predictors. If you're stuck on how to run the regression then read Chapter 5 again (see, these chapters are ordered for a reason!). The resulting analysis is shown in SPSS Output 8.1. It might be a good idea to remind yourself of the group means from Table 8.1. The first thing to notice is that the constant is equal to the mean of the base category (the placebo group). The regression coefficient for the first dummy variable (b_2) is equal to the difference between the means of the high-dose group and the placebo group (5.0 − 2.2 = 2.8). Finally, the regression coefficient for the second dummy variable (b_1) is equal to the difference between the means of the low-dose group and the placebo group (3.2 − 2.2 = 1). This analysis demonstrates how the regression model represents the three-group situation. We can see from the significance values of the *t*-tests that the difference between the high-dose group and the placebo group (b_2) is significant because $p < .05$. The difference between the low-dose and the placebo groups is not, however, significant ($p = .282$).

Coefficients[a]

Model		Unstandardized Coefficients		Standardized Coefficients	t	Sig.
		B	Std. Error	Beta		
1	(Constant)	2.200	.627		3.508	.004
	Dummy Variable 1	2.800	.887	.773	3.157	.008
	Dummy Variable 2	1.000	.887	.276	1.127	.282

a Dependent Variable: Libido

SPSS Output 8.1

Table 8.3 Dummy coding for the four-group experimental design

	Dummy Variable 1	Dummy Variable 2	Dummy Variable 3
Group 1	1	0	0
Group 2	0	1	0
Group 3	0	0	1
Group 4 (base)	0	0	0

A four-group experiment can be described by extending the three-group scenario. I mentioned earlier that you will always need one less dummy variable than the number of groups in the experiment: therefore, this model requires three dummy variables. As before, we need to specify one category that is a base category (a control group). This base category should have a code of 0 for all three dummy variables. The remaining three conditions will have a code of 1 for the dummy variable that described that condition and a code of 0 for the other two dummy variables. Table 8.3 illustrates how the coding scheme would work.

8.2.3. Logic of the *F*-ratio ②

In Chapter 5 we learnt a little about the *F*-ratio and its calculation. To recap, we learnt that the *F*-ratio is used to test the overall fit of a regression model to a set of observed data. I have just explained how ANOVA can be represented as a regression equation, and this should help you to understand what the *F*-ratio tells you about your data. Figure 8.1 shows the Viagra data in graphical form (including the group means, the overall mean and the difference between each case and the group mean). In this example, there were three groups; therefore, we want to test the hypothesis that the means of three groups are different (so, the null hypothesis is that the group means are the same). If the group means were all the same, then we would not expect the placebo group to differ from the low-dose group or the high-dose group, and we would not expect the low-dose group to differ from the

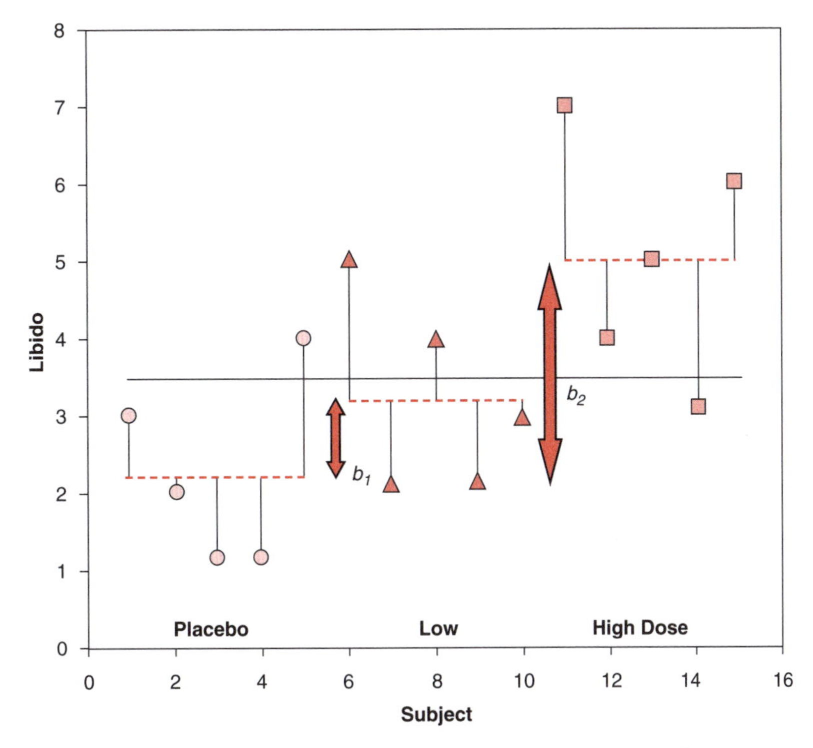

Figure 8.1 The Viagra data in graphical form. The dashed lines represent the mean libido of each group. The shapes represent the libido of individual participants (different shapes indicate different experimental groups). The full horizontal line is the average libido of all subjects

high-dose group. Therefore, on the diagram, the three dashed lines would be in the same vertical position (the exact position would be the grand mean). We can see from the diagram that the group means are actually different because the dashed lines (the group means) are in different vertical positions. We have just found out that in the regression model, b_2 represents the difference between the means of the placebo and the high-dose group, and b_1 represents the difference in means between the low-dose and placebo groups. These two distances are represented in Figure 8.1 by the vertical arrows. If the null hypothesis is true and all the groups have the same means, then these b-coefficients should be 0 (because if the group means are equal then the difference between them will be 0).

The logic of ANOVA follows from what we understand about regression:

- The simplest model we can fit to a set of data is the grand mean (the mean of the outcome variable). When this basic model is fitted there will be a large amount of error between the model and the observed data.
- One of many straight lines can be chosen to model the data collected. If this model fits the data well then it must be better than using the grand mean.

- The intercept and one or more regression coefficients can describe the chosen line.
- The regression coefficients determine the slope of the line; therefore, the bigger the coefficients, the greater the deviation between the line and the grand mean.
- In ANOVA, the regression coefficients are defined by differences between group means.
- The bigger the differences between group means, the greater the difference between the regression line and the grand mean.
- If the differences between group means are large enough, then the resulting line will be a better fit of the data than the grand mean.
- Therefore, if the group means are significantly different, we would expect the regression model to be a better fit of the data than the grand mean.

We can see whether this last statement is true by comparing the improvement in fit due to using the model (rather than the grand mean) with the error that still remains. Another way of saying this is that when the grand mean is used as a model, there will be a certain amount of variation between the data and the grand mean. When a model is fitted it will explain some of this variation but some will be left unexplained. The *F*-ratio is the ratio of the explained to the unexplained variation. Look back at section 5.2.3 to refresh your memory on these concepts before reading on. This may all sound quite complicated, but actually most of it boils down to variations on one simple equation (see Box 8.1).

Box 8.1

You might be surprised to know that ANOVA boils down to one equation (well, sort of) ②

At every stage of ANOVA we're assessing variation (or deviance) from a particular model (be that the most basic model, or the most sophisticated model). We saw back in section 1.4.1 that the extent to which a model deviates from the observed data can be expressed, in general, in the form of equation (8.3). So, in ANOVA, as in regression, we use equation (8.3) to calculate the fit of the most basic model, and then the fit of the best model (the line of best fit). If the best model is any good then it should fit the data significantly better than our basic model:

$$\text{deviation} = \sum(\text{observed} - \text{model})^2 \qquad (8.3)$$

The interesting point is that all of the sums of squares in ANOVA are variations on this one basic equation. All that changes is what we use as the model, and what the corresponding observed data are. Look through the various sections on the sums of squares and compare the resulting equations to equation (8.3); hopefully, you can see that they are all basically variations on this general form of the equation!

8.2.4. Total sum of squares (SS_T) ②

To find the total amount of variation within our data we calculate the difference between each observed data point and the grand mean. We then square these differences and add them together to give us the total sum of squares (SS_T):

$$SS_T = \sum (x_i - \bar{x}_{\text{grand}})^2 \tag{8.4}$$

We also saw in section 1.4.1 that the variance and the sums of squares are related such that variance, $s^2 = SS/(N - 1)$, where N is the number of observations. Therefore, we can calculate the total sums of squares from the variance of all observations (the grand variance) by rearranging the relationship ($SS = s^2(N - 1)$). The grand variance is the variation between all scores, regardless of the experimental condition from which the scores come. Therefore, in Figure 8.1 it would be the sum of the squared distances between each point and the full horizontal line. The grand variance for the Viagra data is given in Table 8.1, and if we count the number of observations we find that there were 15 in all. Therefore, SS_T is calculated as follows:

$$\begin{aligned} SS_T &= s^2_{\text{grand}}(n - 1) \\ &= 3.124(15 - 1) \\ &= 3.124 \times 14 \\ &= 43.74 \end{aligned}$$

Before we move on, it is important to understand degrees of freedom, so have a look at Box 8.2.

Box 8.2

Degrees of freedom ②

The concept of degrees of freedom (*df*) is very difficult to explain, and I have mentioned them several times throughout this book without defining the concept. Most textbooks are bad at explaining degrees of freedom (although Howell, 2002, p. 56 does a good job) and I'm confident that this book will be no exception to this rule! I'll begin with an analogy. Imagine you're manager of a rugby team and you have a team sheet with 15 empty slots relating to the positions on the playing field. There is a standard formation in rugby and so each team has 15 specific positions that must be held constant for the game to be played. When the first player arrives, you have the choice of 15 positions in which to place this player. You place his name in one of the slots and allocate him to a position (e.g. scrum-half) and, therefore, one position on the pitch is now occupied. When the next player arrives, you have the choice of 14 positions but you still have the freedom to choose which position

(Continued)

Box 8.2 (Continued)

this player is allocated. However, as more players arrive, you will reach the point at which 14 positions have been filled and the final player arrives. With this player you have no freedom to choose where they play—there is only one position left. Therefore there are 14 degrees of freedom; that is, for 14 players you have some degree of choice over where they play, but for 1 player you have no choice. The degrees of freedom are one less than the number of players.

In statistical terms the degrees of freedom relate to the number of observations that are free to vary. If we take a sample of four observations from a population, then these four scores are free to vary in any way (they can be any value). However, if we then use this sample of four observations to calculate the standard deviation of the population, we have to use the mean of the sample as an estimate of the population's mean. Thus, we hold one parameter constant. Say that the mean of the sample was 10; then we assume that the population mean is 10 also and we keep this value constant. With this parameter fixed, can all four scores from our sample vary? The answer is no because to keep the mean constant only three values are free to vary. For example, if the values in the sample were 8, 9, 11, 12 (mean = 10) and we changed three of these values to 7, 15 and 8, then the final value *must* be 10 to keep the mean constant. Therefore, if we hold one parameter constant then the degrees of freedom must be one less than the sample size. This fact explains why, when we use a sample to estimate the standard deviation of a population (as we did in section 1.4.1), we have to divide the sums of squares by $N-1$ rather than N alone. For SS_T, we used the entire sample to calculate the sums of squares and so the total degrees of freedom (df_T) are one less than the total sample size ($N-1$). For the Viagra data, this value is 14.

8.2.5. Model sum of squares (SS_M) ②

So far, we know that the total amount of variation within the data is 43.74 units. We now need to know how much of this variation the regression model can explain. In the ANOVA scenario, the model is based upon differences between group means and so the model sums of squares tell us how much of the total variation can be explained by the fact that different data points come from different groups. In short, the model represents the effect of the experimental manipulation.

In section 5.2.3 we saw that the model sum of squares is calculated by taking the difference between the values predicted by the model and the grand mean (see Figure 5.3). In ANOVA, the values predicted by the model are the group means (therefore, in Figure 8.1 the dashed lines represented the values of libido predicted by the model). For each participant the value predicted by the model is the mean for the group to which the participant belongs. In the Viagra example, the predicted value for the five participants in the placebo group will be 2.2, for the five participants in the low-dose condition it will be 3.2, and for the five participants in the high-dose condition it will be 5. The model sum of squares requires us to calculate the differences between each participant's predicted value and the grand mean. These differences are then squared and added together (for reasons that should be clear in your mind by now). We know that the predicted value for participants in a particular group is the mean of that group. Therefore, the easiest way to calculate SS_M is to:

1. Calculate the difference between the mean of each group and the grand mean.
2. Square each of these differences.
3. Multiply each result by the number of participants within that group (n_k).
4. Add the values for each group together.

The mathematical expression of this process is shown in equation (8.5):

$$SS_M = \sum n_k(\bar{x}_k - \bar{x}_{grand})^2 \quad (8.5)$$

Using the means from the Viagra data, we can calculate SS_M as follows:

$$\begin{aligned} SS_M &= 5(2.200 - 3.467)^2 + 5(3.200 - 3.467)^2 + 5(5.000 - 3.467)^2 \\ &= 5(-1.267)^2 + 5(-0.267)^2 + 5(1.533)^2 \\ &= 8.025 + 0.355 + 11.755 \\ &= 20.135 \end{aligned}$$

For SS_M, the degrees of freedom (df_M) will always be one less than the number of parameters estimated. In short, this value will be the number of groups minus 1 (which you'll see denoted as $k - 1$). So, in the three-group case the degrees of freedom will always be 2 (because the calculation of the sums of squares is based on the group means, two of which will be free to vary in the population if the third is held constant).

8.2.6. Residual sum of squares (SS_R) ②

We now know that there are 43.74 units of variation to be explained in our data, and that our model can explain 20.14 of these units (nearly half). The final sum of squares is the residual sum of squares (SS_R), which tells us how much of the variation cannot be explained by the model. This value is the amount of variation caused by extraneous factors such as individual differences in weight, testosterone or whatever. Knowing SS_T and SS_M already, the simplest way to calculate SS_R is to subtract SS_M from SS_T ($SS_R = SS_T - SS_M$); however, telling you to do this provides little insight into what is being calculated and, of course, if you've messed up the calculations of either SS_M or SS_T (or indeed both!) then SS_R will be incorrect also. We saw in section 5.2.3 that the residual sum of squares is the difference between what the model predicts and what was actually observed. We already know that for a given participant, the model predicts the mean of the group to which that person belongs. Therefore, SS_R is calculated by looking at the difference between the score obtained by a person and the mean of the group to which the person belongs. In graphical terms the vertical lines in Figure 8.1 represent this sum of squares. These distances between each data point and the group mean are squared and then added together to give the residual sum of squares, SS_R (see equation (8.6)):

$$SS_R = \sum (x_{ik} - \bar{x}_k)^2 \tag{8.6}$$

Now, the sum of squares for each group represents the sum of squared differences between each participant's score in that group and the group mean. Therefore, we can express SS_R as $SS_R = SS_{group1} + SS_{group2} + SS_{group3} \ldots$ and so on. Given that we know the relationship between the variance and the sums of squares, we can use the variances for each group of the Viagra data to create an equation like we did for the total sum of squares. As such, SS_R can be expressed as equation (8.7):

$$SS_R = \sum s_k^2 (n_k - 1) \tag{8.7}$$

This just means, take the variance from each group (s_k^2), and multiply it by one less than the number of people in that group ($n_k - 1$). When you've done this for each group, add them all up. For the Viagra data, this gives us:

$$\begin{aligned} SS_R &= s^2_{group1}(n_1 - 1) + s^2_{group2}(n_2 - 1) + s^2_{group3}(n_3 - 1) \\ &= (1.70)(5 - 1) + (1.70)(5 - 1) + (2.50)(5 - 1) \\ &= (1.70 \times 4) + (1.70 \times 4) + (2.50 \times 4) \\ &= 6.8 + 6.8 + 10 \\ &= 23.60 \end{aligned}$$

The degrees of freedom for SS_R (df_R) are the total degrees of freedom minus the degrees of freedom for the model ($df_R = df_T - df_M = 14 - 2 = 12$). Put another way, it's $N - k$: the total sample size, N, minus the number of groups, k.

8.2.7. Mean squares ②

SS_M tells us how much variation the regression model (e.g. the experimental manipulation) explains and SS_R tells us how much variation is due to extraneous factors. However, because both of these values are summed values they will be influenced by the number of scores that were summed (e.g. SS_M used the sum of only three different values (the group means) compared to SS_R and SS_T, which used the sum of 14 different values). To eliminate this bias we can calculate the average sum of squares (known as the *mean squares*, MS), which is simply the sum of squares divided by the degrees of freedom. The reason why we divide by the degrees of freedom rather than the number of parameters used to calculate the SS is because we are trying to extrapolate to a population and so some parameters within that population will be held constant. So, for the Viagra data we find the following mean squares:

$$\mathrm{MS_M} = \frac{\mathrm{SS_M}}{df_\mathrm{M}} = \frac{20.135}{2} = 10.067$$
$$\mathrm{MS_R} = \frac{\mathrm{SS_R}}{df_\mathrm{R}} = \frac{23.60}{12} = 1.967$$

$\mathrm{MS_M}$ represents the average amount of variation explained by the model (e.g. the systematic variation), whereas $\mathrm{MS_R}$ is a gauge of the average amount of variation explained by extraneous variables (the unsystematic variation).

8.2.8. The *F*-ratio ②

The *F*-ratio is a measure of the ratio of the variation explained by the model and the variation explained by unsystematic factors. It can be calculated by dividing the model mean squares by the residual mean squares:

$$F = \frac{\mathrm{MS_M}}{\mathrm{MS_R}} \tag{8.8}$$

As with the independent *t*-test, the *F*-ratio is, therefore, a measure of the ratio of systematic variation to unsystematic variation. As such, it is the ratio of the experimental effect to the individual differences in performance. An interesting point about the *F*-ratio is that because it is the ratio of systematic variance to unsystematic variance, if its value is less than 1 then it must, by definition, represent a non-significant effect. The reason why this statement is true is because if the *F*-ratio is less than 1 it means that $\mathrm{MS_R}$ is greater than $\mathrm{MS_M}$, which in real terms means that there is more unsystematic than systematic variance. You can think of this in terms of the effect of natural differences in ability being greater than differences bought about by the experiment. In this scenario, we can, therefore, be sure that our experimental manipulation has been unsuccessful (because it has brought about less change than if we left our participants alone!). For the Viagra data, the *F*-ratio is:

$$F = \frac{\mathrm{MS_M}}{\mathrm{MS_R}} = \frac{10.067}{1.967} = 5.12$$

This value is greater than 1, which indicates that the experimental manipulation had some effect above and beyond the effect of individual differences in performance. However, it doesn't yet tell us whether the *F*-ratio is large enough not to be a chance result. To discover this we can compare the obtained value of *F* against the maximum value we would expect to get by chance alone in an *F*-distribution with the same degrees of freedom (these values can be found in Appendix A.3); if the value we obtain exceeds this critical value we can be confident that this reflects an effect of our independent variable. In this case, with $2(df_m)$ and $12(df_R)$ degrees of freedom the critical values are 3.89 ($p = .05$) and 6.93 ($p = .01$) respectively. The observed value, 5.12, is, therefore, significant at a .05 level of significance but not significant at a .01 level. The exact significance produced by SPSS should, therefore, fall somewhere between .05 and .01 (which, incidentally, it does).

8.2.9. Assumptions of ANOVA ②

The assumptions under which ANOVA is reliable are the same as for all parametric tests based on the normal distribution (see section 3.2). That is, data should be from a normally distributed population, the variances in each experimental condition are fairly similar, observations should be independent and the dependent variable should be measured on at least an interval scale. Now, although I am always banging on about how important assumptions are, they are not completely inflexible. For example, Lunney (1970) investigated the use of ANOVA when the dependent variable was dichotomous (it could have values of only 0 or 1). The results showed that when the group sizes were equal ANOVA was accurate when there were at least 20 degrees of freedom and the smallest response category contained at least 20% of all responses. If the smaller response category contained less than 20% of all responses then ANOVA performed accurately only when there were 40 or more degrees of freedom. This study shows that ANOVA can be quite a robust procedure.

In terms of violations of the assumption of homogeneity of variance, ANOVA is fairly robust when sample sizes are equal. However, when sample sizes are unequal ANOVA is not robust to violations of homogeneity of variance (this is why earlier on I said it's worth trying to collect equal-sized samples of data across conditions!). When groups with larger sample sizes have larger variances than the groups with smaller sample sizes, the resulting *F*-ratio tends to be conservative. That is, it's more likely to produce a non-significant result when a genuine difference does exist in the population. Conversely, when the groups with larger sample sizes have smaller variances than the groups with smaller samples sizes, the resulting *F*-ratio tends to be liberal. That is, it is more likely to produce a significant result when there is no difference between groups in the population (put another way, the Type I error rate is not controlled)—see Glass, Peckham & Saunders (1972) for a review.

Problems resulting from violations of the homogeneity of variance assumption can be corrected (see Box 8.3) and so by far the most serious problem is violations of the assumption of independence. Scariano & Davenport (1987) showed that when this assumption is broken (i.e. observations across groups are correlated) the Type I error rate is substantially inflated. For example, using the conventional .05 Type I error rate when observations are independent, if these observations are made to correlate moderately (say, with a Pearson coefficient of .5) when comparing three groups with 10 observations per group, the actual Type I error rate is .74 (a substantial inflation!). Therefore, if observations are correlated you might think that you are working with the accepted .05 error rate (i.e. you'll incorrectly find a significant result only 5% of the time) when in fact your error rate is closer to .75 (i.e. you'll find a significant result on 75% of occasions when, in reality, there is no effect in the population!).

8.2.10. Planned contrasts ②

The F-ratio tells us only whether the model fitted to the data accounts for more variation than extraneous factors, but it doesn't tell us where the differences between groups lie. So, if the F-ratio is large enough to be statistically significant, then we know only that one or more of the differences between means is statistically significant (e.g. either b_2 or b_1 is statistically significant). It is, therefore, necessary after conducting an ANOVA to carry out further analysis to find out which groups differ. In multiple regression, each b-coefficient is tested individually using a t-test and we could do the same for ANOVA. However, we would need to carry out two t-tests, which would inflate the familywise error rate (see section 8.2.1). Therefore, we need a way to contrast the different groups without inflating the Type I error rate. There are two ways in which to achieve this goal. The first is to break down the variance accounted for by the model into component parts, the second is to compare every group (as if conducting several t-tests) but to use a stricter acceptance criterion such that the familywise error rate does not rise above .05. The first option can be done using planned comparisons (also known as planned contrasts)[3] whereas the latter option is done using *post hoc* comparisons (see next section). The difference between planned comparisons and *post hoc* tests can be likened to the difference between one- and two-tailed tests in that planned comparisons are done when you have specific hypotheses that you want to test, whereas *post hoc* tests are done when you have no specific hypotheses. Let's first look at planned contrasts.

8.2.10.1. Choosing which contrasts to do ②

In the Viagra example we could have had very specific hypotheses. For one thing, we would expect any dose of Viagra to change libido compared to the placebo group. As a second hypothesis we might believe that a high dose should increase libido more than a low dose. To do planned comparisons, these hypotheses must be derived *before* the data are collected. It is fairly standard in the social sciences to want to compare experimental conditions to the control conditions as the first contrast, and then to see where the differences lie between the experimental groups. ANOVA is based upon splitting the total variation into two component parts: the variation due to the experimental manipulation (SS_M) and the variation due to unsystematic factors (SS_R) (see Figure 8.2). Planned comparisons take this logic a step further by breaking down the variation due to the experiment into component parts (see Figure 8.3). The exact comparisons that are carried out depend upon the hypotheses you want to test. Figure 8.3 shows a situation in which the experimental variance is broken down to look at how much variation is created by the two drug conditions compared to the placebo condition (*contrast 1*). Then the variation explained by taking Viagra is broken down to see how much is explained by taking a high dose relative to a low dose (*contrast 2*).

---

3 The terms 'comparison' and 'contrast' are used interchangeably.

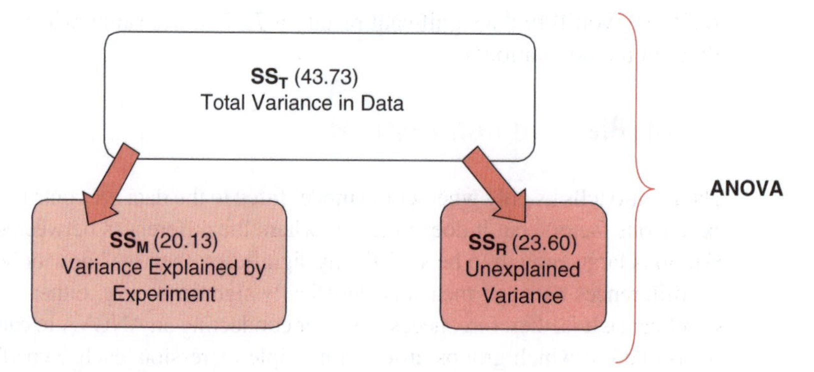

Figure 8.2 Partitioning variance for ANOVA

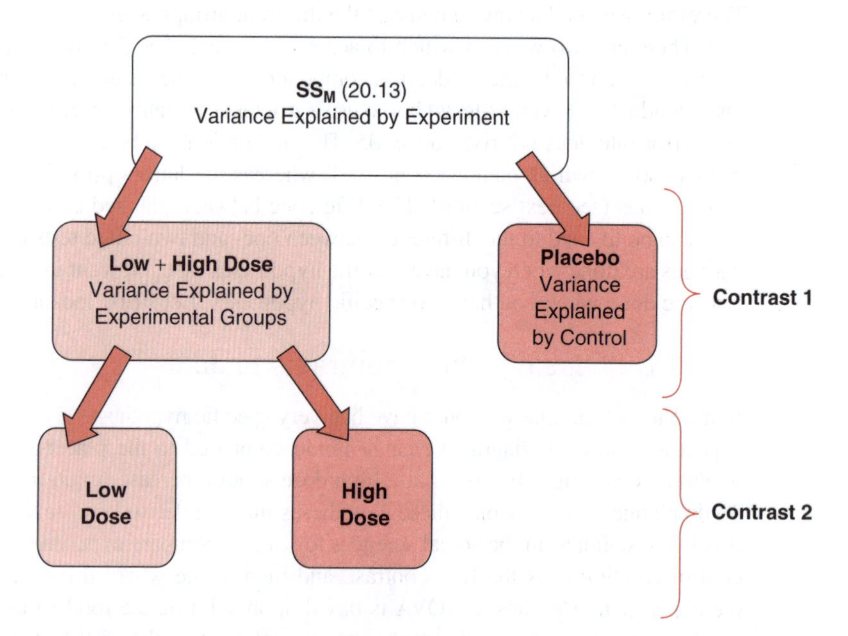

Figure 8.3 Partitioning of experimental variance into component comparisons

Typically, students struggle with the notion of planned comparisons, but there are several rules that can help you to work out what to do. The important thing to remember is that we are breaking down one chunk of variation into smaller independent chunks. This means several things. First, if a group is singled out in one comparison, then it should not reappear in another comparison. So, in Figure 8.3 contrast 1 involved comparing the placebo group to the experimental groups; because the placebo group is singled out, it should not be incorporated into any other contrasts. You can think of partitioning variance as being similar to slicing up a cake. You begin with a cake (the total sum of squares) and you then cut this cake into two pieces (SS_M and SS_R). You then take the piece of cake that represents SS_M and divide this up into smaller pieces. Once you have cut off a piece of cake you cannot stick that piece back on to the original slice, and you cannot stick it onto other pieces of cake, but you can divide it into smaller pieces of cake. Likewise, once a slice of variance has been split from a larger chunk, it cannot be attached to any other pieces of variance, it can only be subdivided into smaller chunks of variance. All of this talk of cake is making me hungry, but hopefully it illustrates a point.

Now, as a second hint to selecting contrasts, the independence of contrasts rule that I've just explained (the cake slicing!) means that you should always have one less contrast than the number of groups (so, there will be $k - 1$ contrasts, where k is the number of conditions you're comparing). Finally, each contrast must compare only two chunks of variance. This final rule is so that we can draw firm conclusions about what the contrast tells us. The F-ratio tells us that some of our means differ, but not which ones, and if we were to perform a contrast on more than two chunks of variance we would have the same problem. By comparing only two chunks of variance we can be sure that a significant result represents a difference between these two portions of experimental variation.

In most social science research we use at least one control condition, and in the vast majority of experimental designs we predict that the experimental conditions will differ from the control condition (or conditions). As such, **one big hint when planning comparisons is to compare all of the experimental groups with the control group or groups as your first comparison**. Once you have done this first comparison, any remaining comparisons will depend upon which of the experimental groups you predict will differ.

To illustrate these principles, Figure 8.4 and Figure 8.5 show the contrasts that might be done in a four-group experiment. The first thing to notice is that in both scenarios there are three possible comparisons (one less than the number of groups). Also, every contrast compares only two chunks of variance. What's more, in both scenarios the first contrast is the same: the experimental groups are compared against the control group or groups. In Figure 8.4 there was only one control condition and so this portion of variance is used only in the first contrast (because it cannot be broken down any further). In Figure 8.5 there were two control groups, and so the portion of variance due to the control conditions (contrast 1) can be broken down again so as to see whether or not the scores in the control groups differ from each other (contrast 3).

In Figure 8.4, the first contrast contains a chunk of variance that is due to the three experimental groups and this chunk of variance is broken down by first looking at whether groups E1 and E2 differ from E3 (contrast 2). It is equally valid to use contrast 2 to compare groups E1 and E3 with E2, or to compare groups E2 and E3 with E1. The exact comparison that you choose depends upon your hypotheses. For contrast 2 in Figure 8.4 to be valid we need to have a good reason to expect group E3 to be different from the other two groups. The third comparison in Figure 8.4 depends on the comparison chosen for contrast 2. Contrast 2 necessarily had to involve comparing two experimental groups against a third, and the experimental groups chosen to be combined must be separated in the final comparison. As a final point, you'll notice that in Figure 8.4 and Figure 8.5, once a group has been singled out in a comparison, it is never used in any subsequent contrasts.

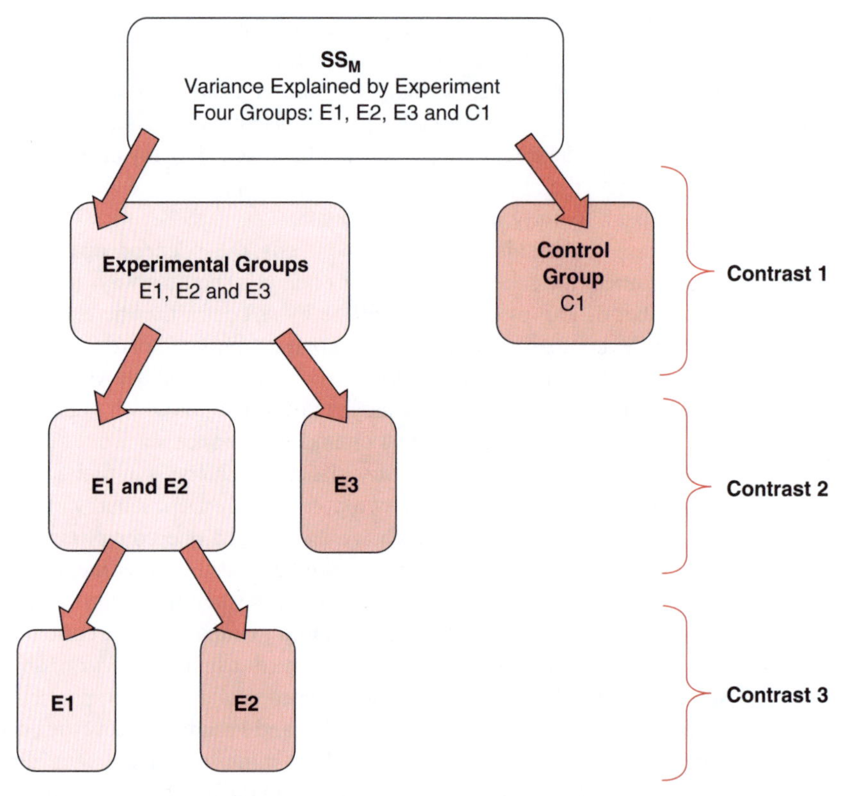

Figure 8.4 Partitioning variance for planned comparisons in a four-group experiment using one control group

When we carry out a planned contrast, we compare 'chunks' of variance and these chunks often consist of several groups. It is perhaps confusing to understand exactly what these contrasts tell us. Well, when you design a contrast that compares several groups to one other group, you are comparing the means of the groups in one chunk with the mean of the group in the other chunk. As an example, for the Viagra data I suggested that an appropriate first contrast would be to compare the two dose groups with the placebo group. The means of the groups are 2.20 (placebo), 3.20 (low dose) and 5.00 (high dose) and so the first comparison, which compared the two experimental groups to the placebo, is comparing 2.20 (the mean of the placebo group) to the average of the other two groups ($(3.2 + 5.0)/2 = 4.10$). If this first contrast turns out to be significant, then we can conclude that 4.10 is significantly greater than 2.20, which in terms of the experiment tells us that the average of the experimental groups is significantly different to the average of the controls. You can probably see that logically this means that, if the standard errors are the same, the experimental group with the highest mean (the high-dose group) will be significantly different from the mean of the placebo group. However, the experimental group with the lower mean

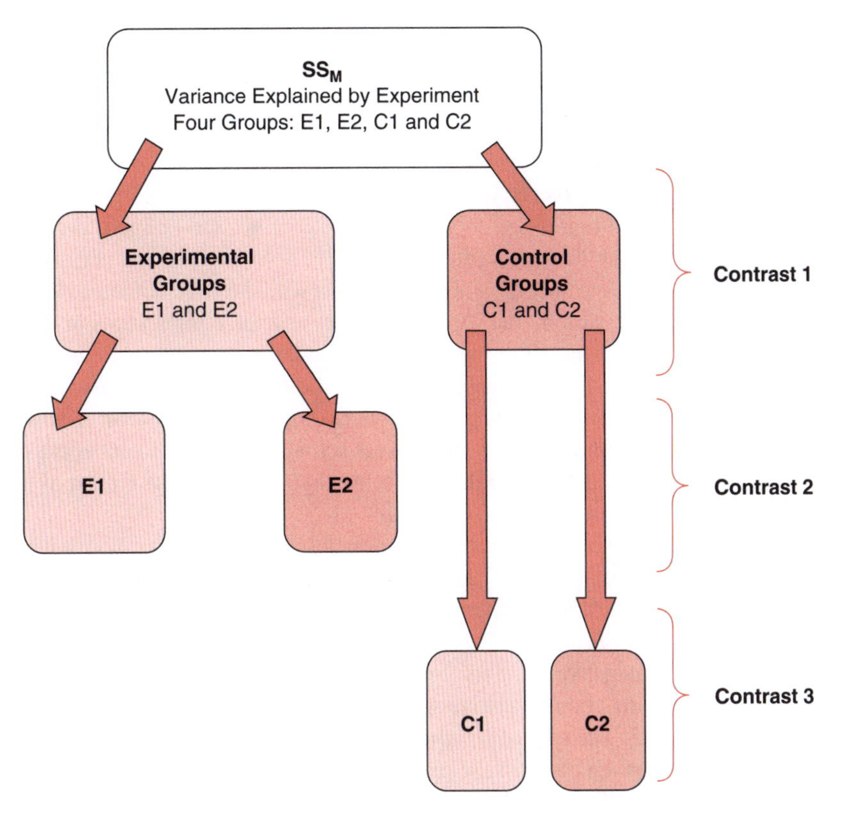

Figure 8.5 Partitioning variance for planned comparisons in a four-group experiment using two control groups

(the low-dose group) might not necessarily differ from the placebo group; we have to use the final comparison to make sense of the experimental conditions. For the Viagra data the final comparison looked at whether the two experimental groups differ (i.e. is the mean of the high-dose group significantly different from the mean of the low-dose group?). If this comparison turns out to be significant then we can conclude that having a high dose of Viagra significantly affected libido compared to having a low dose. If the comparison is non-significant then we have to conclude that the dosage of Viagra made no difference to libido. In this latter scenario it is likely that both doses affect libido more than placebo, whereas the former case implies that having a low dose may be no different to having a placebo. However, the word *implies* is important here: it is possible that the low-dose group might not differ from the placebo. To be completely sure we must carry out *post hoc* tests.

8.2.10.2. Defining contrasts using weights ③

Hopefully by now you have got some idea of how to plan which comparisons to do (if your brain hasn't exploded!). Much as I'd love to tell you that all of the hard work is now over and SPSS will magically carry out the comparisons that you've selected, it won't. To get SPSS to carry out planned comparisons we need to tell it which groups we would like to compare and doing this can be quite complex. In fact, when we carry out contrasts we assign values to certain variables in the regression model (sorry, I'm afraid that I have to start talking about regression again)—just as we did when we used dummy coding for the main ANOVA. To carry out contrasts we simply assign certain values to the dummy variables in the regression model. Whereas before, we defined the experimental groups by assigning the dummy variables values of 1 or 0, when we perform contrasts we use different values to specify which groups we would like to compare. The resulting coefficients in the regression model (b_2 and b_1) represent the comparisons in which we are interested. The values assigned to the dummy variables are known as weights.

This procedure is horribly confusing, but there are a few basic rules for assigning values to the dummy variables to obtain the comparisons you want. I will explain these simple rules before showing how the process actually works. Remember the previous section when you read through these rules, and remind yourself of what I mean by a 'chunk' of variation!

- **Rule 1**: Choose sensible comparisons. Remember that you want to compare only two chunks of variation and that if a group is singled out in one comparison, that group should be excluded from any subsequent contrasts.
- **Rule 2**: Groups coded with positive weights will be compared against groups coded with negative weights. So, assign one chunk of variation positive weights and the opposite chunk negative weights.
- **Rule 3**: The sum of weights for a comparison should be zero: if you add up the weights for a given contrast the result should be zero.
- **Rule 4**: If a group is not involved in a comparison, automatically assign it a weight of 0. If we give a group a weight of 0 then this eliminates that group from all calculations.
- **Rule 5**: For a given contrast, the weights assigned to the group(s) in one chunk of variation should be equal to the number of groups in the opposite chunk of variation.

OK, let's follow some of these rules to derive the weights for the Viagra data. The first comparison we chose was to compare the two experimental groups against the control:

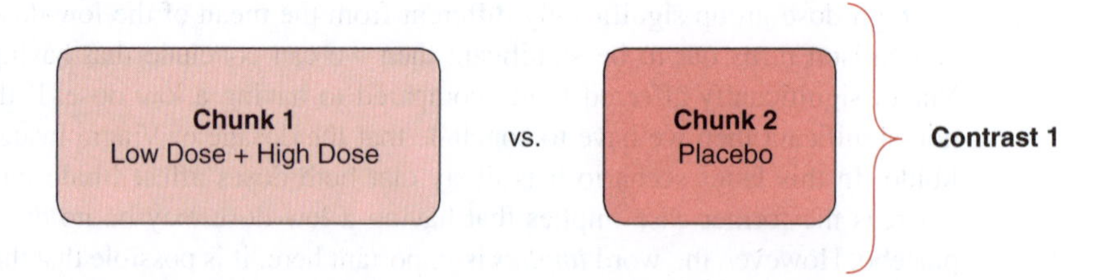

Therefore, the first chunk of variation contains the two experimental groups, and the second chunk contains only the placebo group. Rule 2 states that we should assign one chunk positive weights, and the other negative. It doesn't matter which way round we do this, but for convenience let's assign chunk 1 positive weights, and chunk 2 negative weights:

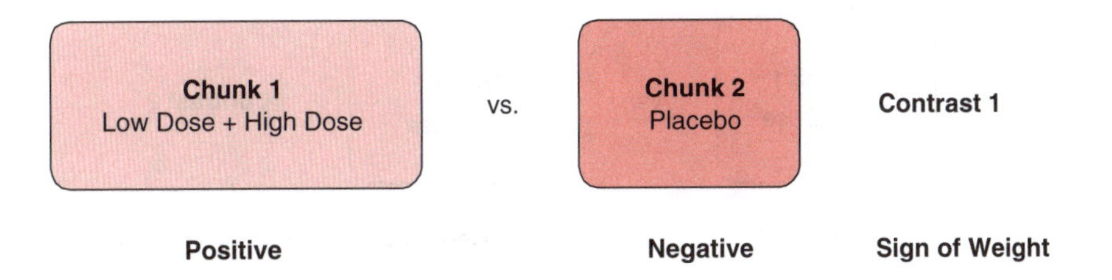

Using rule 5, the weight we assign to the groups in chunk 1 should be equivalent to the number of groups in chunk 2. There is only one group in chunk 2 and so we assign each group in chunk 1 a weight of 1. Likewise, we assign a weight to the group in chunk 2 that is equal to the number of groups in chunk 1. There are two groups in chunk 1 so we give the placebo group a weight of 2. Then we combine the sign of the weights with the magnitude to give us weights of –2 (placebo), 1 (low dose) and 1 (high dose):

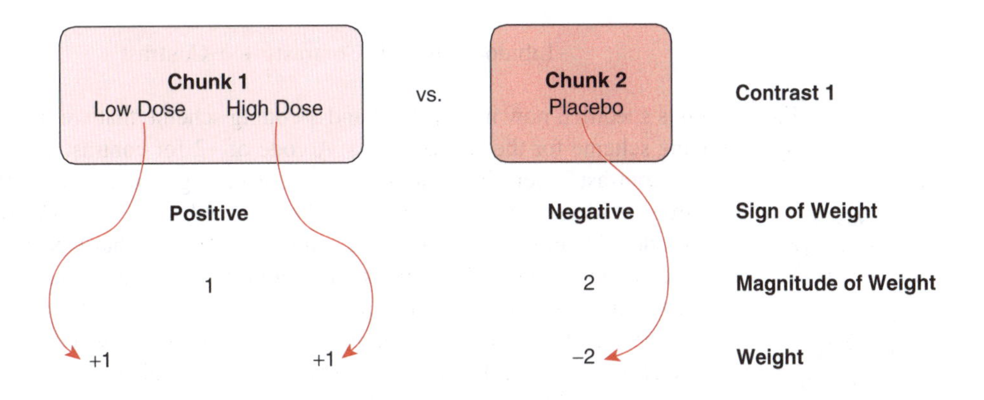

Rule 3 states that for a given contrast, the weights should add up to zero, and by following rules 2 and 5 this rule will always be followed (if you haven't followed these rules properly then this will become clear when you add the weights). So, let's check by adding the weights: sum of weights = 1 + 1 – 2 = 0:

The second contrast was to compare the two experimental groups and so we want to ignore the placebo group. Rule 4 tells us that we should automatically assign this group a weight of 0 (because this will eliminate this group from any calculations). We are left with two chunks of variation: chunk 1 contains the low-dose group and chunk 2 contains the high-dose group. By following rules 2 and 5 it should be obvious that one group is assigned a weight of +1 while the other is assigned a weight of –1. The control group is ignored (and so given a weight of 0). If we add the weights for contrast 2 we should find that they again add up to zero: sum of weights = 1 – 1 + 0 = 0.

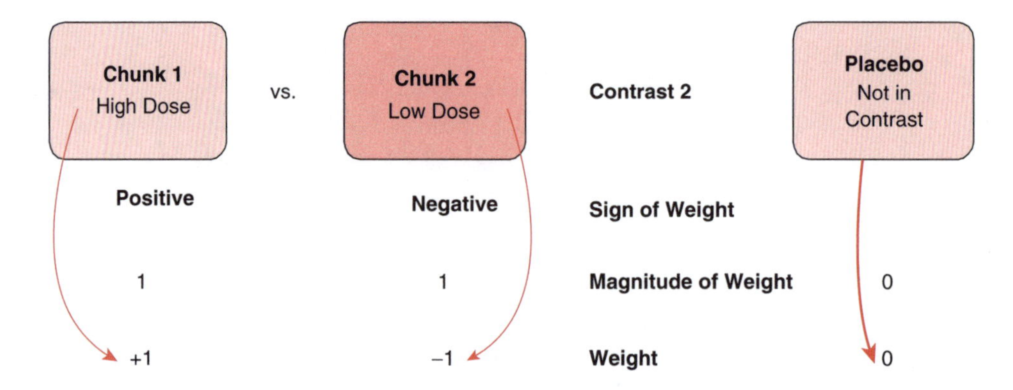

The weights for each contrast are codings for the two dummy variables in equation (8.2). Hence, these codings can be used in a multiple regression model in which b_2 represents contrast 1 (comparing the experimental groups to the control), b_1 represents contrast 2 (comparing the high-dose group with the low-dose group), and b_0 is the grand mean (equation (8.9)):

$$\text{Libido}_i = b_0 + b_1\text{Contrast}_1 + b_2\text{Contrast}_2 \tag{8.9}$$

Each group is specified now not by the 0 and 1 coding scheme that we initially used, but by the coding scheme for the two contrasts. A code of −2 for contrast 1 and a code of 0 for contrast 2 identify participants in the placebo group. Likewise, the high-dose group is identified by a code of 1 for both variables, and the low-dose group has a code of 1 for one contrast and a code of −1 for the other (see Table 8.4).

What are orthogonal contrasts?

It is important that the weights for a comparison sum to zero because it ensures that you are comparing two unique chunks of variation. Therefore, SPSS can perform a *t*-test. A more important consideration is that when you multiply the weights for a particular group, these products should also add up to zero (see final column of Table 8.4). If the products add to zero then we can be sure that the contrasts are *independent* or orthogonal. It is important for interpretation that contrasts are orthogonal. When we used dummy variable coding and ran a regression on the Viagra data, I commented that we couldn't look at the individual *t*-tests done on the regression coefficients because the familywise error rate is inflated (see section 8.2.10 and

Table 8.4 Orthogonal contrasts for the Viagra data

Group	Dummy Variable 1 (Contrast_1)	Dummy Variable 2 (Contrast_2)	Product $\text{Contrast}_1 \times \text{Contrast}_2$
Placebo	−2	0	0
Low Dose	1	−1	−1
High Dose	1	1	1
Total	0	0	0

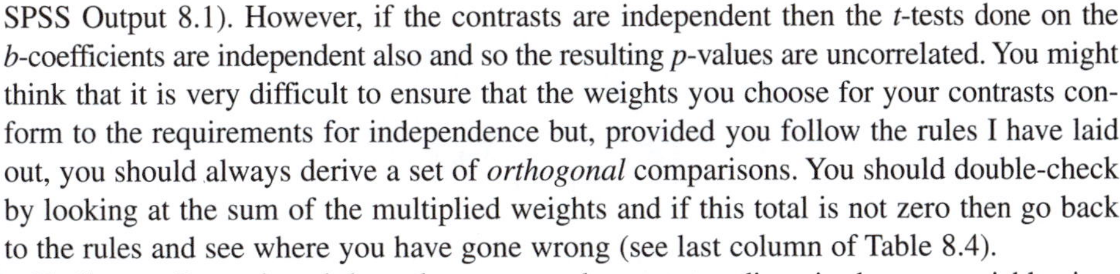

SPSS Output 8.1). However, if the contrasts are independent then the *t*-tests done on the *b*-coefficients are independent also and so the resulting *p*-values are uncorrelated. You might think that it is very difficult to ensure that the weights you choose for your contrasts conform to the requirements for independence but, provided you follow the rules I have laid out, you should always derive a set of *orthogonal* comparisons. You should double-check by looking at the sum of the multiplied weights and if this total is not zero then go back to the rules and see where you have gone wrong (see last column of Table 8.4).

Earlier on, I mentioned that when you used contrast codings in dummy variables in a regression model the *b*-values represented the differences between the means that the contrasts were designed to test. Although it is reasonable for you to trust me on this issue, for the more advanced students I'd like to take the trouble to show you how the regression model works—this next part is not for the faint-hearted and so equation-phobes should move on to the next section! When we do planned contrasts, the intercept b_0 is equal to the grand mean (i.e. the value predicted by the model when group membership is not known):

$$b_0 = \text{grand mean} = \frac{\overline{X}_{\text{High}} + \overline{X}_{\text{Low}} + \overline{X}_{\text{Placebo}}}{3}$$

Placebo group: If we use the contrast codings for the placebo group (see Table 8.4), the predicted value of libido equals the mean of the placebo group. The regression equation can, therefore, be expressed as:

$$\text{Libido}_i = b_0 + b_1\text{Contrast}_1 + b_2\text{Contrast}_2$$

$$\overline{X}_{\text{Placebo}} = \left(\frac{\overline{X}_{\text{High}} + \overline{X}_{\text{Low}} + \overline{X}_{\text{Placebo}}}{3}\right) + (-2b_1) + (b_2 \times 0)$$

Now, if we rearrange this equation and then multiply everything by 3 (to get rid of the fraction) we get:

$$2b_1 = \left(\frac{\overline{X}_{\text{High}} + \overline{X}_{\text{Low}} + \overline{X}_{\text{Placebo}}}{3}\right) - \overline{X}_{\text{Placebo}}$$

$$6b_1 = \overline{X}_{\text{High}} + \overline{X}_{\text{Low}} + \overline{X}_{\text{Placebo}} - 3\overline{X}_{\text{Placebo}}$$

$$6b_1 = \overline{X}_{\text{High}} + \overline{X}_{\text{Low}} - 2\overline{X}_{\text{Placebo}}$$

We can then divide everything by 2 to reduce the equation to its simplest form:

$$3b_1 = \left(\frac{\overline{X}_{\text{High}} + \overline{X}_{\text{Low}}}{2}\right) - \overline{X}_{\text{Placebo}}$$

$$b_1 = \frac{1}{3}\left[\left(\frac{\overline{X}_{\text{High}} + \overline{X}_{\text{Low}}}{2}\right) - \overline{X}_{\text{Placebo}}\right]$$

This equation shows that b_1 represents the difference between the average of the two experimental groups and the control group:

$$3b_1 = \left(\frac{\overline{X}_{\text{High}} + \overline{X}_{\text{Low}}}{2}\right) - \overline{X}_{\text{Placebo}}$$
$$= \frac{5 + 3.2}{2} - 2.2$$
$$= 1.9$$

We planned contrast 1 to look at the difference between the average of the experimental groups and the control and so it should now be clear how b_1 represents this difference. The observant among you will notice that rather than being the true value of the difference between experimental and control groups, b_1 is actually a third of this difference (b_1 = 1.9/3 = .633). The reason for this division is that the familywise error is controlled by making the regression coefficient equal to the actual difference divided by the number of groups in the contrast (in this case 3).

High-dose group: For the situation in which the codings for the high-dose group (see Table 8.4) are used the predicted value of libido is the mean for the high-dose group, and so the regression equation becomes:

$$\text{Libido}_i = b_0 + b_1\text{Contrast}_1 + b_2\text{Contrast}_2$$
$$\overline{X}_{\text{High}} = b_0 + (b_1 \times 1) + (b_2 \times 1)$$
$$b_2 = \overline{X}_{\text{High}} - b_1 - b_0$$

We know already what b_1 and b_0 represent so we place these values into the equation and then multiply by 3 to get rid of some of the fractions:

$$b_2 = \overline{X}_{\text{High}} - b_1 - b_0$$
$$b_2 = \overline{X}_{\text{High}} - \left\{\frac{1}{3}\left[\left(\frac{\overline{X}_{\text{High}} + \overline{X}_{\text{Low}}}{2}\right) - \overline{X}_{\text{Placebo}}\right]\right\} - \left(\frac{\overline{X}_{\text{High}} + \overline{X}_{\text{Low}} + \overline{X}_{\text{Placebo}}}{3}\right)$$
$$3b_2 = 3\overline{X}_{\text{High}} - \left[\left(\frac{\overline{X}_{\text{High}} + \overline{X}_{\text{Low}}}{2}\right) - \overline{X}_{\text{Placebo}}\right] - (\overline{X}_{\text{High}} + \overline{X}_{\text{Low}} + \overline{X}_{\text{Placebo}})$$

If we multiply everything by 2 to get rid of the other fraction, expand all of the brackets and then simplify the equation we get:

$$6b_2 = 6\overline{X}_{\text{High}} - (\overline{X}_{\text{High}} + \overline{X}_{\text{Low}} - 2\overline{X}_{\text{Placebo}}) - 2(\overline{X}_{\text{High}} + \overline{X}_{\text{Low}} + \overline{X}_{\text{Placebo}})$$
$$6b_2 = 6\overline{X}_{\text{High}} - \overline{X}_{\text{High}} - \overline{X}_{\text{Low}} + 2\overline{X}_{\text{Placebo}} - 2\overline{X}_{\text{High}} - 2\overline{X}_{\text{Low}} - 2\overline{X}_{\text{Placebo}}$$
$$6b_2 = 3\overline{X}_{\text{High}} - 3\overline{X}_{\text{Low}}$$

Finally, we can divide the equation by 6 to find out what b_2 represents (remember that 3/6 = 1/2):

$$b_2 = \frac{1}{2}(\overline{X}_{\text{High}} - \overline{X}_{\text{Low}})$$

We planned contrast 2 to look at the difference between the experimental groups:

$$\overline{X}_{\text{High}} - \overline{X}_{\text{Low}} = 5 - 3.2 = 1.8$$

It should now be clear how b_2 represents this difference. Again, rather than being the absolute value of the difference between the experimental groups, b_2 is actually half of this difference (1.8/2 = 0.9). The familywise error is again controlled, by making the regression coefficient equal to the actual difference divided by the number of groups in the contrast (in this case 2).

To illustrate these principles, I have created a file called **Contrast.sav** in which the Viagra data are coded using the contrast coding scheme used in this section. Run multiple regression analyses on these data using **libido** as the outcome and using **dummy1** and **dummy2** as the predictor variables (leave all default options). The main ANOVA for the model is the same as when dummy coding was used; however, the regression coefficients have now changed.

SPSS Output 8.2 shows the result of this regression. The first thing to notice is that the intercept is the grand mean, 3.467 (see, I wasn't telling lies). Second, the regression coefficient for contrast 1 is one-third of the difference between the average of the experimental conditions and the control condition (see above). Finally, the regression coefficient for contrast 2 is half of the difference between the experimental groups (see above). So, when a planned comparison is done in ANOVA a *t*-test is conducted comparing the mean of one chunk of variation with the mean of a different chunk. From the significance values of the *t*-tests we can see that our experimental groups were significantly different from the control ($p < .05$) but that the experimental groups were not significantly different ($p > .05$).

Coefficients[a]

Model		Unstandardized Coefficients		Standardized Coefficients		
		B	Std. Error	Beta	t	Sig.
1	(Constant)	3.467	.362		9.574	.000
	Dummy Variable 1	.633	.256	.525	2.474	.029
	Dummy Variable 2	.900	.443	.430	2.029	.065

a Dependent Variable: Libido

SPSS Output 8.2

8.2.10.3. Non-orthogonal comparisons ②

I have spent a lot of time labouring over how to design appropriate orthogonal comparisons without mentioning the possibilities that non-orthogonal contrasts provide. Non-orthogonal contrasts are comparisons that are in some way related and the best way to get them is to disobey rule 1 in the previous section. Using my cake analogy again, non-orthogonal comparisons are where you slice up your cake and then try to stick slices of cake together again! So, for the Viagra data a set of non-orthogonal contrasts might be to have the same initial contrast (comparing experimental groups against the placebo), but then to compare the low-dose group with the placebo. This disobeys rule 1 because the placebo group is singled out in the first contrast but used again in the second contrast. The coding for this set of contrasts is shown in Table 8.5 and by looking at the last column it is clear that when you multiply and add the codings from the two contrasts the sum is not zero. This tells us that the contrasts are not orthogonal.

Table 8.5 Non-orthogonal contrasts for the Viagra data

Group	Dummy Variable 1 ($Contrast_1$)	Dummy Variable 2 ($Contrast_2$)	Product $Contrast_1 \times Contrast_2$
Placebo	−2	−1	2
Low Dose	1	0	0
High Dose	1	1	1
Total	0	0	3

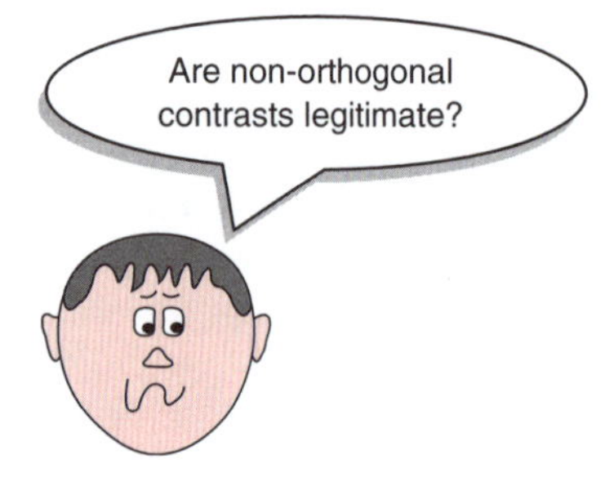

There is nothing intrinsically wrong with performing non-orthogonal contrasts. However, if you choose to perform this type of contrast you must be very careful in how you interpret the results. With non-orthogonal contrasts, the comparisons you do are related and so the resulting test statistics and *p*-values will be correlated to some extent. For this reason you should use a more conservative probability level to accept that a given contrast is statistically meaningful (see section 8.2.11). For this reason I would advise conducting orthogonal planned comparisons whenever possible.

8.2.10.4. Standard contrasts ②

Although under most circumstances you will design your own contrasts, there are special contrasts that have been designed to compare certain situations. Some of these contrasts are orthogonal whereas others are non-orthogonal. Many procedures in SPSS allow you to choose to carry out the contrasts mentioned in this section.

Table 8.6 shows the contrasts that are available in SPSS for procedures such as logistic regression (see section 6.4.3), factorial ANOVA and repeated-measures ANOVA (see Chapters 9 and 10). Although the exact codings are not provided in Table 8.6, examples of the comparisons done in a three- and four-group situation are given (where the groups are

Table 8.6 Standard contrasts available in SPSS

Name	Definition	Contrast	Three Groups	Four Groups
Deviation (first)	Compares the effect of each category (except the first) to the overall experimental effect	1 2 3	2 vs.(1,2,3) 3 vs. (1,2,3)	2 vs. (1,2,3,4) 3 vs. (1,2,3,4) 4 vs. (1,2,3,4)
Deviation (last)	Compares the effect of each category (except the last) to the overall experimental effect	1 2 3	1 vs. (1,2,3) 2 vs. (1,2,3)	1 vs. (1,2,3,4) 2 vs. (1,2,3,4) 3 vs. (1,2,3,4)
Simple (first)	Each category is compared to the first category	1 2 3	1 vs. 2 1 vs. 3	1 vs. 2 1 vs. 3 1 vs. 4
Simple (last)	Each category is compared to the last category	1 2 3	1 vs. 3 2 vs. 3	1 vs. 4 2 vs. 4 3 vs. 4
Repeated	Each category (except the first) is compared to the previous category	1 2 3	1 vs. 2 2 vs. 3	1 vs. 2 2 vs. 3 3 vs. 4
Helmert	Each category (except the last) is compared to the mean effect of all subsequent categories	1 2 3	1 vs. (2,3) 2 vs. 3	1 vs. (2,3,4) 2 vs. (3,4) 3 vs. 4
Difference (reverse Helmert)	Each category (except the first) is compared to the mean effect of all previous categories	1 2 3	3 vs. (2,1) 2 vs. 1	4 vs. (3,2,1) 3 vs. (2,1) 2 vs. 1

labelled 1, 2, 3 and 1, 2, 3, 4 respectively). When you code variables in the data editor, SPSS will treat the lowest-value code as group 1, the next highest code as group 2 and so on. Therefore, depending on which comparisons you want to make you should code your grouping variable appropriately (and then use Table 8.6 as a guide to which comparisons SPSS will carry out). One thing that clever readers might notice about the contrasts in Table 8.6 is that some are orthogonal (i.e. Helmert and difference contrasts) whilst the others are non-orthogonal (deviation, simple and repeated). You might also notice that the comparisons calculated using simple contrasts are the same as those given by using the dummy variable coding described in Table 8.2.

8.2.10.5. Polynomial contrasts: trend analysis ②

One type of contrast deliberately omitted from Table 8.6 is the polynomial contrast. This contrast tests for trends in the data and in its most basic form it looks for a linear trend (i.e. that the group means increase proportionately). However, there are more complex trends such as quadratic, cubic and quartic trends that can be examined. Figure 8.6 shows diagrams of the types of trend that can exist in data sets. The *linear* trend should be familiar to you all by now and represents a simply proportionate change in the value of the dependent variable

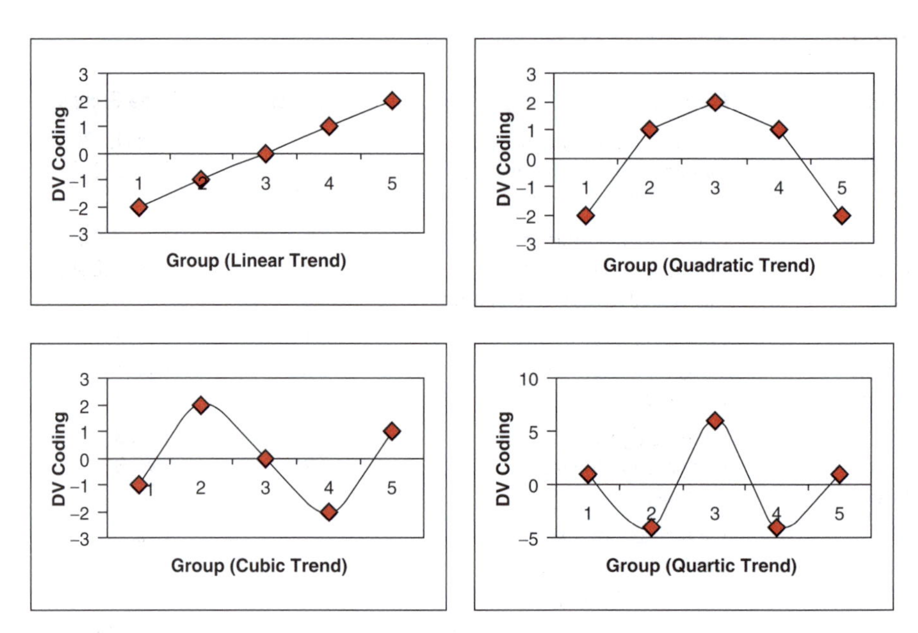

Figure 8.6 Linear, quadratic, cubic and quartic trends across five groups

across ordered categories. A quadratic trend is where there is one change in the direction of the line (e.g. the line is curved in one place). An example of this might be a situation in which a drug enhances performance on a task at first but then as the dose increases the performance drops again. It should be fairly obvious that to find a quadratic trend you need at least three groups (because in the two-group situation there are not enough categories of the independent variable for the means of the dependent variable to change one way and then another). A cubic trend is where there are two changes in the direction of the trend. So, for example, the mean of the dependent variable at first goes up across the first couple of categories of the independent variable, then across the succeeding categories the means go down, but then across the last few categories the means rise again. To have two changes in the direction of the mean you must have at least four categories of the independent variable. The final trend that you are likely to come across is the quartic trend, and this trend has three changes of direction (and so you need at least five categories of the independent variable).

Polynomial trends should be examined in data sets in which it makes sense to order the categories of the independent variable (so, for example, if you have administered five doses of a drug it makes sense to examine the five doses in order of magnitude). For the Viagra data there are only three groups and so we can expect to find only a linear or quadratic trend (and it would be pointless to test for any higher-order trends).

Each of these trends has a set of codes for the dummy variables in the regression model, so we are doing the same thing that we did for planned contrasts except that the codings have already been devised to represent the type of trend of interest. In fact, the graphs in Figure 8.6 have been constructed by plotting the coding values for the five groups. In fact, if you add the codes for a given trend the sum will equal zero and if you multiply the codes you will find that the sum of the products also equals zero. Hence, these contrasts are orthogonal. The great thing about these contrasts is that you don't need to construct your own coding values to do them, because the codings already exist.

8.2.11. *Post hoc* procedures ②

Often it is the case that you have no specific *a priori* predictions about the data you have collected and instead you are interested in exploring the data for any differences between means that exist. This procedure is sometimes called *data mining* or *exploratory data analysis*. Now, personally I have always thought that these two terms have certain 'rigging the data' connotations and so I prefer to think of these procedures as 'finding the differences that I should have predicted if only I'd been clever enough'.

Post hoc tests consist of pairwise comparisons that are designed to compare all different combinations of the treatment groups. So, it is rather like taking every pair of groups and then performing a *t*-test on each pair of groups. This might seem like a particularly stupid thing to say (but then again, I am particularly stupid) in the light of what I have already told you about the problems of inflated familywise error rates. However, pairwise comparisons control the familywise error by correcting the level of significance for each test such that the overall Type I error rate (α) across all comparisons remains at .05. There are several ways in which the familywise error rate can be controlled. The most popular (and easiest) way is to divide α by the number of comparisons, thus ensuring that the cumulative Type I error is below .05. Therefore, if we conduct 10 tests, we use .005 as our criterion for significance. This method is known as the Bonferroni correction, after Carlo Bonferroni (Figure 8.7). There is a trade-off for controlling the familywise error rate and that is a loss of statistical power. This means that the probability of rejecting an effect that

Figure 8.7 Carlo Bonferroni before the celebrity of his correction led to drink, drugs and statistics groupies

does actually exist is increased (a Type II error). By being more conservative in the Type I error rate for each comparison, we increase the chance that we will miss a genuine difference in the data.

Therefore, when considering which *post hoc* procedure to use we need to consider three things: (1) does the test control the Type I error rate; (2) does the test control the Type II error rate (i.e. does the test have good statistical power); and (3) is the test reliable when the test assumptions of ANOVA have been violated?

Although I would love to go into tedious details about how all of the various *post hoc* tests work, there is really very little point. For one thing there are some excellent texts already available for those who wish to know (e.g. Toothaker, 1993; Klockars & Sax, 1986), and for another SPSS provides no less than 18 *post hoc* procedures, so it would use up several square miles of rainforest to explain them. What is important is that you know which *post hoc* tests perform the best according to the three criteria mentioned above.

8.2.11.1. *Post hoc* procedures and Type I (α) and Type II error rates ②

The Type I error rate and the statistical power of a test are linked. Therefore, there is always a trade-off: if a test is conservative (the probability of a Type I error is small) then it is likely to lack statistical power (the probability of a Type II error will be high). Thus, it is important that multiple comparison procedures control the Type I error rate but without a substantial loss in power. If a test is too conservative then we are likely to reject differences between means that are, in reality, meaningful.

The least-significant difference (*LSD*) pairwise comparison makes no attempt to control the Type I error and is equivalent to performing multiple *t*-tests on the data. The only difference is that the LSD requires the overall ANOVA to be significant. The Studentized Newman–Keuls (*SNK*) procedure is also a very liberal test and lacks control over the familywise error rate. *Bonferroni's* and *Tukey's* tests both control the Type I error rate very well but are conservative tests (they lack statistical power). Of the two, Bonferroni has more power when the number of comparisons is small, whereas Tukey is more powerful when testing large numbers of means. Tukey generally has greater power than *Dunn* and *Scheffé*. The Ryan, Einot, Gabriel and Welsch *Q* procedure (*REGWQ*) has good power and tight control of the Type I error rate. In fact, when you want to test all pairs of means this procedure is probably the best. However, when group sizes are different this procedure should not be used.

8.2.11.2. *Post hoc* procedures and violations of test assumptions ②

Most research on *post hoc* tests has been looked at whether the test performs well when the group sizes are different (an unbalanced design), when the population variances are very different, and when data are not normally distributed. The good news is that most multiple comparison procedures perform relatively well under small deviations from normality. The bad news is that they perform badly when group sizes are unequal and when population variances are different.

Hochberg's GT2 and *Gabriel's* pairwise test procedure were designed to cope with situations in which sample sizes are different. Gabriel's procedure is generally more powerful but can become too liberal when the sample sizes are very different. Also, Hochberg's GT2 is very unreliable when the population variances are different and so should be used only when you are sure that this is not the case. There are several multiple comparison procedures that have been specially designed for situations in which population variances differ. SPSS provides four options for this situation: *Tamhane's T2, Dunnett's T3, Games–Howell* and *Dunnett's C*. Tamhane's T2 is conservative and Dunnett's T3 and C keep very tight Type I error control. The Games–Howell procedure is the most powerful but can be liberal when sample sizes are small. However, Games–Howell is also accurate when sample sizes are unequal.

8.2.11.3. Summary of *post hoc* procedures ②

The choice of comparison procedure will depend on the exact situation you have and whether it is more important for you to keep strict control over the familywise error rate or to have greater statistical power. However, some general guidelines can be drawn (see Toothaker, 1993). When you have equal sample sizes and you are confident that your population variances are similar then use REGWQ or Tukey as both have good power and tight control over the Type I error rate. Bonferroni is generally conservative, but if you want guaranteed control over the Type I error rate then this is the test to use. If sample sizes are slightly different then use Gabriel's procedure because it has greater power, but if sample sizes are very different use Hochberg's GT2. If there is any doubt that the population variances are equal then use the Games–Howell procedure because this generally seems to offer the best performance. I recommend running the Games–Howell procedure in addition to any other tests you might select because of the uncertainty of knowing whether the population variances are equivalent.

Although these general guidelines provide a convention to follow, be aware of the other procedures available and when they might be useful to use (e.g. Dunnett's test is the only multiple comparison that allows you to test means against a control mean).

8.3. RUNNING ONE-WAY ANOVA ON SPSS ②

Hopefully you should all have some appreciation for the theory behind ANOVA, so let's put that theory into practice by conducting an ANOVA on the Viagra data. As with the independent *t*-test we need to enter the data into the data editor using a coding variable to specify to which of the three groups the data belong. So, the data must be entered in two columns (one called **dose** which specifies how much Viagra the participant was given and one called **libido** which indicates the person's libido over the following week). The data are in the file **Viagra.sav** but I recommend entering them by hand to gain practice in data entry. I have coded the grouping variable so that 1 = placebo, 2 = low dose and 3 = high dose (see section 2.4.4).

To conduct one-way ANOVA we have first to access the main dialog box using the **Analyze⇒Compare Means⇒One-way ANOVA** menu path (Figure 8.8). This dialog box

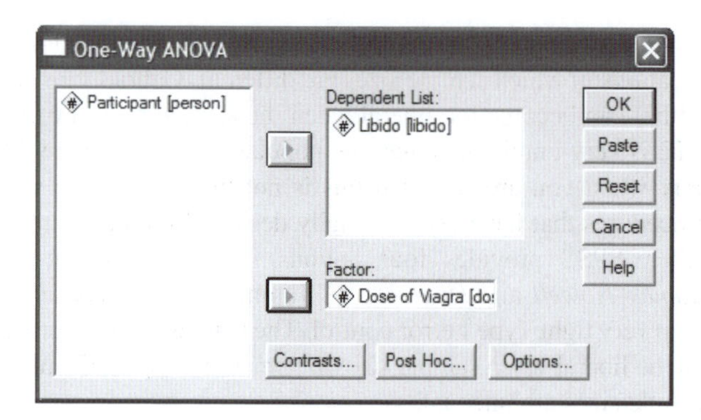

Figure 8.8 Main dialog box for one-way ANOVA

has a space in which you can list one or more dependent variables and a second space to specify a grouping variable, or *factor*. Factor is another term for independent variable and should not be confused with the factors that we will come across when we learn about factor analysis. One thing that I dislike about SPSS is that in various procedures, such as one-way ANOVA, the program encourages the user to carry out multiple tests, which as we have seen is not a good thing. For example, in this procedure you are allowed to specify several dependent variables on which to conduct several ANOVAs. In reality, if you had measured several dependent variables (say you had measured not just libido but physiological arousal and anxiety too) it would be preferable to analyse these data using MANOVA rather than treating each dependent measure separately (see Chapter 14). For the Viagra data we need select only **libido** from the variables list and transfer it to the box labelled *Dependent List* by clicking on ▶. Then select the grouping variable **dose** and transfer it to the box labelled *Factor* by clicking on ▶.

8.3.1. Planned comparisons using SPSS ②

If you click on Contrasts... you access the dialog box that allows you to conduct the planned comparisons described in section 8.2.10.

The dialog box is shown in Figure 8.9 and has two sections. The first section is for specifying trend analyses. If you want to test for trends in the data then tick the box labelled *Polynomial*. Once this box is ticked, you can select the degree of polynomial you would like. The Viagra data have only three groups and so the highest degree of trend there can be is a quadratic trend (see section 8.2.10.3). Now, it is important from the point of view of trend analysis that we have coded the grouping variables in a meaningful order. Now, we expect libido to be smallest in the placebo group, to increase in the low-dose group and then to increase again in the high-dose group. To detect a meaningful trend, we need to have coded these groups in ascending order. We have done this by coding the placebo group with the lowest value 1, the low-dose group with the middle value 2, and the high-dose group with the highest coding value of 3. If we had coded the groups differently, this

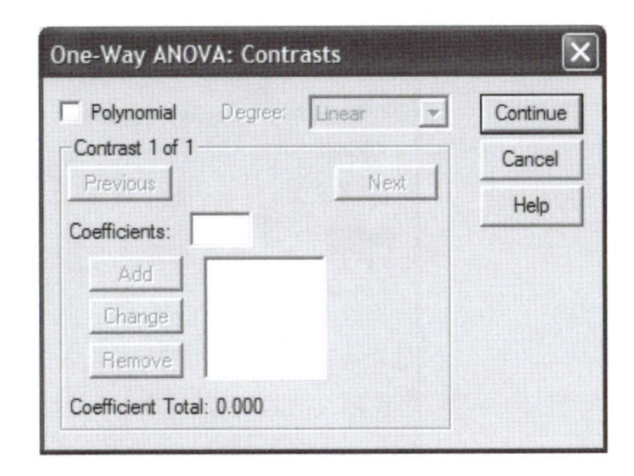

Figure 8.9 Dialog box for conducting planned comparisons

would influence both whether a trend is detected and if a trend is detected whether it is statistically meaningful.

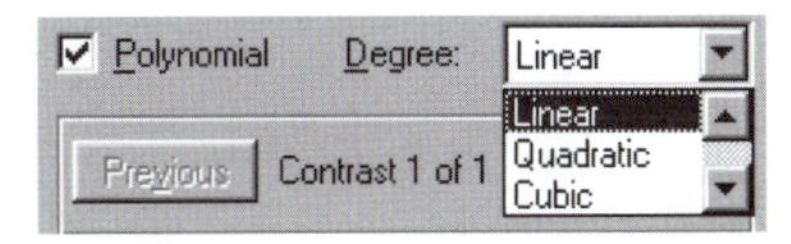

For the Viagra data there are only three groups and so we should select the polynomial option [Polynomial], then select a quadratic degree by clicking on [▼] and then select *Quadratic*. If a quadratic trend is selected SPSS will test for both linear and quadratic trends.

The lower part of the dialog box in Figure 8.9 is for specifying any planned comparisons. To conduct planned comparisons we need to tell SPSS what weights to assign to each group. The first step is to decide which comparisons you want to do and then what weights must be assigned to each group for each of the contrasts. We have already gone through this process in section 8.2.10.2, so we know that the weights for contrast 1 were −2 (placebo group), +1 (low-dose group) and +1 (high-dose group). We will specify this contrast first. It is important to make sure that you enter the correct weight for each group, so you should remember that the first weight that you enter should be the weight for the *first* group (i.e. the group coded with the lowest value in the data editor). For the Viagra data, the group coded with the lowest value was the placebo group (which had a code of 1) and so we should enter the weighting for this group first. Click in the box labelled *Coefficients* with the mouse and then type '−2' in this box and click on [Add]. Next, we need to input the weight for the second group, which for the Viagra data is the low-dose group (because this group was coded in the data editor with the second-highest value). Click in the box labelled *Coefficients* with the mouse and then type '1' in this box and click on [Add]. Finally, we need to input the weight for the last group, which for the Viagra data is the high-dose group (because this group was coded with the highest value in the data editor). Click in the box labelled *Coefficients* with the mouse and then type '1' in this box and click on [Add]. The box should now look like Figure 8.10.

Once you have inputted the weights you can change or remove any one of them by using the mouse to select the weight that you want to change. The weight will then appear in the box labelled *Coefficients* where you can type in a new weight and then click on [Change]. Alternatively, you can click on any of the weights and remove it completely by clicking [Remove]. Underneath

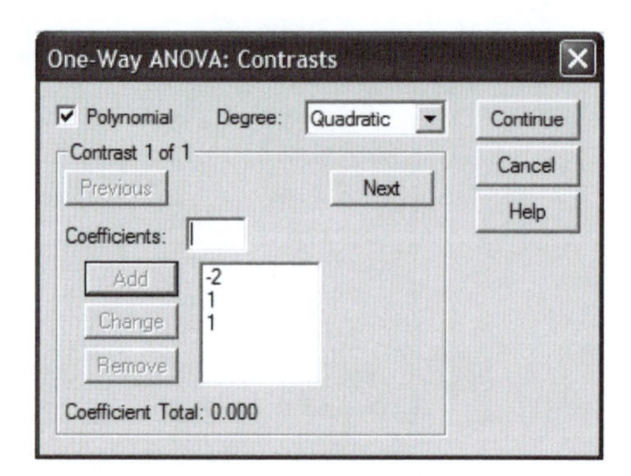

Figure 8.10 Contrasts dialog box completed for the first contrast of the Viagra data

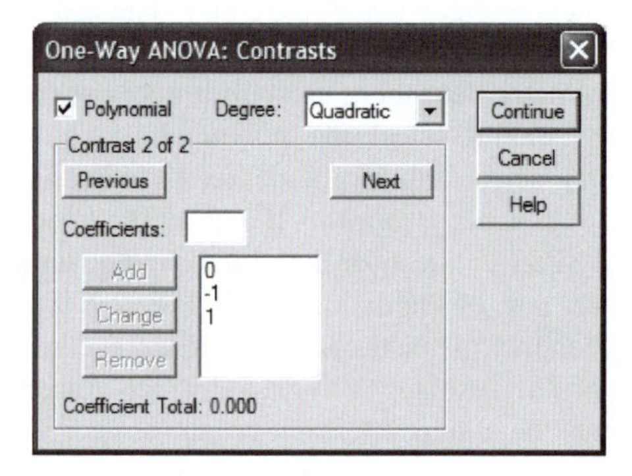

Figure 8.11 Contrasts dialog box completed for the second contrast of the Viagra data

the weights SPSS calculates the coefficient total which, as we saw in section 8.2.10.2, should equal zero. If the coefficient number is anything other than zero you should go back and check that the contrasts you have planned make sense and that you have followed the appropriate rules for assigning weights.

Once you have specified the first contrast, click on Next. The weights that you have just entered will disappear and the dialog box will now read *Contrast 2 of 2*. We know from section 8.2.10.2 that the weights for contrast 2 were: 0 (placebo group), +1 (low-dose group) and −1 (high-dose group). We can specify this contrast as before. Remembering that the first weight we enter will be for the placebo group, we must enter the value 0 as the first weight. Click in the box labelled *Coefficients* with the mouse and then type '0' and click on Add. Next, we need to input the weight for the low-dose group by clicking in the box labelled *Coefficients* and then typing '1' and clicking on Add. Finally, we need to input the weight for the high-dose group by clicking in the box labelled *Coefficients* and then typing '−1' and clicking on Add. The box should now look like Figure 8.11. Notice that the weights add up to zero as they did for contrast 1. It is imperative that you remember to input zero weights for any groups that are not in the contrast. When all of the planned contrasts have been specified click on Continue to return to the main dialog box.

8.3.2. *Post hoc* tests in SPSS ②

Having told SPSS which planned comparisons to do, we can choose to do *post hoc* tests. In theory, if we have done planned comparisons we shouldn't need to do *post hoc* tests (because we have already tested the hypotheses of interest). Likewise, if we choose to conduct *post hoc* tests then we should not need to do planned contrasts (because we have no hypotheses to test). However, for the sake of space we will conduct some *post hoc* tests on the Viagra data. Click on Post Hoc... in the main dialog box to access the *post hoc* tests dialog box (Figure 8.12).

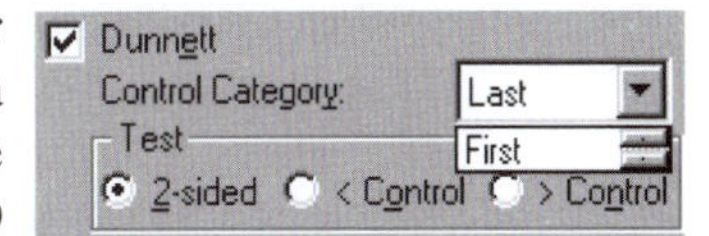

In section 8.2.11.3, I recommended various *post hoc* procedures for various situations. For the Viagra data there are equal sample sizes and so we need not use Gabriel's test. We should use Tukey's test and REGWQ and check the findings with the Games–Howell procedure. We have a specific hypothesis that both the high- and low-dose groups should differ from the placebo group and so we could use Dunnett's test to examine these hypotheses. Once you have selected Dunnett's test, you can change the control category (the default is to use the last category) to specify that the first category be used as the control category (because the placebo group was coded with the lowest value). You can also choose whether to conduct a two-tailed test (2-sided) or a one-tailed test. If you choose a one-tailed test then you must predict whether you believe that the mean of the control group will be less than the mean of a particular experimental group (> Control) or greater than the mean of a particular experimental group (< Control). These are all of the *post hoc* tests that need to be specified and when the completed dialog box looks like Figure 8.12 click on Continue to return to the main dialog box.

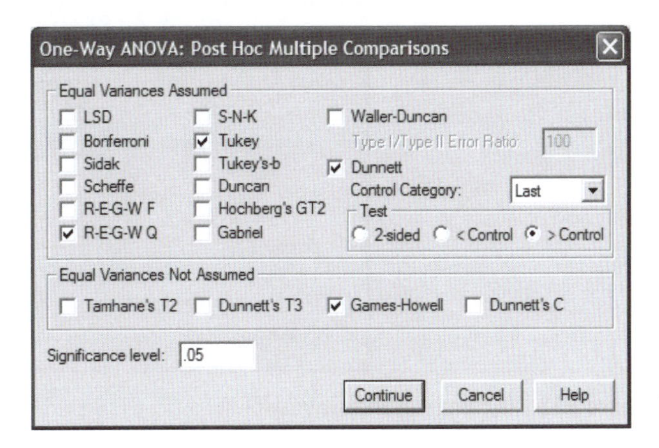

Figure 8.12 Dialog box for specifying *post hoc* tests

8.3.3. Options ②

The options for one-way ANOVA are fairly straightforward. First you can ask for some descriptive statistics which will produce a table of the means, standard deviations, standard errors, ranges and confidence intervals for the means of each group. This option is useful to select because it assists in interpreting the final results. A vital option to select is the homogeneity of variance tests. As with the *t*-test, there is an assumption that the variances of the groups are equal and selecting this option tests this assumption. SPSS uses Levene's test, which tests the hypothesis that the variances of each group are equal (see section 3.6). If the homogeneity of variance assumption is broken, then SPSS offers us two alternative versions of the *F*-ratio: the Brown–Forsythe F (1974) and the Welch F (1951). These two statistics are discussed in Box 8.3 (and if you're really bored, in the file **WelchF.pdf** on the CD-ROM), but suffice it to say they're worth selecting just in case the assumption is broken.

There is also an option to have a *Means plot* (Figure 8.13) and if this option is selected then a line graph of the group means will be produced in the output. Finally, the options let us specify whether we want to exclude cases on a listwise basis or on a per analysis basis. This option is useful only if you are conducting several ANOVAs on different dependent variables. The first option (*Exclude cases analysis by analysis*) excludes any case that has a missing value for either the independent or the dependent variable used in that particular analysis. *Exclude cases listwise* will exclude from *all analyses* any case that has a missing value for the independent variable or any of the dependent variables specified. As such, if you stick to good practice and don't conduct hundreds of ANOVAs on different dependent variables (see Chapter 14 on MANOVA) the default settings are fine. When you have selected the appropriate options, click on Continue to return to the main dialog box and then click on OK to run the analysis.

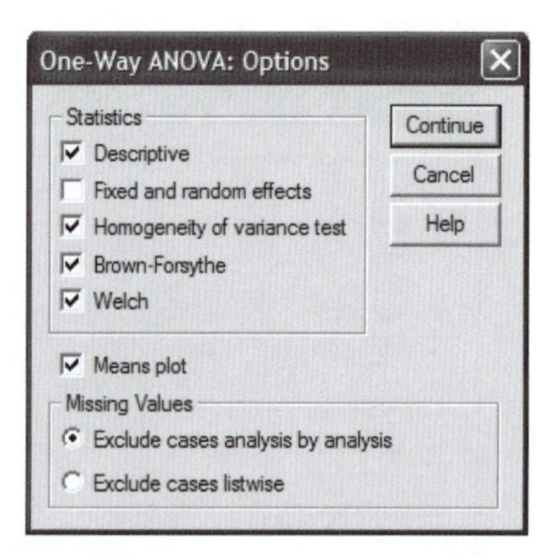

Figure 8.13 Options for one-way ANOVA

Box 8.3

What do I do in ANOVA when the homogeneity of variance assumption is broken? ③

In section 8.2.9 I mentioned that when group sizes are unequal, violations of the assumption of homogeneity of variance can have quite serious consequences. SPSS incorporates options for two alternative *F*-ratios, which have been derived to be robust when homogeneity of variance has been violated. The first is the Brown & Forsythe (1974) *F*-ratio, which is fairly easy to explain. I mentioned earlier that when group sizes are unequal and the large groups have the biggest variance then this biases the *F*-ratio to be conservative. If you think back to equation (8.7) this makes perfect sense because to calculate SS_R variances are multiplied by their sample size (minus 1), so in this situation you get a large sample size cross-multiplied with a large variance, which will inflate the value of SS_R. What effect does this have on the *F*-ratio? Well, the *F*-ratio is proportionate to SS_M/SS_R, so if SS_R is big, then the *F*-ratio gets smaller (which is why it would be more conservative: its value is being overly reduced!). Brown & Forsythe get around this problem by weighting the group variances, not by their sample size, but by the inverse of their sample sizes (actually they use *n/N* so it's the sample size as a proportion of the total sample size). This means that the impact of large sample sizes with large variance is reduced:

$$F_{\mathrm{BF}} = \frac{\mathrm{SS_M}}{\mathrm{SS_{R_{BF}}}} = \frac{\mathrm{SS_M}}{\sum s_k^2(1 - \frac{n_k}{N})}$$

So, for the Viagra data, SS_M is the same as before (20.135), so the equation becomes:

$$\begin{aligned} F_{\mathrm{BF}} &= \frac{20.135}{s^2_{\mathrm{group1}}(1 - \frac{n_{\mathrm{group1}}}{N}) + s^2_{\mathrm{group2}}(1 - \frac{n_{\mathrm{group2}}}{N}) + s^2_{\mathrm{group3}}(1 - \frac{n_{\mathrm{group3}}}{N})} \\ &= \frac{20.135}{1.70(1 - \frac{5}{15}) + 1.70(1 - \frac{5}{15}) + 2.50(1 - \frac{5}{15})} \\ &= \frac{20.135}{3.933} \\ &= 5.119 \end{aligned}$$

This statistic is evaluated using degrees of freedom for the model and error terms. For the model, df_M is the same as before (i.e. $k-1 = 2$), but an adjustment is made to the residual degrees of freedom, df_R. The calculations of this, and the Welch (1951) *F*, are really going beyond the scope of this book, but if you're interested look at Appendix A.5 (or the file **WelchF.pdf** on the CD-ROM), which details the Welch *F*, but for those who lead less tragic lives than me just be aware that these two techniques make adjustments to *F* and the residual degrees of freedom, which combat problems arising from violations of the homogeneity of variance assumption.

(Continued)

Box 8.3 (Continued)

The obvious question is which of the two procedures is best. Tomarken & Serlin (1986) review these and other techniques and seem to conclude that both techniques control the Type I error rate well (i.e. when there's no effect in the population you do indeed get a non-significant *F*). However, in terms of power (i.e. which test is best at detecting an effect when it exists), the Welch test seems to fare best except when there is an extreme mean that has a large variance.

8.4. OUTPUT FROM ONE-WAY ANOVA ②

8.4.1. Output for the main analysis ②

If you load up the Viagra data (or enter it in by hand) and select all of the options I have suggested, you should find that the output looks the same as what follows. If your output is different we should panic because one of us has done it wrong—hopefully not me or more trees have died for nothing. Figure 8.14 shows an error bar chart of the Viagra data with a line superimposed to show the general trend of the means across groups. This graph is not automatically produced by SPSS; however, a line graph will be produced if the *Means plot* option is selected (this graph will look like Figure 8.14 but without the error bars). I have chosen to display an error bar graph because it is slightly more interesting than the line graph alone. It's clear from this chart that all of the error bars overlap, indicating that, at face value, there are no between-group differences (although this measure is only approximate). The line that joins the means seems to indicate a linear trend in that as the dose of Viagra increases so does the mean level of libido.

SPSS Output 8.3 shows the table of descriptive statistics from the one-way procedure for the Viagra data. The first thing to notice is that the means and standard deviations correspond to those shown in Table 8.1. In addition we are told the standard error. You should remember that the standard error is the standard deviation of the sampling distribution of these data (so, for the placebo group, if you took lots of samples from the population from which these data come, the means of these samples would have a standard deviation of .5831). We are also given confidence intervals for the mean. By now, you should be familiar with what a confidence interval tells us, and that is that if we took 100 samples from the population from which the placebo group came and constructed confidence intervals for the mean, then 95 of these intervals would contain the true value of the mean: in other words, the true value of the mean is likely to be between .5811 and 3.8189. Although these diagnostics are not immediately important, we will refer back to them throughout the analysis.

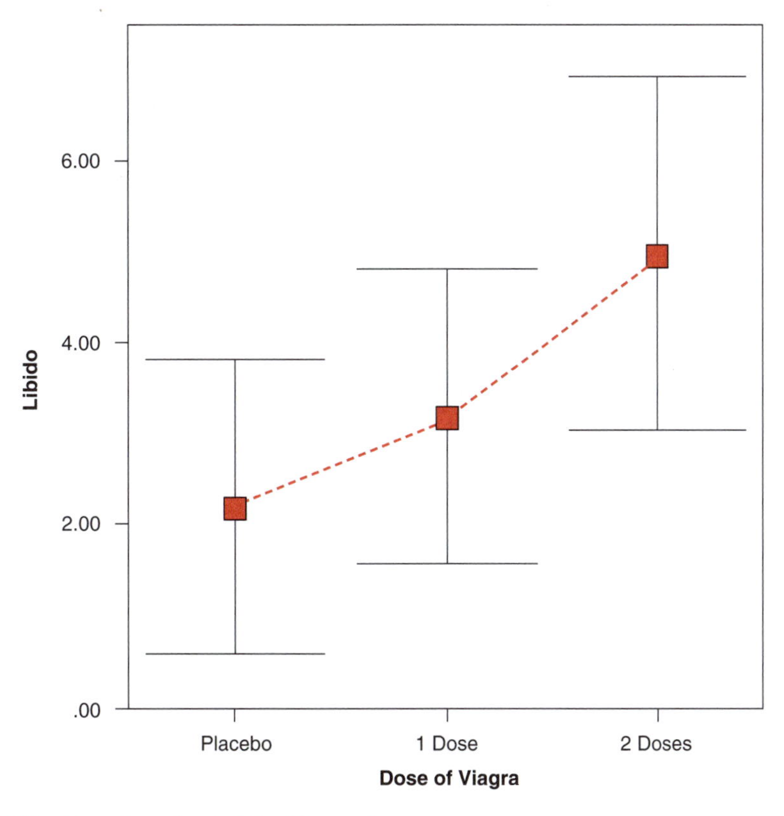

Figure 8.14 Error bar chart of the Viagra data

Descriptives

			N	Mean	Std. Deviation	Std. Error	95% Confidence Interval for Mean		Minimum	Maximum
							Lower Bound	Upper Bound		
Libido	Dose of Viagra	Placebo	5	2.2000	1.3038	.5831	.5811	3.8189	1.00	4.00
		Low Dose	5	3.2000	1.3038	.5831	1.5811	4.8189	2.00	5.00
		High Dose	5	5.0000	1.5811	.7071	3.0368	6.9632	3.00	7.00
		Total	15	3.4667	1.7674	.4563	2.4879	4.4454	1.00	7.00

SPSS Output 8.3

The next part of the output is a summary table of Levene's test (SPSS Output 8.4). This test is designed to test the null hypothesis that the variances of the groups are the same. It is an ANOVA conducted on the absolute differences between the observed data and the mean from which the data came.[4] In this case, Levene's test is, therefore, testing whether the variances of the three groups are significantly different. If Levene's test is significant (i.e. the value of *Sig.* is less than .05) then we can say that the variances are significantly different. This would mean that we had violated one of the assumptions of ANOVA and we would have to take steps to rectify this matter. As we saw in Chapter 3, one common way to rectify differences between group variances is to transform all of the data: if the variances are unequal, they can sometimes be stabilized by taking the square root of every value of the dependent variable and then reanalysing these transformed values (see Chapter 3 and Howell, 2002, pp. 342–349). However, given the apparent utility of Welch's F, or the Brown–Forsythe F, and the fact that transformations often don't help at all, you can instead report the Welch F (and I'd probably suggest reporting this instead of the Brown–Forsythe F unless you have an extreme mean that is also causing the problem with the variances). Luckily, for these data the variances are very similar (hence the high probability value); in fact, if you look at SPSS Output 8.3 you'll see that the variances of the placebo and low-dose groups are identical.

Test of Homogeneity of Variances

	Levene Statistic	df1	df2	Sig.
Libido	.092	2	12	.913

SPSS Output 8.4

SPSS Output 8.5 shows the main ANOVA summary table. The table is divided into between-group effects (effects due to the model—the experimental effect) and within-group effects (this is the unsystematic variation in the data). The between-group effect is further broken down into a linear and quadratic component and these components are the trend analyses described in section 8.2.10.5. The between-group effect labelled *Combined* is the overall experimental effect. In this row we are told the sums of squares for the model (SS_M = 20.13) and this value corresponds to the value calculated in section 8.2.5. The degrees of freedom are equal to 2 and the mean square for the model corresponds to the value calculated in section 8.2.7 (10.067). The sum of squares and mean squares represent the experimental effect. This overall effect is then broken down because we asked SPSS to conduct trend analyses of these data (we will return to these trends in due course). Had we not specified this in section 8.3.1, then these two rows of the summary table would not

4 The interested reader might like to try this out. Simply create a new variable called **diff** (short for difference), which is each score subtracted from the mean of the group to which that score belongs. Then remove all of the minus signs (so, take the absolute value of **diff**) and conduct a one-way ANOVA with **dose** as the independent variable and **diff** as the dependent variable. You'll find that the F-ratio for this analysis is .092, which is significant at $p = .913$!

be produced. The row labelled *Within Groups* gives details of the unsystematic variation within the data (the variation due to natural individual differences in libido and different reactions to Viagra). The table tells us how much unsystematic variation exists (the residual sum of squares, SS_R) and this value (23.60) corresponds to the value calculated in section 8.2.6. The table then gives the average amount of unsystematic variation, the mean squares (MS_R), which corresponds to the value (1.967) calculated in section 8.2.7. The test of whether the group means are the same is represented by the *F*-ratio for the combined between-group effect. The value of this ratio is 5.12, which is the same as was calculated in section 8.2.8. Finally SPSS tells us whether this value is likely to have happened by chance. The final column labelled *Sig.* indicates the likelihood of an *F*-ratio the size of the one obtained occurring by chance. In this case, there is a probability of .025 that an *F*-ratio of this size would occur by chance (that's only a 2.5% chance!). We have seen in previous chapters that social scientists use a cut-off point of .05 as their criterion for statistical significance. Hence, because the observed significance value is less than .05 we can say that there was a significant effect of Viagra.[5] However, at this stage we still do not know exactly what the effect of Viagra was (we don't know which groups differed). One thing that is interesting here is that we obtained a significant experimental effect yet our error bar plot indicated that no difference would be found. This contradiction illustrates how the error bar chart can act only as a rough guide to the data.

ANOVA

				Sum of Squares	df	Mean Square	F	Sig.
Libido	Between Groups	(Combined)		20.133	2	10.067	5.119	.025
		Linear Term	Contrast	19.600	1	19.600	9.966	.008
			Deviation	.533	1	.533	.271	.612
		Quadratic Term	Contrast	.533	1	.533	.271	.612
	Within Groups			23.600	12	1.967		
	Total			43.733	14			

SPSS Output 8.5

Knowing that the overall effect of Viagra was significant, we can now look at the trend analysis. The trend analysis breaks down the experimental effect to see whether it can be explained by either a linear or a quadratic relationship in the data. First, let's look at the linear component. This comparison tests whether the means increase across groups in a

5 A question I get asked a lot by students is 'is the significance of the ANOVA one- or two-tailed, and if it's two-tailed can I divide by 2 to get the one-tailed value?' The answer is that to do a one-tailed test you have to be making a directional hypothesis (i.e. the mean for cats is greater than that for dogs). ANOVA is a non-specific test, so it just tells us generally whether there is a difference or not and because there are several means you can't possibly make a directional hypothesis. As such, it's invalid to halve the significance value.

linear way. Again the sum of squares and mean squares are given, but the most important things to note are the value of the *F*-ratio and the corresponding significance value. For the linear trend the *F*-ratio is 9.97 and this value is significant at a .008 level. Therefore, we can say that as the dose of Viagra increased from nothing to a low dose to a high dose, libido increased proportionately. Moving on to the quadratic trend, this comparison is testing whether the pattern of means is curvilinear (i.e. is represented by a curve that has one bend). The error bar graph of the data strongly suggests that the means cannot be represented by a curve and the results for the quadratic trend bear this out. The *F*-ratio for the quadratic trend is non-significant (in fact, the value of *F* is less than 1, which immediately indicates that this contrast will not be significant).

Finally, SPSS Output 8.6 shows the Welch and Brown–Forsythe *F*-ratios. As it turned out we didn't need these because our Levene's test was not significant, indicating that our variances were equal. However, when homogeneity of variance has been violated you should look at these *F*-ratios *instead* of the ones in the main table. If you're interested in how these values are calculated then look at Box 8.3 and file **Welch F.pdf** or the CD-ROM, but to be honest it's just downright confusing if you ask me and you're much better off just looking at the values in SPSS Output 8.6 and trusting that they do what they're supposed to do (you should also note that the error degrees of freedom have been adjusted and you should remember this when you report the values!).

Robust Tests of Equality of Means

Libido

	Statistic[a]	df1	df2	Sig.
Welch	4.320	2	7.943	.054
Brown–Forsythe	5.119	2	11.574	.026

a Asymptotically F distributed

SPSS Output 8.6

8.4.2. Output for planned comparisons ②

In section 8.3.1 we told SPSS to conduct two planned comparisons: one to test whether the control group was different to the two groups who received Viagra, and one to see whether the two doses of Viagra made a difference to libido. SPSS Output 8.7 shows the results of the planned comparisons that we requested for the Viagra data. The first table displays the contrast coefficients; these values are the ones that we entered in section 8.3.1 and it is well worth looking at this table to double-check that the contrasts are comparing what they are supposed to! As a quick rule of thumb, remember that when we do planned comparisons we arrange the weights such that we compare any group with a positive weight against any group with a negative weight. Therefore, the table of weights shows that contrast 1 compares the placebo group against the two experimental groups, and contrast 2 compares the low-dose group with the high-dose group. It is useful to check this table to make sure that the weights that we entered into SPSS correspond to the weights we intended to enter into SPSS!

Contrast Coefficients

Contrast	Dose of Viagra		
	Placebo	Low Dose	High Dose
1	−2	1	1
2	0	−1	1

Contrast Tests

		Contrast	Value of Contrast	Std. Error	t	df	Sig. (2-tailed)
Libido	Assume equal variances	1	3.8000	1.5362	2.474	12	.029
		2	1.8000	.8869	2.029	12	.065
	Does not assume equal variances	1	3.8000	1.4832	2.562	8.740	.031
		2	1.8000	.9165	1.964	7.720	.086

SPSS Output 8.7

The second table gives the statistics for each contrast. The first thing to notice is that statistics are produced for situations in which the group variances are equal, and when they are unequal. If Levene's test was significant then you should read the part of the table labelled *Does not assume equal variances*. However, for these data Levene's test was not significant and we can, therefore, use the part of the table labelled *Assume equal variances*. The table tells us the value of the contrast itself, which is the weighted sum of the group means. This value is obtained by taking each group mean, multiplying it by the weight for the contrast of interest, and then adding these values together.[6] The table also gives the standard error of each contrast and a *t*-statistic. The *t*-statistic is derived by dividing the contrast value by the standard error ($t = 3.8/1.5362 = 2.47$) and is compared against critical values of the *t*-distribution. The significance value of the contrast is given in the final column and this value is two-tailed. Using the first contrast as an example, if we had used this contrast to test the general hypothesis that the experimental groups would differ from the placebo group, then we should use this two-tailed value. However, in reality we tested the hypothesis that the experimental groups would increase libido above the levels seen in the placebo group: this hypothesis is one-tailed. Provided the means for the groups bear out the hypothesis we can divide the significance values by 2 to obtain the one-tailed probability. Hence, for contrast 1, we can say that taking Viagra significantly increased libido compared to the control group ($p = .0145$). For contrast 2 we also had a one-tailed hypothesis (that a high dose of Viagra would increase libido significantly more than a low dose) and the means bear this hypothesis out. The significance of contrast 2 tells us that a high dose of Viagra

6 For the first contrast this value is:

$$\sum(\overline{X}W) = [(2.2 \times -2) + (3.2 \times 1) + (5.0 \times 1)] = 3.8$$

increased libido significantly more than a low dose (p(one-tailed) = .065/2 = .0325). Notice that had we not had a specific hypothesis regarding which group would have the highest mean then we would have had to conclude that the dose of Viagra had no significant effect on libido. For this reason it can be important as scientists that we generate hypotheses before collecting any data because this method of scientific discovery is more powerful.

In summary, there is an overall effect of Viagra on libido. Furthermore, the planned contrasts revealed that having Viagra significantly increased libido compared to a control group, $t(12) = 2.47$, $p < .05$, and having a high dose significantly increased libido compared to a low dose, $t(12) = 2.03$, $p < .05$.

8.4.3. Output for *post hoc* tests ②

If we had no specific hypotheses about the effect that Viagra might have on libido then we could carry out *post hoc* tests to compare all groups of participants with each other. In fact, we asked SPSS to do this (see section 8.3.2) and the results of this analysis are shown in SPSS Output 8.8. This table shows the results of Tukey's test (known as Tukey's HSD)[7], the Games–Howell procedure and Dunnett's test, which were all specified earlier on. If we look at Tukey's test first (because we have no reason to doubt that the population variances are unequal) it is clear from the table that each group of participants are compared to all of the remaining groups. For each pair of groups the difference between group means is displayed, the standard error of that difference, the significance level of that difference and a 95% confidence interval. First of all, the placebo group are compared to the low-dose group and reveal a non-significant difference (*Sig.* is greater than .05), but when compared to the high-dose group there is a significant difference (*Sig.* is less than .05). This finding is interesting because our planned comparison showed that any dose of Viagra produced a significant increase in libido, yet these comparisons indicate that a low dose does not. Why is there this contradiction? (Have a think about this question before you read on.)

In section 8.2.10.2 I explained that the first planned comparison would compare the experimental groups with the placebo group. Specifically, it would compare the average of the two group means of the experimental groups ((3.2 + 5.0)/2 = 4.1) with the mean of the placebo group (2.2). So, it was assessing whether the difference between these values (4.1 − 2.2 = 1.9) was significant. In the *post hoc* tests, when the low dose is compared to the placebo, the contrast is testing whether the difference between the means of these two groups is significant. The difference in this case is only 1, compared to a difference of 1.9 for the planned comparison. This explanation illustrates how it is possible to have apparently contradictory results from planned contrasts and *post hoc* comparisons. More important, it illustrates how careful we must be in interpreting planned contrasts.

The low-dose group is then compared to both the placebo group and the high-dose group. The first thing to note is that the contrast involving the low-dose and placebo groups is identical to the one just described. The only new information is the comparison between the two experimental conditions. The group means differ by 1.8 which is not

7 The HSD stands for 'honestly significant difference', which has a slightly dodgy ring to it if you ask me!

significant. This result also contradicts the planned comparisons (remember that contrast 2 compared these groups and found a significant difference). Think why this contradiction might exist. For one thing, *post hoc* tests by their very nature are two-tailed (you use them when you have made no specific hypotheses and you cannot predict the direction of hypotheses that don't exist!) and contrast 2 was significant only when considered as a one-tailed hypothesis. However, even at the two-tailed level the planned comparison was closer to significance than the *post hoc* test and this fact illustrates that *post hoc* procedures are more conservative (i.e. have less power to detect true effects) than planned comparisons.

The rest of the table describes the Games–Howell test and a quick inspection reveals the same pattern of results: the only groups that differed significantly were the high-dose and placebo groups. These results give us confidence in our conclusions from Tukey's test because even if the populations' variances are not equal (which seems unlikely given that the sample variances are very similar), then the profile of results still holds true. Finally, Dunnett's test is described and you'll hopefully remember that we asked the computer to compare both experimental groups against the control using a one-tailed hypothesis that the mean of the control group would be smaller than that of both experimental groups. Even as a one-tailed hypothesis, levels of libido in the low-dose group are equivalent to the placebo group. However, the high-dose group has a significantly higher libido than the placebo group.

Multiple Comparisons

Dependent Variable: Libido

	(I) Dose of Viagra	(J) Dose of Viagra	Mean Difference (I-J)	Std. Error	Sig.	95% Confidence Interval	
						Lower Bound	Upper Bound
Turkey HSD	Placebo	Low Dose	−1.0000	.88694	.516	−3.3662	1.3662
		High Dose	−2.8000*	.88694	.021	−5.1662	−.4338
	Low Dose	Placebo	1.0000	.88694	.516	−1.3662	3.3662
		High Dose	−1.8000	.88694	.147	−4.1662	.5662
	High Dose	Placebo	2.8000*	.88694	.021	.4338	5.1662
		Low Dose	1.8000	.88694	.147	−.5662	4.1662
Games–Howell	Placebo	Low Dose	−1.0000	.82462	.479	−3.3563	1.3563
		High Dose	−2.8000*	.91652	.039	−5.4389	−.1611
	Low Dose	Placebo	1.0000	.82462	.479	−1.3563	3.3563
		High Dose	−1.8000	.91652	.185	−4.4389	.8389
	High Dose	Placebo	2.8000*	.91652	.039	.1611	5.4389
		Low Dose	1.8000	.91652	.185	−.8389	4.4389
Dunnett *t* (> control)[a]	Low Dose	Placebo	1.0000	.88694	.227	−.8697	
	High Dose	Placebo	2.8000*	.88694	.008	.9303	

* The mean difference is significant at the .05 level.
a Dunnett *t*-tests treat one group as a control, and compare all other groups against it

SPSS Output 8.8

The table in SPSS Output 8.9 shows the results of Tukey's test and the REGWQ test. These tests display subsets of groups that have the same means. Therefore, the Tukey test creates two subsets of groups with statistically similar means. The first subset contains the placebo and low-dose groups (indicating that these two groups have similar means) whereas the second subset contains the high- and low-dose groups. These results demonstrate that the placebo group has a similar mean to the low-dose group but not the high-dose group, and that the low-dose group has a similar mean to both the placebo and high-dose groups. In other words, the only groups that have significantly different means are the high-dose and placebo groups. The tests provide a significance value for each subset and it's clear from these significance values that the groups in subsets have non-significant means (as indicated by values of *Sig.* that are greater than .05).

These calculations use the harmonic mean sample size. The harmonic mean is a weighted version of the mean that takes account of the relationship between variance and sample size (see Howell, 2002, p. 234). Although you don't need to know the intricacies of the harmonic mean, it is useful that the harmonic sample size is used because it eliminates any bias that might be introduced through having unequal sample sizes. As such, one-way ANOVA provides reliable results even in unbalanced designs (although as I keep pointing out, balanced designs do have many advantages!).

Libido

			Subset for alpha = .05	
	Dose of Viagra	N	1	2
Tukey HSD[a]	Placebo	5	2.2000	
	Low Dose	5	3.2000	3.2000
	High Dose	5		5.0000
	Sig.		.516	.147
Ryan-Einot-Gabriel-Welsch Range	Placebo	5	2.2000	
	Low Dose	5	3.2000	3.2000
	High Dose	5		5.0000
	Sig.		.282	.065

Means for groups in homogeneous subsets are displayed.
a Uses Harmonic Mean Sample Size = 5.000

SPSS Output 8.9

Cramming Samantha's Tips

- The one-way independent ANOVA compares several means, when those means have come from the different groups of people; for example, if you have several experimental conditions and have used different participants in each condition.
- When you have generated specific hypotheses before the experiment use *planned comparisons*, but if you don't have specific hypotheses use *post hoc* tests.

- There are lots of different *post hoc* tests: when you have equal sample sizes and homogeneity of variance is met use *REGWQ* or *Tukey* HSD. If sample sizes are slightly different then use *Gabriel's* procedure, but if sample sizes are very different use *Hochberg's GT2*. If there is any doubt about homogeneity of variance use the *Games–Howell* procedure.
- Test for homogeneity of variance using *Levene's test*. Find the table with this label: if the value in the column labelled *Sig.* is less than .05 then the assumption is violated. If this is the case go to the table labelled *Robust Tests of Equality of Means*. If homogeneity of variance has been met (the significance of Levene's test is greater than .05) go to the table labelled *ANOVA*.
- In the table labelled *ANOVA* (or *Robust Tests of Equality of Means*—see above), look at the column labelled *Sig.*: if the value is less than .05 then the means of the groups are significantly different.
- For contrasts and *post hoc* tests, again look at the columns labelled *Sig.* to discover if your comparisons are significant (they will be if the significance value is less than .05).

8.5. CALCULATING THE EFFECT SIZE ②

One thing you will notice is that SPSS doesn't routinely provide an effect size for one-way independent ANOVA. However, we saw in equation (5.4) that:

$$R^2 = \frac{SS_M}{SS_T}$$

Of course we know these values from the SPSS output. So we can simply calculate r^2 using the between-group effect (SS_M), and the total amount of variance in the data (SS_T)—although for some bizarre reason it's usually called eta squared, η^2. It is then a simple matter to take the square root of this value to give us the effect size r:

$$\begin{aligned} r^2 &= \frac{SS_M}{SS_T} \\ &= \frac{20.13}{43.73} \\ &= .42 \\ r &= \sqrt{.42} \\ &= .65 \end{aligned}$$

Using the benchmarks for effect sizes, this represents a large effect (it is above the .5 threshold for a large effect). Therefore, the effect of Viagra on libido is a substantive finding.

However, this measure of effect size is slightly biased because it is based purely on sums of squares from the sample and no adjustment is made for the fact that we're trying to estimate the effect size in the population. Therefore, we often use a slightly more complex measure called omega squared (ω^2). This effect size estimate is still based on the sums of squares that we've met in this chapter, but like the F-ratio it uses the variance explained by the model, and the error variance (in both cases the average variance, or mean squared error, is used):

$$\omega^2 = \frac{\mathrm{MS_M} - \mathrm{MS_R}}{\mathrm{MS_M} + ((n-1) \times \mathrm{MS_R})}$$

The n in the equation is the number of people in a group (in this case 5). So, in this example we'd get:

$$\begin{aligned}\omega^2 &= \frac{10.07 - 1.97}{10.07 + ((5-1) \times 1.97)} \\ &= \frac{8.10}{10.07 + 7.88} \\ &= .45 \\ \omega &= \sqrt{.45} = .67\end{aligned}$$

As you can see this has led to a slightly higher estimate to using r, and in general ω is a more accurate measure. The one obvious problem is that you can't calculate ω when you have unequal-sized samples (see, I told you to keep them equal!). Although in the sections on ANOVA I will use ω as my effect size measure, think of it as you would r (because it's basically an unbiased estimate of r anyway) and apply the same benchmarks for deciding how substantial an effect is.

Most of the time it isn't that interesting to have effect sizes for the overall ANOVA because it's testing a general hypothesis. Instead, we really want effect sizes for the contrasts (because these compare only two things and so the effect size is considerably easier to interpret). Planned comparisons are tested with the t-statistic and, therefore, we can use the same equation as in section 7.5.5:

$$r_{\text{contrast}} = \sqrt{\frac{t^2}{t^2 + df}}$$

We know the value of t and the df from SPSS Output 8.7 and so we can compute r as follows:

$$\begin{aligned}r_{\text{contrast1}} &= \sqrt{\frac{2.474^2}{2.474^2 + 12}} \\ &= \sqrt{\frac{6.12}{18.12}} \\ &= .58\end{aligned}$$

If you think back to our benchmarks for effect sizes this represents a large effect (it is above .5, the threshold for a large effect). Therefore, as well as being statistically significant, this effect is large and so represents a substantive finding. For contrast 2 we get:

$$r_{\text{contrast2}} = \sqrt{\frac{2.029^2}{2.029^2 + 12}} = \sqrt{\frac{4.12}{16.12}} = .51$$

This too is a substantive finding and represents a large effect size.

8.6. REPORTING RESULTS FROM ONE-WAY INDEPENDENT ANOVA ②

When we report an ANOVA, we have to give details of the *F*-ratio and the degrees of freedom from which it was calculated. For the experimental effect in these data the *F*-ratio was derived from dividing the mean squares for the effect by the mean squares for the residual. Therefore, the degrees of freedom used to assess the *F*-ratio are the degrees of freedom for the effect of the model ($df_M = 2$) and the degrees of freedom for the residuals of the model ($df_R = 12$). Therefore, the correct way to report the main finding would be:

✓ There was a significant effect of Viagra on levels of libido, $F(2, 12) = 5.12$, $p < .05$, $\omega = .67$.

Notice that the value of the *F*-ratio is preceded by the values of the degrees of freedom for that effect. Also, we rarely state the exact significance value of the *F*-ratio: instead we report that the significance value, *p*, was less than the criterion value of .05 and include an effect size measure. The linear contrast can be reported in much the same way:

✓ There was a significant linear trend, $F(1, 12) = 9.97$, $p < .01$, $\omega = .80$, indicating that as the dose of Viagra increased, libido increased proportionately.

Notice that the degrees of freedom have changed to reflect how the *F*-ratio was calculated. I've also included an effect size measure (have a go at calculating this as we did for the main *F*-ratio and see if you get the same value). Also, we have now reported that the *F*-value was significant at a value less than the criterion value of .01. We can also report our planned contrasts:

✓ Planned contrasts revealed that having any dose of Viagra significantly increased libido compared to having a placebo, $t(12) = 2.47$, $p < .05$ (one-tailed), $r = .58$, and that having a high dose significantly increased libido compared to having a low dose, $t(12) = 2.03$, $p < .05$ (one-tailed), $r = .51$.

Note that in both cases I've stated that we used a one-tailed probability.

8.7. VIOLATIONS OF ASSUMPTIONS IN ONE-WAY INDEPENDENT ANOVA ②

I've mentioned several times in this chapter that ANOVA is robust to violations of its assumptions. We also saw that there are measures that can be taken when there is heterogeneity of variance (Box 8.3). However, there is another alternative. There are a group of tests (often called assumption-free, distribution-free and non-parametric tests, none of which are particularly accurate names!). Well, the one-way independent ANOVA also has a non-parametric counterpart called the Kruskal–Wallis test. If you have non-normally distributed data, or have violated some other assumption, then this test can be a useful way around the problem. This test is described in Chapter 13.

8.8. WHAT HAVE WE DISCOVERED ABOUT STATISTICS? ①

This chapter has introduced you to analysis of variance (ANOVA), which is the topic of the next few chapters also. One-way independent ANOVA is used in situations where you want to compare several means, and you've collected your data using different participants in each condition. I started off explaining that if we just do lots of *t*-tests on the same data then our Type I error rate becomes inflated. Hence, we use ANOVA instead. I looked at how ANOVA can be conceptualized as a general linear model (GLM) and so is in fact the same as multiple regression. Like multiple regression, there are three important measures that we use in ANOVA: the total sum of squares, SS_T (a measure of the variability in our data), the model sum of squares, SS_M (a measure of how much of that variability can be explained by our experimental manipulation), and SS_R (a measure of how much variability can't be explained by our experimental manipulation). We discovered that, crudely speaking, the *F*-ratio is just the ratio of variance that we can explain against the variance that we can't. We also discovered that a significant *F*-ratio tells us only that our groups differ, not how they differ. To find out where the difference lies we have two options: specify specific contrasts to test hypotheses (*planned contrasts*), or test every group against every other group (*post hoc tests*). The former are used when we have generated hypotheses before the experiment, whereas the latter are for exploring data when no hypotheses have been made. Finally we discovered how to implement these procedures on SPSS.

8.9. KEY TERMS THAT WE'VE DISCOVERED

- Analysis of variance (ANOVA)
- Bonferroni correction
- Brown–Forsythe *F*
- Cubic trend
- Degrees of freedom
- Deviation contrast
- Difference contrast (reverse Helmert contrast)
- Experimentwise error rate
- Familywise error rate
- Grand variance
- Harmonic mean

- Helmert contrast
- Independent ANOVA
- Omega squared
- Orthogonal
- Pairwise comparisons
- Planned contrasts
- Polynomial contrast
- *Post hoc* tests
- Quadratic trend
- Quartic trend
- Repeated contrast
- Simple contrast
- Harmonic mean
- Helmert contrast
- Weights
- Welch *F*

8.10. SMART ALEX'S TASKS

- **Task 1**: Imagine that I was interested in how different teaching methods affected students' knowledge. I noticed that some lecturers were aloof and arrogant in their teaching style and humiliated anyone who asked them a question, while others were encouraging and supporting of questions and comments. I took three statistics courses where I taught the same material. For one group of students I wandered around with a large cane and beat anyone who asked daft questions or got questions wrong (*punish*). In the second group I used my normal teaching style which is to encourage students to discuss things that they find difficult and to give anyone working hard a nice sweet (*reward*). The final group I remained indifferent to and neither punished nor rewarded their efforts (*indifferent*). As the dependent measure I took the students' exam marks (percentage). Based on theories of operant conditioning, we expect punishment to be a very unsuccessful way of reinforcing learning, but we expect reward to be very successful. Therefore, one prediction is that reward will produce the best learning. A second hypothesis is that punishment should actually retard learning such that it is worse than an indifferent approach to learning. The data are in the file **Teach.sav**. Carry out a one-way ANOVA and use planned comparisons to test the hypotheses that (1) reward results in better exam results than either punishment or indifference; and (2) indifference will lead to significantly better exam results than punishment.②
- **Task 2**: In Chapter 13 (section 13.4) there are some data looking at whether eating soya meals reduces your sperm count. Have a look at this section, access the data for that example, but analyse them with ANOVA. What's the difference between what you find and what is found in section 13.4? Why do you think this difference has arisen?②
- **Task 3**: Students (and lecturers for that matter) love their mobile phones, which is rather worrying given some recent controversy about links between mobile phone use and brain tumours. The basic idea is that mobile phones emit microwaves, and so holding one next to your brain for large parts of the day is a bit like sticking your brain in a microwave oven and selecting the 'cook until well done' button. If we wanted to test this experimentally, we could get six groups of people and strap a mobile phone to their heads (that they can't remove). Then, by remote control, we turn the phones on for a certain amount of time each day. After 6 months, we measure the size of any tumour (in mm^3) close to the site of the phone antenna (just behind the ear). The six groups experienced 0, 1, 2, 3,

4 or 5 hours per day of phone microwaves for 6 months. The data are in **Tumour.sav**. (From Field & Hole, 2003, so there is a very detailed answer in there).②

Answers are in the file on the CD-ROM called **Answers(Chapter 8).pdf**

8.11. FURTHER READING

Howell, D. C. (2002). *Statistical methods for psychology* (5th edition). Belmont, CA: Duxbury. Chapters 11 and 12 provide very detailed coverage of ANOVA and Chapter 16 covers the GLM approach.

Iversen, G. R. & Norpoth, H. (1987). *ANOVA* (2nd edition). Sage university paper series on quantitative applications in the social sciences, 07–001. Newbury Park, CA: Sage. Quite high level, but a good read for those with a mathematical brain.

Klockars, A. J. & Sax, G. (1986). *Multiple comparisons*. Sage university paper series on quantitative applications in the social sciences, 07–061. Newbury Park, CA: Sage. High level but thorough coverage of multiple comparisons—in my view this book is better than Toothaker for planned comparisons.

Rosenthal, R., Rosnow, R. L. & Rubin, D. B. (2000). *Contrasts and effect sizes in behavioural research: a correlational approach*. Cambridge: Cambridge University Press. Fantastic book on planned comparisons by three of the great writers on statistics.

Rosnow, R. L. & Rosenthal, R. (2005). *Beginning behavioural research: a conceptual primer* (5th edition). Englewood Cliffs, NJ: Pearson/Prentice Hall. Look, they wrote another great book!

Toothaker, L. E. (1993). *Multiple comparison procedures*. Sage university paper series on quantitative applications in the social sciences, 07–089. Newbury Park, CA: Sage. Also high level, but gives an excellent précis of *post hoc* procedures.

Wright, D. B. (2002). *First steps in statistics*. London: Sage. If this chapter is too complex then Chapter 7 of Wright is a very readable basic introduction to ANOVA.

CHAPTER 9

ANALYSIS OF COVARIANCE, ANCOVA (GLM 2)

9.1. WHAT WILL THIS CHAPTER TELL US? ②

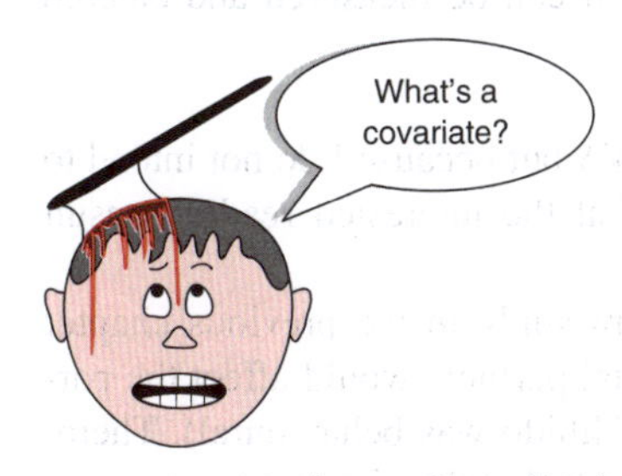

In the previous chapter we saw how one-way ANOVA could be characterized in terms of a multiple regression equation that used dummy variables to code group membership. In addition, in Chapter 5 we saw how multiple regression could incorporate several continuous predictor variables. It should, therefore, be no surprise that the regression equation for ANOVA can be extended to include one or more continuous variables that predict the outcome (or dependent variable). Continuous variables such as these, that are not part of the main experimental manipulation but have an influence on the dependent variable, are known as *covariates* and they can be included in an ANOVA analysis. When we measure covariates and include them in an ANOVA we call it analysis of covariance (or ANCOVA for short). This chapter focuses on this technique. We'll start by having a brief look at the theory of ANCOVA (although don't worry—we're not going into too much detail) and at the fact that there is an additional assumption we need to think about (*homogeneity of regression slopes*). We'll have a look at an example on SPSS and see how to interpret and report results using this technique.

9.2. WHAT IS ANCOVA? ②

In the last chapter, we used an example about looking at the effects of Viagra on libido. Let's think about other things than Viagra that might influence libido: we'll, the obvious one is the libido of the participant's sexual partner (after all 'it takes two to tango'!), but there are other things too such as other medication that suppresses libido (such as antidepressants or the contraceptive pill, both of which lower libido) and fatigue. If these variables (called covariates) are measured, then it is possible to control for the influence they have on the dependent variable by including them in the regression model. From what we know of hierarchical regression (see Chapter 5) it should be clear that if we enter the covariate into the regression model first, and then enter the dummy variables representing the experimental manipulation, we can see what effect an independent variable has *after* the effect of the covariate. As such, we control for (or partial out) the effect of the covariate. There are two reasons for including covariates in ANOVA:

- **To reduce within-group error variance**: In the discussion of ANOVA and *t*-tests we got used to the idea that we assess the effect of an experiment by comparing the amount of variability in the data that the experiment can explain against the variability that it cannot explain. If we can explain some of this 'unexplained' variance (SS_R) in terms of other variables (covariates), then we reduce the error variance, allowing us to assess more accurately the effect of the independent variable (SS_M).
- **Elimination of confounds**: In any experiment, there may be unmeasured variables that confound the results (i.e. variables that vary systematically with the experimental manipulation). If any variables are known to influence the dependent variable being measured, then ANCOVA is ideally suited to remove the bias of these variables. Once a possible confounding variable has been identified, it can be measured and entered into the analysis as a covariate.

There are other reasons for including covariates in ANOVA but because I do not intend to describe the computation of ANCOVA I recommend that the interested reader consult Wildt & Ahtola (1978) or Stevens (1992, Chapter 9).

Imagine that the researcher who conducted the Viagra study in the previous chapter suddenly realized that the libido of the participants' sexual partners would affect the participants' own libido (especially because the measure of libido was behavioural). Therefore, they repeated the study on a different set of participants, but this time took a measure of the partner's libido. The partner's libido was measured in terms of how often they tried to initiate sexual contact. In the previous chapter, we saw that this experimental scenario could be characterized in terms of equation (8.2). Think back to what we know about multiple regression (Chapter 5) and you can hopefully see that this equation can be extended to include this covariate (see equation (9.1)):

$$\text{Libido}_i = b_0 + b_3\text{Covariate}_i + b_2\text{High}_i + b_1\text{Low}_i + \varepsilon_i$$
$$\text{Libido}_i = b_0 + b_3\text{Partner's Libido}_i + b_2\text{High}_i + b_1\text{Low}_i + \varepsilon_i \tag{9.1}$$

9.3. CONDUCTING ANCOVA ON SPSS ②

The data for this example are in Table 9.1 and can be found in the file **ViagraCovariate. sav**. Table 9.1 shows the participant's libido, their partner's libido, and the means (and standard deviations in brackets) of the various scores. I recommend putting these data into the data

Table 9.1 Data from **ViagraCovariate.sav**

Dose	Participant's Libido	Partner's Libido
	3	5
	2	2
	5	6
	2	2
Placebo	2	3
	2	3
	7	7
	2	4
$\bar{X}$	4	5
S	3.22	4.11
	(1.79)	(1.76)
	7	10
	5	8
	3	6
Low Dose	4	7
	4	7
	7	11
	5	9
	4	7
$\bar{X}$	4.88	8.13
S	(1.46)	(1.73)
	9	2
	2	3
	6	5
	3	4
	4	3
	4	3
High Dose	4	2
	6	0
	4	1
	6	3

(Continued)

Table 9.1 (Continued)

	2		0	
	8		1	
	5		0	
	4.85		2.08	
	(2.12)		(1.61)	

editor by hand. This can be done in much the same way as the Viagra data from the previous chapter except that an extra variable must be created in which to place the values of the covariate.

9.3.1. Inputting data ①

In essence, the data should be laid out in the data editor as they are in Table 9.1 (excluding the rows for the means and standard deviations). So, create a coding variable called **dose** and use the *Labels* option to define value labels (as in Chapter 8 I recommend 1 = placebo, 2 = low dose, 3 = high dose). There were different numbers of participants in each condition, so you need to enter nine values of 1 into this column (so that the first nine rows contain the value 1), followed by eight rows containing the value 2, and followed by 14 rows containing the value of 3. At this point, you should have one column with 30 rows of data entered. Next, create a second variable called **libido** and enter the 30 scores that correspond to the person's libido. Finally, create a third variable called **partner** and use the *Labels* option to give this variable a more descriptive title of 'Partner's libido'. Then, enter the 30 scores that correspond to the partner's libido. Remember that the means (and standard deviations) that I've included in the table are not required by SPSS!

9.3.2. Main analysis ②

Most of the factorial ANOVA procedures in SPSS include the facility to include one or more covariates. However, for simpler designs (most designs that don't involve repeated measures) it is probably best to conduct ANCOVA via the general factorial procedure. To access the main dialog box follow the menu path **Analyze⇒General Linear Model ⇒Univariate...** (see Figure 9.1).[1] The main dialog box is similar to that for one-way ANOVA, except that there is a space to specify covariates. Select **libido** and place this in the box labelled *Dependent Variable* by clicking on ▶. Select **dose** and transfer it to the box labelled *Fixed Factor(s)* and then select **partner** and transfer it to the box labelled *Covariate(s)*.

1 **Statistics⇒General Linear Model⇒GLM—General Factorial...** in version 8.0 and earlier.

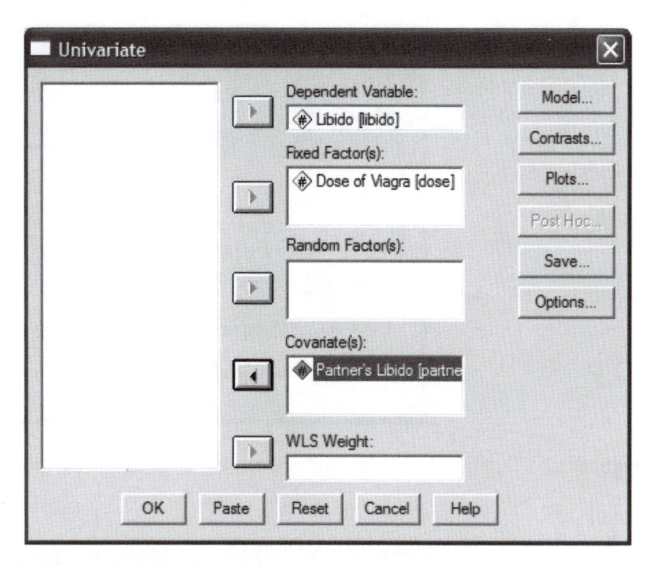

Figure 9.1 Main dialog box for GLM univariate

9.3.3. Contrasts and other options ②

There are various dialog boxes that can be accessed from the main dialog box. The first thing to notice is that if a covariate is selected, the *post hoc* tests are disabled (you cannot access this dialog box). *Post hoc* tests are not designed for situations in which a covariate is specified; however, some comparisons can still be done using contrasts.

Click on Contrasts... to access the *contrasts* dialog box. This dialog box is different to the one we met in Chapter 8 in that you cannot enter codes to specify particular contrasts. Instead, you can specify one of several standard contrasts. These standard contrasts were listed in Table 8.6. In this example, there was a placebo control condition (coded as the first group), so a sensible set of contrasts would be simple contrasts comparing each experimental group with the control. The default contrast in SPSS is a deviation contrast and to change this we must first click on ▾ next to the box labelled *Contrast*. A list of contrasts will drop down and you should select a type of contrast (in this case *Simple*) from this list and the list will automatically disappear. For simple contrasts you have the option of specifying a reference category (which is the category against which all other groups are compared). By default the reference category is the last category: because in this case the control group was the first category (assuming that you coded placebo as 1) we need to change this option by selecting ⊙ First. When you have selected a new contrast, you must click on Change to register this change. The final dialog box should look like Figure 9.2 Click on Continue to return to the main dialog box.

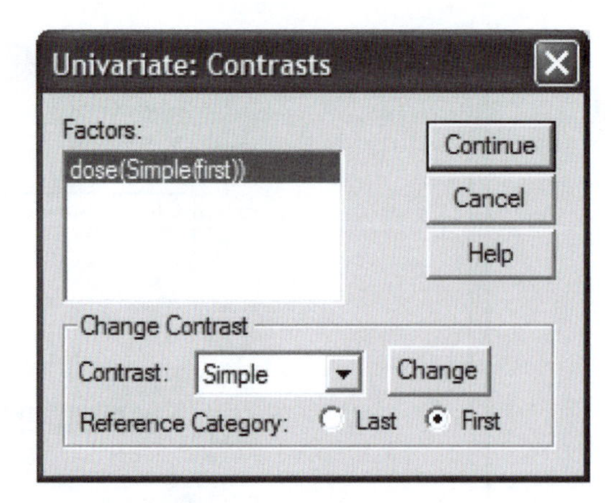

Figure 9.2 Options for standard contrasts in GLM univariate

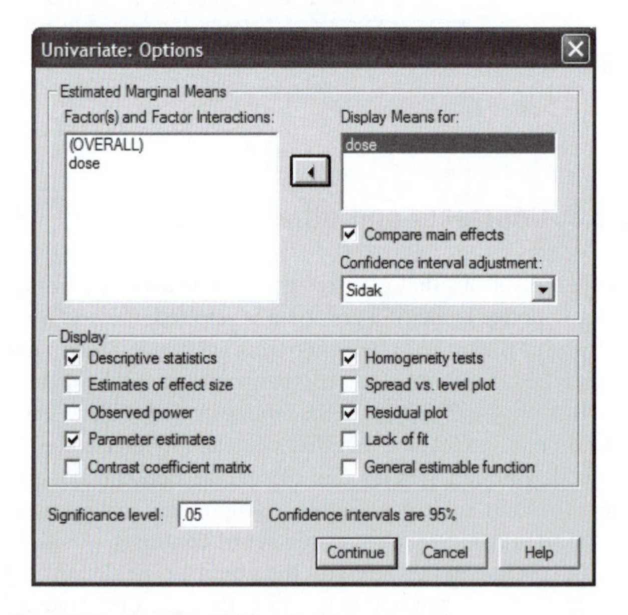

Figure 9.3 *Options* dialog box for GLM univariate

Another way to get *post hoc* tests is by clicking on Options... to access the *options* dialog box (see Figure 9.3). To specify *post hoc* tests, select the independent variable (in this case **dose**) from the box labelled *Estimated Marginal Means: Factor(s) and Factor Interactions* and transfer it to the box labelled *Display Means for* by clicking on ▶. Once a variable has been transferred, the box labelled *Compare main effects* becomes active and you should select this option (☑ Compare main effects). If this option is selected, the box labelled

Confidence interval adjustment becomes active and you can click on ▾ to see a choice of three adjustment levels. The default is to have no adjustment and simply perform a Tukey LSD *post hoc* test (this option is not recommended); the second is to ask for a Bonferroni correction (recommended); the final option is to have a Sidak correction. The Sidak correction is similar to the Bonferroni correction but is less conservative and so should be selected if you are concerned about the loss of power associated with Bonferroni corrected values. For this example use the Sidak correction (we will use Bonferroni later in the chapter). As well as producing *post hoc* tests for the **dose** variable, placing **dose** in the *Display Means for* box will result in a table of estimated marginal means for this variable. These means provide an estimate of the *adjusted* group means (i.e. the means after the covariate has been accounted for). When you have selected the options required (see Box 9.1), click on Continue to return to the main dialog box. There are other options available from the main dialog box. For example, if you have several independent variables you can plot them against each other (which is useful for interpreting interaction effects—see section 10.3.3). For this analysis, there is only one independent variable and so we can click on OK to run the analysis.

Box 9.1

Options for ANCOVA ②

The remaining options in this dialog box are as follows:

- **Descriptive statistics**: This option produces a table of means and standard deviations for each group.
- **Estimates of effect size**: This option produces the value of eta squared (η^2) described in section 8.5, which is a measure of the size of experimental effect. In fact, eta squared is the regression coefficient (R^2) for a non-linear regression line (i.e. a curve) assumed to pass through all group means. In a population this assumption is true, but in samples it is not: therefore, eta squared is usually biased (see section 8.5 and Howell, 2002, section 11.11). For this reason there is little to recommend this option, and the effect size may be estimated more productively using omega squared (ω^2)—see section 9.7.
- **Observed power**: This option provides an estimate of the probability that the statistical test could detect the difference between the observed group means (see section 1.8.5). This measure is of little use because if the *F*-test is significant then the probability that the effect was detected will, of course, be high. Likewise, if group differences were small, the observed power will be low. Observed power is of little use and I would advise that power calculations (with regard to sample size) are made before the experiment is conducted (see Cohen, 1988, 1992; Howell, 2002, for ideas on how to do this by hand; Field, 1998b, for ideas on doing it using a computer; or use the free software G*Power available from the CD-ROM of this book and http://www.psycho.uni-duesseldorf.de/aap/projects/gpower/).

(Continued)

Box 9.1 (Continued)

- **Parameter estimates**: This option produces a table of regression coefficients and their tests of significance for the variables in the regression model (see section 9.5).
- **Contrast coefficient matrix**: This option produces matrices of the coding values used for any contrasts in the analysis. This option is useful only for checking which groups are being compared in which contrast.
- **Homogeneity tests**: This option produces Levene's test of the homogeneity of variance assumption (see sections 3.6 and 8.4.1).
- **Spread vs. level plot**: This option produces a chart that plots the mean of each group of a factor (*X*-axis) against the standard deviation of that group (*Y*-axis). This is a useful plot to check that there is no relationship between the mean and standard deviation. If a relationship exists then the data may need to be stabilized using a logarithmic transformation (see Chapter 3).
- **Residual plot**: This option produces plots of observed-by-predicted-by-standardized residual values. These plots can be used to assess the assumption of equality of variance.

9.4. INTERPRETING THE OUTPUT FROM ANCOVA ②

9.4.1. Main analysis ②

SPSS Output 9.1 shows (for illustrative purposes) the ANOVA table for these data when the covariate is not included. It is clear from the significance value, which is greater than .05, that there are no differences in libido between the three groups; therefore Viagra seems to have no significant effect on libido. It should also be noted that the total amount of variation to be explained (SS_T) was 110.97 (*Corrected Total*), of which the experimental manipulation accounted for 16.84 units (SS_M), whilst 94.12 were unexplained (SS_R).

Tests of Between-Subjects Effects

Dependent Variable: Libido

Source	Type III Sum of Squares	df	Mean Square	F	Sig.
Corrected Model	16.844[a]	2	8.422	2.416	.108
Intercept	535.184	1	535.184	153.522	.000
DOSE	16.844	2	8.422	2.416	.108
Error	94.123	27	3.486		
Total	683.000	30			
Corrected Total	110.967	29			

a R Squared = .152 (Adjusted R Squared = .089)

SPSS Output 9.1

SPSS Output 9.2 shows the results of Levene's test and the ANOVA table when partner's libido is included in the model as a covariate. Levene's test is significant, indicating that the group variances are not equal (hence the assumption of homogeneity of variance has been violated). However, as I've mentioned in section 3.6, Levene's test is not necessarily the best way to judge whether variances are unequal enough to cause problems. A good double-check is to look at the highest and lowest variances. For our three groups we have standard deviations of 1.79 (placebo), 1.46 (low dose) and 2.12 (high dose)—see Table 9.1. If we square these values we get variances of 3.20 (placebo), 2.13 (low dose) and 4.49 (high dose). We then take the largest variance and divide it by the smallest: in this case 4.49/2.13 = 2.11. If the resulting value is less than 2 then we probably don't need to worry too much; if it's greater than 2 (as it is here) then we probably do! However, for the time being don't worry too much about the differences in variances.

The format of the ANOVA table is largely the same as without the covariate, except that there is an additional row of information about the covariate (**partner**). Looking first at the significance values, it is clear that the covariate significantly predicts the dependent variable, because the significance value is less than .05. Therefore, the person's libido is influenced by their partner's libido. What's more interesting is that when the effect of partner's libido is removed, the effect of Viagra becomes significant (*p* is .016 which is less than .05). The amount of variation accounted for by the model (SS_M) has increased to 34.75 units (*corrected model*) of which Viagra accounts for 28.34 units. Most important, the large amount of variation in libido that is accounted for by the covariate has meant that the unexplained variance (SS_R) has been reduced to 76.22 units. Notice that SS_T has not changed; all that has changed is how that total variation is explained.

This example illustrates how ANCOVA can help us to exert stricter experimental control by taking account of confounding variables to give us a 'purer' measure of effect of the experimental manipulation. Without taking account of the libido of the participants' partners we would have concluded that Viagra had no effect on libido, yet clearly it does. However, the effect of the partner's libido seems stronger than that of Viagra. Looking back at the group means from Table 9.1 for the libido data, it seems pretty clear that the significant ANOVA reflects a difference between the placebo group and the two experimental groups (because the low- and high-dose groups have very similar means—4.88 and 4.85—whereas the placebo group mean is much lower at 3.22). However, we'll need to check some contrasts to verify this.

SPSS Output 9.3 shows the parameter estimates selected in the *options* dialog box. These estimates are calculated using a regression analysis with **dose** split into two dummy coding variables (see section 8.2.2 and section 9.5). SPSS codes the two dummy variables such that the last category (the category coded with the highest value in the data editor—in this case the high dose group) is the reference category. This reference category (labelled DOSE=3 in the output) is coded with a 0 for both dummy variables (see section 8.2.2 for a reminder of how dummy coding works). DOSE=2, therefore, represents the difference between the group coded as 2 (low dose) and the reference category

Levene's Test of Equality of Error Variances[a]

Dependent Variable: Libido

F	df1	df2	Sig.
5.525	2	27	.010

Tests the null hypothesis that the error variance of the dependent variable is equal across groups.
a Design: Intercept+ PARTNER+DOSE

Tests of Between-Subjects Effects

Dependent Variable: Libido

Source	Type III Sum of Squares	df	Mean Square	F	Sig.
Corrected Model	34.750[a]	3	11.583	3.952	.019
Intercept	12.171	1	12.171	4.152	.052
PARTNER	17.906	1	17.906	6.109	.020
DOSE	28.337	2	14.169	4.833	.016
Error	76.216	26	2.931		
Total	683.000	30			
Correlated Total	110.967	29			

a R Squared = .313 (Adjusted R Squared = .234)

SPSS Output 9.2

(high dose), and DOSE=1 represents the difference between the group coded as 1 (placebo) and the reference category (high dose). The *b*-values literally represent the differences between the means of these groups and so the significances of the *t*-tests tell us whether the group means differ significantly. The degrees of freedom for these *t*-tests can be calculated as in normal regression (see section 5.2.4) as $N - p - 1$ in which N is the total sample size (in this case 30), and p is the number of predictors (in this case 3, the two dummy variables and the covariate—see equation (9.1)). For these data we get $df = 30 - 3 - 1 = 26$.

Therefore, from these estimates we could conclude that the high dose differs significantly from the placebo group (DOSE=1 in the table) but that the high dose group also differs significantly from the low dose groups (DOSE = 2 in the table). This last conclusion is slightly odd because it contradicts what we initially concluded from the ANOVA (remember the means of the low- and high-dose groups were virtually identical)—can you think why? (All will be revealed in due course!). The final thing to notice is the value of *b* for the covariate (0.483). This value tells us that, other things being equal, if a partner's libido increases by one unit, then the person's libido should increase by just under half a unit (although there is nothing to suggest a causal link between the two). The sign of this coefficient tells us the direction of the relationship between the covariate and the outcome. So, in this example, because the coefficient is positive it means that the partner's libido has

Parameter Estimates

Dependent Variable: Libido

Parameter	B	Std. Error	t	Sig.	95% Confidence Interval	
					Lower Bound	Upper Bound
Intercept	3.843	.625	6.150	.000	2.558	5.127
PARTNER	.483	.196	2.472	.020	.081	.885
[DOSE = 1]	−2.607	.842	−3.095	.005	−4.338	−.876
[DOSE = 2]	−2.894	1.411	−2.051	.050	−5.794	.006
[DOSE = 3]	0[a]	.	.	.	.	.

a This parameter is set to zero because it is redundant

SPSS Output 9.3

a positive relationship with the participant's libido: as one increases so does the other. A negative coefficient would mean the opposite: as one increases, the other decreases.

9.4.2. Contrasts ②

SPSS Output 9.4 shows the result of the contrast analysis specified in Figure 9.2 and compares level 2 (low dose) against level 1 (placebo) as a first comparison, and level 3 (high dose)

Contrast Results (K Matrix)

Dose of Viagra Simple Contrast[a]			Dependent Variable Libido
Level 2 vs. Level 1	Contrast Estimate		−.287
	Hypothesized Value		0
	Difference (Estimate—Hypothesized)		−.287
	Std. Error		1.144
	Sig.		**.804**
	95% Confidence Interval for Difference	Lower Bound	−2.638
		Upper Bound	2.064
Level 3 vs. Level 1	Contrast Estimate		2.607
	Hypothesized Value		0
	Difference (Estimate—Hypothesized)		2.607
	Std. Error		.842
	Sig.		**.005**
	95% Confidence Interval for Difference	Lower Bound	.876
		Upper Bound	4.338

a Reference category = 1

SPSS Output 9.4

Estimates

Dependent Variable: Libido

Dose of Viagra	Mean	Std. Error	95% Confidence Interval	
			Lower Bound	Upper Bound
Placebo	3.313[a]	.572	2.138	4.489
Low Dose	3.027[a]	.962	1.049	5.004
High Dose	5.920[a]	.644	4.597	7.244

a Covariates appearing in the model are evaluated at the following values: Partner's Libido = 4.30

SPSS Output 9.5

against level 1 (placebo) as a second comparison. These contrasts are consistent with what was specified: all groups are compared to the first group. The group differences are displayed: a difference value, standard error, significance value and 95% confidence interval. These results show that the low-dose group did not have a significantly different libido than the placebo group (contrast 1, $p = .804$), but that the high-dose group did differ significantly from the placebo group ($p = .005$). These results are consistent with the regression parameter estimates (in fact, note that contrast 2 is identical to the regression parameters for DOSE=1 in the previous section).

Again, this all seems very odd because, at face value, the significant effect of libido seemed to reflect a difference between the placebo group and the two Viagra groups (which have similar means), yet the contrasts so far contradict these conclusions. The reason for this inconsistency is that the initial conclusion was based on group means that had not been adjusted for the effect of the covariate. These values tell us nothing about the group differences reflected by the significant ANCOVA. SPSS Output 9.5 gives the adjusted values of the group means and it is these values that should be used for interpretation (this is the main reason for selecting the *Display Means for* option). The adjusted means show a very different pattern of responses: it looks as though the significant ANCOVA reflects a difference between the high-dose group and both the low-dose group and the placebo group. The low-dose and placebo groups appear to have fairly similar adjusted means indicating that not enough Viagra does not increase libido above normal levels—a high dose is required! These conclusions support what we know from the contrasts and regression parameters but can be verified with the *post hoc* tests specified in the options menu.

SPSS Output 9.6 shows the results of the Sidak corrected *post hoc* comparisons that were requested as part of the *options* dialog box. The significant difference between the high-dose and placebo groups remains, but interestingly the significant difference between the high- and low-dose groups shown by the regression parameters (SPSS Output 9.3) is gone (p is only .14). This contradiction might result from a loss of power in the *post hoc* tests (remember that planned comparisons have greater power to detect effects than *post hoc* procedures). However, there could be other reasons why these comparisons are non-significant and we should be very cautious in our interpretation of the significant ANCOVA and subsequent comparisons.

Pairwise Comparisons

Dependent Variable: Libido

(I) Dose of Viagra	(J) Dose of Viagra	Mean Difference (I–J)	Std. Error	Sig.[a]	95% Confidence Interval for Difference[a]	
					Lower Bound	Upper Bound
Placebo	Low Dose	.287	1.144	.992	−2.631	3.205
	High Dose	−2.607*	.842	.014	−4.756	−.458
Low Dose	Placebo	−.287	1.144	.992	−3.205	2.631
	High Dose	−2.894	1.411	.144	−6.493	.706
High Dose	Placebo	2.607*	.842	.014	.458	4.756
	Low Dose	2.894	1.411	.144	−.706	6.493

Based on estimated marginal means
*The mean difference is significant at the .05 level.
a Adjustment for multiple comparisons: Sidak

SPSS Output 9.6

9.4.3. Interpreting the covariate ②

I've already mentioned that the parameter estimates (SPSS Output 9.3) tell us how to interpret the covariate. If the *b*-value for the covariate is positive then it means that the covariate and the outcome variable have a positive relationship (as the covariate increases, so does the outcome). If the *b*-value is negative it means the opposite: that the covariate and the outcome variable have a negative relationship (as the covariate increases, the outcome decreases). For these data the *b*-value was positive, indicating that as the partner's libido increases, so does the participant's libido. Another way to discover the same thing is simply to draw a scatterplot of the covariate against the outcome. We came across scatterplots in section 4.4 so have a look back there to find out how to do one. Figure 9.4 shows the resulting scatterplot for these data and confirms what we already know: the effect of the covariate is that as the partner's libido increases, so does the participant's libido (as shown by the slope of the regression line).

9.5. ANCOVA RUN AS A MULTIPLE REGRESSION ②

Although the ANCOVA is essentially done, it may be of interest to rerun the analysis as a hierarchical multiple regression. As an exercise, enter these data and run the analysis yourself by adding two dummy variables to the file **ViagraCovariate.sav** that we've used in this chapter (see section 8.2.2 for help with the dummy coding). If you get stuck then I've included a completed data set called **ViagraCovariate Dummy.sav** on the CD-ROM. To run the analysis, we use the regression procedure (see Chapter 5) with **libido** as the

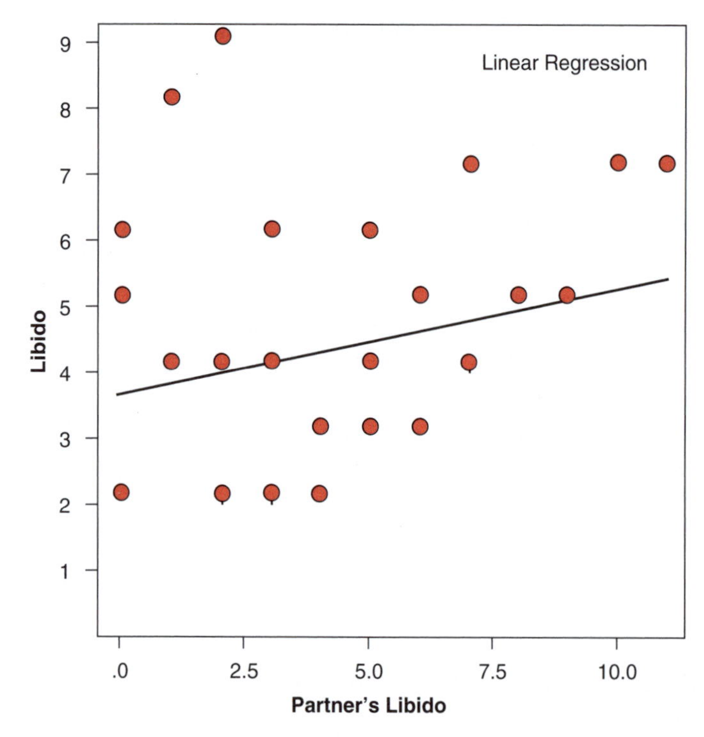

Figure 9.4 Scatterplot of libido against partner's libido

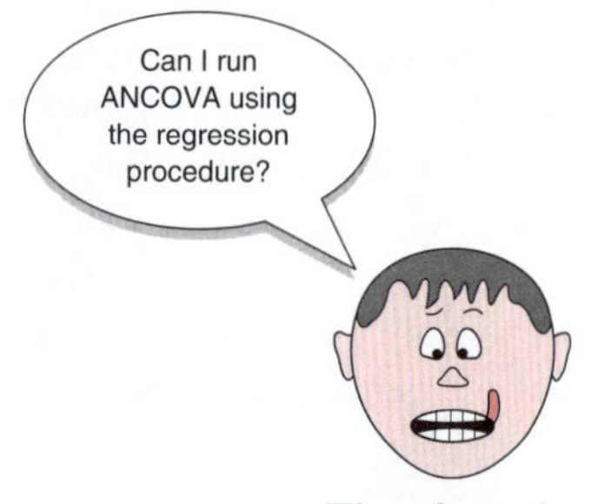

outcome (dependent variable in SPSS terminology) and then in the first block we enter partner's libido (**partner**) as a predictor, and then in a second block we enter both dummy variables. In all cases the entry method should be *Enter* (see section 5.7 for how to enter variables for a hierarchical regression but remember to use the *Enter* method for both blocks). The summary of the resulting regression model (SPSS Output 9.7) shows us the goodness-of-fit of the model first when only the covariate is used in the model, and second when both the covariate and the dummy variables are used.

Therefore, the difference between the values of R^2 (.313 − .058 = .255) represents the individual contribution of the dose of Viagra. Therefore, we can say that the dose of Viagra accounted for 25.5% of the variation in libido, whereas the partner's libido accounted for only 5.8%. This additional information provides some insight into the substantive importance of Viagra. The next table is the ANOVA table, which is again divided into two sections. The top half represents the effect of the covariate alone, whereas the bottom half represents the whole model (i.e. covariate and dose of Viagra included). Notice at the bottom of the ANOVA table (the bit for Model 2) that the entire model (partner's libido and the dummy variables) accounts for 34.57 units of variance (SS_M), there are 110.97 units in total (SS_T) and the unexplained variance (SS_R) is 76.22. The bottom half, therefore,

Model Summary

Model	R	R Square	Adjusted R Square	Std. Error of the Estimate
1	.240[a]	.058	.024	1.932
2	.560[b]	.313	.234	1.712

a Predictors: (Constant), Partner's Libido
b Predictors: (Constant), Partner's Libido, Dummy Variables 1 (Placebo vs. High), Dummy Variable 1
c (Placebo vs. Low)

ANOVA[c]

Model		Sum of Squares	df	Mean Square	F	Sig.
1	Regression	6.413	1	6.413	1.717	.201[a]
	Residual	104.554	28	3.734		
	Total	110.967	29			
2	Regression	34.750	3	11.583	3.952	.019[b]
	Residual	76.216	26	2.931		
	Total	110.967	29			

a Predictors: (Constant), Partner's Libido
b Predictors: (Constant), Partner's Libido, Dummy Variables 1 (Placebo vs. High), Dummy Variable 1 (Placebo vs. Low)
c Dependent Variable: Libido

SPSS Output 9.7

contains the same values as the ANCOVA summary table in SPSS Output 9.2. These values are the same as the row labelled 'corrected model' row of the ANCOVA summary table we encountered when run the analysis as ANCOVA (see SPSS Output 9.2).

SPSS Output 9.8 shows the remainder of the regression analysis. This table of regression coefficients is more interesting. Again, this table is split into two and so the bottom of the table looks at the whole model. When the dose of Viagra is considered with the covariate, the value of *b* for the covariate is .483, which corresponds to the value in the ANCOVA parameter estimates (SPSS Output 9.3). The *b*-values for the dummy variables represent the difference between the means of the low-dose group and the placebo group (**dummy1**) and the high-dose group and the placebo group (**dummy2**)—see section 8.2.2 for an explanation of why. The means of the low- and high-dose groups were 4.88 and 4.85 respectively, and the mean of the placebo group was 3.22. Therefore, the *b*-values for the two dummy variables should be roughly the same (4.88 – 3.22 = 1.66 for dummy 1 and 4.85 – 3.22 = 1.63 for dummy 2). The astute among you might notice from the SPSS output that, in fact, the *b*-values are not only very different from each other (which shouldn't be the case because the high- and low-dose group means are virtually the same) but are different from the values I've just calculated. So, does this mean I've been lying to you for the past 50 pages about what the beta values represent? Well, even I'm not that horrible:

the reason for this apparent anomaly is because the b-values in this regression represent the differences between the adjusted means, not the original means; that is, the difference between the mean of each group and the placebo when these means have been adjusted for the partner's libido. The adjusted values were given in SPSS Output 9.5 and from this table we can see that:

$$
\begin{aligned}
b_{\text{Dummy 1}} &= \overline{X}_{\text{Low(adjusted)}} - \overline{X}_{\text{Placebo(adjusted)}} = 3.027 - 3.313 = -0.286 \\
b_{\text{Dummy 2}} &= \overline{X}_{\text{High(adjusted)}} - \overline{X}_{\text{Placebo(adjusted)}} = 5.920 - 3.313 = -2.607
\end{aligned}
\tag{9.12}
$$

These are the values you can see in the SPSS table. The t-tests conducted on these values show that the significant ANCOVA reflected a significant difference between the high-dose and placebo groups.[2] There was no significant difference between the low-dose and placebo groups. You should also notice that the significances of the t-values are the same as we saw in the contrasts table of the original ANCOVA (see SPSS Output 9.4). As a final point, we obviously don't know whether there was a difference between the low-dose and high-dose groups: to find this out we would need to use different dummy coding (perhaps comparing the high and low to the placebo and then comparing high to low just like we used for the planned comparisons in Chapter 8—see Box 9.2).

Coefficients[a]

Model		Unstandardized Coefficients		Standardized Coefficients	t	Sig.
		B	Std. Error	Beta		
1	(Constant)	3.689	.626		5.894	.000
	Partner's Libido	.158	.120	.240	1.311	.201
2	(Constant)	1.236	.986		1.253	.221
	Partner's Libido	.483	.196	.737	2.472	.020
	Dummy Variable 1 (Placebo vs. Low)	-.287	1.144	-.066	-.251	.804
	Dummy Variable 2 (Placebo vs. High)	2.607	.842	.672	3.095	.005

a Dependent Variable: Libido

SPSS Output 9.8

2 As I mentioned earlier in this chapter the degrees of freedom for these t-tests are $N - p - 1$, as in any regression analysis (see section 5.2.4). N is the total sample size (in this case 30), and p is the number of predictors (in this case 3, the two dummy variables and the covariate—see equation (9.1)). For these data we get $df = 30 - 3 - 1 = 26$.

Box 9.2

Planned contrasts for ANCOVA ③

You may have noticed that although we can ask SPSS to do certain standard contrasts, there is no option for specifying planned contrasts as there was with one-way independent ANOVA (see section 8.3.1). However, these contrasts can be done if we run the ANCOVA through the regression menu. Imagine you chose some planned contrasts as in Chapter 8, in which the first contrast compared the placebo group with all doses of Viagra, and the second contrast then compared the high and low doses (see section 8.2.10). We saw in sections 8.2.10 and 8.3.1 that to do this in SPSS we had to enter certain numbers to code these contrasts. For the first contrast we discovered an appropriate set of codes would be –2 for the placebo group and then 1 for both the high- and low-dose groups. For the second contrast the codes would be 0 for the placebo group, –1 for the low-dose group and 1 for the high-dose group (see Table 8.4). If you want to do these contrasts for ANCOVA, then you enter these values as two dummy variables. So, taking the data in this example, we'd add a column called **Dummy1** and in that column we'd put the value –2 for every person who was in the placebo group, and the value 1 for all other participants. We'd then add a second column called **Dummy2**, in which we'd place a 0 for everyone in the placebo group, a –1 for everyone in the low-dose group and the value 1 for those in the high dose group. The completed data would be like the file **ViagraCovariateContrasts.sav** on the CD-ROM.

Run the analysis as described in section 9.5. The resulting output will begin with a model summary and ANOVA table that should be identical to those in SPSS Output 9.7 (because we've done the same thing as before, the only difference is how the model variance is subsequently broken down with the contrasts). The regression coefficients for the dummy variables will be different, though, because we've now specified different codes.

Coefficients[a]

Model		Unstandardized Coefficients		Standardized Coefficients	t	Sig.
		B	Std. Error	Beta		
1	(Constant)	3.689	.626		5.894	.000
	Partner's Libido	.158	.120	.240	1.311	.201
2	(Constant)	2.009	.986		2.038	.052
	Partner's Libido	.483	.196	.737	2.472	.020
	Dummy Variable 1 (Placebo vs. Low & High)	.387	.238	.276	1.623	.117
	Dummy Variable 2 (Low vs. High)	1.447	.705	.617	2.051	.050

a Dependent Variable: Libido

(Continued)

Box 9.2 (Continued)

The first dummy variable compares the placebo group with the low- and high-dose groups. As such, it compares the adjusted mean of the placebo group (3.313) with the average of the adjusted means for the low- and high-dose groups ((3.027 + 5.920)/2 = 4.474). The *b*-value for the first dummy variable should therefore be the difference between these values: 4.474 − 3.313 = 1.16. However, we also discovered in a rather complex and boring bit of section 8.2.10.2 that this value gets divided by the number of groups within the contrast (i.e. 3) and so will be 1.16/3 = .387 (as it is in the output). The associated *t*-statistic is not significant, indicating that the placebo group was not significantly different from the combined mean of the Viagra groups.

The second dummy variable compares the low- and high-dose groups, and so the *b*-value should be the difference between the adjusted means of these groups: 5.920 − 3.027 = 2.89. We again discovered in section 8.2.10.2 that this value also gets divided by the number of groups within the contrast (i.e. 2) and so will be 2.89/2 = 1.447 (as in the output). The associated *t*-statistic is significant (its significance is exactly .05), indicating that the high-dose group produced a significantly higher libido than the low-dose group after controlling for the effect of the partner's libido.

This illustrates how you can apply the principles from section 8.2.10 to ANCOVA: although SPSS doesn't provide an easy interface to do planned contrasts, they can be done if you use the regression menus rather than the ANCOVA ones!

9.6. ADDITIONAL ASSUMPTIONS IN ANCOVA ③

9.6.1. Homogeneity of regression slopes ③

When an ANCOVA is conducted we look at the overall relationship between the outcome (dependent variable) and the covariate: we fit a regression line to the entire data set, ignoring to which group a person belongs. In fitting this model we therefore assume that this overall relationship is true for all groups of participants. For example, if there's a positive relationship between the covariate and the outcome in one group, we assume that there is a positive relationship in all of the other groups too. If, however, the relationship between the outcome (dependent variable) and covariate differs across the groups then the overall regression model is inaccurate (it does not represent all of the groups). This assumption is very important and is called the assumption of homogeneity of regression slopes. The best way to think of this assumption is to imagine plotting a scatterplot for each experimental condition with the covariate on one axis and the outcome on the other. If you then calculated, and drew, the regression line for each of these scatterplots you should find that the regression lines look more or less the same (i.e. the values of b in each group should be equal).

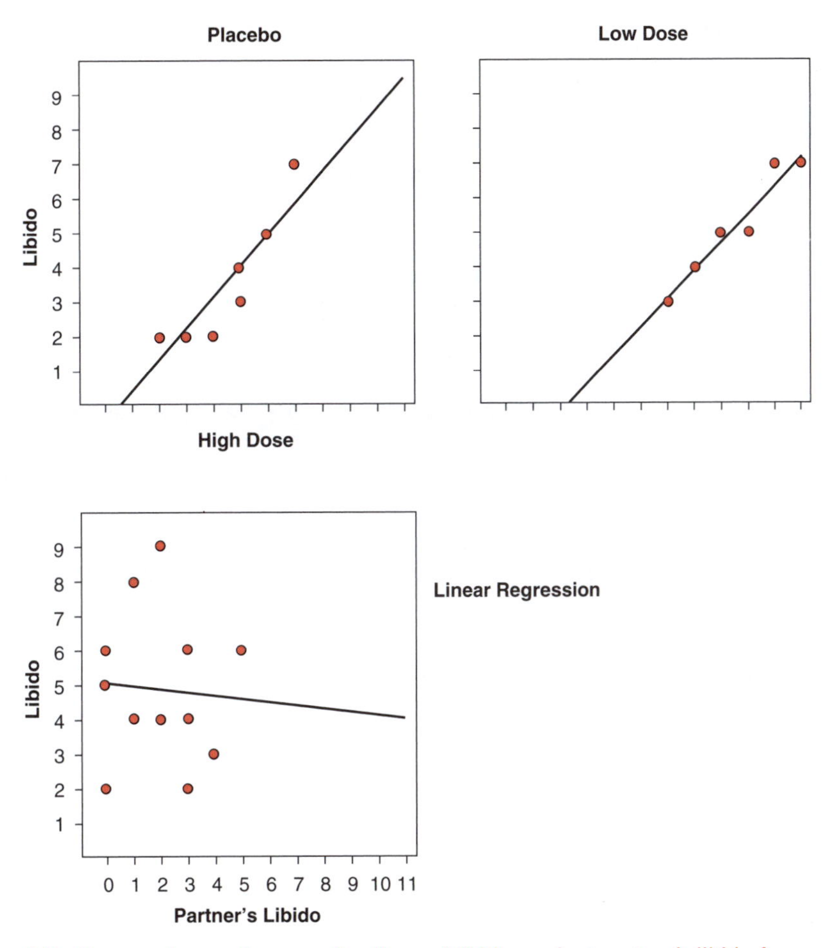

Figure 9.5 Scatterplots and regression lines of libido against partner's libido for each of the experimental conditions

Figure 9.5 shows scatterplots that display the relationship between partner's libido (the covariate) and the outcome (participant's libido) for each of the three experimental conditions. In each scatterplot a dot represents the data from a particular participant and the lines are the regression slopes for the particular group (i.e. they summarize the relationship between libido and partner's libido shown by the dots). It should be clear that there is a positive relationship (the regression line slopes upwards from left to right) between partner's libido and participant's libido in both the placebo and low-dose conditions. However, in the high-dose condition there appears to be no relationship at all between participant's libido and that of their partner (the dots are fairly randomly scattered and, in fact, the regression line slopes downwards from left to right, indicating a slightly negative relationship). This observation gives us cause to doubt whether there is homogeneity of regression slopes (because the relationship between participant's libido and that of their partner is not consistent across the three experimental groups).

9.6.2. Testing for homogeneity of regression slopes in SPSS ③

To test the assumption of homogeneity of regression slopes we need to rerun the ANCOVA but this time use a customized model. Access the main dialog box as before and place the variables in the same boxes as before (the finished box should look like Figure 9.1). To customize the model we need to access the *model* dialog box (Figure 9.6) by clicking on Model... .

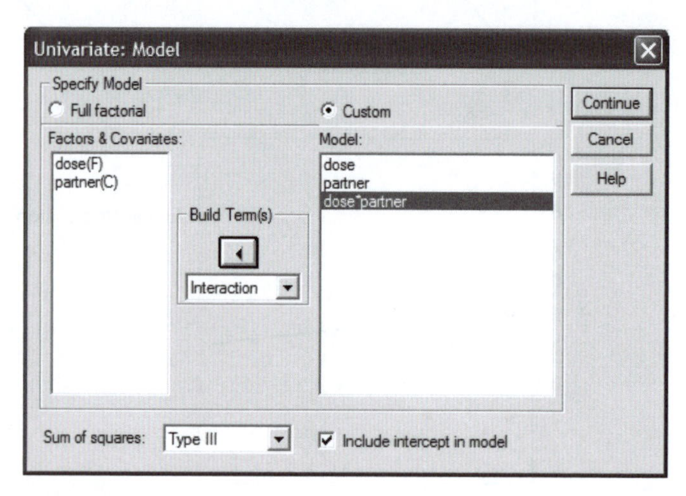

Figure 9.6 GLM univariate *model* dialog box

To customize your model, click on the circle labelled *Custom* to activate the dialog box (Figure 9.6). The variables specified in the main dialog box are listed on the left-hand side and are followed by a letter indicating the variable type (F = fixed factor, C = covariate). To test the assumption of homogeneity of regression slopes, we need to specify a model that includes the interaction between the covariate and dependent variable. Ordinarily, ANCOVA includes only the main effect of dose and partner and does not include this interaction term. To test this interaction term it's important still to include the main effects of dose and partner so that the interaction term is tested controlling for these main effects.

Hence, to begin with you should select **dose** and **partner** (you can select both of them at the same time). Then, where it says *Build Term(s)* there is a drop-down menu. Click on ▾ to access this drop-down menu and then click on *Main effects*. Having selected this, click on ▸ to move the main effects of **dose** and **partner** to the box labelled *Model*. Next we need to specify the interaction term. To do this, select **dose** and **partner** simultaneously and then click on ▾ to access the drop-down menu again, but this time select *Interaction*. Having selected this, click on ▸ to move the interaction of **dose** and **partner** to the box labelled *Model*. The

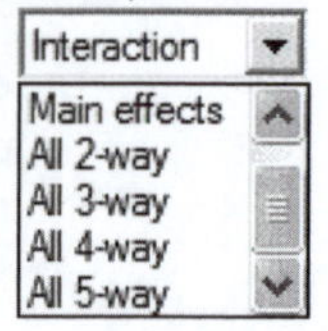

Tests of Between-Subjects Effects

Dependent Variable: Libido

Source	Type III Sum of Squares	df	Mean Square	F	Sig.
Corrected Model	52.000[a]	5	10.400	4.233	.007
Intercept	1.706	1	1.706	.694	.413
DOSE	40.068	2	20.034	8.154	.002
PARTNER	22.430	1	22.430	9.129	.006
DOSE * PARTNER	17.250	2	8.625	3.510	.046
Error	58.967	24	2.457		
Total	683.000	30			
Corrected Total	110.967	29			

a R Squared = .469 (Adjusted R Squared = .358)

SPSS Output 9.9

finished dialog box should look like Figure 9.6. You can find out more about specifying effects in section 10.3.2. Having specified the two main effects and the interaction term, click on Continue to return to the main dialog box and then click on OK to run the analysis.

SPSS Output 9.9 shows the main summary table for the ANCOVA including the interaction term. The effects of the dose of Viagra and the partner's libido are still significant, but the main thing in which we're interested is the interaction term, so look at the significance value of the covariate by outcome interaction (**dose*partner**): if this effect is significant then the assumption of homogeneity of regression slopes has been broken. The effect here is significant ($p < .05$); therefore the assumption is not tenable. Although this finding is not surprising given the pattern of relationships shown in Figure 9.5 it does raise concern about the main analysis, especially in the light of the contradictory findings of the multiple comparisons. This example illustrates why it is important to test assumptions and not just blindly to accept the results of an analysis.

Cramming Samantha's Tips

- Analysis of covariance (ANCOVA) compares several means, but controlling for the effect of one or more other variables (called covariates); for example, if you have several experimental conditions and want to control for the age of the participants.
- In the table labelled *Tests of between-subjects effects*, look at the column labelled *Sig.* for both the covariate and the independent variable (the grouping variable); if the value is less than .05 then for the covariate it means that this variable has a significant relationship to the outcome variable (the dependent variable); for the grouping variable it means that the means of the groups are significantly different after controlling for the effect that the covariate has on the outcome.

(Continued)

Cramming Samantha's Tips (Continued)

- As with ANOVA, if you have generated specific hypotheses before the experiment use *planned comparisons*, but if you don't have specific hypotheses use *post hoc* tests. Although SPSS will let you specify certain standard contrasts, other planned comparisons will have to be done by analysing the data using the regression procedure in SPSS.
- For contrasts and *post hoc* tests, again look to the columns labelled *Sig.* to discover if your comparisons are significant (they will be if the significance value is less than .05).
- Test the same assumptions as for ANOVA, but in addition you should test the assumption of *homogeneity of regression slopes*. This has to be done by customizing the ANCOVA model in SPSS.

9.7. CALCULATING THE EFFECT SIZE ②

We saw in Box 9.1 that we can get SPSS to produce eta squared (η^2), which is just r^2 calculated from the between-group effect, SS_M, divided by the total amount of variance in the data, SS_T. However, this measure of effect size is slightly biased and I recommended not using it. Therefore, as with ANOVA you're well advised to use omega squared (ω^2), which is a less biased version of eta squared (see section 8.5). However, as we saw in section 8.5 this can only be measured when you have equal numbers of participants in each group (and we don't have that here!). So, we're a bit stumped!

However, not all is lost because as I've said many times already, the overall effect size is not nearly as interesting as the effect size for more focused comparisons. These are easy to calculate because we selected regression parameters (see SPSS Output 9.3) and so we have *t*-statistics for the covariate and comparisons between the low- and high-dose groups and the placebo and high-dose groups. These *t*-statistics have $N - 2$ degrees of freedom (see Chapter 5), where N is the total sample size (in this case 30). We can use the same equation as in section 7.5.5:[3]

$$r_{\text{contrast}} = \sqrt{\frac{t^2}{t^2 + df}}$$

3 Strictly speaking, we have to use a slightly more elaborate procedure when groups are unequal. It's a bit beyond the scope of theis book but Rosnow, Rosenthal & Rubin (2000) give a very clear account.

Therefore we get (*t*s from SPSS output 9.3):

$$
\begin{aligned}
r_{\text{Covariate}} &= \sqrt{\frac{2.47^2}{2.47^2 + 28}} \\
&= \sqrt{\frac{6.10}{34.10}} \\
&= .42
\end{aligned}
$$

$$
\begin{aligned}
r_{\text{High Dose vs. Placebo}} &= \sqrt{\frac{-3.095^2}{-3.095^2 + 28}} \\
&= \sqrt{\frac{9.58}{37.57}} \\
&= .50
\end{aligned}
$$

$$
\begin{aligned}
r_{\text{High vs. Low Dose}} &= \sqrt{\frac{-2.051^2}{-2.051^2 + 28}} \\
&= \sqrt{\frac{4.21}{32.21}} \\
&= .36
\end{aligned}
$$

If you think back to our benchmarks for effect sizes these all represent medium to large effect sizes (they're all between .3 and .5). Therefore, as well as being statistically significant, these effects are substantive findings.

9.8. REPORTING RESULTS ②

Reporting ANCOVA is much the same as reporting an ANOVA except we now have to report the effect of the covariate as well. For the covariate and the experimental effect we give details of the *F*-ratio and the degrees of freedom from which it was calculated. In both cases, the *F*-ratio was derived from dividing the mean squares for the effect by the mean squares for the residual. Therefore, the degrees of freedom used to assess the *F*-ratio are the degrees of freedom for the effect of the model ($df_M = 1$ for the covariate and 2 for the experimental effect) and the degrees of freedom for the residuals of the model ($df_R = 26$ for both the covariate and the experimental effect)—see SPSS Output 9.2. Therefore, the correct way to report the main findings would be:

- The covariate, partner's libido, was significantly related to the participant's libido, $F(1, 26) = 6.11$, $p < .05$, $r = .42$. There was also significant effect of Viagra on levels of libido after controlling for the effect of partner's libido, $F(2, 26) = 4.83$, $p < .05$.

We can also report some contrasts:

- Planned contrasts revealed that having a high dose of Viagra significantly increased libido compared to having both a placebo, $t(26) = -3.10$, $p < .05$, $r = .50$, and a low dose, $t(26) = -2.05$, $p < .05$, $r = .36$.

9.9. WHAT HAVE WE DISCOVERED ABOUT STATISTICS? ②

This chapter has shown you how the general linear model (GLM) that was described in Chapter 8 can be extended to include additional variables. The advantages of doing so are that we can control for factors other than our experimental manipulation that might influence our outcome measure. This gives us tighter experimental control, and may also help us to explain some of our error variance, and, therefore, give us a purer measure of the experimental manipulation. We didn't go into too much theory about ANCOVA, just looked conceptually at how the regression model can be expanded to include these additional variables (*covariates*). Instead we jumped straight into an example, which was to look at the effect of Viagra on libido (as in Chapter 8) but including partner's libido as a covariate. I explained how to do the analysis on SPSS and interpret the results but also showed how the same output could be obtained by running the analysis as a regression. This was to try to get the message home that ANOVA and ANCOVA are merely forms of regression! Anyway, we finished off by looking at an additional assumption that has to be considered when doing ANCOVA: the assumption of homogeneity of regression slopes. This just means that the relationship between the covariate and the outcome variable should be the same in all of your experimental groups. We also had a look at how to test this assumption on SPSS. We'll now move on to look at situations in which you've got more than one experimental manipulation.

9.10. KEY TERMS THAT WE'VE DISCOVERED

- Adjusted means
- Analysis of covariance (ANCOVA)
- Homogeneity of regression slopes
- Partial out
- Sidak correction
- Covariates

9.11. SMART ALEX'S TASKS

- **Task 1**: Stalking is a very disruptive and upsetting (for the person being stalked) experience in which someone (the stalker) constantly harasses or obsesses about another person. It can take many forms, from sending intensely disturbing letters

threatening to boil your cat if you don't reciprocate the stalker's undeniable love for you, to literally following you around your local area in a desperate attempt to see which CD you buy on a Saturday (as if it would be anything other than Fugazi!). A psychologist, who'd had enough of being stalked by people, decided to try two different therapies on different groups of stalkers (25 stalkers in each group—this variable is called **Group**). The first group of stalkers he gave what he termed cruel-to-be-kind therapy. This therapy was based on punishment for stalking behaviours; in short every time the stalker followed him around, or sent him a letter, the psychologist attacked them with a cattle prod until they stopped their stalking behaviour. It was hoped that the stalkers would learn an aversive reaction to anything resembling stalking. The second therapy was psychodyshamic therapy, which was a recent development on Freud's psychodynamic therapy that acknowledges what a sham this kind of treatment is (so, you could say it's based on Fraudian theory!). The stalkers were hypnotized and regressed into their childhood, the therapist would also discuss their penis (unless it was a woman in which case they discussed their lack of penis), the penis of their father, their dog's penis, the penis of the cat down the road, and anyone else's penis that sprang to mind. At the end of therapy, the psychologist measured the number of hours in the week that the stalker spent stalking their prey (this variable is called **stalk2**). Now, the psychologist believed that the success of therapy might well depend on how bad the problem was to begin with, so before therapy he measured the number of hours that the patient spent stalking as an indicator of how much of a stalker the person was (this variable is called **stalk1**). The data are in the file **Stalker.sav.** Analyse the effect of therapy on stalking behaviour after therapy, controlling for the amount of stalking behaviour before therapy.②

- **Task 2**: A marketing manager for a certain well-known drinks manufacturer was interested in the therapeutic benefit of certain soft drinks for curing hangovers. He took 15 people out on the town one night and got them drunk. The next morning as they awoke, dehydrated and feeling as though they'd licked a camel's sandy feet clean with their tongue, he gave five of them water to drink, five of them Lucozade (in case this isn't sold outside of the UK, it's a very nice glucose-based drink), and the remaining five a leading brand of cola (this variable is called **drink**). He then measured how well they felt (on a scale from 0 = I feel like death to 10 = I feel really full of beans and healthy) 2 hours later (this variable is called **well**). He wanted to know which drink produced the greatest level of wellness. However, he realized it was important to control for how drunk the person got the night before, and so he measured this on a scale of 0 = as sober as a nun to 10 = flapping about like a haddock out of water on the floor in a puddle of their own vomit. The data are in the file **HangoverCure.sav**. Conduct an ANCOVA to see whether people felt better after different drinks when controlling for how drunk they were the night before.②

The answers are in the file **Answers(Chapter9).pdf** and task 1 has a full interpretation in Field & Hole (2003).

9.12. FURTHER READING

Howell, D. C. (2002). *Statistical methods for psychology* (5th edition). Belmont, CA: Duxbury. Chapter 16.

Rutherford, A. (2000). *Introducing ANOVA and ANCOVA: A GLM Approach*. London: Sage.

Wildt, A. R. & Ahtola, O. (1978). *Analysis of covariance*. Sage university paper series on quantitative applications in the social sciences, 07–012. Newbury Park, CA: Sage. This text is pretty high level but very comprehensive if you want to know the maths behind ANCOVA.

CHAPTER 10

FACTORIAL ANOVA (GLM 3)

10.1. WHAT WILL THIS CHAPTER TELL US? ②

In the preceding two chapters we have looked at situations in which we've tried to test for differences between groups when there has been a single independent variable (i.e. one variable has been manipulated). However, at the beginning of Chapter 8 I said that one of the advantages of ANOVA was that we could look at the effects of more than one independent variable (and how these variables interact). This chapter extends what we already know about ANOVA to look at situations where there are two independent variables. We've already seen in the previous section that it's very easy to incorporate a second variable into the ANOVA framework when that variable is a continuous variable (i.e. not split into groups), but now we'll move on to situations where there is a second independent variable that has been systematically manipulated by assigning people to different conditions.

10.2. THEORY OF FACTORIAL ANOVA (BETWEEN GROUPS) ②

10.2.1. Factorial designs ②

Factorial ANOVA is used when you have two or more independent variables (these variables are sometimes called **factors**, hence the name *factorial* ANOVA). There are several types of factorial design:

- Independent factorial design: This type of experiment is where there are several independent variables or predictors and each has been measured using different participants (between groups).
- Related factorial design: An experiment in which several independent variables or predictors have been measured, but the same participants have been used in all conditions (repeated measures).

- Mixed design: A design in which several independent variables (or predictors) have been measured; some have been measured with different participants whereas others used the same participants.

This section extends the one-way ANOVA model to the factorial case and the subsequent chapters look at repeated-measures designs, factorial repeated-measures designs and finally mixed designs.

Box 10.1

Naming ANOVAs ②

ANOVAs can be quite confusing because there are lots of them. You'll quite often come across things like 'a two-way independent ANOVA was conducted', or 'a two-way repeated-measures ANOVA was conducted' in journals. These statements may look confusing but they're quite easy if you break them down. All ANOVAs have two things in common: they involve some quantity of independent variables and these variables can be measured using either the same or different participants. If the same participants were used we typically use the phrase *repeated measures* and if different participants were used we use the word *independent*. When there are two or more independent variables, it's possible that some variables use the same participants whereas others use different participants. In this case we use the word *mixed*. When we name an ANOVA, we are simply telling the reader how many independent variables we used and how they were measured. In general terms we could write the name of an ANOVA as:

- A (*number of independent variables*) way *how these variables were measured* ANOVA

By remembering this you can understand the name of any ANOVA you come across. Look at these examples and try to work out how many variables were used and how they were measured:

- One-way independent ANOVA
- Two-way repeated-measures ANOVA
- Two-way mixed ANOVA
- Three-way independent ANOVA

The answers you should get are:

- One independent variable measured using different participants
- Two independent variables both measured using the same participants
- Two independent variables: one measured using different participants and the other measured using the same participants
- Three independent variables all of which are measured using different participants

Table 10.1 Data for the beer–goggles effect

Alcohol	None		2 Pints		4 Pints	
Gender	Female	Male	Female	Male	Female	Male
	65	50	70	45	55	30
	70	55	65	60	65	30
	60	80	60	85	70	30
	60	65	70	65	55	55
	60	70	65	70	55	35
	55	75	60	70	60	20
	60	75	60	80	50	45
	55	65	50	60	50	40
Total	**485**	**535**	**500**	**535**	**460**	**285**
Mean	**60.625**	**66.875**	**62.50**	**66.875**	**57.50**	**35.625**
Variance	**24.55**	**106.70**	**42.86**	**156.70**	**50.00**	**117.41**

10.2.2. An example with two independent variables ②

Throughout this chapter we'll use an example that has two independent variables. This is known as a two-way ANOVA. You'll often see ANOVAs referred to as one-way, two-way, three-way and so on, and the number before the word 'way' tells you how many independent variables there were: one-way has one independent variable, three-way has three (see Box 10.1). Anyway, we'll stick with two. An anthropologist was interested in the effects of alcohol on mate selection at night-clubs. Her rationale was that after alcohol had been consumed, subjective perceptions of physical attractiveness would become more inaccurate (the well-known 'beer–goggles effect'). She was also interested in whether this effect was different for men and women. She picked 48 students: 24 male and 24 female. She then took groups of eight participants to a night-club and gave them no alcohol (participants received placebo drinks of alcohol-free lager), 2 pints of strong lager, or 4 pints of strong lager. At the end of the evening she took a photograph of the person that the participant was chatting up. She then got a pool of independent judges to assess the attractiveness of the person in each photograph (out of 100). The data are in Table 10.1 and **goggles.sav**.

10.2.3. Total sum of squares (SS_T) ②

Two-way ANOVA is conceptually very similar to one-way ANOVA. Basically, we still find the total sum of squared errors (SS_T) and break this variance down into variance that can be explained by the experiment (SS_M) and variance that cannot be explained (SS_R). However, in two-way ANOVA, the variance explained by the model is made up of not one experimental manipulation but two. Therefore, we break the model sum of squares down into variance explained by the first independent variable (SS_A), variance explained by the second independent variable (SS_B) and variance explained by the interaction of these two variables ($SS_{A \times B}$).

Basically we start off in the same way as we did for a one-way ANOVA. That is, we calculate how much variability there is between scores when we ignore the experimental condition to which they belong. Remember from one-way ANOVA (equation (8.4)) that SS_T is calculated using the following equation:

$$SS_T = s^2_{grand}(N - 1)$$

The grand variance is simply the variance of all scores when we ignore the group to which they belong. So if we treated the data as one big group it would look as follows:

65	50	70	45	55	30
70	55	65	60	65	30
60	80	60	85	70	30
60	65	70	65	55	55
60	70	65	70	55	35
55	75	60	70	60	20
60	75	60	80	50	45
55	65	50	60	50	40

Grand Mean = 58.33

If we calculate the variance of all of these scores, we get 190.78 (try this on your calculators if you don't trust me). We used 48 scores to generate this value, and so N is 48. As such the equation becomes:

$$\begin{aligned} SS_T &= s^2_{grand}(N - 1) \\ &= 190.78(48 - 1) \\ &= 8966.66 \end{aligned}$$

The degrees of freedom for this SS will be $N - 1$, or 47.

10.2.4. The model sum of squares (SS_M) ②

The next step is to work out the model sum of squares. As I suggested earlier, this sum of squares is then further broken into three components: variance explained by the first independent variable (SS_A), variance explained by the second independent variable (SS_B) and variance explained by the interaction of these two variables ($SS_{A \times B}$).

Before we break down the model sum of squares into its component parts, we must first calculate its value. We know we have 8966.66 units of variance to be explained and our first step is to calculate how much of that variance is explained by our experimental manipulations overall (ignoring which of the two independent variables is responsible). When we did one-way ANOVA we worked out the model sum of squares by looking at the difference between each group mean and the overall mean (see section 8.2.5). We can do the same here. We effectively have six experimental groups if we combine all levels of the two independent variables (three doses for the male participants and three doses for the females). So, given that we have six groups of different people we can then apply the equation for the model sum of squares that we used for one-way ANOVA (equation (8.5)):

$$\mathrm{SS_M} = \sum n_k(\bar{x}_k - \bar{x}_{\text{grand}})^2$$

The grand mean is the mean of all scores (we calculated this above as 58.33), and n is the number of scores in each group (i.e. the number of participants in each of the six experimental groups; eight in this case). Therefore, the equation becomes:

$$\begin{aligned}\mathrm{SS_M} &= 8(60.625 - 58.33)^2 + 8(66.875 - 58.33)^2 + 8(62.5 - 58.33)^2 \\ &\quad + 8(66.875 - 58.33)^2 + 8(57.5 - 58.33)^2 + 8(35.625 - 58.33)^2 \\ &= 8(2.295)^2 + 8(8.545)^2 + 8(4.17)^2 + 8(8.545)^2 + 8(-0.83)^2 + 8(-22.705)^2 \\ &= 42.1362 + 584.1362 + 139.1112 + 584.1362 + 5.5112 + 4124.1362 \\ &= 5479.167\end{aligned}$$

The degrees of freedom for this SS will be the number of groups used, k, minus 1. We used six groups and so $df = 5$.

At this stage we know that the model (our experimental manipulations) can explain 5479.167 units of variance out of the total of 8966.66 units. The next stage is to break down this model sum of squares further to see how much variance is explained by our independent variables separately.

10.2.4.1. The main effect of gender ($\mathrm{SS_A}$) ②

To work out the variance accounted for by the first independent variable (in this case Gender) we need to group the scores in the data set according to which gender they belong. So, basically we ignore the amount of drink that has been drunk, and we just place all of the male scores into one group and all of the female scores into another. So, the data will look like this (note that the first box contains the three female columns from our original table and the second box contains the male columns):

A_1: Female		
65	70	55
70	65	65
60	60	70
60	70	55
60	65	55
55	60	60
60	60	50
55	50	50

Mean Female = 60.21

A_2: Male		
50	45	30
55	60	30
80	85	30
65	65	55
70	70	35
75	70	20
75	80	45
65	60	40

Mean Male = 56.46

We can then apply the equation for the model sum of squares that we used to calculate the overall model sum of squares:

$$SS_A = \sum n_k(\bar{x}_k - \bar{x}_{\text{grand}})^2$$

The grand mean is the mean of all scores (above), and n is the number of scores in each group (i.e. the number of males and females; 24 in this case). Therefore, the equation becomes:

$$\begin{aligned} SS_{\text{Gender}} &= 24(60.21 - 58.33)^2 + 24(56.46 - 58.33)^2 \\ &= 24(1.88)^2 + 24(-1.87)^2 \\ &= 84.8256 + 83.9256 \\ &= 168.75 \end{aligned}$$

The degrees of freedom for this SS will be the number of groups used, k, minus 1. We used two groups (males and females) and so $df = 1$.

10.2.4.2. The main effect of alcohol (SS_B) ②

To work out the variance accounted for by the second independent variable (in this case Alcohol) we need to group the scores in the data set according to how much alcohol was consumed. So, basically we ignore the gender of the participant, and we just place all of the scores after no drinks in one group, the scores after 2 pints in another group, and the scores after 4 pints in a third group. So, the data will look like this:

B_1: None		B_2: 2 Pints		B_3: 4 Pints	
65	50	70	45	55	30
70	55	65	60	65	30
60	80	60	85	70	30
60	65	70	65	55	55
60	70	65	70	55	35
55	75	60	70	60	20
60	75	60	80	50	45
55	65	50	60	50	40
Mean None = 63.75		**Mean 2 Pints = 64.6875**		**Mean 2 Pints = 46.5625**	

We can then apply the same equation for the model sum of squares that we used for the overall model sum of squares and for the main effect of gender:

$$SS_B = \sum n_k(\bar{x}_k - \bar{x}_{\text{grand}})^2$$

The grand mean is the mean of all scores (58.33 as before), and n is the number of scores in each group (i.e. the number of scores in each of the boxes above, in this case 16). Therefore, the equation becomes:

$$\begin{aligned}SS_{\text{alcohol}} &= 16(63.75 - 58.33)^2 + 16(64.6875 - 58.33)^2 + 16(46.5625 - 58.33)^2 \\ &= 16(5.42)^2 + 16(6.3575)^2 + 16(-11.7675)^2 \\ &= 470.0224 + 646.6849 + 2215.5849 \\ &= 3332.292\end{aligned}$$

The degrees of freedom for this SS will be the number of groups used minus 1 (see section 8.2.5 on one-way ANOVA). We used three groups and so $df = 2$.

10.2.4.3. The interaction effect ($SS_{A \times B}$) ②

The final stage is to calculate how much variance is explained by the interaction of the two variables. The simplest way to do this is to remember that the SS_M is made up of three components (SS_A, SS_B and $SS_{A \times B}$); therefore, given that we know SS_A and SS_B we can calculate the interaction term using subtraction:

$$SS_{A\times B} = SS_M - SS_A - SS_B$$

Therefore, for these data, the value is:

$$\begin{aligned}SS_{A\times B} &= SS_M - SS_A - SS_B \\ &= 5479.167 - 168.75 - 3332.292 \\ &= 1978.125\end{aligned}$$

The degrees of freedom can be calculated in the same way, but are also the product of the degrees of freedom for the main effects (either method works):

$$\begin{aligned}df_{A\times B} &= df_M - df_A - df_B \\ &= 5 - 1 - 2 \\ &= 2\end{aligned} \qquad \begin{aligned}df_{A\times B} &= df_A \times df_B \\ &= 1 \times 2 \\ &= 2\end{aligned}$$

10.2.5. The residual sum of squares (SS_R) ②

The residual sum of squares is calculated in the same way as for one-way ANOVA (see section 8.2.6) and again represents individual differences in performance or the variance that can't be explained by factors that were systematically manipulated. We saw in one-way ANOVA that the value is calculated by taking the squared error between each data point and its corresponding group mean. An alternative way to express this was as (see equation (8.7)):

$$SS_R = s^2_{\text{group1}}(n_1 - 1) + s^2_{\text{group2}}(n_2 - 1) + s^2_{\text{group3}}(n_3 - 1) + \cdots + s^2_{\text{group}\,n}(n_n - 1)$$

So, we use the individual variances of each group and multiply them by one less than the number of people within the group (n). We have the individual group variances in our original table of data (Table 10.1) and there were eight people in each group (therefore $n = 8$) and so the equation becomes:

$$\begin{aligned} SS_R &= s^2_{\text{group1}}(n_1 - 1) + s^2_{\text{group2}}(n_2 - 1) + s^2_{\text{group3}}(n_3 - 1) + s^2_{\text{group4}}(n_4 - 1) \\ &\quad + s^2_{\text{group5}}(n_5 - 1) + s^2_{\text{group6}}(n_6 - 1) \\ &= (24.55)(8 - 1) + (106.7)(8 - 1) + (42.86)(8 - 1) + (156.7)(8 - 1) \\ &\quad + (50)(8 - 1) + (117.41)(8 - 1) \\ &= (24.55 \times 7) + (106.7 \times 7) + (42.86 \times 7) + (156.7 \times 7) + (50 \times 7) \\ &\quad + (117.41 \times 7) \\ &= 171.85 + 746.9 + 300 + 1096.9 + 350 + 821.87 \\ &= 3487.52 \end{aligned}$$

The degrees of freedom for each group will be one less than the number of scores per group (i.e. 7). Therefore, if we add the degrees of freedom for each group, we get a total of $6 \times 7 = 42$.

10.2.6. The *F*-ratios ②

Each effect in a two-way ANOVA (the two main effects and the interaction) has its own F-ratio. To calculate these we have first to calculate the mean squares for each effect by taking the sum of squares and dividing by the respective degrees of freedom (think back to section 8.2.7). We also need a mean squares for the residual term. So, for this example we'd have four mean squares calculated as follows:

$$\begin{aligned} MS_A &= \frac{SS_A}{df_A} = \frac{168.75}{1} = 168.75 \\ MS_B &= \frac{SS_B}{df_B} = \frac{3332.292}{2} = 1666.146 \\ MS_{A\times B} &= \frac{SS_{A\times B}}{df_{A\times B}} = \frac{1978.125}{2} = 989.062 \\ MS_R &= \frac{SS_R}{df_R} = \frac{3487.52}{42} = 83.036 \end{aligned}$$

The F-ratios for the two independent variables and their interactions are then calculated by dividing their mean squares by the residual mean squares. Again, if you think back to one-way ANOVA this is exactly the same process!

$$F_A = \frac{MS_A}{MS_R} = \frac{168.75}{83.036} = 2.032$$
$$F_B = \frac{MS_B}{MS_R} = \frac{1666.146}{83.036} = 20.065$$
$$F_{A \times B} = \frac{MS_{A \times B}}{MS_R} = \frac{989.062}{83.036} = 11.911$$

Each of these F-ratios can be compared against critical values (based on their degrees of freedom, which can be different for each effect) to tell us whether these effects are likely to reflect data that have arisen by chance, or reflect an effect of our experimental manipulations (these critical values can be found in Appendix A.3). If an observed F exceeds the corresponding critical values then it is significant. SPSS will calculate these F-ratios and exact significance for each, but what I hope to have shown you in this section is that two-way ANOVA is basically the same as one-way ANOVA except that the model sum of squares is partitioned into three parts: the effect of each of the independent variables, and the effect of how these variables interact.

10.3. FACTORIAL ANOVA USING SPSS ②

10.3.1. Entering the data and accessing the main dialog box ②

To enter these data into the SPSS data editor, remember this golden rule: **levels of a between-group variable go in a single column**. Applying this rule to these data, we need to create two different coding variables in the data editor. These columns will represent gender and alcohol consumption. So, create a variable called **gender** on the data editor and activate the *labels* dialog box. You should define value labels to represent the two genders. We have had a lot of experience with coding values, so you should be fairly happy about assigning numerical codes to different groups. I recommend using the codes male = 0 and female = 1. Once you have done this, you can enter a code of 0 or 1 in this column indicating to which group the person belonged. Create a second variable called **alcohol** and assign group codes by using the *labels* dialog box. I suggest that you code this variable with three values: placebo (no Alcohol) = 1, 2 pints = 2 and 4 pints = 3. You can now enter 1, 2 or 3 into this column to represent the amount of alcohol consumed by the participant. Remember that if you turn the value labels option on you will see text in the data editor rather than the numerical codes. Now, the way this coding works is as follows:

Gender	Alcohol	Participant was
0	1	Male who consumed no alcohol
0	2	Male who consumed 2 pints
0	3	Male who consumed 4 pints
1	1	Female who consumed no alcohol
1	2	Female who consumed 2 pints
1	3	Female who consumed 4 pints

Once you have created the two coding variables, you can create a third variable in which to place the values of the dependent variable. Call this variable **attract** and use the *labels* option to give it the fuller name of 'attractiveness of date'. In this example, there are two independent variables and different participants were used in each condition: hence, we can use the general factorial ANOVA procedure in SPSS. This procedure is designed for analysing between-group factorial designs.

To access the main dialog box for a general factorial ANOVA use the file path **Analyze⇒General Linear Model⇒Univariate…**. The resulting dialog box is shown in Figure 10.1. First, select the dependent variable **attract** from the variables list on the left-hand side of the dialog box and transfer it to the space labelled *Dependent Variable* by clicking on ▶. In the space labelled *Fixed Factor(s)* we need to place any independent variables relevant to the analysis. Select **alcohol** and **gender** in the variables list (these

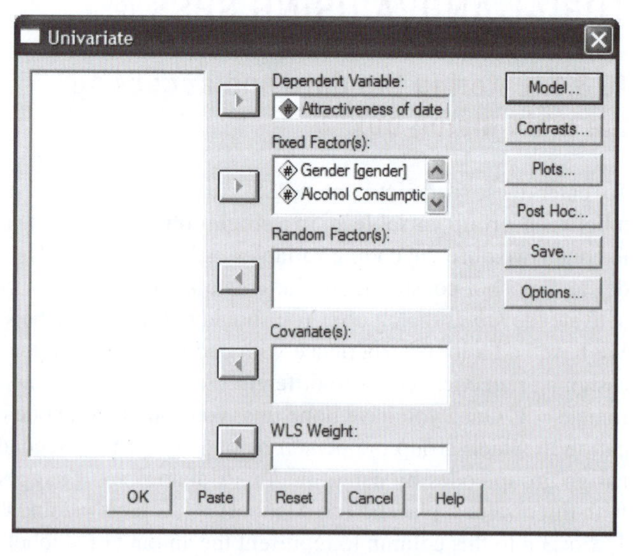

Figure 10.1 Main dialog box for univariate ANOVA

variables can be selected simultaneously by clicking on one, holding the mouse button down and dragging the on-screen pointer over the second variable) and transfer them to the *Fixed Factor(s)* box by clicking on ▶. There are various other spaces that are available for conducting more complex analyses such as random factors ANOVA and factorial ANCOVA. Random factors ANOVA is beyond the scope of this book (interested readers should consult Jackson & Brashers, 1994) and factorial ANCOVA simply extends the principles described in Chapter 9.

10.3.2. Custom models ②

By default SPSS conducts a full factorial analysis (i.e. it includes all of the main effects and interactions of all independent variables specified in the main dialog box). However, there may be times when you want to customize the model that you use to test for certain things. To access the *model* dialog box, click on Model... in the main dialog box (see Figure 10.2). You will notice that, by default, the full factorial model is selected. Even with this selected, there is an option at the bottom to change the type of sums of squares that are used in the analysis. Although we have learnt about sums of squares and what they represent, I haven't talked about different ways of calculating sums of squares. It isn't necessary to understand the computation of the different forms of sums of squares, but it is important that you know the uses of some of the different types. By default, SPSS uses Type III sums of squares, which have the advantage that they are invariant to the cell frequencies. As such, they can be used with both balanced and unbalanced (i.e. different numbers of participants in different groups) designs, which is why they are the default option. Type IV sums of squares are like Type III except that they can be used with data in which there are missing values. So, if you have any missing data in your design, you should change the sums of squares to Type IV.

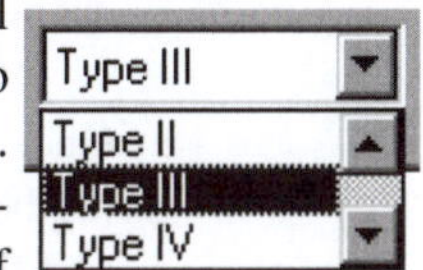

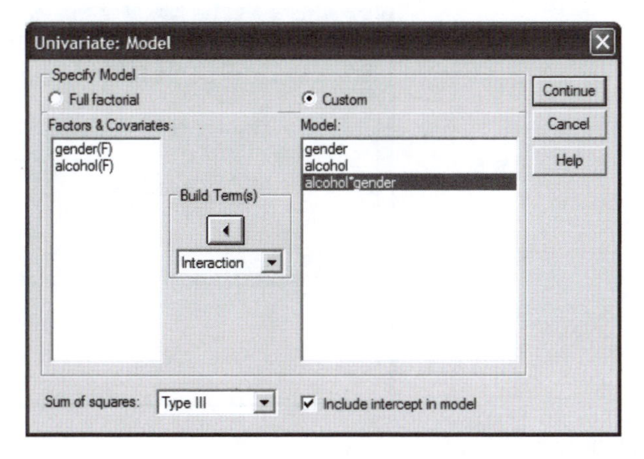

Figure 10.2 Custom models in ANOVA

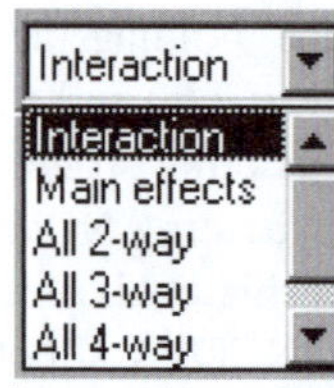

To customize a model, click on the circle labelled *Custom* to activate the dialog box. The variables specified in the main dialog box will be listed on the left-hand side and will be followed, in brackets, by a letter indicating the type of variable it is (F = fixed factor, R = random factor, C = covariate). You can select one or several variables from this list and transfer them to the box labelled *Model* as either main effects or interactions. By default, SPSS transfers variables as interaction terms, but there are several options that allow you to enter main effects, or all two-way, three-way or four-way interactions. These options save you the trouble of having to select lots of combinations of variables (because, for example, you can select three variables, transfer them as all two-way interactions and it will create all three combinations of variables for you). Although model selection has important uses (see section 9.6), it is likely that you'd want to run the full factorial analysis on most occasions.

10.3.3. Graphing interactions ②

Once the relevant variables have been selected, you should click on [Plots] to access the dialog box in Figure 10.3. The *plots* dialog box allows you to select line graphs of your data and these graphs are very useful for interpreting interaction effects. We have only two independent variables, and so there is only one plot worth looking at (the plot that displays levels of one independent variable against the other). Select **alcohol** from the variables list on the left-hand side of the dialog box and transfer it to the space labelled *Horizontal Axis* by clicking on ▶. In the space labelled *Separate Lines* we need to place the remaining independent variable: **gender**. It doesn't actually matter which way round the variables are plotted; you should use your discretion as to which way produces the most sensible graph. When you have moved the two independent variables to the appropriate box, click on

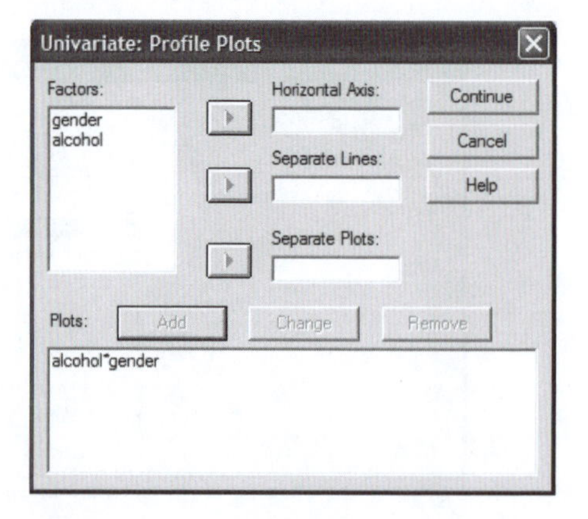

Figure 10.3 *Plots* dialog box

Add and this plot will be added to the list at the bottom of the box. It should be clear that you can plot a whole variety of graphs, and if you had a third independent variable, you would have the option of plotting different graphs for each level of that third variable. The plot that has been selected will help us to interpret any interaction between gender and alcohol consumption. When you have finished specifying graphs, click on Continue to return to the main dialog box.

10.3.4. Contrasts ②

We saw in Chapter 8 that it's useful to follow up ANOVA with contrasts that break down the main effects and tell us where the differences between groups lie. For one-way ANOVA, SPSS had a procedure for entering codes that define the contrasts we want to do. However, for two-way ANOVA no such facility exists and instead we are restricted to doing one of several standard contrasts. These standard contrasts are described in Table 8.6. To be fair, these contrasts will give you what you want in many different situations, but if they don't and you want to define your own contrasts then this has to be done using syntax. Doing contrasts with syntax is probably not going to be that interesting to most people so I've not included it here; however, because I do get asked about it a fair bit and I'm a complete masochist I've prepared a fairly detailed guide on how to do these called **ContrastsUsingSyntax.pdf** that's on the CD-ROM. So, if you want to know more have a look at this additional material.

Anyway, back to the point, we can use standard contrasts for this example. The effect of gender has only two levels, and so we don't really need contrasts for the main effect. The effect of alcohol has three levels: none, 2 pints and 4 pints. We could again select a simple contrast for this variable, and use the first category as a reference category. This would compare the 2-pint group to the no-alcohol group, and then compare the 4-pint category to the no-alcohol group. As such, the alcohol groups would get compared to the no-alcohol group. We could also select a *repeated* contrast. This would compare the 2-pint group to the no-alcohol group, and then the 4-pint group to the 2-pint group (so it moves through the groups comparing each group to the one before). Again, this might be useful. We could also do a Helmert contrast which compares each category against all subsequent categories, so in this case would compare the no-alcohol group to the remaining categories (i.e. all of the groups that had some alcohol) and then would move on to the 2-pint category and compare this to the 4-pint category. Any of these would be fine, but they only give us contrasts for the main effects. In reality, most of the time we actually want contrasts for our interaction term, and they can be obtained only through syntax (oh well, looks like you might have to look at that file after all!).

To get contrasts for the main effect of alcohol click on Contrasts in the main dialog box. We have used the *contrasts* dialog box before in section 9.3.3 and so refer back to that section to help you select a Helmert contrast for the alcohol variable. Once the contrasts have been selected (Figure 10.4), click on Continue to return to the main dialog box.

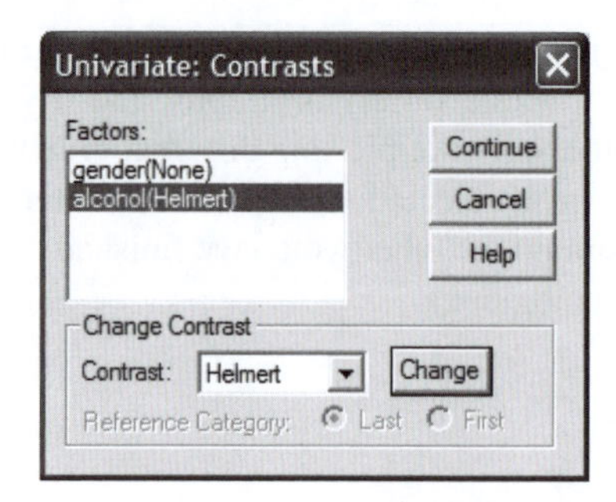

Figure 10.4

10.3.5. *Post hoc* tests ②

Post hoc tests are obtained by clicking on Post Hoc... in the main dialog box to access the *post hoc* tests dialog box (Figure 10.5). The variable **gender** has only two levels and so we don't need to select *post hoc* tests for that variable (because any significant effects can only reflect differences between males and females). However, there were three levels of the **alcohol** variable (no alcohol, 2 pints and 4 pints): hence it is necessary to conduct *post hoc* tests. First, you should select the variable **alcohol** from the box labelled *Factors* and transfer it to the box labelled *Post Hoc Tests for*. My recommendations for which *post hoc* procedures to use are in section 8.2.11 (and I don't want to repeat myself). Suffice to say, you should select the ones in Figure 10.5! Click on Continue to return to the main dialog box.

Univariate: Post Hoc Multiple Comparisons for Observed Means
Factor(s):
gender
alcohol
Post Hoc Tests for:
alcohol
Continue
Cancel
Help
Equal Variances Assumed
LSD
Bonferroni
Sidak
Scheffe
R-E-G-W F
R-E-G-W Q
S-N-K
Tukey
Tukey's-b
Duncan
Hochberg's GT2
Gabriel
Waller-Duncan
Type I/Type II Error Ratio: 100
Dunnett
Control Category: Last
Test
2-sided
< Control
> Control
Equal Variances Not Assumed
Tamhane's T2
Dunnett's T3
Games-Howell
Dunnett's C

Figure 10.5 Dialog box for *post hoc* tests

10.3.6. Options ②

Click on [Options...] to activate the dialog box in Figure 10.6. The options for factorial ANOVA are fairly straightforward. First you can ask for some descriptive statistics, which will display a table of the means, standard deviations, standard errors, ranges and confidence intervals for the means of each group. This is a useful option to select because it assists in interpreting the final results. A vital option to select is the homogeneity of variance tests. As with the *t*-test, there is an assumption that the variances of the groups are equal and selecting this option tests that this assumption has been met. SPSS uses Levene's test, which tests the hypothesis that the variances of each group are equal. Once these options have been selected click on [Continue] to return to the main dialog box, and then click on [OK] to run the analysis.

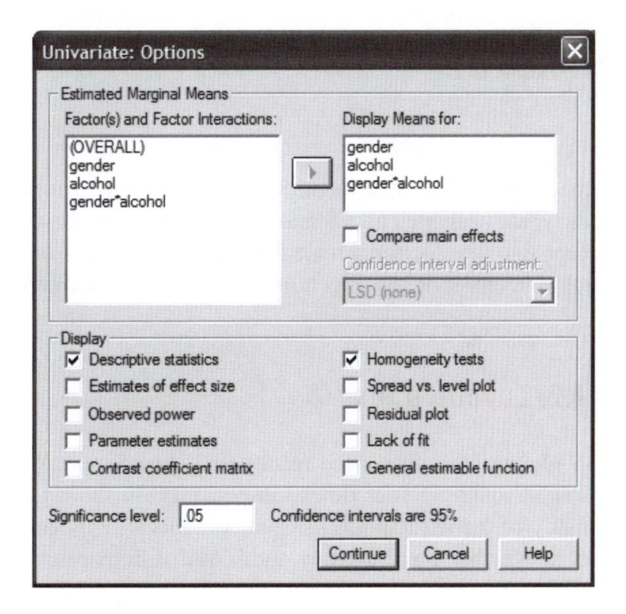

Figure 10.6 *Options* dialog box

10.4. OUTPUT FROM FACTORIAL ANOVA ②

10.4.1. Output for the preliminary analysis ②

SPSS Output 10.1 shows the initial output from factorial ANOVA. This table of descriptive statistics is produced because we asked for descriptives in the *options* dialog box (see Figure 10.6) and it displays the means, standard deviations and number of participants in

Descriptive Statistics

	Alcohol Consumption	Gender	Mean	Std. Deviation	N
Attractiveness of Date	Placebo	Male	66.8750	10.3294	8
		Female	60.6250	4.9552	8
		Total	63.7500	8.4656	16
	2 Pints	Male	66.8750	12.5178	8
		Female	62.5000	6.5465	8
		Total	64.6875	9.9111	16
	4 Pints	Male	35.6250	10.8356	8
		Female	57.5000	7.0711	8
		Total	46.5625	14.3433	16
	Total	Male	56.4583	18.5026	24
		Female	60.2083	6.3381	24
		Total	58.3333	13.8123	48

SPSS Output 10.1

all conditions of the experiment. So, for example, we can see that in the placebo condition, males typically chatted up a female that was rated at about 67% on the attractiveness scale, whereas females selected a male that was rated as 61% on that scale. These means will be useful in interpreting the direction of any effects that emerge in the analysis.

10.4.2. Levene's test ②

SPSS Output 10.2 shows the results of Levene's test. We came across Levene's test in section 3.6 and numerous times subsequently! In short, Levene's test is used to assess the tenability of the assumption of equal variances (homogeneity of variance). Levene's test looks at whether there are any significant differences between group variances and so

Levene's Test of Equality of Error Variances[a]

	F	df1	df2	Sig.
Attractiveness of Date	1.527	5	42	.202

Tests the null hypothesis that the error variance of the dependent variable is equal across groups.
a Design: Intercept+ALCOHOL+GENDER+ALCOHOL *GENDER

SPSS Output 10.2

a non-significant result (as found here) is indicative of the assumption being met. If Levene's test is significant then steps must be taken to equalize the variances through data transformation (taking the square root of all values of the dependent variable can sometimes achieve this goal—see Chapter 3 and Howell, 2002, p. 347).

10.4.3. The main ANOVA table ②

SPSS Output 10.3 is the most important part of the output because it tells us whether any of the independent variables have had an effect on the dependent variable. The important things to look at in the table are the significance values of the independent variables. The first thing to notice is that there is a significant effect of alcohol (because the significance value is less than .05). The *F*-ratio is highly significant, indicating that the amount of alcohol consumed significantly affected who the participant would try to chat up. What this means is that overall, when we ignore whether the participant was male or female, the amount of alcohol influenced their mate selection. The effect of a variable taken in isolation is known as the main effect. The best way to see what this means is to look at a bar chart of the average attractiveness at each level of alcohol (ignore gender completely). This graph can be plotted by using the means in SPSS Output10.1 (see, I told you that those values would come in useful!) and the *Bar...* option of the *Graphs* menu.

Tests of Between-Subjects Effects

Dependent Variable: Attractiveness of Date

Source	Type III Sum of Squares	df	Mean Square	F	Sig.
Corrected Model	5479.167[a]	5	1095.833	13.197	.000
Intercept	163333.333	1	163333.333	1967.025	.000
ALCOHOL	3332.292	2	1666.146	**20.065**	**.000**
GENDER	168.750	1	168.750	**2.032**	**.161**
GENDER * ALCOHOL	1978.125	2	989.062	**11.911**	**.000**
Error	3487.500	42	83.036		
Total	172300.000	48			
Corrected Total	8966.667	47			

a R Squared = .611 (Adjusted R Squared = .565)

SPSS Output 10.3

Figure 10.7 clearly shows that when you ignore gender the overall attractiveness of the selected mate is very similar when no alcohol has been drunk, and when 2 pints have been drunk (the means of these groups are approximately equal). Hence, this significant main effect is *likely* to reflect the drop in the attractiveness of the selected mates when 4 pints have been drunk. This finding seems to indicate that a person is willing to accept a less attractive mate after 4 pints.

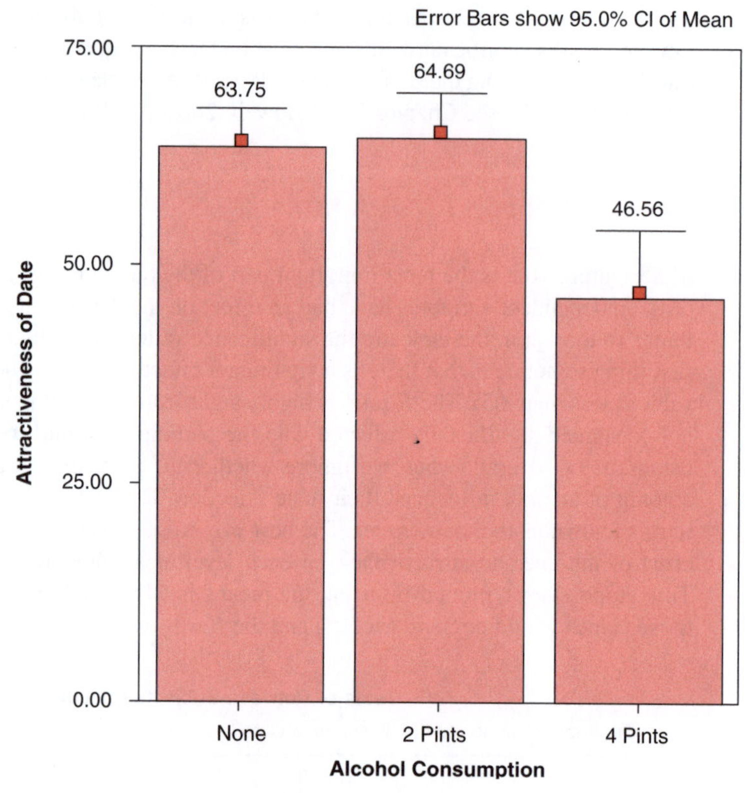

Figure 10.7 Graph showing the main effect of alcohol

The next part of SPSS Output 10.3 tells us about the main effect of gender. This time the *F*-ratio is not significant ($p = .161$, which is larger than .05). What this effect means is that overall, when we ignore how much alcohol had been drunk, the gender of the participant did not influence the attractiveness of the partner that the participant selected. In other words, other things being equal, males and females selected equally attractive mates. Drawing a bar chart of the average attractiveness of mates for men and women (ignoring how much alcohol had been consumed) reveals the meaning of this main effect. Figure 10.8 shows that the average attractiveness of the partners of male and female participants was fairly similar (the means are different by only 4%). Therefore, this non-significant effect reflects the fact that the mean attractiveness was similar. We can conclude from this that, *ceteris paribus*, men and women chose equally attractive partners.

Finally, SPSS Output 10.3 tells us about the interaction between the effect of gender and the effect of alcohol. The *F*-value is highly significant (because the *p*-value is less than .05). What this actually means is that the effect of alcohol on mate selection was different for male participants than it was for females. The SPSS output includes a plot that we asked

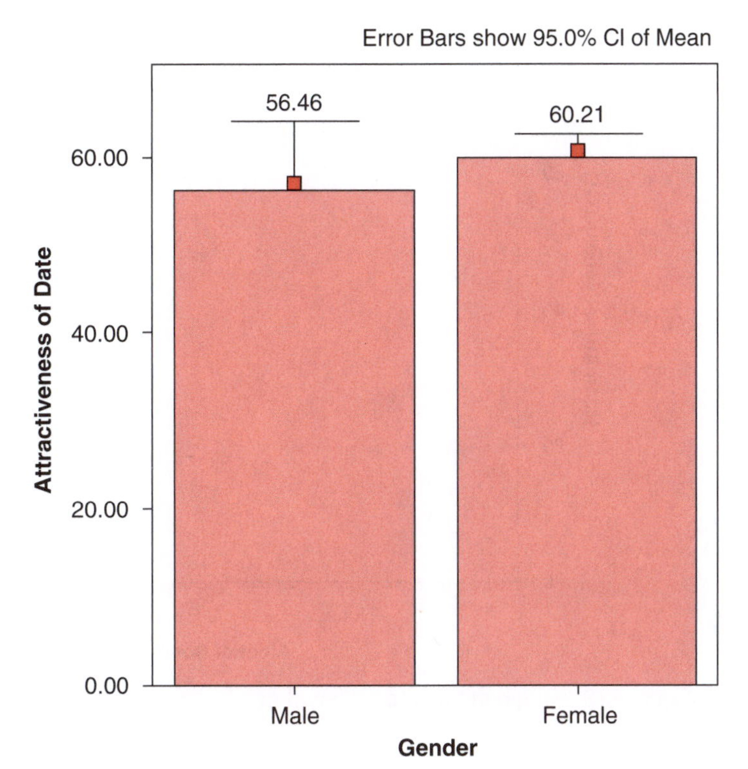

Figure 10.8 Graph to show the effect of gender on mate selection

for (see Figure 10.3) which tells us something about the nature of this interaction effect (see Figure 10.9).

How do I interpret interactions?

Figure 10.9 clearly shows that for women, alcohol has very little effect: the attractiveness of their selected partners is quite stable across the three conditions (as shown by the near-horizontal line). However, for the men, the attractiveness of their partners is stable when only a small amount has been drunk, but rapidly declines when more is drunk. Non-parallel lines usually indicate a significant interaction effect. In this particular graph the lines actually cross, which indicates a fairly large interaction between independent variables. The interaction tells us that alcohol has little effect on mate selection until 4 pints have been drunk and that the effect of alcohol is prevalent only in male participants. In short, the results show that women maintain high standards in their mate selection regardless of alcohol, whereas men have a few beers and then try to get off with anything on legs! One interesting point that these data demonstrate is that we earlier concluded that alcohol significantly affected how attractive a mate was selected (the

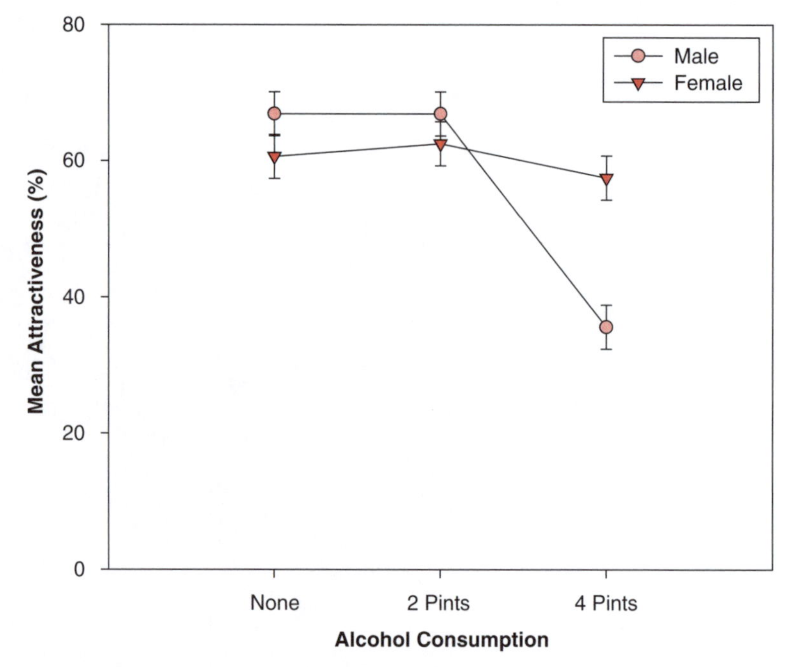

Figure 10.9 Graph of the interaction of gender and alcohol consumption in mate selection

alcohol main effect); however, the interaction effect tells us that this is true only in males (females appear unaffected). This shows how misleading main effects can be: it is usually the interactions between variables that are most interesting in a factorial design.

10.4.4. Contrasts ②

SPSS Output 10.4 shows the results of our Helmert contrast on the effect of alcohol. This helps us to break down the effect of alcohol. The top of the table shows the contrast for *Level 1 vs. Later*, which in this case means the no-alcohol group compared to the two alcohol groups. This tests whether the mean of the no-alcohol group (63.75) is different to the mean of the 2-pint and 4-pint groups combined ((64.69 + 46.56)/2 = 55.625). This is a difference of 8.125 (63.75 – 55.63), which we're told by both the *Contrast Estimate* and the *Difference* in the table. The important thing to look at is the value of *Sig.*, which tells us if this difference is significant. It is, because *Sig.* is .006, which is smaller than .05. We're also told the confidence interval for this difference and because it doesn't cross zero we can be safe in the knowledge that assuming this sample is one of the 95 out of 100 that produces a confidence interval containing the true value of the difference, then the real difference is more than zero (between 2.49 and 13.76 to be precise). So we could conclude that the effect of alcohol is that any amount of alcohol reduces the attractiveness of the dates selected compared to when no alcohol is drunk. Of course this is misleading

Contrast Results (K Matrix)

Alcohol Consumption Helmert Contrast			Dependent Variable: Attractiveness of Date
Level 1 vs. Later	Contrast Estimate		8.125
	Hypothesized Value		0
	Difference (Estimate—Hypothesized)		8.125
	Std.Error		2.790
	Sig.		.006
	95% Confidence Interval for Difference	Lower Bound	2.494
		Upper Bound	13.756
Level 2 vs. Level 3	Contrast Estimate		18.125
	Hypothesized Value		0
	Difference (Estimate—Hypothesized)		18.125
	Std. Error		3.222
	Sig.		.000
	95% Confidence Interval for Difference	Lower Bound	11.623
		Upper Bound	24.627

SPSS Output 10.4

because, in fact, the means for the no-alcohol and 2-pint groups are fairly similar (63.75 and 64.69), so 2 pints of alcohol don't reduce the attractiveness of selected dates! The reason why the comparison is significant is because it's testing the combined effect of 2 and 4 pints, and because 4 pints have such a drastic effect, it drags down the mean. This shows why you need to be careful about how you interpret these contrasts: you need to have a look at the remaining contrast as well.

The bottom of the table shows the contrast for *Level 2 vs. Level 3*, which in this case means the 2-pint group compared to the 4-pint group. This tests whether the mean of the 2-pint group (64.69) is different to the mean of the 4-pint groups combined (46.56). This is a difference of 18.13 (64.69–46.56), which we're told by both the *Contrast Estimate* and the *Difference* in the table. Again, the important thing to look at is the value of *Sig.*, which tells us if this difference is significant. It is, because *Sig.* is .000, which is smaller than .05. We're also told the confidence interval for this difference and because it doesn't cross zero we can be safe in the knowledge that assuming this confidence interval is one of the 95 out of 100 that contains the true value of the difference, then the real difference is more than zero (between 11.62 and 24.63 to be precise). This tells us that having 4 pints significantly reduced the attractiveness of selected dates compared to having only 2 pints.

10.4.5. *Post hoc* analysis ②

The *post hoc* tests (SPSS Output 10.5) break down the main effect of alcohol and can be interpreted as if a one-way ANOVA had been conducted on the **alcohol** variable (i.e. the

Multiple Comparisons

Dependent Variable: Attractiveness of Date

	(I) Alcohol Consumption	(J) Alcohol Consumption	Mean Difference (I – J)	Std. Error	Sig.	95% Confidence Interval	
						Lower Bound	Upper Bound
Bonferroni	None	2 Pints	–.9375	3.22172	1.000	–8.9714	7.0964
		4 Pints	17.1875*	3.22172	.000	9.1536	25.2214
	2 Pints	None	.9375	3.22172	1.000	–7.0964	8.9714
		4 Pints	18.1250*	3.22172	.000	10.0911	26.1589
	4 Pints	None	–17.1875*	3.22172	.000	–25.2214	–9.1536
		2 Pints	–18.1250*	3.22172	.000	–26.1589	–10.0911
Games-Howell	None	2 Pints	–.9375	3.25860	.955	–8.9809	7.1059
		4 Pints	17.1875*	4.16380	.001	6.7981	27.5769
	2 Pints	None	.9375	3.25860	.955	–7.1059	8.9809
		4 Pints	18.1250*	4.35860	.001	7.3104	28.9396
	4 Pints	None	–17.1875*	4.16380	.001	–27.5769	–6.7981
		2 Pints	–18.1250*	4.35860	.001	–28.9396	–7.3104

Based on observed means.
*The mean difference is significant at the .05 level

Attractiveness of Date

	Alcohol Consumption	N	Subset	
			1	2
Ryan-Einot-Gabriel-Welsch Range[a]	4 Pints	16	46.5625	
	None	16		63.7500
	2 Pints	16		64.6875
	Sig.		1.000	.772

Means for groups in homogeneous subsets are displayed.
Based on Type III Sum of Squares
The error term is Mean Square (Error) = 83.036.
a Alpha = .05

SPSS Output 10.5

reported effects for alcohol are collapsed with regard to gender). The Bonferroni and Games–Howell tests show the same pattern of results: when participants had drunk no alcohol or 2 pints of alcohol, they selected equally attractive mates. However, after 4 pints had been consumed, participants selected significantly less attractive mates than after both 2 pints ($p < .001$) and no alcohol ($p < .001$). It is interesting to note that the means of attractiveness of partners after no alcohol and 2 pints were so similar that the probability of the obtained difference between those means is 1 (i.e. completely probable!). The REGWQ test confirms that the means of the placebo and 2-pint conditions were equal whereas the mean of the 4-pint group was different. It should be noted that these *post hoc* tests ignore the interactive effect of gender and alcohol.

In summary, we should conclude that alcohol has an effect on the attractiveness of selected mates. Overall, after a relatively small dose of alcohol (2 pints) humans are still in control of their judgements and the attractiveness levels of chosen partners are consistent with a control group (no alcohol consumed). However, after a greater dose of alcohol, the attractiveness of chosen mates decreases significantly. This effect is what is referred to as the 'beer–goggles effect'! More interestingly, the interaction shows a gender difference in the beer–goggles effect. Specifically, it looks as though men are significantly more likely to pick less attractive mates when drunk. Women, in comparison, manage to maintain their standards despite being drunk. What we still don't know is whether women will become susceptible to the beer–goggles effect at higher doses of alcohol.

Cramming Samantha's Tips

- Two-way independent ANOVA compares several means when there are two independent variables, and different participants have been used in all experimental conditions; for example, if we wanted to know whether different teaching methods worked better for different subjects. You could take students from four courses (Psychology, Geography, Management and Statistics) and assign them to either lecture-based or book-based teaching.
- The two variables are course and method of teaching. The outcome might be the end-of-year mark (as a percentage).
- Test for homogeneity of variance using *Levene's test*. Find the table with this label: if the value in the column labelled *Sig.* is less than .05 then the assumption is violated.
- In the table labelled *Tests of Between-Subjects Effects*, look at the column labelled *Sig.* for all three of your effects: there should be a main effect of each variable, and an effect of the interaction between the two variables; if the value is less than .05 then the effect is significant. For main effects consult *post hoc* tests to see which groups differ, and for the interaction look at an interaction graph or conduct simple effects analysis (Box 10.2).
- For *post hoc* tests, again look to the columns labelled *Sig.* to discover if your comparisons are significant (they will be if the significance value is less than .05).
- Test the same assumptions as for ANOVA.

Box 10.2

Simple effects analysis ③

One popular way to break down an interaction term is to use a technique called simple effects analysis. This analysis basically looks at the effect of one independent variable at individual levels of the other independent variable. So, for example, in our beer–goggles data we could do a simple effects analysis looking at the effect of gender at each level of alcohol. This would mean taking the average attractiveness of the date selected by men and comparing it to that for women after the placebo, then making the same comparison for 2 pints, and then finally for 4 pints. Another way of looking at this is to say we would compare each triangle to the corresponding circle in Figure 10.9: based on the graph, we might expect to find no difference after no alcohol, and after 2 pints (in both cases the triangle and circle are located in about the same position), but we would expect a difference after 4 pints (because the circle and triangle are quite far apart). The alternative would be to compare the mean attractiveness after the placebo, 2 pints and 4 pints for men and then in a separate analysis do the same, but for women. (This would be a bit like doing a one-way ANOVA on the effect of alcohol in men, and then doing a different one-way ANOVA for the effect of alcohol in women.) You don't need to know how these simple effects are calculated (although if you're interested it is explained in Appendix A.6 or the file **Calculating SimpleEffects.pdf** on the CD-ROM), but it is useful to know how to do them on SPSS. Unfortunately, they can't be done through the dialog boxes and instead you have to use SPSS syntax (see section 2.6 to remind yourself about the syntax window). The syntax you need to use in this example is:

```
MANOVA
attract BY gender (0 1) alcohol (1 3)
```

*This initiates the ANOVA by specifying the outcome or dependent variable (**attract**) and then the BY command is followed by our independent variables (**gender** and **alcohol**)—the numbers in brackets are the minimum and maximum group codes that were used to define these variables.

```
/DESIGN = gender WITHIN alcohol (1) gender WITHIN alcohol (2) gender WITHIN alcohol (3)
```

*This specifies the simple effects. For example, 'gender WITHIN alcohol (1)' asks SPSS to analyse the effect of gender at level 1 of alcohol (i.e. when no alcohol was used). The number in brackets should relate to the level of the variable you want to look at (level 1 being the level having the lowest code in the data editor). If we wanted to compare alcohol at levels of gender, then we'd write this the opposite way around:

```
/DESIGN = alcohol WITHIN gender(1) alcohol WITHIN gender(2)
/PRINT
CELLINFO
SIGNIF ( UNIV MULT AVERF HF GG ).
```

(Continued)

Box 10.2 (Continued)

*These final lines just ask for some descriptives for each cell of the analysis (CELLINFO) and for the main ANOVA to be printed (SIGNIF). The syntax for looking at the effect of gender at different levels of alcohol is stored in a file called **GogglesSimpleEffects.sps** for you to look at. Open this file (make sure you also have **goggles.sav** loaded into the data editor) and run the syntax. The output you get will be in the form of text (rather than nice tables). Part of it will replicate the main ANOVA results from SPSS Output 10.3. The simple effects are presented like this:

```
* * * * * A n a l y s i s  o f  V a r i a n c e – design 1 * * * * *

Tests of Significance for ATTRACT using UNIQUE sums of squares

Source of Variation              SS      DF      MS          F     Sig of F

WITHIN+RESIDUAL               6819.79    44    155.00

GENDER WITHIN ALCOHOL(1)       156.25     1    156.25      1.01      .321

GENDER WITHIN ALCOHOL(2)        76.56     1     76.56       .49      .486

GENDER WITHIN ALCOHOL(3)      1914.06     1   1914.06     12.35      .001

(Model)                       2146.87     3    715.62      4.62      .007

(Total)                       8966.67    47    190.78

R-Squared =                       .239

Adjusted R-Squared =              .188
```

Looking at the significance values for each simple effect it appears that there was no significant difference between men and women at level 1 of alcohol (i.e. no alcohol), or at level 2 of alcohol (2 pints), but there was a very significant difference at level 3 of alcohol (which judging from the graph reflects the fact that the mean for men is considerably lower than for women).

Another useful thing to follow up interaction effects is to run contrasts for the interaction term. Unfortunately these can be done only using syntax, and it's a fairly involved process, so if this sounds like something you might want to do then look at the file **ContrastsUsingSyntax.pdf** on the CD-ROM: this is a file I've prepared specially to go through an example of specifying contrasts across an interaction.

10.5. INTERPRETING INTERACTION GRAPHS ②

We've already had a look at one interaction graph when we interpreted the analysis in this chapter. However, interactions are very important, and the key to understanding them is being able to interpret interaction graphs. In the example in this chapter we used Figure 10.9 to conclude that the interaction probably reflected the fact that men and women chose equally attractive dates after no alcohol and 2 pints, but that at 4 pints men's standards

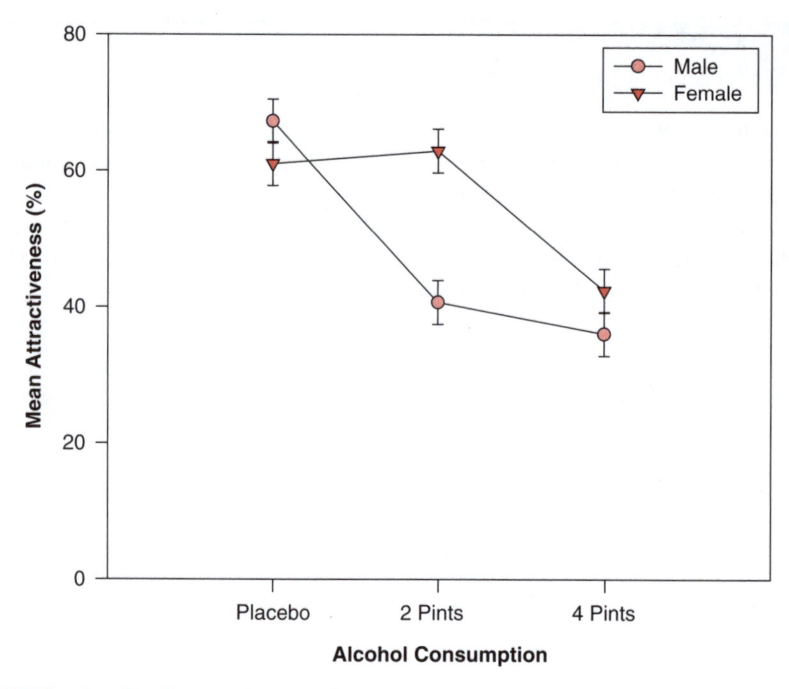

Figure 10.10 Another interaction graph

dropped significantly more than women's. Imagine we'd got the profile of results shown in Figure 10.10; do you think we would've still got a significant interaction effect?

This profile of data probably would also give rise to a significant interaction term because although the attractiveness of men and women's dates are similar after no alcohol and 4 pints of alcohol, there is a big difference after 2 pints. This reflects a scenario in which the beer–goggles effect is equally big in men and women after 4 pints (and doesn't exist after no alcohol) but kicks in quicker for men: the attractiveness of their dates plummets after 2 pints, whereas women maintain their standards until 4 pints (at which point they'd happily date an unwashed gibbon). Let's try another example. Is there a significant interaction in Figure 10.11?

For the data in Figure 10.11 there is unlikely to be a significant interaction because the effect of alcohol is the same for men and women. So, for both men and women, the attractiveness of their dates after no alcohol is quite high, but after 2 pints they drop by a similar amount (the slope of the male and female lines is about the same). After 4 pints there is a further drop, and again this drop is about the same in men and women (the lines again slope at about the same angle). The fact that the line for males is lower than that for females just reflects the fact that across all conditions, men have lower standards than their female counterparts: this reflects a main effect of gender (i.e. males generally chose less attractive dates than females at all levels of alcohol). Two general points that we can make from this are that:

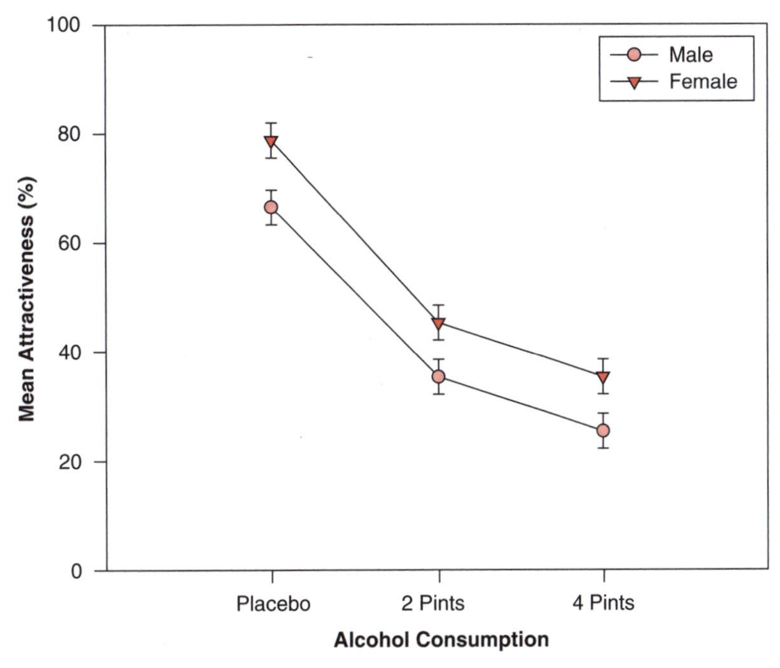

Figure 10.11 A 'lack of' interaction graph

- Significant interactions are shown up by non-parallel lines on an interaction graph. However, it's important to remember that this doesn't mean that non-parallel lines automatically mean that the interaction is significant: whether the interaction is significant will depend on the degree to which the lines are not parallel!
- If the lines on an interaction graph cross over then obviously they are not parallel and this can be a dead giveaway that you have a possible significant interaction. However, contrary to popular belief it isn't *always* the case that if the lines of the interaction graph cross then the interaction is significant.

A further complication is that sometimes people draw bar charts rather than line charts. Figure 10.12 shows some bar charts of interactions between two independent variables. Panels (a) and (b) actually display the data from the example used in this chapter (in fact, why not have a go at plotting them?). As you can see, there are two ways to present the same data: panel (a) shows the data when levels of alcohol are placed along the *x*-axis and different coloured bars are used to show means for males and females, and panel (b) shows the opposite scenario where gender is plotted on the *x*-axis and different colours distinguish the dose of alcohol. Both of these graphs show an interaction effect. What you're looking for are the differences between coloured bars to be different at different points along the *x*-axis. So, for panel (a) you'd look at the difference between the dark and light

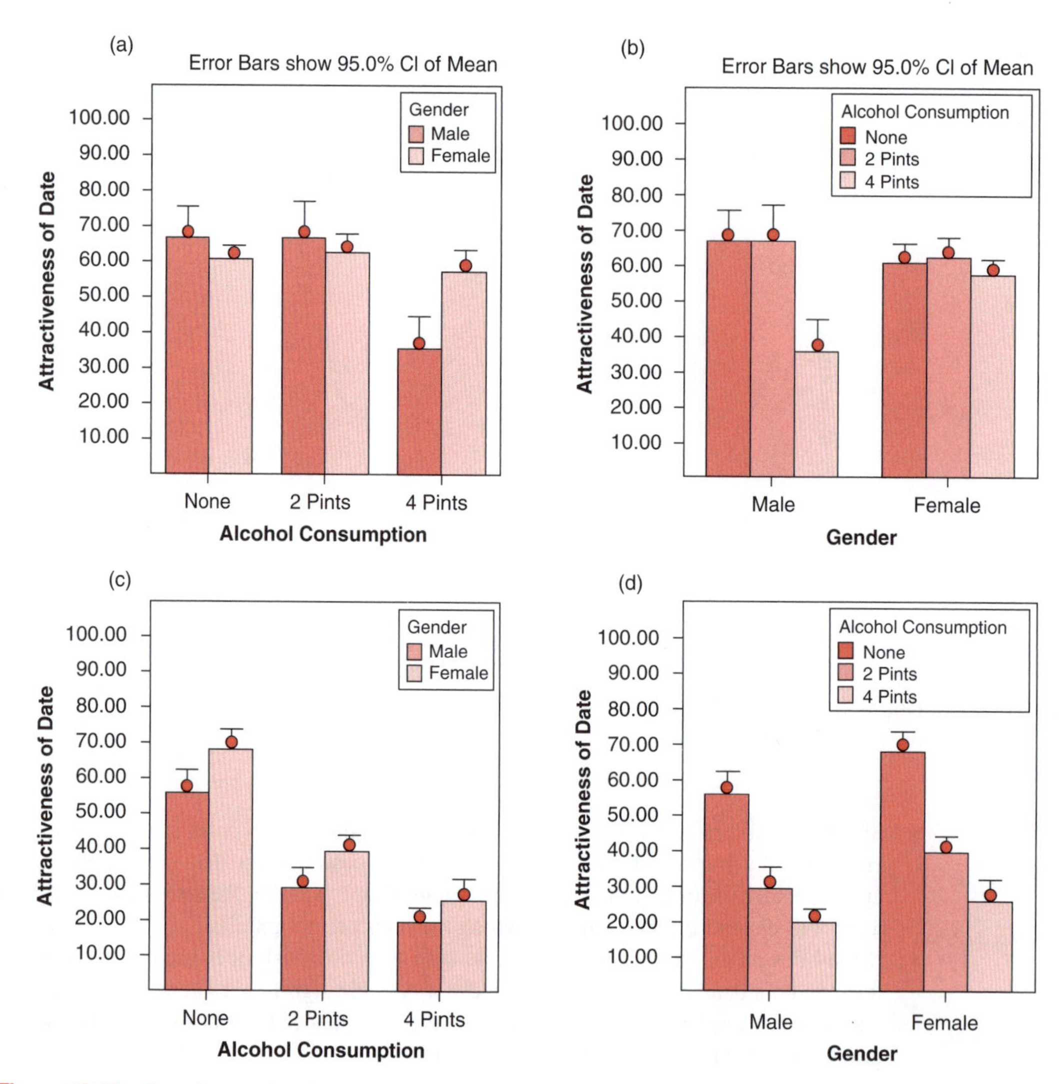

Figure 10.12 Bar charts showing interactions between two variables

bars for no alcohol, and then look to 2 pints and ask 'is the difference between the bars different to when I looked at no alcohol?' In this case the bars look the same at no alcohol as they do at 2 pints: hence, no interaction. However, we'd then move on to look at 4 pints, and we'd again ask 'is the difference between the bars different to what it has been in any of the other conditions?' In this case the answer is yes: for no alcohol and 2 pints, the bars were about the same height, but at 4 pints the lighter bar is much higher than the darker one. This shows an interaction: the pattern of responses changes at 4 pints. Panel (b) shows the same thing but plotted the other way around. Again you look at the pattern of responses. So, first we look at the men and see that the pattern is that the first two bars are the same height, but the last bar is much shorter. The interaction effect is shown up by the fact that for the women there is a different pattern: all three bars are about the same height.

What about panels (c) and (d): do you think there is an interaction? Again, they display the same data in two different ways, but it's different data to what we've used in this chapter. First let's look at panel (c): for the no alcohol data, the lighter bar is a bit bigger than the darker one; moving on to the 2 pint data, the lighter bar is also a bit taller than the darker one, and finally for the 4 pint data, the lighter bar is a little bit higher than the darker one. In all conditions the same pattern is shown—the lighter bar is a bit higher than the darker one (i.e. females pick more attractive dates than men regardless of alcohol consumption)—therefore, there is no interaction. Looking at panel (d) we see a similar result. For men, the pattern is that attractiveness ratings fall as more alcohol is drunk (the bars decrease in height) and then for the women we see the same pattern: ratings fall as more is drunk. This again is indicative of no interaction: the change in attractiveness due to alcohol is similar in men and women.

10.6. CALCULATING EFFECT SIZES ③

As we saw in previous chapters (sections 8.5 and 9.7), we can get SPSS to produce eta squared, η^2 (which is just r^2 by another name). However, you're well advised, for reasons explained in these other sections, to use omega squared (ω^2). The calculation of omega squared becomes somewhat more cumbersome in factorial designs (somewhat being one of my characteristic understatements!). Howell (2002), as ever, does a wonderful job of explaining the complexities of it all (and has a nice table summarizing the various components for a variety of situations). Condensing all of this down, I'll just say that we need first to compute a variance component for each of the effects (the two main effects and the interaction term) and the error, and then use these to calculate effect sizes for each. If we call the first main effect A, and the second main effect B, and the interaction effect A × B, then the variance components for each of these is based on the mean squares of each effect and the sample sizes on which they're based:

$$\hat{\sigma}_{\alpha}^2 = \frac{(a-1)(\text{MS}_\text{A} - \text{MS}_\text{R})}{nab}$$

$$\hat{\sigma}_{\beta}^2 = \frac{(b-1)(\text{MS}_\text{B} - \text{MS}_\text{R})}{nab}$$

$$\hat{\sigma}_{\alpha\beta}^2 = \frac{(a-1)(b-1)(\text{MS}_{\text{A}\times\text{B}} - \text{MS}_\text{R})}{nab}$$

In these equations, a is the number of levels of the first independent variable, b is the number of levels of the second independent variable and n is the number of people per condition. Let's calculate these for our data. We need to look at SPSS Output 10.3 to find out the mean squares for each effect, and for the error term. Our first independent variable was alcohol. This had three levels (hence $a = 3$) and had a mean squares of 1666.146. Our second independent variable was gender, which had two levels (hence $b = 2$) and a mean squares of 168.75. The number of people in each group was eight and the residual mean squares were 83.036. Therefore, our equations become:

$$\hat{\sigma}^2_{\alpha} = \frac{(3-1)(1666.146-83.036)}{8 \times 3 \times 2} = 65.96$$

$$\hat{\sigma}^2_{\beta} = \frac{(2-1)(168.75-83.036)}{8 \times 3 \times 2} = 1.79$$

$$\hat{\sigma}^2_{\alpha\beta} = \frac{(3-1)(2-1)(989.062-83.036)}{8 \times 3 \times 2} = 37.75$$

We also need to estimate the total variability and this is just the sum of these other variables plus the residual mean squares:

$$\begin{aligned}\hat{\sigma}^2_{\text{total}} &= \hat{\sigma}^2_{\alpha} + \hat{\sigma}^2_{\beta} + \hat{\sigma}^2_{\alpha\beta} + \text{MS}_\text{R} \\ &= 65.96 + 1.79 + 37.75 + 83.04 \\ &= 188.54\end{aligned}$$

The effect size is then simply the variance estimate for the effect in which you're interested divided by the total variance estimate:

$$\omega^2_{\text{effect}} = \frac{\hat{\sigma}^2_{\text{effect}}}{\hat{\sigma}^2_{\text{total}}}$$

As such, for the main effect of alcohol we get:

$$\omega^2_{\text{alcohol}} = \frac{\hat{\sigma}^2_{\text{alcohol}}}{\hat{\sigma}^2_{\text{total}}} = \frac{65.96}{188.54} = .35$$

For the main effect of gender we get:

$$\omega^2_{\text{gender}} = \frac{\hat{\sigma}^2_{\text{gender}}}{\hat{\sigma}^2_{\text{total}}} = \frac{1.79}{188.54} = .009$$

For the interaction of gender and alcohol we get:

$$\omega^2_{\text{alcohol} \times \text{gender}} = \frac{\hat{\sigma}^2_{\text{alcohol} \times \text{gender}}}{\hat{\sigma}^2_{\text{total}}} = \frac{37.75}{188.54} = .20$$

To make these values comparable to r we should take the square root, which gives us effect sizes of .59 for alcohol, .09 for gender, and .45 for the interaction term. As such, the effects of alcohol and the interaction are fairly large, but the effect of gender, which was non-significant in the main analysis, is very small indeed (close to zero in fact).

It's also possible to calculate effect sizes for our simple effects analysis (if you read Box 10.2). These effects have 1 degree of freedom for the model (which means they're comparing only two things) and in these situations F can be converted to r using the following equation (which just uses the F-ratio and the residual degrees of freedom):[1]

$$r = \sqrt{\frac{F(1, df_{\mathrm{R}})}{F(1, df_{\mathrm{R}}) + df_{\mathrm{R}}}}$$

Looking at Box 10.2, we can see that we got F-ratios of 1.01, .49 and 12.35 for the effects of gender at no alcohol, 2 pints and 4 pints respectively. For each of these, the degrees of freedom were 1 for the model and 44 for the residual. Therefore, we get the following effect sizes:

$$r_{\mathrm{Gender(No\ Alcohol)}} = \sqrt{\frac{1.01}{1.01 + 44}} = .15$$

$$r_{\mathrm{Gender(2Pints)}} = \sqrt{\frac{0.49}{0.49 + 44}} = .10$$

$$r_{\mathrm{Gender(4Pints)}} = \sqrt{\frac{12.35}{12.35 + 44}} = .47$$

Therefore, the effect of gender is very small at both no alcohol and 2 pints, but becomes large at 4 pints of alcohol.

10.7. REPORTING THE RESULTS OF TWO-WAY ANOVA ②

As with the other ANOVAs we've encountered we have to report the details of the F-ratio and the degrees of freedom from which it was calculated. For the various effects in these data the F-ratios will be based on different degrees of freedom: it was derived from dividing the mean squares for the effect by the mean squares for the residual. For the effects of alcohol and the alcohol × gender interaction, the model degrees of freedom were 2 ($df_{\mathrm{M}} = 2$), but for the effect of gender the degrees of freedom were only 1 ($df_{\mathrm{M}} = 1$).

1 If your F compares more than two things then a different equation is needed (see Rosenthal, Rosnow & Rubin (2000), p. 44), but I prefer to try to keep effect sizes to situations in which only two things are being compared because interpretation is easier.

For all effects, the degrees of freedom for the residuals were 42 (df_R = 84). We can, therefore, report the three effects from this analysis as follows:

- ✓ There was a significant main effect of the amount of alcohol consumed at the night-club, on the attractiveness of the mate they selected, $F(2, 42) = 20.07$, $p < .001$, $\omega^2 = .35$. The Games–Howell *post hoc* test revealed that the attractiveness of selected dates was significantly lower after 4 pints than both after 2 pints and no alcohol (both $p < .001$). The attractiveness of dates after 2 pints and no alcohol were not significantly different.
- ✓ There was a non-significant main effect of gender on the attractiveness of selected mates, $F(1, 42) = 2.03$, $p = .161$, $\omega^2 = .009$.
- ✓ There was a significant interaction effect between the amount of alcohol consumed and the gender of the person selecting a mate, on the attractiveness of the partner selected, $F(2, 42) = 11.91$, $p < .001$, $\omega^2 = .20$. This indicates that male and female genders were affected differently by alcohol. Specifically, the attractiveness of partners was similar in males ($M = 66.88$, $SD = 10.33$) and females ($M = 60.63$, $SD = 4.96$) after no alcohol; the attractiveness of partners was also similar in males ($M = 66.88$, $SD = 12.52$) and females ($M = 62.50$, $SD = 6.55$) after 2 pints; however, attractiveness of partners selected by males ($M = 35.63$, $SD = 10.84$) was significantly lower than those selected by females ($M = 57.50$, $SD = 7.07$) after 4 pints.

10.8. FACTORIAL ANOVA AS REGRESSION ③

We saw in section 8.2.2 that one-way ANOVA could be conceptualized as a regression equation (a general linear model). In this section we'll consider how we extend this linear model to incorporate two independent variables. To keep things as simple as possible I just want you to imagine that we have only two levels of the alcohol variable in our example (none and 4 pints). As such, we have two variables, each with two levels. All of the general linear models we've considered in this book take the general form of:

$$\text{Outcome}_i = (\text{Model}_i) + \text{error}_i$$

For example, when we encountered multiple regression in Chapter 5 we saw that this model was written as (see equation (5.9)):

$$Y_i = (b_0 + b_1X_1 + b_2X_2 + \cdots + b_nX_n) + \varepsilon_i$$

Also, when we came across one-way ANOVA, we adapted this regression model to conceptualize our Viagra example as (see equation (8.2)):

$$\text{Libido}_i = (b_0 + b_2\text{High}_i + b_1\text{Low}_i) + \varepsilon_i$$

In this model, the high and low variables were dummy variables (i.e. variables that can take only values of 0 or 1). In our current example, we have two variables: gender (male or female) and alcohol (none and 4 pints). We can code each of these with zeros and ones: for example, we could code gender as male = 0, female = 1; and we could code the alcohol variable as 0 = none, 1 = 4 pints. We could then directly copy the model we had in one-way ANOVA:

$$\text{Attractive}_i = (b_0 + b_1\text{Gender}_i + b_2\text{Alcohol}_i) + \varepsilon_i$$

Now the astute among you might say 'where has the interaction term gone?' Well, of course, we have to include this too, and so the model simply extends to become (first expressed generally and then in terms of this specific example):

$$\begin{aligned} \text{Outcome}_i &= (b_0 + b_1\text{A}_i + b_2\text{B}_i + b_3\text{AB}_i) + \varepsilon_i \\ \text{Attractive}_i &= (b_0 + b_1\text{Gender}_i + b_2\text{Alcohol}_i + b_3\text{Interaction}_i) + \varepsilon_i \end{aligned} \tag{10.1}$$

The question is, how do we code the interaction term? The interaction term represents the combined effect of alcohol and gender and in fact to get any interaction term in regression you simply multiply the variables involved in that interaction term. This is why you see interaction terms written as gender × alcohol, because in regression terms the interaction variable literally is the two variables multiplied by each other. Table 10.2 shows the resulting variables for the regression (note that the interaction variable is simply the value of the gender dummy variable multiplied by the value of the alcohol dummy variable). So, for example, a male receiving 4 pints of alcohol would have a value of 0 for the gender variable, 1 for the alcohol variable and 0 for the interaction variable. The group means for the various combinations of gender and alcohol are also included because they'll come in useful in due course.

Table 10.2 Coding scheme for factorial ANOVA

Gender	Alcohol	Dummy (Gender)	Dummy (Alcohol)	Interaction	Mean
Male	None	0	0	0	66.875
Male	4 Pints	0	1	0	35.625
Female	None	1	0	0	60.625
Female	4 Pints	1	1	1	57.500

To work out what the *b*-values represent in this model we can do the same as we did for the *t*-test and one-way ANOVA; that is, look at what happens when we insert values of our predictors (gender and alcohol)! To begin with, let's see what happens when we look at men who had no alcohol. In this case, the value of gender is 0, the value of alcohol is 0 and the value of the interaction is also 0. The outcome we predict (as with one-way ANOVA) is the mean of this group (66.875), so our model becomes:

$$\begin{aligned}
\text{Attractive}_i &= b_0 + b_1\text{Gender}_i + b_2\text{Alcohol}_i + b_3\text{Interaction}_i \\
\overline{X}_{\text{Men,None}} &= b_0 + (b_1 \times 0) + (b_2 \times 0) + (b_3 \times 0) \\
b_0 &= \overline{X}_{\text{Men,None}} \\
b_0 &= 66.875
\end{aligned}$$

So, the constant b_0 in the model represents the mean of the group for which all variables are coded as 0. As such it's the mean value of the base category (in this case men who had no alcohol). Now, let's see what happens when we look at females who had no alcohol. In this case, the gender variable is 1 and the alcohol and interaction variables are still 0. Also remember that b_0 is the mean of the men who had no alcohol. The outcome is the mean for women who had no alcohol; therefore, the equation becomes:

$$\begin{aligned}
\text{Attractive}_i &= b_0 + b_1\text{Gender}_i + b_2\text{Alcohol}_i + b_3\text{Interaction}_i \\
\overline{X}_{\text{Women,None}} &= b_0 + (b_1 \times 1) + (b_2 \times 0) + (b_3 \times 0) \\
\overline{X}_{\text{Women,None}} &= \overline{X}_{\text{Men,None}} + b_1 \\
b_1 &= \overline{X}_{\text{Women,None}} - \overline{X}_{\text{Men,None}} \\
&= 60.625 - 66.875 \\
&= -6.25
\end{aligned}$$

So, b_1 in the model represents the difference between men and women for those that had no alcohol. More generally we can say it's the effect of gender for the base category of alcohol (the base category being the one coded with 0, in this case no alcohol). Now let's look at males who had 4 pints of alcohol. In this case, the gender variable is 0, the alcohol variable is 1 and the interaction variable is still 0. We can also replace the b_0 with the mean of the men who had no alcohol. The outcome is the mean for men who had 4 pints; therefore, the equation becomes:

$$\begin{aligned}
\overline{X}_{\text{Men,4 Pints}} &= b_0 + (b_1 \times 0) + (b_2 \times 1) + (b_3 \times 0) \\
\overline{X}_{\text{Men,4 Pints}} &= b_0 + b_2 \\
\overline{X}_{\text{Men,4 Pints}} &= \overline{X}_{\text{Men,None}} + b_2 \\
b_2 &= \overline{X}_{\text{Men,4 Pints}} - \overline{X}_{\text{Men,None}} \\
&= 35.625 - 66.875 \\
&= -31.25
\end{aligned}$$

So, b_2 in the model represents the difference between having no alcohol and 4 pints in men. Put more generally, it's the effect of alcohol in the base category of gender (i.e. the category of gender that was coded with a 0, in this case men). Finally, we can look at females who had 4 pints of alcohol. In this case, the gender variable is 1, the alcohol variable is 1 and the interaction variable is also 1. We can also replace b_0, b_1 and b_2 with what we now know they represent. The outcome is the mean for women who had 4 pints; therefore, the equation becomes:

$$\begin{aligned}
\overline{X}_{\text{Women,4 Pints}} &= b_0 + (b_1 \times 1) + (b_2 \times 1) + (b_3 \times 1) \\
\overline{X}_{\text{Women,4 Pints}} &= b_0 + b_1 + b_2 + b_3 \\
\overline{X}_{\text{Women,4 Pints}} &= \overline{X}_{\text{Men,None}} + (\overline{X}_{\text{Women,None}} - \overline{X}_{\text{Men,None}}) + (\overline{X}_{\text{Men,4 Pints}} - \overline{X}_{\text{Men,None}}) + b_3 \\
\overline{X}_{\text{Women,4 Pints}} &= \overline{X}_{\text{Women,None}} + \overline{X}_{\text{Men,4 Pints}} - \overline{X}_{\text{Men,None}} + b_3 \\
b_3 &= \overline{X}_{\text{Men,None}} - \overline{X}_{\text{Women,None}} + \overline{X}_{\text{Women,4 Pints}} - \overline{X}_{\text{Men,4 Pints}} \\
&= 66.875 - 60.625 + 57.500 - 35.625 \\
&= 28.125
\end{aligned}$$

So, b_3 in the model really compares the difference between men and women in the no-alcohol condition with the difference between men and women in the 4-pint condition. Put another way, it compares the effect of gender after no alcohol to the effect of gender after 4 pints.[2] If you think about it in terms of an interaction graph this makes perfect sense. For example, the top left-hand side of Figure 10.13 shows the interaction graph for these data. Now imagine we calculated the difference between men and women for the no-alcohol groups. This would be the difference between the lines on the graph for the no-alcohol group (the difference between group means, which is 6.25). If we then do the same for the 4-pint group, we find that the difference between men and women is 21.875. If we plotted these two values as a new graph we'd get a line connecting 6.25 to 21.875 (see the bottom left-hand side of Figure 10.13). This reflects the difference between the effect of gender after no alcohol compared to after 4 pints. We know that beta values represent gradients of lines and in fact b_3 in our model is the gradient of this line! Let's also see what happens if there isn't an interaction effect: the right-hand side of Figure 10.13 shows the same data except that the mean for the females who had 4 pints has been changed to 30. If we calculate the difference between men and women after no alcohol we get the same as before: 6.25. If we calculate the difference between men and women after 4 pints we now get 5.625. If we again plot these differences on a new graph, we find a virtually horizontal line. So, when there's no interaction, the line connecting the effect of gender after no alcohol and after 4 pints is flat and the resulting b_3 in our model would be close to 0 (remember that a zero gradient means a flat line). In fact its actual value would be 6.25 – 5.625 = .625.

2 In fact if you rearrange the terms in the equation you'll see that you can also phrase the interaction the opposite way around: it represents the effect of alcohol in men compared to women.

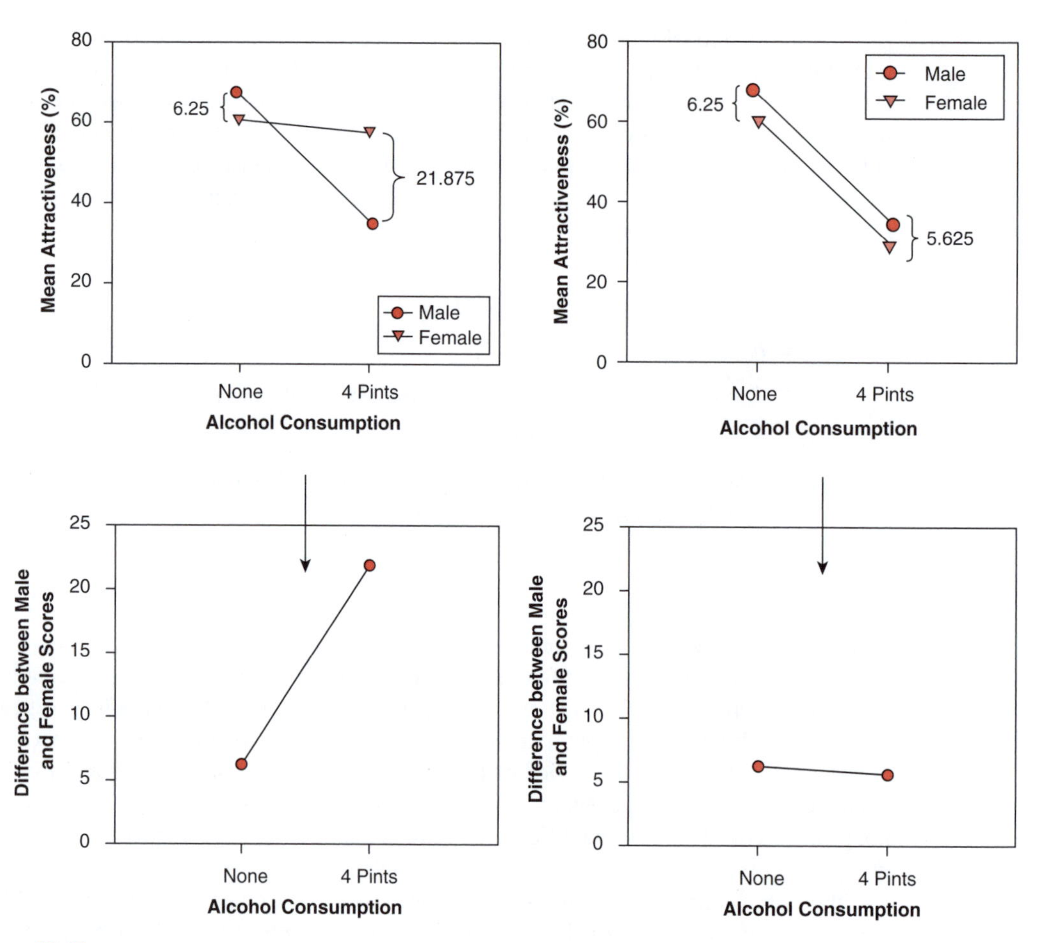

Figure 10.13

Coefficients[a]

Model	Unstandardized Coefficients		Standardized Coefficients	t	Sig.
	B	Std. Error	Beta		
1 (Constant)	66.875	3.055		21.890	.000
Gender	−6.250	4.320	−.219	−1.447	.159
Alcohol Consumption	−31.250	4.320	−1.094	−7.233	.000
Interaction	28.125	6.110	.853	4.603	.000

a Dependent Variable: Attractiveness of Date

SPSS Output 10.6

The file **GogglesRegression.sav** contains the dummy variables used in this example, and just to prove that all of this works, use this file and run a multiple regression on the data. The resulting table of coefficients is in SPSS Output 10.6. The important thing to note is that the various beta values are the same as we've just calculated, which should hopefully convince you that factorial ANOVA is, as is everything it would seem, just regression dressed up in a different costume!

What I hope to have shown you in this example is how even complex ANOVAs are just forms of regression (a GLM). You'll be pleased to know (I'll be pleased to know for that matter) that this is the last I'm going to say about ANOVA as GLM. I hope I've given you enough background so that you get a sense of the fact that we can just keep adding independent variables into our model. All that happens is these new variables just get added into a multiple regression equation with an associated beta value (just like in the regression chapter). Interaction terms can also be added simply by multiplying the variables that interact. These interaction terms will also have an associated beta value. So, any ANOVA (no matter how complex) is just a form of multiple regression.

10.9. WHAT HAVE WE DISCOVERED ABOUT STATISTICS? ②

This chapter has been a whistle-stop tour of factorial ANOVA. In fact we'll come across more factorial ANOVAs in the next two chapters, but for the time being we've just looked at the situation where there are two independent variables, and different people have been used in all experimental conditions. We started off by looking at how to calculate the various sums of squares in this analysis, but most important we saw that we get three effects: two main effects (the effect of each of the independent variables) and an interaction effect. We moved on to see how this analysis is done on SPSS and how the output is interpreted. Much of this was similar to the ANOVAs we've come across in previous chapters, but one big difference was the interaction term. We spent a bit of time exploring interactions (and especially interaction graphs) to see what an interaction looks like and how to spot it! The brave readers also found out how to follow up an interaction with simple effects analysis and also discovered that even complex ANOVAs are simply regression analyses in disguise. Finally we discovered that calculating effect sizes in factorial designs is a complete headache and should be attempted only by the criminally insane. So far we've steered clear of repeated-measures designs, but in the next chapter I have to resign myself to the fact that I can't avoid explaining them for the rest of my life☹

10.10. KEY TERMS THAT WE'VE DISCOVERED

- Beer–goggles effect
- Factorial ANOVA
- Factors
- Independent factorial design
- Interaction effect
- Interaction graphs
- Main effect
- Mixed design
- Related factorial design
- Simple effects analysis

10.11. SMART ALEX'S TASKS

- **Task 1**: People's musical taste tends to change as they get older (my parents, for example, after years of listening to relatively cool music when I was a kid in the 1970s, subsequently hit their mid-forties and developed a worrying obsession with country and western music—or maybe it was the stress of having me as a teenage son!). Anyway, this worries me immensely as the future seems incredibly bleak if it is spent listening to Garth Brooks and thinking 'oh boy, did I underestimate Garth's immense talent when I was in my 20s'. So, I thought I'd do some research to find out whether my fate really was sealed, or whether it's possible to be old and like good music too. First, I got myself two groups of people (45 people in each group): one group contained young people (which I arbitrarily decided was under 40 years of age), and the other group contained more mature individuals (above 40 years of age). This is my first independent variable, **age**, and it has two levels (less than or more than 40 years old). I then split each of these groups of 45 into three smaller groups of 15 and assigned them to listen to Fugazi,[3] ABBA or Barf Grooks (who is a lesser known country and western musician not to be confused with anyone who has a similar name). This is my second independent variable, **music**, and has three levels (Fugazi, ABBA or Barf Grooks). There were different participants in all conditions, which means that of the 45 under forties, 15 listened to Fugazi, 15 listened to ABBA and 15 listened to Barf Grooks; likewise of the 45 over forties, 15 listened to Fugazi, 15 listened to ABBA and 15 listened to Barf Grooks. After listening to the music I got each person to rate it on a scale ranging from –100 (I hate this foul music) through 0 (I am completely indifferent) to +100 (I love this music so much). This variable is called **liking**. The data are in the file **Fugazi.sav**. Conduct a two-way independent ANOVA on them.②
- **Task 2**: Using Box 10.2, change the syntax in **GogglesSimpleEffects.sps** to look at the effect of alcohol at different levels of gender.③

The answers are in the file **Answers(Chapter10).pdf**, and task 1 is an example from Field & Hole (2003) and so has a more detailed answer in there if you feel like you want more detail.

10.12. FURTHER READING

Howell, D. C. (2002). *Statistical methods for psychology* (5th edition). Belmont, CA: Duxbury. Chapter 13.

Rosenthal, R., Rosnow, R. L. & Rubin, D. B. (2000). *Contrasts and effect sizes in behavioural research: a correlational approach*. Cambridge: Cambridge University Press. This is quite advanced but really cannot be bettered for contrasts and effect size estimation.

Rosnow, R. L. & Rosenthal, R. (2005). *Beginning behavioural research: a conceptual primer* (5th edition). Englewood Cliffs, NJ: Pearson/Prentice Hall. Has some wonderful chapters on ANOVA, with a particular focus on effect size estimation, and some very insightful comments on what interactions actually mean.

---

3 See http://www.dischord.com

CHAPTER 11

REPEATED-MEASURES DESIGNS (GLM 4)

11.1. WHAT WILL THIS CHAPTER TELL US? ②

Over the previous three chapters we have looked at a procedure called ANOVA which is used for testing differences between several means. So far we've concentrated on situations in which different people contribute to different means; put another way, different people take part in different experimental conditions. Actually, it doesn't have to be different people (I tend to say people because I'm a psychologist and so spend my life torturing, I mean testing, children in the name of science), it could be different plants, different companies, different plots of land, different viral strains, different goats or even different duck-billed platypuses (or whatever the plural is). Anyway, the point is I've completely ignored situations in which the same people (plants, goats, hamsters, seven-eyed green galactic leaders from space, or whatever) contribute to the different means. I've put it off long enough, and now I'm going to take you through what happens when we do ANOVA on repeated-measures data. To begin with we'll have a fairly detailed look at an additional assumption that we need to consider when using repeated measured designs (*sphericity*). We'll then have a look at the theory behind repeated-measures ANOVA when we have a single independent variable. We'll then go through an example on SPSS before extending things to the situation where we've measured two independent variables. So, by the end of it all you should have almost everything you need to know about repeated-measures ANOVA.

11.2. INTRODUCTION TO REPEATED-MEASURES DESIGNS ②

'Repeated measure' is a term used when the same people participate in all conditions of an experiment. For example, you might test the effects of alcohol on enjoyment of a party. Some people can drink a lot of alcohol without really feeling the consequences, whereas others, like myself, only have to sniff a pint of lager to feel its effects. Therefore, it is important to control for individual differences in tolerance to alcohol. To control these individual differences we can test the same people in all conditions of the experiment: we would test each person after they had consumed 1 pint, 2 pints, 3 pints and 4 pints of lager. After each drink the participant could be given a questionnaire assessing their enjoyment of the party. As such, every participant provides a score representing their enjoyment before the experimental manipulation (no alcohol consumed), after 1 pint, after 2 pints and so on. This design is said to use a repeated measure.

This type of design has several advantages. Most important, it reduces the unsystematic variability in the design (see Chapter 7) and so provides greater power to detect effects. Repeated measures are also more economical because fewer participants are required. However, there is a disadvantage too. In between-group ANOVA the accuracy of the F-test depends upon the assumption that scores in different conditions are independent (Scariano & Davenport, 1987, have documented some of the consequences of violating this assumption). When repeated measures are used this assumption is violated: scores taken under different experimental conditions are likely to be related because they come from the same participants. As such, the conventional F-test will lack accuracy. The relationship between scores in different treatment conditions means that an additional assumption has to be made and, put simplistically, we assume that the relationship between pairs of experimental conditions is similar (i.e. the level of dependence between experimental conditions is roughly equal). This assumption is called the assumption of sphericity, which, trust me, is a pain in the neck to try to pronounce when you're giving statistics lectures at 9 a.m.

11.2.1. The assumption of sphericity ②

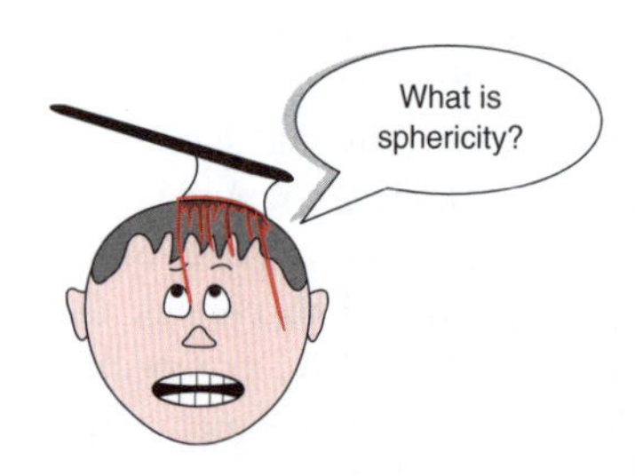

Most of us are taught (especially if you read this book) that it is crucial to have homogeneity of variance between conditions when analysing data from *different* participants, but often we are left to assume that this problem 'goes away' in repeated-measures designs. This is not so, and the assumption of sphericity can be likened to the assumption of homogeneity of variance in between-group ANOVA. Sphericity (denoted by ε and sometimes referred to as *circularity*) is a more general condition of compound symmetry. Compound symmetry holds true when both the variances across conditions are equal (this is the same as the homogeneity of variance assumption in between-group designs) and the covariances between pairs of conditions are equal. So, we assume that the variation within experimental conditions is fairly similar and that no two conditions are any more dependent than any other two. Although compound symmetry has been shown to be a sufficient condition for ANOVA using repeated-measures data, it is not a necessary condition. Sphericity is a less restrictive form of compound symmetry (in fact much of the early research into repeated-measures ANOVA confused compound symmetry with sphericity). Sphericity refers to the equality of variances of the *differences* between treatment levels. So, if you were to take each pair of treatment levels, and calculate the differences between each pair of scores, then it is necessary that these differences have equal variances. As such, **you need at least three conditions for sphericity to be an issue**.

11.2.2. How is sphericity measured? ②

The simplest way to see whether the assumption of sphericity has been met is to calculate the differences between pairs of scores in all combinations of the treatment levels. Once this has been done, you can calculate the variance of these differences. Table 11.1 shows data from an experiment with three conditions. The differences between pairs of scores are computed for each participant and the variance for each set of differences is calculated. We saw above that sphericity is met when these variances are roughly equal. For these data, sphericity will hold when:

$$\text{variance}_{\text{A}-\text{B}} \approx \text{variance}_{\text{A}-\text{C}} \approx \text{variance}_{\text{B}-\text{C}}$$

In these data there is some deviation from sphericity because the variance of the differences between conditions A and B (17.0) is greater than the variance of the differences between A and C and between B and C (10.3). However, these data have *local circularity* (or local sphericity) because two of the variances of differences are identical. Therefore, the sphericity assumption has been met for any multiple comparisons involving these conditions (for a discussion of local circularity see Rouanet & Lépine, 1970). The deviation from sphericity in the data in Table 11.1 does not seem too severe (all variances are *roughly* equal), but can we assess whether a deviation is severe enough to warrant action?

Table 11.1 Hypothetical data to illustrate the calculation of the variance of the differences between conditions

Group A	Group B	Group C	A – B	A – C	B – C
9	12	7	−3	2	5
15	15	12	0	3	3
25	30	20	−5	5	10
35	30	28	5	7	2
30	27	20	3	10	7
		Variance:	**17.0**	**10.3**	**10.3**

11.2.3. Assessing the severity of departures from sphericity ②

SPSS produces a test known as Mauchly's test, which tests the hypothesis that the variances of the differences between conditions are equal. Therefore, if Mauchly's test statistic is significant (i.e. has a probability value less than .05) we should conclude that there are significant differences between the variances of differences; ergo the condition of sphericity is not met. If, however, Mauchly's test statistic is non-significant (i.e. $p > .05$) then it is reasonable to conclude that the variances of differences are not significantly different (i.e. they are roughly equal). So, in short, if Mauchly's test is significant then we must be wary of the F-ratios produced by the computer.

11.2.4. What is the effect of violating the assumption of sphericity? ③

Rouanet & Lépine (1970) provided a detailed account of the validity of the F-ratio under violations of the sphericity assumption. They argued that there are two different F-ratios that can be used to assess treatment comparisons, labelled F' and F'' respectively. F' refers to an F-ratio derived from the mean squares of the comparison in question and the specific error term for the comparison of interest—this is the F-ratio normally used. F'' is derived not from the specific error mean square but from the total error mean squares for *all* repeated-measures comparisons. Rouanet & Lépine (1970) argued that F' is less powerful than F'' and so it may be the case that this test statistic misses genuine effects. In addition, they showed that for F' to be valid sphericity must hold for the *specific comparison in question* (see also Mendoza, Toothaker & Nicewander, 1974). F'' requires only *overall* circularity (i.e. the whole data set must be circular) but because of the non-reciprocal nature of circularity and compound symmetry, F'' does not require compound symmetry, though F' does. So, given that F' is the statistic generally used, the effect of violating sphericity is a loss of power (compared to when F'' is used) and a test statistic (F-ratio) which simply cannot be compared to tabulated values of the F-distribution (for more details see Field, 1998a).

11.2.5. What do we do if we violate sphericity? ②

If data violate the sphericity assumption there are several corrections that can be applied to produce a valid F-ratio. SPSS produces three corrections based upon the estimates of sphericity advocated by Greenhouse & Geisser (1959) and Huynh & Feldt (1976). Both of these estimates give rise to a correction factor that is applied to the degrees of freedom used to assess the observed F-ratio. The calculation of these estimates is beyond the scope of this book (interested readers should consult Girden, 1992); we need to know only that the three estimates differ. The Greenhouse–Geisser correction (usually denoted as $\hat{\varepsilon}$) varies between $1/k - 1$ (where k is the number of repeated-measures conditions) and 1. The closer that $\hat{\varepsilon}$ is to 1.00, the more homogeneous the variances of differences, and hence the closer the data are to being spherical. For example, in a situation in which there are five conditions the lower limit of $\hat{\varepsilon}$ will be $1/(5 - 1)$, or .25 (known as the lower-bound estimate of sphericity).

Huynh & Feldt (1976) reported that when the Greenhouse–Geisser estimate is greater than .75 too many false null hypotheses fail to be rejected (i.e. the correction is too conservative) and Collier, Baker, Mandeville & Hayes (1967) showed that this was also true with a sphericity estimate as high as .90. Huynh & Feldt, therefore, proposed their own less conservative correction (usually denoted as $\tilde{\varepsilon}$). However, Maxwell & Delaney (1990) report that $\tilde{\varepsilon}$ overestimates sphericity. Stevens (1992) therefore recommends taking an average of the two and adjusting the *df* by this averaged value. Girden (1992) recommends that when estimates of sphericity are greater than .75 then the Huynh–Feldt correction should be used, but when sphericity estimates are less than .75 or nothing is known about sphericity at all, then the Greenhouse–Geisser correction should be used instead. We shall see how these values are used in due course.

A final option, when you have data that violate sphericity, is to use multivariate test statistics (MANOVA—see Chapter 14), because they are not dependent upon the assumption of sphericity (see O'Brien & Kaiser, 1985). MANOVA avoids the assumption of sphericity (and all the corresponding considerations about appropriate *F*-ratios and corrections) by using a specific error term for contrasts with 1 *df*, and hence each contrast is only ever associated with its specific error term (rather than the pooled error terms used in ANOVA). MANOVA is covered in depth in Chapter 14, but the repeated-measures procedure in SPSS automatically produces multivariate test statistics.

There is a trade-off in test power between univariate and multivariate approaches (although some authors argue that this can be overcome with suitable mastery of the techniques—O'Brien & Kaiser, 1985). Davidson (1972) compared the power of adjusted univariate techniques with those of Hotelling's T^2 (a MANOVA test statistic) and found that the univariate technique was relatively powerless to detect small reliable changes between highly correlated conditions when other less correlated conditions were also present. Mendoza *et al.* (1974) conducted a Monte Carlo study comparing univariate and multivariate techniques under violations of compound symmetry and normality and found that 'as the degree of violation of compound symmetry increased, the empirical power for the multivariate tests also increased. In contrast, the power for the univariate tests generally decreased' (p. 174). Maxwell & Delaney (1990) noted that the univariate test is relatively more powerful than the multivariate test as *n* decreases and proposed that 'the multivariate approach should probably not be used if *n* is less than $a + 10$ (*a* is the number of levels for repeated measures)' (p. 602). As a rule it seems that when you have a large violation of sphericity ($\varepsilon < .7$) and your sample size is greater than ($a + 10$) then multivariate procedures are more powerful, whilst with small sample sizes or when sphericity holds ($\varepsilon > .7$) the univariate approach is preferred (Stevens, 1992). It is also worth noting that the power of MANOVA increases and decreases as a function of the correlations between dependent variables (Cole, Maxwell, Arvey & Salas, 1994) and so the relationship between treatment conditions must be considered also.

11.3. THEORY OF ONE-WAY REPEATED-MEASURES ANOVA ②

In a repeated-measures ANOVA the effect of our experiment is shown up in the within-participant variance (rather than in the between-group variance). Remember that in independent ANOVA (section 8.2) the within-participant variance is our residual variance

(SS_R); it is the variance created by individual differences in performance. This variance is not contaminated by the experimental effect, because whatever manipulation we've carried out has been done on different people. However, when we carry out our experimental manipulation on the same people then the within-participant variance will be made up of two things: the effect of our manipulation and, as before, individual differences in performance. So, some of the within-subjects variation comes from the effects of our experimental manipulation: we did different things in each experimental condition to the participants, and so variation in an individual's scores will partly be due to these manipulations. For example, if everyone scores higher in one condition than another, it's reasonable to assume that this happened not by chance, but because we did something different to the participants in one of the conditions compared to any other one. *Because* we did the *same* thing to everyone within a particular condition, any variation that cannot be explained by the manipulations we've carried out must be due to random factors outside our control, unrelated to our experimental manipulations (we could call this 'error'). As in independent ANOVA, we use an *F*-ratio that compares the size of the variation due to our experimental manipulations with the size of the variation due to random factors, the only difference being how we calculate these variances. If the variance due to our manipulations is big relative to the variation due to random factors, we get a big value of *F*, and we can conclude that the observed results are unlikely to have arisen by chance.

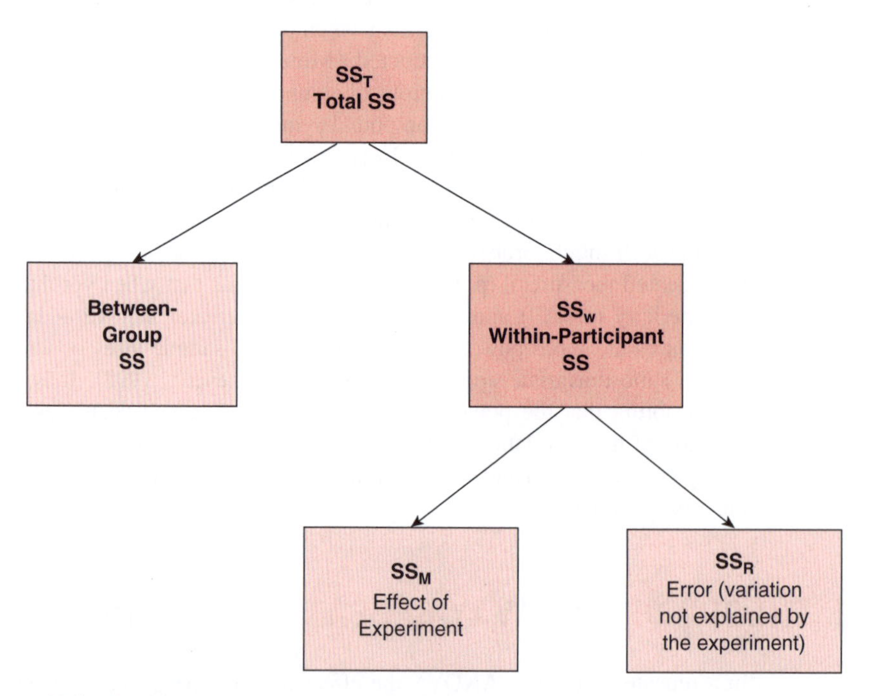

Figure 11.1 Partitioning variance for repeated-measures ANOVA

Figure 11.1 shows how the variance is partitioned in a repeated-measures ANOVA. The important thing to note is that we have the same types of variances as in independent ANOVA: we have a total sum of squares (SS_T), a model sum of squares (SS_M) and a residual sum of squares (SS_R). The *only* difference between repeated-measures and independent ANOVA is where those sums of squares come from: in repeated-measures ANOVA the model and residual sums of squares are both part of the within-participant variance. Let's have a look at an example.

There is often concern among students as to the consistency of marking between lecturers. It is common that lecturers obtain reputations for being 'hard' or 'light' markers (or to use the students' terminology, 'evil manifestations from Beelzebub's bowels' and 'nice people') but there is often little to substantiate these reputations. A group of students investigated the consistency of marking by submitting the same essays to four different lecturers. The mark given by each lecturer was recorded for each of the eight essays. It was important that the same essays were used for all lecturers because this eliminated any individual differences in the standard of work that each lecturer marked. This design is repeated-measures because every lecturer marked every essay. The independent variable was the lecturer who marked the report and the dependent variable was the percentage mark given.

Table 11.2 shows the data for this example. There were eight essays, each marked by four different lecturers. Their marks are shown in the table. In addition, the mean mark given by each lecturer is shown in the table, and also the mean mark that each essay received and the variance of marks for a particular essay. Now, the total variance within essays will in part be caused by the fact that different lecturers are harder or softer markers (the manipulation), and will, in part, be caused by the fact that the essays themselves will differ in quality (individual differences).

Table 11.2 Data for essay marks example

Essay	Tutor 1 (Dr Field)	Tutor 2 (Dr Smith)	Tutor 3 (Dr Scrote)	Tutor 4 (Dr Death)	Mean	S^2
1	62	58	63	64	**61.75**	**6.92**
2	63	60	68	65	**64.00**	**11.33**
3	65	61	72	65	**65.75**	**20.92**
4	68	64	58	61	**62.75**	**18.25**
5	69	65	54	59	**61.75**	**43.58**
6	71	67	65	50	**63.25**	**84.25**
7	78	66	67	50	**65.25**	**132.92**
8	75	73	75	45	**67.00**	**216.00**
Mean	**68.875**	**64.25**	**65.25**	**57.375**		

11.3.1. The total sum of squares (SS_T) ②

Remember from one-way independent ANOVA that SS_T is calculated using the following equation (see equation (8.4)):

$$SS_T = s^2_{grand}(N - 1)$$

Well, in repeated-measures designs the total sum of squares is calculated in exactly the same way. The grand variance in the equation is simply the variance of all scores when we ignore the group to which they belong. So if we treated the data as one big group they would look as follows:

62	58	63	64
63	60	68	65
65	61	72	65
68	64	58	61
69	65	54	59
71	67	65	50
78	66	67	50
75	73	75	45

Grand Mean = 63.9375

The variance of these scores is 55.028 (try this on your calculators). We used 32 scores to generate this value, and so N is 32. As such the equation becomes:

$$\begin{aligned} SS_T &= s^2_{grand}(N - 1) \\ &= 55.028(32 - 1) \\ &= 1705.868 \end{aligned}$$

The degrees of freedom for this sum of squares, as with the independent ANOVA, will be $N - 1$, or 31.

11.3.2. The within-participant (SS_W) ②

The crucial variation in this design is that there is a variance component called the within-participant variance (this arises because we've manipulated our independent variable within each participant). This is calculated using a sum of squares. Generally speaking, when we calculate any sum of squares we look at the squared difference between the mean and individual scores. This can be expressed in terms of the variance across a number of scores and the number of scores on which the variance is based. For example, when we calculated the residual sum of squares in independent ANOVA (SS_R) we used the following equation (look back to equation (8.7)):

$$SS_R = \sum (x_{ik} - \bar{x}_k)^2$$
$$SS_R = \sum s_k^2 (n_k - 1)$$

This equation gave us the variance between individuals within a particular group, and so is an estimate of individual differences within a particular group. Therefore, to get the total value of individual differences we have to calculate the sum of squares within each group and then add them up:

$$SS_R = s^2_{\text{group 1}}(n_1 - 1) + s^2_{\text{group 2}}(n_2 - 1) + s^2_{\text{group 3}}(n_3 - 1)$$

This is all well and good when we have different people in each group, but in repeated-measures designs we've subjected people to more than one experimental condition, and therefore, we're interested in the variation not within a group of people (as in independent ANOVA) but within an actual person. That is, how much variability is there within an individual? To find this out we actually use the same equation but we adapt it to look at people rather than groups. So, if we call this sum of squares SS_W (for within-participant SS) we could write it as:

$$SS_W = s^2_{\text{person 1}}(n_1 - 1) + s^2_{\text{person 2}}(n_2 - 1) + s^2_{\text{person 3}}(n_3 - 1) + \cdots + s^2_{\text{person } n}(n_n - 1)$$

This equation simply means that we are looking at the variation in an individual's scores and then adding these variances for all the people in the study. Some of you may have noticed that in our example, we're using essays rather than people, and so to be pedantic we'd write this as:

$$SS_W = s^2_{\text{essay 1}}(n_1 - 1) + s^2_{\text{essay 2}}(n_2 - 1) + s^2_{\text{essay 3}}(n_3 - 1) + \cdots + s^2_{\text{essay } n}(n_n - 1)$$

The *ns* simply represent the number of scores on which the variances are based (i.e. the number of experimental conditions, or in this case the number of lecturers). All of the variances we need are in Table 11.2, so we can calculate SS_W as:

$$\begin{aligned}
SS_W &= s^2_{\text{essay 1}}(n_1 - 1) + s^2_{\text{essay 2}}(n_2 - 1) + s^2_{\text{essay 3}}(n_3 - 1) + \cdots + s^2_{\text{essay } n}(n_n - 1) \\
&= (6.92)(4 - 1) + (11.33)(4 - 1) + (20.92)(4 - 1) + (18.25)(4 - 1) \\
&\quad + (43.58)(4 - 1) + (84.25)(4 - 1) + (132.92)(4 - 1) + (216)(4 - 1) \\
&= 20.76 + 34 + 62.75 + 54.75 + 130.75 + 252.75 + 398.75 + 648 \\
&= 1602.5
\end{aligned}$$

The degrees of freedom for each person are $n - 1$ (i.e. the number of conditions minus 1). To get the total degrees of freedom we add the *df* for all participants. So, with eight participants (essays) and four conditions (i.e. $n = 4$) we get $8 \times 3 = 24$ degrees of freedom.

11.3.3. The model sum of squares (SS_M) ②

So far, we know that the total amount of variation within the data is 1705.868 units. We also know that 1602.5 of those units are explained by the variance created by individuals' (essays') performances under different conditions. Now some of this variation is the result of our experimental manipulation and some of this variation is simply random fluctuation. The next step is to work out how much variance is explained by our manipulation and how much is not.

In independent ANOVA, we worked out how much variation could be explained by our experiment (the model SS) by looking at the means for each group and comparing these to the overall mean. So, we measured the variance resulting from the differences between group means and the overall mean (see equation (8.5)). We do exactly the same thing with a repeated-measures design. First we calculate the mean for each level of the independent variable (in this case the mean mark given by each lecturer), and compare these values with the overall mean of all marks.

So, we calculate this SS in the same way as for independent ANOVA:

1. Calculate the difference between the mean of each group and the grand mean.
2. Square each of these differences.
3. Multiply each result by the number of entities within that group (n_i).
4. Add the values for each group together.

$$SS_M = \sum n_k(\bar{x}_k - \bar{x}_{grand})^2$$

Using the means from the essay data (see Table 11.2), we can calculate SS_M as follows:

$$\begin{aligned} SS_M &= 8(68.875 - 63.9375)^2 + 8(64.25 - 63.9375)^2 + 8(65.25 - 63.9375)^2 \\ &\quad + 8(57.375 - 63.9375)^2 \\ &= 8(4.9375)^2 + 8(.3125)^2 + 8(1.3125)^2 + 8(-6.5625)^2 \\ &= 554.125 \end{aligned}$$

For SS_M, the degrees of freedom (df_M) are again 1 less than the number of things used to calculate the sum of squares. For the model sums of squares we calculated the sum of squared errors between the four means and the grand mean. Hence, we used four things to calculate these sums of squares. So, the degrees of freedom will be 3. So, as with independent ANOVA the model degrees of freedom are always the number of conditions (k) minus 1:

$$df_M = k - 1 = 3$$

11.3.4. The residual sum of squares (SS_R) ②

We now know that there are 1706 units of variation to be explained in our data, and that the variation across our conditions accounts for 1602 units. Of these 1602 units, our experimental manipulation can explain 554 units. The final sum of squares is the residual sum of squares (SS_R), which tells us how much of the variation cannot be explained by the model. This value is the amount of variation caused by extraneous factors outside of experimental control (such as natural variation in the quality of the essays). Knowing SS_W and SS_M already, the simplest way to calculate SS_R is to subtract SS_M from SS_W ($SS_R = SS_W - SS_M$):

$$\begin{aligned} SS_R &= SS_W - SS_M \\ &= 1602.5 - 554.125 \\ &= 1048.375 \end{aligned}$$

The degrees of freedom are calculated in a similar way:

$$\begin{aligned} df_R &= df_W - df_M \\ &= 24 - 3 \\ &= 21 \end{aligned}$$

11.3.5. The mean squares ②

SS_M tells us how much variation the model (e.g. the experimental manipulation) explains and SS_R tells us how much variation is due to extraneous factors. However, because both of these values are summed values, the number of scores that were summed influences them. As with independent ANOVA we eliminate this bias by calculating the average sum of squares (known as the *mean squares*, MS), which is simply the sum of squares divided by the degrees of freedom:

$$\begin{aligned} MS_M &= \frac{SS_M}{df_M} = \frac{554.125}{3} = 184.708 \\ MS_R &= \frac{SS_R}{df_R} = \frac{1048.375}{21} = 49.923 \end{aligned}$$

MS_M represents the average amount of variation explained by the model (e.g. the systematic variation), whereas MS_R is a gauge of the average amount of variation explained by extraneous variables (the unsystematic variation).

11.3.6. The *F*-ratio ②

The *F*-ratio is a measure of the ratio of the variation explained by the model and the variation explained by unsystematic factors. It can be calculated by dividing the model mean squares by the residual mean squares. You should recall that this is exactly the same as for independent ANOVA:

$$F = \frac{\mathrm{MS_M}}{\mathrm{MS_R}}$$

So, as with the independent ANOVA, the F-ratio is still the ratio of systematic variation to unsystematic variation. As such, it is the ratio of the experimental effect to the effect on performance of unexplained factors. For the marking data, the F-ratio is:

$$F = \frac{\mathrm{MS_M}}{\mathrm{MS_R}} = \frac{184.708}{49.923} = 3.70$$

This value is greater than 1, which indicates that the experimental manipulation had some effect above and beyond the effect of extraneous factors. As with independent ANOVA this value can be compared against a critical value based on its degrees of freedom (df_{M} and df_{R}, which are 3 and 21 in this case).

11.4. ONE-WAY REPEATED-MEASURES ANOVA USING SPSS ②

11.4.1. The main analysis ②

Sticking with the essay marks example, in Chapter 2 we came across the golden rule of the data editor: **each row represents data from one participant while each column represents a level of a variable**. Therefore, separate columns represent levels of a repeated-measures variable. As such, there is no need for a coding variable (as with between-group designs). The data are in Table 11.2 and can be entered into the SPSS data editor in the same format as this table (you don't need to include the columns labelled *Essay*, *mean* or s^2 as they were included only to clarify that the tutors marked the same pieces of work and to help explain how this ANOVA is calculated). To begin with, create a variable called **tutor1** and use the *labels* dialog box to give this variable a full title of 'Dr Field'. In the next column, create a variable called **tutor2**, and give this variable a full title of 'Dr Smith'. The principle should now be clear: so, apply it to create the remaining variables called **tutor3** and **tutor4**. These data can also be found in the file **TutorMarks.sav**.

To conduct an ANOVA using a repeated-measures design, select the *define factors* dialog box by following the menu path **Analyze⇒General Linear Model⇒Repeated Measures…**. In the *define factors* dialog box (Figure 11.2), you are asked to supply a name for the within-subject (repeated-measures) variable. In this case the repeated-measures

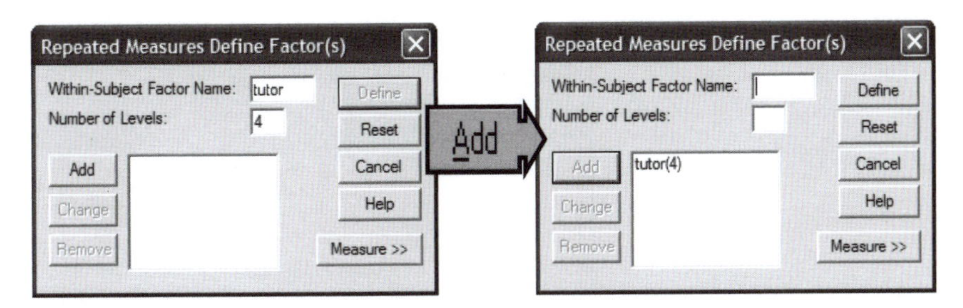

Figure 11.2 *Define factors* dialog box for repeated-measures ANOVA

variable was the lecturer who marked the report, so replace the word *factor1* with the word *tutor*. The name you give to the repeated-measures variable is restricted to eight characters. When you have given the repeated-measures factor a name, you have to tell the computer how many levels there were to that variable (i.e. how many experimental conditions there were). In this case, there were four tutors, so we have to enter the number 4 into the box labelled *Number of Levels*. Click on [Add] to add this variable to the list of repeated-measures variables. This variable will now appear in the white box at the bottom of the dialog box and appears as *tutor(4)*. If your design has several within-subject variables then you can add more factors to the list. When you have entered all of the within-subject factors that were measured click on [Define] to go to the main dialog box.

The main dialog box (Figure 11.3) has a space labelled *Within-Subjects Variables* that contains a list of four question marks followed by a number. These question marks are for the variables representing the four levels of the independent variable. The variables corresponding to these levels should be selected and placed in the appropriate space. We have only four variables in the data editor, so it is possible to select all four variables at once (by clicking on the variable at the top, holding the mouse button down and dragging down over the other variables). The selected variables can then be transferred by clicking on [▸]. When all four variables have been transferred, you can select various options for the analysis. There are several options that can be accessed with the buttons at the bottom of the main dialog box. These options are similar to the ones we have already encountered.

11.4.2. Defining contrasts for repeated measures ②

It is not possible to specify user-defined planned comparisons for repeated-measures designs in SPSS.[1] However, there is the option to conduct one of the many standard contrasts that

1 Actually, as I mentioned in the previous chapter, you can, but only using SPSS syntax. Those who are not already feeling like sticking their head in a mincing machine can read the file **ContrastsUsingSyntax.pdf** on the CD-ROM. Those who do feel like sticking their head in the mincing machine can read the file as well: it will have much the same effect (at least it did on me)!

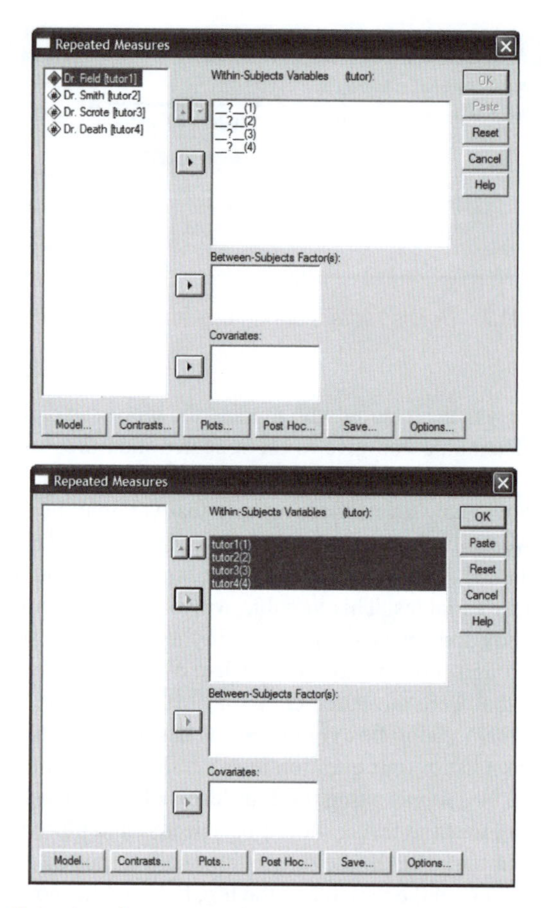

Figure 11.3 Main dialog box for repeated-measures ANOVA (before and after completion)

we have come across previously (see section 9.3.3 for details of changing contrasts). If you click on Contrasts... in the main dialog box you can access the *contrasts* dialog box (Figure 11.4). The default contrast is a polynomial contrast, but to change this default select a variable in the box labelled *Factors*, click on ▾ next to the box labelled *Contrast*, select a contrast from the list and then click on Change. If you choose to conduct a simple contrast then you can specify whether you would like to compare groups against the first or last category. The first category would be the one entered as (1) in the main dialog box and, for these data, the last category would be the one entered as (4). Therefore, the order in which you enter variables in the main dialog box is important for the contrasts you choose.

There is no particularly good contrast for the data we have (the simple contrast is not very useful because we have no control category) so I suggest using the *repeated* contrast, which will compare each tutor against the previous tutor. This contrast can be useful in repeated-measure designs in which the levels of the independent variable have a meaningful order. An example is if you have measured the dependent variable at successive

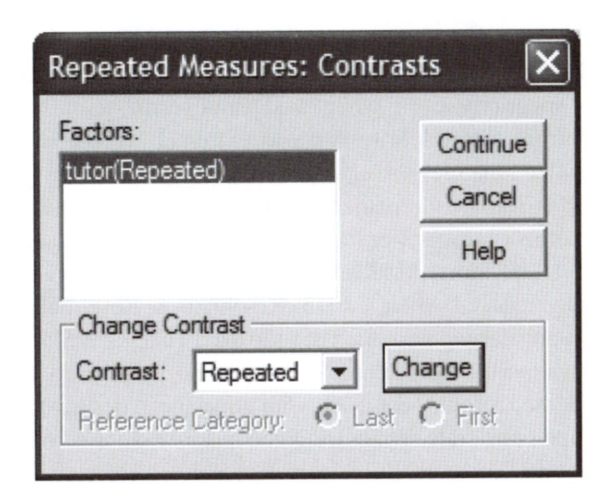

Figure 11.4 Repeated-measures contrasts

points in time, or administered increasing doses of a drug. When you have selected this contrast, click on Continue to return to the main dialog box.

11.4.3. *Post hoc* tests and additional options ③

There are important considerations when thinking about which *post hoc* tests to use. The violation of sphericity has implications for multiple comparisons. Boik (1981) provided an estimable account of the effects of non-sphericity on *post hoc* tests in repeated-measures designs, and concluded that even very small departures from sphericity produce large biases in the *F*-test. He recommends against using these tests for repeated-measures contrasts. When experimental error terms are small, the power to detect relatively strong effects can be as low as .05 (when sphericity = .80). Boik argues that the situation for *multiple* comparisons cannot be improved and concludes by recommending a multivariate analogue. Mitzel & Games (1981) found that when sphericity does not hold ($\varepsilon < 1$) the pooled error term conventionally employed in pairwise comparisons resulted in non-significant differences between two means declared as significant (i.e. a lenient Type I error rate) or undetected differences (a conservative Type I error rate). Mitzel & Games, therefore, recommended the use of separate error terms for each comparison. Maxwell (1980) systematically tested the power and alpha levels for five *post hoc* tests under repeated-measures conditions. The tests assessed were Tukey's wholly significant difference (WSD) test which uses a pooled error term; Tukey's procedure but with a separate error term with either $(n - 1)$ *df* (labelled SEP1) or $(n - 1)(k - 1)$ *df* (labelled SEP2); Bonferroni's procedure (BON); and a multivariate approach—the Roy–Bose simultaneous confidence interval (SCI). Maxwell tested these *a priori* procedures varying the sample size, number of levels of the repeated factor and departure from sphericity. He found that the multivariate approach was always 'too conservative for practical use' (p. 277) and this was most extreme when n (the number of subjects) is small relative to k (the number of conditions). Tukey's test inflated the alpha rate unacceptably with increasing departures from sphericity even when a separate error term was used (SEP1 and SEP2). The Bonferroni method, however, was extremely robust (although *slightly* conservative) and controlled alpha levels regardless of the manipulation. Therefore, in terms of Type I error rates the Bonferroni method was best.

In terms of test power (the Type II error rate) for a small sample ($n = 8$) Maxwell found WSD to be most powerful under conditions of non-sphericity, but this advantage was severely reduced when $n = 15$. Keselman & Keselman (1988) extended Maxwell's work within unbalanced designs. They too used Tukey's WSD, a modified WSD (with non-pooled error variance), Bonferroni t-statistics and a multivariate approach, and found that when unweighted means were used (with unbalanced designs) none of the four tests could control the Type I error rate. When weighted means were used only the multivariate tests could limit alpha rates although Bonferroni t-statistics were considerably better than the two Tukey methods. In terms of power Keselman & Keselman concluded that 'as the number of repeated treatment levels increases, BON is substantially more powerful than SCI' (p. 223).

In summary, when sphericity is violated the Bonferroni method seems to be generally the most robust of the univariate techniques, especially in terms of power and control of the Type I error rate. When sphericity is definitely not violated, Tukey's test can be used. In either case, the Games–Howell procedure, which uses a pooled error term, is preferable to Tukey's test.

For readers using versions of SPSS before version 7.0, this discussion is academic, because as those readers will discover there is no facility for producing *post hoc* tests for repeated-measures designs in these earlier versions! So, why have I included this discussion of which techniques are best? Well, for one thing it is possible to rerun the analysis as a between-group design and make use of the *post hoc* procedures. However, as noted above, most procedures perform very badly with related data (especially if sphericity has been violated) and so I strongly recommend against this approach. However, there are syntax files available for conducting repeated-measures *post hoc* tests (available at http://www.spss.com/tech/macros/). Of these macros the Dunn–Sidak method is probably best because it is less conservative than Bonferroni corrected comparisons.

To those readers who do not have access to the Internet, or find the syntax window puzzling, you can apply Bonferroni comparisons by using the paired t-test procedure. Conduct t-tests on all pairs of levels of the independent variable, then apply a Bonferroni correction to the probability at which you accept any of these tests. This correction is achieved by dividing the probability value (.05) by the number of t-tests conducted. The resulting probability value should be used as the criterion for statistical significance. So, for example, if we compared all levels of the independent variable of the essay data, we would make six comparisons in all and the appropriate significance level would be $.05/6 = .0083$. Therefore, we would accept t-tests that had a significance value less than .0083. One way to salvage what power you can from this procedure is to compare only groups between which you expect differences to arise (rather than comparing all pairs of treatment levels). The fewer tests you perform, the less you have to correct the significance level, and the more power you retain.

The good news for people using SPSS version 7.5 and beyond is that some *post hoc* procedures are available for repeated measures. However, they are not accessed through the usual *post hoc* test dialog box. Instead, they can be found as part of the additional options. These options can be accessed by clicking on [Options...] in the main dialog box to open the *GLM repeated measures: options* dialog box (see Figure 11.5). To specify *post hoc* tests, select the repeated-measures variable (in this case **tutor**) from the box labelled *Estimated Marginal Means: Factor(s) and Factor Interactions* and transfer it to the box labelled *Display Means for* by clicking on [▸]. Once a variable has been transferred, the box labelled *Compare main effects* becomes active and you should select this option ([✓ Compare main effects]). If this option is selected, the box labelled *Confidence interval adjustment* becomes active and you can click on [▾] to see a choice of three adjustment levels. The default is to have no adjustment and simply perform a Tukey LSD *post hoc* test (this is not recommended). The second option is a Bonferroni correction (recommended for the reasons mentioned above), and the final option is a Sidak correction, which should be selected if you are concerned about the loss of power associated with Bonferroni corrected values.

The *options* dialog box (Figure 11.5) has other useful options too. You can ask for descriptive statistics, which will provide the means, standard deviations and number of participants for each level of the independent variable. You can also ask for a transformation matrix, which provides the coding values for any contrast selected in the *contrasts* dialog box (Figure 11.4) and is very useful for interpreting the contrasts in more complex designs. SPSS can also be asked to print out the hypothesis, error, and residual sum of squares and cross-product matrices (SSCPs) and we shall learn about the importance of these matrices in Chapter 14. The option for homogeneity of variance tests will be active

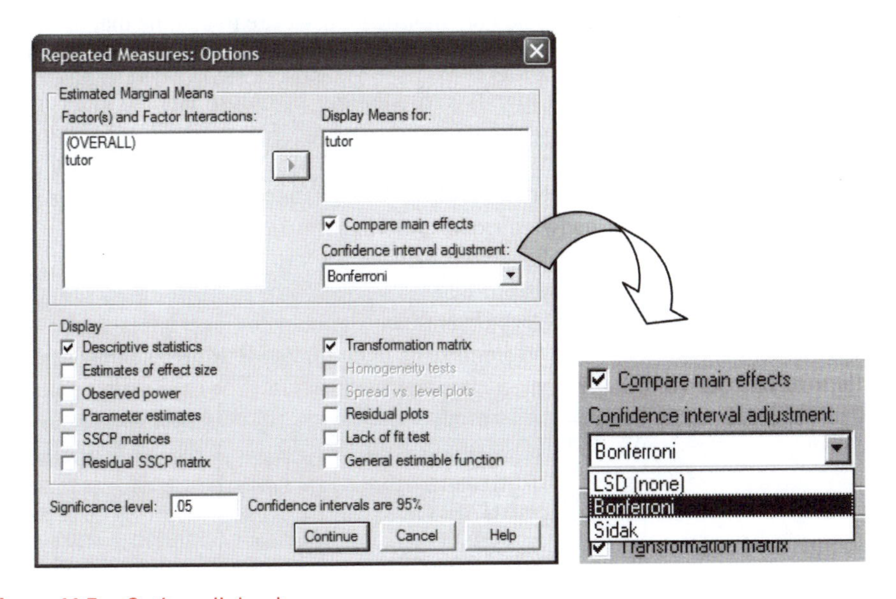

Figure 11.5 *Options* dialog box

only when there is a between-group factor as well (mixed designs—see the next chapter). You can also change the level of significance at which to test any *post hoc* tests; generally, the .05 level is acceptable. When you have selected the options of interest, click on Continue to return to the main dialog box, and then click on OK to run the analysis.

11.5. OUTPUT FOR ONE-WAY REPEATED-MEASURES ANOVA ②

11.5.1. Descriptives and other diagnostics ①

SPSS Output 11.1 shows the initial diagnostic statistics. First, we are told the variables that represent each level of the independent variable. This box is useful to check that the variables were entered in the correct order. The next table provides basic descriptive statistics for the four levels of the independent variable. From this table we can see that, on average, Dr Field gave the highest marks to the essays (that's because I'm nice you see … or it could be because I'm stupid and so have low academic standards!). Dr Death, on the other hand, gave very low grades. These mean values are useful for interpreting any effects that may emerge from the main analysis.

11.5.2. Assessing and correcting for sphericity ②

In section 11.2.3 you were told that SPSS produces a test of whether the data violate the assumption of sphericity. The next part of the output contains information about this test. Mauchly's test should be non-significant if we are to assume that the condition of sphericity has been met. SPSS Output 11.2 shows Mauchly's test for the tutor data, and the important column is the one containing the significance value. The significance value (.043) is less than the critical value of .05, so we accept that the variances of the differences between levels are significantly different. In other words, the assumption of sphericity has been violated. Knowing that we have violated this assumption a pertinent question is: how should we proceed?

Within-Subjects Factors

Measure: MEASURE_1

TUTOR	Dependent Variable
1	TUTOR1
2	TUTOR2
3	TUTOR3
4	TUTOR4

Descriptive Statistics

	Mean	Std. Deviation	N
Dr. Field	68.8750	5.6426	8
Dr. Smith	64.2500	4.7132	8
Dr. Scrote	65.2500	6.9230	8
Dr. Death	57.3750	7.9091	8

SPSS Output 11.1

Mauchly's Test of Sphericity[a]

Measure: MEASURE_1

Within Subjects Effect	Mauchly's W	Approx. Chi-Square	df	Sig.	Epsilon[b]		
					Greenhouse-Geisser	Huynh-Feldt	Lower-bound
TUTOR	.131	11.628	5	.043	.558	.712	.333

Tests the null hypothesis that the error covariance matrix of the orthonormalized transformed dependent variables is proportional to an identity matrix.
a Design: Intercept Within Subjects Design: TUTOR
b May be used to adjust the degrees of freedom for the averaged tests of significance. Corrected tests are displayed in the layers (by default) of the Tests of Within Subjects Effects table

SPSS Output 11.2

We discovered in section 11.2.5 that SPSS produces three corrections based upon the estimates of sphericity advocated by Greenhouse & Geisser (1959) and Huynh & Feldt (1976). Both of these estimates give rise to a correction factor that is applied to the degrees of freedom used to assess the observed F-ratio. The *Greenhouse–Geisser correction* varies between $1/k - 1$ (where k is the number of repeated-measures conditions) and 1. The closer that $\hat{\varepsilon}$ is to 1.00, the more homogeneous the variances of differences, and hence the closer the data are to being spherical. In a situation in which there are four conditions (as with our data) the lower limit of $\hat{\varepsilon}$ will be $1/(4 - 1)$, or .33 (known as the lower-bound estimate of sphericity). SPSS Output 11.2 shows that the calculated value of $\hat{\varepsilon}$ is .558. This is closer to the lower limit of .33 than it is to the upper limit of 1 and it therefore represents a substantial deviation from sphericity. We shall see how these values are used in the next section.

11.5.3. The main ANOVA ②

SPSS Output 11.3 shows the results of the ANOVA for the within-subjects variable. This table can be read much the same as for one-way between-group ANOVA (see Chapter 8). There is a sum of squares for the repeated-measures effect of **tutor**, which tells us how much of the total variability is explained by the experimental effect. Note the value is 554.125, which is the model sum of squares (SS_M) that we calculated in section 11.3.3. There is also an error term, which is the amount of unexplained variation across the conditions of the repeated-measures variable. This is the residual sum of squares (SS_R) that was calculated in section 11.3.4; note that the value is 1048.375 (which is the same value as calculated). As I explained earlier, these sums of squares are converted into mean squares by dividing by the degrees of freedom. As we saw before, the df for the effect of **tutor** is simply $k - 1$, where k is the number of levels of the independent variable. The error df is $(n - 1)(k - 1)$, where n is the number of participants (or in this case, the number of essays) and *k* is as before. The *F*-ratio is obtained by dividing the mean squares for the experimental effect (184.708) by the error mean squares (49.923). As with between-group ANOVA, this test statistic represents the ratio of systematic variance to unsystematic variance. The value of *F* (184.71/49.92 = 3.70) is then compared against a critical value for 3 and 21 degrees of freedom. SPSS displays the exact significance level for the *F*-ratio. The significance of *F* is .028, which is significant because it is less than the criterion value of .05. We can, therefore, conclude that there was a significant difference between the marks awarded by the four lecturers. However, this main test does not tell us which lecturers differed from each other in their marking.

Although this result seems very plausible, we have learnt that the violation of the sphericity assumption makes the *F*-test inaccurate. We know from SPSS Output 11.2 that these data were non-spherical and so we need to make allowances for this violation. The table in SPSS Output 11.3 shows the *F*-ratio and associated degrees of freedom when sphericity is assumed and the significant *F*-statistic indicated some difference(s) between the mean marks given by the four lecturers. In versions of SPSS after version 8, this table

Tests of Within-Subjects Effects

Measure: MEASURE_1

Source		Type III Sum of Squares	df	Mean Square	F	Sig.
TUTOR	Sphericity Assumed	554.125	3	184.708	3.700	.028
	Greenhouse-Geisser	554.125	1.673	331.245	3.700	.063
	Huynh-Feldt	554.125	2.137	259.329	3.700	.047
	Lower-bound	554.125	1.000	554.125	3.700	.096
Error(TUTOR)	Sphericity Assumed	1048.375	21	49.923		
	Greenhouse-Geisser	1048.375	11.710	89.528		
	Huynh-Feldt	1048.375	14.957	70.091		
	Lower-bound	1048.375	7.000	149.768		

a Computed using alpha = .05

Tests of Within-Subjects Effects

Measure: MEASURE_1

Sphericity Assumed

Source	Type III Sum of Squares	df	Mean Square	F	Sig.
TUTOR	554.125	3	184.708	3.700	.028
Error(TUTOR)	1048.375	21	49.923		

a Computed using alpha = .05

SPSS Output 11.3 Repeated-measures ANOVA for versions 8.0 and later (top) and versions 7.0 and 7.5 (bottom)

Tests of Within-Subjects Effects

Measure		Source	Type III Sum of Squares	df	Mean Square	F	Sig.
MEASURE_1	Sphericity Assumed	TUTOR	554.125	3	184.708	3.700	.028
		Error(TUTOR)	1048.375	21	49.923		
	Greenhouse-Geisser	TUTOR	554.125	1.673	331.245	3.700	.063
		Error(TUTOR)	1048.375	11.710	89.528		
	Huynh-Feldt	TUTOR	554.125	2.137	259.329	3.700	.047
		Error(TUTOR)	1048.375	14.957	70.091		
	Lower-bound	TUTOR	554.125	1.000	554.125	3.700	.096
		Error(TUTOR)	1048.375	7.000	149.768		

a Computed using alpha = .05

SPSS Output 11.4

also contains several additional rows giving the corrected values of *F* for the three different types of adjustment (Greenhouse–Geisser, Huynh–Feldt and lower-bound). In versions 7.0 and 7.5 you have to adjust the ANOVA table to see these corrected values. Although most people won't have these versions, it's worth noting that if you use the mouse to double-click on the main ANOVA table, then the table will open up for editing and a new set of menu labels should appear at the top of the window. One of these menus is labelled *Pivot*. Click on this menu and then select the option labelled *Move Layers to Rows*. The corrected *F* values should now be displayed.

SPSS Output 11.4 shows the expanded ANOVA table with the corrected values for each of the three estimates of sphericity and this table now looks like the table from more recent versions of SPSS (as in SPSS Output 11.3). Notice that in all cases the *F*-ratios remain the same; it is the degrees of freedom that change (and hence the critical value against which the obtained *F*-statistic is compared). The degrees of freedom have been adjusted using the estimates of sphericity calculated in SPSS Output 11.2. The adjustment is made by multiplying the degrees of freedom by the estimate of sphericity (see Field, 1998a).[2] The new degrees of freedom are then used to ascertain the significance of *F*. For these data the corrections result in the observed *F* being non-significant when using the Greenhouse–Geisser correction (because $p > .05$). However, it was noted earlier that this correction is quite conservative, and so can miss effects that genuinely exist. It is, therefore, useful to consult the Huynh–Feldt corrected *F*-statistic. Using this correction, the *F*-value is still significant because the probability value of .047 is just below the criterion value of .05.

---

2 For example, the Greenhouse–Geisser estimate of sphericity was .558. The original degrees of freedom for the model were 3; this value is corrected by multiplying by the estimate of sphericity ($3 \times .558 = 1.674$). Likewise the error *df* was 21; this value is corrected in the same way ($21 \times .558 = 11.718$). The *F*-ratio is then tested against a critical value with these new degrees of freedom (1.674, 11.718). The other corrections are applied in the same way.

So, by this correction we would accept the hypothesis that the lecturers differed in their marking. However, it was also noted earlier that this correction is quite liberal and so tends to accept values as significant when, in reality, they are not significant.

This leaves us with the puzzling dilemma of whether or not to accept this *F*-statistic as significant. I mentioned earlier that Stevens (1992) recommends taking an average of the two estimates, and certainly when the two corrections give different results (as is the case here) this is wise advice. If the two corrections give rise to the same conclusion it makes little difference which you choose to report (although if you accept the *F*-statistic as significant it is best to report the conservative Greenhouse–Geisser estimate to avoid criticism!). Although it is easy to calculate the average of the two correction factors and to correct the degrees of freedom accordingly, it is not so easy then to calculate an exact probability for those degrees of freedom. Therefore, should you ever be faced with this perplexing situation (and to be honest that's fairly unlikely) I recommend taking an average of the two significance values to give you a rough idea of which correction is giving the most accurate answer. In this case, the average of the two *p*-values is $(.063 + .047)/2 = .055$. Therefore, we should probably go with the Greenhouse–Geisser correction and conclude that the *F*-ratio is non-significant.

These data illustrate how important it is to use a valid critical value of *F*: it can mean the difference between a statistically significant result and a non-significant result. More important it can mean the difference between making a Type I error and not. Had we not used the corrections for sphericity we would have concluded erroneously that the markers gave significantly different marks. However, I should quantify this statement by saying that this example also highlights how arbitrary it is that we use a .05 level of significance. These two corrections produce significance values only marginally less than or more than .05, and yet they lead to completely opposite conclusions! So, we might be well advised to look at an effect size to see whether the effect is substantive regardless of its significance.

Multivariate Tests[a]

Effect		Value	F	Hypothesis df	Error df	Sig.
TUTOR	Pillai's Trace	.741	4.760[c]	3.000	5.000	.063
	Wilks' Lambda	.259	4.760[c]	3.000	5.000	.063
	Hotelling's Trace	2.856	4.760[c]	3.000	5.000	.063
	Roy's Largest Root	2.856	4.760[c]	3.000	5.000	.063

a Design: Intercept Within Subjects Design: TUTOR
b Computed using alpha = .05
c Exact statistic

SPSS Output 11.5

We also saw earlier that a final option when you have data that violate sphericity is to use multivariate test statistics, because they do not make this assumption (see O'Brien & Kaiser, 1985). SPSS Output 11.5 shows the multivariate test statistics for this example (details of these test statistics can be found in section 14.3.4). The column displaying the

TUTOR[a]

Measure: MEASURE_1

Dependent Variable	TUTOR		
	Level 1 vs. Level 2	Level 2 vs. Level 3	Level 3 vs. Level 4
Dr. Field	1	0	0
Dr. Smith	−1	1	0
Dr. Scrote	0	−1	1
Dr. Death	0	0	−1

a The contrasts for the within subjects factors are:
TUTOR: Repeated contrast

SPSS Output 11.6

significance values clearly shows that the multivariate tests are non-significant (because p is .063, which is greater than the criterion value of .05). Bearing in mind the loss of power in these tests (see section 11.2.5) this result supports the decision to accept the null hypothesis and conclude that there are no significant differences between the marks given by different lecturers. The interpretation of these results should stop now because the main effect is non-significant. However, we will look at the output for contrasts to illustrate how these tests are displayed in the SPSS viewer.

11.5.4. Contrasts ②

The transformation matrix requested in the options is shown in SPSS Output 11.6 and we have to draw on our knowledge of contrast coding (see Chapter 8) to interpret this table. The first thing to remember is that a code of 0 means that the group is not included in a contrast. Therefore, contrast 1 (labelled *Level 1 vs. Level 2* in the table) ignores Dr Scrote and Dr Death. The next thing to remember is that groups with a negative weight are compared to groups with a positive weight. In this case this means that the first contrast compares Dr Field against Dr Smith. Using the same logic, contrast 2 (labelled *Level 2 vs. Level 3*) ignores Dr Field and Dr Death and compares Dr Smith and Dr Scrote. Finally, contrast three (*Level 3 vs. Level 4*) compares Dr Death with Dr Scrote. This pattern of contrasts is consistent with what we expect to get from a repeated contrast (i.e. all groups except the first are compared to the preceding category). The transformation matrix, which appears at the bottom of the output, is used primarily to confirm what each contrast represents.

Above the transformation matrix, we should find a summary table of the contrasts (SPSS Output 11.7). Each contrast is listed in turn, and as with between-group contrasts, an F-test is performed that compares the two chunks of variation. So, looking at the significance values from the table, we could say that Dr Field marked significantly more highly than Dr Smith (*Level 1 vs. Level 2*), but that Dr Smith's marks were roughly equal to Dr Scrote's (*Level 2 vs. Level 3*) and Dr Scrote's marks were roughly equal to Dr Death's (*Level 3 vs. Level 4*). However, the significant contrast should be ignored because of

Tests of Within-Subjects Contrasts

Measure: MEASURE_1

Source	TUTOR	Type III Sum of Squares	df	Mean Square	F	Sig.
TUTOR	Level 1 vs. Level 2	171.125	1	171.125	18.184	.004
	Level 2 vs. Level 3	8.000	1	8.000	.152	.708
	Level 3 vs. Level 4	496.125	1	496.125	3.436	.106
Error(TUTOR)	Level 1 vs. Level 2	65.875	7	9.411		
	Level 2 vs. Level 3	368.000	7	52.571		
	Level 3 vs. Level 2	1010.875	7	144.411		

SPSS Output 11.7

the non-significant main effect (remember that the data did not obey sphericity). The important point to note is that the sphericity in our data has led to some important issues being raised about correction factors, and about applying discretion to your data (it's comforting to know that the computer does not have all of the answers—but it's slightly alarming to realize that this means we have actually to know some of the answers ourselves). In this example we would have to conclude that no significant differences existed between the marks given by different lecturers. However, the ambiguity of our data might make us consider running a similar study with a greater number of essays being marked.

11.5.5. *Post hoc* tests ②

If you selected *post hoc* tests for the repeated-measures variable in the *options* dialog box (see section 11.4.3), then the table in SPSS Output 11.8 will be produced in the output viewer window.

The arrangement of the table in SPSS Output 11.8 is similar to the table produced for between-group *post hoc* tests: the difference between group means is displayed, as well as the standard error, the significance value and a confidence interval for the difference between means. By looking at the significance values we can see that the only difference between group means is between Dr Field and Dr Smith. Looking at the means of these groups (SPSS Output 11.1) we can see that I give significantly higher marks than Dr Smith. However, there is a rather anomalous result in that there is no significant difference between the marks given by Dr Death and myself even though the mean difference between our marks is higher (11.5) than the mean difference between myself and Dr Smith (4.6). The reason for this result is the sphericity in the data. You might like to run some correlations between the four tutors' grades. You will find that there is a very high positive correlation between the marks given by Dr Smith and myself (indicating a low level of variability in our data). However, there is a very low correlation between the marks given by Dr Death and myself (indicating a high level of variability between our marks). It is this large variability between Dr Death and myself that has produced the non-significant result despite the average marks being very different (this observation is also evident from the standard errors).

Pairwise Comparisons

Measure: MEASURE_1

(I) TUTOR	(J) TUTOR	Mean Difference (I–J)	Std. Error	Sig.[a]	95% Confidence Interval for Difference[a]	
					Lower Bound	Upper Bound
1	2	4.625*	1.085	.022	.682	8.568
	3	3.625	2.841	1.000	−6.703	13.953
	4	11.500	4.675	.261	−5.498	28.498
2	1	−4.625*	1.085	.022	−8.568	−.682
	3	−1.000	2.563	1.000	−10.320	8.320
	4	6.875	4.377	.961	−9.039	22.789
3	1	−3.625	2.841	1.000	−13.953	6.703
	2	1.000	2.563	1.000	−8.320	10.320
	4	7.875	4.249	.637	−7.572	23.322
4	1	−11.500	4.675	.261	−28.498	5.498
	2	−6.875	4.377	.961	−22.789	9.039
	3	−7.875	4.249	.637	−23.322	7.572

Based on estimated marginal means

* The mean difference is significant at the .05 level

a Adjustment for multiple comparisons: Bonferroni

SPSS Output 11.8

Cramming Samantha's Tips

- The one-way repeated-measures ANOVA compares several means, when those means have come from the same participants; for example, if you measured people's statistical ability each month over a year-long course.
- In repeated-measures ANOVA there is an additional assumption: *sphericity*. This only needs to be considered when you have three or more repeated-measures conditions. Test for sphericity using *Mauchly's test*. Find the table with this label: if the value in the column labelled *Sig.* is less than .05 then the assumption is violated. If the significance of Mauchly's test is greater than .05 then the assumption of sphericity has been met.
- The table labelled *Tests of Within-Subjects Effects* shows the main result of your ANOVA. If the assumption of sphericity has been met then look at the row labelled *Sphericity Assumed*. If the assumption was violated then read the row labelled *Greenhouse–Geisser* (you can also look at *Huynh–Feldt* but you'll have to read this chapter to find out the relative merits of the two procedures). Having selected the appropriate row, look at the column labelled *Sig.*: if the value is less than .05 then the means of the groups are significantly different.
- For contrasts and *post hoc* tests, again look at the columns labelled *Sig.* to discover if your comparisons are significant (they will be if the significance value is less than .05).

11.6. EFFECT SIZES FOR REPEATED-MEASURES ANOVA ③

As with independent ANOVA the best measure of the overall effect size is omega squared (ω^2). However, just to make life even more complicated than it already is, the equations we've previously used for omega squared can't be used for repeated-measures data! If you do use the same equation on repeated-measures data it will slightly overestimate the effect size. For the sake of simplicity some people do use the same equation for one-way independent and repeated-measures ANOVAs (and I'm guilty of this in another book), but I'm afraid that in this book we're going to hit simplicity in the face, and embrace complexity like our long-lost twin from whom we've been separated for many years.

In repeated-measures ANOVA, the equation for omega squared is (hang onto your hats…):

$$\omega^2 = \frac{[\frac{k-1}{nk}(\text{MS}_\text{M} - \text{MS}_\text{R})]}{\text{MS}_\text{R} + \frac{\text{MS}_\text{BG} - \text{MS}_\text{R}}{k} + [\frac{k-1}{nk}(\text{MS}_\text{M} - \text{MS}_\text{R})]} \tag{11.1}$$

OK, this looks hellish, and, to be fair, it is. But let's break it down. First, there are some mean squares that we've come across before (and calculated before). There's the mean square for the model (MS_M) and the residual mean square (MS_R), both of which can be obtained from the ANOVA table that SPSS produces (SPSS Output 11.3). There's also k, the number of conditions in the experiment, which for these data would be 4 (there were four lecturers), and there's n, the number of people that took part (in this case, the number of essays, 8). The main problem is this term MS_BG. Back at the beginning of section 11.3 (Figure 11.1) I mentioned that the total variation is broken down into a within-participant variation and a between-group variation. In all the subsequent calculations the between-group variation sort of got forgotten about (because we don't actually need it to calculate the F-ratio). Then, lo and behold, it rears its ugly head in the equation for omega squared. The easiest way to calculate this term is by subtraction, because we know from Figure 11.1 that:

$$\text{SS}_\text{T} = \text{SS}_\text{BG} + \text{SS}_\text{W}$$

Now, SPSS doesn't give us SS_W in the output, but we know that this is made up of SS_M and SS_R, which we are given. By substituting these terms and rearranging the equation we get:

$$\text{SS}_\text{T} = \text{SS}_\text{BG} + \text{SS}_\text{M} + \text{SS}_\text{R}$$
$$\text{SS}_\text{BG} = \text{SS}_\text{T} - \text{SS}_\text{M} - \text{SS}_\text{R}$$

The next problem is that SPSS, which is clearly trying to hinder us at every step, doesn't give us SS_T and I'm afraid (unless I've missed something in the output) you're just going to have to calculate it by hand (see section 11.3.1). From the values we calculated earlier, you should get:

$$\begin{aligned} SS_{BG} &= 1705.868 - 554.125 - 1048.375 \\ &= 103.37 \end{aligned}$$

The next step is to convert this to a mean squares by dividing by the degrees of freedom, which in this case are the number of people in the experiment minus 1 ($n - 1$):

$$\begin{aligned} MS_{BG} &= \frac{SS_{BG}}{df_{BG}} = \frac{SS_{BG}}{n-1} \\ &= \frac{103.37}{8-1} \\ &= 14.77 \end{aligned}$$

Having done all this and probably died of boredom in the process, we must now resurrect ourselves with renewed vigour for the effect size equation, which becomes:

$$\begin{aligned} \omega^2 &= \frac{[\frac{4-1}{8\times 4}(184.71 - 49.92)]}{49.92 + \frac{14.77-49.92}{4} + [\frac{4-1}{8\times 4}(184.71 - 49.92)]} \\ &= \frac{12.64}{53.77} \\ &= .24 \end{aligned}$$

So, we get an omega squared of .24. If you calculate it the same way as for the independent ANOVA (check back to section 8.5) you should get a slightly bigger answer (.25 in fact).

I've mentioned at various other points that it's actually more useful to have effect size measures for focused comparisons anyway (rather than the main ANOVA), and so a slightly easier approach to calculating effect sizes is to calculate them for the contrasts we did (see SPSS Output 11.7). For these we can use the equation that we've seen before to convert the F-values (because they all have 1 degree of freedom for the model) to r:

$$r = \sqrt{\frac{F(1, df_R)}{F(1, df_R) + df_R}}$$

For the three comparisons we did, we would get:

$$r_{\text{Field vs. Smith}} = \sqrt{\frac{18.18}{18.18 + 7}} = .85$$

$$r_{\text{Smith vs. Score}} = \sqrt{\frac{.15}{.15 + 7}} = .14$$

$$r_{\text{Scrote vs. Death}} = \sqrt{\frac{3.44}{3.44 + 7}} = .57$$

Therefore, the differences between Drs Field and Smith and Scrote and Death were both large effects, but the differences between Drs Smith and Scrote were small.

11.7. REPORTING ONE-WAY REPEATED-MEASURES ANOVA ②

When we report repeated-measures ANOVA, we give the same details as with an independent ANOVA. The only additional thing we should concern ourselves with is reporting the corrected degrees of freedom if sphericity was violated. Personally, I'm also keen on reporting the results of sphericity tests as well. As with the independent ANOVA the degrees of freedom used to assess the *F*-ratio are the degrees of freedom for the effect of the model (df_M = 1.67) and the degrees of freedom for the residuals of the model (df_R = 11.71). Remember that in this example we corrected both using Greenhouse–Geisser corrected estimates of sphericity. Therefore, we could report the main finding as:

✓ The results show that the mark of an essay was not significantly affected by the lecturer that marked it, $F(1.67, 11.71) = 3.70, p > .05$.

If you chose to report the sphericity test as well, you should report the chi-square approximation, its degrees of freedom and the significance value. It's also nice to report the degree of sphericity by reporting the epsilon value. We'll also report the effect size in this improved version:

✓ Mauchly's test indicated that the assumption of sphericity had been violated ($\chi^2(5) = 11.63, p < .05$); therefore degrees of freedom were corrected using Greenhouse–Geisser estimates of sphericity ($\varepsilon = .56$). The results show that the mark of an essay was not significantly affected by the lecturer that marked it, $F(1.67, 11.71) = 3.70, p > .05, \omega^2 = .24$.

Remember that because the main ANOVA was not significant we shouldn't report any further analysis.

11.8. REPEATED MEASURES WITH SEVERAL INDEPENDENT VARIABLES ②

We have seen already that simple between-group designs can be extended to incorporate a second (or third) independent variable. It is equally easy to incorporate a second, third or even fourth independent variable into a repeated-measures analysis. As an example, some social scientists were asked to research whether imagery could influence public attitudes towards alcohol. There is evidence that attitudes towards stimuli can be changed using positive and negative imagery (e.g. Stuart, Shimp & Engle, 1987; but see Field & Davey, 1999) and these researchers were interested in answering two questions. On the one hand, the government had funded them to look at whether negative imagery in advertising could be used to change attitudes towards alcohol. Conversely, an alcohol company had provided funding to see whether positive imagery could be used to improve attitudes towards alcohol. The scientists designed a study to address both issues. Table 11.3 illustrates the experimental design and contains the data for this example (each row represents a single participant).

Participants viewed a total of nine mock adverts over three sessions. In one session, they saw three adverts: (1) a brand of beer (Brain Death) presented with a negative image (a dead body with the slogan 'drinking Brain Death makes your liver explode'); (2) a brand of wine (Dangleberry) presented in the context of a positive image (a sexy naked man or woman—depending on the participant's gender—and the slogan 'drinking Dangleberry wine makes you a horny stud muffin'); and (3) a brand of water (Puritan) presented alongside a neutral image (a person watching television accompanied by the slogan 'drinking Puritan water makes you behave completely normally'). In a second session (a week later), the participants saw the same three brands, but this time Brain Death was accompanied by the positive imagery, Dangleberry by the neutral image, and Puritan by the negative. In a third session, the participants saw Brain Death accompanied by the neutral image, Dangleberry by the negative image, and Puritan by the positive. After each advert participants were asked to rate the drinks on a scale ranging from −100 (dislike very much) through 0 (neutral) to 100 (like very much). The order of adverts was randomized, as was the order in which people participated in the three sessions. This design is quite complex. There are two independent variables: the type of drink (beer, wine or water) and the type of imagery used (positive, negative or neutral). These two variables completely cross over, producing nine experimental conditions.

11.8.1. The main analysis ②

To enter these data into SPSS we need to recap the golden rule of the data editor, which states that each row represents a single participant's data. If a person participates in all experimental conditions (in this case they see all types of stimuli presented with all types of imagery) then each experimental condition must be represented by a column in the data editor. In this experiment there are nine experimental conditions and so the data need to be entered in nine columns (so, the format is identical to Table 11.3). You should create the following nine variables in the data editor with the names as given. For each one, you should also enter a full variable name (see section 2.4.2) for clarity in the output.

beerpos	Beer	+	Sexy Person
beerneg	Beer	+	Corpse
beerneut	Beer	+	Person in Armchair
winepos	Wine	+	Sexy Person
wineneg	Wine	+	Corpse
wineneut	Wine	+	Person in Armchair
waterpos	Water	+	Sexy Person
waterneg	Water	+	Corpse
waterneu	Water	+	Person in Armchair

Table 11.3 Data from **Attitude.sav**

Drink	Beer			Wine			Water		
Image	+ve	−ve	Neut	+ve	−ve	Neut	+ve	−ve	Neut
	1	6	5	38	−5	4	10	−14	−2
	43	30	8	20	−12	4	9	−10	−13
	15	15	12	20	−15	6	6	−16	1
	40	30	19	28	−4	0	20	−10	2
Male	8	12	8	11	−2	6	27	5	−5
	17	17	15	17	−6	6	9	−6	−13
	30	21	21	15	−2	16	19	−20	3
	34	23	28	27	−7	7	12	−12	2
	34	20	26	24	−10	12	12	−9	4
	26	27	27	23	−15	14	21	−6	0
	1	−19	−10	28	−13	13	33	−2	9
	7	−18	6	26	−16	19	23	−17	5
	22	−8	4	34	−23	14	21	−19	0
	30	−6	3	32	−22	21	17	−11	4
Female	40	−6	0	24	−9	19	15	−10	2
	15	−9	4	29	−18	7	13	−17	8
	20	−17	9	30	−17	12	16	−4	10
	9	−12	−5	24	−15	18	17	−4	8
	14	−11	7	34	−14	20	19	−1	12
	15	−6	13	23	−15	15	29	−1	10

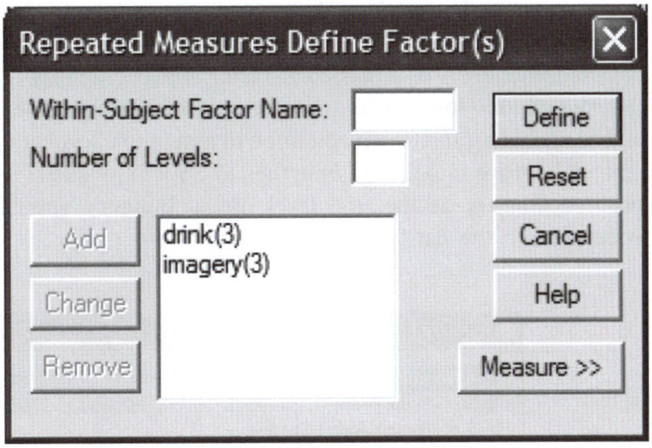

Figure 11.6 *Define factors* dialog box for factorial repeated-measures ANOVA

Once these variables have been created, enter the data as in Table 11.3. If you have problems entering the data then use the file **Attitude.sav**. To access the *define factors* dialog box use the menu path **Analyze⇒General Linear Model⇒Repeated Measures…**. In the *define factors* dialog box you are asked to supply a name for the within-subject (repeated measures) variable. In this case there are two within-subject factors: **drink** (beer, wine or water) and **imagery** (positive, negative and neutral). Replace the word *factor1* with the word *drink*. When you have given this repeated-measures factor a name, you have to tell the computer how many levels there were to that variable. In this case, there were three types of drink, so we have to enter the number 3 into the space labelled *Number of Levels*. Click on Add to add this variable to the list of repeated-measures variables. This variable will now appear in the white box at the bottom of the dialog box and appears as *drink(3)*. We now have to repeat this process for the second independent variable. Enter the word *imagery* into the space labelled *Within-Subject Factor Name* and then, because there were three levels of this variable, enter the number 3 into the space labelled *Number of Levels*. Click on Add to include this variable in the list of factors; it will appear as *imagery(3)*. The finished dialog box is shown in Figure 11.6. When you have entered both of the within-subject factors click on Define to go to the main dialog box.

The main dialog box is essentially the same as when there is only one independent variable except that there are now nine question marks (Figure 11.7). At the top of the *Within-Subjects Variables* box, SPSS states that there are two factors: **drink** and **imagery**. In the box below there is a series of question marks followed by bracketed numbers. The numbers in brackets represent the levels of the factors (independent variables):

In this example, there are two independent variables and so there are two numbers in the brackets. The first number refers to levels of the first factor listed above the box (in this case **drink**). The second number in the bracket refers to levels of the second factor listed above the box (in this case **imagery**). As with one-way repeated-measures ANOVA, you are required to replace these question marks with variables from the list on the left-hand side of the dialog box. With between-group designs, in which coding variables are used, the levels of a particular factor are specified by the codes assigned to them in the data editor. However, in repeated measures designs, no such coding scheme is used and so we determine which condition to assign to a level at this stage. For example, if we entered **beerpos** into the list first, then SPSS will treat beer as the first level of **drink**, and positive imagery as the first level of the **imagery** variable. However, if we entered **wineneg** into the list first, SPSS would consider wine as the first level of **drink**, and

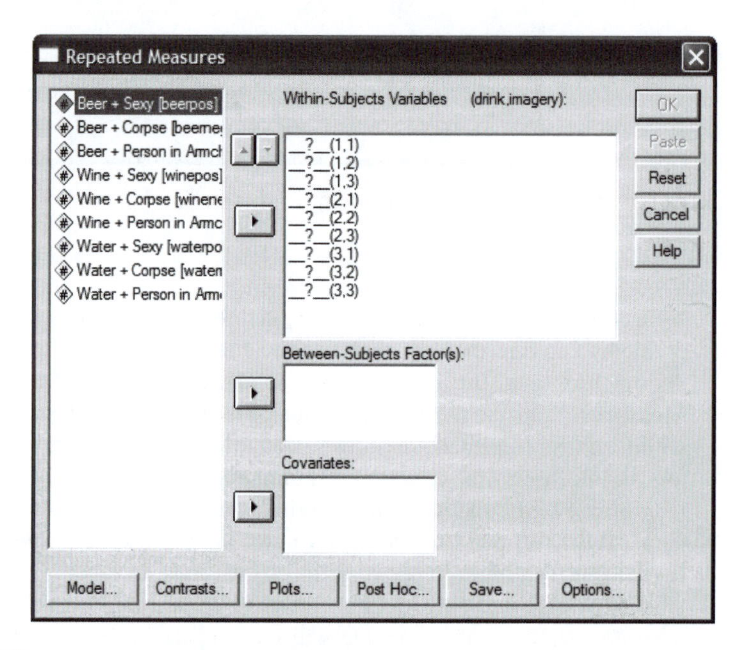

Figure 11.7

_?_(1, 1)	variable representing 1st level of drink and 1st level of imagery
_?_(1, 2)	variable representing 1st level of drink and 2nd level of imagery
_?_(1, 3)	variable representing 1st level of drink and 3rd level of imagery
_?_(2, 1)	variable representing 2nd level of drink and 1st level of imagery
_?_(2, 2)	variable representing 2nd level of drink and 2nd level of imagery
_?_(2, 3)	variable representing 2nd level of drink and 3rd level of imagery
_?_(3, 1)	variable representing 3rd level of drink and 1st level of imagery
_?_(3, 2)	variable representing 3rd level of drink and 2nd level of imagery
_?_(3, 3)	variable representing 3rd level of drink and 3rd level of imagery

negative imagery as the first level of **imagery**. For this reason, it is imperative that we think about the type of contrasts that we might want to do *before* entering variables into this dialog box. In this design, if we look at the first variable, **drink**, there were three conditions, two of which involved alcoholic drinks. In a sense, the water condition acts as a control to whether the effects of imagery are specific to alcohol. Therefore, for this variable we might want to compare the beer and wine conditions with the water condition. This comparison could be done by either specifying a simple contrast (see Table 8.6) in which the beer and wine conditions are compared to the water, or using a difference contrast in which both alcohol conditions are compared to the water condition before being compared to each other. In either case it is essential that the water condition be entered as either the first or last level of the independent variable **drink** (because you can't specify the middle level as the reference category in a simple contrast). Now, let's think about the second factor. The imagery factor also has a control category that was not expected to change attitudes (neutral imagery). As before, we might be interested in using this category as a reference category in a simple contrast[3] and so it is important that this neutral category is entered as either the first or last level.

Based on what has been discussed about using contrasts, it makes sense to have water as level 3 of the **drink** factor and neutral as the third level of the imagery factor. The remaining levels can be decided arbitrarily. I have chosen beer as level 1 and wine as level 2 of the **drink** factor. For the **imagery** variable I chose positive as level 1 and negative as level 2. These decisions mean that the variables should be entered as follows:

beerpos ▶ _?_(1, 1)

beerneg ▶ _?_(1, 2)

beerneut ▶ _?_(1, 3)

winepos ▶ _?_(2, 1)

wineneg ▶ _?_(2, 2)

wineneut ▶ _?_(2, 3)

waterpos ▶ _?_(3, 1)

waterneg ▶ _?_(3, 2)

waterneut ▶ _?_(3, 3)

3 We expect positive imagery to improve attitudes, whereas negative imagery should make attitudes more negative. Therefore, it does not make sense to do a Helmert or difference contrast for this factor because the effects of the two experimental conditions will cancel each other out.

Coincidentally, this order is the order in which variables are listed in the data editor; the coincidence occurred simply because I thought ahead about what contrasts would be done, and then entered variables in the appropriate order! When these variables have been transferred, the dialog box should look exactly like Figure 11.8. The buttons at the bottom of the screen have already been described for the one independent variable case and so I will describe only the most relevant.

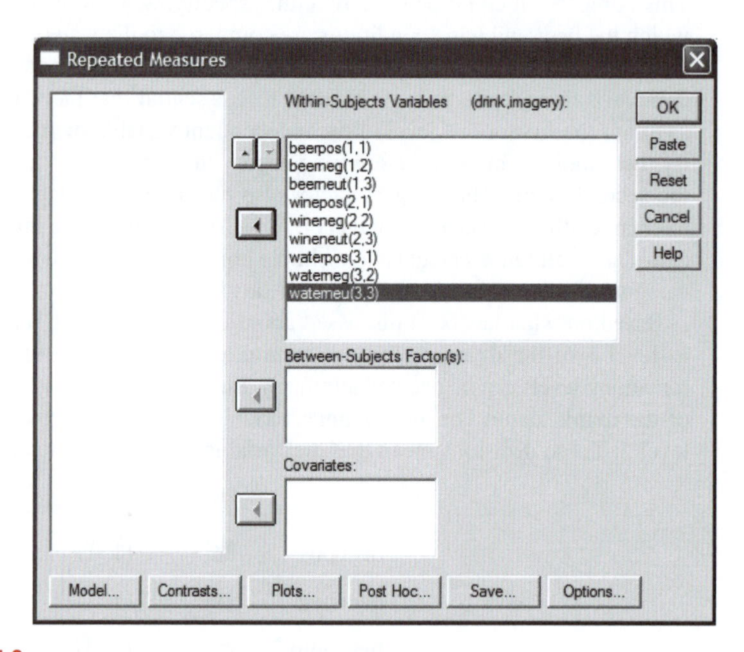

Figure 11.8

11.8.2. Contrasts ②

Following the main analysis it is interesting to compare levels of the independent variables to see whether they differ. As we've seen, there's no facility for entering contrast codes (unless you use syntax) so we have to rely on the standard contrasts available (see Table 8.6). Figure 11.9 shows the dialog box for conducting contrasts and is obtained by clicking on Contrasts... in the main dialog box. In the previous section I described why it might be interesting to use the water and neutral conditions as base categories for the drink and imagery factors respectively. We have used the *contrasts* dialog box before in sections 9.3.3 and 11.4.2 and so all I shall say is that you should select a simple contrast for each independent variable. For both independent variables, we entered the variables such that the control category was the last one; therefore, we need not change the reference category for the simple contrast. Once the contrasts have been selected, click on Continue to return to the main dialog box. An alternative to the contrasts available here is to do a simple effects analysis (see Box 11.1).

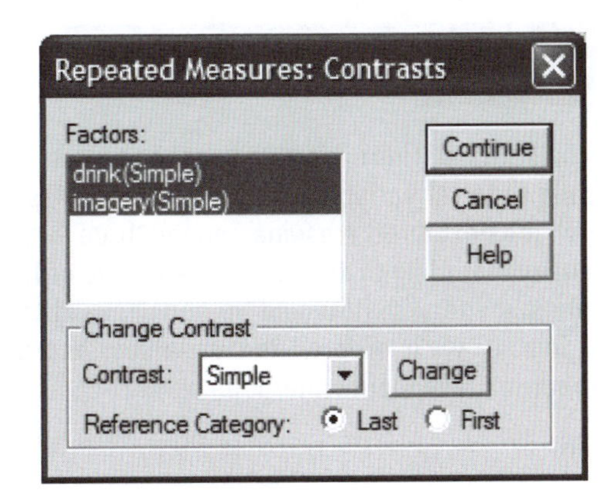

Figure 11.9

Box 11.1

Simple effects analysis ③

We saw in the previous chapter that another way to break down an interaction term is to use a technique called 'simple effects' analysis. This analysis looks at the effect of one independent variable at individual levels of the other independent variable. So, for this example, we could look at the effect of drink for positive imagery, then for negative imagery and then for neutral imagery. Alternatively, we could analyse the effect of imagery separately for beer, wine and water. With repeated-measures designs we can still do simple effects through SPSS syntax, but the syntax we use is slightly different. The syntax you need to use in this example is:

```
MANOVA
beerpos beerneg beerneut winepos wineneg wineneut waterpos waterneg waterneu
/WSFACTORS drink(3) imagery(3)
```

*This initiates the ANOVA by specifying the variables in the data editor that relate to the levels of our repeated-measures variables. The /WSFACTORS command then defines the two repeated-measures variables that we have. The order in which we list the variables from the data editor is important. So, because we've defined drink(3) imagery(3) SPSS starts at level 1 of drink, and then because we've specified three levels of imagery, it uses the first three variables listed as the levels of imagery at level 1 of drink. It then moves on to level 2 of drink and again looks to the next three variables in the list to be the relevant levels of imagery. Finally it moves to level 3 of drink and uses

(Continued)

Box 11.1 (Continued)

the next three variables (the last three in this case) to be the levels of imagery. This is hard to explain, but look at the order of variables, and see that the first three relate to beer (and differ according to imagery), the next three are wine and the three levels of imagery, and the final three are water ordered again according to imagery. Because we ordered them in this way we have to define drink(3) and then imagery(3). (It would be equally valid to write /WSFACTORS imagery(3) drink(3), but only if initially we'd ordered the variables: beerpos winepos waterpos beerneg wineneg waterneg beerneut wineneut waterneu.)

```
/WSDESIGN = MWITHIN drink(1) MWITHIN drink(2) MWITHIN drink(3)
```

*This specifies the simple effects. For example, 'MWITHIN drink(1)' asks SPSS to analyse the effect of imagery at level 1 of drink (i.e. when beer was used). If we wanted to compare drink at levels of imagery, then we'd write this the opposite way around:

```
/WSDESIGN = MWITHIN imagery(1) MWITHIN imagery(2) MWITHIN imagery(3)
/PRINT
SIGNIF( UNIV MULT AVERF HF GG ).
```

*These final lines just ask for the main ANOVA to be printed (SIGNIF). The syntax for looking at the effect of imagery at different levels of drink is stored in a file called **SimpleEffectsAttitude.sps** for you to look at. Open this file (make sure you also have **Attitude.sav** loaded into the data editor) and run the syntax. The output you get will be in the form of text (rather than nice tables). Part of it will replicate the main ANOVA results. The simple effects are presented like this:

```
* * * * * * * * A n a l y s i s   o f   V a r i a n c e - - design 1 * * * * * * * *

Tests involving 'MWITHIN DRINK(1)' Within-Subject Effect.

Tests of Significance for T1 using UNIQUE sums of squares

Source of Variation          SS      DF       MS          F      Sig of F

WITHIN+RESIDUAL          7829.67     19     412.09

MWITHIN DRINK(1)         8401.67      1    8401.67     20.39       .000

- - - - - - - - - - - - - - - - - - - - - - - - - - - - - - - - - - - - - - - - - -

* * * * * * A n a l y s i s   o f   V a r i a n c e - - design 1 * * * * * *
```

Box 11.1 (Continued)

```
Tests involving 'MWITHIN DRINK(2)' Within-Subject Effect.

Tests of Significance for T2 using UNIQUE sums of squares

Source of Variation          SS      DF        MS         F     Sig of F

WITHIN+RESIDUAL            376.00    19      19.79

MWITHIN DRINK(2)          4166.67     1    4166.67    210.55      .000

- - - - - - - - - - - - - - - - - - - - - - - - - - - - - - - - - - - - -

******A n a l y s i s  o f  V a r i a n c e -- design 1 ******

Tests involving 'MWITHIN DRINK(3)' Within-Subject Effect.

Tests of Significance for T3 using UNIQUE sums of squares

Source of Variation          SS      DF        MS         F     Sig of F

WITHIN+RESIDUAL           1500.32    19      78.96

MWITHIN DRINK(3)           742.02     1     742.02      9.40      .006

- - - - - - - - - - - - - - - - - - - - - - - - - - - - - - - - - - - - -
```

The table labelled 'MWITHIN DRINK(1)' gives us an ANOVA of the effect of imagery for beer and the subsequent tables are for wine and water respectively. Looking at the significance values for each simple effect it appears that there were significant effects of imagery at all levels of drink!

We can also follow up interaction effects with specially defined contrasts for the interaction term. Unfortunately these can be done only using syntax, and it's a fairly involved process, so if this sounds like something you might want to do then look at the file **ContrastsUsingSyntax.pdf** on the CD-ROM.

11.8.3. Graphing interactions ②

When we had only one independent variable, we ignored the *plots* dialog box; however, if there are two or more factors, the *plots* dialog box is a convenient way to plot the means

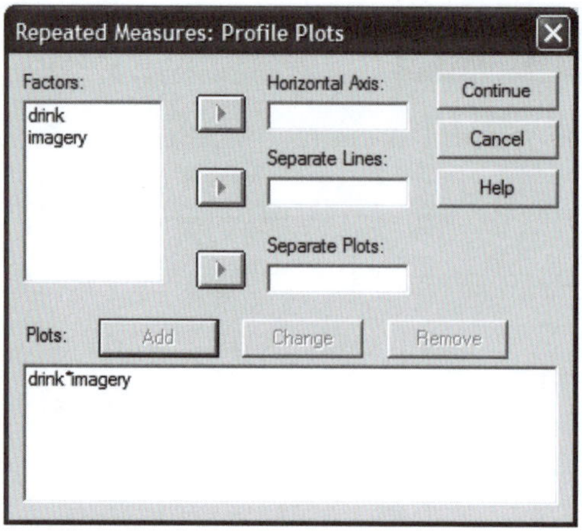

Figure 11.10

for each level of the factors. To access this dialog box click on [Plots...]. Select **drink** from the variables list on the left-hand side of the dialog box and transfer it to the space labelled *Horizontal Axis* by clicking on [▶]. In the space labelled *Separate Lines* we need to place the remaining independent variable: **imagery**. As before, it is down to your discretion which way round the graph is plotted. When you have moved the two independent variables to the appropriate box, click on [Add] and this interaction graph will be added to the list at the bottom of the box (see Figure 11.10). When you have finished specifying graphs, click on [Continue] to return to the main dialog box.

11.8.4. Other options ②

You should notice that *post hoc* tests are disabled for solely repeated-measures designs. Therefore, the only remaining options are in the *options* dialog box, which is accessed by clicking on [Options...]. The options here are the same as for the one-way ANOVA. I recommend selecting some descriptive statistics and you might also want to select some multiple comparisons by selecting all factors in the box labelled *Factor(s) and Factor Interactions* and transferring them to the box labelled *Display Means for* by clicking on [▶] (see Figure 11.11). Having selected these variables, you should tick the box labelled *Compare main effects* ([☑ Compare main effects]) and select an appropriate correction (I chose Bonferroni). The only remaining option of particular interest is to select the *Transformation matrix* option. This option produces a lot of extra output but is important for interpreting the output from the contrasts.

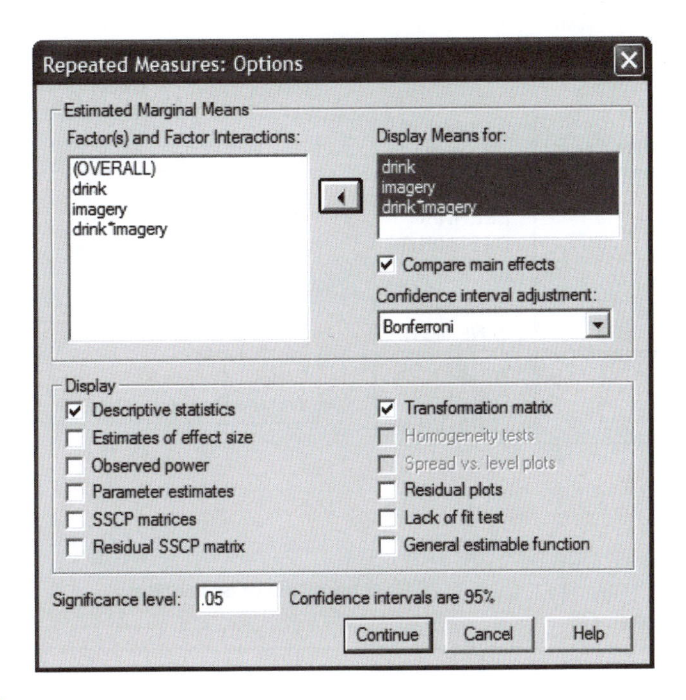

Figure 11.11

11.9. OUTPUT FOR FACTORIAL REPEATED-MEASURES ANOVA ②

11.9.1. Descriptives and main analysis ②

SPSS Output 11.9 shows the initial output from this ANOVA. The first table merely lists the variables that have been included from the data editor and the level of each independent variable that they represent. This table is more important than it might seem, because it enables you to verify that you entered the variables in the correct order for the comparisons that you want to do. The second table is a table of descriptives and provides the mean and standard deviation for each of the nine conditions. The names in this table are the names I gave the variables in the data editor (therefore, if you didn't give these variables full names, this table will look slightly different).

The descriptives are interesting in that they tell us that the variability among scores was greatest when beer was used as a product (compare the standard deviations of the beer variables against the others). Also, when a corpse image was used, the ratings given to the products were negative (as expected) for wine and water but not for beer (so, for some

Within-Subjects Factors

Measure: MEASURE_1

DRINK	IMAGERY	Dependent Variable
1	1	BEERPOS
	2	BEERNEG
	3	BEERNEUT
2	1	WINEPOS
	2	WINENEG
	3	WINENEUT
3	1	WATERPOS
	2	WATERNEG
	3	WATERNEU

Descriptive Statistics

	Mean	Std. Deviation	N
Beer + Sexy	21.0500	13.0080	20
Beer + Corpse	4.4500	17.3037	20
Beer + Person in Armchair	10.0000	10.2956	20
Wine + Sexy	25.3500	6.7378	20
Wine + Corpse	−12.0000	6.1815	20
Wine + Person in Armchair	11.6500	6.2431	20
Water + Sexy	17.4000	7.0740	20
Water + Corpse	−9.2000	6.8025	20
Water + Person in Armchair	2.3500	6.8386	20

SPSS Output 11.9

Mauchly's Test of Sphericity[b]

Measure: MEASURE_1

Within Subjects Effect	Mauchly's W	Approx. Chi-Square	df	Sig.	Epsilon[a]		
					Greenhouse-Geisser	Huynh-Feldt	Lower-bound
DRINK	.267	23.753	2	.000	.577	.591	.500
IMAGERY	.662	7.422	2	.024	.747	.797	.500
DRINK * IMAGERY	.595	9.041	9	.436	.798	.979	.250

Tests the null hypothesis that the error covariance matrix of the orthonormalized transformed dependent variables is proportional to an identity matrix.
a May be used to adjust the degrees of freedom for the averaged tests of significance. Corrected tests are displayed in the layers (by default) of the Tests of Within Subjects Effects table
b Design: Intercept – Within Subjects Design: DRINK–IMAGERY + DRINK*IMAGERY

SPSS Output 11.10

reason negative imagery didn't seem to work when beer was used as a stimulus). The values in this table will help us later to interpret the main effects of the analysis.

SPSS Output 11.10 shows the results of Mauchly's sphericity test (see section 11.2.3) for each of the three effects in the model (two main effects and one interaction). The significance values of these tests indicate that both the main effects of **drink** and **imagery** have violated this assumption and so the *F*-values should be corrected (see section 11.5.2). For the interaction the assumption of sphericity is met (because $p > .05$) and so we need not correct the *F*-ratio for this effect.

SPSS Output 11.11 shows the results of the ANOVA (with corrected *F*-values—if any of you are using version 7.5 or earlier expand the table produced using the method described in section 11.5.3). The output is split into sections that refer to each of the effects

Tests of Within-Subjects Effects

Measure: MEASURE_1

Source		Type III Sum of Squares	df	Mean Square	F	Sig.
DRINK	Sphericity Assumed	2092.344	2	1046.172	5.106	.011
	Greenhouse-Geisser	2092.344	**1.154**	1812.764	**5.106**	**.030**
	Huynh-Feldt	2092.344	1.181	1770.939	5.106	.029
	Lower-bound	2092.344	1.000	2092.344	5.106	.036
Error(DRINK)	Sphericity Assumed	7785.878	38	204.892		
	Greenhouse-Geisser	7785.878	**21.930**	355.028		
	Huynh-Feldt	7785.878	22.448	346.836		
	Lower-bound	7785.878	19.000	409.783		
IMAGERY	Sphericity Assumed	21628.678	2	10814.339	122.565	.000
	Greenhouse-Geisser	21628.678	**1.495**	14468.490	**122.565**	**.000**
	Huynh-Feldt	21628.678	1.594	13571.496	122.565	.000
	Lower-bound	21628.678	1.000	21628.678	122.565	.000
Error(IMAGERY)	Sphericity Assumed	3352.878	38	88.234		
	Greenhouse-Geisser	3352.878	**28.403**	118.048		
	Huynh-Feldt	3352.878	30.280	110.729		
	Lower-bound	3352.878	19.000	176.467		
DRINK * IMAGERY	Sphericity Assumed	2624.422	4	656.106	**17.155**	.000
	Greenhouse-Geisser	2624.422	3.194	821.778	17.155	.000
	Huynh-Feldt	2624.422	3.914	670.462	17.155	.000
	Lower-bound	2624.422	1.000	2624.422	17.155	.001
Error(DRINK*IMAGERY)	Sphericity Assumed	2906.689	76	38.246		
	Greenhouse-Geisser	2906.689	60.678	47.903		
	Huynh-Feldt	2906.689	74.373	39.083		
	Lower-bound	2906.689	19.000	152.984		

SPSS Output 11.11

in the model and the error terms associated with these effects. By looking at the significance values it is clear that there is a significant effect of the type of drink used as a stimulus, a significant main effect of the type of imagery used, and a significant interaction between these two variables. I will examine each of these effects in turn.

11.9.2. The effect of drink ②

The first part of SPSS Output 11.11 tells us the effect of the type of drink used in the advert. For this effect we must look at one of the corrected significance values because sphericity was violated (see above). All of the corrected values are significant and so we should report the conservative Greenhouse–Geisser corrected values of the degrees of freedom. This effect tells us that if we ignore the type of imagery that was used, participants still rated some types of drink significantly differently.

Estimates

Measure: MEASURE_1

DRINK	Mean	Std. Error	95% Confidence Interval Lower Bound	Upper Bound
1	11.833	2.621	6.348	17.319
2	8.333	.574	7.131	9.535
3	3.517	1.147	1.116	5.918

SPSS Output 11.12

In section 11.8.4 we requested that SPSS display means for all of the effects in the model (before conducting *post hoc* tests) and if you scan through your output you should find the table in SPSS Output 11.12 in a section headed *Estimated Marginal Means*.[4] SPSS output 11.12 is a table of means for the main effect of drink with the associated standard errors. The levels of this variable are labelled 1, 2 and 3 and so we must think back to how we entered the variable to see which row of the table relates to which condition. We entered this variable with the beer condition first and the water condition last. Figure 11.12 uses this information to display the means for each condition. It is clear from this graph that beer and wine are naturally rated higher than water (with beer being rated most highly). To see the nature of this effect we can look at the *post hoc* tests (see below) and the contrasts (see section 11.9.5).

SPSS output 11.13 shows the pairwise comparisons for the main effect of drink corrected using a Bonferroni adjustment. This table indicates that the significant main effect reflects a significant difference ($p < .01$) between levels 2 and 3 (wine and water). Curiously, the difference between the beer and water conditions is larger than that for wine and water yet this effect is non-significant ($p > .05$). This inconsistency can be explained by looking at the standard error in the beer condition compared to the wine condition. The standard error for the wine condition is incredibly small and so the difference between means is relatively large (see Chapter 6). Try rerunning these *post hoc* tests but select the uncorrected values (LSD) in the *options* dialog box (see section 11.8.4). You should find that the difference between beer and water is now significant ($p = .02$). This finding highlights the importance of controlling the error rate by using a Bonferroni correction. Had we not used this correction we could have concluded erroneously that beer was rated significantly more highly than water.

4 These means are obtained by taking the average of the means in SPSS Output 11.9 for a given condition. For example, the mean for the beer condition (ignoring imagery) is:

$$\begin{aligned}\overline{X}_{\text{beer}} &= (\overline{X}_{\text{beer+sexy}} + \overline{X}_{\text{beer+corpse}} + \overline{X}_{\text{beer+neutral}})/3 \\ &= (21.05 + 4.45 + 10.00)/3 = 11.83\end{aligned}$$

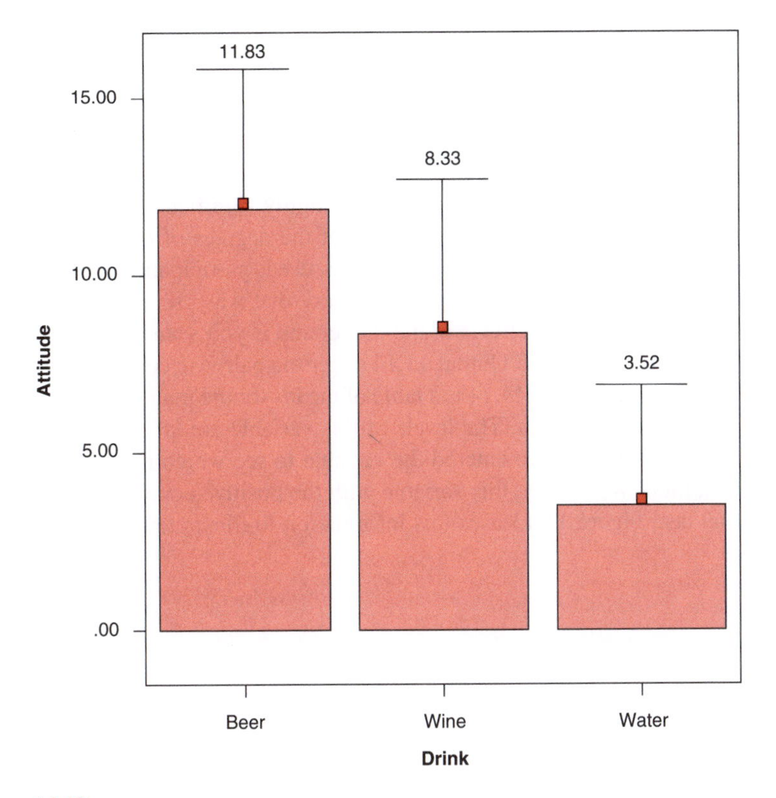

Figure 11.12

Pairwise Comparisons

Measure: MEASURE_1

(I) DRINK	(J) DRINK	Mean Difference (I–J)	Std. Error	Sig.[a]	95% Confidence Interval for Difference[a] Lower Bound	Upper Bound
1	2	3.500	2.849	.703	–3.980	10.980
	3	8.317	3.335	.066	–.438	17.072
2	1	–3.500	2.849	.703	–10.980	3.980
	3	4.817*	1.116	.001	1.886	7.747
3	1	–8.317	3.335	.066	–17.072	.438
	2	–4.817*	1.116	.001	–7.747	–1.886

Based on estimated marginal means

* The mean difference is significant at the .05 level

a Adjustment for multiple comparisons: Bonferroni

SPSS Output 11.13

11.9.3. The effect of imagery ②

SPSS Output 11.11 also indicates that the effect of the type of imagery used in the advert had a significant influence on participants' ratings of the stimuli. Again, we must look at one of the corrected significance values because sphericity was violated (see above). All of the corrected values are highly significant and so we can again report the Greenhouse–Geisser corrected values of the degrees of freedom. This effect tells us that if we ignore the type of drink that was used, participants' ratings of those drinks were different according to the type of imagery that was used. In section 11.8.4 we requested means for all of the effects in the model and if you scan through your output you should find the table in SPSS Output 11.14 (after the pairwise comparisons for the main effect of drink). SPSS Output 11.14 is a table of means for the main effect of imagery with the associated standard errors. The levels of this variable are labelled 1, 2 and 3 and so we must think back to how we entered the variable to see which row of the table relates to which condition. We entered this variable with the positive condition first and the neutral condition last. Figure 11.13 uses this information to illustrate the means for each condition. It

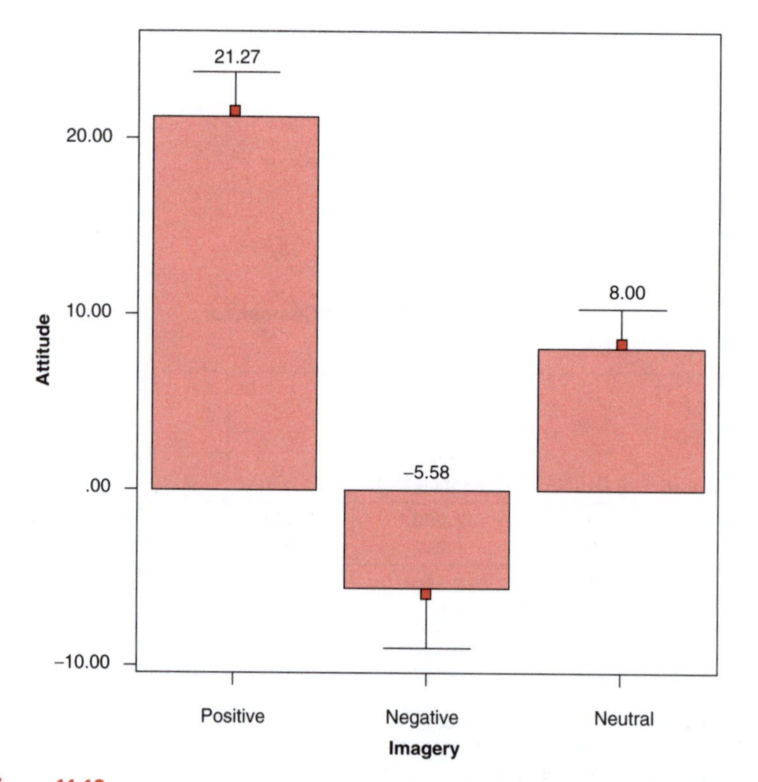

Figure 11.13

Estimates

Measure: MEASURE_1

IMAGERY	Mean	Std. Error	95% Confidence Interval	
			Lower Bound	Upper Bound
1	21.267	.977	19.222	23.312
2	−5.583	1.653	−9.043	−2.124
3	8.000	.969	5.972	10.028

SPSS Output 11.14

Pairwise Comparisons

Measure: MEASURE_1

(I) IMAGERY	(J) IMAGERY	Mean Difference (I–J)	Std. Error	Sig.[a]	95% Confidence Interval for Difference[a]	
					Lower Bound	Upper Bound
1	2	26.850*	1.915	.000	21.824	31.876
	3	13.267*	1.113	.000	10.346	16.187
2	1	−26.850*	1.915	.000	−31.876	−21.824
	3	−13.583*	1.980	.000	−18.781	−8.386
3	1	−13.267*	1.113	.000	−16.187	−10.346
	2	13.583*	1.980	.000	8.386	18.781

Based on estimated marginal means

* The mean difference is significant at the .05 level

a Adjustment for multiple comparisons: Bonferroni

SPSS Output 11.15

is clear from this graph that positive imagery resulted in very positive ratings (compared to the neutral imagery) and negative imagery resulted in negative ratings (especially compared to the effect of neutral imagery). To see the nature of this effect we can look at the *post hoc* tests (see above) and the contrasts (see section 11.9.5).

SPSS Output 11.15 shows the pairwise comparisons for the main effect of imagery corrected using a Bonferroni adjustment. This table indicates that the significant main effect reflects significant differences (all $p < .01$) between levels 1 and 2 (positive and negative), between levels 1 and 3 (positive and neutral), and between levels 2 and 3 (negative and neutral).

11.9.4. The interaction effect (drink × imagery) ②

SPSS Output 11.11 indicated that imagery interacted in some way with the type of drink used as a stimulus. From that table we should report that there was a significant interaction

Estimates

Measure: MEASURE_1

DRINK	IMAGERY	Mean	Std. Error	95% Confidence Interval	
				Lower Bound	Upper Bound
1	1	21.050	2.909	14.962	27.138
	2	4.450	3.869	−3.648	12.548
	3	10.000	2.302	5.181	14.819
2	1	25.350	1.507	22.197	28.503
	2	−12.000	1.382	−14.893	−9.107
	3	11.650	1.396	8.728	14.572
3	1	17.400	1.582	14.089	20.711
	2	−9.200	1.521	−12.384	−6.016
	3	2.350	1.529	−.851	5.551

SPSS Output 11.16

between the type of drink used and imagery associated with it, $F(4, 76) = 17.16$, $p < .001$. This effect tells us that the type of imagery used had a different effect depending on which type of drink it was presented alongside. As before, we can use the means that we requested in section 11.8.4 to determine the nature of this interaction (this table should be below the pairwise comparisons for imagery and is shown in SPSS Output 11.16). The table of means in SPSS Output 11.16 is essentially the same as the initial descriptive statistics in SPSS Output 11.9 except that the standard errors are displayed rather than the standard deviations.

The means in SPSS Output 11.16 are used to create the plot that we requested in section 11.8.3 and this graph is essential for interpreting the interaction. Figure 11.14 shows the interaction graph (slightly modified to make it look prettier!) and we are looking for non-parallel lines. The graph shows that the pattern of responding across drinks was similar when positive and neutral imagery were used. That is, ratings were positive for beer, they were slightly higher for wine and then they went down slightly for water. The fact that the line representing positive imagery is higher than the neutral line indicates that positive imagery gave rise to higher ratings than neutral imagery across all drinks. The bottom line (representing negative imagery) shows a different effect: ratings were lower for wine and water but not for beer. Therefore, negative imagery had the desired effect on attitudes towards wine and water, but for some reason attitudes towards beer remained fairly neutral. Therefore, the interaction is likely to reflect the fact that negative imagery has a different effect to both positive and neutral imagery (because it decreases ratings rather than increasing them). This interaction is completely in line with the experimental predictions. To verify the interpretation of the interaction effect, we need to look at the contrasts that we requested in section 11.8.2.

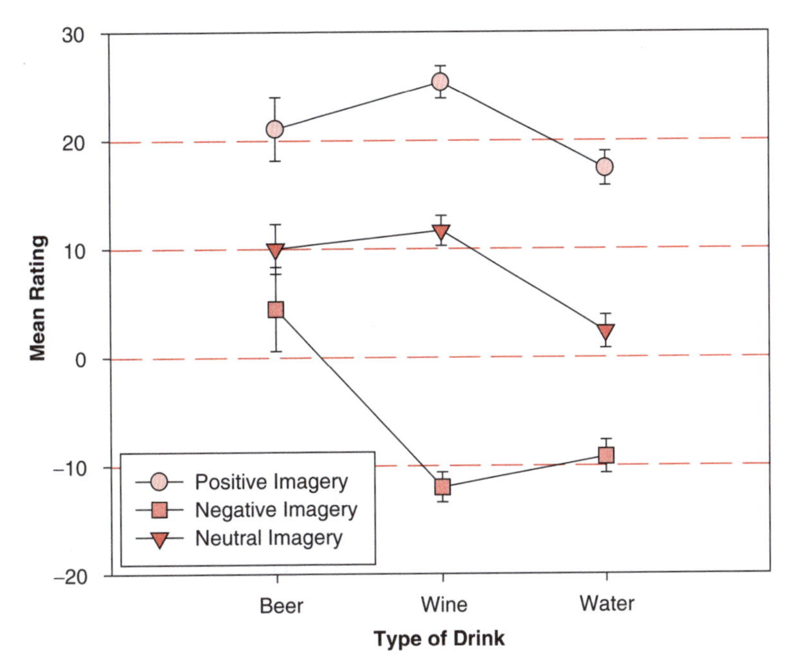

Figure 11.14 Interaction graph for **Attitude.sav**. The type of imagery is represented by the three lines: positive imagery (circles), negative imagery (squares) and neutral imagery (triangles)

11.9.5. Contrasts for repeated-measures variables ②

In section 11.8.2 we requested simple contrasts for the **drink** variable (for which water was used as the control category) and for the **imagery** category (for which neutral imagery was used as the control category). SPSS Output 11.17 shows the summary results for these contrasts. The table is split up into main effects and interactions, and each effect is split up into components of the contrast. So, for the main effect of drink, the first contrast compares level 1 (beer) against the base category (in this case, the last category: water). If you are confused as to which level is which you are reminded that SPSS Output 11.9 lists them for you. This result is significant, $F(1, 19) = 6.22, p < .05$, which contradicts what was found using *post hoc* tests (see SPSS Output 11.13)—why do you think this is? The next contrast compares level 2 (wine) with the base category (water) and confirms the significant difference found with the *post hoc* tests, $F(1, 19) = 18.61, p < .001$. For the imagery main effect, the first contrast compares level 1 (positive) to the base category (the last category: neutral) and verifies the significant difference found with the *post hoc* tests, $F(1, 19) = 142.19, p < .001$. The second contrast confirms the significant difference in ratings found in the negative imagery condition compared to the neutral, $F(1, 19) = 47.07$,

Tests of Within-Subjects Contrasts

Measure: MEASURE_1

Source	DRINK	IMAGERY	Type III Sum of Squares	df	Mean Square	F	Sig.
DRINK	Level 1 vs. Level 3		1383.339	1	1383.339	6.218	.022
	Level 2 vs. Level 3		464.006	1	464.006	18.613	.000
Error(DRINK)	Level 1 vs. Level 3		4226.772	19	222.462		
	Level 1 vs. Level 3		473.661	19	24.930		
IMAGERY		Level 1 vs. Level 3	3520.089	1	3520.089	142.194	.000
		Level 2 vs. Level 3	3690.139	1	3690.139	47.070	.000
Error(IMAGERY)		Level 1 vs. Level 3	470.356	19	24.756		
		Level 2 vs. Level 3	1489.528	19	78.396		
DRINK * IMAGERY	Level 1 vs. Level 3	Level 1 vs. Level 3	320.000	1	320.000	1.576	.225
		Level 2 vs. Level 3	720.000	1	720.000	6.752	.018
	Level 2 vs. Level 3	Level 1 vs. Level 3	36.450	1	36.450	.235	.633
		Level 2 vs. Level 3	2928.200	1	2928.200	26.906	.000
Error(DRINK*IMAGERY)	Level 1 vs. Level 3	Level 1 vs. Level 3	3858.000	19	203.053		
		Level 2 vs. Level 3	2026.000	19	106.632		
	Level 2 vs. Level 3	Level 1 vs. Level 3	2946.550	19	155.082		
		Level 2 vs. Level 3	2067.800	19	108.832		

SPSS Output 11.17

$p < .001$. These contrasts are all very well, but they tell us only what we already knew (although note the increased statistical power with these tests shown by the higher significance values). The contrasts become much more interesting when we look at the interaction term.

11.9.5.1. Beer vs. water, positive vs. neutral imagery ②

The first interaction term looks at level 1 of drink (beer) compared to level 3 (water), when positive imagery (level 1) is used compared to neutral (level 3). This contrast is non-significant. This result tells us that the increased liking found when positive imagery is used (compared to neutral imagery) is the same for both beer and water. In terms of the interaction graph (Figure 11.14) it means that the distance between the circle and the triangle in the beer condition is the same as the distance between the circle and the triangle in the water condition. If we just plot this section of the interaction graph then it's easy to see that the lines are approximately parallel, indicating no interaction effect. We could conclude that the improvement of ratings due to positive imagery compared to neutral is not affected by whether people are evaluating beer or water.

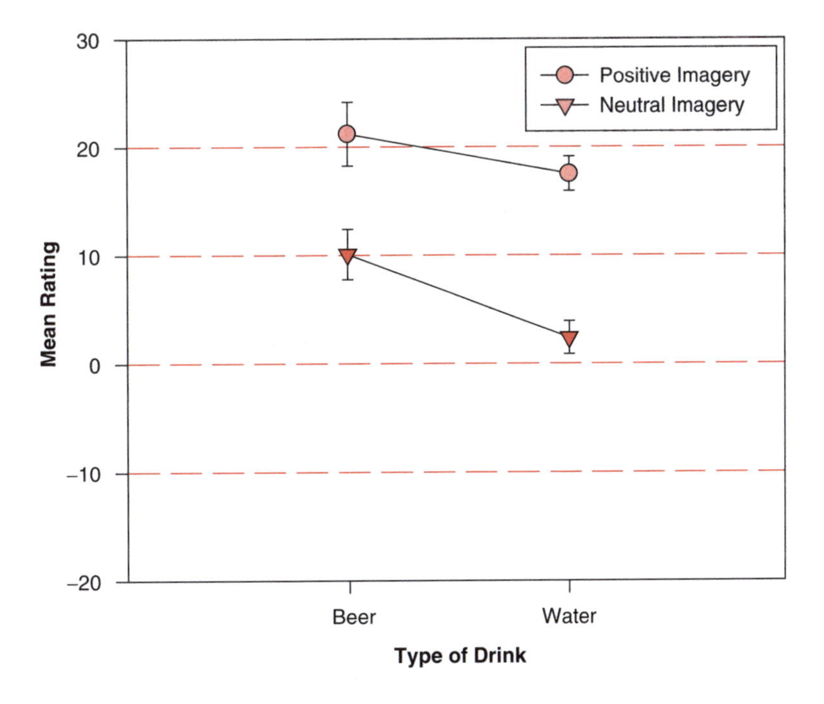

11.9.5.2. Beer vs. water, negative vs. neutral imagery ②

The second interaction term looks at level 1 of drink (beer) compared to level 3 (water), when negative imagery (level 2) is used compared to neutral (level 3). This contrast is

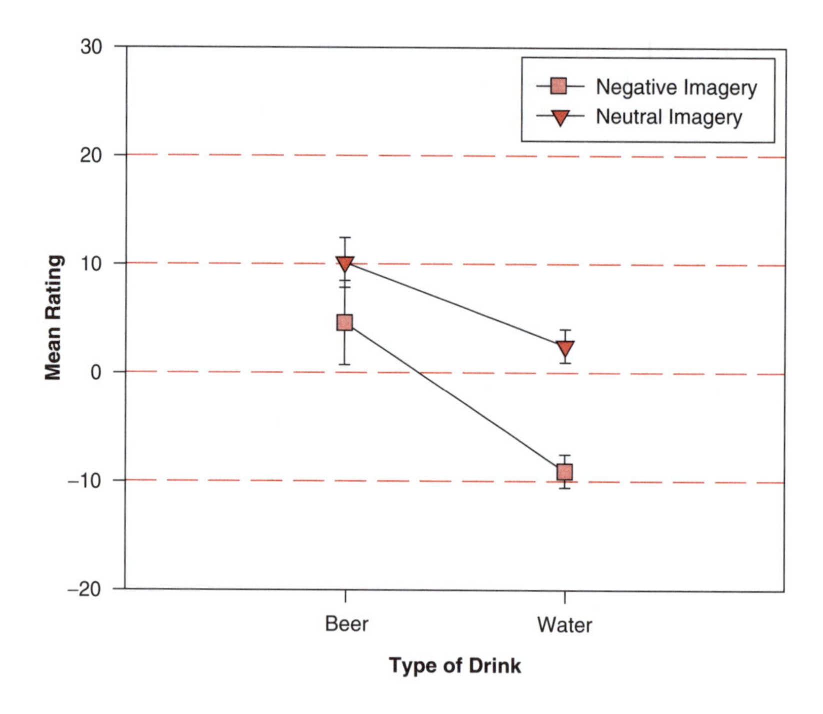

significant, $F(1, 19) = 6.75$, $p < .05$. This result tells us that the decreased liking found when negative imagery is used (compared to neutral imagery) is different when beer is used compared to when water is used. In terms of the interaction graph (Figure 11.14) it means that the distance between the square and the triangle in the beer condition (a small difference) is significantly smaller than the distance between the square and the triangle in the water condition (a larger difference). We could conclude that the decrease in ratings due to negative imagery (compared to neutral) found when water is used in the advert is smaller when beer is used.

11.9.5.3. Wine vs. water, positive vs. neutral imagery ②

The third interaction term looks at level 2 of drink (wine) compared to level 3 (water), when positive imagery (level 1) is used compared to neutral (level 3). This contrast is non-significant, indicating that the increased liking found when positive imagery is used (compared to neutral imagery) is the same for both wine and water. In terms of the interaction graph (Figure 11.14) it means that the distance between the circle and the triangle in the wine condition is the same as the distance between the circle and the triangle in the water condition. If we just plot this section of the interaction graph then it's easy to see that the lines are parallel, indicating no interaction effect. We could conclude that the improvement of ratings due to positive imagery compared to neutral is not affected by whether people are evaluating wine or water.

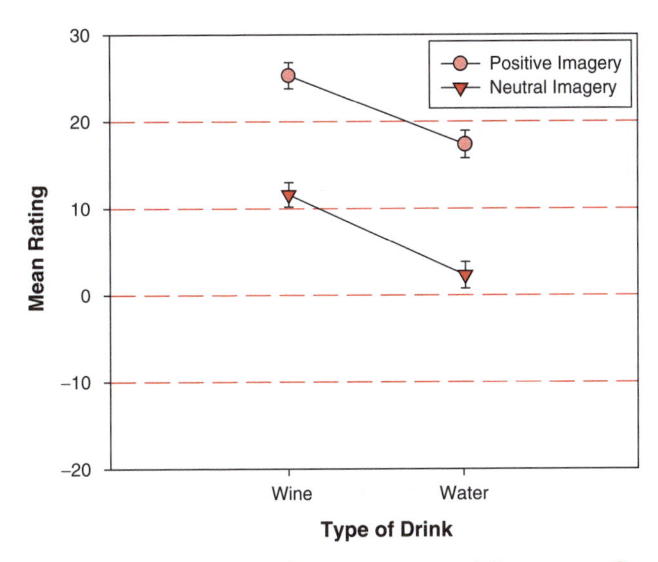

11.9.5.4. Wine vs. water, negative vs. neutral imagery ②

The final interaction term looks at level 2 of drink (wine) compared to level 3 (water), when negative imagery (level 2) is used compared to neutral (level 3). This contrast is significant, $F(1, 19) = 26.91$, $p < .001$. This result tells us that the decreased liking found when negative imagery is used (compared to neutral imagery) is different when wine is used compared to when water is used. In terms of the interaction graph (Figure 11.14) it means that the distance between the square and the triangle in the wine condition (a big difference) is significantly larger than the distance between the square and the triangle in the water condition (a smaller difference). We could conclude that the decrease in ratings due to negative imagery (compared to neutral) is significantly greater when wine is advertised than when water is advertised.

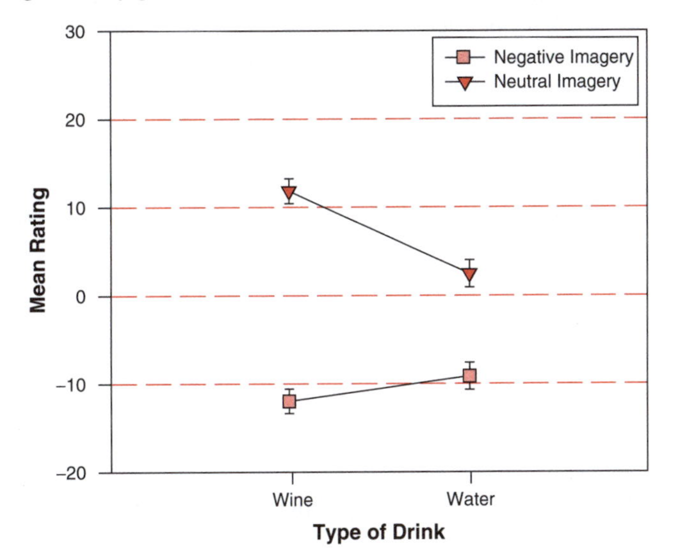

11.9.5.5. Limitations of these contrasts ②

These contrasts, by their nature, tell us nothing about the differences between the beer and wine conditions (or the positive and negative conditions) and different contrasts would have to be run to find out more. However, what is clear so far is that relative to the neutral condition, positive images increased liking for the products more or less regardless of the product; however, negative imagery had a greater effect on wine and a lesser effect on beer. These differences were not predicted. Although it may seem tiresome to spend so long interpreting an analysis so thoroughly, you are well advised to take such a systematic approach if you want truly to understand the effects that you obtain. Interpreting interaction terms is complex, and I can think of a few well-respected researchers who still struggle with them, so don't feel disheartened if you find them hard. Try to be thorough, and break each effect down as much as possible using contrasts and hopefully you will find enlightenment.

Cramming Samantha's Tips

- Two-way repeated-measures ANOVA compares several means when there are two independent variables, and the same participants have been used in all experimental conditions.
- Test the assumption of *sphericity* when you have three or more repeated-measures conditions. Test for sphericity using *Mauchly's test*. Find the table with this label: if the value in the column labelled *Sig.* is less than .05 then the assumption is violated. If the significance of Mauchly's test is greater than .05 then the assumption of sphericity has been met. You should test this assumption for all effects (in a two-way ANOVA this means you test it for the effect of both variables and the interaction term).
- The table labelled *Tests of Within-Subjects Effects* shows the main result of your ANOVA. In a two-way ANOVA you will have three effects: a main effect of each variable, and the interaction between the two. For *each* effect, if the assumption of sphericity has been met then look at the row labelled *Sphericity Assumed*. If the assumption was violated then read the row labelled *Greenhouse–Geisser* (you can also look at *Huynh–Feldt* but you'll have to read this chapter to find out the relative merits of the two procedures). Having selected the appropriate row, look at the column labelled *Sig.*: if the value is less than .05 then the means of the groups are significantly different.
- Break down the main effects and interaction terms using contrasts. These contrasts appear in the table labelled *Tests of Within-Subjects Contrasts*; again look at the columns labelled *Sig.* to discover if your comparisons are significant (they will be if the significance value is less than .05).

11.10. EFFECT SIZES FOR FACTORIAL REPEATED-MEASURES ANOVA ③

Calculating omega squared for a one-way repeated-measures ANOVA was hair-raising enough, and as I keep saying, effect sizes are really more useful when they describe a

focused effect, so I'd advise calculating effect sizes for your contrasts when you've got a factorial design (and any main effects that compare only two groups). SPSS Output 11.17 shows the values for several contrasts, all of which have 1 degree of freedom for the model (i.e. they represent a focused and interpretable comparison) and have 19 residual degrees of freedom. We can use these *F*-ratios and convert them to an effect size *r*, using a formula we've come across before:

$$r = \sqrt{\frac{F(1, df_{\text{R}})}{F(1, df_{\text{R}}) + df_{\text{R}}}}$$

For the two comparisons we did for the drink variable (SPSS Output 11.17), we would get:

$$r_{\text{Beer vs. Water}} = \sqrt{\frac{6.22}{6.22 + 19}} = .50$$

$$r_{\text{Wine vs. Water}} = \sqrt{\frac{18.61}{18.61 + 19}} = .70$$

Therefore, both comparisons yielded very large effect sizes. For the two comparisons we did for the imagery variable (SPSS Output 11.17), we would get:

$$r_{\text{Positive vs. Neutral}} = \sqrt{\frac{142.19}{142.19 + 19}} = .94$$

$$r_{\text{Negative vs. Neutral}} = \sqrt{\frac{47.07}{47.07 + 19}} = .84$$

Again, both comparisons yield very large effect sizes. For the interaction term, we had four contrasts, but again we can convert them to *r* because they all have 1 degree of freedom for the model (SPSS Output 11.17):

$$r_{\text{Beer vs. Water, Positive vs. Neutral}} = \sqrt{\frac{1.58}{1.58 + 19}} = .28$$

$$r_{\text{Beer vs. Water, Negative vs. Neutral}} = \sqrt{\frac{6.75}{6.75 + 19}} = .51$$

$$r_{\text{Wine vs. Water, Positive vs. Neutral}} = \sqrt{\frac{.24}{.24 + 19}} = .11$$

$$r_{\text{Wine vs. Water, Negative vs. Neutral}} = \sqrt{\frac{26.91}{26.91 + 19}} = .77$$

As such the two effects that were significant (beer vs. water, negative vs. neutral and wine vs. water, negative vs. neutral) yield large effect sizes. The two effects that were not significant yielded a medium effect size (beer vs. water, positive vs. neutral) and a small effect size (wine vs. water, positive vs. neutral).

11.11. REPORTING THE RESULTS FROM FACTORIAL REPEATED-MEASURES ANOVA ②

We can report a factorial repeated-measures ANOVA in much the same way as any other ANOVA. Remember that we've got three effects to report, and these effects might have different degrees of freedom. For the main effects of drink and imagery, the assumption of sphericity was violated so we'd have to report the Greenhouse–Geisser corrected degrees of freedom. We can, therefore, begin by reporting the violation of sphericity:

✓ Mauchly's test indicated that the assumption of sphericity had been violated for the main effects of drink, $\chi^2(2) = 23.75$, $p < .001$, and imagery, $\chi^2(2) = 7.42$, $p < .05$. Therefore degrees of freedom were corrected using Greenhouse–Geisser estimates of sphericity ($\varepsilon = .58$ for the main effect of drink and .75 for the main effect of imagery).

We can then report the three effects from this analysis as follows:

✓ All effects are reported as significant at $p < .05$. There was a significant main effect of the type of drink on ratings of the drink, $F(1.15, 21.93) = 5.11$. Contrasts revealed that ratings of beer, $F(1, 19) = 6.22$, $r = .50$, and wine, $F(1, 19) = 18.61$, $r = .70$, were significantly higher than water.

✓ There was also a significant main effect of the type of imagery on ratings of the drinks, $F(1.50, 28.40) = 122.57$. Contrasts revealed that ratings after positive imagery were significantly higher than after neutral imagery, $F(1, 19) = 142.19$, $r = .94$. Conversely, ratings after negative imagery were significantly lower than after neutral imagery, $F(1, 19) = 47.07$, $r = .84$.

✓ There was a significant interaction effect between the type of drink and the type of imagery used, $F(4, 76) = 17.16$. This indicates that imagery had different effects on people's ratings depending on which type of drink was used. To break down this interaction, contrasts were performed comparing all drink types to their baseline (water) and all imagery types to their baseline (neutral imagery). These revealed significant interactions when comparing negative imagery to neutral imagery both for beer compared to water, $F(1, 19) = 6.75$, $r = .51$, and wine compared to water, $F(1, 19) = 26.91$, $r = .77$. Looking at the interaction graph, these effects reflect that negative imagery (compared to neutral) lowered scores significantly more in water than it did for beer, and lowered scores significantly more for wine than it did for water. The remaining contrasts revealed no significant interaction term when comparing positive imagery to neutral imagery both for beer compared to water, $F(1, 19) = 1.58$, $r = .28$, and wine compared to water, $F(1, 19) < 1$, $r = .11$. However, these contrasts did yield small to medium effect sizes.

11.12. WHAT HAVE WE DISCOVERED ABOUT STATISTICS? ②

This chapter has helped us to walk through the murky rabid leg-eating-crocodile infested swamp of repeated-measures designs. The first thing we learnt was that with repeated-measures designs there is yet another assumption to worry about: *sphericity*. Having recovered from this shock revelation, we were fortunate to discover that this assumption, if violated, can be easily remedied. Sorted! We then moved on to look at the theory of repeated-measures ANOVA for one independent variable. Although not essential by any stretch of the imagination, this was a useful exercise to demonstrate that basically it's exactly the same as when we have an independent design (well, there are a few subtle differences but I was trying to emphasize the similarities). We then worked through an example on SPSS, before tackling the particularly foul-tempered, starving hungry, and mad as 'stabby' the mercury-sniffing hatter, piranha fish of omega squared. That's a road I kind of regretted going down after I'd started, but stubborn as ever, I persevered. This led us ungracefully on to factorial repeated-measures designs and specifically the situation where we have two independent variables. We learnt that as with other factorial designs we have to worry about interaction terms. But we also discovered some useful ways to break these terms down using contrasts. If you thought all of that was bad, well, I'm about to throw an added complication into the mix. We can combine repeated-measures and independent designs, and the next chapter looks at this situation. As if this wasn't bad enough, I'm also going to use this as an excuse to show you a design with three independent variables (at this point you should imagine me leaning back in my chair cross-eyed dribbling and laughing maniacally).

11.13. KEY TERMS THAT WE'VE DISCOVERED

- Compound symmetry
- Greenhouse–Geisser correction
- Huynh–Feldt correction
- Lower-bound estimate
- Mauchly's test
- Repeated-measures ANOVA
- Sphericity

11.14. SMART ALEX'S TASKS

- **Task 1**: Imagine I wanted to look at the effect alcohol has on the roving eye. The 'roving eye' effect is the propensity of people in relationships to 'eye-up' members of the opposite sex. I took 20 men and fitted them with incredibly sophisticated glasses that could track their eye movements and record both the movement and the object being observed (this is the point at which it should be apparent that I'm making it up as I go along). Over four different nights I plied these poor souls with 1, 2, 3 or 4 pints of strong lager in a night-club. Each night I measured how many different women they eyed up (a woman was categorized as having been eyed up if the man's eye moved from her head to toe and back up again). To validate this measure we also collected the amount of dribble on the man's chin while looking at a woman. The data are in the file **RovingEye.sav**; analyse them with a one-way ANOVA.②

- **Task 2**: In the previous chapter we came across the beer–goggles effect: a severe perceptual distortion after imbibing vast quantities of alcohol. The specific visual distortion is that previously unattractive people suddenly become the hottest thing since Spicy Gonzalez's extra-hot tabasco-marinated chillies. In short, one minute you're standing in a zoo admiring the orang-utans, and the next you're wondering why someone would put Gail Porter (or whatever her surname is now) into a cage. Anyway, in that chapter, a blatantly fabricated data set demonstrated that the beer–goggles effect was much stronger for men than women, and took effect only after 2 pints. Imagine we wanted to follow up this finding to look at what factors mediate the beer–goggles effect. Specifically, we thought that the beer–goggles effect might be made worse by the fact that it usually occurs in clubs which have dim lighting. We took a sample of 26 men (because the effect is stronger in men) and gave them various doses of alcohol over four different weeks (0 pints, 2 pints, 4 pints and 6 pints of lager). This is our first independent variable, which we'll call alcohol consumption, and it has four levels. Each week (and, therefore, in each state of drunkenness) participants were asked to select a mate in a normal club (that had dim lighting) and then select a second mate in a specially designed club that had bright lighting. As such, the second independent variable was whether the club had dim or bright lighting. The outcome measure was the attractiveness of each mate as assessed by a panel of independent judges. To recap, all participants took part in all levels of the alcohol consumption variable, and selected mates in both brightly and dimly lit clubs. The data are in the file **BeerGogglesLighting.sav**; analyse them with a two-way repeated-measures ANOVA.②
- **Task 3**: Using Box 11.1, change the syntax in **SimpleEffectsAttitude.sps** to look at the effect of drink at different levels of imagery.③

11.15. FURTHER READING

Field, A. P. (1998). A bluffer's guide to sphericity. *Newsletter of the Mathematical, Statistical and Computing section of the British Psychological Society*, *6*(1), 13–22 (available on the CD-ROM as **sphericity.pdf**).

Howell, D. C. (2002). *Statistical methods for psychology* (5th edition). Belmont, CA: Duxbury.

Rosenthal, R., Rosnow, R. L. & Rubin, D. B. (2000). *Contrasts and effect sizes in behavioural research: a correlational approach.* Cambridge: Cambridge University Press. This is quite advanced but really cannot be bettered for contrasts and effect size estimation.

CHAPTER 12

MIXED DESIGN ANOVA (GLM 5)

12.1. WHAT WILL THIS CHAPTER TELL US? ②

The final design that I need to talk about is one in which you have a mixture of between-group and repeated-measures variables: a mixed design. It should be obvious that you need at least two independent variables for this type of design to be possible, but you can have more complex scenarios too (e.g. two between-group and one repeated-measures, one between-group and two repeated-measures, or even two of each). SPSS allows you to test almost any design you might want to, and of virtually any degree of complexity. However, interaction terms are difficult enough to interpret with only two variables so imagine how difficult they are if you include four! The best advice I can offer is to stick to three or fewer independent variables if you want to be able to interpret your interaction terms,[1] and certainly don't exceed four unless you want to give yourself a migraine.

This chapter will go through an example of a mixed ANOVA. There won't be any theory because really and truly you've probably had enough ANOVA theory by now to have a good idea of what's going on (you can read this as 'it's too complex for me and I'm going to cover up my own incompetence by pretending you don't need to know about it'). So, we look at an example using SPSS, and then interpret the output. In the process you'll hopefully develop your understanding of interactions and how to break them down using contrasts.

---

1 Fans of irony will enjoy the four-way ANOVA that I conducted in Field & Davey (1999) and many other publications!

12.2. WHAT DO MEN AND WOMEN LOOK FOR IN A PARTNER? ②

The example we're going to use in this chapter stays with the dating theme. It seems that lots of magazines go on all the time about how men and women want different things from relationships (or perhaps it's just my girlfriend's copies of *Marie Clare*, which I don't read—honestly). The big question to which we all want to know the answer is: are looks or personality more important? Imagine you wanted to put this to the test. You devised a cunning plan whereby you'd set up a speed-dating night.[2] Little did the people who came along know that you'd got some of your friends to act as the dates. Specifically you found nine men and nine women to act as the date. In each of these groups three people were extremely attractive but differed in their personality: one had tonnes of charisma,[3] one had some charisma, and the third person was as dull as this book. Another three people were of average attractiveness, and again differed in their personality: one was highly charismatic, one had some charisma and the third was a dullard. The final three were, not wishing to be unkind in any way, pig-ugly and again one was charismatic, one had some charisma and the final poor soul was mind-numbingly tedious. The participants were the people who came to the speed-dating night, and over the course of the evening they speed-dated all nine members of the opposite sex that you'd set up for them. After their 5 minute date, they rated how much they'd like to have a proper date with the person as a percentage (100% = 'I'd pay large sums of money for your phone number', 0% = 'I'd pay a large sum of money for a plane ticket to get me as far away as possible from you'). As such, each participant rated nine different people who varied in their attractiveness and personality. So, there are two repeated-measures variables: **looks** (with three levels because the person could be attractive, average or ugly) and **personality** (again with three levels because the person could have lots of charisma, have some charisma, or be a dullard). Of course the people giving the ratings could be male or female, so we should also include the gender of the person making the ratings (male or female), and this, of course, will be a between-group variable. The data are in Table 12.1.

---

2 In case speed dating goes out of fashion and no one knows what I'm going on about, the basic idea is lots of men and women turn up to a venue (or just men or just women if it's a gay night), one-half of the group sit individually at small tables and the remainder choose a table, get 3 minutes to impress the other person at the table with their tales of heteroscedastic data, then a bell rings and they get up and move to the next table. Having worked around all of the tables, participants end the evening either stalking the person they fancied or avoiding the hideous weirdo going on about hetro… something or other.

3 The highly attractive people with tonnes of charisma were, of course, taken to a remote clifftop and shot after the experiment because life is hard enough without having people like that floating around making you feel inadequate.

Table 12.1 Data from **LooksOrPersonality.sav** (Att = Attractive, Av = Average, Ug = Ugly)

	High Charisma			Some Charisma			Dullard		
Looks	**Att**	**Av**	**Ugly**	**Att**	**Av**	**Ug**	**Att**	**Av**	**Ug**
	86	84	67	88	69	50	97	48	47
	91	83	53	83	74	48	86	50	46
	89	88	48	99	70	48	90	45	48
	89	69	58	86	77	40	87	47	53
	80	81	57	88	71	50	82	50	45
Male	80	84	51	96	63	42	92	48	43
	89	85	61	87	79	44	86	50	45
	100	94	56	86	71	54	84	54	47
	90	74	54	92	71	58	78	38	45
	89	86	63	80	73	49	91	48	39
	89	91	93	88	65	54	55	48	52
	84	90	85	95	70	60	50	44	45
	99	100	89	80	79	53	51	48	44
	86	89	83	86	74	58	52	48	47
	89	87	80	83	74	43	58	50	48
Female	80	81	79	86	59	47	51	47	40
	82	92	85	81	66	47	50	45	47
	97	69	87	95	72	51	45	48	46
	95	92	90	98	64	53	54	53	45
	95	93	96	79	66	46	52	39	47

12.3. MIXED ANOVA ON SPSS ②

12.3.1. The main analysis ②

To enter these data into SPSS we use the same procedure as the two-way repeated-measures ANOVA that we came across in the previous chapter. Remember that each row in the data editor represents a single participant's data. If a person participates in all experimental conditions (in this case they date all of the people who differ in attractiveness and all of the people who differ in their charisma) then each experimental condition must be represented by a column in the data editor. In this experiment there are nine experimental conditions and so the data need to be entered in nine columns (the format is identical to

Table 12.1). Therefore, create the following nine variables in the data editor with the names as given. For each one, you should also enter a full variable name (see section 2.4.2) for clarity in the output.

att_high Attractive + High Charisma
av_high Average Looks + High Charisma
ug_high Ugly + High Charisma
att_some Attractive + Some Charisma
av_some Average Looks + Some Charisma
ug_some Ugly + Some Charisma
att_none Attractive + Dullard
av_none Average Looks + Dullard
ug_none Ugly + Dullard

Once these variables have been created, enter the data as in Table 12.1. If you have problems entering the data then use the file **LooksOrPersonality.sav**. First we have to define our repeated-measures variables, so access the *define factors* dialog box using the menu path **Analyze⇒General Linear Model⇒Repeated Measures…**. As with two-way repeated-measures ANOVA (see the previous chapter) we need to give names to our repeated-measures variables and specify how many levels they have. In this case there are two within-subject factors: **looks** (attractive, average or ugly) and **charisma** (high charisma, some charisma and dullard). In the *define factors* dialog box replace the word *factor1* with the word *looks*. When you have given this repeated-measures factor a name, tell the computer that this variable has three levels by typing the number 3 into the box labelled *Number of Levels*. Click on Add to add this variable to the list of repeated-measures variables. This variable will now appear in the white box at the bottom of the dialog box and appears as *looks(3)*. Now repeat this process for the second independent variable. Enter the word *charisma* into the space labelled *Within-Subject Factor Name* and then, because there were three levels of this variable, enter the number 3 into the space labelled *Number of Levels*. Click on Add to include this variable in the list of factors; it will appear as *charisma(3)*. The finished dialog box is shown in Figure 12.1. When you have entered both of the within-subject factors click on Define to go to the main dialog box.

The main dialog box is the same as when we did a factorial repeated-measures ANOVA in the previous chapter (see Figure 12.2). At the top of the *Within-Subjects Variables* box, SPSS states that there are two factors: **looks** and **charisma**. In the box below there is a series of question marks followed by bracketed numbers. The numbers in brackets represent the levels of the factors (independent variables)—see the previous chapter for a more detailed explanation.

In this example, there are two independent variables and so there are two numbers in the brackets. The first number refers to levels of the first factor listed above the box (in this case **looks**). The second number in the brackets refers to levels of the second factor listed

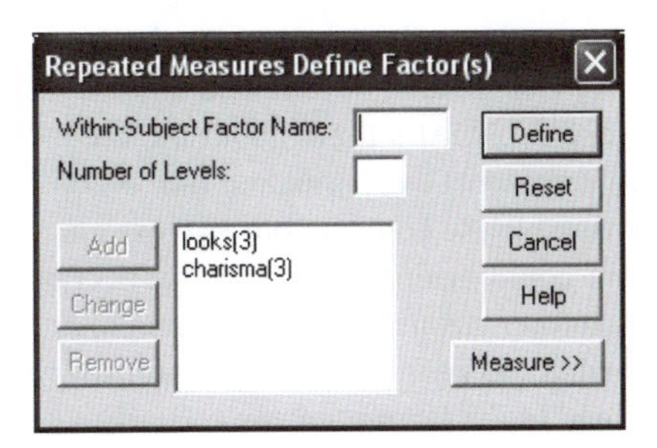

Figure 12.1 *Define factors* dialog box for factorial repeated-measures ANOVA

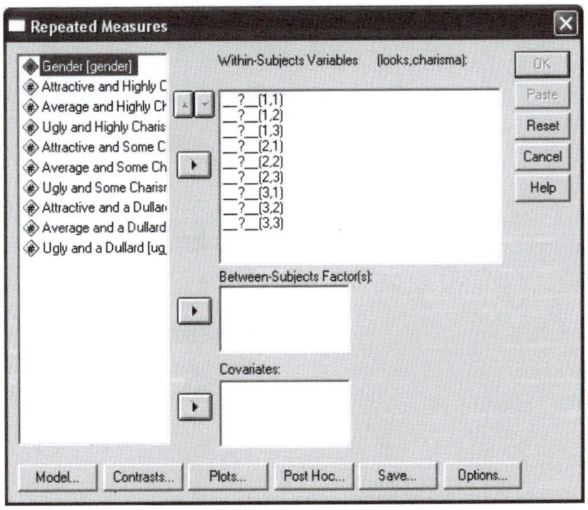

Figure 12.2

above the box (in this case **charisma**). As with the other repeated-measures ANOVAs we've come across, we have to replace the question marks with variables from the list on the left-hand side of the dialog box. With between-group designs, in which coding variables are used, the levels of a particular factor are specified by the codes assigned to them in the data editor. However, in repeated-measures designs, no such coding scheme is used and so we determine which condition to assign to a level at this stage (again look back to the previous chapter for more about this). For this reason, it is imperative that we think about the type of contrasts that we might want to do *before* entering variables into this dialog box. In this experiment, if we look at the first variable, **looks**, there were three conditions: attractive, average and ugly. In many ways it makes sense to compare the attractive and ugly conditions to the average because the average person represents the norm (although it wouldn't be wrong to compare, for example, attractive and average to ugly). This comparison could be done by specifying a simple contrast (see Table 8.6) provided that we make sure that average is coded as our first or last category. Now, let's think about the second factor. The **charisma** factor also has a category that represents the norm, and that is some charisma. Again we could use this as a control against which to compare our two extremes (lots of charisma and none whatsoever). Therefore, we could again conduct a simple contrast comparing everything against some charisma; therefore, this must be entered as either the first or last level.

Based on what has been discussed about using contrasts, it makes sense to have average as level 3 of the **looks** factor and some charisma as the third level of the **charisma** factor. The remaining levels can be decided arbitrarily. I have chosen attractive as level 1 and ugly as level 2 of the **looks** factor. For the **charisma** variable I chose high charisma as level 1 and none as level 2. These decisions mean that the variables should be entered as follows:

att_high	▸	_?_(1,1)
att_none	▸	_?_(1,2)
att_some	▸	_?_(1,3)
ug_high	▸	_?_(2,1)
ug_none	▸	_?_(2,2)
ug_some	▸	_?_(2,3)
av_high	▸	_?_(3,1)
av_none	▸	_?_(3,2)
av_some	▸	_?_(3,3)

Unlike in the previous chapter, I've deliberately made this order different to how the variables are entered into the data editor. This is simply to illustrate that we can enter the variables in any order we like. So far the procedure has been similar to other factorial repeated-measures designs. However, we have a mixed design here, and so we also need to specify our between-group factor as well. We do this by selecting **gender** in the variables list and clicking on ▸ to transfer it to the box labelled *Between-Subjects Factors*. The completed dialog box should look exactly like Figure 12.3. I've already discussed the options for the buttons at the bottom of this dialog box, so I'll talk only about the ones of particular interest for this example.

12.3.2. Other options ②

Following the main analysis it is interesting to compare levels of the independent variables to see whether they differ. As we've seen, there's no facility for entering contrast codes (unless you use syntax) so we have to rely on the standard contrasts available (see Table 8.6). Figure 12.4 shows the dialog box for conducting contrasts and is obtained by clicking on

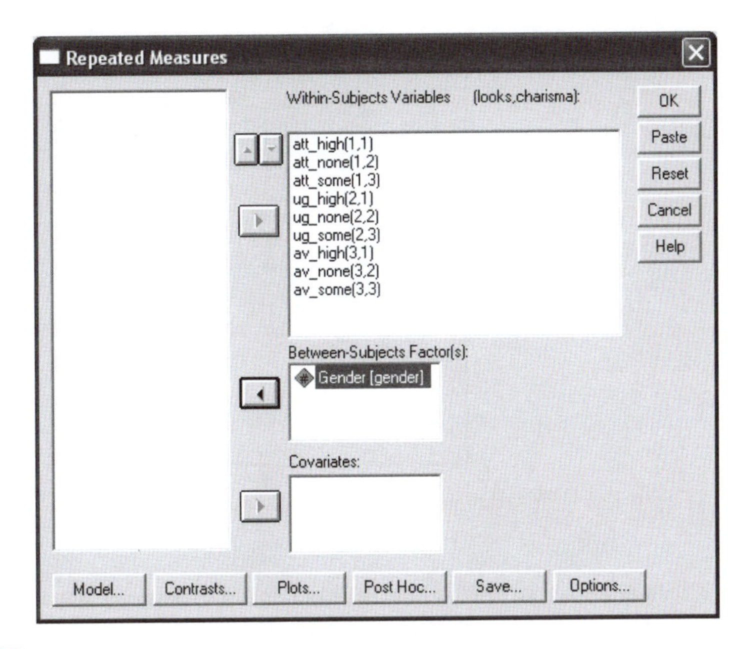

Figure 12.3

Figure 12.4

Contrasts... in the main dialog box. In the previous section I described why it might be interesting to use the average attractiveness and some charisma conditions as base categories for the looks and charisma factors respectively. We have used the *contrasts* dialog box before in sections 9.3.3 and 11.4.2 and so all I shall say is that you should select a simple contrast for each independent variable. For both independent variables, we entered the variables such that the control category was the last one; therefore, we need not change the reference category for the simple contrast. Once the contrasts have been selected, click on Continue to return to the main dialog box.

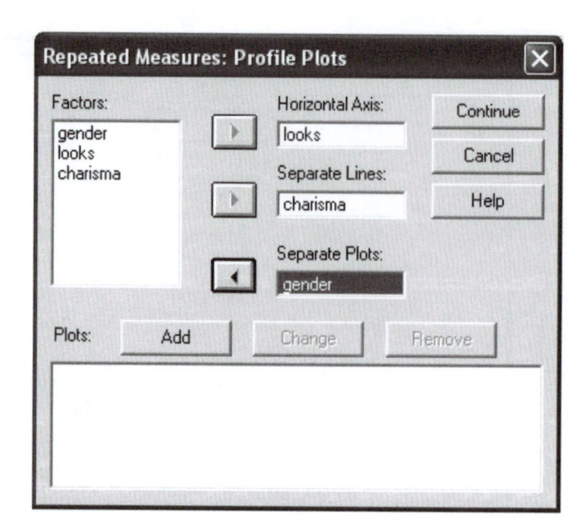

Figure 12.5 *Plots* dialog box for a three-way mixed ANOVA

Gender has only two levels (male or female) so we don't actually need to specify contrasts for this variable. The addition of a between-group factor also means that we can select *post hoc* tests for this variable by clicking on [Post Hoc...]. This action brings up the *post hoc* test dialog box (see section 8.3.2), which can be used as previously explained. However, we need not specify any *post hoc* tests here because the between-group factor has only two levels.

The addition of an extra variable makes it necessary to choose a different graph to the one in the previous chapter's example. Click on [Plots...] to access the dialog box in Figure 12.5. Place **looks** in the space labelled *Horizontal Axis* and **charisma** in space labelled *Separate Line*; finally, place **gender** in the space labelled *Separate Plots*. When all three variables have been specified, don't forget to click on [Add] to add this combination to the list of plots. By asking SPSS to plot the looks × charisma × gender interaction, we should get the interaction graph for looks and charisma, but a separate version of this graph will be produced for male and female participants.

As far as other options are concerned, you should select the same ones that were chosen for the previous example (see section 11.8.4). It is worth selecting estimated marginal means for all effects (because these values will help you to understand any significant effects), but to save space I did not ask for confidence intervals for these effects because we have considered this part of the output in some detail already. When all of the appropriate options have been selected, run the analysis.

12.4. OUTPUT FOR MIXED FACTORIAL ANOVA: MAIN ANALYSIS ③

The initial output is the same as the two-way ANOVA example: there is a table listing the repeated-measures variables from the data editor and the level of each independent variable that they represent. The second table contains descriptive statistics (mean and standard deviation) for each of the nine conditions split according to whether participants were male or female (see SPSS Output 12.1). The names in this table are the names I gave the variables in the data editor (therefore, your output may differ slightly). These descriptive statistics are interesting because they show us the pattern of means across all experimental conditions (so, we use these means to produce the graphs of the three-way interaction).

SPSS Output 12.2 shows the results of Mauchly's sphericity test for each of the three repeated-measures effects in the model. None of the effects violate the assumption of sphericity because all of the values in the column labelled *Sig.* are above .05; therefore, we can assume sphericity when we look at our F-statistics.

SPSS Output 12.3 shows the summary table of the repeated-measures effects in the ANOVA with corrected F-values. As with factorial repeated-measures ANOVA the output is split into sections for each of the effects in the model and their associated error terms. The only difference is that the interactions between our between-group variable of gender and the repeated-measures effects are included also.

Again, we need to look at the column labelled *Sig.* and if the values in this column are less than .05 for a particular effect then it is statistically significant. Working down from the top of the table we find a significant effect of looks, which means that if we ignore whether the date was charismatic, and whether the rating was from a man or a woman, then the attractiveness of a person significantly affected the ratings they received. The looks × gender interaction is also significant, which means that although the ratings were affected by whether the date was attractive, average or ugly, the way in which ratings were affected by attractiveness was different in male and female raters.

Next we find a significant effect of charisma, which means that if we ignore whether the date was attractive, and whether the rating was from a man or a woman, then the charisma of a person significantly affected the ratings they received. The charisma × gender interaction is also significant, so although the ratings were affected by whether the date had high charisma, some charisma or was a dullard, the way in which ratings were affected by charisma was different in male and female raters.

Next we find a significant interaction between looks and charisma, which means that if we ignore the gender of the rater, the profile of ratings across different levels of attractiveness was different for highly charismatic dates, charismatic dates and dullards. (It is equally true to say this the opposite way around: the profile of ratings across different levels of charisma was different for attractive, average and ugly dates.) Just to add to the mounting confusion, the looks × charisma × gender interaction is also significant, meaning that the looks × charisma interaction was significantly different in men and women participants!

This is all a lot to take in, so we'll look at each of these effects in turn in subsequent sections. First, though, we need to see what has happened to our main effect of gender.

12.4.1. The effect of gender ②

The main effect of gender is listed separately from the repeated-measures effects in a table labelled *Tests of Between-Subjects Effects*. Before looking at this table it is important to check the assumption of homogeneity of variance using Levene's test (see section 3.6).

Within-Subjects Factors

Measure: MEASURE_1

LOOKS	CHARISMA	Dependent Variable
1	1	ATT_HIGH
	2	ATT_NONE
	3	ATT_SOME
2	1	UG_HIGH
	2	UG_NONE
	3	UG_SOME
3	1	AV_HIGH
	2	AV_NONE
	3	AV_SOME

Descriptive Statistics

	Gender	Mean	Std. Deviation	N
Attractive and Highly Charismatic	Male	88.30	5.697	10
	Female	89.60	6.637	10
	Total	88.95	6.057	20
Attractive and a Dullard	Male	87.30	5.438	10
	Female	51.80	3.458	10
	Total	69.55	18.743	20
Attractive and Some Charisma	Male	88.50	5.740	10
	Female	87.10	6.806	10
	Total	87.80	6.170	20
Ugly and Highly Charismatic	Male	56.80	5.731	10
	Female	86.70	5.438	10
	Total	71.75	16.274	20
Ugly and a Dullard	Male	45.80	3.584	10
	Female	46.10	3.071	10
	Total	45.95	3.252	20
Ugly and Some Charisma	Male	48.30	5.376	10
	Female	51.20	5.453	10
	Total	49.75	5.476	20
Average and Highly Charismatic	Male	82.80	7.005	10
	Female	88.40	8.329	10
	Total	85.60	8.022	20
Average and a Dullard	Male	47.80	4.185	10
	Female	47.00	3.742	10
	Total	47.40	3.885	20
Average and Some Charisma	Male	71.80	4.417	10
	Female	68.90	5.953	10
	Total	70.35	5.314	20

SPSS Output 12.1

Mauchly's Test of Sphericity[b]

Measure: MEASURE_1

Within Subjects Effect	Mauchly's W	Approx. Chi-Square	df	Sig.	Epsilon[a]		
					Greenhouse-Geisser	Huynh-Feldt	Lower-bound
LOOKS	.960	.690	2	.708	.962	1.000	.500
CHARISMA	.929	1.246	2	.536	.934	1.000	.500
LOOKS*CHARISMA	.613	8.025	9	.534	.799	1.000	.250

Tests the null hypothesis that the error covariance matrix of the orthonormalized transformed dependent variables is proportional to an identity matrix.
a May be used to adjust the degrees of freedom for the averaged tests of significance. Corrected tests are displayed in the Tests of Within-Subjects Effects table.
b Design: Intercept + GENDER
Within Subjects Design: LOOKS+CHARISMA+LOOKS*CHARISMA

SPSS Output 12.2

SPSS produces a table listing Levene's test for each of the repeated-measures variables in the data editor, and we need to look for any variable that has a significant value. SPSS Output 12.4 shows both tables. The table showing Levene's test indicates that variances are homogeneous for all levels of the repeated-measures variables (because all significance values are greater than .05). If any values were significant, then this would compromise the accuracy of the *F*-test for gender, and we would have to consider transforming all of our data to stabilize the variances between groups (see Chapter 3). Fortunately, in this example a transformation is unnecessary. The second table shows the ANOVA summary table for the main effect of gender, and this reveals a non-significant effect (because the significance of .946 is greater than the standard cut-off point of .05).

We can report that there was a non-significant (*ns*) main effect of gender, $F(1, 18) < 1$, *ns*. This effect tells us that if we ignore all other variables, male participants' ratings were basically the same as females'. If you requested SPSS to display means for the gender effect you should scan through your output and find the table in a section headed *Estimated Marginal Means*. SPSS Output 12.5 is a table of means for the main effect of gender with the associated standard errors. This information is plotted in Figure 12.6. It is clear from this graph that men and women's ratings were generally the same.

12.4.2. The effect of looks ②

We came across the main effect of looks in SPSS Output 12.3. Now we're going to have a look at what this effect means. We can report that there was a significant main effect of looks, $F(2, 36) = 423.73$, $p < .001$. This effect tells us that if we ignore all other variables, ratings were different for attractive, average and unattractive dates. If you requested SPSS

Tests of Within-Subjects Effects

Measure: MEASURE_1

Source		Type III Sum of Squares	df	Mean Square	F	Sig.
LOOKS	Sphericity Assumed	20779.633	2	10389.817	423.733	.000
	Greenhouse-Geisser	20779.633	1.923	10803.275	423.733	.000
	Huynh-Feldt	20779.633	2.000	10389.817	423.733	.000
	Lower-bound	20779.633	1.000	20779.633	423.733	.000
LOOKS * GENDER	Sphericity Assumed	3944.100	2	1972.050	80.427	.000
	Greenhouse-Geisser	3944.100	1.923	2050.527	80.427	.000
	Huynh-Feldt	3944.100	2.000	1972.050	80.427	.000
	Lower-bound	3944.100	1.000	3944.100	80.427	.000
Error(LOOKS)	Sphericity Assumed	882.711	36	24.520		
	Greenhouse-Geisser	882.711	34.622	25.496		
	Huynh-Feldt	882.711	36.000	24.520		
	Lower-bound	882.711	18.000	49.040		
CHARISMA	Sphericity Assumed	23233.600	2	11616.800	328.250	.000
	Greenhouse-Geisser	23233.600	1.868	12437.761	328.250	.000
	Huynh-Feldt	23233.600	2.000	11616.800	328.250	.000
	Lower-bound	23233.600	1.000	23233.600	328.250	.000
CHARISMA * GENDER	Sphericity Assumed	4420.133	2	2210.067	62.449	.000
	Greenhouse-Geisser	4420.133	1.868	2366.252	62.449	.000
	Huynh-Feldt	4420.133	2.000	2210.067	62.449	.000
	Lower-bound	4420.133	1.000	4420.133	62.449	.000
Error(CHARISMA)	Sphericity Assumed	1274.044	36	35.390		
	Greenhouse-Geisser	1274.044	33.624	37.891		
	Huynh-Feldt	1274.044	36.000	35.390		
	Lower-bound	1274.044	18.000	70.780		
LOOKS * CHARISMA	Sphericity Assumed	4055.267	4	1013.817	36.633	.000
	Greenhouse-Geisser	4055.267	3.197	1268.295	36.633	.000
	Huynh-Feldt	4055.267	4.000	1013.817	36.633	.000
	Lower-bound	4055.267	1.000	4055.267	36.633	.000
LOOKS * CHARISMA * GENDER	Sphericity Assumed	2669.667	4	667.417	24.116	.000
	Greenhouse-Geisser	2669.667	3.197	834.945	24.116	.000
	Huynh-Feldt	2669.667	4.000	667.417	24.116	.000
	Lower-bound	2669.667	1.000	2669.667	24.116	.000
Error(LOOKS * CHARISMA)	Sphericity Assumed	1992.622	72	27.675		
	Greenhouse-Geisser	1992.622	57.554	34.622		
	Huynh-Feldt	1992.622	72.000	27.675		
	Lower-bound	1992.622	18.000	110.701		

SPSS Output 12.3

Levene's Test of Equality of Error Variances[a]

	F	df1	df2	Sig.
Attractive and Highly Charismatic	1.131	1	18	.302
Attractive and a Dullard	1.949	1	18	.180
Attractive and Some Charisma	.599	1	18	.449
Ugly and Highly Charismatic	.005	1	18	.945
Ugly and a Dullard	.082	1	18	.778
Ugly and Some Charisma	.124	1	18	.729
Average and Highly Charismatic	.102	1	18	.753
Average and a Dullard	.004	1	18	.950
Average and Some Charisma	1.763	1	18	.201

Tests the null hypothesis that the error variance of the dependent variable is equal across groups.
a Design: Intercept+GENDER
Within Subjects Design: LOOKS + CHARISMA + LOOKS*CHARISMA

Tests of Between-Subjects Effects

Meaure: MEASURE_1
Transformed Variable: Average

Source	Type III Sum of Squares	df	Mean Square	F	Sig.
Intercept	94027.756	1	94027.756	20036.900	.000
GENDER	.022	1	.022	.005	.946
Error	84.469	18	4.693		

SPSS Output 12.4

1. Gender

Measure: MEASURE_1

Gender	Mean	Std. Error	95% Confidence Interval	
			Lower Bound	Upper Bound
Male	68.600	.685	67.161	70.039
Female	68.533	.685	67.094	69.973

SPSS Output 12.5

to display means for the looks effect (I'll assume you did from now on) you will find the table in a section headed *Estimated Marginal Means*. SPSS Output 12.6 is a table of means for the main effect of looks with the associated standard errors. The levels of looks are labelled simply 1, 2 and 3, and it's down to you to remember how you entered the variables (or you can look at the summary table that SPSS produces at the beginning of the

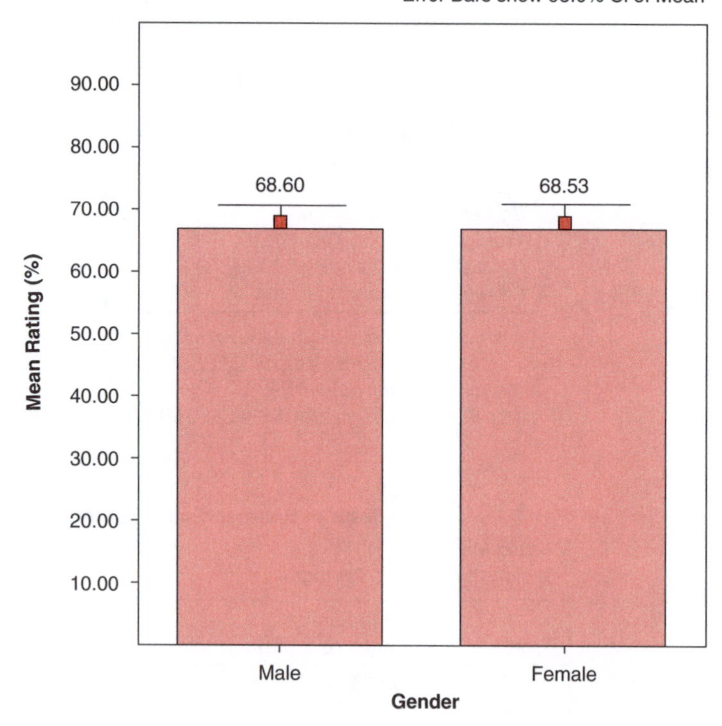

Figure 12.6

2. Looks

Measure: MEASURE_1

Looks	Mean	Std. Error	95% Confidence Interval	
			Lower Bound	Upper Bound
1	82.100	.652	80.729	83.471
2	55.817	.651	54.449	57.184
3	67.783	.820	66.061	69.505

SPSS Output 12.6

output—see SPSS Output 12.1). If you followed what I did then level 1 is attractive, level 2 is ugly and level 3 is average. To make things easier, this information is plotted in Figure 12.7. You can see that as attractiveness falls, the mean rating falls too. So this main effect seems to reflect that the raters were more likely to express a greater interest in going out with attractive people than average or ugly people. However, we really need to look at some contrasts to find out exactly what's going on.

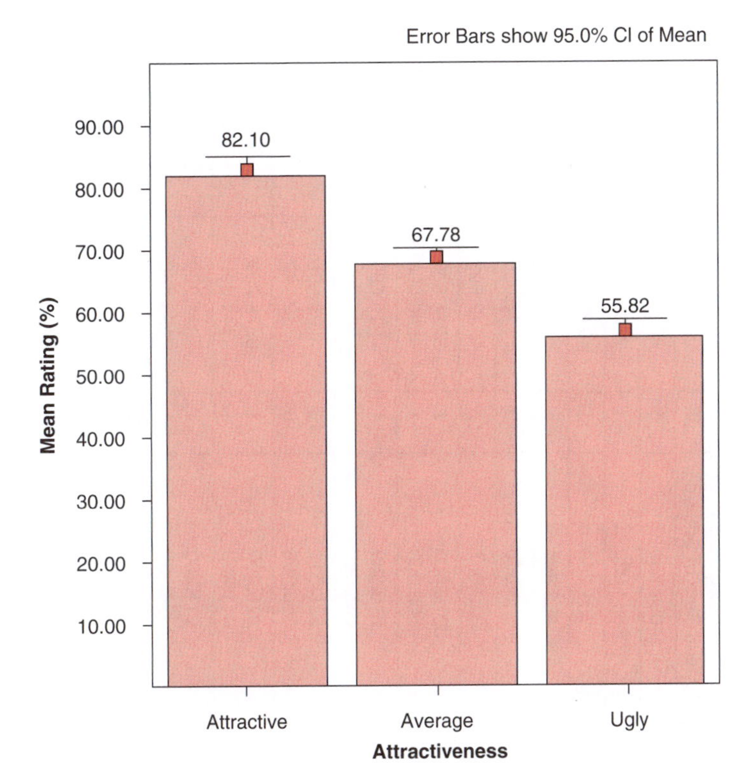

Figure 12.7

SPSS Output 12.7 shows the contrasts that we requested. For the time being, just look at the row labelled *LOOKS*. Remember that we did a simple contrast, and so we get a contrast comparing level 1 with level 3, and then comparing level 2 with level 3; because of the order in which we entered the variables these contrasts represent attractive compared to average (level 1 vs. level 3) and ugly compared to average (level 2 compared to level 3). Looking at the values of F for each contrast, and their related significance values, this tells us that the effect of attractiveness represented the fact that attractive dates were rated significantly higher than average dates, $F(1, 18) = 226.99$, $p < .001$, and average dates were rated significantly higher than ugly ones, $F(1, 18) = 160.07$, $p < .001$.

12.4.3. The effect of charisma ②

The main effect of charisma is in SPSS Output 12.3. We can report that there was a significant main effect of charisma, $F(2, 36) = 328.25$, $p < .001$. This effect tells us that if we ignore all other variables, ratings were different for highly charismatic, a bit charismatic

Tests of Within-Subjects Contrasts

Measure: MEASURE_1

Source	LOOKS	CHARISMA	Type III Sum of Squares	df	Mean Square	F	Sig.
LOOKS	Level 1 vs. Level 3		4099.339	1	4099.339	226.986	.000
	Level 2 vs. Level 3		2864.022	1	2864.022	160.067	.000
LOOKS * GENDER	Level 1 vs. Level 3		781.250	1	781.250	43.259	.000
	Level 2 vs. Level 3		540.800	1	540.800	30.225	.000
Error(LOOKS)	Level 1 vs. Level 3		325.078	18	18.060		
	Level 2 vs. Level 3		322.067	18	17.893		
CHARISMA		Level 1 vs. Level 3	3276.800	1	3276.800	109.937	.000
		Level 2 vs. Level 3	4500.000	1	4500.000	227.941	.000
CHARISMA * GENDER		Level 1 vs. Level 3	810.689	1	810.689	27.199	.000
		Level 2 vs. Level 3	665.089	1	665.089	33.689	.000
Error(CHARISMA)		Level 1 vs. Level 3	536.511	18	29.806		
		Level 2 vs. Level 3	355.356	18	19.742		
LOOKS * CHARISMA	Level 1 vs. Level 3	Level 1 vs. Level 3	3976.200	1	3976.200	21.944	.000
		Level 2 vs. Level 3	441.800	1	441.800	4.091	.058
	Level 2 vs. Level 3	Level 1 vs. Level 3	911.250	1	911.250	6.231	.022
		Level 2 vs. Level 3	7334.450	1	7334.450	88.598	.000
LOOKS * CHARISMA * GENDER	Level 1 vs. Level 3	Level 1 vs. Level 3	168.200	1	168.200	.928	.348
		Level 2 vs. Level 3	6552.200	1	6552.200	60.669	.000
	Level 2 vs. Level 3	Level 1 vs. Level 3	1711.250	1	1711.250	11.701	.003
		Level 2 vs. Level 3	110. 450	1	110. 450	1.334	.263
Error(LOOKS*CHARISMA)	Level 1 vs. Level 3	Level 1 vs. Level 3	3261.600	18	181.200		
		Level 2 vs. Level 3	1944.000	18	108.000		
	Level 2 vs. Level 3	Level 1 vs. Level 3	2632.500	18	146.250		
		Level 2 vs. Level 3	1490.100	18	82.783		

SPSS Output 12.7

3. CHARISMA

Measure: MEASURE_1

CHARISMA	Mean	Std. Error	95% Confidence Interval	
			Lower Bound	Upper Bound
1	82.100	1.010	79.978	84.222
2	54.300	.573	53.096	55.504
3	69.300	.732	67.763	70.837

SPSS Output 12.8

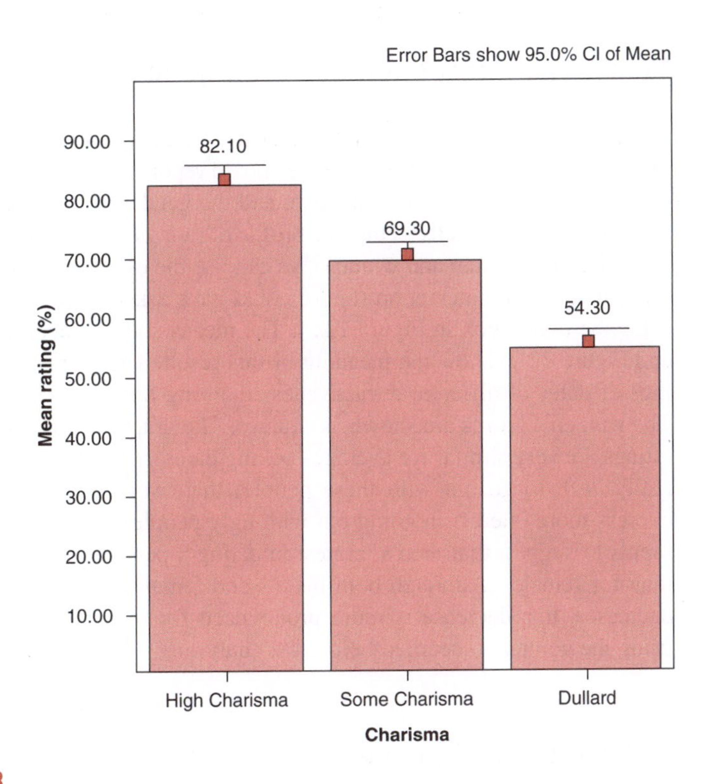

Figure 12.8

and dullard people. The table labelled *CHARISMA* in the section headed *Estimated Marginal Means* tells us what this effect means (SPSS Output 12.8). Again, the levels of charisma are labelled simply 1, 2 and 3. If you followed what I did then level 1 is high charisma, level 2 is no charisma and level 3 is some charisma. This information is plotted in Figure 12.8: as charisma declines, the mean rating falls too. So this main effect seems to reflect that the raters were more likely to express a greater interest in going out with charismatic people than average people or dullards.

SPSS Output 12.7 shows the contrasts that we requested. Looking at the row labelled *CHARISMA* and remembering that we requested simple contrasts, we get a contrast comparing level 1 with level 3, and then comparing level 2 with level 3. How we interpret these contrasts depends on the order in which we entered the repeated-measures variables: in this case these contrasts represent high charisma compared to some charisma (level 1 vs. level 3) and no charisma compared to some charisma (level 2 compared to level 3). The contrasts tell us that the effect of charisma represented the fact that highly charismatic dates were rated significantly higher than dates with some charisma, $F(1, 18) = 109.94$, $p < .001$, and average dates were rated significantly higher than ugly ones, $F(1, 18) = 227.94$, $p < .001$.

12.4.4. The interaction between gender and looks ②

SPSS Output 12.3 indicated that gender interacted in some way with the attractiveness of the date. From the summary table we should report that there was a significant interaction between the attractiveness of the date and the gender of the participant, $F(2, 36) = 80.43$, $p < .001$. This effect tells us that the profile of ratings across dates of different attractiveness was different for men and women. We can use the estimated marginal means to determine the nature of this interaction (or we could have asked SPSS for a plot of gender × looks using the dialog box in Figure 12.5). The means and interaction graph (SPSS output 12.9 and Figure 12.9) show the meaning of this result. The graph shows the average male ratings of dates of different attractiveness ignoring how charismatic the date was (circles). The women's scores are shown as squares. The graph clearly shows that male and female ratings are very similar for average-looking dates, but men give higher ratings (i.e. they're really keen to go out with these people) than women for attractive dates, but women express more interest in going out with ugly people than men. In general this interaction seems to suggest that men's interest in dating a person is more influenced by their looks than for females. Although both male's and female's interest decreases as attractiveness decreases, this decrease is more pronounced for men. This interaction can be clarified using the contrasts specified before the analysis.

12.4.4.1. Looks × gender interaction 1: attractive vs. average, male vs. female ②

The first interaction term looks at level 1 of looks (attractive) compared to level 3 (average), comparing male and female scores. This contrast is highly significant, $F(1, 18) = 43.26$, $p < .001$. This result tells us that the increased interest in attractive dates compared to average-looking dates found for men is significantly more than for women. So, in Figure 12.9 the slope of the line between the circles representing male ratings of attractive dates and average dates is steeper than the line joining the squares representing female ratings of attractive dates and average dates. We can conclude that the preferences for attractive dates, compared to average-looking dates, are greater for males than females.

4. Gender* LOOKS

Measure: MEASURE_1

Gender	LOOKS	Mean	Std. Error	95% Confidence Interval	
				Lower Bound	Upper Bound
Male	1	88.033	.923	86.095	89.972
	2	50.300	.921	48.366	52.234
	3	67.467	1.159	65.031	69.902
Female	1	76.167	.923	74.228	78.105
	2	61.333	.921	59.399	63.267
	3	68.100	1.159	65.665	70.535

SPSS Output 12.9

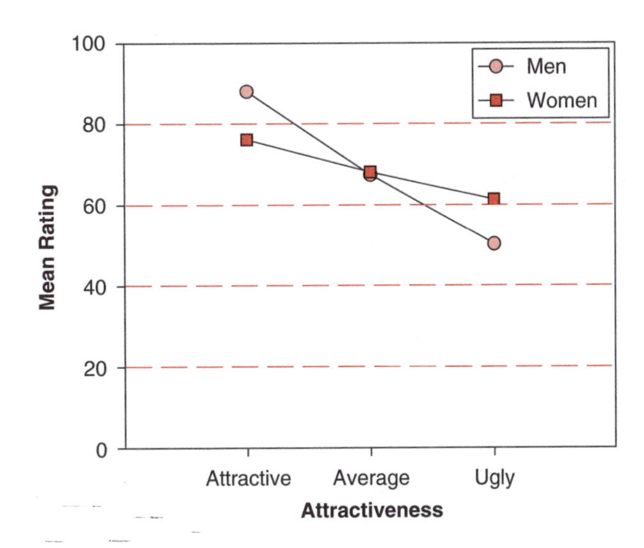

Figure 12.9

12.4.4.2. Looks × gender interaction 2: ugly vs. average, male vs. female ②

The second interaction term compares level 2 of looks (ugly) to level 3 (average), comparing male and female scores. This contrast is highly significant, $F(1, 18) = 30.23$, $p < .001$. This tells us that the decreased interest in ugly dates compared to average-looking dates found for men is significantly more than for women. So, in Figure 12.9 the slope of the line between the circles representing male ratings of ugly dates and average dates is steeper than the line joining the squares representing female ratings of ugly dates and average dates. We can conclude that the preferences for average-looking dates, compared to ugly dates, is greater for males than females.

12.4.5. The interaction between gender and charisma ②

SPSS Output 12.3 indicated that gender interacted in some way with how charismatic the date was. From the summary table we should report that there was a significant interaction between the attractiveness of the date and the gender of the participant, $F(2, 36) = 62.45$, $p < .001$. This effect tells us that the profile of ratings across dates of different levels of charisma was different for men and women. The estimated marginal means (or a plot of gender × charisma using the dialog box in Figure 12.5) tell us the meaning of this interaction (see SPSS Output 12.10 and Figure 12.10, which show the meaning of this result). The graph shows the average male ratings of dates of different levels of charisma ignoring how attractive they were (circles). The women's scores are shown as squares. The graph shows almost the reverse pattern as for the attractiveness data; again male and female ratings are very similar for dates with normal amounts of charisma, but this time men show more interest in dates who are dullards than women do, and women show slightly more interest in very charismatic dates than men do. In general this interaction seems to suggest that women's interest in dating a person is more influenced by their charisma than for men. Although both male's and female's interest decreases as charisma decreases, this decrease is more pronounced for females. This interaction can be clarified using the contrasts specified before the analysis.

12.4.5.1 Charisma × gender interaction 1: high vs. some charisma, male vs. female ②

The first interaction term looks at level 1 of charisma (high charisma) compared to level 3 (some charisma), comparing male and female scores. This contrast is highly significant, $F(1, 18) = 27.20$, $p < .001$. This result tells us that the increased interest in highly charismatic dates compared to averagely charismatic dates found for women is significantly more than for men. So, in Figure 12.10 the slope of the line between the squares representing

5. Gender* CHARISMA

Measure: MEASURE_1

				95% Confidence Interval	
Gender	CHARISMA	Mean	Std. Error	Lower Bound	Upper Bound
Male	1	75.967	1.428	72.966	78.967
	2	60.300	.810	58.598	62.002
	3	69.533	1.035	67.360	71.707
Female	1	88.233	1.428	85.233	91.234
	2	48.300	.810	46.598	50.002
	3	69.067	1.035	66.893	71.240

SPSS Output 12.10

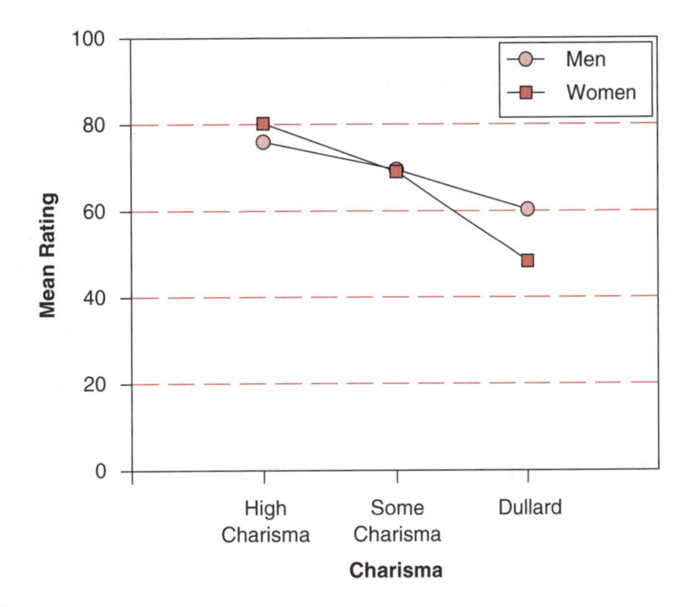

Figure 12.10

female ratings of very charismatic dates and dates with some charisma is steeper than the line joining the circles representing male ratings of very charismatic dates and dates with some charisma. We can conclude that the preferences for very charismatic dates, compared to averagely charismatic dates, are greater for females than males.

12.4.5.2. Charisma × gender interaction 2: dullard vs. some charisma, male vs. female ②

The second interaction term looks at level 2 of charisma (dullard) compared to level 3 (some charisma), comparing male and female scores. This contrast is highly significant, $F(1, 18) = 33.69$, $p < .001$. This result tells us that the decreased interest in dullard dates compared to averagely charismatic dates found for women is significantly more than for men. So, in Figure 12.10 the slope of the line between the squares representing female ratings of dates with some charisma and dullard dates is steeper than the line joining the circles representing male ratings of dates with some charisma and dullard dates. We can conclude that the preferences for dates with some charisma over dullards are greater for females than males.

12.4.6. The interaction between attractiveness and charisma ②

SPSS Output 12.3 indicated that the attractiveness of the date interacted in some way with how charismatic the date was. From the summary table we should report that there was a significant interaction between the attractiveness of the date and the charisma of the date,

6. LOOKS * CHARISMA

Measure: MEASURE_1

LOOKS	CHARISMA	Mean	Std. Error	95% Confidence Interval	
				Lower Bound	Upper Bound
1	1	88.950	1.383	86.045	91.855
	2	69.550	1.019	67.409	71.691
	3	87.800	1.408	84.842	90.758
2	1	71.750	1.249	69.126	74.374
	2	45.950	.746	44.382	47.518
	3	49.750	1.211	47.206	52.294
3	1	85.600	1.721	81.985	89.215
	2	47.400	.888	45.535	49.265
	3	70.350	1.172	67.888	72.812

SPSS Output 12.11

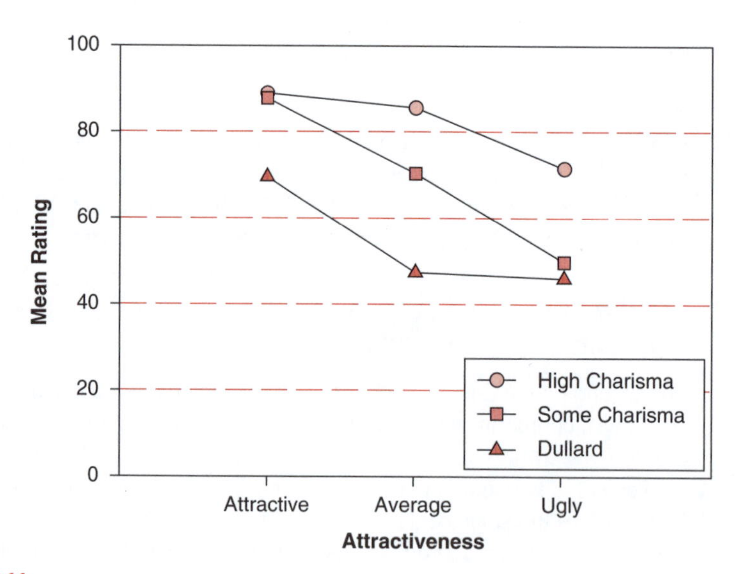

Figure 12.11

$F(4, 72) = 36.63$, $p < .001$. This effect tells us that the profile of ratings across dates of different levels of charisma was different for attractive, average and ugly dates. The estimated marginal means (or a plot of looks × charisma using the dialog box in Figure 12.5) tell us the meaning of this interaction (see SPSS Output 12.11 and Figure 12.11, which show the meaning of this result).

The graph shows the average ratings of dates of different levels of attractiveness when the date also had high levels of charisma (circles), some charisma (squares) and no charisma (triangles). Look first at the difference between attractive and average-looking dates. The interest in highly charismatic dates doesn't change (the line is more or less flat between these two points), but for dates with some charisma or no charisma interest levels decline. So, if you have lots of charisma you can get away with being average looking and people will still want to date you. Now, if we look at the difference between average-looking and ugly dates, a different pattern is observed. For dates with no charisma (triangles) there is no difference between ugly and average people (so if you're a dullard you have to be really attractive before people want to date you). However, for those with charisma, there is a decline in interest if you're ugly (so, if you're ugly, having charisma won't help you much). This interaction is very complex, but we can break it down using the contrasts specified before the analysis.

12.4.6.1. Looks × charisma interaction 1: attractive vs. average, high charisma vs. some charisma ②

The first interaction term investigates level 1 of looks (attractive) compared to level 3 (average looking), comparing level 1 of charisma (high charisma) to level 3 of charisma (some charisma). This is like asking 'is the difference between high charisma and some charisma the same for attractive people and average-looking people?' The best way to understand what this contrast is testing is to extract the relevant bit of the interaction graph in Figure 12.11. If you look at this you can see that the interest (as indicated by high ratings) in attractive dates was the same regardless of whether they had high or low charisma. However, for average-looking dates, there was more interest when that person had high charisma rather than low. The contrast is highly significant, $F(1, 18) = 21.94$, $p < .001$, and tells us that as dates become less attractive there is a greater decline in interest when charisma is average compared to when charisma is high.

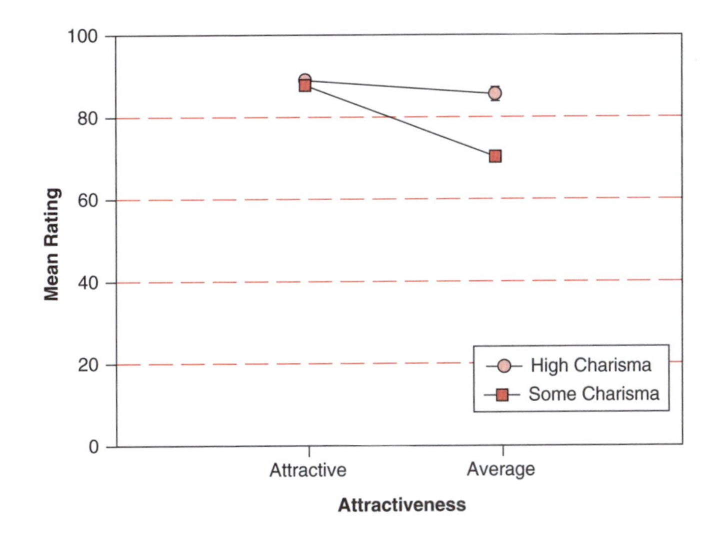

12.4.6.2. Looks × charisma interaction 2: attractive vs. average, dullard vs. some charisma ②

The second interaction term investigates level 1 of looks (attractive) compared to level 3 (average looking), when comparing level 2 of charisma (dullard) to level 3 of charisma (some charisma). This is like asking 'is the difference between no charisma and some charisma the same for attractive people and average-looking people?' Again, the best way to understand what this contrast is testing is to extract the relevant bit of the interaction graph in Figure 12.11. If you look at this you can see that the interest (as indicated by high ratings) in attractive dates was higher when they had some charisma than when they were a dullard. The same is also true for average-looking dates. In fact the two lines are fairly parallel. The contrast is not significant, $F(1, 18) = 4.09$, *ns*, and tells us that as dates become less attractive there is a decline in interest both when charisma is low and when there is no charisma at all.

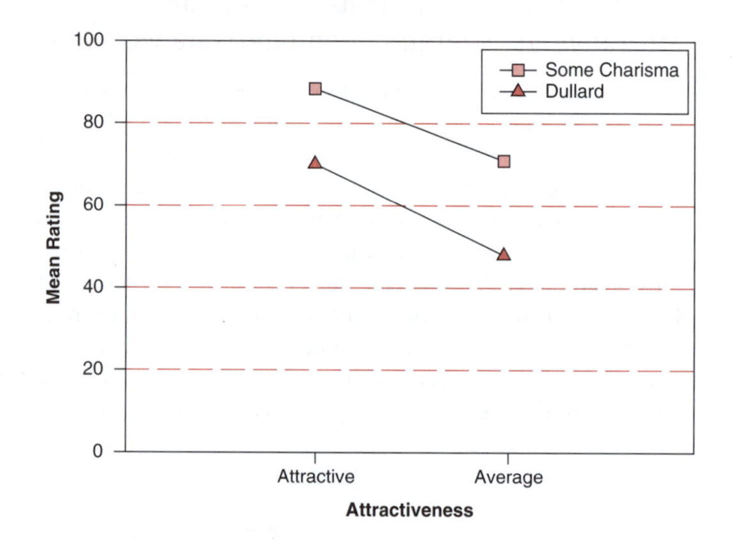

12.4.6.3. Looks × charisma interaction 3: ugly vs. average, high charisma vs. some charisma ②

The third interaction term investigates level 2 of looks (ugly) compared to level 3 (average looking), comparing level 1 of charisma (high charisma) to level 3 of charisma (some charisma). This is like asking 'is the difference between high charisma and some charisma the same for ugly people and average-looking people?' If we again extract the relevant bit of the interaction graph in Figure 12.11 you can see that the interest (as indicated by high ratings) decreases from average-looking dates to ugly ones in both high- and low-charisma dates; however, this fall is slightly greater in the low-charisma dates (the line connecting the squares is slightly steeper). The contrast is significant, $F(1, 18) = 6.23$, $p < .05$, and tells us that as dates become less attractive there is a greater decline in interest when charisma is low compared to when charisma is high.

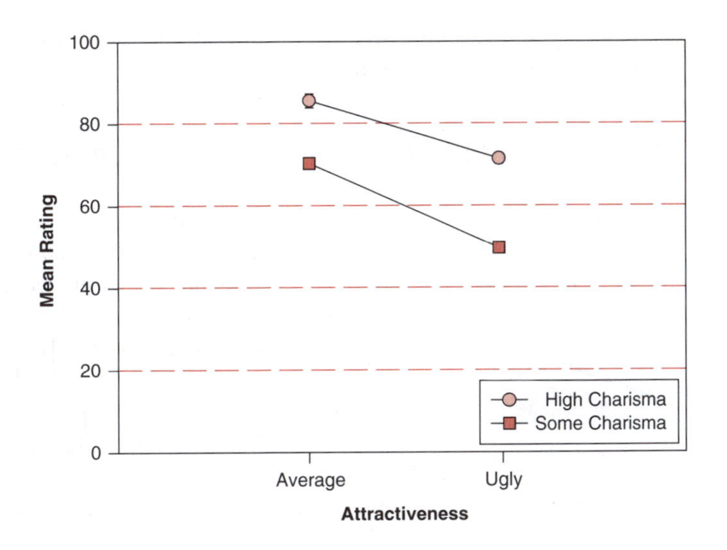

12.4.6.4. Looks × charisma interaction 4: ugly vs. average, dullard vs. some charisma ②

The final interaction term investigates level 2 of looks (ugly) compared to level 3 (average looking), when comparing level 2 of charisma (dullard) to level 3 of charisma (some charisma). This is like asking 'is the difference between no charisma and some charisma the same for ugly people and average-looking people?' If we extract the relevant bit of the interaction graph in Figure 12.11 you can see that the interest (as indicated by high ratings) in average-looking dates was higher when they had some charisma than when they were a dullard, but for ugly dates the ratings were roughly the same regales of the level of charisma. This contrast is highly significant, $F(1, 18) = 88.60$, $p < .001$, and tells us that

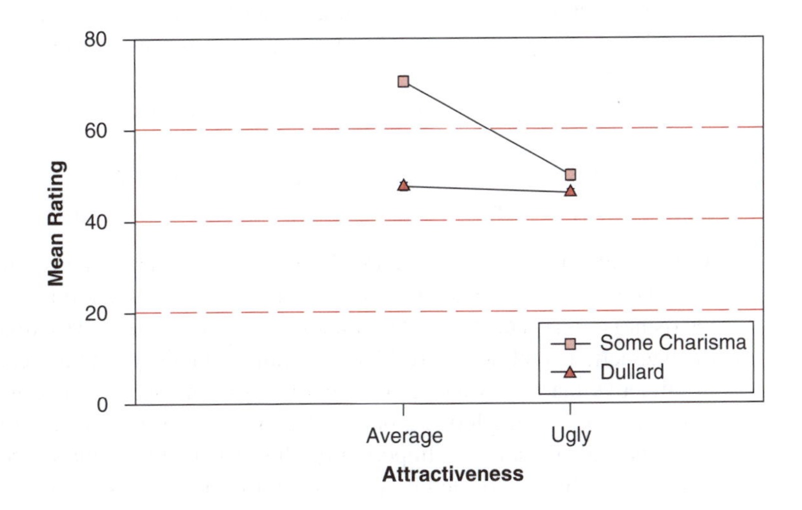

as dates become less attractive the decline in interest in dates with a bit of charisma is significantly greater than for dullards.

12.4.7. The interaction between looks, charisma and gender ③

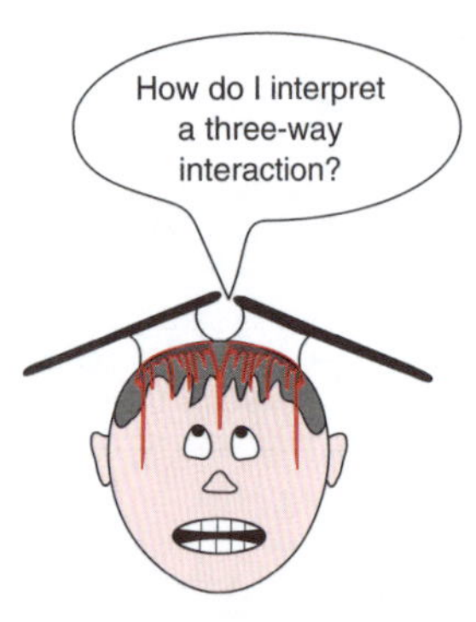

The three-way interaction tells us whether the looks × charisma interaction described above is the same for men and women (i.e. whether the combined effect of attractiveness of the date and their level of charisma is the same for male participants as for female subjects). SPSS Output 12.3 tells us that there is a significant three-way looks × charisma × gender interaction, $F(4, 72) = 24.12$, $p < .001$. The nature of this interaction is revealed in Figure 12.12, which shows the looks × charisma interaction for men and women separately (the means on which the graph is based appear in SPSS Output 12.12). The male graph shows that when dates are attractive, men will express a high interest regardless of charisma levels (the circle, square and dot all overlap). At the opposite end of the attractiveness scale, when a date is ugly, regardless of charisma men will express very little interest (ratings are all low). The only time charisma makes any difference to a man is if the date is average looking, in which case high charisma boosts interest, being a dullard reduces interest, and having a bit of charisma leaves things somewhere in between. The take-home message is that men are superficial cretins who are more interested in physical attributes. The picture for women is very different. If someone has high levels of charisma then it doesn't really matter what they look like, women will express an interest in them (the line of circles is relatively flat). At the other extreme, if the date is a dullard, then they will express no interest in them, regardless of how attractive they are (the line of triangles is relatively flat). The only time attractiveness makes a difference is when someone has an average amount of charisma, in which case being attractive boosts interest, and being ugly reduces it. Put another way, women prioritize charisma over physical appearance. Again, we can look at some contrasts further to break this interaction down. These contrasts are similar to those for the looks × charisma interaction, but they now also take into account the effect of gender as well!

12.4.7.1. Looks × charisma × gender interaction 1: attractive vs. average, high charisma vs. some charisma, male vs. female ③

The first interaction term compares level 1 of looks (attractive) to level 3 (average looking), when level 1 of charisma (high charisma) is compared to level 3 of charisma (some charisma) in males compared to females, $F(1, 18) < 1$, *ns*. If we extract the relevant bits of the interaction graph in Figure 12.12, we can see that interest (as indicated by high ratings) in attractive dates was the same regardless of whether they had high or low charisma. However, for average-looking dates, there was more interest when that person had high charisma rather than low. Importantly, this pattern of results is the same in males and females and this is reflected in the non-significance of this contrast.

7. Gender* LOOKS * CHARISMA

Measure: MEASURE_1

Gender	LOOKS	CHARISMA	Mean	Std. Error	95% Confidence Interval	
					Lower Bound	Upper Bound
Male	1	1	88.300	1.956	84.191	92.409
		2	87.300	1.441	84.273	90.327
		3	88.500	1.991	84.317	92.683
	2	1	56.800	1.767	53.089	60.511
		2	45.800	1.055	43.583	48.017
		3	48.300	1.712	44.703	51.897
	3	1	82.800	2.434	77.687	87.913
		2	47.800	1.255	45.163	50.437
		3	71.800	1.657	68.318	75.282
Female	1	1	89.600	1.956	85.491	93.709
		2	51.800	1.441	48.773	54.827
		3	87.100	1.991	82.917	91.283
	2	1	86.700	1.767	82.989	90.411
		2	46.100	1.055	43.883	48.317
		3	51.200	1.712	47.603	54.797
	3	1	88.400	2.434	83.287	93.513
		2	47.000	1.255	44.363	49.637
		3	68.900	1.657	65.418	72.382

SPSS Output 12.12

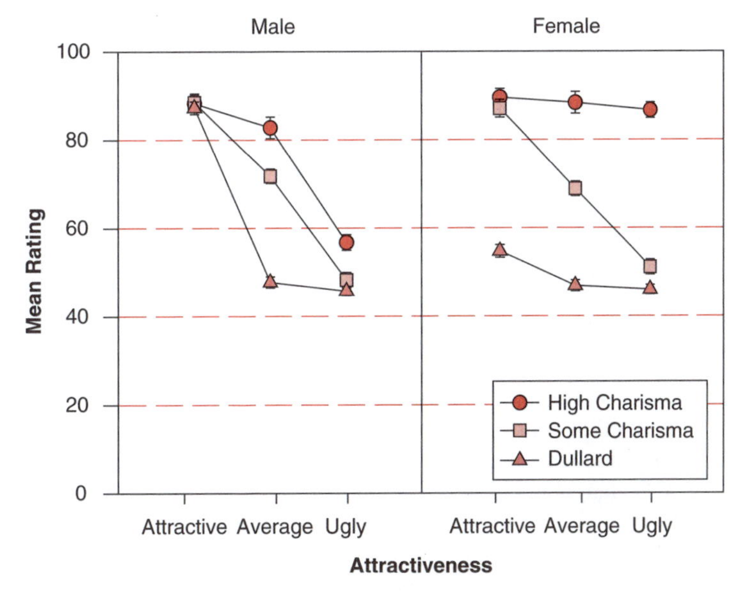

Figure 12.12 Graphs showing the looks × charisma interaction for men and women. Lines represent high charisma (circles), some charisma (squares) and no charisma (triangles)

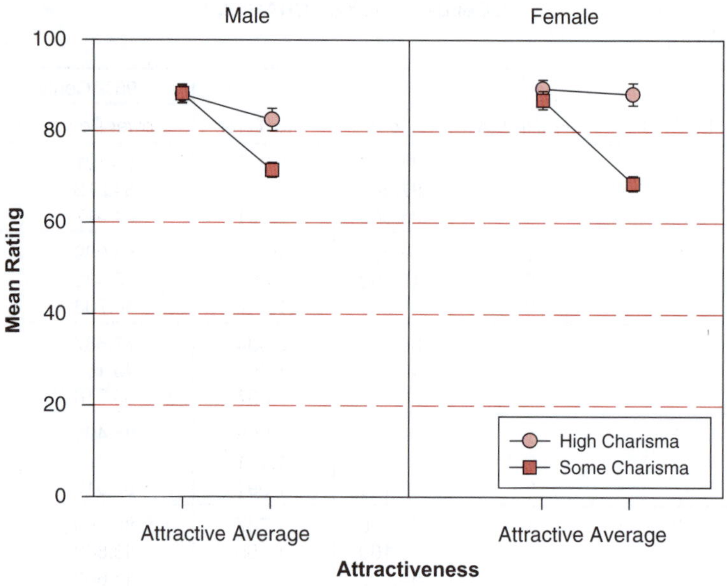

12.4.7.2. Looks × charisma × gender interaction 2: attractive vs. average, dullard vs. some charisma, male vs. female ③

The second interaction term compares level 1 of looks (attractive) to level 3 (average looking), when level 2 of charisma (dullard) is compared to level 3 of charisma (some charisma), in men compared to women. Again, we extract the relevant bit of the interaction graph in Figure 12.12 and you can see that the patterns are different for men and women. This is reflected by the fact that the contrast is significant, $F(1, 18) = 60.67$, $p < .001$. To unpick this we need to look at the graph. First, if we look at average-looking dates, for both men and women more interest is expressed when the date has some charisma than when they have none (and the distance between the square and the triangle is about the same). So the difference doesn't appear to be here. If we now look at attractive dates, we see that men are equally interested in their dates regardless of their charisma, but for women, they're much less interested in an attractive person if they are a dullard. Put another way, for attractive dates, the distance between the square and the triangle is much smaller for men than it is for women. Another way to look at it is that for dates with some charisma, the reduction in interest as attractiveness goes down is about the same in men and women (the lines with squares have the same slope). However, for dates who are dullards, the decrease in interest if these dates are average looking rather than attractive is much more dramatic in men than women (the line with triangles is much steeper for men than it is for women). This is what the significant contrast is telling us.

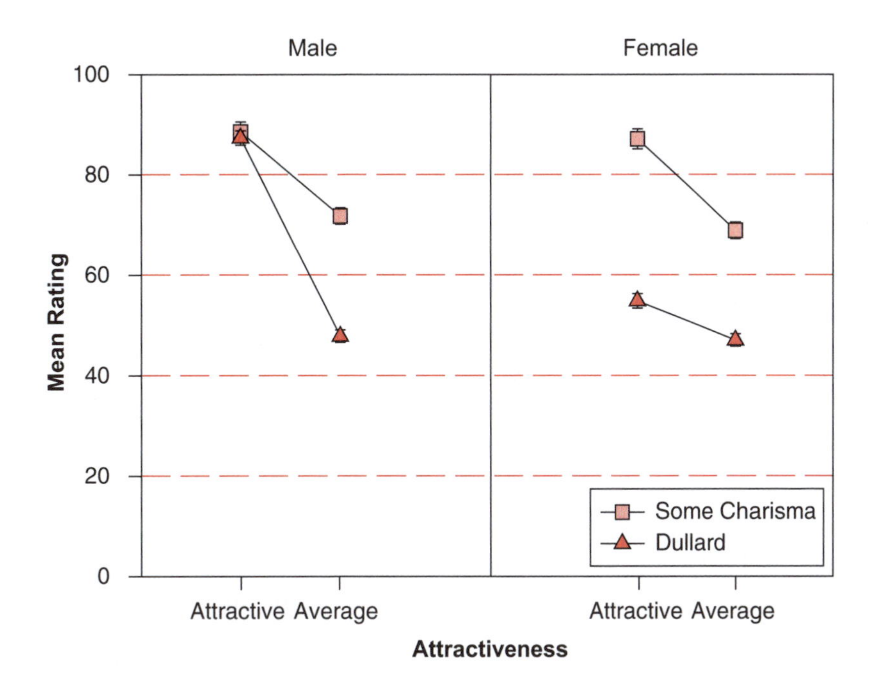

12.4.7.3. Looks × charisma × gender interaction 3: ugly vs. average, high charisma vs. some charisma, males vs. females ③

The third interaction term compares level 2 of looks (ugly) to level 3 (average looking), when level 1 of charisma (high charisma) is compared to level 3 of charisma (some charisma), in men compared to women. Again, we extract the relevant bit of the interaction graph in Figure 12.12 and you can see that the patterns are different for men and women. This is reflected by the fact that the contrast is significant, $F(1, 18) = 11.70$, $p < .01$. To unpick this we need to look at the graph. First, let's look at the men. For men, as attractiveness goes down, so does interest when the date has high charisma and when they have low charisma. In fact the lines are parallel. So, regardless of charisma, there is a similar reduction in interest as attractiveness declines. For women the picture is quite different: when charisma is high, there is no decline in interest as attractiveness falls (the line connecting the circles is flat), but when charisma is lower, the attractiveness of the date does matter and interest is lower in an ugly date than in an average-looking date. Another way to look at it is that for dates with some charisma, the reduction in interest as attractiveness goes down is about the same in men and women (the lines with squares have the same slope). However, for dates who have high charisma, the decrease in interest if these dates are ugly rather than average looking is much more dramatic in men than women (the line with circles is much steeper for men than it is for women). This is what the significant contrast is telling us.

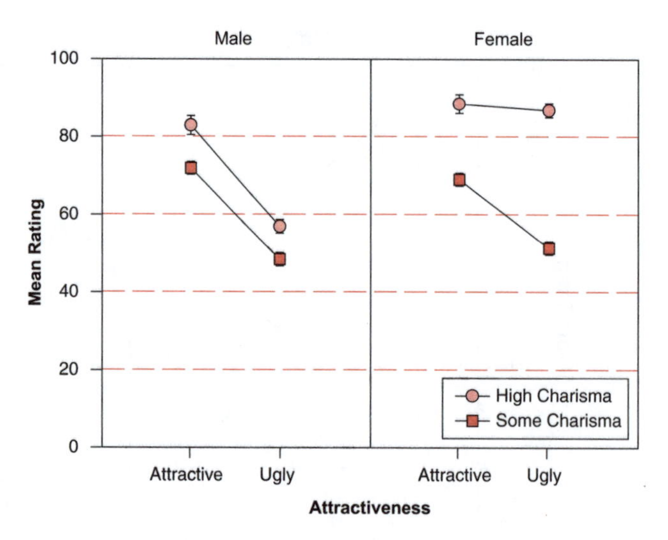

12.4.7.4. Looks × charisma × gender interaction 4: ugly vs. average, dullard vs. some charisma, male vs. female ③

The final interaction term compares level 2 of looks (ugly) to level 3 (average looking), when comparing level 2 of charisma (dullard) to level 3 of charisma (some charisma), in men compared to women. If we extract the relevant bits of the interaction graph in Figure 12.12, we can see that interest (as indicated by high ratings) in ugly dates was the same regardless of whether they had high or low charisma. However, for average-looking dates, there was more interest when that person had some charisma rather than if they were a dullard. Importantly, this pattern of results is the same in males and females and this is reflected in the non-significance of this contrast, $F(1, 18) = 1.33$, *ns*.

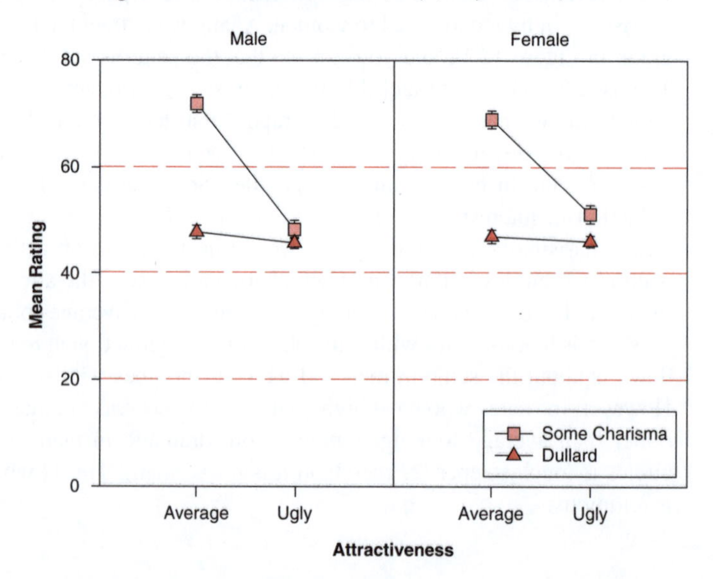

12.4.8. Conclusions ③

These contrasts tell us nothing about the differences between the attractive and ugly conditions, or the high charisma and dullard conditions, because these were never compared. We could rerun the analysis and specify our contrasts differently to get these effects. However, what is clear from our data is that differences exist between men and women in terms of how they're affected by the looks and personality of potential dates. Men appear to be enthusiastic about dating anyone who is attractive regardless of how awful their personality. Women are almost completely the opposite: they are enthusiastic about dating anyone with a lot of charisma, regardless of how they look (and are unenthusiastic about dating people without charisma regardless of how attractive they look). The only consistency between men and women is when there is some charisma (but not lots) in which case for both genders the attractiveness influences how enthusiastic they are about dating the person.

What should be even clearer from this chapter is that when more than two independent variables are used in an ANOVA, it yields complex interaction effects that require a great deal of concentration to interpret (imagine interpreting a four-way interaction!). Therefore, it is essential to take a systematic approach to interpretation and plotting graphs is a particularly useful way to proceed. It is also advisable to think carefully about the appropriate contrasts to use to answer the questions you have about your data. It is these contrasts that will help you to interpret interactions, so make sure you select sensible ones!

Cramming Samantha's Tips

- Mixed ANOVA compares several means when there are two or more independent variables, and at least one of the variables has been measured using the same participants and at least one other variable has been measured using different participants.
- Test the assumption of *sphericity* for the repeated-measures variable(s) when they have three or more conditions using *Mauchly's test*. If the value in the column labelled *Sig.* is less than .05 then the assumption is violated. If the significance of Mauchly's test is greater than .05 then the assumption of sphericity has been met. You should test this assumption for all effects (if there are two or more repeated-measures variables this means you test the assumption for all variables and the corresponding interaction terms).
- The table labelled *Tests of Within-Subjects Effects* shows the results of your ANOVA for the repeated-measures variables and all of the interaction effects. For *each* effect, if the assumption of sphericity has been met then look at the row labelled *Sphericity Assumed*. If the assumption was violated then read the row labelled *Greenhouse–Geisser* (you can also look at *Huynh–Feldt* but you'll have to read the previous chapter to find out the relative merits of the two procedures). Having selected the appropriate row, look at the column labelled *Sig.*: if the value is less than .05 then the means of the groups are significantly different.

(Continued)

Cramming Samantha's Tips (Continued)

- The table labelled *Tests of Between-Subjects Effects* shows the results of your ANOVA for the between-group variables. Look at the column labelled *Sig.*: if the value is less than .05 then the means of the groups are significantly different.
- Break down the main effects and interaction terms using contrasts. These contrasts appear in the table labelled *Tests of Within-Subjects Contrasts*; again look to the columns labelled *Sig.* to discover if your comparisons are significant (they will be if the significance value is less than .05).
- Look at the means, or better still draw graphs, to help you interpret the contrasts.

12.5. CALCULATING EFFECT SIZES ③

I keep emphasizing the fact that effect sizes are really more useful when they summarize a focused effect. This also gives me a useful excuse to circumvent the complexities of omega squared in mixed designs (it's the road to madness I assure you). Therefore, just calculate effect sizes for your contrasts when you've got a factorial design (and any main effects that compare only two groups). SPSS Output 12.7 shows the values for several contrasts, all of which have 1 degree of freedom for the model (i.e. they represent a focused and interpretable comparison) and have 18 residual degrees of freedom. We can use these *F*-ratios and convert them to an effect size *r*, using a formula we've come across before:

$$r = \sqrt{\frac{F(1, df_{\text{R}})}{F(1, df_{\text{R}}) + df_{\text{R}}}}$$

First, we can deal with the main effect of gender because this compares only two groups:

$$r_{\text{Gender}} = \sqrt{\frac{0.005}{0.005 + 18}} = .02$$

For the two comparisons we did for the looks variable (SPSS Output 12.7), we would get:

$$r_{\text{Attractive vs. Average}} = \sqrt{\frac{226.99}{226.99 + 18}} = .96$$

$$r_{\text{Ugly vs. Average}} = \sqrt{\frac{160.07}{160.07 + 18}} = .95$$

Therefore, both comparisons yielded massive effect sizes. For the two comparisons we did for the charisma variable (SPSS Output 12.7), we would get:

$$r_{\text{High vs. Some}} = \sqrt{\frac{109.94}{109.94 + 18}} = .93$$

$$r_{\text{Dullard vs. Some}} = \sqrt{\frac{227.94}{227.94 + 18}} = .96$$

Again, both comparisons yield massive effect sizes. For the looks × gender interaction, we again had two contrasts:

$$r_{\text{Attractive vs. Average, Male vs. Female}} = \sqrt{\frac{43.26}{43.26 + 18}} = .84$$

$$r_{\text{Ugly vs. Average, Male vs. Female}} = \sqrt{\frac{30.23}{30.23 + 18}} = .79$$

Again, these are massive effects. For the charisma × gender interaction, the two contrasts give us:

$$r_{\text{High vs. Some, Male vs. Female}} = \sqrt{\frac{27.20}{27.20 + 18}} = .78$$

$$r_{\text{Dullard vs. Some, Male vs. Female}} = \sqrt{\frac{33.69}{33.69 + 18}} = .81$$

Yet again massive effects (can't you tell the data are fabricated!). Moving on to the looks × charisma interaction, we get four contrasts:

$$r_{\text{Attractive vs. Average, High vs. Some}} = \sqrt{\frac{21.94}{21.94 + 18}} = .74$$

$$r_{\text{Attractive vs. Average, Dullard vs. Some}} = \sqrt{\frac{4.09}{4.09 + 18}} = .43$$

$$r_{\text{Ugly vs. Average, High vs. Some}} = \sqrt{\frac{6.23}{6.23 + 18}} = .51$$

$$r_{\text{Ugly vs. Average, Dullard vs. Some}} = \sqrt{\frac{88.60}{88.60 + 18}} = .91$$

All of these effects are in the medium to massive range. Finally, for the looks × charisma × gender interaction we had four contrasts:

$$r_{\text{Attractive vs. Average, High vs. Some, Male vs. Female}} = \sqrt{\frac{0.93}{0.93 + 18}} = .22$$

$$r_{\text{Attractive vs. Average, Dullard vs. Some, Male vs. Female}} = \sqrt{\frac{60.67}{60.67 + 18}} = .88$$

$$r_{\text{Ugly vs. Average, High vs. Some, Male vs. Female}} = \sqrt{\frac{11.70}{11.70 + 18}} = .63$$

$$r_{\text{Ugly vs. Average, Dullard vs. Some, Male vs. Female}} = \sqrt{\frac{1.33}{1.33 + 18}} = .26$$

As such the two effects that were significant (attractive vs. average, dullard vs. some, male vs. female and ugly vs. average, high vs. some, male vs. female) yielded large effect sizes. The two effects that were not significant yielded close to medium effect sizes.

12.6. REPORTING THE RESULTS OF MIXED ANOVA ②

As you've probably gathered, when you have more than two independent variables, there's a hell of a lot of information to report. You have to report all of the main effects, all of the interactions, and any contrasts you may have done. This can take up a lot of space and one good tip is to reserve the most detail for the effects that actually matter (e.g. main effects are usually not that interesting if you've got a significant interaction that includes that variable). I'm a big fan of giving brief explanations of results in the results section to really get the message across about what a particular effect is telling us and so I tend not just to report results, but to offer some interpretation as well. Having said that, some journal editors are big fans of telling me my results sections are too long. So, you should probably ignore everything I say. Assuming we want to report all of our effects, we could do it something like this (although not as a list!):

- ✓ All effects are reported as significant at $p < .05$. There was a significant main effect of the attractiveness of the date on interest expressed by participant, $F(2, 36) = 423.73$. Contrasts revealed that attractive dates were more desirable than average-looking ones, $F(1, 18) = 226.99$, $r = .96$, and ugly dates were less desirable than average-looking ones, $F(1, 18) = 160.07$, $r = .95$.
- ✓ There was also a significant main effect of the amount of charisma the date possessed on the interest expressed in dating them, $F(2, 36) = 328.25$. Contrasts revealed that dates with high charisma were more desirable than dates with some charisma,

$F(1, 18) = 109.94$, $r = .93$, and dullards were less desirable than dates with some charisma, $F(1, 18) = 227.94$, $r = .96$.

- ✓ There was no significant effect of gender indicating that ratings from male and female participants were in general the same, $F(1, 18) < 1$, $r = .02$.
- ✓ There was a significant interaction effect between the attractiveness of the date and the gender of the participant, $F(2, 36) = 80.43$. This indicates that the desirability of dates of different levels of attractiveness differed in men and women. To break down this interaction, contrasts were performed comparing each level of attractiveness to average looking across male and female participants. These revealed significant interactions when comparing male and female scores to attractive dates compared to average-looking dates, $F(1, 18) = 43.26$, $r = .84$, and to ugly dates compared to average dates $F(1, 18) = 30.23$, $r = .79$. Looking at the interaction graph, this suggests that male and female ratings are very similar for average-looking dates, but men rate attractive dates higher than women, whereas women rate ugly dates higher than men. Although both male's and female's interest decreases as attractiveness decreases, this decrease is more pronounced for men, suggesting that when charisma is ignored, men's interest in dating a person is more influenced by their looks than for females.
- ✓ There was a significant interaction effect between the level of charisma of the date and the gender of the participant, $F(2, 36) = 62.45$. This indicates that the desirability of dates of different levels of charisma differed in men and women. To break down this interaction, contrasts were performed comparing each level of charisma to the middle category of 'some charisma' across male and female participants. These revealed significant interactions when comparing male and female scores to highly charismatic dates compared to dates with some charisma, $F(1, 18) = 27.20$, $r = .78$, and to dullards compared to dates with some charisma $F(1, 18) = 33.69$, $r = .81$. The interaction graph shows that men show more interest in dates who are dullards than women do, and women show slightly more interest in very charismatic dates than men do. Both male's and female's interest decreases as charisma decreases, but this decreases is more pronounced for females, suggesting women's interest in dating a person is more influenced by their charisma than for men.
- ✓ There was a significant interaction effect between the level of charisma of the date and the attractiveness of the date $F(4, 72) = 36.63$. This indicates that the desirability of dates of different levels of charisma differed according to their attractiveness. To break down this interaction, contrasts were performed comparing each level of charisma to the middle category of 'some charisma' across each level of attractiveness compared to the category of average attractiveness. The first contrast revealed a significant interaction when comparing attractive dates to average-looking dates when the date had high charisma compared to some charisma, $F(1, 18) = 21.94$, $r = .74$, and tells us that as dates become less attractive there is a greater decline in interest when charisma is low compared to when charisma is high. The second contrast compared attractive dates to average-looking dates when the date was a dullard compared to when they had some charisma. This was not significant, $F(1, 18) = 4.09$, $r = .43$, and tells us that as dates become less attractive there is a decline in interest both when charisma is low and when there is no charisma at all. The third contrast compared ugly dates to average-looking dates when

the date had high charisma compared to some charisma. This was significant, $F(1, 18) = 6.23$, $r = .51$, and tells us that as dates become less attractive there is a greater decline in interest when charisma is low compared to when charisma is high. The final contrast compared ugly dates to average-looking dates when the date was a dullard compared to when they had some charisma. This contrast was highly significant, $F(1, 18) = 88.60$, $r = .91$, and tells us that as dates become less attractive the decline in interest in dates with a bit of charisma is significantly greater than for dullards.

✓ Finally, the looks × charisma × gender interaction was significant, $F(4, 72) = 24.12$. This indicates that the looks × charisma interaction described previously was different in male and female participants. Again, contrasts were used to break down this interaction; these contrasts compared male and female scores each level of charisma compared to the middle category of 'some charisma' across each level of attractiveness compared to the category of average attractiveness. The first contrast revealed a non-significant difference between male and female responses when comparing attractive dates to average-looking dates when the date had high charisma compared to some charisma, $F(1, 18) < 1$, $r = .22$, and tells us that for both males and females as dates become less attractive there is a greater decline in interest when charisma is low compared to when charisma is high. The second contrast investigated differences between males and females when comparing attractive dates to average-looking dates when the date was a dullard compared to when they had some charisma. This was significant, $F(1, 18) = 60.67$, $r = .88$, and tells us that for dates with some charisma, the reduction in interest as attractiveness goes down is about the same in men and women, but for dates who are dullards, the decrease in interest if these dates are average looking rather than attractive is much more dramatic in men than women. The third contrast looked for differences between males and females when comparing ugly dates to average-looking dates when the date had high charisma compared to some charisma. This was significant, $F(1, 18) = 11.70$, $r = .63$, and tells us that for dates with some charisma, the reduction in interest as attractiveness goes down is about the same in men and women, but for dates who have high charisma, the decrease in interest if these dates are ugly rather than average looking is much more dramatic in men than women. The final contrast looked for differences between men and women when comparing ugly dates to average-looking dates when the date was a dullard compared to when they had some charisma. This contrast was not significant, $F(1, 18) = 1.33$, $r = .26$, and tells us that for both men and women, as dates become less attractive the decline in interest in dates with a bit of charisma is significantly greater than for dullards.

12.7. WHAT HAVE WE DISCOVERED ABOUT STATISTICS? ②

Three-way ANOVA is a confusing nut to crack. I've probably done hundreds of three-way ANOVAs in my life and still I kept getting confused throughout writing this chapter (and so if you're confused after reading it, it's not your fault, it's mine). Hopefully, what you should have discovered is that ANOVA is flexible enough that you can mix and match independent variables that are measured using the same or different participants. In addition we've looked at how ANOVA is also flexible enough to go beyond merely including two independent variables. Hopefully, you've also started to realize why there are good reasons to limit the number of independent variables that you include (for the sake of interpretation). Of course, far more interestingly, you've discovered that men are superficial creatures who value looks over charisma, and that women are prepared to date the hunchback of Notre Dame provided he has a sufficient amount of charisma. It's something of a relief (to me at least) that we now slow the pace down a bit to look at tests that we can use when the assumptions required for parametric tests are broken and cannot be corrected.

12.8. KEY TERMS THAT WE'VE DISCOVERED

- Mixed ANOVA
- Mixed design

12.9. SMART ALEX'S TASKS

- I am going to extend the example from the previous chapter (advertising and different imagery) by adding a between-group variable to the design.[4] To recap, in case you haven't read the previous chapter, participants viewed a total of nine mock adverts over three sessions. In these adverts there were three products (a brand of beer, Brain Death, a brand of wine, Dangleberry, and a brand of water, Puritan). These could be presented alongside positive, negative or neutral imagery. Over the three sessions and nine adverts each type of product was paired with each type of imagery (read the last chapter if you need more detail). After each advert participants rated the drinks on a scale ranging from −100 (dislike very much) through 0 (neutral) to 100 (like very much). The design, thus far, has two independent variables: the type of drink (beer, wine or water) and the type of imagery used (positive, negative or neutral). These two variables completely cross over, producing nine experimental conditions. Now imagine that I also took note of each person's gender. Subsequent to the previous analysis it occurred to me that men and women might respond differently to the products (because, in keeping with stereotypes, men might mostly drink lager whereas women might drink wine). Therefore, I wanted to reanalyse the data taking this additional variable into account. Now, gender is a between-group variable because a participant can be only male or female: they cannot participate as a male and then change into a female and participate again!

4 Previously the example contained two repeated-measures variables (drink type and imagery type), but now it will include three variables (two repeated-measures and one between-group).

The data are the same as in the previous chapter (Table 11.3) and can be found in the file **MixedAttitude.sav**. Run a mixed ANOVA on these data.③

- **Task 2**: Text messaging is very popular amongst mobile phone owners, to the point that books have been published on how to write in text speak (BTW, hope u no wat I mean by txt spk). One concern is that children may use this form of communication so much that it will hinder their ability to learn correct written English. One concerned researcher conducted an experiment in which one group of children was encouraged to send text messages on their mobile phones over a 6 month period. A second group was forbidden from sending text messages for the same period. To ensure that kids in this later group didn't use their phones, this group were given armbands that administered painful shocks in the presence of microwaves (like those emitted from phones)[5]. There were 50 different participants: 25 were encouraged to send text messages, and 25 were forbidden. The outcome was a score on a grammatical test (as a percentage) that was measured both before and after the experiment. The first independent variable was, therefore, text message use (text messagers versus controls) and the second independent variable was the time at which grammatical ability was assessed (before or after the experiment). The data are in the file **TextMessages.sav**.③

Answers are in the file **Answers(Chapter12).pdf** on the CD-ROM. Some more detailed comments about task 2 can be found in Field & Hole (2003).

12.10. FURTHER READING

Howell, D. C. (2002). *Statistical methods for psychology* (5th edition). Belmont, CA: Duxbury.

5 Although this punished them for any attempts to use a mobile phone, because other people's phones also emit microwaves, an unfortunate side effect was that these people acquired a pathological fear of anyone talking on a mobile phone!

CHAPTER 13

NON-PARAMETRIC TESTS

13.1. WHAT WILL THIS CHAPTER TELL US? ①

We've seen in the last few chapters how we can use various techniques to look for differences between means. However, all of these tests rely on parametric assumptions (notably normally distributed data). In Chapter 3, we saw that data are unfriendly and don't always turn up in nice normally distributed packages! Just to add insult to injury it's not always possible to correct for problems with the distribution of a data set—so, what do we do in these cases? The answer is that we have to use special kinds of statistical procedures known as non-parametric tests. Non-parametric tests are sometimes known as assumption-free tests because they make fewer assumptions about the type of data on which they can be used.[1] Most of these tests work on the principle of ranking the data: that is, finding the lowest score and giving it a rank of 1, then finding the next highest score and giving it a rank of 2, and so on. This process results in high scores being represented by large ranks, and low scores being represented by small ranks. The analysis is then carried out on the ranks rather than the actual data. This process is an ingenious way around the problem of using data that break the parametric assumptions. Some people believe that non-parametric tests have less power than their parametric counterparts, but as we shall see in Box 13.2 below, this is not always true. In this chapter we'll look at four of the most common non-parametric procedures: the Mann–Whitney test, the Wilcoxon signed-rank test, Friedman's test and the Kruskal–Wallis test. For each of these we'll discover how to carry out the analysis on SPSS and how to interpret and report the results.

1 Non-parametric tests sometimes get referred to as distribution-free tests, with an explanation that they make *no* assumptions about the distribution of the data. Technically, this isn't true: they do make distributional assumptions (e.g. all the ones in this chapter assume a continuous distribution), but they are less restrictive ones than their parametric counterparts.

13.2. COMPARING TWO INDEPENDENT CONDITIONS: THE WILCOXON RANK-SUM TEST AND MANN–WHITNEY TEST ①

When you want to test differences between two conditions and different participants have been used in each condition then you have two choices: the Mann–Whitney test (Mann & Whitney, 1947) and the Wilcoxon rank-sum test (Wilcoxon, 1945). These tests are the non-parametric equivalent of the independent *t*-test. In fact, both tests are equivalent, and there's another, more famous, Wilcoxon test so it gets very confusing for most of us.

For example, a neurologist might carry out an experiment to investigate the depressant effects of certain recreational drugs. She tested 20 clubbers in all: 10 were given an ecstasy tablet to take on a Saturday night and 10 were allowed only to drink alcohol. Levels of depression were measured using the Beck Depression Inventory (BDI) the day after and midweek. The data are in Table 13.1.

13.2.1. Theory ②

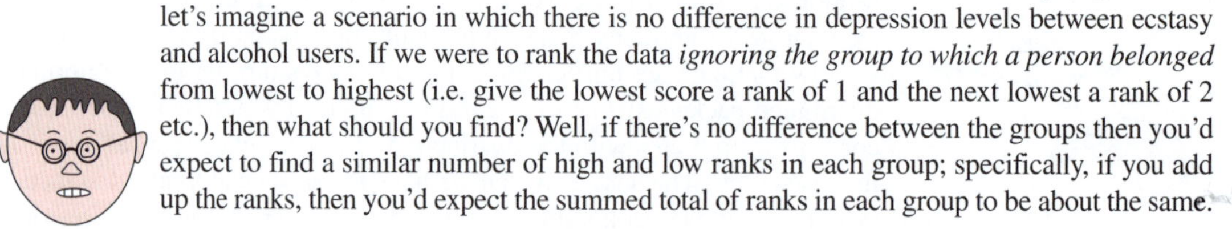

The logic behind the Wilcoxon rank-sum and Mann–Whitney tests is incredibly elegant. First, let's imagine a scenario in which there is no difference in depression levels between ecstasy and alcohol users. If we were to rank the data *ignoring the group to which a person belonged* from lowest to highest (i.e. give the lowest score a rank of 1 and the next lowest a rank of 2 etc.), then what should you find? Well, if there's no difference between the groups then you'd expect to find a similar number of high and low ranks in each group; specifically, if you add up the ranks, then you'd expect the summed total of ranks in each group to be about the same.

Figure 13.1 Frank Wilcoxon

Table 13.1 Data for drug experiment

Participant	Drug	BDI (Sunday)	BDI (Wednesday)
1	Ecstasy	15	28
2	Ecstasy	35	35
3	Ecstasy	16	35
4	Ecstasy	18	24
5	Ecstasy	19	39
6	Ecstasy	17	32
7	Ecstasy	27	27
8	Ecstasy	16	29
9	Ecstasy	13	36
10	Ecstasy	20	35
11	Alcohol	16	5
12	Alcohol	15	6
13	Alcohol	20	30
14	Alcohol	15	8
15	Alcohol	16	9
16	Alcohol	13	7
17	Alcohol	14	6
18	Alcohol	19	17
19	Alcohol	18	3
20	Alcohol	18	10

Now think about what would happen if there was a difference between the groups. Let's imagine that the ecstasy group are more depressed than the alcohol group. If you rank the scores as before, then you would expect the higher ranks to be in the ecstasy group and the lower ranks to be in the alcohol group. Again, if we summed the ranks in each group, we'd expect the sum of ranks to be higher in the ecstasy group than in the alcohol group.

The Mann–Whitney and Wilcoxon rank-sum tests both work on this principle. In fact, when the groups have unequal numbers of participants in them then the test statistic (W_s) for Wilcoxon's rank-sum test is simply the sum of ranks in the group that contains the fewer people; when the group sizes are equal it's the value of the smaller summed rank. Let's have a look at how ranking works in practice.

Figure 13.2 shows the ranking process for both the Wednesday and Sunday data. To begin with, let's use our data for Wednesday, as it's more straightforward. First, just arrange the scores in ascending order, attach a label to remind you which group they came from (I've used A for alcohol and E for ecstasy), then starting at the lowest score assign potential ranks

Wednesday Data

Score	3	5	6	6	7	8	9	10	17	24	27	28	29	30	32	35	35	35	36	39
Rank	1	2	3	4	5	6	7	8	9	10	11	12	13	14	15	16	17	18	19	20
Actual Rank	1	2	3.5	3.5	5	6	7	8	9	10	11	12	13	14	15	17	17	17	19	20
Group	A	A	A	A	A	A	A	A	A	E	E	E	E	A	E	E	E	E	E	E

Sum of Ranks for Alcohol (A) = 59 **Sum of Ranks for Ecstasy (E) = 151**

Sunday Data

Score	13	13	14	15	15	15	16	16	16	16	17	18	18	18	19	19	20	20	27	35
Rank	1	2	3	4	5	6	7	8	9	10	11	12	13	14	15	16	17	18	19	20
Actual Rank	1.5	1.5	3	5	5	5	8.5	8.5	8.5	8.5	11	13	13	13	15.5	15.5	17.5	17.5	19	20
Group	A	E	A	A	A	E	A	A	E	E	E	E	A	A	E	A	E	A	E	E

Sum of Ranks for Alcohol (A) = 90.5 **Sum of Ranks for Ecstasy (E) = 119.5**

Figure 13.2 Ranking the depression scores for Wednesday

starting with 1 and going up to the number of scores you have. The reason why I've called these potential ranks is because sometimes the same score occurs more than once in a data set (e.g. in these data a score of 6 occurs twice, and a score of 35 occurs three times). These are called *tied ranks* and these values need to be given the same rank, so all we do is assign a rank that is the average of the potential ranks for those scores. So, with our two scores of 6, because they would've been ranked as 3 and 4, we take an average of these values (3.5) and use this value as a rank for both occurrences of the score! Likewise, with the three scores of 35, we have potential ranks of 16, 17 and 18; the rank we actually use is the average of these three ranks, $(16 + 17 + 18)/3 = 17$. When we've ranked the data, we add up all of the ranks for the two groups. So, add the ranks for the scores that came from the alcohol group (you should find the sum is 59) and then add the ranks for the scores that came from the ecstasy group (this value should be 151). We take the lowest of these sums to be our test statistic; therefore, the test statistic for the Wednesday data is $W_s = 59$.

Now, have a go with the Sunday data, which has lots of tied ranks and is generally horrible! The answers are in Figure 13.2, and you should find that when you've ranked the data, and added the ranks for the two groups, the sum of ranks for the alcohol group is 90.5

and for the ecstasy group it is 119.5. We take the lowest of these sums to be our test statistic; therefore, the test statistic for the Sunday data is $W_s = 90.5$.

The next issue is: how do we determine whether this test statistic is significant? Well it turns out that the mean ($\overline{W}_S$) and standard error of this test statistic (SE_{W_S}) can be easily calculated from the sample sizes of each group (n_1 is the sample size of group 1, and n_2 is the sample size of group 2):

$$\overline{W}_s = \frac{n_1(n_1 + n_2 + 1)}{2}$$

$$SE_{W_s} = \sqrt{\frac{n_1 n_2(n_1 + n_2 + 1)}{12}}$$

For our data, we actually have equal-sized groups and there are 10 people in each, so n_1 and n_2 are both 10. Therefore, the mean and standard deviation are:

$$\overline{W}_s = \frac{10(10 + 10 + 1)}{2} = 105$$

$$SE_{W_s} = \sqrt{\frac{(10 \times 10)(10 + 10 + 1)}{12}} = 13.23$$

If we know the test statistic, the mean of test statistics, and the standard error, then we can easily convert the test statistic to a z-score using the equation that we came across way back in Chapter 1:

$$z = \frac{X - \overline{X}}{s} = \frac{W_s - \overline{W}_s}{SE_{W_s}}$$

If we calculate this value for the Sunday and Wednesday depression scores we get:

$$z_{\text{Sunday}} = \frac{W_s - \overline{W}_s}{SE_{W_s}} = \frac{90.5 - 105}{13.23} = -1.10$$

$$z_{\text{Wednesday}} = \frac{W_s - \overline{W}_s}{SE_{W_s}} = \frac{59 - 105}{13.23} = -3.48$$

If these values are bigger than 1.96 (ignoring the minus sign) then the test is significant at $p < .05$. So, it looks as though there is a significant difference between the groups on Wednesday, but not on Sunday.

The procedure I've actually described is the Wilcoxon rank-sum test. The Mann–Whitney test, with which many of you may be more familiar, is basically the same. It is based on a test statistic U, which is derived in a fairly similar way to the Wilcoxon procedure (in fact there's a direct relationship between the two). If you're interested, U is calculated using an equation in which n_1 and n_2 are the sample sizes of groups 1 and 2 respectively, and R_1 is the sum of ranks for group 1:

$$U = N_1 N_2 + \frac{N_1(N_1 + 1)}{2} - R_1$$

So, for our data we'd get the following (remember we have 10 people in each group and the sum of ranks for group 1, the ecstasy group, was 119.5 for the Sunday data and 151 for the Wednesday data):

$$U_{\text{Sunday}} = (10 \times 10) + \frac{10(11)}{2} - 119.50 = 35.50$$

$$U_{\text{Wednesday}} = (10 \times 10) + \frac{10(11)}{2} - 151.00 = 4.00$$

SPSS produces both statistics and there is a direct relationship between the two so it doesn't really matter which one you choose!

13.2.2. Inputting data and provisional analysis ①

When the data are collected using different participants in each group, we need to input the data using a coding variable. So, the data editor will have three columns of data. The first column is a coding variable (called something like **drug**) which, in this case, will have only two codes (for convenience I suggest 1 = ecstasy group, and 2 = alcohol group). The second column will have values for the dependent variable (BDI) measured the day after (call this variable **sunbdi**) and the third will have the midweek scores on the same questionnaire (call this variable **wedbdi**).

When you enter the data into SPSS remember to tell the computer that a code of 1 represents the group that were given ecstasy, and that a code of 2 represents the group that were restricted to alcohol (see section 2.4.4). There were no specific predictions about which drug would have the most effect so the analysis should be two-tailed. First, we would run some exploratory analysis on the data and because we're going to be looking for group differences we need to run these exploratory analyses for each group (see sections 3.5 and 3.6). If you do these analyses you should find the same tables shown in SPSS Output 13.1. These tables show first of all that for the Sunday data the distribution for ecstasy, $D(10) = .28$, $p < .05$, appears to be non-normal whereas that for the alcohol data, $D(10) = .17$, *ns*, was normal: we can tell this by whether the significance of the K–S and Shapiro–Wilk tests is less than .05 (and, therefore, significant) or greater than .05 (and, therefore, non-significant (*ns*)). For the Wednesday data, although the data for ecstasy were normal, $D(10) = .24$, *ns*, the data for alcohol appeared to be significantly non-normal, $D(10) = .31$, $p < .01$. This finding would alert us to the fact that a non-parametric test should be used for the Sunday and Wednesday data because one of the variables is not normally distributed for each. You should note that the Shapiro–Wilk statistic yields exact

Tests of Normality

	Type of Drug	Kolmogorov-Smirnov[a]			Shapiro-Wilk		
		Statistic	df	Sig.	Statistic	df	Sig.
Beck Depression Inventory (Sunday)	Ecstasy	.276	10	.030	.811	10	.020
	Alcohol	.170	10	.200*	.959	10	.780
Beck Depression Inventory (Wednesday)	Ecstasy	.235	10	.126	.941	10	.566
	Alcohol	.305	10	.009	.753	10	.004

* This is a lower bound of the true significance.
a Lilliefors Significance Correction

Test of Homogeneity of Variance

		Levene Statistic	df1	df2	Sig.
Beck Depression Inventory (Sunday)	Based on Mean	3.644	1	18	.072
	Based on Median	1.880	1	18	.187
	Based on Median and with adjusted df	1.880	1	10.076	.200
	Based on trimmed mean	2.845	1	18	.109
Beck Depression Inventory (Wednesday)	Based on Mean	.508	1	18	.485
	Based on Median	.091	1	18	.766
	Based on Median and with adjusted df	.091	1	11.888	.768
	Based on trimmed mean	.275	1	18	.606

SPSS Output 13.1

significance values whereas the K–S test sometimes gives an approximation of .2 for the significance (see the Sunday data for the alcohol group) because SPSS cannot calculate exact significances. This finding highlights an important difference between the K–S test and the Shapiro–Wilk test: in general the Shapiro–Wilk test is more accurate. The second table in SPSS Output 13.1 shows the results of Levene's test. For the Sunday data, $F(1, 18) = 3.64$, *ns*, and for Wednesday, $F(1, 18) = .51$, *ns*, the variances are not significantly different, indicating that the assumption of homogeneity has been met. Nevertheless, we need to use non-parametric statistical procedures on the Sunday and Wednesday data because of non-normally distributed data.

13.2.3. Running the analysis ①

First, access the main dialog box by using the **Analyze⇒Nonparametric Tests⇒2 Independent Samples...** menu pathway (see Figure 13.3). Once the dialog box is activated, select both dependent variables from the list (click on **Beck Depression Inventory Sunday [sunbdi]** then, holding the mouse button down, drag over **Beck Depression Inventory Wednesday [wedbdi]**) and transfer them to the box labelled *Test Variable List*

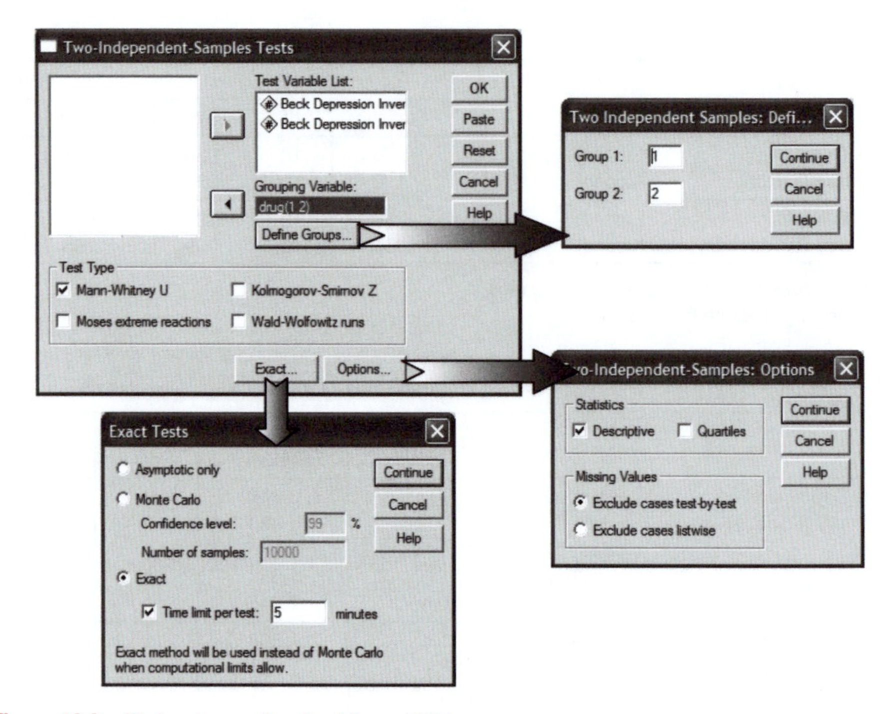

Figure 13.3 Dialog boxes for the Mann–Whitney test

by clicking on . Next, select the independent variable (the grouping variable), in this case **Type of Drug [drug]**, and transfer it to the box labelled *Grouping Variable*. When the grouping variable has been selected the Define Groups... button becomes active and you should click on it to activate the *define groups* dialog box. SPSS needs to know what numeric codes you assigned to your two groups, and there is a space for you to type the codes. In this example we coded our ecstasy group as 1 and our alcohol group as 2, and so you should type these two values in the appropriate space. When you have defined the groups, click on Continue to return to the main dialog box. The main dialog box also provides the facility to do tests other than the Mann–Whitney tests, and these alternatives are explained in Box 13.1.

If you click on Exact... then another dialog box appears.[2] Now, by default SPSS calculates the significance of the Mann–Whitney test using a method that is accurate with large samples (called the *Asymptotic Method*); however, when samples are smaller, or the data are particularly poorly distributed, then more accurate methods are available. The most accurate method is to ask for an *Exact* test, which calculates the significance of the Kruskal–Wallis test exactly. However, to get this precision, there is a price, and because of the complexities of the computation SPSS can take some time to find a

2 This button will appear only if you have the *Exact tests* module of SPSS installed. Remember this in future sections too.

solution—especially in large samples. A slightly less labour-intensive method is to estimate the significance using the Monte Carlo method. This basically involves creating a distribution similar to that found in the sample and then taking several samples (the default is 10,000) from this distribution and from those samples the mean significance value and the confidence interval around it can be created. If that didn't make any sense to you then fear not: the rule of thumb is when samples are large you should probably opt for the Monte Carlo method; when you have small samples (as we have) it's worth opting for the exact test (as I have done in this example). Finally, clicking on Options... opens another dialog box that gives you options for the analysis. These options are not particularly useful because, for example, the option that provides descriptive statistics does so for the entire data set (so doesn't break down values according to group membership). For this reason, I recommend obtaining descriptive statistics using the methods we learnt about in sections 3.4 and 3.5. To run the analyses return to the main dialog box and click on OK.

Box 13.1

Other options for the Mann–Whitney test ②

In the main dialog box there are some other tests that can be selected:

- **Kolmogorov–Smirnov Z**: In Chapter 3 we met a Kolmogorov–Smirnov test that tested whether a sample was from a normally distributed population. This is a different test! In fact, it tests whether two groups have been drawn from the same population (regardless of what that population may be). In effect, this means it does much the same as the Mann–Whitney test! However, the K–S test tends to have better power than the Mann–Whitney test when sample sizes are less than about 25 per group, and so is worth selecting if that's the case.
- **Moses Extreme Reactions**: Great name—makes me think of a bearded man standing on Mount Sinai reading a stone tablet and then suddenly bursting into a wild rage, smashing the tablet and screaming 'what do you mean do not worship any other god?' Sadly, this test isn't as exciting as my mental image. It basically compares the variability of scores in the two groups, so it's a bit like a non-parametric Levene test.
- **Wald–Wolfowitz runs**: Despite sounding like a particularly bad case of diarrhoea this is another variant on the Mann–Whitney test. In this test the scores are rank ordered as in the Mann–Whitney test, but rather than analysing the ranks, this test looks for 'runs' of scores from the same group within the ranked order. Now, if there's no difference between groups then obviously ranks from the two groups should be randomly interspersed. However, if the groups are different then you should see more ranks from one group at the lower end, and more ranks from the other group at the higher end. By looking for clusters of scores in this way the test can determine if the groups differ.

13.2.4. Output from the Mann–Whitney test ①

I explained in section 13.2.1 that the Mann–Whitney test works by looking at differences in the ranked positions of scores in different groups. Therefore, the first part of the output summarizes the data after it has been ranked. Specifically, SPSS tells us the average and total ranks in each condition (see SPSS Output 13.2). Remember that the Mann–Whitney test relies on scores being ranked from lowest to highest: therefore, the group with the lowest mean rank is the group with the greatest number of lower scores in it. Similarly, the group that has the highest mean rank should have a greater number of high scores within it. Therefore, this initial table can be used to ascertain which group had the highest scores, which is useful in case we need to interpret a significant result. You should note that the sums of ranks are the same as those calculated in section 13.2.1 (which is something of a relief to me!)

Ranks

	Type of Drug	N	Mean Rank	Sum of Ranks
Beck Depression Inventory (Sunday)	Ecstasy	10	11.95	119.50
	Alcohol	10	9.05	90.50
	Total	20		
Beck Depression Inventory (Wednesday)	Ecstasy	10	15.10	151.00
	Alcohol	10	5.90	59.00
	Total	20		

SPSS Output 13.2

The second table (SPSS Output 13.3) provides the actual test statistics for the Mann–Whitney test, the Wilcoxon procedure and the corresponding z-score (see section 13.2.1). SPSS Output 13.3 has a column for each variable (one for **sunbdi** and one for **wedbdi**) and in each column there is the value of Mann–Whitney's U-statistic, the value of Wilcoxon's statistic and the associated z-approximation. Note that the values of U, W_s and the associated z-score are the same as we calculated in section 13.2.1!

The important part of the table is the significance value of the test, which gives the two-tailed probability that the magnitude of the test statistic is a chance result. This significance value can be used as it is when no prediction has been made about which group will differ from which. However, if a prediction has been made (e.g. if we said that ecstasy users would be more depressed than alcohol users the day after taking the drug) then we need to calculate the one-tailed probability by taking the two-tailed value and dividing it by 2. For these data, the Mann–Whitney test is non-significant (two-tailed) for the depression scores taken on the Sunday. This finding indicates that ecstasy is no more of a depressant, the day after taking it, than alcohol: both groups report comparable levels of depression. However, for the midweek measures the results are highly significant ($p < .001$). The value of the mean rankings indicates that the ecstasy group had significantly higher levels of depression midweek than the alcohol group. This conclusion is reached by noting that for the Wednesday scores, the average rank is higher in the ecstasy users (15.10) than in the alcohol users (5.90).

Test Statistics[b]

	Beck Depression Inventory (Sunday)	Beck Depression Inventory (Wednesday)
Mann-Whitney U	35.500	4.000
Wilcoxon W	90.500	59.000
Z	−1.105	−3.484
Asymp. Sig. (2-tailed)	.269	.000
Exact Sig. [2* (1-tailed Sig.)]	.280[a]	.000[a]

a Not corrected for ties
b Grouping Variable: Type of Drug

SPSS Output 13.3 (without Monte Carlo exact significance)

Test Statistics[b]

	Beck Depression Inventory (Sunday)	Beck Depression Inventory (Wednesday)
Mann-Whitney U	35.500	4.000
Wilcoxon W	90.500	59.000
Z	−1.105	−3.484
Asymp. Sig. (2-tailed)	.269	.000
Exact Sig. [2* (1-tailed Sig.)]	.280[a]	.000[a]
Exact Sig. (2-tailed)	.288	.000
Exact Sig. (1-tailed)	.144	.000
Point Probability	.013	.000

a Not corrected for ties
b Grouping Variable: Type of Drug

SPSS Output 13.4 (with Monte Carlo exact significance)

SPSS Output 13.4 shows the output for the Mann–Whitney test when exact significance is selected for. I've just included this to show you that we get some extra lines that give us exact significance values (both one- and two-tailed). These don't actually change our conclusions, but be aware that you should probably consult these values in preference to the asymptotic value, especially when sample sizes are small.

13.2.5. Calculating an effect size ②

As we've seen throughout this book it's important to report effect sizes so that people have a standardized measure of the size of the effect you observed, which they can compare to other studies. SPSS doesn't calculate an effect size for us, but we can calculate approximate effect sizes really easily thanks to the fact that SPSS converts the test statistics into a *z*-score. The equation to convert a *z*-score into the effect size estimate, *r*, is as follows (from Rosenthal, 1991, p. 19):

$$r = \frac{Z}{\sqrt{N}}$$

in which *z* is the *z*-score that SPSS produces, and *N* is the size of the study (i.e. the number of total observations) on which *z* is based. In this case SPSS Output 13.3 tells us that *z* is −1.11 for the Sunday data and −3.48 for the Wednesday data. In both cases we had 10 ecstasy users and 10 alcohol users and so the total number of observations was 20. The effect sizes are, therefore:

$$r_{\text{Sunday}} = \frac{-1.11}{\sqrt{20}} = -.25$$

$$r_{\text{Wednesday}} = \frac{-3.48}{\sqrt{20}} = -.78$$

This represents a small to medium effect for the Sunday data (it is below the .3 criterion for a medium effect size) and a huge effect for the Wednesday data (the effect size is well above the .5 threshold for a large effect). Importantly, for the Sunday data it shows how a fairly large effect size can still be non-significant in a small sample!

13.2.6. Writing the results ①

For the Mann–Whitney test, we need only report the test statistic (which is denoted by *U*) and its significance. Of course, we really ought to include the effect size as well. So, we could report something like:

☑ Ecstasy users (*Mdn* = 17.50) didn't seem to differ in depression levels from alcohol users (*Mdn* = 16.00) the day after the drugs were taken, $U = 35.50$, *ns*, $r = -.25$. However, by Wednesday, ecstasy users (*Mdn* = 33.50) were significantly more depressed than alcohol users (*Mdn* = 7.50), $U = 4.00$, $p < .001$, $r = -.78$

Note that I've reported the median for each condition—this statistic is more appropriate than the mean for non-parametric tests. We could also choose to report Wilcoxon's test rather than Mann–Whitney's *U*-statistic and this would be as follows:

☑ Ecstasy users (*Mdn* = 17.50) didn't seem to differ in depression levels to alcohol users (*Mdn* = 16.00) the day after the drugs were taken, $W_s = 90.50$, *ns*, $r = -.25$. However, by Wednesday, ecstasy users (*Mdn* = 33.50) were significantly more depressed than alcohol users (*Mdn* = 7.50), $W_s = 59.00$, $p < .001$, $r = -.78$.

Cramming Samantha's Tips

- The Mann–Whitney test and the Wilcoxon rank-sum test **compare two conditions when different participants take part in each condition and the resulting data are not normally distributed or violate an assumption of the independent *t*-test.**
- Look at the row labelled *Asymp. Sig. (2-tailed)*: if the value is less than .05 then the means of the two groups are significantly different. (If you opted for exact tests then look at the row labelled *Exact Sig. (2-tailed).*)
- Look at the values of the ranks to tell you how the groups differ (the group with the highest scores will have the highest ranks).
- SPSS only provides the two-tailed significance value; if you want the one-tailed significance just divide the value by 2. (If you opted for exact tests then look at the row labelled *Exact Sig. (1-tailed).*)
- Report the *U*-statistic (or W_s if you prefer), the corresponding *z*, and the significance value. Also report the medians and their corresponding ranges (or draw a boxplot).
- You should calculate the effect size and report this too!

Box 13.2

Non-parametric tests and statistical power ②

Ranking the data is a useful way around the distributional assumptions of parametric tests but there is a price to pay: by ranking the data we lose some information about the magnitude of difference between scores. The result is that non-parametric tests can be less powerful than the parametric counterparts. I introduced the notion of statistical power in section 1.8.5: it refers to the ability of a test to find an effect that genuinely exists. So, by saying that non-parametric tests are less powerful, we mean that if there is a genuine effect in our data, then a parametric test is more likely to detect it than a non-parametric one. However, this statement is true only *if the assumptions of the parametric test are met.* So, if we use a parametric test and a non-parametric test on the same data, and those data are normally distributed, then the parametric test will have greater power to detect the effect than the non-parametric test.

The problem is that to define the power of a test we need to be sure that it controls the Type I error rate (the number of times a test will find a significant effect when in reality there is no effect to find—see section 1.8.2). We saw in Chapter 1 that Fisher said that this error rate should be 5%. We know that when data are normally distributed then the Type I error rate of tests based on this distribution is indeed 5%, and so we can work out the power. However, when data are not normal

(Continued)

Box 13.2 (Continued)

the Type I error rate of tests based on this distribution won't be 5% (in fact we don't know what it is for sure as it will depend on the shape of the distribution) and so we have no way of calculating power (because power is linked to the Type I error rate—see section 1.8.5). So, although you often hear (in the first edition of this book, for example!) of non-parametric tests having an increased chance of a Type II error (i.e. more chance of accepting that there is no difference between groups when, in reality, a difference exists), this is true only if the data are normally distributed.

13.3. COMPARING TWO RELATED CONDITIONS: THE WILCOXON SIGNED-RANK TEST ①

The Wilcoxon signed-rank test (Wilcoxon, 1945), not to be confused with the rank-sum test in the previous section, is used in situations in which there are two sets of scores to compare, but these scores come from the same participants. As such, think of it as the non-parametric equivalent of the dependent *t*-test (or a Mann–Whitney test for repeated-measures data). Imagine the experimenter in the previous section was now interested in the *change* in depression levels, within people, for each of the two drugs. We now want to compare the BDI scores on Sunday to those on Wednesday. We still have to use a non-parametric test because the distributions of scores for both drugs were non-normal on one of the two days (see SPSS Output 13.1).

13.3.1. Theory of Wilcoxon's signed-rank test ②

The Wilcoxon signed-rank test works in a fairly similar way to the dependent *t*-test (Chapter 7) in that it is based on the differences between scores in the two conditions you're comparing. Once these differences have been calculated they are ranked (just like in 13.2.1) but the sign of the difference (positive or negative) is assigned to the rank. If we use the same data as before we can compare depression scores on Sunday to those on Wednesday for the two drugs separately.

Table 13.2 shows the ranking for these data. Remember that we're ranking the two drugs separately. First, we calculate the difference between Sunday and Wednesday (that's just Sunday's score subtracted from Wednesday's). If the difference is 0 (i.e. the scores are the same on Sunday and Wednesday) then we exclude these data from the ranking. We make a note of the sign of the difference (was it positive or negative?) and then rank the differences (starting with the smallest) ignoring whether they are positive or negative. The

Table 13.2 Ranking data in the Wilcoxon signed-rank test

BDI Sunday	BDI Wednesday	Difference	Sign	Rank	Positive Ranks	Negative Ranks
			Ecstasy			
15	28	13	+	2.5	2.5	
35	35	0	Exclude			
16	35	19	+	6	6	
18	24	6	+	1	1	
19	39	20	+	7	7	
17	32	15	+	4.5	4.5	
27	27	0	Exclude			
16	29	13	+	2.5	2.5	
13	36	23	+	8	8	
20	35	15	+	4.5	4.5	
				Total =	**36**	**0**
			Alcohol			
16	5	−11	−	9		9
15	6	−9	−	7		7
20	30	10	+	8	+8	
15	8	−7	−	3.5		3.5
16	9	−7	−	3.5		3.5
13	7	−6	−	2		2
14	6	−8	−	5.5		5.5
19	17	−2	−	1		1
18	3	−15	−	10		10
18	10	−8	−	5.5		5.5
				Total =	**8**	**47**

ranking is the same as in section 13.2.1, and we deal with tied scores in exactly the same way. Finally we collect together the ranks that came from a positive difference between the conditions, and add them up to get the sum of positive ranks (T_+). We also add up the ranks that came from negative differences between the conditions to get the sum of negative ranks (T_-). So, for ecstasy, $T_+ = 36$ and $T_- = 0$ (in fact there were no negative ranks), and for alcohol, $T_+ = 8$ and $T_- = 47$. The test statistic, T, is the smaller of the two values, and so is 0 for ecstasy and 8 for alcohol.

To calculate the significance of the test statistic (T), we again look at the mean ($\overline{T}$) and standard error (SE_T), which, like the Mann–Whitney and rank-sum tests in the previous section, are functions of the sample size, n (because we used the same participants, there is only one sample size):

$$\overline{T} = \frac{n(n+1)}{4}$$

$$SE_T = \sqrt{\frac{n(n+1)(2n+1)}{24}}$$

In both groups, n is simply 10 (because that's how many participants were used). However, remember that for our ecstasy group we excluded two people because they had differences of 0; therefore, the sample size we use is 8, not 10. This gives us:

$$\overline{T}_{\text{Ecstasy}} = \frac{8(8+1)}{4} = 18$$

$$SE_{T_{\text{Ecstasy}}} = \sqrt{\frac{8(8+1)(16+1)}{24}} = 7.14$$

For the alcohol group there were no exclusions so we get:

$$\overline{T}_{\text{Alcohol}} = \frac{10(10+1)}{4} = 27.50$$

$$SE_{T_{\text{Alcohol}}} = \sqrt{\frac{10(10+1)(20+1)}{24}} = 9.81$$

As before, if we know the test statistic, the mean of test statistics, and the standard error, then we can easily convert the test statistic to a z-score using the equation that we came across way back in Chapter 1 and the previous section:

$$z = \frac{X - \overline{X}}{s} = \frac{T - \overline{T}}{SE_T}$$

If we calculate this value for the ecstasy and alcohol depression scores we get:

$$z_{\text{Ecstasy}} = \frac{T - \overline{T}}{SE_T} = \frac{0 - 18}{7.14} = -2.52$$

$$z_{\text{Alcohol}} = \frac{T - \overline{T}}{SE_T} = \frac{8 - 27.5}{9.81} = -1.99$$

If these values are bigger than 1.96 (ignoring the minus sign) then it is significant at $p < .05$. So, the test looks as though there is a significant difference between depression scores on Wednesday and Sunday for both ecstasy and alcohol.

13.3.2. Running the analysis ①

To do the same analysis on SPSS we can use the same data as before, but because we want to look at the change for each drug *separately*, we need to use the *split file* command and ask SPSS to split the file by the variable **Type of Drug [drug]**. This process ensures that any subsequent analysis is done for the ecstasy group and the alcohol group separately. Once the file has been split, select the Wilcoxon test dialog box by using the file path **Analyze⇒Nonparametric Tests⇒2 Related Samples…** (Figure 13.4). This dialog box allows you to select other tests too (see Box 13.3).

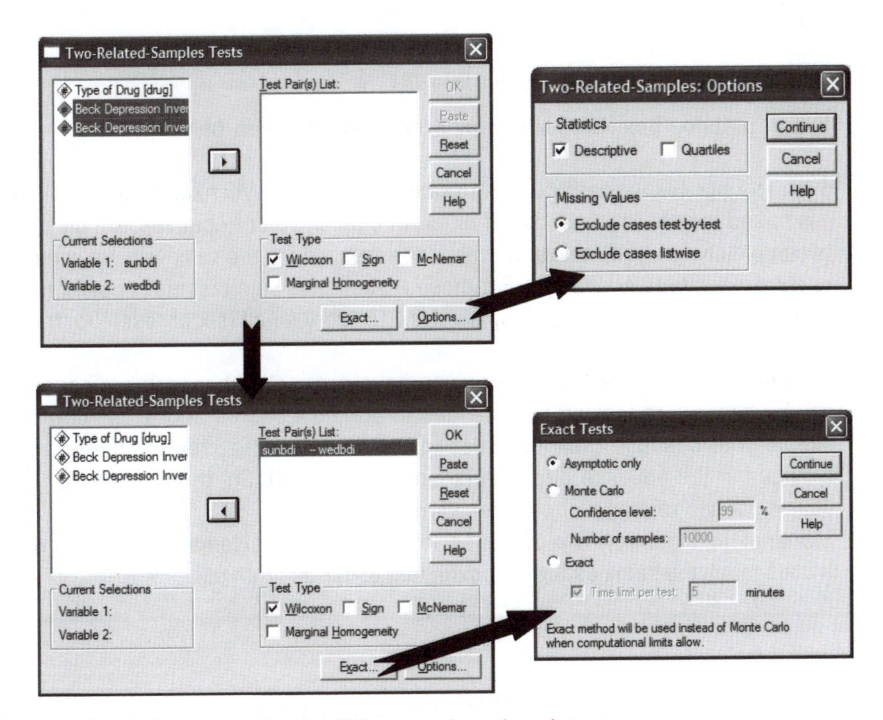

Figure 13.4 Dialog boxes for the Wilcoxon signed-rank test

Once the dialog box is activated, select two variables from the list (click on the first variable with the mouse and then the second). The first variable you select (**Beck Depression Inventory (Sunday) [sunbdi]**) will be named as *Variable 1* in the box labelled *Current Selections*, and the second variable you select (**Beck Depression Inventory**

(Wednesday) wedbdi) appears as *Variable 2*. When you have selected two variables, transfer them to the box labelled *Test Pair(s) List* by clicking on ▶. If you want to carry out several Wilcoxon tests then you can select another pair of variables, transfer them to the variables list, and then select another pair and so on. In this case, we want only one test. If you click on Exact... then another dialog box appears that allows you to select for SPSS to compute exact significance values (see section 13.2.3). I won't go into this again, but suffice it to say when samples are large you should probably opt for the Monte Carlo method, and when you have small samples it's worth opting for the exact test. I haven't opted for either in this example. If you click on Options... then a dialog box appears that gives you the chance to select descriptive statistics. Unlike the Mann–Whitney test, the descriptive statistics here are worth having, because it is the change across variables (columns in the data editor) that is relevant. To run the analysis, return to the main dialog box and click on OK.

Box 13.3

Other options for the Wilcoxon signed-rank test ②

In the main dialog box there are some other tests that can be selected:

- Sign: The sign test basically does the same thing as the Wilcoxon signed-rank test, except that it is based only on the direction of difference (positive or negative). The magnitude of change is completely ignored (unlike in Wilcoxon's test where the rank tells us something about the relative magnitude of change). For these reasons the sign test lacks power (it's not very good at detecting effects) unless samples sizes are very small (six or less). So, frankly, I don't see the point.
- McNemar: This test is useful when you have nominal rather than ordinal data. It's typically used when you're looking for changes in people's scores and it compares the amount of people who changed their response in one direction (i.e. scores increased) to those who changed in the opposite direction (scores decreased). So, this test needs to be used when you've got two related dichotomous variables.
- Marginal Homogeneity: This produces an extension of McNemar's test but for ordinal variables. It does much the same as the Wilcoxon test as far as I can tell.

13.3.3. Output for the ecstasy group ①

If you have split the file, then the first set of results obtained will be for the ecstasy group (SPSS Output 13.5). The first table provides information about the ranked scores. It tells us the number of negative ranks (these are people for whom the Sunday score was greater than the Wednesday score) and the number of positive ranks (people for whom the Wednesday score was greater than the Sunday score). The table shows that for 8 of the 10 participants, their score on Wednesday was greater than on Sunday, indicating greater

Ranks[d]

		N	Mean Rank	Sum of Ranks
Beck Depression Inventory (Wednesday) – Beck Depression Inventory (Sunday)	Negative Ranks	0[a]	.00	.00
	Positive Ranks	8[b]	4.50	36.00
	Tiles	2[c]		
	Total	10		

a Beck Depression Inventory (Wednesday) < Beck Depression Inventory (Sunday)
b Beck Depression Inventory (Wednesday) > Beck Depression Inventory (Sunday)
c Beck Depression Inventory (Wednesday) = Beck Depression Inventory (Sunday)
d Type of Drug = Ecstasy

Test Statistics[b,c]

	Beck Depression Inventory (Wednesday) – Beck Depression Inventory (Sunday)
Z	−2.527[a]
Asymp. Sig. (2-tailed)	.012

a Based on negative ranks
b Wilcoxon Signed Ranks Test
c Type of Drug = Ecstasy

SPSS Output 13.5

depression midweek compared to the morning after. There were two tied ranks (i.e. participants who scored the same on both days). The table also shows the average number of negative and positive ranks and the sum of positive and negative ranks. Below the table are footnotes, which tell us to what the positive and negative ranks relate (so provide the same kind of explanation as I've just made—see, I'm not clever, I just read the footnotes!). In section 13.3.1 I explained that the test statistic, T, is the lowest value of the two types of ranks, so our test value here is the sum of negative ranks (e.g. zero). However, I also showed how this value can be converted to a z-score and this is what SPSS does. The advantage of this approach is that it allows exact significance values to be calculated based on the normal distribution. The second table in SPSS Output 13.5 tells us that the test statistic is based on the negative ranks, that the z-score is −2.53 (which is also the value we calculated in section 13.3.1) and that this value is significant at $p = .012$. Therefore, because this value is based on the *negative* ranks (and because the test statistic is the smaller of the positive and negative ranks the majority of ranks must've been positive), we should conclude that when taking ecstasy there was a significant increase in depression (as measured by the BDI) from the morning after to midweek. If the test statistic had been based on the positive ranks then this would have told us that the results were in the opposite direction (i.e. BDI scores were greater the morning after compared to midweek). Therefore, we can conclude that for ecstasy users there was a significant increase in depression from the next day to midweek ($z = -2.53$, $p < .05$).

13.3.4. Output for the alcohol group ①

The remainder of the output should contain the same two tables but for the alcohol group (if it does not, then you probably forgot to split the file). As before, the first table in SPSS Output 13.6 provides information about the ranked scores. It tells us the number of negative ranks (these are people who were more depressed on Sunday than on Wednesday) and the number of positive ranks (people who were more depressed on Wednesday than on Sunday). The table shows that for 9 of the 10 participants, their score on Sunday was greater than on Wednesday, indicating greater depression the morning after compared to midweek. Unlike the ecstasy takers there were no tied ranks. The table also shows the average number of negative and positive ranks and the sum of positive and negative ranks. Below the table are footnotes that tell us to what the positive and negative ranks relate. As before, the lowest value of ranked scores is converted to a z-score (in this case 8). The second table tells us that the test statistic is based on the positive ranks, that the z-score is −1.99 (again I would point out that this is the value we calculated in 13.3.1—I point this out merely because I'm amazed that my hand calculations actually worked!) and that this value is significant at $p = .047$. Therefore, we should conclude (based on the fact that *positive* ranks were used) that when taking alcohol there was a significant decline in depression (as measured by the BDI) from the morning after to midweek ($z = -1.99$, $p < .05$).

Ranks[d]

		N	Mean Rank	Sum of Ranks
Beck Depression Inventory (Wednesday) – Beck Depression Inventory (Sunday)	Negative Ranks	9[a]	5.22	47.00
	Positive Ranks	1[b]	8.00	8.00
	Tiles	0[c]		
	Total	10		

a Beck Depression Inventory (Wednesday) < Beck Depression Inventory (Sunday)
b Beck Depression Inventory (Wednesday) > Beck Depression Inventory (Sunday)
c Beck Depression Inventory (Wednesday) = Beck Depression Inventory (Sunday)
d Type of Drug = Alcohol

Test Statistics[b,c]

	Beck Depression Inventory (Wednesday) – Beck Depression Inventory (Sunday)
Z	−1.990[a]
Asymp. Sig. (2-tailed)	.047

a Based on positive ranks
b Wilcoxon Signed Ranks Test
c Type of Drug = Alcohol

SPSS Output 13.6

From the results of the two different groups, we can see that there is an opposite effect when alcohol is taken to when ecstasy is taken. Alcohol makes you slightly depressed the morning after but this depression has dropped by midweek. Ecstasy also causes some depression the morning after consumption; however, this depression increases towards the middle of the week. Of course, to see the true effect of the morning after we would have had to take measures of depression before the drugs were administered! This opposite effect between groups of people is known as an interaction (i.e. you get one effect under certain circumstances, and a different effect under other circumstances) and we came across these in Chapters 10–12.

13.3.5. Calculating an effect size ②

The effect size can be calculated in the same way as for the Mann–Whitney test (see the equation in section 13.2.5). In this case SPSS Output 13.6 tells us that for the ecstasy group z is −2.53, and for the alcohol group is −1.99. In both cases we had 20 observations (although we only used 10 people and tested them twice, it is the number of observations, not the number of people, that is important here). The effect size is, therefore:

$$r_{\text{Ecstasy}} = \frac{-2.53}{\sqrt{20}} = -.57$$

$$r_{\text{Alcohol}} = \frac{-1.99}{\sqrt{20}} = -.44$$

This represents a large change in levels of depression when ecstasy is taken (it is above Cohen's benchmark of .5), and a medium to large change in depression when alcohol is taken (it is between Cohen's criteria of .3 and .5 for a medium and large effect respectively).

13.3.6. Writing and interpreting the results ①

For the Wilcoxon test, we need only report the test statistic (which is denoted by the letter T and the smallest of the two sum of ranks), its significance, and preferably an effect size. So, we could report something like:

✓ For Iecstasy users, depression levels were significantly higher on Wednesday (*Mdn* = 33.50) than on Sunday (*Mdn* = 17.50), $T = 0$, $p < .05$, $r = -.57$. However, for alcohol users the opposite was true: depression levels were significantly lower on Wednesday (*Mdn* = 7.50) than on Sunday (*Mdn* = 16.0), $T = 8$, $p < .05$, $r = -.44$.

Alternatively, we could report the values of z:

✓ For ecstasy users, depression levels were significantly higher on Wednesday (*Mdn* = 33.50) than on Sunday (*Mdn* = 17.50), $z = -2.53$, $p < .05$, $r = -.57$. However, for alcohol users the opposite was true: depression levels were significantly lower on Wednesday (*Mdn* = 7.50) than on Sunday (*Mdn* = 16.0), $z = -1.99$, $p < .05$, $r = -.44$.

Cramming Samantha's Tips

- The Wilcoxon signed-rank test compares two conditions when the same participants take part in each condition and the resulting data are not normally distributed or violate an assumption of the dependent *t*-test.
- Look at the row labelled *Asymp. Sig. (2-tailed)*: if the value is less than .05 then the two groups are significantly different.
- Look at positive and negative ranks (and the footnotes explaining what they mean) to tell you how the groups differ (the greater number of ranks in a particular direction tells you the direction of the result).
- As with many tests we've come across, SPSS only provides the two-tailed significance value; if you want the one-tailed significance just divide the value by 2.
- Report the *T*-statistic, the corresponding *z*, the significance value, and an effect size if possible. Also report the medians and their corresponding ranges (or draw a boxplot).

13.4. DIFFERENCES BETWEEN SEVERAL INDEPENDENT GROUPS: THE KRUSKAL–WALLIS TEST ①

In Chapter 8 we discovered a technique called one-way independent ANOVA that could be used to test for difference between several independent groups. I mentioned several times in that chapter that ANOVA is robust to violations of its assumptions. We also saw that there are measures that can be taken when you have heterogeneity of variance (Box 8.3). However, there is another alternative: the one-way independent ANOVA has a non-parametric counterpart called the Kruskal–Wallis test (Kruskal & Wallis, 1952; Figure 13.5). If you have non-normally distributed data, or have violated some other assumption, then this test can be a useful way around the problem.

I read a story in a newspaper recently claiming that scientists had discovered that the chemical genistein, which is naturally occurring in Soya, was linked to lowered sperm counts in western males. In fact, when you read the actual study, it had been conducted on rats, it found no link to lowered sperm counts, but there was evidence of abnormal sexual development in male rats (probably because this chemical acts like oestrogen). The journalist naturally interpreted this as a clear link to apparently declining sperm counts in

Figure 13.5 Joseph Kruskal spotting some more errors in his well-thumbed first edition of *Discovering Statistics...* by that idiot Field

western males (bloody journalists!). Anyway, as a vegetarian who eats lots of soya products and probably would like to have kids one day, I wanted to test this idea in humans rather than rats. I took 80 males and split them into four groups that varied in the number of soya meals they ate per week over a year-long period. The first group was a control group and they had no soya meals at all per week (i.e. none in the whole year); the second group had one soya meal per week (that's 52 over the year); the third group had four soya meals per week (that's 208 over the year) and the final group had seven soya meals a week (that's 364 over the year). At the end of the year, all of the participants were sent away to produce some sperm that I could count (when I say 'I', I mean someone in a laboratory as far away from me as humanly possible).[3]

13.4.1. Theory of the Kruskal–Wallis test ②

The theory of the Kruskal–Wallis test is very similar to that of the Mann–Whitney (and Wilcoxon rank-sum) test, so before reading on look back at section 13.2.1. Like the Mann–Whitney test, the Kruskal–Wallis test is based on ranked data. So, to begin with, you simply order the scores from lowest to highest, ignoring the group to which the score belongs, and then assign the lowest score a rank of 1, the next highest a rank of 2 and so on (see section 13.2.1 for more detail). When you've ranked the data you collect the scores

3 In case any medics are reading this, these data are made up and as I have absolutely no idea what a typical sperm count is they're probably ridiculous, so I apologise and you can laugh at my ignorance!

Table 13.3 Data for the soya example with ranks

No Soya		1 Soya Meal		4 Soya Meals		7 Soya Meals	
Sperm (Millions)	Rank	Sperm (Millions)	Rank	Sperm (Millions)	Rank	Sperm (Millions)	Rank
0.35	4	0.33	3	0.40	6	0.31	1
0.58	9	0.36	5	0.60	10	0.32	2
0.88	17	0.63	11	0.96	19	0.56	7
0.92	18	0.64	12	1.20	21	0.57	8
1.22	22	0.77	14	1.31	24	0.71	13
1.51	30	1.53	32	1.35	27	0.81	15
1.52	31	1.62	34	1.68	35	0.87	16
1.57	33	1.71	36	1.83	37	1.18	20
2.43	41	1.94	38	2.10	40	1.25	23
2.79	46	2.48	42	2.93	48	1.33	25
3.40	55	2.71	44	2.96	49	1.34	26
4.52	59	4.12	57	3.00	50	1.49	28
4.72	60	5.65	61	3.09	52	1.50	29
6.90	65	6.76	64	3.36	54	2.09	39
7.58	68	7.08	66	4.34	58	2.70	43
7.78	69	7.26	67	5.81	62	2.75	45
9.62	72	7.92	70	5.94	63	2.83	47
10.05	73	8.04	71	10.16	74	3.07	51
10.32	75	12.10	77	10.98	76	3.28	53
21.08	80	18.47	79	18.21	78	4.11	56
Total (R_i)	**927**		**883**		**883**		**547**

back into their groups and simply add up the ranks for each group. The sum of ranks for each group is denoted by R_i (where i is used to denote the particular group). Table 13.3 shows the raw data for this example along with the ranks (have a go at ranking the data and see if you get the same results as me!).

Once the sum of ranks has been calculated for each group, the test statistic, H, is calculated as in equation (13.1):

$$H = \frac{12}{N(N+1)} \sum_{i=1}^{k} \frac{R_i^2}{n_i} - 3(N+1) \tag{13.1}$$

In this equation, R_i is the sum of ranks for each group, N is the total sample size (in this case 80) and n_i is the sample size of a particular group (in this case we have equal sample sizes and they are all 20). Therefore, all we really need to do for each group is square the sum of ranks and divide this value by the sample size for that group. We then add up these values. That deals with the middle part of the equation; the rest of it involves calculating various values based on the total sample size. For these data we get:

$$
\begin{aligned}
H &= \frac{12}{80(81)}\left(\frac{927^2}{20} + \frac{883^2}{20} + \frac{883^2}{20} + \frac{547^2}{20}\right) - 3(81) \\
&= \frac{12}{6480}(42{,}966.45 + 38{,}984.45 + 38{,}984.45 + 14{,}960.45) - 243 \\
&= 0.0019(135{,}895.8) - 243 \\
&= 251.66 - 243 \\
&= 8.659
\end{aligned}
$$

This test statistic has a special kind of distribution known as the chi-square distribution (see Chapter 16) and for this distribution there is one value for the degrees of freedom, which is one less than the number of groups ($k - 1$): in this case 3.

13.4.2. Inputting data and provisional analysis ①

When the data are collected using different participants in each group, we need to input the data using a coding variable. So, the data editor will have two columns of data. The first column is a coding variable (called something like **Soya**) which, in this case, will have four codes (for convenience I suggest 1 = no soya, 2 = one soya meal per week, 3 = four soya meals per week, and 4 = seven soya meals per week). The second column will have values for the dependent variable (sperm count) measured at the end of the year (call this variable **sperm**). When you enter the data into SPSS remember to tell the computer which group is represented by which code (see section 2.4.4). The data can be found in the file **Soya.sav**.

First, we run some exploratory analyses on the data and because we're going to be looking for group differences we need to run these exploratory analyses for each group (see sections 3.5 and 3.6). If you do these analyses you should find the same tables shown in SPSS Output 13.7. The first table shows that the K–S test (see section 3.5) was not significant for the control group ($D(20) = 0.181$, $p > .05$) but the Shapiro–Wilk test is significant and this test is actually more accurate (though less widely reported) than the K–S test (see Chapter 3). Data for the group that ate one soya meal per week were significantly different from normal ($D(20) = 0.207$, $p < .05$), as were the data for those that ate four ($D(20) = 0.267$, $p < .01$) and seven ($D(20) = 0.204$, $p < .05$). The second table shows the results of Levene's test. The assumption of homogeneity of

Tests of Normality

	Number of Soya Meals Per Week	Kolmogorov-Smirnov[a]			Shapiro-Wilk		
		Statistic	df	Sig.	Statistic	df	Sig.
Sperm Count (Millions)	No Soya Meals	.181	20	.085	.805	20	.001
	1 Soya Meal Per Week	.207	20	.024	.826	20	.002
	4 Soyal Meals Per Week	.267	20	.001	.743	20	.000
	7 Soya Meals Per Week	.204	20	0.028	.912	20	.071

a Lilliefors Significance Correction

Test of Homogeneity of Variance

		Levene Statistic	df1	df2	Sig.
Sperm Count (Millions)	Based on Mean	5.117	3	76	.003
	Based on Median	2.860	3	76	.042
	Based on Median and with adjusted df	2.860	3	58.107	.045
	Based on trimmed mean	4.070	3	76	.010

SPSS Output 13.7

variance has been violated, $F(3, 76) = 5.12$, $p < .01$, and this is shown by the fact that the significance of Levene's test is less than .05. As such, these data violate two important assumptions: they are not normally distributed, and the groups have heterogeneous variances!

13.4.3. Doing the Kruskal–Wallis test on SPSS ①

First, access the main dialog box by using the **Analyze⇒Nonparametric Tests⇒K Independent Samples...** menu pathway (see Figure 13.6). Once the dialog box is activated, select the dependent variable from the list (click on **Sperm Count (Millions)**) and transfer it to the box labelled *Test Variable List* by clicking on ▶. Next, select the independent variable (the grouping variable), in this case **Soya**, and transfer it to the box labelled *Grouping Variable*. When the grouping variable has been selected the Define Range button becomes active and you should click on it to activate the *define range* dialog box. SPSS needs to know the range of numeric codes you assigned to your groups, and there is a space for you to type the minimum and maximum code. If you followed my coding scheme, then the minimum code we used was 1, and the maximum was 4, so type these

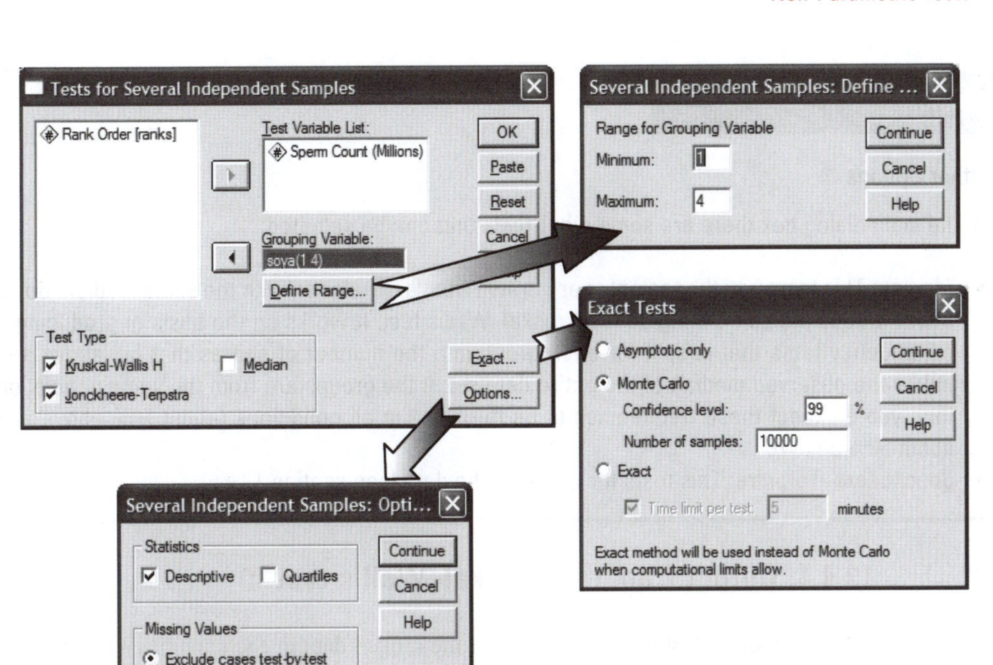

Figure 13.6 Dialog boxes for the Kruskal–Wallis test

numbers into the appropriate spaces. When you have defined the groups, click on Continue to return to the main dialog box. The main dialog box also provides Options... to conduct some tests similar to the Kruskal–Wallis test (see Box 13.4).

If you click on Exact... then you get a dialog box for selecting exact significance values for the Kruskal–Wallis test. I've explained this option in section 13.2.3, so I won't repeat myself here. I'll just recap by saying that when samples are large you should probably opt for the Monte Carlo method (as I have done in this example), and when you have small samples it's worth opting for the exact test. Finally, if you click on Options... then another dialog box appears that gives you options for the analysis. These options are not particularly useful because, for example, the option that provides descriptive statistics does so for the entire data set (so doesn't break down values according to group membership). For this reason, I recommend obtaining descriptive statistics using the methods we learnt about in sections 3.4 and 3.5. To run the analyses return to the main dialog box and click on OK. The final option you can ask for is for the Jonckheere–Terpstra trend test (select Jonckheere-Terpstra). This is useful if you want to look for a linear trend in the data (see section 8.2.10.5).

Box 13.4

Other Options ②

In the main dialog box there are some other tests that can be selected:

- **Median**: This tests whether samples are drawn from a population with the same median. So, in effect it does the same thing as the Kruskal–Wallis test. It works on the basis of producing a contingency table that is split for each group into the number of scores that fall above and below the observed median of the entire data set. If the groups are from the same population then you'd expect these frequencies to be the same in all conditions (about 50% above and about 50% below).
- **Jonckheere–Terpstra**: This tests for trends in the data (see section 13.4.6).

13.4.4. Output from the Kruskal–Wallis test ①

SPSS Output 13.8 shows a summary of the ranked data in each condition and we'll need these for interpreting any effects.

Ranks

	Number of Soya Meals	N	Mean Rank
Sperm Count (Millions)	No Soya Meals	20	46.35
	1 Soya Meal Per Week	20	44.15
	4 Soya Meals Per Week	20	44.15
	7 Soya Meals Per Week	20	27.35
	Total	80	

SPSS Output 13.8

SPSS Output 13.9 shows the test statistic, *H*, for the Kruskal–Wallis test (although SPSS labels it chi-square, because of its distribution, rather than *H*), its associated degrees of freedom (in this case we had four groups so the degrees of freedom are 4 – 1, or 3), and the significance. The crucial thing to look at is the significance value, which is .034; because this value is less than .05 we could conclude that the amount of soya meals eaten per week does significantly affect sperm counts. Note also, the Monte Carlo estimate of significance, which is slightly lower (.033). This is the value we ought to look to rather than the asymptotic value if they yield different results. The confidence interval for significance is also useful: it is .028–.037 and the fact that the boundary does not cross .05 is important because it means that assuming this confidence interval is one of the 99 out of 100 that contains the true value of the significance of the test statistics, the true value is less than .05. This gives us a lot of confidence that the significant effect is genuine. Like

Test Statistics[b,c]

			Sperm Count (Millions)
Chi-Square			8.659
df			3
Asymp. Sig.			.034
Monte Carlo Sig.	Sig.		.033[a]
	99% Confidence Interval	Lower Bound	.028
		Upper Bound	.037

a Based on 10000 sampled tables with starting seed 846668601
b Kruskal–Wallis Test
c Grouping Variable: Number of Soya Meals Per Week

SPSS Output 13.9

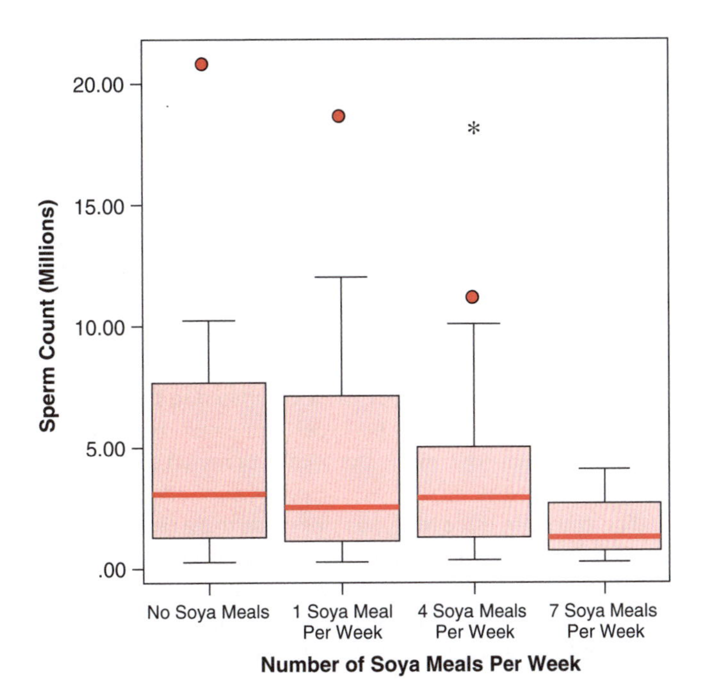

Figure 13.7 Boxplot for the sperm count of individuals eating different number of soya meals per week

a one-way ANOVA, though, this test tells us only that a difference exists; it doesn't tell us exactly where the differences lie.

One way to see which groups differ is to look at a boxplot (see section 3.3.2) of the groups (see Figure 13.7). The first thing to note is that there are some outliers (note the circles and asterisks that lie above the top whiskers)—these are men who produced a particularly rampant amount of sperm. Using the control as our baseline, the medians of the first three groups seem quite similar; however, the median of the group who ate seven soya meals per week does seem a little lower, so perhaps this is where the difference lies. However, these conclusions are subjective. What we really need are some contrasts or *post hoc* tests like those we used in ANOVA (see sections 8.2.10 and 8.2.11).

13.4.5. *Post hoc* tests for the Kruskal–Wallis test ②

There are two ways to do non-parametric *post hoc* procedures, the first being to use Mann–Whitney tests (section 13.2). However, if we use lots of Mann–Whitney tests we will inflate the Type I error rate (section 8.2.1) and this is precisely why we don't begin by doing lots of Mann–Whitney tests! However, if we want to use lots of Mann–Whitney tests to follow up a Kruskal–Wallis test, we can if we make some kind of adjustment to ensure that the Type I errors don't build up to more than .05. The easiest method is to use a *Bonferroni correction*, which in its simplest form just means that instead of using .05 as the critical value for significance for each test, you use a critical value of .05 divided by the number of tests you've conducted. If you do this, you'll soon discover that you quickly end up using a critical value for significance that is so small that it is very restrictive. Therefore, it's a good idea to be selective about the comparisons you make. In this example, we have a control group who had no soya meals. As such, a nice succinct set of comparisons would be to compare each group against the control:

- Test 1: one soya meal per week compared to no soya meals
- Test 2: four soya meals per week compared to no soya meals
- Test 3: seven soya meals per week compared to no soya meals

This results in three tests, so rather than use .05 as our critical level of significance, we'd use .05/3 = .0167. If we didn't use focused tests and just compared all groups with all other groups we'd end up with six tests rather than three (no soya vs. 1 meal, no soya vs. 4 meals, no soya vs. 7 meals, 1 meal vs. 4 meals, 1 meal vs. 7 meals, 4 meals vs. 7 meals) meaning that our critical value would fall to .05/6 = .0083.

SPSS Output 13.10 shows the test statistics from doing Mann–Whitney tests on the three focused comparisons that I suggested. The easiest way to do these in SPSS is to change the values of codes for *Group 1* and *Group 2* that you enter in the *define groups* dialog box (see Figure 13.3). Remember that we are now using a critical value of .0167, so the only comparison that is significant is when comparing those that had seven soya meals a week to those that had none (because the observed significance value of .009 is less than .0167). The other two comparisons produce significance values that are greater than .0167 so we'd have to say they're non-significant. So the effect we got seems mainly to reflect the fact that eating soya seven times per week lowers (I know this from the

No soya vs. 1 meal per week:

Test Statistics[b]

	Sperm Count (Millions)
Mann-Whitney U	191.000
Wilcoxon W	401.000
Z	−.243
Asymp. Sig. (2-tailed)	.808
Exact Sig. [2* (1-tailed Sig.)]	.820[a]

a Not corrected for ties
b Grouping Variable: Number of Soya Meals Per Week

No soya vs. 4 meals per week:

Test Statistics[b]

	Sperm Count (Millions)
Mann-Whitney U	188.000
Wilcoxon W	398.000
Z	−.325
Asymp. Sig. (2-tailed)	.745
Exact Sig. [2* (1-tailed Sig.)]	.758[a]

a Not corrected for ties
b Grouping Variable: Number of Soya Meals Per Week

No soya vs. 7 meals per week:

Test Statistics[b]

	Sperm Count (Millions)
Mann-Whitney U	104.000
Wilcoxon W	314.000
Z	−2.597
Asymp. Sig. (2-tailed)	.009
Exact Sig. [2* (1-tailed Sig.)]	009[a]

a Not corrected for ties
b Grouping Variable: Number of Soya Meals Per Week

SPSS Output 13.10

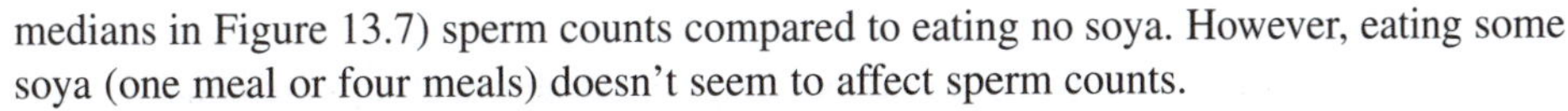

medians in Figure 13.7) sperm counts compared to eating no soya. However, eating some soya (one meal or four meals) doesn't seem to affect sperm counts.

The second way to do *post hoc* tests is essentially the same as doing Mann–Whitney tests on all possible comparisons, but for completeness sake I'll run you through it! It is described by Siegel & Castellan (1988) and involves taking the difference between the mean ranks of the different groups and comparing this to a value based on the value of z (corrected for the number of comparisons being done) and a constant based on the total sample size and the sample size in the two groups being compared. The inequality is:

$$|\overline{R}_u - \overline{R}_v| \geq z_{\alpha/k(k-1)}\sqrt{\frac{N(N+1)}{12}\left(\frac{1}{n_u}+\frac{1}{n_v}\right)} \tag{13.2}$$

The left-hand side of this equation is just the difference between the mean rank of the two groups being compared, but ignoring the sign of the difference (so the two vertical lines that enclose the difference between mean ranks just mean that if the difference is negative then we ignore the negative sign and treat the difference as positive). For the rest of the equation, k is the number of groups (in the soya example 4), N is the total sample size (in this case 80), n_u is the number of people in the first group that's being compared (we have equal group sizes in the soya example so it will be 20 regardless of which groups we compare), and n_v is the number of people in the second group being compared (again this will be 20 regardless of which groups we compare because we have equal group sizes in the soya example). The only other thing we need to know is $z_{\alpha/k(k-1)}$, and to get this value, we need to decide a level for α, which is the level of significance at which we want to work. You should know by now that in the social sciences we traditionally work at a .05 level of significance, so α will be .05. We then calculate $k(k-1)$, which for these data will be $4(4-1) = 12$. Therefore, $\alpha/k(k-1) = .05/12 = .00417$. So, $z_{\alpha/k(k-1)}$ just means 'the value of z for which only $\alpha/k(k-1)$ other values of z are bigger' (or in this case 'the value of z for which only .00417 other values of z are bigger'). In practical terms this means we go to the table in Appendix A.1, look at the column labelled *Smaller Portion* and find the number .00417 (or the nearest value to this, which if you look at the table is .00415), and we then look in the same row at the column labelled z. In this case, you should find that the value of z is 2.64. The next thing to do is to calculate the right-hand side of equation (13.2):

$$
\begin{aligned}
\text{critical difference} &= z_{\alpha/k(k-1)}\sqrt{\frac{N(N+1)}{12}\left(\frac{1}{n_u}+\frac{1}{n_v}\right)} \\
&= 2.64\sqrt{\frac{80(80+1)}{12}\left(\frac{1}{20}+\frac{1}{20}\right)} \\
&= 2.64\sqrt{540(0.1)} \\
&= 2.64\sqrt{54} \\
&= 19.40
\end{aligned}
$$

For this example, because the sample sizes across groups are equal this critical difference can be used for all comparisons. However, when samples sizes differ across groups, the critical difference will have to be calculated for each comparison individually. The next step is simply to calculate all of the differences between the mean ranks of all of the groups (the mean ranks can be found in SPSS Output 13.8) (Table 13.4).

Equation (13.2) basically means that if the differences between mean ranks are bigger than or equal to the critical difference for that comparison, then that difference is significant. In this case, because we have only one critical difference, it means that if any difference is bigger than 19.40, then it is significant. As you can see, all differences are below this value so we would have to conclude that none of the groups were significantly different! This contradicts our earlier findings where the Mann–Whitney test for the no-meals group compared to the seven meals group was deemed significant; why do you think that is? Well, for our Mann–Whitney tests, we did only three comparisons and so only corrected the significance value for the three tests we'd done (.05/3 = .0167). Earlier on in this section I

Table 13.4 Differences between mean ranks for the soya data

Comparison	$\bar{R}_u$	$\bar{R}_v$	$\bar{R}_u - \bar{R}_v$	$\lvert\bar{R}_u - \bar{R}_v\rvert$
No Meals–1 Meal	46.35	44.15	2.20	2.20
No Meals–4 Meals	46.35	44.15	2.20	2.20
No Meals–7 Meals	46.35	27.35	19.00	19.00
1 Meal–4 Meals	44.15	44.15	0.00	0.00
1 Meal–7 Meals	44.15	27.35	16.80	16.80
4 Meals–7 Meals	44.15	27.35	16.80	16.80

said that if we compared all groups against all other groups, that would be six comparisons, so we could accept a difference as being significant only if the significance value was less than .05/6 = .0083. If we go back to our one significant Mann–Whitney test (SPSS Output 13.10) the significance value was .009; therefore, if we had done all six comparisons this would've been non-significant (because .009 is bigger than .0083)! This illustrates what I said earlier about the benefits of choosing selective comparisons.

13.4.6. Testing for trends: the Jonckheere–Terpstra test ②

Back in section 13.4.3 we selected an option for the Jonckheere–Terpstra test, ☑ Jonckheere-Terpstra (Jonckheere, 1954; Terpstra, 1952). This statistic tests for an ordered pattern to the medians of the groups you're comparing. Essentially it does the same thing as the Kruskal–Wallis test (i.e. test for a difference between the medians of the groups) but it incorporates information about whether the order of the groups is meaningful. As such, you should use this test when you expect the groups you're comparing to produce a meaningful order of medians. So, in the current example we expect that the more soya a person eats, the more their sperm count will go down. Therefore, the control group should have the highest sperm count, those having one soya meal per week should have a lower sperm count, the sperm count in the four meals per week group should be smaller still, and the seven meals per week group should have the lowest sperm count. Therefore, there is an order to our medians: they should decrease across the groups. Conversely there might be situations where you expect your medians to increase. For example, there's a phenomenon in psychology known as the 'mere exposure effect', which basically means that the more you're exposed to something, the more you'll like it. Record companies use this to good effect by making sure songs are played on radio for about 2 months prior to their release, so on the day of release, everyone loves the song and is dying to have it and rushes out and buys it sending it to number one.[4] Anyway, if you took three groups and exposed them to

4 Although in most cases the mere exposure effect seems to have the reverse effect on me: the more I hear the manufactured rubbish that gets into the charts the more I want to rid my brain of it.

a song 10 times, 20 times and 30 times respectively and then measured how much people liked the song, you'd expect the medians to increase. Those who heard it 10 times would like it a bit, but those who heard it 20 times would like it more, and those who heard it 30 times would like it the most.

The Jonckheere–Terpstra test (actually referred to more often just as the Jonckheere test) was designed for these situations. In SPSS, it works on the principle that your coding variable (the one that defines the groups) specifies the order in which you expect the medians to change (it doesn't matter whether you expect them to increase or decrease). So, for our soya example, we coded our groups as 1 = no soya, 2 = one soya meal per week, 3 = four soya meals per week, and 4 = seven soya meals per week, so it would test whether the median sperm count increases or decreases across the groups when they're ordered in that way. Obviously we could change the coding scheme and test whether the medians were ordered in a different way. The exact workings of the test are described on the CD-ROM in a file called **Jonckheere.pdf**, but you don't need to know them to use the test. The important thing is to test whether the medians of the groups ascend or descend in the *order specified by the coding variable*.

We saw how to specify the test in section 13.4.3 and SPSS Output 13.11 shows the output from the test for the soya data. This table tells you the number of groups being compared, 4 (just in case you hadn't noticed). It also tells you the value of test statistic J, which is 912. In large samples (more than about eight per group) this test statistic has a sampling distribution that is normal, and a mean and standard deviation that are easily defined and calculated (see the file called **Jonckheere.pdf** on the CD-ROM and you'll

Jonckheere-Terpstra Test[b]

			Sperm Count (Millions)
Number of Levels in Number of Soya Meals Per Week			4
N			80
Observed J-T Statistic			912.000
Mean J-T Statistic			1200.000
Std. Deviation of J-T Statistic			116.333
Std. J-T Statistic			−2.476
Asymp. Sig. (2-tailed)			.013
Monte Carlo Sig. (2-tailed)	Sig.		.013[a]
	99% Confidence Interval	Lower Bound	.010
		Upper Bound	.016
Monte Carlo Sig. (1-tailed)	Sig.		.006[a]
	99% Confidence Interval	Lower Bound	.004
		Upper Bound	.008

a Based on 10000 sampled tables with starting seed 846668601
b Grouping Variable: Number of Soya Meals Per Week

SPSS Output 13.11

find as the output tells you that the mean is 1200 and the standard deviation is 116.33). Knowing these things, we can convert to a z-score by taking the test statistic, subtracting the mean of the sampling distribution from it and then dividing the result by the standard deviation ($z = (912 - 1200)/116.33 = -2.476$). This is much the same as what we did for the Mann–Whitney and Wilcoxon tests. This z-score can then be compared against the values of a normal distribution, and because Jonckheere's test should always be one-tailed (we specify before the experiment the order of the medians) we're looking for a value greater than 1.65 (when we ignore the sign). A value of 2.47 is, therefore, significant. The sign of the z-value does tell us something useful though: if it is positive then it indicates a trend of ascending medians (i.e. the medians get bigger as the values of the coding variable get bigger) but if it is negative (as it is here) it indicates a trend of descending medians (the medians get smaller as the value of the coding variable gets bigger). In this example, because we coded the variables as 1 = no soya, 2 = one soya meal per week, 3 = four soya meals per week, and 4 = seven soya meals per week, it means that the medians get smaller as we go from no soya, to one soya meal, to four soya meals and on to seven soya meals.

You'll also notice that there are one- and two-tailed significance values estimated using Monte Carlo methods. These have appeared because we chose that option for the Kruskal–Wallis test. They confirm what we have already found.

13.4.7. Calculating an effect size ②

Unfortunately there isn't an easy way to convert a chi-square statistic that has more than 1 degree of freedom to an effect size r. You could use the significance value of the Kruskal–Wallis test statistic to find an associated value of z from a table of probability values for the normal distribution (like that in Appendix A.1). From this you could use the conversion to r that we used in section 13.2.5. However, this kind of effect size is rarely that useful (because it's summarizing a general effect). In most cases it's more interesting to know the effect size for a focused comparison (such as when comparing two things). For this reason, I'd suggest just calculating effect sizes for the Mann–Whitney tests that we used to follow up the main analysis.

For the first comparison (no soya vs. 1 meal) SPSS Output 13.10 shows us that z is −0.243, and because this was based on comparing two groups, each containing 20 observations, we had 40 observations in total. The effect size is, therefore:

$$\begin{aligned} r_{\text{No Soya}-1\text{ Meal}} &= \frac{-0.243}{\sqrt{40}} \\ &= -.04 \end{aligned}$$

This represents a very small effect because it is close to zero, which tells us that the effect on sperm counts of having one soya meal per week compared to no soya meals was negligible.

For the second comparison (no soya vs. 4 meals) SPSS Output 13.10 shows us that z is –0.325, and this was again based on 40 observations. The effect size is, therefore:

$$r_{\text{No Soya}-1\text{ Meal}} = \frac{-0.325}{\sqrt{40}} = -.05$$

This again represents a negligible effect because it is close to zero, which tells us that the effect on sperm counts of having four soya meals per week compared to no soya meals was negligible.

For the final comparison (no soya vs. 7 meals) SPSS Output 13.10 shows us that z is –2.597, and this was again based on 40 observations. The effect size is, therefore:

$$r_{\text{No Soya}-1\text{ Meal}} = \frac{-2.597}{\sqrt{40}} = -.41$$

This represents a medium effect, which tells us that the effect of seven soya meals a week lowering sperm counts (compared to having none) was a fairly substantive finding.

We can also calculate an effect size for Jonckheere's test if we want to by using the same equation. This test involves all data, so we have to use 80 as our value of N:

$$r_{\text{Jonckheere}} = \frac{-2.476}{\sqrt{80}} = -.28$$

13.4.8. Writing and interpreting the results ①

For the Kruskal–Wallis test, we need only report the test statistic (which, as we saw earlier, is denoted by H), its degrees of freedom and its significance. So, we could report something like:

✓ Sperm counts were significantly affected by eating soya meals ($H(3) = 8.66$, $p < .05$).

However, we need to report the follow-up tests as well (including their effect sizes):

✓ Sperm counts were significantly affected by eating soya meals ($H(3) = 8.66$, $p < .05$). Mann–Whitney tests were used to follow up this finding. A Bonferroni correction was applied and so all effects are reported at a .0167 level of significance. It appeared that sperm counts were no different when one soya meal ($U = 191$, $r = -.04$) or four soya meals ($U = 188$, $r = -.05$) were eaten per week compared to none. However, when seven soya meals were eaten per week, sperm counts were significantly lower than when no soya was eaten ($U = 104$, $r = -.41$). We can conclude that if soya is eaten every day it significantly reduces sperm counts compared to eating none; however, eating soya less than every day has no significant effect on sperm counts ('phew!' says the vegetarian man!).

Or, we might want to report our trend:

✓ All effects are reported at $p < .05$. Sperm counts were significantly affected by eating soya meals ($H(3) = 8.66$). Jonckheere's test revealed a significant trend in the data: as more soya was eaten, the median sperm count decreased, $J = 912$, $z = -2.48$, $r = -.28$.

Cramming Samantha's Tips

- The Kruskal–Wallis test compares several conditions when different participants take part in each condition and the resulting data are not normally distributed or violate an assumption of one-way independent ANOVA.
- Look at the row labelled *Asymp. Sig.*: if the value is less than .05 then the groups are significantly different.
- You can follow up the main analysis with Mann–Whitney tests between pairs of conditions, but only accept them as significant if they're significant below .05/number of tests.
- If you predict that the means will increase or decrease across your groups in a certain order then do Jonckheere's trend test.
- Report the *H*-statistic, the degrees of freedom and the significance value for the main analysis. For any *post hoc* tests, report the *U*-statistic, and an effect size if possible (you can also report the corresponding *z* and the significance value). Also report the medians and their corresponding ranges (or draw a boxplot).

13.5. DIFFERENCES BETWEEN SEVERAL RELATED GROUPS: FRIEDMAN'S ANOVA ①

In Chapter 11 we discovered a technique called one-way related ANOVA that could be used to test for difference between several related groups. Although, as we've seen, ANOVA is robust to violations of its assumptions, there is another alternative to the repeated-measures case: Friedman's ANOVA (1937). As such, it is used for testing differences between experimental conditions when there are more than two conditions and the same participants have been used in all conditions (each person contributes several scores to the data). If you have non-normally distributed data, or have violated some other assumption, then this test can be a useful way around the problem.

Young people (women especially) can become obsessed with body weight and diets, and because the media are insistent on ramming ridiculous images of stick-thin celebrities down our throats (or should that be 'into our eyes'?) and brainwashing us into believing that these emaciated corpses are actually attractive, we all end up terribly depressed that we're not perfect. So, then other corporate parasites jump on our vulnerability by making loads of money on diets that will help us attain the body beautiful! Well, not wishing to miss out on this great opportunity to exploit people's insecurities I came up with my own diet called the Andikins diet.[5] The basic principle is that you eat like me: you eat no meat, drink lots of Darjeeling tea, eat shed-loads of lovely European cheese, lots of fresh crusty bread, pasta, and chocolate at every available opportunity (especially when writing books), enjoy a few beers at the weekend, play football and rugby at least twice a week and play your drum kit for at least an hour a day or until your girlfriend threatens to saw your arms off and beat you around the head with them for making so much noise. To test the efficacy of my wonderful new diet, I took 10 women who considered themselves to be in need of losing weight and put them on this diet for 2 months. Their weight was measured in kilograms at the start of the diet and then after 1 month and 2 months.

13.5.1. Theory of Friedman's ANOVA ②

The theory for Friedman's ANOVA is much the same as the other tests we've seen in this chapter: it is based on ranked data. To begin with, you simply place your data for different conditions into different columns (in this case there were three conditions so we have three columns). The data for the diet example are in Table 13.5; note that the data are in different columns and so each row represents the weight of a different person. The next thing we have to do is rank the data *for each person*. So, we start with person 1, we look at their scores (in this case person 1 weighed 63.75 kg at the start, 65.38 kg after 1 month on the diet, and 81.34 kg after 2 months on the diet), and then we give the lowest one a rank of 1, the next highest a rank of 2 and so on (see section 13.2.1 for more detail). When you've ranked the data for the first person, you move on to the next person and, starting at 1 again, rank their lowest score, then rank the next highest as 2 and so on. You do this for all people from whom you've collected data. You then simply add up the ranks for each condition (R_i, where i is used to denote the particular group). Table 13.5 shows the raw data for this example along with the ranks (have a go at ranking the data and see if you get the same results as me!).

Once the sum of ranks has been calculated for each group, the test statistic, F_r, is calculated as in equation (13.3):

$$F_r = \left[\frac{12}{Nk(k+1)} \sum_{i=1}^{k} R_i^2 \right] - 3N(k+1) \tag{13.3}$$

5 Not to be confused with the Atkins diet obviously ☺

Table 13.5 Data for the diet example with ranks

	Weight			Weight		
	Start	Month 1	Month 2	Start (Ranks)	Month 1 (Ranks)	Month 2 (Ranks)
Person 1	63.75	65.38	81.34	1	2	3
Person 2	62.98	66.24	69.31	1	2	3
Person 3	65.98	67.70	77.89	1	2	3
Person 4	107.27	102.72	91.33	3	2	1
Person 5	66.58	69.45	72.87	1	2	3
Person 6	120.46	119.96	114.26	3	2	1
Person 7	62.01	66.09	68.01	1	2	3
Person 8	71.87	73.62	55.43	2	3	1
Person 9	83.01	75.81	71.63	3	2	1
Person 10	76.62	67.66	68.60	3	1	2
			R_i	19	20	21

In this equation, R_i is the sum of ranks for each group, N is the total sample size (in this case 10) and k is the number of conditions (in this case 3). This equation is very similar to that of the Kruskal–Wallis test (compare equations (13.1) and (13.3)). All we need to do for each condition is square the sum of ranks and then add up these values. That deals with the middle part of the equation; the rest of it involves calculating various values based on the total sample size and the number of conditions. For these data we get:

$$
\begin{aligned}
F_r &= \left[\frac{12}{(10 \times 3)(3+1)}(19^2 + 20^2 + 21^2)\right] - (3 \times 10)(3+1) \\
&= \frac{12}{120}(361 + 400 + 441) - 120 \\
&= 0.1(1202) - 120 \\
&= 120.2 - 120 \\
&= 0.2
\end{aligned}
$$

When the number of people tested is large (bigger than about 10) this test statistic, like the Kruskal–Wallis test in the previous section, has a chi-square distribution (see Chapter 16) and for this distribution there is one value for the degrees of freedom, which is one less than the number of groups ($k - 1$): in this case 2.

13.5.2. Inputting data and provisional analysis ①

When the data are collected using the same participants in each condition, we need to input the data using different columns. So, the data editor will have three columns of data. The first column is for the data from the start of the diet (called something like **Start**), the second column will have values for the weights after 1 month (called **month1**) and the final column will have the weights at the end of the diet (called **month2**). The data can be found in the file **Diet.sav**.

First, we run some exploratory analyses on the data (see Chapter 3). With a bit of luck you'll get the same table shown in SPSS Output 13.12, which shows that the K–S test (see section 3.5) was not significant for the initial weights at the start of the diet ($D(10) = 0.23$, $p > .05$) but the Shapiro–Wilk test is significant and this test is actually more accurate than the K–S test (see Chapter 3). The data 1 month into the diet were significantly different from normal ($D(10) = .34$, $p < .01$), though the data at the end of the diet do appear to be normal ($D(10) = .20$, $p > .05$). Some of these data are not normally distributed, and so we should use a non-parametric test.

Tests of Normality

	Kolmogorov-Smirnov[a]			Shapiro-Wilk		
	Statistic	df	Sig.	Statistic	df	Sig.
Weight at Start (Kg)	.228	10	.149	.784	10	.009
Weight after 1 month (Kg)	.335	10	.002	.685	10	.001
Weight after 2 month (Kg)	.203	10	.200*	.877	10	.121

*This is a lower bound of the true significance.
a Lilliefors Significance Correction

SPSS Output 13.12

13.5.3. Doing Friedman's ANOVA on SPSS ①

First, access the main dialog box by using the **Analyze⇒Nonparametric Tests⇒K Related Samples...** menu pathway (see Figure 13.8). Once the dialog box is activated, select the three dependent variables from the list (click on **start** and then, keeping the mouse button pressed down, drag the mouse over **month1** and **month2**). Transfer them to the box labelled *Test Variables* by clicking on ▸. If you click on Exact... then you get a dialog box for selecting exact significance values for Friedman's ANOVA (see section 13.2.3). Remember that I said that when samples are large you should probably opt for the Monte Carlo method but with small samples it's worth opting for the exact test. Well, we've got a relatively small sample here so we can select the exact tests. Finally, click on Statistics... to select some descriptive statistics. To run the analyses return to the main dialog box and click on OK. You can also conduct some related tests using the same dialog box as for Friedman's ANOVA (see Box 13.5).

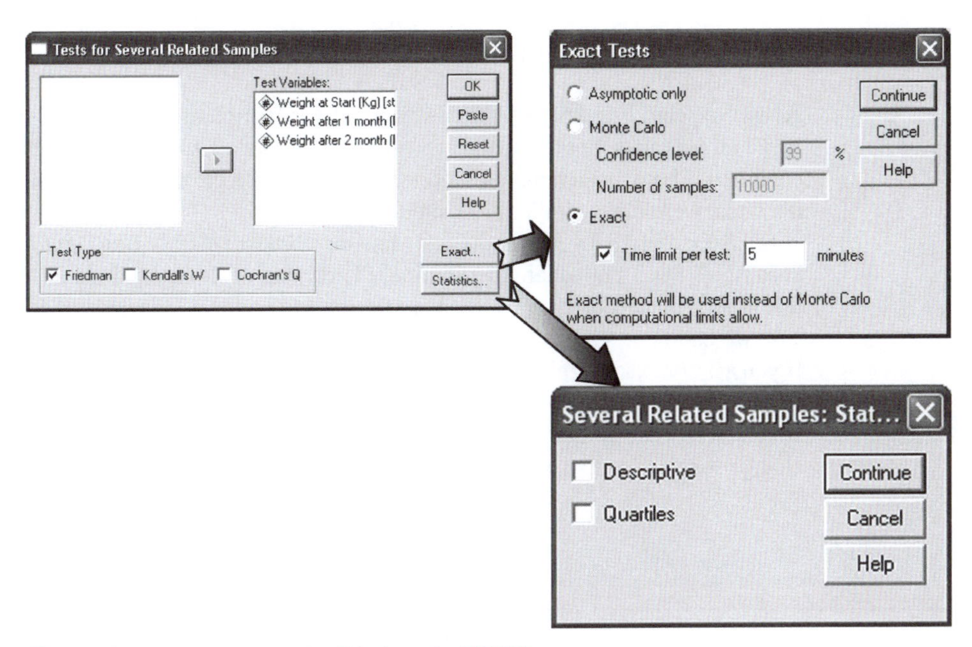

Figure 13.8 Dialog boxes for Friedman's ANOVA

Box 13.5

Other options ②

In the main dialog box there are some other tests that can be selected:

- **Kendall's W**: This is much the same as Friedman's ANOVA but is used specifically for looking at the agreement between raters. So, if, for example, we asked 10 different women to rate the attractiveness of Justin Timberlake, David Beckham and Tony Blair we could use this test to look at the extent to which they agree. This test is particularly useful because, like the correlation coefficient, Kendall's *W* has a limited range: it ranged from 0 (no agreement between judges) to 1 (complete agreement between judges).
- **Cochran's Q**: This test is an extension of McNemar's test (see Box 13.3) and is basically a Friedman test for when you have dichotomous data. So imagine you asked 10 women whether they'd like to snog Justin Timberlake, David Beckham or Tony Blair and they could answer only yes or no. If we coded responses as 0 (no) and 1 (yes) we could do the Cochran test on these data.

13.5.4. Output from Friedman's ANOVA ①

As we've seen, Friedman's ANOVA, like all the non-parametric tests in this chapter, is based on the ranks, not the actual scores. SPSS Output 13.13 shows the mean rank in each condition. These mean ranks are important later for interpreting any effects; they show that the ranks were fairly similar across the conditions.

SPSS Output 13.14 shows the test statistic (SPSS calls this *Chi-Square* rather that F_r because F_r has a chi-square distribution). The value of this statistic is .2, the same value that we calculated earlier. We're also told the test statistic's degrees of freedom (in this case we had three groups so the degrees of freedom are 3 – 1, or 2), and the significance. If you ask for exact significance then this value is given too. The significance value is .905 (or .974 if you read the exact significance), which is well above .05; therefore, we could conclude that the Andikins diet does not have any effect: the weights didn't significantly change over the course of the diet.

Descriptive Statistics

	N	Mean	Std. Deviation	Minimum	Maximum	Percentiles		
						25th	50th (Median)	75th
Weight at Start (Kg)	10	78.0543	20.23008	62.01	120.46	63.5549	69.2288	89.0709
Weight after 1 month (Kg)	10	77.4635	18.61502	65.38	119.96	66.2065	68.5728	82.5385
Weight after 2 month (Kg)	10	77.0668	16.10612	55.43	114.26	68.4525	72.2493	83.8365

Ranks

	Mean Rank
Weight at Start (Kg)	1.90
Weight after 1 month (Kg)	2.00
Weight after 2 month (Kg)	2.10

SPSS Output 13.13

Test Statistics[a]

N	10
Chi-Square	.200
df	2
Asymp. Sig.	.905
Exact Sig.	.974
Point Probability	.143

a Friedman Test

SPSS Output 13.14

13.5.5. *Post hoc* tests for Friedman's ANOVA ②

As with the Kruskal–Wallis test, there are two ways to do non-parametric *post hoc* procedures, which are in essence the same. The first is to use Wilcoxon signed-rank tests (section 13.3) but correcting for the number of tests we do (see sections 8.2.1 and 13.4.5 for the reasons why). The way we correct for the number of tests is to accept something as significant only if its significance is less than α/number of comparisons (the *Bonferroni correction*). In the social sciences this usually means .05/number of comparisons. In this example, we have only three groups, so if we compare all of the groups we simply get three comparisons:

- Test 1: Weight at the start of the diet compared to weight at 1 month.
- Test 2: Weight at the start of the diet compared to weight at 2 months.
- Test 3: Weight at 1 month compared to weight at 2 months.

Therefore, rather than use .05 as our critical level of significance, we'd use .05/3 = .0167. In fact we wouldn't bother with *post hoc* tests at all for this example because the main ANOVA was non-significant, but I'll go through the motions to illustrate what to do. You can do the Wilcoxon tests by selecting the pairs of variables for each comparison in turn and transferring them across to the box labelled *Test Pair(s) List* (see Figure 13.4).

SPSS Output 13.15 shows the test statistics from doing Wilcoxon signed-rank tests on the three comparisons. Remember that we are now using a critical value of .0167, and in fact none of the comparisons are significant as they have one-tailed significance values of .500, .423 and .461 (this isn't surprising because the main analysis was non-significant).

The second way to do *post hoc* tests is very similar to what we did for the Kruskal–Wallis test in section 13.4.5 and is, likewise, described by Siegel & Castellan (1988). Again, we take the difference between the mean ranks of the different groups and compare these differences to a value based on the value of z (corrected for the number of comparisons being done) and a constant based on the total sample size, N (10 in this example) and the number of conditions, k (3 in this case). The inequality is:

$$|\overline{R}_u - \overline{R}_v| \geq z_{\alpha/k(k-1)}\sqrt{\frac{k(k+1)}{6N}} \tag{13.4}$$

The left-hand side of this equation is just the difference between the mean rank of the two groups being compared, but ignoring the sign of the difference. As with Kruskal–Wallis, we need to know $z_{\alpha/k(k-1)}$, and if we stick to tradition and use an α-level of .05, knowing that k is 3, we get $\alpha/k(k-1) = .05/3(3-1) = .05/6 = .00833$. So, $z_{\alpha/k(k-1)}$ just means 'the value of z for which only $\alpha/k(k-1)$ other values of z are bigger' (or in this case 'the value of z for which only .00833 other values of z are bigger'). Therefore, we go to the table in Appendix A.1, look at the column labelled *Smaller Portion* and find the number .00833 and then find the value in the same row in the column labelled z.

Ranks

		N	Mean Rank	Sum of Ranks
Weight after 1 month (Kg) – Weight at Start (Kg)	Negative Ranks	4[a]	7.00	28.00
	Positive Ranks	6[b]	4.50	27.00
	Ties	0[c]		
	Total	10		
Weight after 2 month (Kg) – Weight at Start (Kg)	Negative Ranks	5[d]	6.00	30.00
	Positive Ranks	5[e]	5.00	25.00
	Ties	0[f]		
	Total	10		
Weight after 2 month (Kg) – Weight after 1 month (Kg)	Negative Ranks	4[g]	7.25	29.00
	Positive Ranks	6[h]	4.33	26.00
	Ties	0[i]		
	Total	10		

a Weight after 1 month (Kg) < Weight at Start (Kg)
b Weight after 1 month (Kg) > Weight at Start (Kg)
c Weight after 1 month (Kg) = Weight at Start (Kg)
d Weight after 2 month (Kg) < Weight at Start (Kg)
e Weight after 2 month (Kg) > Weight at Start (Kg)
f Weight after 2 month (Kg) = Weight at Start (Kg)
g Weight after 2 month (Kg) < Weight after 1 month (Kg)
h Weight after 2 month (Kg) > Weight after 1 month (Kg)
i Weight after 2 month (Kg) = Weight after 1 month (Kg)

Test Statistics[b]

	Weight after 1 month (Kg) – Weight at Start (Kg)	Weight after 2 month (Kg) – Weight at Start (Kg)	Weight after 2 month (Kg) – Weight after 1 month (Kg)
Z	−.051[a]	−.255[a]	−.153[a]
Asymp. Sig. (2-tailed)	.959	.799	.878
Exact Sig. (2-tailed)	1.000	.846	.922
Exact Sig. (1-tailed)	.500	.423	.461
Point Probability	.039	.038	.038

a Based on positive ranks
b Wilcoxon Signed Ranks Test

SPSS Output 13.15

In this case there are values of .00842 and .00820, which give *z*-values of 2.39 and 2.40 respectively; because .00833 lies about midway between the values we found we could just take the midpoint of the two *z*-values, 2.395, or we could err on the side of caution and use 2.40. I'll err on the cautious side and use 2.40. We can now calculate the right-hand side of equation (13.4):

$$\begin{aligned}\text{critical difference} &= z_{\alpha/k(k-1)}\sqrt{\frac{k(k+1)}{6N}} \\ &= 2.40\sqrt{\frac{3(3+1)}{6(10)}} \\ &= 2.40\sqrt{\frac{12}{60}} \\ &= 2.40\sqrt{0.2} \\ &= 1.07\end{aligned}$$

When the same people have been used, the same critical difference can be used for all comparisons. The next step is simply to calculate all of the differences between the mean ranks of all of the groups (the mean ranks can be found in SPSS Output 13.13):

Table 13.6 Differences between mean ranks for the diet data

Comparison	$\bar{R}_u$	$\bar{R}_v$	$\bar{R}_u - \bar{R}_v$	$\lvert\bar{R}_u - \bar{R}_v\rvert$
Start–1 Month	1.90	2.00	−0.10	0.10
Start–2 Months	1.90	2.10	−0.20	0.20
1 Month–2 Months	2.00	2.10	−0.10	0.10

Equation (13.4) means that if the differences between mean ranks are bigger than or equal to the critical difference, then that difference is significant. In this case, it means that if any difference is bigger than 1.07, then it is significant. All differences are below this value (Table 13.6) so we could conclude that none of the groups were significantly different and this is consistent with the non-significance of the initial ANOVA.

13.5.6. Calculating an effect size ②

As I mentioned before, there isn't an easy way to convert a chi-square statistic that has more than 1 degree of freedom to an effect size *r* and in any case, it's not always that helpful to have an effect size for a general effect like that tested by Friedman's ANOVA.[6] Therefore, it's more sensible (in my opinion at least) to calculate effect sizes for any comparisons you've done after the ANOVA. As we saw in section 13.3.5 it's straightforward to get an effect size *r* from Wilcoxon's signed-rank test. Alternatively, we could just

6 If you really want to, though, you can (as with the Kruskal–Wallis test) use the significance value of the chi-square test statistic to find an associated value of *z* from a table of probability values for the normal distribution (see Appendix A.1) and then use the conversion to *r* that we've seen throughout this chapter.

calculate effect sizes for the Wilcoxon tests that we used to follow up the main analysis. These effect sizes will be very informative in their own right.

For the first comparison (start weight vs. 1 month) SPSS Output 13.15 shows us that z is –0.051, and because this was based on comparing two conditions each containing 10 observations, we had 20 observations in total (remember that it isn't important that the observations come from the same people). The effect size is, therefore:

$$r_{\text{Start}-1\,\text{Month}} = \frac{-0.051}{\sqrt{20}} = -.01$$

For the second comparison (start weight vs. 2 months) SPSS Output 13.15 shows us that z is –0.255, and this was again based on 20 observations. The effect size is, therefore:

$$r_{\text{Start}-2\,\text{Months}} = \frac{-0.255}{\sqrt{20}} = -.06$$

For the final comparison (1 month vs. 2 months) SPSS Output 13.15 shows us that z is –0.153, and this was again based on 20 observations. The effect size is, therefore:

$$r_{1\text{Month}-2\,\text{Months}} = \frac{-0.153}{\sqrt{20}} = -.03$$

Unsurprisingly, given the lack of significance of the Wilcoxon tests, these all represent a virtually non-existent effect: they are all very close to zero.

13.5.7. Writing and interpreting the results ①

For Friedman's ANOVA we need only report the test statistic (which we saw earlier is denoted by χ^2),[7] its degrees of freedom and its significance. So, we could report something like:

✓ The weight of participants did not significantly change over the 2 months of the diet ($\chi^2(2) = 0.20$, $p < .05$).

Although with no significant initial analysis we wouldn't report *post hoc* tests for these data, in case you need to, you should say something like this (remember that the test

7 The test statistic is often denoted as χ^2_F but the official APA style guide doesn't recognize this term.

statistic T is the smaller of the two sums of ranks for each test and these values are in SPSS Output 13.15):

✓ The weight of participants did not significantly change over the 2 months of the diet ($\chi^2(2) = 0.20$, $p < .05$). Wilcoxon tests were used to follow up this finding. A Bonferroni correction was applied and so all effects are reported at a .0167 level of significance. It appeared that weight didn't significantly change from the start of the diet to 1 month, $T = 27$, $r = -.01$, from the start of the diet to 2 months, $T = 25$, $r = -.06$, or from 1 month to 2 months, $T = 26$, $r = -.03$. We can conclude that the Andikins diet, like its creator, is a complete failure.

Cramming Samantha's Tips

- Friedman's ANOVA compares several conditions when the same participants take part in each condition and the resulting data are not normally distributed or violate an assumption of one-way repeated-measures ANOVA.
- Look at the row labelled *Asymp. Sig.*: if the value is less than .05 then the conditions are significantly different.
- You can follow up the main analysis with Wilcoxon signed-rank tests between pairs of conditions, but only accept them as significant if they're significant below .05/number of tests.
- Report the χ^2-statistic, its degrees of freedom and significance. For any *post hoc* tests report the T-statistic, and an effect size if possible. You can also report the z- and significance value. Also report the medians and their corresponding ranges (or draw a boxplot).

13.6. WHAT HAVE WE DISCOVERED ABOUT STATISTICS? ①

Tests like the t-test and ANOVA can be robust to violations of assumptions, and corrections can be made to ensure these parametric tests are accurate when assumptions are broken. This chapter has dealt with an alternative approach to violations of parametric assumptions, which is to use tests based on ranking the data. We started with the Wilcoxon rank-sum test and the Mann–Whitney test, which is used for comparing two independent groups. This test allowed us to look in some detail at the process of ranking data. We then moved on to look at the Wilcoxon signed-rank test, which is used to compare two related conditions. We examined more complex situations in which there are several conditions (the Kruskal–Wallis test for independent conditions and Friedman's ANOVA for related conditions). For each of these tests we looked at the theory of the test (although these sections could be ignored) and then focused on how to conduct them on SPSS, how to interpret the results and how to report the results of the test. In the process we discovered that drugs make you depressed, soya reduces your sperm count, and my lifestyle is not conducive to losing weight! This ends our look at comparing groups and the final section of the book now focuses on some really hard stuff. Time for a stiff drink I think.

13.7. KEY TERMS THAT WE'VE DISCOVERED

- Cochran's *Q*
- Friedman's ANOVA
- Jonckheere–Terpstra test
- Kendall's *W*
- Kolmogorov–Smirnov *Z*
- Kruskal–Wallis test
- Mann–Whitney test
- McNemar test
- Median test
- Monte Carlo method
- Moses Extreme Reactions
- Non-parametric tests
- Ranking
- Sign test
- Wald–Wolfowitz runs
- Wilcoxon rank-sum test
- Wilcoxon signed-rank test

13.8. SMART ALEX'S TASKS

- **Task 1**: A psychologist was interested in the cross-species differences between men and dogs. She observed a group of dogs and a group of men in a naturalistic setting (20 of each). She classified several behaviours as being dog-like (urinating against trees and lampposts, attempts to copulate with anything that moved, and attempts to lick their own genitals). For each man and dog she counted the number of dog-like behaviours displayed in a 24-hour period. It was hypothesized that dogs would display more dog-like behaviours than men. The data are in the file **MenLikeDogs.sav;** analyse them with a Mann–Whitney test.①
- **Task 2**: There's been much speculation over the years about the influence of subliminal messages on records. To name a few cases, both Ozzy Osbourne and Judas Priest have been accused of putting backward masked messages on their albums that subliminally influence poor unsuspecting teenagers into doing things like blowing their heads off with shotguns. A psychologist was interested in whether backward masked messages really did have an effect. He took the master tapes of Britney Spears' 'Baby one more time' and created a second version that had the masked message 'deliver your soul to the dark lord' repeated in the chorus. He took this version, and the original, and played one version (randomly) to a group of 32 people. He took the same group 6 months later and played them whatever version they hadn't heard the time before. So each person heard both the original and the version with the

masked message, but at different points in time. The psychologist measured the number of goats that were sacrificed in the week after listening to each version. It was hypothesized that the backward message would lead to more goats being sacrificed. The data are in the file **DarkLord.sav**; analyse them with a Wilcoxon signed-rank test.①

- **Task 3**: A psychologist was interested in the effects of television programmes on domestic life. She hypothesized that through 'learning by watching', certain programmes might actually encourage people to behave like the characters within them. This in turn could affect the viewer's own relationships (depending on whether the programme depicted harmonious or dysfunctional relationships). She took episodes of three popular TV shows, and showed them to 54 couples after which the couples were left alone in the room for an hour. The experimenter measured the number of times the couple argued. Each couple viewed all three of the TV programmes at different points in time (a week apart) and the order in which the programmes were viewed was counterbalanced over couples. The TV programmes selected were *Eastenders* (which typically portrays the lives of extremely miserable, argumentative, London folk who like nothing more than to beat each other up, lie to each other, sleep with each other's wives and generally show no evidence of any consideration to their fellow humans!), *Friends* (which portrays a group of unrealistically considerate and nice people who love each other oh so very much—but for some reason I love it anyway!), and a National Geographic programme about whales (this was supposed to act as a control). The data are in the file **Eastenders.sav**; access them and conduct Friedman's ANOVA on the data.①
- **Task 4**: A researcher was interested in trying to prevent coulrophobia (fear of clowns) in children. She decided to do an experiment in which different groups of children (15 in each) were exposed to different forms of positive information about clowns. The first group watched some adverts for McDonald's in which their mascot Ronald McDonald is seen cavorting about with children going on about how they should love their mum. A second group was told a story about a clown who helped some children when they got lost in a forest (although what on earth a clown was doing in a forest remains a mystery). A third group was entertained by a real clown, who came into the classroom and made balloon animals for the children.[8] A final group acted as a control condition and they had nothing done to them at all. The researcher took self-report ratings of how much the children liked clowns resulting in a score for each child that could range from 0 (not scared of clowns at all) to 5 (very scared of clowns). The data are in the file **coulrophobia.sav**; access the data and conduct a Kruskal–Wallis test.①

8 Unfortunately, the first time they attempted the study the clown accidentally burst one of the balloons. The noise frightened the children and they associated that fear response with the clown. All 15 children are currently in therapy for coulrophobia!

The answers for these questions are in the file **Answers(Chapter13).pdf** and because these examples are used in Field & Hole (2003), you could steal this book or photocopy Chapter 7 to get some very detailed answers.

13.9. FURTHER READING

Siegel, S. & Castellan, N. J. (1988). *Nonparametric statistics for the behavioral sciences* (2nd edition). New York: McGraw-Hill. This has become the definitive text on non-parametric statistics, and is the only book seriously worth recommending as 'further' reading. It is probably not a good book for stats-phobes, but if you've coped with my chapter then this book will be an excellent next step.

CHAPTER 14

MULTIVARIATE ANALYSIS OF VARIANCE (MANOVA)

14.1. WHAT WILL THIS CHAPTER TELL US? ②

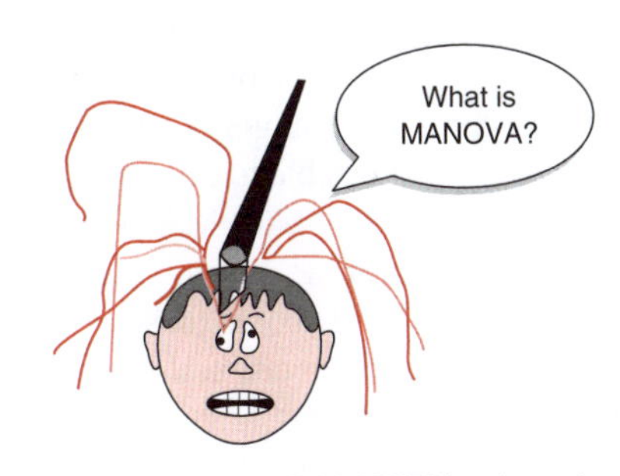

Over the last five chapters, we have seen how the general linear model (GLM) can be used to detect group differences on a single dependent variable. However, there may be circumstances in which we are interested in several dependent variables and in these cases the simple ANOVA model is inadequate. Instead, we can use an extension of this technique known as multivariate analysis of variance (or MANOVA). MANOVA can be thought of as an ANOVA for situations in which there are several dependent variables. The principles of ANOVA extend to MANOVA in that we can use MANOVA when there is only one independent variable or when there are several, we can look at interactions between independent variables, and we can even do contrasts to see which groups differ from each other. ANOVA can be used only in situations in which there is one dependent variable (or outcome) and so is known as a univariate test (univariate quite obviously means 'one variable'); MANOVA is designed to look at several dependent variables (outcomes) simultaneously and so is a multivariate test (multivariate means 'many variables'). This chapter will explain some basics about MANOVA for those of you who want to skip the fairly tedious theory sections and just get on with the test. However, for those who want to know more there is a fairly lengthy theory section to try to explain the workings of MANOVA. We then look at an example using SPSS and see how the output from MANOVA can be interpreted. This leads us to look at another statistical test known as *discriminant function analysis*.

14.2. INTRODUCTION: SIMILARITIES AND DIFFERENCES TO ANOVA ②

If we have collected data about several dependent variables then we could simply conduct a separate ANOVA for each dependent variable (and if you read research articles you'll find that it is not unusual for researchers to do this!). Think back to Chapter 8 and you should remember that a similar question was posed regarding why ANOVA was used in preference to multiple *t*-tests. The answer to why MANOVA is used instead of multiple ANOVAs is the same: the more tests we conduct on the same data, the more we inflate the familywise error rate (see section 8.2.1). The more dependent variables that have been measured, the more ANOVAs would need to be conducted and the greater the chance of making a Type I error.

However, there are other reasons for preferring MANOVA to several ANOVAs. For one thing, there is important additional information that is gained from a MANOVA. If separate ANOVAs are conducted on each dependent variable, then any relationship between dependent variables is ignored. As such, we lose information about any correlations that might exist between the dependent variables. MANOVA, by including all dependent variables in the same analysis, takes account of the relationship between outcome variables. Related to this point, ANOVA can tell us only whether groups differ along a single dimension whereas MANOVA has the power to detect whether groups differ along a combination of dimensions. For example, ANOVA tells us how scores on a single dependent variable distinguish groups of participants (so, for example, we might be able to distinguish drivers, non-drivers and drunk drivers by the number of pedestrians they kill). MANOVA incorporates information about several outcome measures and, therefore, informs us of whether groups of participants can be distinguished by a combination of scores on several dependent measures. For example, it may not be possible to distinguish drivers, non-drivers and drunk drivers only by the number of pedestrians that they kill, but they might be distinguished by *a combination* of the number of pedestrians they kill, the number of lampposts they hit, and the number of cars they crash into. So, in this sense MANOVA has greater power to detect an effect, because it can detect whether groups differ along a combination of variables, whereas ANOVA can detect only if groups differ along a single variable (see Huberty & Morris, 1989, and section 14.2.2). For these reasons, MANOVA is preferable to conducting several ANOVAs.

14.2.1. Words of warning ②

From my description of MANOVA it is probably looking like a pretty groovy little test that allows you to measure hundreds of dependent variables and then just sling them into the analysis. This is not the case. It is not a good idea to lump all of your dependent variables together in a MANOVA unless you have a good theoretical or empirical basis for doing so.

I mentioned way back at the beginning of this book that statistical procedures are just a way of number crunching and so even if you put rubbish into an analysis you will still reach conclusions that are statistically meaningful, but are unlikely to be empirically meaningful. In circumstances where there is a good theoretical basis for including some but not all of your dependent variables, you should run separate analyses: one for the variables being tested on a heuristic basis and one for the theoretically meaningful variables. The point to take on board here is not to include lots of dependent variables in a MANOVA just because you have measured them.

14.2.2. Current controversies ②

Like ANOVA, MANOVA is a two-stage test in which an overall (or omnibus) test is first performed before more specific procedures are applied to tease apart group differences. Unfortunately, there are debates regarding both stages of the MANOVA procedure. For the main analysis there are four commonly used ways of assessing the overall significance of a MANOVA and debate exists about which method is best in terms of power and sample size considerations. In addition, there is substantial debate over how best to analyse and interpret group differences further when the overall MANOVA is significant. There are two main approaches to follow-up analysis: univariate ANOVAs and discriminant analysis. I will consider both.

Another controversy surrounds the power of MANOVA (i.e. the ability of MANOVA to detect an effect that genuinely exists). I mentioned in the previous section that MANOVA had greater power than ANOVA to detect effects because it could take account of the correlations between dependent variables. However, the issue of power is more complex than alluded to by my simple statement. Ramsey (1982) found that as the correlation between dependent variables increased, the power of MANOVA decreased. This led Tabachnick & Fidell (2001) to recommend that MANOVA 'works best with highly negatively correlated DVs, and acceptably well with moderately correlated DVs in either direction' and that 'MANOVA is also wasteful when DVs are uncorrelated' (p. 357). In contrast, Stevens's (1980) investigation of the effect of dependent variable correlations on test power revealed that 'the power with high intercorrelations is in most cases greater than that for moderate intercorrelations, and in some cases it is dramatically higher' (p. 736). These findings are slightly contradictory, which leaves us with the puzzling conundrum of what, exactly, is the relationship between power and intercorrelation of the dependent variables? Luckily, Cole et al. (1994) have done a great deal to illuminate this relationship. They found that the power of MANOVA depends on a combination of the correlation between dependent variables and the *effect size*. In short, if you are expecting to find a large effect, then MANOVA will have greater power if the measures are somewhat different (even negatively correlated) and if the group differences are in the same direction for each measure. If you have two dependent variables, one of which exhibits a large group difference, and one of which exhibits a small, or no, group difference then power will be increased if these variables are highly correlated. The take-home message from Cole et al.'s work is that if you are interested in how powerful the MANOVA is likely to be you should consider not just the intercorrelation of dependent variables but also the size and pattern of group differences that you expect to get. However, it should be noted that Cole et al.'s work is limited to the case of where two groups are being compared and power considerations are more complex in multiple-group situations.

14.2.3. The example for this chapter ②

Throughout the rest of this chapter we're going to use a single example to look at how MANOVA works and then how to conduct one on SPSS. Imagine that we were interested in the effects of cognitive behaviour therapy on obsessive compulsive disorder (OCD). OCD is a disorder characterized by intrusive images or thoughts that the sufferer finds abhorrent (in my case this might be the thought of someone carrying out a *t*-test on data that are not normally distributed, or imagining your parents have died). These thoughts lead the sufferer to engage in activities to neutralize the unpleasantness of these thoughts (these activities can be mental, such as doing a MANOVA in my head to make me feel better about the *t*-test thought, or physical, such as touching the floor 23 times so that your parents won't die). Now, we could compare a group of OCD sufferers after cognitive behaviour therapy (CBT) and after behaviour therapy (BT) with a group of OCD sufferers who are still awaiting treatment (a no-treatment condition, NT).[1] Now, most psychopathologies have both behavioural and cognitive elements to them. For example, in OCD if someone had an obsession with germs and contamination, this disorder might manifest itself in obsessive hand-washing and would influence not just how many times they actually wash their hands (behaviour) but also the number of times they think about washing their hands (cognitions). Similarly, someone with an obsession about bags won't just think about bags a lot, but they might carry out bag-related behaviours (such as saying 'bag' repeatedly, or buying lots of bags). If we are interested in seeing how successful a therapy is, it is not enough to look only at behavioural outcomes (such as whether obsessive behaviours are reduced); it is important to establish whether cognitions are being changed also. Hence, in this example two dependent measures were taken: the occurrence of obsession-related behaviours (**actions**) and the occurrence of obsession-related cognitions (**thoughts**). These dependent variables were measured on a single day and so represent the number of obsession-related behaviours/thoughts in a normal day.

The data are in Table 14.1 and can be found in the file **OCD.sav**. Participants belonged to group 1 (CBT), group 2 (BT) or group 3 (NT) and within these groups all participants had both actions and thoughts measured.

---

1 The non-psychologists out there should note that behaviour therapy works on the basis that if you stop the maladaptive behaviours the disorder will go away, whereas cognitive therapy is based on the idea that treating the maladaptive cognitions will stop the disorder. If you're interested in knowing more then I shamelessly recommend my book on clinical psychology (Field, 2003), which is available from the 'crap books that we can't sell' section of any bookshop.

Table 14.1 Data from **OCD.sav**

	DV 1: Actions			DV 2: Thoughts		
Group:	CBT (1)	BT (2)	NT (3)	CBT (1)	BT (2)	NT (3)
	5	4	4	14	14	13
	5	4	5	11	15	15
	4	1	5	16	13	14
	4	1	4	13	14	14
	5	4	6	12	15	13
	3	6	4	14	19	20
	7	5	7	12	13	13
	6	5	4	15	18	16
	6	2	6	16	14	14
	4	5	5	11	17	18
$\bar{X}$	4.90	3.70	5.00	13.40	15.20	15.00
s	1.20	1.77	1.05	1.90	2.10	2.36
s^2	1.43	3.12	1.11	3.60	4.40	5.56
		$\bar{X}_{\text{grand (Actions)}} = 4.53$			$\bar{X}_{\text{grand(Thoughts)}} = 14.53$	
		$S^2_{\text{grand(Actions)}} = 2.1195$			$S^2_{\text{grand(Thoughts)}} = 4.8780$	

14.3. THEORY OF MANOVA ③

The theory of MANOVA is very complex to understand without knowing matrix algebra, and, frankly, matrix algebra is way beyond the scope of this book (those with maths brains can consult Namboodiri, 1984; or Stevens, 1992). However, I intend to give a flavour of the conceptual basis of MANOVA, using matrices, without requiring you to understand exactly how those matrices are used. Those interested in the exact underlying theory of MANOVA should read Bray & Maxwell's (1985) superb monograph on the subject.

14.3.1. Introduction to matrices ③

First, I should explain what a matrix is: a matrix is simply a collection of numbers arranged in columns and rows. In fact, throughout this book you have been using a matrix without even realizing it: the SPSS data editor. In the SPSS data editor we have numbers arranged in columns and rows and this is exactly what a matrix is. A matrix can have many columns and many rows and we usually specify the dimensions of the matrix using numbers. So, a 2 × 3 matrix is a matrix with two rows and three columns, and a 5 × 4 matrix is one with five rows and four columns. For example:

$$\begin{pmatrix} 2 & 5 & 6 \\ 3 & 5 & 8 \end{pmatrix} \qquad \begin{pmatrix} 2 & 4 & 6 & 8 \\ 3 & 4 & 6 & 7 \\ 4 & 3 & 5 & 8 \\ 2 & 5 & 7 & 9 \\ 4 & 6 & 6 & 9 \end{pmatrix}$$

2×3 matrix $\qquad$ 5×4 matrix

You can think of these matrices in terms of each row representing the data from a single participant and each column as representing data relating to a particular variable. So, for the 5×4 matrix we can imagine a situation where five participants were tested on four variables: so, the first participant scored 2 on the first variable and 8 on the fourth variable. The values within a matrix are typically referred to as *components* or *elements*.

A square matrix is one in which there are an equal number of columns and rows. In this type of matrix it is sometimes useful to distinguish between the diagonal components (i.e. the values that lie on the diagonal line from the top left component to the bottom right component) and the off-diagonal components (the values that do not lie on the diagonal). In the matrix below, the diagonal components are 5, 12, 2 and 6 because they lie along the diagonal line. The off-diagonal components are all of the other values. A square matrix in which the diagonal elements are equal to 1 and the off-diagonal elements are equal to 0 is known as an identity matrix:

$$\begin{pmatrix} 5 & 3 & 6 & 10 \\ 3 & 12 & 4 & 6 \\ 6 & 4 & 2 & 7 \\ 10 & 6 & 7 & 6 \end{pmatrix} \qquad \begin{pmatrix} 1 & 0 & 0 \\ 0 & 1 & 0 \\ 0 & 0 & 1 \end{pmatrix}$$

square matrix $\qquad$ identity matrix

Hopefully, the concept of a matrix should now be slightly less scary than it was previously: it is not some magical mathematical entity, merely a way of representing a data set—just like a spreadsheet.

Now, there is a special case of a matrix where there are data from only one person, and this is known as a *row vector*. Likewise, if there is only one column in a matrix this is known as a *column vector*. In the examples below, the row vector can be thought of as a single person's score on four different variables, whereas the column vector can be thought of as five participants' scores on one variable:

$$\begin{pmatrix} 2 & 6 & 4 & 8 \end{pmatrix} \qquad \begin{pmatrix} 8 \\ 6 \\ 10 \\ 15 \\ 6 \end{pmatrix}$$

row vector $\qquad$ column vector

Armed with this knowledge of what vectors are, we can have a brief look at how they are used to conduct a MANOVA.

14.3.2. Some important matrices and their functions ③

As with ANOVA, we are primarily interested in how much variance can be explained by the experimental manipulation (which in real terms means how much variance is explained by the fact that certain scores appear in certain groups). Therefore, we need to know the sum of squares due to the grouping variable (the systematic variation, SS_M), the sum of squares due to natural differences between participants (the residual variation, SS_R) and of course the total amount of variation that needs to be explained (SS_T): for more details about these sources of variation reread Chapters 5 and 8. However, I mentioned that MANOVA also takes into account several dependent variables simultaneously and it does this by using a matrix that contains information about the variance accounted for by each dependent variable.

For the univariate *F*-test (e.g. ANOVA) we calculated the ratio of systematic variance to unsystematic variance for a single dependent variable. In MANOVA the test statistic is derived by comparing the ratio of systematic to unsystematic variance for several dependent variables. This comparison is made by using the ratio of a matrix representing the systematic variance of all dependent variables to a matrix representing the unsystematic variance of all dependent variables. The matrix that represents the systematic variance (or the model sum of squares for all variables) is denoted by the letter H and is called the hypothesis sum of squares and cross-products matrix (or hypothesis SSCP). The matrix that represents the unsystematic variance (the residual sums of squares for all variables) is denoted by the letter E and is called the error sum of squares and cross-products matrix (or *error SSCP*). Finally, there is a matrix that represents the total amount of variance present for each dependent variable (the total sums of squares for each dependent variable) and this is denoted by T and is called the total sum of squares and cross-products matrix (or total SSCP).

Later, I will show how these matrices are used in exactly the same way as the simple sums of squares (SS_M, SS_R and SS_T) in ANOVA to derive a test statistic representing the ratio of systematic to unsystematic variance in the model. The observant amongst you may have noticed that the matrices I have described are all called sum of squares and cross-products (SSCP) matrices. It should be obvious why these matrices are referred to as sum of squares matrices, but why is there a reference to cross-products in their name? We came across cross-product deviations in Chapter 4 and saw that they represented a total value for the combined error between two variables (so, in some sense they represented an unstandardized estimate of the total correlation between two variables). As such, whereas the sum of squares of a variable is the total squared difference between the observed values and the mean value, the cross-product is the total combined error between two variables. I mentioned earlier that MANOVA had the power to account for any correlation between dependent variables and it does this by using these cross-products.

14.3.3. Calculating MANOVA by hand: a worked example ③

To begin with let's carry out univariate ANOVAs on each of the two dependent variables in our OCD example (see Table 14.1). A description of the ANOVA model can be found in Chapter 8 and I will draw heavily on the assumption that you have read this chapter; if you are hazy on the details of Chapter 8 then now would be a good time to (re)read sections 8.2.4 to 8.2.8.

14.3.3.1. Univariate ANOVA for DV 1 (actions) ②

There are three sums of squares that need to be calculated. First we need to assess how much variability there is to be explained within the data (SS_T), next we need to see how much of this variability can be explained by the model (SS_M), and finally we have to assess how much error there is in the model (SS_R). From Chapter 8 we can calculate each of these values:

- $\mathbf{SS_{T(Actions)}}$: The total sum of squares is obtained by calculating the difference between each of the 20 scores and the mean of those scores, then squaring these differences and adding these squared values up. Alternatively, you can get SPSS to calculate the variance for the action data (regardless of which group the score falls into) and then multiply this value by the number of scores minus 1:

$$\begin{aligned} SS_T &= s^2_{\text{grand}}(n-1) \\ &= 2.1195(30-1) \\ &= 2.1195 \times 29 \\ &= 61.47 \end{aligned}$$

- $\mathbf{SS_{M(Actions)}}$: This value is calculated by taking the difference between each group mean and the grand mean and then squaring them. Multiply these values by the number of scores in the group and then add them together:

$$\begin{aligned} SS_M &= 10(4.90-4.53)^2 + 10(3.70-4.53)^2 + 10(5.00-4.53)^2 \\ &= 10(0.37)^2 + 10(-0.83)^2 + 10(0.47)^2 \\ &= 1.37 + 6.89 + 2.21 \\ &= 10.47 \end{aligned}$$

- $\mathbf{SS_{R(Actions)}}$: This value is calculated by taking the difference between each score and the mean of the group from which it came. These differences are then squared and then added together. Alternatively we can get SPSS to calculate the variance within each group, multiply each group variance by the number of scores minus one and then add them together:

$$
\begin{aligned}
SS_R &= s^2_{CBT}(n_{CBT} - 1) + s^2_{BT}(n_{BT} - 1) + s^2_{NT}(n_{NT} - 1) \\
&= (1.433)(10 - 1) + (3.122)(10 - 1) + (1.111)(10 - 1) \\
&= (1.433 \times 9) + (3.122 \times 9) + (1.111 \times 9) \\
&= 12.9 + 28.1 + 10.0 \\
&= 51.00
\end{aligned}
$$

The next step is to calculate the average sums of squares (the mean square) of each by dividing by the degrees of freedom (see section 8.2.7):

SS	*df*	MS
$SS_{M(Actions)} = 10.47$	2	5.235
$SS_{R(Actions)} = 51.00$	27	1.889

The final stage is to calculate *F* by dividing the mean squares for the model by the mean squares for the error in the model:

$$F = \frac{MS_M}{MS_R} = \frac{5.235}{1.889} = 2.771$$

This value can then be evaluated against critical values of *F*. The point to note here is the calculation of the various sums of squares and what each one relates to.

14.3.3.2. Univariate ANOVA for DV 2 (thoughts) ②

As with the data for dependent variable 1 there are three sums of squares that need to be calculated as before:

$SS_{T\ (Thoughts)}$:

$$
\begin{aligned}
SS_T &= s^2_{grand}(n - 1) \\
&= 4.878(30 - 1) \\
&= 4.878 \times 29 \\
&= 141.46
\end{aligned}
$$

$SS_{M\ (Thoughts)}$:

$$
\begin{aligned}
SS_M &= 10(13.40 - 14.53)^2 + 10(15.2 - 14.53)^2 + 10(15.0 - 14.53)^2 \\
&= 10(-1.13)^2 + 10(0.67)^2 + 10(0.47)^2 \\
&= 12.77 + 4.49 + 2.21 \\
&= 19.47
\end{aligned}
$$

$\mathbf{SS}_{\mathbf{R\ (Thoughts)}}$:

$$\begin{aligned}\mathrm{SS_R} &= s^2_{\mathrm{CBT}}(n_{\mathrm{CBT}} - 1) + s^2_{\mathrm{BT}}(n_{\mathrm{BT}} - 1) + s^2_{\mathrm{NT}}(n_{\mathrm{NT}} - 1)\\ &= (3.6)(10 - 1) + (4.4)(10 - 1) + (5.56)(10 - 1)\\ &= (3.6 \times 9) + (4.4 \times 9) + (5.56 \times 9)\\ &= 32.4 + 39.6 + 50.0\\ &= 122\end{aligned}$$

The next step is to calculate the average sums of squares (the mean square) of each by dividing by the degrees of freedom (see section 8.2.7):

SS	*df*	MS
$SS_{M(Thoughts)} = 19.47$	2	9.735
$SS_{R(Thoughts)} = 122.00$	27	4.519

The final stage is to calculate *F* by dividing the mean squares for the model by the mean squares for the error in the model:

$$F = \frac{\mathrm{MS_M}}{\mathrm{MS_R}} = \frac{9.735}{4.519} = 2.154$$

This value can then be evaluated against critical values of *F*. Again, the point to note here is the calculation of the various sums of squares and what each one relates to.

14.3.3.3. The relationship between DVs: cross-products ②

We know already that MANOVA uses the same sums of squares as an ANOVA, and in the next section we shall see exactly how it uses these values. However, I have also mentioned that MANOVA takes account of the relationship between dependent variables and that it does this by using the cross-products. To be precise, there are three different cross-products that are of interest and these three cross-products relate to the three sums of squares that we calculated for the univariate ANOVAs: that is, there is a total cross-product, a cross-product due to the model, and a residual cross-product. Let's look at the total cross-product ($\mathrm{CP_T}$) first.

I mentioned in Chapter 4 that the cross-product was the difference between the scores and the mean in one group multiplied by the difference between the scores and the mean in the other group. In the case of the total cross-product, the mean of interest is the grand mean for each dependent variable (see Table 14.2). Hence, we can adapt the cross-product equation described in Chapter 4 using the two dependent variables. The resulting equation for the total cross-product is described as in equation (14.1). Therefore, for each dependent variable you take each score and subtract from it the grand mean for that variable. This leaves you with two values per participant (one for each dependent variable) which should

Table 14.2 Calculation of the total cross-product

Group	Actions	Thoughts	Actions $-\bar{X}_{grand(Actions)}$ (D_1)	Thoughts $-\bar{X}_{grand(Thoughts)}$ (D_2)	$D_1 \times D_2$
	5	14	0.47	−0.53	−0.25
	5	11	0.47	−3.53	−1.66
	4	16	−0.53	1.47	−0.78
	4	13	−0.53	−1.53	0.81
CBT	5	12	0.47	−2.53	−1.19
	3	14	−1.53	−0.53	0.81
	7	12	2.47	−2.53	−6.25
	6	15	1.47	0.47	0.69
	6	16	1.47	1.47	2.16
	4	11	−0.53	−3.53	1.87
	4	14	−0.53	−0.53	0.28
	4	15	−0.53	0.47	−0.25
	1	13	−3.53	−1.53	5.40
	1	14	−3.53	−0.53	1.87
BT	4	15	−0.53	0.47	−0.25
	6	19	1.47	4.47	6.57
	5	13	0.47	−1.53	−0.72
	5	18	0.47	3.47	1.63
	2	14	−2.53	−0.53	1.34
	5	17	0.47	2.47	1.16
	4	13	−0.53	−1.53	0.81
	5	15	0.47	0.47	0.22
	5	14	0.47	−0.53	−0.25
	4	14	−0.53	−0.53	0.28
NT	6	13	1.47	−1.53	−2.25
	4	20	−0.53	5.47	−2.90
	7	13	2.47	−1.53	−3.78
	4	16	−0.53	1.47	−0.78
	6	14	1.47	−0.53	−0.78
	5	18	0.47	3.47	1.63
$\bar{X}_{grand}$	4.53	14.53		$CP_T = \Sigma(D_1 \times D_2) = 5.47$	

Table 14.3 Calculating the model cross-product

	$\bar{X}_{group}$ Actions	$\bar{X}_{group}-\bar{X}_{grand}$ (D_1)	$\bar{X}_{group}$ Thoughts	$\bar{X}_{group}-\bar{X}_{grand}$ (D_2)	$D_1 \times D_2$	$N(D_1 \times D_2)$
CBT	4.9	0.37	13.4	−1.13	−0.418	−4.18
BT	3.7	−0.83	15.2	0.67	−0.556	−5.56
NT	5.0	0.47	15.0	0.47	0.221	2.21
$\bar{X}_{grand}$	**4.53**		**14.53**			$CP_M = \Sigma N(D_1 \times D_2) = -7.53$

be multiplied together to get the cross-product for each participant. The total can then be found by adding the cross-products of all participants. Table 14.2 illustrates this process.

$$\mathrm{CP_T} = \sum (x_{i(\text{Actions})} - \overline{X}_{\text{grand(Actions)}})(x_{i(\text{Thoughts})} - \overline{X}_{\text{grand(Thoughts)}}) \tag{14.1}$$

The total cross-product is a gauge of the overall relationship between the two variables. However, we are also interested in how the relationship between the dependent variables is influenced by our experimental manipulation and this relationship is measured by the model cross-product ($\mathrm{CP_M}$). The $\mathrm{CP_M}$ is calculated in a similar way to the model sum of squares. First, the difference between each group mean and the grand mean is calculated for each dependent variable. The cross-product is calculated by multiplying the differences found for each group. Each product is then multiplied by the number of scores within the group (as was done with the sum of squares). This principle is illustrated in equation (14.2) and Table 14.3:

$$\mathrm{CP_M} = \sum n[(\bar{x}_{\text{group(Actions)}} - \overline{X}_{\text{grand(Actions)}})(\bar{x}_{\text{group(Thoughts)}} - \overline{X}_{\text{grand(Thoughts)}})] \tag{14.2}$$

Finally, we also need to know how the relationship between the two dependent variables is influenced by individual differences in participants' performances. The residual cross-product ($\mathrm{CP_R}$) tells us about how the relationship between the dependent variables is affected by individual differences, or error in the model. The $\mathrm{CP_R}$ is calculated in a similar way to the total cross-product except that the group means are used rather than the grand mean (see equation (14.3)). So, to calculate each of the difference scores, we take each score and subtract from it the mean of the group to which it belongs (see Table 14.4):

$$\mathrm{CP_R} = \sum (x_{i(\text{Actions})} - \overline{X}_{\text{group(Actions)}})(x_{i(\text{Thoughts})} - \overline{X}_{\text{group(Thoughts)}}) \tag{14.3}$$

The observant among you may notice that the residual cross-product can also be calculated by subtracting the model cross-product from the total cross-product:

$$\begin{aligned}\mathrm{CP_R} &= \mathrm{CP_T} - \mathrm{CP_M} \\ &= 5.47 - (-7.53) = 13\end{aligned}$$

Table 14.4 Calculation of CP_R

Group	Actions	Actions $-\bar{X}_{group(Actions)}$ (D_1)	Thoughts	Thoughts $-\bar{X}_{group(Thoughts)}$ (D_2)	$D_1 \times D_2$
	5	0.10	14	0.60	0.06
	5	0.10	11	−2.40	−0.24
	4	−0.90	16	2.60	−2.34
	4	−0.90	13	−0.40	0.36
CBT	5	0.10	12	−1.40	−0.14
	3	−1.90	14	0.60	−1.14
	7	2.10	12	−1.40	−2.94
	6	1.10	15	1.60	1.76
	6	1.10	16	2.60	2.86
	4	−0.90	11	−2.40	2.16
$\bar{X}_{CBT}$	**4.9**		**13.4**		$\Sigma =$ **0.40**
	4	0.30	14	−1.20	−0.36
	4	0.30	15	−0.20	−0.06
	1	−2.70	13	−2.20	5.94
	1	−2.70	14	−1.20	3.24
BT	4	0.30	15	−0.20	−0.06
	6	2.30	19	3.80	8.74
	5	1.30	13	−2.20	−2.86
	5	1.30	18	2.80	3.64
	2	−1.70	14	−1.20	2.04
	5	1.30	17	1.80	2.34
$\bar{X}_{BT}$	**3.7**		**15.2**		$\Sigma =$ **22.60**
	4	−1.00	13	−2.00	2.00
	5	0.00	15	0.00	0.00
	5	0.00	14	−1.00	0.00
	4	−1.00	14	−0.00	1.00
NT	6	1.00	13	−2.00	−2.00
	4	−1.00	20	5.00	−5.00
	7	2.00	13	−2.00	−4.00
	4	−1.00	16	1.00	−1.00
	6	1.00	14	−1.00	−1.00
	5	0.00	18	3.00	0.00
$\bar{X}_{NT}$	5		**15**		$\Sigma =$ **−10.00**
				$CP_R = \Sigma(D_1 \times D_2) = 13$	

However, it is useful to calculate the residual cross-product manually in case of mistakes in the calculation of the other two cross-products. The fact that the residual and model cross-products should sum to the value of the total cross-product can be used as a useful double-check.

Each of the different cross-products tells us something important about the relationship between the two dependent variables. Although I have used a simple scenario to keep the maths relatively simple, these principles can be easily extended to more complex scenarios. For example, if we had measured three dependent variables then the cross-products between pairs of dependent variables are calculated (as they were in this example) and entered into the appropriate SSCP matrix (see next section). As the complexity of the situation increases, so does the amount of calculation that needs to be done. At times such as these the benefit of software like SPSS becomes ever more apparent!

14.3.3.4. The total SSCP matrix (*T*) ③

In this example we have only two dependent variables and so all of the SSCP matrices will be 2 × 2 matrices. If there had been three dependent variables then the resulting matrices would all be 3 × 3 matrices. The total SSCP matrix, *T*, contains the total sums of squares for each dependent variable and the total cross-product between the two dependent variables. You can think of the first column and first row as representing one dependent variable and the second column and row as representing the second dependent variable:

	Column 1 Actions	**Column 2 Thoughts**
Row 1 Actions	$SS_{T(Actions)}$	CP_T
Row 1 Thoughts	CP_T	$SS_{T(Thoughts)}$

We calculated these values in the previous sections and so we can simply place the appropriate values in the appropriate cell of the matrix:

$$T = \begin{pmatrix} 61.47 & 5.47 \\ 5.47 & 141.47 \end{pmatrix}$$

From the values in the matrix (and what they represent) it should be clear that the total SSCP represents both the total amount of variation that exists within the data and the total co-dependence that exists between the dependent variables. You should also note that the off-diagonal components are the same (they are both the total cross-product) because this value is equally important for both of the dependent variables.

14.3.3.5. The residual SSCP matrix (*E*) ③

The residual (or error) SSCP matrix, *E*, contains the residual sums of squares for each dependent variable and the residual cross-product between the two dependent variables. This SSCP matrix is similar to the total SSCP except that the information relates to the error in the model:

	Column 1 Actions	Column 2 Thoughts
Row 1 Actions	$SS_{R(Actions)}$	CP_R
Row 1 Thoughts	CP_R	$SS_{R(Thoughts)}$

We calculated these values in the previous sections and so we can simply place the appropriate values in the appropriate cell of the matrix:

$$E = \begin{pmatrix} 51 & 13 \\ 13 & 122 \end{pmatrix}$$

From the values in the matrix (and what they represent) it should be clear that the residual SSCP represents both the unsystematic variation that exists for each dependent variable and the co-dependence between the dependent variables that is due to chance factors alone. As before the off-diagonal elements are the same (they are both the residual cross-product).

14.3.3.6. The model SSCP matrix (*H*) ③

The model (or hypothesis) SSCP matrix, *H*, contains the model sums of squares for each dependent variable and the model cross-product between the two dependent variables:

	Column 1 Actions	Column 2 Thoughts
Row 1 Actions	$SS_{M(Actions)}$	CP_M
Row 1 Thoughts	CP_M	$SS_{M(Thoughts)}$

These values were calculated in the previous sections and so we can simply place the appropriate values in the appropriate cell of the matrix (see below). From the values in the matrix (and what they represent) it should be clear that the model SSCP represents both the systematic variation that exists for each dependent variable and the co-dependence between the dependent variables that is due to the model (i.e. is due to the experimental manipulation). As before the off-diagonal components are the same (they are both the model cross-product):

$$H = \begin{pmatrix} 10.47 & -7.53 \\ -7.53 & 19.47 \end{pmatrix}$$

Matrices are additive, which means that you can add (or subtract) two matrices together by adding (or subtracting) corresponding components. Now, when we calculated univariate ANOVA we saw that the total sum of squares was the sum of the model sum of squares and the residual sum of squares (i.e. $SS_T = SS_M + SS_R$). The same is true in MANOVA except that we are adding matrices rather than single values:

$$\begin{aligned} T &= H + E \\ T &= \begin{pmatrix} 10.47 & -7.53 \\ -7.53 & 19.47 \end{pmatrix} + \begin{pmatrix} 51 & 13 \\ 13 & 122 \end{pmatrix} \\ &= \begin{pmatrix} 10.47 + 51 & -7.53 + 13 \\ -7.53 + 13 & 19.47 + 122 \end{pmatrix} \\ &= \begin{pmatrix} 61.47 & 5.47 \\ 5.47 & 141.47 \end{pmatrix} \end{aligned}$$

The demonstration that these matrices add up should (hopefully) help you to understand that the MANOVA calculations are conceptually the same as for univariate ANOVA, the difference being that matrices are used rather than single values.

14.3.4. Principle of the MANOVA test statistic ④

In univariate ANOVA we calculate the ratio of the systematic variance to the unsystematic variance (i.e. we divide SS_M by SS_R).[2] The conceptual equivalent would therefore be to divide the matrix *H* by the matrix *E*. There is, however, a problem in that matrices are not divisible by other matrices! However, there is a matrix equivalent to division, which is to multiply by what's known as the inverse of a matrix. So, if we want to divide *H* by *E* we

2 In reality we use the mean squares but these values are merely the sums of squares corrected for the degrees of freedom.

have to multiply H by the inverse of E (denoted as E^{-1}). So, therefore, the test statistic is based upon the matrix that results from multiplying the model SSCP with the inverse of the residual SSCP. This matrix is called HE^{-1}.

Calculating the inverse of a matrix is incredibly difficult and there is no need for you to understand how it is done because SPSS will do it for you. However, the interested reader should consult either Stevens (1992) or Namboodiri (1984)—these texts provide very accessible accounts of how to derive an inverse matrix. For readers who do consult these sources, the calculations for this example are included in a file on the CD-ROM called **Appendix Chapter 14.pdf** (you might like to check my maths!). For the uninterested reader, you'll have to trust me on the following:

$$E^{-1} = \begin{pmatrix} 0.0202 & -0.0021 \\ -0.0021 & 0.0084 \end{pmatrix}$$

$$HE^{-1} = \begin{pmatrix} 0.2273 & -0.0852 \\ -0.1930 & 0.1794 \end{pmatrix}$$

Remember that $\mathbf{HE^{-1}}$ represents the ratio of systematic variance in the model to the unsystematic variance in the model and so the resulting matrix is conceptually the same as the F-ratio in univariate ANOVA. There is another problem, though. In ANOVA, when we divide the systematic variance by the unsystematic variance we get a single figure: the F-ratio. In MANOVA, when we divide the systematic variance by the unsystematic variance we get a matrix containing several values. In this example, the matrix contains four values, but had there been three dependent variables the matrix would have had nine values. In fact, the resulting matrix will always contain p^2-values, where p is the number of dependent variables. The problem is how to convert these matrix values into a meaningful single value. This is the point at which we have to abandon any hope of understanding the maths behind the test and talk conceptually instead.

14.3.4.1. Discriminant function variates ④

The problem of having several values with which to assess statistical significance can be simplified considerably by converting the dependent variables into underlying dimensions or factors (this process will be discussed in more detail in Chapter 15). In Chapter 5, we saw how multiple regression worked on the principle of fitting a linear model to a set of data to predict an outcome variable (the dependent variable in ANOVA terminology). This linear model was made up of a combination of predictor variables (or independent variables) each of which had a unique contribution to this linear model. We can do a similar thing here, except that we are interested in the opposite problem (i.e. predicting an independent variable from a set of dependent variables). So, it is possible to calculate underlying linear dimensions of the dependent variables. These linear combinations of the dependent variables are known as *variates* (or sometimes called *latent variables* or *factors*). In this context we wish to use these linear variates to predict which group a person belongs to (i.e. whether they were given CBT, BT or no treatment), so we are using them to discriminate groups of people. Therefore, these variates are called *discriminant functions* or discriminant function variates. Although I have drawn a parallel between these discriminant functions and the model in multiple regression, there is a difference in that we can extract several discriminant functions from a set of dependent variables, whereas in multiple regression all independent variables are included in a single model.

That's the theory in simplistic terms, but how do we discover these discriminant functions? Well, without going into too much detail, we use a mathematical procedure of maximization, such that the first discriminant function (V_1) is the linear combination of dependent variables that maximizes the differences between groups.

It follows from this that the ratio of systematic to unsystematic variance (SS_M/SS_R) will be maximized for this first variate, but subsequent variates will have smaller values of this ratio. Remember that this ratio is an analogue of what the *F*-ratio represents in univariate ANOVA, and so in effect we obtain the maximum possible value of the *F*-ratio when we look at the first discriminant function. This variate can be described in terms of a linear regression equation (because it is a linear combination of the dependent variables):

$$
\begin{aligned}
Y &= b_0 + b_1 X_1 + b_2 X_2 \\
V_1 &= b_0 + b_1 \mathrm{DV}_1 + b_2 \mathrm{DV}_2 \qquad (14.4) \\
V_1 &= b_0 + b_1 \mathrm{Actions} + b_2 \mathrm{Thoughts}
\end{aligned}
$$

Equation (14.4) shows the multiple regression equation for two predictors and then extends this to show how a comparable form of this equation can describe discriminant functions. The *b*-values in the equation are weights (just as in regression) that tell us something about the contribution of each dependent variable to the variate in question. In regression, the values of *b* are obtained by the method of least squares; in discriminant function analysis the *b*-values are obtained from the *eigenvectors* (see Box 5.2) of the matrix HE^{-1}. We can actually ignore b_0 as well because this serves only to locate the variate in geometric space, which isn't necessary when we're using it to discriminate groups.

In a situation in which there are only two dependent variables and two groups for the independent variable, there will be only one variate. This makes the scenario very simple: by looking at the discriminant function of the dependent variables, rather than looking at the dependent variables themselves, we can obtain a single value of SS_M/SS_R for the discriminant function, and then assess this value for significance. However, in more complex cases where there are more than two dependent variables or more than three levels of the independent variable (as is the case in our example), there will be more than one variate. The number of variates obtained will be the smaller of *p* (the number of dependent variables) or $k-1$ (where *k* is the number of levels of the independent variable). In our example, both *p* and $k-1$ are 2, so we should be able to find two variates. I mentioned earlier that the *b*-values that describe the variates are obtained by calculating the eigenvectors of the matrix HE^{-1} and, in fact, there will be two eigenvectors derived from this matrix: one with the b-values for the first variate, and one with the b of the second variate. Conceptually speaking, eigenvectors are the vectors associated with a given matrix that are unchanged by transformation of that matrix to a diagonal matrix (look back to Box 5.2 for a visual explanation of eigenvectors and eigenvalues). A diagonal matrix is simply a matrix in which the off-diagonal elements are zero and by changing HE^{-1} to a diagonal matrix we eliminate all of the off-diagonal elements (thus reducing the number of values that we must consider for significance testing). Therefore, by calculating the eigenvectors and eigenvalues, we still end up with values that represent the ratio of systematic to unsystematic variance (because they are unchanged by the transformation), but there are considerably less of them. The calculation of eigenvectors is extremely complex (insane students can consider reading Strang, 1980, or Namboodiri, 1984), so you can trust me that for the matrix HE^{-1} the eigenvectors obtained are:

$$\text{eigenvector}_1 = \begin{pmatrix} 0.603 \\ -0.335 \end{pmatrix}$$

$$\text{eigenvector}_2 = \begin{pmatrix} 0.425 \\ 0.339 \end{pmatrix}$$

Replacing these values into the two equations for the variates and bearing in mind we can ignore b_0, we obtain the models described in equation (14.5)

$$\begin{aligned} V_1 &= 0.603\,\text{Actions} - 0.335\,\text{Thoughts} \\ V_2 &= 0.425\,\text{Actions} + 0.339\,\text{Thoughts} \end{aligned} \tag{14.5}$$

It is possible to use the equations for each variate to calculate a score for each person on the variate. For example, the first participant in the CBT group carried out 5 obsessive actions, and had 14 obsessive thoughts. Therefore, their score on variate 1 would be −1.675:

$$V_1 = (0.603 \times 5) + (0.335 \times 14) = -1.675$$

The score for variate 2 would be 6.87:

$$V_2 = (0.425 \times 5) + (0.339 \times 14) = 6.871$$

If we calculated these variate scores for each participant and then calculated the SSCP matrices (e.g. H, E, T and HE^{-1}) that we used previously, we would find that all of them have cross-products of zero. The reason for this is because the variates extracted from the data are orthogonal, which means that they are uncorrelated. In short, the variates extracted are independent dimensions constructed from a linear combination of the dependent variables that were measured.

This data reduction has a very useful property in that if we look at the matrix HE^{-1} calculated from the variate scores (rather than the dependent variables) we find that all of the off-diagonal elements (the cross-products) are zero. The diagonal elements of this matrix represent the ratio of the systematic variance to the unsystematic variance (i.e. SS_M/SS_R) for each of the underlying variates. So, for the data in this example, this means that instead of having four values representing the ratio of systematic to unsystematic variance, we now have only two. This reduction may not seem a lot. However, in general if we have p dependent variables, then ordinarily we would end up with p^2-values representing the ratio of systematic to unsystematic variance; by looking at discriminant functions, we reduce this number to p. If there were four dependent variables we would end up with four values rather than 16 (which highlights the benefit of this process).

For the data in our example, the matrix HE^{-1} calculated from the variate scores is:

$$HE^{-1}_{\text{variates}} = \begin{pmatrix} 0.335 & 0.000 \\ 0.000 & 0.073 \end{pmatrix}$$

It is clear from this matrix that we have two values to consider when assessing the significance of the group differences. It probably seems like a complex procedure to reduce the data down in this way: however, it transpires that the values along the diagonal of the matrix for the variates (namely, 0.335 and 0.073) are the *eigenvalues* of the original HE^{-1} matrix. Therefore, these values can be calculates directly from the data collected without first forming the eigenvectors. The calculation of these eigenvalues is included in the appendices but it is not necessary that you understand how they are derived. These eigenvalues are conceptually equivalent to the F-ratio in ANOVA and so the final step is to assess how large these values are compared to what we would expect by chance alone. There are four ways in which the values are assessed.

14.3.4.2. Pillai–Bartlett trace (*V*) ④

The Pillai–Bartlett trace (also known as Pillai's trace) is given by equation (14.6) in which λ represents the eigenvalues for each of the discriminant variates, and s represents the number of variates. Pillai's trace is the sum of the proportion of explained variance on the discriminant functions. As such, it is similar to the ratio of SS_M/SS_T:

$$V = \sum_{i=1}^{s} \frac{\lambda_i}{1 + \lambda_i} \tag{14.6}$$

For our data, this trace turns out to be 0.319, which can be transformed to a value that has an approximate F-distribution.

$$V = \frac{0.335}{1 + 0.335} + \frac{0.073}{1 + 0.073} = 0.319$$

Figure 14.1 Harold Hotelling enjoying my favourite activity of drinking tea

14.3.4.3. Hotelling's T^2 ④

The *Hotelling–Lawley trace* or Hotelling's T^2 (Figure 14.1) is simply the sum of the eigenvalues for each variate (see equation (14.7)) and so for these data its value is 0.408 (0.335 + 0.073). This test statistic is the sum of SS_M/SS_R for each of the variates and so it compares directly to the *F*-ratio in ANOVA:

$$T = \sum_{i=1}^{s} \lambda_i \tag{14.7}$$

14.3.4.4. Wilks's lambda (Λ) ④

Wilks's lambda (Λ) is the product of the *unexplained* variance on each of the variates (see equation (14.8)–the $\prod$ symbol is similar to the summation symbol ($\sum$) that we have encountered already except that it means *multiply* rather than add up). So, Wilks's lambda represents the ratio of error variance to total variance (SS_R/SS_T) for each variate:

$$\Lambda = \prod_{i=1}^{s} \frac{1}{1 + \lambda_i} \tag{14.8}$$

For the data in this example the value is .698, and it should be clear that large eigenvalues (which in themselves represent a large experimental effect) lead to small values of Wilks's lambda: hence statistical significance is found when Wilks's lambda is small:

$$\Lambda = \left(\frac{1}{1+.335}\right)\left(\frac{1}{1+.073}\right) = .698$$

14.3.4.5. Roy's largest root ④

Roy's largest root always makes me think of some bearded statistician with a garden spade digging up an enormous parsnip (or similar root vegetable); however, it isn't a parsnip but, as the name suggests, simply the eigenvalue for the first variate. So, in a sense it is the same as the Hotelling–Lawley trace but for the first variate only (see equation (14.9)):

$$\text{largest root} = \lambda_{\text{largest}} \tag{14.9}$$

As such, Roy's largest root represents the proportion of explained variance to unexplained variance (SS_M/SS_R) for the first discriminant function.[3] For the data in this example, the value of Roy's largest root is simply 0.335 (the eigenvalue for the first variate). So this value is conceptually the same as the *F*-ratio in univariate ANOVA. It should be apparent, from what we have learnt about the maximizing properties of these discriminant variates, that Roy's root represents the maximum possible between-group difference given the data collected. Therefore, this statistic should in many cases be the most powerful.

14.4. ASSUMPTIONS OF MANOVA ③

MANOVA has similar assumptions to ANOVA but extended to the multivariate case:

- **Independence**: Observations should be statistically independent.
- **Random sampling**: Data should be randomly sampled from the population of interest and measured at an interval level.
- **Multivariate normality**: In ANOVA, we assume that our dependent variable is normally distributed within each group. In the case of MANOVA, we assume that the dependent variables (collectively) have multivariate normality with groups.
- **Homogeneity of covariance matrices**: In ANOVA, it is assumed that the variances in each group are roughly equal (homogeneity of variance). In MANOVA we must assume that this is true for each dependent variable, but also that the correlation

3 This statistic is sometimes characterized as $\lambda_{\text{largest}}/(1+\lambda_{\text{largest}})$ but this is not the statistic reported by SPSS.

between any two dependent variables is the same in all groups. This assumption is examined by testing whether the population variance–covariance matrices of the different groups in the analysis are equal.[4]

When these assumptions are broken the accuracy of the resulting multivariate test statistics is compromised. So, what do you do? Well, SPSS doesn't offer a non-parametric version of MANOVA; however, some ideas have been put forward based on ranked data (much like the non-parametric tests we saw in Chapter 13). Although discussion of these tests is well beyond the scope of this book, Zwick (1985) suggests some ideas which can be beneficial when multivariate normality or homogeneity of covariance matrices cannot be assumed.

14.4.1. Checking assumptions ③

Most of the assumptions can be checked in the same way as for univariate tests (see Chapter 8); the additional assumptions of multivariate normality and equality of covariance matrices require different procedures. The assumption of multivariate normality cannot be tested on SPSS and so the only practical solution is to check the assumption of univariate normality for each dependent variable in turn (see Chapter 3). This solution is practical (because it is easy to implement) and useful (because univariate normality is a necessary condition for multivariate normality), but it does not *guarantee* multivariate normality. So, although this approach is the best we can do, I urge readers to consult Stevens (1992, Chapter 6) who provides some alternative solutions.

The assumption of equality of covariance matrices is more easily checked. First, for this assumption to be true the univariate tests of equality of variances between groups should be met. This assumption is easily checked using Levene's test (see section 3.6). As a preliminary check, Levene's test should not be significant for any of the dependent variables. However, Levene's test does not take account of the covariances and so the variance–covariance matrices should be compared between groups using Box's test. This test should be non-significant if the matrices are the same. However, Box's test is very susceptible to deviations from multivariate normality and so can be non-significant not because the matrices are similar, but because the assumption of multivariate normality is not tenable. Hence, it is vital to have some idea of whether the data meet the multivariate normality assumption before interpreting the result of Box's test.

14.4.2. Choosing a test statistic ③

Only when there is one underlying variate will the four test statistics necessarily be the same. Therefore, it is important to know which test statistic is best in terms of test power

4 For those of you who read about SSCP matrices, if you think about the relationship between sums of squares and variance, and cross-products and correlations, it should be clear that a variance–covariance matrix is basically a standardized form of an SSCP matrix.

and robustness. Both Olson (1974; 1976; 1979) and Stevens (1979) have done extensive work on the power of the four MANOVA test statistics. Olson (1974) observed that for small and moderate sample sizes the four statistics differ little in terms of power. If group differences are concentrated on the first variate (as will often be the case in social science research) Roy's statistic should prove most powerful (because it takes account of only that first variate), followed by Hotelling's trace, Wilks's lambda and Pillai's trace. However, when groups differ along more than one variate, the power ordering is the reverse (i.e. Pillai's trace is most powerful and Roy's root is least). One final issue pertinent to test power is that of sample size and the number of dependent variables. Stevens (1980) recommends using fairly small numbers of dependent variables (less than 10) unless sample sizes are large.

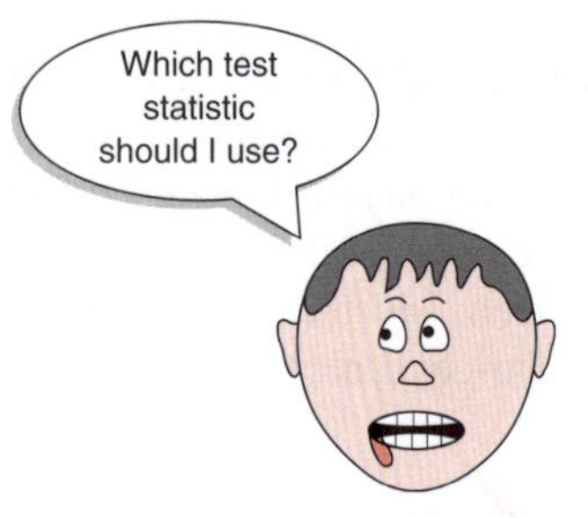

In terms of robustness, all four test statistics are relatively robust to violations of multivariate normality (although Roy's root is affected by platykurtic distributions—see Olson, 1976). Roy's root is also not robust when the homogeneity of covariance matrix assumption is untenable (Stevens, 1979). The work of Olson & Stevens led Bray & Maxwell (1985) to conclude that when sample sizes are equal the Pillai–Bartlett trace is the most robust to violations of assumptions. However, when sample sizes are unequal this statistic is affected by violations of the assumption of equal covariance matrices. As a rule, with unequal group sizes, check the assumption of homogeneity of covariance matrices using Box's test; if this test is non-significant, *and if the assumption of multivariate normality is tenable* (which allows us to assume that Box's test is accurate), then assume that Pillai's trace is accurate.

14.4.3. Follow-up analysis ③

In section 14.2.2 I mentioned that there was some controversy over how best to follow up the main MANOVA. The traditional approach is to follow a significant MANOVA with separate ANOVAs on each of the dependent variables. If this approach is taken, you might well wonder why we bother with the MANOVA in the first place (earlier on I said that multiple ANOVAs were a bad thing to do). Well, the ANOVAs that follow a significant MANOVA are said to be 'protected' by the initial MANOVA (Bock, 1975). The idea is that the overall multivariate test protects against inflated Type I error rates because if that initial test is non-significant (i.e. the null hypothesis is true) then any subsequent tests are ignored (any significance must be a Type I error because the null hypothesis is true). However, the notion of protection is somewhat fallacious because a significant MANOVA, more often than not, reflects a significant difference for one, but not all, of the dependent variables. Subsequent ANOVAs are then carried out on all of the dependent variables, but the MANOVA protects only the dependent variable for which group differences genuinely exist (see Bray & Maxwell, 1985, pp. 40–41). Therefore, you might want to consider applying a Bonferroni correction to the subsequent ANOVAs (Harris, 1975).

The ANOVA approach to following up a MANOVA implicitly assumes that the significant MANOVA is not due to the dependent variables representing a set of underlying dimensions that differentiate the groups. Therefore, some researchers advocate the use of discriminant analysis, which finds the linear combination(s) of the dependent variables that best *separates* (or discriminates) the groups. This procedure is more in keeping with the ethos of MANOVA because it embraces the relationships that exist between dependent variables and it is certainly useful for illuminating the relationship between the dependent variables and group membership. The major advantage of this approach over multiple ANOVAs is that it reduces and explains the dependent variables in terms of a set of underlying dimensions thought to reflect substantive theoretical or psychological dimensions. By default the standard GLM procedure in SPSS provides univariate ANOVAs, but not the discriminant analysis.[5] However, the discriminant analysis can be accessed via different menus and in the remainder of this chapter we will use the OCD data to illustrate how these analyses are done (those of you who skipped the theory section should refer to Table 14.1).

14.5. MANOVA ON SPSS ②

14.5.1. The main analysis ②

Either load the data in the file **OCD.sav**, or enter the data manually. If you enter the data manually you need three columns: one column must be a coding variable for the **group** variable (I used the codes CBT = 1, BT = 2, NT = 3), and in the remaining two columns enter the scores for each dependent variable respectively. Once the data have been entered, access the main MANOVA dialog box by using the **Analyze⇒General Linear Model⇒Multivariate...** menu path (see Figure 14.2).

The ANOVAs (and various multiple comparisons) carried out after the main MANOVA are identical to running separate ANOVA procedures in SPSS for each of the dependent variables. Hence, the main dialog box and options for MANOVA are very similar to the factorial ANOVA procedure we met in Chapter 10. The main difference from the main dialog box is that the space labelled *Dependent Variables* has room for several variables. Select the two dependent variables from the variables list (i.e. **actions** and **thoughts**) and transfer them to the *Dependent Variables* box by clicking on ▸. Select **group** from the variables list and transfer it to the *Fixed Factor(s)* box by clicking on ▸. There is also a box in which you can place covariates. For this analysis there are no covariates; however, you can apply the principles of ANCOVA to the multivariate case and conduct multivariate analysis of covariance (MANCOVA). Once you have specified the variables in the analysis, you can select any of the other dialog boxes by clicking on the buttons on the right-hand side:

---

5 Users of versions before 7.5 should note that the discriminant analysis can be accessed using the **Statistics⇒General Linear Models** (or **ANOVA Models** as it was known in Version 6) ⇒ **Multivariate...** menu.

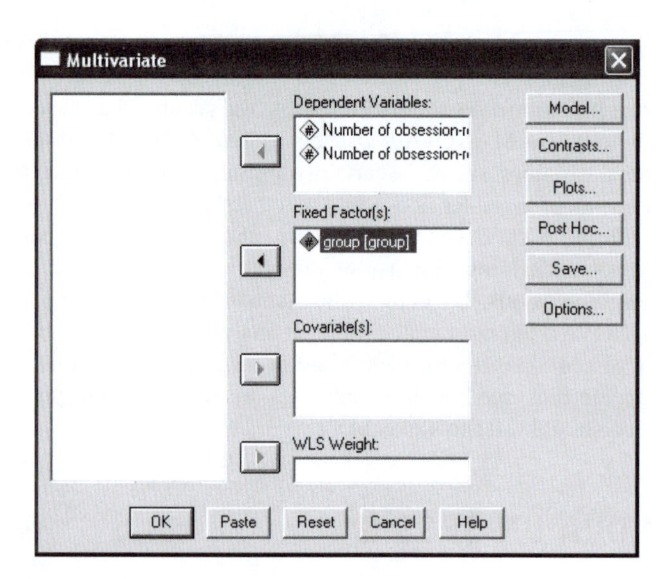

Figure 14.2 Main dialog box for MANOVA

Model... This button opens a dialog box for customizing your analysis and selecting the type of sums of squares used (see section 10.3.2).

Plots... This button opens a dialog box for selecting interaction graphs. This option is useful only when more than two independent variables have been measured (see section 10.3.3).

Save... This button opens a dialog box for saving residuals of the GLM (i.e. regression diagnostics). These options are useful for checking how well the model fits the data (see Chapter 5).

14.5.2. Multiple comparisons in MANOVA ②

The default way to follow up a MANOVA is to look at individual univariate ANOVAs for each dependent variable. For these tests, SPSS has the same options as in the univariate ANOVA procedure (see Chapter 8). The Contrasts... button opens a dialog box for specifying one of several standard contrasts for the independent variable(s) in the analysis. Table 8.6 describes what each of these tests compares, but for this example it makes sense to use a *simple* contrast that compares each of the experimental groups to the no-treatment control group. The no-treatment control group was coded as the last category (it had the highest code in the data editor), so we need to select the group variable and change the contrast to a simple contrast using the last category as the reference category (see Figure 14.3). For more details about contrasts see section 8.2.10.

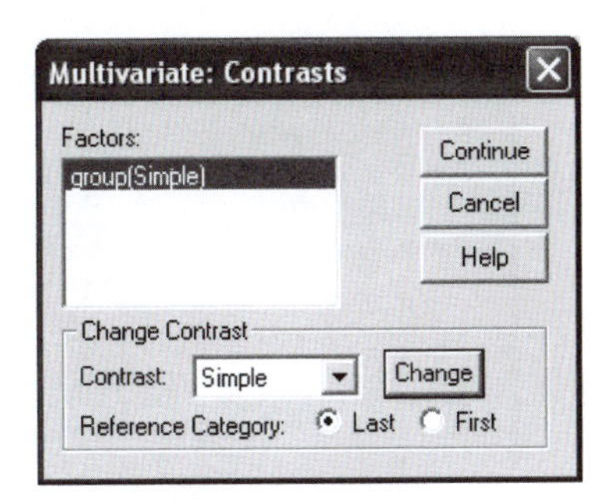

Figure 14.3 Contrasts for independent variable(s) in MANOVA

Instead of running a contrast, we could carry out *post hoc* tests on the independent variable to compare each group to all other groups. To access the *post hoc* tests dialog box click on Post Hoc... . The dialog box is the same as that for factorial ANOVA (see Figure 10.5) and the choice of test should be based on the same criteria as outlined in section 8.2.11. For the purposes of this example, I suggest selecting two of my usual recommendations: REGWQ and Games–Howell. Once you have selected *post hoc* tests return to the main dialog box.

14.5.3. Additional options ③

To access the *options* dialog box, click on Options... in the main dialog box (see Figure 14.4). The resulting dialog box is fairly similar to that of factorial ANOVA (see section 10.3.6); however, there are a few additional options that are worth mentioning.

- **SSCP matrices**: If this option is selected, SPSS will produce the model SSCP matrix, the error SSCP matrix and the total SSCP matrix. This option can be useful for understanding the computation of the MANOVA. However, if you didn't read the theory section you might be happy not to select this option and not worry about these matrices!
- **Residual SSCP matrix**: If this option is selected, SPSS produces the error SSCP matrix, the error variance–covariance matrix and the error correlation matrix. The error variance–covariance matrix is the matrix upon which Bartlett's test of sphericity is based. Bartlett's test examines whether this matrix is proportional to an identity matrix (i.e. that the covariances are zero and the variances—the values along the diagonal—are roughly equal).

The remaining options are the same as for factorial ANOVA (and so have been described in Chapter 10); I recommend rereading that chapter before deciding which options are useful.

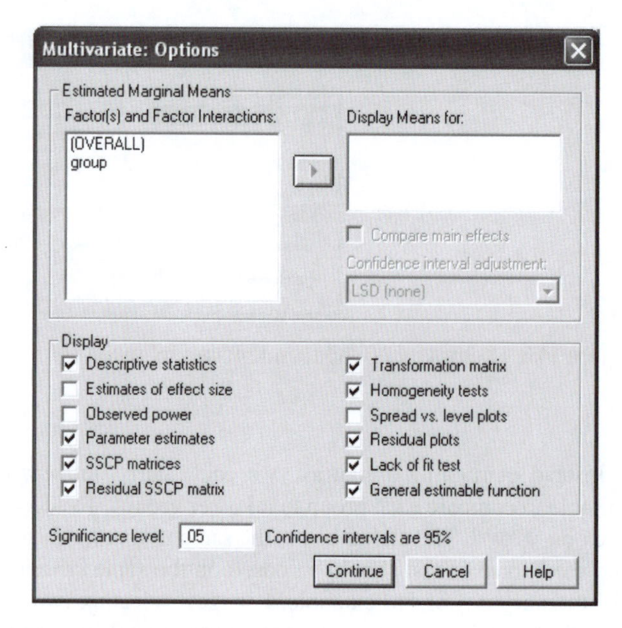

Figure 14.4 Additional options in MANOVA

14.6. OUTPUT FROM MANOVA ③

14.6.1. Preliminary analysis and testing assumptions ③

SPSS Output 14.1 shows an initial table of descriptive statistics that is produced by clicking on the descriptive statistics option in the *options* dialog box (Figure 14.4). This table contains the overall and group means and standard deviations for each dependent variable in turn. These values correspond to those calculated by hand in Table 14.1 and by looking at that table it should be clear what this part of the output tells us. It is clear from the means that participants had many more obsession-related thoughts than behaviours.

Descriptive Statistics

	group	Mean	Std. Deviation	N
Number of obsession-related thoughts	CBT	13.40	1.90	10
	BT	15.20	2.10	10
	No Treatment Control	15.00	2.36	10
	Total	14.53	2.21	30
Number of obsession-related behaviours	CBT	4.90	1.20	10
	BT	3.70	1.77	10
	No Treatment Control	5.00	1.05	10
	Total	4.53	1.46	30

SPSS Output 14.1

Box's Test of Equality of Covariance Matrices[a]

Box's M	9.959
F	1.482
df1	6
df2	18169
Sig.	.180

a Design: Intercept + GROUP

Bartlett's Test of Sphericity[a]

Likelihood Ratio	.042
Approx. Chi-Square	5.511
df	2
Sig.	.064

a Design: Intercept + GROUP

SPSS Output 14.2

SPSS Output 14.2 shows Box's test of the assumption of equality of covariance matrices (see section 14.4.1). This statistic tests the null hypothesis that the variance–covariance matrices are the same in all three groups. Therefore, if the matrices are equal (and therefore the assumption of homogeneity is met) this statistic should be *non-significant*. For these data $p = .18$ (which is greater than .05): hence, the covariance matrices are roughly equal and the assumption is tenable.

If the value of Box's test was significant ($p < .05$) then the covariance matrices are significantly different and so the homogeneity assumption would have been violated. The effect of violating this assumption is unclear. Hakstian, Roed & Lind (1979) report that Hotelling's T^2 is robust in the two-group situation when sample sizes are equal. As a general rule of thumb, if sample sizes are equal then disregard Box's test, because it is highly unstable, and assume Hotelling's and Pillai's statistics to be robust (see section 14.4.2). However, if group sizes are different, then robustness cannot be assumed (especially if Box's test is significant at $p < .001$). The more dependent variables you have measured, and the greater the differences in sample sizes, the more distorted the probability values produced by SPSS become. Tabachnick & Fidell (2001) therefore suggest that if the larger samples produce greater variances and covariances then the probability values will be conservative (and so significant findings can be trusted). However, if it is the smaller samples that produce the larger variances and covariances then the probability values will be liberal and so significant differences should be treated with caution (although non-significant effects can be trusted). As such, Box's test need only really be examined when sample sizes differ: it should not be trusted when multivariate normality cannot be assumed (or is in question), and the variance–covariance matrices for samples should be inspected to assess whether the printed probabilities are likely to be conservative or liberal. In the event that you cannot trust the printed probabilities, there is little you can do except equalize the samples by randomly deleting cases in the larger groups (although with this loss of information comes a loss of power).

Bartlett's test of sphericity tests whether the assumption of sphericity has been met and is useful only in univariate repeated-measures designs because MANOVA does not require this assumption.

14.6.2. MANOVA test statistics ③

SPSS Output 14.3 shows the main table of results. Test statistics are quoted for the intercept of the model (even MANOVA can be characterized as a regression model although

Multivariate Tests[a]

Effect		Value	F	Hypothesis df	Error df	Sig.
Intercept	Pillai's Trace	.983	745.230[c]	2.000	26.000	.000
	Wilks' Lambda	.017	745.230[c]	2.000	26.000	.000
	Hotelling's Trace	57.325	745.230[c]	2.000	26.000	.000
	Roy's Largest Root	57.325	745.230[c]	2.000	26.000	.000
GROUP	Pillai's Trace	.318	2.557	4.000	54.000	.049
	Wilks' Lambda	.699	2.555[c]	4.000	52.000	.050
	Hotelling's Trace	.407	2.546	4.000	50.000	.051
	Roy's Largest Root	.335	4.520	2.000	27.000	.020

a Design: Intercept + GROUP
b Computed using alpha = .05
c Exact statistic

SPSS Output 14.3

how this is done is beyond the scope of this text) and for the group variable. For our purposes, the group effects are of interest because they tell us whether or not the therapies had an effect on the OCD clients. You'll see that SPSS lists the four multivariate test statistics and their values correspond to those calculated in sections 14.3.4.2 to 14.3.4.5. In the next column these values are transformed into an *F*-ratio. The column of real interest, however, is the one containing the significance values of these *F*-ratios. For these data, Pillai's trace ($p = .049$), Wilks's lambda ($p = .050$) and Roy's largest root ($p = .020$) all reach the criterion for significance of .05. However, Hotelling's trace ($p = .051$) is non-significant by this criterion. This scenario is interesting, because the test statistic we choose determines whether or not we reject the null hypothesis that there are no between-group differences. However, given what we know about the robustness of Pillai's trace when sample sizes are equal, we might be well advised to trust the result of that test statistic, which indicates a significant difference. Interestingly, this example highlights the additional power associated with Roy's root (you should note how this statistic is considerably more significant than all others) when the test assumptions have been met.

From this result we should probably conclude that the type of therapy employed had a significant effect on OCD. The nature of this effect is not clear from the multivariate test statistic. First, it tells us nothing about which groups differed from which, and second it tells us nothing about whether the effect of therapy was on the obsession-related thoughts, the obsession-related behaviours, or a combination of both. To determine the nature of the effect, SPSS provides us with univariate tests.

14.6.3. Univariate test statistics ②

SPSS Output 14.4 initially shows a summary table of Levene's test of equality of variances for each of the dependent variables. These tests are the same as would be found if a one-way

Levene's Test of Equality of Error Variances[a]

	F	df1	df2	Sig.
Number of obsession-related thoughts	.076	2	27	.927
Number of obsession-related behaviours	1.828	2	27	.180

Tests the null hypothesis that the error variance of the dependent variable is equal across groups.
a Design: Intercept + GROUP

Tests of Between-Subjects Effects

Source	Dependent Variable	Type III Sum of Squares	df	Mean Square	F	Sig.
Corrected Model	Number of obsession-related thoughts	19.467[b]	2	9.733	2.154	.136
	Number of obsession-related behaviours	10.467[c]	2	5.233	2.771	.080
Intercept	Number of obsession-related thoughts	6336.533	1	6336.533	1402.348	.000
	Number of obsession-related behaviours	616.533	1	616.533	326.400	.000
GROUP	Number of obsession-related thoughts	19.467	2	9.733	2.154	.136
	Number of obsession-related behaviours	10.467	2	5.233	2.771	.080
Error	Number of obsession-related thoughts	122.000	27	4.519		
	Number of obsession-related behaviours	51.000	27	1.889		
Total	Number of obsession-related thoughts	6478.000	30			
	Number of obsession-related behaviours	678.000	30			
Corrected Total	Number of obsession-related thoughts	141.467	29			
	Number of obsession-related behaviours	61.467	29			

a Computed using alpha = .05
b R Squared = .138 (Adjusted R Squared = .074)
c R Squared = .170 (Adjusted R Squared = .109)

SPSS Output 14.4

ANOVA had been conducted on each dependent variable in turn (see Chapter 8). Levene's test should be non-significant for all dependent variables if the assumption of homogeneity of variance has been met. The results for these data clearly show that the assumption has been met. This finding not only gives us confidence in the reliability of the univariate tests to follow, but also strengthens the case for assuming that the multivariate test statistics are robust.

The next part of the output contains the ANOVA summary table for the dependent variables. The row of interest is that labelled *GROUP* (you'll notice that the values in this row are the same as for the row labelled *Corrected Model*; this is because the model fitted to the data contains only one independent variable: **group**). The row labelled *GROUP* contains an ANOVA summary table for each of the dependent variables, and values are given for the sums of squares for both actions and thoughts (these values correspond to the values of SS_M calculated in sections 14.3.3.1 and 14.3.3.2 respectively). The row labelled *Error* contains information about the residual sums of squares and mean squares for each of the dependent variables: these values of SS_R were calculated in sections 14.3.3.1 and 14.3.3.2 and I urge the reader to look back to these sections to consolidate what these values mean. The row labelled *Corrected Total* contains the values of the total sums of squares for each dependent variable (again, these values of SS_T were calculated in sections 14.3.3.1 and 14.3.3.2). The important parts of this table are the columns labelled *F* and *Sig.* in which the *F*-ratios for each univariate ANOVA and their significance values are listed. What should be clear from SPSS Output 14.4 and the calculations made in sections 14.3.3.1 and 14.3.3.2 is that the values associated with the univariate ANOVAs conducted after the MANOVA are *identical* to those obtained if one-way ANOVA was conducted on each dependent variable. This fact illustrates that MANOVA offers only hypothetical protection of inflated Type I error rates: there is no real-life adjustment made to the values obtained.

The values of p in SPSS Output 14.4 indicate that there was a non-significant difference between therapy groups in terms of both obsession-related thoughts ($p = .136$) and obsession-related behaviours ($p = .080$). These two results should lead us to conclude that the type of therapy has had no significant effect on the levels of OCD experienced by clients. Those of you who are still awake may have noticed something odd about this example: the multivariate test statistics led us to conclude that therapy had had a significant impact on OCD, yet the univariate results indicate that therapy has not been successful. Before reading any further, have a think about why this anomaly has occurred.

The reason for the anomaly in these data is simple: the multivariate test takes account of the correlation between dependent variables and so for these data it has more power to detect group differences. With this knowledge in mind, the univariate tests are not particularly useful for interpretation, because the groups differ along a combination of the dependent variables. To see how the dependent variables interact we need to carry out a discriminant function analysis, which will be described in section 14.7.

14.6.4. SSCP matrices ③

If you selected the two options to display SSCP matrices (section 14.5.3), then SPSS will produce the tables in SPSS Output 14.5 and SPSS Output 14.6. The first table (SPSS

Between-Subjects SSCP Matrix

			Number of obsession-related behaviours	Number of obsession-related thoughts
Hypothesis	Intercept	Number of obsession-related behaviours	616.533	1976.533
		Number of obsession-related thoughts	1976.533	6336.533
	GROUP	Number of obsession-related behaviours	**10.467**	**−7.533**
		Number of obsession-related thoughts	**−7.533**	**19.467**
Error		Number of obsession-related behaviours	**51.000**	**13.000**
		Number of obsession-related thoughts	**13.000**	**122.000**

Based on Type III Sum of Squares

SPSS Output 14.5

Output 14.5) displays the model SSCP (H), which is labelled *Hypothesis GROUP* (I have used light shading for this matrix) and the error SSCP (E) which is labelled *Error* (I have used dark shading for this matrix). The matrix for the intercept is displayed also, but this matrix is not important for our purposes. It should be pretty clear that the values in the model and error matrices displayed in SPSS Output 14.5 correspond to the values we calculated in sections 14.3.3.6 and 14.3.3.5 respectively. These matrices are useful, therefore, for gaining insight into the pattern of the data, and especially in looking at the values of the cross-products to indicate the relationship between dependent variables. In this example, the sums of squares for the error SSCP matrix are substantially bigger than in the model (or group) SSCP matrix, whereas the absolute value of the cross-products is fairly similar. This pattern suggests that if the MANOVA is significant then it might be the relationship between dependent variables that is important rather than the individual dependent variables themselves.

SPSS Output 14.6 shows the residual SSCP matrix again, but this time it includes the variance–covariance matrix and the correlation matrix. These matrices are all related. If you look back to Chapter 4, you should remember that the covariance is calculated by dividing the cross-product by the number of observations (i.e. the covariance is the average cross-product). Likewise, the variance is calculated by dividing the sums of squares by the degrees of freedom (and so similarly represents the average sum of squares). Hence, the variance–covariance matrix represents the average form of the SSCP matrix. Finally, we saw in Chapter 4 that the correlation was a standardized version of the covariance (where the standard deviation is also taken into account) and so the correlation matrix represents the standardized form of the variance–covariance matrix. As with the SSCP matrix, these other matrices are useful for assessing the extent of the error in the model.

Residual SSCP Matrix

		Number of obsession-related behaviours	Number of obsession-related thoughts
Sum-of-Squares and Cross-Products	Number of obsession-related behaviours	51.000	13.000
	Number of obsession-related thoughts	13.000	122.000
Covariance	Number of obsession-related behaviours	1.889	.481
	Number of obsession-related thoughts	.481	4.519
Correlation	Number of obsession-related behaviours	1.000	.165
	Number of obsession-related thoughts	.165	1.000

Based on Type III Sum of Squares

SPSS Output 14.6

The variance–covariance matrix is especially useful because Bartlett's test of sphericity is based on this matrix. Bartlett's test examines whether this matrix is proportional to an identity matrix. In section 14.3.1 we saw that an identity matrix was one in which the diagonal elements were 1, and the off-diagonal elements were 0. Therefore, Bartlett's test effectively tests whether the diagonal elements of the variance–covariance matrix are equal (i.e. group variances are the same), and that the off-diagonal elements are approximately zero (i.e. the dependent variables are not correlated). In this case, the variances are quite different (1.89 compared to 4.52) and the covariances slightly different from zero (.48) and so Bartlett's test has come out as nearly significant (see SPSS Output 14.2). Although this discussion is irrelevant to the multivariate tests, I hope that by expanding upon them here you can relate these ideas back to the issues of sphericity raised in Chapter 11, and see more clearly how this assumption is tested.

14.6.5. Contrasts ③

In section 14.5.2 I suggested carrying out a *simple* contrast that compares each of the therapy groups to the no-treatment control group. SPSS Output 14.7 shows the results of these contrasts. The table is divided into two sections conveniently labelled *Level 1 vs. Level 3* and *Level 2 vs. Level 3* where the numbers correspond to the coding of the group variable (i.e. 1 represents the lowest code used in the data editor and 3 the highest). If you coded the group variable using the same codes as I did, then these contrasts represent CBT vs. NT and BT vs. NT respectively. Each contrast is performed on both dependent variables separately and so they are identical to the contrasts that would be obtained from a univariate

ANOVA. The table provides values for the contrast estimate, and the hypothesized value (which will always be zero because we are testing the null hypothesis that the difference between groups is zero). The observed estimated difference is then tested to see whether it is significantly different from zero based on the standard error (it might help to reread Chapters 1 and 7 for some theory on this kind of hypothesis testing). A 95% confidence interval is produced for the estimated difference.

The first thing that you might notice (from the values of *Sig.*) is that when we compare CBT with NT there are no differences in thoughts ($p=.104$) or behaviours ($p=.872$) because both values are above the .05 threshold. However, comparing BT with NT, there is no significant difference in thoughts ($p=.835$) but there is a significant difference in behaviours between the groups ($p=.044$, which is less than .05). The confidence intervals confirm these findings: we have seen before that a 95% confidence interval is an interval that contains the true value of the difference between groups 95% of the time. If these boundaries cross zero (i.e the lower is a minus number and the upper a positive value), then this tells us that the true value of the group difference could be zero (i.e. there will be no difference between the groups). Therefore, we cannot be confident that the observed group difference is meaningful because the true group difference in the population could be zero. If, however, the confidence interval does not cross zero (i.e both values are positive or negative), then we can be confident that the true value of the group difference is different from zero. As such, we can be confident that genuine group differences exist.

Contrast Results (K Matrix)

group Simple Contrast[a]			Dependent Variable: Number of obsession-related thoughts	Dependent Variable: Number of obsession-related behaviours
Level 1 vs. Level 3	Contrast Estimate		−1.600	−.100
	Hypothesized Value		0	0
	Difference (Estimate − Hypothesized)		−1.600	−.100
	Std. Error		.951	.615
	Sig.		.104	.872
	95% Confidence Interval for Difference	Lower Bound	−3.551	−1.361
		Upper Bound	.351	−1.161
Level 1 vs. Level 3	Contrast Estimate		.200	−1.300
	Hypothesized Value		0	0
	Difference (Estimate − Hypothesized)		.200	−1.300
	Std. Error		.951	.615
	Sig.		.835	.044
	95% Confidence Interval for Difference	Lower Bound	−1.751	−2.561
		Upper Bound	2.151	−.039

a Reference category = 3

SPSS Output 14.7

For these data all confidence intervals include zero (the lower bounds are negative whereas the upper bounds are positive) except for the BT vs. NT contrast for behaviours and so only this contrast is significant. This is a little unexpected because the univariate ANOVA for behaviours was non-significant and so we would not expect there to be group differences.

Cramming Samantha's Tips

- MANOVA is used to test the difference between groups across several dependent variables simultaneously.
- Box's test looks at the assumption of equal covariance matrices. This test can be ignored when sample sizes are equal because when they are some MANOVA test statistics are robust to violations of this assumption. If group sizes differ this test should be inspected. If the value of *Sig.* Is less than .001 then the results of the analysis should not be trusted (see section 14.6.1).
- The table labelled *Multivariate Tests* gives us the results of the MANOVA. There are four test statistics listed (*Pillai's Trace, Wilks's Lambda, Hotelling's Trace* and *Roy's Largest Root*). I recommend using Pillai's trace. If the value of *Sig.* for this statistic is less than .05 then the groups differ significantly with respect to the dependent variables.
- ANOVAs can be used to follow up the MANOVA (a different ANOVA for each dependent variable). The results of these are listed in the table entitled *Tests of Between-Subjects Effects.* These ANOVAs can in turn be followed up using contrasts (see Chapters 8 to 12). Personally I don't recommend this approach and suggest conducting a *discriminant function analysis.*

14.7. FOLLOWING UP MANOVA WITH DISCRIMINANT ANALYSIS ③

I mentioned earlier on that a significant MANOVA could be followed up using either univariate ANOVA or discriminant analysis. In the example in this chapter, the univariate ANOVAs were not a useful way of looking at what the multivariate tests showed because the relationship between dependent variables is obviously having an effect. However, these data were designed especially to illustrate how the univariate ANOVAs should be treated cautiously and in real life a significant MANOVA is likely to be accompanied by at least one significant ANOVA. However, this does not mean that the relationship between dependent variables is not important, and it is still vital to investigate the nature of this relationship. Discriminant analysis is the best way to achieve this, and I strongly recommend that you follow up a MANOVA with both univariate tests and discriminant analysis if you want to understand your data fully.

Discriminant analysis is quite straightforward in SPSS: to access the main dialog box simply follow the menu path **Analyze⇒Classify⇒Discriminant...** (see Figure 14.5).

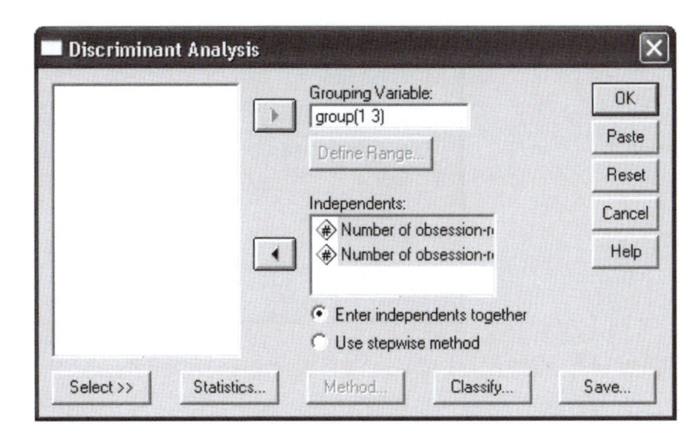

Figure 14.5 Main dialog box for discriminant analysis

The main dialog box will list the variables in the data editor on the left-hand side and provides two spaces on the right: one for the group variable and one for the predictors. In discriminant analysis we look to see how we can best separate (or discriminate) a set of groups using several predictors (so it is a little like logistic regression but where there are several groups rather than two).[6] It might be confusing to think of actions and thoughts as independent variables (after all, they were dependent variables in the MANOVA!) which is why I refer to them as predictors—this is another example of why it is useful not to refer to variables as independent variables and dependent variables in correlational analysis.

To run the analysis, select the variable **group** and transfer it to the box labelled *Grouping Variable* by clicking on ▶. Once this variable has been transferred, the Define Range... button will become active and you should click on this button to activate a dialog box in which you can specify the value of the highest and lowest coding values (1 and 3 in this case). Once you have specified the codings used for the grouping variable, you should select the variables **actions** and **thoughts** and transfer them to the box labelled *Independents* by clicking on ▶. There are two options available to determine how the predictors are entered into the model. The default is that both predictors are entered together and this is the option we require (because in MANOVA the dependent variables are analysed simultaneously). It is possible to enter the dependent variables in a stepwise manner and if this option is selected the Method... button becomes active, which opens a dialog box for specifying the criteria upon which predictors are entered. For the purpose of following up MANOVA, we need only be concerned with the remaining options.

6 In fact, I could have just as easily described discriminant analysis rather than logistic regression in Chapter 6 because they are different ways of achieving the same end result. However, logistic regression has far fewer restrictive assumptions and is generally more robust, which is why I have restricted the coverage of discriminant analysis to this chapter.

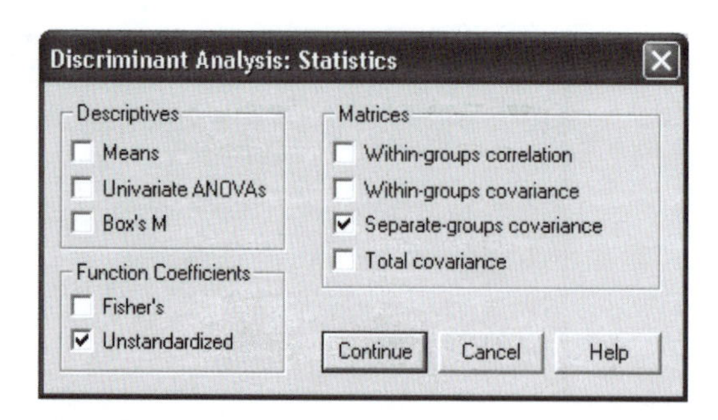

Figure 14.6 Statistics options for discriminant analysis

Click on [Statistics...] to activate the dialog box in Figure 14.6. This dialog box allows us to request group means, univariate ANOVAs and Box's test of equality of covariance matrices, all of which have already been provided in the MANOVA output (so we need not ask for them again). Furthermore, we can ask for the within-group correlation and covariance matrices, which are the same as the residual correlation and covariance matrices seen in SPSS Output 14.6. There is also an option to display a separate-groups covariance matrix, which can be useful for gaining insight into the relationships between dependent variables for each group (this matrix is something that the MANOVA procedure doesn't display and I recommend selecting it). Finally, we can ask for a total covariance matrix, which displays covariances and variances of the dependent variables overall. Another useful option is to select *Unstandardized* function coefficients. This option will produce the unstandardized *b*s for each variate (see equation (14.5)). When you have finished with this dialog box, click on [Continue] to return to the main dialog box.

If you click on [Classify...] you will access the dialog box in Figure 14.7. In this dialog box there are several options available. First, you can select how prior probabilities are determined: if your group sizes are equal then you should leave the default setting as it is; however, if you have an unbalanced design then it is beneficial to base prior probabilities on the observed group sizes. The default option for basing the analysis on the within-group covariance matrix is fine (because this is the matrix upon which the MANOVA is based). You should also request a combined-groups plot, which will plot the variate scores for each participant grouped according to the therapy they were given. The separate-groups plots show the same thing but using different graphs for each of the groups; when the number of groups is small it is better to select a combined plot because it is easier to interpret. The remaining options are of little interest when using discriminant analysis to follow up MANOVA. The only option that is useful is the summary table, which provides an overall gauge of how well the discriminant variates classify the actual participants. When you have finished with the options click on [Continue] to return to the main dialog box.

The final options are accessed by clicking on [Save...] to access the dialog box in Figure 14.8. There are three options available, two of which relate to the predicted group memberships and probabilities of group memberships from the model. These values are comparable to those obtained from a logistic regression analysis (see Chapter 6). The final option is to

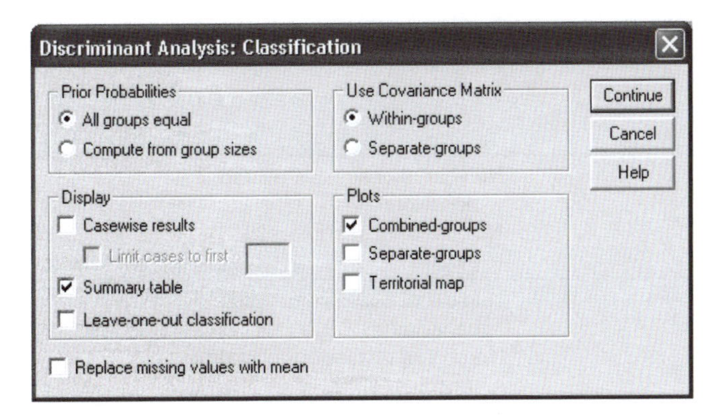

Figure 14.7 Discriminant analysis classification options

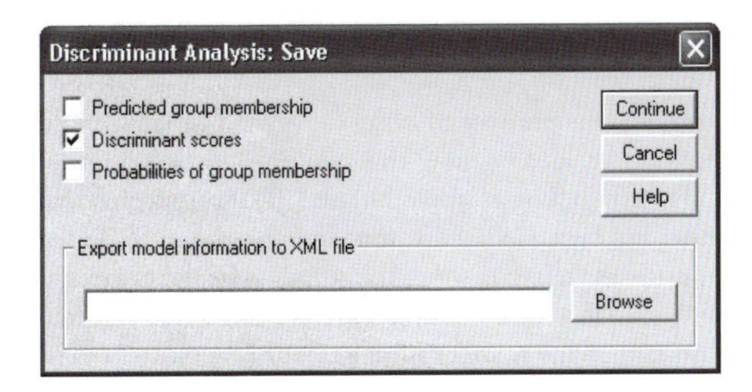

Figure 14.8 *Save new variables* dialog box in discriminant analysis

provide the discriminant scores. These are the scores for each person, on each variate, obtained from equation (14.5). These scores can be useful because the variates that the analysis identifies may represent underlying social or psychological constructs. If these constructs are identifiable, then it is useful for interpretation to know what a participant scores on each dimension.

14.8. OUTPUT FROM THE DISCRIMINANT ANALYSIS ④

SPSS Output 14.8 shows the covariance matrices for separate groups (selected in Figure 14.6). These matrices are made up of the variances of each dependent variable for each group (in fact these values are shown in Table 14.1). The covariances are obtained by taking the cross-products between the dependent variables for each group (shown in Table 14.4 as 0.40, 22.6 and −10) and dividing each by 9—the degrees of freedom, $N - 1$

Covariance Matrices

group		Number of obsession-related behaviours	Number of obsession-related thoughts
CBT	Number of obsession-related behaviours	1.433	4.444E-02
	Number of obsession-related thoughts	4.444E-02	3.600
BT	Number of obsession-related behaviours	3.122	2.511
	Number of obsession-related thoughts	2.511	4.400
No Treatment Control	Number of obsession-related behaviours	1.111	−1.111
	Number of obsession-related thoughts	−1.111	5.556

SPSS Output 14.8

(where N is the number of observations). The values in this table are useful because they give us some idea of how the relationship between dependent variables changes from group to group. For example, in the CBT group behaviours and thoughts have virtually no relationship because the covariance is almost zero. In the BT group thoughts and actions are positively related, so as the number of behaviours decrease, so does the number of thoughts. In the NT condition there is a negative relationship, so if the number of thoughts increases then the number of behaviours decrease. It is important to note that these matrices don't tell us about the substantive importance of the relationships because they are unstandardized (see Chapter 4), they merely give a basic indication.

SPSS Output 14.9 shows the initial statistics from the discriminant analysis. At first we are told the eigenvalues for each variate and you should note that the values correspond to the values of the diagonal elements of the matrix HE^{-1} (calculated in the file **AppendixChapter14.pdf** on the CD-ROM). These eigenvalues are converted into percentage of variance accounted for, and the first variate accounts for 82.2% of variance compared to the second variate, which accounts for only 17.8%. The next part of the output shows Wilks's lambda, which has the same value (0.699), degrees of freedom (4) and significance value (.05) as in the MANOVA (see SPSS Output 14.3). The important point to note from this table is that only one of the variates is significant (the second variate is non-significant, $p = .173$). Therefore, the group differences shown by the MANOVA can be explained in terms of *one* underlying dimension.

The tables in SPSS Output 14.10 are the most important for interpretation. The first table shows the standardized discriminant function coefficients for the two variates. These values are standardized versions of the values in the eigenvectors calculated in section 14.3.4.1. If you recall that the variates can be expressed in terms of a linear regression equation (see equation (14.4)), the standardized discriminant function coefficients are equivalent to the standardized betas in regression. Hence, these coefficients tell us the relative contribution of each variable to the variates. It is clear from the size of the values for these data that the num-

Eigenvalues

Function	Eigenvalue	% of Variance	Cumulative %	Canonical Correlation
1	.335[a]	82.2	82.2	.501
2	.073[a]	17.8	100.0	.260

a First 2 canonical discriminant functions were used in the analysis

Wilks's Lambda

Test of Function(s)	Wilks's Lambda	Chi-square	df	Sig.
1 through 2	.699	9.508	4	.050
2	.932	1.856	1	.173

SPSS Output 14.9

Standardized Canonical Discriminant Function Coefficients

	Function	
	1	2
Number of obsession-related behaviours	.829	.584
Number of obsession-related thoughts	−.713	.721

Structure Matrix

	Function	
	1	2
Number of obsession-related behaviours	.711*	.703
Number of obsession-related thoughts	−.576	.817*

Pooled within-groups correlations between discriminating variables and standardized canonical discriminant functions
Variables ordered by absolute size of correlation within function.
*Largest absolute correlation between each variable and any discriminant function

SPSS Output 14.10

ber of obsessive behaviours has a greater contribution to the first variate than the number of thoughts, but that the opposite is true for variate 2. Also, remembering that standardized beta coefficients vary within ±1, it is noteworthy that both variables have a large contribution to the first variate (i.e. they are both important) because their values are quite close to 1 and −1 respectively. Bearing in mind that only the first variate is important, we can conclude that

Canonical Discriminant Function Coefficients

	Function	
	1	2
Number of obsession-related behaviours	.603	.425
Number of obsession-related thoughts	−.335	.339
(Constant)	2.139	−6.857

Unstandardized coefficients

Functions at Group Centroids

	Function	
group	1	2
CBT	.601	−.229
BT	−.726	−.128
No Treatment Control	.125	.357

Unstandardized canonical discriminant functions evaluated at group means

SPSS Output 14.11

it is necessary to retain both dependent variables in the set of discriminators (because their standardized weights are of a similar magnitude). The fact that one dependent variable has a negative weight and one a positive weight indicates that group differences are explained by the difference between dependent variables.

Another way of looking at the relationship between dependent variables and discriminant variates is to look at the structure matrix, which gives the canonical variate correlation coefficients. These values are comparable to factor loadings (see Chapter 15) and indicate the substantive nature of the variates. Bargman (1970) argues that when some dependent variables have high canonical variate correlations while others have low ones, then the ones with high correlations contribute most to group separation. As such they represent the relative contribution of each dependent variable to group separation (see Bray & Maxwell, 1985, pp. 42–45). We are again interested only in the first variate (because the second was non-significant) and looking at the structure matrix we can conclude that the number of behaviours was slightly more important in differentiating the three groups (because .711 is greater than .576). However, the number of thoughts is still very important because the value of the correlation is quite large. As with the standardized weights, the fact that one dependent variable has a positive correlation, whereas the other has a negative one, indicates that group separation is determined by the difference between the dependent variables.

The next part of the output (SPSS Output 14.11) tells us first the canonical discriminant function coefficients, which are the unstandardized versions of the standardized coefficients described above. As such, these values are the values of b in equation (14.4) and you'll notice that these values correspond to the values in the eigenvectors derived in sec-

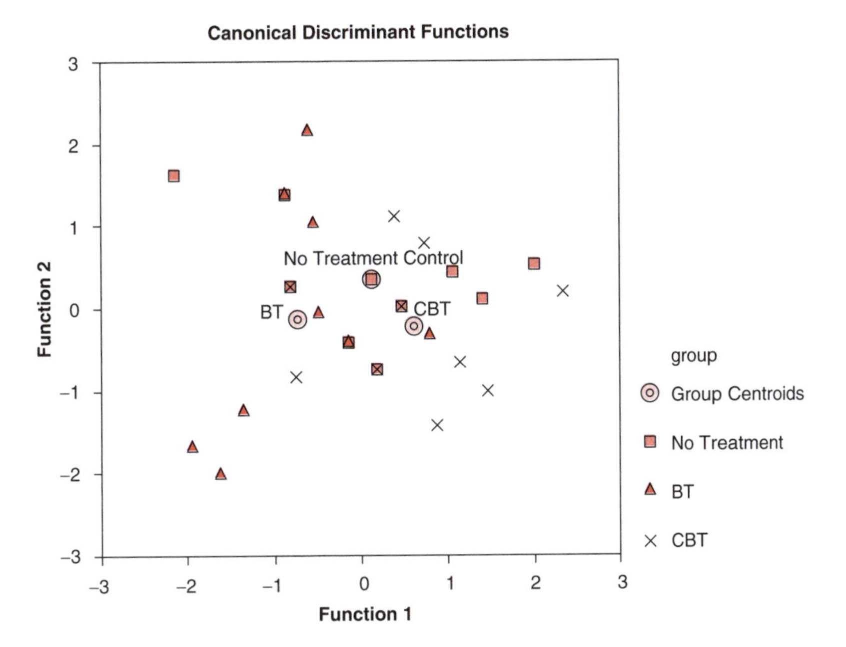

Figure 14.9 Combined-groups plot

tion 14.3.4.1 and used in equation (14.5). The values are less useful than the standardized versions, but do demonstrate from where the standardized versions come. The next table gives the values of the variate centroids for each group. The centroids are simply the mean variate scores for each group. For interpretation we should look at the sign of the centroid (positive or negative), and from these data it looks as if variate 1 discriminates the BT group from the other two (notably the CBT group because the difference between centroids is greatest for these groups). The second variate (which was non-significant) seems to discriminate the NT group from the two experimental groups (but not significantly so).

The relationship between the variates and the groups is best illuminated using a combined-groups plot (selected using the dialog box in Figure 14.7). This graph plots the variate scores for each person, grouped according to the experimental condition to which that person belonged. In addition, the group centroids are indicated, which are the average variate scores for each group.

Figure 14.9 shows this plot for the OCD data, and what is clear from the position of the centroids (the big circles labelled with the group initials) is that variate 1 discriminates the BT group from the CBT (look at the horizontal distance between these centroids). The second variate does not differentiate any groups: we know this already because it was non-significant, but the plot shows that the vertical distances between group centroids is very small, which indicates no group separation on this variate.

Cramming Samantha's Tips

- Discriminant function analysis (DFA) can be used after MANOVA to see how the dependent variables discriminate the groups.
- DFA identifies variates (combinations of the dependent variables) and to find out how many variates are significant look at the tables labelled *Wilks's Lambda*: if the value of *Sig.* is less than .05 then the variate is significantly discriminating the groups.
- Once the significant variates have been identified, use the table labelled *Standardized Canonical Discriminant Function Coefficients* to find out how the dependent variables contribute to the variates. High scores indicate that a dependent variable is important for a variate, and variables with positive and negative coefficients are contributing to the variate in opposite ways.
- Finally, to find out which groups are discriminated by a variate look at the table labelled *Functions at Group Centroids*: for a given variate, groups with values opposite in sign are being discriminated by that variate.

14.9. SOME FINAL REMARKS ④

14.9.1. The final interpretation ④

So far we have gathered an awful lot of information about our data, but how can we bring all of it together to answer our research question: can therapy improve OCD and if so which therapy is best? Well, the MANOVA tells us that therapy can have a significant effect on OCD symptoms, but the non-significant univariate ANOVAs suggested that this improvement is not simply in terms of either thoughts or behaviours. The discriminant analysis suggests that the group separation can be best explained in terms of one underlying dimension. In this context the dimension is likely to be OCD itself (which we can realistically presume is made up of both thoughts and behaviours). So, therapy doesn't necessarily change behaviours or thoughts *per se*, but it does influence the underlying dimension of OCD. So, the answer to the first question seems to be: yes, therapy can influence OCD, but the nature of this influence is unclear.

The next question is more complex: which therapy is best? Figure 14.10 shows graphs of the relationships between the dependent variables and the group means of the original data. The graph of the means shows that for actions, BT reduces the number of obsessive behaviours, whereas CBT and NT do not. For thoughts, CBT reduces the number of obsessive thoughts, whereas BT and NT do not (check the pattern of the bars). Looking now at the relationships between thoughts and actions, in the BT group there is a near-linear positive relationship between thoughts and actions, so the more obsessive thoughts a person has, the more obsessive behaviours they carry out. In the CBT group there is no relationship at all (thoughts and actions vary quite independently). In the no-treatment group there

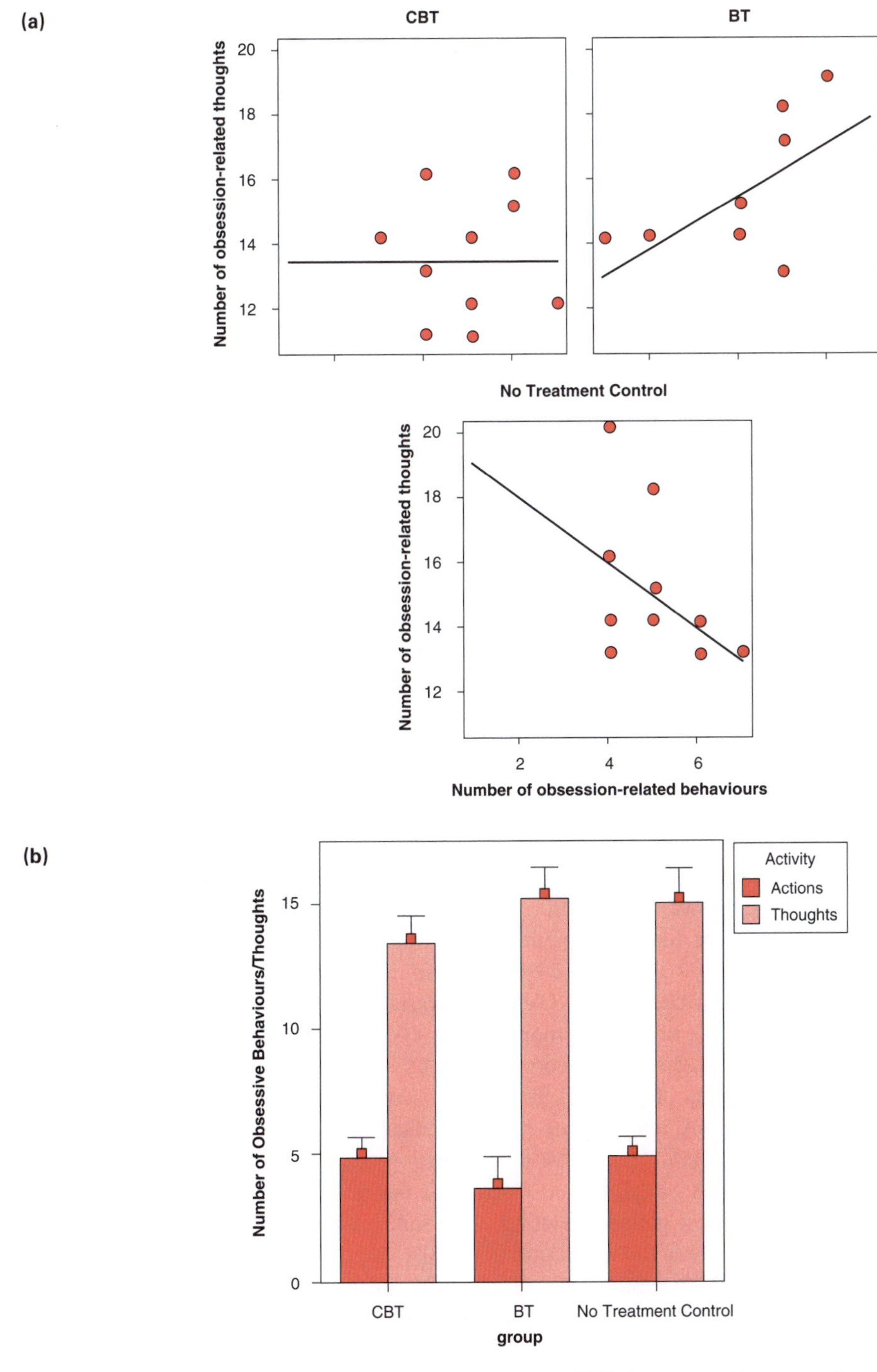

Figure 14.10 Graphs showing (a) the relationships and (b) the means between the dependent variables in each therapy group

is a negative (and non-significant incidentally) relationship between thoughts and actions. What we have discovered from the discriminant analysis is that behaviours are more important in terms of OCD (as a construct) than thoughts (we know this from the canonical variate correlations). Therefore, behaviour therapy seems to be best because it addresses behaviours rather than thoughts (and so compared to CBT is preferable—hence the distance between their group centroids in Figure 14.9). However, the significance of the discriminant function does not necessarily tell us that the BT group was significantly lower than the NT group (so therapy isn't necessarily better than no therapy). So, in short BT has the most influence on OCD as a construct, because of the relative importance of behaviours in that construct compared to cognitions.

14.9.2. Univariate ANOVA or discriminant analysis? ③

This example should have made clear that univariate ANOVA and discriminant analysis are ways of answering different questions arising from a significant MANOVA. If univariate ANOVAs are chosen, Bonferroni corrections should be applied to the level at which you accept significance. The truth is that you should run both analyses to get a full picture of what is happening in your data. The advantage of discriminant analysis is that it tells you something about the underlying dimensions within your data (which is especially useful if you have employed several dependent measures in an attempt to capture some social or psychological construct). Even if univariate ANOVAs are significant, the discriminant analysis provides useful insight into your data and should be used. I hope that this chapter will convince you of this recommendation!

14.10. WHAT HAVE WE DISCOVERED ABOUT STATISTICS? ②

In this chapter we've cackled maniacally in the ear of MANOVA, forced discriminant function analysis to swallow cod-liver oil, and discovered to our horror that Roy has a large root. There are sometimes situations in which several outcomes have been measured in different groups and we discovered that in these situations the ANOVA technique can be extended and is called MANOVA (multivariate analysis of variance). The reasons for using this technique rather than running lots of ANOVAs is that we retain control over the Type I error rate, and we can incorporate the relationships between outcome variables into the analysis. Some of you will have then discovered that MANOVA works in very similar ways to ANOVA, but just with matrices rather than single values. Others will have discovered that it's best to ignore the theory sections of this book. We had a look at an example of MANOVA on SPSS and discovered that just to make life as confusing as possible we get four test statistics relating to the same effect! Of these, I tried to convince you that Pillai's trace was the safest option. Finally, we had a look at the two options for following up MANOVA: running lots of ANOVAs, or discriminant function analysis. Of these, discriminant function analysis gives us the most information, but can be a bit of a nightmare to interpret.

14.11. KEY TERMS THAT WE'VE DISCOVERED

- Bartlett's test of sphericity
- Box's test
- Discriminant analysis
- Discriminant function variates
- Discriminant scores
- Error SSCP (E)
- HE^{-1}
- Homogeneity of covariance matrices
- Hotelling–Lawley trace (T^2)
- Hypothesis SSCP (H)
- Identity matrix
- Matrix
- Multivariate
- Multivariate analysis of variance (or MANOVA)
- Multivariate normality
- Pillai–Bartlett trace (V)
- Roy's largest root
- Square matrix
- Sum of squares and cross-products matrix (SSCP)
- Total SSCP (T)
- Univariate
- Variance–covariance matrix
- Wilks's lambda (Λ)

14.12. SMART ALEX'S TASKS

- A clinical psychologist noticed that several of his manic psychotic patients did chicken impersonations in public. He wondered whether this behaviour could be used to diagnose this disorder and so decided to compare his patients against a normal sample. He observed 10 of his patients as they went through a normal day. He also needed to observe 10 of the most normal people he could find: naturally he chose to observe lecturers at the University of Sussex. He measured all participants using two dependent variables: first, how many chicken impersonations they did in the streets of Brighton over the course of a day, and second, how good their impersonations were (as scored out of 10 by an independent farmyard noise expert). The data are in the file **chicken.sav**; use MANOVA and DFA to find out whether these variables could be used to distinguish manic psychotic patients from those without the disorder.③
- I was interested in whether students' knowledge of different aspects of psychology improved throughout their degree. I took a sample of first years, second years and third years and gave them five tests (scored out of 15) representing different aspects of psychology: **Exper** (experimental psychology such as cognitive and neuropsychology); **Stats** (statistics); **Social** (social psychology); **Develop** (developmental psychology); **Person** (personality). Your task is (1) to carry out an appropriate general analysis to determine whether there are overall group differences along these five measures, (2) to look at the scale-by-scale analyses of group differences produced in the output and interpret the results accordingly, (3) to select contrasts that test the hypothesis that second and third years will score higher than first years on all scales; (4) to select tests that compare all groups to each other—briefly compare these results with the contrasts; and (5) to carry out a separate analysis in which you test whether a combination of the measures can successfully discriminate the groups (comment only briefly on this

analysis). Include only those scales that revealed group differences for the contrasts. How do the results help you to explain the findings of your initial analysis? The data are in the file **psychology.sav**.④

14.13. FURTHER READING

Bray, J. H. & Maxwell, S. E. (1985). *Multivariate analysis of variance*. Sage university paper series on quantitative applications in the social sciences, 07–054. Newbury Park, CA: Sage. This monograph on MANOVA is superb; I cannot recommend anything better.

CHAPTER 15

EXPLORATORY FACTOR ANALYSIS

15.1. WHAT WILL THIS CHAPTER TELL US? ②

In the social sciences we are often trying to measure things that cannot directly be measured (so-called latent variables). For example, management students (or psychologists even) might be interested in measuring 'burnout', which is when someone who has been working very hard on a project (a book, for example) for a prolonged period of time suddenly finds themselves devoid of motivation and inspiration, and wants repeatedly to head-butt their computer. You can't measure burnout directly: it has many facets. However, you can measure different aspects of burnout: you could get some idea of motivation, stress levels, whether the person has any new ideas and so on. Having done this, it would be helpful to know whether these measures really do reflect a single variable. Put another way, are these different variables driven by the same underlying variable? This chapter will look at factor analysis (and *principal component analysis*), a technique for identifying groups or clusters of variables. This technique has three main uses: (1) to understand the structure of a set of variables (e.g. pioneers of intelligence such as Spearman and Thurstone used factor analysis to try to understand the structure of the latent variable 'intelligence'); (2) to construct a questionnaire to measure an underlying variable (e.g you might design a questionnaire to measure burnout); and (3) to reduce a data set to a more manageable size while retaining as much of the original information as possible (e.g. we saw in Chapter 5 that multicollinearity can be a problem in multiple regression, and factor analysis can be used to solve this problem by combining variables that are collinear). Through this chapter we'll discover what factors are, how we find them, and what they tell us (if anything) about the relationship between the variables we've measured.

15.2. FACTORS ②

If we measure several variables, or ask someone several questions about themselves, the correlation between each pair of variables (or questions) can be arranged in what's known as an *R-matrix*. An *R*-matrix is just a correlation matrix: a table of correlation coefficients between variables (in fact, we saw small versions of these matrices in Chapter 4). The diagonal elements of an *R*-matrix are all 1 because each variable will correlate perfectly with itself. The off-diagonal elements are the correlation coefficients between pairs of variables, or questions.[1] The existence of clusters of large correlation coefficients between subsets of variables suggests that those variables could be measuring aspects of the same underlying dimension. These underlying dimensions are known as factors (or *latent variables*). By reducing a data set from a group of interrelated variables into a smaller set of factors, factor analysis achieves parsimony by explaining the maximum amount of common variance in a correlation matrix using the smallest number of explanatory concepts.

There are numerous examples of the use of factor analysis in the social sciences. The trait theorists in psychology used factor analysis endlessly to measure personality traits. Most readers will be familiar with the extraversion–introversion and neuroticism traits measured by Eysenck (1953). Most other personality questionnaires are based on factor analysis (notably Cattell's, 1966a, 16 personality factors questionnaire) and these inventories are frequently used for recruiting purposes in industry and even by some religious groups. However, although factor analysis is probably most famous for being adopted by psychologists, its use is by no means restricted to measuring dimensions of personality. Economists, for example, might use factor analysis to see whether productivity, profits and workforce can be reduced down to an underlying dimension of company growth, and Jeremy Miles told me of a biochemist who used it to analyse urine samples!

Let's put some of these ideas into practice by imagining that we wanted to measure different aspects of what might make a person popular. We could administer several measures that we believe tap different aspects of popularity. So, we might measure a person's social skills (Social Skills), their selfishness (Selfish), how interesting others find them (Interest), the proportion of time they spend talking about the other person during a conversation (Talk 1), the proportion of time they spend talking about themselves (Talk 2), and their propensity to lie to people (the Liar scale). We can then calculate the correlation coefficients for each pair of variables and create an *R*-matrix. Table 15.1 shows this matrix. Any significant correlation coefficients are shown in bold type. It is clear that there are two clusters of interrelating variables. Therefore, these variables might be measuring some common underlying dimension. The amount that someone talks about the other person during a conversation seems to correlate highly with both the level of social skills and how

1 This matrix is called an *R*-matrix, or *R*, because it contains correlation coefficients and *r* usually denotes Pearson's correlation (see Chapter 4)—the *r* turns into a capital letter when it denotes a matrix.

Table 15.1 An *R*-matrix

	Talk 1	Social Skills	Interest	Talk 2	Selfish	Liar
Talk 1	1.000					
Social Skills	**0.772**	1.000				
Interest	**0.646**	**0.879**	1.000			
Talk 2	0.074	−0.120	0.054	1.000		
Selfish	−0.131	0.031	−0.101	**0.441**	1.000	
Liar	0.068	0.012	0.110	**0.361**	**0.277**	1.000

Factor 1

Factor 2

interesting the other finds that person. Also, social skills correlate well with how interesting others perceive a person to be. These relationships indicate that the better your social skills the more interesting and talkative you are likely to be. However, there is a second cluster of variables. The amount that people talk about themselves within a conversation correlates with how selfish they are and how much they lie. Being selfish also correlates with the degree to which a person tells lies. In short, selfish people are likely to lie and talk about themselves.

In factor analysis we strive to reduce this *R*-matrix to its underlying dimensions by looking at which variables seem to cluster together in a meaningful way. This data reduction is achieved by looking for variables that correlate highly with a group of other variables, but do not correlate with variables outside of that group. In this example, there appear to be two clusters that fit the bill. The first factor seems to relate to general sociability, whereas the second factor seems to relate to the way in which a person treats others socially (we might call it Consideration). It might, therefore, be assumed that popularity depends not only on your ability to socialize, but also on whether you are genuine towards others.

15.2.1. Graphical representation of factors ②

Factors (not to be confused with independent variables in factorial ANOVA) are statistical entities that can be visualized as classification axes along which measurement variables can be plotted. In plain English, this statement means that if you imagine factors as being the axis of a graph, then we can plot variables along these axes. The co-ordinates of variables along each axis represent the strength of relationship between that variable and each factor. Figure 15.1 shows such a plot for the popularity data (in which there were only two factors). The first thing to notice is that for both factors, the axis line ranges from −1 to 1, which are the outer limits of a correlation coefficient. Therefore, the position of a given variable depends on its correlation to the two factors. The circles represent the three variables that correlate highly with factor 1 (sociability: horizontal axis)

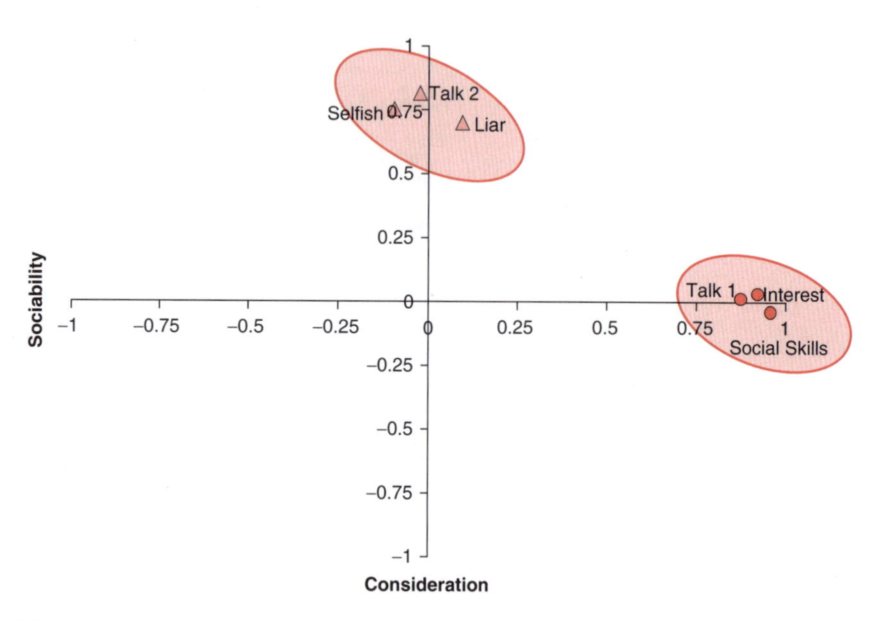

Figure 15.1 Example of a factor plot

but have a low correlation with factor 2 (consideration to others: vertical axis). Conversely, the triangles represent variables that correlate highly with consideration to others but have a low correlation to sociability. From this plot, we can tell that selfishness, the amount a person talks about themselves and their propensity to lie all contribute to a factor which could be called consideration of others. Conversely, how much a person takes an interest in other people, how interesting they are and their level of social skills contribute to a second factor, sociability. This diagram therefore supports the structure that was apparent in the *R*-matrix. Of course, if a third factor existed within these data it could be represented by a third axis (creating a 3-D graph). It should also be apparent that if more than three factors exist in a data set, then they cannot all be represented by a two-dimensional drawing.

If each axis on the graph represents a factor, then the variables that go to make up a factor can be plotted according to the extent to which they relate to a given factor. The co-ordinates of a variable, therefore, represent its relationship to the factors. In an ideal world a variable should have a large co-ordinate for one of the axes, and low co-ordinates for any other factors. This scenario would indicate that this particular variable related to only one factor. Variables that have large co-ordinates on the same axis are assumed to measure different aspects of some common underlying dimension. The co-ordinate of a variable along a classification axis is known as a factor loading. The factor loading can be thought of as the Pearson correlation between a factor and a variable (see Box 15.1). From what we know about interpreting correlation coefficients (see section 4.5.3) it should be clear that if we square the factor loading we obtain a measure of the substantive importance of a particular variable to a factor.

15.2.2. Mathematical representation of factors ②

The axes drawn in Figure 15.1 are straight lines and so can be described mathematically by the equation of a straight line. Therefore, factors can also be described in terms of this equation. We have used the equation of a straight line numerous times throughout this text and so it should be familiar to you. Equation (15.1) reminds us of the equation describing a linear model and then applies this to the scenario of describing a factor. You'll notice that there is no intercept in the equation, the reason being that the lines intersect at zero (hence the intercept is also zero). The *b*s in the equation represent the factor loadings:

$$\begin{aligned} Y_i &= b_1X_1 + b_2X_2 + \cdots + b_nX_n + \varepsilon_i \\ \text{Factor}_i &= b_1\text{Variable}_1 + b_2\text{Variable}_2 + \cdots + b_n\text{Variable}_n + \varepsilon_i \end{aligned} \tag{15.1}$$

Sticking with our example of popularity, we found that there were two factors underlying this construct: general sociability and consideration. We can, therefore, construct an equation that describes each factor in terms of the variables that have been measured. The equations are shown in equation (15.2):

$$\begin{aligned} Y_i &= b_1X_1 + b_2X_2 + \cdots + b_nX_n + \varepsilon_i \\ \text{Sociability}_i &= b_1\,\text{Talk 1}_i + b_2\,\text{Social Skills}_i + b_3\,\text{Interest}_i \\ &\quad + b_4\,\text{Talk 2}_i + b_5\,\text{Selfish}_i + b_6\,\text{Liar}_i + \varepsilon_i \\ \text{Consideration}_i &= b_1\,\text{Talk 1}_i + b_2\,\text{Social Skills}_i + b_3\,\text{Interest}_i \\ &\quad + b_4\,\text{Talk 2}_i + b_5\,\text{Selfish}_i + b_6\,\text{Liar}_i + \varepsilon_i \end{aligned} \tag{15.2}$$

First, notice that the equations are identical in form: they both include all of the variables that were measured. However, the values of *b* in the two equations will be different (depending on the relative importance of each variable to the particular factor). In fact, we can replace each value of *b* with the co-ordinate of that variable on the graph in Figure 15.1 (i.e. replace the values of *b* with the factor loading). The resulting equations are shown in equation (15.3):

$$\begin{aligned} Y_i &= b_1X_1 + b_2X_2 + \cdots + b_nX_n + \varepsilon_i \\ \text{Sociability}_i &= .87\,\text{Talk 1}_i + .96\,\text{Social Skills}_i + .92\,\text{Interest}_i \\ &\quad + .00\,\text{Talk 2}_i - .10\,\text{Selfish}_i + .09\,\text{Liar}_i + \varepsilon_i \\ \text{Consideration}_i &= .01\,\text{Talk 1}_i - .03\,\text{Social Skills}_i + .04\,\text{Interest}_i \\ &\quad + .82\,\text{Talk 2}_i + .75\,\text{Selfish}_i + .70\,\text{Liar}_i + \varepsilon_i \end{aligned} \tag{15.3}$$

Notice that, for the sociability factor, the values of *b* are high for Talk 1, Social Skills and Interest. For the remaining variables (Talk 2, Selfish and Liar) the values of *b* are very low (close to zero). This tells us that three of the variables are very important for that factor (the ones with high values of b) and three are very unimportant (the ones with low values of b). We saw that this point is true because of the way that three variables clustered highly on the factor plot. The point to take on board here is that the factor plot and these equations represent the same thing: the factor loadings in the plot are simply the b-values in these equations (but see box 15.1). For the second factor, inconsideration to others, the opposite pattern can be seen in that Talk 2, Selfish and Liar all have high values of b whereas the remaining three variables have b-values close to zero. In an ideal world, variables would have very high b-values for one factor and very low b-values for all other factors.

These factor loadings can be placed in a matrix in which the columns represent each factor and the rows represent the loadings of each variable onto each factor. For the popularity data this matrix would have two columns (one for each factor) and six rows (one for each variable). This matrix, usually denoted A, can be seen below. To understand what the matrix means try relating the elements to the loadings in equation (15.3). For example, the top row represents the first variable, Talk 1, which had a loading of .87 for the first factor (sociability) and a loading of .01 for the second factor (consideration). This matrix is called the factor matrix or component matrix (if doing principal component analysis)—see box 15.1 to find out about the different forms of this matrix:

$$A = \begin{pmatrix} .87 & .01 \\ .96 & -.03 \\ .92 & .04 \\ .00 & .82 \\ -.10 & .75 \\ .09 & .70 \end{pmatrix}$$

The major assumption in factor analysis is that these algebraic factors represent real-world dimensions, the nature of which must be *guessed at* by inspecting which variables have high loads on the same factor. So, psychologists believe that factors represent dimensions of the psyche, education researchers believe they represent abilities, and sociologists might believe they represent races, or social classes. However, it is an extremely contentious point whether this assumption is tenable and some believe that the dimensions derived from factor analysis are real only in the statistical sense—and are real-world fictions.

Box 15.1

What's the difference between a pattern matrix and a structure matrix? ③

Throughout my discussion of factor loadings I've been quite vague. Sometimes I've said that these loadings can be thought of as the correlation between a variable and a given factor, then at other times I've described these loadings in terms of regression coefficients (*b*). Now, it should be

Box 15.1 (Continued)

obvious from what we discovered in Chapters 4 and 5 that correlation coefficients and regression coefficients are quite different things, so what am I going on about? Shouldn't I make up my mind what the factor loadings actually are?

Well, in vague terms (the best terms for my brain) both correlation coefficients and regression coefficients represent the relationship between a variable and linear model in a broad sense, so the key take-home message is that factor loadings tell us about the relative contribution that a variable makes to factor. As long as you understand that much, you have no problems.

However, the factor loadings in a given analysis can be both correlation coefficients and regression coefficients. In a few sections time we'll discover that the interpretation of factor analysis is helped greatly by a technique known as *rotation*. Without going into details, there are two types: orthogonal and oblique rotation (see section 15.3.6). When orthogonal rotation is used, any underlying factors are assumed to be independent, and the factor loading *is* the correlation between the factor and the variable, but is also the regression coefficient. Put another way, the values of the correlation coefficients are the same as the values of the regression coefficients. However, there are situations in which the underlying factors are assumed to be related or correlated to each other. In these situations, oblique rotation is used and the resulting correlations between variables and factors will differ from the corresponding regression coefficients. In this case, there are, in effect, two different sets of factor loadings: the correlation coefficients between each variable and factor (which are put in the *factor* structure matrix) and the regression coefficients for each variable on each factor (which are put in the *factor* pattern matrix). These coefficients can have quite different interpretations (see Graham, Guthrie & Thompson, 2003).

15.2.3. Factor scores ②

A factor can be described in terms of the variables measured and the relative importance of them for that factor (represented by the value of b). Therefore, having discovered which factors exist, and estimated the equation that describes them, it should be possible also to estimate a person's score on a factor, based on their scores for the constituent variables. As such, if we wanted to derive a score of sociability for a particular person, we could place their scores on the various measures into equation (15.3). This method is known as a *weighted average*. In fact, this method is overly simplistic and rarely used, but it is probably the easiest way to explain the principle. For example, imagine the six scales all range from 1 to 10 and that someone scored the following: Talk 1 (4), Social Skills (9), Interest (8), Talk 2 (6), Selfish (8) and Liar (6). We could replace these values into equation (15.3) to get a score for this person's sociability and their consideration to others (see equation (15.4)). The resulting scores of 19.22 and 15.21 reflect the degree to which this person is sociable and their inconsideration to others respectively. This person scores higher on sociability than inconsideration. However, the scales of measurement used will influence the resulting scores, and if different variables used different measurement scales, then factor scores for different factors cannot be compared. As such, this method of calculating factor scores is poor and more sophisticated methods are usually used.

$$\begin{aligned}
\text{Sociability} &= .87\,\text{Talk 1} + .96\,\text{Social Skills} + .92\,\text{Interest} \\
&\quad + .00\,\text{Talk 2} - .10\,\text{Selfish} + .09\,\text{Liar} \\
\text{Sociability} &= (.87 \times 4) + (.96 \times 9) + (.92 \times 8) + (.00 \times 6) \\
&\quad - (.10 \times 8) + (.09 \times 6) \\
&= 19.22 \\
\text{Consideration} &= .01\,\text{Talk 1} - .03\,\text{Social Skills} + .04\,\text{Interest} \\
&\quad + .82\,\text{Talk 2} + .75\,\text{Selfish} + .70\,\text{Liar} \\
\text{Consideration} &= (.01 \times 4) - (.03 \times 9) + (.04 \times 8) + (.82 \times 6) \\
&\quad + (.75 \times 8) + (.70 \times 6) \\
&= 15.21
\end{aligned} \tag{15.4}$$

15.2.3.1. The regression method ④

There are several sophisticated techniques for calculating factor scores that use factor score coefficients as weights in equation (15.1) rather than using the factor loadings. The form of the equation remains the same, but the *b*s in the equation are replaced with these factor score coefficients. Factor score coefficients can be calculated in several ways. The simplest way is the regression method. In this method the factor loadings are adjusted to take account of the initial correlations between variables; in doing so, differences in units of measurement and variable variances are stabilized.

To obtain the matrix of factor score coefficients (B) we multiply the matrix of factor loadings by the inverse (R^{-1}) of the original correlation or R-matrix. You might remember from the previous chapter that matrices cannot be divided (see section 14.3.4.1). Therefore, if we want to divide by a matrix it cannot be done directly and instead we multiply by its inverse. Therefore, by multiplying the matrix of factor loadings by the inverse of the correlation matrix we are, conceptually speaking, dividing the factor loadings by the correlation coefficients. The resulting factor score matrix, therefore, represents the relationship between each variable and each factor taking into account the original relationships between pairs of variables. As such, this matrix represents a purer measure of the *unique* relationship between variables and factors.

The matrices for the popularity data are shown below. The resulting matrix of factor score coefficients, B, comes from SPSS. The matrices R^{-1} and A can be multiplied by hand to get the matrix B and those familiar with matrix algebra (or who have consulted Namboodiri, 1984, or Stevens, 1992) might like to verify the result (calculations are in the file **FactorScores.pdf** on the CD-ROM). To get the same degree of accuracy as SPSS you should work to at least 5 decimal places.

$$B = R^{-1}A$$

$$B = \begin{pmatrix} 4.76 & -7.46 & 3.91 & -2.35 & 2.42 & -.49 \\ -7.46 & 18.49 & -12.42 & 5.45 & -5.54 & 1.22 \\ 3.91 & -12.42 & 10.07 & -3.65 & 3.79 & -.96 \\ -2.35 & 5.45 & -3.65 & 2.97 & -2.16 & .02 \\ 2.42 & -5.54 & 3.79 & -2.16 & 2.98 & -.56 \\ -.49 & 1.22 & -.96 & .02 & -.56 & 1.27 \end{pmatrix} \begin{pmatrix} .87 & .01 \\ .96 & -.03 \\ .92 & .04 \\ .00 & .82 \\ -.10 & .75 \\ .09 & .70 \end{pmatrix}$$

$$= \begin{pmatrix} .343 & .006 \\ .376 & -.020 \\ .362 & .020 \\ .000 & .473 \\ -.037 & .437 \\ .039 & .405 \end{pmatrix}$$

The pattern of the loadings is the same for the factor score coefficients; that is, the first three variables have high loadings for the first factor and low loadings for the second, whereas the pattern is reversed for the last three variables. The difference is only in the actual value of the weightings, which are smaller because the correlations between variables are now accounted for. These factor score coefficients can be used to replace the b-values in equation (15.4):

$$\begin{aligned} \text{Sociability} &= .343\,\text{Talk 1} + .376\,\text{Social Skills} + .362\,\text{Interest} \\ &\quad + .000\,\text{Talk 2} - .037\,\text{Selfish} + .039\,\text{Liar} \\ \text{Sociability} &= (.343 \times 4) + (.376 \times 9) + (.362 \times 8) + (.000 \times 6) \\ &\quad - (.037 \times 8) + (.039 \times 6) \\ &= 7.59 \\ \text{Consideration} &= .006\,\text{Talk 1} - .020\,\text{Social Skills} + .020\,\text{Interest} \\ &\quad + .473\,\text{Talk 2} + .437\,\text{Selfish} + .405\,\text{Liar} \\ \text{Consideration} &= (.006 \times 4) - (.020 \times 9) + (.020 \times 8) + (.473 \times 6) \\ &\quad + (.437 \times 8) + (.405 \times 6) \\ &= 8.768 \end{aligned} \tag{15.5}$$

Equation (15.5) shows how these coefficient scores are used to produce two factor scores for each person. In this case, the participant had the same scores on each variable as were used in equation (15.4). The resulting scores are much more similar than when the factor loadings were used as weights because the different variances among the six variables has now been controlled for. The fact that the values are very similar reflects that this person not only scores highly on variables relating to sociability, but is also inconsiderate (i.e. they score equally highly on both factors). This technique for producing factor scores ensures that the resulting scores have a mean of 0 and a variance equal to the squared multiple correlation between the estimated factor scores and the true factor values. However, the downside of the regression method is that the scores can correlate not only with factors other than the one on which they are based, but also with other factor scores from a different orthogonal factor.

15.2.3.2. Other methods ②

To overcome the problems associated with the regression technique, two adjustments have been proposed: *the Bartlett method* and *the* Anderson–Rubin method. SPSS can produce factor scores based on any of these methods. The Bartlett method produces scores that are unbiased and that correlate only with their own factor. The mean and standard deviation of the scores is the same as for the regression method. However, factor scores can still correlate with each other. The final method is the Anderson–Rubin method, which is a modification of the Bartlett method that produces factor scores that are uncorrelated and standardized (they have a mean of 0, a standard deviation of 1). Tabachnick & Fidell (2001) conclude that the Anderson–Rubin method is best when uncorrelated scores are required but that the regression method is preferred in other circumstances simply because it is most easily understood. Although it isn't important that you understand the maths behind any of the methods, it is important that you understand what the factor scores represent: namely, a composite score for each individual on a particular factor.

15.2.3.3. Uses of factor scores ②

There are several uses of factor scores. First, if the purpose of the factor analysis is to reduce a large set of data into a smaller subset of measurement variables, then the factor scores tell us an individual's score on this subset of measures. Therefore, any further analysis can be carried out on the factor scores rather than the original data. For example, we could carry out a *t*-test to see whether females are significantly more sociable than males using the factor scores for *sociability*. A second use is in overcoming collinearity problems in regression. If, following a multiple regression analysis, we have identified sources of multicollinearity then the interpretation of the analysis is questioned (see section 5.6.2.3). In this situation, we can simply carry out a factor analysis on the predictor variables to reduce them to a subset of uncorrelated factors. The variables causing the multicollinearity will combine to form a factor. If we then rerun the regression but using the factor scores as predictor variables then the problem of multicollinearity should vanish (because the variables are now combined into a single factor). There are ways in which we can ensure that the factors are uncorrelated (one way is to use the Anderson–Rubin method—see above). By using uncorrelated factor scores as predictors in the regression we can be confident that there will be no correlation between predictors: hence, no multicollinearity!

15.3. DISCOVERING FACTORS ②

By now, you should have some grasp of the concept of what a factor is, how it is represented graphically, how it is represented algebraically, and how we can calculate composite scores representing an individual's 'performance' on a single factor. I have deliberately restricted the discussion to a conceptual level, without delving into how we actually find these mythical beasts known as factors. This section will look at how we find factors. Specifically we will examine different types of method, look at the maths behind one method (principal components), investigate the criteria for determining whether factors are important, and discover how to improve the interpretation of a given solution.

15.3.1. Choosing a method ②

The first thing you need to know is that there are several methods for unearthing factors in your data. The method you choose will depend on what you hope to do with the analysis. Tinsley & Tinsley (1987) give an excellent account of the different methods available. There are two things to consider: whether you want to generalize the findings from your sample to a population and whether you are exploring your data or testing a specific hypothesis. This chapter describes techniques for exploring data using factor analysis. Testing hypotheses about the structures of latent variables and their relationships to each other requires considerable complexity and can be done with computer programs such as AMOS. Those interested in hypothesis testing techniques (known as confirmatory factor analysis) are advised to read Pedhazur & Schmelkin (1991, Chapter 23) for an introduction. Assuming we want to explore our data we then need to consider whether we want to apply our findings to the sample collected (descriptive methods) or to generalize our findings to a population (inferential methods). When factor analysis was originally developed it was assumed that it would be used to explore data to generate future hypotheses. As such, it was assumed that the technique would be applied to the entire population of interest. Therefore, certain techniques assume that the sample used is the population, and so results cannot be extrapolated beyond that particular sample. Principal component analysis is an example of one of these techniques, as is principal factors analysis (*principal axis factoring*) and image covariance analysis (*image factoring*). Of these, principal component analysis and principal factors analysis are the preferred methods and usually result in similar solutions (see section 15.3.3). When these methods are used conclusions are restricted to the sample collected and generalization of the results can be achieved only if analysis using different samples reveals the same factor structure.

Another approach has been to assume that participants are randomly selected and that the variables measured constitute the population of variables in which we're interested. By assuming this, it is possible to develop techniques from which the results can be generalized from the sample participants to a larger population. However, a constraint is that any findings hold true only for the set of variables measured (because we've assumed this set constitutes the entire population of variables). Techniques in this category include the maximum-likelihood method (see Harman, 1976) and Kaiser's alpha factoring. The choice of method depends largely on what generalizations, if any, you want to make from your data.[2]

---

2 It's worth noting at this point that principal component analysis is not in fact the same as factor analysis. This doesn't stop idiots like me from discussing them as though they are, but more on that later.

15.3.2. Communality ②

Before continuing it is important that you understand some basic things about the variance within an *R*-matrix. It is possible to calculate the variability in scores (the variance) for any given measure (or variable). You should be familiar with the idea of variance by now and comfortable with how it can be calculated (if not see Chapter 1). The total variance for a particular variable will have two components: some of it will be shared with other variables or measures (common variance) and some of it will be specific to that measure (unique variance). We tend to use the term 'unique variance' to refer to variance that can be reliably attributed to only one measure. However, there is also variance that is specific to one measure but not reliably so; this variance is called *error* or random variance. The proportion of common variance present in a variable is known as the communality. As such, a variable that has no specific variance (or random variance) would have a communality of 1; a variable that shares none of its variance with any other variable would have a communality of 0.

In factor analysis we are interested in finding common underlying dimensions within the data and so we are primarily interested only in the common variance. Therefore, when we run a factor analysis it is fundamental that we know how much of the variance present in our data is common variance. This presents us with a logical impasse: to do the factor analysis we need to know the proportion of common variance present in the data, yet the only way to find out the extent of the common variance is by carrying out a factor analysis! There are two ways to approach this problem. The first is to assume that all of the variance is common variance. As such, we assume that the communality of every variable is 1. By making this assumption we merely transpose our original data into constituent linear components (known as principal component analysis). The second approach is to estimate the amount of common variance by estimating communality values for each variable. There are various methods of estimating communalities but the most widely used (including alpha factoring) is to use the squared multiple correlation (SMC) of each variable with all others. So, for the popularity data, imagine you ran a multiple regression using one measure (Selfish) as the outcome and the other five measures as predictors: the resulting multiple R^2 (see section 5.5.2) would be used as an estimate of the communality for the variable Selfish. This second approach is what is done in factor analysis. These estimates allow the factor analysis to be done. Once the underlying factors have been extracted, new communalities can be calculated that represent the multiple correlation between each variable and the factors extracted. Therefore, the communality is a measure of the proportion of variance explained by the extracted factors.

15.3.3. Factor analysis vs. principal component analysis ②

I have just explained that there are two approaches to locating underlying dimensions of a data set: factor analysis and principal component analysis. These techniques differ in the communality estimates that are used. Simplistically, though, factor analysis derives a mathematical model from which factors are estimated, whereas principal component analysis merely decomposes the original data into a set of linear variates (see Dunteman, 1989, Chapter 8, for more detail on the differences between the procedures). As such, only factor analysis can estimate the underlying factors and it relies on various assumptions for these estimates to be accurate. Principal component analysis is concerned only with establishing which linear components exist within the data and how a particular variable might contribute to that component. In terms of theory, this chapter is dedicated to principal component analysis rather than factor analysis. The reasons are that principal component analysis is a psychometrically sound procedure, it is conceptually less complex than factor analysis, and it bears numerous similarities to discriminant analysis (described in the previous chapter).

However, we should consider whether the techniques provide different solutions to the same problem. Based on an extensive literature review, Guadagnoli & Velicer (1988) concluded that the solutions generated from principal component analysis differ little from those derived from factor analytic techniques. In reality, there are some circumstances for which this statement is untrue. Stevens (1992) summarizes the evidence and concludes that with 30 or more variables and communalities greater than 0.7 for all variables, different solutions are unlikely; however, with fewer than 20 variables and any low communalities (< 0.4) differences can occur.

The flip-side of this argument is eloquently described by Cliff (1987) who observed that proponents of factor analysis 'insist that components analysis is at best a common factor analysis with some error added and at worst an unrecognizable hodgepodge of things from which nothing can be determined' (p. 349). Indeed feeling is strong on this issue with some arguing that when principal component analysis is used it should not be described as a factor analysis and that you should not impute substantive meaning to the resulting components. However, to non-statisticians the difference between a principal component and a factor may be difficult to conceptualize (they are both linear models), and the differences arise largely from the calculation.[3]

15.3.4. Theory behind principal component analysis ③

Principal component analysis works in a very similar way to MANOVA and discriminant function analysis (see the previous chapter). Although it isn't necessary to understand the mathematical principles in any detail, readers of the previous chapter may benefit from some comparisons between the two techniques. For those who haven't read the previous chapter, I suggest you flick through it before moving ahead!

In MANOVA, various SSCP matrices were calculated that contained information about the relationships between dependent variables. I mentioned before that these SSCP matrices could be easily converted to variance–covariance matrices, which represent the same information but in averaged form (i.e. taking account of the number of observations). I also said that by dividing each element by the relevant standard deviation the variance–covariance matrices becomes standardized. The result is a correlation matrix.

3 For this reason I have used the terms *components* and *factors* interchangeably throughout this chapter. Although this use of terms will reduce some statisticians (and psychologists) to tears, I'm banking on these people not needing to read this book! I acknowledge the methodological differences, but I think it's easier for students if I dwell on the similarities between the techniques and not the differences.

In factor analysis we usually deal with correlation matrices (although it is possible to analyse a variance–covariance matrix too) and the point to note is that this matrix pretty much represents the same information as an SSCP matrix in MANOVA. The difference is just that the correlation matrix is an averaged version of the SSCP that has been standardized.

In MANOVA, we used several SSCP matrices that represented different components of experimental variation (the model variation and the residual variation). In principal component analysis the covariance (or correlation) matrix cannot be broken down in this way (because all data come from the same group of participants). In MANOVA, we ended up looking at the variates or components of the SSCP matrix that represented the ratio of the model variance to the error variance. These variates were linear dimensions that separated the groups tested, and we saw that the dependent variables mapped onto these underlying components. In short, we looked at whether the groups could be separated by some linear combination of the dependent variables. These variates were found by calculating the eigenvectors of the SSCP. The number of variates obtained was the smaller of p (the number of dependent variables) or $k - 1$ (where k is the number of groups). In component analysis we do something similar (I'm simplifying things a little, but it will give you the basic idea). That is, we take a correlation matrix and calculate the variates. There are no groups of observations, and so the number of variates calculated will always equal the number of variables measured (p). The variates are described, as for MANOVA, by the eigenvectors associated with the correlation matrix. The elements of the eigenvectors are the weights of each variable on the variate (see equation (14.5)). These values are the factor loadings described earlier. The largest eigenvalue associated with each of the eigenvectors provides a single indicator of the substantive importance of each variate (or component). The basic idea is that we retain factors with relatively large eigenvalues, and ignore those with relative small eigenvalues.

In summary, component analysis works in a similar way to MANOVA. We begin with a matrix representing the relationships between variables. The linear components (also called variates, or factors) of that matrix are then calculated by determining the eigenvalues of the matrix. These eigenvalues are used to calculate eigenvectors, the elements of which provide the loading of a particular variable on a particular factor (i.e. they are the b-values in equation (15.1)). The eigenvalue is also a measure of the substantive importance of the eigenvector with which it is associated.

15.3.5. Factor extraction: eigenvalues and the scree plot ②

Not all factors are retained in an analysis, and there is debate over the criterion used to decide whether a factor is statistically important. I mentioned above that eigenvalues associated with a variate indicate the substantive importance of that factor. Therefore, it seems logical that we should retain only factors with large eigenvalues. How do we decide whether or not an eigenvalue is large enough to represent a meaningful factor? Well, one technique advocated by Cattell (1966b) is to plot a graph of each eigenvalue (Y-axis) against the factor with which it is associated (X-axis). This graph is known as a scree plot. I mentioned earlier

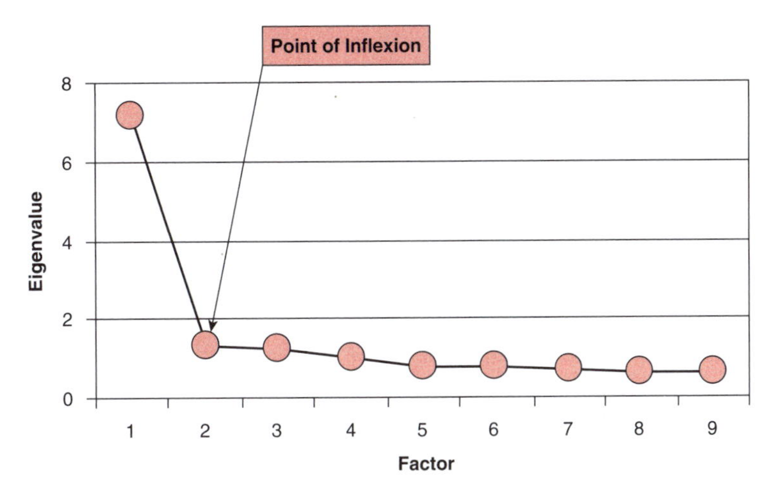

Figure 15.2 Example of a scree plot for data that have only one underlying factor

that it is possible to obtain as many factors as there are variables and that each has an associated eigenvalue. By graphing the eigenvalues, the relative importance of each factor becomes apparent. Typically there will be a few factors with quite high eigenvalues, and many factors with relatively low eigenvalues, and so this graph has a very characteristic shape: there is a sharp descent in the curve followed by a tailing off (see Figure 15.2). Cattell (1966b) argued that the cut-off point for selecting factors should be at the point of inflexion of this curve. With a sample of more than 200 participants, the scree plot provides a fairly reliable criterion for factor selection (Stevens, 1992).

Although scree plots are very useful, factor selection should not be based on this criterion alone. Kaiser (1960) recommended retaining all factors with eigenvalues greater than 1. This criterion is based on the idea that the eigenvalues represent the amount of variation explained by a factor and that an eigenvalue of 1 represents a substantial amount of variation. Jolliffe (1972, 1986) reports that Kaiser's criterion is too strict and suggests the third option of retaining all factors with eigenvalues more than .7. The difference between how many factors are retained using Kaiser's methods compared to Jolliffe's can be dramatic.

You might well wonder how the three methods compare. Generally speaking, research has indicated that Kaiser's criterion is accurate when the number of variables is less than 30 and the resulting communalities (after extraction) are all greater than .7. Kaiser's criterion is also accurate when the sample size exceeds 250 and the average communality is greater than or equal to .6. In any other circumstances you are best advised to use a scree plot provided the sample size is greater than 200 (see Stevens, 1992, pp. 378–380 for more detail). By default, SPSS uses Kaiser's criterion to extract factors. Therefore, if you use the scree plot to determine how many factors are retained you may have to rerun the analysis specifying that SPSS extracts the number of factors you require.

However, as is often the case in statistics, the three criteria often provide different solutions! In these situations the communalities of the factors need to be considered.

In principal component analysis we begin with communalities of 1 with all factors retained (because we assume that all variance is common variance). At this stage all we have done is to find the linear variates that exist in the data—so we have just transformed the data without discarding any information. However, to discover what common variance *really* exists between variables we must decide which factors are meaningful and discard any that are too trivial to consider. Therefore, we discard some information. The factors we retain will not explain all of the variance in the data (because we have discarded some information) and so the communalities after extraction will always be less than 1. The factors retained do not map perfectly onto the original variables—they merely reflect the common variance present in the data. If the communalities represent a loss of information then they are important statistics. The closer the communalities are to 1, the better our factors are at explaining the original data. It is logical that the more factors retained, the greater the communalities will be (because less information is discarded); therefore, the communalities are good indices of whether too few factors have been retained. In fact, with *generalized least-squares factor analysis* and *maximum-likelihood factor analysis* you can get a statistical measure of the goodness-of-fit of the factor solution (see the next chapter for more on goodness-of-fit tests). This basically measures the proportion of variance that the factor solution explains (so, can be thought of as comparing communalities before and after extraction).

As a final word of advice, your decision on how many factors to extract will also depend on why you're doing the analysis in the first place (e.g. if you're trying to overcome multicollinearity problems in regression, then it might be better to extract too many factors than too few).

15.3.6. Improving interpretation: factor rotation ③

Once factors have been extracted, it is possible to calculate to what degree variables load onto these factors (i.e. calculate the loading of the variable on each factor). Generally, you will find that most variables have high loadings on the most important factor, and small loadings on all other factors. This characteristic makes interpretation difficult, and so a technique called factor rotation is used to discriminate between factors. If a factor is a classification axis along which variables can be plotted, then factor rotation effectively rotates these factor axes such that variables are loaded maximally to only one factor. Figure 15.3 demonstrates how this process works using an example in which there are only two factors. Imagine that a sociologist was interested in classifying university lecturers as a demographic group. She discovered that two underlying dimensions best describe this group: alcoholism and achievement (go to any academic conference and you'll see that academics drink heavily!). The first factor, alcoholism, has a cluster of variables associated with it (dark dots) and these could be measures such as the number of units drunk in a week, dependency and obsessive personality. The second factor, achievement, also has a cluster of variables associated with it (light dots) and these could be measures relating to salary, job status and number of research publications. Initially, the full lines represent the factors, and by looking at the co-ordinates it should be clear that the light dots have high

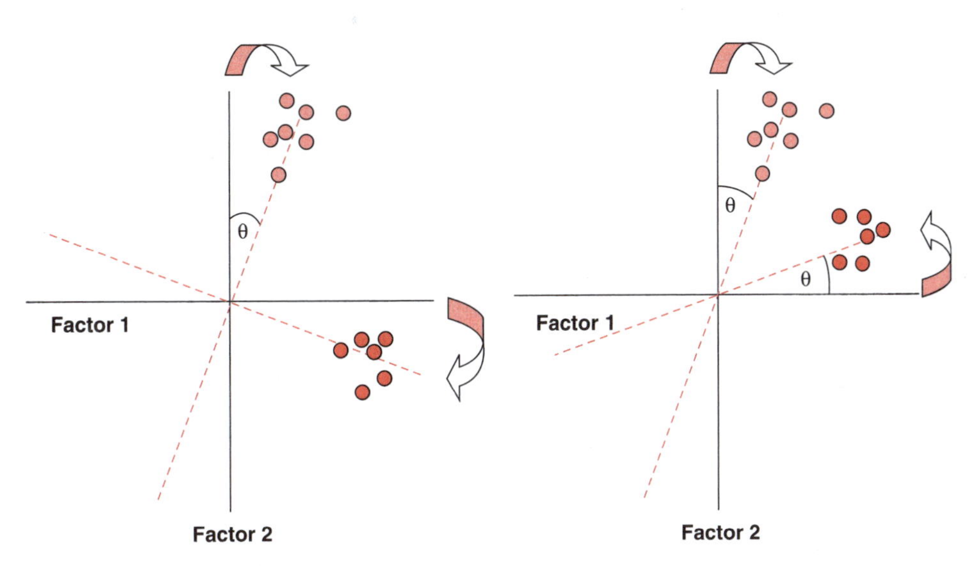

Figure 15.3 Schematic representations of factor rotation. The left graph displays orthogonal rotation whereas the right graph displays oblique rotation (see text for more details). θ is the angle through which the axes are rotated

loadings for factor 2 (they are a long way up this axis), and medium loadings on factor 1 (they are not very far up this axis). Conversely, the dark dots have high loadings for factor 1 and medium loadings on factor 2. By rotating the axes (dotted lines), we ensure that both clusters of variables are intersected by the factor to which they relate most. So, after rotation, the loadings of the variables are maximized onto one factor (the factor that intersects the cluster) and minimized on the remaining factor(s). If an axis passes through a cluster of variables, then these variables will have a loading of approximately 0 on the opposite axis. If this idea is confusing, then look at Figure 15.3 and think about the values of the co-ordinates before and after rotation (this is best achieved by turning the book when you look at the rotated axes).

There are two types of rotation that can be done. The first is orthogonal rotation, and the left-hand side of Figure 15.3 represents this method. In Chapter 8 we saw that the term 'orthogonal' means unrelated, and in this context it means that we rotate factors while keeping them independent. Before rotation, all factors are independent (i.e. they do not correlate at all) and orthogonal rotation ensures that the factors remain uncorrelated. That is why in Figure 15.3 the axes are turned while remaining perpendicular.[4] The other form of rotation is oblique rotation. The difference with oblique rotation is that the factors are allowed to correlate (hence, the axes of the right-hand diagram of Figure 15.3 do not remain perpendicular).

---

4 This term means that the axes are at right angles to one another.

The choice of rotation depends on whether there is a good theoretical reason to suppose that the factors should be related or independent (but see my later comments on this), and also how the variables cluster on the factors before rotation. On the first point, we might not expect alcoholism to be completely independent of achievement (after all, high achievement leads to high stress, which can lead to the drinks cabinet!). Therefore, on theoretical grounds, we might choose oblique rotation. On the second point, Figure 15.3 demonstrates how the positioning of clusters is important in determining how successful the rotation would be (note the position of the dark dots). Specifically, if an orthogonal rotation was carried out on the right-hand diagram it would be considerably less successful in maximizing loadings than the oblique rotation that is displayed. One approach is to run the analysis using both types of rotation. Pedhazur & Schmelkin (1991) suggest that if the oblique rotation demonstrates a negligible correlation between the extracted factors then it is reasonable to use the orthogonally rotated solution. If the oblique rotation reveals a correlated factor structure, then the orthogonally rotated solution should be discarded. In any case, an oblique rotation should be used only if there are good reasons to suppose that the underlying factors *could* be related in theoretical terms.

The mathematics behind factor rotation is complex (especially oblique rotation). However, in oblique rotation, because each factor can be rotated by different amounts a factor transformation matrix, Λ, is needed. The factor transformation matrix is a square matrix and its size depends on how many factors were extracted from the data. If two factors were extracted then it will be a 2×2 matrix, but if four factors were extracted then it becomes a 4×4 matrix. The values in the factor transformation matrix consist of sines and cosines of the angle of axis rotation (θ). This matrix is multiplied by the matrix of unrotated factor loadings, A, to obtain a matrix of rotated factor loadings.

For the case of two factors the factor transformation matrix would be:

$$\Lambda = \begin{pmatrix} \cos\theta & -\sin\theta \\ \sin\theta & \cos\theta \end{pmatrix}$$

Therefore, you should think of this matrix as representing the angle through which the axes have been rotated, or the degree to which factors have been rotated. The angle of rotation necessary to optimize the factor solution is found in an iterative way and different methods can be used.

15.3.6.1. Choosing a method of factor rotation ③

SPSS has three methods of orthogonal rotation (varimax, quartimax and equamax) and two methods of oblique rotation (direct oblimin and promax). These methods differ in how they rotate the factors and, therefore, the resulting output depends on which method you select. Quartimax rotation attempts to maximize the spread of factor loadings for a variable across all factors. Therefore, interpreting variables becomes easier. However, this often results in lots of variables loading highly onto a single factor. Varimax is the opposite in that it attempts to maximize the dispersion of loadings within factors. Therefore, it tries to load a smaller number of variables highly onto each factor resulting in more interpretable clusters of factors. Equamax is a hybrid of the other two approaches and is reported to behave fairly erratically (see Tabachnick & Fidell, 2001). For a first analysis, you should probably select varimax because it is a good general approach that simplifies the interpretation of factors.

The case with oblique rotations is more complex because correlation between factors is permitted. In the case of direct oblimin, the degree to which factors are allowed to correlate is determined by the value of a constant called delta. The default value in SPSS is zero and this ensures that high correlation between factors is not allowed (this is known as direct quartimin rotation). If you choose to set delta greater than zero (up to .8), then you can expect highly correlated factors; if you set delta less than zero (down to −.8) you can expect less correlated factors. The default setting of zero is sensible for most analyses and I don't recommend changing it unless you know what you are doing (see Pedhazur & Schmelkin, 1991, p. 620). Promax is a faster procedure designed for very large data sets.

In theory, the exact choice of rotation will depend largely on whether or not you think that the underlying factors should be related. If you expect the factors to be independent then you should choose one of the orthogonal rotations (I recommend varimax). If, however, there are theoretical grounds for supposing that your factors might correlate then direct oblimin should be selected. In practice, there are strong grounds to believe that orthogonal rotations are a complete nonsense for naturalistic data, and certainly for any data involving humans (can you think of any psychological construct that is not in any way correlated with some other psychological construct?). As such, some argue that orthogonal rotations should never be used.

15.3.6.2. Substantive importance of factor loadings ②

Once a factor structure has been found, it is important to decide which variables make up which factors. Earlier I said that the factor loadings were a gauge of the substantive importance of a given variable to a given factor. Therefore, it makes sense that we use these values to place variables with factors. It is possible to assess the statistical significance of a factor loading (after all, it is simply a correlation coefficient or regression coefficient); however, there are various reasons why this option is not as easy as it seems (see Stevens, 1992, p. 382). Typically, researchers take a loading of an absolute value of more than .3 to be important. However, the significance of a factor loading will depend on the sample size. Stevens (1992) produced a table of critical values against which loadings can be compared. To summarize, he recommends that for a sample size of 50 a loading of .722 can be considered significant, for 100 the loading should be greater than .512, for 200 it should be greater than .364, for 300 it should be greater than .298, for 600 it should be greater than .21, and for 1000 it should be greater than .162. These values are based on an alpha level of .01 (two-tailed), which allows for the fact that several loadings will need to be tested (see Stevens, 1992, pp. 382–384 for further detail). Therefore, in very large samples, small loadings can be considered statistically meaningful. SPSS does not provide significance tests of factor loadings but by applying Stevens's guidelines you should gain some insight into the structure of variables and factors.

The significance of a loading gives little indication of the substantive importance of a variable to a factor. This value can be found by squaring the factor loading to give an estimate of the amount of variance in a factor accounted for by a variable (like R^2). In this respect Stevens (1992) recommends interpreting only factor loadings with an absolute value greater than .4 (which explain around 16% of the variance in the variable).

15.4. RESEARCH EXAMPLE ②

One of the uses of factor analysis is to develop questionnaires: after all if you want to measure an ability or trait, you need to ensure that the questions asked relate to the construct that you intend to measure.[5] I have noticed that a lot of students become very stressed about SPSS. Therefore I wanted to design a questionnaire to measure a trait that I termed 'SPSS anxiety'. I decided to devise a questionnaire to measure various aspects of students' anxiety towards learning SPSS. I generated questions based on interviews with anxious and non-anxious students and came up with 23 possible questions to include. Each question was a statement followed by a five-point Likert scale ranging from 'strongly disagree' through 'neither agree nor disagree' to 'strongly agree'. The questionnaire is printed in Figure 15.4.

The questionnaire was designed to predict how anxious a given individual would be about learning how to use SPSS. What's more, I wanted to know whether anxiety about SPSS could be broken down into specific forms of anxiety. So, in other words, what latent variables contribute to anxiety about SPSS? With a little help from a few lecturer friends I collected 2571 completed questionnaires (at this point it should become apparent that this example is fictitious!). The data are stored in the file **SAQ.sav**. Load the data file into the SPSS data editor and have a look at the variables and their properties. The first thing to note is that each question (variable) is represented by a different column. We know that in SPSS cases (or people's data) are stored in rows and variables are stored in columns and so this layout is consistent with past chapters. The second thing to notice is that there are 23 variables labelled **q1** to **q23** and that each has a label indicating the question. By labelling my variables I can be very clear about what each variable represents (this is the value of giving your variables full titles rather than just using restrictive column headings).

15.4.1. Initial considerations ②

15.4.1.1. Sample size ②

Correlation coefficients fluctuate from sample to sample, much more so in small samples than in large. Therefore, the reliability of factor analysis is also dependent on sample size. Much has been written about the necessary sample size for factor analysis resulting in many 'rules of thumb'. The common rule is to suggest that a researcher has at least 10–15 participants per variable. Although I've heard this rule bandied about on numerous

5 I've included a file on the CD-ROM called **DesigningQuestionnaires.pdf**, which gives some advice on the process of developing questionnaires.

SD = Strongly Disagree, D = Disagree, N = Neither, A = Agree, SA = Strongly Agree

		SD	D	N	A	SA
1	Statistics make me cry	O	O	O	O	O
2	My friends will think I'm stupid for not being able to cope with SPSS	O	O	O	O	O
3	Standard deviations excite me	O	O	O	O	O
4	I dream that Pearson is attacking me with correlation coefficients	O	O	O	O	O
5	I don't understand statistics	O	O	O	O	O
6	I have little experience of computers	O	O	O	O	O
7	All computers hate me	O	O	O	O	O
8	I have never been good at mathematics	O	O	O	O	O
9	My friends are better at statistics than me	O	O	O	O	O
10	Computers are useful only for playing games	O	O	O	O	O
11	I did badly at mathematics at school	O	O	O	O	O
12	People try to tell you that SPSS makes statistics easier to understand but it doesn't	O	O	O	O	O
13	I worry that I will cause irreparable damage because of my incompetence with computers	O	O	O	O	O
14	Computers have minds of their own and deliberately go wrong whenever I use them	O	O	O	O	O
15	Computers are out to get me	O	O	O	O	O
16	I weep openly at the mention of central tendency	O	O	O	O	O
17	I slip into a coma whenever I see an equation	O	O	O	O	O
18	SPSS always crashes when I try to use it	O	O	O	O	O
19	Everybody looks at me when I use SPSS	O	O	O	O	O
20	I can't sleep for thoughts of eigenvectors	O	O	O	O	O
21	I wake up under my duvet thinking that I am trapped under a normal distribution	O	O	O	O	O
22	My friends are better at SPSS than I am	O	O	O	O	O
23	If I am good at statistics people will think I am a nerd	O	O	O	O	O

Figure 15.4 The SPSS anxiety questionnaire (SAQ)

occasions its empirical basis is unclear (although Nunnally, 1978, did recommend having 10 times as many participants as variables). Kass & Tinsley (1979) recommended having between 5 and 10 participants per variable up to a total of 300 (beyond which test parameters tend to be stable regardless of the participant to variable ratio). Indeed, Tabachnick & Fidell (2001) agree that 'it is comforting to have at least 300 cases for factor analysis' (p. 640) and Comrey & Lee (1992) class 300 as a good sample size, 100 as poor and 1000 as excellent.

Fortunately, recent years have seen empirical research done in the form of experiments using simulated data (so-called Monte Carlo studies). Arrindell & van der Ende (1985) used real-life data to investigate the effect of different participant to variable ratios. They concluded that changes in this ratio made little difference to the stability of factor solutions. More recently, Guadagnoli & Velicer (1988) found that the most important factors in determining reliable factor solutions was the absolute sample size and the absolute magnitude of factor loadings. In short, they argue that if a factor has four or more loadings greater than .6 then it is reliable regardless of sample size. Furthermore, factors with 10 or more loadings greater than .40 are reliable if the sample size is greater than 150. Finally, factors with a few low loadings should not be interpreted unless the sample size is 300 or more. More recently MacCallum, Widaman, Zhang & Hong (1999) have shown that the minimum sample size or sample to variable ratio depends on other aspects of the design of the study. In short, their study indicated that as communalities become lower the importance of sample size increases. With all communalities above .6, relatively small samples (less than 100) may be perfectly adequate. With communalities in the .5 range, samples between 100 and 200 can be good enough provided there are relatively few factors each with only a small number of indicator variables. In the worst scenario of low communalities (well below .5) and a larger number of underlying factors they recommend samples above 500.

What's clear from this work is that a sample of 300 or more will probably provide a stable factor solution but that a wise researcher will measure enough variables to measure adequately all of the factors that theoretically they would expect to find.

Another alternative is to use the Kaiser–Meyer–Olkin measure of sampling adequacy (KMO) (see Kaiser, 1970). The KMO can be calculated for individual and multiple variables and represents the ratio of the squared correlation between variables to the squared partial correlation between variables. The KMO statistic varies between 0 and 1. A value of 0 indicates that the sum of partial correlations is large relative to the sum of correlations, indicating diffusion in the pattern of correlations (hence, factor analysis is likely to be inappropriate). A value close to 1 indicates that patterns of correlations are relatively compact and so factor analysis should yield distinct and reliable factors. Kaiser (1974) recommends accepting values greater than .5 as barely acceptable (values below this should lead you either to collect more data or to rethink which variables to include). Furthermore, values between .5 and .7 are mediocre, values between .7 and .8 are good, values between .8 and .9 are great and values above .9 are superb (see Hutcheson & Sofroniou, 1999, pp. 224–225 for more detail).

15.4.1.2. Data screening ②

In this book I have used the expression 'if you put garbage in, you get garbage out'. This saying applies particularly to factor analysis because SPSS will always find a factor solution to a set of variables. However, the solution is unlikely to have any real meaning if the variables analysed are not sensible. The first thing to do when conducting a factor analysis is to look at the intercorrelation between variables. If our test questions measure the same underlying dimension (or dimensions) then we would expect them to correlate with each other (because they are measuring the same thing). Even if questions measure different aspects of the same things (e.g. we could measure overall anxiety in terms of sub-components such as worry, intrusive thoughts and physiological arousal), there should still be high intercorrelations between the variables relating to these sub-traits. If you find any variables that do not correlate with any other variables (or very few) then you should consider excluding these variables before the factor analysis is run. One extreme of this problem is when the correlation matrix resembles an identity matrix (see section 14.3.2). In this case, variables correlate only with themselves and all other correlation coefficients are close to zero. SPSS tests this using Bartlett's test of sphericity (see next section). The correlations between variables can be checked using the *correlate* procedure (see Chapter 4) to create a correlation matrix of all variables. This matrix can also be created as part of the main factor analysis.

The opposite problem is when variables correlate too highly. Although mild multicollinearity is not a problem for factor analysis it is important to avoid extreme multicollinearity (i.e. variables that are very highly correlated) and *singularity* (variables that are perfectly correlated). As with regression, singularity causes problems in factor analysis because it becomes impossible to determine the unique contribution to a factor of the variables that are highly correlated (as was the case for multiple regression). Therefore, at this early stage we look to eliminate any variables that don't correlate with any other variables or that correlate very highly with other variables ($R < .9$). Multicollinearity can be detected by looking at the determinant of the *R*-matrix (see next section).

As well as looking for interrelations, you should ensure that variables have roughly normal distributions and are measured at an interval level (which Likert scales are, perhaps wrongly, assumed to be!). The assumption of normality is most important if you wish to generalize the results of your analysis beyond the sample collected.

15.5. RUNNING THE ANALYSIS ②

Access the main dialog box (Figure 15.5) by using the **Analyze⇒Data Reduction⇒Factor...** menu path. Simply select the variables you want to include in the analysis (remember to exclude any variables that were identified as problematic during the data screening) and transfer them to the box labelled *Variables* by clicking on .

There are several options available, the first of which can be accessed by clicking on Descriptives... to access the dialog box in Figure 15.6. The *Univariate descriptives* option provides means and standard deviations for each variable. Most of the other options relate to the correlation matrix of variables (the *R*-matrix described earlier). The *Coefficients* option produces the *R*-matrix, and the *Significance levels* option will produce a matrix indicating the significance value of each correlation in the *R*-matrix. You can also ask for the *Determinant* of this matrix and this option is vital for testing for multicollinearity or singularity. The determinant of the *R*-matrix should be greater than .00001; if it is less than this value then look through the correlation matrix for variables that correlate very highly ($R > .8$) and consider eliminating one of the variables (or more depending on the extent of the problem) before proceeding. As mentioned in section 6.8, the choice of which of the

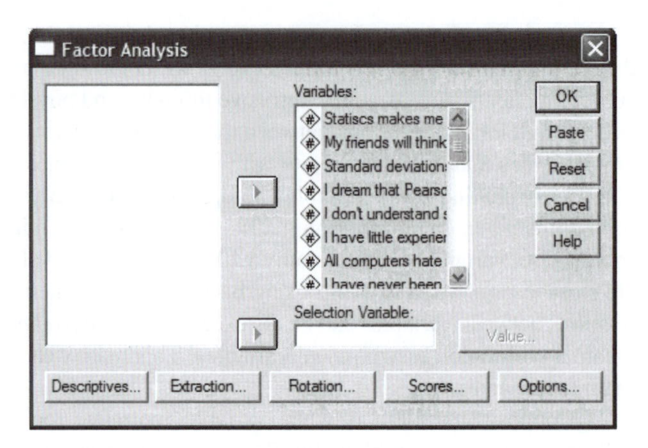

Figure 15.5 Main dialog box for factor analysis

two variables to eliminate will be fairly arbitrary and finding multicollinearity in the data should raise questions about the choice of items within your questionnaire.

KMO and Bartlett's test of sphericity produces the KMO and Bartlett's test. With a sample of 2571 we shouldn't have cause to worry about the sample size (see section 15.4.1.1). We have already stumbled across KMO (see section 15.4.1.1) and seen the various criteria for adequacy. We also came across Bartlett's test in Chapter 14: it examines whether the population correlation matrix resembles an identity matrix (i.e. it tests whether the off-diagonal components are zero). If the population correlation matrix resembles an identity matrix then it means that every variable correlates very badly with all other variables (i.e. all correlation coefficients are close to zero). If it *were* an identity matrix then it would mean that all variables are perfectly independent from one another (all correlation coefficients are zero). Given that we are looking for clusters of variables that measure similar things, it should be obvious why this scenario is problematic: if no variables correlate then there are no clusters to find.

The *Reproduced* option produces a correlation matrix based on the model (rather than the real data). Differences between the matrix based on the model and the matrix based on the observed data indicate the residuals of the model (i.e. differences). SPSS produces these residuals in the lower table of the reproduced matrix and we want relatively few of these values to be greater than .05. Luckily, to save us scanning this matrix, SPSS produces a summary of how many residuals lie above .05. The *Reproduced* option should be selected to obtain this summary. The *Anti-image* option produces an anti-image matrix of covariances and correlations. These matrices contain measures of sampling adequacy for each variable along the diagonal and the negatives of the partial correlation/covariances on the off-diagonals. The diagonal elements, like the KMO, should all be greater than .5 at a bare minimum if the sample is adequate for a given pair of variables. If any pair of variables has a value less than this, consider dropping one of them from the analysis. The off-diagonal elements should all be very small (close to zero) in a good model. When you have finished with this dialog box click on Continue to return to the main dialog box.

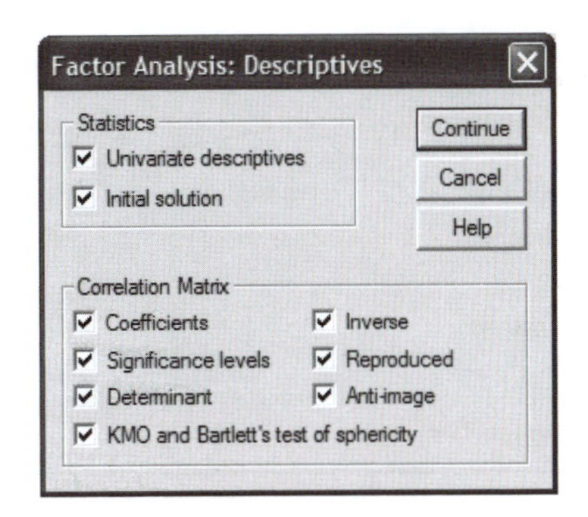

Figure 15.6 Descriptives in factor analysis

15.5.1. Factor extraction on SPSS ②

To access the *extraction* dialog box (Figure 15.7), click on Extraction... in the main dialog box. There are several ways of conducting a factor analysis (see section 15.3.1) and when and where you use the various methods depend on numerous things. For our purposes we will use *principal component* analysis which strictly speaking isn't factor analysis; however, the two procedures may often yield similar results (see section 15.3.3).

In the *Analyze* box there are two options: to analyse the *Correlation matrix* or to analyse the *Covariance matrix*. You should be happy with the idea that these two matrices are actually different versions of the same thing: the correlation matrix is the standardized version of the covariance matrix. Analysing the correlation matrix is a useful default method because it takes the standardized form of the matrix; therefore, if variables have been measured using different scales this will not affect the analysis. In this example, all variables have been measured using the same measurement scale (a five-point Likert scale), but often you will want to analyse variables that use different measurement scales. Analysing the correlation matrix ensures that differences in measurement scales are accounted for. In addition, even variables measured using the same scale can have very different variances and this too creates problems for principal component analysis. Using the correlation matrix eliminates this problem also. There are statistical reasons for preferring to analyse the covariance matrix[6] and generally the results will differ from analysis on the correlation matrix. However, the covariance matrix should be analysed only when your variables are commensurable.

The *Display* box has two options within it: to display the *Unrotated factor solution* and a *Scree plot*. The scree plot was described earlier and is a useful way of establishing how

6 The reason being that correlation coefficients are insensitive to variations in the dispersion of data whereas covariance is and so produces better-defined factor structures (see Tinsley & Tinsley, 1987).

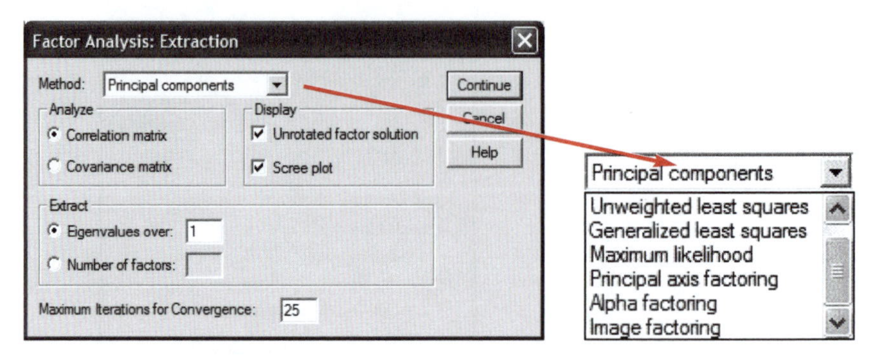

Figure 15.7 Dialog box for factor extraction

many factors should be retained in an analysis. The unrotated factor solution is useful in assessing the improvement of interpretation due to rotation. If the rotated solution is little better than the unrotated solution then it is possible that an inappropriate (or less optimal) rotation method has been used.

The *Extract* dialog box provides options pertaining to the retention of factors. You have the choice of either selecting factors with eigenvalues greater than a user-specified value or retaining a fixed number of factors. For the *Eigenvalues over* option the default is Kaiser's recommendation of eigenvalues over 1, but you could change this to Jolliffe's recommendation of .7 or any other value you want. It is probably best to run a primary analysis with the *Eigenvalues over* 1 option selected, select a scree plot, and compare the results. If looking at the scree plot and the eigenvalues over 1 lead you to retain the same number of factors then continue with the analysis and be happy. If the two criteria give different results then examine the communalities and decide for yourself which of the two criteria to believe. If you decide to use the scree plot then you may want to redo the analysis specifying the number of factors to extract. The number of factors to be extracted can be specified by selecting *Number of factors* and then typing the appropriate number in the space provided (e.g. 4).

15.5.2. Rotation ②

We have already seen that the interpretability of factors can be improved through rotation. Rotation maximizes the loading of each variable on one of the extracted factors whilst minimizing the loading on all other factors. This process makes it much clearer which variables relate to which factors. Rotation works through changing the absolute values of the variables whilst keeping their differential values constant. Click on Rotation... to access the dialog box in Figure 15.8. I've discussed the various rotation options in section 15.3.6.1, but to summarize, the exact choice of rotation will depend on whether or not you think that the underlying factors should be related. If there are theoretical grounds to think that the factors are independent (unrelated) then you should choose one of the orthogonal

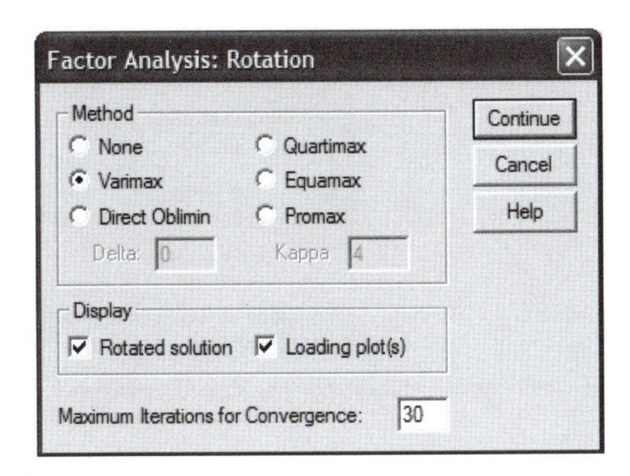

Figure 15.8 Factor analysis: *rotation* dialog box

rotations (I recommend varimax). However, if theory suggests that your factors might correlate then one of the oblique rotations (direct oblimin or promax) should be selected. In this example I've selected varimax.

The dialog box also has options for displaying the *Rotated solution* and a *Loading plot*. The rotated solution is displayed by default and is essential for interpreting the final rotated analysis. The loading plot will provide a graphical display of each variable plotted against the extracted factors up to a maximum of three factors (unfortunately SPSS cannot produce four- or five-dimensional graphs!). This plot is basically similar to Figure 15.1 and it uses the factor loading of each variable for each factor. With two factors these plots are fairly interpretable, and you should hope to see one group of variables clustered close to the *X*-axis and a different group of variables clustered around the *Y*-axis. If all variables are clustered between the axes, then the rotation has been relatively unsuccessful in maximizing the loading of a variable onto a single factor. With three factors these plots can become quite messy and certainly put considerable strain on the visual system! However, they can still be a useful way to determine the underlying structures within the data.

A final option is to set the *Maximum Iterations for Convergence*, which specifies the number of times that the computer will search for an optimal solution. In most circumstances the default of 25 is more than adequate for SPSS to find a solution for a given data set. However, if you have a large data set (like we have here) then the computer might have difficulty finding a solution (especially for oblique rotation). To allow for the large data set we are using, change the value to 30.

15.5.3. Scores ②

The *factor scores* dialog box (Figure 15.9) can be accessed by clicking on [Scores...] in the main dialog box. This option allows you to save factor scores (see section 15.2.3) for each person in the data editor. SPSS creates a new column for each factor extracted and then places the factor score for each person within that column. These scores can then be used

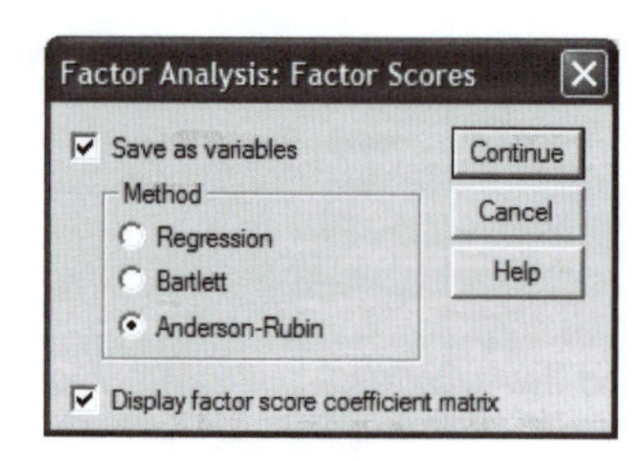

Figure 15.9 Factor analysis: *factor scores* dialog box

for further analysis, or simply to identify groups of participants who score highly on particular factors. There are three methods of obtaining these scores, all of which were described in sections 15.2.4 and 15.2.4.1. If you want to ensure that factor scores are uncorrelated then select the *Anderson-Rubin* method; if correlations between factor scores are acceptable then choose the *Regression* method.

As a final option, you can ask SPSS to produce the factor score coefficient matrix. This matrix was the matrix *B* described in section 15.2.4. This matrix can be useful if, for whatever reason, you wish to construct factor equations such as those in equation (15.5), because it provides you with the values of *b* for each of the variables.

15.5.4. Options ②

This set of options (Figure 15.10) can be obtained by clicking on Options... in the main dialog box. Missing data are a problem for factor analysis just like most other procedures and SPSS provides a choice of excluding cases or estimating a value for a case. Tabachnick & Fidell (2001) have an excellent chapter on data screening (Chapter 4—see also the rather less excellent Chapter 3 of this book). Based on their advice, you should consider the distribution of missing data. If the missing data are non-normally distributed or the sample size after exclusion is too small then estimation is necessary. SPSS uses the mean as an estimate (*Replace with mean*). These procedures lower the standard deviation of variables and so can lead to significant results that would otherwise be non-significant. Therefore, if missing data are random, you might consider excluding cases. SPSS allows you either to *Exclude cases listwise*, in which case any participant with missing data for any variable is excluded, or to *Exclude cases pairwise*, in which case a participant's data are excluded only from calculations for which a datum is missing. If you exclude cases pairwise your estimates can go all over the place so it's probably safest to opt to exclude cases listwise unless this results in a massive loss of data.

The final two options relate to how coefficients are displayed. By default SPSS will list variables in the order in which they are entered into the data editor. Usually, this format is most convenient. However, when interpreting factors it is sometimes useful to list variables by size. By selecting *Sorted by size*, SPSS will order the variables by their factor

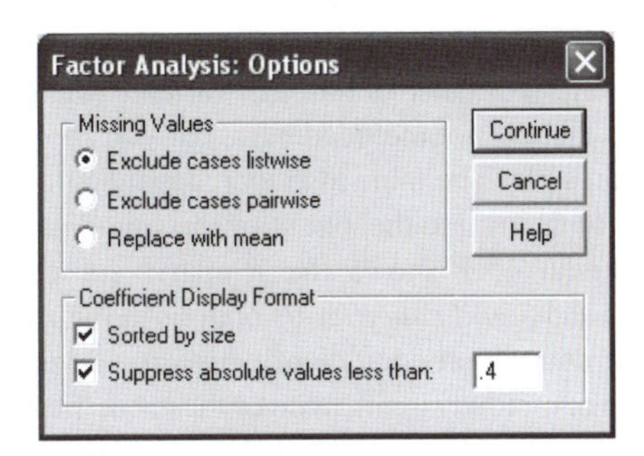

Figure 15.10 Factor analysis: *options* dialog box

loadings. In fact, it does this sorting fairly intelligently so that all of the variables that load highly onto the same factor are displayed together. The second option is to *Suppress absolute values less than* a specified value (by default .1). This option ensures that factor loadings within ±.1 are not displayed in the output. Again, this option is useful for assisting in interpretation. The default value is probably sensible, but on your first analysis I recommend changing it either to .4 (for interpretation purposes) or to a value reflecting the expected value of a significant factor loading given the sample size (see section 15.3.6.2). This will make interpretation simpler. You can, if you like, rerun the analysis and set this value lower just to check you haven't missed anything (like a loading of 0.39). For this example set the value at .4.

15.6. INTERPRETING OUTPUT FROM SPSS ②

Select the same options as I have in the screen diagrams and run a factor analysis with orthogonal rotation. Having done this, select the *Direct Oblimin* option in Figure 15.8 and repeat the analysis. You should obtain two outputs identical in all respects except that one used an orthogonal rotation and the other an oblique one.

For the purposes of saving space in this section I set the default SPSS options such that each variable is referred to only by its label on the data editor (e.g. Q12). On the output *you* obtain, you should find that SPSS uses the value label (the question itself) in all of the output. When using the output in this chapter just remember that Q1 represents question 1, Q2 represents question 2 and Q17 represents question 17. By referring back to Figure 15.4 and matching the question number to the variable name you can identify each question.

15.6.1. Preliminary analysis ②

The first body of output concerns data screening, assumption testing and sampling adequacy. You'll find several large tables (or matrices) that tell us interesting things about our data.

If you selected the *Univariate descriptives* option in Figure 15.6 then the first table will contain descriptive statistics for each variable (the mean, standard deviation and number of cases). This table is not included here, but you should have enough experience to be able to interpret this table. The table also includes the number of missing cases; this summary is a useful way to determine the extent of missing data.

SPSS Output 15.1 shows the *R*-matrix (or correlation matrix)[7] produced using the *Coefficients* and *Significance levels* options in Figure 15.6. The top half of this table contains the Pearson correlation coefficient between all pairs of questions whereas the bottom half contains the one-tailed significance of these coefficients. You should be comfortable with the idea that to do a factor analysis we need to have variables that correlate fairly well, but not perfectly. Also, any variables that correlate with no others should be eliminated. Therefore, we can use this correlation matrix to check the pattern of relationships. The easiest way to do this is by scanning the significance values and looking for any variable for which the majority of values are greater than .05. Then scan the correlation coefficients themselves and look for any greater than .9. If any are found then you should be aware that a problem could arise because of singularity in the data: check the determinant of the correlation matrix and, if necessary, eliminate one of the two variables causing the problem. The determinant is listed at the bottom of the matrix (blink and you'll miss it). For these data its value is 5.271E-04 (which is .0005271) which is greater than the necessary value of .00001 (see section 15.5).[8] Therefore, we can be confident that multicollinearity is not a problem for these data. In summary, all questions in the SAQ correlate fairly well with all others (this is partly because of the large sample) and none of the correlation coefficients are particularly large; therefore, there is no need to consider eliminating any questions at this stage.

SPSS Output 15.2 shows the inverse of the correlation matrix (R^{-1}), which is used in various calculations (including factor scores—see section 15.2.4). This matrix is produced using the *Inverse* option in Figure 15.6 but in all honesty is useful only if you want some insight into the calculations that go on in a factor analysis. Most of us have more interesting things to do than gain insight into the workings of factor analysis and the practical use of this matrix is minimal—so ignore it!

SPSS Output 15.3 shows several very important parts of the output: the KMO, Bartlett's test of sphericity and the anti-image correlation and covariance matrices (note that these matrices have been edited down to contain only the first and last five variables). The anti-image correlation and covariance matrices provide similar information (remember the relationship between covariance and correlation) and so only the anti-image correlation matrix need be studied in detail as it is the most informative. These tables are obtained using the *KMO and Bartlett's test of sphericity* and the *Anti-image* options in Figure 15.6.

We came across the KMO statistic in section 15.4.1.1 and saw that Kaiser (1974) recommends a bare minimum of .5 and that values between .5 and .7 are mediocre, values

---

7 To save space I have edited out several columns of data from the large tables; only data for the first and last five questions in the questionnaire are included.

8 Strictly speaking, the determinant or the correlation matrix should be checked only in factor analysis: in pure principal component analysis it isn't relevant.

Correlation Matrix[a]

		Q01	Q02	Q03	Q04	Q05	Q19	Q20	Q21	Q22	Q23
Correlation	Q01	1.000	−.099	−.337	.436	.402	−.189	.214	.329	−.104	−.004
	Q02	−.099	1.000	.318	−.112	−.119	.203	−.202	−.205	.231	.100
	Q03	−.337	.318	1.000	−.380	−.310	.342	−.325	−.417	.204	.150
	Q04	.436	−.112	−.380	1.000	.401	−.186	.243	.410	−.098	−.034
	Q05	.402	−.119	−.310	.401	1.000	−165	.200	.335	−.133	−.042
	Q06	.217	−.074	−.227	.278	.257	−.167	.101	.272	−.165	−.069
	Q07	.305	−.159	−.382	.409	.339	−.269	.221	.483	−.168	−.070
	Q08	.331	−.050	−.259	.349	.269	−.159	.175	.296	−.079	−.050
	Q09	−.092	.315	.300	−.125	−.096	.249	−.159	−.136	.257	.171
	Q10	.214	−.084	−.193	.216	.258	−.127	.084	.193	−.131	−.062
	Q11	.357	−.144	−.351	.369	.298	−.200	.255	.346	−.162	−.086
	Q12	.345	−.195	−.410	.442	.347	−.267	.298	.441	−.167	−.046
	Q13	.355	−.143	−.318	.344	.302	−.227	.204	.374	−.195	−.053
	Q14	.338	−.165	−.371	.351	.315	−.254	.226	.399	−.170	−.048
	Q15	.246	−.165	−.312	.334	.261	−.210	.206	.300	−.168	−.062
	Q16	.499	−.168	−.419	.416	.395	−.267	.265	.421	−.156	−.082
	Q17	.371	−.087	−.327	.383	.310	−.163	.205	.363	−.126	−.092
	Q18	.347	−.164	−.375	.382	.322	−.257	.235	.430	−.160	−.080
	Q19	−.189	.203	.342	−.186	−.165	1.000	−.249	−.275	.234	.122
	Q20	.214	−.202	−.325	.243	.200	−.249	1.000	.468	−.100	−.035
	Q21	.329	−.205	−.417	.410	.335	−.275	.468	1.000	−.129	−.068
	Q22	−.104	.231	.204	−.098	−.133	.234	−.100	−.129	1.000	.230
	Q23	−.004	.100	.150	−.034	−.042	.122	−.035	−.068	.230	1.000
Sig. (1-tailed)	Q01		.000	.000	.000	.000	.000	.000	.000	.000	.410
	Q02	.000		.000	.000	.000	.000	.000	.000	.000	.000
	Q03	.000	.000		.000	.000	.000	.000	.000	.000	.000
	Q04	.000	.000	.000		.000	.000	.000	.000	.000	.043
	Q05	.000	.000	.000	.000		.000	.000	.000	.000	.017
	Q06	.000	.000	.000	.000	.000	.000	.000	.000	.000	.000
	Q07	.000	.000	.000	.000	.000	.000	.000	.000	.000	.000
	Q08	.000	.000	.000	.000	.000	.000	.000	.000	.000	.005
	Q09	.000	.000	.000	.000	.000	.000	.000	.000	.000	.000
	Q10	.000	.000	.000	.000	.000	.000	.000	.000	.000	.001
	Q11	.000	.000	.000	.000	.000	.000	.000	.000	.000	.000
	Q12	.000	.000	.000	.000	.000	.000	.000	.000	.000	.009
	Q13	.000	.000	.000	.000	.000	.000	.000	.000	.000	.004
	Q14	.000	.000	.000	.000	.000	.000	.000	.000	.000	.007
	Q15	.000	.000	.000	.000	.000	.000	.000	.000	.000	.001
	Q16	.000	.000	.000	.000	.000	.000	.000	.000	.000	.000
	Q17	.000	.000	.000	.000	.000	.000	.000	.000	.000	.000
	Q18	.000	.000	.000	.000	.000	.000	.000	.000	.000	.000
	Q19	.000	.000	.000	.000	.000		.000	.000	.000	.000
	Q20	.000	.000	.000	.000	.000	.000		.000	.000	.039
	Q21	.000	.000	.000	.000	.000	.000	.000		.000	.000
	Q22	.000	.000	.000	.000	.000	.000	.000	.000		.000
	Q23	.410	.000	.000	.043	.017	.000	.039	.000	.000	

a Determinant = 5.271E-04

SPSS Output 15.1

Inverse of Correlation Matrix

	Q01	Q02	Q03	Q04	Q05	Q19	Q20	Q21	Q22	Q23
Q01	1.595	–.028	.087	–.268	–.233	.017	–.024	.011	.002	–.078
Q02	–.028	1.232	–.224	–.057	.013	–.037	.076	.062	–.148	–.003
Q03	.087	–.224	1.661	.138	.057	–.175	.118	.122	–.009	–.103
Q04	–.268	–.057	.138	1.626	–.203	–.049	–.006	–.149	–.045	–.023
Q05	–.233	.013	.057	–.203	1.410	–.024	–.016	–.074	.045	–.006
Q06	.034	–.078	–.072	–.011	–.055	–.023	.080	.069	.058	.025
Q07	.039	.025	.127	–.152	–.072	.105	.077	–.386	.019	–.012
Q08	–.087	–.051	–.013	–.134	–.045	.074	.034	–.039	–.035	.003
Q09	–.023	–.242	–.208	.043	–.027	–.141	.050	–.047	–.156	–.110
Q10	–.017	–.015	–.023	.009	–.124	–.012	.056	.026	.023	.017
Q11	–.075	.061	.121	–.041	.000	–.010	–.140	–.009	.055	.015
Q12	–.011	.046	.147	–.259	–.091	.060	–.100	–.141	.026	–.038
Q13	–.145	–.011	–.055	.040	.007	.014	.028	–.061	.077	–.042
Q14	–.064	.033	.115	–.007	–.040	.063	.002	–.110	.041	–.034
Q15	.138	.050	.013	–.098	.021	.013	–.054	.058	.034	–.030
Q16	–.454	–.017	.142	–.063	–.155	.071	–.008	–.158	–.005	.033
Q17	–.084	–.045	.063	–.064	–.030	–.074	.025	–.077	.015	.080
Q18	–.041	.028	.070	–.044	.004	.047	–.004	–.136	–.037	.033
Q19	.017	–.037	–.175	–.049	–.024	1.264	.120	.048	–.141	–.045
Q20	–.024	.076	.118	–.006	–.016	.120	1.370	–.511	–.014	–.034
Q21	.011	.062	.122	–.149	–.074	.048	–.511	1.830	–.036	.018
Q22	.002	–.148	–.009	–.045	.045	–.141	–.014	–.036	1.200	–.202
Q23	–.078	–.003	–.103	–.023	–.006	–.045	–.034	.018	–.202	1.094

SPSS Output 15.2

between .7 and .8 are good, values between .8 and .9 are great and values above .9 are superb (see Hutcheson & Sofroniou, 1999, pp. 224–225 for more detail). For these data the value is .93, which falls into the range of being superb: so, we should be confident that factor analysis is appropriate for these data.

I mentioned that KMO can be calculated for multiple and individual variables. The KMO values for individual variables are produced on the diagonal of the anti-image correlation matrix (I have highlighted the values in bold). These values make the anti-image correlation matrix an extremely important part of the output (although the anti-image covariance matrix can be ignored). As well as checking the overall KMO statistic, it is important to examine the diagonal elements of the anti-image correlation matrix: the value should be above the bare minimum of .5 for all variables (and preferably higher). For these data all values are well above .5, which is good news! If you find any variables with values below .5 then you should consider excluding them from the analysis (or run the analysis with and without that variable and note the difference). Removal of a variable affects the KMO statistics, so if you do remove a variable be sure to re-examine the new anti-image correlation matrix. As for the rest of the anti-image correlation matrix, the off-diagonal elements represent the partial correlations between variables. For a good factor analysis we want these correlations to be very small (the smaller the better). So, as a final

KMO and Bartlett's Test

Kaiser-Meyer-Olkin Measure of Sampling Adequacy.		.930
Bartlett's Test of Sphericity	Approx. Chi-Square	19334.492
	df	253
	Sig.	.000

Anti-image Matrices

Anti-image Correlation

	Q01	Q02	Q03	Q04	Q05	Q19	Q20	Q21	Q22	Q23
Q01	**.930**	−2.00E-02	5.320E-02	−.167	−.156	1.231E-02	−1.61E-02	6.436E-03	1.459E-03	−5.92E-02
Q02	−2.00E-02	**.875**	−.157	−4.05E-02	1.019E-02	−2.93E-02	5.877E-02	4.139E-02	−.121	−2.39E-03
Q03	5.320E-02	−.157	**.951**	8.381E-02	3.725E-02	−.121	7.790E-02	6.983E-02	−6.57E-03	−7.63E-02
Q04	−.167	−4.05E-02	8.381E-02	**.955**	−.134	−3.42E-02	−4.07E-03	−8.64E-02	−3.25E-02	−1.71E-02
Q05	−.156	1.019E-02	3.725E-02	−.134	**.960**	−1.77E-02	−1.14E-02	−4.59E-02	3.455E-02	−4.75E-03
Q06	2.016E-02	−5.33E-02	−4.23E-02	−6.77E-03	−3.52E-02	−1.53E-02	5.144E-02	3.867E-02	3.984E-02	1.846E-02
Q07	2.264E-02	1.648E-02	7.159E-02	−8.66E-02	−4.43E-02	6.802E-02	4.796E-02	−.208	1.278E-02	−8.07E-03
Q08	−4.91E-02	−3.28E-02	−7.21E-03	−7.49E-02	−2.72E-02	4.696E-02	2.094E-02	−2.05E-02	−2.28E-02	2.202E-03
Q09	−1.64E-02	−.193	−.142	2.960E-02	−1.98E-02	−.111	3.803E-02	−3.06E-02	−.126	−9.25E-02
Q10	−1.22E-02	−1.21E-02	−1.62E-02	6.031E-03	−9.32E-02	−9.16E-03	4.272E-02	1.699E-02	1.867E-02	1.462E-02
Q11	−4.07E-02	3.757E-02	6.429E-02	−2.22E-02	−3.27E-05	−5.81E-03	−8.18E-02	−4.58E-03	3.425E-02	9.600E-03
Q12	−6.62E-03	3.129E-02	8.667E-02	−.154	−5.83E-02	4.030E-02	−6.47E-02	−7.93E-02	1.802E-02	−2.79E-02
Q13	−8.51E-02	−7.60E-03	−3.16E-02	2.303E-02	4.238E-03	8.994E-03	1.785E-02	−3.33E-02	5.233E-02	−2.99E-02
Q14	−3.98E-02	2.313E-02	6.946E-02	−4.26E-03	−2.63E-02	4.399E-02	1.124E-03	−6.32E-02	2.896E-02	−2.56E-02
Q15	8.860E-02	3.680E-02	8.159E-03	−6.21E-02	1.413E-02	9.384E-03	−3.74E-02	3.483E-02	2.485E-02	−2.36E-02
Q16	−.264	−1.14E-02	8.057E-02	−3.62E-02	−9.57E-02	4.659E-02	−5.30E-03	−8.54E-02	−3.26E-03	2.298E-02
Q17	−4.74E-02	−2.87E-02	3.467E-02	−3.55E-02	−1.80E-02	−4.68E-02	1.547E-02	−4.06E-02	1.004E-02	5.457E-02
Q18	−2.31E-02	1.825E-02	3.880E-02	−2.45E-02	2.489E-03	2.980E-02	−2.54E-03	−7.17E-02	−2.40E-02	2.257E-02
Q19	1.231E-02	−2.93E-02	−.121	−3.42E-02	−1.77E-02	**.941**	9.081E-02	3.145E-02	−.115	−3.84E-02
Q20	−1.61E-02	5.877E-02	7.790E-02	−4.07E-03	−1.14E-02	9.081E-02	**.889**	−.323	−1.12E-02	−2.81E-02
Q21	6.436E-03	4.139E-02	6.983E-02	−8.64E-02	−4.59E-02	3.145E-02	−.323	**.929**	−2.41E-02	1.275E-02
Q22	1.459E-03	−.121	−6.57E-03	−3.25E-02	3.455E-02	−.115	−1.12E-02	−2.41E-02	**.878**	−.176
Q23	−5.92E-02	−2.39E-03	−7.63E-02	−1.71E-02	−4.75E-03	−3.84E-02	−2.81E-02	1.275E-02	−.176	**.766**

SPSS Output 15.3

check you can just look through to see that the off-diagonal elements are small (they should be for these data).

Bartlett's measure tests the null hypothesis that the original correlation matrix is an identity matrix. For factor analysis to work we need some relationships between variables and if the *R*-matrix were an identity matrix then all correlation coefficients would be zero. Therefore, we want this test to be *significant* (i.e. have a significance value less than .05). A significant test tells us that the *R*-matrix is not an identity matrix; therefore, there are some relationships between the variables we hope to include in the analysis. For these data, Bartlett's test is highly significant ($p < .001$), and therefore factor analysis is appropriate.

Cramming Samantha's Tips: Preliminary Analysis

- Scan the *Correlation Matrix*: look for variables that don't correlate with any other variables, or correlate very highly ($r = .9$) with one or more other variables. In factor analysis, check that the determinant of this matrix is bigger than .00001; if it is then multicollinearity isn't a problem.
- In the table labelled *KMO and Bartlett's Test* the KMO statistic should be greater than .5 as a bare minimum; if it isn't collect more data. Bartlett's test of sphericity should be significant (the value of *Sig.* should be less than .05). You can also check the KMO statistic for individual variables by looking at the diagonal of the *Anti-Image Matrices*: again, these values should be above .5 (this is useful for identifying problematic variables if the overall KMO is unsatisfactory).

15.6.2. Factor extraction ②

The first part of the factor extraction process is to determine the linear components within the data set (the eigenvectors) by calculating the eigenvalues of the *R*-matrix (see section 15.3.4). We know that there are as many components (eigenvectors) in the *R*-matrix as there are variables, but most will be unimportant. To determine the importance of a particular vector we look at the magnitude of the associated eigenvalue. We can then apply criteria to determine which factors to retain and which to discard. By default SPSS uses Kaiser's criterion of retaining factors with eigenvalues greater than 1 (see Figure 15.7).

SPSS Output 15.4 lists the eigenvalues associated with each linear component (factor) before extraction, after extraction and after rotation. Before extraction, SPSS has identified 23 linear components within the data set (we know that there should be as many eigenvectors as there are variables and so there will be as many factors as variables—see section 15.3.4). The eigenvalues associated with each factor represent the variance explained by that particular linear component and SPSS also displays the eigenvalue in terms of the percentage of variance explained (so, factor 1 explains 31.696% of total variance). It should be clear that the first few factors explain relatively large amounts of variance (especially factor 1) whereas subsequent factors explain only small amounts of variance. SPSS then extracts all factors with eigenvalues greater than 1, which leaves us with four

Total Variance Explained

Component	Initial Eigenvalues			Extraction Sums of Squared Loadings			Rotation Sums of Squared Loadings		
	Total	% of Variance	Cumulative %	Total	% of Variance	Cumulative %	Total	% of Variance	Cumulative %
1	7.290	31.696	31.696	7.290	31.696	31.696	3.730	16.219	16.219
2	1.739	7.560	39.256	1.739	7.560	39.256	3.340	14.523	30.742
3	1.317	5.725	44.981	1.317	5.725	44.981	2.553	11.099	41.842
4	1.227	5.336	50.317	1.227	5.336	50.317	1.949	8.475	50.317
5	.988	4.295	54.612						
6	.895	3.893	58.504						
7	.806	3.502	62.007						
8	.783	3.404	65.410						
9	.751	3.265	68.676						
10	.717	3.117	71.793						
11	.684	2.972	74.765						
12	.670	2.911	77.676						
13	.612	2.661	80.337						
14	.578	2.512	82.849						
15	.549	2.388	85.236						
16	.523	2.275	87.511						
17	.508	2.210	89.721						
18	.456	1.982	91.704						
19	.424	1.843	93.546						
20	.408	1.773	95.319						
21	.379	1.650	96.969						
22	.364	1.583	98.552						
23	.333	1.448	100.000						

Extraction Method: Principal Component Analysis.

SPSS Output 15.4

factors. The eigenvalues associated with these factors are again displayed (and the percentage of variance explained) in the columns labelled *Extraction Sums of Squared Loadings*. The values in this part of the table are the same as the values before extraction, except that the values for the discarded factors are ignored (hence, the table is blank after the fourth factor). In the final part of the table (labelled *Rotation Sums of Squared Loadings*), the eigenvalues of the factors after rotation are displayed. Rotation has the effect of optimizing the factor structure and one consequence for these data is that the relative importance of the four factors is equalized. Before rotation, factor 1 accounted for considerably more variance than the remaining three (31.696% compared to 7.560 5.725 and 5.336%); however, after extraction it accounts for only 16.219% of variance (compared to 14.523, 11.099 and 8.475% respectively).

SPSS Output 15.5 shows the table of communalities before and after extraction. Remember that the communality is the proportion of common variance within a variable (see section 15.3.1). Principal component analysis works on the initial assumption that all variance is common; therefore, before extraction the communalities are all 1 (see column labelled *Initial*). In effect, all of the variance associated with a variable is assumed to be

Communalities

	Initial	Extraction
Q01	1.000	.435
Q02	1.000	.414
Q03	1.000	.530
Q04	1.000	.469
Q05	1.000	.343
Q06	1.000	.654
Q07	1.000	.545
Q08	1.000	.739
Q09	1.000	.484
Q10	1.000	.335
Q11	1.000	.690
Q12	1.000	.513
Q13	1.000	.536
Q14	1.000	.488
Q15	1.000	.378
Q16	1.000	.487
Q17	1.000	.683
Q18	1.000	.597
Q19	1.000	.343
Q20	1.000	.484
Q21	1.000	.550
Q22	1.000	.464
Q23	1.000	.412

Extraction Method: Principal Component

Component Matrix[a]

	Component			
	1	2	3	4
Q18	.701			
Q07	.685			
Q16	.679			
Q13	.673			
Q12	.669			
Q21	.658			
Q14	.656			
Q11	.652			−.400
Q17	.643			
Q04	.634			
Q03	−.629			
Q15	.593			
Q01	.586			
Q05	.556			
Q08	.549	.401		−.417
Q10	.437			
Q20	.436		−.404	
Q19	−.427			
Q09		.627		
Q02		.548		
Q22		.465		
Q06	.562		.571	
Q23				.507

Extraction Method: Principal Component Analysis.
a 4 components extracted

SPSS Output 15.5

common variance. Once factors have been extracted, we have a better idea of how much variance is, in reality, common. The communalities in the column labelled *Extraction* reflect this common variance. So, for example, we can say that 43.5% of the variance associated with question 1 is common, or shared, variance. Another way to look at these communalities is in terms of the proportion of variance explained by the underlying factors. Before extraction, there are as many factors as there are variables, so all variance is explained by the factors and communalities are all 1. However, after extraction some of the factors are discarded and so some information is lost. The retained factors cannot explain all of the variance present in the data, but they can explain some. The amount of variance in each variable that can be explained by the retained factors is represented by the communalities after extraction.

SPSS Output 15.5 also shows the component matrix before rotation. This matrix contains the loadings of each variable onto each factor. By default SPSS displays all loadings; however, we requested that all loadings less than .4 be suppressed in the output (see Figure 15.10) and so there are blank spaces for many of the loadings. This matrix is not particularly important for interpretation, but it is interesting to note that before rotation most variables load highly onto the first factor (this is why this factor accounts for most of the variance in SPSS Output 15.4).

At this stage SPSS has extracted four factors. Factor analysis is an exploratory tool and so it should be used to guide the researcher to make various decisions: you shouldn't leave the computer to make them. One important decision is the number of factors to extract. In section 15.3.5 we saw various criteria for assessing the importance of factors. By Kaiser's criterion we should extract four factors and this is what SPSS has done. However, this criterion is accurate when there are less than 30 variables and communalities after extraction are greater than .7 or when the sample size exceeds 250 and the average communality is greater than .6. The communalities are shown in SPSS Output 15.5, and none exceed .7. The average of the communalities can be found by adding them up and dividing by the number of communalities (11.573/23 = .503). So, on both grounds Kaiser's rule may not be accurate. However, you should consider the huge sample that we have, because the research into Kaiser's criterion gives recommendations for much smaller samples. By Jolliffe's criterion (retain factors with eigenvalues greater than .7) we should retain 10 factors (see SPSS Output 15.4), but there is little to recommend this criterion over Kaiser's. As a final guide we can use the scree plot which we asked SPSS to produce by using the option in Figure 15.7. The scree plot is shown in SPSS Output 15.6 with a thunderbolt indicating the point of inflexion on the curve. This curve is difficult to interpret because the curve begins to tail off after three factors, but there is another drop after four factors before a stable plateau is reached. Therefore, we could probably justify retaining either two or four factors. Given the large sample, it is probably safe to assume Kaiser's criterion; however, you might like to rerun the analysis specifying that SPSS extract only two factors (see Figure 15.7) and compare the results.

SPSS Output 15.7 shows an edited version of the reproduced correlation matrix that was requested using the option in Figure 15.6. The top half of this matrix (labelled *Reproduced Correlations*) contains the correlation coefficients between all of the questions based on the factor model. The diagonal of this matrix contains the communalities after extraction for each variable (I've changed these values to bold type; you can check the values against SPSS Output 15.5).

The correlations in the reproduced matrix differ from those in the R-matrix because they stem from the model rather than the observed data. If the model were a perfect fit of the data then we would expect the reproduced correlation coefficients to be the same as the original correlation coefficients. Therefore, to assess the fit of the model we can look at the differences between the observed correlations and the correlations based on the model. For example, if we take the correlation between questions 1 and 2, the correlation based on the observed data is −.099 (taken from SPSS Output 15.1). The correlation based on the model is −.112, which is slightly higher. We can calculate the difference as follows:

$$\begin{aligned} \text{residual} &= r_{\text{observed}} - r_{\text{from model}} \\ \text{residual}_{Q_1 Q_2} &= (-.099) - (-.112) \\ &= .013 \text{ or } 1.3\text{E-02} \end{aligned}$$

You should notice that this difference is the value quoted in the lower half of the reproduced matrix (labelled *Residual*) for the questions 1 and 2. Therefore, the lower half of the

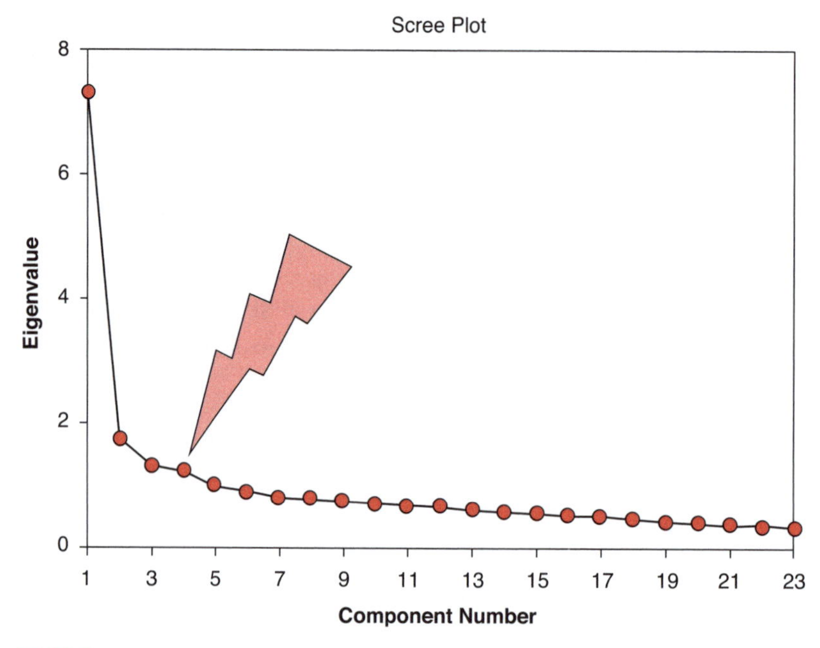

SPSS Output 15.6

reproduced matrix contains the differences between the observed correlation coefficients and the ones predicted from the model. For a good model these values will all be small. In fact, we want most values to be less than .05. Rather than scan this huge matrix, SPSS provides a footnote summary, which states how many residuals have an absolute value greater than .05. For these data there are 91 residuals (35%) that are greater than .05. There are no hard and fast rules about what proportion of residuals should be below .05; however, if more than 50% are greater than .05 you probably have grounds for concern.

Cramming Samantha's Tips: Factor Extraction

- To decide how many factors to extract look at the table labelled *Communalities* and the column labelled *Extraction*. If these values are all .7 or above and you have less than 30 variables then the SPSS default option for extracting factors is fine (Kaiser's criterion of retaining factors with eigenvalues greater than 1). Likewise, if your sample size exceeds 250 and the average of the communalities is .6 or greater then the default option is fine. Alternatively, with 200 or more participants the scree plot can be used.
- Check the bottom of the table labelled *Reproduced Correlations* for the percentage of 'nonredundant residuals with absolute values > .05'. This percentage should be less than 50% and the smaller it is, the better.

Reproduced Correlations

		Q01	Q02	Q03	Q04	Q05	Q19	Q20	Q21	Q22	Q23
Reproduced Correlation	Q01	**.435**[b]	–.112	–.372	.447	.376	–.204	.342	.449	–2.54E-02	4.463E-02
	Q02	–.112	**.414**[b]	.380	–.134	–.122	.357	–.301	–.254	.333	.246
	Q03	–.372	.380	**.530**[b]	–.399	–.345	.403	–.440	–.488	.275	.158
	Q04	.447	–.134	–.399	**.469**[b]	.399	–.231	.353	.480	–4.97E-02	4.179E-02
	Q05	.376	–.122	–.345	.399	**.343**[b]	–.207	.292	.412	–6.03E-02	2.794E-02
	Q06	.218	–3.34E-02	–.200	.278	.273	–.147	–2.06E-02	.244	–.209	–8.17E-02
	Q07	.366	–.148	–.373	.419	.380	–.254	.219	.430	–.179	–3.72E-02
	Q08	.412	2.192E-03	–.270	.390	.312	–.104	.164	.282	–9.89E-02	–.136
	Q09	–4.25E-02	.430	.352	–7.31E-02	–7.97E-02	.363	–.218	–.191	.417	.323
	Q10	.172	–6.12E-02	–.181	.212	.205	–.137	5.763E-03	.188	–.197	–.110
	Q11	.423	–9.74E-02	–.357	.419	.348	–.198	.200	.342	–.209	–.202
	Q12	.402	–.219	–.440	.448	.397	–.302	.354	.503	–.136	1.064E-02
	Q13	.347	–.122	–.342	.395	.360	–.231	.163	.384	–.203	–7.88E-02
	Q14	.362	–.155	–.373	.411	.370	–.254	.241	.431	–.159	–2.16E-02
	Q15	.311	–.158	–.337	.343	.306	–.236	.175	.336	–.230	–.141
	Q16	.440	–.217	–.458	.466	.400	–.299	.373	.494	–.152	–4.93E-02
	Q17	.439	–4.80E-02	–.331	.434	.359	–.162	.196	.347	–.145	–.140
	Q18	.368	–.149	.376	.424	.388	–.259	.215	.439	–.183	–3.16E-02
	Q19	–.204	.357	.403	–.231	–.207	**.343**[b]	–.308	–.324	.294	.196
	Q20	.342	–.301	–.440	.353	.292	–.308	**.484**[b]	.457	–6.78E-02	2.125E-02
	Q21	.449	–.254	–.488	.480	.412	–.324	.457	**.550**[b]	–9.59E-02	3.196E-02
	Q22	–2.54E-02	.333	.275	–4.97E-02	–6.03E-02	.294	–6.78E-02	–9.59E-02	**.464**[b]	.408
	Q23	4.463E-02	.246	.158	4.179E-02	2.794E-02	.196	2.125E-02	3.196E-02	.408	**.412**[b]
Residual [a]	Q01		1.291E-02	3.503E-02	–1.13E-02	2.694E-02	1.500E-02	–.128	–.120	–7.90E-02	–4.91E-02
	Q02	1.291E-02		–6.15E-02	2.169E-02	2.930E-03	–.153	9.905E-02	4.922E-02	–.102	–.147
	Q03	3.503E-02	–6.15E-02		1.893E-02	3.485E-02	–6.11E-02	.115	7.125E-02	–7.09E-02	–8.01E-03
	Q04	–1.13E-02	2.169E-02	1.893E-02		1.785E-03	4.510E-02	–.110	–6.98E-02	–4.87E-02	–7.56E-02
	Q05	2.694E-02	2.930E-03	3.485E-02	1.785E-03		4.136E-02	–9.24E-02	–7.76E-02	–7.22E-02	–6.96E-02
	Q06	–8.72E-04	–4.08E-02	–2.66E-02	1.197E-04	–1.55E-02	–2.01E-02	.122	2.875E-02	4.344E-02	1.304E-02
	Q07	–6.11E-02	–1.12E-02	–9.13E-03	–1.00E-02	–4.07E-02	–1.46E-02	1.628E-03	5.261E-02	1.040E-02	–3.31E-02
	Q08	–8.11E-02	–5.18E-02	1.106E-02	–4.07E-02	–4.36E-02	–5.59E-02	1.145E-02	1.407E-02	1.976E-02	8.616E-02

(Continued)

Reproduced Correlations (Continued)

	Q01	Q02	Q03	Q04	Q05	Q19	Q20	Q21	Q22	Q23
Q09	–4.99E–02	–.115	–5.19E–02	–5.15E–02	–1.60E–02	–.114	5.965E–02	5.467E–02	–.161	–.152
Q10	4.194E–02	–2.28E–02	–1.29E–02	3.359E–03	5.299E–02	9.852E–03	7.830E–02	4.732E–03	6.599E–02	4.844E–02
Q11	–6.60E–02	–4.64E–02	6.094E–03	–5.05E–02	–5.01E–02	–2.07E–03	5.579E–02	4.747E–03	4.741E–02	.116
Q12	–5.67E–02	2.375E–02	3.036E–02	–6.06E–03	–5.01E–02	3.574E–02	–5.63E–02	–6.21E–02	–3.10E–02	–5.71E–02
Q13	7.634E–03	–2.12E–02	2.367E–02	–5.11E–02	–5.81E–02	4.134E–03	4.082E–02	–9.81E–03	7.402E–03	2.583E–02
Q14	–2.39E–02	–9.23E–03	1.909E–03	–5.97E–02	–5.51E–02	–2.55E–04	–1.52E–02	–3.16E–02	–1.08E–02	–2.69E–02
Q15	–6.49E–02	–7.26E–03	2.494E–02	–8.76E–03	–4.49E–02	2.658E–02	3.088E–02	–3.59E–02	6.223E–02	7.911E–02
Q16	5.880E–02	4.966E–02	3.929E–02	–4.99E–02	–5.04E–03	3.159E–02	–.108	–7.36E–02	–3.59E–03	–3.22E–02
Q17	–6.86E–02	–3.90E–02	3.147E–03	–5.16E–02	–4.89E–02	–1.23E–03	9.384E–03	1.601E–02	1.900E–02	4.880E–02
Q18	–2.04E–02	–1.49E–02	9.980E–04	–4.24E–02	–6.59E–02	2.821E–03	1.985E–02	–8.65E–03	2.343E–02	–4.88E–02
Q19	1.500E–02	–.153	–6.11E–02	4.510E–02	4.136E–02		5.985E–02	4.882E–02	–6.01E–02	–7.31E–02
Q20	–.128	9.905E–02	.115	–.110	–9.24E–02	5.985E–02		1.034E–02	–3.19E–02	–5.59E–02
Q21	–.120	4.922E–02	7.125E–02	–6.98E–02	–7.76E–02	4.882E–02	1.034E–02		–3.31E–02	–9.96E–02
Q22	–7.90E–02	–.102	–7.09E–02	–4.87E–02	–7.22E–02	–6.01E–02	–3.19E–02	–3.31E–02		–.177
Q23	–4.91E–02	–.147	–8.01E–03	–7.56E–02	–6.96E–02	–7.31E–02	–5.59E–02	–9.96E–02	–.177	

Extraction Method: Principal Component Analysis.

a Residuals are computed between observed and reproduced correlations. There are 91 (35.0%) nonredundant residuals with absolute values > .05

b Reproduced communalities

SPSS Output 15.7

15.6.3. Factor rotation ②

The first analysis I asked you to run was using an orthogonal rotation. However, you were asked to rerun the analysis using oblique rotation too. In this section the results of both analyses will be reported so as to highlight the differences between the outputs. This comparison will also be a useful way to show the circumstances in which one type of rotation might be preferable to another.

15.6.3.1. Orthogonal rotation (varimax) ②

SPSS Output 15.8 shows the rotated component matrix (also called the rotated factor matrix in factor analysis) which is a matrix of the factor loadings for each variable onto each factor. This matrix contains the same information as the component matrix in SPSS Output 15.5 except that it is calculated *after* rotation. There are several things to consider about the format of this matrix. First, factor loadings less than .4 have not been displayed because we asked for these loadings to be suppressed using the option in Figure 15.10. If you didn't select this option, or didn't adjust the criterion value to .4, then your output will differ. Second, the variables are listed in the order of size of their factor loadings. By default, SPSS orders the variables as they are in the data editor; however, we asked for the output to be *Sorted by size* using the option in Figure 15.10. If this option was not selected your output will look different. Finally, for all other parts of the output I suppressed the variable labels (for reasons of space) but for this matrix I have allowed the variable labels to be printed to aid interpretation.

The original logic behind suppressing loadings less than .4 was based on Stevens's (1992) suggestion that this cut-off point was appropriate for interpretative purposes (i.e. loadings greater than .4 represent substantive values). However, this means that we have suppressed several loadings that are undoubtedly significant (see section 15.3.6.2). However, significance itself is not important.

Compare this matrix with the unrotated solution (SPSS Output 15.5). Before rotation, most variables loaded highly onto the first factor and the remaining factors didn't really get a look in. However, the rotation of the factor structure has clarified things considerably: there are four factors and variables load very highly onto only one factor (with the exception of one question). The suppression of loadings less than .4 and ordering variables by loading size also make interpretation considerably easier (because you don't have to scan the matrix to identify substantive loadings).

The next step is to look at the content of questions that load onto the same factor to try to identify common themes. If the mathematical factor produced by the analysis represents some real-world construct then common themes among highly loading questions can help us identify what the construct might be. The questions that load highly on factor 1 all seem to relate to using computers or SPSS. Therefore we might label this factor *fear of computers*. The questions that load highly on factor 2 all seem to relate to different aspects of statistics; therefore, we might label this factor *fear of statistics*. The three questions that load highly on factor 3 all seem to relate to mathematics; therefore, we might label this factor *fear of mathematics*. Finally, the questions that load highly on factor 4 all contain some component of social evaluation from friends; therefore, we might label this factor *peer evaluation*. This analysis seems to reveal that the initial questionnaire, in reality, is composed of four subscales: fear of computers, fear of statistics, fear of maths and fear of negative peer evaluation. There are two possibilities here. The first is that the SAQ failed to measure what it set out to (namely, SPSS anxiety) but does measure some related constructs. The second is that these four constructs are sub-components of SPSS anxiety; however, the factor analysis does not indicate which of these possibilities is true.

The final part of the output is the factor transformation matrix (see section 15.3.6). This matrix provides information about the degree to which the factors were rotated to obtain a solution. If no rotation were necessary, this matrix would be an identity matrix. If orthogonal rotation were completely appropriate then we would expect a symmetrical matrix (same values above and below the diagonal). However, in reality the matrix is not easy to interpret, although very unsymmetrical matrices might be taken as a reason to try oblique rotation. For the inexperienced factor analyst you are probably best advised to ignore the factor transformation matrix.

15.6.3.2. Oblique rotation ②

When an oblique rotation is conducted, the factor matrix is split into two matrices: the *pattern matrix* and the *structure matrix* (see Box 15.1). For orthogonal rotation these matrices are the same. The pattern matrix contains the factor loadings and is comparable to the factor matrix that we interpreted for the orthogonal rotation. The structure matrix takes into account the relationship between factors (in fact it is a product of the pattern matrix and the matrix containing the correlation coefficients between factors). Most researchers interpret the pattern matrix, because it is usually simpler; however, there are situations in which values in the pattern matrix are suppressed because of relationships between the factors. Therefore, the structure matrix is a useful double-check and Graham et al. (2003) recommend reporting both (with some useful examples of why this can be important).

For the pattern matrix for these data (SPSS Output 15.9) the same four factors seem to have emerged (although for some variables the factor loadings are too small to be displayed). Factor 1 seems to represent fear of statistics, factor 2 represents fear of peer evaluation, factor 3 represents fear of computers and factor 4 represents fear of mathematics. The structure matrix (SPSS Output 15.10) differs in that shared variance is not ignored. The picture becomes more complicated because with the exception of factor 2, several variables load highly onto more than one factor. This has occurred because of the relationship between factors 1 and 3 and factors 3 and 4. This example should highlight why the pattern matrix is preferable for interpretative reasons: because it contains information about the *unique* contribution of a variable to a factor.

The final part of the output is a correlation matrix between the factors (SPSS Output 15.11). This matrix contains the correlation coefficients between factors. As predicted from the structure matrix, factor 2 has little or no relationship with any other

Rotated Component Matrix[a]

	Component			
	1	2	3	4
I have little experience of computers	.800			
SPSS always crashes when I try to use it	.684			
I worry that I will cause irreparable damage because of my incompetence with computers	.647			
All computers hate me	.638			
Computers have minds of their own and deliberately go wrong whenever I use them	.579			
Computers are useful only for playing games	.550			
Computers are out to get me	.459			
I can't sleep for thoughts of eigenvectors		.677		
I wake up under my duvet thinking that I am trapped under a normal distribution		.661		
Standard deviations excite me		−.567		
People try to tell you that SPSS makes statistics easier to understand but it doesn't	.473	.523		
I dream that Pearson is attacking me with correlation coefficients		.516		
I weep openly at the mention of central tendency		.514		
Statistics makes me cry		.496		
I don't understand statistics		.429		
I have never been good at mathematics			.833	
I slip into a coma whenever I see an equation			.747	
I did badly at mathematics at school			.747	
My friends are better at statistics than me				.648
My friends are better at SPSS than I am				.645
If I'm good at statistics my friends will think I'm a nerd				.586
My friends will think I'm stupid for not being able to cope with SPSS				.543
Everybody looks at me when I use SPSS				.427

Extraction Method: Principal Component Analysis.
Rotation Method: Varimax with Kaiser Normalization.
a Rotation converged in 9 iterations

Component Transformation Matrix

Component	1	2	3	4
1	.635	.585	.443	−.242
2	.137	−.168	.488	.846
3	.758	−.513	−.403	.008
4	.067	.605	−.635	.476

Extraction Method: Principal Component Analysis.
Rotation Method: Varimax with Kaiser Normalization.

SPSS Output 15.8

Pattern Matrix[a]

	Component			
	1	2	3	4
I can't sleep for thoughts of eigenvectors	.706			
I wake up under my duvet thinking that I am trapped under a normal distribution	.591			
Standard deviations excite me	−.511			
I dream that Pearson is attacking me with correlation coefficients	.405			
I weep openly at the mention of central tendency	.400			
Statistics makes me cry				
I don't understand statistics				
My friends are better at SPSS than I am		.643		
My friends are better at statistics than me		.621		
If I'm good at statistics my friends will think I'm a nerd		.615		
My friends will think I'm stupid for not being able to cope with SPSS		.507		
Everybody looks at me when I use SPSS				
I have little experience of computers			.885	
SPSS always crashes when I try to use it			.713	
All computers hate me			.653	
I worry that I will cause irreparable damage because of my incompetence with computers			.650	
Computers have minds of their own and deliberately go wrong whenever I use them			.588	
Computers are useful only for playing games			.585	
People try to tell you that SPSS makes statistics easier to understand but it doesn't	.412		.462	
Computers are out to get me			.411	
I have never been good at mathematics				−.902
I slip into a coma whenever I see an equation				−.774
I did badly at mathematics at school				−.774

Extraction Method: Principal Component Analysis.
Rotation Method: Oblimin with Kaiser Normalization.
a Rotation converged in 29 iterations

SPSS Output 15.9

factors (correlation coefficients are low), but all other factors are interrelated to some degree (notably factors 1 and 3, and factors 3 and 4). The fact that these correlations exist tells us that the constructs measured can be interrelated. If the constructs were independent then we would expect oblique rotation to provide an identical solution to an orthogonal rotation and the component correlation matrix should be an identity matrix (i.e. all factors have correlation coefficients of 0). Therefore, this final matrix gives us a guide to whether it is reasonable to assume independence between factors: for these data it appears that we cannot assume independence. Therefore, the results of the

Structure Matrix

	Component			
	1	2	3	4
I wake up under my duvet thinking that I am trapped under a normal distribution	.695		.477	
I can't sleep for thoughts of eigenvectors	.685			
Standard deviations excite me	−.632		−.407	
I weep openly at the mention of central tendency	567		.516	−.491
I dream that Pearson is attacking me with correlation coefficients	.548		.487	−.485
Statistics makes me cry	.520		.413	−.501
I don't understand statistics	.462		.453	
My friends are better at SPSS than I am		.660		
My friends are better at statistics than me		.653		
If I'm good at statistics my friends will think I'm a nerd		.588		
My friends will think I'm stupid for not being able to cope with SPSS		.546		
Everybody looks at me when I use SPSS	−.435	.446		
I have little experience of computers			.777	
SPSS always crashes when I try to use it	.404		.761	
All computers hate me	.401		.723	
I worry that I will cause irreparable damage because of my incompetence with computers			.723	−.429
Computers have minds of their own and deliberately go wrong whenever I use them	.426		.671	
People try to tell you that SPSS makes statistics easier to understand but it doesn't	.576		.606	
Computers are out to get me			.561	−.441
Computers are useful only for playing games			.556	
I have never been good at mathematics				−.855
I slip into a coma whenever I see an equation			.453	−.822
I did badly at mathematics at school			.451	−.818

Extraction Method: Principal Component Analysis.
Rotation Method: Oblimin with Kaiser Normalization.

SPSS Output 15.10

orthogonal rotation should not be trusted: the obliquely rotated solution is probably more meaningful.

On a theoretical level the dependence between our factors does not cause concern; we might expect a fairly strong relationship between fear of maths, fear of statistics and fear of computers. Generally, the less mathematically and technically minded people struggle with statistics. However, we would not expect these constructs to correlate with fear of peer evaluation (because this construct is more socially based). In fact, this factor is the one that correlates fairly badly with all others—so on a theoretical level, things have turned out rather well!

Cramming Samantha's Tips: Interpretation

- If you've conduced orthogonal rotation then look at the table labelled *Rotated Component Matrix*. For each variable, note the component for which the variable has the highest loading. Also, for each component, note the variables that load highly onto it (by high I'd say loadings should be above .4 when you ignore the plus or minus sign). Try to make sense of what the factors represent by looking for common themes in the items that load onto them.
- If you've conduced oblique rotation then look at the table labelled *Pattern Matrix*. For each variable, note the component for which the variable has the highest loading. Also, for each component, note the variables that load highly onto it (by high I'd say loadings should be above .4 when you ignore the plus or minus sign). Double-check what you find by doing the same thing for the *Structure Matrix*. Try to make sense of what the factors represent by looking for common themes in the items that load onto them.

Component Correlation Matrix

Component	1	2	3	4
1	1.000	−.154	.364	−.279
2	−.154	1.000	−.185	8.155E-02
3	.364	−.185	1.000	−.464
4	−.279	8.155E-02	−.464	1.000

Extraction Method: Principal Component Analysis.
Rotation Method: Oblimin with Kaiser Normalization.

SPSS Output 15.11

15.6.4. Factor scores ②

Having reached a suitable solution and rotated that solution we can look at the factor scores. SPSS Output 15.12 shows the component score matrix *B* (see section 15.2.4) from which the factor scores are calculated and the covariance matrix of factor scores. The component score matrix is not particularly useful in itself. It can be useful in understanding how the factor scores have been computed, but with large data sets like this one you are unlikely to want to delve into the mathematics behind the factor scores. However, the covariance matrix of scores is useful. This matrix in effect tells us the relationship between factor scores (it is an unstandardized correlation matrix). If factor scores are uncorrelated then this matrix should be an identity matrix (i.e. diagonal elements will be 1 but all other elements are 0). For these data the covariances are all zero (remembering that 4.37E-16 is actually 0.000000000000000437) indicating that the resulting scores are uncorrelated.

In the original analysis we asked for scores to be calculated based on the Anderson–Rubin method (hence why they are uncorrelated). You will find these scores in the data editor. There should be four new columns of data (one for each factor) labelled *FAC1_1*, *FAC2_1*, *FAC3_1* and *FAC4_1* respectively. If you asked for factor scores in the oblique

Component Score Coefficient Matrix

	Component			
	1	2	3	4
Q01	−.053	.173	.089	.110
Q02	.102	−.129	.086	.281
Q03	.087	−.195	.013	.137
Q04	−.011	.170	.045	.107
Q05	.021	.131	.014	.083
Q06	.383	−.211	−.088	.014
Q07	.213	.004	−.078	.038
Q08	−.129	−.074	.460	.013
Q09	.025	−.029	.108	.354
Q10	.244	−.161	−.021	−.036
Q11	−.066	−.087	.379	−.059
Q12	.097	.161	−.116	.051
Q13	.224	−.065	−.019	.013
Q14	.180	.040	−.084	.043
Q15	.114	−.055	.061	−.058
Q16	−.015	.146	.046	.014
Q17	−.057	−.067	.372	.005
Q18	.242	−.001	−.104	.043
Q19	.048	−.115	.061	.199
Q20	−.195	.359	−.061	−.002
Q21	−.039	.270	−.064	.059
Q22	−.036	.162	−.048	.382
Q23	.032	.211	−.162	.379

Extraction Method: Principal Component Analysis.
Rotation Method: Varimax with Kaiser Normalization.

Component Score Covariance Matrix

Component	1	2	3	4
1	1.000	1.093E-16	.000	.000
2	1.093E-16	1.000	4.373E-16	.000
3	.000	4.373E-16	1.000	.000
4	.000	.000	.000	1.000

Extraction Method: Principal Component Analysis.
Rotation Method: Varimax with Kaiser Normalization.

SPSS Output 15.12

rotation then these scores will appear in the data editor in four other columns labelled *FAC2_1* and so on. These factor scores can be listed in the output viewer using the **Analyze⇒Reports⇒Case Summaries...** command path (see section 5.8.6). Given that there are over 1500 cases you might like to restrict the output to the first 10 or 20. SPSS Output 15.13 shows the factor scores for the first 10 participants. It should be pretty clear that participant 9 scored highly on all four factors and so this person is very anxious about statistics, computing and maths, but less so about peer evaluation (factor 4). Factor scores can be used in this way to assess the relative fear of one person compared to another, or

Case Summaries[a]

	Case Number	FAC1_1	FAC2_1	FAC3_1	FAC4_1
1	1	.10584	−.92797	−1.82768	−.45958
2	2	−.58279	−.18934	−.04137	.29330
3	3	−.54761	.02968	.19913	−.97142
4	4	.74664	.72272	−.68650	−.18533
5	5	.25167	−.51497	−.63353	.68296
6	6	1.91613	−.27351	−.68205	−.52710
7	7	−.26055	−1.40960	−.00435	.90986
8	8	−.28574	−.91775	−.08948	1.03732
9	9	1.71753	1.15063	3.15671	.81083
10	10	−.69153	−.73340	.18379	1.49529

a Limited to first 10 cases

SPSS Output 15.13

we could add the scores up to obtain a single score for each participant (that we might assume represents SPSS anxiety as a whole). We can also use factor scores in regression when groups of predictors correlate so highly that there is multicollinearity.

15.6.5. Summary ②

To sum up, the analyses revealed four underlying scales in our questionnaire that may, or may not, relate to genuine sub-components of SPSS anxiety. It also seems as though an obliquely rotated solution was preferred owing to the interrelationships between factors. The use of factor analysis is purely exploratory; it should be used only to guide future hypotheses, or to inform researchers about patterns within data sets. A great many decisions are left to the researcher using factor analysis and I urge you to make informed decisions, rather than basing decisions on the outcomes you would like to get. The next question is whether or not our scale is reliable.

15.7. RELIABILITY ANALYSIS ②

15.7.1. Measures of reliability ③

If you're using factor analysis to validate a questionnaire, it is useful to check the reliability of your scale. Reliability just means that a scale should consistently reflect the construct it is measuring. One way to think of this is that, other things being equal, a person should get the same score on a questionnaire if they complete it at two different points in time (this is called test–retest reliability). So, someone who is terrified of statistics and who scores highly on our SAQ should score similarly highly if we tested them a month later (assuming they hadn't gone into some kind of statistics-anxiety therapy in that month). Another way to look at reliability is to say that two people who are the same in terms of the construct being measured should get the same score. So, if we took two people who were equally statistics-phobic, then they should get more or less identical scores on the SAQ. Likewise, if we took two people who loved statistics, they should both get equally low scores. It should be apparent that if we took someone who loved statistics and someone who was terrified of it, and they got the same score on our questionnaire, then it wouldn't be an accurate measure of statistical anxiety! In statistical terms, the usual way to look at reliability is based on the idea that individual items (or sets of items) should produce results consistent with the overall questionnaire. So, if we take someone scared of statistics, then their overall score on the SAQ will be high; if the SAQ is reliable then if we randomly select some items from the SAQ, then the person's score on those items should also be high.

The simplest way to do this in practice is to use split-half reliability. This method randomly splits the data set into two. A score for each participant is then calculated based on each half of the scale. If a scale is very reliable a person's score on one half of the scale should be the same (or similar) to their score on the other half: therefore, across several participants scores from the two halves of the questionnaire should correlate perfectly (well, very highly). The correlation between the two halves is the statistic computed in the split-half method, with large correlations being a sign of reliability. The problem with this method is that there are several ways in which a set of data can be split into two and so the results could be a product of the way in which the data were split. To overcome this problem, Cronbach (1951) came up with a measure that is loosely equivalent to splitting data in two in every possible way and computing the correlation coefficient for each split. The average of these values is equivalent to Cronbach's alpha, α, which is the most common measure of scale reliability.[9]

Cronbach's α is

$$\alpha = \frac{N^2\overline{\text{Cov}}}{\sum s^2_{\text{item}} + \sum \text{Cov}_{\text{item}}} \tag{15.6}$$

which may look complicated, but actually isn't. The first thing to note is that for each item on our scale we can calculate two things: the variance within the item, and the covariance between a particular item and any other item on the scale. Put another way, we can construct a variance–covariance matrix of all items. In this matrix the diagonal elements will be the

9 Although this is the easiest way to conceptualize Cronbach's α, whether or not it is exactly equal to the average of all possible split-half reliabilities depends on exactly how you calculate the split-half reliability (see the glossary for computational details). If you use the Spearman–Brown formula, which takes no account of item standard deviations, then Cronbach's α will be equal to the average split-half reliability only when the item standard deviations are equal; otherwise α will be smaller than the average. However, if you use a formula for split-half reliability that does account for item standard deviations (such as Flanagan, 1937; Rulon, 1939) then α will always equal the average split-half reliability (see Cortina, 1993).

variance within a particular item, and the off-diagonal elements will be covariances between pairs of items. The top half of the equation is simply the number of items (N) squared multiplied by the average covariance between items (the average of the off-diagonal elements in the aforementioned variance–covariance matrix). The bottom half is just the sum of all the item variances and item covariances (i.e. the sum of everything in the variance–covariance matrix).

There is a standardized version of the coefficient too, which essentially uses the same equation except that correlations are used rather than covariances, and the bottom half of the equation uses the sum of the elements in the correlation matrix of items (including the ones that appear on the diagonal of that matrix). The normal α is appropriate when items on a scale are summed to produce a single score for that scale (the standardized α is not appropriate in these cases). The standardized α is useful, though, when items on a scale are standardized before being summed.

15.7.2. Interpreting Cronbach's α (some cautionary tales...) ②

You'll often see in books or journal articles, or be told by people, that a value of .7–.8 is an acceptable value for Cronbach's α; values substantially lower indicate an unreliable scale. Kline (1999) notes that although the generally accepted value of .8 is appropriate for cognitive tests such as intelligence tests, for ability tests a cut-off point of .7 is more suitable. He goes on to say that when dealing with psychological constructs, values below even .7 can, realistically, be expected because of the diversity of the constructs being measured.

However, Cortina (1993) notes that such general guidelines need to be used with caution because the value of α depends on the number of items on the scale. You'll notice that the top half of the equation for α includes the number of items squared. Therefore, as the number of items on the scale increases, α will increase. Therefore, it's possible to get a large value of α because you have a lot of items on the scale, and not because your scale is reliable! For example, Cortina reports data from two scales, both of which have $\alpha = .8$. The first scale has only three items, and the average correlation between items was a respectable .57; however, the second scale had 10 items with an average correlation between these items of a less respectable .28. Clearly the internal consistency of these scales differs enormously, yet according to Cronbach's α they are both equally reliable!

A second common interpretation of α is that it measures 'unidimensionality', or the extent to which the scale measures one underlying factor or construct. This interpretation stems from the fact that when there is one factor underlying the data, α is a measure of the strength of that factor (see Cortina, 1993). However, Grayson (2004) demonstrates that data sets with the same α can have very different structures. He showed that an α of .8 can be achieved in a scale with one underlying factor, with two moderately correlated factors and with two uncorrelated factors. Cortina (1993) has also shown that with more than 12 items, and fairly high correlations between items ($r > .5$), α can reach values around and above .7 (.65 to .84). These results compellingly show that α should not be used as a measure of 'unidimensionality'. Indeed, Cronbach (1951) suggested that if several factors exist then the formula should be applied separately to items relating to different factors. In other words, if your questionnaire has subscales, α should be applied separately to these subscales.

The final warning is about items that have a reverse phrasing. For example, in our SAQ that we used in the factor analysis part of this chapter, we had one item (question 3) that was phrased the opposite way around to all other items. The item was 'standard deviations excite me'. Compare this to any other item and you'll see it requires the opposite response. For example, item 1 is 'statistics make me cry'. Now, if you don't like statistics then you'll strongly agree with this statement and so will get a score of 5 on our scale. For item 3, if you hate statistics then standard deviations are unlikely to excite you so you'll strongly disagree and get a score of 1 on the scale. These reverse-phrased items are important for reducing response bias; participants will actually have to read the items in case they are phrased the other way around. For factor analysis, this reverse phrasing doesn't matter, all that happens is you get a negative factor loading for any reversed items (in fact, look at SPSS Output 15.8 and you'll note that item 3 has a negative factor loading). However, in reliability analysis these reverse-scored items do make a difference. To see why, think about the equation for Cronbach's α. In this equation, the top half incorporates the *average* covariance between items. If an item is reverse phrased then it will have a negative relationship with other items; hence the covariances between this item and other items will be negative. The average covariance is obviously the sum of covariances divided by the number of covariances, and by including a bunch of negative values we reduce the sum of covariances, and hence we also reduce Cronbach's α, because the top half of the equation gets smaller. In extreme cases, it is even possible to get a negative value for Cronbach's α, simply because the magnitude of negative covariances is bigger than the magnitude of positive ones! A negative Cronbach's α doesn't make much sense, but it does happen, and if it does, ask yourself whether you included any reverse-phrased items!

If you have reverse-phrased items then you have also to reverse the way in which they're scored before you conduct reliability analysis. This is quite easy. To take our SAQ data, we have one item which is currently scored as 1 = strongly disagree, 2 = disagree, 3 = neither, 4 = agree and 5 = strongly agree. This is fine for items phrased in such a way that agreement indicates statistics anxiety, but for item 3 (standard deviations excite me), disagreement indicates statistics anxiety. To reflect this numerically, we need to reverse the scale such that 1 = strongly agree, 2 = agree, 3 = neither, 4 = disagree and 5 = strongly disagree. This way, an anxious person still gets 5 on this item (because they'd strongly disagree with it).

To reverse the scoring find the maximum value of your response scale (in this case 5) and add 1 to it (so you get 6 in this case). Then for each person, you take this value and subtract from it the score they actually got. Therefore, someone who scored 5 originally now scores 6 − 5 = 1, and someone who scored 1 originally now gets 6 − 1 = 5. Someone in the middle of the scale with a score of 3 will still get 6 − 3 = 3! Obviously it would take a long time to do this for each person, but we can get SPSS to do it for us by using **Transform⇒Compute...** (see section 3.3.4). We came across this command in Chapter 3, and what we do is enter the name of the variable we want to change in the space labelled *Target Variable* (see Figure 15.11) (in this case the variable is called **q03**). You can use a different name if you like, but if you do SPSS will create a new variable and you must

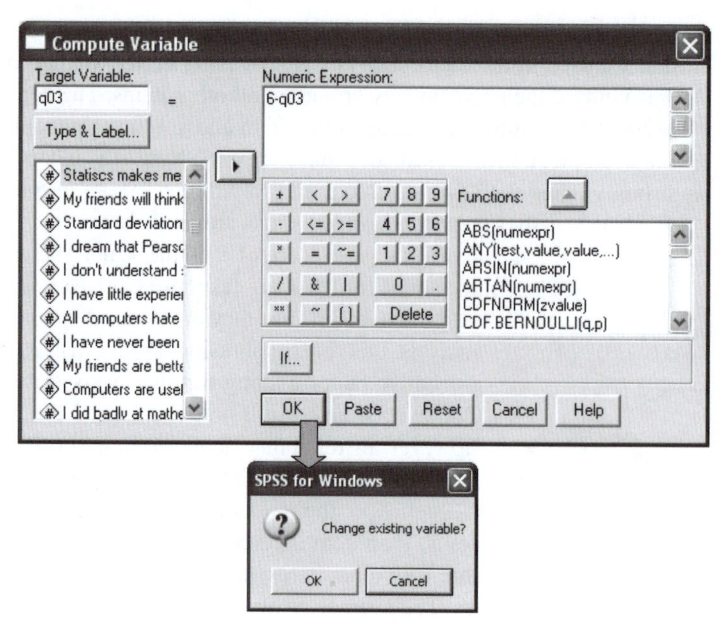

Figure 15.11 Reverse scoring item 3 (**q03**) in SPSS

remember that it's this new variable that you need to use in the reliability analysis. Then, where it says *Numerical Expression* you need to tell SPSS how to compute the new variable. In this case, we want to take each person's original score on item 3, and subtract that value from 6. Therefore, we simply type 6-q03 (which means 6 minus the value found in the column labelled *q03*). If you've used the same name then when you click on OK you'll get a dialog box asking if you want to change the existing variable; just click on OK if you're happy for the new values to replace the old ones. Figure 15.11 shows this process.

15.7.3. Reliability analysis on SPSS ②

Let's test the reliability of the SAQ using the data in **SAQ.sav**. Now, you should have reverse scored item 3 (see above), but if you can't be bothered then load the file **SAQ (Item 3 Reversed).sav** instead. Remember also that I said we should conduct reliability analysis on any subscales individually. If we use the results from our orthogonal rotation (look back at SPSS Output 15.8), then we have four subscales:

1. Subscale 1 (*Fear of computers*): items 6, 7, 10, 13, 14, 15, 18
2. Subscale 2 (*Fear of statistics*): items 1, 3, 4, 5, 12, 14, 10, 21
3. Subscale 3 (*Fear of mathematics*): items 8, 11, 17
4. Subscale 4 (*Fear of Peer evaluation*): items 2, 9, 19, 22, 23

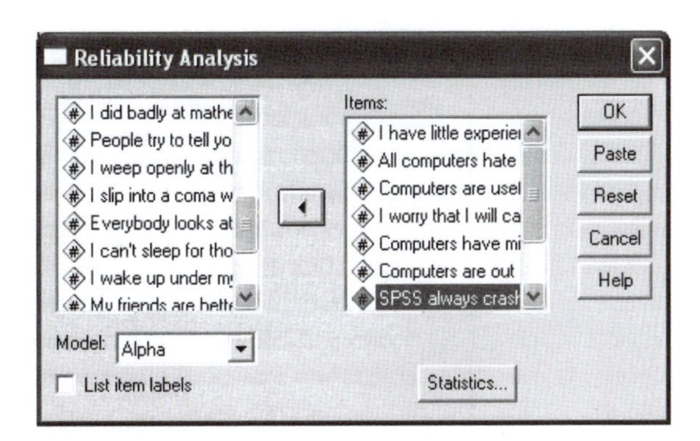

Figure 15.12 Main dialog box for reliability analysis

To conduct each reliability analysis on these data you need to follow the **Analyze⇒Scale⇒Reliability Analysis...** menu path to display the dialog box in Figure 15.12. Select any items from the list that you want to analyse (to begin with let's do the items from the fear of computers subscale) on the left-hand side of the dialog box and transfer them to the box labelled *Items* by clicking on ▸.

Selecting the *List item labels* checkbox will list all of the variable labels for each variable (which can be useful for checking to which items your variables relate). There are several reliability analyses you can run, but the default option is Cronbach's α. You can change the method (for example to the split-half method) by using the drop-down box labelled *Model*, but the default method is a good one to select.

If you click on Statistics... you can access the dialog box in Figure 15.13. In the *statistics* dialog box you can select several things, but the one most important for questionnaire reliability is: *Scale if item deleted*. This option provides a value of Cronbach's α for each item on your scale. It tells us what the value of α would be if that item were deleted. If our questionnaire is reliable then we would not expect any one item to affect the overall reliability greatly. In other words, no item should cause a substantial decrease in α. If it does then we have serious cause for concern and we should consider dropping that item from the questionnaire. As .8 is seen as a good value for α, we would hope that all values of *alpha if item deleted* should be around .8 or higher.

The inter-item correlations and covariances (and summaries) provide us with correlation coefficients and averages for items on our scale. We should already have these values from our factor analysis so there is little point in selecting these options. However, if you haven't already done a factor analysis then it's useful to ask for inter-item correlations because the overall α is affected by the number of items being analysed, and so you might want to check back to see whether the items seem to inter-relate well. Options like the *ANOVA Table* will simply compare the central tendency of different items on the questionnaire using an *F test* (it conducts a one-way repeated-measures ANOVA on the items on the questionnaire), a *Friedman chi-square* (if your data are ranked), or a *Cochran chi-square* (if you're data are dichotomous; for example, if items on the questionnaire had yes/no responses).

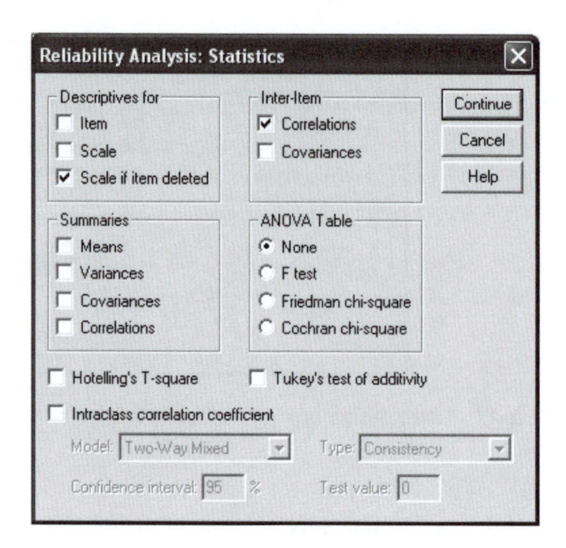

Figure 15.13 Statistics for reliability analysis

The *Hotelling's T-square* does much the same but produces the multivariate equivalent of the *F test*. These tests are useful if you want to check that items have similar distributional properties (i.e. the same average value), but given the large sample sizes you ought to be using for factor analysis, they will inevitably produce significant results even when only small differences exist between the means of questionnaire items.

Use the simple set of options in Figure 15.13 to run a basic reliability analysis. Click on Continue to return to the main dialog box and then click on OK to run the analysis.

15.7.4. Interpreting the output ②

SPSS Output 15.14 shows the results of this basic reliability analysis for the fear of computing subscale. The values in the column labelled *Corrected Item-Total Correlation* are the correlations between each item and the total score from the questionnaire. In a reliable scale all items should correlate with the total. So, we're looking for items that don't correlate with the overall score from the scale: if any of these values are less than about .3 (depends slightly on your sample size—with bigger samples smaller correlation coefficients are acceptable) then we've got problems because it means that a particular item does not correlate very well with the scale overall. Items with low correlations may have to be dropped. For these data, all data have item-total correlations above .3, which is encouraging.

The values in the column labelled *Alpha if Item Deleted* are the values of the overall α if that item isn't included in the calculation. As such, they reflect the change in Cronbach's α that would be seen if a particular item were deleted. The overall α is .823, and so all values in this column should be around that same value. What we're actually looking for

R E L I A B I L I T Y   A N A L Y S I S   -   SCALE   (ALPHA)

Correlation Matrix

	Q06	Q07	Q10	Q13	Q14	Q15	Q18
Q06	1.0000						
Q07	.5136	1.0000					
Q10	.3222	.2837	1.0000				
Q13	.4664	.4421	.3020	1.0000			
Q14	.4022	.4407	.2547	.4498	1.0000		
Q15	.3599	.3914	.2952	.3422	.3801	1.0000	
Q18	.5133	.5009	.2925	.5329	.4983	.3429	1.0000

N of Cases = 2571.0

Item-total Statistics

	Scale Mean if Item Deleted	Scale Variance if Item Deleted	Corrected Item-Total Correlation	Squared Multiple Correlation	Alpha if Item Deleted
Q06	15.8650	17.6141	.6187	.3981	.7906
Q07	15.1684	17.7370	.6190	3949	.7905
Q10	15.8114	20.7360	.3999	.1665	.8239
Q13	15.6429	18.8086	.6067	.3844	.7937
Q14	15.2159	18.7188	.5768	.3504	.7980
Q15	15.3259	19.3217	.4913	.2497	.8119
Q18	15.5235	17.8324	.6474	.4475	.7855

Reliability Coefficients   7 items

Alpha = .8234   Standardized item alpha = .8214

SPSS Output 15.14

is values of α greater than the overall α. If you think about it, if the deletion of an item increases Cronbach's α then this means that the deletion of that item improves reliability. Therefore, any items that result in substantially greater values of α than the overall α may need to be deleted from the scale to improve its reliability. None of the items here would substantially affect reliability if they were deleted. The worst offender is question 10: deleting this question would increase α from .823 to .824. Nevertheless this increase is not dramatic and both values reflect a reasonable degree of reliability.

Finally, and perhaps most importantly, the value of *Alpha* at the very bottom is Cronbach's α: the overall reliability of the scale. To reiterate, we're looking for values in the magnitude of .7 to .8 (or thereabout's) bearing in mind what we've already noted about effects from the number of items. In this case α is slightly above .8, and is certainly in the

R E L I A B I L I T Y   A N A L Y S I S   -   SCALE   (ALPHA)

Item-total Statistics

	Scale Mean if Item Deleted	Scale Variance if Item Deleted	Corrected Item-Total Correlation	Squared Multiple Correlation	Alpha if Item Deleted
Q01	21.7569	21.4417	.5361	.3435	.8017
Q03	20.7165	19.8250	.5492	3093	.7996
Q04	21.3450	20.4105	.5750	.3553	.7955
Q05	21.4088	20.9422	.4944	.2724	.8066
Q12	20.9716	20.6393	.5715	.3370	.7962
Q16	21.2517	20.4507	.5973	.3886	.7928
Q20	20.5068	21.1761	.4185	.2440	.8185
Q21	20.9603	19.9385	.6061	.3988	.7908

Reliability Coefficients   8 items

Alpha = .8208   Standardized item alpha = .8234

SPSS Output 15.15

region indicated by Kline, so this probably indicates good reliability. As a final point, it's worth noting that if items do need to be removed at this stage then you should rerun your factor analysis as well to make sure that the deletion of the item has not affected the factor structure!

OK, let's move on to to do the fear of statistics subscale (items 1, 3, 4, 5, 12, 16, 20 and 21). I won't go through the SPSS again, but SPSS Output 15.15 shows the output from the analysis (to save space I've omitted the inter-item correlations). The values in the column labelled *Corrected Item-Total Correlation* are again all above .3, which is good. The values in the column labelled *Alpha if Item Deleted* are the values of the overall α if that item isn't included in the calculation. The overall α is .821, and none of the items here would increase the reliability if they were deleted. This indicates that all items are positively contributing to the overall reliability. The overall α is also excellent (.821) because it is above .8, and indicates good reliability.

Just to illustrate the importance of reverse-scoring items before running reliability analysis, SPSS Output 15.16 shows the reliability analysis for the fear of statistics subscale but done on the original data (i.e. without item 3 being reverse scored). Note that the overall α is considerably lower (.606 rather than .821). Also, note that this item has a negative item-total correlation (which is a good way to spot if you have a potential reverse-scored item in the data that hasn't been reverse scored!). Finally, note that for item 3, the *Alpha if Item Deleted* is about .8. That is, if this item was deleted then the reliability would improve from about .6 to about .8! This, I hope, illustrates that failing to reverse-score items that have been phrased oppositely to other items on the scale will mess up your reliability analysis!

```
R E L I A B I L I T Y   A N A L Y S I S   -   SCALE   (ALPHA)

Item-total Statistics

              Scale          Scale         Corrected
              Mean           Variance      Item-          Squared        Alpha
              if Item        if Item       Total          Multiple       if Item
              Deleted        Deleted       Correlation    Correlation    Deleted

Q01           20.9277        12.1247        .5051          .3435          .5213
Q03           20.7165        19.8250       -.5492          .3093          .7996
Q04           20.5158        11.4467        .5260          .3553          .5048
Q05           20.5795        11.7138        .4662          .2724          .5230
Q12           20.1424        11.7393        .5006          .3370          .5154
Q16           20.4224        11.5842        .5291          .3886          .5066
Q20           19.6776        12.1073        .3528          .2440          .5576
Q21           20.1311        11.1894        .5410          .3988          .4969

_
Reliability Coefficients      8 items

Alpha = .6055            Standardized item alpha = .6413
```

SPSS Output 15.16

Moving swiftly on to the fear of maths subscale (items 8, 11 and 17), SPSS Output 15.17 shows the output from the analysis. The values in the column labelled *Corrected Item-Total Correlation* are again all above .3, which is good, and the values in the column labelled *Alpha if Item Deleted* indicate that none of the items here would increase the reliability if they were deleted because all values in this column are less than the overall reliability of .819. As with the previous two subscales, the overall α is around .8, which indicates good reliability.

Finally, if you run the analysis for the final subscale of peer evaluation, you should get the output in SPSS Output 15.18. The values in the column labelled *Corrected Item-Total Correlation* are all around .3, and in fact for item 23 it is below .3. This indicates fairly bad internal consistency and identifies item 23 as a potential problem. The values in the column labelled *Alpha if Item Deleted* indicate that none of the items here would increase the reliability if they were deleted because all values in this column are less than the overall reliability of .5699 (or .570 rounded off). Unlike the previous subscales, the overall α is quite low (.57), and although this is in keeping with what Kline says we should expect for this kind of social science data, it is well below the other scales. The scale has five items, compared to seven, eight and three on the other scales, so its reduced reliability is not going to be dramatically affected by the number of items (in fact, it has more items than the fear of maths subscale). If you look at the items on this subscale, they cover quite diverse themes of peer evaluation, and this might explain the relative lack of consistency. This might lead us to rethink this subscale.

R E L I A B I L I T Y   A N A L Y S I S - SCALE (ALPHA)

Item-total Statistics

	Scale Mean if Item Deleted	Scale Variance if Item Deleted	Corrected Item-Total Correlation	Squared Multiple Correlation	Alpha if Item Deleted
Q08	4.7219	2.4701	.6845	.4704	.7396
Q11	4.7036	2.4530	.6818	.4672	.7422
Q17	4.4920	2.5037	.6520	.4251	.7725

Reliability Coefficients   3 items

Alpha = .8194   Standardized item alpha = .8195

SPSS Output 15.17

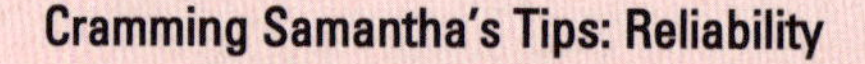

Cramming Samantha's Tips: Reliability

- Reliability is really the consistency of a measure.
- Reliability analysis can be used to measure the consistency of a questionnaire.
- Remember to reverse-score any items that were reverse phrased on the original questionnaire before you run the analysis.
- Run separate reliability analyses for all subscales of your questionnaire.
- Cronbach's α indicates the overall reliability of a questionnaire and values around .8 are good (or .7 for ability tests and such like).
- The *Alpha if Item Deleted* column tells you whether removing an item will improve the overall reliability: values greater than the overall reliability indicate that removing that item will improve the overall reliability of the scale. Look for items that dramatically increase the value of α.
- If you do remove items, rerun your factor analysis to check that the factor structure still holds!

15.8. WHAT HAVE WE DISCOVERED ABOUT STATISTICS? ②

This chapter has made us tiptoe along the craggy rock face that is factor analysis. This is a technique for identifying clusters of variables that relate to each other. One of the difficult things with statistics is realizing that they are subjective: many books (this one included I suspect) create the impression that statistics are like a cookbook and if you follow the instructions you'll get a nice tasty chocolate cake (yum!). Factor analysis perhaps more than any other test in this book illustrates how incorrect this is. The world of statistics is full of arbitrary rules (.05 being the classic example) and nearly all of the time,

```
R E L I A B I L I T Y   A N A L Y S I S   -   SCALE   (ALPHA)

Item-total Statistics

              Scale          Scale        Corrected
               Mean        Variance         Item-          Squared         Alpha
             if Item       if Item          Total          Multiple       if Item
             Deleted       Deleted       Correlation     Correlation      Deleted

Q02          11.4609        8.1186          .3389           .1337          .5153
Q09          10.2380        6.3955          .3907           .1674          .4765
Q19          10.7923        7.3810          .3162           .1060          .5218
Q22          10.1964        7.2824          .3776           .1441          .4870
Q23           9.6499        7.9879          .2389           .0689          .5628

Reliability Coefficients     5 items

Alpha =  .5699          Standardized item alpha =  .5724
```

SPSS Output 15.18

whether you realize it or not, you have to act upon your own discretion. So, if nothing else I hope you've discovered enough to give you sufficient discretion about factor analysis to act upon! We saw that the first stage of factor analysis is to scan your variables to check that they relate to each other to some degree but not too strongly. The factor analysis itself has several stages: check some initial issues (e.g. sample size adequacy), decide how many factors to retain, and finally decide which items load onto which factors (and try to make sense of the meaning of the factors). Having done all that you can consider whether the items you have are reliable measures of what you're trying to measure.

15.9. KEY TERMS THAT WE'VE DISCOVERED

- Alpha factoring
- Anderson–Rubin method
- Common variance
- Communality
- Component matrix
- Confirmatory factor analysis
- Cronbach's α
- Direct oblimin
- Equamax
- Extraction
- Factor
- Factor analysis
- Factor loading
- Factor matrix
- Factor scores
- Factor transformation matrix, Λ
- Kaiser–Meyer–Olkin (KMO) measure of sampling adequacy
- Kaiser's criterion
- Latent variable
- Oblique rotation
- Orthogonal rotation
- Pattern matrix
- Principal component analysis
- Promax
- Quartimax

- Random variance
- Reliability
- Rotation
- Scree plot
- Split-half reliability
- Structure matrix
- Test–retest reliability
- Unique variance
- Varimax

15.10. SMART ALEX'S TASK

- The University of Sussex is constantly seeking to employ the best people possible as lecturers (no, really, it is). Anyway, it wanted to revise a questionnaire based on Bland's theory of research methods lecturers. This theory predicts that good research methods lecturers should have four characteristics: (1) a profound love of statistics; (2) an enthusiasm for experimental design; (3) a love of teaching; and (4) a complete absence of normal interpersonal skills. These characteristics should be related (i.e. correlated). The 'Teaching Of Statistics for Scientific Experiments' (TOSSE) already existed, but the university revised this questionnaire and it became the 'Teaching Of Statistics for Scientific Experiments—Revised' (TOSSE—R). The university gave this questionnaire to 239 research methods lecturers around the world to see if it supported Bland's theory. The questionnaire is in Figure 15.14, and the data are in **TOSSE-R.sav**. Conduct a factor analysis (with appropriate rotation) to see the factor structure of the data.②

SD = Strongly Disagree, D = Disagree, N = Neither, A = Agree, SA = Strongly Agree

		SD	D	N	A	SA
1	I once woke up in the middle of a vegetable patch hugging a turnip that I'd mistakenly dug up thinking it was Roy's largest root	○	○	○	○	○
2	If I had a big gun I'd shoot all the students I have to teach	○	○	○	○	○
3	I memorize probability values for the *F*-distribution	○	○	○	○	○
4	I worship at the shrine of Pearson	○	○	○	○	○
5	I still live with my mother and have little personal hygiene	○	○	○	○	○
6	Teaching others makes me want to swallow a large bottle of bleach because the pain of my burning oesophagus would be light relief in comparison	○	○	○	○	○
7	Helping others to understand sums of squares is a great feeling	○	○	○	○	○
8	I like control conditions	○	○	○	○	○
9	I calculate three ANOVAs in my head before getting out of bed every morning	○	○	○	○	○
10	I could spend all day explaining statistics to people	○	○	○	○	○

SD = Strongly Disagree, D = Disagree, N = Neither, A = Agree, SA = Strongly Agree						
		SD	**D**	**N**	**A**	**SA**
11	I like it when people tell me I've helped them to understand factor rotation	○	○	○	○	○
12	People fall asleep as soon as I open my mouth to speak	○	○	○	○	○
13	Designing experiments is fun	○	○	○	○	○
14	I'd rather think about appropriate dependent variables than go to the pub	○	○	○	○	○
15	I soil my pants with excitement at the mere mention of factor analysis	○	○	○	○	○
16	Thinking about whether to use repeated- or independent measures thrills me	○	○	○	○	○
17	I enjoy sitting in the park contemplating whether to use participant observation in my next experiment	○	○	○	○	○
18	Standing in front of 300 people in no way makes me lose control of my bowels	○	○	○	○	○
19	I like to help students	○	○	○	○	○
20	Passing on knowledge is the greatest gift you can bestow on an individual	○	○	○	○	○
21	Thinking about Bonferroni corrections gives me a tingly feeling in my groin	○	○	○	○	○
22	I quiver with excitement when thinking about designing my next experiment	○	○	○	○	○
23	I often spend my spare time talking to the pigeons… and they die of boredom	○	○	○	○	○
24	I tried to build myself a time machine so that I could go back to the 1930s and follow Fisher around on my hands and knees licking the floor on which he'd just trodden	○	○	○	○	○
25	I love teaching	○	○	○	○	○
26	I spend lots of time helping students	○	○	○	○	○
27	I love teaching because students have to pretend to like me or they'll get bad marks	○	○	○	○	○
28	My cat is my only friend	○	○	○	○	○

Figure 15.14 The Teaching of Statistics for Scientific Experiments—Revised (TOSSE—R)

15.11. FURTHER READING

Cortina, J. M. (1993). What is coefficient alpha? An examination of theory and applications. *Journal of Applied Psychology*, 78, 98–104. A very readable paper on Cronbach's α.

Dunteman, G. E. (1989). *Principal components analysis*. Sage university paper series on quantitative applications in the social sciences, 07–069. Newbury Park, CA: Sage. This monograph is quite high level but comprehensive.

Field, A. P. (2000). *Designing a questionnaire*. This is a handout giving advice on how to go about designing questionnaires, and is on the CD-ROM in a file called **DesigningQuestionnaires.pdf**. It might be worth reading if you're thinking of designing a questionnaire.

Pedhazur, E. & Schmelkin, L. (1991). *Measurement, design and analysis*. Hillsdale, NJ: Erlbaum. Chapter 22 is an excellent introduction to the theory of factor analysis.

Tabachnick, B. G. & Fidell, L. S. (2001). *Using multivariate statistics* (4th edition). Boston: Allyn & Bacon. Chapter 13 is a technical but wonderful overview of factor analysis.

CHAPTER 16

CATEGORICAL DATA

16.1. WHAT WILL THIS CHAPTER TELL US? ①

At last, the final chapter! From my own perspective I've saved the worst until last because I rarely collect categorical data, and so know remarkably little about analysing it! Still, a complete absence of knowledge about a topic has never stopped me writing about it in the past. Sometimes, we are interested not in test scores, or continuous measures, but in categorical variables. These are not variables involving cats (although the examples in this chapter might convince you otherwise), but are what we have so far termed grouping variables. They are variables that describe categories of entities. We've come across these types of variables in virtually every chapter of this book. There are different types of categorical variable (see section 4.5.6), but in theory a person, or case, should fall into only one category. Good examples of categorical variables are gender (with few exceptions people can be only biologically male or biologically female)[1], pregnancy (a woman can be only pregnant or not pregnant), and voting in an election (as a general rule you are allowed to vote for only one candidate). In all cases (except logistic regression) so far, we've used such categorical variables to predict some kind of continuous outcome, but there are times when we want to look at relationships between lots of categorical variables. This chapter looks at two techniques for doing this. We begin with the simple case of two categorical variables, and discover the chi-square statistic (which we're not really discovering because we've unwittingly come across it countless times before). We then extend this model to look at relationships between several categorical variables.

---

1 Before anyone objects violently, I am aware that numerous chromosomal and hormonal conditions exist that complicate the matter. Also, people can have a different gender identity to their biological gender.

16.2. THEORY OF ANALYSING CATEGORICAL DATA ①

We will begin this chapter by looking at the simplest situation that you could encounter: that is, analysing two categorical variables. If we want to look at the relationship between two categorical variables then we can't use the mean or any similar statistic because we don't have any variables that have been measured continuously. Trying to calculate the mean of a categorical variable is completely meaningless because the numeric values we attach to different categories are arbitrary, and the mean of those numeric values will depend on how many members each category has. Therefore, when we've measured only categorical variables, we analyse frequencies. That is, we analyse the number of things that fall into each combination of categories. If we take an example, a researcher was interested in whether animals could be trained to line-dance. He took 200 cats and tried to train them to line-dance by giving them either food or affection as a reward for dance-like behaviour. At the end of the week he counted how many animals could line dance and how many could not. There are two categorical variables here: **training** (the animal was trained using either food or affection, not both), and **dance** (it either learnt to line-dance or it did not). By combining categories, we end up with four different categories. All we then need to do is to count how many cats fall into each category. We can tabulate these frequencies as in Table 16.1 (which shows the data for this example) and this is known as a contingency table.

Table 16.1 Contingency table showing how many cats will line-dance after being trained with different rewards

		Training		
		Food as Reward	Affection as Reward	Total
Could They Dance?	Yes	28	48	76
	No	10	114	124
	Total	38	162	200

16.2.1. Pearson's chi-square test ①

If we want to see whether there's a relationship between two categorical variables (i.e. does the amount of cats that line-dance relate to the type of training used?) we can use the Pearson's chi-square test. This is an extremely elegant statistic based on the simple idea of comparing the frequencies you observe in certain categories to the frequencies you might expect to get in those categories by chance. All the way back in Chapters 1, 5 and 8 we saw that if we fit a model to any set of data we can evaluate that model using a very simple equation (or some variant of it):

$$\text{deviation} = \sum(\text{observed} - \text{model})^2$$

This equation was the basis of our sums of squares in regression and ANOVA. Now, when we have categorical data we can use the same equation. There is a slight variation in that we divide by the model scores as well, which is actually much the same process as dividing the sum of squares by the degrees of freedom in ANOVA. The resulting statistic is Pearson's chi-square (χ^2) and is given in equation (16.1):

$$\chi^2 = \sum \frac{(\text{Observed}_{ij} - \text{Model}_{ij})^2}{\text{Model}_{ij}} \tag{16.1}$$

The observed data are, obviously, the frequencies in Table 16.1, but we need to work out what the model is. In ANOVA the model we use is group means, but as I've mentioned we can't work with means when we have only categorical variables. Therefore, we use 'expected values'. One way to estimate the expected values would be to say 'well, we've got 200 cats in total, and four categories, so the expected value is simply 200/4 = 50'. This would be fine if, for example, we had the same number of cats that had affection as a reward and food as a reward; however, we didn't: 38 got food and 162 got affection as a reward. Likewise there are not equal numbers that could and couldn't dance. To take account of this, we calculate expected values for each of the cells in the table (in this case there are four cells) and we use the column and row totals for a particular cell to calculate the expected value:

$$\text{Model}_{ij} = E_{ij} = \frac{\text{Row Total}_i \times \text{Column Total}_j}{n}$$

n is simply the total number of observations (in this case 200). We can calculate these expected values for the four cells within our table (row total and column total are abbreviated to RT and CT respectively):

$$\text{Model}_{\text{Food, Yes}} = \frac{\text{RT}_{\text{Yes}} \times \text{CT}_{\text{Food}}}{n} = \frac{76 \times 38}{200} = 14.44$$

$$\text{Model}_{\text{Food, No}} = \frac{\text{RT}_{\text{No}} \times \text{CT}_{\text{Food}}}{n} = \frac{124 \times 38}{200} = 23.56$$

$$\text{Model}_{\text{Affection, Yes}} = \frac{\text{RT}_{\text{Yes}} \times \text{CT}_{\text{Affection}}}{n} = \frac{76 \times 162}{200} = 61.56$$

$$\text{Model}_{\text{Affection, No}} = \frac{\text{RT}_{\text{No}} \times \text{CT}_{\text{Affection}}}{n} = \frac{124 \times 162}{200} = 100.44$$

Given that we now have these model values, all we need to do is take each value in each cell of our data table, subtract from it the corresponding model value, square the result, and then divide by the corresponding model value. Once you've done this for each cell in the table, just add them up!

$$
\begin{aligned}
\chi^2 &= \frac{(28-14.44)^2}{14.44} + \frac{(10-23.56)^2}{23.56} + \frac{(48-61.56)^2}{61.56} + \frac{(114-100.44)^2}{100.44} \\
&= \frac{(13.56)^2}{14.44} + \frac{(-13.56)^2}{23.56} + \frac{(-13.568)^2}{61.56} + \frac{(13.56)^2}{100.44} \\
&= 12.73 + 7.80 + 2.99 + 1.83 \\
&= 25.35
\end{aligned}
$$

This statistic can then be checked against a distribution with known properties. All we need to know is the degrees of freedom and these are calculated as $(r-1)(c-1)$ in which r is the number of rows and c is the number of columns. Another way to think of it is the number of levels of each variable minus 1 multiplied. In this case we get $(2-1)(2-1) = 1$ *df*. If you were doing the test by hand, you would find a critical value for the chi-square distribution with 1 *df* and if the observed value was bigger than this critical value you would say that there was a significant relationship between the two variables. These critical values are produced in Appendix A4, and for 1 *df* the critical values are 3.84 ($p = .05$) and 6.63 ($p = .01$) and so because the observed chi squared is bigger than these values it is significant at $p < .01$. However, if you use SPSS, it will simply produce an estimate of the precise probability of obtaining a chi-square statistic as big as (in this case) 25.35 by chance.

16.2.2. The likelihood ratio ②

An alternative to Pearson's chi-square is the likelihood ratio statistic, which is based on maximum-likelihood theory. The general idea behind this theory is that you collect some data and create a model for which the probability of obtaining the observed set of data is maximized, and then compare this model to the probability of obtaining those data under the null hypothesis. The resulting statistic is, therefore, based on comparing observed frequencies with those predicted by the model:

$$
L\chi^2 = 2\sum \text{Observed}_{ij} \ln\left(\frac{\text{Observed}_{ij}}{\text{Model}_{ij}}\right) \qquad (16.2)
$$

in which ln is the natural logarithm (this is a standard mathematical function that we came across in Chapter 6; you can find it on your calculator usually labelled as ln or $\log_e$). Using the same model and observed values as in the previous section, this would give us:

$$L\chi^2 = 2\left[28 \times \ln\left(\frac{28}{14.44}\right) + 10 \times \ln\left(\frac{10}{23.56}\right) + 48 \times \ln\left(\frac{48}{61.56}\right) + 114 \times \ln\left(\frac{114}{100.44}\right)\right]$$
$$= 2[(28 \times .662) + (10 \times -.857) + (48 \times -.249) + (114 \times .127)]$$
$$= 2[18.54 - 8.57 - 11.94 + 14.44]$$
$$= 24.94$$

As with Pearson's chi-square, this statistic has a chi-square distribution with the same degrees of freedom (in this case 1). As such it is tested in the same way: that is, we could look up the critical value of chi-square for the number of degrees of freedom we have. As before, the value we have here will be significant because it is bigger than the critical values of 3.84 ($p = .05$) and 6.63 ($p = .01$). For large samples this statistic will be roughly the same as Pearson's chi-square, but is preferred when samples are small.

16.2.3. Yates's correction ②

When you have a 2 × 2 contingency table (i.e. two categorical variables each with two categories) then Pearson's chi-square tends to produce significance values that are too small (in other words, it tends to make a Type I error). Therefore, Yates suggested a correction to the Pearson formula (usually referred to as Yates's continuity correction. The basic idea is that when you calculate the deviation from the model (the $\text{Observed}_{ij} - \text{Model}_{ij}$ in equation (16.1)) you subtract .5 from the absolute value of this deviation before you square it. In plain English this means you calculate the deviation, ignore whether it is positive or negative, subtract .5 from the value and then square it. Pearson's equation then becomes:

$$\chi^2 = \sum \frac{(|\text{Observed}_{ij} - \text{Model}_{ij}| - .5)^2}{\text{Model}_{ij}}$$

For the data in our example this just translates into:

$$\begin{aligned}\chi^2 &= \frac{(13.56 - .5)^2}{14.44} + \frac{(13.56 - .5)^2}{23.56} + \frac{(13.56 - .5)^2}{61.56} + \frac{(13.56 - .5)^2}{100.44} \\ &= 11.81 + 7.24 + 2.77 + 1.70 \\ &= 23.52\end{aligned}$$

The key thing to note is that it lowers the value of the chi-square statistic and, therefore, makes it less significant. Although this seems like a nice solution to the problem there is a fair bit of evidence that this over-corrects and produces chi-square values that are too small!

Howell (2002) provides an excellent discussion of the problem with Yates's correction for continuity if you're interested; all I will say is that although it's worth knowing about, it's probably best ignored!

16.3. ASSUMPTIONS OF THE CHI-SQUARE TEST ①

It should be obvious that the chi-square test does not rely on assumptions such as having continuous normally distributed data like most of the other tests in this book (categorical data cannot be normally distributed because they aren't continuous). However, the chi-square test still has two important assumptions:

1. For the test to be meaningful it is imperative that each person, item or entity contributes to only one cell of the contingency table. Therefore, you cannot use a chi-square test on a repeated measures design (e.g. if we had trained some cats with food to see if they would dance and then trained the same cats with affection to see if they would dance, we couldn't analyse the resulting data with Pearson's chi-square test).
2. The second important point is that the expected frequencies should be greater than 5. Although it is acceptable in larger contingency tables to have up to 20% of expected frequencies below 5, the result is a loss of statistical power (so, the test may fail to detect a genuine effect). Even in larger contingency tables no expected frequencies should be below 1. Howell (2002) gives a nice explanation of why violating this assumption creates problems.

Finally, although not an assumption, it seems fitting to mention in a section in which a gloomy and foreboding tone is being used that proportionately small differences in cell frequencies can result in statistically significant associations between variables. Therefore, we must look at row and column percentages to interpret any effects we get. These percentages will reflect the patterns of data far better than the frequencies themselves (because these frequencies will be dependent on the sample sizes in different categories).

16.4. DOING CHI-SQUARE ON SPSS ①

There are two ways in which categorical data can be entered: enter the raw scores, or enter weighted cases. We'll look at both in turn.

16.4.1. Entering data: raw scores ①

If we input the raw scores, it means that every row of the data editor represents each entity about which we have data (in this example, each row represents a cat). So, you would create two coding variables (**training** and **dance**) and specify appropriate numeric codes for each.

The **training** could be coded with a 0 to represent a food reward and 1 to represent affection and **dance** could be coded with 0 to represent an animal that danced and 1 to represent one that did not. For each animal, you put the appropriate numeric code into each column. So a cat that was trained with food that did not dance would have 0 in the training column and 1 in the dance column. The data in the file **Cats.sav** are entered in this way and you should be able to identify the variables described. There were 200 cats in all and so there are 200 rows of data.

16.4.2. Entering data: weight cases ①

An alternative method of data entry is to create the same coding variables as before, but to have a third variable that represents the number of animals that fell into each combination of categories. In other words, we input the frequency data (the number of cases that fall into a particular category). We could call this variable **frequent**. Figure 16.1 shows the data editor with this third variable added. Now, instead of having 200 rows, each one representing a different animal, we have one row representing each combination of categories and a variable telling us how many animals fell into this category combination. So, the first row represents cats that had food as a reward and who then danced. The variable **frequent** tells us that there were 28 cats that had food as a reward and then danced. This information was previously represented by 28 different rows in the file **Cats.sav** and so you can see how this method of data entry saves you a lot of time! Extending this principle, we can see that when affection was used as a reward 114 cats did not dance.

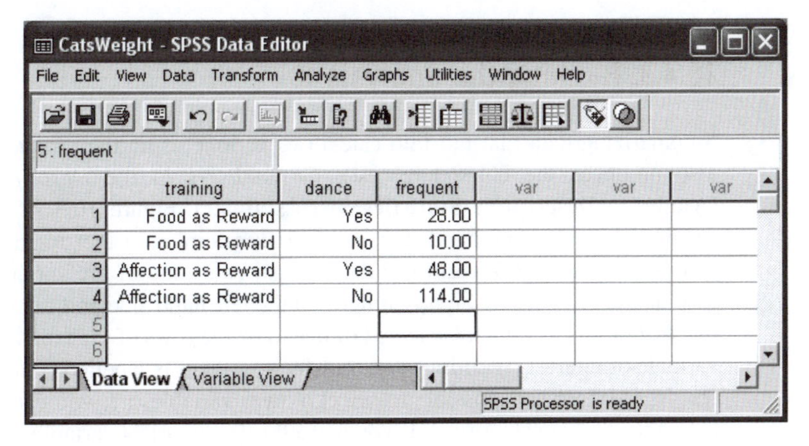

	training	dance	frequent	var	var	var
1	Food as Reward	Yes	28.00			
2	Food as Reward	No	10.00			
3	Affection as Reward	Yes	48.00			
4	Affection as Reward	No	114.00			
5						
6						

Figure 16.1 Data entry using weighted cases

Entering data using a variable representing the number of cases that fall into a combination of categories can be quite labour saving. However, to analyse data entered in this way we must tell the computer that the variable **frequent** represents the number of cases

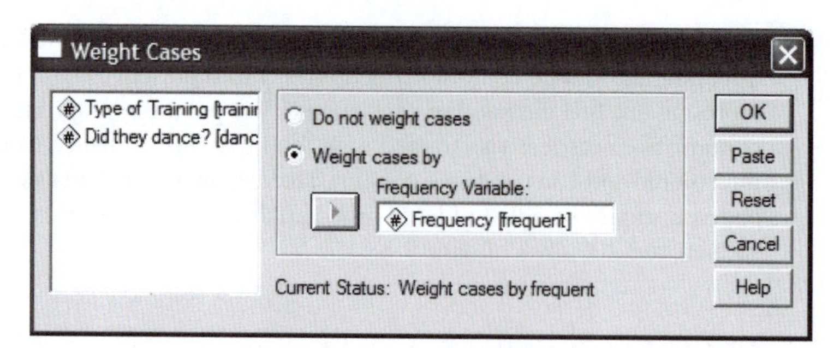

Figure 16.2 The dialog box for the *weight cases* command

that fell into a particular combination of categories. To do this, access the *weight cases* function by using the menu path **Data⇒Weight Cases...** to access the dialog box in Figure 16.2. Select the *Weight cases by* option and then select the variable in which the number of cases is specified (in this case **frequent**) and transfer it to the box labelled *Frequency Variable* by clicking on ▶. This process tells the computer that it should weight each category combination by the number in the column labelled **frequent**. Therefore, the computer will pretend, for example, that there are 28 rows of data that have the category combination 0, 0 (representing cats trained with food who danced). Data entered in this way are in the file **CatsWeight.sav** and if you use this file you must remember to weight the cases as described.

16.4.3. Running the analysis ①

Summarizing data that fall into categories is done using the *crosstabs* command (which also produces the chi-square test). *Crosstabs* is in the *Descriptive Statistics* menu (**Analyze⇒Descriptive Statistics⇒Crosstabs...**). Figure 16.3 shows the dialog boxes for the *crosstabs* command (the variable **frequent** is in the diagram because I ran the analysis on the **CatsWeight.sav** data). First, enter one of the variables of interest in the box labelled *Row(s)* by highlighting it on the left-hand side and transferring it by clicking on ▶. For this example, I selected **dance** to be the rows of the table. Next, select the other variable of interest (**training**) and transfer it to the box labelled *Column(s)* by clicking on ▶. In addition, it is possible to select a layer variable (i.e. you can split the rows of the table into further categories). If you had a third categorical variable (as we will later in this chapter) you could split the contingency table by this variable (so layers of the table represent different categories of this third variable).

If you click on Statistics... a dialog box appears in which you can specify various statistical tests. The most important options under the statistics menu for categorical data are described in Box 16.1.

Box 16.1

Statistical options for *crosstabs*

- **Chi-square**: This performs the basic Pearson chi-square test (section 16.2.1). The chi-square test detects whether there is a significant association between two categorical variables. However, it does not say anything about how strong that association might be.
- **Phi and Cramer's V**: These are measures of the strength of association between two categorical variables. Phi is used with 2 × 2 contingency tables (tables in which you have two categorical variables and each variable has only two categories). Phi is calculated by taking the chi-square value and dividing it by the sample size and then taking the square root of this value. If one of the two categorical variables contains more than two categories then Cramer's *V* is preferred to phi because phi fails to reach its minimum value of zero (indicating no association) in these circumstances.
- **Lambda**: Goodman and Kruskal's λ measures the proportional reduction in error that is achieved when membership of a category of one variable is used to predict category membership on the other variable. A value of 1 means that one variable perfectly predicts the other, whereas a value of 0 indicates that one variable in no way predicts the other.
- **Kendall's statistic**: This statistic is discussed in section 4.5.5.

Select the chi-square test, the continuity correction, phi and lambda and then click on Continue. If you click on Cells... a dialog box appears in which you can specify the type of data displayed in the crosstabulation table. It is important that you ask for expected counts because for chi-square to be accurate these expected counts must exceed certain values. The basic rule of thumb is that with 2 × 2 contingency tables no expected values should be below 5. In larger tables the rule is that all expected counts should be greater than 1 and no more than 20% of expected counts should be less than 5. It is also useful to have a look at the row, column and total percentages because these values are usually more easily interpreted than the actual frequencies and provide some idea of the origin of any significant effects. Once these options have been selected click on Continue to return to the main dialog box. From here you can click on Exact... (if you have exact tests installed). I've explained this option in section 13.2.3 so I won't dwell on it here, but select the *Exact* test option. Click on Continue to return to the main dialog box and then click on OK to run the analysis.

16.4.4. Output for the chi-square test ①

The crosstabulation table produced by SPSS (SPSS Output 16.1) contains the number of cases that falls into each combination of categories and is rather like our original contingency table. We can see that in total 76 cats danced (38% of the total) and of these 28 were

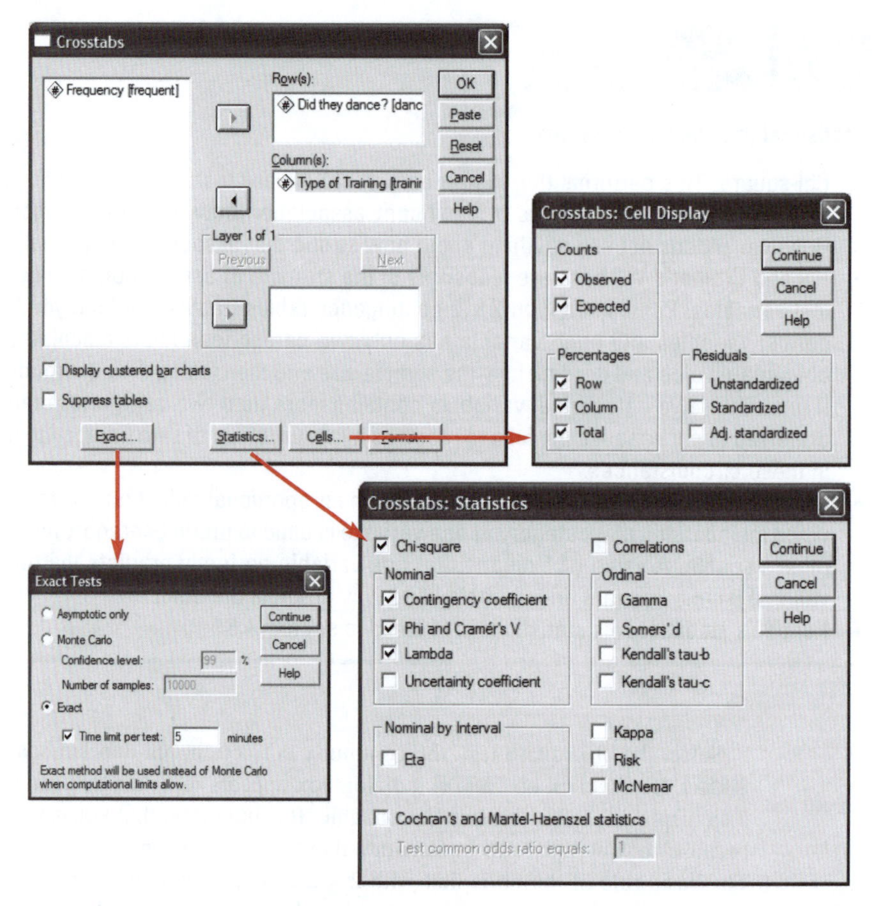

Figure 16.3 Dialog boxes for the *crosstabs* command

trained using food (36.8% of the total that danced) and 48 were trained with affection (63.2% of the total that danced). One hundred and twenty-four cats didn't dance at all (62% of the total) and of those that didn't dance, 10 were trained using food as a reward (8.1% of the total that didn't dance) and a massive 114 were trained using affection (91.9% of the total that didn't dance). The numbers of cats can be read from the rows labelled *Count* and the percentages are read from the rows labelled *% within Did they dance?* We can also look at the percentages within the training categories by looking at the rows labelled *% within Type of Training*. This tells us, for example, that of those trained with food as a reward, 73.7% danced, and 26.3% did not. Similarly, for those trained with affection only 29.6% danced compared to 70.4% that didn't. In summary, when food was used as a reward most cats would dance, but when affection was used most cats refused to dance.

Did they dance? * Type of Training Crosstabulation

			Type of Training		
			Food as Reward	Affection as Reward	Total
Did they dance?	Yes	Count	28	48	76
		Expected Count	14.4	61.6	76.0
		% within Did they dance?	36.8%	63.2%	100.0%
		% within Type of Training	73.7%	29.6%	38.0%
		% of Total	14.0%	24.0%	38.0%
	No	Count	10	114	124
		Expected Count	23.6	100.4	124.0
		% within Did they dance?	8.1%	91.9%	100.0%
		% within Type of Training	26.3%	70.4%	62.0%
		% of Total	5.0%	57.0%	62.0%
Total		Count	38	162	200
		Expected Count	38.0	162.0	200.0
		% within Did they dance?	19.0%	81.0%	100.0%
		% within Type of Training	100.0%	100.0%	100.0%
		% of Total	19.0%	81.0%	100.0%

SPSS Output 16.1

Before moving on to look at the test statistics itself it is vital that we check that the assumption for chi-square has been met. The assumption is that in 2×2 tables (which is what we have here), all expected frequencies should be greater than 5. If you look at the expected counts in the crosstabulation table (which incidentally are the same as we calculated earlier), it should be clear that the smallest expected count is 14.4 (for cats that were trained with food and did dance). This value exceeds 5 and so the assumption has been met. If you found an expected count lower than 5 the best remedy would be to collect more data to try to boost the proportion of cases falling into each category.

As we saw earlier, Pearson's chi-square test examines whether there is an association between two categorical variables (in this case the type of training and whether the animal danced or not). As part of the *crosstabs* procedure SPSS produces a table (SPSS Output 16.2) that includes the chi-square statistic and its significance value. The Pearson chi-square statistic tests whether the two variables are independent. If the significance value is small enough (conventionally *Sig.* must be less than .05) then we reject the hypothesis that the variables are independent and accept the hypothesis that they are in some way related. The value of the chi-square statistic is given in the table (and the degrees of freedom) as is the significance value. The value of the chi-square statistic is 25.356, which is within rounding error of what we calculated in section 16.2.1. This value is highly significant ($p < .001$), indicating that the type of training used had a significant effect on whether an animal would dance.

A series of other statistics are also included in the table (many of which have to be specifically requested using the options in the dialog box in Figure 16.3). *Continuity Correction* is Yates's continuity corrected chi-square (see section 16.2.3) and its value

Chi-Square Tests

	Value	df	Asymp. Sig. (2-sided)	Exact Sig. (2-sided)	Exact Sig. (1-sided)
Pearson Chi-Square	25.356[b]	1	.000		
Continuity Correction[a]	23.520	1	.000		
Likelihood Ratio	24.932	1	.000		
Fisher's Exact Test				.000	.000
Linear-by-Linear Association	25.229	1	.000		
N of Valid Cases	200				

a Computed only for a 2 × 2 table
b 0 cells (.0%) have expected count less than 5. The minimum expected count is 14.44

SPSS Output 16.2

is the same as the value we calculated earlier (23.52). As I mentioned earlier, this test is probably best ignored anyway, but it does confirm the result from the main chi-square test. The *Likelihood Ratio* is the statistic we encountered in section 16.2.2 (and is again within rounding error of the value we calculated: 24.93). Again this confirms the main chi-square result, but this statistic would be preferred in smaller samples.

Underneath the chi-square table there are some footnotes relating to the assumption that expected counts should be greater than 5. If you forgot to check this assumption yourself, SPSS kindly gives a summary of the number of expected counts below 5. In this case, there were no expected frequencies less than 5 so we know that the chi-square statistic should be accurate.

The highly significant result indicates that there is an association between the type of training and whether the cat danced or not. What we mean by an association is that the pattern of responses (i.e. the proportion of cats that danced to the proportion that did not) in the two training conditions is significantly different. This significant finding reflects the fact that when food is used as a reward, about 74% of cats learn to dance and 26% do not, whereas when affection is used, the opposite is true (about 70% refuse to dance and 30% do dance). Therefore, we can conclude that the type of training used significantly influences the cats: they will dance for food but not for love! Having lived with a lovely cat for many years now, this supports my cynical view that cats will do nothing unless there is a bowl of cat food waiting for them at the end of it!

If requested, SPSS will produce another table of output containing some additional statistical tests. Most of these tests are measures of the strength of association. These measures are based on modifying the chi-square statistic to take account of sample size and degrees of freedom and they try to restrict the range of the test statistic from 0 to 1 (to make them similar to the correlation coefficient described in Chapter 4). These are shown in SPSS Output 16.3.

Symmetric Measures

		Value	Approx. Sig.
Nominal by Nominal	Phi	.356	.000
	Cramer's V	.356	.000
	Contingency Coefficient	.335	.000
N of Valid Cases		200	

a Not assuming the null hypothesis
b Using the asymptotic standard error assuming the null hypothesis

SPSS Output 16.3

- **Phi**: This statistic is accurate for 2 × 2 contingency tables. However, for tables with greater than two dimensions the value of phi may not lie between 0 and 1 because the chi-square value can exceed the sample size. Therefore, Pearson suggested the use of the coefficient of contingency.
- **Cramer's *V***: When both variables have only two categories, phi and Cramer's *V* are identical. However, when variables have more than two categories Cramer's statistic can attain its maximum of 1—unlike the other two—and so it is the most useful.
- **Contingency coefficient**: This coefficient ensures a value between 0 and 1 but, unfortunately, it seldom reaches its upper limit of 1 and for this reason Cramer devised Cramer's V.

For these data, Cramer's statistic is .36 out of a possible maximum value of 1. This represents a medium association between the type of training and whether the cats danced or not (if you think of it like a correlation coefficient then this represents a medium effect size). This value is highly significant ($p < .001$) indicating that a value of the test statistic that is this big is unlikely to have happened by chance, and therefore the strength of the relationship is significant. These results confirm what the chi-square test already told us but also give us some idea of the size of effect.

16.4.5. Calculating an effect size ②

Although Cramer's *V* is an adequate effect size (in the sense that it is constrained to fall between 0 and 1 and is, therefore, easily interpretable), a more common, and possibly more useful, measure of effect size for categorical data is the odds ratio. Odds ratios are particularly useful in 2 × 2 contingency tables because for these tables the interpretation of the odds ratio is very clear. However, this isn't as restrictive as you might think because, as I've said more times than I care to recall in the GLM chapters, effect sizes are only ever useful when they summarize a focused comparison. A 2 × 2 contingency table is the categorical data equivalent of a focused comparison!

The odds ratio is simple enough to calculate. If we look at our example, we can first-calculate the odds that a cat danced given that it had food as a reward. This is simply the number of cats that were given food and danced, divided by the number of cats that were given food but didn't dance:

$$\begin{aligned}\text{odds}_{\text{dancing after food}} &= \frac{\text{number that had food and danced}}{\text{number that had food but didn't dance}}\\ &= \frac{28}{10}\\ &= 2.8\end{aligned}$$

Next we calculate the odds that a cat danced given that it had affection as a reward. This is simply the number of cats that were given affection and danced, divided by the number of cats that were given affection but didn't dance:

$$\begin{aligned}\text{odds}_{\text{dancing after affection}} &= \frac{\text{number that had affection and danced}}{\text{number that had affection but didn't dance}}\\ &= \frac{48}{114}\\ &= 0.421\end{aligned}$$

The odds ratio is simply the odds of dancing after food divided by the odds of dancing after affection:

$$\begin{aligned}\text{odds ratio} &= \frac{\text{odds}_{\text{dancing after food}}}{\text{odds}_{\text{dancing after affection}}}\\ &= \frac{2.8}{0.421}\\ &= 6.65\end{aligned}$$

What this tells us is that if a cat was trained with food it was 6.65 times more likely to dance. As you can see this is an extremely elegant and easily understood metric for expressing the effect you've got!

16.4.6. Reporting the results of chi-square ①

When reporting Pearson's chi-square we simply report the value of the test statistic with its associated degrees of freedom and the significance value. The test statistic, as we've seen, is denoted by χ^2. The SPSS output tells us that the value of χ^2 was 25.36, that the degrees of freedom on which this was based were 1, and that it was significant at $p < .001$. It's also useful to reproduce the contingency table and my vote would go to quoting the odds ratio too. As such, we could report:

✓ There was a significant association between the type of training and whether or not cats would dance $\chi^2(1) = 25.36$, $p < .001$. This seems to represent the fact that based on the odds ratio cats were 6.65 times more likely to dance if trained with food than if trained with affection.

Cramming Samantha's Tips

- If you want to test the relationship between two categorical variables you can do this with *Pearson's chi-square test* or the *likelihood ratio statistic.*
- Look at the table labelled *Chi-Square Tests.* If the *Exact Sig.* Value is less than .05 for the row labelled *Pearson Chi-Square* then there is a significant relationship between your two variables.
- Check underneath this table to make sure that no expected frequencies are less than 5.
- Look at the crosstabulation table to work out what the relationship between the variables is. Better still, calculate the *odds ratio.*
- Report the χ^2-statistic, the degrees of freedom and the significance value. Also report the contingency table.

16.5. SEVERAL CATEGORICAL VARIABLES: LOGLINEAR ANALYSIS ③

So far we've looked at situations in which there are only two categorical variables. However, often we want to analyse more complex contingency tables in which there are three or more variables. For example, what about if we took the example we've just used but also collected data from a sample of 70 dogs? We might want to compare the behaviour in dogs with that in cats. We would now have three variables: **animal** (dog or cat), **training** (food as reward or affection as reward) and **dance** (did they dance or not?). This couldn't be analysed with the Pearson chi-square and instead has to be analysed with a technique called loglinear analysis.

16.5.1. Chi-square as regression ④

To begin with let's have a look at how our simple chi-square example can be expressed as a regression model. Although we already know about as much as we need to about the chi-square test, if we want to understand more complex situations life becomes considerably easier if we consider our model as a general linear model (i.e. regression). All of the general linear models we've considered in this book take the general form of:

$$\text{Outcome}_i = (\text{Model}_i) + \text{error}_i$$

For example, when we encountered multiple regression in Chapter 5 we saw that this model was written as (see equation (5.9)):

$$Y_i = (b_0 + b_1 X_1 + b_2 X_2 + \cdots + b_n X_n) + \varepsilon_i$$

Table 16.2 Coding scheme for dancing cats

Training	Dance	Dummy (Training)	Dummy (Dance)	Interaction	Frequency
Food	Yes	0	0	0	28
Food	No	0	1	0	10
Affection	Yes	1	0	0	48
Affection	No	1	1	1	114

Also, when we came across one-way ANOVA, we adapted this regression model to conceptualize our Viagra example as (see equation (8.2)):

$$\text{Libido}_i = b_0 + b_2\text{High}_i + b_1\text{Low}_i + \varepsilon_i$$

The *t*-test was conceptualized in a similar way. In all cases the same basic equation is used; it's just the complexity of the model that changes. With categorical data we can use the same model in much the same way as with regression to produce a linear model. In our current example we have two categorical variables: training (food or affection) and dance (yes they did dance or no they didn't dance). Both variables have two categories and so we can represent each one with a single dummy variable (see section 5.10.1) in which one category is coded as 0 and the other as 1. So for training, we could code 'food' as 0 and 'affection' as 1, and we could code the dancing variable as 0 for 'yes' and 1 for 'no' (see Table 16.2).

This situation might be familiar if you think back to factorial ANOVA (section 10.8) in which we also had two variables as predictors. In that situation we saw that when there are two variables the general linear model became (think back to equation (10.1)):

$$\text{Outcome}_i = (b_0 + b_1\text{A}_i + b_2\text{B}_i + b_3\text{AB}_i) + \varepsilon_i$$

in which A represents the first variable, B represents the second, and AB represents the interaction between the two variables. Therefore, we can construct a linear model using these dummy variables that is exactly the same as the one we used for factorial ANOVA (above). The interaction term will simply be the training variable multiplied by the dance variable (look at Table 16.2 and if it doesn't make sense look back to section 10.8 because the coding is exactly the same as this example):

$$\begin{aligned}\text{Outcome}_i &= (\text{Model}_i) + \text{error}_i \\ \text{Outcome}_{ij} &= (b_0 + b_1\text{Training}_i + b_2\text{Dance}_j + b_3\text{Interaction}_{ij}) + \varepsilon_{ij} \qquad (16.3)\end{aligned}$$

However, because we're using categorical data, to make this model linear we have actually to use log values (see Chapter 6) and so the actual model becomes:[2]

$$\begin{aligned} \ln(O_i) &= \ln(\text{Model}_i) + \ln(\varepsilon_i) \\ \ln(O_{ij}) &= (b_0 + b_1\text{Training}_i + b_2\text{Dance}_j + b_3\text{Interaction}_{ij}) + \ln(\varepsilon_{ij}) \end{aligned} \quad (16.4)$$

The training and dance variables and the interaction can take the values 0 and 1, depending on which combination of categories we're looking at (Table 16.2). Therefore, to work out what the *b*-values represent in this model we can do the same as we did for the *t*-test and ANOVA and look at what happens when we replace training and dance with values of 0 and 1. To begin with, let's see what happens when we look at when training and dance are both zero. This represents the category of cats that got food reward and did line-dance. When we used this sort of model for the *t*-test and ANOVA the outcomes we used were taken from the observed data: we used the group means (e.g. see sections 7.8 and 8.2.2). However, with categorical variables means are rather meaningless because we haven't measured anything on an ordinal or interval scale; instead we merely have frequency data. Therefore, we simply use the observed frequencies (rather than observed means) as our outcome instead. In Table 16.1 we saw that there were 28 cats that had food for a reward and did line-dance; if we use this as the observed outcome then the model can be written as (if we ignore the error term for the time being):

$$\ln(O_{ij}) = b_0 + b_1\text{Training}_i + b_2\text{Dance}_j + b_3\text{Interaction}_{ij}$$

For cats that had food reward and did dance, the training and dance variables and the interaction will all be 0 and so the equation reduces to:

$$\begin{aligned} \ln(O_{\text{Food, Yes}}) &= b_0 + (b_1 \times 0) + (b_2 \times 0) + (b_3 \times 0) \\ \ln(O_{\text{Food, Yes}}) &= b_0 \\ \ln(28) &= b_0 \\ b_0 &= 3.332 \end{aligned}$$

Therefore, b_0 in the model represents the log of the observed value when all of the categories are zero. As such it's the log of the observed value of the base category (in this case cats that got food and danced). Now, let's see what happens when we look at cats that had affection as a reward and danced. In this case, the training variable is 1 and the dance

2 Actually, the convention is to denote b_0 as θ and the *b*-values as λ, but I think these notational changes serve only to confuse people so I'm sticking with *b* because I want to emphasize the similarities to regression and ANOVA.

variable and the interaction are still 0. Also, our outcome now changes to be the observed value for cats that received affection and danced (from Table 16.1 we can see that the value is 48). Therefore, the equation becomes:

$$\begin{aligned}\ln(O_{\text{Affection, Yes}}) &= b_0 + (b_1 \times 1) + (b_2 \times 0) + (b_3 \times 0)\\ \ln(O_{\text{Affection, Yes}}) &= b_0 + b_1\\ b_1 &= \ln(O_{\text{Affection, Yes}}) - b_0\end{aligned}$$

Remembering that b_0 is the expected value for cats that had food and danced, we get:

$$\begin{aligned}b_1 &= \ln(O_{\text{Affection, Yes}}) - \ln(O_{\text{Food, Yes}})\\ &= \ln(48) - \ln(28)\\ &= 3.871 - 3.332\\ &= .539\end{aligned}$$

The important thing is that b_1 is the difference between the log of the observed frequency for cats that received affection and danced and the log of the observed values for cats that received food and danced. Put another way, within the group of cats that danced it represents the difference between those trained using food and those trained using affection.

Now, let's see what happens when we look at cats that had food as a reward and did not dance. In this case, the training variable is 0, the dance variable is 1 and the interaction is again 0. Our outcome now changes to be the observed frequency for cats who received food but did not dance (from Table 16.1 we can see that the value is 10). Therefore, the equation becomes:

$$\begin{aligned}\ln(O_{\text{Food,No}}) &= b_0 + (b_1 \times 0) + (b_2 \times 1) + (b_3 \times 0)\\ \ln(O_{\text{Food,No}}) &= b_0 + b_2\\ b_2 &= \ln(O_{\text{Food,No}}) - b_0\end{aligned}$$

Remembering that b_0 is the expected value for cats that had food and danced, we get:

$$\begin{aligned}b_2 &= \ln(O_{\text{Food,No}}) - \ln(O_{\text{Food, Yes}})\\ &= \ln(10) - \ln(28)\\ &= 2.303 - 3.332\\ &= -1.029\end{aligned}$$

The important thing is that b_2 is the difference between the log of the observed frequency for cats that received food and danced and the log of the observed frequency for cats that received food and didn't dance. Put another way, within the group of cats that received food as a reward it represents the difference between cats that didn't dance and those that did. Finally, we can look at cats that had affection and danced. In this case, the training and dance variables are both 1, and the interaction (which is the value of training multiplied by the value of dance) is also 1. We can also replace b_0, b_1 and b_2 with what we now know they represent. The outcome is the log of the observed frequency for cats that received affection but didn't dance (this expected value is 114—see Table 16.1); therefore, the equation becomes (I've used the shorthand of A for affection, F for food, Y for yes and N for No):

$$
\begin{aligned}
\ln(O_{\text{A,N}}) &= b_0 + (b_1 \times 1) + (b_2 \times 1) + (b_3 \times 1) \\
\ln(O_{\text{A,N}}) &= b_0 + b_1 + b_2 + b_3 \\
\ln(O_{\text{A,N}}) &= \ln(O_{\text{F,Y}}) + [\ln(O_{\text{A,Y}}) - \ln(O_{\text{F,Y}})] + [\ln(O_{\text{F,N}}) - \ln(O_{\text{F,Y}})] + b_3 \\
\ln(O_{\text{A,N}}) &= \ln(O_{\text{A,Y}}) + \ln(O_{\text{F,N}}) - \ln(O_{\text{F,Y}}) + b_3 \\
b_3 &= \ln(O_{\text{A,N}}) - \ln(O_{\text{F,N}}) + \ln(O_{\text{F,Y}}) - \ln(O_{\text{A,Y}}) \\
&= \ln(114) - \ln(10) + \ln(28) - \ln(48) \\
&= 1.895
\end{aligned}
$$

So, b_3 in the model really compares the difference between affection and food when the cats didn't dance to the difference between food and affection when the cats did dance. Put another way, it compares the effect of training when cats didn't dance to the effect of training when they did dance.

The final model is, therefore:

$$\ln(O_{ij}) = 3.332 + .539\text{Training} - 1.029\text{Dance} + 1.895\text{Interaction} + \ln(\varepsilon_{ij})$$

The important thing to note here is that everything is exactly the same as with factorial ANOVA except that we dealt with log transformed values (in fact compare this section with section 10.8 to see just how similar everything is). In case you still don't believe me that this works as a GLM, I've prepared a file called **CatRegression.sav** which contains the two variables **dance** and **training** (both dummy coded with zeros and ones as described above) and the interaction (**int**). There is also a variable called **observed** which contains the observed frequencies in Table 16.1 for each combination of **dance** and **training**. Finally, there is a variable called **lno**, which is the natural logarithm of these observed frequencies (remember that throughout this section we've dealt with the log observed values). Run a multiple regression analysis using this file with **lno** as the outcome, and **training**, **dance** and **interaction** as your three predictors. SPSS Output 16.4 shows the resulting coefficients table from this regression. The important thing to note is that the constant, b_0, is 3.332 as calculated above, the beta value for type of training, b_1, is .539 and for dance, b_2, is −1.030, both of which are within rounding error of what was calculated above. Also the coefficient for the interaction, b_3, is 1.895 as predicted. There is one interesting point, though: all of the standard errors are zero, or put differently there is *no* error at all in this

Coefficients[a]

Model		Unstandardized Coefficients		Standardized Coefficients		
		B	Std. Error	Beta	t	Sig.
1	(Constant)	3.332	.000		6.3E+08	.000
	Type of Training	.539	.000	.307	8.1E+07	.000
	Did they dance?	−1.030	.000	−.725	−1.0E+08	.000
	Interaction	1.895	.000	1.361	1.7E+08	.000

a Dependent Variable: LN (Observed values)

SPSS Output 16.4

model. This is because the various combinations of coding variables completely explain the observed values. This is known as a saturated model and I shall return to this point later so bear it in mind. For the time being, I hope this convinces you that chi-square can be conceptualized as a linear model.

OK, this is all very well, but the heading of this section did rather imply that I would show you how the chi-square test can be conceptualized as a linear model. Well, basically, the chi-square test looks at whether two variables are independent; therefore, it has no interest in the combined effect of the two variables, only their unique effect. So we can conceptualize chi-square in much the same way as the saturated model, except that we don't include the interaction term. If we remove the interaction term, our model becomes:

$$\ln(\text{Model}_{ij}) = b_0 + b_1\text{Training}_i + b_2\text{Dance}_j$$

With this new model, we cannot predict the observed values as we did for the saturated model because we've lost some information (namely, the interaction term). Therefore, the outcome from the model changes, and thus the betas change too. We saw earlier that the chi-square test is based on 'expected frequencies'. Therefore, if we're conceptualizing the chi-square test as a linear model, our outcomes will be these expected values. If you look back to the beginning of this chapter you'll see we already have the expected frequencies based on this model. We can recalculate the betas based on these expected values:

$$\ln(E_{ij}) = b_0 + b_1\text{Training}_i + b_2\text{Dance}_j$$

For cats that had food reward and did dance, the training and dance variables will be 0 and so the equation reduces to:

$$\begin{aligned}
\ln(E_{\text{Food, Yes}}) &= b_0 + (b_1 \times 0) + (b_2 \times 0) \\
\ln(E_{\text{Food, Yes}}) &= b_0 \\
b_0 &= \ln(14.44) \\
&= 2.67
\end{aligned}$$

Therefore, b_0 in the model represents the log of the expected value when all of the categories are 0.

When we look at cats that had affection as a reward and danced, the training variable is 1 and the dance variable and the interaction are still 0. Also, our outcome now changes to be the expected value for cats that received affection and danced:

$$\begin{aligned}
\ln(E_{\text{Affection, Yes}}) &= b_0 + (b_1 \times 1) + (b_2 \times 0) \\
\ln(E_{\text{Affection, Yes}}) &= b_0 + b_1 \\
b_1 &= \ln(E_{\text{Affection, Yes}}) - b_0 \\
&= \ln(E_{\text{Affection, Yes}}) - \ln(E_{\text{Food, Yes}}) \\
&= \ln(61.56) - \ln(14.44) \\
&= 1.45
\end{aligned}$$

The important thing is that b_1 is the difference between the log of the expected frequency for cats that received affection and danced and the log of the expected values for cats that received food and danced. In fact, the value is the same as the column marginal; that is, the difference between the total number of cats getting affection and the total number of cats getting food: ln(162) – ln(38) = 1.45.

When we look at cats that had food as a reward and did not dance, the training variable is 0, and the dance variable is 1. Our outcome now changes to be the expected frequency for cats who received food but did not dance:

$$\begin{aligned}
\ln(E_{\text{Food,No}}) &= b_0 + (b_1 \times 0) + (b_2 \times 1) \\
\ln(E_{\text{Food,No}}) &= b_0 + b_2 \\
b_2 &= \ln(O_{\text{Food,No}}) - b_0 \\
&= \ln(O_{\text{Food,No}}) - \ln(O_{\text{Food, Yes}}) \\
&= \ln(23.56) - \ln(14.44) \\
&= .49
\end{aligned}$$

Therefore, b_2 is the difference between the log of the expected frequencies for cats that received food and didn't or did dance. In fact, the value is the same as the row marginal; that is, the difference between the total number of cats that did and didn't dance: ln(124) – ln(76) = 0.49.

We can double-check all of this by looking at the final cell:

$$\begin{aligned}
\ln(E_{\text{Affection,No}}) &= b_0 + (b_1 \times 1) + (b_2 \times 1) \\
\ln(E_{\text{Affection,No}}) &= b_0 + b_1 + b_2 \\
\ln(100.44) &= 2.67 + 1.45 + .49 \\
4.61 &= 4.61
\end{aligned}$$

The final chi-square model is, therefore:

$$\ln(O_{ij}) = \text{Model}_{ij} + \ln(\varepsilon_{ij})$$
$$\ln(O_{ij}) = 2.67 + 1.45\text{Training} + .49\text{Dance} + \ln(\varepsilon_{ij})$$

We can rearrange this to get some residuals (the error term):

$$\ln(\varepsilon_{ij}) = \ln(O_{ij}) - \text{Model}_{ij}$$

In this case, the model is merely the expected frequencies that were calculated for the chi-square test, so the residuals are the differences between the observed and expected frequencies.

This demonstrates how chi-square can work as a linear model, just like regression and ANOVA, in which the betas tell us something about the relative differences in frequencies across categories of our two variables. If nothing else made sense I want you to leave this section aware that chi-square (and analysis of categorical data generally) can be expressed as a linear model (although we have to use log values). We can express categories of a variable using dummy variables, just as we did with regression and ANOVA, and the resulting beta values can be calculated in exactly the same way as for regression and ANOVA. In ANOVA, these beta values represented the difference between the means of a particular category compared to a baseline category. With categorical data, the beta values represent the same thing, the only difference being that rather than dealing with means, we're dealing with expected values. Grasping this idea (that regression, *t*-tests, ANOVAs and categorical data analysis are basically the same) will help (me) considerably in the next section.

16.5.2. Loglinear analysis ③

In the previous section, after nearly reducing my brain to even more of a rotting vegetable than it already is in trying to explain how categorical data analysis is just another form of regression, I ran the data through an ordinary regression on SPSS to prove that I wasn't talking complete gibberish. At the time I rather glibly said 'oh, by the way, there's no error in the model, that's odd isn't it?' and sort of passed this off by telling you that it was a 'saturated' model and not to worry too much about it because I'd explain it all later just as soon as I'd worked out what the hell was going on. That seemed like a good avoidance tactic at the time but unfortunately I now have to explain what I was going on about.

To begin with, I hope you're now happy with the idea that categorical data can be expressed in the form of a linear model provided that we use log values (this, incidentally, is why the technique we're discussing is called loglinear analysis). From what you hopefully already know about ANOVA and linear models generally, you should also be cosily tucked up in bed with the idea that we can extend any linear model to include any amount of predictors and any resulting interaction terms between predictors. Therefore, if we can represent a simple two-variable categorical analysis in terms of a linear model, then it shouldn't amaze you to discover that if we have more than two variables this is no problem: we can extend the simple model by adding whatever variables and the resulting interaction terms. This is all you really need to know. So, just as in multiple regression and ANOVA, if we think of things in terms of a linear model, then conceptually it becomes very easy to understand how the model expands to incorporate new variables. So, for example, if we have three predictors (A, B and C) in ANOVA (think back to section 12.3) we end up with three two-way interactions (AB, AC, BC) and one three-way interaction (ABC). Therefore, the resulting linear model of this is just:

$$\text{Outcome}_i = (b_0 + b_1\text{A} + b_2\text{B} + b_3\text{C} + b_4\text{AB} + b_5\text{AC} + b_6\text{BC} + b_7\text{ABC}) + \varepsilon_i$$

In exactly the same way, if we have three variables in a categorical data analysis we get an identical model, but with an outcome in terms of logs:

$$\ln(\text{O}_{ijk}) = (b_0 + b_1\text{A}_i + b_2\text{B}_j + b_3\text{C}_k + b_4\text{AB}_{ij} + b_5\text{AC}_{ik} + b_6\text{BC}_{jk} + b_7\text{ABC}_{ijk}) + \ln(\varepsilon_{ijk})$$

Obviously the calculation of beta values and expected values from the model becomes considerably more cumbersome and confusing but that's why we invented computers: so that we don't have to worry about it! Loglinear analysis works on these principles. However, as we've seen in the two-variable case, when our data are categorical and we include all of the available terms (main effects and interactions) we get no error: our predictors can perfectly predict our outcome (the expected values). So, if we start with the most complex model possible, we will get no error. The job of loglinear analysis is to try to fit a simpler model to the data without any substantial loss of predictive power. Therefore, loglinear analysis typically works on a principle of backward elimination (yes, the same kind of backward elimination that we can use in multiple regression—see section 5.5.3.3). So we begin with the saturated model, and then we remove a predictor from the model and using this new model we predict our data (calculate expected frequencies, just like in the chi-square test), and then see how well the model fits the data (i.e. are the expected frequencies close to the observed frequencies?). If the fit of the new model is not very different from the more complex model, then we abandon the complex model in favour of the new one. Put another way, we assume the term we removed was not having a significant impact on the ability of our model to predict the observed data.

However, the analysis doesn't just remove terms randomly, it does it hierarchically. So, we start with the saturated model and then remove the highest-order interaction, and assess the effect that this has. If removing the interaction term has no effect on the model then it's obviously not having much of an effect; therefore, we get rid of it and move on to remove any lower-order interactions. If removing these interactions has no effect then we carry on to any main effects until we find an effect that does affect the fit of the model if it is removed.

To put this in more concrete terms, at the beginning of the section on loglinear analysis I asked you to imagine we'd extended our training and line-dancing example to incorporate a sample of dogs. So, we now have three variables: animal (dog or cat), training (food or affection) and dance (did they dance or not?). Just as in ANOVA this results in three main effects, three interactions involving two variables and one interaction involving all three variables. So we get the following effects: animal, training, dance, animal × training, animal × dance, training × dance, and animal × training × dance. When I talk about backward elimination all I mean is that loglinear analysis starts by including all of these effects, it then takes the highest-order interaction (in this case the three-way interaction of animal × training × dance) and removes it. It constructs a new model without this interaction, and from the model calculates expected frequencies. It then compares these expected frequencies (or model frequencies) to the observed frequencies using the standard equation for the likelihood ratio statistic (see section 16.2.2). If the new model significantly changes the likelihood ratio statistic, then removing this interaction term has a significant effect on the fit of the model and we know that this effect is statistically important. If this is the case SPSS will stop there and tell you that you have a significant three-way interaction! It won't test any other effects because with categorical data all lower-order effects are consumed within higher-order effects. If, however, removing the three-way interaction doesn't significantly affect the fit of the model then SPSS moves on to lower-order interactions. Therefore, it looks at the animal × training, animal × dance, and training × dance interactions in turn and constructs models in which these terms are not present. For each model it again calculates expected values and compares them to the observed data using a likelihood ratio.[3] Again, if any one of these models does result in a significant change in the likelihood ratio then the term is retained and SPSS won't move on to look at any main effects involved in that interaction (so, if the animal × training interaction is significant it won't look at the main effects of animal or training). However, if the likelihood ratio is unchanged then the analysis removes the offending interaction term and moves on to look at main effects.

I mentioned that the likelihood statistics (see section 16.2.2) are used to assess each model. From the equation it should be clear how this equation can be adapted to fit any model: the observed values are the same throughout, and the model frequencies are simply the expected frequencies from the model being tested. For the saturated model, this statistic will always be 0 (because the observed and model frequencies are the same so the ratio of observed to model frequencies will be 1, and $\ln(1) = 0$), but as we've seen, in other cases it will provide a measure of how well the model fits the observed frequencies. To test whether a new model has changed the likelihood ratio, all we need do is to take the likelihood ratio for a model, and subtract from it the likelihood statistic for the previous model (provided the models are hierarchically structured):

$$L\chi^2_{\text{Change}} = L\chi^2_{\text{Current Model}} - L\chi^2_{\text{Previous Model}}$$

I've tried in this section to give you a flavour of how loglinear analysis works, without actually getting too much into the nitty-gritty of the calculations. I've tried to show you how we can conceptualize a chi-square analysis as a linear model and then relied on what I've previously told you about ANOVA to hope that you can extrapolate these conceptual ideas to understand roughly what's going on. The curious among you might want to know

3 It's worth mentioning that for every model, the computation of expected values differs, and as the designs get more complex, the computation gets increasingly tedious and incomprehensible (at least to me); however, you don't need to know the calculations to get a feel for what is going on.

exactly how everything is calculated and to these people I have two things to say: 'I don't know' and 'I know a really good place where you can buy a straitjacket'. If you're that interested then Tabachnick & Fidell (2001) have, as ever, written a wonderfully detailed and lucid chapter on the subject which frankly puts this feeble attempt to shame. Still, assuming you're happy to live in relative ignorance we'll now have a look at how to do a loglinear analysis.

16.6. ASSUMPTIONS IN LOGLINEAR ANALYSIS ②

Loglinear analysis is an extension of the chi-square test and so has similar assumptions; that is, an entity should fall into only one cell of the contingency table (i.e. cells of the table must be independent) and the expected frequencies should be large enough for a reliable analysis. In loglinear analysis with more than two variables it's all right to have up to 20% of cells with expected frequencies less than 5; however, all cells must have expected frequencies greater than 1. If this assumption is broken the result is a radical reduction in test power—so dramatic in fact that it may not be worth bothering with the analysis at all. Remedies for problems with expected frequencies are (1) collapse the data across one of the variables (preferably the one you least expect to have an effect!); (2) collapse levels of one of the variables; (3) collect more data; or (4) accept the loss of power.

If you want to collapse data across one of the variables then certain things have to be considered:

1. The highest-order interaction should be non-significant.
2. At least one of the lower-order interaction terms involving the variable to be deleted should be non-significant.

Let's take the example we've been using. Say we wanted to delete the animal variable; then for this to be valid, the animal × training × dance variable should be non-significant, and either the animal × training or the animal × dance interaction should also be significant. You can also collapse categories within a variable. So, if you had a variable of 'season' relating to spring, summer, autumn and winter, and you had very few observations in winter, you could consider reducing the variable to three categories: spring, summer, autumn/winter perhaps. However, you should really only combine categories that it makes theoretical sense to combine.

Finally, some people overcome the problem by simply adding a constant to all cells of the table, but there really is no point in doing this as it doesn't address the issue of power.

16.7. LOGLINEAR ANALYSIS USING SPSS ②

16.7.1. Initial considerations ②

Data are entered for loglinear analysis in the same way as for the chi-square test (see sections 16.4.1 and 16.4.2). The data for the cat and dog example are in the file **CatsandDogs.sav**.

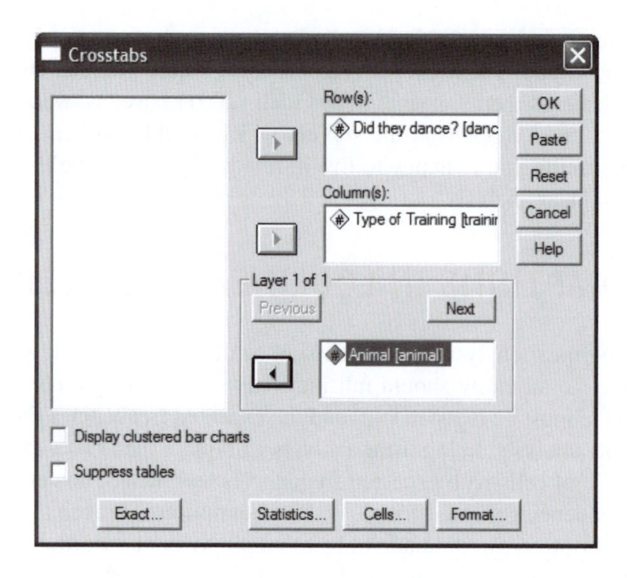

Figure 16.4

Open this file. Notice that this file has three variables (**animal**, **training** and **dance**) and each one contains codes representing the different categories of these variables. To begin with we should use the *crosstabs* command to produce a contingency table of the data. Use section 16.4.3 to help you to do this. The finished dialog box is in Figure 16.4.

The crosstabulation table produced by SPSS (SPSS Output 16.5) contains the number of cases that fall into each combination of categories. The top half of this table is the same as SPSS Output 16.1 because the data are the same (we've just added some dogs) and if you look back in this chapter there's a summary of what this tells us. For the dogs we can summarize the data in a similar way. In total 49 dogs danced (70% of the total) and of these 20 were trained using food (40.8% of the total that danced) and 29 were trained with affection (59.2% of the total that danced). Twenty-one dogs didn't dance at all (30% of the total) and of those that didn't dance, 14 were trained using food as a reward (66.7% of the total that didn't dance) and 7 were trained using affection (33.3% of the total that didn't dance). The numbers of dogs can be read from the rows labelled *Count* and the percentages are read from the rows labelled *% within Did they dance*? In summary, a lot more dogs danced (70%) than didn't (30%). About half of those that danced were trained with affection and about half with food as a reward. In short, dogs seem more willing to dance than cats (70% compared to 38%), and they're not too worried about what training method is used.

Before moving on to look at the test statistics it is vital that we check that the assumptions of loglinear analysis have been met: specifically, there should be no expected counts less than 1, and no more than 20% less than 5. If you look at the expected counts in the crosstabulation table, it should be clear that the smallest expected count is 10.2 (for dogs that were trained with food but didn't dance). This value still exceeds 5 and so the assumption has been met.

Did they dance? * Type of Training * Animal Crosstabulation

Animal				Type of Training		Total
				Food as Reward	Affection as Reward	
Cat	Did they dance?	Yes	Count	28	48	76
			Expected Count	14.4	61.6	76.0
			% within Did they dance?	36.8%	63.2%	100.0%
			% within Type of Training	73.7%	29.6%	38.0%
			% of Total	14.0%	24.0%	38.0%
		No	Count	10	114	124
			Expected Count	23.6	100.4	124.0
			% within Did they dance?	8.1%	91.9%	100.0%
			% within Type of Training	26.3%	70.4%	62.0%
			% of Total	5.0%	57.0%	62.0%
	Total		Count	38	162	200
			Expected Count	38.0	162.0	200.0
			% within Did they dance?	19.0%	81.0%	100.0%
			% within Type of Training	100.0%	100.0%	100.0%
			% of Total	19.0%	81.0%	100.0%
Dog	Did they dance?	Yes	Count	20	29	49
			Expected Count	23.8	25.2	49.0
			% within Did they dance?	40.8%	59.2%	100.0%
			% within Type of Training	58.8%	80.6%	70.0%
			% of Total	28.6%	41.4%	70.0%
		No	Count	14	7	21
			Expected Count	10.2	10.8	21.0
			% within Did they dance?	66.7%	33.3%	100.0%
			% within Type of Training	41.2%	19.4%	30.0%
			% of Total	20.0%	10.0%	30.0%
	Total		Count	34	36	70
			Expected Count	34.0	36.0	70.0
			% within Did they dance?	48.6%	51.4%	100.0%
			% within Type of Training	100.0%	100.0%	100.0%
			% of Total	48.6%	51.4%	100.0%

SPSS Output 16.5

16.7.2. The loglinear analysis ②

Having established that the assumptions have been met we can move on to the main analysis. The best way to run a loglinear analysis is using the **Analyze⇒Loglinear Statistics⇒ Model Selection…** menu to access the main dialog box in Figure 16.5. Select any variable that you want to include in the analysis by selecting them with the mouse and then transferring them to the box labelled *Factor(s)* by clicking on ▸. When there is a variable in this box the Define Range... button becomes active. Just like with the *t*-test and several of the non-parametric tests we encountered in Chapter 13 we have to tell SPSS the codes that we've

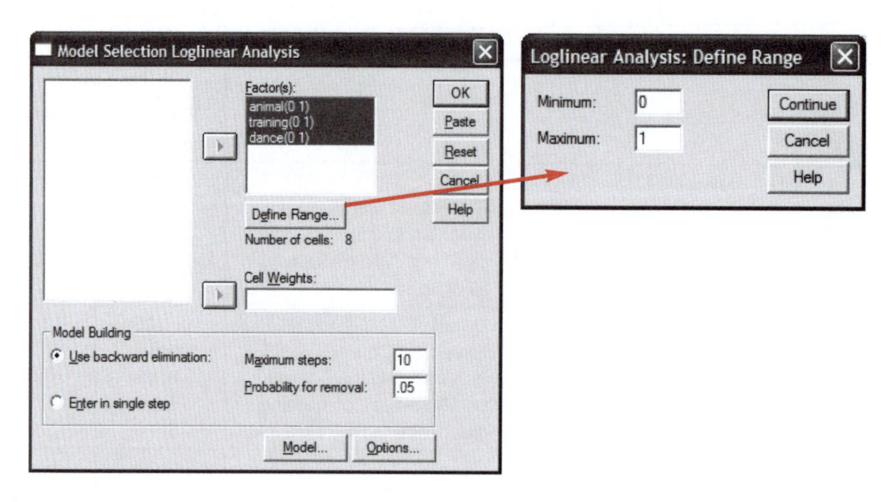

Figure 16.5 Main dialog box for loglinear analysis

used to define our categorical variables. Select a variable in the *Factor(s)* box and then click on [Define Range...] to activate a dialog box that allows you to specify the value of the minimum and maximum code that you've used for that variable. In fact all three variables in this example have the same codes (they all have two categories and I coded them all with 0 and 1) so we can select all three; then click on [Define Range...] and type 0 in the *Minimum* box and 1 in the *Maximum* box. When you've done this click on [Continue] to return to main dialog box.

The default options in this main box are fine; the main thing to note is that by default SPSS uses backward elimination (as I've described elsewhere). You can actually select *Enter in a single step*, which is a non-hierarchical method (in which all effects are entered and evaluated, like forced entry in multiple regression). In loglinear analysis the combined effects take precedence over lower-order effects and so there is little to recommend non-hierarchical methods.

If you click on [Model...] then this will open a dialog box very similar to those we saw in ANOVA (e.g. see Figure 10.2). By default SPSS fits the saturated model, and this is what we should be fitting. However, you can define your own model if you like by specifying individual main effects and interaction terms. Unless you have a very good reason for not fitting the saturated model then leave well alone!

Clicking on [Options...] opens the dialog box in Figure 16.6. There's few options to play around with really (the default options are fine). The only two things you can select are *Parameter estimates*, which will produce a table of parameter estimates for each effect (by parameter estimates I mean just a *z*-score and associated confidence interval), and an *Association table*, which will produce chi-square statistics for all of the effects in the model. This may be useful in some situations, but as I've said before, if the higher-order interactions are significant then we shouldn't really be interested in the lower-order effects because they're confounded with the higher-order effects. When you've finished with the options click on [Continue] to return to the main dialog box and then click on [OK] to run the analysis.

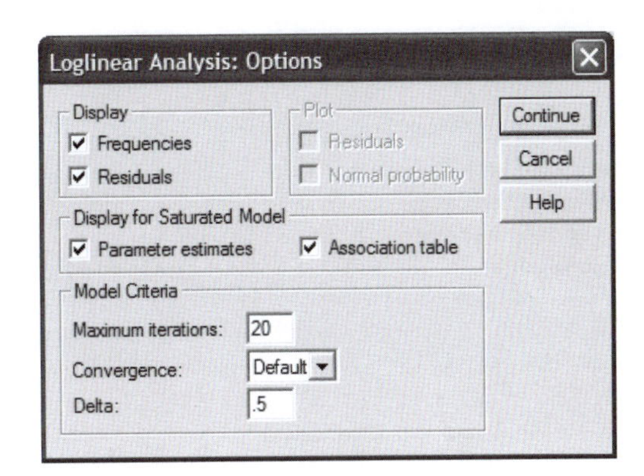

Figure 16.6 Options for loglinear analysis

16.8. OUTPUT FROM LOGLINEAR ANALYSIS ③

SPSS Output 16.6 shows the initial output from the loglinear analysis. To begin with the output tells us that we have 270 cases (remember that we had 200 cats and 70 dogs and this is a useful check that no cases have been lost!). SPSS then lists all of the factors in the model and the number of levels they have (in this case all have two levels). To begin with SPSS fits the saturated model (all terms are in the model including the highest-order interaction, in this case the animal × training × dance interaction). SPSS then gives us the observed and expected counts for each of the combinations of categories in our model. These values should be the same as the original contingency table except that each cell has .5 added to it (this value is the default and is fine, but if you want to change it you can do so by changing delta in Figure 16.6). The final bit of this initial output gives us two goodness-of-fit statistics (Pearson's chi-square and the likelihood ratio statistic, both of which we came across at the beginning of this chapter). In this context these tests are testing the hypothesis that the frequencies predicted by the model (the expected frequencies) are significantly different from the actual frequencies in our data (the observed frequencies). Now, obviously, if our model is a good fit of the data then the observed and expected frequencies should be very similar (i.e. not significantly different). Therefore, we want these statistics to be non-significant. A significant result would mean that our model was significantly different from our data (i.e. the model is a bad fit of the data). In large samples these statistics should give the same results but the likelihood ratio statistic is preferred in small samples. In this example, both statistics are 0 and yield a probability value, *p*, of-INF, which is a rather confusing way of saying that the probability is very high. Put another way, at this stage the model *perfectly* predicts the data. If you read the theory section this shouldn't surprise you, as I showed there that the saturated model is a perfect fit of the data and I also mentioned that the resulting likelihood ratio would be zero. What's interesting in loglinear analysis is what bits of the model we can then remove without significantly affecting the fit of the model.

```
* * * * * * * * H I E R A R C H I C A L   L O G   L I N E A R * * * * * * * *

DATA Information

      270 unweighted cases accepted.
        0 cases rejected because of out-of-range factor values.
        0 cases rejected because of missing data.
      270 weighted cases will be used in the analysis.

FACTOR Information

   Factor      Level    Label
   ANIMAL        2                          Animal
   TRAINING      2      Type of Training
   DANCE         2      Did they dance?

- - - - - - - - - - - - - - - - - - - - - - - - - - - - - - - - - - - - - - -
DESIGN 1 has generating class

   ANIMAL*TRAINING*DANCE

 Note: For saturated models .500 has been added to all observed cells.
 This value may be changed by using the CRITERIA = DELTA subcommand.

The Iterative Proportional Fit algorithm converged at iteration 1.
The maximum difference between observed and fitted marginal totals is .000
and the convergence criterion is .250

- - - - - - - - - - - - - - - - - - - - - - - - - - - - - - - - - - - - - - -
Observed, Expected Frequencies and Residuals.

   Factor      Code          OBS count     EXP count     Residual     Std Resid
ANIMAL      Cat
 TRAINING      Food as
  DANCE         Yes              28.5          28.5          .00          .00
  DANCE         No               10.5          10.5          .00          .00
 TRAINING     Affectio
  DANCE         Yes              48.5          48.5          .00          .00
  DANCE         No              114.5         114.5          .00          .00

ANIMAL      Dog
 TRAINING      Food as
  DANCE         Yes              20.5          20.5          .00          .00
  DANCE         No               14.5          14.5          .00          .00
 TRAINING     Affectio
  DANCE         Yes              29.5          29.5          .00          .00
  DANCE         No                7.5           7.5          .00          .00

- - - - - - - - - - - - - - - - - - - - - - - - - - - - - - - - - - - - - - -
Goodness-of-fit test statistics

   Likelihood ratio chi square =     .00000    DF = 0  P = -INF
            Pearson chi square =     .00000    DF = 0  P = -INF
- - - - - - - - - - - - - - - - - - - - - - - - - - - - - - - - - - - - - - -
```

SPSS Output 16.6

The next part of the output (SPSS Output 16.7) tells us something about which components of the model can be removed. The first bit of the output is labelled *Tests that K-way and higher order effects are zero* and underneath there is a table showing likelihood ratio and chi-square statistics when $K = 3$, 2 and 1 (as we go down the rows of the table). The first row ($K = 3$) is testing whether removing the three-way effect *and* higher-order effects will significantly affect the fit of the model. Now of course, the three-way interaction is the highest-order effect that we have so this is simply testing whether removal of the three-way interaction (i.e. the animal × training × dance interaction) will significantly affect the fit of the model. If you look at the two columns labelled *Prob* then you can see that both the chi-square and likelihood ratio tests agree that removing this interaction will significantly affect the fit of the model (because the probability value is less than .05). The next row of the table ($K = 2$) tells us whether removing the two-way interactions (i.e. the animal × training, animal × dance, and training × dance interactions) and any higher-order effects will affect the model. In this case there is a higher-order effect (the three-way interaction) so this is testing whether removing the two-way interactions *and* the three-way interaction would affect the fit of the model. This is also highly significant (which shouldn't surprise us because the test contains the three-way interaction, which as we've already seen is highly significant) indicating that if we removed the two-way interactions and the three-way interaction then this would have a significant detrimental effect on the model. The final row ($K = 1$) tells us whether removing the one-way effects (i.e. the main effects of animal, training and dance) and any higher-order effects will significantly affect the fit of the model. There are lots of higher-order effects here—there are the two-way interactions and the three-way interaction—and so this is basically testing whether if we remove everything from the model there will be a significant effect on the fit of the model. This again is highly significant (which we would expect because as we've already seen the three-way interaction is highly significant and this test includes that interaction). If this test was non-significant (if the values of *Prob* were above .05) then this would tell you that removing everything from your model would not affect the fit of the model (in other words, overall, the combined effect of your variables and interactions is not significant).

The next part of the table expresses the same thing but without including the higher-order effects. It's labelled *Tests that K-way effects are zero* and then lists tests for when $K = 1$, 2 and 3. The first row ($K = 1$), therefore, tests whether removing the main effects (the one-way effects) has a significant detrimental effect on the model. The probability values are smaller than .05 indicating that if we removed the main effects of animal, training and dance from our model it would significantly affect the fit of the model (in other words, these effects are significant predictors of the data). The second row ($K = 2$) tests whether removing the two-way interactions has a significant detrimental effect on the model. The probability values are less than .05 indicating that if we removed the animal × training, animal × dance, and training × dance interactions then this would significantly reduce how well the model fits the data. In other words, one or more of these two-way interactions is a significant predictor of the data. The final row ($K = 3$) tests whether removing the three-way interaction has a significant detrimental effect on the model. The probability values are less than .05 indicating that if we removed the animal × training × dance interaction then this would significantly reduce how well the model fits the data. In other words, this three-way

```
Tests that K-way and higher order effects are zero.

   K    DF   L.R. Chisq    Prob    Pearson Chisq    Prob    Iteration

   3     1      20.305    .0000          20.778    .0000            4
   2     4      72.267    .0000          67.174    .0000            2
   1     7     200.163    .0000         253.556    .0000            0

Tests that K-way effects are zero.

   K    DF   L.R. Chisq    Prob    Pearson Chisq    Prob    Iteration

   1     3     127.896    .0000         186.382    .0000            0
   2     3      51.962    .0000          46.396    .0000            0
   3     1      20.305    .0000          20.778    .0000            0
```

SPSS Output 16.7

interaction is a significant predictor of the data. This row should be identical to the first row of the previous table (the *Tests of K-way and higher order effects are zero*) because it is the highest-order effect and so in the previous table there were no higher-order effects to include in the test (look at the output and you'll see the results are identical).

What this is actually telling us is that the three-way interaction is significant: removing it from the model has a significant effect on how well the model fits the data. We also know that removing all two-way interactions has a significant effect on the model, but you have to remember that loglinear analysis should be done hierarchically and so these two-way interactions aren't of interest to us because the three-way interaction is significant (we'd only look at these effects if the three-way interaction was non-significant).

If you selected an *Association table* in Figure 16.6 then you'll get the table in SPSS Output 16.8. This simply breaks down the table that we've just looked at into its component parts. So, for example, although we know from the previous output that removing all of the two-way interactions significantly affects the model, we don't know which of the two-way interactions is having the effect. This table tells us. We get a Pearson chi-square test for each of the two-way interactions and the main effects and the column labelled *Prob* tells us which of these effects is significant (values less than .05 are significant). We can tell from this that the animal × dance, training × dance, and animal × training interactions are all significant. Likewise, we saw in the previous output that removing the one-way effects (the main effects of animal, training and dance) also significantly affected the fit of the model, and these findings are confirmed here because the main effects of animal and training are both significant. However, the main effect of dance is not (the probability value is greater than .05). Interesting as these findings are we should ignore them because of the hierarchical nature of loglinear analysis: these effects are all confounded with the higher-order interaction of animal × training × dance.

If you selected *Parameter estimates* in Figure 16.6 then you'll get the table in SPSS Output 16.9. This simply tells us the same thing as the previous table (i.e. it

```
* * * * * * * * H I E R A R C H I C A L   L O G   L I N E A R * * * * * * * *
Tests of PARTIAL associations.

  Effect Name                       DF   Partial Chisq   Prob    Iter

  ANIMAL*TRAINING                    1       13.760      .0002     2
  ANIMAL*DANCE                       1       13.748      .0002     2
  TRAINING*DANCE                     1        8.611      .0033     2
  ANIMAL                             1       65.268      .0000     2
  TRAINING                           1       61.145      .0000     2
  DANCE                              1        1.483      .2233     2
- - - - - - - - - - - - - - - - - - - - - - - - - - - - - - - - - - - - - - - -
```

SPSS Output 16.8

```
Note: For saturated models .500 has been added to all observed cells.
 This value may be changed by using the CRITERIA = DELTA subcommand.

 Estimates for Parameters.

 ANIMAL*TRAINING*DANCE

 Parameter       Coeff.       Std. Err.  Z-Value Lower   95 CI Upper    95 CI
     1        .3600938577      .08335      4.32010         .19672        .52347

 ANIMAL*TRAINING

 Parameter       Coeff.       Std. Err.  Z-Value Lower   95 CI Upper    95 CI
     1       -.4020174410      .08335     -4.82306        -.56539       -.23865

 ANIMAL*DANCE

 Parameter       Coeff.       Std. Err.  Z-Value Lower   95 CI Upper    95 CI
     1       -.1970307093      .08335     -2.36381        -.36040       -.03366

 TRAINING*DANCE

 Parameter       Coeff.       Std. Err.  Z-Value Lower   95 CI Upper    95 CI
     1        .1042911061      .08335      1.25120        -.05908        .26766

 ANIMAL

 Parameter       Coeff.       Std. Err.  Z-Value Lower   95 CI Upper    95 CI
     1        .4036938932      .08335      4.84318         .24032        .56707

 TRAINING

 Parameter       Coeff.       Std. Err.  Z-Value Lower   95 CI Upper    95 CI
     1       -.3281973781      .08335     -3.93743        -.49157       -.16483

 DANCE

 Parameter       Coeff.       Std. Err.  Z-Value Lower   95 CI Upper    95 CI
     1        .2319101606      .08335      2.78226         .06854        .39528
- - - - - - - - - - - - - - - - - - - - - - - - - - - - - - - - - - - - - - - -
```

SPSS Output 16.9

provides individual estimates for each effect) but it does so using a z-score rather than a chi-square test. This can be useful because we get confidence intervals, and also because the value of z gives us a useful comparison between effects (if you ignore the plus or minus sign, the bigger the z, the more significant the effect is). So, if you look at the z-values you can see that the main effect of animal is the most important effect in the model ($z = 4.84$) followed by the animal × training interaction ($z = -4.82$) and then the animal × training × dance interaction ($z = 4.32$) and so on. However, it's worth reiterating that in this case we don't need to concern ourselves with anything other than the three-way interaction.

The final bit of output (SPSS Output 16.10) deals with the backward elimination. SPSS will begin with the highest-order effect (in this case the animal × training × dance interaction), it removes it from the model, sees what effect this has, and if it doesn't have a significant effect then it moves on to the next highest effects (in this case the two-way interactions). However, we've already seen that removing the three-way interaction will have a significant effect and this is confirmed at this stage by the table labelled *If Deleted Simple Effect is*, which confirms that removing the three-way interaction has a significant effect on the model. Therefore, the analysis stops here: the three-way interaction is not removed and SPSS evaluates this final model. The bit of the output labelled *The best model has generating class* tells us that the model that best fits the data includes the three-way interaction, and because of the hierarchical nature of the analysis this is the only effect that we need to interpret. Finally SPSS evaluates this final model with the likelihood ratio statistic and we're looking for a non-significant test statistic which indicates that the expected values generated by the model are not significantly different from the observed data (put another way, the model is a good fit of the data). In this case the result is very non-significant, indicating that the model is a good fit of the data.[4]

The next step is to try to interpret this interaction. The first useful thing we can do is to plot the frequencies across all of the different categories. You should plot the frequencies in terms of the percentage of the total (these values can be found in the crosstabulation table in SPSS Output 16.5 in the rows labelled *% of Total*). The resulting graph is shown in Figure 16.7 and this shows what we already know about cats: they will dance (or do anything else for that matter) when there is food involved but if you train them with affection they're not interested. Dogs on the other hand will dance when there's affection involved (actually more dogs danced than didn't dance, regardless of the type of reward, but the effect is more pronounced when affection was the training method). In fact, both animals show similar responses to food training; it's just that

4 The fact that the analysis has stopped here is unhelpful because I can't show you how it would proceed in the event of a non-significant three-way interaction. However, it does keep things simple and if you're interested in exploring loglinear analysis further, the tasks at the end of the chapter and the answers on the CD-ROM do show you what happens when the highest-order interaction is not significant.

```
* * * * * * * * H I E R A R C H I C A L   L O G   L I N E A R * * * * * * * *

Backward Elimination (p = .050) for DESIGN 1 with generating class

 ANIMAL*TRAINING*DANCE

 Likelihood ratio chi square = .00000                   DF = 0 P = -INF
 - - - - - - - - - - - - - - - - - - - - - - - - - - - - - - - - - - - - - - - -
If Deleted Simple Effect is          DF  L.R. Chisq Change  Prob   Iter
 ANIMAL*TRAINING*DANCE                1        20.305       .0000    4

Step 1

 The best model has generating class
    ANIMAL*TRAINING*DANCE
 Likelihood ratio chi square = .00000 DF = 0 P = -INF
- - - - - - - - - - - - - - - - - - - - - - - - - - - - - - - - - - - - - - - -
The final model has generating class

  ANIMAL*TRAINING*DANCE

The Iterative Proportional Fit algorithm converged at iteration 0.
The maximum difference between observed and fitted marginal totals is .000
and the convergence criterion is .250
- - - - - - - - - - - - - - - - - - - - - - - - - - - - - - - - - - - - - - - -
 Observed, Expected Frequencies and Residuals.

      Factor   Code    OBS count  EXP count  Residual  Std Resid

ANIMAL     Cat
 TRAINING     Food as
  DANCE          Yes       28.0       28.0       .00       .00
  DANCE          No        10.0       10.0       .00       .00
 TRAINING     Affectio
  DANCE          Yes       48.0       48.0       .00       .00
  DANCE          No       114.0      114.0       .00       .00

ANIMAL     Dog
 TRAINING     Food as
  DANCE          Yes       20.0       20.0       .00       .00
  DANCE          No        14.0       14.0       .00       .00
 TRAINING     Affectio
  DANCE          Yes       29.0       29.0       .00       .00
  DANCE          No         7.0        7.0       .00       .00
- - - - - - - - - - - - - - - - - - - - - - - - - - - - - - - - - - - - - - - -
 Goodness-of-fit test statistics

Likelihood ratio chi square =     .00000     DF = 0 P = -INF
        Pearson chi square =     .00000     DF = 0 P = -INF
- - - - - - - - - - - - - - - - - - - - - - - - - - - - - - - - - - - - - - - -
```

SPSS Output 16.10

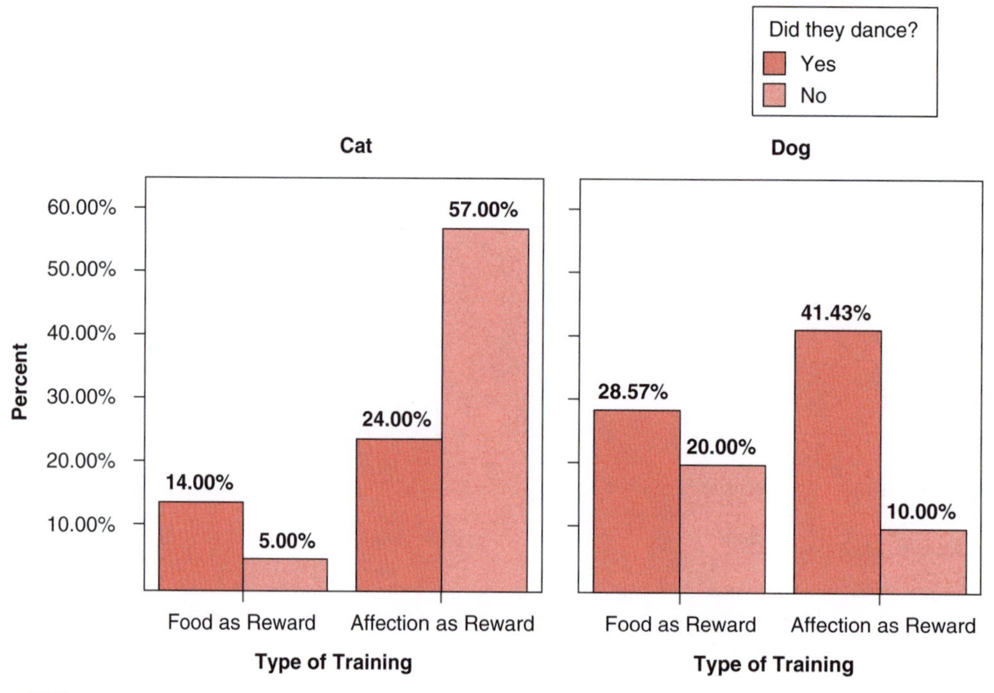

Figure 16.7

cats won't do anything for affection. So cats are sensible creatures who only do stupid stuff when there's something in it for them (i.e. food), whereas dogs are just plain stupid!

16.9. FOLLOWING UP LOGLINEAR ANALYSIS ②

An alternative way to interpret a three-way interaction is to conduct chi-square analysis at different levels of one of your variables. For example, to interpret our animal × training × dance interaction, we could perform a chi-square test on training and dance but do this separately for dogs and cats (in fact the analysis for cats will be the same as the example we used for chi-square). You can then compare the results in the different animals. Try doing this using the split file command (see section 3.4). The results and interpretation for cats are in SPSS Output 16.2 and for dogs the output is shown in SPSS Output 16.11. For dogs there is still a significant relationship between the types of training and whether they danced but it is less strong (the chi-square is 3.93 compared to 25.2 in the cats).[5] This reflects the fact that dogs are more likely to dance if given affection than if given food: the opposite of cats!

5 The chi-square statistic depends on the sample size, so really you need to calculate effect sizes and compare them to make this kind of statement (unless you had equal numbers of dogs and cats!).

Chi-Square Tests[c]

	Value	df	Asymp. Sig. (2-sided)	Exact Sig. (2-sided)	Exact Sig. (1-sided)
Pearson Chi-Square	3.932[b]	1	.047		
Continuity Correction[a]	2.966	1	.085		
Likelihood Ratio	3.984	1	.046		
Fisher's Exact Test				.068	.042
Linear-by-Linear Association	3.876	1	.049		
N of Valid Cases	70				

a Computed only for a 2 × 2 table

b 0 cells (.0%) have expected count less than 5. The minimum expected count is 10.20

c Animal = Dog

SPSS Output 16.11

16.10. EFFECT SIZES IN LOGLINEAR ANALYSIS ②

As with Pearson's chi-square, one of the most elegant ways to report your effects is in terms of odds ratios. Odds ratios are easiest to understand for 2 × 2 contingency tables and so if you have significant higher-order interactions, or your variables have more than two categories, it is worth trying to break these effects down into logical 2 × 2 tables and calculating odds ratios that reflect the nature of the interaction. So, for example, in this example we could calculate odds ratios for dogs and cats separately. We have the odds ratios for cats already (section 16.4.5), and for dogs we would get:

$$
\begin{aligned}
\text{odds}_{\text{dancing after food}} &= \frac{\text{number that had food and danced}}{\text{number that had food but didn't dance}} \\
&= \frac{20}{14} \\
&= 1.43 \\
\text{odds}_{\text{dancing after affection}} &= \frac{\text{number that had affection and danced}}{\text{number that had affection but didn't dance}} \\
&= \frac{29}{7} \\
&= 4.14 \\
\text{odds ratio} &= \frac{\text{odds}_{\text{dancing after food}}}{\text{odds}_{\text{dancing after affection}}} \\
&= \frac{1.43}{4.14} \\
&= .35
\end{aligned}
$$

This tells us that if a dog was trained with food it was .35 times more likely to dance (i.e. it was less likely to dance). Another way to say this is that it was 1/.35 = 2.90 times *less* likely to dance. Compare this to cats that were 6.65 times more likely to dance. As you can see, comparing the odds ratios for dogs and cats is an extremely elegant way to present the three-way interaction term in the model.

16.16.11. Reporting the results of loglinear analysis ②

When reporting loglinear analysis you need to report the likelihood ratio statistic for the final model, usually denoted just by χ^2. For any terms that are significant you should report the chi-square change, or you could consider reporting the *z*-score for the effect and its associated confidence interval. If you break down any higher-order interactions in subsequent analyses then obviously you need to report the relevant chi-square statistics (and odds ratios). For this example we could report:

✓ The three-way loglinear analysis produced a final model that retained all effects. The likelihood ratio of this model was $\chi^2(0) = 0$, $p = 1$. This indicated that the highest-order interaction (the animal × training × dance interaction) was significant, $\chi^2(1) = 20.31$, $p < .001$. To break down this effect, separate chi-square tests on the training and dance variables were performed separately for dogs and cats. For cats, there was a significant association between the type of training and whether or not cats would dance, $\chi^2(1) = 25.36$, $p < .001$; this was true in dogs also, $\chi^2(1) = 3.93$, $p < .05$. Odds ratios indicated that cats were 6.65 times more likely to dance if trained with food than if trained with affection, but dogs were only .35 times more likely to dance (or 2.90 times less likely to dance). Therefore, the analysis seems to reveal a fundamental difference between dogs and cats: cats are more likely to dance for food rather than affection, whereas dogs are more likely to dance for affection than food.

Cramming Samantha's Tips

- If you want to test the relationship between more than two categorical variables you can do this with *loglinear analysis*.
- Loglinear analysis is hierarchical: the initial model contains all main effects and interactions. Starting with the highest-order interaction, terms are removed to see whether their removal significantly affects the fit of the model. If it does then this term is not removed and all lower-order effects are ignored.
- Look at the table labelled *The best model has generating class* to see which effects have been retained in the final model. Then look at the table labelled *If Deleted Simple Effect is* to see the individual significance of the retained effects (look at the column labelled *Prob*: values less than .05 indicate significance).
- Look at the *Goodness-of-fit test statistics* for the final model: if this model is a good fit of the data then this statistic should be non-significant (*Prob* should be bigger than .05).
- Look at the crosstabulation table to interpret any significant effects (% of total for cells is the best thing to look at).

16.12. WHAT HAVE WE DISCOVERED ABOUT STATISTICS? ①

When I wrote the first edition of this book I had always intended to do a chapter on loglinear analysis, but by the time I got to that chapter I had already written 300 pages more than I was contracted to do, and had put so much effort into the rest of it that, well, the thought of that extra chapter was giving me thoughts of large cliffs and jumping. This time around, I wanted to make sure that at the very least I did a loglinear chapter. However, when I came to it, I'd already written 200 pages more than I was supposed to for this new edition, and with deadlines fading into the distance, history was repeating itself. It won't surprise you to know then that I'm really happy to have written the damn thing! This chapter has taken a very brief look at analysing categorical data. What I've tried to do is to show you how really we approach categorical data in much the same way as any other kind of data: we fit a model, we calculate the deviation between our model and the observed data, and we use that to evaluate the model we've fitted. I've also tried to show that the model we fit is the same one that we've come across throughout this book: it's a linear model (regression). When we have only two variables we can use Pearson's chi-square test or the likelihood ratio test to look at whether those two variables are associated. In more complex situations, we simply extend these models into something known as a loglinear model. This is a bit like an ANOVA for categorical data: for every variable we have, we get a main effect but we also get interactions between variables. Loglinear analysis simply evaluates all of these effects hierarchically to tell us which ones best predict our outcome.

16.13. KEY TERMS THAT WE'VE DISCOVERED

- Categorical variable
- Chi-square test
- Contingency table
- Cramer's *V*
- Goodman and Kruskal's λ
- Loglinear analysis
- Odds ratio
- Phi
- Saturated model
- Yates's continuity correction

16.14. SMART ALEX'S TASKS

- Certain editors at Sage Publications like to think they're a bit of a whiz at football (soccer if you prefer). To see whether they are better than Sussex lecturers and postgraduates we invited various employees of Sage to join in our football matches (oh, sorry, I mean we invited them down for important meetings about books). Every player was only allowed to play in one match. Over many matches, we counted the number of players that scored goals. The data are in the file **SageEditorsCan'tPlayFootball.sav**; do a chi-square test to see whether more publishers or academics scored goals. We predict that Sussex people will score more than Sage people.③

- I was interested in whether horoscopes are just a figment of people's minds. Therefore, I got 2201 people, made a note of their star sign (this variable, obviously, has 12 categories: Capricorn, Aquarius, Pisces, Aries, Taurus, Gemini, Cancer, Leo, Virgo, Libra, Scorpio and Sagittarius) and whether they believed in horoscopes (this variable has two categories: believer or unbeliever). I then sent them a horoscope in the post of what would happen over the next month: everybody, regardless of their star sign, received the same horoscope, which read: 'August is an exciting month for you. You will make friends with a tramp in the first week of the month and cook him a cheese omelette. Curiosity is your greatest virtue, and in the second week, you'll discover knowledge of a subject that you previously thought was boring, statistics perhaps. You might purchase a book around this time that guides you towards this knowledge. Your new wisdom leads to a change in career around the third week, when you ditch your current job and become an accountant. By the final week you find yourself free from the constraints of having friends, your boy/girlfriend has left you for a Russian ballet dancer with a glass eye, and you now spend your weekends doing loglinear analysis by hand with a pigeon called Hephzibah for company.' At the end of August I interviewed all of these people depending on how closely their lives matched the fictitious horoscope and I classified the horoscope as having come true, or not. The data are in the file **Horoscope.sav**. Conduct a loglinear analysis to see whether there is a relationship between the person's star sign, whether they believe in horoscopes, and whether the horoscope came true.③

The answers can be found on the CD-ROM in **Answers(Chapter16).pdf**.

16.15. FURTHER READING

Hutcheson, G. & Sofroniou, N. (1999). *The multivariate social scientist*. London: Sage.

Tabachnick, B. G. & Fidell, L. S. (2001). *Using multivariate statistics* (4th edition). Boston: Allyn & Bacon. Chapter 7 is a fantastic account of loglinear analysis.

CHAPTER 17

EPILOGUE

Here's some questions that the writer sent

Can an observer be a participant?

Have I seen too much?

Does it count if it doesn't touch?

If the view is all I can ascertain,

Pure understanding is out of range.

(Fugazi, 2001)

So, you've reached the end of the book, eh? If you've read the book from cover to cover then you've invested a large portion of your life (possibly even your whole life unless you're a very fast reader) in this book and I'm grateful. You might have learnt something about statistics (and if not, I certainly learnt a lot through writing it), but more important we've had our fair share of sex, drugs and rock'n'roll (in the examples, obviously).

One of the big problems with being passionate about teaching is that you're constantly noticing all of the ways in which you're a terrible teacher. The day I got my copy of the first edition of this book I was very excited: I opened it up to admire my handiwork and the first thing I spotted was a mistake! Admittedly it was a small one, but the rot set in and I've been finding annoying imperfections ever since. So, when the publishers decided they wanted a new version I thought 'great', now I can do it properly and add in all those things that people always ask me about. Obviously I went into obsessive mode (am rarely out of it, in fact) and 700 pages later I'm hoping I never have to write another word about statistics ever again! So, if you could just cut out the following form and send it to Michael Carmichael, c/o Sage Publications, I'd be really grateful.

Dear Mr Carmichael

I am just writing to say that I have bought Andy Field's book *Discovering Statistics Using SPSS* and I have found it to be a work of near perfection. Should you ever get the urge to publish a third edition, then I must insist you not change a single word, simply slap on a new cover and pay someone other than Andy to update some of the SPSS windows. Changing any word, even a small one, would be a heinous crime and would jeopardize all future sales of the book. Also, it's not fair that nice man Andy works so hard just so that you and your editor chums can jolly around London in your limousines having champagne breakfasts and snorting cocaine; you should give him some more money and pay for a very long holiday, which he surely deserves after all of his toil.

Yours sincerely

[*insert your name here*]
Valued Sage Customer

GLOSSARY

This glossary contains brief definitions of all key terms in this book. Words in italics refer to items that are defined elsewhere in the glossary.

–2LL: the *log-likelihood* multiplied by minus 2. This version of the likelihood is used in logistic regression.

α-level: the probability of making a *Type I error* (usually this value is .05).

A Life: what you don't have when writing statistics textbooks.

Adjusted mean: in the context of *analysis of covariance* this is the value of the group mean adjusted for the effect of the *covariate*.

Adjusted predicted value: a measure of the influence of a particular case of data. It is the predicted value of a case from a model estimated without that case included in the data. The value is calculated by re-estimating the model without the case in question, then using this new model to predict the value of the excluded case. If a case does not exert a large influence over the model then its predicted value should be similar regardless of whether the model was estimated including or excluding that case. The difference between the predicted value of a case from the model when that case was included and the predicted value from the model when it was excluded is the *DFFit*.

Adjusted R^2: a measure of the loss of predictive power or *shrinkage* in regression. The adjusted R^2 tells us how much variance in the outcome would be accounted for if the model had been derived from the population from which the sample was taken.

Alpha factoring: a method of *factor analysis*.

Analysis of covariance: a statistical procedure that uses the *F*-ratio to test the overall fit of a linear model controlling for the effect that one or more *covariates* have on the *outcome variable*. In experimental research this linear model tends to be defined in terms of group means and the resulting ANOVA is therefore an overall test of whether group means differ after the variance in the *outcome variable* explained by any *covariates* has been removed.

Analysis of variance: a statistical procedure that uses the F-ratio to test the overall fit of a linear model. In experimental research this linear model tends to be defined in terms of group means and the resulting ANOVA is therefore an overall test of whether group means differ.

ANCOVA: acronym for *analysis of covariance*.

Anderson–Rubin method: a way of calculating *factor scores* which produces scores that are uncorrelated and *standardized* with a mean of 0 and a standard deviation of 1.

ANOVA: acronym for *analysis of variance*.

Autocorrelation: when the *residuals* of two observations in a regression model are correlated.

b_i: unstandardized regression coefficient. Indicates the strength of relationship between a given predictor, i, and an outcome in the units of measurement of the predictor. It is the change in the outcome associated with a unit change in the predictor.

β_i: standardized regression coefficient. Indicates the strength of relationship between a given predictor, i, and an outcome in a *standardized* form. It is the change in the outcome (in standard deviations) associated with a one standard deviation change in the predictor.

β-level: the probability of making a *Type II error* (Cohen, 1992, suggests a maximum value of .2).

Bartlett's test of sphericity: unsurprisingly, this is a test of the assumption of *sphericity*. The test examines whether a *variance–covariance matrix* is proportional to an *identity matrix*. Therefore, it effectively tests whether the diagonal elements of the *variance–covariance matrix* are equal (i.e. group variances are the same), and that the off-diagonal elements are approximately zero (i.e. the *dependent variables* are not *correlated*). Jeremy Miles, who does a lot of multivariate stuff, claims he's never ever seen a matrix that reached non-significance using this test and, come to think of it, I've never seen one either (although I do less multivariate stuff) so you've got to wonder about it's practical utility.

Beer–goggles effect: the phenomenon that people of the opposite gender (or the same depending on your sexual orientation) appear much more attractive after a few alcoholic drinks.

Between-group design: another name for *independent design*.

Between-subject design: another name for *independent design*.

Bimodal: a description of a distribution of observations that has two *modes*.

Biserial correlation: a standardized measure of the strength of relationship between two variables when one of the two variables is *dichotomous*. The biserial correlation coefficient is used when one variable is a continuous dichotomy (e.g. has an underlying continuum between the categories).

Bivariate correlation: a correlation between two variables.

Blockwise regression: another name for *hierarchical regression*.

Bonferroni correction: a correction applied to the *α-level* to control the overall *Type I error rate* when multiple significance tests are carried out. Each test conducted should use a criterion of significance of the α-level (normally .05) divided by the number of tests conducted. This is a simple but effective correction, but tends to be too strict when lots of tests are performed.

Boredom effect: refers to the possibility that performance in tasks may be influenced (the assumption is a negative influence) by boredom/lack of concentration if there are many tasks, or the task goes on for a long period of time.

Box test: a test of the assumption of *homogeneity of covariance matrices*. This test should be non-significant if the matrices are the same. Box's test is very susceptible to deviations from *multivariate normality* and so can be non-significant not because the *variance–covariance matrices* are similar across groups, but because the assumption of multivariate normality is not tenable. Hence, it is vital to have some idea of whether the data meet the multivariate normality assumption (which is extremely difficult) before interpreting the result of Box's test.

Boxplot (a.k.a. box–whisker diagram): a graphical representation of some important characteristics of a set of observations. At the centre of the plot is the *median*, which is surrounded by a box, the top and bottom of which are the limits within which the middle 50% of observations fall (the *interquartile range*). Sticking out of the top and bottom of the box are two whiskers which extend to the most and least extreme scores respectively.

Box–whisker plot: see *boxplot*.

Brown–Forsythe *F*: a version of the *F*-ratio designed to be accurate when the assumption of *homogeneity of variance* has been violated.

Categorical variable: any variable made up of categories of objects/entities. The UK degree classifications are a good example because a degree is classified as a 1, 2:1, 2:2, 3, pass or fail. Therefore, graduates form a categorical variable because they will fall into only one of these categories (hopefully the category of students receiving a 1st!).

Chi-square distribution: a *probability distribution* of the sum of squares of several normally distributed variables. It tends to be used to test hypotheses about categorical data.

Chi-square test: although this term can apply to any *test statistic* having a *chi-square distribution*, it generally refers to Pearson's chi-square test of the independence of two categorical variables. Essentially it tests whether two categorical variables forming a *contingency table* are associated.

Cochran's *Q*: this test is an extension of *McNemar's test* and is basically a *Friedman's ANOVA* for *dichotomous* data. So imagine you asked 10 people whether they'd like to shoot Justin Timberlake, David Beckham and Tony Blair and they could answer only yes or no. If we coded responses as 0 (no) and 1 (yes) we could do the Cochran test on these data.

Coefficient of determination: the proportion of variance in one variable explained by a second variable. It is the *Pearson correlation coefficient* squared.

Common variance: variance shared by two or more variables.

Communality: the proportion of a variable's variance that is *common variance*. This term is used primarily in *factor analysis*. A variable that has no *unique variance* (or *random variance*) would have a communality of 1 whereas a variable that shares none of its variance with any other variable would have a communality of 0.

Complete separation: a situation in *logistic regression* when the outcome variable can be perfectly predicted by one predictor or a combination of predictors! Suffice it to say this situation makes your computer have the equivalent of a nervous breakdown: it'll start gibbering, weeping, and saying it doesn't know what to do.

Component matrix: general term for the *structure matrix* in SPSS *principal component analysis*.

Compound symmetry: a condition that holds true when both the variances across conditions are equal (this is the same as the *homogeneity of variance* assumption) and the *covariances* between pairs of conditions are also equal.

Confidence interval: for a given statistic calculated for a sample of observations (e.g. the mean), the confidence interval is a range of values around that statistic that are believed to contain, with a certain probability (e.g. 95%), the true value of that statistic (i.e. the population value).

Confirmatory factor analysis (CFA): a version of *factor analysis* in which specific hypotheses about structure and relations between the *latent variables* that underlie the data are tested.

Confounding variable: a variable (that we may or may not have measured) other than the *predictor variables* in which we're interested that potentially affects an *outcome variable*.

Contingency table: a table representing the cross-classification of two or more *categorical variables*. The levels of each variable are arranged in a grid, and the number of observations falling into each category is noted in the cells of the table. For example, if we took the categorical variables of **glossary** (with two categories: whether an author was made to write a glossary or not) and **mental state** (with three categories: normal, sobbing uncontrollably and utterly psychotic), we could construct a table as below. This tells us that 127 authors who were made to write a glossary ended up as utterly psychotic, compared to only 2 who did not write a glossary.

		Glossary		
		Author made to write glossary	No glossary	Total
Mental State	Normal	5	423	428
	Sobbing uncontrollably	23	46	69
	Utterly psychotic	127	2	129
	Total	155	471	626

Cook's distance: a measure of the overall influence of a case on the model. Cook & Weisberg (1982) have suggested that values greater than 1 may be cause for concern.

Counterbalancing: a process of systematically varying the order in which experimental conditions are conducted. In the simplest case of there being two conditions (A and B), counterbalancing simply implies that half of the participants complete condition A followed by condition B, whereas the remainder do condition B followed by condition A. The aim is to remove systematic bias caused by *practice effects* or *boredom effects*.

Covariance: a measure of the 'average' relationship between two variables. It is average *cross-product deviation* (i.e. the cross-product divided by one less than the number of observations).

Covariance ratio (CVR): a measure of whether a case influences the variance of the parameters in a *regression model*. When this ratio is close to 1 the case has very little influence on the variances of the model parameters. Belsey et al. (1980) recommend the following: if the CVR of a case is greater than $1 + [3(k + 1)/n]$ then deleting that case will damage the precision of some of the model's parameters, but if it is less than $1 - [3(k + 1)/n]$ then deleting the case will improve the precision of some of the model's parameters (k is the number of predictors and n is the sample size).

Covariate: a variable that has a relationship with (in terms of *covariance*), or has the potential to be related to, the *outcome variable* we've measured.

Cox and Snell's R^2_{CS}: a version of the *coefficient of determination* for logistic regression. It is based on the log-likelihood of a model (*LL(New)*) and the log-likelihood of the original model (*LL(baseline)*), and the sample size, n. However, it is notorious for not reaching its maximum value of 1 (see *Nagelkerke's* R^2_N).

Cramer's V: a measure of the strength of association between two *categorical variables* used when one of these variables has more than two categories. It is a variant of *phi* used because when one or both of the categorical variables contain more than two categories phi fails to reach its minimum value of 0 (indicating no association).

Cronbach's α: a measure of the reliability of a scale defined by:

$$\alpha = \frac{N^2\,\overline{\text{Cov}}}{\sum s^2_{\text{item}} + \sum \text{Cov}_{\text{item}}}$$

in which the top half of the equation is simply the number of items (*N*) squared multiplied by the average covariance between items (the average of the off-diagonal elements in the *variance–covariance matrix*). The bottom half is the sum of all the elements in the *variance–covariance matrix*.

Cross-product deviations: a measure of the 'total' relationship between two variables. It is the deviation of one variable from its mean multiplied by the other variable's deviation from its mean.

Cross-validation: Assessing the accuracy of a model across different samples. This is an important step in *generalization*. In a *regression model* there are two main methods of cross-validation: *adjusted* R^2 or data splitting, in which the data are split randomly into two halves, and a regression model is estimated for each half and then compared.

Crying: what you feel like doing after writing statistics textbooks.

Cubic trend: if you connected the means in ordered conditions with a line then a cubic trend is shown by two changes in the direction of this line. You must have at least four ordered conditions.

Data editor: the main window in SPSS in which you enter data and carry out statistical functions.

Data view: there are two ways to view the contents of the *data editor* window. The data view shows you a spreadsheet and can be used for entering raw data (see also *variable view*).

Degrees of freedom: an impossible thing to define in a few pages let alone a few lines. Essentially the number of 'entities' that are free to vary when estimating some kind of statistical parameter. In a more practical sense, it has a bearing on significance tests for many commonly used *test statistics* (such as the *F-ratio*, *t-test*, *chi-square statistic*) and determines the exact form of the *probability distribution* for these *test statistics*. The explanation involving rugby players in Chapter 8 is far more interesting…

Deleted residual: a measure of the influence of a particular case of data. It is the difference between the *adjusted predicted value* for a case and the original observed value for that case.

Dependent *t*-test: a test using the *t-statistic* that establishes whether two means collected from the same sample (or related observations) differ significantly.

Dependent variable: another name for *outcome variable*. This name is usually associated with experimental methodology (which is the only time it really makes sense) and is so called because it is the variable that is not manipulated by the experimenter and so its value depends on the variables that have been manipulated. To be honest I just use outcome variable all the time—it makes more sense (to me) and is less confusing.

Deviance: the difference between the observed value of a variable and the value of that variable predicted by a statistical model.

Deviation contrast: a non-orthogonal *planned contrast* that compares the mean of each group (except first or last depending on how the contrast is specified) to the overall mean.

DFA: acronym for discriminant function analysis (see *discriminant analysis*).

DFBeta: a measure of the influence of a case on the values of b_i in a *regression model*. If we estimated a regression parameter b_i and then deleted a particular case and re-estimated the same regression parameter b_i then the difference between these two estimates would be the DFBeta for the case that was deleted. By looking at the values of the DFBetas, it is possible to identify cases that have a large influence on the parameters of the regression model; however, the size of DFBeta will depend on the units of measurement of the regression parameter.

DFFit: a measure of the influence of a case. It is the difference between the *adjusted predicted value* and the original predicted value of a particular case. If a case is not influential then its DFFit should be zero—hence, we expect non-influential cases to have small DFFit values. However, we have the problem that this statistic depends on the units of measurement of the outcome and so a DFFit of 0.5 will be very small if the outcome ranges from 1 to 100, but very large if the outcome varies from 0 to 1.

Dichotomous: description of a variable that consists of only two categories (e.g. the variable gender is dichotomous because it consists of only two categories: male and female).

Difference contrast: a non-orthogonal *planned contrast* that compares the mean of each condition (except the first) to the overall mean of all previous conditions combined.

Direct oblimin: a method of *oblique rotation*.

Discriminant analysis: also known as discriminant function analysis. This analysis identifies and describes the *discriminant function variates* of a set of variables and is useful as a follow-up test to *MANOVA* as a means of seeing how these variates allow groups of cases to be discriminated.

Discriminant function variate: a linear combination of variables created such that the differences between group means on the transformed variable is maximized. It takes the general form $Variate_1 = b_1X_1 + b_2X_2 + \cdots + b_nX_n$.

Discriminant score: a score for an individual case on a particular *discriminant function variate* obtained by replacing that case's scores on the measured variables in the equation that defines the variate in question.

Dummy variables: a way of recoding a categorical variable with more than two categories into a series of variables all of which are *dichotomous* and can take on values of only 0 or 1. There are seven basic steps to create such variables: (1) count the number of groups you want to recode and subtract 1; (2) create as many new variables as the value you calculated in step 1 (these are your dummy variables); (3) choose one of your groups as a baseline (i.e. a group against which all other groups should be compared, such as a control group); (4) assign that baseline group values of 0 for all of your dummy variables; (5) for your first dummy variable, assign the value 1 to the first group that you want to compare against the baseline group (assign all other groups 0 for this variable); (6) for the second dummy variable assign the value 1 to the second group that you want to compare against the baseline group (assign all other groups 0 for this variable); (7) repeat this process until you run out of dummy variables.

Durbin–Watson test: tests for serial correlations between errors in *regression models*. Specifically, it tests whether adjacent residuals are correlated, which is useful in assessing the assumption of *independent errors*. The test statistic can vary between 0 and 4 with a value of 2 meaning that the residuals are uncorrelated. A value greater than 2 indicates a negative correlation between adjacent residuals, whereas a value below 2 indicates a positive correlation. The size of the Durbin–Watson statistic depends upon the number of predictors in the model and the number of observations. For accuracy, you should look up the exact acceptable values in Durbin & Watson's (1951) original paper. As a very conservative rule of thumb, values less than 1 or greater than 3 are definitely cause for concern; however, values closer to 2 may still be problematic depending on your sample and model.

Effect size: an objective and standardized measure of the magnitude of an observed effect. Measures include Cohen's *d*, Glass' *g* and Pearson's correlations coefficient *r*.

Equamax: a method of *orthogonal rotation* that is a hybrid of *quartimax* and *varimax*. It is reported to behave fairly erratically (see Tabachnick & Fidell, 2001) and so is probably best avoided.

Error bar chart: a graphical representation of the mean of a set of observations that includes the 95% confidence interval of the mean. The mean is usually represented as a dot, square or rectangle extending to the value of the mean. The confidence interval is represented by a line protruding from the mean (upwards, downwards or both) to a short horizontal line representing the limits of the confidence interval. Error bars can be drawn using the standard error or standard deviation instead of the 95% confidence interval.

Error SSCP (*E*): the error sum of squares and cross-product matrix. This is a *sum of squares and cross-product matrix* for the error in a predictive *linear model* fitted to *multivariate* data. It represents the *unsystematic variance* and is the *multivariate* equivalent of the *residual sum of squares*.

Eta squared: an *effect size* measure that is the ratio of the *model sum of squares* to the *total sum of squares*—so, in essence, *the coefficient of determination* by another name. Frankly it's a bit of a waste of time: not only is it biased, but it also measures the overall effect of the ANOVA and so can't be interpreted in a meaningful way.

Exp(B): an indicator of the change in *odds* resulting from a unit change in the predictor in *logistic regression*. If the value is greater than 1 then it indicates that as the predictor increases, the odds of the outcome occurring increase. Conversely, a value less than 1 indicates that as the predictor increases, the odds of the outcome occurring decrease.

Experimental hypothesis: the prediction that your experimental manipulation will have some effect or that certain variables will relate to each other.

Experimentwise error rate: The probability of making a *Type I error* in an experiment involving one or more statistical comparisons when the null hypothesis is true in each case.

Extraction: a term used for the process of deciding whether a *factor* in *factor analysis* is statistically important enough to 'extract' from the data and interpret. The decision is based on the magnitude of the eigenvalue associated with the factor. See *scree plot*, *Kaiser's criterion*.

***F*-ratio**: a test statistic with a known *probability distribution* (the *F*-distribution). It is the ratio of the average variability in the data that a given model can explain to the average variability unexplained by that same model. It is used to test the overall fit of the model in *simple regression* and *multiple regression*, and to test for overall differences between group means in experiments.

Factor: another name for an *independent variable* or *predictor* that's typically used when describing experimental designs. However, to add to the confusion, it is also used synonymously with *latent variable* in factor analysis.

Factor analysis: a *multivariate* technique for identifying whether the correlations between a set of observed variables stem from their relationship to one or more *latent variables* in the data, each of which takes the form of a *linear model*.

Factor loading: the *regression coefficient* of a variable for the *linear model* that describes a *latent variable* or *factor* in *factor analysis*.

Factor matrix: general term for the *structure matrix* in SPSS *factor analysis*.

Factor scores: a single score from an individual entity representing its performance on some *latent variable*. The score can be crudely conceptualized as follows: take an entity's score on each of the variables that make up the factor and multiply it by the corresponding *factor loading* for the variable, then add these values up (or average them).

Factor transformation matrix, Λ: a matrix used in *factor analysis*. It can be thought of as containing the angles through which factors are rotated in factor *rotation*.

Factorial ANOVA: an analysis of variance involving two or more *independent variables* or *predictors*.

Familywise error rate: The probability of making a *Type I error* in any family of tests when the null hypothesis is true in each case. The 'family of tests' can be loosely defined as a set of tests conducted on the same data set and addressing the same empirical question.

Frequency distribution: a graph plotting values of observations on the horizontal axis, and the frequency with which each value occurs in the data set on the vertical axis (a.k.a. *histogram*).

Friedman's ANOVA: a non-parametric test of whether more than two related groups differ. It is the non-parametric version of one-way *repeated-measures ANOVA*.

Generalization: the ability of a statistical model to say something beyond the set of observations that spawned it. If a model generalizes, it is assumed that predictions from that model can be applied not just to the sample on which it is based, but to a wider population from which the sample came.

Glossary: a collection of grossly inaccurate definitions (written late at night when you really ought to be asleep) of things that you thought you understood until some evil book publisher forced you to try to define them.

Goodman and Kruskal's λ: measures the proportional reduction in error that is achieved when membership of a category of one variable is used to predict category membership of the other variable. A value of 1 means that one variable perfectly predicts the other, whereas a value of 0 indicates that one variable in no way predicts the other.

Goodness-of-fit: an index of how well a model fits the data from which it was generated. It's usually based on how well the data predicted by the model correspond with the data that were actually collected.

Grand mean: the *mean* of an entire set of observations.

Grand variance: the *variance* within an entire set of observations.

Greenhouse–Geisser correction: an estimate of the departure from *sphericity*. The maximum value is 1 (the data completely meet the assumption of *sphericity*) and the minimum is the *lower bound*. Values below 1 indicate departures from sphericity and are used to correct the *degrees of freedom* associated with the corresponding *F-ratios* by multiplying them by the value of the estimate. Some say the Greenhouse–Geisser correction is too conservative (strict) and recommend the *Huynh–Feldt correction* instead.

Harmonic mean: a weighted version of the *mean* that takes account of the relationship between variance and sample size. It is calculated by summing the reciprocal of all observations, then dividing by the number of observations. The reciprocal of the end product is the harmonic mean:

$$H = \frac{1}{\frac{1}{n}\sum_{i=1}^{n}\frac{1}{x_i}}$$

Hat values: another name for *leverage*.

***HE*$^{-1}$**: a matrix that is functionally equivalent to the *hypothesis SSCP* divided by the *error SSCP* in *MANOVA*. Conceptually it represents the ratio of *systematic* to *unsystematic variance*, so is a *multivariate* analogue of the *F-ratio*.

Helmert contrast: a non-orthogonal *planned contrast* that compares the mean of each condition (except the last) to the overall mean of all subsequent conditions combined.

Heteroscedasticity: the opposite of *homoscedasticity*. This occurs when the residuals at each level of the predictor variable(s) have unequal variances. Put another way, at each point along any predictor variable, the spread of residuals is different.

Hierarchical regression: a method of *multiple regression* in which the order in which predictors are entered into the regression model is determined by the researcher based on previous research: variables already known to be predictors are entered first, new variables are entered subsequently.

Histogram: see *frequency distribution*.

Homogeneity of covariance matrices: an assumption of some *multivariate* tests such as *MANOVA*. It is an extension of the *homogeneity of variance assumption* in *univariate* analyses. However, as well as assuming that *variances* for each *dependent variable* are the same across groups, it also assumes that relationships (*covariances*) between these *dependent variables* are roughly equal. It is tested by comparing the population *variance–covariance matrices* of the different groups in the analysis.

Homogeneity of regression slopes: an assumption of *analysis of covariance*. This is the assumption that the relationship between the *covariate* and *outcome variable* is constant across different treatment levels. So, for three treatment conditions, if there's a positive relationship between the covariate and the outcome in one group, we assume that there is a similar-sized positive relationship between the covariate and outcome in the two other groups too.

Homogeneity of variance: the assumption that the variance of one variable is stable (i.e. relatively similar) at all levels of another variable.

Homoscedasticity: an assumption in regression analysis that the residuals at each level of the predictor variables(s) have similar variances. Put another way, at each point along any predictor variable, the spread of residuals should be fairly constant.

Hosmer and Lemeshow's R^2_L: A version of the *coefficient of determination* for logistic regression. It is a fairly literal translation in that it is the –2LL for the model divided by the original –2LL; in other words, it's the ratio of what the model can explain compared to what there was to explain in the first place!

Hotelling–Lawley trace (T^2): a *test statistic* in *MANOVA*. It is the sum of the eigenvalues for each *discriminant function variate* of the data and so is conceptually the same as the *F-ratio* in *ANOVA*: it is the sum of the ratio of *systematic* and *unsystematic variance* (SS_M/SS_R) for each of the variates.

Huynh–Feldt correction: an estimate of the departure from *sphericity*. The maximum value is 1 (the data completely meet the assumption of sphericity). Values below this indicate departures from sphericity and are used to correct the *degrees of freedom* associated with the corresponding *F-ratios* by multiplying them by the value of the estimate. It is less conservative than the *Greenhouse–Geisser* estimate, but some say it is too liberal.

Hypothesis: a prediction about the state of the world (see *experimental hypothesis* and *null hypothesis*).

Hypothesis SSCP (*H*): the hypothesis sum of squares and cross-product matrix. This is a *sum of squares and cross-product matrix* for a predictive *linear model* fitted to *multivariate* data. It represents the *systematic variance* and is the *multivariate* equivalent of the *model sum of squares*.

Identity matrix: a square matrix (i.e. has the same number of rows and columns) in which the diagonal elements are equal to 1, and the off-diagonal elements are equal to 0. The following are all examples:

$$\begin{pmatrix} 1 & 0 \\ 0 & 1 \end{pmatrix} \quad \begin{pmatrix} 1 & 0 & 0 \\ 0 & 1 & 0 \\ 0 & 0 & 1 \end{pmatrix} \quad \begin{pmatrix} 1 & 0 & 0 & 0 \\ 0 & 1 & 0 & 0 \\ 0 & 0 & 1 & 0 \\ 0 & 0 & 0 & 1 \end{pmatrix}$$

Independence: the assumption that one data point does not influence another. When data come from people, it basically means that the behaviour of one participant does not influence the behaviour of another.

Independent ANOVA: *analysis of variance* conducted on any design in which all *independent variables* or *predictors* have been manipulated using different participants (i.e. all data come from different entities).

Independent design: an experimental design in which different treatment conditions utilize different organisms (e.g. in psychology, this would mean using different people in different treatment conditions) and so the resulting data are independent (a.k.a. between-group or between-subject designs).

Independent errors: for any two observations in regression the *residuals* should be uncorrelated (or independent).

Independent factorial design: an experimental design incorporating two or more *predictors* (or *independent variables*) all of which have been manipulated using different participants (or whatever entities are being tested).

Independent *t*-test: a test using the *t-statistic* that establishes whether two means collected from independent samples differ significantly.

Independent variable: another name for a *predictor variable*. This name is usually associated with experimental methodology (which is the only time it makes sense) and is so called because it is the variable that is manipulated by the experimenter and so its value does not depend on any other variables (just on the experimenter). I just use *predictor variable* all the time because the meaning of the term is not constrained to a particular methodology.

Interaction effect: the combined effect of two or more *predictor variables* on an *outcome variable*.

Interaction graph: a graph showing the means of two or more *independent variables* in which the means of one variable are shown at different levels of the other variable. Unusually the means are connected with lines, or are displayed as bars. These graphs are used to help understand *interaction effects*.

Interquartile range: the limits within which the middle 50% of an ordered set of observations fall.

Interval data: data measured on a scale along the whole of which intervals are equal. For example, people's ratings of this book on Amazon.com can range from 1 to 5; for these data to be interval it should be true that the increase in appreciation for this book represented by a change from 3 to 4 along the scale should be the same as the change in appreciation represented by a change from 1 to 2, or 4 to 5.

Jonckheere–Terpstra test: this statistic tests for an ordered pattern of medians across independent groups. Essentially it does the same thing as the *Kruskal–Wallis test* (i.e. tests for a difference between the medians of the groups) but it incorporates information about whether the order of the groups is meaningful. As such, you should use this test when you expect the groups you're comparing to produce a meaningful order of medians.

Kaiser–Meyer–Olkin measure of sampling adequacy (KMO): this can be calculated for individual and multiple variables and represents the ratio of the squared correlation between variables to the squared *partial correlation* between variables. It varies between 0 and 1: a value of 0 indicates that the sum of partial correlations is large relative to the sum of correlations, indicating diffusion in the pattern of correlations (hence, *factor analysis* is likely to be inappropriate); a value close to 1 indicates that patterns of correlations are relatively compact and so factor analysis should yield distinct and reliable factors. Values between 0.5 and 0.7 are mediocre, values between 0.7 and 0.8 are good, values between 0.8 and 0.9 are great, and values above 0.9 are superb (see Hutcheson & Sofroniou, 1999).

Kaiser's criterion: a method of *extraction* in *factor analysis* based on the idea of retaining factors with associated eigenvalues greater than 1. This method appears to be accurate when the number of variables in the analysis is less than 30 and the resulting *communalities* (after *extraction*) are all greater than 0.7; or when the sample size exceeds 250 and the average communality is greater than or equal to 0.6.

Kendall's tau: a non-parametric correlation coefficient similar to *Spearman's correlation coefficient*, but should be preferred when you have a small data set with a large number of tied ranks.

Kendall's *W*: This is much the same as *Friedman's ANOVA* but is used specifically for looking at the agreement between raters. So, if, for example, we asked 10 different women to rate the attractiveness of Justin Timberlake, David Beckham and Tony Blair we could use this test to look at the extent to which they agree. Kendall's *W* ranges from 0 (no agreement between judges) to 1 (complete agreement between judges).

Kolmogorov–Smirnov test: a test of whether a distribution of scores is significantly different from a *normal distribution*. A significant value indicates a deviation from normality, but this test is notoriously affected by large samples in which small deviations from normality yield significant results.

Kolmogorov–Smirnov Z: not to be confused with the *Kolmogorov–Smirnov test* that tests whether a sample comes from a normally distributed population. This tests whether two groups have been drawn from the same population (regardless of what that population may be). It does much the same as the *Mann–Whitney test* and *Wilcoxon rank-sum test*! This test tends to have better power than the Mann-Whitney test when sample sizes are less than about 25 per group.

Kruskal–Wallis test: non-parametric test of whether more than two independent groups differ. It is the non-parametric version of one-way *independent ANOVA*.

Kurtosis: measures the degree to which scores cluster in the tails of a *frequency distribution*. A *platykurtic* distribution is one that has many scores in the tails and so is quite flat. In contrast, *leptokurtic* distributions are relatively thin in the tails and so look quite pointy.

Latent variable: a variable that cannot be directly measured, but is assumed to be related to several variables that can be measured.

Leptokurtic: see *kurtosis*.

Levene's test: tests the hypothesis that the variances in different groups are equal (i.e. the difference between the variances is zero). A significant result indicates that the variances are significantly different—therefore, the assumption of *homogeneity of variances* has been violated. When sample sizes are large, small differences in group variances can produce a significant Levene's test and so the *variance ratio* is a useful double check.

Leverage: leverage statistics (or hat values) gauge the influence of the observed value of the outcome variable over the predicted values. The average leverage value is $(k + 1)/n$ in which k is the number of predictors in the model and n is the number of participants. Leverage values can lie between 0 (the case has no influence whatsoever) and 1 (the case has complete influence over prediction). If no cases exert undue influence over the model then we would expect all of the leverage values to be close to the average value. Hoaglin & Welsch (1978) recommend investigating cases with values greater than twice the average $(2(k + 1)/n)$ and Stevens (1992) recommends using three times the average $(3(k + 1)/n)$ as a cut-off point for identifying cases having undue influence.

Likelihood: the probability of obtaining a set of observations given the parameters of a model fitted to those observations.

Linear model: a model that is based upon a straight line.

Logistic regression: a version of *multiple regression* in which the outcome is *dichotomous*.

Log-likelihood: a measure of error, or unexplained variation, in categorical models. It is based on summing the probabilities associated with the predicted and actual outcomes and is analogous to the *residual sum of squares* in multiple regression in that it is an indicator of how much unexplained information there is after the model has been fitted. Large values of the log-likelihood statistic indicate poorly fitting statistical models, because the larger the value of the log-likelihood, the more unexplained observations there are. The log-likelihood is the logarithm of the *likelihood*.

Loglinear analysis: a procedure used as an extension of the *chi-square test* to analyse situations in which you have more than two *categorical variables* and you want to test for relationships between these variables. Essentially, a *linear model* is fitted to the data that predict expected frequencies (i.e. the number of cases expected in a given category). In this respect it is much the same as *analysis of variance* but for entirely categorical data.

Lower Bound: the name given to the lowest possible value of the *Greenhouse–Geisser* estimate of *sphericity*. Its value is $1/k - 1$, in which k is the number of treatment conditions.

Mahalanobis distances: measures the influence of a case by examining the distance of cases from the mean(s) of the predictor variable(s). You need to look for the cases with the highest values. It is not easy to establish a cut-off point at which to worry, although Barnett & Lewis (1978) have produced a table of critical values dependent on the number of predictors and the sample size. From their work it is clear that even with large samples ($N = 500$) and five predictors, values above 25 are cause for concern. In smaller samples ($N = 100$) and with fewer predictors (namely, three) values greater than 15 are problematic, and in very small samples ($N = 30$) with only two predictors values greater than 11 should be examined. However, for more specific advice, refer to Barnett & Lewis's (1978) table.

Main effect: the unique effect of a *predictor variable* (or *independent variable*) on an *outcome variable*. The term is usually used in the context of *ANOVA*.

Mann–Whitney test: a *non-parametric test* that looks for differences between two independent samples. That is, it tests whether the populations from which two samples are drawn have the same location. It is functionally the same as *Wilcoxon's rank-sum test*, and both tests are non-parametric equivalents of the *independent t-test*.

MANOVA: acronym for *multivariate analysis of variance*.

Matrix: a collection of numbers arranged in columns and rows. The values within a matrix are typically referred to as *components* or *elements*.

Mauchly's test: a test of the assumption of *sphericity*. If this test is significant then the assumption of sphericity has not been met and an appropriate correction must be applied to the *degrees of freedom* of the *F-ratio* in *repeated-measures ANOVA*. The test works by comparing the *variance–covariance matrix* of the data to an *identity matrix*, if the variance–covariance matrix is a scalar multiple of an *identity matrix* then sphericity is met.

Maximum-likelihood estimation: a way of estimating statistical parameters by choosing the parameters that make the data most likely to have happened. Imagine for a set of parameters we calculated the probability (or likelihood) of getting the observed data: if this probability was high then these particular parameters yield a good fit of the data; conversely, if the probability is low, these parameters are a bad fit of our data. Maximum-likelihood estimation chooses the parameters that maximize the probability.

McNemar's test: this tests differences between two related groups (see *Wilcoxon signed-rank test* and *sign test*), when you have *nominal data*. It's typically used when you're looking for changes in people's scores and it compares the proportion of people who changed their response in one direction (i.e. scores increased) to those who changed in the opposite direction (scores decreased). So, this test needs to be used when you've got two related dichotomous variables.

Mean: A simple statistical model of the centre of a distribution of scores. A hypothetical estimate of the 'typical' score.

Mean squares: a measure of average variability. For every *sum of squares* (which measure the total variability) it is possible to create a mean squares by dividing by the number of things used to calculate the sum of squares (or some function of it).

Median: The middle score of a set of ordered observations. When there is an even number of observations the median is the average of the two scores that fall either side of what would be the middle value.

Median test: a non-parametric test of whether samples are drawn from a population with the same median. So, in effect, it does the same thing as the *Kruskal–Wallis test*. It works on the basis of producing a contingency table that is split for each group into the number of scores that fall above and below the observed median of the entire data set. If the groups are from the same population then you'd expect these frequencies to be the same in all conditions (about 50% above and about 50% below).

Mixed ANOVA: *analysis of variance* used when you have a *mixed design*.

Mixed design: an experimental design incorporating two or more *predictors* (or *independent variables*) at least one of which has been manipulated using different participants (or whatever entities are being tested) and at least one of which has been manipulated using the same participants (or entities). Also known as a split-plot design because Fisher developed ANOVA for analysing agricultural data involving 'plots' of land containing crops.

Mode: the most frequently occurring score in a set of data.

Model sum of squares: a measure of the total amount of variability for which a model can account. It is the difference between the *total sum of squares* and the *residual sum of squares*.

Monte Carlo method: a term applied to the process of using data simulations to solve statistical problems.

Moses Extreme Reactions: a non-parametric test that compares the variability of scores in two groups, so it's a bit like a non-parametric *Levene's test*.

Multicollinearity: a situation in which two or more variables are very closely linearly related.

Multimodal: a description of a distribution of observations that has more than two *modes*.

Multiple *R*: the multiple correlation coefficient. It is the correlation between the observed values of an outcome and the values of the outcome predicted by a multiple regression model.

Multiple regression: an extension of *simple regression* in which an outcome is predicted by a linear combination of two or more predictor variables. The form of the model is $Y_i = (b_0 + b_1X_1 + b_2X_2 + \cdots + b_nX_n) + \varepsilon_i$ in which the outcome is denoted as Y, and each predictor is denoted as X. Each predictor has a regression coefficient b_i associated with it, and b_0 is the value of the outcome when all predictors are zero.

Multivariate: means 'many variables' and is usually used when referring to analyses in which there are more than one *outcome variable* (e.g. *MANOVA*, *principal component analysis*, etc.).

Multivariate analysis of variance: family of tests that extend the basic *analysis of variance* to situations in which more than one *outcome variable* has been measured.

Multivariate normality: an extension of a normal distribution to multiple variables. It is a *probability distribution* of a set of variables $v' = [v_1, v_2, \cdots, v_n]$ given by:

$$f(v_1, v_2, \ldots, v_n) = 2\pi^{-\frac{n}{2}} \left| \sum \right|^{-\frac{1}{2}} \exp\left[-\frac{1}{2}(v-\mu)' \sum^{-1} (v-\mu) \right]$$

in which μ is the vector of means of the variables, and Σ is the *variance–covariance matrix*. If that made any sense to you then you're cleverer than I am.

Nagelkerke's R^2_N: A version of the *coefficient of determination* for logistic regression. It is a variation on *Cox and Snell's* R^2_{CS} which overcomes the problem that this statistic has of not being able to reach its maximum value.

Negative skew: see *skew*.

Nominal data: where numbers merely represent names, for example the numbers on sports players shirts: a player with the number 1 on her back is not necessarily worse than a player with a 2 on her back. The numbers have no meaning other than denoting the type of player (i.e. fullback, centre forward, etc.).

Non-parametric tests: a family of statistical procedures that do not rely on the restrictive assumptions of parametric tests. In particular they do not assume that data come from a normal distribution.

Normal distribution: a *probability distribution* of a random variable that is known to have certain properties. It is perfectly symmetrical (has a *skew* of 0), and has a *kurtosis* of 0.

Null hypothesis: reverse of the *experimental hypothesis* that your prediction is wrong and that the predicted effect doesn't exist.

Numeric variables: variables involving numbers.

Oblique rotation: a method of *rotation* in *factor analysis* that allows the underlying factors to be correlated.

Odds: the probability of an event occurring divided by the probability of that event not occurring.

Odds ratio: the ratio of the *odds* of an event occurring in one group compared to another. So, for example, if the odds of dying after writing a glossary are 4, and the odds of dying after not writing a glossary are .25, then the odds ratio is 4/.25 = 16. This means that if you write a glossary you are 16 times more likely to die than if you don't. An odds ratio of 1 would indicate that the *odds* of a particular outcome are equal in both groups.

Omega squared: an *effect size* measure associated with ANOVA that is less bias than *eta squared.* It is a (sometimes hideous) function of the *model sum of squares* and the *residual sum of squares* and isn't actually much use because it measures the overall effect of the ANOVA and so can't be interpreted in a meaningful way. In all other respects it's great, though.

One-tailed test: a test of a directional hypothesis. For example, the hypothesis 'the longer I write this glossary, the more I want to place my editor's genitals in a starved crocodile's mouth' requires a one-tailed test because I've stated the direction of the relationship (see also *two-tailed test*).

Ordinal data: data that tell us not only that things have occurred, but also the order in which they occurred. These data tell us nothing about the differences between values. For example, gold, silver and bronze medals are ordinal: they tell us that the gold medallist was better than the silver medallist; however, they don't tell us how much better (was gold a lot better than silver, or were gold and silver very closely competed?).

Orthogonal: this means perpendicular (at right angles) to something. It tends to be equated to *independence* in statistics because of the connotation that perpendicular *linear models* in geometric space are completely independent (one is not influenced by the other).

Orthogonal rotation: a method of *rotation* in *factor analysis* that keeps the underlying factors independent (i.e. not correlated).

Outcome variable: a variable whose values we are trying to predict from one or more *predictor variables.*

Outlier: an observation very different from most others. Outliers can bias statistics such as the mean.

Pairwise comparisons: comparisons of pairs of means.

Parametric test: a test that requires data from one of the large catalogue of distributions that statisticians have described. Normally this term is used for parametric tests based on the *normal distribution*, which require four basic assumptions that must be met for the test to be accurate: normally distributed data (see *normal distribution*), *homogeneity of variance*, *interval* or *ratio data*, and *independence*.

Part correlation: another name for a *semi-partial correlation.*

Partial correlation: a measure of the relationship between two variables while 'controlling' the effect one or more additional variables has on both.

Partial out: to partial-out the effect of a variable is to control for the effect that the variable has on the relationship between two other variables (see *partial correlation*).

Pattern matrix: a matrix in *factor analysis* containing the *regression coefficients* for each variable on each *factor* in the data (see also *structure matrix).*

Pearson's correlation coefficient: or Pearson's product-moment correlation coefficient, to give it its full name, is a *standardized* measure of the strength of relationship between two variables. It can take any value from −1 (as one variable changes, the other changes in the opposite direction by the same amount), through 0 (as one variable changes the other doesn't change at all), to +1 (as one variable changes, the other changes in the same direction by the same amount).

Perfect collinearity: exists when at least one predictor in a *regression model* is a perfect linear combination of the others (the simplest example being two predictors that are perfectly correlated—they have a correlation coefficient of 1).

Phi: a measure of the strength of association between two *categorical variables*. Phi is used with 2 × 2 *contingency tables* (tables in which there are two categorical variables and each variable has only two categories). Phi is a variant of the *chi-square test*, χ^2: $\phi = \sqrt{\chi^2/n}$, in which n is the total number of observations.

Pillai–Bartlett trace (*V*): a *test statistic* in *MANOVA*. It is the sum of the proportion of explained variance on the *discriminant function variates* of the data. As such, it is similar to the ratio of SS_M/SS_T.

Planned comparisons: another name for *planned contrasts*.

Planned contrasts: a set of comparisons between group means that are constructed before any data are collected. These are theory-led comparisons and are based on the idea of partitioning the variance created by the overall effect of group differences into gradually smaller portions of variance. These tests have more power than *post hoc tests*.

Platykurtic: see *kurtosis*.

Point–biserial correlation: a standardized measure of the strength of relationship between two variables when one of the two variables is *dichotomous*. The point–biserial correlation coefficient is used when the dichotomy is discrete, or true (i.e. one for which there is no underlying continuum between the categories). An example of this is pregnancy: you can be either pregnant or not, there is no in between.

Polynomial contrast: a contrast that tests for trends in the data. In its most basic form it looks for a linear trend (i.e. that the group means increase proportionately).

Population: in statistical terms this usually refers to the collection of units (be they people, plankton, plants, cities, suicidal authors, etc.) to which we want to generalize a set of findings or a statistical model.

Positive skew: see *skew*.

***Post hoc* tests**: a set of comparisons between group means that were not thought of before data were collected. Typically these tests involve comparing the means of all combinations of pairs of experimental conditions. To compensate for the number of tests conducted, each test uses a strict criterion for significance. As such, they tend to have less power than *planned contrasts*. They are usually used for exploratory work for which no firm hypotheses were available on which to base planned contrasts.

Power: the ability of a test to detect an effect of a particular size (a value of 0.8 is a good level to aim for).

Practice effect: refers to the possibility that a participant's performance in a task may be influenced (positively or negatively) if they repeat the task because of familiarity with the experimental situation and/or the measures being used.

Predictor variable: a variable that is used to try to predict values of another variable known as an *outcome variable*.

Principal component analysis (PCA): a *multivariate* technique for identifying the linear components of a set of variables.

Probability distribution: a curve describing an idealized *frequency distribution* of a particular variable from which it is possible to ascertain the probability with which specific values of that variable will occur. For categorical variables it is simply a formula yielding the probability with which each category occurs.

Promax: a method of *oblique rotation* that is computationally faster than *direct oblimin* and so useful for large data sets.

Quadratic trend: if you connect the means in ordered conditions with a line then a quadratic trend is shown by one change in the direction of this line (e.g. the line is curved in one place): the line is, therefore, U-shaped. You must have at least three ordered conditions.

Quartic trend: if you connect the means in ordered conditions with a line then a quartic trend is shown by three changes in the direction of this line. You must have at least five ordered conditions.

Quartimax: a method of *orthogonal rotation*. It attempts to maximize the spread of factor loadings for a variable across all *factors*. This often results in lots of variables loading highly onto a single factor.

Random variance: variance that is unique to a particular variable but not reliably so.

Randomization: the process of doing things in an unsystematic or random way. In the context of experimental research the word usually applies to the random assignment of participants to different treatment conditions.

Ratio data: *interval data* but with the additional property that ratios are meaningful. For example, people's ratings of this book on Amazon.com can range from 1 to 5; for these data to be ratio not only must they have the properties of *interval data*, but in addition a rating of 4 should genuinely represent someone who enjoyed this book twice as much as someone who rated it as 2. Likewise, someone who rated it as 1 should be half as impressed as someone who rated it as 2.

Regression coefficient: see b_i and β_i.

Regression model: see *simple regression* and *multiple regression*.

Related design: another name for a *repeated-measures design*.

Related factorial design: an experimental design incorporating two or more *predictors* (or *independent variables*) all of which have been manipulated using the same participants (or whatever entities are being tested).

Reliability: the ability of a measure to produce consistent results when the same entities are measured under the same conditions.

Repeated contrast: a non-orthogonal *planned contrast* that compares the mean in each condition (except the first) to the mean of the preceding condition.

Repeated-measures ANOVA: an *analysis of variance* conducted on any design in which the *independent variable* (*predictor*) or *variables* (*predictors*) have all been measured using the same participants in all conditions.

Repeated-measures design: an experimental design in which different treatment conditions utilize the same organisms (i.e. in psychology, this would mean the same people take part in all experimental conditions) and so the resulting data are related (a.k.a. related or within-subject designs).

Residual: The difference between the value a model predicts and the value observed in the data on which the model is based. When the residual is calculated for each observation in a data set the resulting collection is referred to as the *residuals*.

Residual sum of squares: a measure of the variability that cannot be explained by the model fitted to the data. It is the total squared *deviance* between the observations and the value of those observations predicted by whatever model is fitted to the data.

Residuals: see *residual*.

Reverse Helmert contrast: another name for a *difference contrast*.

Roa's efficient score statistic: a statistic measuring the same thing as the *Wald statistic* but which is computationally easier to calculate.

Rotation: a process in *factor analysis* for improving the interpretability of *factors*. In essence, an attempt is made to transform the factors that emerge from the analysis in such a way as to maximize *factor loadings* that are already large, and minimize factor loadings that are already small. There are two general approaches: *orthogonal rotation* and *oblique rotation*.

Roy's largest root: a *test statistic* in *MANOVA*. It is the eigenvalue for the first *discriminant function variate* of a set of observations. So, it is the same as the *Hotelling–Lawley trace* but for the first variate only. It represents the proportion of explained variance to unexplained variance (SS_M/SS_R) for the first discriminant function.

Sample: a smaller (but hopefully representative) collection of units from a *population* used to determine truths about that population (e.g. how a given population behaves in certain conditions).

Sampling distribution: the *probability distribution* of a statistic. You can think of this as follows: if we take a *sample* from a *population* and calculate some statistic (e.g. the *mean*), the value of this statistic will depend somewhat on the sample we took. As such the statistic will vary slightly from sample to sample. If, hypothetically, we took lots and lots of samples from the population and calculated the statistic of interest we could create a frequency distribution of the values we get. The resulting distribution is what the sampling distribution represents: the distribution of possible values of a given statistic that we could expect to get from a given population.

Saturated model: a model that perfectly fits the data and, therefore, has no error. It contains all possible *main effects* and *interactions* between variables.

Scatterplot: a graph that plots values of one variable against the corresponding value of another variable (and the corresponding value of a third variable can also be included on a three-dimensional scatterplot).

Scree plot: a graph plotting each *factor* in a *factor analysis* (*X*-axis) against its associated eigenvalue (*Y*-axis). It shows the relative importance of each factor. This graph has a very characteristic shape (there is a sharp descent in the curve followed by a tailing off) and the point of inflexion of this curve is often used as a means of *extraction*. With a sample of more than 200 participants, this provides a fairly reliable criterion for *extraction* (Stevens, 1992).

Semi-partial correlation: a measure of the relationship between two variables while 'controlling' the effect that one or more additional variables has on one of those variables. If we call our variables x and y, it gives us a measure of the variance in y that x alone shares.

Shapiro–Wilk test: a test of whether a distribution of scores is significantly different from a *normal distribution*. A significant value indicates a deviation from normality, but this test is notoriously affected by large samples in which small deviations from normality yield significant results.

Shrinkage: the loss of predictive power of a regression model if the model had been derived from the population from which the sample was taken, rather than the sample itself.

Sidak correction: a slightly less conservative variant of a *Bonferroni correction*.

Sign test: tests whether two related samples are different. It does the same thing as the *Wilcoxon signed-rank test*. Differences between the conditions are calculated and the sign of the difference (positive or negative) is analysed as it indicates the direction of differences. The magnitude of change is completely ignored (unlike in Wilcoxon's test where the rank tells us something about the relative magnitude of change), and for this reason it lacks *power*. However, its computational simplicity makes it a nice party trick if ever anyone drunkenly accosts you needing some data quickly analysed without the aid of a computer—doing a sign test in your head really impresses people. Actually it doesn't, they just think you're a sad gimboid.

Simple contrast: a non-orthogonal *planned contrast* that compares the mean in each condition to the mean of either the first or last condition depending on how the contrast is specified.

Simple effects analysis: This analysis looks at the effect of one *independent variable* (categorical *predictor variable*) at individual levels of another *independent variable*.

Simple regression: a *linear model* in which one variable or outcome is predicted from a single predictor variable. The model takes the form $Y_i = (b_0 + b_1X_i) + \varepsilon_i$, in which Y is the outcome variable, X is the predictor, b_1 is the regression coefficient associated with the predictor and b_0 is the value of the outcome when the predictor is zero.

Skew: measure of the symmetry of a *frequency distribution*. Symmetrical distributions have a skew of 0. When the frequent scores are clustered at the lower end of the distribution and the tail points towards the higher or more positive scores, the value of skew is positive. Conversely, when the frequent scores are clustered at the higher end of the distribution and the tail points towards the lower more negative scores, the value of skew is negative.

Spearman's correlation coefficient: a standardized measure of the strength of relationship between two variables that does not rely on the assumptions of a *parametric test*. It is *Pearson's correlation coefficient* performed on data that have been converted into ranked scores.

Sphericity: a less restrictive form of *compound symmetry* which assumes that the variances of the differences between data taken from the same participant (or other entity being tested) are equal. This assumption is most commonly found in *repeated-measures ANOVA* but applies only where there are more than two points of data from the same participant. (see also *Greenhouse–Geisser correction*, *Huynh–Feldt correction*).

Split-half reliability: a measure of *reliability* obtained by splitting items on a measure into two halves (in some random fashion) and obtaining a score from each half of the scale. The correlation between the two scores, corrected to take account of the fact that the correlations are based on only half of the items, is used as a measure of reliability. There are two popular ways to do this. Spearman (1910) and Brown (1910) developed a formula that takes no account of the standard deviation of items:

$$r_{sh} = \frac{2r_{12}}{1 + r_{12}}$$

in which r_{12} is the correlation between the two halves of the scale. Flanagan (1937) and Rulon (1939), however, proposed a measure that does account for item variance:

$$r_{sh} = \frac{4r_{12} \times s_1 \times s_2}{s_T^2}$$

in which s_1 and s_2 are the standard deviations of each half of the scale, and S_T^2 is the variance of the whole test. See Cortina (1993) for more detail.

Square matrix: a *matrix* that has an equal number of columns and rows.

Standard deviation: an estimate of the average variability (spread) of a set of data measured in the same units of measurement as the original data. It is the square root of the variance.

Standard error: the standard deviation of the *sampling distribution* of a statistic. For a given statistic (e.g. the *mean*) it tells us how much variability there is in this statistic across *samples* from the same *population*. Large values, therefore, indicate that a statistic from a given sample may not be an accurate reflection of the population from which the sample came.

Standard error of differences: if you were to take several pairs of samples from a population and calculate their means, then you could also calculate the difference between their means. If you plotted these differences between sample means as a *frequency distribution*, you would have the *sampling distribution* of differences. The standard deviation of this sampling distribution is the *standard error of differences*. As such it is a measure of the variability of differences between sample means.

Standardization: the process of converting a variable into a standard unit of measurement. The unit of measurement typically used is *standard deviation* units (see also *z-scores*). Standardization allows us to compare data when different units of measurement have been used (we could compare weight measured in kilograms with height measured in inches).

Standardized: see *standardization*.

Standardized DFBeta: a *standardized* version of DFBeta. These standardized values are easier to use than *DFBeta* because universal cut-off points can be applied. Stevens (1992) suggests looking at cases with absolute values greater than 2.

Standardized DFFit: a *standardized* version of *DFFit*.

Standardized residuals: the *residuals* of a model expressed in standard deviation units. Standardized residuals with an absolute value greater than 3.29 (actually we usually just use 3) are cause for concern because in an average sample a value this high is unlikely to happen by chance; if more than 1% of our observations have standardized residuals with an absolute value greater than 2.58 (we usually just say 2.5) there is evidence that the level of error within our model is unacceptable (the model is a fairly poor fit of the sample data); and if more than 5% of observations have standardized residuals with an absolute value greater than 1.96 (or 2 for convenience) then there is also evidence that the model is a poor representation of the actual data.

Stepwise regression: a method of multiple regression in which variables are entered into the model based on a statistical criterion (the semi-partial correlation with the outcome variable). Once a new variable is entered into the model, all variables in the model are assessed to see whether they should be removed.

String variables: variables involving words (i.e. letter strings). Such variables could include responses to open-ended questions such as 'how much do you like writing glossary entries?'; the response might be 'about as much as like placing my gonads on hot coals'.

Structure matrix: a matrix in *factor analysis* containing the *correlation coefficients* for each variable on each *factor* in the data. When *orthogonal rotation* is used this is the same as the *pattern matrix*, but when oblique rotation is used these matrices are different.

Studentized deleted residual: a measure of the influence of a particular case of data. This is a standardized version of the *deleted residual*.

Studentized residuals: a variation on *standardized residuals*. Studentized residuals are the *unstandardized residual* divided by an estimate of its standard deviation that varies point by point. These residuals have the same properties as the standardized residuals but usually provide a more precise estimate of the error variance of a specific case.

Sum of squared errors: another name for the *sum of squares*.

Sum of squares (SS): an estimate of total variability (spread) of a set of data. First the *deviance* for each score is calculated, and then this value is squared. The SS is the sum of these squared deviances.

Sum of squares and cross-products (SSCP) matrix: a *square matrix* in which the diagonal elements represent the *sum of squares* for a particular variable, and the off-diagonal elements represent the *cross-products* between pairs of variables. The SSCP matrix is basically the same as the *variance–covariance matrix*, except that it expresses variability and between-variable relationships as total values, whereas the variance–covariance matrix expresses them as average values.

Suppressor effects: when a predictor has a significant effect but only when another variable is held constant.

Syntax: pre-defined written commands that instruct SPSS what you would like it to do (writing 'bugger off and leave me alone' doesn't seem to work ...).

Syntax editor: a window in SPSS for writing and editing *syntax*.

Systematic variation: variation due to some genuine effect (be that the effect of an experimenter doing something to all of the participants in one sample but not in other samples, or natural variation between sets of variables). You can think of this as variation that can be explained by the model that we've fitted to the data.

***T*-statistic**: Student's t is a *test statistic* with a known *probability distribution* (the t-distribution). In the context of regression it is used to test whether a regression coefficient b is significantly different from zero; in the context of experimental work it is used to test whether the differences between two means are significantly different from zero (see also *independent t-test* and *dependent t-test*).

Test–retest reliability: the ability of a measure to produce consistent results when the same entities are tested at two different points in time.

Test statistic: a statistic for which we know how frequently different values occur. The observed value of such a statistic is typically used to test *hypotheses*.

Tolerance: tolerance statistics measure *multicollinearity* and are simply the reciprocal of the *variance inflation factor* (1/VIF). Values below 0.1 indicate serious problems, although Menard (1995) suggests that values below 0.2 are worthy of concern.

Total SSCP matrix (*T*): the total sum of squares and cross-product matrix. This is a *sum of squares and cross-product matrix* for an entire set of observations. It is the *multivariate* equivalent of the *total sum of squares*.

Total sum of squares: a measure of the total variability within a set of observations. It is the total squared *deviance* between each observation and the overall mean of all observations.

Transformation: the process of applying a mathematical function to all observations in a data set, usually to correct some distributional abnormality such as *skew* or *kurtosis*.

Two-tailed test: a test of a non-directional hypothesis. For example, the hypothesis 'writing this glossary has some effect on what I want to do with my editor's genitals' requires a two-tailed test because it doesn't suggest the direction of the relationship (see also *one-tailed test*).

Type I error: occurs when we believe that there is a genuine effect in our population, when in fact there isn't.

Type II error: occurs when we believe that there is no effect in the population when, in reality, there is.

Unique variance: variance that is specific to a particular variable (i.e. is not shared with other variables). We tend to use the term 'unique variance' to refer to variance that can be reliably attributed to only one measure, otherwise it is called *random variance*.

Univariate: this means 'one variable' and is usually used to refer to situations in which only one *outcome variable* has been measured (i.e. *ANOVA*, *t-tests*, *Mann–Whitney tests,* etc.).

Unstandardized residuals: the *residuals* of a model expressed in the units in which the original outcome variable was measured.

Unsystematic variation: variation that isn't due to the effect in which we're interested (so, it could be due to natural differences between people in different samples, such as differences in intelligence or motivation). You can think of this variation as one that can't be explained by whatever model we've fitted to the data.

Variable view: there are two ways to view the contents of the *data editor* window. The variable view allows you to define properties of the variables for which you wish to enter data (see also *data view*).

Variance: an estimate of average variability (spread) of a set of data. It is the sum of squares divided by the number of values on which the sum of squares is based minus 1.

Variance–covariance matrix: a square matrix (i.e. same number of columns and rows) representing the variables measured. The diagonals represent the *variances* within each variable, whereas the off-diagonals represent the *covariances* between pairs of variables.

Variance inflation factor (VIF): a measure of *multicollinearity*. The VIF indicates whether a predictor has a strong linear relationship with the other predictor(s). Myers (1990) suggests that a value of 10 is a good value at which to worry. Bowerman and O'Connell (1990) suggest that if the average VIF is greater than 1, then multicollinearity may be biasing the regression model.

Variance ratio: the variance of the group with the biggest variance divided by the variance of the group with the smallest variance. If this ratio is less than 2, then it's safe to assume *homogeneity of variance*.

Variance sum law: this states that the variance of a difference between two independent variables is equal to the sum of their variances.

Varimax: a method of *orthogonal rotation*. It attempts to maximize the dispersion of *factor loadings* within *factors*. Therefore, it tries to load a smaller number of variables highly onto each factor resulting in more interpretable clusters of factors.

Viewer: SPSS window in which the output of any analysis is displayed.

VIF: see *variance inflation factor*.

Wald statistic: a *test statistic* with a known *probability distribution* (a *chi-square distribution*) that is used to test whether the *b*-coefficient for a predictor in a *logistic* regression model is significantly different from zero. It is analogous to the *t-statistic* in a *regression model* in that it is simply the *b*-coefficient divided by its standard error. The Wald statistic is inaccurate when the regression coefficient (*b*) is large, because the standard error tends to become inflated, resulting in the Wald statistic being underestimated.

Wald–Wolfowitz runs: another variant on the *Mann-Whitney test*. Scores are rank ordered as in the Mann–Whitney test, but rather than analysing the ranks, this test looks for 'runs' of scores from the same group within the ranked order. Now, if there's no difference between groups then obviously ranks from the two groups should be randomly interspersed. However, if the groups are different then you should see more ranks from one group at the lower end, and more ranks from the other group at the higher end. By looking for clusters of scores in this way the test can determine if the groups differ.

Weights: a number by which something (usually a variable in statistics) is multiplied. The weight assigned to a variable determines the influence that variable has within a mathematical equation: large weights give the variable a lot of influence.

Welch *F*: a version of the *F*-ratio designed to be accurate when the assumption of *homogeneity of variance* has been violated. Not to be confused with the sqwelch test which is where you shake your head around after writing statistics books to see if you still have a brain.

Wilcoxon signed-rank test: a *non-parametric test* that looks for differences between two related samples. It is the non-parametric equivalent of the *related t-test*.

Wilcoxon's rank-sum test: a *non-parametric test* that looks for differences between two independent samples. That is, it tests whether the populations from which two samples are drawn have the same location. It is functionally the same as the *Mann–Whitney test*, and both tests are non-parametric equivalents of the *independent t-test*.

Wilks's lambda (Λ): a *test statistic* in *MANOVA*. It is the product of the unexplained variance on each of the *discriminant function variates* so it represents the ratio of error variance to total variance (SS_R/SS_T) for each variate.

Within-subject design: another name for a *repeated-measures design*.

Yates's continuity correction: an adjustment made to the *chi-square test* when the *contingency table* is two rows by two columns (i.e. there are two categorical variables both of which consist of only two categories). In large samples the adjustment makes little difference and is slightly dubious anyway (see Howell, 2002).

z-score: the value of an observation expressed in standard deviation units. It is calculated by taking the observation, subtracting from it the mean of all observations, and dividing the result by the standard deviation of all observations. By converting a distribution of observations into z-scores you create a new distribution that has a mean of 0 and a standard deviation of 1.mm

APPENDIX

A.1 TABLE OF THE STANDARD NORMAL DISTRIBUTION

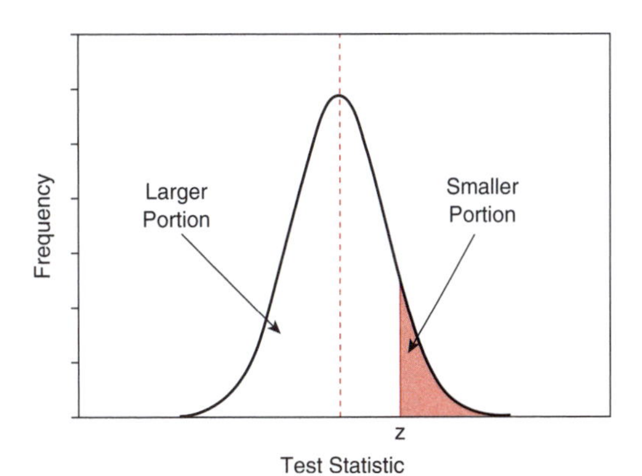

z	Larger Portion	Smaller Portion	y
.00	.50000	.50000	.3989
.01	.50399	.49601	.3989
.02	.50798	.49202	.3989
.03	.51197	.48803	.3988
.04	.51595	.48405	.3986
.05	.51994	.48006	.3984
.06	.52392	.47608	.3982
.07	.52790	.47210	.3980
.08	.53188	.46812	**.3977**
.09	.53586	.46414	.3973
.10	.53983	.46017	.3970
.11	.54380	.45620	.3965
.12	.54776	.45224	.3961
.13	.55172	.44828	.3956
.14	.55567	.44433	.3951
.15	.55962	.44038	.3945
.16	.56356	.43644	.3939
.17	.56749	.43251	.3932

z	Larger Portion	Smaller Portion	y
.18	.57142	.42858	.3925
.19	.57535	.42465	.3918
.20	.57926	.42074	.3910
.21	.58317	.41683	.3902
.22	.58706	.41294	.3894
.23	.59095	.40905	.3885
.24	.59483	.40517	.3876
.25	.59871	.40129	.3867
.26	.60257	.39743	.3857
.27	.60642	.39358	.3847
.28	.61026	.38974	.3836
.29	.61409	.38591	.3825
.30	.61791	.38209	.3814
.31	.62172	.37828	.3802
.32	.62552	.37448	.3790
.33	.62930	.37070	.3778
.34	.63307	.36693	.3765
.35	.63683	.36317	.3752

(Continued)

(Continued)

z	Larger Portion	Smaller Portion	y
.36	.64058	.35942	.3739
.37	.64431	.35569	.3725
.38	.64803	.35197	.3712
.39	.65173	.34827	.3697
.40	.65542	.34458	.3683
.41	.65910	.34090	.3668
.42	.66276	.33724	.3653
.43	.66640	.33360	.3637
.44	.67003	.32997	.3621
.45	.67364	.32636	.3605
.46	.67724	.32276	.3589
.47	.68082	.31918	.3572
.48	.68439	.31561	.3555
.49	.68793	.31207	.3538
.50	.69146	.30854	.3521
.51	.69497	.30503	.3503
.52	.69847	.30153	.3485
.53	.70194	.29806	.3467
.54	.70540	.29460	.3448
.55	.70884	.29116	.3429
.56	.71226	.28774	.3410
.57	.71566	.28434	.3391
.58	.71904	.28096	.3372
.59	.72240	.27760	.3352
.60	.72575	.27425	.3332
.61	.72907	.27093	.3312
.62	.73237	.26763	.3292
.63	.73565	.26435	.3271
.64	.73891	.26109	.3251
.65	.74215	.25785	.3230
.66	.74537	.25463	.3209
.67	.74857	.25143	.3187
.68	.75175	.24825	.3166
.69	.75490	.24510	.3144
.70	.75804	.24196	.3123
.71	.76115	.23885	.3101
.72	.76424	.23576	.3079
.73	.76730	.23270	.3056
.74	.77035	.22965	.3034
.75	.77337	.22663	.3011
.76	.77637	.22363	.2989
.77	.77935	.22065	.2966
.78	.78230	.21770	.2943
.79	.78524	.21476	.2920
.80	.78814	.21186	.2897

z	Larger Portion	Smaller Portion	y
.81	.79103	.20897	.2874
.82	.79389	.20611	.2850
.83	.79673	.20327	.2827
.84	.79955	.20045	.2803
.85	.80234	.19766	.2780
.86	.80511	.19489	.2756
.87	.80785	.19215	.2732
.88	.81057	.18943	.2709
.89	.81327	.18673	.2685
.90	.81594	.18406	.2661
.91	.81859	.18141	.2637
.92	.82121	.17879	.2613
.93	.82381	.17619	.2589
.94	.82639	.17361	.2565
.95	.82894	.17106	.2541
.96	.83147	.16853	.2516
.97	.83398	.16602	.2492
.98	.83646	.16354	.2468
.99	.83891	.16109	.2444
1.00	.84134	.15866	.2420
1.01	.84375	.15625	.2396
1.02	.84614	.15386	.2371
1.03	.84849	.15151	.2347
1.04	.85083	.14917	.2323
1.05	.85314	.14686	.2299
1.06	.85543	.14457	.2275
1.07	.85769	.14231	.2251
1.08	.85993	.14007	.2227
1.09	.86214	.13786	.2203
1.10	.86433	.13567	.2179
1.11	.86650	.13350	.2155
1.12	.86864	.13136	.2131
1.13	.87076	.12924	.2107
1.14	.87286	.12714	.2083
1.15	.87493	.12507	.2059
1.16	.87698	.12302	.2036
1.17	.87900	.12100	.2012
1.18	.88100	.11900	.1989
1.19	.88298	.11702	.1965
1.20	.88493	.11507	.1942
1.21	.88686	.11314	.1919
1.22	.88877	.11123	.1895
1.23	.89065	.10935	.1872
1.24	.89251	.10749	.1849
1.25	.89435	.10565	.1826

(Continued)

z	Larger Portion	Smaller Portion	y
1.26	.89617	.10383	.1804
1.27	.89796	.10204	.1781
1.28	.89973	.10027	.1758
1.29	.90147	.09853	.1736
1.30	.90320	.09680	.1714
1.31	.90490	.09510	.1691
1.32	.90658	.09342	.1669
1.33	.90824	.09176	.1647
1.34	.90988	.09012	.1626
1.35	.91149	.08851	.1604
1.36	.91309	.08691	.1582
1.37	.91466	.08534	.1561
1.38	.91621	.08379	.1539
1.39	.91774	.08226	.1518
1.40	.91924	.08076	.1497
1.41	.92073	.07927	1476
1.42	.92220	.07780	.1456
1.43	.92364	.07636	.1435
1.44	.92507	.07493	.1415
1.45	.92647	.07353	.1394
1.46	.92785	.07215	.1374
1.47	.92922	.07078	.1354
1.48	.93056	.06944	.1334
1.49	.93189	.06811	.1315
1.50	.93319	.06681	.1295
1.51	.93448	.06552	.1276
1.52	.93574	.06426	.1257
1.53	.93699	.06301	.1238
1.54	.93822	.06178	.1219
1.55	.93943	.06057	.1200
1.56	.94062	.05938	.1182
1.57	.94179	.05821	.1163
1.58	.94295	.05705	.1145
1.59	.94408	.05592	.1127
1.60	.94520	.05480	.1109
1.61	.94630	.05370	.1092
1.62	.94738	.05262	.1074
1.63	.94845	.05155	.1057
1.64	.94950	.05050	.1040
1.65	.95053	.04947	.1023
1.66	.95154	.04846	.1006
1.67	.95254	.04746	.0989
1.68	.95352	.04648	.0973
1.69	.95449	.04551	.0957
1.70	.95543	.04457	.0940

z	Larger Portion	Smaller Portion	y
1.71	.95637	.04363	.0925
1.72	.95728	.04272	.0909
1.73	.95818	.04182	.0893
1.74	.95907	.04093	.0878
1.75	.95994	.04006	.0863
1.76	.96080	.03920	.0848
1.77	.96164	.03836	.0833
1.78	.96246	.03754	.0818
1.79	.96327	.03673	0804
1.80	.96407	.03593	.0790
1.81	.96485	.03515	.0775
1.82	.96562	.03438	.0761
1.83	.96638	.03362	.0748
1.84	.96712	.03288	.0734
1.85	.96784	.03216	.0721
1.86	.96856	.03144	.0707
1.87	.96926	.03074	.0694
1.88	.96995	.03005	.0681
1.89	.97062	.02938	.0669
1.90	.97128	.02872	.0656
1.91	.97193	.02807	.0644
1.92	.97257	.02743	.0632
1.93	.97320	.02680	.0620
1.94	.97381	.02619	.0608
1.95	.97441	.02559	.0596
1.96	.97500	.02500	.0584
1.97	.97558	.02442	.0573
1.98	.97615	.02385	.0562
1.99	.97670	.02330	.0551
2.00	.97725	.02275	.0540
2.01	.97778	.02222	.0529
2.02	.97831	.02169	.0519
2.03	.97882	.02118	.0508
2.04	.97932	.02068	.0498
2.05	.97982	.02018	.0488
2.06	.98030	.01970	.0478
2.07	.98077	.01923	.0468
2.08	.98124	.01876	.0459
2.09	.98169	.01831	.0449
2.10	.98214	.01786	.0440
2.11	.98257	.01743	.0431
2.12	.98300	.01700	.0422
2.13	.98341	.01659	.0413
2.14	.98382	.01618	.0404
2.15	.98422	.01578	.0396

(Continued)

(Continued)

z	Larger Portion	Smaller Portion	y
2.16	.98461	.01539	.0387
2.17	.98500	.01500	.0379
2.18	.98537	.01463	.0371
2.19	.98574	.01426	.0363
2.20	.98610	.01390	.0355
2.21	.98645	.01355	.0347
2.22	.98679	.01321	.0339
2.23	.98713	.01287	.0332
2.24	.98745	.01255	.0325
2.25	.98778	.01222	.0317
2.26	.98809	.01191	.0310
2.27	.98840	.01160	.0303
2.28	.98870	.01130	.0297
2.29	.98899	.01101	.0290
2.30	.98928	.01072	.0283
2.31	.98956	.01044	.0277
2.32	.98983	.01017	.0270
2.33	.99010	.00990	.0264
2.34	.99036	.00964	.0258
2.35	.99061	.00939	.0252
2.36	.99086	.00914	.0246
2.37	.99111	.00889	.0241
2.38	.99134	.00866	.0235
2.39	.99158	.00842	.0229
2.40	.99180	.00820	.0224
2.41	.99202	.00798	.0219
2.42	.99224	.00776	.0213
2.43	.99245	.00755	.0208
2.44	.99266	.00734	.0203
2.45	.99286	.00714	.0198
2.46	.99305	.00695	.0194
2.47	.99324	.00676	.0189
2.48	.99343	.00657	.0184
2.49	.99361	.00639	.0180
2.50	.99379	.00621	.0175
2.51	.99396	.00604	.0171
2.52	.99413	.00587	.0167
2.53	.99430	.00570	.0163
2.54	.99446	.00554	.0158
2.55	.99461	.00539	.0154
2.56	.99477	.00523	.0151
2.57	.99492	.00508	.0147
2.58	.99506	.00494	.0143
2.59	.99520	.00480	.0139
2.60	.99534	.00466	.0136
2.61	.99547	.00453	.0132

z	Larger Portion	Smaller Portion	y
2.62	.99560	.00440	.0129
2.63	.99573	.00427	.0126
2.64	.99585	.00415	.0122
2.65	.99598	.00402	.0119
2.66	.99609	.00391	.0116
2.67	.99621	.00379	.0113
2.68	.99632	.00368	.0110
2.69	.99643	.00357	.0107
2.70	.99653	.00347	.0104
2.71	.99664	.00336	.0101
2.72	.99674	.00326	.0099
2.73	.99683	.00317	.0096
2.74	.99693	.00307	.0093
2.75	.99702	.00298	.0091
2.76	.99711	.00289	.0088
2.77	.99720	.00280	.0086
2.78	.99728	.00272	.0084
2.79	.99736	.00264	.0081
2.80	.99744	.00256	.0079
2.81	.99752	.00248	.0077
2.82	.99760	.00240	.0075
2.83	.99767	.00233	.0073
2.84	.99774	.00226	.0071
2.85	.99781	.00219	.0069
2.86	.99788	.00212	.0067
2.87	.99795	.00205	.0065
2.88	.99801	.00199	.0063
2.89	.99807	.00193	.0061
2.90	.99813	.00187	.0060
2.91	.99819	.00181	.0058
2.92	.99825	.00175	.0056
2.93	.99831	.00169	.0055
2.94	.99836	.00164	.0053
2.95	.99841	.00159	.0051
2.96	.99846	.00154	.0050
2.97	.99851	.00149	.0048
2.98	.99856	.00144	.0047
2.99	.99861	.00139	.0046
3.00	.99865	.00135	.0044
⋮	⋮	⋮	⋮
3.25	.99942	.00058	.0020
⋮	⋮	⋮	⋮
3.50	.99977	.00023	.0009
⋮	⋮	⋮	⋮
4.00	.99997	.00003	.0001

Values calculated by author using SPSS 11.

A.2. CRITICAL VALUES OF THE *T*-DISTRIBUTION

	Two-Tailed Test		One-Tailed Test	
df	*0.05*	*0.01*	*0.05*	*0.01*
1	12.71	63.66	6.31	31.82
2	4.30	9.92	2.92	6.96
3	3.18	5.84	2.35	4.54
4	2.78	4.60	2.13	3.75
5	2.57	4.03	2.02	3.36
6	2.45	3.71	1.94	3.14
7	2.36	3.50	1.89	3.00
8	2.31	3.36	1.86	2.90
9	2.26	3.25	1.83	2.82
10	2.23	3.17	1.81	2.76
11	2.20	3.11	1.80	2.72
12	2.18	3.05	1.78	2.68
13	2.16	3.01	1.77	2.65
14	2.14	2.98	1.76	2.62
15	2.13	2.95	1.75	2.60
16	2.12	2.92	1.75	2.58
17	2.11	2.90	1.74	2.57
18	2.10	2.88	1.73	2.55
19	2.09	2.86	1.73	2.54
20	2.09	2.85	1.72	2.53
21	2.08	2.83	1.72	2.52
22	2.07	2.82	1.72	2.51
23	2.07	2.81	1.71	2.50
24	2.06	2.80	1.71	2.49
25	2.06	2.79	1.71	2.49
26	2.06	2.78	1.71	2.48
27	2.05	2.77	1.70	2.47
28	2.05	2.76	1.70	2.47
29	2.05	2.76	1.70	2.46
30	2.04	2.75	1.70	2.46
35	2.03	2.72	1.69	2.44
40	2.02	2.70	1.68	2.42
45	2.01	2.69	1.68	2.41
50	2.01	2.68	1.68	2.40
60	2.00	2.66	1.67	2.39
70	1.99	2.65	1.67	2.38
80	1.99	2.64	1.66	2.37
90	1.99	2.63	1.66	2.37
100	1.98	2.63	1.66	2.36
∞ (z)	1.96	2.58	1.64	2.33

All values computed by the author using SPSS 11.

A.3. CRITICAL VALUES OF THE *F*-DISTRIBUTION

		df (numerator)									
df (denominator)	*p*	1	2	3	4	5	6	7	8	9	10
1	.05	161.45	199.50	215.71	224.58	230.16	233.99	236.77	238.88	240.54	241.88
	.01	4052.18	4999.50	5403.35	5624.58	5763.65	5858.99	5928.36	5981.07	6022.47	6055.85
2	.05	18.51	19.00	19.16	19.25	19.30	19.33	19.35	19.37	19.38	19.40
	.01	98.50	99.00	99.17	99.25	99.30	99.33	99.36	99.37	99.39	99.40
3	.05	10.13	9.55	9.28	9.12	9.01	8.94	8.89	8.85	8.81	8.79
	.01	34.12	30.82	29.46	28.71	28.24	27.91	27.67	27.49	27.35	27.23
4	.05	7.71	6.94	6.59	6.39	6.26	6.16	6.09	6.04	6.00	5.96
	.01	21.20	18.00	16.69	15.98	15.52	15.21	14.98	14.80	14.66	14.55
5	.05	6.61	5.79	5.41	5.19	5.05	4.95	4.88	4.82	4.77	4.74
	.01	16.26	13.27	12.06	11.39	10.97	10.67	10.46	10.29	10.16	10.05
6	.05	5.99	5.14	4.76	4.53	4.39	4.28	4.21	4.15	4.10	4.06
	.01	13.75	10.92	9.78	9.15	8.75	8.47	8.26	8.10	7.98	7.87
7	.05	5.59	4.74	4.35	4.12	3.97	3.87	3.79	3.73	3.68	3.64
	.01	12.25	9.55	8.45	7.85	7.46	7.19	6.99	6.84	6.72	6.62
8	.05	5.32	4.46	4.07	3.84	3.69	3.58	3.50	3.44	3.39	3.35
	.01	11.26	8.65	7.59	7.01	6.63	6.37	6.18	6.03	5.91	5.81
9	.05	5.12	4.26	3.86	3.63	3.48	3.37	3.29	3.23	3.18	3.14
	.01	10.56	8.02	6.99	6.42	6.06	5.80	5.61	5.47	5.35	5.26
10	.05	4.96	4.10	3.71	3.48	3.33	3.22	3.14	3.07	3.02	2.98
	.01	10.04	7.56	6.55	5.99	5.64	5.39	5.20	5.06	4.94	4.85
11	.05	4.84	3.98	3.59	3.36	3.20	3.09	3.01	2.95	2.90	2.85
	.01	9.65	7.21	6.22	5.67	5.32	5.07	4.89	4.74	4.63	4.54
12	.05	4.75	3.89	3.49	3.26	3.11	3.00	2.91	2.85	2.80	2.75
	.01	9.33	6.93	5.95	5.41	5.06	4.82	4.64	4.50	4.39	4.30
13	.05	4.67	3.81	3.41	3.18	3.03	2.92	2.83	2.77	2.71	2.67
	.01	9.07	6.70	5.74	5.21	4.86	4.62	4.44	4.30	4.19	4.10
14	.05	4.60	3.74	3.34	3.11	2.96	2.85	2.76	2.70	2.65	2.60
	.01	8.86	6.51	5.56	5.04	4.69	4.46	4.28	4.14	4.03	3.94
15	.05	4.54	3.68	3.29	3.06	2.90	2.79	2.71	2.64	2.59	2.54
	.01	8.68	6.36	5.42	4.89	4.56	4.32	4.14	4.00	3.89	3.80
16	.05	4.49	3.63	3.24	3.01	2.85	2.74	2.66	2.59	2.54	2.49
	.01	8.53	6.23	5.29	4.77	4.44	4.20	4.03	3.89	3.78	3.69
17	.05	4.45	3.59	3.20	2.96	2.81	2.70	2.61	2.55	2.49	2.45
	.01	8.40	6.11	5.18	4.67	4.34	4.10	3.93	3.79	3.68	3.59
18	.05	4.41	3.55	3.16	2.93	2.77	2.66	2.58	2.51	2.46	2.41
	.01	8.29	6.01	5.09	4.58	4.25	4.01	3.84	3.71	3.60	3.51

(Continued)

df (denominator)	*p*	*df* (numerator) 1	2	3	4	5	6	7	8	9	10
19	.05	4.38	3.52	3.13	2.90	2.74	2.63	2.54	2.48	2.42	2.38
	.01	8.18	5.93	5.01	4.50	4.17	3.94	3.77	3.63	3.52	3.43
20	.05	4.35	3.49	3.10	2.87	2.71	2.60	2.51	2.45	2.39	2.35
	.01	8.10	5.85	4.94	4.43	4.10	3.87	3.70	3.56	3.46	3.37
22	.05	4.30	3.44	3.05	2.82	2.66	2.55	2.46	2.40	2.34	2.30
	.01	7.95	5.72	4.82	4.31	3.99	3.76	3.59	3.45	3.35	3.26
24	.05	4.26	3.40	3.01	2.78	2.62	2.51	2.42	2.36	2.30	2.25
	.01	7.82	5.61	4.72	4.22	3.90	3.67	3.50	3.36	3.26	3.17
26	.05	4.23	3.37	2.98	2.74	2.59	2.47	2.39	2.32	2.27	2.22
	.01	7.72	5.53	4.64	4.14	3.82	3.59	3.42	3.29	3.18	3.09
28	.05	4.20	3.34	2.95	2.71	2.56	2.45	2.36	2.29	2.24	2.19
	.01	7.64	5.45	4.57	4.07	3.75	3.53	3.36	3.23	3.12	3.03
30	.05	4.17	3.32	2.92	2.69	2.53	2.42	2.33	2.27	2.21	2.16
	.01	7.56	5.39	4.51	4.02	3.70	3.47	3.30	3.17	3.07	2.98
35	.05	4.12	3.27	2.87	2.64	2.49	2.37	2.29	2.22	2.16	2.11
	.01	7.42	5.27	4.40	3.91	3.59	3.37	3.20	3.07	2.96	2.88
40	.05	4.08	3.23	2.84	2.61	2.45	2.34	2.25	2.18	2.12	2.08
	.01	7.31	5.18	4.31	3.83	3.51	3.29	3.12	2.99	2.89	2.80
45	.05	4.06	3.20	2.81	2.58	2.42	2.31	2.22	2.15	2.10	2.05
	.01	7.23	5.11	4.25	3.77	3.45	3.23	3.07	2.94	2.83	2.74
50	.05	4.03	3.18	2.79	2.56	2.40	2.29	2.20	2.13	2.07	2.03
	.01	7.17	5.06	4.20	3.72	3.41	3.19	3.02	2.89	2.78	2.70
60	.05	4.00	3.15	2.76	2.53	2.37	2.25	2.17	2.10	2.04	1.99
	.01	7.08	4.98	4.13	3.65	3.34	3.12	2.95	2.82	2.72	2.63
80	.05	3.96	3.11	2.72	2.49	2.33	2.21	2.13	2.06	2.00	1.95
	.01	6.96	4.88	4.04	3.56	3.26	3.04	2.87	2.74	2.64	2.55
100	.05	3.94	3.09	2.70	2.46	2.31	2.19	2.10	2.03	1.97	1.93
	.01	6.90	4.82	3.98	3.51	3.21	2.99	2.82	2.69	2.59	2.50
150	.05	3.90	3.06	2.66	2.43	2.27	2.16	2.07	2.00	1.94	1.89
	.01	6.81	4.75	3.91	3.45	3.14	2.92	2.76	2.63	2.53	2.44
300	.05	3.87	3.03	2.63	2.40	2.24	2.13	2.04	1.97	1.91	1.86
	.01	6.72	4.68	3.85	3.38	3.08	2.86	2.70	2.57	2.47	2.38
500	.05	3.86	3.01	2.62	2.39	2.23	2.12	2.03	1.96	1.90	1.85
	.01	6.69	4.65	3.82	3.36	3.05	2.84	2.68	2.55	2.44	2.36
1000	.05	3.85	3.00	2.61	2.38	2.22	2.11	2.02	1.95	1.89	1.84
	.01	6.66	4.63	3.80	3.34	3.04	2.82	2.66	2.53	2.43	2.34

(Continued)

(Continued)

df (denominator)	*p*	*df* (numerator) 15	20	25	30	40	50	1000
1	.05	245.95	248.01	249.26	250.10	251.14	251.77	254.19
	.01	6157.31	6208.74	6239.83	6260.65	6286.79	6302.52	6362.70
2	.05	19.43	19.45	19.46	19.46	19.47	19.48	19.49
	.01	99.43	99.45	99.46	99.47	99.47	99.48	99.50
3	.05	8.70	8.66	8.63	8.62	8.59	8.58	8.53
	.01	26.87	26.69	26.58	26.50	26.41	26.35	26.14
4	.05	5.86	5.80	5.77	5.75	5.72	5.70	5.63
	.01	14.20	14.02	13.91	13.84	13.75	13.69	13.47
5	05	4.62	4.56	4.52	4.50	4.46	4.44	4.37
	.01	9.72	9.55	9.45	9.38	9.29	9.24	9.03
6	.05	3.94	3.87	3.83	3.81	3.77	3.75	3.67
	.01	7.56	7.40	7.30	7.23	7.14	7.09	6.89
7	.05	3.51	3.44	3.40	3.38	3.34	3.32	3.23
	.01	6.31	6.16	6.06	5.99	5.91	5.86	5.66
8	.05	3.22	3.15	3.11	3.08	3.04	3.02	2.93
	.01	5.52	5.36	5.26	5.20	5.12	5.07	4.87
9	.05	3.01	2.94	2.89	2.86	2.83	2.80	2.71
	.01	4.96	4.81	4.71	4.65	4.57	4.52	4.32
10	.05	2.85	2.77	2.73	2.70	2.66	2.64	2.54
	.01	4.56	4.41	4.31	4.25	4.17	4.12	3.92
11	.05	2.72	2.65	2.60	2.57	2.53	2.51	2.41
	.01	4.25	4.10	4.01	3.94	3.86	3.81	3.61
12	.05	2.62	2.54	2.50	2.47	2.43	2.40	2.30
	.01	4.01	3.86	3.76	3.70	3.62	3.57	3.37
13	.05	2.53	2.46	2.41	2.38	2.34	2.31	2.21
	.01	3.82	3.66	3.57	3.51	3.43	3.38	3.18
14	.05	2.46	2.39	2.34	2.31	2.27	2.24	2.14
	.01	3.66	3.51	3.41	3.35	3.27	3.22	3.02
15	.05	2.40	2.33	2.28	2.25	2.20	2.18	2.07
	.01	3.52	3.37	3.28	3.21	3.13	3.08	2.88
16	.05	2.35	2.28	2.23	2.19	2.15	2.12	2.02
	.01	3.41	3.26	3.16	3.10	3.02	2.97	2.76
17	.05	2.31	2.23	2.18	2.15	2.10	2.08	1.97
	.01	3.31	3.16	3.07	3.00	2.92	2.87	2.66
18	.05	2.27	2.19	2.14	2.11	2.06	2.04	1.92
	.01	3.23	3.08	2.98	2.92	2.84	2.78	2.58

(Continued)

df (denominator)	p	df (numerator) 15	20	25	30	40	50	1000
19	0.05	2.23	2.16	2.11	2.07	2.03	2.00	1.88
	0.01	3.15	3.00	2.91	2.84	2.76	2.71	2.50
20	0.05	2.20	2.12	2.07	2.04	1.99	1.97	1.85
	0.01	3.09	2.94	2.84	2.78	2.69	2.64	2.43
22	0.05	2.15	2.07	2.02	1.98	1.94	1.91	1.79
	0.01	2.98	2.83	2.73	2.67	2.58	2.53	2.32
24	0.05	2.11	2.03	1.97	1.94	1.89	1.86	1.74
	0.01	2.89	2.74	2.64	2.58	2.49	2.44	2.22
26	0.05	2.07	1.99	1.94	1.90	1.85	1.82	1.70
	0.01	2.81	2.66	2.57	2.50	2.42	2.36	2.14
28	0.05	2.04	1.96	1.91	1.87	1.82	1.79	1.66
	0.01	2.75	2.60	2.51	2.44	2.35	2.30	2.08
30	0.05	2.01	1.93	1.88	1.84	1.79	1.76	1.63
	0.01	2.70	2.55	2.45	2.39	2.30	2.25	2.02
35	0.05	1.96	1.88	1.82	1.79	1.74	1.70	1.57
	0.01	2.60	2.44	2.35	2.28	2.19	2.14	1.90
40	0.05	1.92	1.84	1.78	1.74	1.69	1.66	1.52
	0.01	2.52	2.37	2.27	2.20	2.11	2.06	1.82
45	0.05	1.89	1.81	1.75	1.71	1.66	1.63	1.48
	0.01	2.46	2.31	2.21	2.14	2.05	2.00	1.75
50	0.05	1.87	1.78	1.73	1.69	1.63	1.60	1.45
	0.01	2.42	2.27	2.17	2.10	2.01	1.95	1.70
60	0.05	1.84	1.75	1.69	1.65	1.59	1.56	1.40
	0.01	2.35	2.20	2.10	2.03	1.94	1.88	1.62
80	0.05	1.79	1.70	1.64	1.60	1.54	1.51	1.34
	0.01	2.27	2.12	2.01	1.94	1.85	1.79	1.51
100	0.05	1.77	1.68	1.62	1.57	1.52	1.48	1.30
	0.01	2.22	2.07	1.97	1.89	1.80	1.74	1.45
150	0.05	1.73	1.64	1.58	1.54	1.48	1.44	1.24
	0.01	2.16	2.00	1.90	1.83	1.73	1.66	1.35
300	0.05	1.70	1.61	1.54	1.50	1.43	1.39	1.17
	.01	2.10	1.94	1.84	1.76	1.66	1.59	1.25
500	.05	1.69	1.59	1.53	1.48	1.42	1.38	1.14
	.01	2.07	1.92	1.81	1.74	1.63	1.57	1.20
1000	.05	1.68	1.58	1.52	1.47	1.41	1.36	1.11
	.01	2.06	1.90	1.79	1.72	1.61	1.54	1.16

All values computed by the author using SPSS 11.

A.4. CRITICAL VALUES OF THE CHI-SQUARE DISTRIBUTION

	p				*p*	
df	.05	.01		*df*	.05	.01
1	3.84	6.63		25	37.65	44.31
2	5.99	9.21		26	38.89	45.64
3	7.81	11.34		27	40.11	46.96
4	9.49	13.28		28	41.34	48.28
5	11.07	15.09		29	42.56	49.59
6	12.59	16.81		30	43.77	50.89
7	14.07	18.48		35	49.80	57.34
8	15.51	20.09		40	55.76	63.69
9	16.92	21.67		45	61.66	69.96
10	18.31	23.21		50	67.50	76.15
11	19.68	24.72		60	79.08	88.38
12	21.03	26.22		70	90.53	100.43
13	22.36	27.69		80	101.88	112.33
14	23.68	29.14		90	113.15	124.12
15	25.00	30.58		100	124.34	135.81
16	26.30	32.00		200	233.99	249.45
17	27.59	33.41		300	341.40	359.91
18	28.87	34.81		400	447.63	468.72
19	30.14	36.19		500	553.13	576.49
20	31.41	37.57		600	658.09	683.52
21	32.67	38.93		700	762.66	789.97
22	33.92	40.29		800	866.91	895.98
23	35.17	41.64		900	970.90	1001.63
24	36.42	42.98		1000	1074.68	1106.97

A.5. THE WELCH *F*-TEST

See the file **Welch F.pdf** on the CD-ROM.

A.6. CALCULATING SIMPLE EFFECTS

See the file **Calculating Simple Effects.pdf** on the CD-ROM.

A.7. JONCKHEERE'S TREND TEST

See the file **Jonckheere.pdf** on the CD-ROM.

A.8. CHAPTER 14

See the file **Appendix Chapter 14.pdf** on the CD-ROM.

A.9. CALCULATION OF FACTOR SCORE COEFFICIENTS

See the file **Factor Scores.pdf** on the CD-ROM.

REFERENCES

Agresti, A. & Finlay, B. (1986). *Statistical methods for the social sciences* (2nd edition). San Francisco: Dellen.

American Psychological Association (2001). *Publication manual of the American Psychological Association* (5th edition). Washington, DC: APA Books.

Arrindell, W. A. & van der Ende, J. (1985). An empirical test of the utility of the observer-to-variables ratio in factor and components analysis. *Applied Psychological Measurement, 9*, 165–178.

Bargman, R. E. (1970). Interpretation and use of a generalized discriminant function. In R. C. Bose et al. (eds.), *Essays in probability and statistics*. Chapel Hill: University of North Carolina Press.

Barnett, V. & Lewis, T. (1978). *Outliers in statistical data*. New York: Wiley.

Belsey, D. A., Kuh, E. & Welsch, R. (1980). *Regression diagnostics: identifying influential data and sources of collinearity*. New York: Wiley.

Berry, W. D. (1993). *Understanding regression assumptions*. Sage university paper series on quantitative applications in the social sciences, 07–092. Newbury Park, CA: Sage.

Berry, W. D. & Feldman, S. (1985). *Multiple regression in practice*. Sage university paper series on quantitative applications in the social sciences, 07–050. Beverly Hills, CA: Sage.

Bock, R. D. (1975). *Multivariate statistical methods in behavioural research*. New York: McGraw-Hill.

Boik, R. J. (1981). A priori tests in repeated-measures designs: effects of nonsphericity. *Psychometrika, 46*(3), 241–255.

Bowerman, B. L. & O'Connell, R. T. (1990). *Linear statistical models: an applied approach* (2nd edition). Belmont, CA: Duxbury.

Bray, J. H. & Maxwell, S. E. (1985). *Multivariate analysis of variance*. Sage university paper series on quantitative applications in the social sciences, 07–054. Newbury Park, CA: Sage.

Brown, M. B. & Forsythe, A. B. (1974). The small sample behaviour of some statistics which test the equality of several means. *Technometrics, 16*, 129–132.

Brown, W. (1910). Some experimental results in the correlation of mental abilities. *British Journal of Psychology, 3*, 296–322.

Cattell, R. B. (1966a). *The scientific analysis of personality*. Chicago: Aldine.

Cattell, R. B. (1966b). The scree test for the number of factors. *Multivariate Behavioral Research, 1*, 245–276.

Cliff, N. (1987). *Analyzing multivariate data*. New York: Harcourt, Brace Jovanovich.

Cohen, J. (1968). Multiple regression as a general data-analytic system. *Psychological Bulletin, 70*(6), 426–443.

Cohen, J. (1988). *Statistical power analysis for the behavioural sciences* (2nd edition). New York: Academic Press.

Cohen, J. (1990). Things I have learned (so far). *American Psychologist, 45*(12), 1304–1312.

Cohen, J. (1992). A power primer. *Psychological Bulletin, 112*(1), 155–159.

Cohen, J. (1994). The earth is round ($p < .05$). *American Psychologist, 49*(12), 997–1003.

Cole, D. A., Maxwell, S. E., Arvey, R. & Salas, E. (1994). How the power of MANOVA can both increase and decrease as a function of the intercorrelations among the dependent variables. *Psychological Bulletin, 115*(3), 465–474.

Collier, R. O., Baker, F. B., Mandeville, G. K. & Hayes, T. F. (1967). Estimates of test size for several test procedures based on conventional variance ratios in the repeated-measures design. *Psychometrika, 32*(2), 339–352.

Comrey, A. L. & Lee, H. B. (1992). *A first course in factor analysis* (2nd edition). Hillsdale, NJ: Erlbaum.

Cook, R. D. & Weisberg, S. (1982). *Residuals and influence in regression*. New York: Chapman & Hall.

Cooper, C. L., Sloan, S. J. & Williams, S. (1988). *Occupational Stress Indicator Management Guide*. Windsor, Berks: NFER-Nelson.

Cortina, J. M. (1993). What is coefficient alpha? An examination of theory and applications. *Journal of Applied Psychology, 78*, 98–104.

Cox, D. R. & Snell, D. J. (1989). *The analysis of binary data* (2nd edition). London: Chapman & Hall.

Cronbach, L. J. (1951). Coefficient alpha and the internal structure of tests. *Psychometrika*, *16*, 297–334.

Cronbach, L. J. (1957). The two disciplines of scientific psychology. *The American Psychologist*, *12*, 671–684.

Davidson, M. L. (1972). Univariate versus multivariate tests in repeated-measures experiments. *Psychological Bulletin*, *77*, 446–452.

Domjan, M., Blesbois, E. & Williams, J. (1998). The adaptive significance of sexual conditioning: Pavlovian conditioning of sperm release. *Psychological Science*, *9*, 411–415.

Dunteman, G. E. (1989). *Principal components analysis*. Sage university paper series on quantitative applications in the social sciences, 07–069. Newbury Park, CA: Sage.

Durbin, J. & Watson, G. S. (1951). Testing for serial correlation in least squares regression, II. *Biometrika*, *30*, 159–178.

Eriksson, S.-G., Beckham, D. & Vassell, D. (2004). Why are the English so shit at penalties? a review. *Journal of Sporting Ineptitude*, *31*, 231–1072.

Erlebacher, A. (1977). Design and analysis of experiments contrasting the within- and between-subjects manipulations of the independent variable. *Psychological Bulletin*, *84*, 212–219.

Eysenck, H. J. (1953). *The structure of human personality*. New York: Wiley.

Field, A. P. (1998a). A bluffer's guide to sphericity. *Newsletter of the Mathematical, Statistical and Computing Section of the British Psychological Society*, *6*(1), 13–22 (available on the CD-ROM as **sphericity.pdf**).

Field, A. P. (1998b). Review of nQuery Adviser Release 2.0. *British Journal of Mathematical and Statistical Psychology*, *52*(2), 368–369.

Field, A. P. (2000). *Discovering statistics using SPSS for Windows: advanced techniques for the beginner*. London: Sage.

Field, A. P. (2001). Meta-analysis of correlation coefficients: a Monte Carlo comparison of fixed- and random-effects methods. *Psychological Methods*, *6*, 161–180.

Field, A. P. (2003). *Clinical psychology*. Crucial: Exeter. (See http://www.learningmatters.co.uk)

Field, A. P. & Davey, G. C. L. (1999). Reevaluating evaluative conditioning: a nonassociative explanation of conditioning effects in the visual evaluative conditioning paradigm. *Journal of Experimental Psychology: Animal Processes*, *25*(2), 211–224.

Field, A. P. & Hole, G. J. (2003). *How to design and report experiments*. London: Sage.

Fisher, R. A. (1925/1991). *Statistical methods, experimental design, and scientific inference*. Oxford: Oxford University Press. (This reference is for the 1991 reprint.)

Flanagan, J. C. (1937). A proposed procedure for increasing the efficiency of objective tests. *Journal of Educational Psychology*, *28*, 17–21.

Foster, J. J. (2001). *Data analysis using SPSS for Windows: a beginner's guide* (2nd edition). London: Sage.

Friedman, M. (1937). The use of ranks to avoid the assumption of normality implicit in the analysis of variance. *Journal of the American Statistical Association*, *32*, 675–701.

Girden, E. R. (1992). *ANOVA: repeated measures*. Sage university paper series on quantitative applications in the social sciences, 07–084. Newbury Park, CA: Sage.

Glass, G. V., Peckham, P. D. & Sanders, J. R. (1972). Consequences of failure to meet assumptions underlying analysis of variance and covariance. *Educational Research*, *42*, 237–288.

Gould, S. J. (1981). *The Mis measure of Man*. London: Penguin.

Graham, J. M., Guthrie, A. C. & Thompson, B. (2003). Consequences of not interpreting structure coefficients in published CFA research: a reminder. *Structural Equation Modeling*, *10*(1), 142–153.

Grayson, D. (2004). Some myths and legends in quantitative psychology. *Understanding Statistics*, *3*(1), 101–134.

Green, S. B. (1991). How many subjects does it take to do a regression analysis? *Multivariate Behavioural Research*, *26*, 499–510.

Greenhouse, S. W. & Geisser, S. (1959). On methods in the analysis of profile data. *Psychometrika*, *24*, 95–112.

Guadagnoli, E. & Velicer, W. (1988). Relation of sample size to the stability of component patterns. *Psychological Bulletin*, *103*, 265–275.

Hakstian, A. R., Roed, J. C. & Lind, J. C. (1979). Two-sample T^2 procedure and the assumption of homogeneous covariance matrices. *Psychological Bulletin*, *86*, 1255–1263.

Hardy, M. A. (1993). *Regression with dummy variables*. Sage university paper series on quantitative applications in the social sciences, 07–093. Newbury Park, CA: Sage.

Harman, B. H. (1976). *Modern factor analysis* (3rd edition, revised). Chicago: University of Chicago Press.

Harris, R. J. (1975). *A primer of multivariate statistics*. New York: Academic Press.

Hoaglin, D. & Welsch, R. (1978). The hat matrix in regression and ANOVA. *American Statistician*, *32*, 17–22.

Hoddle, G., Batty, D. & Ince, P. (1998). How not to take penalties in important soccer matches. *Journal of Cretinous Behaviour*, *1*, 1–2.

Hosmer, D. W. & Lemeshow, S. (1989). *Applied logistic regression*. New York: Wiley.

Howell, D. C. (1997). *Statistical methods for psychology* (4th edition). Belmont, CA: Duxbury.

Howell, D. C. (2002). *Statistical methods for psychology* (5th edition). Belmont, CA: Duxbury.

Huberty, C. J. & Morris, J. D. (1989). Multivariate analysis versus multiple univariate analysis. *Psychological Bulletin*, *105*, 302–308.

Hutcheson, G. & Sofroniou, N. (1999). *The multivariate social scientist*. London: Sage.

Huynh, H. & Feldt, L. S. (1976). Estimation of the Box correction for degrees of freedom from sample data in randomised block and split-plot designs. *Journal of Educational Statistics*, *1*(1), 69–82.

Jackson, S. & Brashers, D. E. (1994). *Random factors in ANOVA*. Sage university paper series on quantitative applications in the social sciences, 07–098. Thousand Oaks, CA: Sage.

Jolliffe, I. T. (1972). Discarding variables in a principal component analysis, I: artificial data. *Applied Statistics*, *21*, 160–173.

Jolliffe, I. T. (1986). *Principal component analysis*. New York: Springer-Verlag.

Jonckheere, A. R. (1954). A distribution-free *k*-sample test against ordered alternatives. *Biometrika*, *41*, 133–145.

Kaiser, H. F. (1960). The application of electronic computers to factor analysis. *Educational and Psychological Measurement*, *20*, 141–151.

Kaiser, H. F. (1970). A second-generation little jiffy. *Psychometrika*, *35*, 401–415.

Kaiser, H. F. (1974). An index of factorial simplicity. *Psychometrika*, *39*, 31–36.

Kass, R. A. & Tinsley, H. E. A. (1979). Factor analysis. *Journal of Leisure Research*, *11*, 120–138.

Keselman, H. J. & Keselman, J. C. (1988). Repeated measures multiple comparison procedures: effects of violating multisample sphericity in unbalanced designs. *Journal of Educational Statistics*, *13*(3), 215–226.

Kinnear, P. R. & Gray, C. D. (2000). *SPSS for Windows made simple (Release 10)*. Hove: Psychology Press.

Kline, P. (1999). *The handbook of psychological testing* (2nd edition). London: Routledge.

Klockars, A. J. & Sax, G. (1986). *Multiple comparisons*. Sage university paper series on quantitative applications in the social sciences, 07–061. Newbury Park, CA: Sage.

Kruskal, W. H. & Wallis, W. A. (1952). Use of ranks in one-criterion variance analysis. *Journal of the American Statistical Association*, *47*, 583–621.

Loftus, G. R. & Masson, M. E. J. (1994). Using confidence intervals in within-subject designs. *Psychonomic Bulletin and Review*, *1*(4), 476–490.

Lunney, G. H. (1970). Using analysis of variance with a dichotomous dependent variable: an empirical study. *Journal of Educational Measurement*, *7*(4), 263–269.

MacCallum, R. C., Widaman, K. F., Zhang, S. & Hong, S. (1999). Sample size in factor analysis. *Psychological Methods*, *4*(1), 84–99.

Mann, H. B. & Whitney, D. R. (1947). On a test of whether one of two random variables is stochastically larger than the other. *Annals of Mathematical Statistics*, *18*, 50–60.

Maxwell, S. E. (1980). Pairwise multiple comparisons in repeated-measures designs. *Journal of Educational Statistics*, *5*(3), 269–287.

Maxwell, S. E. & Delaney, H. D. (1990). *Designing experiments and analyzing data*. Belmont, CA: Wadsworth.

Menard, S. (1995). *Applied logistic regression analysis*. Sage university paper series on quantitative applications in the social sciences, 07–106. Thousand Oaks, CA: Sage.

Mendoza, J. L., Toothaker, L. E. & Crain, B. R. (1976). Necessary and sufficient conditions for F ratios in the $L \times J \times K$ factorial design with two repeated factors. *Journal of the American Statistical Association*, *71*, 992–993.

Mendoza, J. L., Toothaker, L. E. & Nicewander, W. A. (1974). A Monte Carlo comparison of the univariate and multivariate methods for the groups by trials repeated-measures design. *Multivariate Behavioural Research*, *9*, 165–177.

Miles, J. & Shevlin, M. (2001). *Applying regression and correlation: a guide for students and researchers*. London: Sage.

Mitzel, H. C. & Games, P. A. (1981). Circularity and multiple comparisons in repeated-measures designs. *British Journal of Mathematical and Statistical Psychology, 34*, 253–259.

Myers, R. (1990). *Classical and modern regression with applications* (2nd edition). Boston, MA: Duxbury.

Nagelkerke, N. J. D. (1991). A note on a general definition of the coefficient of determination. *Biometrika, 78*, 691–692.

Namboodiri, K. (1984). *Matrix algebra: an introduction*. Sage university paper series on quantitative applications in the social sciences, 07–38. Beverly Hills, CA: Sage.

Norušis, M. J. (1997). *SPSS® for Windows™ Base system user's guide, release 7.5*. SPSS Inc.

Nunnally, J. C. (1978). *Psychometric theory*. New York: McGraw-Hill.

O'Brien, M. G. & Kaiser, M. K. (1985). MANOVA method for analyzing repeated-measures designs: an extensive primer. *Psychological Bulletin, 97*(2), 316–333.

Olson, C. L. (1974). Comparative robustness of six tests in multivariate analysis of variance. *Journal of the American Statistical Association, 69*, 894–908.

Olson, C. L. (1976). On choosing a test statistic in multivariate analysis of variance. *Psychological Bulletin, 83*, 579–586.

Olson, C. L. (1979). Practical considerations in choosing a MANOVA test statistic: a rejoinder to Stevens. *Psychological Bulletin, 86*, 1350–1352.

Pedhazur, E. & Schmelkin, L. (1991). *Measurement, design and analysis*. Hillsdale, NJ: Erlbaum.

Ramsey, P. H. (1982). Empirical power of procedures for comparing two groups on *p* variables. *Journal of Educational Statistics, 7*, 139–156.

Rosenthal, R. (1991). *Meta-analytic procedures for social research* (revised). Newbury Park, CA: Sage.

Rosenthal, R., Rosnow, R. L. & Rubin, D. B. (2000). *Contrasts and effect sizes in behavioural research: a correlational approach*. Cambridge: Cambridge University Press.

Rosnow, R. L. & Rosenthal, R. (2005). *Beginning behavioural research: a conceptual primer* (5th edition). Englewood Cliffs, NJ: Pearson/Prentice Hall.

Rosnow, R. L., Rosenthal, R. & Rubin, D. B. (2000). Contrasts and correlations in effect-size estimation. *Psychological Science, 11*, 446–453.

Rouanet, H. & Lépine, D. (1970). Comparison between treatments in a repeated-measurement design: ANOVA and multivariate methods. *The British Journal of Mathematical and Statistical Psychology*, *23*, 147–163.

Rowntree, D. (1981). *Statistics without tears: a primer for non-mathematicians*. London: Penguin.

Rulon, P. J. (1939). A simplified procedure for determining the reliability of a test by split-halves. *Harvard Educational Review*, *9*, 99–103.

Scariano, S. M. & Davenport, J. M. (1987). The effects of violations of independence in the one-way ANOVA. *The American Statistician*, *41*(2), 123–129.

Siegel, S. & Castellan, N. J. (1988). *Nonparametric statistics for the behavioral sciences* (2nd edition). New York: McGraw-Hill.

Spearman, C. (1910). Correlation calculated with faulty data. *British Journal of Psychology*, *3*, 271–295.

SPSS Inc. (1997). *SPSS® Base 7.5 syntax reference guide*. SPSS Inc.

Stevens, J. P. (1979). Comment on Olson: choosing a test statistic in multivariate analysis of variance. *Psychological Bulletin*, *86*, 355–360.

Stevens, J. P. (1980). Power of the multivariate analysis of variance tests. *Psychological Bulletin*, *88*, 728–737.

Stevens, J. P. (1992). *Applied multivariate statistics for the social sciences* (2nd edition). Hillsdale, NJ: Erlbaum.

Strang, G. (1980). *Linear algebra and its applications* (2nd edition). New York: Academic Press.

Stuart, E. W., Shimp, T. A. & Engle, R. W. (1987). Classical conditioning of consumer attitudes: four experiments in an advertising context. *Journal of Consumer Research*, *14*, 334–349.

Studenmund, A. H. & Cassidy, H. J. (1987). *Using econometrics: a practical guide*. Boston: Little, Brown.

Tabachnick, B. G. & Fidell, L. S. (2001). *Using multivariate statistics* (4th edition). Boston: Allyn & Bacon.

Terpstra, T. J. (1952). The asymptotic normality and consistency of Kendall's test against trend, when ties are present in one ranking. *Indagationes Mathematicae*, *14*, 327–333.

Tinsley, H. E. A. & Tinsley, D. J. (1987). Uses of factor analysis in counseling psychology research. *Journal of Counseling Psychology*, *34*, 414–424.

Tomarken, A. J. & Serlin, R. C. (1986). Comparison of ANOVA alternatives under variance heterogeneity and specific noncentrality structures. *Psychological Bulletin*, *99*, 90–99.

Toothaker, L. E. (1993). *Multiple comparison procedures*. Sage university paper series on quantitative applications in the social sciences, 07–089. Newbury Park, CA: Sage.

Welch, B. L. (1951). On the comparison of several mean values: an alternative approach. *Biometrika*, *38*, 330–336.

Wilcoxon, F. (1945). Individual comparisons by ranking methods. *Biometrics*, *1*, 80–83.

Wildt, A. R. & Ahtola, O. (1978). *Analysis of covariance*. Sage university paper series on quantitative applications in the social sciences, 07–012. Newbury Park, CA: Sage.

Williams, J. M. G. (2001). *Suicide and attempted suicide*. London: Penguin.

Wright, D. B. (1997). *Understanding statistics: an introduction for the social sciences*. London: Sage.

Wright, D. B. (2002). *First steps in statistics*. London: Sage.

Wright, D. B. (2003). Making friends with your data: improving how statistics are conducted and reported. *British Journal of Educational Psychology*, *73*, 123–136.

Zwick, R. (1985). Nonparametric one-way multivariate analysis of variance: a computational approach based on the Pillai–Bartlett trace. *Psychological Bulletin*, *97*, 148–152.

INDEX

To help you find what you need, page numbers in red indicate key entries (definitions, glossary entries or detailed explanations of the term)

D

S

W

X

Y

Z

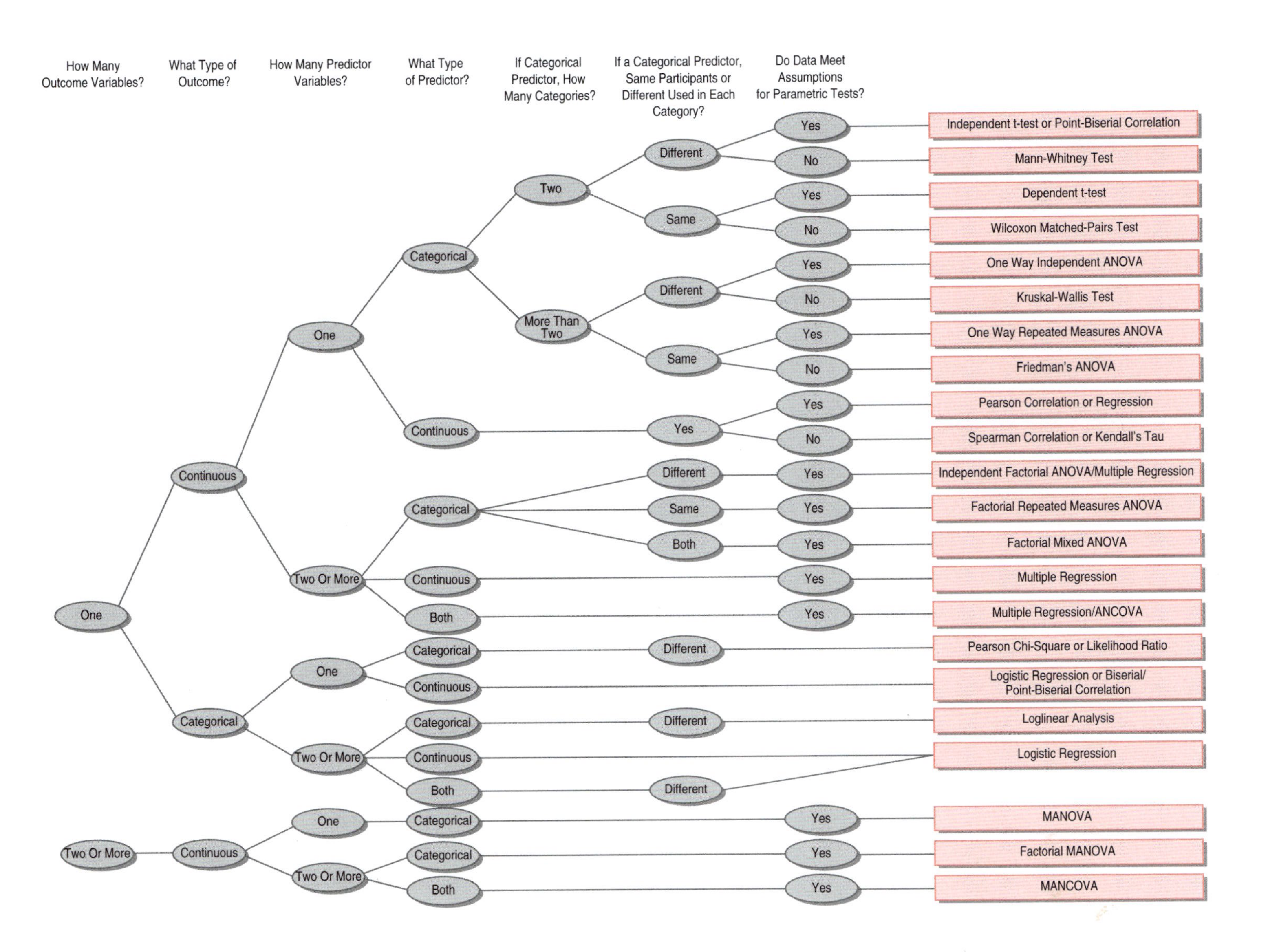
How Many Outcome Variables?
What Type of Outcome?
How Many Predictor Variables?
What Type of Predictor?
If Categorical Predictor, How Many Categories?
If a Categorical Predictor, Same Participants or Different Used in Each Category?
Do Data Meet Assumptions for Parametric Tests?
One
Two Or More
Continuous
Categorical
One
Two Or More
Categorical
Continuous
Both
Two
More Than Two
Different
Same
Both
Yes
No
Independent t-test or Point-Biserial Correlation
Mann-Whitney Test
Dependent t-test
Wilcoxon Matched-Pairs Test
One Way Independent ANOVA
Kruskal-Wallis Test
One Way Repeated Measures ANOVA
Friedman's ANOVA
Pearson Correlation or Regression
Spearman Correlation or Kendall's Tau
Independent Factorial ANOVA/Multiple Regression
Factorial Repeated Measures ANOVA
Factorial Mixed ANOVA
Multiple Regression
Multiple Regression/ANCOVA
Pearson Chi-Square or Likelihood Ratio
Logistic Regression or Biserial/ Point-Biserial Correlation
Loglinear Analysis
Logistic Regression
MANOVA
Factorial MANOVA
MANCOVA

workers for the beet fields and bean fields and strawberry fields and celery fields and tomato fields and orchards. What could be done? Someone in Sacramento had an idea.

I had worked with Braceros much of my life, starting by hoeing beans at nine years old and graduating to weeding and thinning sugar beets by the time I was thirteen and fourteen. My brother and I had connections because our father did odd jobs for quite a few farmers and ranchers. As a favor to our dad, the ranchers would hire us to build a fence or work cattle or inoculate turkeys, and the farmers would hire us in the fields along with the Braceros, knowing we'd work just as hard and just as long. Side by side, with *el cortito,* the short-handled hoe with a grip at twelve or fifteen inches that was later made illegal, I had spent several summers every day bent over the sugar-beet rows next to men and women from Mexico. You arrive at the field early, when everything is still wet from night irrigation and morning fog; you get out of your vehicle and take out the hoe you sharpened at home and look at the first row for the day. The row lines out toward the horizon, curving away to infinity, and you start chopping weeds or thinning beets, row after row after row after row after row as the day warms and grows hot and becomes a mean flame on your curved back. Crop dusters buzz overhead and sometimes you get nailed when they dust, but you laugh and wipe your eyes. Beside you people sing and talk while bent over. You learn good things to say, usually involving the key word *tu* and often *madre. Hijo del puta,* you learn, and *pendejo* and *chingate.* You learn when such words are funny and when they are absolutely not funny. You learn that the name of your town, Atascadero, means "mud hole" in Spanish—not "place of standing waters" as you'd been told—and everyone laughs. More than anything else, you learn to defer all gratification not just to the end of the row or field or day, but to the end of the summer, maybe the end of years. But all the time you're thinking you won't do this forever. You fantasize about the perfect job. Someday someone will pay you as much money as you can possibly imagine, say $3.50 an hour, to backpack in a mountain wilderness with meadows and streams and snowfields, like the pictures you've seen. But the men and women in the rows beside you don't think like that. They think of their families across the border, babies and kids in school, and they imagine working many years in the fields and sending or taking the money back home. One day you forget when *tu* is not funny, and you get in a fight and think you've won, but when you turn your back the other guy swings a ten-pound sugar beet by its stalk and lays you out cold. Many years later it still hurts in frigid weather.

The men in Sacramento couldn't just say they were taking young Black men out of the cities to put them into the fields of the San Joaquín Valley. Black people had an inherited, not exactly pleasant, collective memory of being put to work in American fields, and provoking that particular memory wasn't likely to relax inner-city tensions. The government men decided to call it an economic opportunity work program. Any lucky person with a sufficiently low income, they announced, could qualify to work in the fields for minimum wage. They advertised the program heavily and recruited in Los Angeles, Stockton, Compton, East Palo Alto, Oakland—those places where summer jobs for Black teens had never existed and where young Black males with time on their hands posed potential complications for the coming summer. Somehow we heard about it in Atascadero. It sounded like fun.

It sounded like fun. That's the kind of town we lived in. About twenty of us signed up to have fun in the tomato fields of the San Joaquín Valley during the summer of 1965. We got on a bus and headed for a former Bracero camp in the Valley, stopping en route for soft drinks at a gas station in the tiny town of Soledad, a hundred yards from the maximum security prison there, and then cruising right on through the golden Gabilan Mountains and over to the Central Valley where temperatures doubled minute by minute all day long.

I at least knew what field labor was. My high-school buddies did not. And no one had explained to us that there would be two or three hundred Black guys from the meanest, hardest, nastiest inner cities in California already waiting in the camp, guys who, by the time we arrived, were already intensely unhappy about what the white folks had talked them into once again.

The Bracero camp was on the outskirts of Merced, an agricultural town whose name, which means "mercy," was very misleading. The camp consisted of former military barracks from World War II that had housed thousands of Mexican workers over the years since the war, row upon row of narrow buildings with faint remnants of flaking gray-white paint and roofs of sun-curled asphalt sheeting. Between the barracks was hard-packed dirt, and standing around on the dirt were hundreds of young Black men in knots and clumps and clusters that turned with one motion toward our bus. A new-looking chain-link fence ten feet high surrounded the labor camp, with three rows of barbed wire slanting inward along the top, similar to what we'd seen at Soledad. The gate closed behind us, and those of us not staring at the scene in front of us looked back to see a guard take his position in front of the locked gate.

The bus stopped a few yards from one of the barracks and we filed off the bus and through the open doorway quickly, trying to look but not look like we were looking at those already in camp as we slid into the brown-gray light inside. Metal cots, like in those old military and prison movies and in the bedroom I'd shared with my brothers at home a couple of houses back, lined both sides of the long, narrow room. On the cots, mattresses had been flattened to thin, hard layers by decades of bone-tired, sweating workers whose mass must have increased with exponential despair until some mattresses were mere wafers. A pair of grayish, rust-marked cotton sheets and a bristly green wool army blanket were stacked at the foot of each bed, and gloom rose off the stained mattresses and swirled like tule fog through the bad air. Guys who'd never been quiet in their lives, who'd bullied everybody they'd ever met with words and feet and fists, stood and just looked around, duffle bags dropped to the floor beside them.

It took a few minutes before the first to flop down on the cots began to notice the bedbugs. None of us had ever seen or been intimate with a bedbug before. Cans of insecticide powder were issued that afternoon, and we sprinkled it liberally under and over our mattresses, into the sheets, between sheets and blankets. It took a couple of days for the lice to surface, but within a week we'd cut each other's hair to quarter-inch stubs with a shared pair of scissors someone dug up.

A skinny white man with a tiny mustache, wire-rim sunglasses, a new straw cowboy hat and polished cowboy boots, jeans, and a pearl-snap shirt came in and explained things, standing just inside the open door. We would start work the next day. We would line up for breakfast at 4:30 and be at the fields by 6:30 because the afternoons got hot. Real hot. He luxuriated on the vowels, stretching the words out on his slow tongue the way a snake might uncoil on a warm floor. We would pay a modest fee for room and board, he explained, for transportation to and from the fields, for health and hazard insurance. All would be deducted from our checks in advance to make things easier for everyone. We would remain inside the compound after work. Basketball courts had been constructed for our recreational needs. He showed his teeth. "I bet you all can find a few basketball players among them Black boys," he said, keeping the teeth exposed in a straight, thin line. "I bet them boys'd jest really love to play with you all."

"We'll whip their asses. Shit," Shane Turner started to say, but when everyone just turned and looked at him, his voice kind of died and he scratched a spot on his forearm. Shane Turner was tougher than anybody any of us knew, but even he wasn't that tough.

That's when Doc Dupree stepped out from the back of the group and slapped Shane on the shoulder. "Thas right," Doc said, grinning and looking at the cowboy. "We'll whip those Black boys' asses."

The cowboy took a step backward. Doc was our one Black guy, what we called "colored," from the next town over, Paso Robles. None of us knew Doc very well, but he'd regaled us with stories of his white women all the way over on the bus, and not even Shane had scraped together the balls to question Doc's veracity or the right of a Black boy to white girls. Or maybe Shane just enjoyed the stories too much to stop them. Shane liked to be screwed with and to see others screwed with. He probably liked Doc's stories.

Tall and skinny, with a huge grin and muscles you could see work whenever he lifted an arm to scratch his belly, Doc looked and smelled like trouble. Fun trouble maybe, but some kind for sure. He explained on the bus that he had about sixteen kinds of black belts in sixteen kinds of judo and karate, the sort of stuff we only knew through James Bond movies. Doc sounded like somebody who'd be at home in James Bond's world of Pussy Galore and Odd Job and all.

The cowboy apparently understood Doc the same way we did. His thin smile got thinner. "I bet you will," he nearly whispered. He took another step toward the door. "Supper's at six. Better be quick." He turned and reached the open doorway and looked back over his shoulder. "Or them black boys'll get it all."

When the cowboy was gone, Doc spread his large hands and studied his palms. "Sheeit," he said. "Shee double it." He looked up. "You leave these cowboys and niggers to me."

Even Shane and Ty Cook, the toughest and meanest Atascadero produced—so tough and mean they'd both end up in Soledad a few years later, where one would kill the other in a fight—were clearly willing to let Doc handle cowboys and niggers. The crowds of black faces that had turned toward the bus and the bedbugs that looked big enough to turn mattresses had together clearly shaken everyone's confidence.

No one could eat the food. Foul scraps from bone yards and rotten, discarded fruit from the packing sheds, orange water they called juice, brown water they called coffee, commodity powdered eggs cooked on a twenty-foot grill and chopped into six-inch squares elastic enough to hold aloft by one corner. Lunch was worse. Crowded onto flatbeds, we'd be driven in the dark to fields many miles away. We'd pick green tomatoes in bitter cold until the sun rose and fired flaming howitzers at us. The first day at lunchtime another flatbed truck pulled up with several large, perhaps ten-gallon, thermoses that we were told contained soup. A Black kid from Los Angeles announced that he was hungry enough to eat something none of the rest of us would even say as he vaulted up onto the truck and pulled the lid off one of the thermoses and immediately vomited. The liquid in the container was rotten, whatever it was. Not once did anyone even taste it. Lugs of smashed and rotted peaches and pears sat beside the thermoses. We found out later we were paying a handsome sum for such food. We also learned an important lesson when back in camp we raised the top of the old-fashioned soft-drink dispenser, the kind where you put your money in and slid the bottles along a row until they could be lifted out. When we lifted the cooler lid, we found that the top of each bottle had a neat hole in it precisely the size of a straw. Those running the camp took one look and never refilled the pop dispenser.

They took us out on flatbed trucks before daylight and brought us back when the workday was over, locking the gate behind us. We were told not to leave the camp, so we dug a hole under the back fence and took turns sneaking out in small groups and going into Merced to buy food we could eat and shoot pool until it became obvious to even the slowest of us that we were going collectively broke. We called home from the pay phone outside the camp office, asking for money. We pooled resources and bought cans of Spam, bologna, white bread, mustard, Twinkies. Doc reported that the others in the camp were getting angry, but not at us.

We made peace on the first day when, on the way back from the first dinner we could not eat, we walked into a wall of black faces with expressions intended to cause worry. A large crowd had assembled to ensure that those who were not Black would regret their genetic inheritance. But all was defused when Doc stepped forward and picked up the largest fellow and tossed him around for a while and dropped him. That's how everyone learned Doc really did have sixteen black belts in sixteen kinds of judo or karate. And, we then suspected but later learned for sure, he also had, as he liked to say, truly "laid more pipe than a plumber."

So delighted with Doc's exhibition was everyone but the victim, a kind of truce was formed around the icon of Doc Dupree. After that we played basketball in the evening—black, white, and other—and I met a guy from Compton shorter than me at about five-feet-six who could dunk with both hands. The Black guys graciously took a white guy or two on each team—being part Choctaw and Cherokee did not exempt me from whiteness—the way in grade school each team had to accept a girl or a kid with glasses who couldn't throw a ball.

The men contracted to run the work camp made one effort to ensure our happiness. The first Saturday we were all bussed out to a lake where we swam and lay in the sun and drank watery Kool-Aid. It was a recreation lake, and the Mercedites already there recreating when we arrived vacated the sandy beach posthaste, leaving it to the couple hundred of us. We never went back to the lake, but the next day, Sunday, the gates were opened and we were allowed to walk into town, where some went to a movie, some shot pool, and most just wandered around staring at those staring at us and then headed back to the labor camp. Men driving by in pickups shouted as we walked along the county road, and a beer bottle spun through the air once to shatter against a fence post. The small group of us who were not Black seemed for some reason to infuriate the locals the most. We'd all heard what they shouted before in movies, but we'd never had those things directed at us. Shane challenged several pickups to stop and fight, but no one stopped.

The whole thing lasted less than three weeks. At the end of two weeks we lined up at the camp office to get paid. After expenses deducted for room and board, transportation from home to the camp and from the camp to the fields each day, insurance, and the money I'd spent on food from town, I found I had lost twelve dollars in the transaction. Like nearly everyone else, I was broke. Halfway through the third week, the citizens of Merced—not, I'm certain, the most upstanding citizens of Merced—decided to burn the labor camp. As night fell, a mob approached, howling like mobs in movies we'd all seen, as if we were Frankenstein's monster in the camp castle. From our point of view inside the fence and between barracks, the mob was an indistinct dark mass characterized by obscure wavelike motion and a strange guttural and shrill wail. County sheriff's vehicles with blinding spotlights drove between the mob and the front gate. We all waited between the barracks, prepared for the big fight. Steel legs had been twisted off cots, and an arsenal of previously secret knives appeared, even among our group, shadowy objects in fists. Cars were turned over and burned in the vague distance, and I later heard that in the madness outside a woman had been raped and killed, but the mob was stopped beyond the gate.

We sat up that night waiting for a second assault that didn't come and listening to the crackle of voices over police radios, and in the morning we all emerged slowly from the barracks. No adults remained in the camp, no county sheriff's vehicles were in sight. The office was deserted, the supervisors gone, the front gate open, no one there but us.

People wandered around with questioning eyes, gathered in small groups, discussed possibilities, watched warily the roads and fields outside the fence. Until finally we began to leave. Those who could do so lined up with strange patience and oddly delicate courtesies to call home from the pay phone outside the office, to ask for money or a ride. Those of us who knew such calls were futile began to slowly walk out the front gate, peering into the distance up and down the narrow county road and across fields. A couple hundred of us, mostly Black, began dispersing along the narrow roads in small groups, thin streams of young men walking slowly west and south and north, looking up quickly when a car or truck approached but not bothering to stick out a thumb.

It took two days to hitchhike home, maybe three hundred miles, with two friends from Atascadero. We slept in the grass beneath an oak tree near Creston the first night and then caught a ride the rest of the way the next morning with a man delivering sacks of chicken feed. On hearing our story, he stopped outside of Paso Robles and bought us all breakfast. How long, I wonder today, did it take those Black guys from Watts to get home, and did anyone buy them breakfast? What were they thinking as they walked along those Central Valley farm roads, beside the slow, dangerous-looking canals, between the immense rows of vegetables and stacked irrigation pipes, occasionally sticking a thumb out, most likely without even looking up, knowing no one would stop for them. How long did it take for the streams of young Black men to disperse along that alluvial pattern of roads, and what must the scene have looked like from up above? Finally, I wonder if they went home and tried to burn Los Angeles, or if they just sat back in astonishment at what those white people had done to them yet again and let other, less worldly men and women do the burning. And where are those fellows today, the ones I picked tomatoes and played basketball and watched a mob with? Do they sit in midlife and wonder, as I do, whether it really happened at all? Whether their memories, like mine, are warped and shadowed far beyond reliability. Whether even trying to put such a thing into words is an absurd endeavor, as if such things are best left to turn and drift in inarticulate memory like those river pebbles that get worn more and more smooth over time until there are no edges.

CHAPTER THREE

Mushroom Nights

If Tommy Outhouse hadn't worked at the mushroom farm, things probably would have turned out differently for him. But you never know. It wasn't the name, since Tommy had pretty well come to terms with his last name long before he arrived at the mushroom farm. Son of an Okinawan mother and a GI father who had somehow emerged from the American melting pot named for a toilet, Tommy had come to the U.S. as a skinny little boy whose slanty eyes and hilarious name had instantly made life pure hell. Seeing his son come home from school bloody every day, Tommy's father enrolled him in martial arts classes at a young age and, as soon as he could, Tommy took up weight-lifting and nutrition. By age seventeen, when he transferred into our school for his senior year, he looked like a very handsome Asian Mister World, with muscles on top of muscles, and he had several advanced belts in ways of fighting that were mysterious and awe-inspiring to the rest of us. Tommy was invincible, and aside from the time three drunks tried to mug him in an alley in Santa Barbara and paid dearly for their foolishness, no one was ever self-destructive enough to propose violence to Tommy. Because Tommy was himself nonviolent, was in fact the nicest human being a person could ever meet, he led a peaceful existence—that is, except for inner demons you could see working at him every minute of the day, things that would flare just behind his brown eyes and stretch his constant smile to what looked like a fracture line. We worried about Tommy, some of us, because as self-absorbed as we were in our own teenage torments, we knew that Tommy suffered more and that Tommy was unlucky. Take the case of the three men who attacked him in an alley. All three ended up in the hospital, and Tommy was charged with assault. Three drunks from a bar try to rob a sixteen-year-old boy, and he ends up with a criminal record for assault because of his martial-arts training. Growing up as we did, we knew that some people were just born unlucky and that Tommy was one of those, more than most.

Hobie Joe owned the mushroom farm, growing mainly fancy creams, though once in a great while he would get a wild hair and briefly experiment with another, darker kind of mushroom. Mushrooms are called "hobies" in the business. Art the Wop was the foreman, and Gordon, who suffered from encephalitis, or sleeping sickness, was the only full-time employee. The regular part-time labor force consisted of myself and Tommy; my childhood friend and fishing companion, Glen, who would grow up to be a staff writer for the *San Francisco Chronicle;* Jim, who would later join the radical Weathermen and whose parents bought a cow's head from the slaughter house each month and ate the whole thing; and Joe's worthless son Pat, who in the year 2000 would turn up on national television as the scientist flunky for right-wingers denying the reality of global warming. It surprised none of us when it turned out that Pat, the boss's kid, had betrayed Tommy.

We weren't exactly a close-knit group at the mushroom farm, but we shared a portion of camaraderie, doing killingly hard work together with the joy of sixteen- and seventeen-year-olds in the prime of physical health and confident that greater adventures were on the horizon. Except, that is, for Pat, who did not like manual labor and preferred to sit and smoke and make caustic remarks when his father wasn't around. Pat's life was planned for him, and, seeing his life stretched out before him, he was dreadfully bored. Unlike the rest of us, who were not thought to be college material, he knew he would be successful. His older brother was a gay graduate student at Berkeley, author of a brilliant parody of *The Graduate* titled "The Grad You Ate." Our older brothers and friends were already part of a war in a place called Vietnam, a war that had not become ugly yet for most of us, and we looked forward to our turns at such excitement as soon as we graduated.

Spring in the Salinas Valley was warming up with orange poppies, blue lupines, red and white paintbrush, and a vast sea of wild oats on the hillsides. Flocks of blackbirds rose from alfalfa fields to darken the sky and settle again, and mourning doves cooed in the shade all day long. Nearly every morning on my way to school, if I looked just right, I could see a bobcat sitting well back in a clump of poison-oak brush beside the road, waiting for rabbits too slow to evade a speeding car or truck. A dazed black bear had wandered down from Pine Mountain earlier in the spring, when there weren't supposed to be any bears left, and the fact that he'd been promptly shot and killed by Sage Barrow, our part-time deputy sheriff and part-time high-school janitor, had not lessened our amazement at his brief existence. The days were getting long as the solstice drew near, and the nights were already hot and crackling with the lyrics of Bob Dylan and Wolfman Jack's raspy howl from Bakersfield. When we came out of the mushroom houses in the evening, we were greeted by bats swooping and dodging all about us, and as we climbed into our souped-up cars and pickups we felt exhausted and happy. It was the best time of year, the best time of life.

The three mushroom houses, perhaps a hundred feet long and twenty-five wide, were split-level, thick-walled concrete caves half-buried in a southwestern slope of Pine Mountain on the side opposite the town graveyard and just a couple of miles downriver from the state hospital for the criminally insane. Dark and cold, the houses remained the uniform sixty-two degrees loved by mushrooms. Blazing sunlight slanted down over the green-black Santa Lucia Mountains in the west, over the narrowing southern end of the golden Salinas Valley where our town lay, over the bright sand of the wide river bed, and over the great white oaks between Joe's house and the mushroom houses to flood the side of Pine Mountain, flaming on the wild oats and erasing the Ponderosa pine shadows higher up the slope, but the sun never penetrated Joe's mushroom caves.

Hobie Joe was a retired chemistry professor from the University of Chicago. One day while traversing the shabby neighborhoods surrounding the university Joe had the idea of quitting his job and raising mushrooms, an idea he acted upon at once. He knew and loved mushrooms, and from the day he moved to California and bought his small farm, mushrooms were his life. "Goddamn, let's get a viggle on!" was his unvarying exhortation to his foreman and minuscule labor force. Joe's transition from professor to mushroom farmer made sense to Art, who used to announce with regularity that students were just like mushrooms: "Keep 'em in the dark and feed 'em shit."

Behind the houses, up against the cut-back hillside, Art kept the mushroom compost cooking in enormous piles thirty feet long and ten feet high. Consisting of racehorse manure trucked from the Santa Anita track down south mixed with straw, the cooking compost gave off a rich, steaming air that permeated everything. The idea and reality of compost lay heavy over our lives. Mushrooms grow in compost. Joe's houses were two-story affairs, with two wooden mushroom beds on each level in two long rows, an alley on either side and down the middle between the beds. The upper floor consisted of two-by-ten running boards laid down over two-by-four cross-pieces. To fill the upper mushroom beds with compost, we ran wheelbarrows heaped with the burning-hot, steaming stuff into doors that opened onto the upper level in the back of each house, balancing along the running boards, trying desperately not to steer the too-heavy loads off the boards and onto the cement floor ten feet below. Each run with a compost-filled wheelbarrow took every bit of teenage strength we had. Each day left our bodies wrung, our clothes and hair and pores dense with the rich feel and smell of horse shit and fermented straw.

We pitchforked the compost into the beds, leveled it, broadcast mushroom spawn over it, and then covered that with an inch of sterile dirt. Almost at once, it seemed, mushrooms emerged, and when they were firm, creamy, unopened buttons, those of us with the proper skill harvested them. Harvesting took place whenever a crop came on, whether two o'clock in the afternoon or two o'clock in the morning. To harvest, you worked the beds in the dark, carrying a small trouble light with a bare bulb that you hung over the bed where you were working and snipping the mushroom stalks with a little curved knife. The rest of the mushroom house was an unlit, drenched blackness in which, when you were picking alone in the predawn, sounds whispered and rustled. In those solitary hours, of which I spent many, the mushrooms would sometimes begin to dance, to tiptoe back and forth in mushroom chorus lines at the edges of the light and scurry about in the dark. I was one of the only part-timers allowed to harvest alone, a distinction that weighed heavily upon my young shoulders.

When a house was harvested and the mushrooms trucked off to markets, we pitchforked all the dirt and compost back into wheelbarrows and wheeled it out of the houses. We scraped the boards of each bed and cleaned and hosed each house spotless, and then Art steamed the house at frantically high temperatures to kill everything so that no mycogyny would corrupt the next crop. When the house was sterile, we brought new compost in and began the cycle again.

Art was a tough, brash, loud, friendly, trouble-seeking, short, dark, big-bellied guy in his mid-twenties from Bakersfield, proudly of Italian extraction. At frequent intervals he would pull up his filthy white T-shirt to show us a knife scar that angled up the width of his prominent hirsute belly. The other guy, he'd swear, ended up at the bottom of a Central Valley irrigation canal. He'd laugh, his black eyes twinkling with the memory. "You pussies think you're tough?" he'd shout when Joe wasn't around, looking from one to the other of us, though never at Tommy. For Art it was as if Tommy didn't really exist. Only Tommy was exempt from the teasing and endless banter. When someone else spilled a wheelbarrow, Art's outrage knew no bounds and we'd hear about it all day. When Tommy lost control of a load, which often happened, Art looked the other way. Our families and ancestry were fair game for Art's obscene critiques, but not Tommy's. It was as if the foreman and Tommy walked through different dimensions of mushroom-farm time and space.

Gordon was a soundless giant in his forties, six-feet-four and broad shouldered, who lived in a separate, slow world from the rest of us. He'd spent time in prison for accidentally killing his brother when he was much younger and had in some mysterious way contracted sleeping sickness, which Art preferred to call "brain fever." Immune to Art's sarcasm and most other stimuli from the outside world, Gordon walked and worked with a focus far away from us, speaking only once in a great while to utter a sudden, shocking obscenity apropos of nothing and then lapse back into himself. If Gordon forgot to take his pills, a frequent occurrence, he would fall asleep in unanticipated places and at odd times. Often, I would arrive at the farm very early to find Gordon asleep in a rickety chair in the tiny office–coffee room, his forehead resting on a coffee mug so that when I leaned and shouted in his ear, "Gordon, wake up and take your pills!" he would jerk upward with a grunt and a deeply indented circle in the middle of his forehead. One day Gordon fell asleep leaning unnoticed in the dark against a mushroom bed and was steamed by Art along with the rest of the house, coming out hours later a deep red but otherwise seemingly unaffected.

Art liked to keep things stirred up if he could. After I got into a fight at a high-school dance, the last real fistfight of my life, Art went out of his way to seek out and hire Larry, my antagonist in the brawl. Catching me alone, Art would whisper, "Larry says you fight like a pussy." Then he'd say the same to Larry. To Art's infinite disgust, he found that Larry and I had already formed one of those manly teenage friendships that grow out of bloodying each other's noses. The fact that we'd both been dragged from the gym and pounded by deputy sheriffs' nightsticks right after pounding each other cemented the new bond. When Art saw his plan fail, he fired Larry since he hadn't needed an additional worker anyway.

Tommy and I were friends to a degree. "Bananas," he would solemnly advise me as he studied my skinny frame. Bananas, he promised, would provide bulk. Under his tutelage, I ate bananas and drank concoctions of raw eggs and protein powder and milk—laying a cholesterol foundation to last a lifetime—and lifted weights obsessively at school. But despite raw eggs and bananas and weights, I couldn't convince my muscles to look like Tommy's, and despite our tenuous friendship, there was a distance between him and the rest of the world that, like everyone else, I couldn't bridge. Tommy didn't trust the big world and he didn't fit into the small world of our town. The harder he tried the more obvious it was. Some of us liked him and worried about him and watched over him as much as we were capable of, shackled by our innate self-interest, but he was a conspicuous fish in the wrong pond. At work, able to lift or carry twice as much as any of us, he would stumble and trip, drop his pitchfork to the floor below, spill his wheelbarrow full of compost, and in the end get less accomplished than anyone except Pat, the boss's son. School was worse. Atascadero was a white and brown town. Hispanic families had been there first, generations before when the whole valley had been stolen from the Chumash Indians and given out as land grants by Spanish rulers. Those families remained, though landless and impoverished now, their descendants blending more or less harmoniously with the whites who had come later. Darker-skinned men and women from Mexico worked in the fields and lived like familiar passing shadows on the verges of our community. But Tommy was neither white nor brown, and we, his peers, had grown up throwing mock bombs while shouting "Bombs over Tokyo!" on the playground. The town had no context for Tommy. And there was an edge to him that people couldn't get around, a tautness behind his constant smile and friendly eyes that made you think of the way a golf ball snaps apart when you cut to its tightly wound center. In a school with a graduating class of seventy, our coaches were desperate for anyone who could bounce or carry or throw a ball or pin an opponent. Things were so desperate that the ninety-five-pound state wrestling champ was pressed into service as football quarterback and promptly got smashed by a two-hundred-pound linebacker from Paso Robles. Basketball players were expected to also be wide receivers and third basemen. But the coaches ignored Tommy, the finest physical specimen to ever grace our school. He didn't volunteer and he wasn't invited. He went to classes and to work at the mushroom farm and home, friendly to everyone but friends with almost none. At lunch he sat with one or two of us from the mushroom farm and said nothing. He was never taunted or teased.

Once Joe pressed Tommy into service picking mushrooms. No one else was available and the job was too big for just me; a whole house had come on at once and the hobies had to be harvested and shipped between evening and morning. It was to be one of those all-night jobs and Joe was desperate.

Harvesting mushrooms is a delicate task. Day or night, you work in the dark, carrying your trouble light with you to hang over the immediate spot where you're picking so that it illuminates a few round feet of shadowy forms. Beyond the small circle of light everything is black. Within that circle you must pick only those mushrooms that are ready: full capped but not yet open, exactly as they're found in the market. You must not touch the cap of the mushroom but rather slide your small, curved knife to the base of the stem and snip it off, lifting the mushroom from below in a single motion and placing it gently in the same little box in which it will travel to the store. If you touch the mushroom cap you bruise it and it's worthless. You must do all this at a frenetic pace.

Tommy and I worked opposite each other on different sides of the same mushroom beds, the circles of our lights just intersecting. In that surreal light, Tommy's face took on angular shadows and sharp planes that today make me think of an actor in a Noh play, a comparison I did not have back then. Back then his face was just strange and a bit frightening each time I glanced up, and mine must have been the same to him. We worked for a long time in silence. From his fumbling, I knew that many of Tommy's mushrooms would be bruised and Joe would be furious. Realizing that Tommy was rushing to keep up with me, I slowed my pace, knowing Joe would be livid about that as well.

Around one in the morning I paused and looked up to find Tommy staring at me across the silhouetted tops of mushrooms. "You're different," he said.

I didn't reply, waiting for him to explain himself. Difference was the ultimate fear of my teen years, of everyone's as far as I could tell. None of us wanted to be different. Difference marked you as a target, neutralized defenses.

"Your eyes," he said finally. "They're like mine."

It took a moment for me to realize he meant that my eyes were almond shaped like his, not as extreme perhaps but definitely different. We had known each other for seven or eight months, and he had never mentioned it before. I'd been harassed throughout grade school with chants of "Ching chong, Chinaman," but my schoolmates had long since stopped noticing or at least commenting.

"How come?" he asked.

I hesitated and then explained that I was part American Indian, that my eyes came from my father's Choctaw world. Choctaw or Cherokee meant nothing to Tommy, but "Indian" registered. Tommy hadn't known that about me because, like almost everyone else, he had never been to my house to meet my family, and since no one talked to him at school, he hadn't heard it through the grapevine. We were a rather notoriously poor family of nine kids living on the very edge of the town, next to the river, and of my friends only Glen, who was also part Indian, had ever come to my house. I had suffered for years from a tormented consciousness of my family's "difference" from everyone else in the town, and had worked hard to erase that difference in the world I lived in each morning when I left our squalid shack of a home. I did not want to hear my difference articulated by the shadowy mask of my friend. Except for the muscles, I realize now, I did not want to be like him.

"But you're okay," he said. And in his statement I heard more question and wonder than confirmation. He wasn't telling me I was okay; he was asking how I managed to fit in, to live with it, to fend off the kinds of razor-edged doubts that worked beneath his skin.

I had no response. I didn't feel "okay," but I also knew that, for reasons I couldn't fathom, Tommy lived in a world infinitely removed from whatever that word implied. In not being okay, I was a rank amateur compared to Tommy, and as I saw him lower his gaze and reach awkwardly for another mushroom I felt saved in some strange kind of way. Somehow my being saved seemed connected to Tommy's not being.

The last thing he said that night was, "I've screwed up a lot, but I'm not going to screw up any more." I stopped waiting for him after that and was soon cutting mushrooms several yards away, seeing his fumbling light in the distance.

When one day at the mushroom farm Tommy announced he had a date for the Girl's League Formal, we were ecstatic for him. His date was a year younger and very pretty. I never knew if Tommy had been too shy to ask a girl out or if he'd tried without success, but as far as any of us knew he had not even talked to a girl in our school since transferring in. We thought this was a breakthrough. For the Girl's League Formal, girls asked boys. I was going with Becky Hogue, a cheerleader who had been foolish enough to ask me, and my thoughts were on the old, yellowing dinner jacket my brother had left behind. We were all happy for Tommy on the day of the formal, when he came by to show off his rented tuxedo. Miles behind Tommy in the ways of the world, I was planning on knocking off work in time to shower and put on my brother's too-big jacket and slacks before picking up poor Becky, and had just come out of House Number Two when Tommy drove up in the mushroom farm's van. Because he didn't have a car, he had arranged with Pat to borrow the Volkswagen van in which Hobie Joe delivered mushrooms. With several hours to go before the formal, Tommy was obviously enjoying his fine clothes and wheels. Joe was out of town for a couple of days, and what neither Tommy nor the rest of us knew was that Pat had not told his father about the loan of the van.

None of us called our boss Hobie Joe to his face. Art, who roared with delight at being called Art the Wop, had originated the owner's name, but even he was careful to keep it *sotto voce.* None of us would have wanted to hurt Joe's feelings, because despite his bitterly low wages and constant exhortations to work harder and get a wiggle on, he was a wonderful boss. On hot days he would lug out ice-cold six-packs of Brew 102, the cheapest beer on the market but a gift that made us his slaves since none of us was old enough to drink legally. And one day he took me aside and offered to pay for my entire college education if I would apply to a major university, an offer that left me fully speechless and stuns me even to this day—one I didn't know how to accept at the time since college wasn't in my plans. But like Art, Joe treated Tommy differently, becoming cautious around him in a way everyone could see, as if Tommy might do something unpredictable and therefore dangerous. It was clear that Joe didn't trust Tommy, and perhaps Pat was right in not telling his father about the loan of the van. Maybe Pat deserved credit we never gave him for an act of generosity none of us would have expected or perhaps matched, even a bit of courage in deceiving his father for Tommy's sake. Tommy had not approached any of the rest of us for the loan of a car, probably because he suspected he'd be disappointed. I know that, despite how much I liked Tommy, I would not have handed over to him the '57 Chevy my brother had left in my care even if I had not planned to use it myself that night. Today, many years later, the thought of Tommy in that too-powerful car is frightening.

When he came by on the day of the formal, Tommy was more excited and animated than I had ever seen him, and he was careful to keep his distance from our compost aromas while he laughed at our banter. The last vision any of us had of Tommy was of him in his tuxedo laughing and waving through the Volkswagen window as he drove away from the great oak trees surrounding the farm, heading in the late afternoon sunlight to pick up his date for dinner. Whereas I was taking Becky to a local chicken restaurant (where I would not have enough money to pay the bill and we would flee in ignominy), Tommy was taking his date to Morro Bay, a dashing seafood town twenty miles away over a pass in the coast range.

What happened next is just a story, third-hand and therefore even more unreliable than memory. They say Tommy had bolstered his courage with Southern Comfort before picking up his date, though none of us suspected that when he came by the mushroom farm. On the twisting highway over the coast range, he lost control of the van and rolled it onto its side. With the girl unhurt but screaming hysterically, Tommy clambered out a window and lifted the smashed vehicle back onto its wheels. Leaving the terrified girl in the van, he hitchhiked to Morro Bay and called Pat. Fearing his father's wrath, Pat phoned the police to report the van stolen.

A Highway Patrol car had already found the damaged van with the sobbing girl and was taking her home when the report went out. Upon coming home, Joe was properly outraged and justified in all his previous suspicions of Tommy. Pat, to my knowledge, never confessed and left the next year for the University of Chicago. Becky went to Brigham Young University, married a Mormon, and had many children, while I went to work in a can factory in Hayward. We never saw or heard of Tommy Outhouse again.

CHAPTER FOUR

The Hunter's Dance

I was halfway across the Black River on a perfect, early summer day, thigh deep in the easy current and letting my vision drift thoughtlessly up the grass of the opposite bank. Ponderosa pines laid dark, wavering lines on the canyon slope, and in the bright noon sunlight between two oblong shadows my eyes stopped on a doe and a coyote fighting for the life of a new-born fawn. With the breeze in my direction and the utter focus of this life-battle, neither deer nor coyote noticed me a couple hundred feet away in the water.

Upright on precarious hind legs, the whitetail doe sparred with the coyote, jabbing spindly forelegs at his retreating face. Behind the deer, a miniscule spotted fawn lay tightly against a big log, pushing itself as far into the crease between log and earth as it could get, doing precisely what its mother had ordered. The coyote feinted and dodged and retreated as the doe advanced in her awkward, fencing stance. And then, when he had lured her sufficiently far from the fawn, he suddenly raced around her and slashed at the baby, ripping a gash along the tiny neck and shoulder before the mother could wheel and drive him away. Once more he teased the doe until she again was too far from her newborn, whereupon he dove and slashed and retreated.

I knew I should do nothing, told myself sternly to just turn away. A voyeur, I should not even be watching this casual death, should not transform the uncomplicated reality of this thing into a dreadful aesthetic. This was a reality beyond anything I could judge or affect. For more time than I could comprehend, coyotes had taken new fawns from their mothers. The winter and spring had been particularly wet, and forage was rich and thick. Nature had responded to plenty with plenty, and rabbits were everywhere, coveys of quail exploding from every chokecherry and wild-grape thicket, wild turkeys flocking with crowds of scurrying chicks. All over the White Mountain Apache reservation on this trip I'd seen does with late-born fawns, sometimes twins. The abundant coyotes, like the large, dark male on the hillside above, were fat and sleek. The vignette I was witnessing had taken place uncountable times and would, God willing, take place uncountable more times long after I was compost.

Maybe it was my anthropocentric conviction that the coyote was grinning, enjoying the game it surely could not lose as it teased the doe away from the fawn. Maybe it was the fact of the mother's desperate courage in facing the formidable coyote, that complication by a mother's instinct of the clean line between predator and prey. Or perhaps it was just an inexcusable human weakness, blindness, or selfishness that converted everything into personal significance. For whatever reason, I banged my aluminum fly-rod case loudly against a rock in the middle of the stream, leaving a sharp dent in the tube and a crass metallic noise echoing in the otherwise silent canyon.

Coyote and deer paused and turned toward me and then in an instant both were gone, vanished over opposite sides of the low ridge. As quickly as I could with a heavy backpack, I waded out of the river and climbed up to the log. Slashed deeply in several places, the fawn, perhaps hours old, was in shock and doomed there in the hot sun. Already, flies buzzed around the hardening blood that covered its spotted sides. Already, the eyes were glazed.

I walked away, feeling utterly stupid and impotent. The doe, I knew, would never return, traumatized as she was not only by her battle but perhaps even more by my sudden, awful presence. The coyote, who like all coyotes considered man merely a temporary inconvenience, would return and enjoy the meal he'd earned. I would hike out to my truck and drive to the little town of Alpine, an elk-hunting town two hours away, for the burger I always bought in the same café on my way back to Albuquerque after each trip down to the river. The doe, with such fawn loss undoubtedly programmed into her genetic code, would very likely bear again and try again to preserve the small life given to her. More fawns would fall prey to the same coyote, who looked to be in his prime.

The day before, my fourth day on the small river, I had fished upstream from my camp, walking, wading, and casting, all thought and energy focused on reading water and catching fish, so that when I finally noticed the sun declining above the deep canyon I was three or four miles from my tent. I turned and began to retrace my steps, casting indifferently here and there just to see my brown and gray fly touch perfect pools and eddies. I'd already kept two rainbow trout for dinner and released many others, and I had no great interest in catching more fish.

I'd walked maybe a quarter of a mile when I saw the first lion tracks. Nearly as large as my outstretched hand, they were superimposed over the boot tracks I'd left as I meandered upstream, their softly rounded and surprisingly deep imprints looking precise and quite deliberate. As I tracked myself back toward camp, it became clear that the cougar had tracked me for over two miles, stepping where I'd stepped. Since I'd moved very slowly for those miles, stopping and casting and catching and releasing trout, studying currents and shadowy riffles, the cat must have moved just as slowly, must have stalked and toyed with me from a discreet distance, pausing where I paused, watching me with the same intensity I had watched the dark water, perhaps cocking an ear toward the splashing fish I magically drew from the stream. All cats are curious, including mountain lions, and the cougar might have just found me an interesting anomaly, an entertaining spectacle worth wasting a couple of hours on. It might have been a fairly young cat who had not seen a man before and was intrigued.

But most likely, given the realpolitik of nature, the cougar was contemplating me as dinner. Much evidence has taught us that the great cats have little difficulty placing humans within their spectrum of viable meals. Mountain lions, a naturalist friend had told me, have evolved canine teeth designed to fit precisely into the interstitial spaces between the vertebrae of deer just over the medulla oblongata, bringing a quick and efficient death. Would the same fit work the same inevitable magic upon my neck, I wondered? I felt strangely flattered to imagine myself thus accepted into a cougar's world, to be granted such a place of respect along with the deer and elk. But then, had the lion elected to sample my flavor I undoubtedly would have felt less flattered, and perhaps, I had to admit, I was still walking toward my camp precisely because I was not sufficiently part of that world. Perhaps I had been spared because unlike those other creatures I had lost my rightful place. Or, finally, maybe it was nothing more complicated or subtle than the nine-foot graphite fly rod that saved me, a strange, waving antenna the cat was not predisposed to mess with.

Two hours after I left the dying fawn, I was driving back toward New Mexico, up the spine of the White Mountains, pondering what had taken place. What kind of god, I wondered, would decree that coyotes should live by the horrific death of a creature just hours old, by the pain of a mother incapable of protecting her young? What kind of moral universe decreed that countless millions of animals must experience terror and shrieking pain in the beaks and jaws of other animals, every day, every hour, everywhere, until the natural state of the earth, it seemed, must be a single unending shriek of pain? The same god that caused the cougar, that magnificent creature of sinewy tendon and killing instincts, to contemplate me as sustenance, potential energy to drive those killing muscles. There's nothing like a sudden awakening to the fact that you are part of the food chain to cast a new light on things. And not a few human beings have had the privilege of actually being eaten by mountain lions.

In long retrospect, I ask other questions. What kind of god would decree that on my way back from the river I should stop in Alpine for a hamburger made of the ground flesh of a cow raised for that purpose, hormone fed and fattened in muddy feedlots, crowded into stockyards with eyes gouged out by jostling horns, and finally lined into slaughterhouse factories to be killed with the sounds and smells of myriad other deaths in its ears and nostrils? For five days I had camped on the river, catching trout every day and releasing most, but each day selecting two or three fish to kill and cook for dinner. What god gave those trout lives into my hands? What kind of angry god has put so many lives at the disposal of others, the multitudinous rabbits mere grain for the predators? Is there a code to govern such infinite and necessary deaths, values by which we live and make die?

It is in the end the awful beauty of the dance of deer and coyote that I remember best from that summer day. The magnificent, futile courage of the doe, precarious on hind feet and jabbing hooves, and the luring, masterful control of the coyote. A pas de deux of death on one hand and a mere coyote lunch on the other. A pain so great it seems even today to rend the air I breathe, but whose pain? A fearful, perfect symmetry. My own minuet with mountain lion strikes me as a passing grace, a brief moment of near acceptance, a question unanswered.

I grew up hunting. We were a poor family of nine children, and the little income our father could make as a ranch hand or truck driver did not go far. The year I entered the University of California, 1968, his total income was three thousand dollars. Before I was old enough to carry a rifle, he made his living and ours hunting in the forests and swamps along the Yazoo River in Mississippi the way his Choctaw ancestors always had. We children would wake up each morning to coon skins tacked to a shed wall to dry, and there would be alligator sandwiches about which no one spoke. Later, in California at eight years old, I killed my first bird with a slingshot, watching as a small, round stone crushed the back of a thrush's head and the bird fell with a soft, dead weight. I had no conscious purpose in killing the thrush and could never, to save my childish soul, have explained why I did so, tossing the bird carelessly into the brush afterwards. Hindsight today tells me that my purpose that day was the same purpose that drives a kitten to stalk and tumble a feather or strand of yarn. Like the kitten, I was practicing to be a serious killer when my time came. I had selected my ammunition carefully and stalked the thrush with all the skill I could muster, making no sound, blending with the surrounding manzanita and scrub oak until I was close. Not long after the thrush, I owned a BB gun and brought death to sparrows, linnets, bluejays, quail, and doves and a sad number of other birds and lizards. We ate the quail and doves, and sometimes the other birds as well, but most of my killing was a terrible, harsh wastefulness for which I am ashamed these years later, practice killing that would enrage me were I to witness it today. Then my brother, two years older, and I both acquired .22 rifles, and we hunted rabbits, putting to use the skills we had honed on small birds. We were now real hunters, and every day during the proper season we brought home rabbits, often five each day—shot through the head, gutted and skinned where they died—and our family ate them, often having no other food in the house. After school and on weekends and during summers, we worked the tomato and beet fields and bucked hay and built fences on ranches and with our earnings bought shotguns and deer rifles. Then we hunted quail and doves the right way, and bandtail pigeons, and deer, bringing everything home to be eaten. A handful of quail was a snack, a harvest of pigeons a family dinner; a deer, of which there were many, lasted longer even in our hungry house. We stalked and devoured the world around us, hunting and killing and eating, wasting nothing during those more mature years. Not once did I feel the killing I did was wrong; it was as thoughtless and natural as plucking an apple from a tree. Furthermore, my brother and I had crucial roles within the family. We provided sustenance from hillside and brushy river bottom.

For a while as a teen I hunted with a bow, a fifty-five-pound Bear recurve that constituted my only gesture toward sport hunting. I stopped, however, when one day I pinned a rabbit to a tree and watched the little thing spin in crying circles around the arrow shaft. Rabbits do cry in pain, and deer do scream. If you've heard either, you do not forget, and you never want to hear such a thing again. To shoot and eat a rabbit was proper and necessary. But to inflict unnecessary pain and the horror that came with it was wrong and immoral.

I have not hunted, have not shot an animal, since I left home for college and, while I was away, my mother sold all my guns. It wasn't malicious or even thoughtless; the family, with six children still at home, simply needed money in a hurry as was often the case. But my mother's act changed my life, no doubt. Perhaps I would have stopped hunting anyway, given the different world I was entering. In college I knew no one else who hunted, knew no one who did not think hunting was a barbaric act, no one who could understand how a family might not be able to live without killing and eating the wild animals around them. This was the late '60s, early '70s, and we were all supposed to love life in a flowery sort of way, to enjoy bean curd and sprouts, to abjure killing and embrace free love, an exchange I was quite willing to make. A small-town kid cast into an alien world, I would not have tried to impress my values upon my new friends even if I had thought success possible. Who was I to defend a way of life I had never questioned or even pondered? Very likely I would have drifted away from hunting anyway, but without guns that possible drift became a definite destination.

After college I became a wilderness ranger for the U.S. Forest Service in the North Cascades, where, during the fall, I patrolled the high hunt, that anomalous season in which men climbed on horseback into the high country, above timberline, to kill deer and bear and mountain goats. During the high hunt I saw the dark underbelly of killing for sport, killing done by men for no other purpose than the killing itself. I was awakened in my tent one morning by a too-close rifle shot and crawled outside to find a gut-shot bear cub squalling heart-breakingly a few yards beyond my camp, a Seattle hunter nearby with his expensive scoped rifle. The bear cub was not legal game, but the cub had to be dispatched. The man had wanted a bearskin, and the little cub had looked immense to his urban eyes, the same eyes that had failed to notice my blue tent near his target. In Seattle the man dined on much better fare than bear meat, and the fine he paid for his error did not affect his life-style. On another day I found two hunters at the end of a trail, preparing to climb into their truck with the head, horns, and cape of a mountain goat. The rest of the goat had been left behind, in violation of the law, because the hike out had been difficult and they had no interest in the meat. They wanted the trophy and could easily purchase much better cuisine than stringy goat flesh. Nobody, one of the men explained, really eats that stuff. Eons before, a man might have greatly valued a bearskin or a mountain goat's head and horns and might have honored and even danced with such a signifier—in an act much like love—to demonstrate his ability to provide for his family and community, and his gratitude to the animal that had given itself to him so generously. But somewhere along the generations, the thing that signaled a man's ability to provide, and that man's kinship with the natural world, had become merely a trophy, a meaningless signifier of destruction.

One afternoon during that same season, deep inside that same Cascade wilderness, I had looked up to find a cougar poised on the crest of a snow ridge above me, her soft gold outlined against a deep blue sky. The cat studied me a hundred feet below her, ponderous and awkward with my seventy-pound pack, her face a study in concentration. Obviously she had paused in midstride, having seen me just before I saw her, and it seemed equally obvious that certain questions were playing in her cat mind. Aware of possible problematic answers to such questions and in no mood to be dined upon, I raised my ice axe high above my head to suggest that I was someone of significance, a creature of dimension too great to be toyed with. After holding me in her quiet regard for what seemed minutes but must have been seconds, she turned and vanished with great dignity over the ridgeline. Unlike those city hunters, with their gut-shot bear cub and wasted mountain goat, the cougar had no need for a trophy and understood her world perfectly. She had granted me the consideration due something foreign and potentially valuable or dangerous, and after due regard had gone on with the serious business of life, leaving me standing absurdly alone in the mountains with ice axe aloft. If the encounter had gone differently, she might well have killed and eaten me with only the slightest difficulty, and her action would have been proper and correct from any objective point of view not my own. Unlike that lion, the trophy hunters had no place in the moral world of the Cascade wilderness, or in any moral world. In her regard such killers would have been a mere corruption of the earth.

♦ ♦ ♦

Today I live in the Manzano Mountains in New Mexico, on an edge of the Cibola National Forest. A dozen yards beyond our kitchen window hangs a bird feeder beneath which cottontail rabbits come to graze on bluejay spillings. Some mornings four or five of the plump cottontails crowd one another beneath the feeder, and I tease my daughters with the possibility that one weekend when I'm alone I'll go into competition with the owls and hawks and coyotes and bobcats and foxes and I'll eat those juicy bunnies. My daughters, who have had pet bunnies playing in their bedrooms, become only mock angry because they know I won't really kill and eat the rabbits. They don't know me as a hunter. That was another life. In all the time they've known me, I've killed nothing, not even a gopher. But the truth is that I simply cannot look at those cottontails without imagining them frying in butter, lightly floured, delicious. It's hard to shake old habits of thought, old instincts born of necessity.

My daughters are right, of course; I won't eat the wild pets that feed and frolic outside our windows. I don't need to. We have chicken in the freezer, ground beef for the burgers we have once a month or so, a round steak to chop up when green chile stew is necessary, tofu for those difficult dinners—all from a local grocery. When not backpacking, I fly-fish with barbless hooks in trophy trout waters and release every fish because I'm told I must (not because I think it is a morally superior act), and I buy cod and salmon from the market. But still, I have a running argument with my wife and daughters. If we are not to be vegetarians, I argue, we should raise chickens so that we don't have to buy chickens or eggs from a store. Perhaps I should take up deer hunting again, I suggest, maybe elk or antelope as well. We are irresponsible, I argue, letting others do the killing for us. Worse than irresponsible, immoral, for by subsidizing the meat industry we subsidize the most horrific forms of cruelty. As a child and teen I worked on chicken ranches and I know those places where chickens are raised a hundred thousand at a time in long, dark buildings, fed hormones and antibiotics, never allowed sunlight, never allowed to know any life except a brief, dank prelude to death. With cattle—creatures I have helped brand, castrate, dehorn, and butcher—it is worse. And with the intelligent pig it is worse still on the massive hog farms where they live in squalor and vilely pollute the environment prior to their squealing factory deaths. If we raised chickens the way my family did when I was a child, they could roam and scratch in the dirt, flirt with one another, fight, worry about hawks and bobcats, make love, and live full chicken lives until the day they became food for us. We would grant them dignity and they would feed us, a satisfactory exchange. If I hunted deer, the animals would live their lives in full dimension up to the moment those lives ceased, and I would once again join the ranks of the cougar and coyote, and now, in the Southwest, the reintroduced Mexican gray wolf, as a predator. Together we would ensure our mutual dignity, the deep regard of hunter and hunted. Surely my bullet would be quicker than the coyote's slashing canines, and while it would be more human, it would certainly be no more humane, for is it possible to think that nature has not, over countless millennia, prepared the deer to meet its coyote or cougar death, or readied the rabbit for the owl? Such death must be sacramental, a matter of mutual knowledge and profound respect. How else may the owl or bobcat find the rabbit except by way of the most complete knowledge possible, and what is knowledge if not respect? What is such complete knowledge and respect but a kind of love?

There is no angry god out among the trees and beside the streams where deer walk and wait for the cougar. The Creator can have no shame even in the face of a futile, heroic struggle of a doe for the life of her fawn. True understanding does not bring a metal fly-rod case down on a rock to shatter the most natural reciprocity between predator and prey. This is my body, nature says, this my blood. A necessary agony and resurrection in sinew and leaf. He who loves this world is, in heart, a tracker and hunter, can be nothing else.

CHAPTER FIVE

In the Service of Forests

Moonless midnight on a project fire in the mountains outside of Tucson. Two guys in yellow Nomex shirts and hard hats studying the uphill slope, watching for flaming pineapples; fifteen hotshots with piss-pumps, shovels, and pulaskis spread out up and down the south side of the canyon chasing spots in the dark and pretending to mop up, their headlamps bobbing; three of us with chain saws cutting down burning oaks, flamethrowers, taking turns. The night is utterly black except for the dim, erratic headlamps and the red eyes of glowing hotspots here and there on the canyon sides where the coals of an oak will burn all night. In its fifth day, smoke from the fire has cut off the stars and lies in a heavy blanket just above our heads. The raw, burned-over black mountains are so steep you can't stand except on game trails and in the bowls around trees, each canyon dropping to a dry sand wash. Vietnam vet helicopter pilots, with grinning, sweating faces focused on something far away, have been screaming up the canyons each evening just before dark, banking as close as possible to rock faces and the blackened skeletons of trees, to wheel and spin and drop us and vanish, laughing. To them it's just a new kind of war with nobody shooting, and soon the casualty rate will climb so high from crashes that the Forest Service will consider banning vets from the controls of their helicopters and borate bombers. At midnight it's blistering hot and the fire that sprinted through here during daylight still lingers in the scorched air and sizzling, blackened dirt and rocks that burn through gloves. Hot spots flare now and then in intermittent red and yellow flame against the black canyon side when the night breeze licks at them. The live oaks are hollow, and the fire has gotten inside many of them to burn invisibly, just waiting for all the hotshot crews and prison crews and smoke jumpers and helattack crews to go home so it can come out secretly and blast the desert mountains yet again.

It's our job to probe for such secret fires, expose their bright hearts and subdue them. In a normal world, a world not overrun by proliferating humanity with all its needs, fires would burn through these dry mountains every few years, keeping the brush and small growths down so that such trees as these once lovely oaks were spared. But a century of fire suppression has built up a massive fuel load everywhere, in every forest and brushy desert in America, so that any fire will explode lethally, instantly surfeited and monstrous, paradoxically burning everything because there is so much to burn. Fires are greedy and will take what is there unless kept small. Nature's fires, given freely and often by lightning, were meant to be delicate things, just lapping at the fragile growth that had appeared since the previous delicate fire, their tongues too frail to do more than touch an oak or pine trunk before passing on. But man has created forests where delicate fires gorge and grow immense within minutes or hours and then howl like banshees from canyon to ridge, ridge to valley.

Desert mountain fires have surprises not found elsewhere. Rattlesnakes have crawled deep into the earth to survive but climb back to the warm surface once the fire passes. They wait in cool spots under the edges of rocks where the flame hasn't reached, the kinds of places that look good for sitting during breaks and the kinds of places also favored by the small, poisonous amber scorpions of this country. Yuccas burn through, the first explosion of wildfire searing off the leaves, and then, once the fire has moved to a distant ridgeline, the slow moldering burn that remains eating steadily through the roots until what is left is the round body of the yucca that, once freed, becomes a rolling, bounding, leaping ball of flame careening down the mountainside. We call them pineapples. The mountains outside of Tucson grow many yuccas, and the night was spectacular with reckless balls of flame.

The finest moment that night, the picture you never forget, came when Rich Dent caught fire. When you cut into a hollow oak, letting a rush of oxygen into the burning center, flame explodes into the night, sometimes shooting four-foot tongues of fire into the dark. It was Rich's turn, and Joe and I stood back, our saws on the ground beside us, watching him cut. We called ourselves spotters, watching for an unexpected shift or twist in the tree, for anything surprising and therefore dangerous. In reality we'd all been cutting so long we didn't need a spotter, but it was better if one cut and two watched. Good enough for government work, we said.

The live oaks were small, seldom more than a foot through and twenty feet high. Felling them was a simple affair complicated by only two factors. One was balance and, therefore, trajectory. The twisting branches and the black night made it difficult to calculate which way a tree's weight lay and therefore which way it might want to fall, regardless of where we put the cuts. The other factor was internal fire. Some trees held no fire, while others were flaming bombs waiting to explode. The only way to find out was to cut. Rich cut into fire and for a moment was outlined in flame, holding a chainsaw stitched in flame where the gas and oil had ignited, his wiry body torqued and etched in a brilliant border of flame. For an instant he was made a dazzling star against the utterly black night, and then he threw the saw and stepped out of the fire sign into darkness, unburned and unhurt.

The Prescott Hotshots had been on quite a few fires that season, and each one was special. The Tucson fire, for example, was an experiment in optics and heat. Our crew got there late, after the first lines were in, and we were sent out at night to mop up, primarily to fell the burning oaks and watch for line jumps. Because we were hotshots, an initial attack fire crew, we felt we were above mopping up. We'd been pulled off a mesa fire south of Flagstaff and rushed down to Tucson, and we resented being badly used. By night we dodged balls of incandescent fire that leapt down the canyons with trampoline-like skill, and with our chainsaws we turned blackened oaks into flaming candles that scorched the desert sky. Sometime after daylight, one of the small, fast helicopters would swoop in suddenly over a ridge top, invariably scaring the crap out of everyone, and scoop us up a few at a time to go back to the main fire camp, where we would eat breakfast and then try furiously to sleep. The temperature stayed a consistent 110°, scarcely diminishing at night and hammering the desert by day, and the mesquite and palo verde trees cast a frail shade on the burning sand. A couple of hundred firefighters remained from the army that had rushed in from all over the West when the fire was new and threatening Tucson, and a third of these would be sprawled during the day in myriad impossible shapes on tarps beneath the sparse and spindly bushes.

What we called a "convict" crew from an Arizona prison was cooking and serving meals on the Tucson fire. We worked with prison fire crews fairly often, and for the most part liked them. An inmate on such a crew had saved Joe and me from incineration on a big fire below Winslow. Our hotshot crew foreman, whose father was politically connected in the Forest Service, didn't know much about fires and compensated by strutting and bluffing. Joe was the other sawyer, a powerfully built kid of twenty, Golden Gloves champ of Arizona, gentle soul, vastly experienced on fires because his father had been a fire-control officer for the Forest Service. When we arrived in the middle of the night, the Winslow fire was crowning out, racing across ponderosa stands on the high, flat surface of the Mogollon Rim like a dancer just touching the tops of the trees in eighty-foot walls of flame. We'd backed off, waiting for new orders and watching a big Caterpillar close by trying to cut line at least a quarter of a mile from the fire. Even that far back, the fire illuminated everything with a bloody light that flared on sweaty faces and bright, teary eyes. The wind had picked up, and flaming cinders were landing everywhere, each one starting a new fire in the stone-dry forest. We all knew we had to do something pretty soon. In front of us was a stand of doghair pine, the kind of stunted, thin trees that grow close together like grass, each stem maybe three to six inches in diameter. When the foreman ordered us to cut a line through the doghair, Joe and I looked at each other. Dried out by the summer heat and the approaching fire, such a stand could go up like gasoline. Anybody who'd fought forest fires in the West very much had seen it happen. The foreman had not.

Joe shook his head. "I don't think that's a good idea," he said softly. Behind us the rest of the twenty-man crew stood watching, knowing they'd be expected to follow us through the doghair and knowing they didn't want to do that.

"Did I ask you what you thought?" the fat foreman shouted. "Or did I tell you to start cutting?" He looked at me, his face red in the light from my headlamp and distant flames. "You take lead," he said.

I was older, not the oldest on the crew, as that honor went to another sawyer who had spent twenty years riding motorcycles and dropping acid before finding his true calling, but older than Joe and the rest of the crew. I had a graduate degree from a university and was supposed to be smart.

"I don't think it's a good idea either," I said. "You ever seen what happens in that kind of stuff?"

The foreman began to wave his hands. We could hear a helicopter not far away, and the Caterpillar roaring at the earth while fallers from other crews were dropping big trees a few hundred yards ahead of the tractor. Each falling tree seemed to suck the oxygen out of the air as it went down with a dreadful muted sound, and the smoke was thickening around us. We were a long way from the vehicles that had brought us in. Trying to decide what to say to my boss, who could easily get me fired, I glanced at the prison crew on one side of the small clearing, leaning on pulaskis and hazel hoes and shovels and alternately watching the bulldozer and our foreman. Some of the prisoners had never been on a big fire before and looked nervously interested in the outcome of the disagreement. Others looked like they were having a very good time. The head of the crew, an imposing fellow with massive chest and arms and a pale, crewcut-topped face, was watching Joe and me and the foreman, his hardhat dangling in one hand and his expression curious.

The foreman was opening his mouth to shout when the head of the prison crew stepped forward. He winked at me and Joe and, without looking at the foreman, said, "If this motherfucker gives you another order of any kind, you tell me." He turned to the foreman. "These boys don't go in there, asshole," he said.

The foreman gave no more orders. Five minutes after the confrontation, the stand of doghair exploded like a napalm bomb, the fireline was breached, the Caterpillar was abandoned to the flames and everyone evacuated.

I was therefore predisposed to like convict crews, and the biggest, most tattooed, shave-headed, undershirted convict of all had stood at the window of the cook trailer throughout the Tucson fire, ladling out very good food with an impassive expression. Project fires were like that: lots of food, steaks and baked potatoes and fresh salads, fruit juices and sodas. After eating we spread tarps under the skeletal trees and saguaro shadows and listened to insects hum and the infinite sunlight sizzle while trying to sleep. For most of us that was a hopeless endeavor, but a few could sleep anywhere in any conditions. Joe was one of those. I'd seen him fall asleep while standing in chow line on the Winslow fire and drop like a downed snag.

I lay there the second day of the preposterous Tucson fire, cursing night-shift mop-up and wishing I could sleep in preparation for the night work ahead. Exactly overhead, the sun was an enormous circle of white flame taking up most of the sky, and insects snapped and buzzed in the palo verde above me. I was thinking of scorpions, because a guy on the Flagstaff crew had been stung by one that morning while trying to sleep. And I was thinking of the waves of sweat running down my face onto my neck and bare arm where my cheek rested. Absentmindedly, I let my eyes roam across the scores of bodies sprawled everywhere in the faint shade. I noticed the giant convict cook standing outside the food trailer, watching us, his eyes focused on Joe, who appeared to be sleeping heavily, his head on his arm and his hard hat tilted against his face to shield his eyes. I glanced lightly at Joe and back to the cook, wondering if Joe was the cook's type. Something about the cook's expression caught me. He was transfixed upon Joe like a hunting hawk, his enormous shaved head glistening with sweat and his pale eyes unblinking. He did not look amorous as he reached a hand down beside the trailer and came up with a pulaski, that half-axe-half-hoe fireline tool. Pausing for a moment, his eyes still fixed on my fellow sawyer, the giant cook, undoubtedly incarcerated for a crime beyond words and allowed out like this by a bookkeeping error, began to walk slowly toward us, aiming directly for Joe. I froze for an instant and then began to rise to shout a warning, but I was too late. The cook took three quick, silent steps and swung the pulaski. Joe woke and leapt back with a shout. Beside where Joe's hard hat lay, a headless diamondback rattler was flipping and writhing.

Sometimes I think the Prescott Hotshot crew was the best time of my life. Twenty of us worked out, played horseshoes, drank great quantities of beer, and fought forest fires together. We were supposedly the only technical rock-climbing hotshot crew in the country, and therefore for three days a week, when not on fires, we practiced rock-climbing and rappelling, going to the beautiful rock of the Granite Dells outside of Prescott to climb and slide. A small, isolated tavern sat alongside the road to Jerome not far from the Dells, and at lunchtime we'd coil our ropes and repair to the bar for burgers. At first the foreman imposed rules. One beer maximum with lunch. But the foreman, I'll call him Ralph, was fond of beer. After the first couple of days he relaxed his rule. Two beers with lunch wouldn't hurt anything. And soon it became apparent that several beers with lunch were quite harmless. Rock-climbing after lunch began to be more exciting, so much so that most of the crew preferred to remain at the tavern with the foreman while just a few of us went back out with ropes, chocks, bongs, beeners, and climbing shoes. (My shoes had belonged to a guy who died climbing in Chile and were a gift to me from his friends.) We never used our skills on a fire, but if the occasion had arisen we could have been heroes.

Firefighting was fun. Today I see yellow-shirted hotshots on the news, always on a big, dramatic wildfire, and I hear the media voices describing the miserable, dangerous work while the camera focuses on a smoke-stained and filthy face, and I laugh. Just off camera are a bunch of men and women having the time of their lives and making good money to boot. In hotshot thinking, a good fire year is one in which all the forests burn. Sure, it's dangerous, and people have died, including some of my friends, but that's not what it's all about. As a sawyer I was up front, cutting the first swath of line while a puller stood behind my shoulder pulling away the cut brush and limbs and tossing them to the side for someone else. Behind the three sawyers came the cutting and scraping squads. We could cut a ten-foot line down to mineral earth in minutes, working in controlled mania. We were efficient, good at what we did, young, excited. I went on to fight fires on a California crew and was even accepted into smoke-jumper's school in Hamilton, Montana, at which time my wife made it clear that I could be a smoke jumper or be married. Prescott, however, was the best. Twenty-five years later, I have my Prescott Hotshots fire shirt hanging in a closet, as if I might one day need it.

But being a hotshot in Arizona and California wasn't really the finest experience the Forest Service offered. The finest was in the North Cascades.

I came to the Cascades in 1969 to work on a trail crew. I'd been a seasonal ranger for the State Park Service in California, registering campers at San Simeon State Beach a few miles south of Hearst Castle, driving up and down the Big Sur coastline to watch sea otters surf and play, tossing bags of dye into the surf to measure the speed of the tide, and studying wind and berm crests. In the evening a couple of the other seasonal rangers and I would alter consciousnesses and swim in the outdoor pool up at the so-called Castle, gliding underwater to stare at the marble sculptures and exotic tiles and then filling my '49 Buick with illicit oranges on the way down the hill. No one was permitted to pick the fruit, by orders of William Randolph Hearst's ghost. We stole it.

My college roommate had worked on a trail crew for the Forest Service in the Glacier Peak Wilderness and suggested that I join him the summer of my junior year. Joined by another friend with whom I'd grown up and gone to junior college and the University of California, I took a Greyhound bus north from Santa Barbara in late spring of '69 to work trail eighty miles or so south of the Canadian border. We carefully cut off our long hair before heading north into logging country. My father had logged in the redwoods of Northern California many years before, right after coming out from Mississippi, but I still didn't trust loggers. We'd heard stories designed to cause apprehension. I'd been in the Northwest only once before, when the same friend and I had hitched to British Columbia right out of high school on an excellent adventure. This would be different, though, because Larry, my roommate, had told us stories of meadows and glaciers, bears and eagles, waterfalls and rivers so fast they rolled boulders all night. He'd shown photographs of the Cascade high country, photos in which the sun always shone on lupine meadows and glaciated peaks and sparkling alpine lakes. I'd never been in such country, the kind of world I'd seen in illustrations for *Heidi,* and that I'd dreamed about as a kid.

The bus pulled off the highway an hour north of Seattle, and the driver opened the doors. Darrington was still thirty-five miles away up the Stillaguamish River, but no buses went any closer to the town. It had been raining steadily since we'd crossed the Oregon border the night before, and with every mile northward the rain doubled and tripled in intensity until it was pounding on the Greyhound and turning the windows into aquarium glass. We got off the bus and retrieved our duffle bags from the belly of the machine and stood rather forlornly watching the bus pull away toward Canada, flipping the corduroy collars of our Lee Storm Rider jean jackets up against the waves of rain sweeping across the valley. A hamburger stand and a gas station crouched on opposite sides of the road near the freeway ramps, and the clouds appeared to be resting like indigent whales right on the two roofs and spewing rain with casual vehemence straight at us. Above the roofs we could see only soggy clouds. The floor of the world surrounding us appeared to be a sodden green-black rising along the river to the trunks of trees eaten by clouds.

I had decided on the hamburger stand when Glen stuck out his thumb and flagged down an empty log truck that came humping down off the highway. The driver was motioning to us even before the truck stopped, and when Glen climbed up and pulled the passenger door open with me on his heels, the driver grinned widely and said, "Crazier'n *Catch-22,* ain't it?" We'd met our first logger, Tom, who would become a good and close friend.

I watched the valley narrow through the truck's windows. The clouds seemed to ride the rising slopes of mountains on either side as we moved closer to the river and climbed. Soon the trunks of big trees lined both sides of the road, the tops of the trees appearing to tear the clouds into tatters that hung in rags.

Darrington was at the end of the paved road, a town of seven hundred lost souls hidden as far back as roads went in the North Cascades, a minuscule logging town hunched at the foot of Whitehorse Peak between two wild rivers, the Sauk and the Stillaguamish. More than pavement ended in Darrington. Our new friend dropped us off where he turned toward the mill a half-mile from the ranger station. We'd shouldered our duffles and begun to walk when almost at once a pickup truck stopped and two giant human beings got out and, without a pause, asked us if we would prefer to have our hair burned off with white gas or cut with a chainsaw. The bigger one lifted a saw with a four-foot bar out of the back of the truck and jerked it to howling life for emphasis. It would be nice to say we responded with clever and biting repartee, but the truth is that we just picked up our duffles and ran toward the ranger station, thankful that the loggers chose not to pursue.

Such scenes were common in Darrington, a town whose majority population shared two surnames and where outsiders, usually called "flatlanders," were an endangered species. We learned to live by stealth while in town, but since our job was to spend ten days out of every two weeks camped miles back in the Glacier Peak Wilderness, we saw little of Darrington and therefore little of the kinds of trouble possible. What we saw a great deal of was the most extravagant beauty the natural world has to offer.

We lived in the attic of the district ranger station bunkhouse on days off, and the rest of the time we headed into the wilderness as part of a four-man trail crew. A string of highly intelligent and sensitive mules carried our tent, tools, and provisions, and a highly intelligent and vastly more sensitive mule skinner guided the pack string. Anyone seeing Chuck, the skinner, knew at once what was expected. No foolish talk, foolish behavior, or thoughtless expressions of any values not embraced within Chuck's philosophy. A person did his work, whatever that might be, and ate pancakes for breakfast. Diamond hitches were used to lash pack boxes to mules, and diamond hitches were always tied correctly. College education might not condemn a person to hell, but such corruption required a great deal of overcoming. A trail crew member walked lightly, spoke rarely, and did what was expected without unnecessary expenditure of spoken words. When Chuck was around.

Chuck, however, with his Lummi wife Wilma, packed us into wherever we were going and left us there, returning with the mules—Kansas City, Rosco, Kitty, and the rest—ten days later to pack us out. Within the ten days we were on our own, in a world more marvelous than any I could have imagined or invented.

The rain stopped and stayed away for four months. The high country was at first an infinite space of snow-covered ridges and meadows rising up to icy black granite and glacier-bound peaks, all beneath a blue, blue sky. When the snow melted, the same country was glacier lilies and lupines and heather and then, later, huckleberries and blueberries unending. Blue lakes hid in every meadowed crease and granite fold. Marmots whistled and golden eagles rode updrafts out of the converging drainages. We worked hard, cutting windfalls with a two-man crosscut and double-bit axe, grading trail and chopping roots with pulaskies, building cedar bridges over small and large streams out of red and yellow cedar logs that might have been lying on the forest floor for fifty years. When we scraped the moss off and cut rounds with the crosscut and split the rounds with mallets and wedges, we found beauty each time. Bridges of the fresh-split cedar emitted a sacred aroma and shone like gold.

We worked all day and climbed nearby peaks after our eight hours were over. Black bears grazed on blueberries near our camps and never once touched the steak and chicken and fruit we kept in snowbanks close to our tent. They were wild bears, satisfied with our presence if we behaved properly, happy to ignore and be ignored. Grouse humphed in every stand of alpine fir and mountain hemlock, and white-splashed ptarmigan raced from heather to heather. Never having heard of giardia, we drank from every stream and spring. We ate vast breakfasts, lunches, and dinners and grew into the greatest physical condition any of us would ever experience in our lifetimes.

We all returned a second summer. I was working on a masters degree in English. Glen, who quickly came to be called Whistle Pig due to habits shared with the sedentary marmot, was completing a B.A in English. Digby already had his M.A. in English, and Mick had a graduate degree in biology. The other two crews contained a Ph.D. in political philosophy, a masters in urban planning, a degree in journalism, and an artist of great talent with a college degree testifying to his talent. We were a highly educated lot who drank Jack Daniels around a fire at night and read aloud Gary Snyder's poems about grubbing trail and logging in and near the very same wilderness. We exulted in the life we had chosen, even momentarily.

After two seasons on trail crew I wanted more. I wanted to be able to keep my eyes elevated and not have to focus on the log I was cutting, the bridge I was building, the trail I was grading. I became a wilderness ranger in my third season with the Forest Service. I had passed through one perfect world into another. I put on a uniform and a badge and a pack and began to walk through the Glacier Peak Wilderness, on trail and off trail, for days and weeks and, it turned out, years. Out five days and in two, out ten days and in four—I patrolled the river drainages and the high country, going where I chose and camping wherever I found myself. Sometimes for days on end I saw no one, walking the high ridges above the clouds, watching a hundred Cascade peaks north into Canada and south to Mount Rainer dodge in and out of sight, camping each night by a different alpine lake. A child with eight brothers and sisters, I was utterly alone for the first time in my life for day after day and night after night in the backcountry. Solitude and silence became nonnegotiable commodities, essential ingredients in the life I expected. I climbed glaciated peaks to radio back reports of frozen lakes to the district station forty or fifty miles away, and I sat high above the surrounding world, watching and breathing.

And then one day I was ready for a different life. I got married and moved to Arizona to fight fires. After the wedding my wife's well-positioned relatives bent their heads and asked, solemnly, what I planned to *dooo*? In my J. C. Penney's wedding suit and my work boots, I solemnly replied that I thought I would fight forest fires for the rest of my life. And all these many years later, in a different life, I find myself dreaming of dark forests that crown out in racing flames, of small oak trees that explode like comets into the night sky, and of alpine ridges a million miles from the world. In my dreams I seek out wildfire and search for eagles that soar over granite and glacier. In my next life I will do it all over again, every single thing.

CHAPTER SIX

Ringtail Moon

I'd arrived at the Grand Canyon just at sunset, in those ancient days when you could drive along the south rim all the way to the Hermit Trail and park your '64 VW Beetle before starting down. Folks up in the park headquarters would have preferred a registered backpacker to an unregistered one like me, but it was late and I was anxious to be on my way. I'd just gotten back from a project fire on the Mogollan Rim below Winslow, the kind of summer fire that crowns out in a ponderosa forest and streaks across the night sky in eighty-foot flames, the kind that draws fire crews from all over the West and thrives on adrenalin from every quarter. The Prescott Hotshots, for whom I ran a chainsaw, had arrived at that fire around midnight and driven right into the middle of the whole damned fire carnival. I'd been vividly aware of driving into the middle of the thing because I was the one who drove.

Five days of endless fireline shifts, exploding doghair-pine clusters, a lost helicopter, an abandoned and burned-over Caterpillar tractor, paper sleeping bags in camp and a few hours' sleep on black and smoking slopes, and I was ready for the quiet isolation of the Canyon. Those night walls of flame were still dancing on the backs of my retinas, the roar of chainsaws still mixed with howling and whipping of fire fed by huge winds. I needed a soundless world, a space in which walls were stable and secure, air a dependable source. With four days off, I headed for the Canyon, getting a late start after a day of postfire cleanup and saw-tuning around the hotshot station, but wanting to be inside the Canyon and well away from chainsaws and the smell of gas by the next sunrise.

I'd parked my car, thrown my pack on, and gotten a mile down the trail by the time the sun was gone. At two miles I was stumbling in the thinning dark, more exhausted than I'd expected, with a near-vertical drop of uncertain depth to my immediate left and sheer rock an arm's length to my right. I decided to camp for the night at a three-sided stone shelter built into the rock wall beside the trail, its open face looking out into the black air above Hermit Creek a long, long way below. I could still taste the Winslow fire in my throat and smell it in my skin, and my eyes felt scorched the way they always did after a big fire.

As I lay in my open mummy bag, feeling the day's heat fading into the surrounding rock, a great disc of moon began to rise over the canyon. I watched it climb until it hung there, brilliant in the cloudless sky, bloated and heavy, just beneath the roof edge of the shelter as if it had elbowed its way directly out of Hermit Creek. Shadows unfurled around me, and each line and crease was abruptly present, hard edged and clear. I studied the growing light with astonishment. And then a black and hairy hand appeared, groping at the end of a sinewy arm slowly down from the shelter roof, enormous, spidery fingers given full, terrifying dimension in the great moon. For what seemed a very long time the sooty hand probed, monstrous fingers flexing toward me, and then it was gone.

I lay still for a moment, listening to my heartbeat, feeling what Emily Dickinson once called "zero at the bone," and then I squirmed out of the sleeping bag and tiptoed from the shelter two steps to the trail's edge, peering up and down, fortified by pure fear. An animal scuttered in the moon's shadows and I jumped. A small, wiry form with a large tail scurried away from me up the trail, and at once I understood: a ringtail, whose delicate, prehensile hand had been made abruptly monstrous by the framing moon and infinitude of canyon.

A curious ringtail, a rising moon, and an exhausted warp of space and time had converged to alter the way I saw my world. For an instant, reality had fractured and the impossible—a grasping black hand as big as the moon—had been conceivable, full of meaning and significance and capable of reshaping a lifetime of knowledge of how the world works into something utterly new and transforming. The next morning the July heat lashed the Canyon by seven A.M., and I continued partway down the Hermit Trail, scattering a small herd of bighorn sheep who seemed to step blithely off into space at my approach, and then I left the Hermit and cut over to the Boucher Trail. Under an already pernicious sun, I squirmed down through a break in the canyon strata, where for fifty vertical feet the trail became a climber's chimney, and eventually descended all the way to the side-canyon camp beside Louie Boucher Creek. The camp was a paradise where a handful of stunted trees cast shade beside the stream, and on the cliff face a few hundred feet above, a hanging waterfall dripped down whatever kind of greenery lived up high in so deep a place.

As I made camp where the wise prospector had made his a long time before, bitterly regretting that in my fire-scorched state I had forgotten to bring shorts for the summer trip, a mutant deer, with antlers jutting in scores of crazy directions and angles, wandered up the creek to graze beside me, friendly and soft eyed, and an inner voice of some kind drew me to a large boulder beside the creek a dozen feet from the deer. The voice told me to turn over the rock, which I did. Underneath the big rock, carefully wrapped in a plastic bag, were a new pair of gym shorts and a copy of Colin Fletcher's book *The Man Who Walked Through Time,* his account of a journey on foot the length of the inner Canyon. Such a Canyon gift is not to be taken lightly, and certainly not refused. For the remainder of my trip I wore the red shorts and, in odd moments, read the book, leaving both, a bit more frayed, beneath a different rock miles away where a different voice would entice someone else to whom the moon's great hand had beckoned.

Whether we come as refugees from fire or merely unexpecting innocents, the Grand Canyon speaks to us and alters our vision of the world, provides surprising gifts and distorts suddenly our sense of proportion and possibility. To see it once is to never be free of it; to step into it even once is to live that step the remainder of one's days. To grasp the immensity of the canyon requires an act of imagination for which our everyday world does not prepare us. To enter the air of the Grand Canyon is to breathe differently, measure one's world by new degrees, see past the earth's skin to neither flesh nor bone but something other, an inner form or spirit that requires radical adjustments. To describe the canyon requires the impossible, as writers have been discovering for half a millennium now, since the first Spanish invaders looked into the chasm and attempted to measure and claim the New World in words.

American writer John C. Van Dyke, cranky and impatient before attempts at the impossible, despaired in 1919 of artists' endeavors at realistic renderings of the Canyon. Product of the cultural convergence that delivered the twentieth century into the realities of Einstein, Freud, a relativity of inner and outer perceptions, and new, shimmering subjectivities, where fragments were being shored against everyone's ruin in the Western world and a phenomenon called World War I had just permanently altered our sense of measurable reality, Van Dyke declares, "Impressionism, in its rightful meaning of giving the realistic or objective impression of the fact, is possibly the better method of procedure." Possibly indeed. Writers, Van Dyke insists, fare no better than painters. The poet comes to the Canyon, he complains, "with a fine frenzy in his eye and a thick feeling in his throat," but by the time he transfers his emotion to paper "it has proved to be merely a disjointed rhapsody." Not just poets find the Canyon impossible to describe: "Even the people who write prose, and are not popularly supposed to be bothered with fine frenzies, have their troubles in describing the Canyon. They have not enough adjectives to go around or to reach up and over. Language fails them" (216–17). Van Dyke, like everyone, surrenders, declaring finally, "The Canyon is impossible." Signification fails; language did not come into being in the presence of such phenomena and remains utterly insufficient before it. No signifier within a footstep of the sudden plunge of the Canyon is answerable to any signified whatsoever. The center does not hold, the blood-dimmed tide floods at sunset, and all ceremonies of innocence are reimagined and revivified. Any falcon swooping those endless and endlessly changing shadows could not possibly hear a falconer. An anarchy of impressions is loosed upon every perception. Only the Native holds the measure of the place, because the Native, the Indian, has had many, many generations in which to learn resistance to the jealousy of measure and voice, the futile desire for words to reach up and over.

Van Dyke, forlornly conscious of the impossibility of his task, set out nonetheless with admirable determination to find words for the wordlessness of the Canyon in his 1919 book, *The Grand Canyon of the Colorado: Recurrent Studies in Impressions and Appearances.* In the first paragraphs on the first page of his book, he grows eloquent, warming up: "The great abyss, without hint or warning, opens before your feet. For the moment the earth seems cleft in twain and you are left standing at the brink. . . . The depth yawns to engulf you." In a charmingly staged ecstasy of horror, Van Dyke exclaims, "Ah! the terror of it!" The problem, he suggests, is dimension: "The mind keeps groping for a scale of proportion—something whereby we can mentally measure. Standards of comparison break down and common experience helps us not at all" (11). The Canyon reaches great, hairy hands out of its bloated moon to grasp your soul, wrenching it out of common experience. Van Dyke nonetheless tries with a kind of shotgun approach, piling verbs and modifiers as if the right ones might spill out by accident, hyphenating his prose into new, almost frantic shapes, desperately trying rhyme, alliteration, anything to keep things from falling apart, to keep the Canyon's anarchy at bay:

> The rock forms are florid, fantastic, flamboyant, and yet planned on so vast a scale that they are impressive and commanding through sheer mass. The colors are hectic, sky-flushed, fire-fused, perhaps leached and bleached by rain or flung off in vivid tones by blazing sunlight. Sometimes a vermilion-red glows

> beside a fire-green, while at other times, so subtle is the blend that you cannot draw a line between gold and orange or purple and mauve. The light shifts almost like the footlights in a ballet, showing a silver, a saffron, a pink, a heliotrope. (6)

Van Dyke finds the Master Planner at work, moving on a vast scale, but the human line cannot hold in his perception.

The problems of proportion and description presented by the Grand Canyon were discovered, of course, long before 1919. In 1540 the first European to record himself doing so teetered on the canyon's rim, much chagrined as he confronted yet another impossibility in an impossible New World that was supposed to deliver up not an immeasurable reality but measurable wealth, not distortions of time and space but infinite youth, Cibola and more. Sent by Coronado to find the great river that Native people had spoken of, García López de Cárdenas found instead a great, gaping crack in the earth at the bottom of which could be seen a brown ribbon of water. Cardenas estimated the distant trickle to be about six feet wide. Three men sent scrambling down toward the river were soon horrified at the shape-shifting obstacles of the canyon. Returning to the rim, Cárdenas's men explained in dismay that rocks that from above appeared as tall as a man were in reality much higher than the great tower of Seville, over three hundred feet.

Two centuries after Cárdenas, while Americans to the east were declaring themselves free of the English crown and solely in charge of despoiling their own portion of the continent, a Spanish priest named Francisco Tomás Garces became, in 1776, the second-recorded European to set eyes on the Canyon and tell about the encounter. Wasting no words in any attempt to fix the grandeur of what he called "Puerto de Bucareli," Garces got right to the point, angrily denouncing this wonder of the world as "that prison [*calabozo*] of cliffs and canyons." Undoubtedly still in a pique, Father Garces would be bludgeoned to death by Natives near the river in 1781. Half a century later, an American trapper named James Ohio Pattie—passing with his small party across New Mexico and Arizona while rescuing fair damsels from Comanches, eating ravens and buzzards and the party's unlucky dogs, and frequently shooting and killing the Native inhabitants of the invaded region like a lethal Monty Python—stumbled upon the Grand Canyon. Like Garces, Pattie was not notably impressed by the beauty of the landscape. Having just found three of his trapping party "cut in pieces and spitted before a great fire" (149) by unknown Natives of the region, Pattie was not predisposed to enjoy dramatic scenery when he stumbled upon the snow-covered Canyon rim. "A march more gloomy and heart-wearing, to people hungry, poorly clad, and mourning the loss of their companions cannot be imagined," he wrote before observing: "We arrived where the river emerges from these horrid mountains, which so cage it up, as to deprive all human beings of the ability to descend to its banks, and make use of its waters" (150–51).

To the early explorers the Grand Canyon where I sought refuge from fire and fire signs was often just a bad joke, a terrible turn of terrain, an obstacle even worse than those irritating people called Indians who were spread inopportunely across the continent, including, irritatingly, along and even in the great Canyon. Once the West was, as euphemists like to put it, "opened up," the Natives, when not still bitterly resisting the inexorable invasion at places like the Little Bighorn, became romantic subjects; at the same time, radical geography became picturesque scenery rather than obstacle. Attitudes changed in respect to the "prison of cliffs and canyons." In 1857 an artist with an early unsuccessful Canyon expedition, F. W. von Egloffstein, wrote, "We paused in wondering delight, surveying this stupendous formation." The intrepid von Egloffstein, after tumbling into the depths of Havasu Canyon with the aid of a broken ladder, took the opportunity to pay a visit to the Havasupai Indian community. Unlike most others, the failed expedition's leader, Lieutenant Joseph Christmas Ives, had little trouble finding words for his reaction to the canyon, declaring succinctly, "The region is, of course, altogether valueless. It can be approached only from the south, and after entering it there is nothing to do but leave. Ours has been the first, and will doubtless be the last, party of whites to visit this profitless locality. It seems intended by nature that the Colorado river, along the greater portion of its lonely and majestic way, shall be forever unvisited and undisturbed" (Krutch, 14).

Ives's prophecy was not quite accurate. In 1869 the Grand Canyon's great explorer, Major John Wesley Powell, would descend the length of the chasm by boat and become the first to officially travel through the depths from end to end. Powell's expedition began at Green River, Wyoming, and lost one of its four boats, a twenty-one-foot craft wisely called *No-Name,* before even coming close to the Grand Canyon. Once in the bowels of the canyon proper, in the midst of hourly terrors, Powell wrote:

> We are three-quarters of a mile in the depths of the earth, and the great river shrinks into insignificance, as it dashes its angry waves against the walls and cliffs, that rise to the world above; they are but puny ripples, and we but pygmies, running up and down the sands, or lost among the boulders. We have an unknown distance yet to run; an unknown river yet to explore. What falls there are, we know not; what rocks beset the channel, we know not; what walls rise over the river, we know not. (83)

All of the world is known, of course, except to the newly arrived, invasive eye. Powell's great unknown distance, river, falls, rocks, and soaring walls were well known to the Native Americans who had long lived and hunted and prayed within the canyon, as Powell himself noted. The Euramerican eye had simply not yet perceived this world, nor possessed it by mapping through word and measured line in the Western tradition. The great unknown resided at the center of Powell's soul, and with his maps and charts he would lay an Ariadne thread down to the heart of the labyrinth, slaying the monster of infinite imaginative possibility and returning alive from the depths. In the Canyon, Powell, the one-armed Civil War veteran, came closer than any other invader to finding words for the impossible. Almost out of provisions, trapped between known terrors behind and unimaginable terrors ahead, Powell wrote in his journal:

> The walls, now, are more than a mile in height. A thousand feet of this is up through granite crags, then steep slopes and perpendicular cliffs rise, one above another, to the summit. The gorge is black and narrow below, red and gray and flaring above, with crags and angular projections on the walls, which, cut in many places by side canons [*sic*], seem to be a vast wilderness of rocks. Down in these grand, gloomy depths we glide, ever listening, for the mad waters keep up their roar; ever watching, ever peering ahead, for the narrow canon is winding, and the river is closed in so that we can see but a few hundred yards, and what there may be ahead we know not; but we listen for falls, and watch for rocks, or stop now and then, in the bay of a recess, to admire the gigantic scenery. (86)

In the midst of his own dark night of the soul, Powell came to echo the earlier Spanish priest, calling the canyon "our granite prison."

Powell opened up the Grand Canyon to waves of humanity that rolled across the continent like a crowned-out forest fire. He made it possible for European-Americans, plunged into a kind of permanent colonial crisis of strangers in strange lands, to imagine the canyon if not to describe it. But a world cannot be appropriated fully into the imagination until it is plumbed, measured, diagramed, and, above all, mapped with the invaders' names superimposed over the indigenous landscape. In 1880 a geologist named Clarence Edward Dutton led a United States Geological Survey expedition into the Grand Canyon and gave us the first of those many odd names that flavor the canyon with a residue borrowed from European empire and metanarrative: Brahma Temple, Vishnu Temple, Shiva Temple, the Tower of Babel, the Hindu Amphitheater. Two decades later, Van Dyke would grumble that "years ago, when this country was young and defenseless, some people more or less broke in on the Canyon and exhausted the pantheon of gods in giving names to the buttes and promontories." Van Dyke concludes, "The buttes and isolated points that have been looming heavenward in majestic isolation for thousands of years before the coming of the Pale Face, are not made more grand or comprehensible by likening them by name to the squat temples of Bhudda [*sic*] or Shiva or Zoroaster" (13–14).

With the Canyon made more human and comprehensible—in European terms—by Powell, settlers began to drift in toward the rim and side canyons, perhaps the most famous of whom was Captain John Hance. A storyteller of the first order, Hance became well known for the many yarns he spun for passersby and guests at his Grandview Point cabin. One of his favorite stories told of the time he tried to leap the canyon on horseback to escape from Indians. His horse naturally failing to make the other side and plummeting toward the rocks below, Hance saved himself only by dismounting at the last moment. He also liked to describe the time the canyon filled so completely with clouds that he was able to walk across by snowshoe, endangered only briefly by a clearing trend.

As the nineteenth century revolved into the twentieth, and romancing the West became safer and thus more popular and necessary as a means of obscuring any less pleasant realities out there—such as what happened to the prototypes for those cartoon Indians chasing Hance—the Grand Canyon began to attain celebrity status in the American imagination as a kind of hyperreal Nature. The great environmentalist John Muir paid a visit in 1898. After observing how approachable the Canyon now was, so much so that children and what he nicely called "tender, pulpy people" might access it in comfort, Muir was, as usual, overcome to the point of Van Dyke's "fine frenzy" when it came time to capture the canyon:

> It seems a gigantic statement for even nature to make, all in one mighty stone word, apprehended at once like a burst of light, celestial color its natural vesture, coming in glory to mind and heart as to a home prepared for it from the very beginning. Wildness so godful, cosmic, primeval, bestows a new sense of earth's beauty and size. Not even from high mountains does the world seem so wide, so like a star in glory of light on its way through the heavens. (100)

Teddy Roosevelt followed Muir to the "godful" Canyon in 1901, in a hiatus from slaughtering big-game animals and Cuban Spaniards, and made an eloquent plea for conservation of the Canyon. Roosevelt put his federal money where his mouth was by creating the Grand Canyon National Monument in 1908.

Writers have consistently approached the Canyon with a "thick feeling" in the throat. Joseph Wood Krutch, one of the last American men of letters with a fine eye for both landscape and the American appetite for landscape destruction, tried his hand in *Grand Canyon: Today and All Its Yesterdays,* writing, "Seen only from the rim and thought of only as spectacle, the 'view' has some of the insubstantiality of a cloudscape changing color and form almost from moment to moment" (20). Zane Grey, poet laureate of the western novel, gave the Canyon his best shot, too, in a swash-and-spur-buckling novel titled *The Call of the Canyon,* burbling, "What a stupendous chasm, gorgeous in sunset color on the heights, purpling into mystic shadows in the depths! There was a wonderful brightness of all the millions of red and yellow and gray surfaces. . . . Deep dark-blue shadows, like purple sails of immense ships, in wonderful contrast with the bright sunlit slopes, grew and rose toward the east" (252–53).

As Van Dyke told us, it isn't just the writer who has difficulty with the Canyon. Baron von Egloffstein, he of the broken ladder, was the first to try capturing the Grand Canyon in sketch or painting, with results described by Krutch as "more like something out of Dante's *Inferno* than like Arizona." California novelist Wallace Stegner also complained about von Egloffstein's attempts: "Nothing here is realistic: stratification is ignored, forms are wildly exaggerated, the rocks might as well be of the texture of clouds." In the baron's wake came Thomas Moran, who accompanied Powell and prepared the engravings for *Powell's Exploration of the Colorado River of the West* in 1875. Moran's paintings and engravings aroused wide interest in the Grand Canyon, so much so that Congress purchased Moran's greatest Canyon painting, *The Grand Chasm of the Colorado,* for the dizzying sum of ten thousand dollars. Van

Dyke puts Grand Canyon painting into perspective for us: "You cannot turn the Canyon into a tone of color, or arrange it as a merely graceful pattern of form, without distorting truth and falling into insipidity. Indeed, there are many difficulties in the way of the individual who would put the Canyon on canvas. More than one painter has come to grief over it" (216).

And it becomes no easier with time, as nature writer Colin Fletcher demonstrated in the book I found beneath a rock, *The Man Who Walked Through Time,* his 1967 account of backpacking the two-hundred-mile length of the canyon. Fletcher wisely demurred from attempts at word-painting, going only so far as to write, "We came to the lip of the Rim. And there, defeating my senses, was the depth. The depth and the distances. . . . In the first moment of shock, with my mind already exploding beyond old boundaries, I knew that something had happened to the way I looked at things" (5–6). That's what the Grand Canyon does. It alters forever the way we look at things, the way we see our world. Even when that world is someone else's and, like Father Garces, we stumble upon it in all ignorance if not innocence.

What we can say of the canyon is this: Between 7 million and 10 million years ago a dome pushed up in the path of a river not yet named Colorado, and the river, left with no choice, began cutting its way through this obstacle. Today, the canyon is more than 200 miles long, from four to eighteen miles wide, and up to 6,000 feet deep. Of the rock it has exposed, the youngest is 200 million years old, the oldest more than 1.7 billion, and upon that rock Native people lived and worshipped long before the first European despaired of perspective and proportion there, and upon that rock Native people still live and pray. The river is young and tough, draining an area of 242,000 square miles and composed for the most part of water you can cut with a knife. Though dammed unforgivably just above the Canyon, the river is not stopped, for a water of such desire that it could cut through millions of years and a mile of rock will not be stopped by a mere twentieth-century work of steel and concrete. We must know that, in the twenty-first century, the Colorado will finish silting up the artificial lake sacrilegiously named for Powell; it will push, strain, erode, crush, collapse, and wear away through sheer will the thin human-made wall. It will never stop wanting the sea with all its power, for it cannot.

Ringtails, small catlike creatures of surpassing beauty and curiosity, related to the raccoon, live along the Grand Canyon of the Colorado and at night may investigate a campsite, reaching down from an ascending moon to probe the air. Such creatures attain proper dimension only in the Canyon, where the world speaks to us in its real form, undisguised, sincere, forever. "We opened our eyes upon the world with awe," Van Dyke concludes in the final lines of his long paean to the Canyon, "and we close them at the last groping our way in starry spaces."

CHAPTER SEVEN

My Criminal Youth

In Mississippi I was no criminal. With my brother I wandered a humid, immaterial paradise, innocent, naming a dark, damp world of wonders, owning nothing and desiring little. In California, after age seven, I learned criminality, to invade and violate others' lives, to bear away others' possessions, to lie, prevaricate, hide. I became predacious, understood the world as made of removable objects and obdurate enemies. In Mississippi the Indian and Black people around us possessed little to be desired unto oneself: a new cotton sack for picking, a head scarf, a broken-bladed pocketknife, a hound pup. Things that anyone might possess and no one would dispute, at least seen from childish eyes.

California, I saw at once, was made of desirable objects and those who possessed such objects. New bicycles leaned on kickstands on front porches of nice houses with picture windows through which television sets flickered. The owners of such bicycles came and went with one brother or sister and two parents in long, sharp-fendered and shiny automobiles and wore mohair sweaters and Boy and Girl Scout uniforms. Basketball hoops edged concrete driveways, and the bicycle riders were to be seen afoot on Saturday afternoons shooting a basketball with their fathers and older brothers. Sprinklers made perfect halos of sparkling drops on lawns, and television antennas stood straight and imposing on roofs as we passed rattling by in a car spilling with children on our way to a different kind of home, our father spitting tobacco juice indiscriminately out the window as he drove.

I have to be fair and exact: my older brother did not steal. Gene was honest, lived by a code he himself invented, trod a straight and narrow path through impossibilities clear eyed and truthful. Many years later Gene would take the same straight path from high school to Vietnam and then, with the same clear-eyed honesty, tell me not to follow. Unlike my brother, I stole and lied, and not from want or even envy. I was not driven to have what I could not have or to right intuited ancient wrongs stemming from the removal of Choctaw and Cherokee peoples or the oppression of my ancestors, or any such truck. The truth is I didn't think much about being Indian, one way or another. Perhaps I unconsciously wanted to dispossess the privileged class of their bourgeois materials, to level the playing field a bit in my eight-year-old imagination. More likely my motivation was the pure pleasure of the act itself, the stealth and secrecy, the cunning, the risk and the perverse triumph. And the comradeship of common larceny.

It happened in San Leandro, California, in third grade. My father had gotten a job as a truck driver not long after we'd moved permanently out from Mississippi, and we abruptly found ourselves living for a short while near San Francisco, on the periphery of a quasi-middle-class community on the edge of a great city. We were the bottom rung, of course, in our little shingle-sided house on a steep hillside surrounded by more affluent residences. Used to roaming the woods along the Yazoo River in Mississippi, with no rules or restrictions, we took to the hills of the Bay Area enthusiastically. It was an oddly rural place, with a creek running down out of the green and lush hills into deep pools that held big carp. The hills, unfortunately, were off limits because from them came drinking water for the city. A great, beautiful reservoir lay up there, the birthplace of the wonderful carp stream, and high barbed-wire-topped fences stood between us and the reservoir.

I might never have become a criminal, certainly not a successful criminal, had our older cousin not come to live with us at that time. Darrell was fifteen and right out of the worst part of Kansas City. Black hair in a shining ducktail, a neat curl to one side of his upper lip reminiscent of Ricky Nelson, white T-shirts with sleeves rolled up, dark jeans with bottoms rolled, and loafers with a dime in each one—Darrell understood and was equipped for life. His father was Uncle Bob, my mother's bad brother who had several wives in different places at the same time, children here and there, a lengthy criminal record and intimate knowledge of various penal systems, and great charm. Darrell's mother was a Cherokee from Arkansas. His mother, Aunt Fern, stayed with us for a year also, and she was always telling Darrell stories about Quanah Parker and Purty Boy Floyd, insisting Purty Boy Floyd was an Indian also and sometimes mixing the two up as far as I could tell. Her method was to sort of talk at him while she was cooking or ironing and he was racing past not even listening, as if she thought that maybe she could hit him in flight with enough stray words over time that eventually a whole story would stick. I'd stop and listen when I wasn't afraid of being left behind, which wasn't often. Darrell was even more Indian than we were, but you couldn't tell. He didn't want to be an Indian; he wanted to be Dillinger, and he looked like a dangerous, darker version of Ricky Nelson who was on television a lot at that time. Though he would have a complete set of false teeth by the time he was twenty-two, at fifteen Darrell still had most of his front teeth and liked to flash a wolfish smile at the world.

Uncle Bob had dropped Darrell and Aunt Fern off one day and vanished. We kids were delighted. There were six of us (three more yet to be born), but none as old and wise as Darrell, and we could see interesting things written all over our cousin. So could my parents, apparently, for they did not act pleased by the way Uncle Bob's brief visit had turned out. Darrell seemed to take the turn of events with equanimity. He was used to making the best of things and settled right in.

Darrell had been in a gang in Kansas City, a gang that arranged street fights for recreational purposes and, if one could believe his stories, used sawed-off shotguns and such in their pursuit of happiness. He looked around and was disappointed at the prospects. "We need a gang," he said to my brother and me. Gene, being older and significantly more mature in fifth grade than I was in third, just looked at Darrell curiously and shook his head. A gang seemed reasonable to me, but I didn't know anyone beyond fifth grade, and the two or three guys I knew were even afraid of my older sister. Darrell put his plans for a gang aside for the time being and settled for teaching us how to make zip guns and various explosive devices. For a while we destroyed inanimate objects while our parents were at work. Johnnie, the boy next door whose father was in the military, joined our gang happily, and we stole ammunition from his father—mostly vast belts of machine gun ammo that magically turned up in the family garage—to take apart for the gunpowder inside, listening to "Purple People Eater" and "Tequila" and the Everly Brothers on a transistor radio Darrell had acquired as we pliered the casings apart. Because Darrell had advanced in sophistication well beyond the requirements of common education, while we spent days in school he stayed home and studied his surroundings.

Darrell got bored blowing up mailboxes and such and began to consider his options. There were a lot of bicycles to be had. Every porch had a bike or two or three. Bicycles were valued property and to all appearances were a staple commodity of the world Darrell had been dropped into. They provided mobility and excitement in our hilly world. He decided we would steal bicycles. Gene would have nothing to do with it, but his private code forbade tattling. I thought it sounded interesting. Johnnie liked the idea, although he already had an expensive Schwinn and therefore no personal need, and did require that we agree not to steal his bike. Darrell enlisted another neighbor, an older guy in eighth grade, who had no bike.

For a few months we stole bicycles. We went out at night, slipping invisibly through alleys and along neighborhood sidewalks until we spotted a bicycle on a porch. No one, it seemed, had thought of locking bikes in the mid-1950s. Wally and Beaver Cleaver didn't lock their bikes; not even Eddie Haskell locked his bike. Timmy always coasted his bike to a stop and leaned it against a porch railing, taking Lassie with him, unconcerned about bicycle thieves. There was a niche to be filled, the times were ripe for our gang, and we plundered San Leandro of its bicycles.

I recall one moment in particular. Eight years old, I crept up to a nice house on a nice street, while my gang of three waited behind in the shadows. A very nice three-speed leaned on its kickstand in front of a big window. Behind the bike, through the window, I saw Mom and Dad and Sis sitting on a light-colored couch and a boy about my brother's age in a comfortable-looking chair. They were laughing together at something on the big television.

I remember standing on the porch, plainly illuminated by the porch light, watching the family watch television. We did not yet have a TV in our house, and the eight of us could never have watched at the same time if we had owned one. There was a wholeness about the scene, a kind of completeness that I had never imagined in a family before. In my experience families were in constant flux, coming and going, messy and uncontained. Strange cousins could appear one day and stay for several years with no evident disruption even as additional cousins showed up and vanished. Never had I imagined a family holding a single image in mind or sight even for a moment. The scene inside the picture window opened onto a world I had never considered, one I knew instinctively did not desire my company.

I tore myself away with great difficulty and stole the bicycle, unthinkingly kicking the kickstand up and rolling the bike down the porch steps with no concern for clatter. In the driveway I paused and looked back, and no one had moved. To this very day that family lives in perfect time, balanced in the flickering gray light of a black-and-white television set, forever, as I steal their bicycle. And I still feel the awfulness of the moment in which I broke into that perfect order, violated whatever it was inside the house that I could not understand, and bore an objective portion away into the night. At the time I knew it marked a point of departure, like a ticket bought and punched for a journey to a place I didn't even suspect existed.

While our father was busy driving trucks across America, a line diver, and our mother was desperately trying to keep up with, feed, and clothe six children, we took bicycles apart and put bicycles together. We ran a bike chop shop in the front yard of our rental home, with piles of rims and handlebars and goosenecks and tires and frames that adults didn't seem to notice. Darrell sold bicycle parts for a quarter, fifty cents, or a dollar to kids who wandered by. We rode hybrid bikes until bored, and then we reconfigured them into new bikes. We gave bicycle parts to friends. And still we were obscenely rich in stolen bicycles, surfeited with bikes, glutted with spokes and rims and brake cables. Our bike-stealing days finally waned due to limited storage and wholesale marketing capacity. Darrell had not counted on difficulties arising from the fact that he could not safely sell stolen bicycles or parts to someone whose bicycle we had recently stolen. Since virtually everyone in need of a bicycle in our region fit that description, he was stymied. Something had to be done.

A half-mile beyond our house a minute creek emerged from the hills. It came down out of the reform-school property in a concrete pipe about two feet in diameter. Once outside the reformatory fence, it spilled out of that pipe into something like a small settling pool, and then about fifty feet farther down it entered another conduit, a concrete pipe so big that even Darrell could walk upright through its entrance. This concrete pipe, technically now an underground stream, ran down and down beneath the city, growing smaller and smaller until a person had to crawl on hands and knees and then wriggle on his belly to plop out a small opening into a fetid pool far down in an urban world. We had made the journey downpipe several times, and we had gazed at the fold in the hills above the reform school from a high road outside the fence, imagining where the creek must be a real stream tumbling from the mountains.

Walking and crawling and wriggling through slimy water in the smelly dark until you could slip like an eel into a worse-smelling pool was great fun, but Darrell had come up with a better use for the drain pipe about the same time we became bicycle entrepreneurs. His idea was to acquire, somehow and without expense, many burlap bags, or gunny sacks as we called them, and fill the bags with sand from the sides of the creek in front of the culvert opening. In this way we enlarged the pool adjacent to the pipe and also filled sandbags that could be used to plug the orifice of the drain. We dammed the culvert, leaving a spillway near the top, and created a rapidly filled pool twenty feet across and perhaps six feet down at its deepest point. It was to be our swimming pool the following summer when the days grew hot.

Darrell realized the days were already growing hot for our gang, so he decided the pool was a perfect place to dump our disturbing mountain of bicycle parts. By night we hauled frames and rims and seats and handlebars and gears and brakes and cables and dumped the whole load into our incipient swimming pool. Everything sank, and our yard was clean and our good names restored. We felt clean and virtuous until men with badges showed up at our door.

Someone had noticed the dammed culvert and spreading pool. Someone had alerted the officials as to the identities of the dam builders. The police had come to offer us an option: dam deconstruction or reform school.

They stood by while we dismantled the dam and emptied the sandbags and carried wet gunnysacks away. They watched with astonishment as a graveyard of dismembered bicycles rose above the falling brown water. They turned inquiring eyes upon us, and we, in turn, studied the dense foliage upstream. Strangely, they allowed us to go home unquestioned that day, but we began to hear rumors about missing bicycles being matched with newly surfaced bicycle parts. Darrell said we had nothing to worry about as long as nobody finked. He made it obvious that finking was not a viable option for anyone.

Even before the heat from the bicycle caper had dissipated, Darrell was looking around. Burglary was always a possibility, and we became a small gang of burglars. I was the smallest, and because of that fact I was the one boosted through windows with the task of tiptoeing to open locked doors. In this way we engaged in what television shows like to call breaking and entering, and I learned an intense measure of fear as I crept through darkened houses alone, navigating strange hallways and negotiating ominous furniture and alien fragrances. Always there was the sharp-edged fear that I would never get out again, that I'd be locked forever in whatever empty, strange-smelling, shadowy home I had penetrated. Today I recognize certain metaphorical dimensions of those black hours in alien rooms, feeling in the dark from dim recesses toward a locked door, but in third grade I knew only the melding of excitement and terror. In those moments after squeezing through an open window, I was alone in a way I've never felt since.

I took nothing material from those foreign homes. I believe, however, that my city cousin did take things. Darrell became easy with money and made frequent trips down the hill to town on his bike. I didn't know the word "fence" back then except as that obstacle between us and the reservoir or the reformatory. And it was about this time that the boys' school, the reformatory for juvenile delinquents, appealed to Darrell's best interests. He announced one evening that it would be worthwhile to break into the "juvie." "Think of it," he said, the curl in his lip intensifying.

We thought of it. Even Gene was attracted. Apparently, breaking into a place everyone else was trying to break out of did not violate his private code. In retrospect, I believe that he found something oddly noble in the endeavor. I thought the idea was crazy, as did Johnnie, who resolutely said "No way." Because my brother and cousin were going, I had no choice, no matter how much I longed to stay home with Johnnie and make more bombs.

We scaled the chain link fence outside the reform-school garden. Darrell showed us how to lay our coats over the slanting barbed wire at the top of the fence and then, once over the top with coats left in place for the return trip, lower ourselves enough to drop the hundred feet or so to the ploughed earth. It hardly hurt and as soon as each of us could breathe again, we scurried for the shadows of trees along the creek that ran through the school. In the shadows we made our first logistical decisions. Darrell wanted to find an open window in one of the buildings, urging us to consider the neat things that would undoubtedly be inside, such as guns and billy clubs. Gene shook his head. He had breached the barrier, but he would not steal no matter how much his bigger cousin muttered and ridiculed as I sat to one side feeling very small.

Finally Darrell gave in. "All right," I could tell he was about to say "chickenshit," but nobody ever said such a thing to Gene, not ever, not even when they were fifteen and experienced in gang warfare and he was only twelve. "All right," Darrell finally whispered. "They've got those bee boxes on the hill. Let's get some honey. Stupid bees'll be asleep."

Darrell just couldn't leave without taking something, apparently anything, and Gene saw no real crime in stealing honey from bees. Bees made more than enough for themselves and would just make more if we took some. Bees were fair game, so we sidled around the trees and brush and waded the shallow little creek and climbed the hillside above the reform-school buildings where we had seen the bee boxes from the high road outside the juvie.

The boxes were exactly where they were supposed to be, six or eight of them lined up on the hillside, small white ghosts in the dark. Darrell, every gesture manifesting his disgust at his lack of a real gang, approached a box and casually lifted the lid off. The bees were apparently having a party, for they were not only up late and wide awake, but immediately irritated. Darrell stood for a moment in apparent shock, and the next instant he was running and slapping at himself and screaming, not shouting but screaming.

Gene and I started to run ahead of Darrell just as the lights came on. Big spotlights like we'd seen in cartoons began to sweep back and forth across the hillside. A siren went off.

We began to run back and forth, in and out of the blinding lights just like cartoon characters, Darrell still squealing with stinging bees. It was Gene who grabbed my arm and said, "Come on."

Gene ran straight down the hill toward the creek just above the juvie buildings, and I followed as fast as I could. As soon as he saw what we were doing, Darrell passed me in a real beeline for the creek.

Gene was already in the water and heading downstream, with Darrell right behind him when I got there. I jumped in and ran splashing after them, the siren palpable at my back, and almost immediately I saw what Gene was up to. A few yards in front of him, the stream vanished into the black maw of a culvert pipe. I was thinking that nobody could fit into that hole when Gene fell to his knees, collapsed to his belly, and disappeared. Darrell hesitated just a few seconds and slithered after him. I followed, imagining the deep pool that used to block the other end of that very pipe, happy I was no longer wearing a coat and intensely thankful for the police who made us take our dam apart. That night we burglarized our own house, slipping through a bedroom window after silently hosing each other down outside.

It was only a couple of weeks later that we decided to build a coaster for the deadly hills that defined our neighborhood. I had agitated for such an undertaking for months, only to be ridiculed, but now suddenly Darrell was excited about the idea. We'd make a real coaster, he said, sneering, not one of those soapbox pieces of shit other boys in the neighborhood had built. First we needed wheels, real ones. Something in Darrell's tone made Gene opt out of the search for wheels, though I could see the idea of a real, serious coaster pulled at him strongly.

A hospital was three or four miles away, an easy walk. We knew it well because we had regularly raided the orange trees of the retirement community attached to the hospital. I followed Darrell, who had a pair of vice grips in his back pocket, to the hospital and through shrubbery to the back of the hospital. There, beside a loading ramp, was his target: a large hand truck, or dolly, with huge, hard rubber wheels.

We watched the doors for ten minutes until he was pretty confident, and then Darrell approached the hand truck and removed the wheels in a matter of seconds, beckoning to me with a jerk of his head to come and help. The wheels were too heavy to carry, too heavy for me to even lift, so we rolled them around the hospital and through the bushes and down a long asphalt side road and onto a main road and all the way home with cars rushing past and people staring.

The heavy, solid rubber tires would be on the back, like racing slicks, Darrell explained. For the front he said we needed lighter balloon tires, precisely the kind found on good-quality wheelbarrows. Then we needed a real automobile steering wheel, which was obtained and attached to a pipe around which we wrapped steel cable that controlled a pivoting front axle. Upon that chassis we built a heavy, indestructible-looking coaster. Johnnie won the coin toss and got to test drive it down Dead Man's Hill.

Dead Man's Hill plunged several hundred yards at a terrifyingly precipitous drop straight into a crossroad. A sharp turn to the left would give the coaster a long, gradual runout, and it was a turn, Darrell decided, that a good driver could easily make. A missed turn, however, would send coaster and driver shooting off a ten-foot drop into someone's house. The someone was an unpleasant man whom I had once brained with a rock while he sat reading his paper in front of a window. I had been trying to knock out a streetlight, and the rock that went through his window was completely unintentional. Nonetheless, the man bore an intense dislike for the lot of us, and me in particular. Johnnie required repeated reassurance that making the corner on Dead Man's Hill would be a piece of cake for a good driver. Johnnie was about to get a learner's permit, so he seemed to think it imperative to be thought a good driver. He climbed into the coaster as we held it. At the bottom of the hill, Gene was waving frantically, signaling, we thought, that the coast was clear.

It became quickly apparent that Gene was signaling the opposite. Only he had accurately remembered our code, and the fact he was trying to communicate was that a large moving van was approaching the intersection.

Johnnie saw the truck in plenty of time. Scarcely halfway down the hill, he had ample leisure to plan his strategy, time his response, and both avoid the truck which had inexplicably stopped right at the bottom of the hill and make the turn. It happened in seconds.

We saw Johnnie wait until the right moment to turn, demonstrating his poise and skill as a good driver. Too soon and he'd roll the coaster, too late and it would be worse.

He spun the wheel with a cool that even at eight years old I admired and envied. But the coaster curved right instead of left. We had wound the steering cable the wrong way.

Johnnie began to frantically try to pull the steering wheel in the opposite direction and he succeeded precisely enough to drive it straight into the truck.

When we got there, Johnnie lay in a pile of splintered wood and tangled cable, halfway beneath the running board of the truck. Blood was flowing, and as I traced it to its origin I discovered that Johnnie's right hand was almost severed just above the knuckles. Otherwise he seemed in good shape.

Johnnie went to the hospital where we had gotten our coaster wheels, and doctors sewed his hand back together. The day after he got home from the hospital we had a ceremonial burning of the coaster's remains in a vacant lot. Neighbors came to the edges of their yards to glare, but no one protested. The next day, Darrell knocked on the doors of two nearby houses, explaining that he'd heard they needed wheelbarrow wheels and that he had extras. The owners of wheel-less barrows looked at the familiar objects in Darrell's hands and set their mouths hard, saying nothing as he magnanimously laid the wheels before them.

Shortly afterwards Uncle Bob appeared one day and took Darrell and Aunt Fern away. A few days after that our family moved to another town, just ahead of the law according to the rumors my older sister confided years later. With Darrell gone and no further guidance, my criminal youth more or less ended. We moved to a small white house with peeling paint in a remote canyon near Paso Robles, at the southern end of the Salinas Valley, where there were no houses to burglarize, no hospitals to ravage, no neighbors to scowl. From that time onward, sometimes with Gene but more often alone, I broke and entered upon nature herself, burgled her hills and oaks, ransacked her solitude for days and weeks and months and years on end, feeling my way toward a door that must exist somewhere. And just as it was during my brief burglary days, I have tried my best to take nothing, exulting instead in the thrill of stealth, the craft of cunning, the successful escape, the knowledge that all is never lost.

One could do worse.

CHAPTER EIGHT

The Syllogistic Mixedblood

How Roland Barthes Saved Me from the *indians*

I'm looking at a photograph of my great-grandfather, Nora Miriam Bailey's father, who would vanish from her life shortly after the moment of this photo. A very young man with an angular face and intelligent, intent expression, he looks distinctly like some kind of "other," and I have diverse reasons for valuing that discovered otherness in my ancestor. With his extraordinarily narrow, slanted eyes and pronounced cheekbones, in a different context he could be the offspring of a Chinese railroad worker in the American West or anyone from a lean and hungry part of Asia. However, I am seeking quite specifically an *indian* in this photo, a Native American ancestor.[1] Seated as he is in front of a log home with a blanket nailed over the door, surrounded by a vaguely "other"-looking family (except for his mother standing behind him and his daughter before him—both of whom look disappointingly "not-other") in what I know to be Oklahoma in 1913, just a few years after it ceased being Indian Territory, I *know* that he is, in fact, a Cherokee-Irish mixedblood from whom I am descended. But this is my ex post facto reading of the images before me, my discovery of what I bring intact to the photograph, ultimately a small part of what I wish to find there. Did I not bring to the photograph a narrative, a complex of stories set in motion by words my mother wrote on the back of the photograph shortly before her death, I could not place or define these people or this place, though there are indeed clues if, like Nabokov's Kinbote searching for his own pale fire, I bend my implacable detective's gaze closely enough. Like the mad Kinbote, I *can* read my own image and story into and thus out of the text before me. Says Roland Barthes: "Those ghostly traces, photographs, supply the token presence of dispersed relatives" (9).

I am disappointed, of course, to find no *indian* in this formal family portrait (taken most likely by an itinerant photographer who had a decent camera, necessary technical knowledge, and no cultural artifacts to secure). This family from whom I am descended wears no recognizably Indian cultural artifacts; nor are they surrounded by any such signifiers. (Though there is possibility in the blanket nailed across the cabin door: what if my great-grandfather had perversely wrapped the blanket around himself for this picture?) The cabin, alas, could be any homesteader's cabin in Sooner country. The nondescript clothing they wear could be any poor frontier family's best clothing. They could, in fact, *be* any frontier family.

I know I am descended from indigenous peoples on both my mother's and father's sides. I know this because they told me so, as their parents told them, and so on. Mississippi Choctaw on my father's side and Oklahoma Cherokee on my mother's. Only recently did we discover the box of photographs from which I have taken this one of my great-grandfather's family. My mother's grandfather (he was—and remains in the photograph—of mixed indigenous American and European extraction) is even included in the 1910 Indian census for the state of Oklahoma and on the Cherokee tribal rolls. I lay the photograph before me and search for absent origins. Such a search, the Anishinaabe writer Gerald Vizenor has informed us all, must seek "a hyperreal simulation and . . . the ironic enactment of a native presence by an absence in a master narrative" (*Fugitive Poses* 27). Extracted ultimately in the picture, my great-grandfather's image takes on the polysemous character Barthes has identified as the essence of such images. To find the Indian in the photographic cupboard, I must narratively construct him out of his missing presence, for my great-grandfather was Indian but not *an Indian.* Like most mixedbloods, I assume, I am aware of the irony of such a wistful enactment, and I have seen it in my own children and brothers and sisters and friends, the familiar tortured syllogism: Indians have Indian ancestors. I have Indian ancestors. Therefore, I am an Indian. Unfortunately, in Indian Territory nothing is so simple.

Clue: *He* looks like me when I was also young, but I am now older than my great-grandfather will ever have been. In a mirror, I have much the same eyes and mouth and perhaps something similar to the vulpine distrust registered in his resistant expression; but Vizenor has warned us that the "*indian* has never been real in the mirror, or a name of presence in the simulations of history" (*Fugitive Poses* 28). Nonetheless, I felt a kind of tremor of recognition when I saw the picture for the first time a couple of years ago. We had heard of this man, my eight brothers and sisters and I, throughout our lives. John Bailey, the mixedblood Cherokee grandfather who, our aunt told us, guided wagon trains in Indian Territory (this last surely a childishly apocryphal story about the man our aunt never met). But it was not until several years after our mother's death that the box of photographs containing this particular photo surfaced. For the first time we had some kind of hard, objective evidence that John Bailey had existed, as well as a photograph of our long-dead grandmother as a three-year-old standing between her longer-dead father's knees with his probably longer-dead-still mother standing behind him and brothers and sisters around him.

It was as if a ghost had ceased its rattling and moaning to finally show itself for the first time. He is concrete now in this image, a mixedblood ancestor, for as Barthes insists, "in Photography I can never deny that the *thing has been there.* . . . Hence it would be better to say that Photography's inimitable feature (its *noeme*) is that someone has seen the referent . . . *in flesh and blood,* or again *in person*" (76, 79). Susan Sontag, anticipating Barthes, explains: "A photograph passes for incontrovertible proof that a given thing happened. The picture may distort; but there is always a presumption that something exists, or did exist, which is like what's in the picture" (5). The key in Sontag's statement, of course, is the "like." Someone "like" my great-grandfather, with his rakishly tilted hat and amused-looking but utterly resistant expression, existed and was seen *in flesh and blood* and *in person* by at least those people around him and behind the other aperture. Existed and etched himself into the features of my rogue uncle, my mother's brother, dead now for many years, and into my own face and most likely other portions of my self. Others among the seven adults and three children in the photograph look on with sullen, uncertain, or indeterminate expressions, but my great-grandfather squints with definite amusement and perhaps a touch of contempt mingling with curiosity and a great withholding, his two hands lightly draped upon the shoulders of his toddler daughter, whose small, pale hands in turn look to be firmly holding to each of his knees. They are oddly interlocked in the photograph, the child who would be my wild-drinking, rootless, and reckless grandmother and her cagey-looking father. Oddly, because I know from family stories that he would vanish from her life almost as soon as the photograph was taken, abandoning her for reasons no one ever seemed to know (maybe death, the way his grandson, my uncle Bob, would be murdered young; maybe mere irresponsibility). What I know is that the fact of absence already haunts this photograph as does a greater mystery for me.

And greater discovery. My grandmother's mother was said by my mother and aunt to have possibly been named Edith or Emily and to have been a "fullblooded Cherokee." But there is no written record of her, and no one really knows her name or anything beyond the originary simulation of fullbloodedness. Thus our great-grandmother would seem to have joined the legion of invisible Cherokee great-grandmothers in American lore, that frontier pageant of ghostly Indian princesses who haunt our metanarrative bloodlines. She vanished, the stories always indicated, as soon as our grandmother was born and bore no other trace in family memory, a storied "fact" I have cited in my own writings. However—and I am stunned right now by this recognition—a very young, pretty, *Indian-looking* woman stands behind my great-grandfather in this recently discovered photograph, her posture straight and dignified and, most significantly, her right hand resting possessively and confidently upon John Bailey's shoulder. They are linked thus by hands, my infant grandmother, her father, and this unnamed woman. No one else in the photograph is touching another; all hands are held to the sides or on knees or together in laps. Only my grandmother, her father, and this young woman touch one another as though intuiting that this intimate and crucial connection will too soon be broken. "The eyes and hands of wounded fugitives in photographs are the sources of stories, the traces of native survivance" says Vizenor, "all the rest is ascribed evidence, surveillance, and the interimage simulations of dominance" (*Fugitive Poses* 158). These hands, forming a signifying chain from child through man to woman, are not Indian but human. *I think she is my great-grandmother.* She does not look like the other, heavily Irished family members in the photograph. She is an other "other" in this grouping. Though no one is alive who can verify the possibility, I believe I have discovered my great-grandmother, who has stood invisibly in this photograph for all the months I have perused it, the invisible Native in the tableau vivant of Indianness I have sought to construct, rendered invisible by the narrative I have brought to this photograph. Told that my grandmother's mother did not exist, I could not therefore see her image before me. "Whatever it grants to vision and whatever its manner," Barthes says, "a photograph is always invisible: it is not it that we see.... The referent adheres" (6). In the case of my ancestral photograph, the fiction has appropriated the referent; looking for traces of an absent origin, I have been blinded to the purely human image here. But I have found her at this moment. Now I believe deeply that this young Cherokee woman had being, *in flesh and blood,* that she flashed *in person* across Indian Territory and Oklahoma for brief years and vanished abruptly. Studying the diminutive, very attractive young woman, I doubt that she is the fullblood I have been told she was. She doesn't look quite *indian* enough, though she looks beautifully Native, someone I would like to know. Suddenly the photograph opens to a new narrative, more poignant, humanly resonant. Off to the side, however, I hear the whispered warning of Barthes, who writes that "essentially the camera makes everyone a tourist in other people's reality, and eventually in one's own" (57).

Barthes, scrutinizing a photograph of his mother as a child, tells himself that "she is going to die." "I shudder," he writes, "*over a catastrophe which has already occurred.*" A moment before, contemplating his response to a photograph of Lewis Payne, taken while Payne was waiting to be hanged for the attempted assassination of U.S. Secretary of State W. H. Seward, Barthes has written: "The photograph is handsome, as is the boy: that is the *studium*. But the *punctum* is: *he is going to die.* I read at the same time: *This will be* and *this has been;* I observe with horror an anterior future of which death is the stake" (96). Examining this picture of my grandmother as a toddler standing between the knees of her father and surrounded by, I am now certain, her vanishing mother, her aunts, uncles, and grandmother, I feel the encroachment of an old catastrophe as well: here are roots of uncertainty and disequilibrium. Already I see in the set of my infant grandmother's tilted mouth the defiance and dis-ease that will mark her and the disasters that she will fling at everyone within the range of her helter-skelter life. Very shortly after the photograph was taken, the family in the photograph would disappear from my grandmother's life, leaving her to carom between homes until, at the brutally early age of thirteen, she would herself be a mother who would in turn abandon her own children. This incipient, catastrophic absence of my grandmother's parents and family haunts the photograph.

All of these people are many years dead, but nonetheless Barthes's catastrophe of the photograph as memento mori is not my catastrophe. Death leaves very little residue, actually, despite Barthes's long pondering upon his vanished mother. ("Natives mourned the presence, not the absence of the dead," Vizenor has written.) For me, the *punctum* in this photograph of my ancestors is not *he is dead,* or they are dead, but *he is me*—or they are me.[2] Already in the photograph he and they (and therefore this portion of me) are disturbing absences, pricking, bruising, poignant; and I observe an anterior future in which absence, not death, is the stake. What catastrophe occurred to sever the connections represented in these images, what sent my grandmother at such a fragile age spinning off into a solitary and damaging life and erased all memories of the mother from family stories? Is this simply an aspect of the familiar story of deracination and loss that marks mixedblood history?

Heritage, history, and story are the real crucibles of catastrophe. Phenotype is phenomenal. "But more insidious, more penetrating than likeness," writes Barthes, "the Photograph sometimes makes appear what we never see in a real face (or in a face reflected in a mirror): a generic feature, the fragment of oneself or of a relative which comes from some ancestor. In a certain photograph, I have my father's sister's 'look.' The Photograph gives a little truth, on condition that it parcels out the body. But this truth is not that of the individual, who remains irreducible; it is the truth of lineage" (103). "Lineage," Barthes adds, "reveals an identity stronger, more interesting than legal status—more reassuring as well, for the thought of origins soothes us, whereas that of the future disturbs us, agonizes us" (105). I have much of my great-grandfather's look, the angular and rather disturbing face of the photograph that will be passed first to his grandson, my uncle Bob, who would be rabbiting across all our lives, disrupting, stealing, laughing, telling wild tales, being taken away to prison again and again, and dying young and violently.

The thought of origins does not soothe all of us. For some, the thought of origins can, in fact, be far more agonizing than the future. If, like a Nabokov creation, I can make of the past a necessary fiction, I must also acknowledge that the past's residence within language makes it inescapably mere fiction. However, the future that recedes infinitely before me, beyond language, is helpless to challenge my construction of it, to repudiate my reading. Despite my most strenuous efforts within language, I can determine nothing of the inexorably absolute past, of origins; the past, with its endless accretions, renders me inauthentic. As Barthes admonishes, "The Photograph does not call up the past" (82). Who were those people in the photograph? I *know* that they were mixedbloods in Indian Territory, living in what my family persisted in calling the Cherokee Nation, with Irish and probably other European roots as well—know this because that is the fiction I have memorized, my history. But as Barthes also points out, quoting Nietszche, "A labyrinthine man never seeks the truth, but only his Ariadne" (73). Winding into the labyrinth of the Territory, I find no thread of consciousness or knowledge, no Native Ariadne. There are fibers out of which a thread might be woven, bits of family story, old photographs with scribbling on the backs, but no coherent narrative. Despite the elusive promise of my newly found great-grandmother in the photograph, who would make a splendid Ariadne indeed, the remaining fibers prove too fragmentary for weaving, too thin to cohere. In the end is only the maze and the monster of hybrid potential at its center. I remain in the labyrinth, puzzled, hearing the approach of my own footsteps.

We know they are mixedbloods. My mother and aunt and uncle and grandmother spoke of these people and of this place. But these people could be any people anywhere. Barthes, captivated by the way subjects in photographs are dressed, including his mother, declares that "clothing is perishable, it makes a second grave for the loved being" (64). Clothing may indeed be perishable, but costumes are not, as nineteenth-century ethnographic photographers understood when they kept "typical" Indian dress on hand for those Indian subjects who might not possess such essential attire.[3] Costumes stop clocks. Had my ancestors in the Territory been dressed up as Indians, they would have escaped this Barthesian grave—never mind that their Cherokee relatives had for many years already been dressing and living much as European Americans dressed and lived. Fringed leather, beads, feathers, braids—such signifiers would tell me and anyone else exactly who these people were and are—for these signs of Indianness escape mutability, evade death. These people, however, were *Indian* but not *Indians*. I search vainly for signifiers of Indian presence in the photograph: if only this or that could pronounce these people Indians, but the log cabin with blanket nailed over the door, the nondescript poor-folks' clothing, the suspicious posture—none of this says "Indian." I look more closely at the faces. The hair is for the most part dark, a good "Indian" aspect. Some of the faces have the unmistakably "Asian" look to them that people descended from Bering Strait immigrants are supposed to have, most pronouncedly my great-grandfather's face. The complexions appear to be of varying hues, from light to dark. Distressingly, my young grandmother looks rather white, though her father and aunts appear suitably dark. One of the uncles, August Edward Bailey (I know from my mother's writing on the back of the photograph), looks very much like a surly Irishman. Aunt Nora, John Bailey's sister, appears quite Cherokee to me, but perhaps that is wish fulfillment on my part, since Cherokees, too, come in all forms. I do not seek Irish signifiers in this family, for the stories that have defined me from birth in my own hearing have never been Irish, and of course I enter the labyrinth in search of what I already know to be at the center: the monster of my own hybridization. Were our family stories about Irish clans and leprechauns, rather than lightning-struck trees and little people, perhaps I would study these photographs with different desire and less difficulty. If it is true, as Sontag insists, that "to photograph is to appropriate the thing photographed" (4), it is more true that to gaze upon a photograph is to appropriate into our own originary history the object of the photograph. In studying this photograph of my ancestors, I cannot deny that I am attempting to appropriate a kind of "Indianness" into my own life.

Writing of Ishi, the famed last survivor of the exterminated Yahi people in California, Vizenor has explained: "The tribes have become the better others, to be sure, and the closer the captured experiences are to the last wild instances in the world, the more valuable are the photographs. . . . The last wild man and other tribal people captured in photographs have been resurrected in a nation eager to create a tragic past" (Lippard 66, 71). In exploring my family photo nearly a century old, I cannot avoid the bothersome sense that I, too, am searching for a "better other," clues to "the last wild instances in the world," that I desire to force my long-dead relatives to enact a kind of cultural striptease to accommodate my desire for lineal contact with the original absent other. But I desire and need something the images in the photograph will not divulge. Though I may have found a great-grandmother, Indianness remains hidden away in the recesses that resist my entrance. These figures repudiate me, and I feel a kind of anger or resentment emanating from their images, as if they anticipate my future larceny. "Nevertheless," Sontag says, "the camera's rendering of reality must always hide more than it discloses" (23).

Sontag declares furthermore that "in America, the photographer is not simply a person who records the past but the one who invents it" (67). But nothing is invented in this photograph of my great-grandfather's family, and nothing remains to be appropriated. Only resistant images are recorded, resisting depth, narrativization, invention, appropriation, assimilation. Faced with this photograph, and others in the family box, I realize that the photographer who supposedly invents the past, like an Edward Curtis photographing "real" Indians, actually invents nothing but rather goes in search of what already exists prior to his subjects, has already been invented by the myth-making consciousness of America, to find, recognize, and verify that prior invention (how otherwise could he bear the signifiers of authenticity with him as props?). "Indians" are thus invented, but conversely, by the same process, mixedbloods are recorded and erased, having no place in the metanarrative of fixed colonial others. The opaqueness rendered toward the camera, the exclusion of the photographer from the implied life, the resistance of the pose, all of this comes from the fact that mixedbloods *cannot* be known—and they know full well that they cannot be known, that the camera will obscure them except within their own vision. Ours, they seem to say, is a life undefined and beyond your reach; we invent such a life each day, sui generis. Robert Young has written about history that "the other is neutralized as a means of encompassing it: ontology amounts to a philosophy of power, as egotism in which the relation with the other is accomplished through its assimilation into the self" (13). Young quotes Hélène Cixous, who, remarking upon what she terms an "annihilating dialectical magic," states, "I saw that the great, noble, 'advanced' countries established themselves by expelling what was 'strange'; excluding it but not dismissing it; enslaving it" (1).

Mixedbloods were and are exempt from such ontology and from this annihilating dialectical magic because they inhabited and still inhabit a boundary zone. This is not the bleeding "open wound" described in Gloria Anzaldúa's *Borderlands,* for no tragic victimage is embraced or celebrated here; though it is, in fact, very much like Anzaldúa's description of the borderland as "a vague and undetermined place created by the emotional residue of an unnatural boundary" (3). Cherokee mixedbloods have had a long and honorable place within post-Columbian America beginning with the first influx of Europeans. A paragraph from Grace Steele Woodward's *The Cherokees* gives us a sense of the early and thorough reality of such miscegenation:

> John Adair, from Ireland, married Mrs. Ge-ho-ga Foster, a fullblood Cherokee of the Deer Clan. George Lowrey married Nannie of the Holly clan, and their son George (born about 1770) figured prominently in the affairs of the [Cherokee] Nation until his death in 1852. . . . From mixed-blood unions consummated in the Colonial period came future Cherokee leaders, saints, and sinners. (85–86)

Today there is an Adair County in Oklahoma, and the list of prominent Cherokee mixedbloods is unending, including such figures as John Ross, one-eighth Cherokee by blood quantum and highly educated but the most prominent of all Cherokee leaders and the most eloquent opponent of Removal in the early nineteenth century; Sequoyah (George Guess), creator of the Cherokee syllabary; John Rollin Ridge, whose father was educated in New England and who in 1854 became the first Native American novelist. Countless mixedbloods made the long and deadly walk to Indian Territory and built their homes and farms there, entering into a new life as resilient and adaptive survivors, the way Native peoples of the Americas always have. No countable cultural artifacts or commodities, these people were simply bent on survival in their place and time. As Vizenor has pointed out, before the twentieth century in America, "Natives had practiced medicine, composed music, published histories, novels, and poetry, won national elections, and traveled around the world" (*Fugitive Poses* 165). But these Natives were not the Indians sought feverishly by archival imagists on the other side of the aperture and therefore not the ones whose images became, as George Steiner has explained about the "past that rules us," imprinted as "symbolic constructs of the past . . . almost in the manner of genetic information, on our sensibility" (3). The very existence of these heroic pragmatists and syncretic survivors was "obscured by the interimage simulations of the *indian,* the antithesis of civilization in photographs and motion pictures" (*Fugitive Poses* 165).[4] Mixedbloods or fullbloods, Native Americans who shunned the static props desired and proffered by cultural artificers, and insisted on living and surviving in the dynamic moment, were erased by those who manufacture kitschy history for acquisitive colonial power.

Mixedbloods cannot be appropriated because they cannot be defined. By the absoluteness of their irrefutable presence, beyond myth and metaphor, stubbornly obstructing the construction of simulated Indians, they are annoying obstacles to white appropriation. Mixedbloods cannot be neutralized or encompassed or assimilated. In *White Mythologies,* Robert Young quotes Said: "In occupying two places at once . . . the depersonalized, dislocated colonial subject can become an incalculable object, quite literally, difficult to place. The demand of [colonial] authority cannot unify its message nor simply identify its subjects" (143). Young also quotes Homi Bhabha to the effect that "in racial stereotyping 'colonial power produces the colonized as a fixed reality which is at once an 'other' and yet entirely knowable and visible'" (143). Mixedbloods are Said's ultimate "incalculable object" or "dislocated colonial subjects"; neither "knowable" nor "visible" in Bhaba's terms, they resist racial stereotyping and fixed realities as they balance within their two sites of Native and Euramerican selves.

"There is a history of darkness in the making of images," the novelist and photographer Wright Morris has pointedly noted. About nineteenth-century photographs of Indians, Morris suggests: "Presented in their full regalia, as if intuiting a final judgement, they faced the photographer as if assembled to be shot" (16). The mixedbloods in our family photographs imitate nothing, nor do they face the camera ready to be shot. They are ambivalent survivors ready to slip away at the first twitch of history's trigger finger. Uncostumed, they defiantly or sullenly or suspiciously regard the camera as a strange diversion to be briefly borne before they get on with their discordant lives, leaving the photographic record of this one moment to flutter on casual winds across generations. They inhabit neither past nor future. They give me nothing.

These photographs taken in Indian Territory are records of invisibility, and it is within the invisible that I locate mixedblood identity. There is no space for mixedbloods within the national fantasy; therefore they remain uninvented.

"Thus the air," says Barthes, "is the luminous shadow which accompanies the body; and if the photograph fails to show this air, then the body moves without a shadow, and once this shadow is severed . . . there remains no more than a sterile body" (110). In Choctaw cosmology a person possesses two shadows, the *shilup* and *shilombish,* two souls as it were. In life and death these two shadows have different responsibilities, different paths.[5] Though Choctaw and Cherokee belief systems are quite different from one another, it strikes me that the generic Native mixedblood resembles this Choctaw figure of two shadows, differences cast by the unitary unnamable coming between perception and light, without which differences there would indeed be only the sterile body. In the photographs these shadows form something like a palimpsest of soul, paradoxically simultaneous shimmerings over the flat surface. Posing with two shadows, mixedbloods split the history of darkness at the heart of image making, withholding what the colonial contriver would take. *Indian* but not *Indians,* Irish but not Irishmen.

Few looking at photos of mixedbloods would be likely to say, "But they don't look like Irishmen," but everyone seems obligated to offer an opinion regarding the degree of Indianness represented. Even my wife, who has borne twenty-five years of my personal mixedblood musings and who knows my knotted family history, peruses these ancestral photographs and says, "They don't look very Indian." I hold my tongue, not responding with what I have said at other times to other people: "Do you mean they don't look Navajo or Lakota or Cherokee or Lummi or Yurok or fullblood or halfblood or quarterblood or *what*?" How, in fact, I want to ask, could anyone in a photograph look like an "Indian," something that never existed. The answer is, of course, that these mixedbloods could look "*indian*" only by being what Vizenor has labeled "cultural ritualists": by dressing in costume, posing as the absent other, stopping the clock. Had these mixedblood Cherokee ancestors posed in traditional Plains Indian headdresses and beadwork, everyone would immediately know they were Indian. A Navajo-Laguna friend (a "real" fullblooded Indian) looks at a photograph of one of my great-aunts who happens to be very dark complexioned, has a pronouncedly large nose, and stares at the camera with sullen suspicion, and says, "*She* looks like *something*." What he means, of course, is that she is dark like his relatives and therefore looks like what he thinks of as *Indian*. In my experience, my fox-faced great-grandfather, though not nearly as darkly hued as the aunt, looks more "Indian."

How do we combat this essentialist discourse when it comes even from those we love and like? How do we teach our own children the meaning of "indianness" in their own blood when they can only look to these photographs for signs they will not find? Where is the thread that will take us to the heart of this maze and back out to the ordered world where paths ostensibly do not turn back upon themselves and monsters ostensibly do not lurk? "How are we to distinguish between the real and the imitation?" asks Wright Morris (4). I see in these mixedblood ancestors the kind of suspicious yet resolute indeterminacy that I feel in my own life and see in my own face, a kind of Native negative capability. The Indian has never been real in the mirror. I tell my children that I am not an Indian in the photographs they preserve.

CHAPTER NINE

In a Sense Abroad

Clowns and Indians, Poodles and Drums— Discoveries in France

The four boys from the Northern Ute and Blackfeet reservations had their drum set up on the right side of the stage, and the Blackfeet girls and one Ute grass dancer were lined up ready to dance on the left side. Amid the bright lights, I could see the cameraman trying to get the best angle or light on the three extraordinarily beautiful young girls in the jingle dress, fancy shawl dress, and women's traditional outfit, and the handsome warrior in the grass dance regalia. Before coming into the studio, Logan, the powerfully built grass dancer, had discovered he'd left his roach headdress on the train from Paris. Calls had been made, but of course the porcupine-quill roach was gone, now the possession of some incredulous Frenchman, and Logan couldn't dance without it. He was silently disconsolate. Luckily, however, one of our French hosts had offered the loan of his own porcupine-quill roach, a nineteenth-century Crow roach he wore when he did traditional Northern Plains dances in France and which he happened to have with other dance paraphernalia in the back of his car at that very moment. Now Logan stood on stage complete, imposing and dignified with his borrowed roach. Of the group, he was the oldest at twenty, a pillar of maturity and responsibility part of the time and obviously saddened by his own human weaknesses the rest of the time. Like all of us.

In the monitor, I watched as the camera panned across the stage and back again, strafing me and the beautiful and urbane hostess on our stools in the middle of the stage. The guy behind the camera looked to be salivating at the exotic vision of *les indiennes* in the TV studio. It was Mexican Pete's Wild West show all over again, the toast of Paree or, in this case, Nancy in eastern France. I regretted not having my Iron Eyes Cody wig, turkey feathers, and bone choker. I knew I looked too much like a professor and not enough like an extra from *Dances with Wolves* to suit French television audiences. The Blackfeet kids looked *très indienne;* fourteen to seventeen years old, they had never been out of Montana before. The three Ute guys had traveled a bit on the powwow circuit back home and though the oldest was not yet twenty-one, they had powwow scuff marks all over them and were men of the Indian world. Of the four at the drum, only the two from the Northern Ute reservation were experienced singers. The two younger Blackfeet boys, fifteen and sixteen, hadn't really sung before, not with a drum and not on French television. But that was okay since they looked Indian as hell, were *beaucoup* handsome, and could keep the beat at the drum and lipsynch while the two Ute guys carried the songs. Back home people would have seen what was up immediately and had a whole lot of fun at the boys' expense, but here in France it didn't matter.

Here in France these 50-percent lipsynching kids would get a standing ovation from twelve hundred delerious *indianistes* in a town called Belfort three days after this television show. For singing a forty-nine with the refrain of "Indian girls, Indian girls—how I love them Indian girls." They could have sung a Lawrence Welk tune for all the French audience would have known. On previous trips I'd experimented with French gullibility, making up wild "Indian" stories to see what they'd buy. Every male Choctaw infant is left alone in the woods for two weeks at six months of age, I'd explained. Not all survive, but the ones who do are forever close to nature and can speak with animals. I was a survivor. Finally, unable to bear the unceasing demand for clichéd romance any longer, I'd told a group at a wine-cellar bookstore reading in Bordeaux that I wasn't Indian at all but was actually descended from Soviet spies parachuted into Mississippi in order to foment revolution among Indians and Blacks. When the attempt failed, my parents had been forced to go underground, posing as mixed-blood Choctaw and Cherokee farm workers and having nine children as a cover. My editor grimaced and chalked up yet another reason not to invite me back to France. Tom King, the noted Cherokee writer who'd been sharing the reading in Bordeaux with me, leaned over and said, "I think it's time we got outta here, Kimo Sabe." Under the onslaught of essentialist, romantic, unshakably confident French pronouncements upon *les indiennes,* I had grown passive aggressive. While they mocked my crude sorties upon the French language, I graciously taught them expressions in Choctaw such as *Halito, chem achukma ishkitini,* which might translate as something like, "Hello, how are you great horned owl?" I was jaded by France by the time of the dancing poodle television event.

None of the kids had been to Europe before. The youngest girl, who was fourteen years old and so lovely with her flashing brown eyes, high cheekbones, bright smile, and long, shiny hair that people tended to stop and stare wherever they encountered her, had never been as far as Missoula or even been on a plane before. She had missed the flight out of Great Falls with the others and caught a later flight by herself, changing planes a couple of times before arriving at Charles de Gaulle Airport alone and taking the train into Paris without a word of French. She arrived grinning and happy as a new songbird about this strange foreign world, chattering at once with the other girls about the weird human beings and cute boys served up by France. All of the Blackfeet kids had flown to Paris alone, since the two adult chaperones who were supposed to accompany them had somehow ended up somewhere else when it came time to fly out of Great Falls. The Ute boys were far too sophisticated to have contemplated chaperones in the first place. They had already taken the metro from Vincennes, where we had been staying in the burbs of Paris, to the Eiffel Tower, luring two of the Blackfeet youngsters along with them. In Paris they had discovered the liberal disposition of French bars to the extent that the metro had shut down by the time when, in the very wee hours, they were ready to return to the hotel. Their response had been to walk miles through Paris, with no map, no French language, only their internal global positioning senses, and arrive back at the hotel sometime after daylight. They were definitely ready for a Gallic vision quest.

I'd been in France a few times, the first time fifteen years earlier when I had a Fulbright Fellowship in Italy. In the spring of that year, 1981, my wife and I had spent two weeks luxuriating anonymously in Paris and living on bread and bottled water. Signora Cipriana Scelba, "La Scelba" as they called her in Rome, head of the Italian Fulbright commission, had made sure I was given a generous stipend of $300 a month for my stay in Italy, and so when we decided to head to Paris after eight months of teaching in Pisa, we were already penniless. We thought that would be okay, since the Italians had convinced us all Parisians were dogs and would treat us like dogs, making a short visit of perhaps two days to see the Louvre the best plan. Of course the Parisians had been delightfully friendly on that trip and Paris had been Paris, so we'd stayed two weeks, starving and having a very fine time. The later trips, all in the '90s, had been to plug my novels published in France. On this particular trip with the kids, I was supposed to promote a book called *Bone Game*, or *Le joueur de ténèbres* in French. I'd warned my French editor not to team me up again with anyone like the well-known Native activist I'd gone through hell and gendarmes with on the first trip, the trip with a real Plains Indian leaning halfway out a cab window at two A.M. screaming extraordinarily creative expletives at surprised tourists on the Left Bank. After dinner and much wine—for the French find it *trop gauche* to not insist everyone drink wine, even when one valiantly attempts to demure—she had very logically come to the conclusion that these European and American tourists were responsible for all the really bad things back home, for five hundred years of genocide and ethnocide and ugly racism, and she wanted them to know it. Like the cartoon light bulb, the unpleasant realization had flashed for her that we had been brought to France for their amusement, that she in particular was there for exotic color. Crushed into my corner of the taxi, with our communal French editor—fresh off the powwow circuit—between us, I was in no mood to disabuse her of her sudden enlightenment.

Another time, the editor, Francis, had teamed me up with a Native American poet and rap artist on a nightmare tour from Brittany on the north coast to Marseilles on the south. The poet and would-be rap star wore tawny makeup and ultra-black hair dye that kept running awkwardly as he sweated and shouted his Indian anger at loving French audiences. On one occasion he cursed and stomped out of a performance when four young "goddamned Lakotas" as he called them showed up unexpectedly with a drum, intent on singing with us. "Traditional enemies" I had explained to the bemused audience in the huge FNAC superstore performance center in Rennes while the Lakota guys set up their drum and began to sing, oblivious to my presence. "I cannot feel your pain," one very *indianized* French lady, replete with ultra-black braids, bones and feathers and turquoise, had shouted at us during that trip. "I want to share your pain," she exclaimed as the management dragged her away. "Lady, you ain't heard a word I said," the poet–rap artist replied, black dye streaking down his face to blur with makeup.

I really should have known what to expect, given the carnival of madness I'd already experienced on previous trips. Indians are so popular in France that anything can happen when you are exposed to the public. It's impossible to be prepared, especially since my publisher makes a point of never telling me beforehand what is expected. Francis tried to anticipate and avoid all problems or confusions by answering all questions about Indian things before any of the indigenous Americans with him could speak. Time and again, I sat with Francis and other Native writers at a book festival or writers' conference or "debate" while a French audience asked questions and Francis replied in French with thirty- or forty-minute-long lectures. To the French audience that was apparently just fine. It became clear to all of us that the French liked to talk to each other while we sat silently looking stoic and oppressed and noble and such.

I had been on national television, on *Canal Plus* in Paris, with Shirley MacLaine, men in drag, and life-sized puppets. That had been during a previous trip to promote my second novel, *The Sharpest Sight,* when the publicity department of my wealthy publisher, Albin Michel, had used their considerable clout to get me on prime time with a big Hollywood name. It was a coup of such magnitude that folks at the giant glass publishing offices couldn't believe it. A nobody like me on national television with a great American actress? *Comment*? How had it been accomplished, they asked one another. It was like being on a version of the *Tonight Show* in which no one spoke English. For me it was stardom qualified by the fact that the night before the show I had to wash my Levi's and best shirt out in the hotel shower and semidry both with intense applications of a hair dryer. Real celebrities didn't do such things, I was certain. As I was being rushed to the television studio in Paris, I kept smelling the mildew aroma of my still-damp clothing while Christine, my publisher's manic, thin, chain-smoking public relations woman, was chanting in the taxi, "Open your eyes, Louee. Smile beeg." Being slanty-and-slitty eyed and naturally dour, when I tried to follow her advice I looked like I had been deflowered only moments before and wasn't sure what to make of the experience. Smiles don't come all that easily to anyone in my family. It's a traditional Indian thing, I explained to Christine. Traditionally, she had to understand, Choctaws did not smile for fear of offending various spirits who might become jealous of such happiness. The fear arose, perhaps, from those weeks alone in the woods as an infant. My *shilup* and *shilombish,* inside and outside shadows, might just quarrel right in front of the television cameras, resulting in permanent schizophrenia should I appear too jolly. But I tried.

I am confident that Shirley MacLaine, who looked fresh and lovely—and I'm certain smelled the same—did not wash her clothes in her hotel shower before the show. I am also confident that I should have known the tall, massive, deep-voiced blonde in the make-up chair next to me as we both got ready for television was not a woman, a fact I did not determine until the cameras were on both of us and Shirley. It was only later, after Shirley had explained to the cameras apropos of a fact neither the distinguished host nor I could ascertain that she had dearly loved the "black mammy" who raised her, and after I was already out of the studio, that I learned my voice had been translated for the national viewing audience across all of France by a young Vietnamese-French woman. What had I looked like, I wondered later, speaking in the voice of a young Vietnamese-French woman? Coyote would have relished the moment, would have continued speaking in the voice of a young Vietnamese-French woman for the remainder of the trip. *Qu'est-ce que c'est,* Coyote? the French would say.

After one reading in Marseilles, Francis and I dined in what appeared to be an actual restaurant set up on long wooden tables backstage in a theater-warehouse near the waterfront. Actors and actresses rushed frantically back and forth, sitting down to eat with makeup that, in the "restaurant" lights, made them all look insanely manic-depressive as they chattered madly and then vanished. Somehow my editor and I ended up being driven back to our posh hotel by two young actresses, both very beautiful and excited and bent on convincing us that skinny-dipping in the marvelously polluted Bay of Marseilles would be a killer thing to do. "Louee," my editor confided when both of the actresses had rushed from the car for a moment to make a mysterious phone call, "Theese can be very dangerous." I concurred.

Then there was the incident in Rennes with Danielle Mitterand, widow of the late president of France. I had attended a large demonstration in support of Leonard Peltier, expecting to hear Madame Mitterand, president of *France Liberté,* give a stirring speech only to be informed precisely at the last moment, as I stood convivially next to her, that it was not the famous Madame Mitterand who would give the speech but rather I, the Indian author. With absolutely no forewarning, I had been pushed onto a stage in front of five hundred people waiting to be educated, challenged, and stirred to action. I glared down at the widow of the late French president, and she smiled back supportively as I tried to summon forth a stirring speech. And the unfortunate events involving the plastic teepees in a small town near the French Alps where I later stood for an hour watching rainbow trout hover in the beautifully clear waters of the river running behind the town hall. Or the trouble with the fascist journalists at a dinner north of Strasbourg. Or the near-riot at the International Writers Parliament, when in typical American fashion I mistook street theater for an assassination attempt. Francis and I had mutually good reasons to distrust one another. *De trop.*

I really should not have been surprised on television that day with the Ute and Blackfeet kids in Nancy, but the clowns and poodle were still a healthy shock since Francis had assured me that this time I would not be accompanied on my promotion tour. "Ah well, Louees," he explained after the fact, "you see I knew you should be very happy to be weeth these wonderful kids. They are very wonderful, don't you think?"

They were very wonderful, indeed. The boys at the drum had their baseball caps on backwards and fingers covering one ear while they mouthed songs to themselves silently and practiced drum strokes without making contact. I was supposed to talk literature with the suave hostess who, seconds before coming onto the stage, had been smoking furiously out in a hallway as her makeup was checked. She was, I had decided as I watched her in the hallway, terrified of Indians and what might happen on her show. It was clear she had bigger career plans than a regional variety show, and wild Indians could have a deleterious effect on such plans, or perhaps the contrary. *Que será será.* On the stool beside me she looked radiant, beautiful to a point of unreality, and stank unmercifully of cigarette smoke, perfume, and fear-sweat, a tragic combination. My French editor sat on a stool to my right, ready to translate. In my mind I could hear Christine, the publisher's p.r. woman, saying, "Open your eyes, Louie. Smile beeg." I deliberately narrowed my eyes and looked dour, an angry, stoic half-Indian just across the ice bridge.

The show began, and the hostess turned to ask me a question about literature. As I opened my mouth to respond with wit and profound intellect, a clown came walking across the stage in front of me, on his hands, an upside-down, tall, lanky clown. Before I could recall what profundity I had been prepared to utter, a second, smaller clown came from the other side of the stage, bulbous nose, white bald pate, red side hair, and all the usual clown stuff, and began to play a clarinet as a small blue poodle started to spin and dance and bark to the music. The hostess smiled at the cute dog, and I watched myself on the monitor as backdrop for two clowns and a dancing poodle. Then at a prearranged signal I must have missed, the boys began to drum and sing, and the four dancers began to dance. Had someone cued the kids in advance that when the clowns and dancing dog appeared they were to go into action? Or had the kids just recognized the perfect time to push chaos to the brink? I heard the singers go into a forty-nine, with the beautifully nasal refrain, "Indian girls, Indian girls. How I love them Indian girls." Logan, the grass dancer, shuffle-stepped powerfully toward center stage, his head down, roach waving, and immense calf muscles flexing. The tall clown, still balancing impressively on his hands, tried to run away from the grass dancer and collapsed noisily, sending the poodle yapping into the arms of the small clown with the clarinet. Logan kept dancing as the hostess shrank toward me, tilting dangerously on her stool, and I shrank toward Francis to escape the hostess's beautiful smell. The tall clown had slithered off the stage and was now watching from below as the girls danced forward and Logan retreated, at which time I looked up to notice that the television monitor was showing a commercial for cosmetics.

As the commercial ran and the singers really got into their shrill forty-nine song and the poodle barked monotonously, the dancers continued, unaware that they had been preempted. At that point a Frenchman in a disheveled black sport coat and wrinkled slacks and blue shirt, with the wild thinning gray hair of a set-piece scientist, came out of a door and stepped up onto the stage. The commercial ended and the dancers realized that something else was going on and they were apparently not expected to dance any longer. The dancers moved slowly back to the edge of the stage, the drum quieted, the dog was silenced in the clown's lap off-camera, and with supreme grace the beautiful, fragrant hostess began to interview the man in the sport coat about a marvelous discovery he had made, something along the lines of time travel or antigravity. I stood up and walked off the stage and out of the studio, my sudden departure fazing no one as far as I could tell.

This was, indeed, not the first time my *très* earnest editor had tricked me into devilishly bizarre touring arrangements and circus shows. I really should have learned long before encountering the clowns and poodle. My first novel to be translated into French, a book called *The Sharpest Sight,* had just come out in Paris in 1995 when my editor, whom I had at that time never met, called to tell me the publisher wanted me to come over for a promotional tour. I had been invited, he added, to attend the International Writers Parliament in Strasbourg as well. As exciting as a trip to Paris and around France was, I politely explained that I didn't think I could afford it. When he replied that they would pay all of my expenses, I was suspicious. Many writers, I knew, experienced such things, but I was not a writer used to being mollycoddled, or even much noticed. I wrote odd fiction with little commercial panache. Had I known that, in fact, I was not to be mollycoddled in France and had I suspected in the least what lay ahead, I would have fled to the Superstition Wilderness of Arizona before venturing to Paris. But Paris is seductive, and I went.

Having traveled the French nation in four cardinal directions a number of times during the past few years, having loitered sensitively in Paris cafes with the notorious writer Thomas King, and having washed living oysters down with better wine than I am accustomed to drinking at book fairs on the French coast, I have learned some things. I have learned that I absolutely do not speak French, not a word of French despite my at one time erroneous belief to the contrary, a lesson others should heed. I have learned that while life in France can occasionally be very good for someone posing as a famous writer, life would be much better for me in France if I looked more appropriately like a Plains warrior, preferably Lakota. I have learned that as an Indian in France one should have a real "Indian" name and be angry as much as possible about things done to one's people, that the average citizen in France knows a great deal more about American Indians and U.S. history than I do, and that one out of two Frenchmen and women have been given Indian names, been adopted into one or more tribes, been to Crow Fair, and been on vision quests, often in the Bois de Boulogne. I have learned that while a surprising number of the French intelligentsia have impressive knowledge about the Leonard Peltier case and are properly outraged over America's brutally racist politics exemplified by the case, they are not at all amenable to discussing the darker aspects of French minority politics. I have also learned that one should never eat duck that smells strange, even in an expensive restaurant even in Paris, and that the French do not take "No Smoking" signs with any degree of seriousness whatsoever.

As we all know, the French do take art very seriously, however. The French read Proust and they have produced Robbe-Grillet. I felt rather vain, therefore, when my second novel, *The Sharpest Sight,* received the Roman Noir prize as the best *noir* novel published in France in 1995. Though I still do not know what a *noir* novel is, I was deeply honored upon learning of the award and even more deeply honored when my literary prize arrived. But I must step back to my editor's phone call informing me of the award. "You have won an award, Louees," he said in his always-happy-but-suspecting-trouble-ahead voice. "The Roman Noir prize." Francis, whose name must be pronounced *Frahncees,* speaks excellent French, having lived nearly every year of his life within spitting distance of the Eiffel Tower. *Roman noir* sounded very French when he said it, and very important.

"But," he added, "zere ees a problehm." Francis's English only begins to sound like an American movie when zere ees a problehm. I waited. "They have given me your prize but I do not know how to get it to you." Assuming the best, I said, "Why not just mail it? I'm sure I can cash it over here." Francis was silent for a proper moment before replying, "I am afraid it is not money." After another silence, he said, "Eet ees a gun, a peestol."

The *roman noir* prize, it turned out, was a 9-millimeter Beretta pistol, despite the fact that the prize-givers call it the "Calibre 38" award. It turned out I had written a book that had something significant in common with Raymond Chandler and Dashiell Hammett's work, which is well beloved in France. Without realizing it, I had written a *noir* Indian novel. The .38-caliber award was a 9-millimeter Beretta. I studied the situation for a moment and finally explained to Francis that I was honored to have a Beretta pistol and would consult with a gun-collector friend who would surely know how to ship such a thing from Paris to Albuquerque; I was sure many guns were shipped from Paris to Albuquerque every day.

I consulted with my friend, a Hemingway specialist, who carefully explained all the proper channels and paperwork and the long, long time involved in such a delicate international transaction. I decided to call Francis and convey the complex information. Before I could do so, however, a delivery man knocked at the door bearing a box from my editor inscribed, in large black letters, *LIVRES*. Books. Inside was a very attractive Beretta pistol which I now have framed in a kind of shadow box in my office. That is the moral of this story.

CHAPTER TEN

Roman Fervor, or Travels in Hypercarnevale

"Pianos Anders met Nella Vesco at Mario's in the Trastevere. It was raining in Rome." That was the beginning of a story I wrote in Italy, a story titled "Roman Fervor." Pianos Anders was the Devil and cruised around Rome on a Moto Guzzi motorcyle. Nella was a naïve American. Rome was seductive. The Coliseum got involved in the plot, with moonlight visits to its malarial environs. Mario's restaurant figured heavily. The names had come to me one day as I was sitting in a tiny and fairly dingy hotel room on the Left Bank in Paris, the first room I'd ever seen with a bidet in it which I mistook, of course, for a different piece of plumbing. I was alternately scanning a newspaper and glancing out the rain-streaked window, feeling like young Hemingway or maybe D'Arcy McNickle direct from the Flathead Reservation of Montana to Paris. I'd been killing time, laboring in my weak French through a short piece in *Le Monde* about a woman named Nella Vesco. I cannot remember why Nella Vesco was newsworthy, but I remember looking up from the article and seeing a store across the street with "Pianos Anders" in gold paint on the window. The names in conjunction on a rainy Paris day struck a chord, and I carried them away with me when my wife and I returned to Rome. I threw the story away in 1987, on a day of despair in the Sandia Mountains of New Mexico, filling a thirty-gallon garbage bag with paper out in my converted tack-room office in the barn. Into the bag went everything I'd ever written up to that point except a novel manuscript that I'd forgotten my agent had safely in New York. The manuscripts were all precomputer constructions, and so were gone forever with, of course, no great loss. I was never going to write again.

Today I regret only that one story, the only portion of which I retain in memory being the first two lines. Those first lines are a souvenir of a wonderful year my wife, Polly, and I spent in Italy, first in Rome and then up north where I held a Fulbright Fellowship at the University of Pisa. During that year we glided around Italy, from Rome to Pisa, Pisa to Florence, the hill country of Tuscany and Umbria, Siena, San Gimignano, Perugia, Bologna, Aosta, Milan, down to Bari, Brindisi, and back up the coast to Venice and Bolzano, over to Rapallo and the Italian Riviera. We had no money, for the Italian Fulbright Commission did not believe in money, but we had plentiful time because Italians believe in time. Italians know how to luxuriate in time, to dwell in it and bathe in it and wear it like rare perfume. We were often desperately poor, but sinfully rich in hours and days and weeks and months, for my job was to teach in an Italian university, and Italian academics long ago deciphered the key to a good life, and the key involves abundant disposable time. From Pisa we went to Paris, to Amsterdam, to Copenhagen, Munich, Salzburg, Vienna, Berne, up the Jungfrau, down to Barcelona and back to Geneva and Zermatt and over the Simplon and to the Matterhorn from two nations. We went to Greece. We ate bread and drank bottled water and I diminished rapidly throughout the year, eventually leaving twenty-five pounds of my meager one hundred and sixty-five behind when we returned home to America. I was taken for Basque crossing the Spanish border and thoroughly searched and interrogated by the formidable Guardia Civil; for Hungarian crossing the Austrian border and searched for illicit canned goods; for a suspicious character everywhere, while Polly provided respectable ballast in difficult waters.

In 1986 the Italian man of extreme letters, Umberto Eco, published *Travels in Hyperreality*, essays recounting in part his journey across America in search of what makes this country tick. "The 'completely real'" in America, he concluded, "becomes identified with the 'completely fake'" (7). Looking, oddly enough, in Hollywood, Disneyland, San Franciso's Fisherman's Wharf, Hearst Castle, and various wax museums, Eco found, not very surprisingly, exactly what he had come in search of, what he calls "an America of furious hyperreality," "horror vacui," the "creche-ification of the bourgeois universe." Eco's recorded journey shows clearly that all travel is inner travel, and that, as Emerson understood, our giants go with us. The America Eco found and illuminated is Eco's inner America, not a world anyone else inhabits, a wilderness of signs decoded by an Italian intellect according to a magic semiotic ring forged and brought along for the quest. The Italy I found in 1980 was and remains similarly my own Italy, a fresh Old World encountered with new and untrained eyes that naively embraced signs and wonders. My wife and I found an Italy of furious subtextuality, where signs were indeed signs and sometimes did not lie but rather fell away into ludic abyss or *carnevale.*

♦ ♦ ♦

Mario's was a solid little restaurant in the Trastevere region of Rome where students often lived, that bohemian quarter across the Tiber River from the more respectable parts of the city. At Mario's one could dine on half a roast chicken, an excellent salad, rice, and the unassuming yet confident house wine for three dollars and fifty cents. True, the confident wine was piped in each morning through a two-inch hose from a tanker truck—a disconcerting event I witnessed more than once—but nonetheless it was a notch above the Red Mountain I'd drunk as a student in Santa Barbara.

We would walk the ten minutes from our *pensione,* the Ponte Rotto, run by an order of nuns, to Mario's almost every night for the couple of months we lived in Rome, strolling on those fine September evenings through alleys carpeted with discarded hypodermic needles where the methadone population shot up, we supposed. Sometimes, we crossed the river on the unbroken other bridge and dined cheaply in other parts of the city, meandering up toward the Castel Sant'Angelo, of Benvenuto Cellini fame, and the Vatican to watch the lights on the river, or taking a bus through the monoxide-shrouded streets to the Piazza di Spagna to get a tourist sandwich and wander yet again through Keats's little apartment at the foot of the Spanish Steps, *la scalinata,* where he wrote and languished to his youthful end. On first looking into Keats's home, seeing fragments of manuscripts in that little room, the bare furniture, and the feeling of intense life outside, I easily imagined the young genius there, felt his sense of isolation in the face of mortality, the thoughts that he would soon indeed cease to be. Being young and melodramatic myself, a country boy with a college exposure to great poetry but no sophistication at all and thus none of the necessary cynicism that protects us, I felt Keats in that room to the very marrow of my bones, and convinced Polly to go with me to the Protestant cemetery to see the great poet's grave along with Shelley's, standing over the stones and musing upon a hidden morrow in midnight, a triple sight in blindness keen, and what great leagues I'd traveled since boyhood in Mississippi. Facing those dead poets in Rome, I truly felt that I'd gotten somewhere, even if I was already older than poor Johnny Keats would ever be. Somewhere I have a photograph of the graves, in a box with other images. (Years later I would similarly seek out the graves of T. S. Eliot and Byron in Westminster Abbey and stand awe-stricken once more.)

We were waiting in Rome for the proper time to go north to Pisa, the way countless legions must have waited in Rome over the many centuries for orders to go north, so I could assume teaching duties at the university. Though I had tried, I'd been unable to get information in advance that would tell me when classes began for the year in Pisa. When I fretted aloud about this state of uncertainty, the Romans would assure me repeatedly that the situation was quite normal and that in any case it would be a dire mistake to go to Pisa too soon. "A very little city," they would say. "Provincial. Nothing to do at all—and those Pisani." They would shake their heads and sometimes sigh, leaving the sordid truths about the Pisani unsaid but deeply inferred. Occasionally someone would look at me closely and ask if all American Indians looked like me. Upon finding that we had lived in Arizona, where I had been a fire fighter, they would look even more closely and ask if the Apaches in Arizona were still dangerous. Only those Indians blended of Choctaw, Cherokee, Irish, and French look like me, I would answer, and yes, the Apaches were still very dangerous, especially if you ended up in the wrong bar at the wrong time. Once in a while I would attempt to call the university in Pisa to discover when I could expect classes to begin, but I never succeeded in reaching anyone other than an elderly sounding secretary who would shout *Pronto* and *Non c'é* and then hang up.

We were not having such a bad time in Rome, meanwhile, and weren't terribly unhappy about not being called north to Pisa. In Rome our responsibilities included having breakfasts of astonishingly good coffee and very hard, very bad rolls with chocolate paste each morning in the *pensione,* brought to our tables by the *sorelle* who whisked silently and disapprovingly about the beautifully tiled, twelfth-century breakfast room. Each morning I would outrage the sisters by asking if I might just have one more cup of the superb coffee, *per favore. Mi piace,* I would say, flexing my new Italian, *molto delizioso.* The *sorella* would frown, mumble indecipherable Italian deprecations very likely calling upon the Deity or at least a Vatican power to witness this American abomination, and shuffle away to bring the second coffee. Caffé Americano it was, heavy and rich and thick with milk and sugar. Years later when I found I had been lactose intolerant all my life, I understood why my stomach was in even more dire pain than usual throughout our stay in Italy, but at the time I didn't put the clues together and probably would have continued to drink that coffee come hell or gastrointestinal revolt even had I known.

After breakfast our tasks in Rome took on a different complexion, for then we felt that duties called for us to venture out into the city, sticking our hands bravely into the Bocca della Veritá each day as we passed, knowing full well that if we were dishonest people our hands would be left in the stony mouth of the sculpture. We learned to cross terrifying Roman streets by burying ourselves inside coveys of scurrying nuns, for not even Roman drivers will run over the *sorelle.* We learned to board the always impossibly crowded busses by wedging ourselves behind the massive, black-clad sisters and following their torpedo phalanxes into the vehicles. The Church was endlessly valuable to us and, like the famed Osage Doctor Revard, I discovered and claimed it in the name of my nation. I also discovered that one out of two Italian women looks like Sophia Loren and the other one out of two looks like her more beautiful sister. Polly thought Italian men were equally beautiful, but I believe she imagined that. Even the bus drivers, however, were obviously dressed by Milanesi fashion designers. Everything was beautiful and all was shrouded in fumes—people, cars, buses, mopeds, and scooters all smoking all the time.

It wasn't all fun or easy. Sometimes we had to pass through underground pedestrian tunnels where signs warned in English of "Bad Atmospheric Conditions," even passing the spot where Caesar had been killed in one pedestrian underpass, or so the sign said. *Et tu, Brute,* I would say each time in passing, to the disgust of my spouse. And sometimes we had to brave the elements in the Medici Gardens, where the sun could be unmercifully hot and sprinkles of rain would occasionally nip us dashing into the Villa Borghese or out of an arbor or museum. We went to a theater and saw *L'Impero Colpisce Ancora* and *Urban Cowboy,* delighting at the sounds of *Ciao cowboy* from the screen. On a television set across the alley from our Ponte Rotto room the theme music from *Bonanza* played every night. We grew restless and nostalgic and needed to force the moment to its crisis. I was, after all, supposed to be in Italy to teach American literature at a university in Pisa, a very little city hours away by train and full of indescribable Pisani. One day we boarded a train and went north to find the truth.

Il Magnifico Dottore Professore Rolando Anzilotti, the extraordinarily distinguished professor of American literature at the University of Pisa, internationally renowned scholar, hero of the Italian resistance in World War II, member of Parliament, and creator of Pinocchio Village outside of Pisa, met us at the train station. With the authoritative zest for life I would come to associate with Professor Anzilotti, he quickly bundled the two of us plus bags into an automobile the size of a microwave oven and accelerated from zero to sixty (kilometers) in five seconds on the medieval streets of Pisa. Seated in front and bitterly envying Polly the relative safety of the back seat, I stared at the ancient buildings and youthful people blurring past on one side, vibrating with the cobblestones and afraid to look ahead. When I did turn frontward, I noticed a sign indicating that we were in an extremely narrow one-way alley and not going in the direction prescribed. "Professor Anzilotti," I murmured, "isn't this a one-way street?" There wasn't even enough room for a bicycle on either side of the car. He downshifted and accelerated even more, the tiny motor screaming. "Yes, I know," he shouted above the motor and cobblestones, "that is why I must go fast." And so it went.

In the hotel later, we discovered a disconcerting odor rising from bathroom fixtures and dissipating throughout our room. "There is a problem," I explained to the young man reading a paper at the desk. "A very strong odor." Before I could finish, the young man looked up from his soccer news and said, "*Non c'é problema*" and looked back down. He was correct, as I was to discover. There was no problem. Every plumbing fixture in Pisa gave off the same fragrance as a normal occurrence, the influence of the beautiful Arno River—and stunningly beautiful it was with the beige and orange city reflected at sunset—that flowed through town and could be recognized by its unique odor several miles away. This, we learned, was one of the many paradoxes of Italy. One of the most beautiful places in the world simply stank to high heaven in particular places at particular times, including much of the time. Tanneries.

The next morning I made my pilgrimage to Professor Anzilotti's office on the second level of a building undoubtedly familiar to

Dante, right next to the Hunger Tower. Being American, I got to the point, asking when classes would begin, when I should expect to start teaching.

"Well," Professor Anzilotti replied, "I am not sure."

I hesitated for a moment, assuming I had asked the question incorrectly. "When, I mean, does the term begin?" I repeated deferentially.

"Well," he said, "you see. I am not sure. Perhaps you should ask Professor Coltelli."

Professor Laura Coltelli was in charge of another university institute in which I would also be teaching. I decided to try once more before seeking Professor Coltelli.

"Can you perhaps give me a ball-park figure?" I said, knowing Professor Anzilotti's good grasp of American slang. "Just an approximate date, like perhaps November fifteenth?" It was already well into October, and our residence was still Rome.

"Well." Professor Anzilotti closed his eyes and slowly nodded his great, silver-maned head. "Perhaps it is possible. See what Professor Coltelli has to say."

I left in search of the other institute, wandering about the medieval part of the city through buildings that constituted a university but looked no different from all the other ancient buildings in the city. I found Professor Coltelli in a smaller version of the office I had left twenty minutes earlier and promptly asked if she could tell me when I would be expected to start teaching.

"Well," she said. "I am not sure."

We looked at each other, taking measure it seemed.

"What," she said at last, "did Professor Anzilotti say?"

I decided on the truth. "He said November 15 sounded possible."

La Professoréssa Coltelli closed her eyes and considered. "That is a good date," she said finally.

Back at Professor Anzilotti's office I found him in the same position I'd left him in. "What," he inquired, "did Professor Coltelli say?"

I closed my eyes for a moment. "Well," I responded. "Professor Coltelli said November fifteenth sounded good."

"Well, let us say classes will begin on November 15."

And, of course, my classes began on November fifteenth, which I discovered much later was close to the official date for the opening of Italian universities. *I professori* had been making pleasant with me, making *uno scherzo*. And my students were allowed an extra ten days. But still there was the matter of textbooks. I had agreed to teach a graduate seminar on the works of John Steinbeck, very popular in Italy, an "American seminar" as Professor Anzilotti insisted, meaning I was expected to force Italian students to actually speak in a class and, even more extraordinarily, to write papers. For Professor Coltelli's institute I would teach an undergraduate survey of American literature.

I had corresponded with Professor Anzilotti at some length about the texts to be required for each class. I was earnest, serious, careful, and after much worry and scrutiny I had decided upon a particular anthology for the undergraduate survey and a proper selection of Steinbeck's fiction and nonfiction for the seminar. Where, I asked Professor Anzilotti, should the students purchase the textbooks?

He raised his great head and looked at me beneath heavy lids. "Well," he said. "You see, the students cannot purchase the textbooks."

I hesitated before asking why, intuiting that the answer would not clarify my situation greatly.

He kept his patient expression fixed upon me, showing that he was willing to go to great lengths for a perhaps dull-witted American. "Well," he replied finally. "The students cannot purchase the textbooks because we did not order the textbooks."

I squinted slightly. "But," I said, "I thought we had agreed that you would order textbooks."

"Well." He smiled gently, obviously showing enormous patience with his slow visitor. "You see, we did not order the textbooks because the students would not have purchased the textbooks."

Classes began on November fifteenth, without textbooks. Earnest to a fault, I put on my gray tweed jacket, blue shirt and tie, and charcoal slacks and, carrying an umbrella like all good Pisani, bussed my way to the university district. At first, for my undergraduates, of whom there might be one hundred or twenty students present on any given day depending upon uncertain forces I knew nothing about, I would type ditto masters on the little portable typewriter I had brought with me, spending hours typing "Crossing Brooklyn Ferry," "The Open Boat," stories by Jewett and Freeman, Eliot's "Prufrock" and even—it still hurts to think of this—"The Waste Land." I'd rush into the American literature offices, throw off my jacket and roll up my sleeves, and, watched inscrutably by the thousand-year-old librarian, run dittoes off on the great hand-cranked machine, staining my hands and arms a brilliant purple.

For the first month I diligently dittoed and handed out copies of poems and stories to the undergraduates, a small crowd or a great sea of disturbingly beautiful people, mostly female and mostly improved versions of Sophia Loren. The students would be very polite and grateful for the poems and stories I gave them, and they would take them away. By the end of the fourth week, however, I realized with a dawning certainty that no one had read anything I had spent so many hours typing and spilling purple ink over. They simply were politely and smilingly accepting the papers and taking them to some mysterious destination where they remained eternally unread. I came to my senses and became a performance artist and storyteller.

Because most of the lovely young students spoke or comprehended only very limited English, I read poems and passages to them slowly and carefully and told stories in my limited Italian. I told them the story of the boy and the runaway slave on the Mississippi River, reading aloud the wonderful descriptions of daybreak and lightning storms and Huck's passionate decision to go to hell for Jim. I told them what errant critics thought and said about the boy and the slave on the river, and together we condemned those critics as too dense to understand the true genius of the novel's ending as well as I, and now my students, did. I recounted the adventures of Humphrey Van Weyden and Wolf Larson, reading Wolf's wonderful philosophy so forcefully that the students became one and all violent materialists for a day. We all decided we, too, would rather reign in hell than serve in heaven. For part of one day we worked excruciatingly together to translate the lobster and squid at the beginning of *The Financier* into a common language. We made our way through a much-abridged oral version of *Moby-Dick* and through Wharton's sad story of Lily Bart doomed by a cruel society, a tale the students understood quickly. Attendance improved dramatically after the first three weeks until the room was packed with students who now had a heightened appreciation for the American system of education. I indicated to them that there were poems and stories and novels written and published by Native Americans, or "Red Indians" as indigenous Americans are often called in Europe. Trickster tales had enormous appeal, and the pathos of *House Made of Dawn,* in my thirty-minute oral version, enraptured them. I recited "Sunday Morning," "The Idea of Order at Key West," "Stopping by Woods on a Snowy Evening," and "The Red Wheelbarrow." Students listened to everything politely, barely breathing, coming away from my class with significant knowledge of and admiration for a very peculiar version of American literature, a kind of knowledge that must have fascinated the Italian professors who directed the students' oral examinations. There were no papers to read and no homework assignments to grade, nothing to do but perform for three hours each week. My profession would never again seem so delightful to me

For the students in the graduate seminar I scoured Tuscany and Umbria and even more distant places. Bookstores in towns like Perugia and Siena, Florence and Rapallo, all had copies of various Steinbeck novels. I found one here and two there, and I made gifts of the novels to the six students. "Communisti," as they called themselves, and to a person great admirers of Ronald Reagan, they sat in my seminar in fine suits and fur coats and very slowly, painfully, internalized the foreign concept of verbalizing personal feelings about a novel.

"So," I would ask brightly, "what do you think of this novel? Was it truly necessary for George to shoot poor Lenny? And what about the fact that it was a German pistol?"

The students, all of whom spoke nearly perfect English, would look at me politely, and after a few minutes I would try a different tack, a different question with the same results. Finally, I explained that we were supposed to discuss the books, that such was the nature of a seminar. They were supposed to tell me what they found in the books, how they felt about them, and we would discuss those things. It was their class, truly, and I was there to guide them perhaps but not to lecture.

On the third day of silence, one young man took pity on me. "*Professore,*" he said, obviously struggling for a proper explanation. "Well, you see, although we are *communisti* we have been produced by a Catholic system. In such a system you are supposed to give us, maybe you would call it the catechism." He smiled, relieved at having found the right concept. "And then, you see, we write it down and tell it to you."

"I give you the truth and you write it down?"

We had a good time. I met the timeless crone who had shouted "*Non c'é!*" at me and hung up during my calls from Rome. Reminiscent of the Sybil hanging in her wicker basket in *The Waste Land,* she was older even than the thirteenth-century university, bent and wizened and untouchable in the enormity of her remoteness. She could neither hear nor see with any certainty, and in her keeping rested the great, massive, invaluable cast-iron key to the university's minuscule library. Such an undertaking was it to persuade her to rise from her mountainous desk and climb, bent and twisted, over a period of what seemed painful hours, the precipitous flight of marble stairs to the little library and insert the key into the glass doors covering whatever segment of literature one desired, that one seldom summoned the courage to do so. Consequently, we read alphabetically, having grabbed an armload of books from whichever library case she had opened. Winter was a season of J's: Helen Hunt Jackson; James, both Henry and William; Randall Jarrell; Jeffers and Jefferson; Jewett and Joyce.

Pisa was a communist stronghold in a conservative region. The train station in Bologna had been bombed the year before, and we all lived in fear of the *fascisti*, the ultra-right-wing terrorists whose practice was to strike randomly, blowing up public places to intimidate the population. The Brigate Rosse, or Red Brigades, on the left, in contrast, selected their targets carefully, preferring assassination to indiscriminate terror. In the very center of Pisa, the national government had, in a display of exceptional insight and miscalculation, placed the training center and barracks for a formidable military division, the feared *paracadutisti*, or paratroopers, the Italian Green Berets. Friction at times developed between the communist students of the university and the right-wing paratroopers, especially when the lonely soldiers were drawn toward the young women who lounged on scooters in the plaza, as happened one April afternoon.

The boyfriend took exception and beat up the paratrooper. The paratrooper rushed to the *paracadutisti* barracks and returned with a hundred friends who came marching through the streets in a menacing phalanx giving fascist salutes. The students responded by beating up all of the paratroopers, who then fled in a wild panic back to the barracks. The next day mothers began arriving from all over Italy, and by the following morning furious and concerned mothers lined the street in front of the entrance to the *paracadutisti* center, protesting the federal government's failure to protect their sons.

Italy was different, I decided. When Polly's purse was stolen, the lesson of difference was driven home. We had been traveling north of the Alps, guarding our meager resources with extreme care, but we had returned to Pisa and felt at home there after several months. She let her guard down, and a deft boy on a scooter snatched the purse right from under her arm as she boarded a bus. Passport, credit cards, a small amount of cash, and a treasured little pocketknife were in the purse, and we were upset.

We reported the theft to a beautifully dressed *poliziotto* standing on the edge of the plaza. He informed us that we must report the theft not to the *polizía* but rather to the *carabinieri* and gave us directions to the proper location. The *carabinieri* dwelt inside a courtyard with an imposing gate guarded by impeccably uniformed men with machine guns. We inquired nervously at a small office as to the correct procedure for reporting a stolen purse and were greeted with muted outrage. One did not report stolen purses to the *carabinieri.* The *carabinieri* were involved in a war against terrorism, not purse snatchers. Stolen purses belonging to foreigners were correctly reported to the Ufficio per Stranieri. We were given directions and set off again. It was getting late in the afternoon, and we worried that the office for foreign victims of crime might be closed.

The office was very small, a cubicle up two flights of stairs, dimly lit but open. Behind a desk sat a disheveled balding man in a white shirt, sleeves rolled up and wrinkled suitcoat on the back of his oak chair. He needed a shave and looked cross but perhaps patient. The room reeked of great quantities of smoke over long duration. He motioned us to sit and said, "Yes."

"Well," I said. "You see, my wife's purse has been stolen. We were told to report it here."

"By whom were you told to do this?" His English was perfect, even to the tired accent that would have fit on the Barney Miller show.

"The *polizía* and the *carabinieri,* both."

He nodded. "Well, yes, that is correct."

We waited. He considered his cluttered desk top very sadly before finally saying, "But I am afraid you cannot report the stolen purse at this time."

"Well," I said. "We were told to do this."

"*Dunque.*" His sadness at the impermeable ignorance of foreigners had caused him to slip back into Italian. "You see, you cannot report this because you do not have the correct form."

"Form?" Polly said very politely.

"You must purchase the correct form, the *carta bollata.*" He seemed resigned to the impossibility of anyone purchasing such a form.

"And do you have such forms?" Polly continued, her voice now hopeful.

He shook his head. "You must go to the *tabacchino* to purchase the form." He explained to us how to find the *tabacchino,* which was not far away.

When we returned with the form, the detective smiled broadly. No longer despairing of the impenetrable ignorance of all *stranieri,* he set to work quickly and efficiently to help us fill the form out properly, at the end of which time he sat back in his swivel chair and took a deep breath, seemingly sad once more.

"Do you think there is any chance of finding the purse?" Polly's voice indicated that she, like me, held no such hope.

He tapped his long fingers on the desk, sighed, and examined the form before him as if it might provide an answer. "Well, you see," he said, "we will find the purse."

My wife and I looked at one another.

"And the purse will contain your passport and credit cards." He sighed. "But we will not recover the money or the little knife."

We left the office shrouded in mystery. Four days later, two nattily attired officers of the *commissariato di polizía* rang the doorbell of our apartment in the postmodern apartment building where we lived amidst vineyards outside of town. Very politely they presented my wife's purse and wished us a fine day. "The little knife," one of them said sadly. "Well, I am afraid it is not there." "And the money," the other officer shrugged and offered a commiserating smile as they departed.

The Scottish woman in the apartment below scurried up the stairs to ask why the coppers had been there. Her Algerian husband was outside, so she could leave her apartment. It was safe because, as she indicated through our fourth-floor window, her husband was at that moment ascending the scaffolding of another apartment building under construction. He was, she explained, upset because some of the workmen had whistled at her as she'd walked by just a few minutes before.

We watched him climb to the level of the half-dozen undershirted workmen. I opened the window. Her husband was waving his arms in front of the men who were watching him with what appeared friendly curiosity. Then the men all began to wave their arms. The husband pushed one of the men, who pushed him back, both almost losing their balance. "You should understand," he shouted in Italian. "Because you, too, are Algerian." All the men grew calm, contemplative. The husband turned and descended the scaffolding, and the neighbor rushed back down the stairs to her apartment. On the morning after their wedding in his place of birth, she had confided to my wife, her husband's family had paraded the red-stained sheets of the nuptial bed through the streets of their little town.

♦ ♦ ♦

For exercise I ran miles through the countryside, two and sometimes three hours at a time, passing through several little villages, a tourist in running shoes. One day eight miles into my run I found a pair of enormous antique, hand-forged pliers. I ran the eight miles home holding the beautiful pliers, and nearly every car that passed me stopped to ask if I required help repairing my automobile. I bought an Italian racing bike, a Legnano *da corsa,* and rode it often up into the mountains outside of Pisa. We had tried to buy such a bike for Polly also, but the stout gentleman who owned the bicycle shop, and had in his time been a great racer, refused to sell such a bike to a woman. It was not proper and he would not commit such a sacrilege even for a handsome profit. I admired his sense of ethics, though Polly did not.

We left Pisa in the spring to travel in Europe before going home, staying one last day, however, for Pisa's grand *carnevale* celebration. Festivities were to begin in a plaza beside the beautiful, fragrant Arno. The newspaper had promised that four marching bands from neighboring towns would wind their way through Pisa from four directions, converging upon the plaza. Furthermore, the newspaper hinted, there was to be a grand surprise.

We arrived early to get view-seats on the high wall along the river. Music blared from a PA system. A tiered wedding cake that appeared to be made of beautifully painted and filigreed cardboard, twenty feet in diameter and at least fifteen feet high, dominated the center of the plaza. An anxious crowd milled around the cake, some dancing to the loudspeaker music. In the distance we could hear the strains of four marching bands already approaching, each band playing a quite different tune from the others and the four blending with an interesting effect as they drew closer.

When the marching bands sounded as if they were each just around a corner, the crowd began to cheer. Suddenly the top of the wedding cake flew off, the sides of the top tier fell down, and seven or eight men in pink tutus, tights, and full ballet regalia appeared. It was the famous Trocadero dance troupe from Monaco, ballerinos in drag. The crowd went wild as the Trocs began pirouetting and the four bands pushed suddenly into sight from the four opposing streets, their disparate tunes now an earsplitting cacophony drowning out the local music.

The crowd, caught between four advancing pincers of the musical phalanxes, grew apprehensive, and then, as the bands converged, the crowd began to panic, spilling against the great cake with its dancing ballerinas. The bands had been told to march but not to stop, and they crashed into one another in a massive collision, crushing and engulfing the crowd of carnival goers, pushing against the cardboard cake, which began to topple. Men in tutus scrambled as everyone shouted at once and the trombones and tubas and French horns and drums played on in a mad tidal swirl. High and safe on our river balustrade, we watched, travelers in a hypercarnevale, not Eco's furious hyperreality but rather a kind of furious reality, manic and rich and real to its unbelievable end. Pianos Anders appeared on his Moto Guzzi, Nella Vesco perched behind, as he revved the powerful engine and blared his powerful horn and grinned, waving to us from the maelstrom.

The mixedblood Cherokee Bailey family in front of their log home near Muldrow, Oklahoma, 1913. The author's maternal grandmother is standing between the knees of her father, John Bailey.

The author, in a sense abroad, posing in Paris in 1998. Photo by Thomas King.

Hanging out at the Pensione Ponte Rotto with Polly, Rome, 1981, ready to go home. Photo by Pearl Jones.

Ranger days in the Glacier Peak Wilderness, Darrington, Washington, 1974. Photo by Bernie Smith.

At the Forum, Rome, imagining *hypercarnevale*, 1980. Photo by Polly Owens.

In the high country, sitting on ice axe, Glacier Peak Wilderness, 1976. Photo by Bernie Smith.

Forlorn introspection on being told to give a speech for France Liberté, with Madame Mitterand in Rennes, France, 1998.

Winning the crosscut-saw championship with Rich (Joe oiling), Prescott National Forest, Arizona, 1975. Photo by Dan Fagre.

With fish on trail crew, Milk Creek, Glacier Peak Wilderness, 1971.

"I hear the train." Great-Great Uncle August Edward Bailey, between the tracks. Near Muldrow, Oklahoma, 1911.

Three trail crews, with the Milk Creek Trail Crew flag, Darrington, Washington, 1971. *Clockwise from right, squatting:* Whistle Pig Martin, G. I. Joe, Owens with whipsaw, Digby, Captain P. G. Fox, Bogmire with chainsaw, Dago, Ralph, and Whitey.

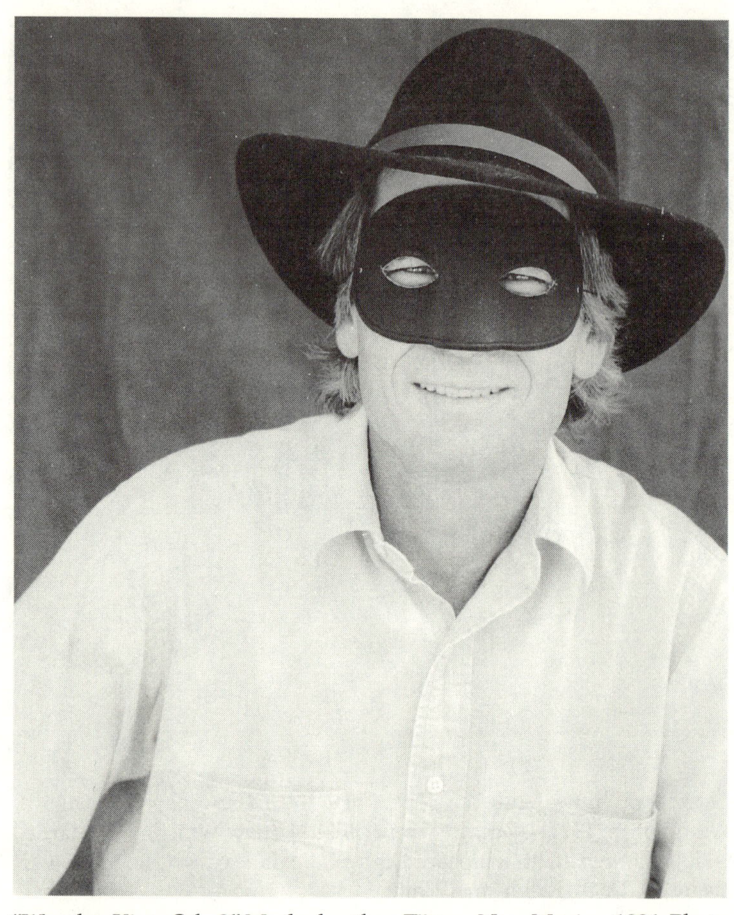

"We who, Kimo Sabe?" Masked author, Tijeras, New Mexico, 1994. Photo by Thomas King.

PART TWO

Inventions

CHAPTER ELEVEN

Coyote Story, or the Birth of a Critic

Coyote was going along, maybe forty miles north of Pietown. About half an hour before sunset on the southwestern edge of the Acoma reservation, the cliffs had deepened to a cabernet red. Bluebirds sat on and fluttered upward from fence posts, pronghorn antelope twitched their tails and stared at the distance, and the flat washes on either side of the road held vanishing monsoon pools, clumps of chamisa and sage, sparse grama, white poppies, and orange paintbrush.

Coyote carried a jackrabbit in his mouth, the head and ass end flopping as Coyote dogtrotted toward the distant mesas. If asked why he seemed to be in a hurry, Coyote would undoubtedly have said, in that patient-but-condescending way, "I got shit to do." What caught my eye this day, however, was the raven that flapped down right in front of Coyote and began to dance, jumping a few inches off the rain-pocked dirt and fluttering his wings like shiny black shawl fringes and bopping his head like a Beat poet.

I stopped the pickup and rolled down the passenger-side window to see and hear better, since I was driving north and the show was on the east side of the empty two-lane road. Raven dipped and dodged and then opened his beak and squawked a high-pitched laugh at Coyote, and that was when I saw the second raven, just off to the side, about the same time that Coyote saw it.

Coyote stopped in his tracks and eyed the raven in front of him, and I could see the anger slide down over his face along with another expression that must have had something to do with the possibility of a raven side dish.

Coyote dropped the rabbit and stood with mouth open, tongue hanging out, eyeing the raven in front of him. He took a step toward the bird, and the bird jumped up and fluttered its wings and cawed another mocking laugh, giving no ground. Coyote took a second step and a third, and the rabbit was behind his back feet now. That was when the other raven ran in with back-folded wings and grabbed at a hind rabbit foot, but Coyote was too quick and spun almost in time for a raven snack.

Raven number two leaped and fluttered back between chamisa bushes a healthy distance away, shaking her feathers until she looked sleek again, while raven number one began his irritating dance once more. Coyote sat down to consider the situation, turning to study raven number two for a long moment with an expression that looked like a cross between amusement and irritation, and then turning back to watch the raven dancer in front of him. Meanwhile, the dancer had dipped and bopped a foot or two closer and stretched out his scrawny neck to caw a definite insult at Coyote, very likely having to do with Coyote's relations. That's when I got out of the pickup and went to lean on the government fence, the kind with reinforced steel corners and tight barbed wire made by people who don't care how much money they spend, not the kind most people make with gnarly juniper posts held off the ground by loose swirls of rusting wire.

♦ ♦ ♦

"Tight spot, enit?" I said, looking carefully over Coyote's head at the wine-colored cliffs where a Spaniard had carved graffiti hundreds of years earlier. *Paso por aquí,* the conquistador had written in the rock. And pass he had, leaving behind a few hundred thousand heirs. The very top of one of the cliff towers had eroded into an exact image of a giant, stylized eagle looking down at the road.

Coyote looked at both ravens, one after another, and then turned his head toward me.

"*Yatahey, wasichu,*" he said.

I nodded.

"How come you can talk to me? Only Indians are supposed to talk with Coyote."

"I am an Indian," I replied.

Coyote glanced at the watching ravens, placing a forefoot on the dead rabbit. "Give me a break," he said. "Cut me some slack. No way. Get outta here."

"You've been hanging out around those truckers at Sky City," I said, "to talk like that."

"No way, fugidaboudit. Those truckers are poets. They use metaphors. Califuckinfornia. Minnefuckinsota. Get this, 'We truckers are a funny bunch.' Want to hear the rest?"

I shook my head, having read that one myself in a bathroom in Arizona.

"Read that in a stall over there at the casino. I know my feminine self should be offended, but I find the sexual bravado rather charming. What part of you is Indian? Must be below the belt. Drop your pants and let's see."

I shook my head. This was a long, long time ago, when two-leggeds and four-leggeds could still talk to one another. Late summer of 1989. I was driving back from a fishing trip on the White Mountain Apache reservation in Arizona.

"*Et tu, Brute.* You ever hear of the trick I played on Vulva Woman?" Coyote grinned, his tongue hanging out. The ravens shuffled closer, one in front and one on the east side, making no sound.

"What kind of juice you run in that thing?" Coyote nodded toward my 1980 Datsun pickup.

"Anybody ever tell you that you suffer from attention deficit disorder?" I asked. "I'm Choctaw and Cherokee, somewhere between a quarter and three-eighths."

"Oh, Cherokee, enit? Ain't we all? About half those white men and buffalo soldiers that got rubbed out with Custer were Cherokee, or so they kept trying to say near the end there. Even some of those Crow scouts started saying they were Cherokee. Didn't help. You ever eat buffalo? Must be tough having only a quarter or a three-eighths. Girls don't like it. Oh them indun girls, indun girls. How I love them indun girls." He hummed to himself and looked dubiously down at the rabbit.

"Stock four-cylinder with a Ford carb and distributor," I said, glancing fondly back toward my high-rise 4-w-d camper-shelled pickup. I'd bought the thing from a private at Kirtland Air Force Base who'd stripped all the emissions control stuff off of it, jacked the body up to put a second gas tank in, put in a lift kit, added big tires, and made it totally illegal to operate in most parts of the U.S. I wasn't surprised Coyote admired it. I noticed that the first raven had inched to within a couple feet of the rabbit.

Coyote looked at the bird and grinned. Both birds jumped backwards simultaneously.

"Simultaneously?" Coyote grinned. "Want to hear some truck-driver poetry? It can be offensive. I go over there and read the bathroom writing. Trucker oral tradition."

I shook my head and Coyote watched me with narrowing expression. Abruptly, the north raven ran in and pecked, coming away with a rabbit eyeball.

"Damn!" Coyote yelled. "You let him abruptly get the best part."

"That's disgusting," I said.

Coyote laughed and picked up the rabbit. Both ravens danced up and down angrily, hopping from foot to foot.

"I abruptly got shit to do," Coyote said around the rabbit in his mouth as he turned and started loping back the direction he'd come from.

CHAPTER TWELVE

Blessed Sunshine

Prologue: Among photos in a cardboard box left behind when my mother died is one of a most strange kind of raft-boat. What follows here is the story I have told myself to explain my child-grandmother's extraordinary, imageless presence in this photograph. Though a fiction leavened with the yeast of memory and inherited story, it seems to me truly no greater fiction than everything I know to be "true" about Nora Miriam Bailey, the hard-drinking, wild, destructive, selfish, tormented, inexhaustible woman I was told to call Grandmother and about whom I learned fragments of story more rare and unbelievable than any fiction I or my brothers and sisters might have invented. Those "true" stories left us with ghost-afflicted images of a woman who gave away her three children, our mother and aunt and uncle, to different farmers at a country barn dance near Tahlequah, Oklahoma. Gave them away and left with a stranger, a bottle in her waving hand being my mother's last image of the mother who bore her at age thirteen. The last image until a year later when three little people came in a dream to tell my mother she would be freed from the sharecropper who abused her, through his death, and her mother appeared from nowhere to reclaim her. The "true" stories tell us that, after our grandmother had reclaimed all three children, the four of them walked for a whole winter across west Texas, hiding in the woods, stealing food, living any way they could, barefoot on frozen ground. Each of those splinters of story leaving a festering sore that in some way must be dealt with.

What made Nora Miriam Bailey that mother and grandmother? Can we doubt that a raft called Blessed Sunshine played its part in carving that tormented adult? The photograph exists.

BLESSED SUNSHINE

Nora Miriam Bailey, six years old, woke in darkness, lost and terrified until she felt the thick river shift beneath her. The raft sighed and settled, and she heard the brittle kree of a kingfisher pass by in the night. She remembered then. In her brief life, she'd known different brown waters, but nothing like this Texas river or this raft with its load of death. On the other side of the cold barrel stove and tin chimney, in a wooden box of a bed, the man breathed so hard she could feel the rancid air moving in the tiny room. The small woman in the bed with him made no sound at all, and Nora thought how the woman had looked like a white-breasted bird in the gray daylight. River dampness rose through wood floor and corn-shuck mattress and pasted the flour-sack nightgown to her skinny thighs. She hugged the patchwork comforter closer to her chin, hoping to hear the kingfisher again in the night but hearing only the man's breathing now and the quiet slap of water. The heavy air, which bore the weight of fried fish, well-used lard, the extinguished kerosene lantern, fleshed animal hides, unwashed human bodies, and the rotting river below them, pressed her to the limp mattress and floor. She understood that her life had been given over to unknown forces and strangeness yet unnamed, and she determined with all the cunning of her years to live through this, too. "*Waddo,*" the man had said to Mr. Timmons, "Thank you" in Indian. She was a thing given, and thus was sown the seed of her own giving scant years later.

Atop the raft-house, beyond her sight even had she been able to penetrate that darkness, were coon skins scraped and stretched on stick frames like death's wings, and the bloated body of a wildcat laid casually on the roof's edge. This had been her first vision when she'd arrived the day before, and it remained with her in the complete darkness of the little room. If a wind came up, she knew the skin sails would carry the raft down that black river of ghosts where her mother and granma waited. This was her terror and her consolation.

They'd traveled a week to get there in the wagon, Mr. Timmons with his thin, serious face coaxing and threatening the mules every step of the way. "This here country's Texas," he'd said on the second afternoon. "Yeller rose." He'd smiled at her with his crooked teeth. "How'd you like to be the yeller rose, little Nora Myrum, the yeller rose o' Texas?" On the splintery seat beside him, she looked away toward the big horseflies swirling about the poor mules. She'd seen a yellow rose in Nakey Waters's yard the week Nakey died of fever. Mr. Timmons began to mutter a song about the "yeller rose o' Texas" that was the only gal for him. She heard "diamunts" and something like "Roserlee." His sudden voice after so much silence was frightening, like a ghost or an owl call at night.

"A owl called that night Granma died," she said, looking at the backs of the mules' heads. "A screech owl."

Mr. Timmons stopped his song and nodded his long head once, acknowledging the gravity of what she'd said. Then they were both quiet for a long time, watching the horseflies and the ragged country they were passing through.

Blackjack oaks ranged the low hills on either side of the road, and ravens kept darting shadows over the wagon so she had to close her eyes not to see. Sometimes, when she opened her eyes and brushed her thin hair back, she saw squirrels moving in the woods, and she thought of her granma's biscuits and squirrel gravy. In the lard pail, behind the seat with the bedrolls, were still half a dozen of Mrs. Timmons's hard biscuits. The Timmonses were neighbors, their plank-walled cabin just a few miles from the Bailey home. Minnie Timmons was the one that found her and brought her home.

After what seemed like days of silence, when they had left the noisy roads and all towns behind, when only jays called from the brush and hawks and those black buzzards circled the pale sky with motionless wings, he had begun to talk. "They's Cherokees down this way, too, passel of 'em." His dark eyes looked out over the mules toward the dry hills ahead of them, and he lifted and dropped the leather straps gently across the animals' backs. "Old settlers that ain't never moved. You got relatives among this here Trinity River bunch. Don't you worry, little girl, these folks'll keer for you." Still, except that one time he never looked at her and never smiled, and even at six years old she understood the mix of relief and guilt in his voice, tones she'd heard plenty of times. Whoever was there, in that place he was taking her, might not care for her at all. Maybe just like the Timmonses they'd have too many children of their own to take in another one. Or maybe they'd all died or just gone away, the way people did.

A ploughed section of reddish-brown earth edged up beside the dirt road, and she watched the furrows sway out of sight toward the woods and wondered what this kind of people planted in the earth, and did such people have children too, and then abruptly they were in a region without ploughed earth or any sign at all of people like them. They had come suddenly into a country like nothing she'd seen before, forests of stumps and splintered brush going away from the road on both sides as far as she could see, all gray and dead, the earth gouged and pocked. No animals moved in the blasted woods, and the sun shone only dimly through the thickened air.

The river, when it rose into view, was starved and brown, its water seeming not to move at all, the broken forest reflected in its surface. Tied to a sliver of stump, in the middle of nothing, the raft with its wood shack looked lost and crazy, and she had to focus on it for a long time to know it was really there. Close to the raft and river, the earth was scraped like a sore, and beyond the scraped place stood dark, splintered trees and heaped brush, like a world that had been smashed and bruised by powerful lightning. Here the air seemed burned, a gray haze that hung like smoke over everything, tangling in the fragments of trees. She knew they had found a place never touched by sun, that they had arrived in the Darkening Land, the place of nightwalkers and death that her granma had told her of. Then she understood the black ravens that had shadowed the wagon for two days, the dark figures that had peopled her nightmares for so long. This was the Nightland described in stories, where the nightwalkers were.

Mr. Timmons had stopped the wagon atop the long slant to the river, set the brake, and stared with her at the scene. She could hear him suck air between his teeth as they sat together, neither moving at all. On the side of the raft-house was a picture of a fish, large and evil, and someone had painted letters and numbers above the picture. Even she could see that some of the letters were backwards. "It's a ignorant people," she said to herself so he couldn't hear. Her granma had taught her letters because ignorance was of the devil.

When Mr. Timmons spoke, his voice was forced. "Blessed sunshine. You see them words painted, Nora Myrum? It's your daddy's uncle owns that boat."

She could feel him taking in the horror of the place, feel even fear emanating from the grown man beside her. She knew that to evoke the sacred Sunland in such a place was what her granma would call blasphemy. The fish painted on the boat would devour sunshine like that whale that ate somebody in the Bible.

"Blessed sunshine. Must be good people would paint that on their house," Mr. Timmons said, and she knew he was talking not to her.

She'd sat stock still on the wagon seat, eyeing the lightning-burned world and seared earth, in the middle of which was the death raft. The big fish on the side of the raft-house scared her. The tall man standing in front of the raft with his moustache, chin whiskers, and fur hat scared her even more, and the little woman stared at her with cocked head, wordless like a bird.

Mr. Timmons had stayed for a supper of fried catfish and corn bread but had begged leave to start back home before dark. She'd gone out with them, down the wood ramp to stand on the shore and watch the wagon go away, his skinny back curved toward home, wanting to call after him but holding to the silence that had mainly kept her all those days. She knew Minnie Timmons would have let her stay. She'd been her mother's friend, and she was Cherokee, too. He wasn't Cherokee. And she'd heard them argue. "Cain't hardly fill the bellies of these six we already got," he'd said. "Them folks down there is her own flesh and blood."

It was like the place after the big flood where all the bones danced. Like everything she had feared since first her mother and then her granma died and for a week she stayed in their cabin alone, eating the corn bread someone had left after they buried Granma Storms.

She had no illusions regarding what lay ahead of her.

♦ ♦ ♦

At twelve years, she was a scrawny sliver of bone and muscle, small for her age and covered over with taut flesh and flour-sack dress and looked at the world with eyes unchanged since her first vision of the raft. At twelve years she knew the river and the world it creased on both banks, knew the man she called Uncle and the woman she called Aunt Telitha who lay now like wrought iron in the bed. She knew to listen and watch the slightest movements or inclinations of man and animal, air and water, to anticipate and move herself without sound or motion, to act before words. Knew the infinite meaning of Blessed Sunshine.

She watched with the eyes of a fishing bird, poised at the cusp of the raft as her father's uncle spoke.

"You got to imagine it, little Nora, got to jest wrap your mind around all that water. Jest water and air and nary a thing beyond. Brown water like this yere, and nothin' but a whole world of it. Look out there." He raised one arm from the pole that pushed the raft upstream and pointed at the sluggish brown river. "The Lord giveth and taketh away. He giveth this world like black woods a night when they ain't no moon nor stars and jest splinters o' lightnin' to see and nary a trail and them branches whippin' a body's face and He takes those we love, before all others. Before all others."

She knew the river like her own breath. Could read it like the Bible he had taught her to parse. She knew when the hue of the current shifted that there'd been a gully washer somewhere far off and it was coming to them, knew when a bar was forming that no one might ever see or when a submerged tree was making a bad reversal that could jamb the raft tight. Knew the dark places under a cutbank where the huge catfish would lay up and could be grappled out, which undercarved bank would likely deposit a big cottonwood in their path soon, knew which tongues of deep woods held the raccoons he needed to hunt for their corn meal and coffee and lard. Knew when this old man who cared for her would come back from the shore world drunk on whiskey and crazy.

Those days and nights she would sometimes hide on shore with Aunt Telitha, rolled in a tarp and blankets to keep the cold and chiggers off, hearing him shouting Bible words and Indian words all mixed up that skipped off the river like flat stones. It was one of those nights from which Aunt Telitha had taken sick.

Uncle Jyker pulled the long pole out of the muddy water, leaving a putrid swirl that she followed with her eyes, and then plunged it down again, pushing the raft against the slow current next to the bank. Rains had spilled the river and gathered trees and brush into its current so that the raft moved next to a floating forest. A bloated newborn calf rocked in tangled branches so close she could have reached out to touch its matted hide, flies rising and settling on the red fur and whitened eyes. She'd seen many dead things along the river, mostly fish like useless gar and an occasional drum, but also dogs and stray livestock. Inside the cabin the old lady, Aunt Telitha, lay sick in bed for the tenth day, saying nothing but only looking at them the way Nora had seen animals sicken and die. Nora thought she would make tea when the raft was still again.

"And Jehovah saw that the wickedness of man was great in the earth, and he said it repenteth me that I have made them." He turned his brown eyes on her, and she looked away toward a big cottonwood that hung over the river, its roots like snakes exposed by the current. Building midday clouds made bulbous shadows like toads' heads on the water, and flies and gnats swarmed in the still air so she had to wave them away.

"And the earth was filled with violence and corrupted flesh, and God said unto Noah this here is the end of all flesh and I will destroy them all with the earth." He turned and spat tobacco juice into the water and then drove the pole back into the river. The raft edged upstream and she studied the new batch of skins stretched to dry on the roof.

"And the waters prevailed and all flesh died that moved on the earth, both birds and cattle and things that creepeth. And Noah only was left. Imagine that. All alone on that water on a boat like this here. And they must've been trash floatin' everwhere since Jehovah drowned all the towns and farms and woods. All them dead people and livestock and wild creatures floatin' and rottin' on the wide corrupt face of the water like this here river. And the Bible says Noah sent forth a raven, old Raven hisself, and he went forth to and fro and dried up all the waters." He poled the raft into a deep eddy and held it in place for a moment. "And then Old Buzzard come out and flew over the earth, making all these mountains and valleys and rivers ever time his wings touched before Noah went out. Now take the rope and tie it on that tree yonder." He nodded toward an oak ten feet above the waterline and continued as she jumped ashore.

"And Jehovah said I ain't going to smite the earth no more for man's sake, for that the imagination of man is evil from his youth. And the fear of you and the dread of you all shall be upon ever beast of the earth and upon ever bird of the heavens. Ever livin' thing shall be food for you, ever livin' thing, and surely your blood, the blood of your lives, I will require; at the hand of ever beast will I require it: and at the hand of man, even at the hand of ever man's brother, will I require the life of man. Whoso sheddeth man's blood, by man shall his blood be shed. Whoso sheddeth man's blood, by man shall his blood be shed. And be fruitful and multiply. Jehovah said all that and a pile more."

As she wrapped the rope around the trunk with a half hitch, she considered Jehovah and imagined Old Buzzard touching mountains into being with the delicate tips of his wings. It was after the flood that someone heard all the bones of the drowned people dancing. She heard the rattle of bones inside the cabin and imagined Aunt Telitha dancing a dance of bones.

"Five hundred years, is how old Noah was." He flipped the plank from the raft to shore and walked onto land, immediately testing the knot she had tied and nodding with satisfaction. "Five hundred years." He stopped and cocked his head like a raven. "I ain't but fifty, and my bones hurt like thunder most mornins. Jest imagine how that old ringworm must of felt." He smiled at her through the beard and reached out to ruffle her short hair. "Ever livin' thing, little Nora. We ain't no more'n a speck of a firefly." She saw his eyes glistening with moisture, and the sight frightened her more than anything ever had.

"I been a preacher and a doctor and a drunk," he said to the river. "And a coon hunter ain't no difference. Skinnin' somebody or somethin' one way or t'other. Your Aunt Telitha is going to die, little Nora."

He sat down suddenly under the oak she'd tied the raft to, and his head fell onto his arms over his overalled knees. She saw the bony body creak and shiver like a snag in a hard current. His voice, when it came, was muffled. "This here is the end of all flesh, little Nora. For out of it thou wast taken. And the eyes of both them was opened."

"You could go for a doctor," Nora said, looking down at him. "A doctor could make her well."

He pushed the fur hat back from his eyes and raised his head and studied the scarred knuckles of both hands, shaking his head. "I been a doctor, little Nora. I seen death come like a thief in the night plenty times. He drove us out from that tree of life, you see, lest we et of it and never died. Me and your aunt been together twenty-three years this summer."

"Maybe a doctor could help Aunt Telitha."

He looked at her, seeming to wake up. "Don't say her name now, little Nora. You can't say that name no more now." His eyes widened. "Ain't you heard it last night? I seen it fly past when I was in the woods comin' down toward the boat, a fiery creature leavin' sparks and soundin' all the world like a raven, but no bird ever made a mark in the air like that thing. I run, but when that comes for a body they ain't nothin' to do. He done took her from this corrupt earth. Like a thief in the night."

Thunder sounded far off above the riparian forest, and she noticed for the first time that black clouds had piled up over the river, casting no reflections at all in the dark water.

When she went into the cabin, her father's aunt lay on the floor as though she had been hurled from the bed, her body as hard and twisted as a piece of knotted stove wood, her face toward the wall. Nora stopped halfway across the tiny room and stared for a moment before she went back outside.

He still sat beneath the oak and watched her. "They move acrost water without no noise nor a ripple, and they get awful lonely."

She stood on the edge of the raft, listening to him and hearing a hawk keening somewhere downriver. Nearby a fish jumped heavily, the sound caught up by a new roll of thunder. She became aware of the lap of water between raft and shore, and she felt the raised, splintery grain of the wood against her bare feet. The forest leaned over the river, darkening the cove they lay in, and a light wind had begun to carry the smell of mud in dank ribbons along the shore.

For the thousandth time since she had seen the raft of skin sails six years before, she thought of her granma and then, helplessly, her mother, feeling as though she must be able to see their faces. She had been loved, by a mother whose hair was black and whose brown eyes smiled over her, whose words now were strange, like words spoken under water. By a grandmother who had held her when they buried a mother and daughter at once and whose thin face never smiled. She had been gifted with a childhood of love that was abruptly gone, and now with a childhood as quickly bereft. When she looked back at the great fish and the strange sails quivering in a growing breeze, she knew she would not set foot on the raft again where, for half her life, she had floated above the immeasurable river.

She would be a traveler by land now for all the days of her life.

CHAPTER THIRTEEN

Yazoo Dusk

1. COLUMBUS BAILEY

"You got to fish ever goddamn day." That walker hound started up again outside, probably smelling last night's coon tracks by the chickens. The old man rubbed two or three white chin hairs and looked up at me under the bill of that Redskins cap. My brain started running like a scared rabbit, heading out toward the river and then doubling back to the woods and thinking about climbing one of those big oaks. But it wasn't any use, just like it hadn't ever been any use, and I let those black eyes catch me.

"It ain't a lot I know," he said. I studied the mason jar of shine he had surrounded by those big, bony hands. "But one thing I can tell you is you got to fish ever goddamned day." One of his knuckles was busted open and raw looking.

Uncle Col Bailey had come as close to doing it as anybody I ever heard of, though these days he only fished in his own head, ever since he'd started hearing fish scream. I'd heard stories about him as a boy being out there on the river when the sun came up and out there when the sun went down. Columbus Bailey was a famous fisherman all over the Yazoo country, up to Yazoo City and down as far as Vicksburg. Everybody said it was because he could think like a fish. He used to talk to them, too, I heard. Once over in

Louisiana, in Catahoula Parish where my grandma was born and where those big Catahoula hounds come from, I had a man ask me if I knew Col Bailey the fisherman. I don't know why I told that man no, I hadn't never heard of him, though I was more or less named for him, if you believe one story. Cole Bailey's my name, and my dad told me he come as close to Uncle Col's name as he could without spelling out the whole name of that Italian Spanish burglar, but truth is my dad likes a good story. Uncle Col is my great uncle on the Choctaw and Cajun and Irish side, you see, and I'd made the mistake this summer day of coming all the way to Mississippi to ask Uncle Col's advice on love. If I'd thought about it at all, I'd have remembered that with Uncle Col it's best to ask a little question, because the old man's answers was always a hundred times bigger than any question, maybe a thousand.

I'd started junior college out in California a year before, because my cousin, second cousin sort of, whose name is Cole also, Cole McCurtain, talked me into it, said go on out to California and work at that mushroom farm he used to work at before he became a professor and then go to junior college. I could've gone to junior college right here in Mississippi of course, but Cole McCurtain thought it'd be good for me to see something else, and he'd liked California pretty well, I guess. The first thing that happened, naturally, was I fell in love and suffered more than I ever thought a person could. Cole McCurtain didn't help a bit when I called him to tell him how bad it was, just grinned over the phone. Adela Camacho was as beautiful as a winter sunrise on the river, when all that new light comes down through the black bones of trees and lays gold on the slow water and you don't hear nothing except maybe a lone dog out there deep in the woods and the air is that kind of sweet cold that makes you dizzy. Skin as smooth and brown as maple syrup, eyes black as the old man's but big and deep and a smile you couldn't forget in a million years. And that way of looking back just when you think she's gone.

"So my nephew come all the way from California to ask me about love." The old man got up slow, unfolding his long body carefully, the way you open a new-sharpened jackknife. Lifting his cap off, he ran a hand through what was left of his white hair so I could see brown fingers right against the same brown skin on his head. He settled the cap back, adjusting the thing two or three times to get it just right the way he always did, while going over to the stove. Being summer and well after daylight, it was already getting warm in his little cabin, but Uncle Col was always cold, always had been, so he opened up the stove and began shoving some more sticks in and stirring the coals around.

"How old is this girl?" He'd closed the stove door and spun around so fast I'd half jumped. "Sit down, boy." He jerked his head toward the other chair, a high-backed wooden thing he'd got from a white woman across the river. Twenty years old and I was still "boy."

"Not so young," I said, thinking if he knew she was eighteen he'd start laughing and not stop like he did sometimes.

He pulled a frying pan off a shelf and set it on the stove, and I realized why he'd poked the fire up. When he reached for the slab of bacon on the same shelf and started cutting, I felt a jab of hunger. I could already smell bacon cooking even before it was in the pan.

"You could make some coffee." With his back to me, he nodded sideways toward the crate shelf with the coffeepot and can of Folgers in it.

"Love." He sliced the bacon thick with a filleting knife and laid eight slices delicately in the iron skillet.

As I reached for the door pull to go outside, I heard him mutter the same word, and then a brittle kind of chuckling followed me out the door. That hound pup came up wriggling and slapping my legs with his bony tail, and I saw a couple of yellow chicken feathers on his snout. I knew he hadn't killed any chickens, because Uncle Col would have known a chicken-killing dog before the first thought of chicken dinner ever crossed the animal's mind. The pup must have been snuffling coon scent by the pen and snuffled the feathers onto his jowls.

I pumped a few times to get the rust taste out and then filled the pot and headed back inside. It was true that I'd driven four days in a '57 Chevy to talk to the old man. I was desperate, crazy in love and ignorant. I'd been stupid enough to propose marriage the fourth time we were together, sitting out in the Pismo dunes with a bottle of wine and a box of fried chicken. When she laughed, I felt like my heart had one of those cracks you get in a windshield, the kind that start out small but unless something's done just keeps going right across until the whole thing comes apart in your face. Six months later, I knew I had to do something. I couldn't live without her, I knew that. One day she'd say she might love me, she didn't know, and then the next she'd act like she hardly knew me. And when she took me to her parents' house I felt like a spaceman setting there with everybody talking Spanish and her sisters giggling behind their hands at me. But that's another story.

Back inside, the old man was turning the bacon with a fork and scratching his flat belly under the overalls with the other hand. I set the pot on the stove and dumped in two fistfuls of coffee, standing so close I could smell smoke and lye soap and age on him. By the time I'd woke up, he'd already been to the river and took a bath like he did every day. Most of his life, he'd check his trot line and reset it before his bath, but he didn't do that any more.

"Ever day." He forked the bacon onto a tin plate and broke six eggs into the deep grease so that they became brown and lacey the second they hit.

Over the years, I'd noticed that Uncle Col was different from most Indians I knew, even mixedbloods like him. For one thing, most Indian elders didn't swear so much. But the old man had a hard time getting a sentence out without at least a damn in it. For another thing, he was kind of a loner. I'd never known him to go down to Philadelphia where a lot of the Choctaws were, and he didn't even cross the river to where some of his own family was scattered. When Cole McCurtain was back here that time, I never even knew it because Cole was down there with Luther Cole, and him and Uncle Col hadn't talked in years. All these names kind of run together. Indian names is like that sometimes.

Columbus Bailey didn't go nowhere to see people, but people come to see him. He had things they wanted. The old man was *alikchi.* He knew the kinds of medicines people needed. Right now I needed for him to tell me how to make Adela Camacho fall head over heels all the way to the bottom in love with me and marry me. I know she didn't think much of me working at the mushroom farm, but why do you think I'm going to junior college, I told her.

He didn't talk again until we were setting at the bare wood plank table with bacon and eggs and boiling hot black coffee. He'd put the moonshine up on the shelf next to the bacon and with some kind of magic produced a skillet of corn bread not more than a day old. He was sopping corn bread in egg and bacon grease and looking doubtfully at his coffee.

"Luther Cole was my friend," he said finally, still not looking up.

I got interested immediately, hooked by the fact the old man hadn't added any swear words to the statement. Luther Cole had lived down the river about five miles in a little cabin. I'd only seen the old Choctaw man a couple of times, and each time I'd been scared to death. There was an old woman with him both times, tall and beautiful with hair silver as a new dollar, shiny dark skin, and a big smile. The old couple had seemed like ghosts to me. Uncle Col never just answered a question, you see.

2. LUTHER COLE

Dusk lay close in the woods along the river, the way it does toward the end of summer just before the sun goes down and the trees lift night up like an old blanket until everything's black. When there ain't any stars and maybe only a new moon, night in those coon woods gets so thick a man can walk right into himself in the dark. Luther Cole lay under a muscadine thicket, the heavy vines drooping all around him and over him heavy as hanging ropes and the big purple fruit kind of sweet and rotten smelling, sixteen years old and watching the girl on the front porch.

Slivers of gray edging toward black in the pecan tree and darker still where the woods sloped down to the river in vine and brush, tree thickened and rancid with August. Luther Cole saw the owl swoop up from the woods to land in the tree, near-white belly and broad, pale wings so soft he felt smothered by the sound where he lay beneath them vines. Damp summer rose hot in leaves and rotten earth through the sleeveless flannel shirt and spread across flat boy-stomach and chest, itching and creeping inside his shirt and overalls. Start of crickets and frogs down toward the water and a poorwill beginning, too, its double voice out over the water the way it had been all of the boy's life. But the boy's eyes were on the owl, *ishkitini,* again. He realized he'd stopped breathing only when he began once more, sucking the thick Yazoo air in like he was drowning, thinking about moccasins and copperheads, too, and stranger violence.

When the owl disappeared through the high gable window of the big house, Luther Cole lay for a few moments and then unbent from the ground, brushing leaves out of the straight black hair that hung past his shoulders. Damn the girl, he thought, and that old grandmother who wouldn't stay dead like she ought to. Fourteen years old and he knew he'd already been in love all his whole life and more, with the Yazoo just a hundred yards down the hill telling him violence would indeed lay hold of every goddamned thing the moment he moved. Hadn't it already taken hold of everybody he came from all the way back to Red Shirt and more?

Running a hand down overalled stomach and muddy knees, he moved out of the woods, pushing aside the muscadines, and angled cautiously across to the pecan tree, watching the darkened window. At the tree he embraced the straight trunk and climbed like an inchworm to the first heavy limb, where he pulled himself up with thin, hard arms to stand and reach his scrawny body upward again and climb until he was at the big branch near the window and he began edging outward, gripping higher limbs with his hands and balancing like a circus clown, bare toes grabbing smooth bark.

A few feet from the open window he stopped and peered inward, determination stronger than fear of the old woman. The *alikchi* had been sending dreams all over the damned place. He'd found himself tumbling down at the bottom of the river, head over heels in a strong current, knowing he was supposed to understand but failing and angry when he couldn't. And the girl wouldn't help because the *alikchi* was her grandmother.

A foot began to itch, and he lifted the other foot to scratch just as the owl appeared in the window, cream-white and big, the yellow eyes locked on him so that he dizzied and fell.

♦ ♦ ♦

"Serves you right for spying on me."

Onatima Blue Wood stood looking down at him in her yellow nightgown, arms folded across her chest, a thick black braid over each shoulder, and a scowl on her extraordinarily beautiful face and in her green eyes. Dark cusps hung beneath sharp cheekbones and shadowed her fine lips.

"She didn't mean to make you fall. We didn't know you were where you shouldn't be, spying on other people's business." She cocked her head and continued to look down at him where he lay folded awkwardly on the grass.

"And don't pretend you're hurt when you're not, Luther Cole. I've seen you fall twice as bad as that a hundred times and never get hurt."

Unmoving, Luther stared up at her thinking that maybe this time he was truly dead and finally she'd be sorry. The dark had spread out from the woods to cover the sky like a blanket full of holes, and around Onatima's head a new constellation was spinning, *fichik luak, lukoli, shubota.* Something like that. *Issuba,* the ones pointing northward. *Tohwikeli.* The old language was spinning in his head, too.

"Your head is on fire," he said, wishing he knew the old language to speak to her.

"Enough hog talk, Luther Cole. Now get up off the ground before my father hears us and runs you off again." She stepped back, never uncrossing her arms from her sixteen-year-old chest, and shook her head.

Luther gathered the pieces of himself and stood up carefully, feeling joints coming back into alignment.

"It's a good thing I grabbed them branches on my way down. Else that old lady would've killed me this time."

"Nonsense. Grandmother would never hurt one of my friends."

A dog began barking in the kennel behind the house, and instantly ten others joined in, all wanting to be out nosing things in the night. They heard an upstairs window open and the old man yell. The dogs stopped, and then Luther was aware of the hounds sulking and the silence that spread over the whole countryside at the sound of the man's voice. The shadow of the big barn reached out at them, and he realized that the moon was there. How long had he lain in the woods watching, he wondered.

"You have no reason to watch me and worry about me, Luther Cole. You know I can take care of myself." She reached out to brush debris off his shoulder and her hand continued to his hair, gently raking twigs out. "You should see yourself."

"One of your friends?" He shook himself and finished raking twigs and leaves from his tangled hair. "I could've been killed by that witch. Why's she coming around so much all of a sudden?"

"You know why. Now go on home before my father comes out here. I wish I'd never ever kissed you, Luther Cole."

"But you did. And not just once, neither. You said we'd be married."

The girl shook her head and then wrapped her arms around her shoulders. "Child's talk, you know that. You're still a boy, Luther Cole."

"And you're growd up all of a sudden. It's because of him, Granger."

She shook her head once more. "I told him to go away. It's because of nobody, and you're going to school, too."

"I ain't going to no school, especially not way up there."

"Any school. Don't say double negatives. You know better."

"I'll kill that white man."

"I told him to go away and he will. Now you have to go, too. We aren't children any longer, Luther. My father might have you whipped if he finds you out here again. You know he said he would."

"You think I'm scared of that? You think that Granger'll go away just because you say so? Didn't you say it before?"

"My father encourages him. Now go home, Luther."

She stepped forward and touched both of his shoulders with her hands and kissed him on the forehead, stepping back again before he could react. "I'm not a child any more, Luther. We can't play in the woods anymore. Everything's different."

"Nothing's different. I'll come back tomorrow. You stay home from church. Say you have a gut ache like you used to and stay home when they go. I'll come."

She shook her head. "Not tomorrow. I can't. You don't understand at all. You have to go to that school. Goodbye, Luther Cole."

"I'll kill that white man."

Onatima turned away. "Go home and sleep, Luther. Granma won't let it happen."

The girl went away, gliding barefooted through the grass around the back of the house without any sound or motion in the air so that abruptly he found himself alone.

3. ONATIMA BLUE WOOD

If you think it's easy being an Indian girl, you're crazy. People think because my father's rich, or richer than most in these parts anyway, that it's all easy. Last night the moon climbed up behind the pecan tree, climbing up out of the woods the way I've seen a coon go up, smooth and slow, until it sat right there on the highest branches and waited for something. I understood because I've been waiting for something all my life. People always look at me. Like they expect something, something that a girl who wasn't half Indian wouldn't do. And school, that's a real barrel of fun. Ever since I was little if I read "See Dick run" the teachers acted like a prize mule had sat up and recited the Lord's Prayer. Like an Indian wasn't supposed to read, like my mother, who was born in Scotland actually, on the Isle of Skye, and whose name I can't say any longer since she passed on, hadn't taught me to read when I was not even three and a half years old and I wasn't writing whole sentences by the time I was four, which is more than a lot of grown-up white folks around here can do. As if it was an accomplishment for a half-Indian girl to read about stupid Dick running when I'd read all of the Brothers Grimm and Hans Christian Andersen by second grade.

Those teachers never knew anything about the real stories, the ones my grandmother, my *apokni,* told me when she came to live with us after Mother's death. And the other things, *alikchi* medicine. She taught me those words, too, and I listened as hard as I could but still only heard a little bit. Deep woods stories, olden time true stories. Chahta things, *fahpo* things, what Granma called *aiukli.* Stories I could not even let my father know I heard, because Father was modern, a fullblood stepping away from Indian stories into this thing called America. Two worlds, two kinds of stories. But those little people in the fairy tales, we knew them, my granma and me. They were *bohpuli,* senders and throwers, hide behinds. Be secret, she said, tell no one, even Father—her own son—and especially teachers. The first time she came after they buried her, to keep the teaching alive, I was frightened for only a moment until I understood. I read *Moby-Dick* at thirteen years of age and understood every ephemeral shadow of that teller's story, Ishmael, outcast like me and Luther Cole, wanderer, friend to Indians. Ha. And Burns's wee cowering timorous beastie from my mother. Having both worlds, I thought, would put me ahead of them all. But that Luther, pulling his head into the Indian world like a snapping turtle into one of those ridged shells, didn't seem to worry about being behind one little bit. Girls mature sooner, Granma said. Those Jobe and Owens girls didn't seem to, however.

♦ ♦ ♦

I lay there in my whitest nightgown, after mothers and grandmothers had gone away, and let the moonlight bathe me in its radiance. I glowed, a princess, a rare and radiant maiden whom the angels named Onatima. My skin was made of light, and I felt like I could just float right out the window. Luther had been here, hiding in the woods and making animal calls that any fool who wasn't deaf and retarded would know wasn't any animal that ever lived. Like an old-time Indian, Luther Cole didn't approve of the ways of white people. "Apple" he called my own father, an Indian who was not just red but red-brown on the outside and white like all rich people on the inside. He would take me away, across the river to that life his kind of Chahta lived, not understanding at all the way these two worlds had steel hooks pulling from both sides until I thought I would be left just bloody muscle and broken bone, all the while thinking, remembering, waiting. All those years from when we'd both run and fall and laugh like babies, knowing Luther had marked me for his world against my father's desire but, and this so strange, precisely within the wishes of my mother who told stories of Highland clans and warriors so in keeping with Granma's Choctaw stories. Massacred MacDonalds. Highland ghosts. Pushed toward an Indian world by my red-haired, freckled, indescribably beautiful mother and denied the same world by my fullblood, dark-as-stone father. Is it any wonder I am a misfit, a castaway, a Gypsy, an Ishmael? Not ignoring what is good, I, too, am quick to perceive an evil, or something like that. Like him a lover of words. English words, because I don't speak the old language at all. Promised while still a child by my father to Granger with his stable of horses and that school in the North that taught him not a fraction of what I learned from two women in my father's house against his knowing and his will. Which is the way women work nowadays, unlike the olden times when, Granma told me, Choctaw women were the ones to make decisions, the ones at the center.

Luther Cole has my heart and will take it away to that raw plank cabin on the other side of the river, but Granger Collins shall have my woman's body, alas. All the world ordains as much. And I should be grateful that a white man of standing would marry a halfbreed, they say. Not for Onatima Blue Wood to leave a big white house and barns and stables and cross a brown river to twelve square feet of darkness. They say. Granma says, Wait. She comes now at night and says the story will be a long and twisted one. She says this is not the real story. All time is one, she says, and I must bide it as well as Luther Cole, who is, she says, the one who will learn. Whom I will teach.

Down there by the river, where I walk sometimes, the water is brown and heavy, moving so intently that I know there is no end to it. Never, I think, have Chahta people been far from such water, and I know I cannot leave it. Granger talks of other places, Saint Louis, Chicago, New York, but he might as well be talking about the bottom of an ocean I've never even seen. Away from this river I shall die. Luther Cole can go off to school and come back, for he carries the river inside him, but this brown water is not inside me. It surrounds me, gives me breath. I'll stay and outlast them all, Granger, my father, all. I will not change my name nor cut my hair for him or anyone.

4. LUTHER COLE

Luther Cole lay under the muscadine thicket, the heavy vines drooping heavy as hanging ropes, sixteen years old and watching the girl on the front porch the way he'd been watching her for the past hour. Something was up, and it didn't involve the fool sitting on the porch with her.

"Chiggers for sure," the boy whispered aloud, thinking how downright stupid it was to be lying in the damp rot of leaves and decayed bark and bird-eaten fruit pulp but unable to lift himself from the earth and tell his body to go elsewhere. The same way he'd been unable to ward off the girl all these years.

On the wide porch Onatima shone in light from a lamp inside the house, and he could see the fine, taut brown cheeks and smiling lips and smooth forehead and long, long black hair that he knew she refused to wear coiled the way she should but insisted on letting fall straight and shining dark to the seat of the chair. Uncut all these years. The white dress rose in a crimped collar halfway up her neck, with buttons down the front and long sleeves that lapped over her hands. He could see her straight shoulders in the wicker chair, the fan in her right hand, and the white man who sat in the other wicker chair facing her. He hated and pitied the man, knowing who he was and why he was and how he was and so much more than he'd ever wanted to know about any man. Rich white folks were like that, living big and wide across everybody's lives, like they owned the full moon and the whole sky of stars and everyone who walked or ran way down below moon and sky. Rich Indians, like the girl's father, were different. They kept their lives quiet, moving slowly, careful to cast no shadows across others' paths or doorways, a way of life both ancient and learned, sometimes painfully. Pitied the man because the man didn't know the truth about the girl and himself, or anyone maybe. *Hatak yushpakamma,* bewitched.

Onatima Blue Wood was the girl's name, and her father was the richest Choctaw on that part of the Yazoo, her mother a white woman who'd come to teach the poor Indians like Luther Cole and married one of them. Like Luther Cole, the girl was a mixedblood, a halfbreed. But she was different because of the big white house that pushed that porch out toward the river, the barns and stables full of blooded horses, the white dress, the old Choctaw man inside the house wearing a suit and vest and, when he wanted to, a fine white hat and driving a buggy and paying Black men and women to do his work.

Luther studied the man in the lamplight. Granger Collins, a name that made him want to cut the suit coat off the man with a rusty can lid. At least fifteen years older than the girl who, Luther knew exactly, had turned sixteen just three weeks before. A fool.

Lying deep in the moldering leaves and stench of Yazoo mud and all the tangled darkness of the Choctaw people who had lived and died on that land for a thousand years, Luther Cole was not his father and mother, for he had never known either, nor his brothers and sisters whom he had never known, nor his Choctaw and French blood relations from across the Mississippi in Louisiana, of whom he had heard only whispers and speculations that forked like rabbit trails only to run out in river brush; nor was he the sixteen-year-old boy who had been in love with Onatima since before either had breathed the wet air of the Yazoo Trace or imbibed the old stories. But he was, steeped in blood and bone, much more than he knew or the white man could know or the girl would desire to know. More than the screech owl bothering the newly risen night not far off, causing the small hairs to stand on the boy's neck, or the hounds baying dinner in one of the barns behind the big house, or the slow slap of the river on its mud bank to his back, or all the lives that had graced and perished the land before and after what they called the Removal.

The white man was still on the porch, the girl had not run off yet with that gambler, followed by the bitter old shadow of her fullblood father. And Luther Cole had not yet been sent up to Haskell.

5. TELLER

Poking around in the woods. Swallowtail coat and plug hat, those black suit pants worn to shine and the white shirt now possum gray, his beard full of moss and twigs, and not finding much at all. I can see the committee, all the way back here from the Territory, looking for Chahta people. *Chahta isht atia. Chahta yakni. Chahta okla. Chahta hatak. Chahta anumpa.* So many words now, throats dry and cracking. And the people staying right back there in the woods, watching the strangers with familiar sounding names, some of them like the bearded one, more white than Choctaw, searching for what they said was relatives.

They been gone so long, you see, more than two generations out there in Indian Territory, that we almost forgot them. Some had relatives, but it's a long way out there, and the trail just disappeared. I remember when it happened, all the people herded up like hogs drove out of the woods into pens and put on boats and wagons, coming in from all over, thousands of them, little kids hiding behind those deep brown eyes, scared, lost, hanging on to folks, men trying to look like they knew but knowing no more than their own children, and the women really knowing the way women always know when home is gone and what they're standing on isn't the same and keeps shifting and they have to somehow keep touching every thing and body so there's a whole somewhere. There was the rich ones, like Pitchlynn, and some stayed in Mississippi, but most was poor and most started walking westward, going the way the stories told us we'd come from so long ago. People clung to that, they were going back where they started in time immemorial walking with the brothers, Chahta and Chikasa, following the leaning pole to this home.

"If you are Choctaw, you are entitled to land out in the Territory, where most of the people are now." That's what the committee said, speaking patiently the way they might of spoke to a child. "You just have to come out there and you'll have land now. It's called allotment. Our land will be given to each one of us, each family, and you are entitled. If you are Choctaw."

If they were Choctaw, as if white men from somewhere north could just come and tell people who they were. The people watched, that Yazoo mud on their feet. Those other Choctaws had been taken away seventy and eighty years before by the government, leaving a handful who wouldn't go, or couldn't. Torn away, like a piece of flesh caught on a briar and just ripped from the body, except this time it was the body itself ripped away, leaving that bloody piece behind. There was some that turned back from that awful walk, straggling home to tell of dying children and old folks, births in frozen mud beside trail and road, white farmers charging ten times the price for any kind of food. A picture of a daybreak that showed a hundred horses standing dead, frozen in mud up to their withers. A hundred Indian horses coming out of the night like that frozen, all facing back the way they'd come. "I druther die right here, right this minute," the coming-back people said. So the ones left behind knew the truth of what they called Removal. And there was some that come back from the Territory now and then, and some that left Mississippi and went out there, some that slipped back and forth during the war when so many Indian people fought and got cut to pieces for something that didn't mean a thing to them.

So we knew about the other place where our relatives went, knew they built up a Choctaw Republic out there with schools and all until that war killed everything. But we didn't know our relatives any more until the committees come back to get us. "It's the law," they explained. "We got to offer every Choctaw some of that land. One hundred and sixty acres for head of household . . ."

They didn't seem Choctaw no more. We didn't go.

That was the year Luther Cole was born across the Yazoo in a cabin, into a Choctaw world of so much magic that it already had possessed Onatima Blue Wood for more than a year in the big white house with Black servants and blooded horses in a stable.

CHAPTER FOURTEEN

The Dancing Poodle of Arles

It began in Barcelona, in the howling tunnel of the North Station, where faces were skulls and no signs told of anything, sooted concrete walls beneath the street and the Spanish trains wailing to a stop and lurching away like the drunks he remembered back home on the reservation, looking for another fix. When the dark-faced man with the briefcase pointed, they leaped aboard the wooden car—he faithless, she stoic before what she knew would be an absurd fate—and found a nun and five schoolgirls with an orange guitar. "La Paloma" the girls sang as the train crawled over the Pyrenees—gray rock and yellow leaf—and into Toulouse, where he foolishly insisted on ordering dinner at six P.M. and was insulted and they boarded a late train for Geneva.

In the train compartment she huddled in a corner near the door, eyes closed but mouth too firm for sleep. He stared out at the bulking shadows and quick, jerky lights of small towns, again thinking of home. Each village, he thought, would be the last one and then the magic would be gone. He remembered the other journey. How many years before? When the night walls of Avignon, crenellations outlined in golden light, had appeared outside their window as they raced toward Paris, like a magic trick. Now, between the fragile illuminations of the towns, the dark lay thick and heavy upon the flat surface of southern France. The train hurried across the land as if it touched nothing, and he thought of another world where old women mouthed timeless stories and grandfathers sat watching stars bend across the sky in the ancient patterns of those same stories. That was a world of order, a different world. How did I get here, he wondered. How have we come so far?

In Arles at midnight the train stopped and they, alone, got off and crossed toward the waiting room in a gray wind strung like barbed wire along the cold tracks. He clenched his jaw and noticed her shivering in her beige jacket, stumbling on the crushed granite, her face turned from him. The thought of a wait till morning in the little station made his bones ache. The connecting train for Geneva, a city as cool and logical as a tea tray, would not arrive until six.

The small rectangle of the room contained only a blue, oblivious old man reading a newspaper behind the bars of the ticket window. She walked quickly to a brown wooden bench along the wall near the door, and he followed, noting the smooth lines of her gray pants and the swirl of her yellow hair upon the wall when she leaned back and shut her eyes against him. He pushed his own dark ponytail back and sat near her, dropping the two bags beside the bench, feeling that the trip was not working and it was his fault they'd left Toulouse at a ridiculous time since he'd been the one to rage against the French, the one to insist upon the cool, restful logic of Geneva. But then, she'd resisted only slightly, too dissatisfied, he thought, to care very much. Beside her, he felt the sooty darkness hanging like a wave over the station, and again he thought of his grandmother and the others back home. They would have stories to explain all of this.

The light dampened the yellow tiles of the walls and soaked into the red floor, and he slid his back down the hard bench in search of less torturous constructions. He tried to burrow inside her mind, the way he'd once imagined he could, but all he found were images of the children, shards of ten years together, what you received finally in exchange for the quick delight, the intake of breath. She would be thinking of the kids at home with her sister, not of the soaring, lovely confusion of Gaudi's temple of the sacred family in Barcelona, La Sagrada Familia. Gaudi had tried to make sense of it all in that edifice of nonsense, a coyote construction that defied logic and law and climbed skyward in spite of everything. Everything should have collapsed but didn't.

And then the red door of the station opened and a young girl entered, followed by a large blue-gray poodle. The girl's black hair was cut straight across her wide forehead and rounded to her neck, and her green eyes flashed at them when she smiled. Her arms hung at her sides and her fingertips curled around the long sleeves of her violet sweater. Behind the girl the blue poodle walked erect on its hind feet, its forefeet curled neatly in front of its chest and its brown eyes bright with intelligence. The girl made a sign with thumb and forefinger, and the poodle yelped once and began to spin and bark, blurring the yellow walls of the room. The man in the ticket window lowered his newspaper and raised it again.

For a moment they watched the blue poodle dance and the young girl smile, and then the girl made a strange nasal sound to the dog in a language they could not understand. The poodle turned to them and then danced back through the open door, followed by the girl in the violet sweater. When the door closed, he saw his wife's blue eyes and knew she was laughing, though she kept the laughter carefully inside.

She touched his arm, startling him. "There's something familiar about this place," she said. "Arles—have we been here before?"

He thought about the question. They hadn't been there before, but "I know what you mean," he replied. Something about the place, like a delightful word one could not quite articulate, or a fragment that, if it fell into place, would connect everything the way his grandmother began each story with pure sound. There was a familiar story here somewhere, but which story?

He got up and approached the door and looked through the dingy glass at the lights outside. "Want to take a walk?" he asked. "There are lights."

She came to the door and stood beside him, looking doubtfully through the small windows like, he thought, someone looking through the walls of an aquarium. At the thought, the images became watery and strange.

She glanced at the two bags by the bench.

"They'll be fine here," he said, opening the door.

♦ ♦ ♦

The streetlights were egg-yolk swirls above the dark paving stones, and here and there lighted windows made yellowish-green rectangles. The blue poplars along the edge of the street wavered smoothly toward the sky, and he saw that the sky itself was a swirl of yellow stars like pinwheels trailing streams of light and nearly obscuring the indigo background. It occurred to him that what he saw was all color, without form—stronger, simpler, more vibrant colors than he'd ever seen, exaggerating the essentials of the street and sky but leaving the obvious town in vague shadow.

They stepped cautiously into the street, balancing on the uneven slate-colored stones. Halfway down the block an outdoor café burned with light, the sulfur-yellow wall and awning framing eight white tables with a scattering of dark forms around them. Near three empty tables in the street, he could see the brown outlines of four figures walking. The blazing light of the café was balanced on the opposite side of the street by mauve walls shading into near black broken by orange slashes suggestive of lighted windows. The buildings on both sides rose to a wedge of purple sky and enormous stars like cream-colored flowers in full petal. A frightening sense of déjà vu swept over him as he stared at the café, and clutching his wife's arm he turned from the scene and fled down a side street, turning again at a small avenue called Rue Lamartine. The tap of her shoes on the polished stones matched the audible tap of his pulse.

They hurried past a window where yellow light made a thick line between green shutters, and she stopped abruptly and turned back, pulling her arm from his. He watched her approach the window and he followed, stepping carefully onto the gray curb behind her. Together they peered through the crack between the green shutters at the small room with its pale violet walls and floor of red tile, yellow chairs and lilac door. On a tumbled bed a scarlet coverlet was turned back from sheets and pillows of light greenish lemon, and on the edge of the bed sat a man with an emaciated, luminous head outlined with tangled red hair, the lines of the hair heavy, as if laid on in an impulse of thick strokes. The man's high, pale forehead angled down to razor-edged cheekbones, hollow cheeks, and a sharp chin under a sparse confusion of orange beard, and wrapped around the head and under the chin so that it covered one ear with a thick pad was a white rag.

He touched his wife's arm. "We can't," he whispered.

The face lifted, and the eyes, like dark whirlpools, stared directly at them through the shutters. The tiny room, with its broad lines of furniture, portraits, and wrinkled clothes hung on pegs, expressed a deep, inviolable rest in painful contrast with the man on the bed.

They shrank like burglars from the familiar face and window. Overhead, the stars tumbled out of control, madly, and they walked quickly back to the station.

♦ ♦ ♦

An hour after daybreak they sat in the train for Geneva. Through the broad windows of their compartment, he watched a world of shocking turbulence, a stormy sea of wheat fields, grain like patterned sand tossed with rolling bushes, and the green flames of cypresses quivering upward against blue, knobbed mountains curling in waves upon themselves. Overhead, a swirling pale-blue sky and thick lines of angular black crows above stubble cornfields and a curving swath of red-green road.

As the train sped toward the cool blue of Geneva, he watched southern France explode in light. "A kingdom of light," he thought.

He turned from the crows like black, folded staples, and absent-mindedly fingering the lobe of one ear, he smiled at his wife. She reached out to touch his shoulder, her hair a sudden chaos of gold against the shimmering green pattern of the seat. Ahead of the train, the Alps waited in tumult.

"Arles," she said, musing on the sound of the word. "There's something familiar about that town."

He knew what she meant, and he remembered the old stories. The strongest colors should be used bravely, he thought, for time will only soften them too much.

CHAPTER FIFTEEN

Winter Rain

Every day now for six months he'd awakened to the river, rising out of sleep the way the winter river rose from its brushy bed, hearing before anything the throbbing of the current under the sand, his eyes opening to the sun caught like blood in the naked branches of the sycamores and cottonwoods out there in the river. He'd watched the river through its changes, taking to his bed in midsummer when the stream had subsided to a thin trickle in the heart of the wide swath of brush and sand. He'd seen the little stream vanish, leaving pools here and there that he knew from past years would be marked by the tracks of coons around the tails and bones of big fish. And then the pools, too, had disappeared, and the river had become its real self, a subterranean river, the largest and most powerful in the world, a current he could feel beneath his feet when he walked the crusty sand and a throbbing he felt every day now lying in the bed before the windows he'd had specially made for the watching.

Through the bigger window he could see the smooth sweep of the sycamore branches rising like cold shadows at the edge of the river, and the islands of brush tangled black and topped by the brittle bones of the cottonwoods out in the middle of the sand. Through the window in the other wall, he could see Pine Mountain, where each evening the sun impaled itself on the black points of the pines and where the old graveyard hung on its flattened bench halfway up, the boy's grave shadowed always by the thick timber of the mountain. Below that must be the black river, *Oka lusa hacha,* the one the Choctaws said a good man could cross, and hadn't he been that, a good man? *Shilup* or *shilombish,* inside or outside shadow? The *shilup* would cross the river to a good place, they said. The outside shadow would remain behind, an owl or a fox. That world was so far away now back there in Mississippi; *sapa iskitini* and *yukpa* they said of that place, whose words though? For a long time words and strange voices had come to him when he didn't expect them, and dreams of a deep canyon not like this river, a chasm with fast, black water at the bottom roaring so loudly it frightened him. In his dreams he had to cross that river but could find no trail down and was afraid. He could see a log across the deep river, like one of the cottonwoods outside his window but different. The bark was gone from the log and it looked too slippery and dangerous.

This morning he watched the early sun flame on the wings of a red-tailed hawk as the bird wheeled and settled onto one of the bare sycamores with a cry that cut through the walls of his room. Below, he knew, the rabbits would huddle in the black brush, made nervous by the scream of the hawk, tensed against the panic that would send them recklessly across the bare sand. With its wings folded the bird turned a deep bronze, and the old man thought of his own thin face and hawklike blade of nose. His vision shifted to the other, smaller window and then curved slowly back to the river. The boy's shadow seemed to stalk the river now, the angle of the cap and thin line of the gun cutting against sand and brush, denying years and distance. *Shilombish,* that was the word. Outside shadow. The good *shilup* would have crossed the river long ago.

The arms and hands with their bleached brown fingers fluttered against the sheets and a cry rose from the wreck of his body and then the door opened and the woman who cared for him entered the room, her face bright and concerned, the silver hair in a careful bun behind her head.

"Good morning," she said. "Did we sleep well last night?" Her voice carried a feel of caution, the old man thought, the voice of one conscious of hidden dangers.

He tried to focus on the swift rise of the sycamore and the dark bird, to pretend to himself that he had not heard, that she did not exist. All of his life he had been lifted and moved about by such voices, and now he felt that he should be beyond that, should have earned a distance like that of the hawk. Or the boy, the child of his middle age. Instead, he felt himself flushed from hiding and heard himself saying in his dry, wintry voice, "Very well, thank you."

But it wasn't true, of course. The dreams had come again, the girl mixed up with the deep, clear pools, the long black stream of her hair flowing, fanning out with the current of the underground river, her brown body and her bright smile with all of its promise turning and tumbling in the insistent current. Waking, he thought of the boy, and sleeping, he dreamed of the pools, the deep pits cut into the river many years before so that the underground stream was freed and exposed to sun and sky, and the cold, clear water ran out of the sand into a shimmering translucence of rainbow-sided fish and shifting pebbles. Why the girl had come back to him after all these years perplexed him. For a long time he had held her away, had insisted to himself that the life he lived was the life he had chosen, that what he had been was what he had imagined himself to be, not merely what he had had to be.

For fifty years now, since trading the dampness of Mississippi for California's light, he had lived beside the river, had matched desire and reflex against the small lives of the dry river, had taken the boy there and taught him the times one must not kill, had learned to let the pulse of the underground current move through and beyond him. But only now, it seemed, was he coming close to the river's meaning.

The woman turned a handle and a part of the bed rose, and he found himself sitting almost upright so that the plane of the river tilted at an awkward angle, the skeletons of trees seeming to grope for balance, to sway outside the glass. The hawk's cry cut through the room, and the bird soared on a backward beating of wings and disappeared into the far reaches of the river, out of the frame of his vision. He imagined the glide of the bird's shadow over the sand, recreated the clumps of brush, the pattern of the river that he had known, and saw deep in his mind the kindling of the sun on the red tail.

Shilup or *shilombish*? Inside or outside shadow? It was important to remember which. The woman began to raise spoonfuls of pale food to his mouth and he felt himself swallowing. Now the sun had risen out of his window, and he shifted his eyes to imagine the splinters of light that fell through the pines of the mountain beyond the other window. He remembered the light on the pine needles and felt the prick of the needles as they lay by the graveyard a lifetime before—before the boy, the war, almost everything—her hair like water on his face as she lay over him, his fingers clutching the steep slope to keep from falling away forever.

"Everyone in town asks about you," she was saying as she raised the spoon. "Everyone thinks so highly of you, you know."

It was true, he knew. Everyone always had. A good boy in that other life, a good man in this, and they'd all been happy when the child came, for he would make a good father. They'd come to the house. To stand with him beside the woodpile and stare out over the river where the silence was like thin winter ice, and to glance with him toward the graveyard on the mountain. To sit in the chairs and tell him about the little town, their words like the dry clay that rattled when you walked the coastal hills. But when they saw that his eyes were on the window and his detachment showed no awareness of their words, they ceased coming. Now they did not come and the old man was relieved. It was not that the town had forgotten him, he knew; it simply relied on him to die the way it had always relied on him to do the right thing. But there was the matter of the dark river and the log he could not identify. He did not think he could get across.

"Mrs. Paaske had her appendix removed last week," the woman was saying. "They said it came close to bursting before they got her to the hospital."

How, he wondered, had words begun to fail with the boy so many years ago, crumbling between them like the inky shale of the hills to the west. Sometime, he couldn't remember how long before, Earl, his oldest friend, had come to sit in the same chair and look at him, seeing, the old man knew, his own end in the other's.

"The band-tails are back," Earl had said, and the old man felt the stir of wing tips brush through, smelled the explosion of pine and oak. Even then he'd been old and Earl's eye too sure when they hunted the pigeons, the boy watching and the old man hating the friend's theft. Perhaps then, that day when frost didn't melt beneath the oak shadows, the boy had begun the slow curve away and toward the war. Perhaps with Earl, and with the smooth gray arc of the 12-gauge and the tumbling birds.

Answered with silence, Earl had risen, disappearing beyond the doorway, and the old man's eyes had opened again to the river and the dark mountain.

The river was predictable in its own way. In the summer and fall it retreated into its bed and became an underground river, a brush and sand and cottonwood river, leaving only long white fingers exposed, and rabbits broke like dark fists over the red and gold and brown leaves and frost coated the dead grasses and small branches snapped like memories. Then in the winter and early spring the rains came pounding down over the coastal mountains, thrashing the final leaves from the white oaks, flattening the wild oats on the hillsides, smashing against the sand and sending the creeks in brown explosions toward the river. The river would rise out of its bed and rage in a dark wall that carried everything before it. And then as quickly as it rose up it would subside, sinking into the sand until only the stream was left at the heart of the river and then only the pools and finally the sand itself.

Almost all of his life he had watched it, and now he felt the tension in the river, the waiting to break out again and sweep everything away. Never had he felt it so powerfully, always that pounding current beneath the smooth sand, that urgency.

In his sleep he fought the dreams, refusing the deeper, darker river with its slippery log, and at the same time denying the girl her presence in the beautiful current of this other stream, dodging away from the deep pools cut into this river. Always in those pools it was summer and the sun was hot on the sand and they dove and swam into the heart of the light for a slashing moment. Her black hair soaked up the light until it burned his hand, and her smile promised a life. A different life.

"We aren't feeling well today," she said.

Outside the window, the light was graying and in the distance he saw clouds gathering. He knew that for weeks now they had lain up there on the crest of the coast range beyond his sight, trying to slip on over into the valley but never quite succeeding. Though he couldn't see them, he knew they had been the thin cirrus that came apart in tatters in the faded air. But now it was time for the real rains, and the clouds he saw were the bigger, heavier cumulus coming together out there, darkening the sun and putting a shadow over the river.

The woman followed his gaze to the window, then took away the tray and rose, drifting silently through the doorway. He could hear her moving things around in another room the way she always had. Now she would read, or knit, while she waited for him to sleep. And later she would come, he knew, to find him and talk to him and feed him again and ask if there was something he needed.

He closed his eyes against the mountain beginning to gray with cloud, and the girl appeared, laughing, her dark hair streaming with rain, the water flowing over her body. The clouds rushed to seal the sky, and immediately he saw the first drops come sweeping in on a new wind to streak his windows so that the sycamores and cottonwoods warped and wavered like old men. Then the real rains came hard and he felt the river responding, felt the current quicken and rise up in its bed until it broke from the white sand in a furious black torrent he knew he could never cross.

The woman who cared for him listened to the rain drumming on the roof and looked down at the old man. She crouched in the chair and reached both hands behind her to free the long silver hair in a sudden flood over her shoulders. She clasped both arms across her chest and began to rock as she remembered so many years before when she had led him to the deepest pools of the river.

CHAPTER SIXTEEN

Shelter

Sleet came like broken glass out of the southwest, cutting the tops off the peaks and riming the alpine fir and hemlock. He worked to burn the shelter in the saddle of the pass, the black granite mountains leaning in over him, their heads in gray cloud obscured further by the blowing ice, a small dark form moving cautiously against the green meadow and gray wind, tearing the cedar shakes from the sides of the shelter as the two women struggled on the trail in the river valley below. Piling the old, dry shakes around the three-sided log structure as they walked the washed-out trail with roots grabbing and holding with each step the way roots always did down there where the ice was rain and the dark forest crowded in upon them and they climbed upward toward the pass at the center point where he worked to burn the shelter, piling and burning and piling as logs fell into logs, raking embers upon embers until the shelter was a pile of glowing coals as big as his mountaineering tent and to the north and south the black mountains moved away in cloud, while east and west the pass fell to timbered river valleys one of which the two women climbed slowly up from, two Indians noting the medicine plants, feeling the familiar rain, singing softly to themselves as he burned the shelter their father had built and they thought they would sleep in that night in spite of the ravens barking from the trees to warn them, their joints cold and their songs warming, the earth drumming, the river talking clearly down the valley below him up there burning the shelter because he was the ranger.

♦ ♦ ♦

The day he burned the White Pass shelter, sleet cut like broken glass out of the southwest, slicing the tops off the granite peaks and riming the alpine fir and mountain hemlock, bending him close over the flames. He worked in the saddle of the pass, using the pulaski to tear cedar shakes from the shelter walls and piling them to get the fire started, then burning and piling as the log structure collapsed, raking embers upon embers until what had been a shelter in the snowy pass for eighty years was a small pile of glowing coals no bigger than the orange tent pitched in the trees nearby. It was his job there, so many miles from even the nearest gravel road and so many more from the little logging town where the ranger station stood. A job winter snow had begun by bowing the ancient shelter low until it sagged over the heather meadow now in early summer like a broken bird wing.

He'd seen at once when he backpacked in that morning from his camp along the river below that the shelter was done for. Three sides and roof bent close to the earth as if inspecting after these eighty years the mark of hoof and boot, the peaks like a jagged wall all around shredding the clouds. A spot of brown from a mile away on the long meadowed ridge as he approached, the broken shelter had brought the pass and the Cascade peaks rushing toward itself, the single point of reference.

A job he felt good about, backpacking in to remove this man-made thing.

♦ ♦ ♦

Rain came, settling on the downswept branches of fir and hemlock, slipping down the shaggy and gleaming needles to. . . . The rain fell onto the downswept branches and collected and fell to the undergrowth with a steady hammering. The two old women climbed steadily upward, out of the thick timber to the switch-backs where the trees were stunted, twisted dwarf things.

♦ ♦ ♦

A hard climb we had of it, my sister and me, slogging up the Northfork trail seven miles, the red earth washed out of the trail leaving those roots that reach up to throw you down whenever they get a chance, and there along the river the salmonberry and blackberry vines grabbing at you when you went by and the nettles swiping at your face. Plants you can make medicine from if you need to, but you don't want them swatting you in the face, my canvas pack with its little straps cutting into my shoulders and that can of beans stabbing me in the back.

The first night we camped seven miles in at Red Creek, just before the trail starts switchbacking up toward the pass. Sarah caught some trout like she always does, like our father taught her when she was a girl all them many years ago and like he tried to teach me, and we put up a tarp between the trees and built a fire and listened to the creek and the river talking where they came together just below our camp while a camp-robber jay came and begged until my sister gave him some bread, which was a mistake. Her sitting against that big cedar at the edge of the tarp, her silver hair in braids and the red bandana on top of her head, the rain and the creek and the river talking about something I almost couldn't understand. Me thinking about how the next night we'd stay in the shelter our father built up there at the pass, where you needed a shelter from the storms that lived up there.

♦ ♦ ♦

A job he felt good about, removing this man-made thing, back-packing in to erase the intrusion, the unnatural structure. He warmed first his front and then his back against the coals, the shel-ter now only a glowing circle the size of his internal-frame pack that rested next to the orange tent and the little camp stove he carried. The steaming parka he wore over his down coat telling him it was time to turn the other side to the fire, his heavy mountaineering boots oblivious to ice and fire, the tent with the sleeping bag inside reassuring him when he looked that way through the ice strung like wire now across the meadow so that he pulled his wool hat lower and tilted the hood of his parka away from the blowing snow. Over his tent the small, stunted trees bent and curled their limbs together thickly. After he finished he would have tea.

♦ ♦ ♦

It was sleet on the switchbacks and ice like a flint knife on the long, high ridge trail across the meadow toward the shelter, with Sarah bent into the wind so that the ice caught on her bandana and the top of her pack and stayed there, me thinking how small she looked hunched into the mountain that way, how like our father when he was alive, before he died, like our father who is in the graveyard on the Stillaguamish who built the shelter we would sleep in that night with the odor of mice and mules and all of the people and animals who had slept there during eighty years. Where when the clouds rose we could look up and see the great mother mountain, Dakobed, that the whites called something else not seeing what the mountain really meant there. Me thinking how far we'd come since the beginning.

Not even looking up, Sarah, when the ptarmigan exploded out of that bunch of heather right beside her to fall away down the valley still like a snowball this early in the summer before his feathers changed, feeling maybe how good it was to come out in the mountains again like we used to with our father and feeling maybe how old she was, as old as him when he died. Not seeing either, when I did, the smoke up there in the saddle of the pass like a little feather blowing away with the ice so that you had to be lucky or look hard to see it at all.

♦ ♦ ♦

"The ice storm swept down from the north," he wrote in his journal as he waited for the tea water to boil, "turning the gnarled trees from black-green to white, riming the hairs of his mustache so that each stood out stiff and brittle." The stunted trees, twisted from three centuries in the high country, poked their spiny branches at the orange rain fly of his tent, testing its tautness, scraping a brittle shaft across the fabric with a little tearing sound. The wind picked up, sweeping down the black ice of the peaks with greater acceleration, probing the clusters of curved trees, seeking him out, plucking the guy lines of the tent to make them sing, experimenting with the fire of the little stove.

"Where the shelter stood is now nothing. Tomorrow he will spade up the earth and transplant until there is no evidence."

♦ ♦ ♦

Close now to the top of the ridge where the heather turned to bare granite carved sharp by wind, the two women bent into the storm and walked toward the pass, the one in front bent more tightly, closely to the thin trail, the one in back unfolded slightly so that she could look across the saddle of the pass for the thing she could not find, not even smoke now but instead a bright spot of color in the small trees.

♦ ♦ ♦

My hands in the mittens cold and hard now, the sharp pains decreasing, a bad sign, and Sarah still looking down so that she does not see the shelter is gone and we are walking toward only a spot of color in the small trees at the edge of the pass. So that she doesn't see the way I do.

♦ ♦ ♦

The ranger squatting beside his small stove balanced on a rock, the two women climbing steadily and quickly toward the pass, the trees touching the tent fly and the wind plucking at the cords. The mountains leaning, clouded, toward the empty meadow. The crooked hemlock branch hooking the nylon and lifting, the wind working at the cords until fly and tent rip and lift into the iced air. Whirled into dark, into silence, he wonders where he is, might be, should be. Dakobed is hidden in clouds.

CHAPTER SEVENTEEN

Soul-Catcher

The old man held the rifle in one hand and walked bent over under the weight of the gunnysack on his back, as if studying the tangle of roots that was the trail. Behind him three lanky brown and black-and-white hounds crowded close to his thin legs and threw nervous glances at the wet forest all around. The only sound was that of the old man's boots and the occasional whine of one of the dogs. The sliver of moon had set, and the trail was very dark. The light from the carbide lamp on his hat cast a phosphorescent glow around the group, so that the old man, with his long, silver hair, might have been one of the Choctaw shadows on the bright path home.

Out of the dark to the old man's right came a scream that cut through the swamp like jagged tin and sent the hounds trembling against his legs.

"Hah! Get back, you!" he scolded, turning to shake his head at the cringing dogs. "That cat ain't going to eat you, not yet."

The dogs whined and pushed closer so that the old man stumbled and caught himself and the light from the headlamp splashed upon the trail. He shook his head again and chuckled, making shadows dance around them. He knew what it was that stalked him. The black *koi* hadn't been seen in the swamps during the old man's lifetime, but as a child he'd heard the stories so often that he knew at once what the *koi* meant. It was an old and familiar story. He'd felt the black one out there in the swamps for a long time. The bird, *falachito,* had called from the trees to warn him, and he had listened and gone on because what else was there to do? All of his life he had been prepared to recognize the soul-catcher when it should come.

The old man also knew that the screamer was probably the panther that the fool white man, Reeves, had wounded near Satartia a couple of weeks before. He could feel the animal's anger there in the darkness, feel the hatred like grit between his teeth. And he felt great pity for the injured cat.

The boar coon in the sack was heavy, and the old man thought that he should have brought the boy along to help, but then the forest opened and he was at the edge of his cabin clearing, seeing the thread of his garden trail between the stubble of the past year's corn and the dried husks of melon and squash vines. Behind him the panther screamed again, this time to his left. The cat had been circling like that for the past hour, never getting any closer or any farther away.

He paused at the edge of the clearing and spoke a few words in a low voice, trying to communicate his understanding and sympathy to the wounded animal and his knowledge of what was there to the soul-catcher. For a moment he leaned the rifle against his leg and reached up to touch a small pouch that hung inside his shirt. All of his life the old man had balanced two realities, two worlds, a feat that had never struck him as particularly noteworthy or difficult. But as the cat called out once more, he felt a shadow fall over him. The animal's cry rose from the dark waters of the swamp to the stars and then fell away like one of the deep, bottomless places in the river.

When the old man pulled the leather thong to open the door, the hounds shot past and went to cower beneath the plank beds. He lowered the bag to the puncheon floor and pushed the door closed. After a moment's thought he dropped the bolt into place before reaching with one hand to hang the twenty-two on nails beneath a much larger rifle. Finally, he looked at the teenage boy sitting on the edge of one of the beds with a book in his lap. The lantern beside the boy left half of his upturned face in shadow, as if two faces met in one, but the old man could see one green eye and the fair skin, and he wondered once more how much Choctaw there was in the boy.

The boy looked up fully and stared at the old uncle, the distinct epicanthic fold of each eye giving the boy's face an oddly Asian quality.

"*Koi,*" the old man said. "A painter. He followed me home."

After a moment's silence, the boy said, "You going to keep him?"

The old man grinned. The boy was getting better.

"Not this one," he replied. "He's no good. A fool shot him, and now he's mad." He studied the air to one side of the boy and seemed to make a decision. "Besides, this black one may be *nalusa-chito,* the soul-catcher. He's best left alone, I think."

The boy's grin died quickly, and the old man saw fear and curiosity mingle in the pale eyes.

"Why do you think it's *nalusachito*?" The word was awkward on the boy's tongue.

"Sometimes you just know these things. He's been out there a while. The bird warned me, and now that fool white man has hurt him."

"*Nalusachito* is just a myth," the boy said.

The old man looked at the book in the boy's lap. "You reading that book again?"

The boy nodded.

"A teacher give that book to your dad one time, so's he could learn all about his people, the teacher said. He used to read that book, too, and tell me about us Choctaws." The old man grinned once more. "After he left, I read some of that book."

The old man reached a hand toward the boy. "Here, let me read you the part I like best about us people." He lifted a pair of wire-rimmed glasses from a shelf above the rifles and slipped the glasses on.

The boy held the book out and the old man took it. Bending so that the lantern light fell across the pages, he thumbed expertly through the volume.

"This is a good book, all right. Tells us all about ourselves. This writer was a smart man. Listen to this." He began to read, pronouncing each word with care, as though it were a foreign language. *"The Choctaw warrior, as I knew him in his native Mississippi forest, was as fine a specimen of manly perfection as I have ever beheld."* He looked up with a wink. *"He seemed to be as perfect as the human form could be. Tall, beautiful in symmetry of form and face, graceful, active, straight, fleet, with lofty and independent bearing, he seemed worthy in saying, as he of Juan Fernandez fame: 'I am monarch of all I survey.' His black piercing eye seemed to penetrate and read the very thoughts of the heart, while his firm step proclaimed a feeling sense of manly independence. Nor did their women fall behind in all that pertains to female beauty."*

The old man looked at the boy. "Now there's a man that hit the nail on the head." He paused for a moment. "You ever heard of this Juan Fernandez? Us Choctaws didn't get along too good with Spanish people in the old days. Remind me to tell you about Tuscalusa some time."

The boy shook his head. "Alabama?"

The old man nodded. "I read this next part to Old Lady Blue Wood that lives crost the river. She says this is the smartest white man she ever heard of." He adjusted the glasses and read again. *"They were of such unnatural beauty that they literally appeared to light up everything around them. Their shoulders were broad and their carriage true to Nature, which has never been excelled by the hand of art, their long, black tresses hung in flowing waves, extending nearly to the ground; but the beauty of the countenances of many of those Choctaw and Chickasaw girls was so extraordinary that if such faces were seen today in one of the parlors of the fashionable world, they would be considered as a type of beauty hitherto unknown."*

He handed the book back to the boy and removed the glasses, grinning all the while. "Now parts of that do sound like Old Lady

Blue Wood. That unnatural part, and that part about broad shoulders. But she ain't never had a carriage that I know of, and she's more likely to light into anybody that's close than to light 'em up."

The boy looked down at the moldy book and then grinned weakly back at the old uncle. Beneath the floppy hat, surrounded by the acrid smell of the carbide headlamp, the old man seemed like one of the swamp shadows come into the cabin. He thought about his father, the old man's nephew, who had been only half Choctaw but looked nearly as dark and indestructible as the uncle. Then he looked down at his own hand in the light from the kerosene lantern. The pale skin embarrassed him, gave him away. The old man, his great-uncle, was Indian, and his father had been Indian, but he wasn't.

There was a thud on the wood shingles of the cabin's roof. Dust fell from each of the four corners of the cabin and onto the pages of the damp book.

"*Nalusachito* done climbed up on the roof," the old man said, gazing at the ceiling with amusement. "He moves pretty good for a cat that's hurt, don't he?"

The boy knew the uncle was watching for his reaction. He steeled himself, and then the panther screamed again and he flinched.

The old man nodded. "Only a fool or a crazy man ain't scared when soul-catcher's walking around on his house," he said.

"You're not afraid," the boy replied, watching as the old man set the headlamp on a shelf and hung the wide hat on a nail beside the rifles. He pulled a piece of canvas from beneath the table and spread it on the floor. As he dumped the coon out onto the canvas, he looked up with a chuckle. "That book says Choctaw boys always respected their elders. I'm scared all right, but I know about that cat, you see, and that's the difference. That cat ain't got no surprises for me because I'm old, and I done heard all the stories."

The boy glanced at the book.

"It don't work that way," the old man said. "You can't read them. A white man comes and he pokes around and pays somebody or maybe somebody feels sorry for him and tells him stuff and he writes it down. But he don't understand, so he can't put it down right, you see."

How do you understand, the boy wanted to ask as he watched the uncle pull a knife from its sheath on his hip and begin to skin the coon, making cuts down each leg and up the belly so delicately that the boy could see no blood at all. The panther shrieked overhead, and the old man seemed not to notice.

"Why don't you shoot it?" the boy asked, looking at the big deer rifle on the wall, the .30-40 Krag from the Spanish-American War.

The old man looked up in surprise.

"You could sell the skin to Mr. Wheeler for a lot of money, couldn't you?" Mr. Wheeler was the Black man who came from across the river to buy the coon skins.

The old man squinted and studied the boy's face. "You can't hunt that cat," he said patiently. "*Nalusachito*'s something you got to accept, something that's just there.

"You see," he continued, "what folks like that fool Reeves don't understand is that this panther has always been out there. We just ain't noticed him for a long time. He's always there, and that's what people forget. You can't kill him." He tapped his chest with the handle of the knife. "*Nalusachito* comes from in here."

The boy watched the old man in silence. He knew about the soul-catcher from the book in his lap. It was an old superstition, and the book didn't say anything about *nalusachito* being a panther. That was something the old man invented. This panther was very real and dangerous. He looked skeptically at the old man and then up at the rifle.

"No," the old man said. "We'll just let this painter be."

He pulled the skin off over the head like a sweater, leaving the naked body of the raccoon shining like a baby in the yellow light. Under the beds the dogs sniffed and whined, and overhead the whispers moved across the roof.

The old man held the skin up and admired it and laid it, fur down, on the bench beside him. "I sure ain't going outside to nail this up right now," he said, the corners of his mouth suggesting a grin. He lifted the bolt and pushed the door open and swung the body of the coon out into the dark. When he closed the door there was a snarl and an impact on the ground. The dogs began to growl and whimper, and the old man said, "You, Yvonne! Hoyo!" and the dogs shivered in silence.

The boy watched the old man wash his hands in the bucket and sit on the edge of the other bed to pull off his boots. Each night and morning since he had come it had been the same. The old uncle would go out at night and come back before daylight with something in the bag. Usually the boy would awaken to find the old man in the other plank bed, sleeping like a small child so lightly that the boy could not see or hear him breathe. But this night the boy had awakened in the very early morning, torn from sleep by a sound he wasn't conscious of hearing, and he had sat up with the lantern and book to await the old man's return. He read the book because there was nothing else to read. The myths reminded him of fairy tales he'd read as a child, and he tried to imagine his father reading them.

The old man was a real Choctaw—*Chahta okla*—a fullblood. Was the ability to believe the myths diluted with the blood, the boy wondered, so that his father could, when he had been alive, believe half as strongly as the old man and he, his father's son, half as much yet? He thought of the soul-catcher and he shivered, but he knew that he was just scaring himself the way kids always did. His mother had told him how they said that when his father was born the uncle had shown up at the sharecropper's cabin and announced that the boy would be his responsibility. That was the Choctaw way, he said, the right way. A man must accept responsibility for and teach his sister's children. Nobody had thought of that custom for a long time, and nobody had seen the uncle for years, and nobody knew how he'd even learned of the boy's birth, but there he was come out of the swamps across the river with his straight black hair hanging to his shoulders under the floppy hat and his face dark as night so that the mother, his sister, screamed when she saw him. And from that day onward the uncle had come often from the swamps to take the boy's father with him, to teach him.

The old man rolled into the bed, pulled the wool blanket to his chin, turned to the wall, and was asleep. The boy watched him and then turned down the lamp until only a dim glow outlined the objects in the room. He thought of Los Angeles, the bone-dry hills and yellow air, the home where he'd lived with his parents before the accident that killed them. It was difficult to be Choctaw, to be Indian there, and he'd seen his father working hard at it, going to urban powwows where the fancy dancers spun like beautiful birds, growing his black hair long. His father had taught him to hunt in the desert hills and to say a few phrases like *Chahta isht ia* and *Chahta yakni* in the old language. The words had remained only sounds, the powwow dancers only another Southern California spectacle for a green-eyed, fair-skinned boy. But the hunting had been real, a testing of desire and reflex he had felt all the way through.

Indians were hunters. Indians lived close to the land. His father had said those things, often. He thought about the panther. The old man would not hunt the black cat, and had probably made up the story about *nalusachito* as an excuse. The panther was dangerous. For a month the boy had been at the cabin and had not ventured beyond the edges of the garden except to go out in the small rowboat onto the muddy Yazoo River that flanked one side of the clearing. The swampy forest around the cabin was like the river, a place in which nothing was ever clear: shadows, swirls, dark forms rising and disappearing again, nothing ever clearly seen. And each night he'd lain in the bed and listened to the booming and cracking of the swamp like something monstrously evil and thought of the old man killing things in the dark, picturing the old man as a solitary light cutting the darkness.

The panther might remain, its soft feet whispering maddeningly on the cabin roof each night while the old man hunted in the swamp. Or it might attack the old man who would not shoot it. For the first time the boy realized the advantage in not being really Choctaw. The old uncle could not hunt the panther, but he could because he knew the cat for what it really was. It would not be any more difficult than the wild pigs he'd hunted with his father in the coast range in California, and it was no different than the cougars that haunted those same mountains. The black one was only a freak of nature.

Moving softly, he lifted the heavy rifle from its nails. In a crate on the floor he found the cartridges and, slipping on his red-plaid mackinaw, he dropped the bullets into his pocket. Then he walked carefully to the door, lifted the bolt, stepped through, and pulled the door silently closed. Outside, it was getting close to dawn and the air had the clean, raw smell of that hour tainted by the sharp odor of the river and swamp. The trees were unsure outlines protruding from the wall of black surrounding the cabin on three sides. Over the river the fog hovered in a gray somewhat lighter than the air, and a kingfisher called in a shrill *kree* out across the water.

He pushed shells into the rifle's magazine and then stepped along the garden trail toward the trees, listening carefully for the sounds of the woods. Where even he knew there should have been the shouting of crickets, frogs, and a hundred other night creatures, there was only silence beating like the heartbeat drum at one of the powwows. At the edge of the clearing he paused.

In the cabin the old man sat up and looked toward the door. The boy had an hour before full daylight, and he would meet *nalusachito* in that transitional time. He fingered the medicine around his neck and wondered about such a convergence. There was a meaning beyond his understanding, something that could not be avoided.

The boy brushed aside a muscadine vine and stepped into the woods, feeling his boots sink into the wet floor. It had all been a singular journey toward this, out of the light of California, across the burning earth of the Southwest, and into the darkness of this place. Beyond the garden, in the uncertain light, the trunks of trees, the brush and vines, were like a curtain closing behind him. Then the panther cried in the damp woods somewhere in front of him, the sound insinuating itself into the night like one of the tendrils of fog that clung to the ground. He began to walk on the faint trail toward the sound, the air so thick he felt as though he were suspended in fluid, his movements like those of a man walking on the floor of a sea. His breathing became torturous and liquid, and his eyes adjusted to the darkness and strained to isolate the watery forms surrounding him.

When he had gone a hundred yards the panther called again, a strange, dreamlike, muted cry different from the earlier screams, and he hesitated a moment and then left the trail to follow the cry. A form slid from the trail beside his boot, and he moved carefully away, deeper into the woods beyond the trail. Now the light was graying, and the leaves and bark of the trees became delicately etched as the day broke.

The close scream of the panther jerked him into full consciousness, and he saw the cat. Twenty feet away, it crouched in a clutter of vines and brush, its yellow eyes burning at him. In front of the panther was the half-eaten carcass of the coon.

He raised the rifle slowly, bringing it to his shoulder and slipping the safety off in the same movement. With his action, the panther pushed itself upright until it sat facing him on its haunches, braced by the front feet. It was then he saw that one of the front feet hung limply, a festering wound on the shoulder. He lined the notched sight of the rifle against the cat's head, and he saw the burning go out of the eyes. The panther watched him calmly, waiting as he pulled the trigger. The animal toppled backwards and kicked for an instant and was still.

He walked to the cat and nudged it with a boot. *Nalusachito* was dead. He leaned the rifle against a tree and lifted the cat by its four feet and swung it onto his back, surprised at how light it was and feeling the sharp edges of the ribs through the fur. He felt sorrow and pity for the hurt animal he could imagine hunting awkwardly in the swamps, and he knew that what he had done was right. He picked up the rifle and turned back toward the cabin, walking bent under the weight of his burden.

♦ ♦ ♦

When he opened the cabin door with the cat on his shoulder, the old man was sitting in the chair facing him. He leaned the rifle against the bench and swung the panther carefully to the floor and looked up at the old man, but the old man's eyes were fixed on the open doorway. Beyond the doorway *nalusachito* crouched, ready to spring.

PART THREE

Refractions

CHAPTER EIGHTEEN

As If an Indian Were Really an Indian

Native American Voices and Postcolonial Theory

It's very convenient, isn't it, that so much of what we perhaps loosely term postcolonial theory today is written by real "Indians"—with such names as Chakrabarty, Chakravorty, Gandhi, Bhabha, Mohanty, and so on—so that, in writing or speaking about indigenous Native American literature, we can, if we desire, quote without even changing noun or modifier, as if an Indian were really an Indian. I am myself neither a real Indian like Gayatri Chakravorty Spivak or Satya Mohanty, nor a theorist, but merely a teacher who also writes, a writer who also teaches—and my reflections should be taken in that context. I also wish to borrow a caution from the seminal postcolonial writer Frantz Fanon, who declares in *Black Skin, White Masks,* "Since I was born in the Antilles, my observations and my conclusions are valid only for the Antilles" (14). Following and narrowing Fanon's wise lead, I must say that in significant measure I am my own Antilles and therefore my observations and conclusions are valid only for myself. As an American of deeply mixed heritage and somewhat unique upbringing, I speak on behalf of and from the perspective of no one else.

Edward Said has written about *"strategic location,* which [he explains] is a way of describing the author's position in a text with regard to the . . . material he writes about" (*Orientalism* 20). My own strategic location vis-à-vis what we have come to call Native

American literature is a complicated and contingent one. Descended from Mississippi Choctaw and Oklahoma Cherokee peoples on both paternal and maternal sides, I am also Euramerican, with Irish and Cajun French ancestors. My strategic location, therefore, may be found in what I think of as a kind of frontier zone, which elsewhere I have referred to as "always unstable, multidirectional, hybridized, characterized by heteroglossia, and indeterminate" (*Mixedblood Messages* 26). In this sense my work, like that of many other writers identified as Native American, might in fact be accurately described as what postcolonial theorists have called "migrant" or "diasporic" writing, for while I spent my earliest years in Choctaw country in Mississippi, I do not write from the heart of a reservation site or community and was not raised within a traditional culture. It would not be incorrect to say, in fact, that today in the U.S., urban centers and academic institutions have come to constitute a kind of diaspora for Native Americans who through many generations of displacement and orchestrated ethnocide are often far from their traditional homelands and cultural communities. Such a frontier/transcultural location is an inherently unstable position, and one from which it is difficult and undoubtedly erroneous to assume any kind of essential stance or strategy, despite many temptations to do so.

A commonality of issues, shared by all of us who work in the field of Native American literature, marks this unsettled zone of frontier, interstitial scholarship. Dipesh Chakrabarty has suggested that non-Western thought is consistently relegated to the status of what Foucault might term a "subjugated knowledge" or what others define as a "minor" or "deterritorialized" knowledge (Gandhi 43). Chakrabarty writes that while Third World historians "feel a need to refer to works in European history; historians of Europe do not feel any need to reciprocate. . . . We cannot even afford an equality or symmetry of ignorance at this level without taking the risk of appearing 'old fashioned' or 'out-dated'" (Chakrabarty 2). To Chakrabarty's lament, I would add that critics and teachers of Native American literature are in a similar position. We are properly expected to have and exhibit a crucial knowledge of canonical European and Euramerican literature; if we fail to be familiar with Shakespeare, Chaucer, Proust, Flaubert, Dickinson, Faulkner, Eliot, Joyce, Pound, Yeats, Keats, Woolf, Tolstoy, Tennyson, and so forth—not to mention the latest poststructuralist theory—we are simply not taken seriously and probably will not earn a university degree in the first place. That, it is presumed, is the foundational knowledge, the "grand narrative of legitimation" in our particular field. However, while I have a great fondness for and some knowledge of canonical Western literature, like Chakrabarty I have often noticed that the majority of my colleagues in various "English" departments around the U.S. know very little if anything about Native American literature, written or oral; nor do they often exhibit any symptom of feeling it to be incumbent upon them to gain such knowledge. It may be pleasant to believe, as the guest editor of a recent issue of *Modern Fiction Studies* declared, that "Native American literature has become part of the mainstream,"[1] but, despite incremental change over the past couple of decades, it would only take a few minutes in the English departments of most universities, even in the Southwest, the heart of supposed "Indian Country," to be disabused of such wishful thinking. In fact, surprisingly, it would not take much time spent browsing through contemporary critical/theoretical texts—including especially those we call postcolonial—to discover an even more complete erasure of Native American voices.

It seems that Chakrabarty's "symmetry of ignorance" applies even less to postcolonial theorists' awareness of Native American voices than it does to mainstream academia. While Native American literature is gradually finding a niche within academia, one discovers, predominantly, an absence of Native American voices in works by major cultural theorists and respected writers. We find, for example, a surprising erasure of Native presence and voice in Toni Morrison's *Playing in the Dark,* a highly praised work in which this Nobel-winning African American writer adopts the rhetoric of imperial discovery and cites Hemingway's "Tontos" as well as

Melville's *Pequod* without ever noting an indigenous Native presence shadowing her figuration of blackness in America. The most extraordinary denigration of Native American voices is found, however, in Said's *Culture and Imperialism,* where this celebrated father of postcolonial theory dismisses Native American writing in a single phrase as "that sad panorama produced by genocide and cultural amnesia which is beginning to be known as 'native American literature'" (304). We can add to this list Homi Bhabha's total silencing of indigenous Native Americans in his influential study, *The Location of Culture.* Bhabha gives the impression of being acutely aware of a wide panoply of minority voices in this book, referencing Hispanic and Black American writers, for instance, and extensively praising Morrison's *Beloved,* but nowhere, not even in a whispered aside, does he note the existence of a resistance literature arising from indigenous, colonized inhabitants of the Americas. How, one wonders, can this student of postcoloniality, difference, liminality, and what he terms "culture as a strategy of survival" (172) be utterly ignorant of or indifferent to such writers as N. Scott Momaday, James Welch, Leslie Silko, Louise Erdrich, Simon Ortiz, or, most incredibly, the Anishinaabe writer and theorist Gerald Vizenor? How, one wonders, can any serious student of the "indigenous or native narrative," the term Bhabha uses to define his subject, not read and deal with the radically indigenous theory of Vizenor? Imagine what lengths Bhabha and other postcolonial theorists would have to go to in order to come to terms with Vizenor's "trickster hermeneutics"—which effectively subsumes poststructuralist and postmodern theory into a Native American paradigm and discursive system—or the work of poet Luci Tapahonso, who writes in both Navajo and English from a position deeply embedded in her Diné culture while moving freely within the academic diaspora, liberating and appropriating villanelles, sestinas, and sonnets to a Navajo voice and epistemology.

Those of us who moan about the "center's" refusing to hear the "margin's" cries of protest, or, worse yet, the metropolitan center's refusing even to acknowledge that there is a voice and a multitudinous language out here, should face certain truths. Yes, it is very difficult—almost impossible, really—for the truly new written work of protest or resistance or "other" narrative even to be published, let alone acknowledged. But why should that not be the case? The American world of books, which is to say New York, with its half-dozen major commercial publishing houses remaining—actually publishing conglomerates reflecting the "globalization" of industry that is merely an extension of the marketing phalanx of Western corporatism across all boundaries—is a for-profit endeavor.[2] Controlled by highly educated persons of European and European American descent, the publishing industry manufactures a product designed primarily for consumers of European and European American origin. It's not simply because, as the noted bank robber said, that's where the money is. As Nabokov's mad reader/editor Kinbote demonstrates so superbly in the novel *Pale Fire,* we all want to read ourselves into and back out of every text and will go to great extremes to ensure that possibility. Naturally, the primarily white, Eurocentrically educated reader will want to find him- and herself inside the covers of any text. Thus Richard Ford wins a Pulitzer Prize for an uninspired, entirely predictable, white middle-class male angst story such as *Independence Day,* or a sophomoric tale of New York such as *Bright Lights, Big City* is made a best-seller by New York. This is unfortunate for those of us who create and desire to read texts containing "other" stories, but it is not inherently an immoral or unpredictable proposition. A novel by and about a Blackfeet Indian growing up among other Blackfeet Indians and the remnants of an indigenous culture in a far corner of America, or a radically displaced Pueblo Indian suffering deracination in a southwestern world of ceremonial stories and kivas—these things are alien to the book-buying public of America except in the broadest human sense. There is no moral imperative that we can bring to bear to coerce publishers into manufacturing such texts or consumers into buying and reading them. Simply put, it is quite natural for a dominant culture to manufacture and consume primarily those artifacts that reflect and reinforce its own sense of self, of culture. We can argue all we want that so-called "marginalized" voices must have a place in American publishing because America's multiculturality is increasing exponentially. We, too, we feel, deserve books that mirror our worlds, our cultures, our values, our particular forms of struggle and enlightenment. Like Nabokov's Kinbote, we want to find our own Zemblas in the texts we read. We can argue that the world of manufactured cultural artifacts called books is controlled by an elite whose consciousness is in turn manufactured in universities controlled by and for the same elite, and we will be correct. But that is, in hard ways, beside the point. Books that appeal to the metropolitan center's desire for the exotic, the colorful fringes of itself—such as the new rush of novels by writers from (formerly imperial) India—and that do so without radically disturbing the center's sense of well-being, will continue to be produced. In America we will continue to have the *Native Sons, Invisible Mans, Beloveds, Woman Warriors,* and *Love Medicines.* But the voices that tell stories too disturbing or too alien will be kept silenced or at best on the publishing margins represented by small presses and, more than ever today, university presses. In the end we should not be surprised by the fact that the dominant culture will continue to produce consumer goods—and therefore cultural capital—for itself, even as we rail about justice from the far margins. Therefore we will be forced to continue to find other venues, other paths toward our audiences, outside of New York and the half-dozen remaining publishing houses—a not necessarily unhealthy situation.

With that being said, we can turn again to those intellectuals whose announced intention it is to contest the hegemony of cultural totalitarianism always already at the center, those individuals and institutions who make a self-satisfied profession of challenging the orthodoxy and monologic authority of the imperial powers. They cannot be excused for the same ignorance, the same deafness, the same blindness, the same refusal to see or hear the margin's protests. In these preening intellectuals, such ignorance does indeed represent a moral failure. It is, in fact, difficult if not impossible to take seriously any cultural/critical theorist who expounds upon "post-coloniality" or liminality and at the same time appears ignorant of the rapidly growing body of Native American literature, or who, if he or she is aware of it, clearly relegates it to a "minor," "subjugated, or "deterritorialized" knowledge worthy of only silence or erasure. Quite ironically, it becomes quickly obvious that these cultural critics are supremely the products of the imperial crucible of the Western university, speaking the languages of Cambridge, Oxford, Harvard, Yale, Princeton, the language of the imperial center. Within this great silence, it comes as more of a surprise than it should to find Trinh Minh-Ha, in *Woman, Native, Other,* paying careful attention to Laguna Pueblo writer Leslie Marmon Silko's storytelling and then at least noting such Native American writers as Silko, Momaday, Joy Harjo, Vizenor, and Linda Hogan in a chapter from her 1991 essay collection, *When the Moon Waxes Red.*

As we argue, nonetheless, for the recognition missing from such texts and, concomitantly, a place for the margin on the literary and academic stage, we must be wary of what Rey Chow has called the language of victimization and "self-subalternisation," which, Chow points out, "has become the assured means to authority and power" in the metropolis, a phenomenon Leela Gandhi refers to succinctly as the "professionalisation of the margin" (Gandhi 128) and Vizenor calls "aesthetic victimry" (*Fugitive Poses* 12). And we need to examine carefully those Native American texts that do make it somehow from the margin to if not the center at least an orbital relationship with the homogenizing gravity of that center. Citing her own frustrating experience in a white-male-dominated academic conference as an example, Gayatri Spivak warns that "The putative center welcomes selective inhabitants of the margin in order to better exclude the margin." Spivak—a "real" Indian in America—quotes Adrienne Rich, who says, "This is the meaning of female tokenism: that power withheld from the vast majority of women is offered to the few, so that it may appear that any truly qualified woman can gain access to leadership, recognition, and reward" (Spivak, *In Other Worlds* 107).

In mentioning postcolonial theory and Native American literature in the same breath, we should also keep carefully in mind the fact that America does not participate in what is sometimes termed the "colonial aftermath" or postcolonial condition. There is no need in the U.S. for what has been called the "therapeutic retrieval of the colonial past" in postcolonial societies (Gandhi 5). In *The Wretched of the Earth,* Fanon wrote, "Two centuries ago, a former European colony decided to catch up with Europe. It succeeded so well that the United States of America became a monster, in which the taints, the sickness, and the inhumanity of Europe have grown to appalling dimensions" (313). Regardless of whether we share Fanon's harsh critique of the U.S., I think it is crucial for us to remember that the American Revolution was not truly a war to throw off the yoke of colonization as is popularly imagined, but rather a family squabble among the colonizers to determine who would be in charge of the colonization of North America, who would control the land and the lives of the indigenous inhabitants. America never became postcolonial. The indigenous inhabitants of North America can stand anywhere on the continent and look in every direction at a home usurped and colonized by strangers who, from the very beginning, laid claim not merely to the land and resources but to the very definition of the natives. While it may be true figuratively, as Fanon writes, that "it is the settler who has brought the native into existence and who perpetuates his existence" (*The Wretched of the Earth* 36), it is even more true, as Vizenor has argued, that the "Indian" is a colonial invention, a hyperreal construction. "The *indians,*" Vizenor writes, "are the romantic absence of natives. . . . The *indian* is a simulation and loan word of dominance. . . . The *indian* is . . . the other in a vast mirror" (*Fugitive Poses* 14, 35, 37).

In imagining the Indian, America imagines itself, and I know of no text that more brilliantly and efficiently illuminates America's self-imagining than Walt Whitman's poem, "Facing West from California's Shores." Whitman speaks in his poem as and for America, and since it is a brief poem, it is worth quoting in its entirety:

Facing west from California's shores,
Inquiring, tireless, seeking what is yet unfound,
I, a child, very old, over waves, towards the house of maternity, the land of migrations, look afar,
Look off the shores of my Western sea, the circle almost circled;
For starting westward from Hindustan, from the vales of Kashmere,
From Asia, from the north, from the God, the sage, and the hero,
From the south, from the flowery peninsulas and the spice islands,
Long having wander'd since, round the earth having wonder'd,
Now I face home again, very pleas'd and joyous,
(But where is what I started for so long ago?
And why is it yet unfound?)

In her introduction to postcolonial theory, drawing upon Heidegger, Gandhi writes that "the all-knowing and self-sufficient Cartesian subject violently negates material and historical alterity/Otherness in its narcissistic desire to always see the world in its own self-image" (39). In this extraordinary poem, Whitman articulates Euramerica's narcissistic desire not merely to negate indigenous Otherness and possess the American continent, but to go far beyond North America to what is implicitly "my" sea, "my" Hindustan, "my" Kashmere, "my" Asia, and so on. In her extraordinarily thorough "Translator's Preface" to Derrida's *Of Grammatology*, Spivak reminds us of Nietzsche's famous "will to power" and his explanation that "the so-called drive for knowledge can be traced back to a drive to appropriate and conquer" (*Of Grammatology* xxii). The imperial, "inquiring" American of Whitman calls to mind not only this Nietzschean will to power but also Spivak's point that "the desire to explain might be a symptom of the desire to have a self that can control knowledge and world that can be known" and that "the possibility of explanation carried the presupposition of an explainable (even if not fully) universe and an explaining (even if imperfectly) subject. These presuppositions assure our being. Explaining, we exclude the possibility of the heterogeneous."

Spivak concludes that "every explanation must secure and assure a certain kind of being-in-the-world, which might as well be called our politics" (*In Other Worlds* 104–6).

The Euramerican "child" of Whitman's remarkable poem claims possession of the womb of humanity in a great universalizing narcissism that subsumes everything in its tireless, inquiring quest after its own image, thereby excluding heterogeneity from the very womb, and it is pleased and joyous about the whole endeavor. Gandhi writes that "in order for Europe to emerge as the site of civilisational plenitude, the colonised world had to be emptied of meaning" (15). Whitman brilliantly comprehends this grand gesture in the Americas, and in his final question he even more brilliantly registers the ultimate, haunting emptiness the colonizing consciousness faces at last once it is forced to confront a world emptied of meaning outside of itself. Whitman's question—"(But where is what I started for so long ago? / And why is it yet unfound?)"—resonates not merely through the heart of America's own metanarrative, leaving the Jay Gatsbys of America alone and painfully perplexed, but through the rest of the world marked by European narcissism and expansion. In Whitman's poem America, having completed a stunningly deadly, self-determinedly innocent waltz across the bodies and cultures of Native inhabitants, seeks to claim not just the territory of the world but also the womb of humanity, seeding its own consciousness alone within that embryonic place of origin. This self-willed, brutal child looks to the "house of maternity" beyond "Hindustan," "Kashmere," and all of Asia to legitimize its claim to primacy and thus legitimacy as the inheritor of all.

The large stride of Whitman's American child moves across and through hundreds of indigenous cultures. They are the Others who must be both subsumed and erased in a strange dance of repulsion and desire that has given rise to both one of the longest sustained histories of genocide and ethnocide in the world as well as a fascinating drama in which the colonizer attempts to empty out and reoccupy not merely the geographical terrain but the constructed space of the indigenous other. And if we doubt the place of the imagined Indian in America's imperial self-construction, we can turn to Scribner's *Concise Dictionary of American History,* which explains: "The experiences of the Indian agencies with people of a race, color, and culture wholly unlike our own have been of great value to United States officials dealing with similar peoples throughout the world" (Andrews 463).

Fanon has written, "Man is human only to the extent to which he tries to impose his existence on another man in order to be recognized by him." He adds, "In its immediacy, consciousness of self is simple being-for-itself. In order to win the certainty of oneself, the incorporation of the concept of recognition is essential" (*Black Skin, White Masks* 216, 217). European America holds a mirror and a mask up to the Native American. The tricky mirror is that Other presence that reflects the Euramerican consciousness back at itself, but the side of the mirror turned toward the Native is transparent, letting the Native see not his or her own reflection but the face of the Euramerican beyond the mirror. For the dominant culture, the Euramerican controlling this surveillance, the reflection provides merely a self-recognition that results in a kind of being-for-itself and, ultimately, as Fanon suggests, an utter absence of certainty of self. The Native, in turn, finds no reflection directed back from the center, no recognition of "being" from that direction.

The mask is one realized over centuries through Euramerica's construction of the "Indian" Other. In order to be recognized, and to thus have a voice that is heard by those in control of power, the Native must step into that mask and BE the Indian constructed by white America. Paradoxically, of course, like the mirror, the mask merely shows the Euramerican to himself, since the masked Indian arises out of the European consciousness, leaving the Native behind the mask unseen, unrecognized for him/herself. In short, to be seen and heard at all by the center—to not share the fate of Ralph Ellison's Invisible Man—the Native must pose as the absolute fake, the fabricated "Indian," like the dancing puppets in Ellison's novel.

If a fear of inauthenticity is the burden of postmodernity, as has been suggested by David Harvey in *The Condition of Postmodernity* among others, it is particularly the burden not only of the Euramerican seeking merely his self-reflection, but even more so that of the indigenous American in the face of this hyperreal "Indian." In *Midnight's Children,* Salman Rushdie's character Saleem Sinai says, "above all things, I fear absurdity." Quoting Rushdie's character and reflecting on Sinai's predicament, Gandhi writes that "the colonial aftermath is also fraught by anxieties and fears of failure which attend the need to satisfy the burden of expectation" (5). America not only contains no "colonial aftermath," it also places no burden of expected achievement on the American Indian. On the contrary, for Native Americans the only burden of expectation is that he or she put on the constructed mask provided by the colonizer, and the mask is not merely a mirror but more crucially a static death mask, fashioned beforehand, to which the living person is expected to conform. He or she who steps behind the mask becomes the Vanishing American, a savage/noble, mystical, pitiable, romantic fabrication of the Euramerican psyche fated to play out the epic role defined by Mikhail Bakhtin: "The epic and tragic hero," Bakhtin writes, "is the hero who by his very nature must perish. . . . Outside his destiny the epic and tragic hero is nothing; he is, therefore, a function of the plot fate assigns him; he cannot become the hero of another destiny or another plot" (*The Dialogic Imagination* 36). While the "Indian" holds a special and crucial place within the American narrative, the Native who looks beyond his or her immediate community and culture for recognition finds primarily irrelevance and absurdity.

Fanon declares that "the town belonging to the colonized people, or at least the native town, the Negro village . . . the reservation, is a place of ill fame, peopled by men of evil repute. They are born there, it matters little where or how; they die there, it matters not where, not how. . . . The native town is a hungry town, starved of bread, of meat, of shoes, of coal, of light" (*The Wretched of the Earth* 39). Could this not be a description of Pine Ridge, South

Dakota, today, a hungry reservation town and the poorest county in America, home to the very Plains Indians so celebrated in America's fantastic construction of Indianness, the aestheticized subjects of such romantic and insidious films as *Dances With Wolves*? Fanon writes: "For a colonized people the most essential value, because the most concrete, is first and foremost the land: the land which will bring them bread and, above all, dignity" (*The Wretched of the Earth* 44). Above all, Native Americans have been deprived of land, and of the dignity that derives from the profound and enduring relationship with homeland.

To understand this contemporary colonialism as it affects Native Americans, we need to recognize what Gandhi, a "real Indian" writer and theorist, has called the "relationship of reciprocal antagonism and desire between coloniser and colonised." Gandhi suggests that "the battle lines *between* native and invader are replicated *within* native and invader" (4, 11–12). America's desire to control knowledge, to exclude the heterogeneous, and to assure a particular kind of being-in-the-world depends upon a total appropriation and internalization of this colonized space, and to achieve that end, America must make the heterogeneous Native somehow assimilable and concomitantly erasable. What better way to achieve that end than to invent the Indian as an Other that springs whole-formed from the Euramerican psyche? In *Black Skin, White Masks,* Fanon says simply that "what is often called the black soul is the white man's artifact" (14). It can be argued, and has been argued effectively by such Native American writers as Gerald Vizenor and Philip Deloria, that what is called the Indian is the white man's invention and artifact. In his book *Playing Indian,* Deloria documents exhaustively and persuasively America's obsession with first constructing the Indian as Other and then inhabiting that constructed Indianness as fully as possible. Deloria describes marvelous "Indianation" ceremonies in which white Americans induct one another into Indianness. In one such ceremony, a "spirit" voice intones, "The red men are my children. Long ago I saw in the future their destruction, and I was very sad."

According to Deloria, the "spirit tells initiates that the only way to placate the mournful Indian shades is to preserve their memory and customs. The society's sachem then replies that the membership will accept the 'delightful task.' The ceremony concludes by offering the initiate complete redemption and a new life through mystic rebirth as an Indian child. 'Spirit,' prays the sachem, 'receive us as your children. Let us fill too the place of those who are gone'" (78).

Deloria comments: "Like all these boundary-crossing movements, however, Indianized quests for authenticity rested upon a contradictory foundation. In order to be authentic, Indians had to be located outside modern American societal boundaries. Because they were outside those boundaries, however, it became more difficult to get at them, to lay claim to the characteristics Indians had come to represent" (115). To extend Deloria's observation, I would also suggest that the artifactual "Indian," unlike the indigenous Native, could be easily gotten at because he or she was a pure product of America. Deloria also examines Natives who step into the Euramerican-constructed mask to play "cultural politics for social and political ends," actual indigenous Native Americans who "found themselves acting Indian, mimicking white mimickings of Indianness" (189). "If being a survivor of the pure, primitive old days meant authenticity, and if that in turn meant cultural power that might be translated to social ends, it made sense for a Seneca man to put on a Plains headdress, white America's marker of that archaic brand of authority" (189).

In the midst of such extraordinary torsions, where is the Native writer to locate her or himself, and how is that writer to find or have a voice in a world articulated through another language or knowledge system? Raja Rao described the challenge faced by the indigenous artist who writes in a colonizer's language by saying, "One has to convey in a language that is not one's own the spirit that is one's own" (Gandhi 150). Rao also said, "We cannot write like the English. We should not. We cannot write only as Indians" (Gandhi 151). Native Americans, with few exceptions, such as Navajo writers like Luci Tapahonso and Rex Lee Jim who write in both Navajo and English, do not have the option of writing and publishing in an indigenous language, and even more than those of whom Rao speaks, Native authors surely cannot "write only as Indians." After five hundred years of war, of colonial infantilization and linguistic erasure, cultural denigration, and more, how and where does the Native writer discover a voice that can be heard at the metropolitan center?

Bhabha, borrowing from V. S. Naipaul's *The Mimic Men,* writes of "colonial mimicry," which, he declares, "conceals no presence or identity behind its mask" (*The Location of Culture* 86). Gandhi labels such mimicry the "sly weapon of anti-colonial civility, an ambivalent mixture of deference and disobedience," which "inaugurates the process of anti-colonial self-differentiation through the logic of inappropriate appropriation" (149–50). In glossing Bhabha's formulation, Gandhi argues that "the most radical anti-colonial writers are 'mimic men'" and that "the paradigmatic moment of anti-colonial counter-textuality is seen to begin with the first indecorous mixing of Western genres with local content" (150). Gandhi's description of such countertextual mimicry would seem to apply well to such a work as Cherokee writer John Rollin Ridge's 1854 novel, *The Life and Adventures of Joaquín Murieta, the Celebrated California Bandit,* in which Ridge appropriates the genre of western romance to tell a deeply encoded story of colonial oppression and brutality. Similarly, the native Okanagon writer Mourning Dove performed a brilliant and clearly subversive act of appropriation in her 1927 novel, *Cogewea, the Half-Blood: A Depiction of the Great Montana Cattle Range,* a novel that mimics the classic western romance with indigenous mixedbloods in the roles always reserved for European American cowboys. But neither Ridge nor Mourning Dove found acceptance in the centers that control textual production in America. Ridge, grandson of traditional Cherokee leader Major Ridge and founding editor of the *Sacramento Bee* newspaper in California, promptly vanished from the landscape of American literature to be rediscovered a century later in a kind of ethnographical salvage operation. Mourning Dove found it took not only collaboration with a white editor, but a dozen years and money earned from her labor in apple orchards and hops fields to get *Cogewea* published and then forgotten for half a century until its discovery by Native American scholars and feminist critics. Acceptance and recognition by the metropolitan center would not come for a Native American writer until 1969, when the Kiowa author N. Scott Momaday won a Pulitzer Prize for his first novel, *House Made of Dawn*. According to a member of the Pulitzer jury, an award to the author of this novel "might be considered as a recognition of the arrival on the American literary scene of a matured, sophisticated literary artist from the original Americans" (Schubnell 93).

I want to focus briefly upon Momaday's historic accomplishment. And in doing so, while keeping in mind that we are discussing a "colonial" rather than "postcolonial" text, it may be helpful here to borrow from the critic Timothy Brennan, who argues that the "privileged postcolonial text is typically accessible and responsive to the aesthetic and political taste of liberal metropolitan readers. The principal pleasures of this cosmopolitan text accrue from its managed exoticism. It is both 'inside' and 'outside' the West." Such texts provide a "familiar strangeness, a trauma by inches" (quoted in Gandhi 162). Bhabha declares that "the desire to emerge as 'authentic' through mimicry—through a process of writing and repetition—is the final irony of partial representation" (88). How does a colonial or postcolonial author—a collective whom I have earlier identified as "migrant" or diasporic in the postcolonial sense—achieve a text so accessible and responsive to metropolitan readers? Aijaz Ahmad suggests that "among the migrants themselves, only the privileged can live a life of constant mobility and surplus pleasure, between Whitman and Warhol as it were. Most migrants tend to be poor and experience displacement not as cultural plenitude but as torment. . . . Postcoloniality is also, like most things, a matter of class" (Gandhi 164). It is not difficult to extrapolate from Ahmad's statement the fact that most Native Americans live lives fashioned by generations of displacement with resulting impoverishment, suffering, and silence. While life for urban Native Americans can be difficult, the poorest and most desperate places in America are Native American reservations, populated by indigenous people who commonly live with as much as 85 percent unemployment, deplorable health care and even worse education, horrifying rates of alcohol and drug addiction, an epidemic of fetal alcohol syndrome, and the highest suicide rates and lowest life expectancy of any ethnic group in the nation—all the effects of profoundly institutionalized racism. It is apparently delightful to caricature Native Americans as sports mascots and in movies, but as long as the real people are hidden from sight on rural lands reserved for their containment, it isn't necessary for the dominant culture even to contemplate the Natives' quality of life. And before everyone protests, let me quote Jean-Paul Sartre, who wrote acutely in his preface to Fanon's *The Wretched of the Earth,* "Our worthiest souls contain racial prejudice" (21). How else could our worthiest souls live with Washington Redskins, Cleveland Indians, Atlanta Braves, Chief Wahoo, the Tomahawk Chop, and so on, all of which are merely the slightest indices of the far more profound currents that impoverish Native Americans in incalculable ways? Would we live so comfortably today with the New Jersey Jews, Newark Negroes, Cleveland Chicanos, Houston Honkies, Atlanta Asians, and so on? Entering the twenty-first century, would the city of Washington cheer a football team whose mascot scampered around at half time in blackface or wearing a yarmulke and carrying a menorah? Again, Bhabha adds an interesting note when he writes that "the discriminatory identities constructed across traditional cultural norms and classifications, the Simian Black, the Lying Asiatic—all these are *metonymies* of presence. They are strategies of desire" (90).

America has in large part at least recognized the impropriety of such racial stereotypes as those Bhabha describes—the "Simian Black" and "lying" or "inscrutable" Asian—even if it has perhaps not relinquished them entirely. However, the same operative strategy of desire that for a very long time governed representation of other minority cultures clearly still controls Euramerican discourse representing the Native American Indian. Given the uncomfortable realities of most contemporary Native American lives, as we look around us in academia, I think it is perhaps time to recognize that what we are calling Native American literature is represented largely, if not exclusively, by the sorts of privileged texts Brennan describes and is created by those migrant or diasporic Natives who live lives of relatively privileged mobility and surplus pleasure. As a group, published Native American authors have an impressively high rate of education, most possessing not merely a university degree but at least some graduate work if not an advanced degree. We may go back to our families and communities periodically or regularly, we may—like N. Scott Momaday—even be initiated into a traditional society within our tribal culture, but we are inescapably both institutionally privileged by access to Euramerican education and distinctly migrant in the sense that we possess a mobility denied to our less privileged relations.

The question remains: what must the colonial, or postcolonial, writer—in this case specifically the indigenous Native American, mixedlood, or fullblood—do to be allowed a voice like Shakespeare's cursing Caliban? Let us consider the Pulitzer Prize juror's words regarding Momaday's award for *House Made of Dawn*. What might such a statement as "the arrival on the American literary scene of a matured, sophisticated literary artist from the original Americans" possibly mean? I would suggest, as I have elsewhere, that these words indicate that an aboriginal writer has finally learned to write like the colonial center that determines legitimate discourse. Momaday so successfully mimicked the aesthetics of the center that he was allowed access. This—the evolution of the colonial subject to *something like* the level of the colonial master—fulfills the idealized, utopian colonial vision of reconstructing the colonial subject to reflect the colonizing gaze. To be heard at the center—to achieve both authenticity and recognition within the imperial gaze that controls dissemination of discourse—the Native must step into the mask and into Whitman's circle; and the more fully we enter that empowering circle, the more successful we may be. The most prosperous of such texts are both accessible to the aesthetic and political tastes of the metropolitan center and, perhaps more significantly, present to such readers a carefully managed exoticism that is entertaining but not discomfiting to the non-Native reader. Elsewhere I have called the most extreme versions of this publishing phenomenon "literary tourism." And, as I have also suggested in my book *Mixedblood Messages,* certainly Momaday's Pulitzer Prize testifies to a Native American author's successful entry into that privileged circle. *House Made of Dawn,* with its visible and pervasive indebtedness to classic modernist texts, is clearly "accessible and responsive to the aesthetic and political taste of liberal metropolitan readers." Just as clearly, Momaday's novel offers what Brennan calls a "managed exoticism." The Pueblo world of the protagonist, with its "familiar strangeness" and "trauma by inches," is never threatening or very discomfiting to its Euramerican readership, while it provides a convincing and colorful tour of Indian country, including titillating glimpses of Native ceremony and even witchcraft. Exoticism packaged in familiar and therefore accessible formulas, carefully managed so as to admit the metropolitan reader while never implying that the painful difficulties illuminated within the text are the responsibility of that reader—clearly *House Made of Dawn* is both "inside" and "outside" the West.

To say all of the above is not to deny the aesthetic or artistic achievement or even the important political ramifications of Momaday's novel or many others that might be similarly described. *House Made of Dawn* is superbly subversive. Seen in this way, Momaday becomes one of the "mimic men," to borrow Naipul's term, who realize a genuinely appropriationist and subversive end by packaging a text in sufficient imperial wrappings to get it past the palace guards into the royal Pulitzer chambers. It is an impressive achievement, yet an achievement requiring extensive education of its author—all the way to a doctorate from Stanford University—within the knowledge system of the metropolitan center as well as careful crafting to conform to that center's expectations. A peripatetic citizen of the U.S. Native American diaspora, a modern "migrant" intellectual and artist, Momaday—like all the rest of us who find ourselves in somewhat similar if less celebrated circumstances—has not only entered the trajectory of Whitman's circular journey but has internalized that circle. Momaday's "strategic location" vis-à-vis the Pueblo world he wrote about in his first novel, like the strategic location of most Native American writers, is surpassingly complex, shifting, hybridized, interstitial, and unstable.

Those of us working in the field of what we call Native American literature can and undoubtedly will chafe at the ignorance and erasure of Native American voices within the metropolitan center and within what at times appears to be the loyal opposition to that center called postcolonial theory. And we can and undoubtedly will continue to try to make our voices heard—to "give voice to the silent," as a recent academic conference termed the endeavor. However, such negotiations are never simple or free of cost. It seems that a necessary, if difficult, lesson for all of us may well be that in giving voice to the silent we unavoidably give voice to the forces that conspire to effect that silence. Finally, as we approach nearer to literate Euramerica, the circle within us may be almost circled, and like Whitman's America, we may be left asking, "But where is what I started for so long ago?"

CHAPTER NINETEEN

Staging indians

Native Sovenance and Survivance in Gerald Vizenor's "Ishi and the Wood Ducks"

"Ishi was never his real name." So begins Gerald Vizenor's "Ishi Obscura," in the *anishinaabe* author's *Manifest Manners* (1994). An essay on the last "wild man" of America captured in a museum and photographs, "Ishi Obscura" interrogates the invention of Ishi, the reputed last of the Yahi people in California, as museum artifact and photographic "other." In "Ishi and the Wood Ducks," Vizenor's four-act play, the author invites the Yahi survivor into the courtroom of contemporary "indianness" and, through Ishi, deconstructs the deadly desire of Euramerica for the simulated Indian. "The *indian,*" Vizenor writes in *Fugitive Poses,* "is a misnomer, a simulation with no referent and with the absence of natives; *indians* are the other, the names of sacrifice and victimry" (27).[1] Ishi, the actual man who walked out of California's Sierra Nevada range to die in a museum, offers tragic visionaries the quintessential material for "sacrifice and victimry," what Vizenor also terms "the metaphor of the native at the littoral" and a history of "aesthetic sacrifice" (*Fugitive Poses* 27). However, Vizenor's aim in giving the "last man of the stone age" a storyteller's voice in his play is to contradict the entropic story of the American "*indian.*"

Vizenor has had a long and intense fascination with Ishi, the latest fruit of which is "Ishi and the Wood Ducks," a work that deploys compassion, humor, and the ironies of trickster to dissect

Euramerica's invention of Indians at the same time that it pays homage to the man called Ishi. At the heart of the play's deconstruction of Ishi as the invented "other" is what Vizenor defines as "trickster hermeneutics": "the interpretation of simulations in the literature of survivance, the ironies of descent and racialism, transmutation, third gender, and themes of transformation in oral tribal stories and written narratives" (*Manifest Manners* 15). As a "postindian" survivor, Vizenor's Ishi steps outside the frame of the nostalgic colonial camera, defies the melancholy and terminal definition called "Indian," and appears on stage as a good-humored signifier of indeterminate human consciousness, a counteragent against the "hyperreal simulation and . . . ironic enactment of a native presence by an absence in a master narrative" (*Fugitive Poses* 27). In the end, despite Euramerica's desperate attempts to fix Ishi as a phenomenon called a "primitive Indian," Vizenor makes it clear that Ishi exists forever in the moment of his stories, reinventing himself within the oral tradition with each utterance.

While the first line in Vizenor's opening "Historical Introduction" to "Ishi and the Wood Ducks" quotes a description of the newly discovered Ishi as "a pathetic figure crouched upon the floor," Ishi's first words in the play are "Have you ever heard the wood duck stories," a telling contradiction. Within his stories Ishi embodies what Vizenor terms "survivance" and "sovenance": "The native stories of survivance are successive and natural estates; survivance is an active repudiation of dominance, tragedy, and victimry. . . . Native sovenance is that sense of reason in native stories . . . not the romance of an aesthetic absence or victimry" (*Fugitive Poses* 15). "Ishi and the Wood Ducks" provides Vizenor with yet another opportunity to dissect the pathological impulse that has driven Euramerica for five centuries to discover an aboriginal something commensurate with its own needs, to call that something "Indian," and to obscure the Native in the process. Imagining the Indian as a natural resource to be mined for some kind of catalyst that will make the "American" alloy so much the stronger, the Euramerican metanarrative peels away the man or woman to find the Indian. Vizenor's art strips the artifactual veneer of Indianness to find the Native humanity at the center, to liberate the original American from the deadening servitude of Indianness.

In the writings of Gerald Vizenor, as his readers know well, "real" life permeates fiction and fiction warps real events and recognizable contemporary and historical figures into trickster's endless and endlessly self-reflexive tropes. Those who know Vizenor's work will also be familiar with the utopian impulse that drives often harsh satires, the same utopian desires that lie at the paradoxical heart of trickster stories. The goal is to illuminate hypocrisies and false positions, to challenge all accepted mores or fixed values, to shock and disturb us and cause us to reexamine and readjust moment-by-moment our conceptions of and relationships with both self and environment. The purpose of "holotropic" trickster,[2] with his/her indeterminable shape-shifting, contradictions, and assaults upon our rules and values, is to make the world better. "Ishi and the Wood Ducks" has its inception in Vizenor's "real life" challenge to the University of California, a challenge arising out of his desire to readjust history, force the university and the rest of society to recognize its false positions and culpability, and do honor to an exceptional man we erroneously call Ishi.

In the fall of 1985, as a visiting professor in Ethnic Studies at the University of California, Berkeley, Gerald Vizenor formally proposed that the north part of the campus's Dwinelle Hall (which housed and still houses the offices of Native American Studies) be renamed Ishi Hall. The student senate at Berkeley voted unanimously to support Vizenor's proposal, praising in their bill Ishi's "intellectual contributions to the University of California in the fields of anthropology, linguistics, and Native American studies."[3] When an administrator in the university's splendidly titled Space Assignment and Capital Improvements Group demurred and offered instead to consider naming a proposed addition to Kroeber Hall after Ishi, Vizenor said no: "We're not about to yield to anthropology and a new intellectual colonialism," he declared.[4] For seven years—that powerful tribal number—the matter rested there.

Who doesn't know the name of Ishi? "He was the last of his tribe," Mary Ashe Miller, a reporter for the *San Francisco Call,* wrote on September 6, 1911, before going on to declare that "[t]he man is as aboriginal in his mode of life as though he inhabited the heart of an African jungle; all of his methods are those of primitive peoples" (*Manifest Manners* 128). Miller, suffering pangs of ethnostalgic despair over a vanishing, generic, and thus entirely invented savagery, obviously saw through Ishi the man to the aesthetic of absence Vizenor describes. Discovered weak and starving in a slaughterhouse in northern California in August 1911, the prototypical "primitive" and ultimate Yahi was immediately placed in jail by the county sheriff because "no one around town could understand his speech or he theirs" (*Manifest Manners* 131). Not the first or last Native to be imprisoned for possessing liminal linguistic identity, the man who would be called Ishi was quickly transferred to the proprietorship of the U.C. Berkeley Museum of Anthropology and the care of noted anthropologist Alfred Kroeber, as Theodora Kroeber explained with undoubtedly unconscious irony: "Within a few days the Department of Indian Affairs authorized the sheriff to release the wild man to the custody of Kroeber and the museum staff" (*Manifest Manners* 131). In a statement that doubles delightfully back upon itself trickster-fashion (the "wild" man "released" into "custody"), Kroeber's words make it clear that the generic "Indian" was immediately and apparently unselfconsciously recognized as property of a federal agency (the BIA), which could "authorize" his transfer, as property, to the "custody" of another institution. Nowhere in this statement is there room for the possibility of the indigenous person's own individual agency; Ishi was, like all "Indians," a "bankable simulation" (*Manifest Manners* 11).

The last known survivor of the Yahis, a division of the Yana Tribe of California, the Native person called Ishi was the perfect Indian for colonial Euramerica, the end result of five hundred years of attempts to create something called "Indian." With his family and entire tribe slaughtered, starved, and decimated by disease like countless thousands of the original inhabitants of California, he became the quintessential last Vanishing American, a romantic, artifactual savage who represented neither threat nor obstacle but instead a benign natural resource to be mined for what white America could learn about itself. And when, after living for five years in the anthropology museum at Berkeley, he did perish as scripted, on March 25, 1916, of tuberculosis contracted from his white captors and custodians, the *Chico Record,* a little newspaper in a minuscule northern California town, saw through the staged drama with cynical insight:

> Ishi, the man primeval, is dead. . . . He furnished amusement and study to the savants at the University of California for a number of years . . . but we do not believe he was the marvel that the professors would have the public believe. He was just a starved-out Indian from the wilds of Deer Creek who, by hiding in its fastnesses, was able to long escape the white man's pursuit. And the white man with his food and clothing and shelter finally killed the Indian just as effectually as he would have killed him with a rifle. ("Ishi and the Wood Ducks" 3)

Perhaps most exceptional in this newspaper account is the writer's recognition that the "white man's pursuit" of the Indian is inexorable. Central to the extensive body of Vizenor's writing is his recognition that the white man's pursuit of the Indian is indeed unending, that it represents an unrelenting desire to capture and inhabit the heart of the Native just as surely as Euramerica has appropriated and occupied the Native landscape. Key to this pursuit is language, the struggle for definition and dominance through authoritative discourse.

The University of California must have thought, with considerable relief, that the irritating matter of Ishi Hall had ended when Visiting Professor Gerald Vizenor left the campus to teach elsewhere. However, in January 1992, having given up an endowed chair at the University of Oklahoma to return to Berkeley as a senior professor, Vizenor raised the banner of Ishi's cause once more, writing to Chancellor Chang-Lin Tien to say: "This is a proposal to change the name of the north part of Dwinelle Hall to Ishi Hall in honor of the first Native American Indian who served with distinction the University of California." This time, a full seven years after his first proposal had vanished into the bowels of a university committee, Vizenor cleverly made sure that his proposal coincided with the hullabaloo surrounding the quincentenary of Columbus's "discovery" of America. "This is the right moment," he explained to the chancellor, "the quincentenary is time to honor this tribal man who served with honor and good humor the academic interests of the University of California."[5]

Vizenor's proposal was forwarded to a committee once again. Two months later, in March, after hearing nothing, he wrote politely to the chancellor: "I understand the administrative burdens at this time, even so, my proposal is urgent because it is tied to an unmistakable historical moment." Three months later, in June 1992, he wrote to the chancellor yet again, this time a bit more heatedly:

> Consider the sense of resentment and anger that you might feel if your reasonable initiatives to honor a tribal name and emend institutional racism were consigned to the mundane commerce of a campus space subcommittee. Sixty-eight years ago today, on June 2, 1924, Native American Indians became citizens for the first time of the United States. Indeed, these historical ironies, and the national sufferance of the quincentenary, could be much too acrimonious to resume; the very institutions and the foundational wealth of this state are based on stolen land and the murder of tribal people.

On August 31, 1992, the Director of the Space Management and Capital Programs Committee on campus circulated a memo informing everyone that "The Naming of Buildings Subcommittee did not approve naming a wing of Dwinelle after Ishi. It suggested instead that the central courtyard of Dwinelle Hall be named *Ishi Court.*"

On May 7, 1993—almost eight years after Vizenor's initial proposal—*Ishi Court* was officially dedicated on the Berkeley campus. In his dedication address, Vizenor said, "There is a wretched silence in the histories of this state and nation: the silence of tribal names. The landscapes are overburdened with untrue discoveries. There are no honorable shadows in the names of dominance. The shadows of tribal names and stories persist, and the shadows are our natural survivance." He concluded his speech by declaring, "Ishi Court, you must remember, is the everlasting center of Almost Ishi Hall."

"Ishi was never his real name," Vizenor writes in "Ishi Obscura." "Ishi is a simulation, the absence of his tribal names. . . . Ishi the obscura is discovered with a bare chest in photographs; the tribal man named in that simulation stared over the camera, into the distance" (*Manifest Manners* 126). A Yahi word meaning "one of the people," the name Ishi masked the Native survivor's sacred name, which he never told to anyone. Ishi became the simulation. "The notion of the aboriginal and the primitive combined both racialism and postmodern speciesism," Vizenor writes in "Ishi Obscura," "a linear consideration that was based on the absence of monotheism, material evidence of civilization, institutional violence, and written words more than on the presence of imagination, oral stories, the humor in trickster stories, and the observation of actual behavior and experience" (*Manifest Manners* 128).

For Euramerica, the Indian is defined by absence, not presence. The determination is to know the "other" as not-European and thus to delineate the European, while paradoxically to find and extract some aboriginal distillation that, like the single drop of black in the white paint of Ralph Ellison's *Invisible Man*, will somehow make white America greater. The goal is, as Vizenor stresses, never to *know* the "other." "The word Indian," Vizenor writes, "is a colonial enactment, not a loan word, and the dominance is sustained by the simulation that has superseded the real tribal names. The Indian was an occidental invention that became a bankable simulation; the word has no referent in tribal languages or cultures. The postindian is the absence of the invention, and the end of representation in literature; the closure of that evasive melancholy of dominance. Manifest manners are the simulations of bourgeois decadence and melancholy" (*Manifest Manners* 11). Finally, once the Native is fully defined as the "absent other" the aboriginal space has been figuratively (and, of course, often quite literally through concerted programs of genocide) emptied so that it may be reoccupied by the colonizer, realizing the invader's wistful dream of achieving an "original" relationship with the invaded space.

For indigenous Americans, the paradox that arises from the ironies of invention is that it seems the Native person must pose as the absent other—the *indian,* as Vizenor writes—in order to be seen or heard, must actually become the simulation at the border. Simply put, the central paradox of Indian identity is that Indians must pose as simulations in order to be seen as "real." Vizenor quotes Umberto Eco on this point: "This is the reason for this journey into hyperreality, in search of instances where the American imagination demands the real thing and, to attain it, must fabricate the absolute fake." Vizenor adds that "Indians, in this sense, must be the simulations of the 'absolute fakes' in the ruins of representation, or the victims in literary annihilation" (*Manifest Manners* 9). "Ishi," he writes, "has become one of the most discoverable tribal names in the world; even so, he has seldom been heard as a real person" (137). In "Ishi and the Wood Ducks," the patient Yahi survivor becomes the "postindian," one of the original people of America freed from the humorless "melancholy of dominance" that requires the invented Indian.

Vizenor's play opens with an introduction that sketches the history of Ishi's discovery in northern California and his transfer from jail to museum. The introduction quotes a number of statements by those who surrounded Ishi in the museum, including Kroeber's declaration that Ishi "has perceptive powers far keener than those of highly educated white men. He reasons well, grasps an idea quickly, has a keen sense of humor, is gentle, thoughtful, and courteous and has a higher type of mentality than most Indians." To Kroeber's dubious praise,Vizenor adds the observation by Thomas Waterman, the museum's linguist, that "this wild man has a better head on him than a good many college men." Following the introduction, a prologue opens with the characters of Ishi and an "old woman" named Boots Story both waiting on a bench outside a federal courthouse, Ishi dressed in "an oversized suit and tie" and carrying a leather briefcase. Boots, a Gypsy, informs Ishi that she must appear in federal court in order to get a "real name." Otherwise she will be deported, or sent "home."

The cast is drawn from the same ten characters in each of the play's four acts, but, with the exception of Ishi himself and his choral accomplice Boots, the roles and identities of the characters change in each act, a strategy the author explains: "The actors are the same but the names of characters are repeated in the prologue and four acts of the play. The sense of time, manifest manners, and historical contradictions are redoubled and enhanced by the mutations of identities in the same characters."

As noted above, Ishi's first words in the prologue, addressed to Boots, are "Have you ever heard the wood duck stories?" A few lines later he says, "Ishi is my museum name, not my real name." Turning to the audience, he explains with nice irony, "Kroeber was an anthropologist and got me out of jail to live in a museum (*pause*), he was one of my very first friends." He adds: "Kroeber named me (*pause*), in my own language, he named me the last man of the stone age."

This prologue foregrounds the archival contradictions and values of names, museums, and stories: "How about a museum of stories?" Ishi asks Boots, invoking "that trace of creation and natural reason in native stories." When Boots says, "Who can remember stories anymore?" Ishi replies, "That's why you need your own museum." "Names without stories are the end," Ishi says with finality. "Ishi and the Wood Ducks" discovers the dead end of the name "Indian," which arises out of a static, monologic story of cultural dominance (and ethnocide) central to the Euramerican metanarrative, but which has no story at all within the hundreds of cultures it supposedly comprehends, or as Vizenor explains above, "has no referent in tribal languages or cultures."

The first act of the play takes place in the Museum of Anthropology at Berkeley, where Ishi is first seen flaking arrowheads (including one out of a piece of broken glass) in front of a "wickiup." Boots Story, silenced on stage, has mutated into a custodial worker who pushes her duster around the room and offers silent facial responses to the words and events of the act. The act centers on a visit to Ishi by characters associated with the museum and attempts by Ashe Miller, the newspaper reporter, to interview Ishi while Prince Chamber, a photographer accompanying her, tries unsuccessfully to take Ishi's photograph. In the opening lines of the act, Saxton Pope, "the medical doctor," praises Ishi as a natural healer and asks him to "show us your home in the mountains" and to teach Pope's son "how to hunt and fish with a bow and arrow, and how to make arrowheads out of broken bottles." Ishi turns to the audience to say, "The mountains are dangerous." When pressed by Pope, he explains that the danger comes from the "savages": the gold miners who "have no stories" and "no culture." Informed of the photographer's visit, Ishi says, "Pictures are never me," and "My stories are lost in pictures." Ishi's words here echo Vizenor's insistence in "Ishi Obscura" that "the miners were the real savages; they had no written language, no books, no manners . . . no stories in the blood. They were the agents of civilization," and "the gaze of those behind the camera haunts the unseen margins of time and scene in the photograph; the obscure presence of witnesses at the simulation of savagism could become the last epiphanies of a chemical civilization" (*Manifest Manners* 127). "Savages land in your pictures," Ishi tells the anthropologist Kroeber. (Roland Barthes wrote that "essentially the camera makes everyone a tourist in other people's reality, and eventually in one's own" [57]).

When Ashe Miller says, "Big Chiep [Ishi's nickname for Kroeber] said you come from a 'puny civilization' and that you are 'unspeakably ignorant' (*pause*). Do you know anything?" Ishi's response is "Too much pina." "Pina," we are quickly told, means "pain," an explanation that embarrasses Kroeber. Act one ends with the photographer's request that Ishi engage in what Vizenor has called "cultural striptease": "Ishi, bare your chest, the light is good." Ishi, content to remain unseen and unheard as a "primitive" and refusing to simulate Euramerica's fake Indian, whispers to Kroeber, who explains (in an instance drawn from Ishi's "real life" in the museum): "Ishi say he not see any other people go without them, without clothes (*pause*), and he say he never take them off no more." Ishi repudiates aesthetic absence and victimry, refuses to stop the clock. Obsessed with the way his mother is dressed in a photograph, Barthes declares that "clothing is perishable, it makes a second grave for the loved being" (64). Costumes are not perishable, however. Early ethnographic photographers clearly understood this when they carted generic "Indian" dress around for those Native subjects who might not possess such essential accoutrements.[6] Ishi perceptively refuses to costume himself in essential savagery for the camera.

Act 2 is set in the Mount Olivet Cemetery columbarium, where Zero Larkin, a "native sculptor with a vision," has come seventy years after Ishi's death in order to receive inspiration from Ishi's ashes. He is accompanied by Ashe Miller, the reporter from act 1, and Prince Chamber, the photographer who asked Ishi to bare his chest in that act. Trope Browne, undoubtedly one of the enormous mixedblood Browne dynasty that populates Vizenor's 1988 work *The Trickster of Liberty* (in which Ishi graduates with Tune Browne from the University of California), serves as attendant at the columbarium. When Zero asks Trope "Are you a brother?" Trope responds that his answer depends on "How much money you want." Observing and commenting upon the scene, unseen by the others, are Ishi and Boots Story.

Act 2 targets both invented Indians and those who would pose as experts on such inventions. When Ishi denies to Boots that he was ever a shaman or that he made the pot in which his ashes are held—both of which "facts" are maintained by the manager of the cemetery, Angel Day—Boots says, "Angel lied then." Ishi replies, "Not really, she's an expert on Indians. . . . Indians are inventions, so what's there to lie about?"

Zero Larkin, the posturing, scathingly satirized Native sculptor, intones, "Ishi is with me, our spirit is one in his sacred name," before adding, "I'm going to blast his sacred signature . . . at the bottom of my stone sculpture, my tribute to his power as an Indian." When asked if he's finished the sculpture, Zero says, "Not yet, but with the inspiration of his name it won't take long." Ashe Miller, the penetrating journalist from act 1, asks, "Zero does it make a difference to anyone that you are not from the same tribe as Ishi?" and Zero, the Vizenoresque "*varionative*,"[7] replies, "We are both tribal artists, and that's our identity." Not aware of the fact that Ishi names not a person but a constructed museum artifact, Zero himself embraces his identity as generic invented Indian, what Vizenor called a "colonial enactment" in "Ishi Obscura." When he touches the false burial urn, Zero chants, "Ishi is with me, his spirit is coming through my fingers." "What does it feel like?" the reporter queries, to which Zero answers, "The greatest power of my life, the spirit of my people." After Ashe asks if Zero means that his "people" are indigenous northern Californians like Ishi's, the dialogue continues to the detriment of the "Native" sculptor:

> ZERO: We are one as tribal people.
>
> Zero trembles, and he smiles as the camera comes closer. Trope moves closer to be sure he does not drop the pot.
>
> ASHE: Which one?
>
> ANGEL: He means a universal tribal spirit.
>
> ASHE (ironically): That explains it.

Act 2 ends with Boots chanting, "Zero, zero and the cemetery liars," and Ishi asking, "Do you think they could hear my duck stories?"

The third act of "Ishi and the Wood Ducks" features a meeting of the Committee on Names and Spaces in Kroeber Hall at the University of California. We are told, "The faculty is seated at a conference table to consider a proposal to rename the building Ishi Hall." In this act Angel Day has mutated into a professor of anthropology, Trope Browne a history professor, Ashe Miller appropriately a professor in mass communication, and Prince Chamber in visual arts. Professor Alfred Kroeber, the distinguished anthropologist, is present as sponsor of the proposal to rename his own hall in honor of Ishi.

"Why would he want to change his own name?" Trope Browne asks just before Kroeber appears on stage. When questioned, Kroeber says simply, "Ishi made my name," and adds, "Ishi's name is not genuine." Angel Day says, "Several years ago, as you know, we received a proposal from an errant faculty member to change the name of Kroeber Hall to Big Chiep Hall (*pause*). Naturally, and in honor of your distinction as an anthropologist, we chose not to discuss the proposal." Kroeber replies, "I supported that proposal." Kroeber goes on to explain that "The Big Chiep is one of his [Ishi's] name stories. . . . Ishi honored me with a nickname, in the same sense that we gave him his name (*pause*); the contradiction, of course, is that our name for him is romantic, while his name for me is a story."

Throughout this act, Ishi and Boots stand at the edge of the stage and comment upon the proceedings. "Kroeber said it," Boots tells Ishi, "you are his anthropologist." "Not a chance," Ishi replies, "he's the subject not the object." When Boots says, "Who are you then?" Ishi's answer is "The last object of their stone age." When the committee, just like the "real" committee that buried Vizenor's proposals for Ishi Hall in 1985 and 1992, decides that Kroeber's proposal is a "sad romantic visitation" and denies the request, Ishi and Boots—still invisible—begin to address the faculty members directly, with Ishi saying, "Doctor Ishi Hearst proposes that no building on campus would bear the same name for more than two years at a time."[8] The buildings, Ishi suggests, "would be known by their nicknames and stories, and no building would hold even a nickname for more than two years."

"Ishi and the Wood Ducks" concludes with act 4 set in the federal courtroom outside of which Ishi and Boots waited in the prologue. Ishi, it turns out, has been brought to court on charges of violating the Indian Arts and Crafts Act of 1990, the charge being that "he sold objects as tribal made, and could not prove that he was in fact a member of a tribe or recognized by a reservation government." In this final act Kroeber is the presiding federal judge, Ashe Miller the prosecutor, Pope the defense attorney. Other characters from the first three acts are court functionaries or witnesses (and a kind of chorus), including Boots Story. The act opens with Ishi placing "several bows, arrows, arrowheads, and fire sticks on the defense table."

Ishi, we learn, has been arrested at the Santa Fe Indian Market because "other artists complained that he was not a real Indian." Ishi's dilemma is twofold: on the one hand, like many contemporary Native Americans Ishi belongs to a tribe that is not federally recognized (in his case because his tribe has been ruled extinct, though a number of extant California tribes are not recognized today); at the same time, Ishi cannot be a real Indian because "the *indian* is a simulation, the absence of natives; the *indian* transposes the real, and the simulation of the real has no referent, memories, or native stories" (*Fugitive Poses* 15). Clearly, other "real" Indian artists saw greater profit for themselves in barring Ishi from the lucrative native arts-and-crafts market and are exploiting the colonial definition of Indian identity to their advantage. The Indian Arts and Crafts Act of 1990 decrees that it is "unlawful to offer or display for sale any good, with or without a Government trademark, in a manner that falsely suggests it is Indian produced, an Indian product, or the product of a particular Indian or Indian tribe or Indian arts and crafts organization, resident within the United States." The Act stipulates that "[f]or a first criminal offense, an individual is fined not more than $250,000 and/or jailed not more than 5 years; subsequent violations are not more than $1,000,000 and/or 15 years."[9]

Clearly, as Vizenor has pointed out, the "occidental invention" of Indianness has become a very lucrative "bankable simulation" in essentialist and ethnostalgic America. Were Judge Kroeber to find him guilty of criminal offense, Ishi would be in serious trouble indeed. Not only has Ishi violated the Indian Arts and Crafts Act, he is also charged with purchasing a false certificate of

Indianness, an enrollment form from the "Dedicednu Indians of California" based in Laguna Beach, a charge that brings great laughter from the courtroom.

Judge Kroeber declares that "Ishi names a very remote tribe with no immediate evidence of his descent and that, clearly, is a burden under the provisions of the new law." In Ishi's defense Saxton Pope argues that "my client is more tribal than anyone in the country" despite the fact that Ishi makes arrowheads out of broken bottles. For support Pope points out the spurious (and romantically racist) evidence that Ishi has wide feet, "would rather not wear shoes," and has leather thongs in his ears, none of which counts for much in court. In desperation Pope then adds, "Ishi never heard of Christmas," but Ashe as prosecutor replies in Vizenoresque language that "The absence is not a presence of character." Finally, Pope introduces Ishi's wood duck stories as the ultimate evidence of his client's tribal identity, arguing "oral stories are none other than a real character, the character of tribal remembrance, the character that heard and remembered the wood duck stories." As Ishi sings "Winotay, winotay, winotay," Pope begins the wood duck stories that (the actual) Ishi liked to tell in his museum home, stories that could take seven hours in the telling and be told only after dark.

Clearly, Judge Kroeber has a humanitarian and philosophical bent. This Kroeber is not the anthropologist who once labeled Ishi "unspeakably ignorant," but rather the man who, at the time of Ishi's death, wrote, "As to disposal of the body, I must ask you . . . to yield nothing at all under any circumstances. If there is any talk about the interests of science, say for me that science can go to hell. We propose to stand by our friends." This philosophical judge, who bears the name of Ishi's inventor, asks rhetorically, "What is criminal in the imagination of a tribal artist?" before adding, "Colonial inventions are criminal, not tribal survivance." A few lines later, the judge ponders, "Consumer fraud? Or is this a case of cultural romance?" to which the court clerk Prince Chambers replies, "Who knows the difference in Santa Fe?" When the prosecutor says, "The government protects the true Indian," Kroeber responds with, "Then the Indian, not the buyer, must beware."

Forced to make a ruling, Judge Kroeber finds that "clearly the performance of wood duck stories, no matter how great the audience response, does not establish tribal character or identity under the provisions of the Indian Arts and Crafts Act of 1990." He then turns directly to the audience to ask, "What would you do under the circumstances?"

Act 4 and the play both end with Kroeber ruling that "Ishi is real and the law is not. Therefore, my decision is to declare that the accused is his own tribe. Ishi is his own sovereign tribal nation. . . . Ishi, the man so named, has established a tribal character in a museum and in his endless wood duck stories. . . . Ishi is an artist, he is our remembrance of justice, and that is his natural character." Ishi becomes "a native presence" in the play, "not the romance of an aesthetic absence or victimry" (*Fugitive Poses* 15).

"Ishi and the Wood Ducks" manages in brief compass to suggest nearly all of the major issues found in Vizenor's significant body of published work. Among these are the colonial invention of Indianness; the "word wars" that characterize the five-hundred-year-old Euramerican program to fix the Indian within what, in *Bearheart,* Vizenor calls a "terminal creed"; the posturing of those "varionatives" who (like Zero Larkin) would pose within the dominant culture's constructed definition of universal Indianness; the "archival" erasure of the Native in photographs of the *indian;* and the "survivance" of the "postindian" within oral stories. In foregrounding the issue of Ishi's undocumentable identity, Vizenor illustrates sharply his contention in *Manifest Manners* that "the Indian is the simulation of the absence, an unreal name; however, the misnomer has a curious sense of legal standing. Some of the definitions are ethnological, racial, literary, and juristic sanctions." Vizenor concludes this discussion of Indian identity in *Manifest Manners* with the observation, "Clearly, the simulations of tribal names, the absence of a presence in a mere tribal misnomer, cannot be sustained by legislation or legal maneuvers" (14).

In Ishi's defense counsel's argument that "oral stories are none other than a real character, the character of tribal remembrance," Vizenor echoes his position in "Ishi Obscura": "The natural development of the oral tradition is not a written language. The notion, in the literature of dominance, that the oral advances to the written, is a colonial reduction of natural sound, heard stories, and the tease of shadows in tribal remembrance" (*Manifest Manners* 72). In "Ishi and the Wood Ducks," the written advances to the oral and the co-constructive audience—brought immediately onto the stage by both Ishi and Judge Kroeber—become the active jury challenged to deconstruct the *indian* to find the lonesome survivor who, in good humor, honored his sacred name and presented himself in a museum as "one of the people."

CHAPTER TWENTY

A Story of a Talk, My Own Private India, or Dorris and Erdrich Remap Columbus

Way back in the twentieth century, during the quincentennial hoopla over Columbus's "discovery" of America, I was invited to be part of a panel discussion of Native American writing at the Postmodern American Literature Association annual conference, or PAL as we in the profession call it. The meeting was in San Francisco, the first city I ever saw as a teen not too far out of Mississippi and right off the chicken farm, and still my favorite city in the world after Rome, and maybe Paris. The San Francisco to which I had come with my high-school journalism class many years before for a conference of sorts in Elu's Basque Restaurant, above which very sophisticated and beautiful women, as I recall, leaned out of windows and beckoned futilely to us confused sixteen-year-old boys. The San Francisco where on the same trip I'd seen my first stage production, something called *The Fantastiks* that still haunts my mind with ethereal confusion and where I ended up backstage amidst much hilarity and outlandish people in a story different from this one.

Because it was San Francisco, I was quick to accept the invitation and fly to what we always called The City and, despite an escalating bout with flu, to prepare for my part in the panel discussion by celebrating the quincentennial extensively with a friend until somewhere well past the midnight hour preceding the morning of my talk. When the sun finally rolled all the way across the continent and lodged against our tall hotel, I had been awake all night, first celebrating Columbus with ironic gusto and then more ironically celebrating the fact that my lodgings came with a convenient bathroom as my flu became more apparent in my unfocused state. When I began to make my way to the official hotel PAL site for the panel on Native writing, I did not feel at all well. My friend Hector, who had dutifully honored the discovery of the New World with me in the City by the Bay but was not engaged to speak professionally that day, had murmured *"vaya con dios"* and *"buena suerte, ese"* sometime around two A.M. as we'd made our way back to the hotel. My head had made the infernally happy sound of trolley cars into an internally persuasive discourse, and my stomach was in no mood for Rice-a-Roni, the authoritative San Francisco Treat. I remember, walking those few blocks, being under the impression that the famous hills of San Francisco appeared each and every one to be clones of the famous Lombard Street with its sinuous and sinister curves.

I trudged, paper in hand, up the winding and folding streets through romantic fog, or illusions of fog, thinking weakly of McTeague, the monster San Francisco dentist who pulled teeth with his thumb and forefinger and nibbled off his wife's fingers. It was the kind of morning, with its inner and outer fogs, ready-made for a monster dentist, and with the pages of my twenty-minute talk my only weapon, I felt vulnerable, like someone with a low survival quotient in the naturalistic universe where teacup tragedies did not exist. It may not be true that Mark Twain said the coldest winter he ever spent was a summer in San Francisco, but if he said it, even apocryphally, he was prevaricating as usual, for December in San Francisco is colder than July. "You don't want to get your hands froze on Christmas, do you?" kept echoing in my head as I navigated like Christopher C. or idiot Benjy on the day of sarsaparilla toward my own private India.

Some time before, in the relative safety of Santa Cruz with its earthquakes and serial killers, I had decided to give a talk on a quincentennial novel, Michael Dorris and Louise Erdrich's *The Crown of Columbus*, a book I had found both extremely entertaining and a clever exploitation and subversion of the whole Columbus schtick. A quincentennial trickster maneuver that had not, I thought, received the respect it was due. I was, to use a favorite expression, somewhat taken aback when I arrived at the place of the panel discussion, however. Like most academics, especially those of us in Native American literature accustomed to seeing our panels situated in the eight A.M. slot of the last day of any conference, I was used to mumbling a paper into a microphone in front of a half-dozen or dozen people who looked like deer in headlights. (Once, at a strange conference in Omaha, no one had shown up at all for a panel I was on, and the other two speakers and I took turns sitting in the audience and applauding wildly.)

I wasn't prepared for this Postmodern Literature Association audience. The room overflowed. Every chair was taken, every standing space filled, people sat in the aisles, leaned in the doorway, tiptoed to look over other people's heads in the doorway. Eager, curious, earnest, intellectual faces proliferated in the room. *Buena suerte, ese,* I thought as I probed and excused myself to the front, where on a small stage two tables were divided by a podium. The crowd had come to see and hear some of my more illustrious cospeakers, who would undoubtedly prefer to remain unnamed in this recounting. The distinguished scholar chairing the panel, whom at a different conference I had once accused of being my illegitimate mother and who has never allowed me to forget that gaucherie, gave me a long, concerned, and somewhat illegitimately motherly look and showed me where to sit at one of the tables. She suspected, I knew, only half of my debilitating state, being ignorant of my raging influenza but well trained to recognize familiar conditions of quincentennial celebrants.

On my left was a friend, a beautiful and magnificent Navajo poet of great dignity and presence who turned and smiled at me with intense sympathy and penetrating understanding. On my right was a strange and substantial person of indefinable identity who wore a beaded headband around his thinning red hair, a bone choker around his ample red neck, and regalia looking vaguely like bits swiped from numerous museums over his more-than-ample torso. This, every signifier shouted, was an Indian. In my chair sat a person without a single cultural signifier who was just coming to an awareness that cause-and-effect dynamics are actual, that celebrations of the quincentennial in combination with escalating flu symptoms the night before one must fulfill one's obligations to the profession constitute a cause, and that one's inability to focus, hear, or think clearly might be labeled an effect. The myriad red person to my right appeared to be pulsating, the way an octopus flushes when angry or upset just before releasing a massive ink cloud. I clutched my paper on *The Crown of Columbus* the way Christopher himself must have clutched his maps of ocean routes to the land of plenty.

♦ ♦ ♦

Anyone identified as a Native American writer was in somewhat of an awkward position during that quincentennial year. We were rather frequently expected to trot or be trotted out as representatives of quintessential quincentennial victimage, to be used like a penitente's flail to provide delicious excoriation for the exquisitely sensitive nerve endings of Euramerica's and the Western world's collective guilt. Suffering for this one year was supposed to be such sweet sorrow, and Native Americans were expected to be not people of color but people of choler like my palpitating colleague on the PAL panel. The Native American was expected to pose as romantic, tragic, and epic, and there were plenty of folks around willing to pose; it was out of rebellion against such manufactured angst that arose my quincentennial celebration with a Hispanic friend, heir himself to not only indigenous Sonoran but also Spanish heritage.

As Nash Twostar explains in *The Crown of Columbus*, today's Native American is "an improbable exception, a survivor of survivors of survivors" (364). To survive has meant to escape the absolute past of the romantic victim and to carve out through cunning and sheer endurance another destiny and another plot. That morning, sitting and listening to my colleagues, I felt like a survivor of a survivor of a survivor, all of the roles being my own. He of the red hair, headband, and bone choker seemed, to my unreliable eyes, to be displaying evidence of being indeed a person of choler as he stood at the podium and spoke wailingly of the travails of his people, until I began to fear an exploding cloud of obfuscating ink that would bring panicked flight from the jammed room and resulting disaster. The throbbing in my head had increased exponentially and my stomach was trying to find allies for a glorious revolution.

♦ ♦ ♦

Earlier that fall I had been asked by the editor of a Third World newspaper, a self-described *Marxist* editor, to write a piece on Columbus. I agreed because the editor was a friend whose politics I admired, but I did so a little reluctantly because I suspected that what he had in mind was not what I intended to do. I wrote an essay for the Marxist newspaper about the four-thousand-year-old man who had been recently discovered frozen in the Austrian Alps. I called my essay, "Columbus and the Tyrolean Traveler." I wrote about the haunting earnestness of this traveler. He had prepared so well for his interrupted journey—with finely crafted leather clothing, "a sort of wooden backpack, a leather pouch hanging from his belt with a fire-flint, probably a bow, a stone necklace, a knife with a stone blade." A newspaper account showed a photograph with the caption, "The frozen Bronze-Age corpse was well preserved."

I imagined this Bronze Age man setting out to walk somewhere that required climbing over a snowy, icy pass, perhaps to visit relatives, bear a communication, maybe sight-see. Clearly a man of stature in whatever community he inhabited, he had given forethought to the journey, sitting up at night to consider distance, difficulty, and need and finally preparing carefully and well. His route was a challenging one through Alpine heights, a world even then of rock and ice not to be taken lightly (we're told by experts today that the Alpine glaciers have diminished by half during the last century or so). All must have gone well until an avalanche caught him unawares just as they catch folks today, and four thousand years later he turned up, surprised but silent. I imagined how this man looked out across the mountains he must have known by a different name, how much he may have marveled at the beauty of alpinglow and perhaps a flock of chamois against black rock, the family he had left behind or was climbing toward, the stories he replayed in his mind as he climbed with his staff up the icy chutes, the songs he sang in his solitary journey.

♦ ♦ ♦

"Imagine he was headed for a literature conference," says Coyote. "Still about as live as most of those professors, enit?"

♦ ♦ ♦

Ahem. The Tyrolean traveler's story had abruptly become history/ourstory. His remains were said to be of "extraordinary scientific significance." He had been chemically preserved, refrigerated, dissected, and subsumed into a modern metanarrative, where he remains today in his twenty-first-century deep freeze. It struck me that this was to a quite disturbing extent exactly the approach Europeans and European Americans have always had and for the most part continue to have toward all indigenous peoples. Like the Vanishing American so dear to Anglo-America's heart, the four-thousand-year-old time-traveler is of great value precisely because he is so very dead, an artifact frozen in time—like the recently discovered brain of Ishi, the last of the Yahi—to be studied, chemically preserved, dissected, and eventually placed in a museum just like hundreds of thousands of Native American remains and subsumed into modern Euramerican history.

As I expected, the editor of the small radical newspaper was not very happy with my essay—was, in fact, quite unhappy with it. It wasn't what he expected or needed to whet the collective anger or guilt of his readers. As Vivian Twostar says so well in *The Crown of Columbus,* "He assumed that [I] would pen a vitriolic lament, an excoriation blaming Columbus for all the Indians' troubles" (51). When I did not do so, he was justifiably disappointed and declined my clever and deftly penned reflection on the Tyrolean Traveler.

♦ ♦ ♦

That morning at the PAL conference, hearing my own voice on something like tape delay, I spoke to the overflow crowd about what I thought and still today think is the most impressive accomplishment in Dorris and Erdrich's Columbus novel, an accomplishment worthy of recognition and applause even these years later with much difficult water beneath many bridges. Like Vivian Twostar, these two mixedblood authors chose not to pen a simplistic fiction of outrage. Instead, Dorris and Erdrich defiantly, even subversively, seized the low ground of American literature with a potboiling adventure story of the discovery of the Discoverer. And within the pages of their page turner, the two authors managed to cleverly encapsulate the political struggle between competing discourses at the heart of five hundred years of America's "discovery," that struggle which began when, as Dorris and Erdrich explain, Columbus "took possession of all that was before him simply by speaking certain formulaic words" (186). Thus was the New World introduced to the authoritative discourse of the Old, a language encountered by New World inhabitants with, as Mikhail Bakhtin says, "its authority already fused to it" (342).

Near the beginning of *The Crown of Columbus,* the authors quote a passage from Las Casas's version of Columbus's diary. In this, one of his first diary entries addressed to the Spanish monarchs, Columbus writes of the Lucayan Natives: "Our Lord pleasing, at the time of my departure I will take six of them from here to Your Highnesses in order that they may learn to speak" (6). Half a millennium later, *The Crown of Columbus* and the rest of the growing canon of literature by Native Americans demonstrate that

Columbus succeeded beyond his most earnest expectations. The people he mistakenly and unrepentingly called "Indians" have indeed learned to speak, and they have appropriated the master discourse, making the invaders' language their language, with a lowercase *e,* and turning it against the center. Because of such often defiant acts of appropriation, as Robert Young points out in *White Mythologies,* "the First World is now having to come to terms with the fact that it is no longer always positioned in the first person with regard to the Second or Third Worlds" (125).

And, like the four-thousand-year-old Bronze Age Tyrolean backpacker, Native Americans were well-prepared for this long journey, for in oral cultures stories are omnipresent and omnipotent. Through stories, as Vivian Twostar's Navajo grandmother, Angeline, knows in *The Crown of Columbus,* truth and history form and change, always growing, shifting ground, always dynamic and syncretic.

In Las Casas's version of Columbus's diary, it is clear that—in spite of Columbus's monologic certainty of linguistic authority—the Guanahanis and all of the indigenous peoples Columbus encounters know very well how to use communication as a tool or a weapon. Again and again, Columbus records trickster narratives such as that of the two men who, through signs, "showed the Spaniards that some pieces of flesh were missing from their bodies, and they gave the Spaniards to understand that the cannibals had eaten them by mouthfuls" (237). "The Admiral did not believe it," Las Casas dutifully writes, but it isn't hard to imagine these two tribal tricksters nudging one another and winking as they tell their tale, just as the Natives must have done as again and again they sent gullible Columbus sailing onward toward those promised mountains of gold always somewhere else, anyplace but here.

With their revision of Columbus—and thus history—Dorris and Erdrich accomplish a similar act of trickery, claiming the iconographic terrain of this nowhere man—who, according to various speculations in this novel might have been Italian, Jewish, Russian, Norwegian, and/or gay or might not have existed at all—claim him as Indian territory, reimagining the image and myth from a

Native American perspective. The paradigm of discovery is cleverly reversed, the course and discourse remapped.

The journey of discovery is undertaken by a Native woman whose priggish New England lover tags along. Roger Williams, the lover, is a poet-dilettante resembling the absurd would-be poet/editor Kinbote in Vladimir Nabokov's *Pale Fire*. Roger is penning his own "unrhymed monologue about Columbus—a reconstructed voice as in Browning's 'My Last Duchess'" (16)—a poem about a duke's thorough colonization of his young wife. Throughout most of the novel Roger is precisely such a monologist, capable of speaking only in reconstructed voices; about his poem he declares, "Content followed form and I bent the facts to my will" (51). Friends must point out to him that it is he and not Vivian's Navajo grandmother who is foreign. His heteroglot lover, Vivian—"Coeur d'Alene-Navajo-Irish-Hispanic-Sioux-by-marriage"—admits all voices into her conceptual horizon, identifying with the Columbus who "spoke all languages with a foreign accent" and "was propelled by alienation" (124); the Columbus who "was a nexus of imaginary lives, of stories" (125). Throughout the novel, Roger and Vivian struggle for control of language to such an extent that at one point Vivian notes that Roger "refused to relinquish command of the language" (223). Rather rapidly, however, Roger is forced to relinquish it, until in the end he is an accessory to Vivian's triumph, described by Nash as "simply tuned in to Mom, as though he had something to learn" (368).

The *Crown of Columbus* begins with a reversal, as "something marvelous" washes up from the sea and is found by an "island girl." At the time of discovery the island girl, Valerie Clock, has come down to rake the sand for flotsam and jetsam, the detritus of Euramerican civilization. The marvelous discovery is an infant, the sea's promise of new life, a new world. Cast up from the sea is Violet, the mixedblood daughter of Vivian and Roger, a symbol of America's heterogeneous future, pure potential and a fruitful merger of old and new worlds. As Valerie's island relatives recognize, the infant is like Moses, a promise of redemption.

The novel is flush with such images of birth and rebirth. Vivian Twostar's search for clues to Columbus's missing diary is paralleled by Violet's preparations for birth. The sea becomes an embryonic space out of which lives—as in Shakespeare's *The Tempest*—are startlingly delivered. First Violet washes up to begin the story. Later, the dynastic entrepreneur, Cobb—Roger Williams's evil twin—washes ashore, naked, with eyes that "spoke . . . the way a baby's do" (8). Cobb is "cradled . . . like a child" by Racine Seelbinder, the intuitive international man comfortable in every country. Much later in the novel, though not much later in the story, Vivian will pilot Cobb's powerful boat to shore, having wrested control of the craft from its owner by kicking Cobb overboard, using a technique learned in Grotz's Academy of Karate.

To complete her voyage of discovery, Vivian will descend into the earth—the earth-navel, emergence place, womb—and join Roger Williams. Like another famous figure in Western mythology, Roger is entombed for three days within the earth—in his own words, "a dark presence in the lower darkness"—where he puts computer disk aside and becomes a part of the oral tradition, reciting Columbus's history in his own unfinished poem, telling to himself the story that is history. And finally, when he is ready for rebirth, he ascends toward the light accompanied by Vivian, his savior, and by a symbol of much older sacrifice and resurrection. When Roger and Vivian emerge, Nash describes them: "Roger was next out of the ground, Mom's pale twin, a tall, skinny yellow chalk ghost. . . . He was a larva. . . . Both of them were dazed and disoriented, like they had forgotten the world, like they had been buried and returned to life" (358). The implication seems to be that Roger, deep in his sepulcher-in-the-sea, has cast off his despotic, power-driven Yankee self who seeks to command language and it is that other, darker self which has washed up in the form of the naked, cradled Cobb. In this twinning with Vivian, Roger has seemingly been reborn, part and particle of the new America represented by his mixedblood lover. When Nash asks how he came to be in the cave, Roger responds, "It is the destiny of Adam" (358). In this emergence

Dorris and Erdrich have conflated Christian and Native American myths, those stories that attempt to explain to us who we are—where we came from and where we are going—just as Vivian's name—Twostar—suggests a twinning of celestial significance.

Clearly, Dorris and Erdrich wanted this novel to represent a new beginning. As mixedblood authors, they use their text to celebrate the potential for a merger of worlds made concrete in their own selves and texts, a reconciliation articulated by Roger, who declares, "I am satisfied in our differences" (376) . Rather than only lamenting the centuries of genocide and injustices done to Native peoples, they have seized the opportunity to look ahead.

But finally we must come to the illusive crown for which the novel is titled. When Vivian and Roger discover the mysterious crown on its pedestal deep in the earth, it turns out to be a crown of thorns, within a crystal box, encased in bat guano. After reciting a portion of the Navajo Blessing Way, Vivian smashes the priceless container, and Nash comments to the reader: "The world has become a small place, all parts connected, where an Indian using an ancient Asian art can break into an old European box, witnessed by someone who grew up in Australia" (365). "What we have here," Vivian explains, "is Europe's gift to America. . . . What we have here was the promise, the pledge, the undiluted intent, the preconceived ideal before any fact was known. . . . And Columbus left it unopened. Never given. Never accepted" (368).

When the crown is exposed to air, it remains mysteriously preserved; only the thorns fall off. Now the celebrated symbol of ultimate suffering and sacrifice, that millenarian promise of redemption from earthly pain in a postapocalyptic paradise that propelled the American Myth from Atlantic to Pacific, has been rendered painless. The crown of nature can be worn without suffering, just as Violet, the daughter of a vital future, is rediscovered in "THE CHURCH OF THE LIVING GOD."

The premise of *The Crown of Columbus* is that the great discoverer brought with him a crown to be given to the New World. In her newly discovered Diario, Vivian finds the sentence, *"I knelt with the Crown and dedicated myself once again, Most Serene Princess, to finding the gold which will pay Spain's task*" (202). Vivian insists that "this crown business was not mentioned by Las Casas: Nowhere in the biographies or surviving letters had there been any description or reference to a crown. Just in my pages" (203). Vivian, however, is not quite accurate. Las Casas does, in fact, mention several crowns, writing of the "five kings" who come with the great "Guacanagarí, all with their crowns displaying their high rank. . . . The king came to receive the Admiral as soon as he reached land," Las Casas writes, "and took him by the arm to the same house as yesterday where he had a dais and chairs in which he seated the Admiral. Next he took off the crown from his own head and put it on the Admiral's" (296–97).

Perhaps, contrary to Vivian's speculation, *this* is the Crown of Columbus. Perhaps the great Guacanagarí reached forth to crown the invader with a symbol of his own sacrifice, a portentous gift from the New World to the Old. Entombed in the bat-cave for half a millennium, the resurrected crown might thus describe a new history: Columbus died for your sins. Now you can begin again. And try to do it right.

The *Crown of Columbus* ends with what Walt Whitman would call "the circle almost circled." Valerie Clock, the island girl, sits watching the ocean, and Dorris-Erdrich write, "It was a while before Valerie started to think of the sea as a place to cross, but once she did, she couldn't stop." With this cleverly subversive, potboiling detective story, these two mixedblood authors seemed to be answering that great representative American, Whitman, who asked the one question that cuts to the heart of five hundred years of history: "But where is what I started for so long ago / And why is it yet unfound?"

A survivor, I left my audience that morning in San Francisco with Robert Young's profound statement: "Until the lonely hour arrives in which the philosophical proof of the truth of history is produced, then history will inevitably continue as a representation of and interpretation of the past" (22). Like other quincentennial writings, including those by Native Americans, *The Crown of Columbus* is just such a representation and interpretation, with no lonely hour yet arrived. Frailly human, prepare as we may, each of us sets out uneasily across glaciers and frozen passes, through winding, cold-shrouded streets for his or her own Alpine destination or private India—celebrate however long and how we dare, arrive how and when we may, call it what we will. We may get lost along the way, may not reach the shores we had in mind, may even discover ourselves embroiled in someone else's story that we then try to call our own. We may surface from ice millennia later, intact, preserved, the mystery of our own private mission yet to be written. If we're lucky, a Las Casas will interpret for us, an audience will with patience wait.

Notes

CHAPTER 8

1. For this specific use, I have borrowed this form of the word from Gerald Vizenor, who explains that the "*indian* is the simulation of a logocentric other," without referent. On most occasions I will retain the familiar "Indian." See Vizenor's *Fugitive Poses* for extensive discussion of the term.

2. "A photograph's punctum is that accident which pricks me (but also bruises me, is poignant to me)," Barthes explains (*Camera Lucida* 27).

3. See Dorothy and Thomas Hoobler, *Photographing the Frontier* (New York: G. P. Putnam's Sons, 1980) 117.

4. Of his own mixedblood father, Vizenor writes: "My father was not a cultural ritualist; he never surrendered to shamans, base traditions, or absence as the source of a native sense of presence. He and many others of his generation moved from reservations to cities some sixty years ago during a severe economic depression in the country. Their concern, and with a much richer sense of humor at the time, was survivance" (*Fugitive Poses* 87).

5. This is not a rare belief. Vizenor cites virtually the same belief in doubled spirits among the Anishinaabes (*Fugitive Poses* 94).

CHAPTER 18

1. Nancy J. Peterson, "Introduction: Native American Literature—From the Margins to the Mainstream," *Modern Fiction Studies* 45: 1, 3.

2. James Colbert, in "Past Piff: In the Narrative Garden of Contemporary American Fiction," quotes Cliff Becker, who announced at the

1998 Associated Writing Programs annual conference that the purchase of Random House by the German conglomerate Bertelsmann meant that "there are now ony six commercial publishers remaining in the United States" (48).

CHAPTER 19

1. For a definition of "indian" vs. Indian, see *Fugitive Poses*, esp. 14–16.

2. Vizenor defines trickster as a "comic holotrope: the whole figuration; an unbroken interior landscape that beams various points of view in temporal reveries." *The Trickster of Liberty: Tribal Heirs to a Wild Baronage* (Minneapolis: University of Minnesota Press, 1988) x.

3. *The Daily Californian*, October 25, 1985, p. 1.

4. Ibid.

5. Vizenor to Chancellor Chang-Lin Tien, January 9, 1992, author's personal papers.

6. See Dorothy and Thomas Hoobler, *Photographing on the Frontier* (New York: G. P. Putnam's Sons, 1980) 117. And Vizenor, *Fugitive Poses* 163.

7. "The *varionative* is an uncertain curve of native antecedence; obscure notions of native sovenance and presence" (*Fugitive Poses* 15).

8. In "Ishi Obscura," Vizenor describes Phoebe Apperson Hearst as the creator of the Department and Museum of Anthropology at the University of California (*Manifest Manners* 131).

9. U.S. Department of the Interior—Indian Arts and Crafts Board, "Summary and Text of Title I, Public Law 101-644 [104 Stat. 4662], Act of 11,29/90."

Works Cited

Andrews, Wayne, ed. *Concise Dictionary of American History.* New York: Charles Scribner's Sons, 1962.

Anzaldúa, Gloria. *Borderlands: La Frontera.* San Francisco: aunt lute books, 1987.

Austin, Mary, and John Muir. *Writing the Western Landscape.* Ed. Ann H. Zwinger. Boston: Beacon Press, 1994.

Bakhtin, Mikhail. *The Dialogic Imagination: Four Essays by M. M. Bakhtin.* Ed. Michael Holquist. Tr. Caryl Emerson and Michael Holquist. Austin: University of Texas Press, 1981.

Barthes, Roland. *Camera Lucida: Reflections on Photography.* New York: Hill and Wang, 1981

Bhabha, Homi K. *The Location of Culture.* London and New York: Routledge, 1994.

Chakrabarty, Dipesh. "Postcoloniality and the Artifice of History: Who Speaks for 'Indian' Pasts?" *Representations* 37 (1992): 1–26.

Colbert, James. "Past Piff: In the Narrative Garden of Contemporary American Fiction." *Revista Canaria de Estudios Ingleses* 39 (1999): 47–69.

Columbus, Christopher. *The Diario of Christopher Columbus's First Voyage to America 1492–1493.* Abstracted by Fray Bartolomé de las Casas. Transcribed and translated into English with notes and a concordance of the Spanish by Oliver Dunn and James E. Kelley, Jr. Norman: University of Oklahoma Press, 1989.

Deloria, Philip. *Playing Indian.* New Haven: Yale University Press, 1998.

Dorris, Michael, and Louise Erdrich. *The Crown of Columbus.* New York: HarperCollins Publishers, 1991.

Fanon, Frantz. *Black Skin, White Masks.* New York: Grove Press, 1967.

———. *The Wretched of the Earth*. New York: Grove Presss, 1963.

Fletcher, Colin. *The Man Who Walked Through Time*. New York: Alfred A. Knopf, 1968.

Gandhi, Leela. *Postcolonial Theory: A Critical Introduction*. New York: Columbia University Press, 1998.

Garces, Father Francisco. *A Record of Travels in Arizona and California, 1775–1776*. Tr. John Galvin. San Francisco: John Howell Books, 1965.

Grey, Zane. *The Call of the Canyon*. New York: Black's Readers Service Company, 1924.

Harvey, David. *The Condition of Postmodernity: An Enquiry into the Origins of Cultural Change*. Oxford: Basil Blackwell, 1989.

Krutch, Joseph Wood. *Grand Canyon: Today and All Its Yesterdays*. New York: William Sloane Associates, 1958.

Lippard, Lucy R. *Partial Recall*. New York: The New Press, 1992.

Morris, Wright. *Time Pieces: Photographs, Writing, and Memory*. New York: Aperture Foundation, 1989.

Owens, Louis. *Mixedblood Messages: Literature, Film, Family, Place*. Norman: University of Oklahoma Press, 1998.

Pattie, James O. *The Personal Narrative of James O. Pattie of Kentucky*. Ed. Timothy Flint. Chicago: The Lakeside Press, 1930.

Powell, J. W. *The Exploration of the Colorado River and Its Canyons*. New York: Dover Publications, 1961.

Rabaté, Jean-Michel. *Writing the Image after Roland Barthes*. Philadelphia: University of Pennsylvania Press, 1997.

Said, Edward W. *Culture and Imperialism*. New York: Knopf, 1993.

———. *Orientalism*. New York: Vintage Books, 1979.

Schubnell, Matthias. *N. Scott Momaday: The Cultural and Literary Background*. Norman: University of Oklahoma Press, 1985.

Sontag, Susan. *On Photography*. New York: Farrar, Straus and Giroux, 1973.

Spivak, Gayatri Chakravorty. *In Other Worlds: Essays in Cultural Politics*. New York: Routledge, 1988.

———. "Translator's Preface." *Of Grammatology*, by Jacques Derrida. Tr. Gayatri Chakravorty Spivak. Baltimore: Johns Hopkins University Press, 1997.

Steiner, George. *In Bluebeard's Castle*. New Haven: Yale University Press, 1971.

Trinh T. Minh-ha. *When the Moon Waxes Red: Representation, Gender and Cultural Politics*. New York and London: Routledge, 1991.

———. *Woman, Native, Other*. Bloomington: Indiana University Press, 1989.

Van Dyke, John C. *The Grand Canyon of the Colorado: Recurrent Studies in Impressions and Appearances*. New York: Charles Scribner's Sons, 1920.

Vizenor, Gerald. *Fugitive Poses: Native American Indian Scenes of Absence and Presence*. Lincoln: University of Nebraska Press, 1998.

———. "Ishi and the Wood Ducks." In *Native American Literature: A Brief Introduction and Anthology*. Ed. Gerald Vizenor. New York: Harper-Collins, 1995. 299–336.

———. *Manifest Manners: Postindian Warriors of Survivance*. Hanover and London: Wesleyan University Press, 1994.

———. *The Trickster of Liberty: Tribal Heirs to a Wild Baronage*. Minneapolis: University of Minnesota Press, 1988.

Womack, Craig S. *Red on Red: Native American Literary Separatism*. Minneapolis: University of Minnesota Press, 1999.

Woodward, Grace Steele. *The Cherokees*. Norman: University of Oklahoma Press, 1963.

Young, Robert. *White Mythologies: Writing History and the West*. London: Routledge, 1990.

Index

Note: This index refers only to those chapters with primarily academic interest or focus.